T0362379

BIOMOLECULAR CRYSTALLOGRAPHY

Principles, Practice, and Application to Structural Biology

BIOMOLECULAR CRYSTALLOGRAPHY

Principles, Practice, and Application to Structural Biology

Bernhard Rupp

Garland Science
Taylor & Francis Group

Garland Science
Vice President: Denise Schanck
Editor: Summers Scholl
Assistant Editor: Alex Engels
Production Editor: Simon Hill
Typesetting: Georgina Lucas
Copyeditor: Sally Huish
Proofreader: Heather Whirlow Cammarn
Cover Design: Matthew McClements, Blink Studio Ltd.
Illustrator: Bernhard Rupp
Index: Indexing Specialists (UK) Ltd

©2010 by Garland Science, Taylor & Francis Group, LLC

ISBN 978-0-8153-4081-2

This book contains information obtained from authentic and highly regarded
sources. Reprinted material is quoted with permission, and sources are indicated.
A wide variety of references are listed. Reasonable efforts have been made to
publish reliable data and information, but the author and the publisher cannot
assume responsibility for the validity of all materials or for the consequences
of their use. All rights reserved. No part of this book covered by the copyright
herein may be reproduced or used in any format in any form or by any means—
graphic, electronic, or mechanical, including photocopying, recording, taping, or
information storage and retrieval systems-without permission of the publisher.

Library of Congress Cataloging-in-Publication Data
Rupp, Bernhard.
  Biomolecular crystallography / Bernhard Rupp.
    p. cm.
  ISBN 978-0-8153-4081-2
  1. X-ray crystallography. 2. Macromolecules--Structure. 3.
Biomolecules--Structure. 4. Proteins--Structure. I. Title.
  QP519.9.X72.R87 2010
  572'.36--dc22

                                                      2009036505

Published by Garland Science
Taylor & Francis Group, LLC, an informa business

711 Third Avenue, 8th floor, New York, NY 10017, USA,
and 3 Park Square, Milton Park, Abingdon, OX14 4RN, UK.

Printed by CPI Group (UK) Ltd, Croydon CR0 4YY

15 14 13 12 11 10 9 8 7 6 5 4 3 2

Visit our web site at http://www.garlandscience.com

Preface

I never teach my pupils; I only attempt to provide the conditions in which they can learn.

Albert Einstein

Undertake difficult tasks by approaching what is easy in them.

Lao Tzu

Biomolecular Crystallography (BMC) describes the science and art of X-ray crystallography of biological macromolecules. Biomolecular crystallography started in earnest around the mid-1930s in Cambridge and Oxford, England. Given the formidable practical and theoretical challenges, few at that time could have predicted its enormous impact on the field of structural biology. Today, X-ray crystallography routinely provides near-atomic resolution details of molecular structures of practically unlimited size.

The fascination and challenge of macromolecular crystallography lies in the wide range of disciplines combined in its practice. A great deal of molecular biology and biochemical laboratory work is performed during protein production, and often protein engineering plays a critical role in obtaining crystallizable material. Protein crystallization itself succeeds by a combination of critical rational analysis and the incorporation of all available information on the biological and cellular context of the material, frequently augmented by a touch of biochemical intuition. Diffraction data collection following the rules of diffraction physics takes place at synchrotron storage rings, some of the most advanced (and expensive) machines accelerator physicists have conceived. Armed with the data, an array of sophisticated computer programs taking advantage of powerful computational algorithms and maximum likelihood statistics goes to work, and often a first structure model results in an hour or less from automated model building programs.

Perhaps the biggest change in modern macromolecular crystallography is that it is now conducted largely by biologists and biomedical scientists interested in a specific structural problem. Such problem-oriented, hypothesis-driven research applies crystallography as a tool in the study of structural biology. With comprehensively trained professional crystallographers mostly engaged in developing increasingly sophisticated methods, algorithms, and automation, the training of future structural biologists has become an important and difficult task. The American Crystallographic Association recognizes the lack of adequate training for structural biologists and in 2006 a special education task force generated a white paper on crystallographic education, which is available at the National Academies' Web site and at the BMC book Web site (www.ruppweb.org). This book attempts to meet these recommendations.

Many early classics of general crystallography were strong on physical principles but not dedicated to the, then young, field of macromolecular crystallography. Since the 1976 seminal publication of *Protein Crystallography* by Sir Thomas Blundell and Dame Louise Johnson, substantial developments in crystallographic techniques have happened. Although a number of texts have appeared that serve certain aspects of the field very well, no comprehensive graduate-level textbook or accessible reference exists that is geared toward the needs of the structural biologist conducting crystallography. Much of the knowledge is dispersed in multi-authored volumes, which despite (or because of) their in-depth treatment and technical excellence are not exactly ideal self-study or course material.

A modern text on biomolecular crystallography should cover not only advanced crystallographic techniques and methods, but also the front-end aspects of protein crystallography: strategy development, protein production and purification, and crystallization, as well as the intense feedback taking place now between crystallization, structure analysis, and protein engineering. The strong interaction between bioinformatics, proteomics, and protein engineering on one hand, and the determination and analysis of protein structures on the other hand, has long been standard practice in the pharmaceutical industry, and it has now increasingly become the norm in any structure-determination project. A textbook should balance rigor and accessibility to meet the needs of the practitioner of protein crystallography, which today (and even more in the future) is the structural biologist well educated in crystallography.

In some parts of the book, the demands of mathematical thoroughness may conflict with the understandable desire of students to get to work as quickly as possible. But I have good reasons not to subscribe to a "mathematics-free" approach to biological crystallography. First, a few lines of equations can save pages of text and prevent semantic arguments. Second, there are certain relations in the physical sciences which evade what Sir Karl Popper calls "*Kübeltheorie des Alltagsverstandes,*" or translated, "evade the common household logic." Although I attempt to provide common sense explanations whenever possible in early sections of a chapter, they are supported by mathematics as necessary. Essential mathematical basics and tools are reviewed in the Appendix. I am sure that the serious student of crystallography will find enough theory in the chapters to be able to move on to advanced specialized literature. I also hope to impress on the casual user of crystallography that at least a qualitative understanding of the theory behind the "black box" programs is essential.

A general philosophy of the entire book is that of a probabilistic Bayesian treatment. In macromolecular crystallography (as well as in science in general) we rarely know anything with absolute certainty. Instead, it is our responsibility to provide an assessment of our confidence in our data and our structure models. It is extremely important to understand that crystallographic structure models are just that: models derived from sometimes very good and sometimes quite weak experimental data. In any case, they must withstand objective scrutiny against the data as well as against established prior information such as fundamental stereochemistry and all laws of nature.

Biomolecular crystallography is rapidly developing and some aspects of its practice described as state-of-the art may be superseded eventually (or turn out to be less elegant or even plain wrong). I have tried to consider the current state of the literature until about mid 2009, and I provide on occasion some necessarily subjective outlook of where I think the field is going. Particularly in the area of protein engineering, new techniques are published at an ever increasing pace and quickly tested and adopted for crystallography. This is one of the most appealing aspects of biomolecular crystallography and structural biology: It never gets boring, there are always new ideas to try, new methods to develop, and ever more daring structure–function hypotheses to test.

With this in mind, I hope you enjoy your study of Biomolecular Crystallography.

Bernhard Rupp, September 2009

Foreword

In 1976 Tom Blundell and Louise Johnson published their classic text *Protein Crystallography*. For many years it was regarded as the definitive reference for practitioners of X-ray structure analysis. Bernhard Rupp's *Biomolecular Crystallography* is, in every way, a worthy successor.

When Blundell and Johnson wrote their book, the field seemed to be reasonably well established. In the words of the authors, "It seemed to us that the subject has now reached a stage of maturity when a detailed review in the form of a book is needed." Looking back, however, synchrotron radiation was very much in its infancy. Computing and computer graphics were primitive by modern standards. The molecular biological techniques that are used to facilitate protein purification and to engineer ready-made proteins that are easier to crystallize and come with their own heavy atoms were yet to be invented. The possibility of using a single small frozen crystal to collect multiple high-resolution data sets at different wavelengths in a few minutes and to determine the structure the same day was unimaginable. The Protein Data Bank had been established but deposition of data was *laissez faire*. (In 1977 there were 74 entries, a third of which were different coordinate sets for hen egg-white lysozyme, hemoglobin, papain and cytochrome c's.)

Given the advances, breadth, and depth of current biomolecular crystallography, the challenge of providing a comprehensive and authoritative overview, starting from first principles, is formidable. Dr. Rupp has, however, succeeded admirably. The book starts with a general overview of the crystallographic method and then provides comprehensive coverage of protein structure, protein crystallization, protein production and purification, followed by details of the crystallographic method.

It is not in the least surprising to learn that the author has spent the last five years working almost exclusively on this volume. He has provided an invaluable resource which will be the reference and textbook of choice for beginner and expert alike.

Brian W. Matthews

A Note to the Reader

Overall design and style of the book

Each chapter starts with an introductory quote, some of historic interest, some humorous—generally setting the scene for the chapter. In the chapters themselves I try to build the subject from basic, yet well-illustrated, concepts. Once conceptual associations have been established, the depth of the treatment increases and the qualitative observations become embedded in the necessary mathematical formalism. Important topics introduced at an introductory level are reinforced in the more rigorous sections and revisited from their practical perspective in examples.

Throughout the text are Boxes containing essential concepts. These are generally short, declarative statements summarizing and emphasizing important insights derived in the surrounding sections. Key Concepts are also collected in a separate section at the end of each chapter and organized by topic. They contain the minimum qualitative information that should be retained after finishing a chapter and are ideal for generating comprehension questions. Glossary terms are highlighted in blue throughout the book.

The Sidebars contain historic information, sometimes protocols, critical remarks, or interesting additional information that does not necessarily fit the flow of the text. Here is a first example (Sidebar 0-1).

Chapters 8, 10, 11, 12, and 13 include explicit practical examples. I have tried to provide a representative sample of scenarios (they include S-SAD, SAD, and MAD phasing, phase extension, density modification and averaging, and molecular replacement phasing, as well as automated model building and maximum likelihood refinement). For each example data and models are available, which can be downloaded either from the PDB itself, or from the Web sites cited in the text and from the supplemental BMC Web site. I have employed a selection of contemporary programs, but for reasons of simplicity, popularity, and my own familiarity, most examples use programs from the CCP4 suite and the *SHELX* suite of programs. The molecular replacement example also uses evolutionary programming (*EPMR*) and torsion angle refinement (*CNS*). The model-building

---

Sidebar 0-1 Why no history chapter? Encyclopedic textbooks tend to contain sections covering the history of their subject in the introduction. I have decided to omit such a separate section. After all, many of the first students of the pioneers involved in the most exciting early periods of protein crystallography—largely taking place in the UK in Cambridge and King's College, and at Caltech in Pasadena, California—are fortunately still with us, and others have left us first-hand reports and reminiscences about their work and their participation in the development of the field. There are no better sources than these primary accounts, and their genuine stories need not be retold by me. The most comprehensive tale about the early years of protein crystallography is an extraordinarily entertaining compilation by Richard Dickerson, now retired from UCLA, retelling and commenting on the early work taking place in

Cambridge. His highly recommended book *Present at the Flood – How Structural Molecular Biology Came About* has served as a source and inspiration for numerous historic sidebars in the main chapters. It is more entertaining and motivating to provide historic information—often tales of concurrent agony and persistence—in sidebars when the subject in question is discussed, instead of in the introduction where the relevance of a historic viewpoint might not yet be fully appreciated. I have also corresponded with crystallographers directly involved in some of the early work; Brian Matthews and George Sheldrick deserve particular credit for discussions of early phasing techniques.

Several volumes of *Methods of Enzymology, Acta Crystallographica* Section D, and *Protein Science* include historic accounts by primary sources. Such specialized reviews are individually cited in the corresponding sidebars.

section employs the automated model-building Web services *TEXTAL* and *ARP/wARP*. This selection does not mean that any other programs and suites such as *PHENIX* or *SHARP* are any less powerful or user friendly; many of them are described or mentioned in the text.

Many of the figures and illustrations in the book are emblematic of figures found in publications describing and analyzing macromolecular structures. I provide briefly in Chapter 13 the tricks, programs, and tools used to prepare appealing and persuasive figures. This is also where the respective programs used in the production of the figures are cited and credited.

How this book is organized

The book is organized into 13 chapters, arranged into 5 parts, and an appendix. The sequence of chapters more or less follows the general course of a crystallographic structure determination project.

Part I, *From Sequence to Crystals*, contains the introductory chapter, which provides some general guidance on what to expect from a crystallographic study, together with an overview of the progression and key stages of structure determination. Proper project planning considering the critical stages of the process helps to balance expectations with the anticipated results necessary to answer your original question, and early consideration of alternative approaches avoids problems and disappointment in the later stages of the project.

Chapter 2 provides an introduction to protein (and DNA) structure. Unlike classical introductions to protein structure, such as the more comprehensive and excellent *Introduction to Protein Structure* by Carl-Ivar Branden and John Tooze, we instead focus on features directly relevant to crystallographic structure determination, and introduce illustrations of electron density as early as possible to train the eye in feature recognition. Properties of basic and modified residues as determinants of protein chemistry are discussed, together with how their interactions give rise to the great variety of structural folds we find in nature. After the introduction of secondary structure and basic structural motifs a brief exploration of fold space follows, illustrated with a variety of structures ranging from drug targets to protein–DNA complexes.

The theory and practice of protein crystallization forms the subject of Chapter 3. There is good reason why we treat protein crystallization *before* protein production: Once we understand the necessary conditions that need to be met to allow for self-assembly of molecules into a regular crystal lattice we can better appreciate and apply proper strategies for protein engineering. We lay out the physicochemical foundation of protein crystallization and focus on generally applicable methods and analysis of crystal growth. Many illustrative examples of crystals and relevant electron density features are shown. The chapter ends with a section on special methods including protein–DNA complexes, membrane protein crystallization, and molecular scaffolding.

With an understanding of the properties that a crystallizable protein should have, we develop in Chapter 4 the strategies to produce proteins using recombinant DNA technology and heterologous over expression in various host systems. Given the vast array of laboratory techniques available, we have to limit the treatment to the techniques most suitable for the structural biologist. Protein engineering techniques at the DNA level as well as some chemical and biochemical modifications useful for obtaining crystallizable protein variants are discussed. Advanced cloning techniques, expression host selection, and fusion constructs are illustrated with examples, and various outlooks of emerging methods are included. The chapter ends with a brief discussion of physicochemical methods for the assessment of conformational purity ranging from light-scattering methods to heteronuclear multidimensional NMR.

In Part II, *Fundamentals of Protein Crystallography*, Chapter 5 develops the foundations of crystal packing and symmetry. The asymmetry of protein molecules imposes significant limitations on the crystal symmetry, which initially allows a somewhat less formal treatment than in conventional crystallography texts. Many important concepts of symmetry and crystal geometry become readily apparent during crystal-building exercises using projections of real molecules. At the end of the chapter, formal treatment of symmetry elements and coordinate system transformations is provided, followed by a first introduction of the reciprocal lattice. Chapter 6 treats the physical principles of X-ray scattering and rapidly extends from atoms to crystals. Working with complex structure factors requires proper handling of computations in the complex plane, which is made more intuitive with numerous illustrations. Once the fundamentals are covered, we can compute the first structure factors and explore their properties. An extensive treatment of anomalous scattering follows, necessary because anomalous methods are central to modern phasing techniques. The formal rigor increases at the end of the chapter and a discussion of the reciprocal lattice and diffraction geometry using the Ewald construction prepares us for data collection.

Before we progress to the practice of data collection, we pause for an interlude on statistical methods. Modern crystallography relies a great deal on probabilistic methods, and in a biosciences curriculum they are generally not treated in a form specific and suitable for crystallography. The treatment of statistics is specifically selected for crystallographic applications and includes, beyond descriptive statistics, the important concepts of Bayesian inference and maximum likelihood. While it may not be necessary to work through Chapter 7 in full depth during the first contact or in an introductory course, it will be nearly inevitable that at a later point one has to refer back to the foundations laid out in Chapter 7.

Part III, *From Crystal to Data*, consists of a single chapter. Data collection is actually the last "real" experiment conducted during a structure determination and considering the efforts that went into a project up to this point, getting the best possible data is prudent. Chapter 8 discusses modern instrumentation as well as the practical aspects of data collection and data processing, including treatment of twinning and re-indexing. Visualization of reciprocal space is provided by many diagrams and actual diffraction images. The descriptive part is followed by a practical example of processing high resolution synchrotron data and a high-redundancy S-SAD data set collected on a laboratory X-ray diffractometer. The ultimate goal of the chapter is to drive across the point that data quality will, without mercy, determine the quality and hence usefulness of the final structure.

Part IV, *Determining Your Structure*, gets to the core of the *in silico* part of structure determination and computational crystallography. The short Chapter 9 builds the foundation with an in-depth discussion of Fourier transforms and their practical application. Illustrated with many examples are the effects of missing data, measurement errors, and phase errors, which immediately leads to the recognition of the phase problem and the phenomenon of phase bias. A brief section about Patterson functions prepares us further for Chapter 10, in which we explore experimental phasing methods. The concept of difference data and the resulting basis for substructure solution are discussed, including the treatment of phase probabilities by classical and maximum likelihood methods. This technically demanding subject is demonstrated and reinforced with multiple plots of 2-dimensional probability distributions. Density modification and phase extension are indispensable tools necessary to obtain quality electron density maps, and are correspondingly treated. Practical examples of S-SAD phasing and MAD based on the data processed in Chapter 8 close the chapter.

Chapter 11 expands the topic of non-crystallographic symmetry with a formal treatment of general transformations in 3-dimensional space. The subject often proves difficult and I have taken great care to present consistent notation. The ability to handle rotations and translations in space leads directly to exploration of molecular replacement (MR) methods, where a structurally similar model provides the initial phases. In addition to the actual MR techniques ranging from evolutionary algorithms to maximum likelihood functions, the recognition and treatment of phase bias, model selection, and automated MR procedures are discussed. As always, practical examples illustrate the theoretical treatment.

Chapter 12 presents the next steps of building an initial structure model into electron density and refining it against the experimental data. Model building and restrained reciprocal space refinement, together with concurrent error correction, are repeated until a reliable structure model is obtained. Following a thorough discussion of stereochemical restraints and the principles of optimization algorithms, the foundations of restrained maximum likelihood refinement are developed. Parameterization, the necessity of cross-validation, and the role of log-(free)-likelihood in optimizing restraint weights are emphasized. A final section illustrates the rebuilding of a molecular replacement solution as well as manual and automated model building of the MAD structure phased in Chapter 10.

Part V, *Making Sense of Your Structure*, continues the structure solution adventure to answer questions of biomedical relevance. Chapter 13 therefore elaborates on the finer details of local structure validation, including several cautionary tales from the literature. The emphasis is on understanding the structure model as a molecular hypothesis with a (hopefully high) probability of being correct, and thus subject to testing against all available information—experimental evidence, biochemical and stereochemical knowledge, and in fact all known laws of nature. Special topics include drug target structures and the subtle art of building correct ligand structures. The validation section is followed by a final subchapter about analysis and presentation of molecular structures in publications and presentations.

The Appendix provides a refresher for mathematics as needed for the derivation of the fundamentals.

Practice makes perfect

The performance of inexpensive desktop or laptop computers today allows even complex crystallographic calculations to be performed in minutes to hours. I recommend you install at least the CCP4 suite of programs (including the model building program *Coot*) and the *SHELXC/D/E* suite of programs on your computer. Although you ultimately should gain familiarity with all available programs, the simplicity of installation, the large and helpful user community, and an excellent bulletin board provide a good starting point.

The CCP4 suite and *SHELX* programs are freely available to academic users, and they are offered as executables for Windows and Linux. CCP4 consists of numerous program modules for practically every conceivable computing task in a crystal structure determination project. As a great advantage to the beginning user, the interactive version, CCP4i, offers a tcl/tk based graphical user interface, which makes navigating through the logical flow of programs and their input quite intuitive. The advanced user still can root through the underlying shell scripts and explore individual programs in depth. Another benefit of the CCP4i GUI is the ease of plotting graphs of critical data and parameters. We will frequently show CCP4i plots in this book. The *SHELX* programs have a lean command line interface with reasonable defaults, and they are powerful and robust.

One of the most valuable assets of CCP4 is an active and immensely supportive user community, which exchanges information, help, and hints on the CCP4 e-mail bulletin board. The topics discussed range from installation problems and usage issues to practical and theoretical aspects of crystallography. Even if you do not have a specific problem, regularly following the discussions is exceptionally educational. The CCP4 bulletin board requires a subscription. Information can be found on the CCP4 Web site or the BMC Web site.

Supplemental Web material

Although BMC is a self-contained text, a substantial amount of supplemental material and resources can be found on the BMC Web site (www.ruppweb.org/). In addition to my venerable Web tutorial, a collection of links to crystallographic and structural biology resources, sites, centers of excellence, other Web teaching material are provided and updated regularly. With the rapid progress in the field, new developments and resources are expected to become available even while the book is in press.

Acknowledgments

This textbook, discussing in hundreds of pages the interdisciplinary science of macromolecular crystallography, could not have been researched and written without the significant involvement, assistance, and also encouragement of the biomolecular crystallography community at large. My grateful acknowledgments therefore are due to numerous individuals providing valuable contributions as follows.

The CCP4 community. Special thanks are due to all the contributors to the CCP4 bulletin board who have responded to my postings and volunteered advice and insights as well as ideas and images of crystals, data files, or model coordinates for perusal in my book. I wish to acknowledge them all, even if their material did not make it into final print. When material from these sources is included, I specifically acknowledge the contributors again in the text or figure caption. Thank you for volunteering images, ideas, data, or literature (in historical sequence):

Andrea Schmidt, EMBL Outstation Hamburg; Debanu Das, University of California, Berkeley; Michael Kolbe, MPI for Infection Biology, Berlin; Bernard Collins, National Jewish Medical and Research Center, Denver; Sebastiano Pasqualato, IFOM, Milan; Erik Debler, The Scripps Research Institute, La Jolla; Flip Hoedemaeker, Kcy Drug Prototyping BV, Amsterdam; Terese Bergfors, Uppsala University; Sergei V. Strelkov, Catholic University of Leuven; Ewa Skrzypczak-Jankun, University of Toledo; Axel Martin, Ruhr-Universität Bochum; Jonathan Grimes, Oxford University; Raquel Lieberman, Harvard University; Juan-Maria Ruiz, Granada; Ehmke Pohl, University of Durham, UK; Dominika Borek, UT Southwestern Medical Center, Dallas; Kenneth Frankel, LBNL; Leonard Banaszak, University of Minnesota, Twin Cities; Graeme Winter, Daresbury Laboratories; Jürgen Bosch, University of Washington; James Whisstock, Monash University, Australia; Mitchell Miller, Stanford University; Andrea Hadfield, University of Bristol; Yvonne Leduc, University of Saskatchewan; Pete Artymiuk, University of Sheffield; Richard Bryant, University of Oxford; Paul Adams, LBNL; Helen Berman, RCSB Rutgers; Frances Bernstein, Bernstein & Sons; Tom Terwilliger, LANL; Jiamu Du, Shanghai Institute for Biological Sciences; Aiping Dong and Xiaohui Xu, University of Toronto; Henry Bellamy,

CAMD, Louisiana State University, Baton Rouge; Alexander McPherson and Aaron Greenberg, University of California, Irvine; Lukasz Salwinski, University of California, Los Angeles; Peter Zwart, LBNL; Han Remaut, Birkbeck College, London; Dimitri Svergun, EMBL Outstation Hamburg; Simon Colebrook, Oxford University; Joanne Nettleship, Oxford Protein Production Facility; Gunter Stier, MPI Heidelberg; Gloria Borgstahl and Jason Porta, University of Nebraska, Omaha; Artem Evdokimov, Pfizer Global R&D, Groton, CT; Mark Wilson, University of Nebraska, Lincoln; Gyorgy Snell, LBNL; Isabel Usón, Instituto de Biologia Molecular de Barcelona, Spain.

I am particularly indebted to all the members of the community of experts who answered, often beyond the scope of the CCP4 bulletin board, the many desperate and poorly articulated questions I have posted, generally at 3 AM California time, on the bulletin board. Amongst those are (in random order) Dale Tronrud, Ian Tickle, Bart Hazes, Kay Diederichs, Randy Read, Garib Murshudov, Eleanor Dodson, Fred Vellieux, Axel Brunger, Alexandre Urzhumtsev, George Sheldrick, Lynn Ten Eyck, Peter Zwart, Gerard Kleywegt, Roberto Steiner, Bernie Santarsiero, Gerard Bricogne, Clemens Vonrhein, Ron Stenkamp, Ethan Merritt, Mark Wilson, James Holton, Chris Hall, Bob Sweet, and Colin Nave.

Commercial vendors. Several commercial vendors have provided trial software, equipment images, and technical drawings for my perusal. I thank Michael Ruf and Sue Byram from Bruker AXS; Ron Hamlin and Chris Nielsen from ADSC Corporation; Jim Pflugrath and Kris Tesh from Rigaku-MSC; and Ruben Abagyan, Molsoft LLC, for the ICM Pro trial license used to produce numerous images of protein structures. Finally, my own company, q.e.d. life science discoveries, has generously supported the online materials and my time devoted during the last four years to the production of this book.

I also wish to acknowledge the University of California, Irvine, and Alexander McPherson for providing me with an adjunct affiliation and Katherine Kantardjieff, CSU Fullerton, for academic library access. During a sabbatical stay in Jim Sacchettini's laboratory at Texas A&M University, some of the ideas for the book evolved from discussions with his lively group of graduate students and postdocs—thank you all, guys!

Reviewers. I am most indebted to the reviewers who were extremely supportive and did a fantastic job in fine-combing through the manuscript and correcting sloppiness, inaccuracies, and many outright errors. As in almost every first edition, there will be remaining mistakes, errors, and omissions. I wish to declare that I alone take the blame for these and would encourage the readers to report any errors or unclear statements as well as any suggestions for improvements directly to me through the BMC Web site (www.ruppweb.org).

I am particularly indebted to Ian Tickle for extensive technical review and personal discussions of Chapters 7, 10, and 12, assuring adequacy of my statistical and likelihood treatments. Ian has gone beyond anything one could possibly expect from a reviewer in numerous emails and by providing extensive discussions as well as literature references about challenging topics. Dale Tronrud critically worked through Chapter 12, and discussions with him were the inspiration for additional ideas for several figures in Chapter 6. James Holton contributed his expertise by reviewing synchrotron physics and data collection in Chapter 8 and also provided a copy of his awesome *MLFSOM* program. Chapters 6 and 8 have also undergone extensive field review by Jim Pflugrath who provided most valuable suggestions based on his experiences from Cold Spring Harbor Laboratory crystallography boot camp. George Sheldrick, Tobias Beck, Tim Grüne, and Brian Matthews greatly contributed to improvements in Chapter 10 and also provided the data for experimental phasing examples. The personal discussions with George Sheldrick proved exceptionally helpful in tightening up the experimental phasing Chapter 10. During review of Chapters 7, 9, and the mathematical appendix Mark Wilson provided valuable suggestions and review. Chapter 4 would not have been possible without the review, corrections, and substantial additions by Artem Evdokimov. Gunter Stier has provided additional corrections and figures for Chapter 4. Sections of Chapter 12 dealing with molecular dynamics and *CNS*-specific features were reviewed and corrected by Axel Brunger. Chapter 1 has been edited, reviewed, and shortened by Katherine Kantardjieff, who also provided comments and instructive additions for the ligand structure analysis section in Chapter 13. For this final chapter I thank Gloria Borgstahl for her useful suggestions, Artem Evdokimov for additional review, and Jim Naismith for permission to reproduce his collected experiences about publishing in high impact journals.

Ron Stenkamp, Michael Chapman, and Ehmke Pohl have served as developmental reviewers of numerous chapters, and I thank them for their copious helpful comments and suggestions. Ehmke Pohl has also contributed a draft for the section on DNA-protein complex crystallization.

Many other colleagues have provided assistance ranging from figure preparation to program modifications. Amongst those I thank Jonathan Grimes, University of Oxford, for the beautiful figure of the PRD1 phage head and the phage head crystal images. Peter Briggs has kindly modified the CCP4 mapslicer interface for me, and Garib Murshudov added for me several options to plot reflection statistics in *REFMAC5*. Kevin Cowtan, YSBL York, UK, kindly allowed me to reproduce selected Fourier transform images from his tutorial. Paul Emsley gave me a personalized introduction to his program *Coot*. He and Bernhard Lohkamp patiently responded to my desperate plotting and graphing questions. Gloria Borgstahl and Jeff Lovelace prepared the special figure of an incommensurately modulated protein structure, and I thank Mark Wilson for the TLS figures made with Tim Fenn's POVscript+. Artem Evdokimov assembled a figure of HSQC spectrum for Chapter 4. Richard Dickerson kindly provided his amusing electron density figure leading into Chapter 9.

Editorial and production staff. The idea to develop my basic web tutorial into a textbook crystallized when Bob Rogers, senior editor at Garland Science convinced me to take on the project of extending the rudimentary Web contents of Crystallography 101 into a full text book. It is largely due to his encouragement and his patient but persistent inquiries about progress that the first chapters were ever written. Little did I know then what to expect, and had I known, I am not sure if I would have been foolish enough to embark on this multi-year project. When Bob retired, assistant editor Alex Engels and editor Summers Scholl, together with vice president Denise Schanck, took over and provided patient but unrelenting guidance throughout the process. I wish to acknowledge the great work of the production staff in England under senior production editor Simon Hill. My copyeditor Sally Huish has done a terrific job sorting out my inconsistent language and grammar, and the great layout and typesetting is by Georgina Lucas. The agonizing task of proofing the text and more than 600 equations was thouroughly accomplished by Heather Whirlow Cammarn.

Louise Jones, publishing editor of the International Union of Crystallography, provided me with permissions and reproductions of space group diagrams from the International Tables for Crystallography and figures from IUCr journal publications.

Contents

Detailed Contents

Chapter 10

Experimental phasing   473

Chapter 11

Non-crystallographic symmetry and molecular replacement   547

Chapter 12

Model building and refinement   607

PART I
From Sequence to Crystals

Introduction: Preparing for your study

> Almost all aspects of life are engineered at the molecular level, and without understanding molecules we can only have a sketchy understanding of life itself.
>
> Francis Crick (1988), *What Mad Pursuit: A Personal View of Scientific Discovery*, p. 61

Chapter 1 provides a brief introduction emphasizing the relation between form and function, followed by an overview of the principles and challenges of macromolecular crystallography. The relevance of protein crystal structures compared with the solution state is discussed. A brief interlude about molecular cryptanalysis leads into an overview of the method, including general guidance on planning and executing a structure determination in order to answer a biological question. A short overview of coordinate files and the role of the Protein Data Bank concludes the introduction. The main purpose of this chapter is to provide a quick overview of the method, also suitable for entry level courses or users of crystallographic models, while preparing the ground for the serious study of the remaining chapters.

1.1 Molecular structure defines function

The molecular structure of matter defines its properties and function. This simple but far-reaching statement is true for all matter. The properties of gasses, liquids, rocks, semiconductors, or small organic molecules are defined by their molecular structure, as are the functions of proteins and their complex macromolecular assemblies.[1] The motivation to use X-ray crystallography as the primary means of macromolecular structure determination is founded on the fact that accurate and precise molecular structure models—often revealing details at the atomic resolution level—can be obtained rapidly and reliably by means of X-ray crystallography. About 90% of all structure models deposited in the Protein Data Bank (PDB) are determined by X-ray crystallography, while the remaining 10% are determined by solution nuclear magnetic resonance (NMR) spectroscopy, largely in the 5 to 25 kDa range of molecular weight.

X-ray crystallography can provide highly detailed molecular structure models of large molecular assemblies. Elucidating the atomic details of molecular interactions is particularly important, for example, in the clarification of enzymatic mechanisms and is essential for drug target structures serving as leads in structure guided drug design.[2] In addition to providing accurate models of molecular structures, another advantage of crystallographic structure determination is that no limits of principle prevent the accurate description of very large molecular structures or molecular complexes, as evidenced in the nearly 2 MDa structure of the 50S ribosomal subunit[3] containing 27 different proteins (~4000 residues) and the ribosomal 5S and 23S RNA, together comprising 2833 nucleotides. The even larger 66 MDa protein capsid of the bacteriophage head PRD1 is assembled from ~2000 subunits of 18 different proteins[4] (illustrated in Figure 2-2). As the emphasis in modern biosciences shifts toward the understanding of entire biological systems, the capability to determine large, multi-unit complex structures, thereby providing a detailed picture of specific protein–protein interactions within their overall interaction networks,[5] will become increasingly important.

Crystallography provides the foundation of modern structural biology and structure-based drug discovery. Exciting structures that increase our understanding of molecular form and function are determined at an ever increasing pace (Figure 1-1). Whole virus particles, the F_1-ATPase,[6] the potassium ion channel,[7] fundamental structural information about cellular recognition by antigen presenting cells,[8] immunity and T-cell signaling,[9] or the functional insight provided by crystal structures of the ribosomal machinery are just a few examples that have tremendously extended our understanding of the structural basis of biomolecular function. At the same time, the quest for new therapeutic drugs has launched massive public and commercial efforts to determine the details of drug–target interactions, and to develop whole structure-guided drug discovery pipelines.[10] The role and use of protein drug target crystallography is steadily increasing, and a growing number of therapeutic drugs, which are either the direct result of structure based discovery or where structure guided lead optimization has played at least a significant role, are reaching the market.[11] Among those drugs are the well-publicized human immunodeficiency virus (HIV) protease inhibitors amprenavir (Agenerase®) and nelfinavir (Viracept®), which were developed with knowledge of the crystal structure of HIV protease;[12] the influenza drug Tamiflu® (Oseltamvir);[13] and the onco-therapeutic Gleevec® (Imatinib).[14] Drug target structures are therefore frequently encountered as illustrative examples in this book (Sidebar 1-1).

X-ray diffraction is a fundamental technique. Analysis of the first X-ray diffraction patterns, obtained in 1912 by Walther Friedrich and Paul Knipping in Max von Laue's laboratory (Sidebar 6-7) from crystals of simple compounds such as diamond, rock salt, or zinc sulfide, confirmed in a fundamental way the atomic constitution of matter and the interactions and bonding of the atoms. Given that the atom theory and particularly quantum mechanics were then in their infancy, these diffraction experiments provided the much needed crucial evidence and support for the development of the atomic theory of matter. Interestingly, shortly after X-rays were discovered by Wilhelm Conrad Röntgen in 1895 (Sidebar 6-1) and before the first diffraction patterns of simple compounds were recorded, it was already known that also proteins could form crystals. It would take more than three decades of technical developments until the first diffraction patterns of protein crystals could be recorded and another two decades until the successful determination of the first macromolecular structure, myoglobin, in 1957 (Sidebar 10-1). Many more relevant historic tidbits and anecdotes are provided in sidebars throughout this book.

Macromolecular crystallography is no trivial pursuit. Although we take modern biomolecular crystallography for granted as a mature tool, the method itself is less than 100 years old and only with the advance of computers has it become possible to tackle molecular structures the size of a protein. While today we are spoiled by recombinant DNA techniques, robotic automation, and easy access to powerful tunable synchrotron X-ray radiation sources,[15] the first protein crystallographers were true pioneers and their efforts to obtain the first protein structure were no less than heroic. Not without reason, the field of crystallography and structural biology is rich with Nobel Laureates.

Sidebar 1-1 List of selected macromolecular structures presented in illustrations.

Bovine pancreatic trypsin inhibitor, Figure 2-1

Ribosome 50S subunit, Figure 2-1

Clostridium botulinum neurotoxin (holotoxin), Figure 2-1

PRD1 bacteriophage head, Figure 2-2

Human apolipoprotein E4, Figure 2-11

Concanavalin A, a plant lectin, Figure 2-15

Ribonuclease A, Figure 2-16

Ornithine decarboxylase complexed with DFMO, Figure 2-30

Ribonuclease inhibitor, Figure 2-35

Cyclooxygenase 2 (COX2), Figure 2-38

Bacterial membrane porin, Figure 2-40, 3-44

GABA$_A$ pentameric neuroreceptor, Figure 2-41

B-DNA decamer, Figure 2-44

Zn-finger complexed with B-DNA, Figure 2-46

p53 tumor suppressor complexed with DNA, Figure 2-47

Mycobacterium tuberculosis InhA complexed with isoniazid, Figure 2-48

Nucleosome core particle, Figure 2-49

Ferritin, thaumatin, Figure 3-16

Trypsin complexed with benzamidine, Figure 3-41

Clostridium tetani neurotoxin, ganglioside binding domain, Figure 3-42

KcsA potassium channel, Figure 3-46, 3-47

Corynebacterium diphtheriae, DtxR repressor complexed with duplex DNA, Figure 3-48

Bacillus subtilis organic hydroperoxide-resistance protein OhrB, Figure 4-4

Green fluorescent protein, engineered, Figure 4-14

Apolipoprotein E3, 22 kDa LDL receptor binding domain, Se-Met mutant, Figure 4-15

Mistic membrane insertion fusion protein, Figure 4-16

OprP, a trimeric bacterial outer membrane protein, Figure 4-18

Bacillus anthracis alanine racemase, Figure 4-19

Quercetin 2,3-dioxygenase, glycosylated, Figure 4-20

Hepatocytic growth factor complex, Figure 4-25

Recombinant glucosylceramidase, defective in Gaucher's disease, Figure 5-4 ff., 6-16

Mycobacterium tuberculosis rhamnose epimerase RmlC, drug target, Figure 5-21, 11-6

Chicken calmodulin, Figure 5-22, 11-7

SARS coronavirus ancillary protein, Figure 5-33

Staphylococcus aureus superantigen toxin-like fragment, Figure 5-34

SH2 domain of the C-terminal human Src kinase (Csk), Figure 5-41

European mistletoe viscotoxin A1, Figure 10-35

Antibody F$_{ab}$ fragment against human LDL receptor, Figure 11-2

GTP cyclohydrolase I complex with its feedback regulatory protein GFRP, Figure 11-12

Rhodobacter sphaeroides cytochrome c' dimer, Figure 11-22

E. coli medium chain length acyl-CoA thioesterase II (TesB), Figure 12-40

Antibody F$_{ab}$ fragment against supersweetener, Figure 13-13

Cellular retinoic acid binding protein, CRABP, Figure 13-16

Sporobolomyces salmonicolor carbonyl reductase, Figure 13-14

Human acetylcholinesterase, AChE, Figure 13-15

E. coli small-conductance mechanosensitive channel, MscS, Figure 13-23

Primary literature references are provided at their respective locations. Many more annotated and illustrated examples for important macromolecular structures may be found in the "Molecule of the Month" column edited by David Goodsell on the Rutgers Protein Data Bank (RCSB-PDB) web site.

Remote facility access. New crystallographic techniques and methods, as well as easy access to remote synchrotron data collection, have made it possible for practically every structural biology laboratory—given proper training—to reliably pursue crystallographic structure determination projects. The apparent ease with which complex biomolecular structures can be obtained, with the help of remote synchrotron data collection from submitted crystals (referred to as "FedEx crystallography") and advanced (but not foolproof) software tools, has occasionally given rise to the casual attitude that macromolecular crystallography is a simple biochemical technique on a par with running an SDS gel or recording an absorption spectrum. This notion is gravely mistaken (Sidebar 1-2), as the preview of the practical and fundamental challenges in the next section, as well as selected examples of cautionary tales from the literature dispersed in appropriate places throughout the text, will testify. The great technical advances

Figure 1-1 Growth of the Protein Data Bank. The cumulative number of entries in the Protein Data Bank (PDB) plotted against the year of deposition follows quite well Richard Dickerson's formula (boxed in the upper left corner) predicting an exponential growth law in analogy of Moore's law for packing density of electronic microchips.[16] Only two protein structures were deposited at the PDB (then hosted in Brookhaven) in 1972, but by the late 1970s, researchers had determined the crystal structures of 132 proteins. In 1978, Richard Dickerson, a pioneer in protein and DNA crystallography and then a professor of physical chemistry at Caltech, examined the number of available protein structures and derived the equation predicting the number of structures to be deposited in the future. You can test for yourself based on current deposition numbers how surprisingly accurate Richard Dickerson's early prediction remains. Figure courtesy of Christine Zardecki,[17] RCSB Rutgers.

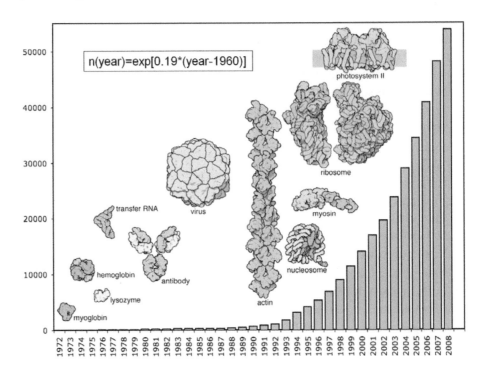

$$n(year)=\exp[0.19*(year-1960)]$$

resulting from efforts of various Protein Structure Initiatives worldwide have even led to the opinion that nowadays one may simply submit the amino acid sequence of the protein to one of the Protein Structure Centers, and in due course structure coordinates will be returned. While this approach may succeed for simple bacterial proteins, the complexity of eukaryotic or human proteins almost always requires significantly more effort. Some of the complications affecting eukaryotic proteins and membrane proteins and how to approach them are discussed in Chapter 4.

A major part of the process of any structure determination, and the key to success, is proper planning; a clear view is needed of the objectives of the structure determination, in particular what information at what level of detail will be necessary to answer your specific question or to test your specific biomedical hypothesis, and how to reach the specified goals efficiently. The following provides a brief introduction to the method of crystallographic structure determination, outlining key stages in a well-planned protein structure determination. The rewards of understanding the technical details of macromolecular crystallography become obvious when the process turns difficult and structure determination is less straightforward. Having command over this most powerful technique of structure determination will enhance your confidence and also allow you to take responsibility for the extraordinarily persuasive power conveyed by a crystallographic structure model.

Box 1-1 Crystallography provides the foundation of modern structural biology. The power of macromolecular crystallography results from the fact that highly accurate models of large molecular structures and molecular complexes can be determined, often at a near atomic level of detail. Crystallographic structure models have provided insight into molecular form and function, establishing the basis for structural biology and structure-guided drug discovery. Non-proprietary protein structure models are made available to the public by deposition in the Protein Data Bank, which holds more than 57 000 entries as of May 2009.

1.2 Principle and challenges of crystallography: A preview

The basic single-crystal diffraction experiment is deceptively simple: A single crystal of the material of interest is placed into a finely focused X-ray beam, and the diffraction images are recorded. The electron density representing the atomic structure of the molecules in the crystal is reconstructed by Fourier methods from the diffraction data, and an atomic model of the structure is built into the electron density. Figure 1-2 shows the basic principle, while photographs of actual modern X-ray diffractometers can be found in Chapter 8. While the above description of the diffraction experiment appears quite straightforward, there are a number of practical and conceptual challenges that need to be addressed. Various nontrivial principles form the basis of the extraordinary power of crystallography, making it not just a most important tool of structural biology but an attractive field of study in its own right.

Practical challenges. The price to be paid for obtaining informative high resolution X-ray structures is that a well-diffracting protein crystal needs to be produced; the obvious practical issue is that without crystals there is no crystallography. Proteins, nucleic acids, or molecular complexes thereof—by their nature as large and flexible macromolecules—seldom self-assemble readily into the regular, periodically repeating arrangements typical for crystals. Indeed, growing well-diffracting protein crystals can prove quite challenging; we dedicate the whole of Chapter 3 to the principle and practice of growing protein crystals. Moreover, even obtaining the material to be crystallized in the first place is nontrivial; more often than not the protein needs to be modified or engineered so that it in fact can crystallize, as elaborated in Chapter 4. An additional complication during the data collection process (Chapter 8) originates from the fact that biological material, such as protein crystals, is highly susceptible to radiation damage by the intense ionizing radiation it is exposed to during the X-ray diffraction experiment. To prevent decay of the crystals during X-ray exposure, cooling of crystals to cryogenic temperatures (somewhat above the boiling point of liquid nitrogen of 77 K or −196°C) is near universal practice. However, many crystals are difficult to flash-cool and it is hard to predict which cryoprotection conditions will work.

Fundamental challenges. The diffraction process is fundamentally different from microscopic imaging; crystallography is not an imaging technique. The crucial difference is that visible light scattered from objects can be focused through refractive lenses to create a magnified image of the object. This is not the

Figure 1-2 The principle of X-ray structure determination. A crystal mounted on a goniostat with at least one rotatable axis is exposed to a finely collimated, intense X-ray beam in the 5–20 keV energy range (~2.3 to 0.6 Å wavelength). Individual diffraction images are recorded on an area detector during small rotation increments of the crystal and combined into a diffraction data set. However, the diffraction images are not direct images of the molecule. Diffraction images are transforms of the molecular shape into reciprocal space—equivalent to being transformed into "secret code." This secret reciprocal space code must be deciphered for each structure by the crystallographer. The basic mathematical tool of back-transformation from reciprocal diffraction space into direct molecular space is the Fourier transform (FT), which together with separately acquired phases for each diffraction spot allows synthesizing or reconstructing the electron density (blue grid) of the molecules self-assembled into the diffracting crystal. An atomic model of the structure, represented in the figure by a ribbon model, is then built into the 3-dimensional electron density. The absence of phases in the diffraction data is the origin of the phase problem in crystallography and a suitable phasing strategy needs to be developed for each structure determination.

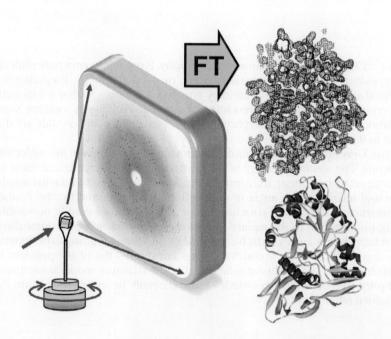

Sidebar 1-2 **So easy a monkey can do it?** "A certain portion of the research community has tended to regard crystal structure analysis as entirely too easy and the value of crystallographic results has been challenged as being irrelevant beyond the solid state, despite countless examples of correlations between structural features observed in the solid state and the chemical, physical and biological properties of the same molecules *in vitro* and *in vivo*, as measured by a wide range of techniques. The rapidity with which new structures are determined and the rate at which crystallographic databases are expanding fuels the argument that crystallography is easy and crystallographic instrument manufacturers reinforce this opinion by claiming that with today's instrument and software anyone can determine a crystal structure. Even Judith Howard, past president of the British Crystallographic Association, was quoted out of context as saying 'instruments have been so advanced that you sometimes feel you can train a monkey to use it'. Unfortunately, a monkey so trained would know about as much about the underlying phenomena of diffraction, the proper use and analysis of diffraction data and the structural information it provides as many other current users who lack formal crystallographic training." Bill Duax, Editor, *International Union of Crystallography Newsletter* **9** (2) 2001.

case for X-rays, which are also electromagnetic radiation but of several orders of magnitude shorter wavelength and correspondingly higher energy (Chapter 6); the refractive index of X-rays in different materials is essentially equal and close to unity, and no refractive lenses can be constructed for X-rays. Instead, the electron density of the scattering molecular structure must be reconstructed by Fourier transform techniques (Chapter 9). In itself, the Fourier reconstruction from reciprocal diffraction space back into direct molecular space poses no difficulties in principle, with the unfortunate qualifier that for this type of reconstruction *two* terms are needed as Fourier coefficients: the structure factor amplitudes, readily accessible in the form of the square root of the measured and corrected diffraction spot intensities; and as a second term for each observed diffraction spot, its relative phase angle. These phase angles are not directly accessible and must be supplied by additional phasing experiments (Chapters 10 and 11). The absence of directly accessible phases constitutes the phase problem in protein crystallography and it is the reason why protein structure determination can be quite difficult; it is also theoretically demanding, which makes protein crystallography in itself a very interesting field of active basic research. Finally, each atom of the structure contributes to each diffraction spot or reflection in a complex *nonlinear* way. As a consequence, the refinement of the structure model—adjusting it so that its model parameters (the positional coordinates and a B-factor for each atom as a measure of atomic mobility or general displacement) best describe the observed data—is nontrivial. The number of measured reflections seldom sufficiently exceeds the number of adjustable model parameters to allow free refinement, and prior stereochemical knowledge must nearly always be incorporated into macromolecular refinement in the form of stereochemical restraints that keep the structure model within physically reasonable bounds. One can readily appreciate why macromolecular crystallography is nontrivial (Sidebar 1-2).

1.3   Protein molecules and the crystalline state

The native location of a protein molecule is not the ordered solid state but generally some fluid environment, often an aqueous solution. The intracellular cytosol and the extracellular environment are generally quite crowded with other molecules, small and large. Cell membranes harboring receptors and channels are also fluid assemblies, with parts of the protein molecules extending into the cytosol and/or extracellular fluid. The question arises of how the fact that proteins have self-assembled into solid crystals affects the interpretation of the resulting crystal structures and their biological relevance with respect to their

Box 1-2 **Challenges of protein crystallography.** Proteins are generally difficult to crystallize and without crystals there is no crystallography. Preparing the material and modifying the protein by protein engineering so that it can actually crystallize is nontrivial. Prevention of radiation damage by ionizing X-ray radiation routinely requires cryocooling of crystals and many crystals are difficult to flash-cool.

The X-ray diffraction patterns do not provide a direct image of the molecular structure. The electron density of the scattering molecular structure must be reconstructed by Fourier transform techniques. Both structure factor amplitude and relative phase angle of each reflection are required for the Fourier reconstruction. While the structure factor amplitudes are readily accessible, being proportional to the square root of the measured reflection intensities, the relative phase angles must be supplied by additional phasing experiments. The absence of directly accessible phases constitutes the phase problem in crystallography. The nonlinear refinement of the structure model is nontrivial and prior stereochemical knowledge must generally be incorporated into the restrained refinement.

native solution state. As explained in great detail in Chapter 3, protein crystals are not rigid bricks. They actually are formed by a loose periodic network of weak, non-covalent interactions (Figure 3-5) and contain large solvent channels which allow relatively free diffusion of small molecules through the crystal. This solvent access maintains a solution environment (which is actually a requirement for protein crystals: once they dry out they almost invariably stop diffracting) that accommodates quite some conformational freedom for surface-exposed side chains or loops as well as some "breathing" motions of the structure core.

Enzyme activity and the solid state. Comparison of many nuclear magnetic resonance (NMR) solution structure ensembles with crystallographic structure models has shown that the core structure of protein molecules remains unchanged[18] compared with the solution state during crystallization. In addition, enzymes packed in crystals even maintain biological activity. The maintained activity in crystals actually creates a challenge for the crystallographer, often necessitating the design of inactive enzyme substrate analogs or substitutes in order to dissect the molecular reaction mechanisms.

Molecular flexibility. The maintenance of the core structure and of enzymatic function shows that crystal structures are a very good approximation of the native protein solution structure. Nonetheless, highly flexible or mobile regions, frequently the amino- or carboxyl-termini of the protein chain or flexible loops connecting secondary structure elements, can be poorly defined or even absent in the electron density and thus can be modeled only with limited confidence. This situation is also reflected in the fact that in NMR structures, highly dynamic and flexible regions exhibit large deviations between the multiple models of an NMR ensemble. In either case, the absence of a well-ordered structure is a genuine reflection of the dynamic behavior of the protein molecules and not a weakness of either technique. NMR and X-ray crystallography in fact complement each other,[19] and refinement of molecular models against data from both of the methods combined can ultimately give a more accurate structure than either method alone.[20] The advantage of crystallography, however, is the more detailed information in well-defined parts of the molecule and in principle no size limitation, which restricts solution NMR to molecules below ~35 kDa.

Crystal packing affects local regions. In certain situations flexible and dynamic regions of a protein molecule can be rigidly fixed in a specific conformation as a result of crystal packing interactions. In most cases this represents just a snapshot of one possible conformation out of many and it must be understood that such a specific conformation may not represent the local protein structure in solution. A simple safeguard against misinterpretation—which is usually assignment of certain biological relevance to regions where that is *de facto* not warranted—is to display all neighboring, symmetry-related molecules in the crystal structure and examine if any intermolecular interactions are present that are a result of crystal packing. Such packing induced artifacts can also hamper for example drug discovery by altering or blocking binding sites and thus preventing an otherwise active substance from binding. Binding properties can be altered either directly through intermolecular contacts of key residues with neighboring protein molecules, or through normally flexible loops covering the binding site and thus preventing access; or indirectly by affecting allosteric binding sites of the protein. It is good practice to examine the possible presence of packing effects in a crystal structure before attempting any ligand docking or drug lead optimization.

Large conformational changes destroy crystals. The fact that protein molecules are periodically packed in a crystal lattice clearly places limitations on the direct observation of processes involving large conformational changes, which invariably destroy the delicate molecular packing arrangement of a protein crystal. Molecular transport processes or interactions involving extended conformational rearrangements therefore require multiple, stepwise "snapshot" structure determinations in order to dissect the details of such inherently dynamic processes. Enzymatic reactions involving limited, local structural changes in protein conformation can sometimes be elucidated by technically challenging time-resolved X-ray diffraction studies.[21]

Dynamic behavior of molecules. Although no large-scale dynamic movements can take place in a crystal, there is in fact some information about the dynamic behavior of the molecules present in the crystal structure. High B-factors for an atom, a generic displacement measure indicating how well the model is defined in that region, are indicative of flexibility or disorder in that local region. Analysis of concerted movements of entire regions of the molecule by TLS parameterization during refinement can give indications for the propensity toward even substantial domain movements. The unqualified critique that protein structures are just a static snapshot is thus oversimplifying and often a manifestation of ignorance on the part of the model user. We will discuss the finer details of the implications and analysis of the dynamic behavior of proteins in Chapter 12.

Protein structure quality is locally defined. The most important fact to understand is that crystallographic protein structure models naturally have very clear and well-defined regions, while disordered parts can reflect dynamic behavior, increased flexibility, or static and displacement up to complete local disorder. Certain structure types, for example helix bundles or globular proteins, tend to form very stable and well-defined structures, while others, no matter how proficient the crystallographer, will always remain less well defined because of their inherent dynamic flexibility. Local analysis of protein crystal structures will be treated throughout the text and more deeply in Chapter 13.

Box 1-3 Crystallographic structure models versus proteins in solution. Protein crystals are formed by a loose periodic network of weak, non-covalent interactions and contain large solvent channels. The solvent channels allow relatively free diffusion of small molecules through the crystal and also provide conformational freedom for surface-exposed side chains or loops. The core structure of protein molecules in solution as determined by NMR is identical to the crystal structure. Even enzymes generally maintain activity in protein crystals. Crystal packing can affect local regions of the structure where surface-exposed side chains or flexible surface loops form intermolecular crystal contacts. Large conformational movements destroy crystals and cannot be directly observed though a single crystal structure. Limited information about the dynamic behavior of molecules can be obtained from analysis of the B-factors as a measure of local displacement or by analysis of correlated displacement by TLS (Translation-Libration-Screw) analysis. The quality of a protein structure is a local property. Surface-exposed residues or mobile loops may not be traceable in electron density, no matter how well defined the rest of the structure is.

1.4   Interlude: Molecular cryptanalysis

The computer-savvy student of today may find it entertaining to view crystallography as the cryptanalysis of reciprocal space—decoding the secret molecular message carried by the diffraction data back into direct molecular space. A number of quite interesting parallels between crystallography and cryptanalysis exist, and in periods of despair it often helps to entertain the perspective of decoding a riddle or secret message. Technically speaking, each molecule is encoded by a single-use encryption key and each structure determination requires one to find its particular key. Sometimes finding the right key or methods may be straightforward, but on occasion it will not be trivial at all.

The first step in code breaking just as in crystallography is to intercept the message. For us this means data collection—without diffraction data, there can be no structure determination. Let us assume we have intercepted the following cipher conveying a text message relevant to macromolecular crystallography:

```
19 17 17 19 14 21 17 16 18 24 16 19 15 18 24 08 22 03 12 18 06 03 04 06 22 12
18 14 07 12 19 08 08 18 24 12 19 14 18 06 18 26 17
```

The first requirement for cracking the code is that we understand its nature and its language. We readily recognize that the code elements are limited to numbers not exceeding 26, so it is reasonable to assume that we are dealing with some simple one-to-one encryption of an alphabetic message. Given that this text is in English, we furthermore assume that the message is also in English. The same necessity to understand the nature and language of the code exists in crystallography and Part II, *Fundamentals of Protein Crystallography*, will establish the basics necessary for the understanding of crystal geometry, reciprocal space, scattering of X-rays by molecular crystals, and the analysis of the resulting X-ray diffraction data.

We can take our first shot at deciphering the code by conducting a simple statistical analysis of the probability distribution of the code elements (Figure 1-3A). We know that the most frequent letter in English text is the letter *e*, and we can assign in a first trial the letter *e* to the corresponding number 18, similar to a trial seeding position in crystallographic substructure solution methods. Notice how important it is to intercept the code correctly (i.e., to collect your data accurately); one simple transcription error swapping one 18 with 19 completely changes the trial assignment (Figure 1-3B). Statistical methods play an important role in crystallographic analysis (Chapter 7) and phasing methods are particularly sensitive to poorly measured data—accurate data collection is important. Note, however, that a code number of say 58 instead of 18 would be easily recognizable as highly improbable value or error, equivalent to the concept of outlier detection commonly applied in crystallographic data processing.

A statistical technique we can apply now to improve our first trial "structure" is based on pattern recognition. Take for example the first four code numbers; they show a very peculiar pattern:

```
19 17 17 19 14 21 17 16  E 24 16 19 15  E 24 08 22 03 12  E 06 03 04 06 22 12
 E 14 07 12 19 08 08  E 24 12 19 14  E 06  E 26 17
```

These symmetric code numbers very likely do not stand for all consonants (C) or all vowels (V) (you may struggle to construct a sensible combination) but rather a combination starting with CVVC or VCCV. We can search now through a dictionary and try all possible solutions—a multi-solution approach—and pick the one that makes the most sense (i.e., scores the highest). One of the first hits is ATTACK, which expands our code as follows:

```
A  T  T  A  C  K  T 16  E 24 16  A 15  E 24 08 22 03 12  E 06 03 04 06 22 12
 E  C 07 12  A 08 08  E 24 12  A  C  E 06  E 26  T
```

Figure 1-3 Statistical analysis of encoded data. The frequency of the code symbols is compared with the expectation values for letters. In panel (A), the most frequent code symbol 18 is correctly identified with the letter e. A simple transcription or measurement error (B) changing 18 to 19 completely changes the relative frequencies and serves as a reminder that diffraction data need to be accurately collected, which is particularly important for experimental phasing methods to succeed.

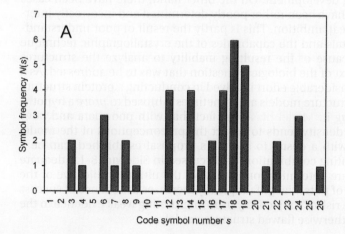

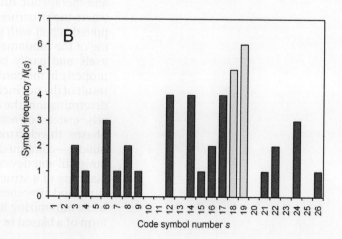

Pattern recognition and image seeking methods are found in various flavors in crystallography. The most common application is to model building, where, based on shape compatibility, the electron density is recognized as belonging to a certain secondary structure element such as an α-helix or β-sheet, or representing a side chain corresponding to a certain kind of amino acid residue. Automated methods, as well as the crystallographer during manual model building, rely on pattern recognition. While the individual often bests the computer in recognizing matching elements, the computer succeeds by brute force through trying many more possible solutions.

Applying pattern recognition techniques was a great step forward for our code cracking and now we apply prior knowledge about the context of the message to further analysis. Prior knowledge or the prior probability of a structure model plays an important role in crystallography because it helps us to judge how well our crystallographic model complies with the body of independently established scientific knowledge. This helps to make sure that despite all its new and exciting features, our structure model remains within physically reasonable bounds, complying with stereochemical restraints and other fundamental laws of nature.

Given that our encrypted message is of some importance for bio*molecular* crystallography, the sequence

```
06 22 12   E   C 07 12   A 08
```

attracts our attention—it probably means MOLECULAR. We further complete our clear text (the initial structure model) now and obtain the following message:

```
A   T   T   A   C   K   T 16   E 24 16   A 15   E 24   R   O 03   L   E   M 03 04   M   O L
E   C   U   L   A   R   R E 24   L   A   C   E   M   E 26   T
```

What remains now is to complete, refine, and polish our model by filling the remaining gaps. It is left up to the reader to finish the cryptanalysis of this message referring to a common method of solving the crystallographic phase problem.

1.5 Planning and executing a protein structure determination

No more than two decades ago, the pursuit and successful determination of a single protein structure by multiple isomorphous replacement phasing could earn a Ph.D. and warrant a high impact publication. Today, the technique has matured to a point that in many cases structural biologists with little specific training in crystallography can quite successfully determine a protein structure. On the one hand, this has tremendously enriched our knowledge about fundamental biochemistry, reaction mechanisms, or the molecular basis for disease and therapeutic drug development. On the other hand, there have been cases where the structures have been rather poorly determined and incorrectly interpreted, albeit with great ambition. This is partly the result of poor understanding of the fundamentals and the capabilities of the crystallographic technique itself, and partly because of the resulting inability to analyze the structures properly in the context of the biological question that was to be addressed. As a result of the not inconsiderable effort involved in conducting a protein structure determination, the structure models are sometimes (ab)used to *prove* a hypothesis, instead of *testing* it. Model bias—the fact that with poor data and poor phases, the electron density tends to reflect the preconceptions of the model builder—combined with a desire to see one's proposal established, can be a powerful and devastating combination, as discussed in Sidebar 13-1. Adequate planning of a structure determination in view of the ultimate objective of the study and awareness of potential pitfalls greatly increases the chance of success while minimizing the risk of subjectively over-interpreting weak evidence in the form of a biased or otherwise flawed structure model.

Project planning and alternatives

Any actual protein crystallography study starts with a considerable amount of information gathering and planning. Just as anywhere in research, "a year in the laboratory can save a day in the library". The information gathering stage includes, besides comprehensive literature searches and bioinformatics, the response to questions that need to be asked in the broader context of maximizing the return on the considerable investment of time and effort your structure determination project presents. Figure 1-4 illustrates the major stages of a protein structure determination project.

There is *a priori* no guarantee that a crystal structure study will succeed. It can fail at many stages and in addition to a sound scientific hypothesis to be proposed or tested by the structure model, any research plan should include alternate options (or projects) if the experiment cannot be completed as planned. While the up-front molecular biology and cloning have quite reasonable success rates, protein production and protein crystallization remain major challenges. In both cases, initial average success rates are around 30% and data from the NIH Protein Structure Initiative have shown that the chances of obtaining a crystal structure of an average globular bacterial protein in one straight path without modification to the initial strategy are only 10–20%. On the other hand, once a diffracting crystal is obtained, the chances are very good that structure determination will succeed given a proper phasing strategy. Ultimately, the analysis of the model will show whether the structure actually answers the initial question. Quite often, however, serendipity plays a significant role and many exciting breakthroughs (and *Nature* papers) in structural biology are the result of unexpected features discovered and supported by the analysis of a well-determined crystal structure model.

Certain categories of structure determination projects are inherently risky. The high impact of an integral membrane protein structure comes at a price and one must be prepared to spend perhaps years to determine just one single structure— if at all successful. It is advisable in such high risk/high impact situations to negotiate with the supervisor or laboratory head an alternate completion target for a thesis or assignment. On the other hand, it cannot be expected that a basic molecular replacement structure of a mutant of a well-known household

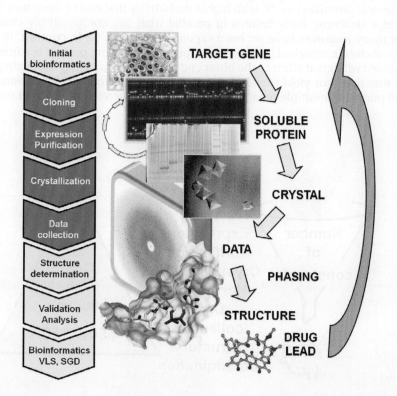

Figure 1-4 Overview of protein structure determination. The bar on the left side of the figure lists major stages of a crystal structure determinaton project. The dark blue shading indicates experimental procedures while the light shading indicates work performed in-silico on computers. The results of the structure analysis frequently feed back into the design of a refined study, particularly in structure guided drug discovery. VLS: virtual ligand screening; SGD: structure guided drug discovery. Consult Figure 1-8 for a more detailed diagram of key steps in structure determination and the corresponding Chapters in this book.

enzyme, of which there are already many structures deposited in the PDB, will yield a high impact publication or a Ph.D. thesis.

Keeping track of the goal

Even before technicalities of cloning, protein production, crystallization, and actual structure determination are detailed, the first step in planning is to assert what qualities the final structure model must have in order to address your underlying question. Not all crystallographic structure models are created equal: A drug discovery study will necessitate a much higher level of detail (see Figure 1-6) than the determination of the novel fold of a newly discovered protein or speculation about conformational rearrangements in a huge complex of cellular secretion machinery.[22] When the minutiae of a binding mechanism are to be determined, cofactors or substrate analogs generally must be present in a high quality structure. Comparison between the apo-structure and the complex with bound ligands may also be informative. Dynamic or flexible structures may need multiple structure determinations in different crystal forms to inform about the range of conformations that can be assumed, and large multi-domain structures may have to be expressed and crystallized as separate domains, while a hypothetical model of the holo-complex can be assembled *in silico*. Protein engineering (Chapter 4) is quite often a necessary prerequisite to facilitate crystallization. Some crystal forms may also exhibit packing contacts that can hinder the binding of ligands or make biological interpretation otherwise difficult.

Automation and parallel approaches

The major preparatory steps of a structural study, namely protein expression and protein crystallization, are experimentally driven and non-deterministic, in the sense that the outcome cannot be predicted from first principles. Multiple expression and crystallization trials under different conditions are inherently suitable for parallelization and automation. Robotic crystallization screening is already an established procedure in industry as well as in academic laboratories and parallel expression and purification strategies are being adopted in structural biology laboratories. The primary benefit of automated *parallel* approaches is the broad basis from which experiments are started, and a failure of an individual protein construct to express or crystallize is not fatal to the entire project. A multi-pronged, parallel approach also allows a quick selection of several promising leads, with higher probability that one of them may finally yield a structure. Early failures in parallel trials are comparatively cheap, as not many resources have yet been expended (Figure 1-5). In contrast, in *serial* approaches, a marginal lead such as a poorly expressing protein is sometimes pursued with great effort to the bitter end only to yield a weakly diffracting crystal that does not yield usable data or sufficient information. Broad screening and pursuit of multiple parallel constructs greatly increase the odds of success.

Figure 1-5 Keeping the funnel full. Given the high attrition rate in early experimental steps of a protein structure determination project, a parallel approach pursuing multiple variants of the protein target significantly increases the chances of success. Parallel approaches are also less susceptible to late stage failures, because the pressure to pursue one single protein construct that already shows warning signs to the bitter end is diminished. The most expensive failure is a perfectly good diffraction data set that cannot be phased, because all the costly and time-consuming laboratory experiments have been conducted already. Developing a suitable phasing strategy early on is thus part of good study planning.

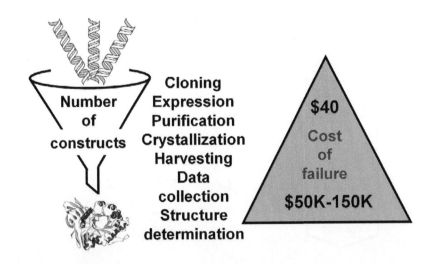

Protein engineering

Parallel approaches beginning with cloning and preparation of different protein constructs have additional advantages. Despite common perception, protein crystallization is not the biggest challenge in crystallography. The challenge begins much earlier, namely to produce a protein that actually *can* be crystallized. The crystallizability of a protein is to an overwhelming degree determined by the properties of the molecule itself. Intermolecular interactions between the molecules must be favorable toward self-assembly into a well-formed, periodic crystal lattice. If the protein either lacks suitably located surface residues, or has structural features such as large termini or flexibly tethered domains that prevent it inherently from crystallizing, then no amount of crystallization screening will ever yield usable crystals. Clearly, starting from a variety of protein constructs, usually with different affinity tags, possibly limited truncation, or stable sub-domains increases the chances of success. In a similar horizontal approach, orthologs from different species provide variants already engineered by nature as an alternate pathway. There is good reason to discuss protein production (Chapter 4) for the purpose of a crystallographic study *after* we have examined the principles of crystallization in Chapter 3. The requirements a protein must meet to be suitable for crystallization—particularly as regards conformational homogeneity—are generally much more stringent than they are for a functional or biochemical assay.

Crystal growth

Once a reasonably pure, soluble, and stable protein is obtained, crystallization screening begins. Despite the fundamental molecular complexity of the crystallization process outlined in Chapter 3, the actual setup of the common crystallization experiment is simple and easy to automate. Numerous crystallization techniques have been developed, each with different merits depending on the specific purpose (initial screening, optimization of growth, ease of harvesting, ease of automation) and type of research environment. Maximizing comprehensive sampling of crystallization space with the least amount of material plus ease of miniaturization favors a robotic setup of nano-drops,[23, 24] and sampling statistics play a significant role in the design of efficient crystallization screening strategies.

Crystallization trials are observed under a microscope at regular time intervals and the experimenter, sometimes assisted by image recognition software, decides whether a potentially suitable crystal has been obtained. In many cases, the initial crystallization conditions need to be further refined or optimized before a well-diffracting crystal is obtained.

Harvesting, cryocooling, and mounting of crystals

Another critical experimental step, and probably the least systematically explored, is harvesting and cryocooling of the crystals. Protein crystals that have grown in the crystallization trials must be harvested from their mother liquor and mounted on a diffractometer for data collection. There are numerous hazards associated with this procedure. Protein crystals are very small, ranging in size between few 100 μm to about 10 μm in size, and because of the high solvent content and weak, non-covalent intermolecular interactions they are very fragile. Like every biological material, they are susceptible to severe radiation damage in the intense X-ray beams needed for data collection. Preventing damage from exposure to the ionizing X-ray radiation is the foremost reason to cool the crystals to cryogenic temperatures, where damage from the ionizing radiation is at least greatly reduced. During the rapid flash-cooling (or quenching) to liquid nitrogen temperature, the formation of crystalline ice in the mother liquor surrounding the crystal must be avoided. Crystalline ice formation invariably destroys the delicate protein crystals.

The reason why cryocooling procedures are rarely systematically investigated is easy to understand: Once a cryocondition is found that yields a diffracting crystal and good data are collected, there is little incentive to mount the remaining crystals or to grow or flash-cool more of them for systematic studies. In

Sidebar 1-3 Checklist of key typical questions to consider during the planning of a structure study.
- What is the primary purpose of the structure determination?
- What degree of detail (resolution) do I need to meet study objectives?
- Does sequence analysis indicate any anomalies, such as intrinsic disorder?
- Does the molecule contain multiple domains? Can they be expressed separately?
- Are structures with significant sequence identity already known, which I may use for molecular replacement phasing?
- Are there any orthologs in other species available?
- Is there a cloning system available with minimal need for redesign and subcloning?
- In what systems can the gene be expressed: bacteria, yeast, insect cells, mammalian cells?
- Is the expression system suitable for selenomethionine labeling for experimental phasing?
- Can posttranslational modifications or decorations be of importance? Can they be removed/prevented?
- Are a significant number of disulfide bridges present that might prevent folding in bacterial expression systems/reducing conditions?
- How can I tag the protein for affinity capture?
- Does the purification strategy introduce reagents that may affect crystallization?

- Do I need to co-purify or co-express in the presence of binding partners, ligands or cofactors, or ions?
- In the absence of any diffracting crystals, are salvage strategies available or is protein engineering possible?
- Do I need to soak metal ions for heavy atom phasing or anomalous phasing? Ligands? Cofactors?
- Do I have prior information to guide crystallization screening?
- Do I have means for physicochemical post-mortem analysis such as light scattering if crystallization fails?
- Do I have synchrotron time available when the crystals are ready?
- Are all the crystallographic programs in place?
- Do I have the analysis tools and programs?
- Are resources available for pursuing parallel approaches?
- Which options exist to redesign the study if the results are inconclusive?
- Are alternate structure determination methods such as NMR or cryo-EM available?
- Can additional biochemical experiments be proposed based on the structure analysis that further support the structure based hypothesis?

Many similar questions will be specifically addressed in the corresponding strategy sections of Chapter 3 (Protein crystallization) and Chapter 4 (Proteins for crystallography).

addition, the delicate micro-manipulations vital to crystal harvesting are hard to automate. Remember that crystals are expensive; losing them amounts to a correspondingly costly late-stage failure.

Diffraction data collection

Once a diffracting crystal has been harvested and mounted in a cryo-loop, the actual process of the structure determination begins. In contrast to expression and crystallization screening of biological material, which more or less depend on trial and error experimentation (regardless of the clever screening strategies deployed), the path from data to structure can be understood from physical first principles. Once the data are collected, all subsequent steps—data processing, phasing calculations, electron density reconstruction, model building, structure refinement, validation and analysis—are conducted *in silico* with computer programs.

Information content of data. The diffraction pattern images recorded on area detectors are indexed, integrated, and scaled, and unit cell and space group are determined as discussed in Chapter 8. A reduced and hopefully complete data set, essentially representing a periodically sampled reciprocal space transform of the molecules in the crystal, is obtained. The extent to which the crystal diffracts (well-ordered crystals diffract to a higher resolution than poorly ordered ones) directly determines how detailed the final reconstruction of the electron density will be, and hence, ultimately and without mercy, how detailed and accurate the resulting model can be. While the exact correlation between diffraction limit and sampling density will be derived in Chapter 9, it is easy to understand the obvious benefits of high resolution. Figure 1-6 emphasizes the qualitative connection between the extent of diffraction and amount of detail discernible in the electron density reconstruction (discussed in detail in Chapter 9). Note again that crystallography does not *image* atoms or molecules; it reveals the

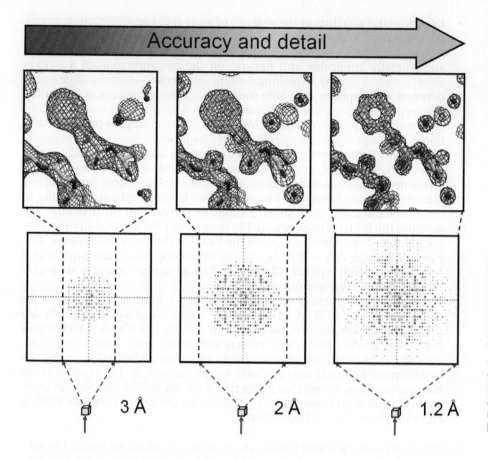

Figure 1-6 **Data quality determines structural detail and accuracy.** The qualitative relation between the extent of X-ray diffraction, the resulting amount of available diffraction data, and the quality and detail of the electron density reconstruction and protein structure model are evident from this figure: The crystals are labeled with the nominal resolution d_{min} given in Å (Ångström) and determined by the highest diffraction angle (corresponding to the closest sampling distance in the crystal, thus termed d_{min}) at which X-ray reflections are observed. Above each crystal is a sketch of the corresponding diffraction pattern, which contains significantly more data at higher resolution, corresponding to a smaller distance between discernible objects of approximately d_{min}. As a consequence, both the reconstruction of the electron density (blue grid) and the resulting structure model (stick model) are much more detailed and accurate. The non-SI unit Å (10^{-8} cm or 0.1 nm = 10^{-10} m) is frequently used in the crystallographic literature, simply because it is of the same order of magnitude as atomic radii (~0.77 Å for carbon) or bond lengths (~1.54 Å for the C–C single bond).

mathematically reconstructed electron density of all the molecules that form the protein crystal.

Phase determination

The measured intensities of the diffraction spots (or reflections) that form the data set provide only the magnitude or amplitude of the diffracted X-rays, but the phase relations between the reflections are absent (Figure 1-7). The phases for each reflection, however, are necessary to reconstruct the electron density. Hence, because of the lack of phase information in the diffraction patterns, direct reconstruction of the electron density of the molecules via Fourier transforms from the intensity or amplitude data alone is generally not possible (known as the phase problem in crystallography). Obtaining the missing phases is the conceptually most challenging part in the *in silico* segment of a structure determination project. In protein crystallography, there are two major avenues to obtain the missing phases:

- **Molecular replacement.** If a previously determined, structurally similar model is available, it can be used to calculate initial phases, which are then applied in the initial reconstruction of the electron density. The task involves finding the correct position of the search molecule in the crystal and thus the method is named molecular replacement, in the sense of repositioning (not substituting) the search molecule. Although the method allows a quick determination of initial phases, it can introduce severe model phase bias: Because phases dominate the reconstruction of the electron density, the initial structure will largely reflect the features of the search model and not the true structure, and phase bias removal methods are extensively used to build the correct model structure. Overall, about three-quarters of all structures deposited in the Protein Data Bank are solved by molecular replacement techniques exploiting a structurally similar model as a source of initial phases. Chapter 11 discusses molecular replacement in depth.

- **Experimental phasing.** In the absence of a suitable known structure model, phases must be determined *de novo* by a separate phasing experiment, hence the name experimental phasing. Experimental phasing methods are generally applicable and depend on the determination of a marker atom substructure by exploiting—often very weak—intensity differences between isomorphous data sets (maintaining the same crystal structure). Isomorphous differences are found between native and derivative crystals that contain a heavy atom soaked into the crystal, forming the basis for traditional isomorphous replacement phasing methods. Anomalous differences in data sets from crystals containing anomalous scatters—generally atoms heavier than H, C, N, and O—are most frequently used for experimental phase determination. Even the minuscule anomalous signal from natively present sulfur can be used in certain cases and, when measured carefully, to determine the sulfur atom substructure. The heavy atom substructure then provides the initial phases required to reconstruct the electron density. One popular phasing technique exploits the fact that the amino acid methionine can be replaced with selenomethionine by overexpressing the protein in a suitable expression system (Chapter 4). The Se atoms then provide a site-specific source of anomalous phasing signal.

  Experimental phasing accounts for about a quarter of all structures deposited in the PDB. A variety of experimental phasing methods are available (Chapter 10), and together with density modification techniques even a single anomalous data set may suffice to solve a protein structure. Anomalous phasing methods dominate, and even traditional isomorphous replacement methods are routinely supplemented by orthogonal phase information from anomalous diffraction data. Table 10-1 contains an overview of the most common phasing techniques (including their bewildering and sometimes MAD abbreviations).

In almost all cases, the initial protein structure phases obtained during the sub-structure phasing stage are further enhanced by various density modification techniques providing substantially improved electron density maps into which the initial protein structure model is built.

Electron density interpretation and model building

Once an interpretable electron density map (a 3-dimensional contour grid of the electron density) is obtained from improved experimental phases, a model of the protein structure must be built into the electron density. The model building is carried out using computer graphics programs that display the electron density and allow placement and manipulation of protein backbone markers and residues. Various electron density fitting and geometry refinement tools as well as automated model building programs greatly accelerate the process. Automated model building programs often provide a quite reasonable starting model, which can then be completed and polished by hand as discussed in Chapter 12.

Figure 1-7 The crystallographic phase problem. In order to reconstruct the electron density of the molecule, two quantities need to be provided for each reflection (data point): the structure factor amplitude, F_{hkl}, which is directly obtained through the experiment and is proportional to the square root of the measured intensity of the diffraction spot or reflection; and the phase angle of each reflection, α_{hkl}, which is not directly observable and must be supplied by additional phasing experiments. The methods and mathematics of electron density reconstruction by Fourier methods are extensively treated in Chapter 9.

$$\rho(x,y,z) = \frac{1}{V} \sum_{-h}^{h} \sum_{-k}^{k} \sum_{-l}^{l} F_{hkl} \cdot \exp[-2\pi i(hx + ky + lz - \alpha_{hkl})]$$

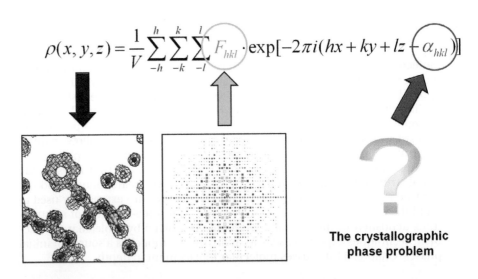

The crystallographic phase problem

Model building is often perceived as the most dreaded part of the structure determination. Model building certainly constitutes the most intense involvement in the protein structure and with poor electron density a large degree of chemical intuition and experience are required. It is here where quality data, good phases, and high resolution deliver substantial payoff: In addition to revealing much higher detail and allowing greater confidence when building the model, the tedium of model building is also greatly reduced.

Restrained reciprocal space refinement

Despite most careful model building in real space, the initial model will, in addition to missing certain parts, contain many small errors such as incorrect bond lengths and angles, poor backbone geometry, or improbable torsion angles. These small errors are corrected during the course of reciprocal space refinement. The refinement program adjusts the atomic position coordinates and B-factors of the model so that the differences between observed and calculated diffraction data are minimized.[25] The diffraction data are a reciprocal space representation of the molecule, hence the name reciprocal space refinement. Various other corrections such as anisotropic scaling and bulk solvent corrections are also applied during the refinement.[26] The global measure of the agreement between calculated and observed structure factor amplitudes is a linear residual, the crystallographic R-value, and its cross-validation equivalent, R(free) (discussed in detail in Chapters 7 and 12).

Restraints and cross-validation. While matching observed and calculated intensities by shifting atomic parameters, the refinement program also restrains the model to conform to certain stereochemical expectation values. Many geometry values, particularly covalent bond lengths and bond angles, are well known *a priori*, and there is no reason to assume that they would be any different in each protein structure. These stereochemical restraints also address a general problem in protein structure refinement, which is the low data to parameter ratio. In this situation, refinement is generally not stable against experimental data alone. Restraint terms creating penalties for deviations from known geometry target values[27] serve as additional observations, and safeguards against overfitting are necessary. Overfitting is the introduction and variation of additional model parameters, which artificially improves the fit between observed and calculated data, but actually does not improve the structure model. Common sources of overfitting are the introduction of too many water molecules or placing various kinds of solvent molecules (or ligands) into spurious density. Cross-validation against a small subset of unused (free) data, monitored by the R(free)-value for the free data set,[28] is standard practice.

Structure validation

Even after careful model correction and refinement, certain errors persist or are not reported by the refinement program's diagnostics, or they are just overlooked. A protein structure can be huge and several thousand residues must be built and refined to obtain the final structure model. It is quite possible (but not good practice) that some less interesting parts of the structure have been neglected. During refinement and rebuilding the model is thus constantly subjected to an array of validation tools, ranging from basic geometry checks to detailed chemical and folding plausibility checks based on prior knowledge (Chapter 13).

We have already mentioned that the quality of a protein structure is a *local property* and global agreement indicators such as the R-values cannot be specific as far as the local quality of a structure model is concerned. Local geometry validation programs evaluate the model geometry on a per-residue basis and flag outliers. Extended stretches of consistently high deviations are indicative of serious "problem zones" within the model. A very powerful method of assessing the local quality of a protein structure is the real space correlation coefficient of the model against a bias minimized electron density map,[29–31] which will be discussed in detail together with additional validation tools in Chapter 13.

To fix model errors, the electron density around the questionable residues is inspected, necessary corrections are made, and the structure is again refined. This process is repeated until all significant errors are corrected and the difference electron density shows no more offending high-level features. It is important that errors are corrected as much as possible (time and commitment permitting) because fixing many small errors can result in a significant overall improvement in structure quality. Once reasonably polished the structure is ready for detailed analysis and the structure factor amplitudes and the model coordinates are deposited with the Protein Data Bank and released to the public.

Analysis and description of the structure

One of the most exciting parts of any crystallographic study is the first analysis of the new structure. One is already quite familiar with the molecule from the preceding model building and refinement steps, and often the story unfolds during model building. The detailed analysis then depends to a large degree on the purpose of the structure study and on which hypothesis the structure model was intended to test. Always remember that the structure model itself is a hypothesis; it must withstand scrutiny against both the experimental evidence in the form of electron density and the entire pool of established prior knowledge and all laws of nature. Unconventional and unusual features always require *strong and convincing* support through experimental evidence, including all the additional supporting biochemical evidence that may exist in addition to the crystallographic evidence.

Figure 1-8 provides a review of the key steps in a protein structure determination, including references to the corresponding book chapters.

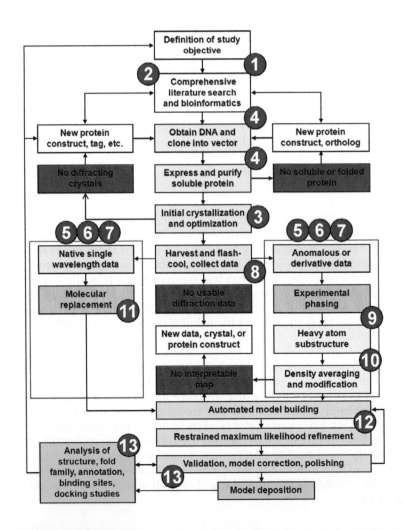

Figure 1-8 Key stages in the structure determination process. The flow diagram provides an overview of the major steps in a structure determination project, labeled with the chapter numbers treating the subject or related general fundamentals. Blue shaded boxes indicate experimental laboratory work, while all steps past data collection are conducted *in silico*. Protein production is discussed in Chapter 4, once we understand the process of crystallization and the requirements a protein must meet to be able to crystallize (Chapter 3).

1.6   Crystallographic models and coordinate files

We have thus far discussed the basic steps to obtain a crystallographic protein structure model, but not how to represent the model. Protein structure models at minimum must contain the atomic coordinates of each atom contained in the asymmetric unit cell, generally provided in Cartesian world coordinates in dimensions of Å. In addition, the basic crystallographic information in the form of cell parameters (unit cell dimensions a, b, c, and angles α, β, γ) and space group (Chapter 5) allows a display program to generate all symmetry related molecules to assemble the packing environment of the molecules in the crystal structure. The coordinates and additional information are deposited in the Protein Data Bank (PDB).

The Protein Data Bank

Models of protein structures created by experimental X-ray crystallography or NMR are saved in the form of coordinate data files, headed by additional method-specific information. The authoritative public repository and archive for these experimentally derived data files is the Protein Data Bank, PDB.[32] The PDB archives contain atomic coordinates, bibliographic citations, primary and secondary structure information, as well as crystallographic structure factors and NMR experimental data. Purely computational models are no longer accepted by the PDB. In addition to collecting, annotating,[33] and curating data files, the PDB provides deposition and validation services,[34] cross-links to other databases, and also provides a variety of analysis tools and structure viewers. The PDB, however, is *not* responsible for the correctness of deposited structures. It is the ultimate responsibility of a proficient crystallographer to deposit a model as free of errors as reasonably achievable (Chapters 12 and 13 will expand in great detail on the practical aspects of what "reasonably achievable" means in the nontrivial context of macromolecular refinement).

Format of protein structure coordinate files

A PDB file is identified by a four character PDB identification code. The first letter is a number, which was originally intended to indicate revisions to the file by incrementing the numbers, followed by three alphanumeric characters. A PDB file has two main sections: a header and the actual atom coordinate records. The two most common file formats are a fixed-length, 80-character per line, key-worded record format descending from the FORTRAN computer programming language standards; and a more modern, variable length record format called the macromolecular Crystallographic Information Format, or mmCIF.

Sidebar 1-4 A brief history of the Protein Data Bank and the Molecular Structure Database. The Protein Data Bank (PDB) was formally created in 1971 and initially maintained at the Brookhaven National Laboratory. In 1999 it moved to the Research Collaboratory for Structural Bioinformatics (RCSB) at Rutgers, under the direction of Helen Berman, John Westbrook, Phil Bourne at UC San Diego, and others, where it has grown into a formidable collection of data and tools.[17] Since 2003 three sites harbor the world wide PDB (wwPDB). In addition to the RCSB, the wwPDB is also mirrored in Japan at the Institute for Protein Research in Osaka and at the European Bioinformatics Institute, EBI, in Hinxton (close to Cambridge), UK. The EBI hosts the Protein Data Bank Europe[35] (PDBe, formerly the Macromolecular Structure Database), which includes an additional array of structural bioinformatics analysis tools and provides deeper annotation and cross-referencing of the PDB files which goes beyond the standard PDB services. Particularly

useful on the EBI/PDBe site is the PDBsum database[36] providing an at-a-glance overview of deposited PDB structures including topology drawings, motif analysis, sequence alignments, binding pocket analysis, interface analysis, and more. The PDBe also hosts a mirror of the Electron Density Server (EDS), a powerful validation tool that, in addition to electron density maps, provides real space correlation plots of structures on a per-residue basis. Such plots, which we will discuss in Chapter 13 in great detail, compare the model with the actual electron density and provide quite accurate information about the local quality of the structure. However, the EDS plots can only be created if the crystallographer has also deposited the corresponding data file in the form of the structure factor amplitudes against which the model was refined. It may come as a surprise that crystallographers were not required until early 2008 to deposit their primary data in the form of measured structure factor amplitudes together with the model coordinates.

Translators between the two common formats and to extendable mark-up language (XML) format[37] are available from the PDB sites. Despite its shortcomings, the fixed record PDB file format is persistent, largely because it is quite easy to read. The file format and the dictionary items are published on the PDB web site, and a brief introduction, with particular emphasis on the fixed-format CRYST and ATOM records and their peculiarities, is provided in Appendix A.1.

Graphical display of structure models

The first low-resolution protein structure models were made from clay displaying only secondary structure elements, later followed by complicated atomic brass ball models on steel rod assemblies (Sidebar 12-1). Given the size and complexity of protein molecules, all serious analysis is now carried out with the assistance of graphical display programs.

All display programs have in common that they can read at least the atom coordinate records and useful ones can also interpret the crystallographic information and generate symmetry related molecules and display crystal packing. Note that displaying only the asymmetric unit as contained in the coordinate file does not suffice for full analysis of crystal contacts; you need to display all neighboring molecules including translationally related symmetry copies (detailed in Chapter 5). As there is no specific connectivity record in the PDB files for standard residues, the display programs generally generate the bond information based on atom type and distance to neighboring atoms. The secondary structure is in most cases computationally assigned by DSSP[38] or a similar program. This means that occasionally a display program cannot interpret and properly draw ligands or non-standard residues, and poor or unrefined models tend to have missing bonds and only partial secondary structure assignment. Note that the deposited model does not always reveal the biologically relevant molecular assembly, as explained in Section 2.9.

Ball and stick models, space filling models, and secondary structure cartoons or ribbons are practically always available in molecular display programs for informative rendering, and depending on the purpose, various other properties can be mapped onto certain mesh objects or onto surface representations. Charge distribution, hydrophobicity, and binding pocket surfaces are common properties to display. Most of the crystallographic web browser plug-ins or Java-based display applets are quite useful for a first overview, but despite some of the applets allowing quite reasonable display options, fully detailed analysis usually requires specific software. We will encounter figures prepared with several programs throughout the text (Sidebar 13-7 describes how the figures were generated) and a summary of popular molecular graphics programs and their availability is given in the online supplement.

1.7   Crystallographic computer programs

Protein crystallography depends heavily on computational methods. Crystallographic computing has made substantial progress, largely as a result of abundant and cheap high performance computing. It is now possible to determine and analyze complex crystal structures entirely on inexpensive laptop or desktop computers with a few GB of memory. Automation and user interfaces have reached a high level of sophistication (although compatibility and integration issues remain). As a result, the actual process of structure solution, although the theoretically most sophisticated part in a structure determination, is commonly not considered a bottleneck in routine structure determination projects. Given reliable data of decent resolution (~2.5 Å or better) and no overly large or complex molecules, many structures can in fact be solved *de novo* and refined (although probably not completely polished) within several hours. Automated model building programs—many of them available as web services—have removed much of the tedium of initial model building. Several of the crystallographic programs and program suites will be used or introduced in the corresponding chapters, and a current compilation can be found in the

Sidebar 1-5 Not all models are created equal. While graphical display programs can generate beautiful pictures of protein structures, these images of great persuasive power lack crucial quality information. The structure representations—commonly ribbon diagrams or ball and stick models—look equally convincing and convey a deceptive sense of precision regardless of whether the model coordinates originate from a poor homology model, a low resolution 3.5 Å X-ray structure, or from a high quality 1.2 Å atomic resolution structure. The true proof is in the image of the electron density and how well it is matched by the model. A real space correlation plot provides a rapid overview of the fit between model and electron density for the entire structure, and it is good practice to show the model together with clear and properly contoured electron density whenever discussing an important feature of the structure. The majority of figures of structure details in this book will thus show the electron density together with parts of the structure model.

online supplement. The supplemental web material also contains links to the program web sites, which often contain useful tutorial materials, and to web sites providing sample data and additional exercises.

1.8   Key concepts

- The power of macromolecular crystallography lies in the fact that highly accurate models of large molecular structures and molecular complexes can be determined at often near atomic level of detail.
- Crystallographic structure models have provided insight into molecular form and function, and provide the basis for structural biology and structure guided drug discovery.
- Non-proprietary protein structure models are made available to the public by deposition in the Protein Data Bank, which holds more than 57 000 entries as of May 2009.
- Proteins are generally difficult to crystallize; without crystals there is no crystallography.
- Preparing the material and modifying the protein by protein engineering so that it can actually crystallize is nontrivial.
- Radiation damage by ionizing X-ray radiation requires cryocooling of crystals, and many crystals are difficult to flash-cool.
- The X-ray diffraction patterns are not a direct image of the molecular structure.
- The electron density of the scattering molecular structure must be reconstructed by Fourier transform techniques.
- Both structure factor amplitude and relative phase angle of each reflection are required for the Fourier reconstruction.
- While the structure factor amplitudes are readily accessible, being proportional to the square root of the measured reflection intensities, the relative phase angles must be supplied by additional phasing experiments.
- The absence of directly accessible phases constitutes the phase problem in crystallography.
- The nonlinear refinement of the structure model is nontrivial and prior stereochemical knowledge must generally be incorporated into the restrained refinement.
- Protein crystals are formed by a loose periodic network of weak, non-covalent interactions and contain large solvent channels.
- The solvent channels allow relatively free diffusion of small molecules through the crystal and also provide conformational freedom for surface-exposed side chains or loops.
- The core structure of protein molecules in solution as determined by NMR is identical to the crystal structure.
- Even enzymes generally maintain activity in protein crystals.
- Crystal packing can affect local regions of the structure where surface-exposed side chains or flexible surface loops form intermolecular crystal contacts.
- Large conformational movements destroy crystals and cannot be directly observed though a single crystal structure.
- Limited information about the dynamic behavior of molecules can be obtained from analysis of the B-factors as a measure of local displacement or by analysis of correlated displacement by TLS (Translation-Libration-Screw) analysis.
- The quality of a protein structure is a local property. Surface-exposed residues or mobile loops may not be traceable in electron density, no matter how well defined the rest of the structure is.

1.9   Additional reading

1.   Rhodes G (2006) *Crystallography Made Crystal Clear*, 3rd Ed, London, UK: Academic Press.

2.   Blow D (2002) *Outline of Crystallography for Biologists*. Oxford, UK: Oxford University Press.

3.   Glusker JP, & Trueblood KN (1990) *Crystal Structure Analysis. A Primer*. New York, NY: Oxford University Press.

1.10 References

1. Schenk H (Ed.) (1998) *Crystallography Across the Sciences.* Chester, UK: International Union for Crystallography.

2. Blundell TL, Jhoti H, & Abell C (2001) High-throughput crystallography for lead discovery in drug design. *Nat. Rev. Drug Discovery* 1, 45–54.

3. Ban N, Nissen P, Hansen J, et al. (2000) The complete atomic structure of the large ribosomal subunit at 2.4 Å resolution. *Science* 289, 905–920.

4. Cockburn JJ, Abrescia NG, Grimes JM, et al. (2004) Membrane structure and interactions with protein and DNA in bacteriophage PRD1. *Nature* 432(7013), 122–125.

5. Xenarios I, & Eisenberg D (2001) Protein interaction databases. *Curr. Opin. Struct. Biol.* 12, 334–339.

6. Abrahams JL, & Leslie AGW (1996) Methods used in the structure determination of the bovine mitochondrial F1 ATPase. *Acta Crystallogr.* D52, 30–42.

7. Zhou Y, Morais-Cabral JH, Kaufman A, et al. (2001) Chemistry of ion coordination and hydration revealed by a K+ channel-Fab complex at 2.0 Å resolution. *Nature* 414(6859), 43–48.

8. Kim J, Urban RG, Strominger JL, et al. (1994) Toxic shock syndrome toxin-1 complexed with a class II major histocompatibility molecule HLA-DR1. *Science* 266(5192), 1870–1874.

9. Garcia KC, Degano M, Stanfield RL, et al. (1996) An $\alpha\beta$ T cell receptor structure at 2.5 Å and its orientation in the TCR-MHC complex. *Science* 274, 209–219.

10. Tickle IJ, Sharff A, Vinkovic M, et al. (2004) High-throughput protein crystallography and drug discovery. *Chem. Soc. Rev.* 33, 558–565.

11. Congreve M, Murray CW, & Blundell TL (2005) Structural biology and drug discovery. *Drug Discovery Today* 10(13), 895–907.

12. Greer J, Erickson JW, Baldwin JJ, et al. (1994) Application of the three-dimensional structures of protein target molecules in structure-based drug design. *J. Med. Chem.* 37, 1035–1054.

13. Kim CU, Lew W, Williams MA, et al. (1998) Structure-activity relationship studies of novel carbocyclic influenza neuraminidase inhibitors. *J. Med. Chem.* 41(14), 2451–2460.

14. Atwell S, Adams JM, Badger J, et al. (2004) A novel mode of Gleevec binding is revealed by the structure of spleen tyrosine kinase. *J. Mol. Biol.* 53(31), 55827–55832.

15. Hendrickson WA (2000) Synchrotron crystallography. *Trends Biochem. Sci.* 25, 637–643.

16. Moore GE (1965) Cramming more components onto integrated circuits. *Electronics* 8(15), 2–5.

17. Berman H (2008) The Protein Data Bank: a historical perspective. *Acta Crystallogr.* A64(1), 88–95.

18. Etter M (1988) NMR and X-ray crystallography: Interfaces and challenges. *ACA Transactions* 24.

19. Brünger AT (1997) X-ray crystallography and NMR reveal complementary views of structure and dynamics. *Nat. Struct. Biol.* 4, 862–865.

20. Shaanan B, Gronenborn AM, Cohen GH, et al. (1992) Combining experimental information from crystal and solution studies: joint X-ray and NMR refinement. *Science* 257(5072), 961–964.

21. Moffat K (1998) Ultrafast time resolved crystallography. *Nat. Struct. Biol. Suppl.* 5, 641–642.

22. Zimmer J, Nam Y, & Rapoport TA (2008) Structure of a complex of the ATPase SecA and the protein-translocation channel. *Nature* 455, 936–943.

23. Santarsiero BD, Yegian DT, Lee CC, et al. (2002) An approach to rapid protein crystallization using nanodroplets. *J. Appl. Crystallogr.* 35(2), 278–281.

24. DeLucas LJ, Bray TL, Nagy L, et al. (2003) Efficient protein crystallization. *J. Struct. Biol.* 142(1), 188–206.

25. TenEyck LF, & Watenpaugh K (2001) Introduction to refinement. *International Tables for Crystallography F,* 369–374.

26. Kostrewa D (1997) Bulk solvent correction: Practical application and effects in reciprocal and real space. *CCP4 News. Protein Crystallogr.* 34, 9–22.

27. Konnert JH, & Hendrickson WA (1980) A restrained-parameter thermal-factor refinement procedure. *Acta Crystallogr.* A36, 110–119.

28. Brünger AT (1992) Free R value: A novel statistical quantity for assessing the accuracy of crystal structures. *Nature* 355, 472–475.

29. Bründén CI, & Jones TA (1990) Between objectivity and subjectivity. *Nature* 343, 687–689.

30. Reddy V, Swanson S, Sacchettini JC, et al. (2003) Effective electron density map improvement and structure validation on a Linux multi-CPU web cluster: The TB Structural Genomics Consortium Bias Removal Web Service. *Acta Crystallogr.* D59, 2200–2210.

31. Vaguine AA, Richelle J, & Wodak SJ (1999) SFCHECK: a unified set of procedures for evaluating the quality of macromolecular structure-factor data and their agreement with the atomic model. *Acta Crystallogr.* D55, 191–20.

32. Berman HM, Westbrook J, Feng Z, et al. (2000) The Protein Data Bank. *Nucleic Acids Res.* 28, 235–242.

33. Burkhardt K, Schneider B, & Ory J (2006) A biocurator perspective: Annotation at the Research Collaboratory for Structural Bioinformatics Protein Data Bank. *PLoS Comput. Biol.* 2(10).

34. Dutta S, Burkhardt K, Swaminathan GJ, et al. (2008) Data deposition and annotation at the Worldwide Protein Data Bank. In Kobe B, Guss M, & Huber T (Eds.), *Structural Proteomics: High-Throughput Methods.* New York, NY: Humana Press/Springer.

35. Velankar S, Best C, Beuth B, et al. (2010) PDBe: Protein Data Bank in Europe. *Nucl. Acids Res.* 38, D308-D317.

36. Laskowski RA (2001) PDBsum: summaries and analyses of PDB structures. *Nucleic Acids Res.* 29, 221–222.

37. Westbrook J, Ito N, Nakamura H, et al. (2004) PDBML: The representation of archival macromolecular structure data in XML. *Bioinformatics*, online.

38. Kabsch W, & Sander C (1983) Dictionary of protein secondary structure: pattern recognition of hydrogen-bonded and geometrical features. *Biopolymers* 22, 2577–2637.

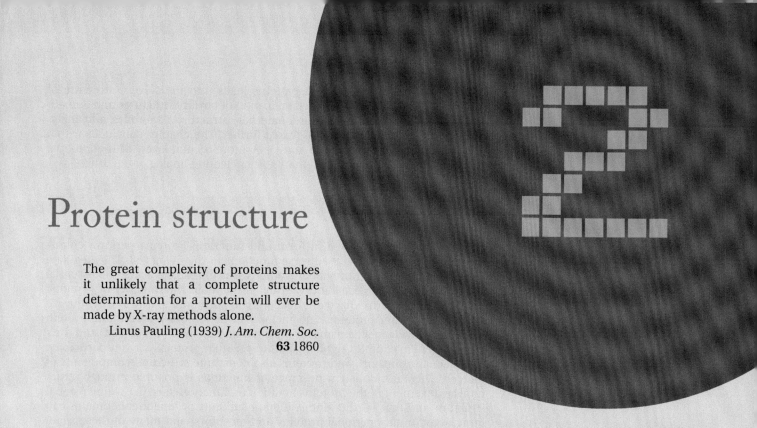

Protein structure

The great complexity of proteins makes it unlikely that a complete structure determination for a protein will ever be made by X-ray methods alone.

Linus Pauling (1939) *J. Am. Chem. Soc.* **63** 1860

The first protein structure models based on direct experimental evidence in the form of crystallographic data were derived in the 1950s by pioneers such as Max Perutz, John Kendrew, and fellow crystallographers at the Medical Research Laboratory of the Cavendish Laboratory in Cambridge and in Oxford by Dorothy Hodgkin — only 15 years after Linus Pauling's famous remark quoted above. Early protein crystallographers had very little guidance from prior knowledge when interpreting their first low resolution electron density maps and were initially quite puzzled by the lack of regular geometric features in protein molecules. Today, the interpretation of sophisticated high-resolution structures is greatly aided by the accumulated knowledge gained from the tens of thousands of protein structures determined, and the fundamentals of protein structure are required basic knowledge for structural biologists as well as protein crystallographers.

A comprehensive and systematic description of protein structures is not possible in this chapter because of the sheer number of structures and the complexity of their structure– function relationships. Books listed in the general reference section are available for the reader interested either in a more biological, function-oriented perspective or in a comprehensive catalog of protein structure. Instead, we focus here on aspects relevant for protein crystallography and the analysis of protein structures. To emphasize the practical aspects, illustrations feature actual electron density whenever possible. At the same time, examples are selected that emphasize a specific structural feature important for biological function or for structure based drug design. This approach serves two purposes: First, we train the eye in the recognition of electron density features and their proper interpretation, and second, we place the occasionally dry subject matter into a directly relevant biological context.

The chapter begins with a brief introduction including the fundamentals of molecular geometry and then follows the hierarchy of protein structure organization, beginning with backbone conformation and the basic secondary structure elements. The descriptive section of secondary structure elements is followed by a quite extensive section on amino acid side chain properties and their interactions. A semi-quantitative discussion of the various side chain interactions and their physical chemistry is necessary to appreciate their importance in tertiary and quaternary structure formation as well as in the interpretation of binding site geometry and specificity.

A discussion of hydrogen bond patterns leads into the solvent structure of proteins, which is followed by a classification of tertiary structures and domain organization. Membrane proteins, quaternary structure assemblies, and analysis of oligomer interfaces are discussed further. The chapter concludes with a brief section on DNA structure and DNA–protein complexes, with additional structural examples and a few remarks on nucleotide-analog based drugs.

2.1   Introduction

Proteins are highly diverse in form and function. Proteins perform countless fundamental and highly diverse functions in the living cell. They regulate DNA transcription and provide, together with ribonucleic acids, the essential ribosomal translation machinery. Proteins maintain the integrity of genomic information, perform enzymatic reactions in metabolic pathways, synthesize and degrade other proteins, and metabolize xenobiotics. Membrane protein receptors provide mechanisms for information transfer between cells and their environment in complex signaling pathways; other proteins actively participate in molecular transport, and proteins also constitute structural components of the cell. This enormously wide range of functions is possible through structural arrangement of the proteins in different, function-specific, 3-dimensional structures or folds. Furthermore, different subunits or functional domains can form larger, multifunctional proteins and receptors, and many of these again can combine into specific cellular machinery such as ATP motors in bacterial flagellates or multi-unit DNA replication or repair machineries.

Specific side chain properties generate diversity. The remarkable variety of protein form and function is rooted in the specific amino acid sequence of each protein. In all organisms, 20 common proteinogenic L-α-amino acids form the basic building blocks of proteins. In addition, selenocysteine is found in a few specialized redox enzymes, while pyrrolysine is present only in enzymes of methanogenic archaea. Posttranslational modifications (such as phosphorylation or methylation) and decorations (mostly glycosylation) add further variety to nature's inventory of protein structure.[1] The D-amino acids found in bacteria and fungi, commonly incorporated as components in cell wall biosynthesis, are created by other proteins (racemases) from their L-stereoisomers. It is also important to realize that not all expressed proteins have a defined structure *per se*; some need chaperoning to assume their fold, or they may assume a specific fold only in context with partners or certain environmental triggers. Similarly, not each and every part of a protein has to have a well-defined structure — large loops and even whole flexible domains can be present in proteins. The number of such intrinsically disordered or partially ordered proteins may be higher than commonly anticipated[2] and seems to constitute a significant fraction of the expressed genome.[3]

Protein flexibility and variability. Flexibility and variable conformation (be it intrinsic or a result of inhomogeneous posttranslational decoration, for example) impose clear difficulties for crystallographic structure determination. The self-assembly of proteins into regular, periodic crystals is rather difficult when the building blocks are flexible or have different conformation, resulting in poor crystals or no crystallization at all. A great deal of protein engineering effort is thus spent to coerce proteins with disordered regions or low intrinsic capability to form crystals into well-behaved — and still functional — molecules.

2.2   Basics of protein structure

Proteins are unbranched polypeptide chains consisting of 22 different proteinogenic L-α-amino acids. Twenty of the amino acids are common building blocks of all proteins (see Table 2-2) while selenocysteine and pyrrolysine are quite rare. Each of the amino acids has a different side chain with distinctly different properties, which combined with posttranslational modifications give rise to the extraordinary variety in protein functionality. Below about 40–50 amino acids (or residues) the term peptide is frequently used, but the definition is not unambiguous and, technically, large proteins are also just polypeptides. A certain number of residues are necessary to perform a particular biochemical function, and functional domains usually contain around 50–150 residues (functional motifs within a domain can be smaller). The size of protein molecules ranges from the lower limit of around 50 residues for single domain proteins to several thousand residues in multi-functional and multi-domain proteins (Figure 2-1). Even larger complexes can be formed from multiple protein subunits. For example, thousands of actin molecules assemble into the actin filaments of muscle fibers. Ribosomes are large molecular complexes formed from many different proteins together with a large fraction of ribosomal RNA (ribosomes are in fact ribozymes). It attests to the power of crystallography and to the persistence of the structural biologists that structures in the mega-dalton (MDa) range such as the 50S ribosomal subunit have been successfully determined at remarkably-high resolution (2.4 Å) for such a large and complex molecular assembly.

Structural hierarchy of proteins. The sequence of the different amino acids in the peptide chain, beginning at the N-terminus, is termed the primary structure of the protein. In the majority of cases, the sequence completely determines how the protein will fold into its 3-dimensional structure. During the folding process, specific secondary structure elements generally form first, such as α-helices and β-strands. The spatial arrangement of the various secondary structure elements then defines the tertiary structure, or the 3-dimensional fold of the protein. Within the conformational space of folded 3-dimensional structures (the fold space), distinct structural motifs and domains common among the structures can be recognized. Finally, the biologically active unit is in many cases a larger homo-oligomer of the same protein, or a hetero-oligomer or complex of different proteins. For example, many enzymes form dimers, while neuroreceptors frequently form hetero-pentamers of different, but structurally highly homologous subunits. The arrangement of subunits defines the quaternary structure. The hierarchy of protein structure organization (visualized in Figure 2-3) will be explored in the following chapters in more detail.

For the crystallographer and structural biologist, the understanding of protein structure must cover a wide range, extending from the global view necessary

Figure 2-1 Protein structures determined by X-ray crystallography. The small protein (top left) is the bovine pancreatic trypsin inhibitor (58 residues, ~8 kDa), one of the first proteins whose structure was refined at atomic resolution.[4] Its two distinct secondary structure elements are shown in ribbon presentation (a red α-helix and two cyan anti-parallel β-strands). The larger molecule (bottom left, ~1300 residues, ~150 kDa) is the neurotoxin secreted by the bacterium *Clostridium botulinum*. Botulinum neurotoxins are the most toxic substances known to man. They consist of three functionally distinct domains: a Zn-protease (cyan ribbon), a translocation domain (orange), and the ganglioside-binding domain (green), which is shown bound to a recognition peptide from the neuronal cell surface (red helix).[5] The right panel shows the large (50S) ribosomal subunit of the bacterium *Haloarcula marismortui*. The huge, nearly 2 MDa large structure[6] contains 27 different proteins (~4000 residues) and the ribosomal 5S and 23S RNA, together having 2833 nucleotides. For their work ultimately leading to the determination of the ribosome 50S subunit structure, V. Ramakrishnan, T. A. Steitz and A. E. Yonah shared the Nobel Prize in Chemistry. PDB entries 1bpi,[4] 3bta,[5] and 1ffk.[6]

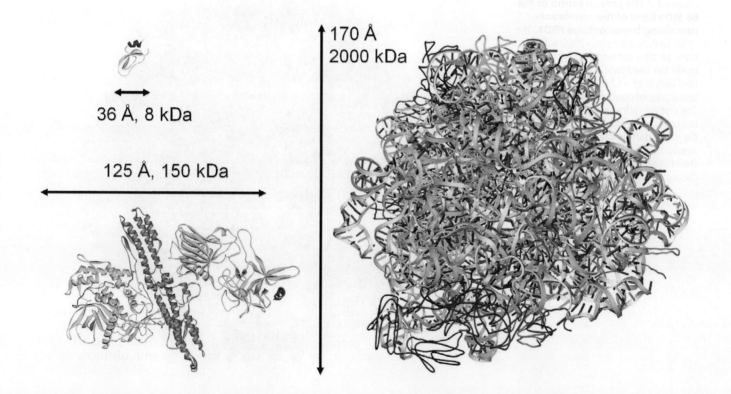

36 Å, 8 kDa

125 Å, 150 kDa

170 Å
2000 kDa

Sidebar 2-1 **Virus capsid structures.** Virus capsids can form assemblies even larger than the ribosome particles, but they are generally highly symmetric and are built from only a few different protein subunits. Nevertheless, their size and large unit cells make virus structure determination a formidable task. The most impressive particles crystallized are the 66 MDa heads of membrane-containing bacteriophage PRD1, containing ~2000 subunits from 18 different proteins, a host-derived lipid bilayer of ~12 500 lipid molecules, and a core of double stranded DNA (Figure 2-2).[7,8] The crystals diffracted to 4 Å with unit cell constants of nearly 1000 Å, and the virion particle itself measures 640 Å across.[9] The structure is so large that it is difficult to show it on the same scale as the other molecules in Figure 2-1.

Box 2-1 **Protein structure hierarchy.** Proteins or polypeptides are polyionic macromolecules assembled from 20 common proteinogenic L-α-amino acids. The sequence of amino acids constitutes the primary structure, the first level of organization of protein structure. Secondary structure elements are formed by distinct *backbone interactions*. The secondary structure elements are organized in small motifs and can form one or more independently folding domains. The arrangement of the protein domains defines the tertiary structure. Several protein molecules can assemble together at the level of quaternary structure.

to describe and analyze new folds, down to the minute analysis of enzymatic mechanisms or receptor–ligand interactions. We begin with some basic definitions of molecular geometry before we examine in detail the building principles of protein structure.

Fundamentals of molecular geometry

For any deeper discussion of protein structure one has to understand the basics of molecular geometry. Bond lengths, bond angles, and torsion (dihedral) angles are frequently used terms in the geometric description of the protein structure, and they are indispensable in crystallographic work.

Bond lengths, bond angles, and torsion angles

The bond length d_{AB} is defined as the distance between the center of two atoms A and B. Between three atoms A, B, and C, a bond angle γ_{ABC} can be measured, and between four consecutive atoms, a torsion angle τ_{ABCD} is defined. The torsion angle is equivalent to the dihedral angle between the planes ABC and BCD and is computed via the angle between the normal vectors of these planes. Viewed from the direction of the plane ABC along BC, a torsion to the right (clockwise) is defined as positive (0 to 180°), and to the left as negative (0 to –180°). The basic stereochemical descriptors are depicted in Figure 2-4, and Table 2-1 lists the observed values of a few common bond lengths and their standard deviation.

Figure 2-2 **The protein capsid of the 66 MDa head of the membrane-containing bacteriophage PRD1.** The virion particle contains ~2000 subunits from 18 different proteins. Enclosed inside the head capsid are a host-derived lipid bilayer of ~12 500 lipid molecules and a core of double-stranded DNA. The crystals diffracted to 4 Å with unit cell constants of nearly 1000 Å, and the virion particle itself measures 640 Å across. Image courtesy of Nicola Abrescia, David Stuart, and Jonathan Grimes, Oxford University. PDB entry 1w8x.[7]

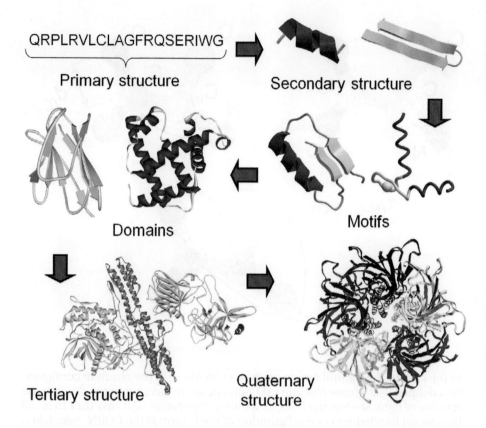

Primary structure
QRPLRVLCLAGFRQSERIWG

Secondary structure

Domains

Motifs

Tertiary structure

Quaternary structure

Figure 2-3 The hierarchy of protein structure organization. From a chain of defined amino acid sequence or primary structure, secondary structures form through distinct backbone interactions. Several secondary structure elements combine through side chain interactions into smaller structural motifs, which together with other secondary structure elements form distinct, independently folding structural domains. From the same protein chain, several domains with different functions can fold, assembling into the complete tertiary structure. Finally, several protein chains can form functionally important homo- or hetero-oligomeric assemblies, defining the quaternary structure.

Sidebar 2-2 Hydrogen atoms in protein structure models. In protein crystallographic work, the hydrogen atoms are generally only visible in electron density at ultra high resolution (about 1.2 Å or better). Because their so-called *riding* positions are known and can be calculated when needed, they are normally omitted in crystallographic models (the presence of hydrogen atoms is, however, implicitly and fully accounted for in crystallographic distance and energy restraints). The remaining panels of Figure 2-4 therefore show alanine in a representation typical for a crystallographic model, *sans* hydrogens. We will generally omit hydrogen atoms in riding positions in protein structure models, unless they are needed to emphasize a structural or stereochemical feature.

Bond lengths and bond angles generally vary only by a small amount from target values, which have been derived from small molecule X-ray structures and high resolution protein structures.[10] Long side chains assume certain energetically preferred conformations defined by specific side chain torsion angles. Torsion angles show much wider distributions than bond angles, but play an important role in macromolecular structure for several reasons: (i) They often assume distinct preferred values. (ii) Torsion about a bond can impact the positions of atoms far away as downstream atoms rotate as a group. (iii) The combined effects of (i) and (ii) accumulated over successive bonds in a polymer give near-infinite combinations of conformations that account for most of the diversity in molecular structures. (iv) The energy barriers to deformation of torsion angles are less imposing than for bond lengths or angles, allowing torsion angles to contribute more to the variation in molecular structure.

Chiral centers
The building blocks of a protein are L-α-amino acids, in which the α-carbon atom bonds to four different ligands and thus forms a chiral center. The notation is derived from the relation to the chiral centers of glyceraldehyde, and happens

Peptide bond	Average length (Å)	Single bond	Average length (Å)	Hydrogen bond	Average length (Å)
Cα–C	1.525 ± 0.021	C–C	1.540 ± 0.027	O–H - - - O–H	2.8 ± 0.2
C–N	1.329 ± 0.014	C–N	1.489 ± 0.030	N–H - - - O=C	2.9 ± 0.2
N–C	1.458 ± 0.019	C–O	1.420 ± 0.020	O–H - - - O=C	2.8 ± 0.2

Table 2-1 Common bond lengths. Mean values and standard deviations of some covalent bond lengths and hydrogen bonds commonly occurring in protein structures. The average or mean values are derived from accurate small molecule X-ray structures and high resolution protein structures. Because they also serve as the basis for stereochemical restraints in macromolecular refinement (Chapter 12), the mean values are also known as restraint target values.[11]

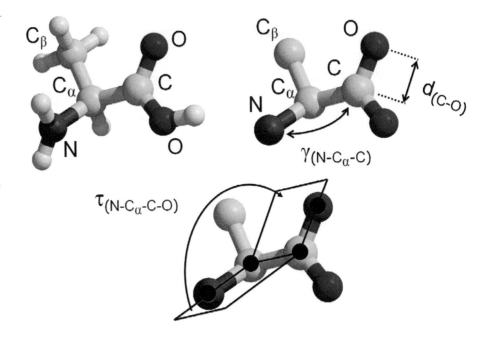

Figure 2-4 Definition of bond length, bond angle, and torsion angle. A ball and stick representation of the amino acid alanine is labeled with atom names and shows the definition of bond lengths, bond angles, and torsion angles. The color scheme applied to distinguish atoms is a common default scheme used by protein model display programs (carbon atoms yellow, oxygen red, nitrogen blue). Hydrogen atoms in riding positions are explicitly shown only in the top left panel (gray atoms) and are normally omitted. Because the torsion angle can be interpreted as the angle between two planes, torsion angles are also called dihedral angles.

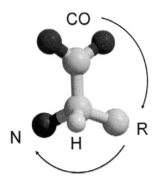

Figure 2-5 The CORN rule. The CORN rule is a practical aid for determining the configuration of chiral Cα centers of amino acids. When the central Cα-atom is viewed with the H atom pointing toward the observer (or out of the paper plane), the ligand sequence reads "CO-R-N" in clockwise rotation for a L-amino acid.

in the case of the natural amino acids to coincide with the absolute configuration defined as 2S (note that the L/D nomenclature has nothing to do with the rotation of light; see Sidebar 6-13, absolute configuration). An easy trick to recall the correct handedness or configuration of the L-form is the CORN rule: when the Cα-atom is viewed with the H atom pointing toward the observer as shown in Figure 2-5, the ligand names read "CO-R-N" in a clockwise rotation. The branched amino acids threonine and isoleucine have a second chiral center at their Cβ carbon atom, and their absolute conformation is (2S, 3R). The chirality of amino acids is an invariable chemical property.

2.3   The geometry of the polypeptide chain

The protein chain emerges during the process of mRNA translation in the ribosomes. The chain begins at the N-terminus and is extended through peptide bond formation between the carboxyl group of the preceding amino acid and the amino group of the following amino acid. The equilibrium lies on the hydrolyzed side of the reaction, and the cell has to expend energy in the process of peptide synthesis. The linear sequence of the amino acids is the first level in the organization of a protein and thus called the primary structure of the peptide or protein. Counting of amino acids (or residues) starts at the N-terminal end (NH₂-group) of the chain. Because most proteins used in crystallographic studies today are created by recombinant techniques, the protein sequence is normally known from the readily available cDNA sequence. Peptide sequences

Box 2-2 **Molecular geometry** The most common descriptors for molecular geometry are bond lengths, bond angles, and torsion angles. The proper chirality or handedness of backbone Cα carbon atoms (L-conformation or 2S configuration in absolute terms) can be checked by the CORN rule. The amino acids threonine and isoleucine have a second chiral center at Cβ with 3R configuration. Specific types of bond distances and angles as well as chirality are highly restrained to their known mean values in macromolecular structure refinement, where they provide important, stabilizing prior knowledge.

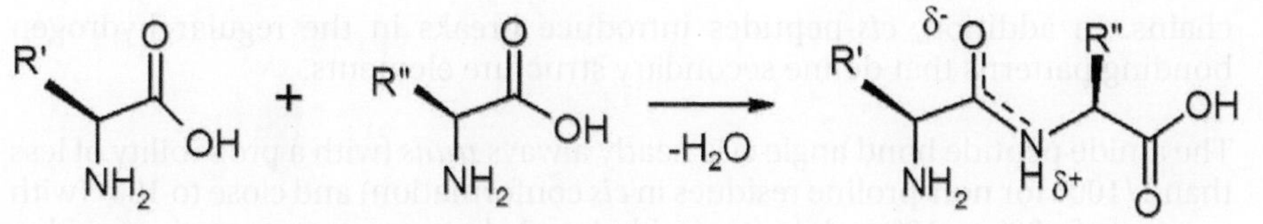

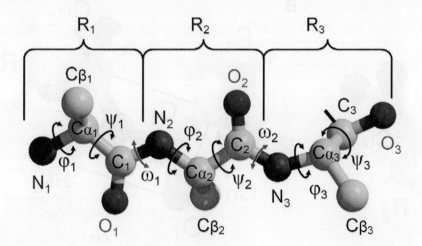

Figure 2-6 **Formation of a peptide bond.** Note that the peptide bond has partial double bond character because of resonance stabilization, and carries a strong dipole moment. The latter makes the CO group a particularly strong hydrogen bond acceptor and the NH a good hydrogen donor group, an important feature in the formation of distinct secondary structure elements.

are usually given in single letter code in sequence databases or in 3-letter code (see Table 2-2). The 3-letter code is used in the SEQRES records of a PDB file.

Peptide bond basics and the protein backbone

A basic sketch illustrating peptide bond formation is shown in Figure 2-6, with R representing any of the different side chains. The peptide bond between two amino acids is formed between the C-terminal carboxyl group of the first amino acid and the amino group of the following amino acid. The peptide bond is planar because of the delocalization of electrons along O_n–C_n–N_{n+1}–H_{n+1}, giving it partial double bond character and creating a dipole moment of ~3.5 debye (note from Table 2-1 that the C–N distance of the peptide bond is shorter than a normal C–N single bond). As a result of the dipole polarization, the peptide bond has a strong hydrogen bond acceptor in the oxygen atom of the CO group and a willing donor in the NH group. The formation of hydrogen bonds between the peptide backbone atoms is crucial for the formation of secondary structure elements. In addition, the planarity of the peptide bond imposes significant restraints on the conformational freedom and flexibility of the protein backbone.

Figure 2-7 shows the N-terminal stretch of a peptide, whose backbone is formed by a sequence of atoms N_1–$C\alpha_1$–C_1–N_2–$C\alpha_2$–C_2–N_3–$C\alpha_3$–C_3–N_4– and so forth. There are three atoms per backbone chain segment, and three torsion angles between three pairs of successive atoms: the peptide torsion angles φ (phi), ψ (psi), and ω (omega). These three torsion angles are the only degrees of freedom for the polypeptide backbone conformation. While the torsion angles φ (phi) and ψ (psi) are restrained to certain combinations by non-bonded van der Waals repulsions (and only to a minor degree by additional specific steric hindrances resulting from collisions between bulky residues extending beyond the Cβ carbons), the planar peptide bond with its partial double bond character is highly restrained (but not completely rigid) to ω-angles around 180° or 0° for *trans*- and *cis*-peptides, respectively. The angle of the torsions is defined as positive when looking in the direction of the peptide chain and twisting to the right.

Trans- and *cis*-peptide bonds

In both the *trans*- and *cis*-peptides, six atoms of each residue ($C\alpha_n$ C_n O_n N_{n+1} HN_{n+1} $C\alpha_{n+1}$) are restrained into a planar conformation. However, as indicated from a comparison of the *trans* versus *cis* conformation in Figure 2-8, *cis*-peptide formation is significantly less favorable because of the limited number of conformations that are possible without collisions between adjacent side

Figure 2-7 **Backbone torsion angles.** The N-terminal 3-residue stretch of a peptide Ala-Ala-Ala- containing three *trans*-peptide bonds is shown. Three torsion angles for each residue, φ (phi), ψ (psi), and ω (omega), define the conformation of the peptide backbone. While combinations of torsion angles φ and ψ are only restrained by van der Waals repulsion and fall into several allowed, energetically favored regions, the *trans* omega-torsion around the partially delocalized, planar peptide bond (short red arrow) is highly restrained to 180°.

chains. In addition, *cis*-peptides introduce breaks in the regular hydrogen bonding patterns that define secondary structure elements.

The amide peptide bond angle ω is nearly always *trans* (with a probability of less than 1/1000 for non-proline residues in *cis* conformation) and close to 180° (with a range of –20° to 10°), while the imide bonds between a non-proline residue and a subsequent proline are somewhat more frequently (~7%) found in *cis* conformation.[12] For the cyclical amino acid proline, the *cis* and *trans* conformation are geometrically very similar, with comparable energies differing by only a few kcal mol[-1]. It is rather crucial to verify the presence of non-proline *cis*-peptides carefully with respect to the electron density (Figure 2-9) during model building, because *cis*-peptide bonds represent local spots of high conformational energy in a polypeptide chain. *Cis*-peptides appear more frequent in turns and loops, but this observation may be partly because loops are difficult to build in poorly defined electron density and their conformation is generally uncertain.

Backbone torsion angles — the Ramachandran plot

The backbone conformation of a polypeptide is completely defined by the sequence of the torsion angles along its peptide chain. A 2-dimensional scatter plot of the φ–ψ torsion angle pairs for each residue shows certain preferred regions where φ–ψ torsion angle combinations cluster. Such a 2-dimensional representation of φ–ψ torsion angle pairs is named the *Ramachandran plot* after its inventor.[14]

Figure 2-10 depicts an empirical Ramachandran plot, displaying the regions of frequent φ–ψ torsion angle combinations. Preferred core regions are highlighted in color, while the surrounding contours signify regions of higher conformational energy, where the probability of finding a residue with such a torsion angle pair is correspondingly lower. Regions outside the contours contain highly unfavorable, strained torsions which are very unlikely to occur because of strong repulsive van der Waals interactions (Figure 12-18 shows a corresponding Lennard-Jones potential curve). The regions of the Ramachandran plot thus represent a 2-dimensional projection of a conformational energy surface, which can be computed from non-bonded van der Waals interactions between backbone atoms. The preferred torsion angles cluster in the minima of the energy surface. The distributions in the practically used φ–ψ torsion angle plots are based on analysis of empirical observations[15] and vary slightly depending on the size and quality of the data set used in their analysis. The backbone torsion angles of secondary structure elements (Section 2.4) cluster in specific regions of the Ramachandran plot.

The Ramachandran plot is a very valuable tool for structure validation because it quickly reveals the plausibility of the geometry of a protein backbone trace. The Ramachandran plot can also be interpreted as displaying the probability

Figure 2-8 *Trans*- and *cis*-peptide bonds. The *trans*-peptide bond, recognizable by the two sequential Cα atoms opposed, has an ω-angle of 180° while the *cis*-peptide bond with an ω-angle of 0° can be easily recognized by the Cα atoms located on the same side of the peptide bond. In both cases, the peptide bond is planar and contains the same atoms, but in *cis* conformation (right panel) the torsional freedom is much more limited because of side chain collisions. The Cα–Cα distance in a *cis*-peptide is ~2.9 Å, while the Cα–Cα distance across a *trans*-peptide bond is ~3.8 Å.

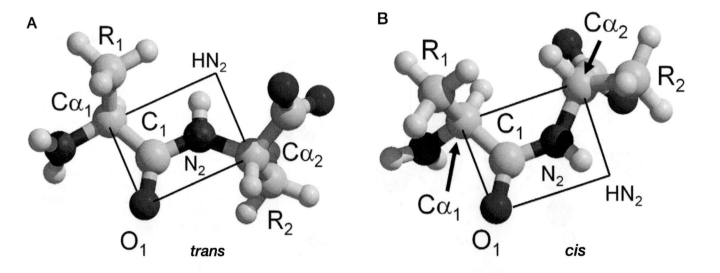

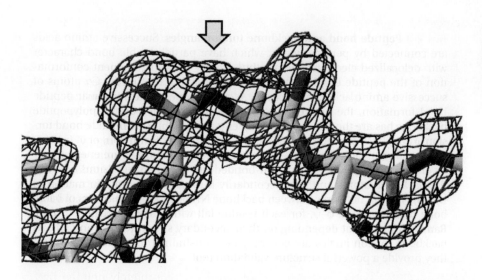

Figure 2-9 *Cis*-peptide bond and its electron density. A *cis*-peptide bond between Gly and Ala in a connecting loop is shown in experimental electron density (blue grid). Both the clear electron density and the fact that the *cis* conformation is structurally conserved in this particular fold family allow making this particular *cis* assignment with confidence. PDB entry 1upi.[13]

of a given certain φ–ψ torsion angle combination to occur, which is highest in the most favorable regions. It is important, however, to understand that in the process of assuming a minimum energy state when the protein is folded, local strain can be induced on the protein backbone, and a small number of residues (no more than approximately 1–2%) can be expected to adopt energetically less favorable, and thus less probable, conformations. A special case are glycine residues, which have no Cβ extending from Cα (only a second hydrogen atom) and thus enjoy much larger conformational freedom. In addition, glycines do not possess a chiral center, which leads to a centrosymmetric distribution of their torsion angle distribution in the φ–ψ plot.

Analysis of the φ–ψ angle distribution is particularly valuable during crystallographic model building, because the φ–ψ torsion angles are generally not or only weakly restrained in crystallographic refinement (Chapter 12) and thus not subject to biased, artificially narrowed distributions (as restrained covalent bond lengths and angles are). We will thus make extensive use of Ramachandran plots in the validation and analysis as discussed in Chapter 13.

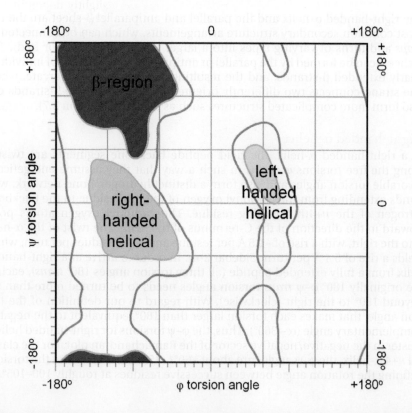

Figure 2-10 Backbone torsion angle distribution. The protein backbone torsion angles are represented in the form of an empirical Ramachandran plot.[15] The colored regions are energetically most preferred, with the surrounding regions having higher torsional energies. The φ–ψ combinations outside the blue contoured regions are so unfavorable that they are usually not observed. The torsion angles of various secondary structure elements cluster in typical regions. The right-handed helical region largely contains residues in right-handed α-helices, and some 3_{10}- and π-helices. The β-region contains β-strands, β-proline helices, and β-turns, while the left-handed helical region contains torsion angle pairs as observed in left-handed helices. Note that the left-handed helical region has opposite signs in the φ–ψ distribution compared with the regular, right-handed helical region. Left-handed helix conformations tend to be short and rare, because a clash between backbone carbonyl oxygen and Cβ atoms makes them energetically less favorable.

Box 2-3 **Peptide bond and backbone torsion angles.** Successive amino acids are connected by peptide bonds, which have partial double bond character with delocalized electrons and are thus planar. The most frequent conformation of the peptide bond is the *trans* conformation, that is, the Cα atoms of successive amino acids are opposed. As a consequence of the planar peptide bond formation, the only degrees of conformational freedom a polypeptide backbone has are the two torsional angles φ and ψ, while the peptide bond torsion ω is highly restrained to +180° (dominant *trans* conformation) or 0° (rare *cis* conformation). Only limited φ–ψ torsion angle combinations are energetically favored, primarily as a result of non-bonded van der Waals repulsions between the backbone atoms and only secondarily through the optimal formation of hydrogen bond networks between backbone NH and CO groups. Pairs of backbone torsion angles, φ–ψ, for each residue fall within specific regions on the Ramachandran plot depending on their secondary structure environment. As backbone torsion angles are generally not constrained in model refinement, they provide a powerful structure validation tool.

2.4   Secondary structure elements

When a polypeptide chain is folded so that it repeatedly contains the same energetically favored backbone torsions, specific secondary structure elements are formed. They assume a typical backbone conformation that optimally satisfies the strong hydrogen bonds between the CO- and NH-groups of peptide bonds in neighboring residues. Only a very limited number of backbone conformations optimally satisfy torsion angle restraints and hydrogen bonding patterns at the same time, which gives rise to the secondary structure elements *helices*, *sheets*, and *turns*. Common to all secondary structure elements is therefore a well-defined hydrogen bonding network between backbone carbonyl oxygen atoms acting as acceptors and the backbone nitrogen atoms as hydrogen donors of subsequent residues (in α-helices and turns) or residues from adjacent strands (in β-sheet structures).

The right-handed α-helix and the parallel and antiparallel β-sheet are the two most common secondary structure arrangements, which can be connected by *loops* and turns of varying types into a larger tertiary structure (or fold). The β-sheets can be formed by the parallel or antiparallel arrangement of individual, nearly extended β-strands, and the resulting sheets can also bifurcate, where one strand connects two differently oriented β-sheets. Multiple β-strands can also form more complicated structures such as β-barrels (Section 2.7).

Right-handed α-helices

In a right-handed α-helix, the rigid peptide backbone segments are twisted along the free torsions φ and ψ in such a way that they assume energetically favorable torsion angles and also form a distinct hydrogen bond network, with bonds extending from the carbonyl oxygen of the *n*th residue to the backbone nitrogen of the *n*+4th following residue. The carbonyl oxygen atoms point upward in the direction of the C-terminus of the helix. The twist of the α-helix is to the right, with a rise of ~1.5 Å per residue and ~3.6 residues per turn, which yields a rise of 5.4 Å per turn (Sidebar 2-4). In order to arrive at a right-handed helix from a fully extended peptide (all three torsion angles 180° *trans*), each of the originally 180° φ–ψ *trans* torsion angles needs to be turned more than 90° beyond 180° to the right (clockwise). With regard to our definition of the torsion angle, that makes each torsion larger than 180°, equivalent to the negative complementary angle ($\alpha - 360°$). Thus, the φ–ψ torsions for right-handed helices cluster in the negative/negative sector of the Ramachandran plot. For the classical *n*+4 α-helix, the φ–ψ angles are about −60°/−45°, with the sum of the torsions defining the rotation angle between successive residues at roughly 100–105°.

Sidebar 2-3 Pre-crystallography protein structure models. That proteins could be crystallized has been known since the mid to late 19th century, and by the 1940s, about 150 different proteins had been crystallized.[16] At the same time, crystallographic techniques, which had developed rapidly since the 1920s, had already yielded insight into the regular and ordered constitution of matter in small molecule crystals, and there were some expectations that proteins would also show regularity in their molecular structure. When the first protein structure was determined by John Kendrew and his model of sperm whale myoglobin was published in 1958,[17] the complexity and complete lack of regularity in protein molecules came as a surprise; regularity and integral symmetry had at least to some degree been expected because protein crystals diffracted X-rays quite well.

Similar to the well-known case of DNA, fiber diffraction patterns of hair and wool (consisting of keratin fibers), collected in the early 1930s by William Astbury, showed characteristic repeats indicative of a helical structure, with a rise of ~5.1 Å per turn. These first patterns were thought to result from a helical conformation, called the α-form. When moist wool fibers were stretched, the diffraction pattern changed and Astbury thus proposed a second, extended β-form. The names of the two most frequent secondary structure elements were thus coined.[18] Although his models of the protein chain as a "molecular centipede" were not correct in detail, the existence of a hydrogen bonding network between backbone atoms was already predicted by Astbury. However, in the absence of crystal structures of peptides and a developed theory of the chemical bond, the partial double bond nature of the — as a consequence rigid — peptide bond was not realized, and a "cyclol theory" proposing a series of covalent six-ring structures for the backbone geometry remained in discussion until the 1940s.

Based on Linus Pauling's theory of the chemical bond and the knowledge of peptide bond distances and angles from small molecule crystallographic structures (including the crucial rigidity of the peptide bond), and aware of the stereochemical advantage of certain φ–ψ torsion angles, Pauling, Corey, and Branson[19] derived, in 1951, theoretical models for two protein secondary structure elements,[20] several years before any protein structures at sufficient resolution confirmed most of their ideas. The first structure, which in reference to Astbury's patterns they appropriately called the α-helix, is correct in detail, with the exception of their guess for the absolute configuration. Their α-helix is left-handed and assembled from D-amino acids. It was again crystallographic evidence — the determination of the absolute configuration of molecules by anomalous X-ray diffraction, pioneered by Bijvoet in the 1950s — which showed that the natural amino acids are L-amino acids and consequently, α-helices in proteins are right-handed. The proposed γ-helix, although in rise similar to the π-helix of today but with reversed carboxyl group direction, is not found in nature. A few months later, Pauling and colleagues published a second series of papers, describing the other most prevalent type of secondary structures, the β-sheet structures.[21]

The personalized historic account of Richard Dickerson[22] includes commented reprints of seminal key publications from this fascinating early period of protein crystallography.

Helices are the most abundant form of secondary structure[26] (~35%), and their lengths span the whole region from the minimum of four residues to more than 50 residues such as in the transcription-regulating leucine zippers (Figure 2-47). Numerous structures exist that contain only α-helices, connected by short turns. Helices assemble frequently into helix bundles (Figure 2-11), from which a wide array of important proteins such as structural coiled coils, cytochromes,

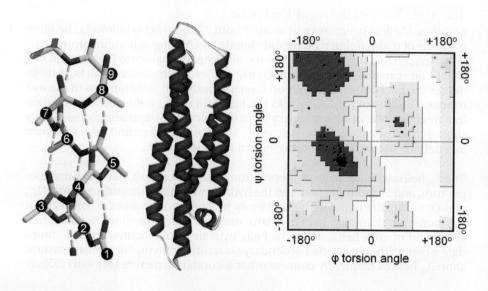

Figure 2-11 Protein α-helices. Left panel shows a polyalanine stretch in α-helical conformation. Note the hydrogen bonds (green dotted lines) from the backbone carbonyl oxygen of residue *n* to the backbone nitrogen of residue *n*+4. The residues are labeled at the Cα carbon atom. Center panel: A typical 4-helix bundle is the 22 kDa fragment of apolipoprotein E4, an allelic isoform of apo-E implicated in late onset familial Alzheimer's disease.[23] Note that in the Ramachandran plot[24] (right panel) practically all residues except those few located in the connecting loops and turns are sharply clustered in the right-handed α-helical region. PDB entry 1b68.[25]

Sidebar 2-4 **Nomenclature of helices and screws.** A helix can be assembled from building blocks by a screw operation that combines each successive rotation around an axis with a translation parallel to that axis. This is true for protein α-helices, the DNA double helix, and also for crystallographic screw axes in general. The geometry of a general helix is described by the number of subunits S needed to complete a full turn, and the rise R, which is the distance required to complete one full turn. The rise of a single unit is then given by R/S. Note that S does not necessarily have to be an integer number (Figure 2-12).

Alternatively, helices can be described in BKP (Bragg–Kendrew–Perutz) nomenclature as S_a helices, where S is again the number of amino acids per turn, and a is the number of backbone atoms (including the H atom) in the *atom ring* closed by the hydrogen bond. The α-helix is in this nomenclature a 3.6_{13} ($3 \times 4 + 1$) helix, and the π-helix is defined as a 4.4_{16} ($3 \times 5 + 1$) helix. It is interesting that Bragg, Kendrew, and Perutz were really close in 1950[27] to finding the proper α-helix. Being crystallographers, however, they assumed that crystallographic, proper 4-fold screw symmetry should be present in their proposed 4_{13} helix. The small difference to the correct non-integer 3.6_{13} α-helix causes some strain and less perfect hydrogen bond alignment, and their 4_{13} helix just does not look as "right" as Pauling, Corey and Branson's correct 3.6_{13} α-helix.[19]

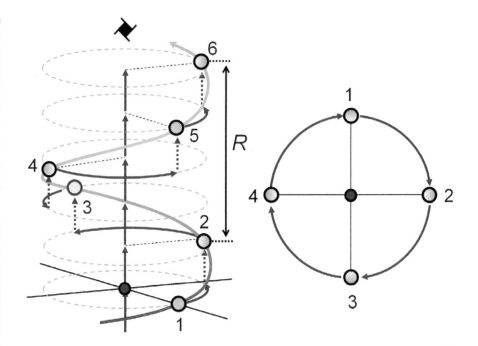

Figure 2-12 **Helix orientation.** A succession of counter clockwise rotations by 90° in a projection with the helical axis pointing upwards plus translation along the helical axis leads to a right-handed helix (left drawing). The sense or handedness of the helix is determined by looking along the helical axis (right panel) and following the sequence of the points (it does not matter in which direction one looks along the helix- or screw axis, the handedness remains the same). In crystallographic plain axes as well as screw axes, the rotation is exactly 360/n degrees, with n = 2, 3, 4, and 6. The depicted helix is compatible with a crystallographic 4-fold (4_1) screw operation, indicated by the symbol ✦ above it. The right-handed protein α-helix in contrast has a non-integer number of ~3.6 residues per turn, corresponding to ~100° counter clockwise rotation between residues. Screw operations are further explained in Chapter 5.

lipoproteins, human growth hormone, or the transmembrane helices of integral membrane proteins are assembled. Because of the accumulated charges and unsaturated hydrogen bonds at both ends of the helix, helices are generally located on the outside of a protein. We postpone the discussion of the assembly of protein structures from secondary structure elements until we have discussed the different chemical properties of the side chains, which play a crucial role in protein folding and stability.

The 3_{10}-helix, π-helix, and PP-II helix

Additional helical structures can be built from L-amino acids following the same principle of maintaining rigid peptide bonds and hydrogen bonding from a carbonyl oxygen to a subsequent backbone nitrogen. In analogy to the classical n+4 α-helix one can define a right-handed n+3 and n+5 helix. However, in both these helices the backbone conformation is energetically not as optimal as in the n+4 α-helix. The core of the helix packs too tightly in the 3_{10} helices, which are thus less frequent (4% of all secondary structure elements) and also considerably shorter than α-helices. The π-helices are too loosely packed and their existence as a defined secondary structure element is questionable.

The 3_{10} helix, with the hydrogen bond from residue n to n+3, has three residues per turn, and 10 atoms lie between the hydrogen bond donor and acceptor (N_{n+3} and O_n), thus the name 3_{10} helix. Its φ–ψ torsions cluster around –49°/–26° in the Ramachandran plot. In crystal structures we observe a short 3_{10} helix often at the end of an α-helix, when the helix twist needs to "tighten" to accommodate a connection to another secondary structure element. For similar reasons, short 3_{10} helices frequently connect other secondary structure elements (Figure

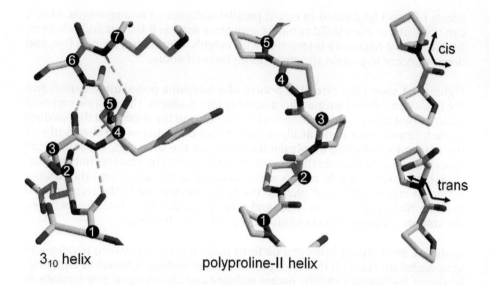

3₁₀ helix   polyproline-II helix

Figure 2-13 The right-handed 3₁₀ helix and the left-handed trans-polyproline-II helix. The residues of the helices are numbered at their Cα positions. Note the *n*+3 backbone hydrogen bonds in the 3₁₀ helix. No hydrogen network can exist in the proline helix because of the absence of a hydrogen atom on the proline nitrogen. A PP-II helix therefore does not comply with the strict definition of secondary structure elements via their backbone hydrogen bonding patterns. The right panel emphasizes the *cis* and *trans* conformation of a proline dipeptide. Note the similar conformation of the proline rings in both *cis* and *trans* conformation. PDB entries 2d2e[30] and 1vzj.[31]

2-13). The 3₁₀ helices probably also play a significant role during the process of protein folding or in conformational molecular switches. In contrast to 3₁₀ helices, π-helices are extremely rare and have been proposed only in a handful of proteins. Nevertheless, it has been proposed that conformations occurring in π-helical conformation have evolved to provide unique structural features within proteins and may have certain functional advantages.[28]

A helical secondary structure element that is frequently found on the surface of globular proteins is the left-handed *n*+2 polyproline-II (PP-II) helix, consisting of a series of *trans*-prolines or at least *trans*-proline rich stretches. The PP-II helix is a flexible structural element, and biologically quite relevant. As an example, binding of the proline-rich region of the Src activating platelet-derived growth factor β (PDGF-β) to the SH3 (Src homology 3) domain of the Src tyrosine kinase triggers a downstream cytoplasmtic signaling cascade responsible for cell proliferation. Defects in the Src tyrosine kinase lead to uncontrolled mitogenic signaling, which in turn causes aberrant cell proliferation in a number of human cancers.[29] Similar to the 3₁₀ helices, PP-II helices can serve as a bridge between different secondary structures. Long PP-II helices containing enzymatically Cγ-hydroxylated prolines are building blocks of the twisted collagen fibers.

β–strand based sheet structures

β-strand based secondary structures are derived from an extended conformation of the polypeptide chain. A fully extended, stretched-out peptide chain (both torsion angles φ and ψ as well as the peptide bond torsion ω in 180° *trans* conformation) would extend in a simple zigzag pattern — not too favorable because of the close proximity of the carbonyl oxygen and the adjacent side chains. The Ramachandran plot (Figure 2-10) confirms that the torsion pairs around +180°/+180° are not a favored combination. However, with a small change in torsion angles, a favorable extended conformation can be achieved. The polypeptide chain in each β-strand is nearly — but not fully — extended with φ–ψ torsions of about −135°/135° clustering in the upper left quadrant of the Ramachandran plot. The incomplete extension induces a *pleat* in β-sheets (Figure 2-14) and results in a *n*+2 Cα-distance of about 6 Å, compared with the 7.6 Å for a fully extended but energetically not favorable +180°/+180° conformation.

In contrast to the helices, where the backbone interactions were sequential (from residue *n* to *n*+4) within the same helix, in β-strand based secondary structures the backbone interactions occur between different β-strands. There are principally two arrangements possible: parallel or antiparallel alignment of the strands, leading to parallel or antiparallel β-sheets. In addition, as the number of individual strands that can form a β-structure is not limited, extended sheets can also be formed in mixed parallel–antiparallel arrangements. Sheets can bifurcate, or even build complex structures such as β-barrels and β-helices. Because of the necessary interaction with neighboring chains, a single β-strand does not occur in protein structures as an isolated entity.

Figure 2-14 shows the extended β-form of a 5-residue polyalanine stretch and the basic parallel and antiparallel assembly into β-sheets. The peptide carbonyl groups point in alternating directions in the plane of the sheet, and the residues in each strand point periodically up above and down below the sheets (this is the reason for selecting a polyalanine chain for the drawings, because larger side chains would obscure the view of the backbone). The cross section emphasizes the location of side chains above and below the sheet, and the distinct up–down zigzag pattern or pleat of the backbone leading to the term *pleated β-sheets*. Sheets are also not rigid, and there is considerable flexibility between the strands. Extended sheets often show a right-handed twist.

In theory, inter-strand hydrogen patterns differ distinctly between parallel and antiparallel sheets, with the close pair of parallel hydrogen bonds in antiparallel sheets leading to a slightly higher stability and clustering of φ–ψ torsions of about –140°/135° compared with the parallel β-sheet with –120°/115° torsions. However, the hydrogen bonding patterns are not as strictly ideal as Figure 2-14 might suggest. Intermediate patterns with an offset of the individual strands against each other are caused by geometric restraints of side chain packing. This is shown in Figure 2-15 for a backbone section of an actual antiparallel β-sheet from concanavalin A, a sturdy plant lectin and hardy perennial of crystallization studies and structural biology.[32]

β-bulges and irregular sheets

Sometimes the regular hydrogen bond pattern observed in β-sheets becomes locally disrupted. This is largely the consequence of residues with helical torsion angles located in the β-strands, and various types of so-called β-bulges have been catalogued.[33] A frequent result of bulges is a distortion of one of the β-strands which falls out of register with respect to the inter-strand hydrogen bonding, leading to an increase in the right-handed twist of the sheet. Almost all β-bulges occur in antiparallel strands, and a classical example is the bulged

Figure 2-14 Formation of β-sheets from individual β-strands. Left pair: the extended conformation of the β-strand viewed in the sheet plane, and viewed across the sheet plane. Note that the carboxyl groups alternate in extending *in* the plane of the sheet, while the residues protrude alternating *above* and *below* each strand. Also visible is the typical pleat of the sheet in cross section. Right two pairs: inter-strand hydrogen bond arrangement in parallel and antiparallel β-sheets.

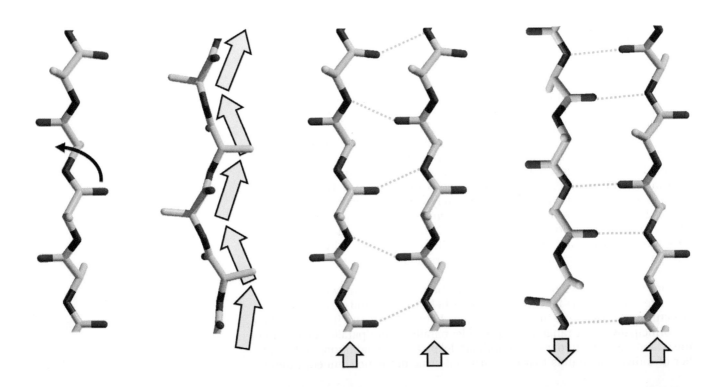

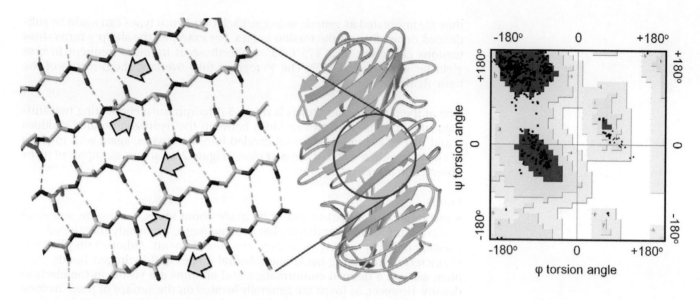

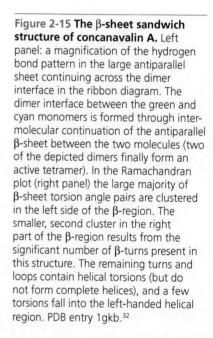

extended hairpin due to a Gly–Ser insertion into a hairpin loop of ribonuclease A[34] shown in Figure 2-16. The model of ribonuclease strongly contrasts the near-perfect antiparallel β-sheets of concanavalin A: in the ribonuclease β-structure elements one recognizes a number of irregular crossovers of strands where the sheet folds back onto itself, including a bifurcation (one strand belongs to two different sheets).

Turns and loops

Turns and loops connect secondary structure elements. While turns are relatively sharp and show internal backbone hydrogen interactions, and thus qualify as *bona fide* secondary structure elements, loops are longer, often flexible sequences of residues that do not have regular backbone interactions and do not belong to any class of regular secondary structure elements. Nevertheless, in both turns and loops, the residues must still have allowable backbone torsion angles. It is a grave mistake to believe that a polypeptide in random coil conformation can have random φ–ψ torsions![35]

Turns

Classification of turns is based on the distance between residues involved in the backbone hydrogen bond interaction.[36] There are $n+2$ turns (γ), $n+3$ turns (β), $n+4$ turns (α), and $n+5$ turns (π). If the turns are longer but still exhibit a defined structure, although without a defined backbone hydrogen network,

Figure 2-15 The β-sheet sandwich structure of concanavalin A. Left panel: a magnification of the hydrogen bond pattern in the large antiparallel sheet continuing across the dimer interface in the ribbon diagram. The dimer interface between the green and cyan monomers is formed through inter-molecular continuation of the antiparallel β-sheet between the two molecules (two of the depicted dimers finally form an active tetramer). In the Ramachandran plot (right panel) the large majority of β-sheet torsion angle pairs are clustered in the left side of the β-region. The smaller, second cluster in the right part of the β-region results from the significant number of β-turns present in this structure. The remaining turns and loops contain helical torsions (but do not form complete helices), and a few torsions fall into the left-handed helical region. PDB entry 1gkb.[32]

Figure 2-16 Ribonuclease A contains irregular β-sheets and a β-bulge. The left panel shows the interruption of the antiparallel β-sheet through an insertion of an additional residue in one strand. In contrast to the near-perfect regularity of the sheets in concanavalin A (Figure 2-15), the β-strands are irregular and show crossovers and a bifurcation. PDB entry 7rsa.[34]

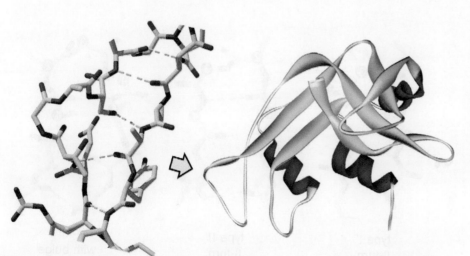

they are annotated as generic ω-loops. Each of the turn types can again be subdivided depending on the torsion angles. For example, the sharp γ-turns show torsions clustering near +75°/–65°, while those of its more frequent inverse γ' cluster around –75°/+65°. The γ' regions thus overlaps the β-region of the basic Ramachandran plot.

One of the most frequent turns is the n+3 β-hairpin turn connecting two antiparallel β-strands (with two residues between the hydrogen bonded residues n and n+3). Hairpin turns are subdivided into three types, again with torsions in characteristic regions of the φ–ψ plot.[36] Figure 2-17 shows examples of turns observed in protein structures.

Loops

A few important remarks need to be made about loops, a constant source of aggravation for the crystallographer. Omega-loops, as already mentioned, are irregular links connecting secondary structure elements. Although they lack the characteristic repeating backbone dihedral angles and hydrogen bonds, they often assume a defined conformation and are thus are visible in the electron density. However, as loops are generally located on the surface of proteins (one may revisit the ribbon drawings of the molecules to verify this fact), loops can also be flexible and as a consequence poorly defined in electron density.

Lack of defined electron density does not necessarily imply the absence of any function. Many loops participate in molecular recognition, where they assume defined conformations upon contact with the binding partner. Similarly, upon ligand binding, some loops undergo rearrangement and open or close a binding pocket, which is often associated with allosteric regulation mechanisms. In such cases, it may not be possible to trace the loop in an apo-form (without ligands, cofactors, or general binding partners) of the protein, and it may be necessary to determine the structure in both free and ligand-bound states. In this sense, loops are in fact important (secondary) structure elements;[37] they just do not fit the rigid requirement of specific torsion angles and hydrogen bond patterns which one can apply to helices, sheets, and shorter turns. Figure 2-18 shows two exposed surface loops, where the electron density becomes increasingly weak in the solvent region. In such loops, the models range from uncertain to completely conjectural, and in the best case represent just one possible conformation of a flexible multi-conformation state. Nevertheless, the torsion angles of loops still must fall in allowed regions of φ–ψ torsions!

It is worth noting that flexible loops are also a problem for NMR, because of the absence of defined NOE restraints between neighboring residues. This makes the study of intrinsically disordered molecules or regions, which often need a certain molecular context to assume defined shapes, generally difficult. Unfortunately, dynamic structural rearrangements play significant roles in very

Figure 2-17 Types I' and II' are the most frequent β-turns. In β-turns, two residues separate the hydrogen bonded residues, while in the α-turn three residues are located between the hydrogen bonded residues. Note that the β-sheet with the α-turn in the right panel has also a bulge after residue number five. PDB entries 1gkb[32] and 1upi.[13]

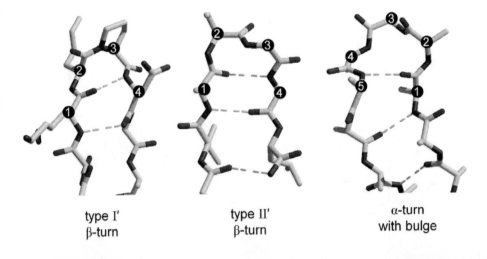

type I'
β-turn

type II'
β-turn

α-turn
with bulge

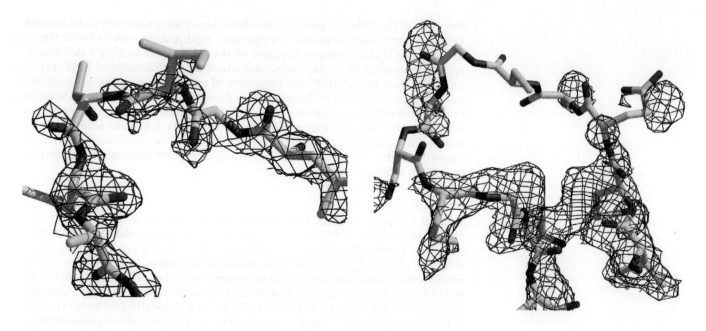

Figure 2-18 **Solvent-exposed loops.** In the left panel, the backbone conformation of the loop can be reasonably well traced, and the general location of the loop is still reasonably well defined. In the right panel, the situation is much worse, and the tracing of the entire loop is uncertain.

important areas such as protein folding, molecular switches, and regulatory mechanisms.

We have now completed the discussion of the basic secondary structure elements — α-helices, sheets, and turns — which are formed through specific hydrogen bonding patterns between the backbone oxygen and nitrogen atoms, leading to characteristic torsion angle distributions. In addition, we have recognized loops as essential secondary structure elements outside of this rigid definition. How these secondary structure elements now interact with each other and form the tertiary structure depends crucially on the properties of the amino acid side chains. So far, we did not have to pay much attention to the nature of the side chains, and it is time to revisit the issue before we proceed to the discussion of structural folds, domains, and motifs.

> Box 2-4 **Secondary structure elements.** The optimal formation of hydrogen bond networks with energetically favored backbone conformations leads to formation of specific secondary structure elements: α-helices, β-sheets, turns, loops, and a variety of other, less frequent secondary structures. The 3.6_{13} α-helices are formed by hydrogen bonds extending from the carbonyl oxygen of the nth residue to the backbone nitrogen of the n+4th following residue of the same helix. The second most frequent secondary structure elements are β-sheets, which can be formed in parallel or antiparallel arrangement between different β-strands. The helices and sheets are connected by turns and loops. Solvent-exposed connecting loops between secondary structure elements are occasionally disordered and only partially traceable in electron density.

2.5   Amino acids

The discussion of secondary structure elements focused on backbone conformations, which were largely independent of the type of side chain of each residue. It is now necessary to examine in greater detail the specific properties of the amino acid side chains. Depending on its side chain properties, a residue will have a certain propensity to be on the surface of the protein, packed inside in the hydrophobic core, or contribute through specific interactions to the stabilization of secondary structure elements. The space requirements and geometric restraints of the different residues also determine their propensity to be located in certain secondary elements. For example, rigid proline with its cyclic structure induces kinks in helices. Glycines on the other hand, lacking a side chain, appear frequently in turns, loops, and wherever conformational flexibility is required. Most importantly, the chemistry of enzymes and receptors is determined by the properties of the residues in the active site pocket through specific interactions with their ligands. As structure-based drug design is a major area in modern structural biology and medicinal chemistry, it is worthwhile focusing some effort on understanding the chemistry of amino acid residues in proteins. Examples of drug target structures and drug–target interactions will appear frequently in subsequent chapters.

Classification of residues

Each of the proteinogenic L-α-amino acids is characterized by specific side chain properties that define via non-covalent interactions the local properties as well as the overall structure of the protein. The different side chains, R, determine the chemical properties of the amino acid residue. In an isolated amino acid, the NH_2 group and the COOH group are charged at physiological pH (7.4), but in a peptide chain only the first, N-terminal residue and the last, C-terminal residue carry the charged NH_3^+ and COO^- groups, respectively. The residues in between carry no ionic charges at the backbone. A residue in our nomenclature is thus a monomer as it appears in the polypeptide chain, including the amino acid side chain plus the peptide backbone. Figure 2-19 clarifies the definition.

A simple primary classification of the residues into *hydrophobic, polar,* and *charged* residues according to side chains properties is common. This classification provides a good starting point for structural aspects, and each group is subdivided further according to the functional chemistry of the residues. A comprehensive compilation of the properties for the common 20 residues is provided in Table 2-2, where the color coding follows the functional classification. In order to train the eye in the recognition of the electron density of residues, we will show stick models of the residues together with their electron density in the following images. Hydrogen atoms are omitted, as they are rarely visible in electron density, and their location can be easily calculated or imagined in the known riding positions.

Hydrophobic residues

The aliphatic residues alanine, valine, leucine, and the branched isoleucine, the aromatic residue phenylalanine, and the cyclic residue proline are hydrophobic (Figure 2-20). Among these six, proline is special insofar as its ring structure

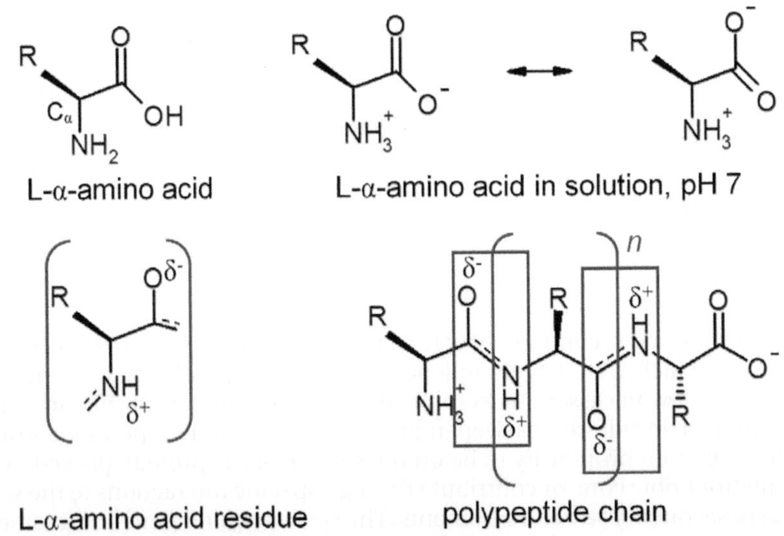

Figure 2-19 Definition of L-α-amino acid, residue, and polypeptide chain. Hydrogen atoms at the chiral Cα atom are omitted. Top left: L-α-amino acid in uncharged state. In solution, the functional terminal groups are charged, and because of resonance the two oxygen atoms are equivalent. Bottom row: a residue of an L-α-amino acid. A residue lacks the second carboxyl oxygen and the second hydrogen atom on the amide group, which are lost during formation of the peptide bonds. The bottom right panel shows the formal composition of a polypeptide chain of *n* residues, with the blue boxes containing the planar peptide bonds.

L-α-amino acid

L-α-amino acid in solution, pH 7

L-α-amino acid residue

polypeptide chain

Residue name	Three letter code	Single letter code	Relative abundance (%) in E.C.	Residue MW	pK$_a$	VdW volume (Å^3)	Residue chemistry	Sec. str. pref.	Hydro- phob.	H- bond D/A
Glycine	GLY	G	7.8	57.05		48		D	0.16	
Alanine	ALA	A	13.0	71.09		67	H	H+	0.25	
Valine	VAL	V	6.0	99.14		105	H	E	0.54	
Proline	PRO	P	4.6	97.12		90	H, cyclic	D	−0.07	
Leucine	LEU	L	7.8	113.16		124	H	H	0.53	
Isoleucine	ILE	I	4.4	113.16		124	H	E	0.73	
Phenylalanine	PHE	F	3.3	147.18		135	H, aryl	E	0.61	
Methionine	MET	M	3.8	131.19		124	H	H	0.26	
Aspartate	ASP	D	9.9	114.11	3.9	91	C, −	D	−0.72	A
Glutamate	GLU	E	10.8	128.14	4.3	109	C, −	H	−0.62	A
Lysine	LYS	K	7.0	129.17	10.5	135	C, +	H	−1.10	D
Arginine	ARG	R	5.3	156.16	12.5	148	C, +	-	−1.76	D
Histidine	HIS	H	0.7	137.14	6.0	118	C, +, P	-	−0.40	D, A
Serine	SER	S	6.0	87.08		73	P, -OH	D	−0.26	D, A
Threonine	THR	T	4.6	101.11		93	P, -OH	E	−0.18	D. A
Tyrosine	TYR	Y	2.2	163.18	10.1	141	P, -OH, aryl	E	0.02	D, A
Asparagine	ASN	N	9.9	114.11		96	P, -NH$_2$	D	−0.64	D, A
Glutamine	GLN	Q	10.8	129.12		114	P, -NH$_2$	H	− 0.69	D, A
Tryptophan	TRP	W	1.0	186.21		163	P, -NH-, aryl	E	0.37	D
Cysteine	CYS	C	1.8	103.15	8.3	86	P, -SH	-	0.04	D
-COO$^-$	-	-	-	44.01	3.9		C, −			A
-NH$_3^+$	-	-	-	17.04	7.4		C, +			D
H$_2$O	HOH	-	-	18.02	7.0		P			D, A

Table 2-2 Properties of amino acid residues. Color-coded for hydrophobic (yellow), acidic (red), basic (blue), polar hydroxyl groups (orange), polar amino groups (light blue) and cysteine (green). Legend for residue chemistry: H: hydrophobic, P: polar, C: charged (positive or negative as indicated). Secondary structure preference: H: helical, E: extended, D: destabilizing. Hydrogen bonds: A: acceptor, D: donor. The listed relative hydrophobicity values are consensus values used in the computation of the hydrophobic moments.[38]

connects back to its own nitrogen atom forming an intramolecular imide, which leads to the special structural properties of proline due to the absence of a free φ-torsion. Methionine, with a methyl group bound to its sulfur atom in the δ-position, also falls into the class of hydrophobic residues.

Hydrophobic core propensity

Hydrophobic residues have a high propensity to be located in the core of the protein, and the formation of a hydrophobic core is an important driving force in the protein folding process (−0.03 kcal mol^{-1} stabilizing energy per Å^2 of buried hydrophobic surface). In structure elements such as helix bundles, the sides of the helices facing each other contain a large number of hydrophobic residues. The situation is even more pronounced in transmembrane helices, which have an overall preference for hydrophobic residues and can thus be quite accurately predicted.[39] So-called *inside–outside* distributions of residues are also based on the statistical preferences of hydrophobic residues to be located inside the protein, and charged residues to be surface-exposed on the outside.

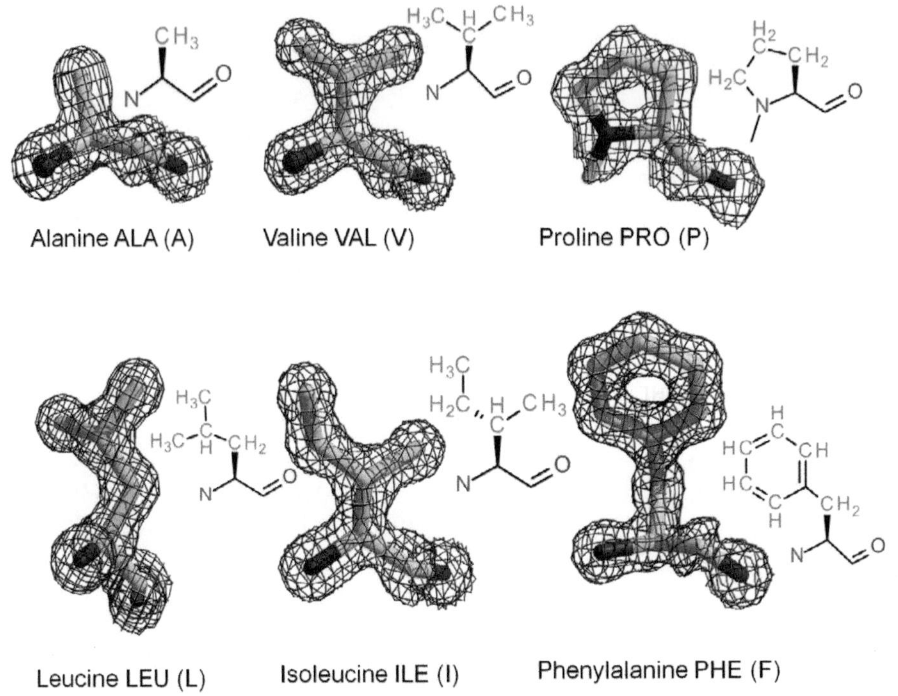

Figure 2-20 **The residues of the six hydrophobic amino acids** alanine, valine, proline, leucine, isoleucine and phenylalanine shown as stick models (*sans* hydrogens) in electron density typical for a structure in the 1.5–1.8 Å range. Note the second chiral center at the isoleucine Cβ-atom.

Figure 2-20 **The residues of the six hydrophobic amino acids** alanine, valine, proline, leucine, isoleucine and phenylalanine shown as stick models (*sans* hydrogens) in electron density typical for a structure in the 1.5–1.8 Å range. Note the second chiral center at the isoleucine Cβ-atom.

The inside–outside distributions are thus useful in protein structure validation to judge the plausibility of a structure model.[40] Although less of an issue in well-defined high resolution structures, unusual inside–outside distributions can be an indication of gross mis-tracing in low resolution X-ray structures. If hydrophobic residues are located in loops, they are frequently the points of high flexibility and disorder. The absence of side chain density of the isoleucine residue in Figure 2-18 may serve as an illustrative example.

Aromatic phenylalanine

In contrast to purely aliphatic residues, the aromatic phenylalanine can in fact form dipole interactions through its π-electrons. This is most frequently observed in π-stacking with other aromatic residues (phenylalanine, tyrosine, or tryptophan, to a lesser degree also histidine) or with the multiple lone pairs of the guanidyl group of arginine. In addition, water molecules are often found above the π-ring of phenylalanine, in agreement with theoretical density functional calculations.[32] As examined later, polar π-stacking interactions also play a role in base pair stacking in DNA and a number of protein–drug or DNA–drug interactions.

Proline

Proline has unique properties as a cyclic imino acid. The ring closure onto its own nitrogen atom creates an imide, which leads to special properties and structural propensities for proline residues. In a peptide chain, the absence of a hydrogen atom at the cyclic proline nitrogen means that no hydrogen bond can be formed to other residues. This particularly disturbs α-helices, where the contact to the residue four positions downstream is broken. Proline thus induces kinks or breaks in α-helices. On the other hand, the highly restrained −70° φ-torsion can provide a good "lead" turn into a helical conformation, and proline is a frequent first residue in an α-helix. In addition, because of the fixed −70° φ-torsion, proline is rare in sheets, but the absence of the amide hydrogen increases its frequency at sheet boundaries, where it may reduce the potential for aggregation as a result of unsaturated hydrogen bonds. In proline, the *cis* and *trans* conformations are geometrically very similar (Figure 2-13), with comparable energies differing by only a few kcal mol[-1], and *cis*-proline bonds are thus more frequent (~7% of all X–Pro bonds) than non-proline *cis*-peptide bonds (~0.1%). The proline ring also can assume a variety of discrete puckering patterns, but they do not seem to be of decisive influence on protein structure.

Methionine

The amino acid methionine (Figure 2-21) also has a higher propensity for hydrophobic environments such as core regions. Its special importance for crystallography lies in the fact that it can be replaced by the chemically practically equivalent seleno-methionine (Se-Met or MSE), which is introduced into proteins during overexpression (Chapter 4). Se produces a strong anomalous signal at wavelengths slightly shorter than 1.0 Å, which is ideally suited for MAD phasing experiments (Chapter 10). The basic structural difference from regular Met is the higher electron density at the Se position, and Se-Met labeled proteins are highly isomorphous to their native counterparts (Sidebar 4-2). Se-Met seems to be more susceptible to chemical oxidation, which is generally not a serious problem in its use as an anomalous scattering label, but it increases susceptibility to X-ray radiation damage compared with native methionine.

Glycine

Glycine (Figure 2-21) is the simplest amino acid, playing a special role in structural elements. Although not strictly a hydrophobic residue, it fits into the series as the member with the smallest side chain, which is a second hydrogen atom. Glycine therefore is achiral, and because of the absence of a side chain Cβ atom, glycine can assume a much wider variety of φ–ψ backbone torsions. As a consequence, glycine has high conformational flexibility and is thus frequently found where sharp turns or flexible hinges between structural elements occur. On the other hand, the high conformational flexibility of glycine leads to entropic losses if it assumes a rigid conformation, and it can disrupt conformationally highly restrained structure elements, particularly α-helices.

Charged residues

In addition to participating in intramolecular interactions crucial for structural integrity, charged and polar residues play a significant role in the chemistry of enzyme reactions and ligand and receptor recognition. Charged residues display a high propensity to be located on the surface of a protein, and they determine the polyionic character of proteins. As a consequence, the charges are located at specific locations on the surface. The distribution of the surface charges changes with pH, and the ability to affect local charge distribution by pH changes[41] is extremely important for protein crystallization (Chapter 3). At the isoelectric point, pI, the total net charge of a protein is zero and its solubility at a minimum.

Acidic residues

We already mentioned that at physiological pH the first and last residue of a polypeptide are ionized and charged (NH_3^+ and COO^-). The two acidic residues, aspartate and glutamate, also contain carboxyl groups with low pK_a values and are thus fully negatively charged at physiological pH.

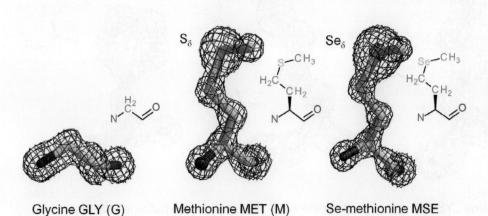

Glycine GLY (G)   Methionine MET (M)   Se-methionine MSE

Figure 2-21 The residues of glycine, methionine, and the non-standard amino acid selenomethionine. Note the higher electron density level at Se position compared with regular methionine. The PDB heterogen code for Se-Met is MSE.

Figure 2-22 The charged acidic residues aspartate and glutamate. The positive charge is delocalized between the equivalent oxygen atoms, corresponding to a description with two tautomeric forms.

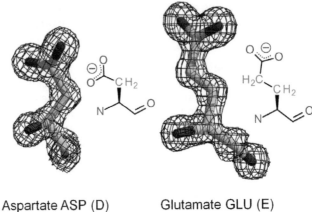

Aspartate ASP (D)   Glutamate GLU (E)

Although the carboxyl groups of Asp and Glu (Figure 2-22) are chemically practically identical, the additional flexibility introduced through the additional methylene group in Glu strongly affects their structural propensities. As an example, Asp is associated with increasing disorder while the glutamate residue has a high propensity to form helices, and both act as capping residues for helices. A significant role of both residues is the formation of salt bridges in intermolecular crystal contacts and the chelating of positively charged metal cations, the rationale for using divalent metal ions as additives in crystallization.[42]

Basic residues

The basic residues lysine and arginine (Figure 2-23) are fully ionized at physiological pH and carry a positive charge because their pK_a values are quite high. As surface side chains, they participate in inter- and intramolecular salt bridges. The high conformational entropy of the four consecutive methyl groups in Lys imposes an entropic penalty upon crystallization, which is the rationale for improvement of crystallization propensity by Lys to Ala mutations[43] via surface entropy reduction (SER) methods (Chapter 4). Under certain conditions the pK_a of Lys can be lowered to a point where its Nζ participates in various reactions; the best example perhaps is the Schiff base formation with pyridoxal phosphate in PLP-dependent enzymes. Lysine has a preference for helical conformations.

In histidine (Figure 2-23), the nitrogen atoms in its imidazole ring are only weakly protonated around physiological pH due to its pK_a only slightly below

Figure 2-23 The basic amino acids lysine, arginine, and histidine. While Lys and Arg are fully ionized at physiological conditions, His is variably protonated depending on pH and can act as a powerful nucleophile in enzymatic reactions.

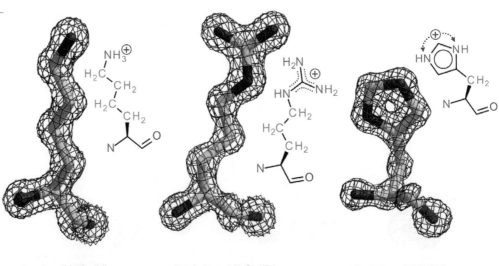

Lysine LYS (K)   Arginine ARG (R)   Histidine HIS (H)

neutral, with the positive charge delocalized between Nδ and Nε. At basic pH, His can be a very strong basic nucleophile, and thus His plays a significant role in enzyme reactions such as protease activity.

Arginine is one of the largest amino acids, but has less entropic freedom than Lys because of the rigidity of the δ-guanidyl group with its delocalized positive charge. The electrons above and below the plane of the guanidyl group allow it to participate in π-stacking with aromatic residues (Figure 2-32). Charged interactions of Arg with carboxyl groups are occasionally found as intermolecular contacts (Figure 2-31). Through these charged interactions (also called salt bridges or ion pairs) Arg also interacts with neighboring residues; a prime example is the loss of LDL receptor binding activity (resulting in type III hyperlipoproteinemia in homozygous individuals) in the apolipoprotein E2 isoform through the Arg112 to Cys112 mutation, which flips a key residue out of the receptor binding site.[44]

Polar residues

The remaining amino acids have polar side chains, which can participate in dipole–dipole interactions or act as hydrogen bond donors and/or acceptors. In this group fall the aliphatic hydroxyl residues serine and threonine as well as the aromatic hydroxyl residue tyrosine (Figure 2-24); the amide residues asparagine and glutamine and the bulkiest residue, tryptophan (Figure 2-25); and finally, the sulfur-containing cysteine residue (Figure 2-26). The hydrogen atoms on their functional groups do not dissociate under physiological conditions. However, a number of important posttranslational modifications and decorations take place at these residues, and the hydroxyl residues play an important part in enzyme reactions, such as in the ubiquitous family of α/β serine hydrolases. The ability to form hydrogen bonds makes most of the polar residues also important for inter- and intramolecular contacts.

Hydroxyl residues

The hydroxyl-group-containing residues participate in hydrogen bonds, and phosphorylation of their hydroxyl groups by kinases is the most ubiquitous mechanism in signal transduction. Structurally, tyrosine can participate in aromatic π-stacking, and all polar hydroxyl residues can act as either hydrogen bond donors or acceptors. The specific handedness of the second chiral center in threonine may be the reason for the helix capping propensity of threonine, which in turn might have induced the selective pressure toward the (2S, 3R) configuration.[45]

Amide residues

The amide residues asparagine and glutamine are frequently found on protein surfaces. Because of their ability to act both as hydrogen bond donors through their NH₂ group and as acceptors via the amide oxygen, they are involved in numerous hydrogen bond and dipole interactions. The major difference

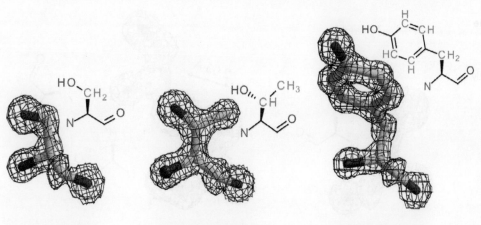

Figure 2-24 The hydroxyl residues serine, threonine, and tyrosine. Note the second chiral center at the Thr Cβ atom. The absolute conformation of threonine is (2S, 3R).

Serine SER (S)   Threonine THR (T)   Tyrosine TYR (Y)

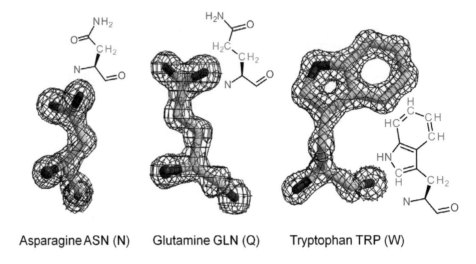

Figure 2-25 The amide residues asparagine and glutamine, and the polar aromatic residue tryptophan. Their large and characteristic planar shape makes Trp residues a valuable positional marker during model building and tracing of a protein chain in electron density.

Asparagine ASN (N)   Glutamine GLN (Q)   Tryptophan TRP (W)

between Asn and Gln is the increased conformational freedom of glutamine, which as a consequence is well tolerated in α-helices, while the more restricted asparagine has increased propensity for inducing disorder. Tryptophan can only donate one hydrogen atom. The large, characteristic shape of tryptophan is helpful in initial model building, where it is readily recognizable because of its size and thus is useful as a sequence anchor.

Cysteine and disulfide bridges

Sulfur-containing cysteine (Figure 2-26) is a particularly interesting residue from a structural point of view. The thiol group of cysteine is relatively reactive, particularly with mercury, gold, and other heavy-metal compounds, which provides for some of the most useful heavy-metal derivatives for isomorphous replacement and anomalous phasing. Solvent-exposed cysteines can be oxidized by disulfide reducing agents such as β-mercaptoethanol, forming S,S-(2-hydroxyethyl)-thiocysteine (Figure 2-27). When electron density beyond the Sγ atom is visible for a cysteine residue, it is good practice to check for the presence of potential reducing agents in the protein stock buffer and the crystallization cocktail.

Under oxidizing conditions, cysteine can form covalent disulfide bonds with neighboring cysteine residues (Figure 2-26) with a bond energy of about 62 kcal mol[-1], a S–S distance of 2.03 Å, and a Cβ–S–S bond angle distribution of ~100–110°, with a mean value of 105°.[46] Disulfide bonds can link secondary

Figure 2-26 The thiol residue cysteine and the S–S bond. Under oxidizing conditions, a disulfide bond can form between two cysteine residues. Proper formation of disulfide bonds is frequently a problem when eukaryotic proteins are expressed in prokaryotic bacterial expression hosts with their reducing cytosolic environment.

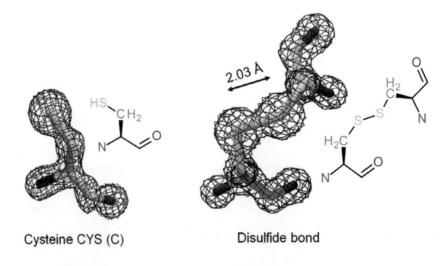

Cysteine CYS (C)   Disulfide bond

structure elements and thereby stabilize tertiary structure, and their formation can be crucial for the proper folding particularly of eukaryotic proteins. Because of the requirement of oxidizing conditions, disulfide bonds are usually found in surface proteins and extracellular proteins, for example in secreted proteases such as trypsin. Under reducing conditions during overexpression in bacteria, S–S bonds generally do not form, and special bacterial strains for the production of disulfide bond containing proteins have been developed (Chapter 4).

As a result of the reactivity of the sulfhydryl group and the increased X-ray absorption of the sulfur atom, cysteines are also the residues most directly affected by radiation damage.[47] On the other hand, because of X-ray absorption by the sulfur atoms of the relatively abundant native methionine and cysteine, these residues are the source of anomalous scattering contributions, which allow phase determination by the sulfur single-wavelength anomalous diffraction (S-SAD) method.[48] Careful and redundant data collection is required, but S-SAD phasing is attractive as it works well at the longer wavelength (1.54178 Å) produced by copper anode lab generators. Chapter 8 (Instrumentation and data collection) and Chapter 10 (Experimental phasing) contain a practical S-SAD phasing example.

Ambiguities in side chain electron density

During model building, the crystallographer (or an automated model building program) places residues in the electron density, with the most distinguishing feature of each residue being the shape of the side chain. Well defined side chain density, together with knowledge of the sequence, generally allows reliable side chain placement based on the observed density shape. Some ambiguity, however, exists that cannot be resolved by relying solely on electron density shape: The orientation of asparagine, glutamine, and histidine cannot be unequivocally assigned from electron density alone. The scattering difference between nitrogen ($7e^-$) and oxygen ($8e^-$) is generally not enough to distinguish the two possible orientations of the amide groups or of the His imidazole ring. The proper conformation of Asn, Gln, and His needs to be assigned according to chemical plausibility, such as satisfying hydrogen bond networks, which is often straightforward in the case of Asn and Gln. The fact that His can have different protonation states depending on its actual local pK_a often makes the assignment of His side chain orientation more subtle.

Surprisingly, a large fraction of protein crystal structures still have incorrectly assigned Asn and Gln side chain amide conformations (Figure 12-32). This is even more astonishing, as checking programs exist, which flag questionable assignments, and a recent one based on self-consistent empirical potentials[49] as well as the validation program *MolProbity*[50] even return "flipped" coordinate files. Note that just swapping the atom names is not correct; the amide groups as well as the imidazole ring need to be rotated around the preceding torsion. Even then, because of the different bond lengths and angles, the flipped coordinates need to be re-refined against the X-ray data (discussed in Chapter 12). Improperly assigned NQH residues seem to be nonchalantly regarded as a nuisance error, but for purposes such as ligand docking and molecular modeling, the proper assignment of hydrogen bond donors and acceptors is crucial.

Side chain torsions and multiple side chain conformations

Just as the backbone torsions are limited in their conformational freedom, the side chains of residues assume certain preferred conformations. Long aliphatic side chains leading to the positively charged groups in lysine and arginine can assume a number of staggered conformations, while valine, for example, has a strong preference for a single Cα–Cβ torsion with a narrow torsion angle distribution. The torsion angle distributions are collected in torsion angle libraries, and are used as restraints — although not as strict as bond angles and distances — in refinement and for structure validation. The local environment defined by intra- or intermolecular contacts determines which of the possible favored torsions is actually assumed. Strong interactions with other residues or ligands can lead to strained torsions, but very unusual side chain torsions certainly warrant closer inspection. A special refinement protocol exploits the fact that fewer parameters are needed to describe a protein through its torsion angles, and torsion angle refinement is frequently used in conjunction with simulated annealing molecular dynamics in early stages of structure determination (Chapter 12).

The fact that the local environment determines which torsion angle combinations are assumed also means that in a local environment that is ambivalent or fluctuating, multiple conformations of the same side chain are possible. This is frequently the case with surface-exposed residues. As shown in Figure 12-33, the conformational split can range from discrete dual conformations to partial side chain density and up to near complete absence of discernible side chain electron density past Cβ in multiple, fluctuating conformations.

Modified residues

Posttranslational modification of residues occurs frequently in eukaryotic systems. A few examples are phosphorylation of hydroxyl residues Ser, Thr, and Tyr via kinases in cell signaling (despite their crucial importance, only about 1% of all kinases are tyrosine kinases; the remainder are Ser and Thr kinases), carboxylation of glutamate, N-glycosylation of amide residue Asn in cell surface recognition and transport proteins, Cγ-hydroxylation of prolines in structural collagen proteins, acetylation of N-termini as protection against degradation, methylation of lysine residues, PLP cofactor binding to lysine in PLP-dependent enzymes, or the covalent link to heme groups in bacterial cytochromes. Various posttranslational modifications are shown in Figure 2-27, and Figure 4-20 illustrates an example of N-glycosylation.

Proteins expressed in bacterial systems generally lack such posttranslational modifications. Nevertheless, occasionally extra density is observed protruding from residues, mostly on the surface of proteins. Here it is advisable to check the protein's preparation and purification history as well as the crystallization cocktails for reagents that could attack reactive groups. A classical example is the previously mentioned formation of S,S-(2-hydroxyethyl)-thiocysteine through reaction of cysteine with the reducing agent β-mercaptoethanol shown in Figure 2-27.

About pK_a values of amino acids

The tabulated pK_a values for amino acids tend to vary widely from source to source. The reason is foremost that pK_a values are perturbed by the local environment defined by the neighboring residues and by solvent polarity, all of which affect the acid–base equilibria.[51] Already the pK_a values for the NH_3^+ and COO^- termini vary for the free amino acids, and they depend further on the position of the termini in the context of the folded protein. The same holds for the other charged residues. For example, low polarity raises the pK_a of an acid residue, because it destabilizes the ionized form. Conversely, if a basic lysine residue is buried in a hydrophobic environment, its pK_a will be lower. We mentioned already that the shifting of the pH dramatically changes the charge distribution on the protein surface, which is therefore a powerful parameter in crystallization screening. Similarly, the chemistry of each residue can be very different

Figure 2-27 Covalently modified residues. Left and center panels: N-trimethyl lysine (PDB heterogen code M3L) and N-acetylated alanine (N-terminal acetylation) in chicken smooth muscle calmodulin. Right panel: S,S-(2-hydroxyethyl)-thiocysteine (PDB heterogen code CME), formed through reaction of the surface-exposed cysteine modified with reducing agent β-mercaptoethanol present in the purification buffer and in the crystallization cocktail. PDB entries 1up5 and 1upi.[13]

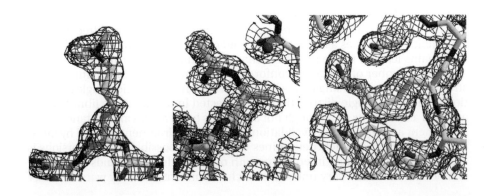

Box 2-5 **Amino acid side chains.** The side chains of amino acid residues in the polypeptide chain are chemically different and cover a wide variety of properties, coarsely divided into hydrophobic, acidic, basic, and polar. They are responsible for intramolecular interactions and self assembly of several secondary structure elements into the tertiary structure and quaternary structure, and they define the chemistry of the active site of a protein.

depending on its pK_a value, and even extreme pK_a values can be found for buried residues. It is important to keep this in mind when interpreting enzyme reaction mechanisms or ligand and drug interactions in crystal structures. Histidine, for example, is a powerful nucleophile in the basic regions above its pK_a, but can also participate in charged interactions in the acidic region.

Figure 2-28 summarizes the acid–base equilibria of the charged amino acid residues and the N- and C-termini. Remember that at the pK_a, the charged and uncharged form are present in equal amounts, and the farther away the pH from the pK_a in the direction of charged form, the more the equilibrium will be on the charged side and *vice versa*.

2.6   The interactions of side chains

Side chain interactions play a crucial role in the stability and assembly of protein folds, in intermolecular crystal contacts, and in any protein–ligand interactions. As crystallographers and structural biologists, we have to interpret and understand in every crystal structure the wide range of interactions in which side chains are involved. Specific side chain properties have been detailed already in the previous section, and a semi-quantitative review of the types of interactions we encounter between side chains and their binding partners follows. It is important to understand that all weak non-covalent interactions such as hydrogen bonds, ionic, polar, and even van der Waals (vdW) interactions are based

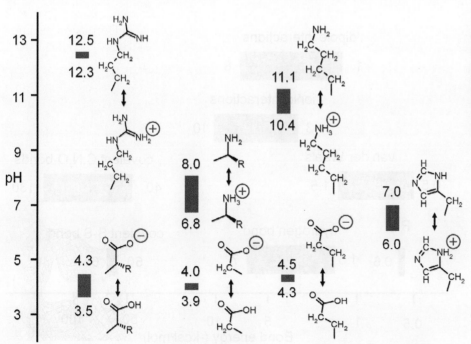

Figure 2-28 Logarithmic acid dissociation constant (pK_a) of charged amino acids. The figure provides a quick overview of the normal range of pK_a values for charged amino acid side chains and of amino- and carboxy-terminal groups. Unusual and extreme local environments can further shift these normal ranges to atypical values.

Sidebar 2-5 **Covalent binding in suicide substrates and crystal cross linking.** In drug–ligand interactions, *suicide* substrates such as α-difluoromethylornithine (DFMO) or 5-fluorouracil permanently block binding sites (Figure 2-30). This is particularly exploited in cancer drugs or antibiotics, while most other interactions between drugs and their protein targets are designed to be reversible[52] and thus non-covalent. Another example for decidedly negative effects of covalent enzyme blocking is the family of organophosphorus nerve gases (e.g., Sarin), which irreversibly block the catalytic serine residue in the ester binding pocket of acetylcholinesterase (see also Sidebar 13-5). Of practical use in crystallization are covalent cross-linking agents such as glutaraldehyde, which can be soaked into crystals to improve their stability.

on electronic interactions, and the distinction between them is thus sometimes floating. Multiple effects can contribute to an interaction between side chains, and it is rare to find pure interactions of precisely one or the other character. Figure 2-29 summarizes the energy range and interaction distance of the most important types of intermolecular interactions.

Covalent interactions

Covalent interactions involve the formation of strong and highly directional covalent bonds (and thus chemical alteration of the side chains), which distinguishes them from the other, weak interactions. Covalent binding energies involving C, N, and O atoms of side chains are generally in the $\sim$–100 kcal mol^{-1} range (more than one order of magnitude stronger than the weak, reversible interactions), and they are often irreversibly formed or require enzyme action for their formation and breakage. Examples we already discussed are the covalent modifications of side chains, and the covalent attachment of cofactors to enzymes. In protein–protein interactions, covalent bonds normally do not play a role, with the exception of the intramolecular S–S bonds of $\sim$–62 kcal mol^{-1} binding energy.

Ionic interactions

This class of weak interactions encompasses charged electrostatic interactions, with the terms *ion pair* or *salt bridge* specific to interactions where both partners are charged. The strongest bonds in this class, with energies up to –10 kcal mol^{-1}, are ion pairs between two side chains of opposite charge. Typical examples are the ion pairs formed between negatively charged acidic residues and positively charged basic side chains (Figure 2-31). Ionic interactions have no directionality. A negatively charged residue can also act as a hydrogen bond acceptor, and negatively charged residues with hydrogen atoms can act as hydrogen donors. Ion pairs also are frequent mediators of strong intermolecular crystal contacts. In protein–drug interactions, ion pairs are not as frequent, because the presence of charged residues decreases the lipophilicity of the drug molecule, which makes it difficult for the drug to cross membrane (blood–brain) barriers.

Hydrogen bonds

Hydrogen bonds are largely dipolar interactions that involve at least one hydrogen atom. Hydrogen bonds require a hydrogen donor and a hydrogen acceptor group, and have binding energies of typically –3 to –7 kcal mol^{-1}. Good

Figure 2-29 Typical ranges for bond energies of side chain interactions. RT denotes the thermal energy at room temperature (293 K). Note the logarithmic energy scale. The numbers in the red boxes give the approximate radial distance dependence for charged and polar interactions, and an approximate interaction range for the directionally dependent hydrogen bonds. The numbers left and right of the red boxes flank the approximate bond energy range. In contrast to the weak non-covalent interactions, covalent bonds have specific and discrete bond distances and bond angles.

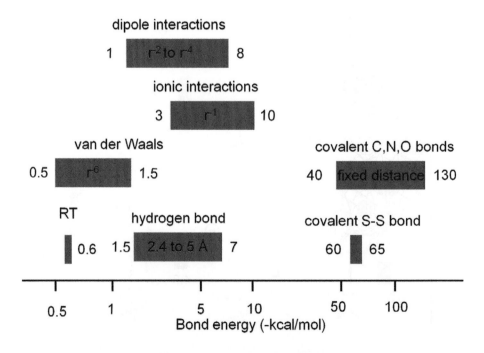

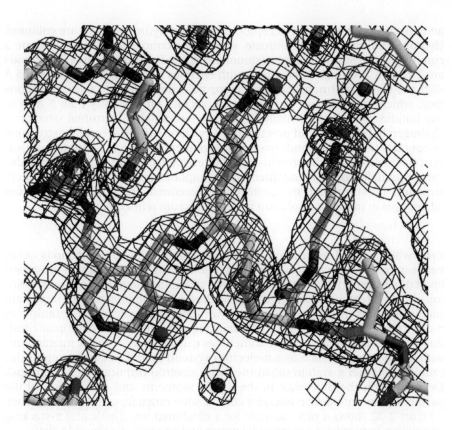

Figure 2-30 Covalent binding of ligands. The suicide substrate α-difluoromethylornithine (DFMO) covalently cross-links a cysteine residue with the PLP cofactor in the active site of ornithine decarboxylase from *Trypanosoma brucei*, the parasite that causes sleeping sickness. PDB file 2tod.[53]

hydrogen acceptors are the acidic residues, the oxygen atoms of the polypeptide backbone, followed by sp^2 hybridized entities with oxygen atoms which may themselves have hydrogen bound, but still have lone pairs to attract hydrogen, such as hydroxyl residues or water molecules. Common hydrogen donors are amide or imide groups, or weaker acceptors with hydrogen bound atoms, sharing their hydrogen with a strong acceptor. Hydrogen bonds are semi-directional,

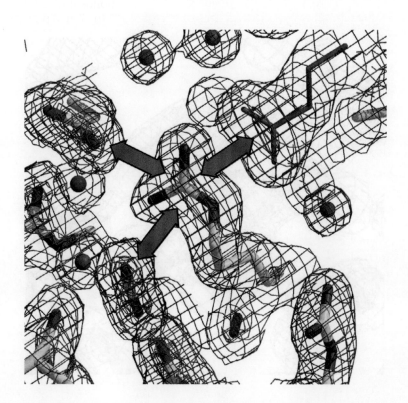

Figure 2-31 Multiple charged intermolecular ion–ion interactions. The positively charged, basic guanidyl group of an Arg side chain is surrounded by three negatively charged, acidic Glu residues. Note that the Arg also participates in a crystal contact to a symmetry-related Glu residue shown in purple. PDB entry 1nfo.[44]

Sidebar 2-6 The ubiquitous role of the hydrogen bond.

"Although the hydrogen bond is not strong it has great significance in determining the properties of substances. Because of its small bond energy and the small activation energy involved in its formation and rupture, the hydrogen bond is especially suited to play a part in reactions occurring at normal temperatures. It has been recognised that hydrogen bonds restrain protein molecules to their native configurations, and I believe that as the methods of structural chemistry are further applied to physiological problems it will be found that the significance of the hydrogen bond for physiology is greater than that of any other single structural feature"

Linus Pauling,[56] in *The Nature of the Chemical Bond*, 1960, Chapter 12, pp. 449–450.

and are strongest when the acceptor, donor, and hydrogen atom are collinear. Hydrogen bonds can also bifurcate, where for example two acceptors share a common hydrogen atom. Typical donor–acceptor distances for hydrogen bonds are 2.7–2.9 Å, although interactions with energies above RT start at around 5 Å distance within a strong donor–acceptor pair. Beyond ~5 Å, the interactions fade, which is the reason why solvent molecules farther away than 5 Å from any bonding partner are usually flagged as questionable in protein structures. Hydrogen bonds are found practically everywhere, within secondary structure, in contacts between secondary structure elements, in crystal contacts, in solvent interactions, and in protein–ligand interactions. Hydrogen bonds can also carry a partially covalent and directional component[54] (often ignored in protein structure literature), and the inclusion of directional components has certain advantages in medium resolution protein structure refinement.[55]

Polar interactions

Even when no hydrogen atom is available, dipole–dipole interactions occur between partners carrying an opposing dipole moment. There are many varieties of dipole (polar) interactions, which are all grouped together in this category. Ion–dipole interactions as well as dipole–dipole interactions play an important role in receptor-drug interactions, and we mentioned already the example of π-electron–dipole interactions in the stacking and interactions of aromatic residues. The latter interactions are sometimes called charge-transfer interactions, with a donor group (such as a π-electron system) paired with an acceptor (for example, another π-system substituted with electron withdrawing (-E) groups). Depending on the difference in the dipole moments and distribution of the partners, polar interaction energies cover a wide range from ~–1 to–8 kcal mol^{-1}. Figure 2-32 shows a nice example for a combined ion–dipole and π-stacking interaction between an Arg guanidyl group and an aromatic Phe side chain.

Van der Waals interactions

Van der Waals (vdW) interactions are weak and nonspecific, and are based on intermolecular London dispersion forces. They are weak attractive forces caused by polarization of the electron clouds when atoms approach each other, up to the point where the electron distributions start to overlap and cause rapidly increasing repulsive forces once the distances are below the sum of the

Figure 2-32 Polar π-stacking interactions between the guanidyl group of Arg (δ⁺ dipole) and the π-electrons of a phenylalanine (δ⁻ dipole). This interaction also has some charged character, and a weak hydrogen donor–acceptor interaction between Arg and the Phe π-system may also contribute to the interaction.

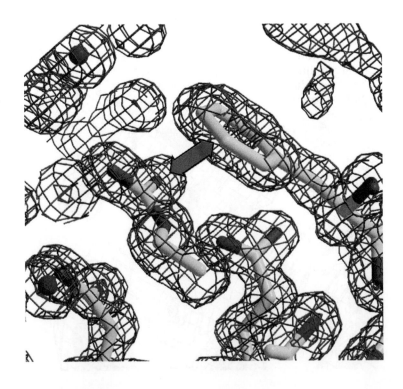

vdW radii of the partners (Figure 12-18 shows the corresponding Lennard-Jones potential curve). Although each individual vdW interaction is quite weak (–0.5 to –1.5 kcal mol^{-1}) and barely compensating for RT energy (0.6 kcal mol^{-1}), the sum of many vdW interactions can reach quite substantial values. Buried surfaces or contact areas between proteins can be large, and vdW forces for example provide a significant part of protein–protein oligomer interface interactions. The same holds for drug–receptor interactions, where good shape complementarity in addition to specific interactions can contribute a significant part of the binding energy.

Hydrophobic interactions

We already mentioned in the discussion of hydrophobic side chains that hydrophobic interactions are the driving force behind the initial folding of proteins. Hydrophobic residues tend to associate in a polar medium such as water, and a hydrophobic protein core is formed. Around hydrophobic residues, water molecules must satisfy their hydrogen bonds with each other, and an ordered water clathrate (or cage) is formed, which decreases the entropy of the solvent. When two hydrophobic groups approach each other, these clathrates "melt" together and fewer water molecules need to be ordered. This gain in solvent entropy is equivalent to a decrease in free energy ($\Delta G = \Delta H - T\Delta S$), and a stabilizing interaction results. As this is not a real attractive force between hydrophobic groups, the energy is hard to quantify for a "mole" of hydrophobicity. As a guide, an additional -CH$_2$- group contributes about – 0.7 kcal mol^{-1} to a hydrophobic interaction, and as in the case of vdW interactions, a large number of hydrophobic interactions can contribute significantly to the overall interaction energy.

Solvent structure

With the properties of hydrogen bonds and side chain interactions fresh in mind, it is timely to consider the structure of the solvent around the protein. Soluble proteins in cellular compartments or intercellular space do not float around freely, but they share — just as they do in crystals — with other proteins a very crowded environment full of water molecules, small organic molecules and metabolites, and ions of every element present in the cellular environment.

On the surface of the protein we find charged and polar residues, as well as unsatisfied backbone hydrogen bond acceptors and donors from terminated secondary structures and connecting loops. It is therefore not unexpected that these exposed charges and dipoles hold on to whatever satisfies and maximizes the network of hydrogen bonds and other possible interactions. The most prevalent molecule available both in protein crystals and in (non-membrane) cellular environments are water molecules. They can act as both hydrogen donors and acceptors, and are thus able to form extensive networks with multiple partners. Figure 2-33 shows hydrogen bonded solvent atoms that form a network between residues of symmetry related molecules (Figures 12-27 and 12-28 provide additional examples).

Hydration shells and water dynamics

High resolution X-ray structures have shown that the narrow boundary between the protein and the solvent indeed exhibits a locally specific and often functionally relevant structure, including ordered and conserved solvent molecules within one or two coordination shells. The solvent within this ~7 Å wide boundary layer is often well ordered in a crystal structure. In fact, the same protein structures determined in different space groups and with different crystal packing show that a large number of the strongly bonded water molecules are conserved, and also form an integral part of the protein's solution structure. During the discussion of hydrophobic core formation we already recognized that the release of ordered water molecules from ordered clathrates around hydrophobic residues is a major driving force in protein folding. Similarly, entropic gains in the change of the surrounding water structure are a most significant factor driving the thermodynamics of crystallization, and we will revisit this subject in the crystallization chapter.

Sidebar 2-7 Binding energy and binding constant. To appreciate the crucial role and effect of different interactions for structure based drug design and lead optimization, it is useful to establish a relation for the binding energy between a generic receptor–ligand complex [RL] and the associated *binding constant*, K_d. The lower the binding constant (which is defined as a dissociation constant), the better the ligand [L] or drug binds to the receptor [R]. With the K_d defined as

$$K_d = \frac{[R][L]}{[RL]} \text{ , and}$$

using $\Delta G^o = -RT \ln \frac{1}{K_d}$

or $\Delta G^o = RT \ln K_d$ (2-1)

we can estimate the effect of any additional favorable interaction on the binding constant of a drug: A change of one order of magnitude in K_d (ln 10) equals 2.302·RT. With RT = 0.62 kcal mol^{-1} at physiological body temperature (310 K), we obtain a gain in binding energy of –1.43 kcal mol^{-1} for each order of magnitude in decrease in K_d. This means that the addition of a single, strong hydrogen bond of ~–5.7 kcal mol^{-1} to a drug–receptor interaction network can improve the binding constant of a drug so much that it binds 4 orders of magnitude (10 000 fold) better! This clearly shows that targeted modification of protein–drug interactions through rational, structure guided drug design holds great promise. Structure based drug design has thus become a standard method in modern drug discovery, complementing assay-based combinatorial high-throughput compound screening.

Figure 2-33 An intermolecular hydrogen bonded solvent network. Water molecules are frequently bound to the surface residues of proteins and are often involved in crystal contacts. The residues shown in purple stick representation belong to a symmetry related molecule. PDB entry 2j9n.[57]

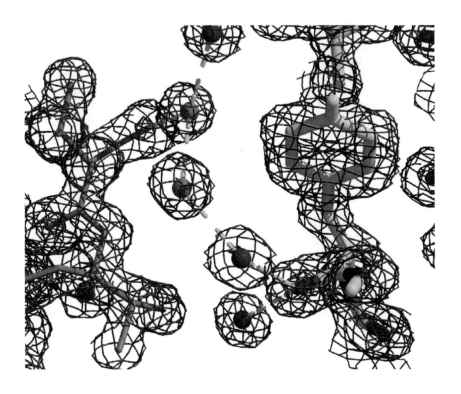

Deuterium exchange studies have demonstrated that strongly bonded water molecules and even buried water atoms still exchange with the continuous, dynamically moving solvent around the protein. This emphasizes the fact that proteins undergo small dynamic fluctuations on a fast time scale which may be not, or only partly, evident from the crystal structure. We will discuss later in more detail how dynamic motion of a protein can be inferred from X-ray structures by anisotropic displacement analysis (Chapter 12).

The fact that the solvent beyond the nearest coordination shells around the protein is completely disordered, can be used to advantage in map improvement techniques such as solvent flattening or solvent flipping. These density modification techniques exploit the fact that partial structure factors for disordered solvent with an average electron density of ~0.4 e⁻ Å⁻³ can be calculated and used to improve the initial experiment phases. Density modification techniques are very powerful and we will treat them in Chapter 10.

2.7   Protein tertiary structure

Secondary structure elements, which are generally defined by their typical backbone hydrogen bond patterns and backbone torsion angles, self-assemble, largely via side chain interactions and unsatisfied backbone interactions, into

Box 2-6 **Side chain interactions.** Side chain interactions are generally noncovalent, weak, and non-directional, with bond energies the range of –1 to –10 kcal mol⁻¹. They also form the basis for enzyme–ligand and drug–receptor interactions. Solvent-exposed side chains are frequently disordered and can be modeled in split, dual conformations. From electron density alone, placement of N, Q, and H side chains is ambiguous except at atomic resolution, and chemical plausibility of the hydrogen bond and interaction network must be considered for correct NQH side chain conformations. Conserved solvent molecules (water) form an integral part of the solution structure of the protein.

a domain or protein fold. Within a folded domain, the core typically is rich in hydrophobic residues, while the surface is decorated with charged and polar residues. The simplest classification of protein domains into folds is according to their constituting secondary structure components.[58] The four fundamental fold classes are thus α-helical structures, pure β-structures, α/β-structures that containing a mixture of α-helices and β-strands, and a fourth α+β class that comprises folds consisting of discrete α-only and β-only elements. We will discuss only a few typical examples in the following sections, and refer for a more extensive description of functionally important protein structure families to the excellent text of Brändén and Tooze.[59] Instead, we focus on aspects of importance for crystallography — such as domain structure, the identification of new folds and structures, and the functional relevance of quaternary structure assemblies.

Structural domains

Domains are independently folding 3-dimensional structures, about 50–250 residues in size, generally with a specific function that often gives the domain its name, such as ATP-binding cassette, LDL receptor binding domain, catalytic domain, and so forth. They can exist as functional single domain proteins of their own right, or they can be part of a larger, multi-functional, multi-domain protein. Domains are used by nature as "cassettes" or building blocks, which can be assembled or swapped in different orders. Within a structure, an exchange of domains (domain swapping) can occur and be exploited for protein engineering purposes.

Multi-domain proteins

Several functionally different and distinct domains can fold from a single protein chain, and such multi-domain proteins are common. For example, the *Clostridium botulinum* neurotoxin depicted in Figure 2-1 has three distinct, functionally different domains: (i) a ganglioside-binding domain that targets and binds the holotoxin to the intersynaptic membrane; (ii) a translocation domain that participates in membrane vesicle fusion and translocation into the neuronal cytosol of its third domain; (iii) the actual Zn-protease neurotoxin. The combination of all the domains of a protein comprises its tertiary structure. Finally, different oligomers can form from single-chain proteins, which constitutes the quaternary structure. Homo-dimers of enzymes and hetero-pentamers of neuroreceptors are examples of common, functionally relevant quaternary structure assemblies.

Domain expression and truncation

For crystallography, the fact that domains fold independently and are generally stable without their partners is of significant relevance — it allows us to express and crystallize single domains in isolation. This is particularly helpful if the multi-domain protein in question is either very large or the domains are flexibly tethered, which generally makes self-assembly into well-ordered, periodic crystal lattices difficult. Figure 2-34 depicts the domain structure of a membrane receptor with a single transmembrane crossing helix, where separately expressed domains have been successfully used for structure determination. However, if a

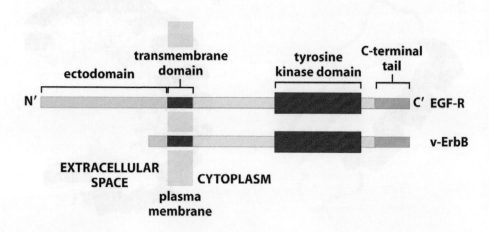

Figure 2-34 **Domain structure of the epidermal growth factor receptor (EGF-R) and the related mutant oncoprotein v-ErbB.** Both the growth factor binding ectodomain of the EGF-R as well as the tyrosine kinase (TK) domains have been separately expressed and the X-ray structures determined. Normal signal transduction occurs via dimerization of receptors upon ligand binding and subsequent transphosphorylation, where one receptor TK domain phosphorylates the other dimer partner. The oncogenic v-ErbB receptor lacks a complete ectodomain but causes persistent mitogenic signaling independent of ligand binding. See also Sidebar 5-2 discussing the role of dimerization. With permission from reference 29 (Garland Science).

domain depends on its surrounding partners for proper folding (as for example seems to be the case for the long, helical translocation domain of *C. botulinum* neurotoxin in Figure 2-1 or for polytopic membrane proteins), the domain-cutting approach does not work. We will examine the different methods of utilizing domain truncation for protein engineering in more detail in Chapter 4.

Structural motifs

Within a domain, we can often distinguish smaller, but conserved combinations of secondary elements that repeatedly appear either in different protein folds or repeat within the same fold. Such secondary structure sub-assemblies are called structural motifs. They are often associated with (and named after) a specific function, but they are not complete functional proteins or domains by themselves. We will briefly discuss two motifs that have interesting features from a crystallographic point of view.

The β-α-β motif

Among the first motifs detected in protein structures is the β-α-β motif. It is interesting insofar as it solves a problem which the observant reader may have discovered during the study of β-sheet structures. In the antiparallel β-sheet structures such as shown in the β-sandwich of concanavalin A in Figure 2-15, forming the antiparallel sheet out of a contiguous polypeptide chain is quite simple: just fold the peptide chain over, and the reverse strand is automatically connected through a short loop. Such an arrangement can be infinitely repeated, and the extended sheets of antiparallel β-strands building the β-barrel domain shown in Figure 2-37 are built this way. For a parallel β-sheet, however, the peptide chain has to trace back to the N-terminal edge of the sheet, across the strands, which can be accomplished by a long flexible loop, or better with a more stable α-helix. The connection of the two parallel β-strands with a helix creates a sturdy motif, which is used to build α/β barrel structures consisting of repeating and overlapping β-α-β motifs, exemplified by the triosephosphate isomerase and the α/β horseshoe domain shown in Figure 2-35. Typically, such α/β barrels contain the active sites in the core walled by the parallel β-strands.

The calcium-binding EF-hand

Another hardy perennial in the garden of structural motifs is the Ca-binding EF-hand (Figure 2-36). It exemplifies both a metal binding site and a very common motif in important Ca^{2+}-dependent signaling proteins, and has recently been discovered in domain-swapped dimers of the pollen allergen PhI p 7.[61] The EF-hand motif has been extensively described in the literature and in protein structure textbooks.[59] For us it is of interest as it shows a prime example for a metal binding site that is an integral part of the protein structure. The EF-hand is a sturdy helix-turn-helix motif with the turn looping around an octahedrally coordinated Ca^{2+} cation. The name EF-hand stems from the numbering of the helices in parvalbumin, where this motif was first described in the

Figure 2-35 The β-α-β motif (A) serving as structural repeat building up the α/β horseshoe domain (B) of the ribonuclease inhibitor. PDB entry 2bnh.[60]

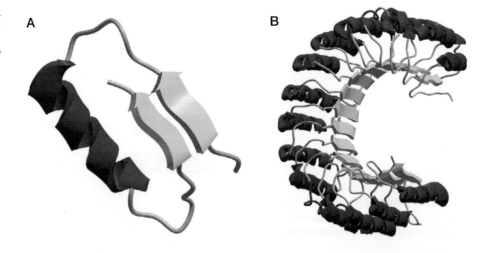

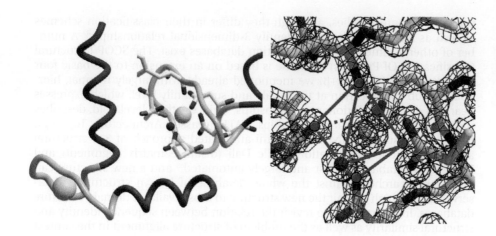

Figure 2-36 The Ca-binding EF-hand motif. Next to the topological overview (left) is the ball and stick representation of the Ca-contacting residues and water molecule. The electron density (right panel) includes in the same orientation a sketch of the slightly distorted octahedral coordination around the Ca²⁺ cation. PDB entry 1up5.

early 1970s, and the shape was perceived as a hand with extended fingers that cups the calcium ion. The calcium ion itself is coordinated by a backbone oxygen, a water atom, two carboxyl oxygen atoms of Asp residues, the amide oxygen of a Gln residue, and the forked carboxyl group of a glutamate residue. The interactions are a quite complex mixture of ionic and polar dipole interactions, and although there are seven atoms around the calcium, distances indicate that the bifurcated contact to the Glu residues has a large ionic component, with the sixth corner of the octahedron presenting the point of highest negative partial charge density. A table of protein–metal coordination distances is provided in Appendix A.11.

Domain structure databases

A crucial question for a crystallographer or structural biologist is how to make sure whether a newly determined protein structure that did not show any significant sequence similarity in a sequence search against genomic data represents a new fold, or whether it belongs to an already known fold family. This quest involves a search of 3-dimensional fold classification databases.

Structure is more conserved than sequence

There are a few important points worth emphasizing: While high sequence similarity practically assures structural similarity, the opposite is not true. Structure is much more conserved than sequence, meaning that structurally (and thus frequently, functionally) similar proteins do not have to share similar sequences. Their structural similarity can be a result of divergent evolution or of very distant homology.

For the crystallographer, this means that despite the fact that a sequence did not yield any alignment with a member of any known structure family, the fold may still turn out to belong to a well-known structure family. It is thus a good idea to carefully check all available structure family databases before engaging in a detailed new fold discussion. Irrespective of the result of these efforts, it needs to be understood that domain recognition or assignment is a very difficult task, and it may well be that a "new" structure qualifies at a lower or different level of structural superposition or connectivity as a member of a known structure family. Variable criteria in domain definition include for example levels of superpositional granularity or how connectivity is treated. Subtleties such as domain swapping and domain structure assembled from different oligomer subunits contribute to the difficulty of precisely defining domains. How many structure families exist, and how discrete (or degenerate) the fold space population is, is thus a subject of debate. At present, at the most granular level of homologous superfamily about 1500 different domain structure types are distinguished.

Major domain structure databases

The three largest fold classification databases, SCOP, CATH, and Dali/FSSP show a reasonable correspondence of about 80% between their 3-dimensional

Sidebar 2-8 Structure prediction will not obsolete experimental structure determination. It is worthwhile noting that *in silico* structure prediction without significant sequence identity by threading and fold recognition methods is difficult, and the chances for a correct *ab initio* predicted structure are relatively low.[64] The difficulty of predicting protein structures without a sequentially similar experimental structure model is one of the reasons that motivated the Protein Structure Initiative (PSI) of the U.S. National Institutes of Health (NIH). The rather optimistic premise is that once all folds occurring in nature have been found, threading or fold recognition methods would be able to predict the 3-dimensional structure for any given sequence. How successful these efforts will finally be, remains to be seen. If one adds to the limited accuracy of molecular modeling at low sequence identity (<20–30%) the importance of conformational changes upon ligand/ partner binding and context-induced folding of intrinsically disordered (or low complexity) regions, we recognize that experimental structure determination will remain crucially important for the foreseeable future.

protein structure families, although they differ in their classification schemes as well as in their methods to identify 3-dimensional relationships.[62] A number of other 3-dimensional classification databases exist. The SCOP (Structural Classification Of Proteins) database is based on an extension to the basic four groups (α, β, α/β, and α+β) we mentioned already, with largely manual, hierarchical sub-classification at the family and superfamily level, which expresses near and distant evolutionary relationships, while the third level, fold, describes the actual 3-dimensional relationships. CATH (Classes, Architecture, Topology, Homologous superfamily) uses a semi-automated approach of 3-d structural and phylogenetic classification, while Dali (distance matrix alignment) and its domain database are fast and largely automated[63] and a new structure can rapidly be searched against the whole database of known structures. Meta-servers exist which submit the new structure to most available domain structure databases. In Chapter 11 we revisit the relation between sequence identity and structural similarity as well as the problem of structure alignment in the context of molecular replacement phasing.

Examples of domain architecture

Instead of cataloging structures by function, as is common in structural biology and protein structure texts, we will give only a few selected examples for fold architectures that are representative of common structure classes, or serve as example cases which we use throughout the text. At the most detailed level of superfamilies, about 1500 distinct fold entries are represented in both the SCOP and CATH databases.

Figure 2-37 shows in the top row two family members of the α-only class we used previously as examples: the up-down α-bundle of apolipoprotein and the orthogonal α-non-bundle of myoglobin, the first protein structure ever determined. The β-barrel is a very interesting fold of the mostly β-class that forms porins which allow passive diffusion through bacterial membranes, form the trepanizing "drill" of bacteriophage stems, and provide the scaffold of the green fluorescent protein (GFP, Figure 4-13). The β-*sandwich* is another β-only architecture, which you may recognize in addition to a β-*trefoil* as one of the two distinct structural β-domains comprising the C-terminal fragment of the *C. botulinum* neurotoxin in Figure 2-1, and in the concanavalin A jelly-roll dimer from Figure 2-15. Important immune system molecules such as the HLA (human leukocyte antigen) family or antibodies are also organized in β-sandwich topologies. A very common architecture in metabolic enzymes is the triosephosphate isomerase (TIM) barrel, an α/β barrel we encountered in the PLP-dependent enzyme ornithine decarboxylase (Figure 2-30) which consists of the β-α-β motif repeats discussed previously. The α/β horseshoe demonstrates a beautiful, leucine-rich repeat architecture with a central β-core and the helices of the repeats lining the outside of the structure.

2.8   Membrane proteins

A structure family that requires special treatment is that of membrane proteins. It is estimated that about 1/3 of all human proteins are membrane proteins, and they constitute the large majority of drug targets including G-protein coupled

Box 2-7 **Domain structure of proteins.** From one polypeptide chain, several structural domains can independently fold. A domain generally has a specific function. Smaller structural motifs can be repeated within a domain, or appear repeatedly in different domains. The structural domains can be classified in a hierarchy of 3-dimensional similarity. About 1500 structurally different folds have been defined so far at the superfamily level. For a given function, structure is generally more conserved than sequence.

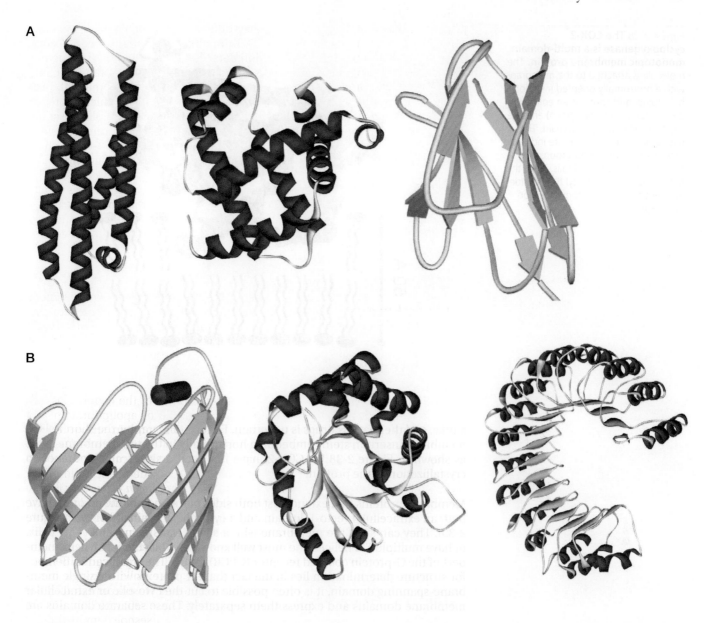

Figure 2-37 **Six common domain structures** at the level of architecture in CATH. (A) α-helix bundle, orthogonal α-non-bundle, and β-sandwich; (B) β-barrel, α/β-barrel, and the α/β horseshoe. PDB entries 1bz4,[23] 1mbn,[17] 1ynl,[65] 2por,[66] 1hkw,[67] and 2bnh.[60]

receptors (GPCRs), growth factors, neuroreceptors, ion channels, proton pumps, ABC transporters, and a variety of other, important cellular machineries such as photoreaction centers or ATPase motors. In contrast to hydrophilic *globular proteins* floating around in intercellular space or in the cytosol, membrane proteins are intimately associated with cellular membranes, which do in fact — as we know from X-ray crystallography — form an integral part of their structure. This intimate membrane association makes intact membrane proteins very difficult to express, solubilize, and crystallize. Less than 2% of all PDB entries are integral membrane proteins.

Classification of membrane proteins

A useful classification of membrane proteins from the structural point of view is to group them by the extent to which they are inserted or integrated into the membrane. The simplest anchoring is via a short, helical tail or a horizontal coil into a single layer of the membrane bilayer, so-called *monotopic* membrane proteins. In this family belong the well-known cyclooxygenases (the prostaglandin synthases COX-1 and COX-2) which are the targets of non-steroidal anti-inflammatory drugs (NSAIDs). Unfortunately, the selective COX-2 inhibitors, which do not suppress the production of gastrointestinally beneficial prostaglandins via the COX-1 isozyme, have been linked to increased risk of heart attack and

Figure 2-38 The COX-2 cyclooxygenase is a multi-domain monotopic membrane protein. The molecule is attached to the membrane with a horizontally oriented insertion helix (blue) and contains an epithelial growth factor domain (green) and the largely helical catalytic domain. The membrane phospholipids are indicated with the red polar head groups and the two black fatty acid tails. Note the detergent molecule (octyl β-glucoside) adjacent to the membrane insertion helix. PDB entry 1cvu.[69]

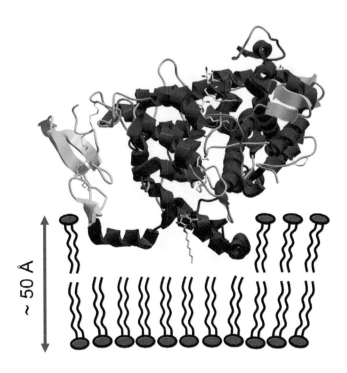

stroke, and their fate as drugs is uncertain. For crystallography, the short helical membrane insertion stems embedding horizontally in the top membrane layer, as shown in Figure 2-38 for COX-2, have not been a significant hindrance to crystallization of the protein.[68]

Membrane proteins that extend past both sides of the membrane and that have both an extracellular (ecto-) domain and a cytosolic domain are *bitopic* (Figure 2-39). They can span the membrane with a single transmembrane (TM) helix, or have multiple TM helices; the most well known are the seven-TM-helix members of the G-protein coupled receptor (GPCR) family. The significant difference for structure determination lies in the fact that for proteins with a single membrane-spanning domain, it is often possible to cut the cytosolic or extracellular membrane domains and express them separately. These separate domains are

Figure 2-39 Common topologies for helical membrane proteins. The leftmost protein is a monotopic membrane protein, with a horizontal insertion helix that anchors it in a single layer of the cell membrane, as seen in the example of COX2. In the middle is a bitopic membrane protein with a single transmembrane crossing helix, typical for the family of growth factor receptors. In the depicted case, both the C-terminal and N-terminal domains might be expressed and isolated separately from the transmembrane helix. The most difficult cases to produce and to crystallize are proteins with multiple transmembrane helices, because the protein chain traces back into the transmembrane part, as shown in the right part of the figure. The 7-TM GPCR receptors belong to this family.

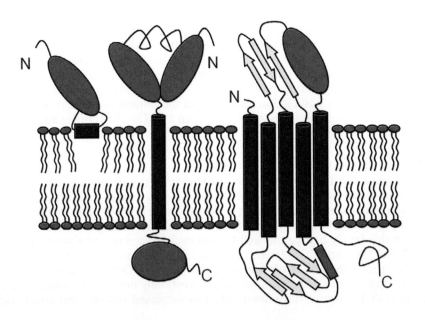

then purified and crystallized just as regular globular proteins. Typical examples are the tyrosine kinases of the epithelial growth factor receptor (EGF-R) family involved in a variety of human cancers, against which selective kinase inhibitors (Iressa®, gefitinib; and Traveca®, erlotinib) and novel receptor domain antagonists have been designed.[29]

Separately expressing the extracellular domains of *polytopic* multi-pass TM proteins is only possible when the protein chain of the ectodomain does not trace back into the membrane and does not form an integral part of the transmembrane domain. A typical example is the dimeric Venus' flytrap (VFT) domain of the metabotropic glutamate receptor,[70] which could be expressed and crystallized, despite the complete glutamate receptor being a type-II, 7-TM GPCR .

Polytopic membrane proteins are difficult to crystallize

The most difficult cases for structure determination are the polytopic, integral, and multi-pass transmembrane proteins, which cannot be domain truncated or cut without loss of structural integrity. Unfortunately, they represent a great number of important membrane proteins from inner membranes of bacteria and mitochondria (bacteriorhodopsins, K⁺ and Ca²⁺ ion channels, Cl⁻ channels, cytochromes) and receptors in neuronal membranes (GPCRs; acetylcholine receptors, AChRs; the pentameric GABA$_A$ neuroreceptors; opioid receptors; and many more). In those cases, the protein is generally unstable when removed from the phospholipid membrane, which poses difficulties for expression, purification, and crystallization. The membrane proteins are then usually gently extracted with detergents that keep the transmembrane stem stable by covering it with their hydrophobic tails, and the protein must remain in various detergents during the entire purification and crystallization process. Once they are successfully solubilized, membrane proteins crystallize generally through normal protein–protein contacts. Complexing with antibodies has also been shown effective by mediating crystal contacts through the antibody molecules (Chapter 3).

The collection of domains illustrated in Figure 2-37 includes a porin structure as a member of the β-stranded membrane proteins from the outer membrane of Gram-negative bacteria (Figure 2-40). In addition to serving as an example for an interesting antiparallel β-strand barrel forming a passive diffusion channel through bacterial membranes, the electron density along the hydrophobic outside of the membrane-inserted β-barrel clearly shows detergent molecules (Figure 3-44). Detergents play an important role in the production and crystallization of membrane proteins (Chapters 3 and 4).

Figure 2-40 The bacterial porin from *Rhodobacter capsulatus*. Shown are the ribbon presentation of the β-barrel and the electron density of a cross section through the barrel parallel to the inner layer of the membrane. Note the large channel in the center, which is gated just above the shown cross section and allows diffusion of molecules up to ~600 Da molecular weight into the bacterial cell. PDB entry 2por.[66]

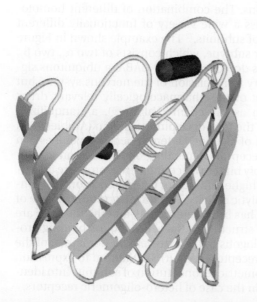
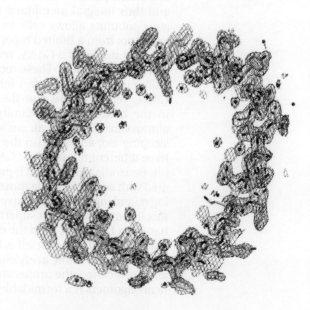

Box 2-8 **Membrane proteins.** Membrane proteins form the most important class of drug receptors. Their isolation, production, and crystallization for structure determination are difficult because of the hydrophobicity of their transmembrane parts. The exposed extracellular and cytosolic domains of monotopic and bitopic membrane proteins can sometimes be crystallized separately from their transmembrane helix. With polytopic membrane proteins with multiple transmembrane helices this is generally not possible. Phospholipids and detergents are tightly integrated with the transmembrane stems of membrane proteins and are sometimes observed in crystal structures.

2.9   Quaternary structure generates functional variety

As discussed in previous sections, a single polypeptide chain can form several structurally and functionally distinct domains, thus creating in an elegant way the enormous functional variety among proteins from a limited number of different domains by swapping or differently combining the subunits. This economic concept can be extended to the combination by non-covalent association of several protein chains into multi-functional, multimeric assemblies. Such assemblies are described as the quaternary structure of the protein or protein complex in question. Quaternary assemblies are frequently homo-oligomers formed from multiple instances of the same polypeptide chain, the simplest case being a homodimer. Such dimers are very common (over 1/3 of all structures in the PDB), and many enzymes form homodimers, where the active site is frequently formed by parts of both molecules of the dimer.

Multimeric assemblies

Modular assembly of functional domains allows sequential processing of substrates in metabolic synthesis pathways. A typical example is tryptophan synthase.[71] The enzyme is a $\alpha_2\beta_2$-tetramer, with different subunits catalyzing separate steps in the reactions from indole-3-glycerolphosphate via indole, which is combined with serine to finally yield tryptophan. The $\alpha_2\beta_2$-tetramer is nearly two orders of magnitude more active than the isolated subunits, because the tetrameric enzyme complex does not release the intermediate indole during the two-step reaction.

An illustrative example for functionally versatile hetero-oligomeric receptors is the superfamily of the GABA$_A$ (γ-aminobutyric acid) neuroreceptors, which are important pentameric transmitter-gated ion channels of the Cys-loop family and thus integral membrane proteins. The combination of different homologous subunits allows cells to express a wide variety of functionally different receptors from a limited repertoire of subunits.[72] The example shown in Figure 2-41 is a model of a GABA$_A$ receptor subtype, which consists of two α_1, two β_2, and one γ_2 subunits. These receptors do not only bind GABA, a ubiquitous signaling molecule in the fast inhibitory transmission of the nervous system, but GABA$_A$ receptors are also the target of the pharmacologically relevant drugs in the benzodiazepine family such as diazepam (Valium®), a tranquilizer; alprazolam (Xanax®), an anxiolytic; the non-diazepine zolpidem (Ambien®), a sleeping aid; and are also the target of alcohol. Different benzodiazepines can have different affinities for GABA$_A$ receptors depending on the subunit assembly. Benzodiazepines which preferably bind to α_1-containing receptors are associated with sedation, while those with higher affinity for GABA$_A$ receptors containing α_2 and/or α_3 subunits have anxiolytic activity. To date, no crystal structure of this important receptor superfamily has been determined, and the models[72] are based on homology with the crystal structure of the acetylcholine binding protein[73] (AChBP), which itself is homologous to the extracellular domain(s) of the pentameric nicotinic acetylcholine receptor (nAChR). In addition to expression and purification, the proper stoichiometric reconstitution of subunits into identical oligomers is a formidable task in the case of hetero-oligomeric receptors.

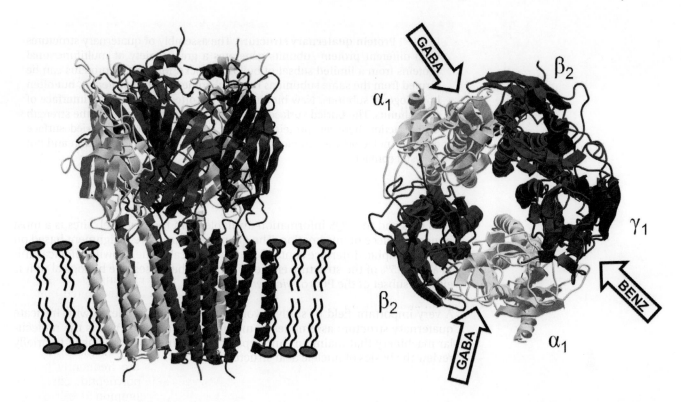

Biological unit and protein quaternary structure

An interesting problem results from the fact that the same weak, non-covalent interaction patterns that assemble the various proteins into biologically relevant, active multimers are also present in intermolecular crystal packing contacts. The question is, then, how true dimers or higher oligomers in crystal structures can be distinguished from merely packing-generated assemblies. The answer is relatively clear if the interaction energy between the monomers is large; in this case the assembly is likely to also persist in solution and in all likelihood has some biological relevance. A typical case for a strong dimer (two of which form the tetrameric, biologically active assembly) is the tight and continuous contact between the two β-sheets in the sandwich structure of concanavalin A shown in Figure 2-15. However, the distinction between true oligomerization contacts and crystal packing mediated interaction is not so clear cut when the intermolecular interactions are weak.

Buried surface area. In the absence of direct experimental evidence, a common first estimate for the strength of the interaction between protein molecules is gained through the analysis of the surface area per molecule that is buried upon contact formation. The buried surface area per molecule is half the difference between the solvent accessible surface area calculated for each of the two separated molecules and the accessible surface area of the complex. Above ~700–800 Å², the interaction is likely a true oligomer interaction that is stable in solution and biologically relevant. A buried surface area below about 400–500 Å² probably indicates only a crystal contact. This leaves a gray area in the range of about 400–700 Å², where additional experimental evidence gathered by light scattering, size exclusion chromatography, or ultracentrifugation may shed light on the solution aggregation state. The buried surface area and other properties defining the quaternary structure can be analyzed using bioinformatics tools such as the Protein Quaternary Structure (PQS) Server[74], its successor PISA, and ProtBuD[75] (Protein Biological Unit database). In general, the assignment of quaternary structure and the likely biological unit from crystal structure data alone is not trivial.[76] First, it is important to remember that the crystallographic asymmetric units in 52% of all PDB entries do not represent the actual biological unit.[75] Just loading the asymmetric unit into a graphic display program does not necessarily provide the biologically relevant oligomeric state, and the BIOMOL record in the

Figure 2-41 Model of the quaternary structure of the pentameric GABA_A neuroreceptor. The GABA_A receptor consists of five homologous subunits: two α_1 (yellow), two β_2 (red), and one γ_2 subunit (blue). Neurotransmitter- and drug-binding sites are located at the interfaces between different subunits (GABA: γ-aminobutyric acid; BENZ: benzodiazepine binding site). The cytosolic domain is missing in the model. Model coordinates[72] courtesy of Margot Ernst and Werner Sieghart, Medical University of Vienna, Austria.

Box 2-9 **Protein quaternary structure.** The assembly of quaternary structures from different protein subunits generates a great variety of multifunctional proteins from a limited subset of building blocks. Multimeric proteins can be formed from the same subunit as homo-oligomers or from different, but often homologous, subunits. New binding sites are often formed at the interface of the subunits. The buried surface area provides a first estimate for the strength of interaction between protein subunits. Above ~700–800 Å^2 buried surface area per molecule, the interaction is likely a true oligomer interaction and not a crystal contact.

PDB files and the PQS information available from the PDB Web sites is a most important piece of information. The quaternary structure assignment based on crystallographic (interface contact) data is identical to the known biological unit in about 85% of the structure entries. In most other cases, the biological unit is either a subset of the PQS or *vice versa*.

A very important field of study involving molecular complexes with intricate quaternary structure assemblies of multiple proteins and DNA is the molecular machinery that maintains and repairs the genome. We thus need to briefly review the basics of nucleic acid structure.

2.10 Nucleic acids and their protein complexes

Many proteins, including transcription factors, regulators, repair enzymes, or DNA-packing proteins, are studied by X-ray crystallography in complex with nucleic acid oligomers. Nucleotides or modified nucleotides are also frequently encountered as cofactors, substrates, and ligands in protein structures. In addition, a number of potent drugs and mutagens bind to, or are incorporated into, DNA (deoxyribonucleic acid) directly and are frequently studied by crystallography (although direct drug–DNA and small protein–DNA interactions are an established area of NMR studies). We already are familiar with the fundamentals of protein structure, and in the following we will limit our discussion of molecular DNA and RNA structure — which are discussed in practically all introductory biochemistry textbooks — to a review of the structural basics and the principal interactions between DNA/RNA and proteins from a crystallographer's point of view.

Early single crystal studies confirmed the model of the hydrated B-form of DNA occurring in physiological conditions, and soon thereafter the molecular structures of dehydrated A-form, plus a new, normally not *in vivo* observed Z-DNA conformation were reported. Figure 2-42 illustrates the basic features of the B-DNA structure.

Nucleic acid structure

The basic building blocks of DNA are deoxynucleotides (or nucleotides in the case of RNA). Their varying and specificity-conferring components are four cyclic, organic bases, either from the purine group (adenine, A, and guanine, G), or one of the pyrimidine bases (thymine, T, and cytosine, C), while uracil, U, replaces T in RNA. The bases are linked through an N-glycosidic bond to β-D-ribose, a cyclic pentose, and the combination of bases plus sugar is called a nucleoside. The C5′ carbon of the nucleoside sugar is linked to one, two, or three phosphate moieties, forming mono-, di- or tri-nucleotides. In the case of DNA, the sugar is 2-deoxyribose lacking the 2′-OH group, and the nucleotide is then a deoxyribonucleotide. Figure 2-43 illustrates the basic nucleotide structure and nomenclature.

Monophosphate nucleotides are connected into a single strand through bonds formed between the 3′ ribose carbon of the downstream nucleotide and the

Sidebar 2-9 **The first single crystal structure of DNA.** Fiber diffraction patterns of DNA in A- and B-form were known since the 1930s, and their interpretation by Watson and Crick led to the correct model of the hydrated B-form of DNA in 1953.[77] Interestingly, it took about 20 more years after the first protein structure had been determined until the first single crystal structure of DNA was obtained in 1979 by Richard Dickerson at Caltech.[78] The main reason for the significantly later date of DNA single crystal structure determinations compared with protein structures is that DNA fragments cannot be isolated in defined size and sequence from natural DNA — in contrast to proteins, which can generally be purified to homogeneity from tissue. For successful crystallization, the DNA fragments generally must have identical length and sequence, which could not be achieved until DNA synthesis methods were developed.

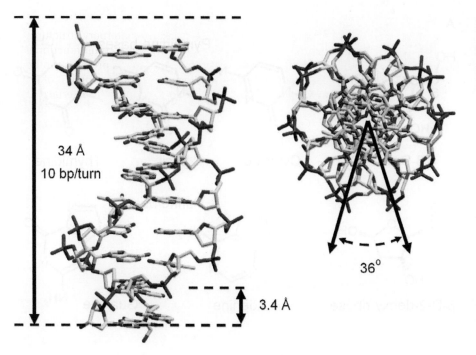

34 Å
10 bp/turn

3.4 Å

36°

Figure 2-42 **Canonical B-DNA structure.** The typical B-DNA double helix has a rise of 3.4 Å per base pair, and 10 base pairs are necessary to complete one turn of the DNA. The negatively charged phosphate–sugar backbone (purple phosphorus atoms) is located on the outside of the helix, and the base pairs, stacked on top of each other, are accessible from the side through the major and minor grooves. The angle offset between successive base pairs (right panel) is 36°. PDB entry 1d49.[79]

phosphate at the 5′ ribose position of the subsequent nucleotide. The direction of a single strand thus is defined as leading from 5′ to 3′. In the case of DNA, the nucleotides are deoxynucleotides, and in the case of RNA nucleotides, thymidine is replaced by uridine. A complementary strand of DNA with matched A–T and C–G base pairs (Watson–Crick base pairs) leads in the opposite direction, satisfying the hydrogen bonding between the keto form of the base pairs. In addition to the hydrogen bonded base pairing, strong π-stacking interactions between the base pairs stabilize the double helical structure with its major and minor groove (Figure 2-44). The backbone on the outside carries one negative charge per phosphate, and the DNA is thus an acidic polyanion. The negative charges distributed along the backbone need to be compensated, either by cations in solution such as Mg^{2+}, or by basic residues of protein binding partners. Protamines (short Arg- and Lys-rich peptides) stabilize DNA in sperm cells, and DNA or RNA binding proteins generally show distinct positively charged, basic patches where the polyanionic nucleic acid backbone interacts.

The fact that natural DNA molecules are for all practical purposes infinitely long, in addition to being highly negatively charged, posed special problems for crystallization. Practically all DNA crystallization cocktails contain Mg^{2+} as counterions, and in the typical orthorhombic B-DNA crystal structure, 10-mers of DNA are arranged so that they stack as if they would form an infinite B-DNA chain. Many of the early DNA crystal structures were obtained from C–G flanked palindromic sequences, simply because only one type of strand needs to be synthesized, and because C–G pairs share three hydrogen bonds and are thus more stable than A–T pairs with only two common hydrogen bonds. In crystal structures, the distinct ladder-like structure of DNA is readily recognizable in electron density.

DNA damage. DNA is susceptible to a number of environmental damages, which need to be repaired by a complex repair-protein machinery in order to keep the genomic information intact. The most common types of damage to DNA by mutagenic events are chemical modification of the bases and cross linking of bases, while severe oxidative events can also cause single- or double-strand breakage. Even in the absence of exogenic mutagens, at a physiological pH of ~7 water still contains 10^{-7} reactive hydrogen and hydroxyl ions. Depurination, depyrimidation, and deamidation of nucleotides are common, as is oxidation of bases, and electrophilic alkylating agents can add alkyl groups to bases.

Sidebar 2-10 Lessons learned from the tale of the DNA structure discovery. The history of the discovery of the DNA structure provides an interesting lesson for the practicing structural biologist or crystallographer. It needed both the decisive high-quality fiber diffraction images recorded by Rosalind Franklin and Maurice Wilkins as well as the correct interpretation of those photographs by James Watson and Francis Crick, who conceived the DNA base-pairing model, to solve the crucial puzzle of the structural basis of genomic information storage. Without the accurate experimental data collected by Rosalind Franklin providing key experimental restraints, the modeling by Watson and Crick would have likely remained far more conjectural. Contemporary biomolecular crystallography and structural biology thus require scientists with the experimental skills to generate adequate material and data, but who are also able to conduct a thorough analysis and interpretation of the resulting structure. A most exciting aspect of structural studies is thus the high degree of collaboration between diverse fields ranging from bioinformatics, molecular biology, protein chemistry, and computational crystallography to structure-based drug design and preclinical research.

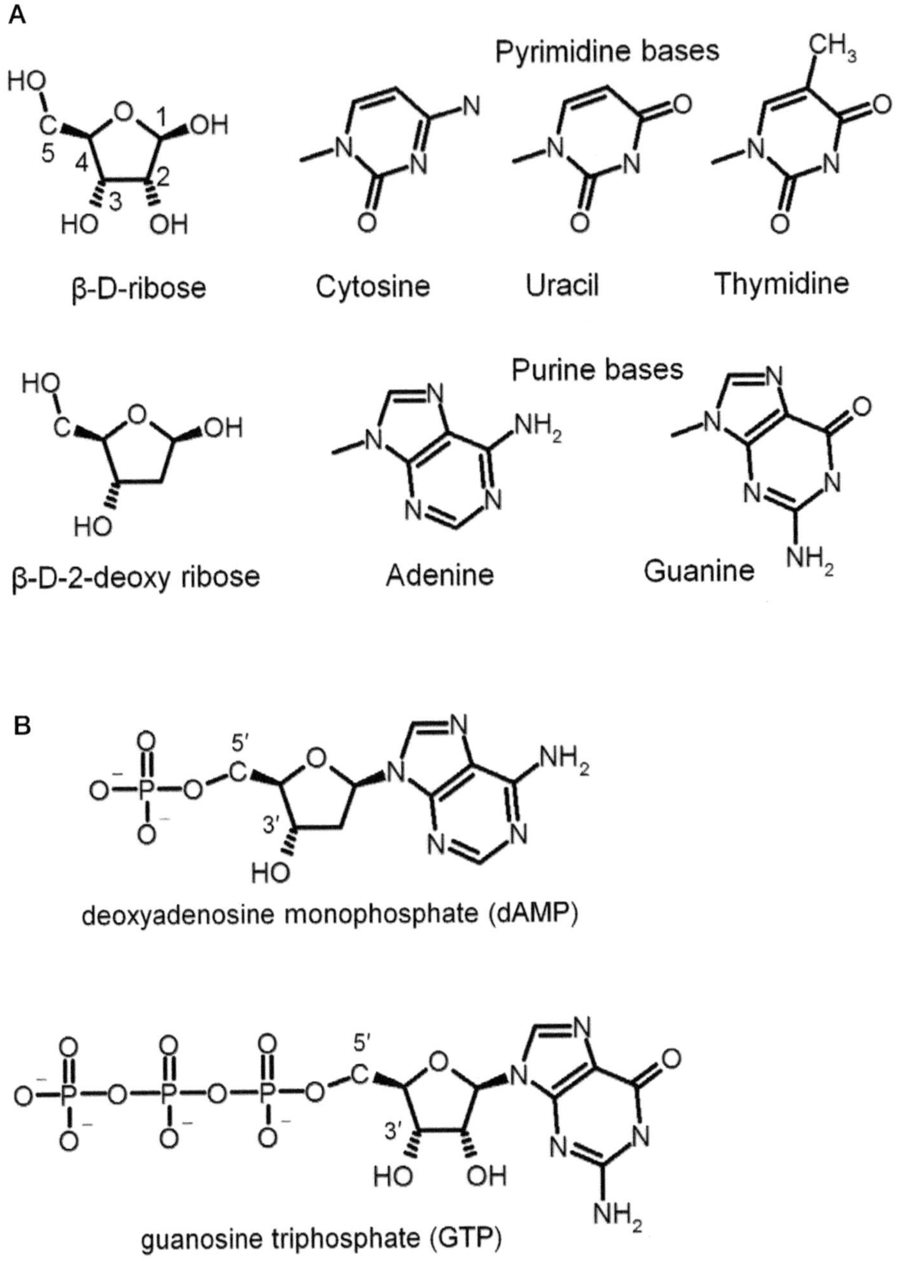

Figure 2-43 The basic building blocks of nucleotides. The pyrimidine and purine bases (A) are joined through an N-glycosidic bond to carbon 1 of the ribose sugar to form a nucleoside. The nucleoside (where the sugar numbering now is primed) is linked through the 5′ ribose carbon to up to three phosphates to form a nucleotide (B). To assemble a DNA or RNA strand, mononucleotides are joined through the 3′ carbon atom of the sugar ring to the phosphate group of the subsequent nucleotide. During synthesis, DNA or RNA strands are extended by adding the 5′ linked phosphate of a nucleotide to the 3′ end of the growing strand.

A

Pyrimidine bases

β-D-ribose   Cytosine   Uracil   Thymidine

Purine bases

β-D-2-deoxy ribose   Adenine   Guanine

B

deoxyadenosine monophosphate (dAMP)

guanosine triphosphate (GTP)

Box 2-10 DNA structure and its complexes with proteins. Natively occurring B-DNA is a polyanionic macromolecule assembled from 4 different deoxynucleotides. Two complementary strands combine in a double helix, with stacked base pairs pointing inside and the phosphate–sugar backbone outside. Fragments of double stranded B-DNA are most frequently complexed with DNA binding proteins involved in maintenance, repair, and transcriptional control of the genome. Specific DNA binding motifs are highly conserved and probe the DNA base pair sequence frequently through recognition helices in the major groove, supported by additional recognition motifs. Nucleotides are also important cofactors, energy transporters, and signaling molecules, and modified nucleotides are used as drugs.

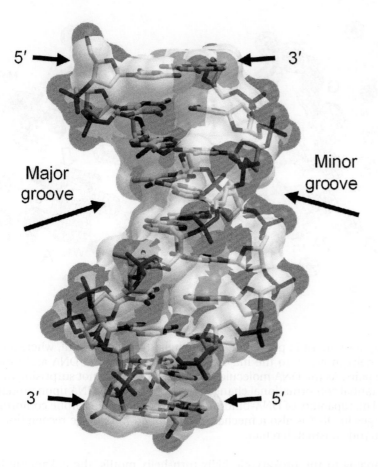

5′ ← → **3′**

Major
groove →

Minor
groove ←

3′ ← → **5′**

Figure 2-44 **Space filling model of B-DNA.** The image shows that the base pairs are most easily accessible in the wide major groove, where they frequently participate in specificity-conveying interactions with DNA-binding proteins.

UV radiation can lead to covalent crosslinking of base pairs, which need to be excised and replaced by nucleotide excision and repair (NER) complexes, and many highly carcinogenic substances generate adducts or intercalate between base pairs. The capability of such agents to halt or impair DNA replication is, on the other hand, also used in cancer chemotherapeutics, which inhibit the DNA replication necessary for the rapid cell division and growth of cancer tissue. Unfortunately, such therapies are not very tumor-specific, and the side effects on other tissues undergoing frequent replication such as gut lining, hematopoietic (blood cell generating) stem cells, or hair follicles are severe. We will discuss nucleotide-based drugs in a following section.

Protein–DNA interactions

While interactions via the negative charges of the phosphate backbone to basic residues are non-specific, the sides of the base pairs exposed in the major and minor groove present distinct patterns of hydrogen bond acceptors, hydrogen donors, and non-polar contacts to ring hydrogen atoms and methyl groups. Clearly, these exposed patterns are highly specific, dependent on the sequence of the base pairs, and provide the basis for the specificity of protein–DNA interaction and recognition. In addition, sequence-specific distortions from the canonical B-DNA structure may also contribute to recognition. Figure 2-45 shows a cross section through B-DNA along a G–C and T–A base pair and the recognition site nomenclature.

DNA recognition motifs

To accomplish any regulatory or repair task, proteins must be able to specifically recognize certain sequences or defects in DNA. A number of specialized DNA recognition motifs have thus evolved in proteins, depending on whether they probe the major groove or minor groove of DNA. The most ubiquitous DNA

Sidebar 2-11 **The p53 protein and DNA repair.** The DNA recognition motifs are in general highly conserved among transcription factors, but unique patterns such as those in the tumor suppressor gene p53 are also found. The p53 protein halts the cell cycle progression if DNA damage is encountered, and both the DNA-binding and protein-binding interactions of this key protein are susceptible to mutational defects leading to a variety of cancers. Even in the absence of mutagenic events, errors are introduced by the replication machinery, but most of them are corrected during strand synthesis by a number of proofreading DNA polymerases, followed by checks via the mismatch repair (MMR) machinery. Together, these corrections by DNA repair proteins lead to an error rate of about one in 10^9 replicated base pairs.

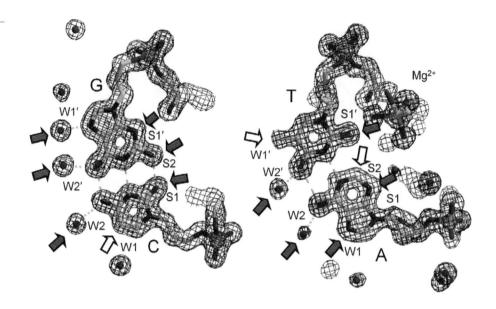

Figure 2-45 Cross section through B-DNA structure. Shown is a thin slab through a B-DNA double helix, with a G–C pair and a T–A pair. The major groove, to the left of the base pairs, is distinctly wider and more accessible than the minor groove (right side of base pairs). Arrows depict contact or recognition points (red, hydrogen bond acceptor; blue, hydrogen bond donor; white, non-polar contact with ring hydrogen or -CH$_3$ group). For C–G and A–T pairs, the recognition site numbering is just swapped (W1 = W1', etc.). Note that ring nitrogen atoms without a bound hydrogen atom can act as hydrogen bond acceptors. In this crystal structure of unbound DNA, most hydrogen bonds are satisfied through contacts with discrete solvent water molecules. Also distinctly visible is an octahedral coordinated Mg^{2+} ion coordinated with water and backbone-phosphate oxygen. PDB entry 1d49.[79]

recognition motif in prokaryotes is the helix-turn-helix motif, where a recognition helix inserts into the major groove of the bacterial DNA and probes the base pairs. As the DNA molecule is rather flexible, it is not surprising that quite substantial conformational changes can be induced in DNA upon protein binding. The capability of DNA-binding proteins to induce or probe conformational changes in DNA is also a mechanism exploited in damage recognition during eukaryotic mismatch repair.

An analog to the prokaryotic helix-turn-helix motifs, the eukaryotic homeodomain also contains a recognition helix probing the major groove of DNA, largely through hydrogen bonds via conserved Asn, Arg, Trp, and Phe residues. In contrast, the eukaryotic basal transcription factor TATA box binding protein (TBP) probes the minor groove across an extended eight base-pair region of the strongly bent DNA, with hydrophobic interactions largely responsible for the protein–DNA association. Very common eukaryotic transcriptional regulator motifs are Zn-fingers, which frequently form modular repeats, with their helices probing the DNA major groove (Figure 2-46).

Leucine zippers consist of two large helices, interacting with each other through Leu-rich stretches, which clamp the DNA like tweezers in the major groove (Figure 2-47), and other zipper-based transcription factors of the b/HLH/zip

Figure 2-46 Z-finger DNA-binding motif. The depicted transcription factor is a Zn-finger, a triple repeat from mouse immediate early protein ZIF268, complexed with a B-DNA 10-mer with a single base pair overhang at the end of each strand. The right panel illustrates how each of the recognition helices "fingers" into the major groove of the DNA, thereby interacting with the exposed sides of three base pairs per finger. The Zn^{2+} ions in each helix-loop motif stabilizing the conformation are shown as light blue spheres. The linker regions between the repeats are flexible and contain a glycine residue, enabling the protein to wrap around the DNA upon binding. PDB entry 1aay.[80]

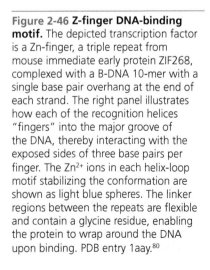

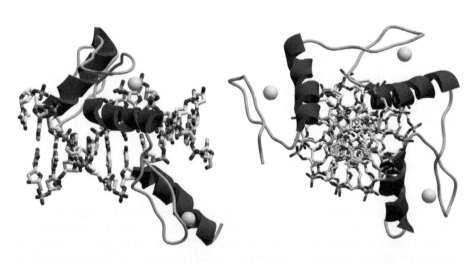

family contain an additional helix-loop-helix motif. For many transcription factors, homodimerization is a common mechanism involved in DNA recognition. In contrast to the major groove dominated binding patterns discussed before, the core domain of the Zn-containing tumor suppression factor p53 binds DNA in a quite complex, sequence-specific form in a AT rich region of the minor groove, in the major groove, and also through an Arg residue to backbone phosphate on the minor groove side of an extended stretch of DNA (Figure 2-47). As expected, mutations in these DNA-binding residues of the core domain are the cause for many human cancers.[29]

Co-crystallization of protein–DNA complexes is not trivial, generally requiring DNA strands of varying lengths to be synthesized, and we will address protein–DNA complex crystallization briefly in a special section of the Chapter 3.

Nucleotide analogs and drugs

Nucleotides are not only important as the building blocks of nucleic acids, but they also are important molecules involved in energy transport and storage (ATP, ADP), electron transfer and redox reactions (NAD, NADP), cellular signaling via G-proteins (GTP, GDP), and as second messengers (cAMP, cGMP); they are also generally important co-factors in proteins. Many crystal structures therefore are determined in complex with a variety of nucleotides.

The fact that these nucleotides can undergo enzymatically catalyzed reactions such as hydrolysis in crystallization solution, or even when soaked in crystals of active enzymes, poses a problem for crystallography: we often need to co-crystallize or soak with non-hydrolyzable analogs in order to obtain stable nucleotide–protein complexes. There are three principal possibilities to modify nucleotides: at the phosphate group, on the 2′ and 3′ positions of the sugar (for NMR and spectroscopic labeling), or at the base itself. Non-hydrolyzable nucleotide analogs for crystallography are generally obtained by modification in the phosphate group region, while the bases are the target of modifications for drug molecules. Phosphate analogs such as non-hydrolyzable vanadates have also been used in assessment of binding energy in the transduction process.

Figure 2-47 The leucine zipper structure and the p53 tumor suppressor core domain. The leucine zipper forms two long helices that probe the DNA major groove (left panel). The p53 tumor suppressor core domain interacts through one recognition helix with base pairs in the minor groove, and the second helix probes the major groove. In addition, an arginine residue interacts with the sugar–phosphate backbone. The residues shown on the p53 molecule are those most frequently involved in tumorigenic mutations. The DNA is a 21-mer stretch with overhangs. Overhangs allow the DNA to continue through crystal packing as if it were an infinitely extended stretch. In this crystal structure, two more p53 molecules bind unspecifically to the same DNA stretch and are omitted. PDB entries 1ysa[81] and 1tup.[82]

Sidebar 2-12 The prodrug isoniazid binds to NAD. Some prodrugs are converted into their active form by host enzymes. A good example is the conversion of isoniazid, a first line drug against *Mycobacterium tuberculosis* (MTB), by the mycobacterial enzyme KatG into an active form bound to nicotinamide adenine dinucleotide (NAD). This modified drug then inhibits InhA, the enoyl-acyl-carrier-protein reductase, which is necessary for the biosynthesis of mycolic fatty acids. These ~50–60 carbon long mycolic fatty acids are required for the assembly of the cell wall of MTB and their absence stops bacterial cell growth and proliferation. Because humans do not synthesize mycolic acids, cell-wall synthesis inhibiting drugs have no comparable human targets and generally fewer side effects. The crystal structure of drug bound InhA, determined in Jim Sacchettini's laboratory, clearly showed how the active drug blocks the NAD binding site of the enzyme and provided proof for the chemical constitution and conformation of the active drug, isonicotinic acyl-NADH (Figure 2-48).

An interesting fact is that many nucleic base derived drugs are enzymatically converted into their active nucleotides *in vivo*. The cancer drug 5-fluorouracil, for example, is converted via several pathways into the 2′ deoxynucleotide, which then inhibits the enzyme thymidylate synthase, which is the sole source of *de novo* synthesis of thymidylate. Cells affected by the drug are thus unable to proliferate because they cannot synthesize new DNA because of the lack of thymidine bases. We have already mentioned that unfortunately other rapidly regenerating, and thus DNA synthesizing, human tissues such as gut lining and hematopoietic cells are also affected, leading to the side-effects of these antineoplastic agents administered against rapidly growing cancers.

Organization of DNA in chromatin

We conclude our section about protein–DNA complex structure with the tale of a crystallographic *tour de force* that took nearly 15 years to accomplish — the determination of the nucleosome core particle by Timothy Richmond and colleagues.

The human genome consists of approximately 3 billion nucleotide pairs organized into 23 chromosomes. Uncoiled and extended end to end, the DNA would extend about 3 meters, which obviously needs some packing to fit into a cell. In eukaryotes, the genetic material is thus neatly organized in complexes of DNA with core histones and other chromosomal proteins that together form nuclear chromatin. The nucleosome core chromatin repeating unit includes two copies each of four core histones H2A, H2B, H3 and H4 (collective molecular mass 206 kDa) wrapped by 146–147 residues of contiguous DNA. The question was how these nucleosome core particles assemble and how exactly the DNA wraps around the core histone proteins.

The crystal structure of the nucleosome core particle was the culmination of more than 15 years of work involving laborious protein preparation, persistent protein–DNA complex crystallization, and solid protein crystallography. The four core histone proteins were individually expressed in milligram quantities in *Escherichia coli*, purified under denaturing conditions, then refolded into histone octamers and assembled into nucleosome core particles using 146 base pair (bp) defined-sequence DNA fragment derived from human α-satellite DNA. The first crystals obtained in 1984 diffracted only to 7 Å, but in 1997 finally a 2.8 Å structure was determined and the crystals have been further optimized in

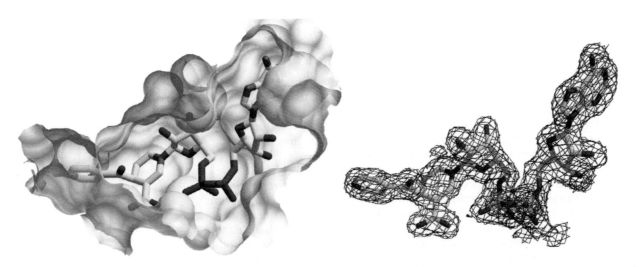

Figure 2-48 The active form of the antitubercular drug isoniazid. Isoniazid (isonicotinic acid hydrazide) is converted by the mycobacterial enzyme KatG into the active drug isonicotinic acyl-NADH, shown here in the binding pocket of the target InhA, an enoyl-acyl-carrier-protein reductase, essential for the biosynthesis of mycolic fatty acids (left panel). The clear electron density (right panel) obtained in the 1.75 Å structure of its complex with InhA confirmed beyond doubt the chemical composition and the inhibiting mode of the modified drug. Nucleotides can be readily localized through the high electron density level at their phosphorus atoms, which also contribute an anomalous signal usable for experimental phasing. PDB entry 2nv6.[83]

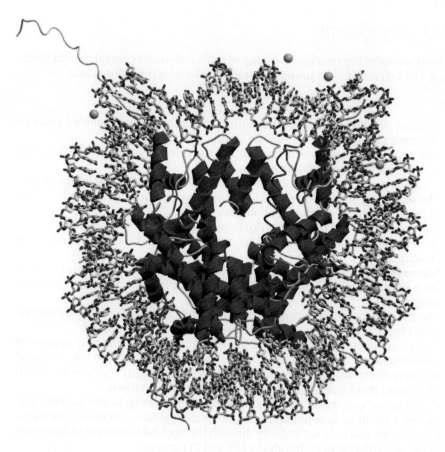

Figure 2-49 The nucleosome core particle. The nucleosome core particle consists of two copies each of the core histone proteins, H2A, H2B, H3, and H4, wrapped by 146 base pairs of DNA. The tails of the histone proteins, some of which are seen protruding from between the base pairs, form strong or weak nucleosomal interactions, depending on their state of acetylation: When hyperacetylated, nucleosomal interactions are weakened; the DNA is not constrained on the surface of the nucleosome and becomes accessible to transcription factors. PDB entry 1aoi.[85]

recent years (partly by using a 147 bp DNA fragment instead of the 146 bp piece) and diffracted finally to ~2 Å. Crystals were grown by vapor diffusion over the course of 1 to 3 weeks and flash cooled in liquid propane at −120 °C prior to data collection at the European Synchrotron Radiation Facility (ESRF) in Grenoble. Three heavy atom derivatives were used for MIR phasing, prepared by soaking with mercury nitrate, methyl mercury nitrate, and tetrakis-acetoxymercuri-methane (TAMM).[84] The initial MIR electron density map was used to trace 120 bp of DNA and 745 amino acids, and iterative rounds of model building and refinement resulted in a model containing the entire DNA and histone octamer (Figure 2-49).

The structure revealed that the histone amino terminal tails pass over and between the gyres of the DNA superhelix, making contact with neighboring core particles. When the histone tails are hypoacetylated, they constrain the wrapping of the DNA on the nucleosome surface by promoting strong nucleosomal interactions. Upon hyperacetylation, the nucleosomal interactions are weakened, the histone tails no longer constrain the DNA on the surface of the nucleosome, and the DNA is now accessible to transcription factors. The bending and supercoiling of the DNA on the nucleosome seems to promote binding of transcription factors and augment interactions between different proteins of the transcriptional machinery. Very recently, the structure of a tetrameric nucleosome complex with DNA has been determined[86] at low resolution (9 Å), and it is likely that a high resolution structure of the nuclear tetrasome will eventually be determined.

The emphasis on the importance and biological relevance of quaternary structure and assemblies of proteins and DNA finishes our chapter on protein structure. A summary of key concepts follows for your review.

2.11 Key concepts

A brief review of key concepts — grouped into sections for general protein structure, DNA structure, and protein crystallography — is provided below.

Protein structure

- Proteins or polypeptides are polyionic macromolecules assembled from 20 common proteinogenic L-α-amino acids.
- The sequence of amino acids constitutes the primary structure, the first level of organization of protein structure.
- Secondary structure elements are formed by distinct backbone interactions.
- The secondary structure elements are organized in small motifs that form one or more independently folding domains.
- The arrangement of the protein domains defines the tertiary structure.
- Several protein molecules can assemble together at the level of quaternary structure.
- Successive amino acids are connected by formation of peptide bonds, which have partial double bond character with delocalized electrons and are thus planar.
- The most frequent conformation of the peptide bond is the *trans* conformation, that is, the Cα atoms of successive amino acids are opposed.
- As a consequence of the planar peptide bond formation, the only degrees of conformational freedom a polypeptide backbone has are the two torsional angles φ and ψ, while the peptide bond torsion ω is highly restrained to +180° (dominant *trans* conformation) or 0° (rare *cis* conformation).
- Only limited φ–ψ torsion angle combinations are energetically favored, primarily a result of non-bonded van der Waals repulsions between the backbone atoms and only secondarily through the optimal formation of hydrogen bond networks between backbone NH and CO groups.
- The optimal arrangement of hydrogen bond networks with energetically favored backbone conformations leads to formation of specific secondary structure elements: α-helices, β-sheets, turns, loops, and a variety of other, less frequent secondary structures.
- The side chains of amino acid residues in the polypeptide chain are chemically different and cover a wide variety of properties, coarsely divided into hydrophobic, acidic, basic, and polar.
- The interactions of side chains lead to the self-assembly of several secondary structure elements into the tertiary structure of the folded protein.
- Side chain interactions are generally non-covalent, weak, and non-directional, with bond energies in the range of −1 to −10 kcal mol^{-1}.
- From one polypeptide chain, several structural domains can independently fold. A domain generally has a specific function.
- Smaller structural motifs can be repeated within a domain or appear repeatedly in different domains.
- The structural domains can be classified in a hierarchy of three-dimensional similarity. About 1500 structurally different folds have been defined so far at the superfamily level.
- Several proteins formed from separate polypeptide chains can associate into a functionally relevant quaternary structure. Quaternary assemblies can be formed from the same protein (homodimers, tetramers, etc.), or from different protein chains (hetero-oligomers).
- Conserved solvent molecules (water) form an integral part of the solution structure of the protein, just as phospholipids are tightly integrated with the transmembrane stems of membrane proteins.

DNA structure

- Natively occurring B-DNA is a polyanionic macromolecule assembled from four different deoxynucleotides. Two complementary strands combine in a double helix, with stacked base pairs pointing inside and the phosphate–sugar backbone outside.

- Fragments of double stranded B-DNA are most frequently complexed with DNA binding proteins involved in maintenance, repair, and transcriptional control of the genome.
- Specific DNA binding motifs are highly conserved and probe the DNA base pair sequence frequently through recognition helices in the major groove, supported by additional recognition motifs.
- Nucleotides are also important cofactors, energy transporters, and signaling molecules, and modified nucleotides are used as drugs.

Protein crystallography

- The most common descriptors for molecular geometry are bond lengths, bond angles, and torsion angles.
- The proper chirality or handedness of backbone Cα carbon atoms (L-configuration, or 2S configuration in absolute terms) can be checked by the CORN rule.
- The amino acids threonine and isoleucine have a second chiral center at Cβ with 3R configuration.
- Specific types of bond distances and angles as well as chirality are highly restrained to their known mean values in macromolecular structure refinement, where they provide important, stabilizing prior knowledge.
- Pairs of backbone torsion angles φ–ψ for each residue fall in specific regions on the Ramachandran plot depending on their secondary structure environment. As backbone torsion angles are generally not constrained in refinement, they provide a powerful validation tool.
- Solvent-exposed side chains are frequently disordered and can be modeled in split conformations.
- Solvent-exposed connecting loops between secondary structure elements can also be disordered and only partially traceable in electron density.
- From electron density alone, placement of N, Q, and H side chains is ambiguous except at atomic resolution, and chemical plausibility of the hydrogen bond and interaction network must be considered for correct NQH side chain conformations.
- Discrete solvent molecules (largely water) form an integral part of the solution structure of the protein and can be modeled into electron density.
- Additional chemical plausibility needs to be applied to infer the nature of non-water solvent molecules and to interpret electron density of unknown ligands or solvent molecules.
- Phospholipids are tightly integrated with the transmembrane stems of membrane proteins, and are sometimes visible in electron density together with detergent molecules.
- Membrane proteins form the most important class of drug receptors. Their isolation, production, and crystallization for structure determination is difficult because of the hydrophobic nature of their transmembrane stems.
- The quaternary structure of a protein is likely associated with its biological unit, but the PDB coordinate file contains only the crystallographic asymmetric unit, which is only in less than half of all cases of oligomeric assemblies identical with the complete biological unit.

2.12 Additional reading

1. Lesk AM (2004) *Introduction to Protein Science: Architecture, Function and Genomics.* Oxford, UK: Oxford University Press.

2. Brändén C, & Tooze J (1999) *Introduction to Protein Structure.* New York, NY: Garland Publishing.

3. Creighton TE (1992) *Proteins: Structures and Molecular Properties.* New York, NY: W.H. Freeman & Company.

4. Alberts B, Johnson A, Lewis J, et al. (2007) *Molecular Biology of the Cell.* New York, NY: Garland Publishing.

5. Silverman RB (2004) *The Organic Chemistry of Drug Design and Drug Action.* New York, NY: Elsevier Academic Press.

6. Weinberg RA (2006) *The Biology of Cancer.* New York, NY: Garland Science.

7. Fersht A (1999) *Structure and Mechanism in Protein Science: A Guide to Enzyme Catalysis and Protein Folding.* New York, NY: W.H. Freeman & Company.

8. Atkins P, & de Paula J (2006) *Physical Chemistry for the Life Sciences.* New York, NY: Freeman & Company.

2.13   References

1. Walsh CT (2006) *Posttranslational Modifications*. Greenwood Village, CO: Roberts & Company.

2. Dunker AK, Lawson JD, Brown CJ, et al. (2001) Intrinsically disordered protein. *J. Mol. Graph. Model.* 19(1), 26–59.

3. Tompa P (2002) Intrinsically unstructured proteins. *Trends Biochem. Sci.* 27(10), 527–533.

4. Parkin S, Rupp B, & Hope H (1996) Structure of bovine pancreatic trypsin inhibitor at 125 K: Definition of carboxyl-terminal residues Gly57 and Ala58. *Acta Crystallogr.* D52(1), 18–29.

5. Lacy DB, Tepp W, Cohen AC, et al. (1998) Crystal structure of botulinum neurotoxin type A and implications for toxicity. *Nat. Struct. Biol.* 5(10), 898–902.

6. Ban N, Nissen P, Hansen J, et al. (2000) The complete atomic structure of the large ribosomal subunit at 2.4 Å resolution. *Science* 289, 905–920.

7. Abrescia NG, Cockburn JJ, Grimes JM, et al. (2004) Insights into assembly from structural analysis of bacteriophage PRD1. *Nature* 432(7013), 68–74.

8. Cockburn JJ, Abrescia NG, Grimes JM, et al. (2004) Membrane structure and interactions with protein and DNA in bacteriophage PRD1. *Nature* 432(7013), 122–125.

9. Bamford JK, Cockburn JJ, Diprose J, et al. (2002) Diffraction quality crystals of PRD1, a 66-MDa dsDNA virus with an internal membrane. *J. Struct. Biol.* 139(2), 103–112.

10. Engh RA, & Huber R (1991) Accurate bond and angle parameters for X-ray structure refinement. *Acta Crystallogr.* A47, 392–400.

11. Engh RA, & Huber R (2001) Structure quality and target parameters. In Rossmann MG & Arnold E (Eds.), *International Tables for Crystallography*. Dordrecht/Boston/London: Kluwer Academic Publishers.

12. Weiss MS, & Hilgenfeld R (1999) A method to detect nonproline *cis* peptide bonds in proteins. *Biopolymers* 50, 536–544.

13. Kantardjieff K, Kim C-Y, Naranjo C, et al. (2004) *Mycobacterium tuberculosis* rmlC epimerase (Rv3465): a promising drug target structure in the rhamnose pathway. *Acta Crystallogr.* D 60, 895–902.

14. Ramachandran GN, Ramakrishnan C, & Sasisekharan V (1963) Stereochemistry of polypeptide chain configurations. *J. Mol. Biol.* 7, 95–99.

15. Lovell SC, Davis IW, Arendall WB, 3rd, et al. (2003) Structure validation by C-alpha geometry: phi, psi and C-beta deviation. *Proteins* 50(3), 437–450.

16. McPherson A (1999) *Crystallization of Biological Macromolecules*. Cold Spring Harbor, NY: Cold Spring Harbor Laboratory Press.

17. Kendrew JC, Bodo G, Dintzis HM, et al. (1958) A three-dimensional model of the myoglobin molecule obtained by X-ray analysis. *Nature* 181, 662–666.

18. Astbury WT, & Sisson WA (1935) X-ray studies of the structures of hair, wool and related fibres. III. The configuration of the keratin molecule and its orientation in the biological cell. *Proc. R. Soc. London.* A150, 533–551.

19. Pauling L, Corey RB, & Branson HR (1951) The structure of proteins; two hydrogen-bonded helical configurations of the polypeptide chain. *Proc. Natl. Acad. Sci. U.S.A.* 37(4), 205–211.

20. Eisenberg D (2003) The discovery of the alpha-helix and beta-sheet, the principal structural features of proteins. *Proc. Natl. Acad. Sci. U.S.A.* 100(20), 11207–11220.

21. Pauling L, & Corey RB (1951) The pleated sheet, a new layer configuration of polypeptide chains. *Proc. Natl. Acad. Sci. U.S.A.* 37(5), 251–256.

22. Dickerson RE (2005) *Present at the Flood*. Sunderland, MA: Sinauer Associates.

23. Segelke BW, Forstner M, Knapp M, et al. (2000) Conformational flexibility in the apolipoprotein E amino-terminal domain structure determined from three new crystal forms: Implication for lipid binding. *Protein Sci.* 9, 886–897.

24. Laskowski RA, MacArthur MW, Moss DS, et al. (1993) PROCHECK: a program to check the stereochemical quality of protein structures. *J. Appl. Crystallogr.* 26(2), 283–291.

25. Dong J, Peters-Libeu CA, Weisgraber KH, et al. (2001) Interaction of the N-terminal domain of apolipoprotein E4 with heparin. *Biochemistry* 40(9), 2826–2834.

26. Kabsch W, & Sander C (1983) Dictionary of protein secondary structure: pattern recognition of hydrogen-bonded and geometrical features. *Biopolymers* 22, 2577–2637.

27. Bragg WL, Kendrew JC, & Perutz MF (1950) Polypeptide chain configurations in crystalline proteins. *Proc. R. Soc. London* A203, 321–357.

28. Fodje MN, & Al-Karadaghi S (2002) Occurrence, conformational features and amino acid propensities for the pi-helix. *Protein Eng.* 15(5), 353–358.

29. Weinberg RA (2006) *The Biology of Cancer*. New York, NY: Garland Science.

30. Watanabe S, Kita A, & Miki K (2005) Crystal structure of atypical cytoplasmic ABC-ATPase SufC from *Thermus thermophilus* HB8. *J. Mol. Biol.* 353(5), 1043–1054.

31. Dvir H, Harel M, Bon S, et al. (2004) The synaptic acetylcholinesterase tetramer assembles around a polyproline II helix. *EMBO J.* 23(22), 4394–4405.

32. Kantardjieff KA, Höchtl P, Segelke BW, et al. (2002) Concanavalin A in a dimeric crystal form: revisiting structural accuracy and molecular flexibility. *Acta Crystallogr.* D58, 735–743.

33. Chan AW, Hutchinson EG, Harris D, et al. (1993) Identification, classification, and analysis of beta-bulges in proteins. *Protein Sci.* 2(10), 1574–1590.

34. Wlodawer A, Svensson LA, Sjolin L, et al. (1988) Structure of phosphate-free ribonuclease A refined at 1.26 Å. *Biochemistry* 27(8), 2705–2717.

35. Rupp B, & Segelke BW (2001) Questions about the structure of the botulinum neurotoxin B light chain in complex with a target peptide. *Nat. Struct. Biol.* 8, 643–664.

36. Rose GD, Gierasch LM, & Smith JA (1985) Turns in peptides and proteins. *Adv. Protein Chem.* 37, 1–109.

37. Fetrow JS (1995) Omega loops: nonregular secondary structures significant in protein function and stability. *FASEB J.* 9(9), 708–717.

38. Eisenberg D, Weiss RM, & Terwilliger TC (1984) The hydrophobic moment detects periodicity in protein hydrophobicity. *Proc. Natl. Acad. Sci. U.S.A.* 81(1), 140–144.

39. Moller S, Croning MD, & Apweiler R (2001) Evaluation of methods for the prediction of membrane spanning regions. *Bioinformatics* 17(7), 646–653.

40. Hoft RRW, Vriend G, Sander C, et al. (1996) Errors in protein structures. *Nature* 381, 272.

41. Kantardjieff K, Jamshidian M, & Rupp B (2004) Distributions of pI vs pH provide strong prior information for the design of crystallization screening experiments. *Bioinformatics* 20(14), 2171–2174.

42. Trakhanov S, Kreimer DI, Parkin S, et al. (1998) Cadmium induced crystallization of proteins: II. Crystallization of the *Salmonella typhimurium* histidine binding protein in complex with L-histidine, L-arginine or L-lysine. *Protein Sci.* 7, 600–604.

43. Derewenda ZS (2004) The use of recombinant methods and molecular engineering in protein crystallization. *Methods* 34(3), 354–363.

44. Dong L-M, Parkin S, Trakhanov SD, et al. (1996) Novel mechanism for defective receptor binding of apolipoprotein E2 in type III hyperlipoproteinemia. *Nat. Struct. Biol.* 3, 718–722.

45. Altschuler EL (2001) Alpha helix capping and the conformation of threonine. *Med. Hypotheses* 56(4), 478–479.

46. Petersen MT, Jonson PH, & Petersen SB (1999) Amino acid neighbours and detailed conformational analysis of cysteines in proteins. *Protein Eng.* 12(7), 535–548.

47. Garman E, & Nave C (2002) Radiation damage to crystalline biological molecules: current view. *J. Synchrotron Radiat.* 9, 327–328.

48. Ramagopal UA, Dauter M, & Dauter Z (2003) Phasing on anomalous signal of sulfurs: what is the limit? *Acta Crystallogr.* D59, 1020–1027.

49. Weichenberger CX, & Sippl MJ (2006) Self-consistent assignment of asparagine and glutamine amide rotamers in protein crystal structures. *Structure* 14(6), 967–972.

50. Chen VB, Arendall WB, III, Headd JJ, et al. (2010) MolProbity: all-atom structure validation for macromolecular crystallography. *Acta Crystallograph.* D66(1), 12–21.

51. Fersht A (1999) *Structure and Mechanism in Protein Science: A Guide to Enzyme Catalysis and Protein Folding*. New York, NY: W.H. Freeman & Company.

52. Silverman RB (2004) *The Organic Chemistry of Drug Design and Drug Action*. New York, NY: Elsevier Academic Press.

53. Grishin NV, Osterman AL, Brooks HB, et al. (1999) X-ray structure of ornithine decarboxylase from *Trypanosoma brucei*: the native structure and the structure in complex with alpha-difluoromethylornithine. *Biochemistry* 38(46), 15174–15184.

54. Morozov A, Kortemme T, Tsemekhman K, et al. (2005) Close agreement between the orientation dependence of hydrogen bonds observed in protein structures and quantum mechanical calculations. *Proc. Natl. Acad. Sci. U.S.A.* 101(18), 6946–6951.

55. Fabiola F, Bertram R, Korostelev A, et al. (2002) An improved hydrogen bond potential: Impact on medium resolution protein structures. *Protein Sci.* 11(6), 1415–1423.

56. Pauling L (1960) *The Nature of the Chemical Bond and the Structure of Molecules and Crystals*. Ithaca, NY: Cornell University Press.

57. Viola R, Carman P, Walsh J, et al. (2007) Operator-assisted harvesting of protein crystals using a universal micromanipulation robot. *J. Appl. Crystallogr.* 40(4), 539–545.

58. Levitt M, & Chothia C (1976) Structural patterns in globular proteins. *Nature* 261(5561), 552–558.

59. Brändén C, & Tooze J (1999) *Introduction to Protein Structure*. New York, NY: Garland Publishing.

60. Kobe B, & Deisenhofer J (1996) Mechanism of ribonuclease inhibition by ribonuclease inhibitor protein based on the crystal structure of its complex with ribonuclease A. *J. Mol. Biol.* 264(5), 1028–1043.

61. Verdino P, Westritschnig K, Valenta R, et al. (2002) The cross-reactive calcium binding pollen allergen, Phl p 7, reveals a novel dimer assembly. *EMBO J.* 21(19), 5007–5016.

62. Hadley C, & Jones DT (1999) A systematic comparison of protein structure classifications: SCOP, CATH and FSSP. *Structure* 7(9), 1099–1112.

63. Holm L, & Sander C (1998) Touring protein fold space with Dali/FSSP. *Nucleic Acids Res.* 26(1), 316–319.

64. Baker D, & Sali A (2001) Protein structure prediction and structural genomics. *Science* 294(5540), 93–96.

65. Gokulan K, Khare S, Ronning D, et al. (2005) Co-crystal structures of NC6.8 Fab identify key interactions for high potency sweetener recognition: Implications for the design of synthetic sweeteners. *Biochemistry* 44, 9889–9898.

66. Weiss MS, & Schulz GE (1992) Structure of porin refined at 1.8 Å resolution. *J. Mol. Biol.* 227(2), 493–509.

67. Gokulan K, Rupp B, Pavelka MS, Jr., et al. (2003) Crystal structure of *Mycobacterium tuberculosis* diaminopimelate decarboxylase, an essential enzyme in bacterial lysine biosynthesis. *J. Biol. Chem.* 278(206), 18588–18596.

68. Filipponi E, Cecchetti V, Tabarrini O, et al. (2000) Chemometric rationalization of the structural and physicochemical basis for selective cyclooxygenase-2 inhibition: Toward more specific ligands. *J. Comput. Aided Mol. Des.* 14(3), 277–291.

69. Kiefer JR, Pawlitz JL, Moreland KT, et al. (2000) Structural insights into the stereochemistry of the cyclooxygenase reaction. *Nature* 405(6782), 97–101.

70. Kunishima N, Shimada Y, Tsuji Y, et al. (2000) Structural basis of glutamate recognition by a dimeric metabotropic glutamate receptor. *Nature* 407, 971–977.

71. Nishio K, Morimoto Y, Ishizuka M, et al. (2005) Conformational changes in the alpha-subunit coupled to binding of the beta 2-subunit of tryptophan synthase from *Escherichia coli*: Crystal structure of the tryptophan synthase alpha-subunit alone. *Biochemistry* 44(4), 1184–1192.

72. Ernst M, Bruckner S, Boresch S, et al. (2005) Comparative models of GABA$_A$ receptor extracellular and transmembrane domains: Important insights in pharmacology and function. *Mol. Pharmacol.* 68(5), 1291–1300.

73. Brejc K, van Dijk WJ, Klaassen RV, et al. (2001) Crystal structure of an ACh-binding protein reveals the ligand-binding domain of nicotinic receptors. *Nature* 411(6835), 269–276.

74. Krissinel E, Henrick K (2007) Inference of Macromolecular Assemblies from Crystalline State. *J. Mol. Biol.* 372, 774–797.

75. Xu , Canutescu A, Obradovic Z, et al. (2006) ProtBuD: A database of biological unit structures of protein families and superfamilies. *Bioinformatics* 22(23), 2876–2882.

76. Kobe B, Guncar G, Buchholz R, et al. (2008) Crystallography and protein-protein interactions: Biological interfaces and crystal contacts. *Biochem. Soc. Trans.* 36(Pt 6), 1438–1441.

77. Watson JD, & Crick FH (1953) Molecular structure of nucleic acids: A structure for deoxyribose nucleic acid. *Nature* 171(4356), 737–738.

78. Wing R, Drew H, Takano T, et al. (1980) Crystal structure analysis of a complete turn of B-DNA. *Nature* 287(5784), 755.

79. Quintana JR, Grzeskowiak K, Yanagi K, et al. (1992) Structure of a B-DNA decamer with a central T-A step: C-G-A-T-T-A-A-T-C-G. *J. Mol. Biol.* 225(2), 379–395.

80. Pavletich NP, & Pabo CO (1991) Zinc finger-DNA recognition: crystal structure of a Zif268-DNA complex at 2.1 Å. *Science* 252(5007), 809–817.

81. Ellenberger TE, Brandl CJ, Struhl K, et al. (1992) The GCN4 basic region leucine zipper binds DNA as a dimer of uninterrupted alpha helices: Crystal structure of the protein-DNA complex. *Cell* 71(7), 1223–1237.

82. Cho Y, Gorina S, Jeffrey PD, et al. (1994) Crystal structure of a p53 tumor suppressor-DNA complex: Understanding tumorigenic mutations. *Science* 265(5170), 346–355.

83. Vilcheze C, Wang F, Arai M, et al. (2006) Transfer of a point mutation in *Mycobacterium tuberculosis* inhA resolves the target of isoniazid. *Nat. Med.* 12(9), 1027–1029.

84. O'Halloran TV, Lippard SJ, Richmond TJ, et al. (1987) Multiple heavy-atom reagents for macromolecular X-ray structure determination. Application to the nucleosome core particle. *J. Mol. Biol.* 194(4), 705–712.

85. Davey CA, Sargent DF, Luger K, et al. (2002) Solvent mediated interactions in the structure of the nucleosome core particle at 1.9 Å resolution. *J. Mol. Biol.* 319(5), 1097–1113.

86. Schalch T, Duda S, Sargent DF, et al. (2005) X-ray structure of a tetranucleosome and its implications for the chromatin fibre. *Nature* 436(7047), 138–141.

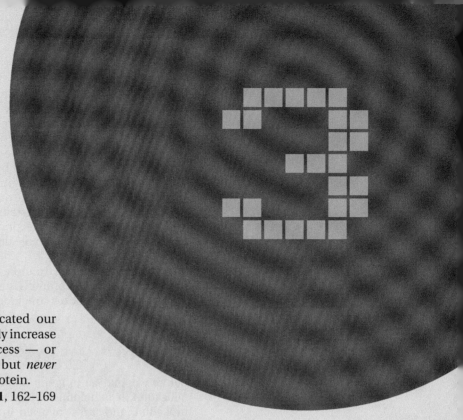

Protein crystallization

> In the end, no matter how sophisticated our reasoning and statistics are, one can only increase the *probability* of crystallization success — or increase the degree of belief in it — but *never guarantee* success for any particular protein.
>
> B. Rupp (2003) *J. Struct. Biol.* **141**, 162–169

Crystallizing the protein is a crucial step during the course of a protein structure determination, and there is no known method of predicting exactly under what conditions a specific protein will form single crystals. The general procedure is to reduce the solubility of the stock protein, in the hope that the protein separates from the solution while self-assembling into diffracting crystals. The reason for the difficulty in controlling crystallization lies foremost in the complexity of the weak interactions that are necessary between the irregularly shaped and flexible protein molecules in order to self-assemble into a regular, periodic crystal lattice. Viewing a protein crystal as a periodic network of specific interactions emphasizes that each protein's intrinsic properties, such as local surface charge distribution, flexibility, and conformational homogeneity, are key factors predetermining the chance for success in protein crystallization.

Many practitioners consider protein crystallization more of an art than a science. Perceiving crystallization as some kind of magic, however, brings with it the temptation to accept a less rational approach toward the subject: Crystallization tips are often based on anecdotal evidence and single events. Mystery surrounds the appearance of crystals (or the lack thereof), and many gadgets and kits promising salvation are commercially offered. There is very little justification for such a creationist view of protein crystallization: Although we have no means of accurately predicting conditions for successful crystallization of a specific protein, and despite the seemingly overwhelming number of possible crystallization conditions, we can tackle the problem in a systematic manner. Understanding the physical chemistry of crystallization, appreciating the crucial role of the protein itself, and properly using sampling statistics all play important roles on the path to success. Statistical analysis also allows the development of a rationale for deciding when it is time to terminate further crystallization trials and instead to begin modifying the protein. Some proteins do have properties that make them intrinsically unsuitable for crystallization, and *protein engineering*, addressed in the next chapter, may be applied to coerce these proteins into a crystallizable form.

Chapter 3 attempts to provide guidance through the process of protein crystallization with emphasis on rational strategies leading to a successful outcome of a structure determination project. After an exploration of the nature of protein crystals, we will briefly review the classic protein crystallization experiment, realizing the need to revisit and potentially modify protein properties in view of the crystallization process. In the following sections, the physical chemistry of crystallization and the chemistry and interactions of precipitants and reagents used in protein crystal growth are examined in more detail.

The remaining sections describe and explain the most common protein crystallization techniques, followed by a discussion of efficient sampling strategies to explore the crystallization parameter space. Inspection of the crystal growth experiments and analysis of the outcomes follows. A dedicated section treats the subject of protein–ligand crystallization as well as heavy atom soaking. Actual crystal harvesting, cryoprotection, and mounting are only briefly introduced, and are treated in considerable detail in Chapter 8.

A section highlighting examples of more difficult crystallization problems including membrane proteins, antibody scaffolding, and the crystallization of protein–DNA complexes concludes the crystallization chapter.

3.1   Overview of protein crystallization experiments

Before we analyze in great detail the fundamental principles of protein crystallization, it is helpful to acquire a brief overview of the basic experimental procedure and tasks involved in practical protein crystallization. In order to form a protein crystal, protein molecules must separate from solution and self-assemble into a periodic crystal lattice. One generally starts from a protein solution of suitably high concentration (usually in the mg ml^{-1} range) to which reagents that reduce protein solubility (precipitants) are added. Once the solubility limit of the protein is exceeded, the solution becomes supersaturated and metastable, and given favorable conditions and the presence of nucleation sites, crystals may start to grow. A useful tool to conceptualize the processes during protein crystallization, the crystallization diagram, will be introduced later in this chapter.

As it is not predictable which precipitants and conditions will in fact yield well-diffracting crystals, many initial crystallization trials with varying reagents, different reagent concentrations, and several pH or temperature levels are necessary. Frequently, only poor crystals are observed in initial trials, and a systematic optimization of the initial crystallization conditions leading to diffraction quality crystals follows the initial screening.

The most popular manual technique for growing protein crystals is the vapor-diffusion technique, which is illustrated by the hanging-drop setup in Figure 3-1: A few µl of the protein solution are mixed with an equal amount of reservoir solution containing the precipitation reagent cocktail. A drop of this mixture is placed on a siliconized glass slide, which covers and seals the reservoir. Since the protein–precipitant mixture in the hanging drop is less concentrated than the

Figure 3-1 Basic hanging-drop vapor diffusion. Hanging-drop vapor diffusion has been in use for over 30 years for the manual setup of protein crystallization. The reservoir (generally one well of a multi-well assay plate) is partially filled with several hundred µl of crystallization cocktail. A small drop (a few µl or less) of this cocktail is set in the center of a siliconized cover slide, and mixed there with an equal volume of protein stock solution (green). The cover slide is then turned over and placed on the greased rim of the reservoir well. The mixing with protein has reduced the precipitant cocktail concentration to half of the original value, and the sealed system thus equilibrates by water vapor diffusion from the drop into the reservoir solution, thus effectively increasing the concentration of all constituents (protein and precipitation cocktail reagents) in the crystallization drop. During this process the drop becomes supersaturated, nucleation can occur, and protein crystals may grow from the supersaturated solution.

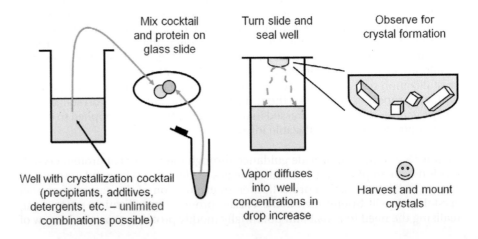

Mix cocktail and protein on glass slide

Turn slide and seal well

Observe for crystal formation

Well with crystallization cocktail (precipitants, additives, detergents, etc. – unlimited combinations possible)

Vapor diffuses into well, concentrations in drop increase

Harvest and mount crystals

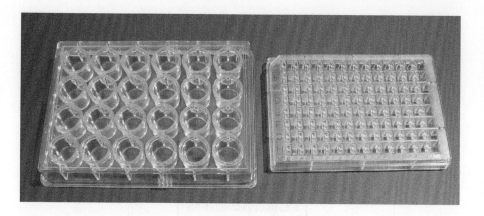

Figure 3-2 Crystallization plates.
A 24-well Linbro plate used for a manual hanging-drop crystallization setup is shown to the left. Originally designed for cell culture work, the Linbro plates made from polystyrene are cheap, and commercially available with pre-greased rims. The wells are covered with siliconized 22 mm circular cover slides. Although not suitable for automated work, the hanging-drop method using Linbro plates is still useful in small laboratories. The crystallization plate to the right is a typical 96-well sitting-drop plate in the smaller, standardized high-throughput SBS (Society for Biomolecular Sciences) format (Intelliplate, Art Robbins Instruments). Crystallization robotics most frequently use sitting-drop plates with a SBS footprint.

reservoir solution (remember: the protein solution was mixed with the reservoir solution at a ratio of approximately 1:1), water evaporates from the drop into the reservoir. As a result, the concentrations of both protein and precipitant in the drop slowly increase until the protein's solubility limit is exceeded. The solution becomes supersaturated, nucleation and phase separation occur, and crystals may form.

The most basic format for the hanging-drop vapor-diffusion technique is a 24-well Linbro plate (Figure 3-2). These clear polystyrene plates are relatively cheap, and each well can be individually sealed with a siliconized glass slide. The drops are easily observed under a microscope. A variety of other crystallization setup techniques exist, each with certain advantages depending on the specific requirements such as suitability for robotic automation or miniaturization (Section 3.7).

Once a suitable single crystal, preferably of regular, blocky shape, is detected, it needs to be harvested under a microscope and mounted on the diffractometer. Crystal sizes vary from a few hundred μm down to about 20–50 μm amenable for synchrotron data collection, and sometimes even as small as ~10 μm for specialized microfocus synchrotron beam lines.[1] Nylon fiber loops or more advanced laser-etched Kapton loops are commonly used to harvest the crystals by scooping them out of their mother liquor. To minimize radiation damage by the intense X-rays, the looped crystals are, in the vast majority of cases, rapidly cooled down (quenched or flash-cooled) to liquid nitrogen temperature. The flash-cooling may require the additional step of cryoprotection, depending on whether the crystallization cocktail already contains cryoprotectants or not. Large molecular weight polyethylene glycols, glycerol, or high salt content in the cocktail may suffice to prevent the formation of crystalline ice—which invariably destroys the crystals—upon cooling. In the absence of cryoprotectants in

Box 3-1 Protein crystallization basics. Protein crystals are periodic self-assemblies of large and often flexible macromolecules, held together by weak intermolecular interactions. Protein crystals are generally fragile and sensitive to environmental changes. In order to form crystals, the protein solution must become supersaturated. In the supersaturated, thermodynamically metastable state, nucleation can occur and crystals may form while the solution equilibrates. The most common technique for protein crystal growth is by vapor diffusion, where water vapor equilibrates from a drop containing protein and a precipitant into a larger reservoir with higher precipitant concentration. Given the large size and inherent flexibility of most protein molecules combined with the complex nature of their intermolecular interactions, crystal formation is an inherently unlikely process, and many trials may be necessary to obtain well-diffracting crystals.

Figure 3-3 Workflow of a crystallization project. The flow diagram shows the basic tasks and their relation and feedback in a crystallization project, starting at the protein level. Color indicates basic liquid handling (magenta); crystallization plate setup and handling (light blue); and mounting and data collation tasks (green).

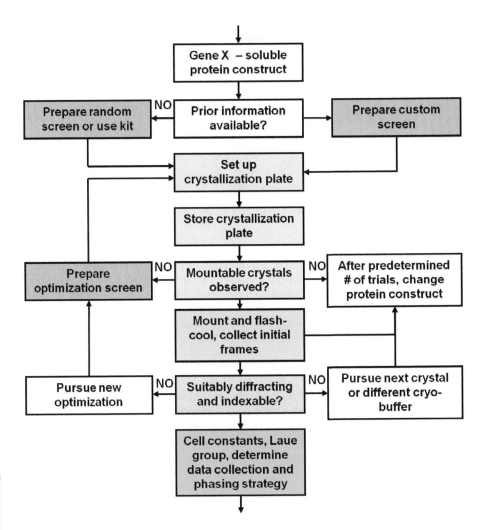

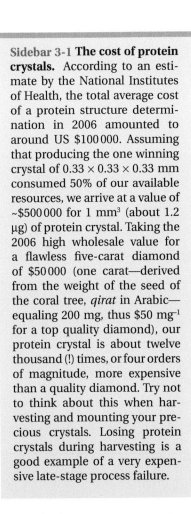

Sidebar 3-1 The cost of protein crystals. According to an estimate by the National Institutes of Health, the total average cost of a protein structure determination in 2006 amounted to around US $100 000. Assuming that producing the one winning crystal of 0.33 × 0.33 × 0.33 mm consumed 50% of our available resources, we arrive at a value of ~$500 000 for 1 mm³ (about 1.2 µg) of protein crystal. Taking the 2006 high wholesale value for a flawless five-carat diamond of $50 000 (one carat—derived from the weight of the seed of the coral tree, *qirat* in Arabic—equaling 200 mg, thus $50 mg⁻¹ for a top quality diamond), our protein crystal is about twelve thousand (!) times, or four orders of magnitude, more expensive than a quality diamond. Try not to think about this when harvesting and mounting your precious crystals. Losing protein crystals during harvesting is a good example of a very expensive late-stage process failure.

the crystallization cocktail, the crystals are swept through (or briefly soaked in) a cryobuffer containing a suitable cryoprotectant before being quenched. Again, no specific prediction of which cryobuffer will be best is possible, and multiple trials may be necessary to obtain a diffracting crystal. The harvested and cryo-cooled crystals are finally mounted on a diffractometer and the diffraction data are collected (Chapter 8). The workflow of protein crystallization is outlined in Figure 3-3.

3.2   Understanding protein crystals

It seems obvious why a major chapter in this book is dedicated to protein crystal growth; after all, without crystals, no crystallography. However, one needs to be more specific here: for the purpose of structure determination, we need to concern ourselves with the growth of single crystals of proteins. A single crystal useful for X-ray diffraction is relatively large, in the range of 50 µm to about 0.5 mm in size. Single crystals generally consist of many growth domains, which are nearly identically aligned and very often—but not necessarily—display sharp edges and well-defined crystal faces. In contrast, polycrystalline material contains many thousands of tiny, randomly oriented microcrystals, each again a single crystal, but as an individual much too small for a crystal structure determination. Polycrystalline material is investigated using powder diffraction methods, which play an important role in the identification of different crystal morphologies of pharmaceutical drug preparations(Sidebar 3-5). For the purpose of structure determination, powder methods based on the Rietveld profile refinement method[2] are not yet powerful enough to allow routinely *de novo* protein structure determination, but progress has been made in applying molecular replacement and isomorphous replacement phasing techniques in combination with powder diffraction data.[3]

The crystals we are familiar with from mineral collections or common small molecule crystals such as salt or sugar crystals create associations with qualities such as *hard*, *durable*, *pretty*, and *precious*. Unfortunately, only the latter two qualities hold true for protein crystals: Protein crystals often do form very beautiful single crystals, with well-defined edges and crystal faces as shown in Figure 3-4. However, external appearance is not necessarily a fail-safe indication of diffraction quality. Perfectly formed crystals may not diffract, while unsightly fragments dissected from crystal clusters can produce excellent diffraction patterns.

Properties and assembly of protein crystals

Crystals are periodic assemblies of fundamental building blocks, which can be atoms, small molecules, or whole proteins and even huge protein–protein or protein–nucleic acid complexes. In general crystallography, the concept that crystals are built from many small identical unit cells by periodically stacking them like solid bricks in a rigid crystal lattice is common. For the formal treatment of crystal geometry and diffraction theory in Chapters 5 and 6, the abstract notion of unit cells and crystal lattices is indeed practical and exceptionally useful.

For the purpose of understanding protein crystal growth as a dynamic process, it is, however, much more instructive to focus on the periodic network of intermolecular interactions between the molecules forming the crystal. Figure 3-5 shows a simple, translationally periodic arrangement of molecules, representing a primitive 2-dimensional crystal. It becomes immediately clear from the figure that only a limited number of interactions exists (indicated by the arrows) which form the network of intermolecular forces that keeps the large protein molecules connected. In addition, these interactions have to take place at specific locations on the surface of our protein molecules in order to allow ordered self-assembly into a crystal.

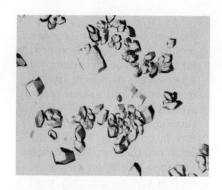

Figure 3-4 Crystals of tetragonal lysozyme. The figure shows tetragonal lysozyme crystals of varying quality, imaged from a hanging drop with a digital microscope camera (Figure 3-36). Hen egg white lysozyme is a hardy perennial of protein crystallization and widely used in practical demonstrations as well as in systematic crystallization studies. It is readily available and can be easily crystallized with common reagents (Sidebar 3-4). Note that some crystals are very well developed single crystals with sharp edges, while others are twinned and inter-grown in clusters. External appearance does not always correlate with diffraction quality. Perfectly formed crystals may not diffract, while unsightly fragments dissected from crystal clusters can produce excellent diffraction patterns.

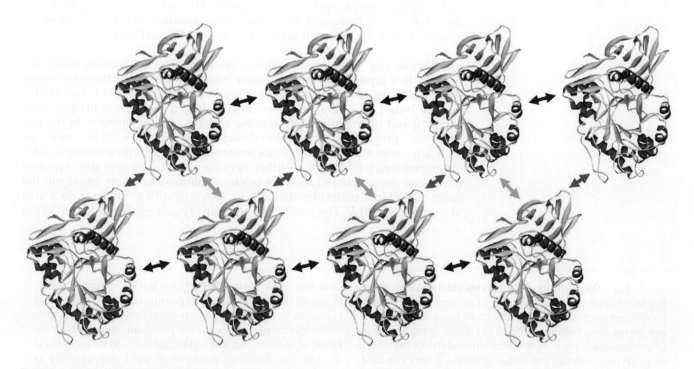

Figure 3-5 Protein crystals are formed by a sparse network of weak intermolecular interactions. The example shows protein molecules assembled into a primitive 2-dimensional lattice, connected by three different types (red, green, blue) of periodically repeating intermolecular interactions. The interactions are both sparse and weak, and as a consequence protein crystals are fragile and sensitive to mechanical stress and environmental changes.

Sidebar 3-2 **Protein crystals require a solvent environment for stability.** The first attempted X-ray studies of protein crystals from pepsin and insulin—initiated in 1934 by young Dorothy Crowfoot in J. D. Bernal's laboratory in Cambridge[5]—led to the insight that dried protein crystals lose their diffraction properties. Protein crystals generally must be kept hydrated at all times, even during X-ray diffraction data collection. Initially carried out in sealed capillaries at room temperature, diffraction data are now routinely collected from flash-cooled crystals at near liquid nitrogen temperature. It took Dorothy Crowfoot Hodgkin and a team of accomplished crystallographers 36 years to finally determine and publish the crystal structure of insulin in 1969.[6] The successful structure determination of cholesterol iodide, penicillin, and vitamin B_{12} led to her unshared 1964 Nobel Prize in Chemistry for "her determination by X-ray techniques of the structures of biologically important molecules."[7]

A number of important points can immediately be concluded from the situation depicted in Figure 3-5. Because the interactions between the irregular shaped, large, and flexible protein molecules are both weak and sparse, one cannot expect that protein crystals will be very hard and sturdy. The voids between molecules will be filled with disordered mother liquor, not contributing any significant interactions. Most importantly, the intermolecular contacts between the protein molecules are very specific, and they are inherent of the particular protein. This means that, no matter how diligently and skillfully we try to crystallize a protein, if a protein is fundamentally unsuitable for crystallization, it cannot, and will not, form good crystals or crystallize at all. Over and over in this text will we emphasize the necessity and importance of protein modification to overcome such fundamental barriers to crystallization,[4] originating from the nature of the protein molecules themselves. We already mentioned in Chapter 2 the examples of independent domain expression and domain truncation as frequent methods of producing smaller and more crystallizable protein constructs, and we will dedicate the whole of Chapter 4 to protein engineering for crystallography.

Crystal stability and intermolecular contacts

Crystal stability is a function of building block size and intermolecular properties: A crystal formed of small, covalently bonded atoms with a high ratio of strong directional bonds per unit volume will be extraordinarily stable (for example, diamond). Even ionic salt crystals, with large coordination numbers and many non-directional, electrostatic interactions are still quite durable. Molecular crystals of organic substances such as sugar still have quite a number of intermolecular hydrogen bonds and van der Waals (vdW) contacts per unit volume, and are reasonably stable.

In contrast, proteins, as building blocks, consist of long macromolecular chains of 20 different amino acids, often many hundred residues long, which fold into a 3-dimensional, irregularly shaped protein molecule (Chapter 2). Because protein molecules have irregular shapes and frequently possess dangling ends such as disordered termini or flexible loops, they are inherently not easily stacked and assembled into a regular, periodic lattice. The lack of internal symmetry of the chiral protein molecules places additional restraints on the possibilities for arranging them into a crystal lattice (discussed further in Chapter 5).

In addition, the intermolecular interactions keeping the protein molecules together in a crystal belong to the same class of weak interactions we recognized in Chapter 2 as being responsible for interactions in and between proteins *in vivo*. Dipole–dipole interactions, hydrogen bonds, salt bridges, vdW contacts, and hydrophobic interactions, all have binding energies in the low $-$kcal mol^{-1} range, in contrast to the strong covalent (about -100 kcal mol^{-1}) or ionic interactions in mineral crystals (return to Figure 2-29 for a review of side-chain mediated interactions). Further, because of the substantial size of protein molecules, the number of contacts per unit volume also is low, usually in the range of 3–12 intermolecular contacts per molecule with a contact surface area of around 200–500 Å^2. The result is that protein crystals—if they can form in the

Sidebar 3-3 **Do proteins dislike crystallizing?** An interesting idea which may shed some light on why it is generally so difficult to crystallize proteins arises from an evolutionary perspective. Soluble proteins in cellular compartments or intercellular space do not float around freely, but—just as in crystals—share, with other proteins, a very crowded environment, full of small molecules, nutrients, and copies of themselves and other proteins. It is conceivable that proteins perhaps had to evolve precisely to *not* aggregate and associate with each other under normal circumstances. Uncontrolled spontaneous crystallization certainly would compromise the viability of a normal cell, and some empirical evidence points toward the possibility of negative evolutionary design.[8] An interesting curiosity in this context is the fact that *Bacillus thuringiensis,* used commercially as a biopesticide, actually stores its insecticidal proteins as perfectly diffracting protein microcrystals.[9]

first place—are generally very fragile, soft (more like a cube of jelly instead of a small brick), and sensitive to all kinds of environmental variations. The contacts between molecules can also be quite anisotropically distributed, similar to minerals such as graphite or mica, exhibiting distinct cleavage planes. For similar reasons, uncooperative protein crystals thus may delaminate or shatter into shards upon touch or other physical disturbance.

Local nature of crystal contacts

The local distribution of surface residues responsible for maintaining the intermolecular contacts between proteins in a crystal is highly specific for each protein. Because proteins are polyions and carry surface charges (even when proteins have no net charge at their isoelectric point), charged residues are often key anchor points for intermolecular contacts. Just as with the other non-covalent interactions—hydrogen bonds and non-bonded dipole–dipole interactions, hydrophobic contacts, and van der Waals interactions—they cannot be predicted unless the structure of the protein is already known. Charged residues are, however, excellent starting candidates for protein engineering, and particularly lysine and glutamate/glutamine residues with high surface entropy have been successfully mutated to alanine for improved crystallization,[14] as we will discuss in Chapter 4.

The fact that particular and protein-specific local contacts are of decisive importance for protein crystal formation does already raise a warning sign for rules and crystallization "tips" based on global protein properties and related classifications: Such predictors may provide coarse guidance, but cannot provide any specific information about local contact formation (we will again, in Chapter 12, encounter the insensitivity of global descriptors regarding local errors).

Detailed packing analysis of crystal structures revealed that protein–protein crystal contact formation indeed appears to be an essentially stochastic process.[15] Moreover, because current protein structure prediction is not accurate enough nor can protein–solvent interactions be modeled with the necessary precision to pinpoint all contributions to the free energy of crystallization, *ab initio* crystallization prediction for proteins is not possible. Notwithstanding the progress in designing molecular assemblies by engineering of packing contacts between known structural templates,[16] the absence of *ab initio* calculations from sequence and physical principles means that we must rely on statistical sampling strategies (Section 3.8) to quantify the likelihood of a certain protein crystallizing under given conditions.

Crystal packing effects, artifacts, and solvent

The fact that specific residues are involved in crystal contacts raises the question about artifacts introduced by crystallization. We mentioned in Chapter 1 that—despite the fact that the core structure and even the enzymatic function of proteins are maintained in crystals—flexible and dynamic regions can be fixed in a specific conformation because of crystal packing interactions, and altered conformations of flexible regions may be induced. As a consequence, such an apparently specific conformation observed in a crystal structure may not actually be a dominant representation of that part of the protein structure in solution. A simple safeguard against misinterpretation—which usually implies assignment of certain biological relevance that is *de facto* not warranted—is to display all neighboring molecules in the crystal structure, and examine contact regions carefully for conformations that are likely a result of crystal packing. Determining the structure from multiple different crystal forms may also help to resolve the question of crystallization artifacts in the structure model.

Protein crystals contain on average around 50% solvent, mostly disordered in large solvent channels between the stacked molecules or along plain rotation axes in the crystal structure. The solvent contains water and all other molecules and ions present in the crystallization cocktail, plus anything carried through from purification into the protein stock solution. Solvent molecules in the nearest coordination shells tend to be well ordered and can form extensive networks,

Sidebar 3-4 The first protein crystals originated from abundant natural sources. Given the scarcity and sensitivity to degradation of many proteins, it is not surprising that the first protein molecules that were crystallized and studied came from abundant natural sources and were also relatively sturdy.[10] Examples are jack bean proteins (lectins such as concanavalin or urease[11] from *Canavalis ensiformis*), pepsin and insulin from various animals, and sperm whale myoglobin[12] (it may seem odd that the now protected *Physeter macrocephalus* falls into the category of abundant raw material, but in the deprivation after World War II, whale meat was considered an alternative to beef in Great Britain; the lasting impact on her cuisine to this date is the subject of debate).

Ferritin crystallizes readily by the addition of a few drops of $CdSO_4$ solution to horse spleen juice extracts. Even lysozyme can be crystallized directly from egg white[13] with a few drops of concentrated NaCl solution. From those and earlier observations on plant proteins, it became obvious that crystallization is also a practical purification technique. A great advantage, not to be underestimated, is that once a single crystal sample is obtained, crystallography is thus seldom plagued with impurity effects. For successful crystallization in the first place, nonetheless, reasonable sample purity of ~95% or higher is generally required.

Sidebar 3-5 Crystal morphology in pharmaceutical preparations. Specific crystal forms are of particular interest in pharmaceutical preparations. Different crystal forms and crystal morphologies of the same drug can have different properties as far as release and absorption are concerned. This holds for polycrystalline material (usually small molecule drugs) as well as for microcrystalline suspensions of peptide and protein drugs such as insulin or interferon-α used for injection. As a result, morphology and crystal forms present in drug formulations are a frequent source of patent disputes and challenges.

as already discussed (Figure 2-33). Other cocktail molecules such as methyl-pentane-diol (MPD, Figure 12-30), glycerol, phosphate, sulfate, and metal cations can often be distinguished in electron density when they are bound to surface residues. Beyond the nearest coordination shells, the solvent rapidly becomes disordered. Some molecules in the cocktail such as bidentate organic acids[17] or metal ions[18] can also participate in intermolecular contacts, and may in fact be essential for crystal formation.[19] While the lack of strong bonds and the presence of large voids explain the fragility of the crystals, the extended solvent channel network also has pivotal advantages, because the solvent channels allow diffusion of heavy metal ions (important for experimental phasing) or even larger substrates or inhibitors (relevant for drug discovery) into the protein crystals.

Crystal forms and morphology

It is not uncommon to observe different crystal forms under varying crystallization conditions, and multiple crystal forms may even be present in the same crystallization drop. This polymorphism can be used to advantage, because different crystal forms may exhibit significantly different diffraction quality. It is worthwhile trying to optimize all the crystal forms present rather than just focusing on the one that looks best by visual assessment in the initial screens, in part because polymorphism can also resolve questions regarding crystallization artifacts. The same crystal form can also appear in different crystal habits, having the same crystal structure, but exhibiting different pronouncement of the various crystal faces.

On some occasions, very rapid crystal growth can lead to inclusion of liquid in the crystals or to formation of skeleton crystals, when the edges grow fastest and not enough material is available to form and fill the crystal faces. Crystallizing at a lower temperature will likely slow kinetics and may improve crystals suffering from rapid-growth related defects.

3.3   Protein properties and crystallization

In the introduction we discussed that the crystallizability of a protein is *a priori* determined by the properties of the protein. Specific intermolecular interactions between the molecules must be formed in order to self-assemble the molecules into a well-formed, periodic crystal lattice. If the protein either lacks suitably located surface residues, or has structural features such as large flexible termini or flexibly tethered domains that inherently prevent it from crystallizing, then no amount of crystallization screening will ever yield usable crystals. Clearly, starting from a variety of protein constructs, usually with different affinity tags, limited truncation constructs, or expression of stable domains, increases the chances of success. In a similar horizontal approach, orthologs from different

Box 3-2 The nature of protein crystals. The protein molecules in a crystal are connected through a network of few and specific intermolecular interactions. Between the packed protein molecules, large voids remain that are filled with solvent. The voids between molecules allow the crystal to exchange liquid with the environment, and small molecules such as ligands or drug molecules can be soaked into crystals. For experimental phase determination, heavy metal ions can be soaked into the crystal, where they may specifically bind to certain residues and form marker atoms for phase determination.

Intermolecular packing interactions can change the conformation of surface residues or flexible loops of a molecule, but the core of a protein maintains its native conformation in the crystalline state; enzymes generally remain active in the crystalline state.

species are already properly engineered by nature and can provide alternate starting proteins for crystallization.

Planning ahead and anticipating the necessity of crystallization early in the protein preparation stages of a structure determination project is of significant advantage. The protein properties can provide powerful restraints directing the screening strategy. Once we are aware of the principles and necessary conditions for protein crystallization presented in this chapter, we will in fact devote the subsequent chapter to protein engineering and "salvage strategies" for crystallization. A short overview is necessary at this point to make us aware of important protein-related properties relevant for crystallization. Chapter 4 will provide a more detailed discussion of protein preparation and protein engineering in view of the requirements for successful crystallization.

Protein constructs and affinity tags

Given that the vast majority of purification protocols involve an affinity capture step, the question arises of what effect leading N-terminal or trailing C-terminal tags can exert on crystallization. There are many examples of successful crystallization with a variety of affinity tags, from mostly short poly-His tags up to substantial polypeptide tags.[20] In those cases, evidence is conclusive that tags were unproblematic. When tags needed to be removed to facilitate crystallization, causality between tag removal and successful crystallization is less well established; a vast pool of other parameters change in an uncontrolled fashion during expression and purification and may be factors influencing crystallization. No convincing statistics or predictions are available at this time about when and whether an expression tag is hurting crystallization, and quite substantial tags—largely N-terminal—have not hindered crystallization. As sub-cloning and expression testing are relatively cheap and easy to perform, pursuit of multiple constructs with varying tags as well as different truncations is a valuable and efficient option to obtain more soluble or better crystallizable protein[4, 21] (Chapter 4).

Only ~10% of all tagged sequences deposited in the PDB do actually show electron density for the tag. In some cases, the residues of the tag participate in crystal contacts as shown in Figure 3-6. Statistical analysis of pairs of tagged and untagged structures have shown[22] that the tags generally do not seem to affect the conformation of the rest of the protein, with the exception of the few terminal residues adjacent to the affinity tag. The most commonly used tags are N-terminal His$_6$-tags.

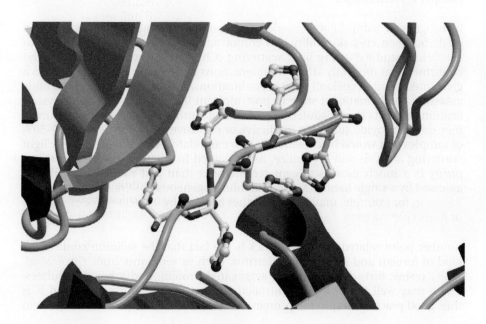

Figure 3-6 **C-terminal histidine affinity purification tag participating in crystal contacts.** The C-terminal His$_6$-tag of the molecule in the lower part of the figure interacts with the symmetry related copy shown in the top part of the figure. Not shown are additional solvent molecules that participate in an intricate network of intermolecular contacts. PDB entry 2bv9 and untagged molecule 2bvn.[23]

Batch variation and contaminants

It is quite common that different batches of the same protein do not show the same crystallization behavior. Thus, a second batch prepared from the same construct may actually crystallize if the first one did not (or *vice versa*!). Proteins also tend to acquire all kinds of "hitch hikers" such as cofactors, detergents, lipids, or membrane components that co-purify and vary from batch to batch, only to be detected later in electron density during model building.[24–26] Ligand binding sites in particular can attract all kind of detritus from the environment, leading on the one hand to spurious and easily misinterpreted electron density in the active sites,[27] or on the other hand to unexpected discoveries.[28] Moreover, batch variation is probably the most common reason for failure to successfully optimize the crystals after obtaining initial screening hits. Keeping some of the same protein batch aside for optimization may be wise.

Protein concentration

A commonly asked question is how concentrated the protein needs to be for successful crystallization. The often quoted rule of "at least 10 mg ml^{-1}" is not sustainable in view of the evidence. Although the average protein concentration extracted from PDB data is around 14 mg ml^{-1}, there are many examples of successful crystallization in the low mg range and even lower.[29] The required concentration depends on the individual protein, and instead of an absolute value, a more rationally defensible guideline is "as high as reasonably achievable" in each respective case. As protein crystallization requires reaching a supersaturated solution, the initial protein concentration must be high enough to actually reach this state during the crystallization experiment. A majority of clear drops observed in crystallization trials thus indicates too low a concentration. A few initial trials of observing a sub-µl drop of protein solution mixed with highly concentrated precipitants such as 30% PEG 5000, 4 M ammonium sulfate, or 30% isopropanol can quickly determine whether precipitation can be achieved. Under such aggressive precipitation conditions, precipitation should readily be achievable if the protein is sufficiently concentrated.

Purity, freshness, and conformational state

Impurities generally impede crystal growth. As with concentration, a clear cut absolute requirement for purity cannot be given. A reasonable single-band appearance in a well-loaded SDS gel (> 95% purity) is certainly a good starting point. For most proteins, degradation occurs over time—sometimes rapidly—and using the protein fresh seems to be of advantage for crystallization. Activity decreases proportionally with the amount of functioning material, but even small amounts of degraded protein or oligomeric aggregates may drastically hamper crystallization.

Investigations using light-scattering methods have shown that a correlation exists between crystallizability and conformational purity,[30] and second virial coefficient studies[31] using light scattering (Chapter 4) have (not surprisingly) confirmed that modestly attractive interactions between protein molecules in a given precipitation cocktail favor crystallization. Although the value of methods assessing conformational state in post-mortem investigation of failed crystallization attempts is undisputed—by revealing a probable cause of the failure—they do not provide sufficient predictive or exclusive guidance.[32] About 20–30% of samples that would be considered poor candidates for crystallization by light scattering analysis still crystallize. As explained in Chapter 4, *conformational purity* is a much more stringent requirement than just molecular purity as assessed by a single band SDS gel, and inhomogeneous posttranslational modifications, for example, may require further analysis by isoelectric focusing (IEF) or mass spectroscopy.

Another point related to protein stock is the fact that the solution contains all kind of foreign and endogenous detritus, such as remnants from chromatography resins, dirt, denatured and aggregated protein, and other particulates. These may well act in an uncontrolled fashion as nucleation sites, and it is thus good practice to spin the protein stock down before aspirating the protein

Box 3-3 **The protein is the most crucial factor in determining crystallization success.** Given that a crystal can only form if specific interactions between molecules can occur in an orderly fashion, the inherent properties of the protein itself are the primary factors determining whether crystallization can occur. A single-residue mutation can make all the difference between successful crystallization and complete failure. Important factors related to the protein that influence crystallization are its purity, the homogeneity of its conformational state, the freshness of the protein, and the additional components that are invariably present—but often unknown or unspecified—in the protein stock solution.

solution. This is particularly advisable if the protein stock has been frozen and thawed, where partial denaturation often occurs.

Proteins can, in rare circumstances, crystallize from quite impure solutions (crystallization, after all, is a purification technique). If a sample of a not-yet-perfect batch does not crystallize, then further purification may indeed be necessary. However, losing the majority of the material in low-recovery polishing steps before trying crystallization with a small sample of the not-yet perfectly pure material is inefficient. Examples for crystallization from crude tissue extracts of proteins such as ferritin, myoglobin, or lysozyme[10] illustrate that point.

Buffers, salts, and additives in protein stock

The protein stock solution should contain as few other reagents as possible, but as many as necessary to keep the protein in solution at a reasonably high concentration relative to its solubility maximum. Generally, a buffer solution is required, and sometimes also low salt concentrations may be necessary to stabilize (or salt-in) the protein. Weak, preferably organic buffers such as 10 mM HEPES are commonly used (in order for the crystallization cocktail to be able to drive the pH of the crystallization drop, the stock buffer must be rather weak). Additives, ligands, specific cofactors, or even detergents may be needed to keep the protein stable and active, and may place additional restraints on the choice of crystallization reagents. Certain cocktail components such as Ca^{2+} ions and phosphate stock buffer—a favorite of protein biochemists but less suitable for crystallization—are incompatible. It is also rather wasteful to screen a protein that is unstable below physiological pH against a screening kit that contains a large number of low pH cocktails.

The various reagents and additives such as lipids, detergents, or cofactors acquired throughout the course of purification, are often only discovered once the structure is determined. Each of these components may play a critical role in the crystallization process. For the success of protein prediction based on statistical analysis and machine learning methods, availability and quality of all protein related information is crucial. Absence of critical purification information will hamper your ability to mine your crystallization data for the prediction of crystallization success as well as make it potentially impossible to reproduce a crystallizing protein batch. Given the increasing importance of exploring multiple constructs to obtain crystallizable proteins, the need for complete and accurate book-keeping during protein production should be evident.

3.4   Thermodynamics of protein solutions

The first necessary step in any crystallization experiment is to coerce the protein gently out of solution, so that in the process of phase separation crystals can form. The solubility of a protein can be reduced by adding precipitants to the solution, by removing solvent (water) from the protein solution, or by a combination of both. Once the solubility limit of the protein is exceeded, the solution becomes supersaturated and, given a nucleation event, the excess protein molecules separate from solution. In favorable conditions the molecules may, in the process, self-assemble into crystals. Solubility reducing agents or precipitants are thus a primary component of any crystallization cocktail. The pH of the crystallization cocktail, generally stabilized by a buffer, also determines the level of protein solubility, but a more important effect of pH shifts on protein crystallization is the variation of the local surface charge distribution. The overall effect of pH changes is much more complex than simple induction of solubility changes. Temperature also affects protein solubility, which, depending on precipitant cocktail composition, may either increase (more frequently) or decrease with temperature.[10] The practical value of temperature selection lies rather in the control of nucleation and growth kinetics. Generally, a lower temperature means slower kinetics, but again, no prediction of the optimal temperature for a specific crystallization experiment is possible. Finally, various kinds of additives drastically affect crystallization behavior by modifying the intermolecular interactions between the protein molecules that are necessary to form crystals.

Solubility diagrams and supersaturation

A practical way to represent the change of protein solubility with precipitant is the solubility diagram (Figure 3-7). In these diagrams, the protein concentration is plotted on the vertical axis, and the precipitant concentration is plotted horizontally. Solubility diagrams are derived from experimental data, and are thus true phase diagrams. In contrast, the "crystallization diagrams" that are generally used are schematic representations that often contain additional "regions" which are actually depicting areas of different kinetic processes, and not representative of true phase equilibria. In these diagrams, the "precipitant" represents an entire cocktail, with the ratio of its individual components constant (i.e. the protein–precipitant system is represented as a pseudo-binary phase diagram).

The solubility line depicted in the solubility diagram separates the region that contains a single phase, the protein solution, from a two-phase region that contains protein and saturated protein solution in thermodynamic equilibrium. However, it is possible to supersaturate the protein solution. Supersaturation creates a metastable state, where the system is not in equilibrium, but the kinetics are hampered and do not allow the system to relax into equilibrium. The most common methods to achieve supersaturation of a protein solution are:

- Addition of a precipitant to the protein solution (batch methods).

- Removal of water from the protein–precipitant solution by vapor diffusion.

- Exchange of solvent by dialysis.

- Free-interface diffusion.

- Change of pH.

- Any combination thereof.

Once the activation barrier toward equilibration is overcome (for example, by spontaneous or external creation of nucleation sites), excess protein will come out of solution, and a protein-rich phase such as a precipitate (or a crystal) in equilibrium with saturated solution will form. It is important to understand that the limits of the metastable zone are thermodynamically clearly defined by the stability condition

$$\left(\frac{\partial^2 \overline{G}}{\partial x_1^2}\right) > 0$$

(3-1)

where $\overline{G}$ is the mean molar Gibbs energy of the system, in principle measurable as excess mixing enthalpy, and x_1 is the molar fraction of the protein. If we further increase the protein concentration, we finally reach a point where the

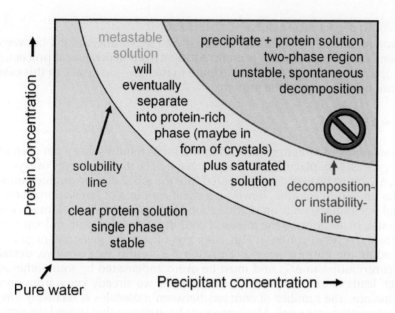

Figure 3-7 A basic solubility phase diagram for a given temperature. The diagram visualizes the general observation that the higher the precipitant concentration in the solution, the lower the maximal achievable protein concentration in the solution and *vice versa*. Between the solubility line and the decomposition line lies the metastable region representing the supersaturated protein solution, which will eventually— given the necessary kinetic nucleation events—equilibrate and separate into a protein-rich phase (such as precipitate or crystals) and saturated protein solution.

curvature (the second derivative) of $\overline{G}$ becomes negative and the highly supersaturated solution becomes unstable and must spontaneously decompose. This spinodal decomposition line delineates the metastable solution from the region of complete instability. Although thermodynamically clearly defined and part of a phase diagram, the decomposition line is experimentally not easily accessible and thus generally omitted or only schematically drawn in crystallization phase diagrams.

Effect of pH on protein solubility

The pH of the solution exerts a very strong effect on protein crystallization (Figure 3-8) . Although the solubility minima correspond well with the isoelectric point (pI, the pH at which where the net charge of the protein is zero), the correlation of pI and the actual pH of crystallization is weak, meaning that proteins do *not* crystallize best or most frequently at their pI. The pH change is nevertheless a key parameter and immensely useful for protein crystallization screening, and will be discussed later in the section on crystallization cocktails. The strong effect of pH on crystallization success is likely more a result of affecting the local charge distribution (and thus creating specific favorable packing interactions) than a net effect of pH on protein solubility.

Effect of temperature on protein solubility

Protein solubility can either increase or decrease with temperature, often varying between precipitants even for the same protein. Statistics show that most proteins are crystallized either at room temperature (a poor definition itself) or at 4°C. This binary choice results from the fact that traditionally proteins are prepared and purified at reduced temperature, commonly in a 4°C cold-room. The reasons for the choice of this specific temperature are largely historic, stemming from the attempt to slow down degradation by proteases, which is a more common problem in proteins isolated from tissue preparations than for those obtained via overexpression. There are presently no systematic investigations that demonstrate a significant overall advantage of one specific temperature for crystallization. However, individual proteins may crystallize better at a certain temperature, and a lower temperature can be helpful to slow crystal growth or to change the dominance of a certain crystal form. That notwithstanding, a reasonably constant temperature level should be of advantage in principle, simply to keep stable whatever crystals may have formed.

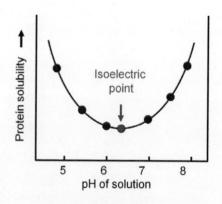

Figure 3-8 Protein solubility versus pH of protein solution. The protein shown in this example has its solubility minimum at its isoelectric point of ~6.3, where the sum of positive and negative charges (the net charge of the protein) is zero. Even at the isoelectric point, there are still numerous (but net compensating) local charges present on the surface of the protein.

Entropic contributions drive crystallization

A protein is stable in solution when net attractive interactions between the solvent and the protein are present. As in any physicochemical process, both entropic and enthalpic terms contribute to the free energy ΔG, in this case the solvation free energy of the protein,

$$\Delta G_s = \Delta H_s - T\Delta S_s \tag{3-2}$$

Self-assembly of protein molecules into crystals requires that a net drop in free energy must take place during the process; that is, the free energy of crystallization, ΔG_c, must be negative. Compared with the relatively unrestrained tumbling of the protein in solution, however, a significant loss of translational and rotational freedom happens upon formation of a regular network of interactions in a crystal, in addition to the losses of conformational freedom of loop residues involved in crystal contacts. This large loss of entropy, given the negative sign preceding the entropy term in Equation 3-2, results in a positive, destabilizing contribution to ΔG_c, and must be overcompensated by some other effect, which leads to an interesting observation: As we already know from previous discussions, the number of contacts between molecules is relatively low, and the interactions are weak. Measurements have shown that indeed the enthalpic contributions ΔH_c (heat of formation) are only weakly negative or practically insignificant (between -17 and ~ 0 kcal mol^{-1}) for protein crystal formation.[33] At the same time, the entropic loss for the protein $\Delta S_{protein}$, estimated just from the loss of translational and rotational freedom, is around -25 to -75 cal mol^{-1} K^{-1}, which leads, for room temperature, to a positive, destabilizing free energy contribution $-T\Delta S$ of about $+7.5$ to $+25$ kcal mol^{-1}.[34]

The large destabilizing entropic contribution overwhelming the enthalpic terms begs the question: where does the remaining stabilizing contribution come from that renders the total free energy of crystallization ΔG_c negative, as required for driving the equilibrium from protein in solution toward protein in a crystalline state? As already discussed in Chapter 2, as a driving force for protein folding, setting free the ordered water molecules forming the clathrate cage around hydrophobic residues during formation of the hydrophobic protein core provides a strong entropic gain. Similarly, the release of water molecules across both hydrophobic and polar surface residues during the formation of a crystal contributes to the entropy gains of the solvent, exceeding the entropy loss largely caused by the loss of motional freedom upon crystallization. The entropic gain in the range of 25–150 cal mol^{-1} K^{-1} corresponds to the release of around 5–30 ordered water molecules from the surface of the protein upon crystal formation.[35]

We can now write down the fundamental equation for the free energy of crystallization ΔG_c with split entropy terms (loss term for the protein, gain for the solvent):

$$\Delta G_c = \Delta H_c - T(\Delta S_{protein} + \Delta S_{solvent}) \tag{3-3}$$

This equation sheds light on several important facts regarding the thermodynamics of crystallization. As large entropy terms of opposite sign but similar magnitude compete in the entropy part of the equation, the difference between them will be small. In other words, relatively small changes in each of the entropy terms can have dramatic effects on the resulting total entropy change. This provides a plausible explanation of why the mutation of a single surface residue can have such drastic effects, and why the mutation of surface residues with high entropic freedom—foremost Lys, Glu, and Gln—to Ala can lead to significant improvements of crystallization:[33] If a high entropy residue is replaced with a low entropy residue, the total motional entropy loss of the mutant protein is smaller compared with the native form. With constant $\Delta S_{solvent}$, the entropy difference in Equation 3-3 becomes larger, and thus an additional negative, stabilizing, contribution to ΔG_c is gained. Targeted surface mutation as a general method of improving crystallization propensity will be discussed in Chapter 4.

Box 3-4 **Fundamental properties of protein solutions.** Crystallization is a special form of phase separation from a homogeneous solution, where the protein-rich phase in equilibrium with the protein solution is an ordered crystal. For phase separation to occur, the solution must become supersaturated, where it is thermodynamically metastable and will upon nucleation equilibrate into a protein-rich phase and protein solution. Supersaturation is a thermodynamic necessity to achieve phase separation. If the nucleation process and other kinetic parameters such as growth kinetics are favorable, the protein-rich phase may form as a protein crystal. Other possible protein-rich phases that can form are various liquid phases (protein "oils") and solid precipitates.

The phase relations in a protein solution can be represented in a pseudo-binary phase diagram with protein concentration and the precipitant concentration as parameters. The pH of a protein solution has a strong effect on the solubility, but crystallization does not preferentially occur at the isoelectric point, where protein solubility is at its minimum. Temperature also affects protein solubility and crystallization, but no general preferences can be predicted. Protein crystallization is an entropy-driven process; the release of water molecules across both hydrophobic and polar surface residues during the formation of a crystal contributes to the entropy gains of the solvent, exceeding the entropy loss largely caused by the loss of motional degrees of freedom upon crystallization. Self-assembly into crystals can be assisted by trial of additives that stabilize the protein, mediate crystal contacts, fine-tune intermolecular interactions, or otherwise modify protein solubility.

Equation 3-3 emphasizes that a fine balance between enthalpic gains and entropic changes during crystallization must be established in order to achieve crystallization. This explains why very minute changes can upset the delicate equilibrium between a crystal and the surrounding solution from which it grew, and even very small changes in environmental and chemical conditions can lead to degradation or disappearance of crystals. The protein–solvent interactions that keep the protein in solution compete with both intramolecular interactions of similar nature (but in much larger number, which keep the protein folded) and with the rather weak and sparse intermolecular interactions that need to be present for a crystal to form. During a crystallization experiment, we thus want to shift the equilibrium toward increasing intermolecular interactions, but to do so gently without denaturing the molecule.

Compounding the difficulties caused by the necessity to achieve a state in which the overall forces balance favorably toward negative ΔG_c—an absolutely necessary thermodynamic requirement—is the fact that whether this scenario can actually be realized depends critically on microscopic kinetic phenomena, foremost nucleation and growth kinetics. Thus, achieving negative ΔG_c is a *necessary* but not *sufficient* condition for crystallization as we will examine further.

3.5   Self-assembly and crystallization kinetics

At this point, we have acquired a good working understanding of how to reduce protein solubility. Reducing the protein solubility allows us to move into the metastable region of the phase diagram, a necessary, but not sufficient, condition for phase separation and thus for crystal formation. Whether the thermodynamically possible scenario of phase separation and self-assembly of molecules into a crystal actually will happen depends on the kinetic parameters governing the system development.

Activation and phase separation

A metastable system generally requires some perturbation or an activation event to overcome the kinetic barriers toward equilibrium. In a metastable, supersaturated protein solution this means that some form of nucleation has to happen before the supersaturated solution, which is still single-phase, can overcome the activation barrier and separate into a protein-rich phase in equilibrium with a saturated protein solution (Figure 3-9). In fortuitous cases, the protein-rich phase separating from the supersaturated solution is a crystal, but solid precipitates or liquid phases (protein "oils") are also, unfortunately common, protein-rich phases.

During crystallization, phase separation and phase transition into a new protein rich phase occurs. Phase transitions are initiated by so-called critical events, which are kinetically driven and hard to predict in complex systems such as protein solutions. Nonetheless, investigations by physical methods such as atomic force microscopy (AFM, pioneered by Alexander McPherson for protein applications)[10] or dynamic light scattering methods (DLS) show that, as expected, all established fundamental physical principles of phase formation apply to protein crystallization, with added complexity because of the intricacy of the multicomponent system which a protein crystallization drop represents.

Supersaturated solutions and nucleation

The supersaturated protein solutions generated by crystallization experiments are of fairly high concentration, and sometimes approach in density the crowded environment of the interior of a cell. Some protein solutions can in fact be supersaturated to extreme values. Lysozyme for example can be supersaturated to a concentration of several hundred mg ml^{-1}. This is remarkable, because a simple estimate, assuming a specific density of protein of about 1.35 g ml^{-1} and 50% solvent content, shows that an average protein crystal contains protein in a nominal concentration of about 600–700 mg ml^{-1}.

Homogeneous and heterogeneous nucleation. Given that the molecules in a supersaturated protein solution are rather crowded, they collide quite frequently with each other. Some of these collisions may happen in orientations that allow the formation of favorable contacts, and the local gain in binding energy may overcome the entropic loss resulting from the increase of order. This leads locally to a lowering of the system energy, a favorable event. However, other intermolecular bounces may disrupt this initial, ordered assembly, and the transient nucleus disappears again. Such events happen repeatedly in the solution, and it is qualitatively understandable that the probability of formation of more stable and larger nuclei increases with supersaturation. At some point, a nucleus may reach a critical size, and the fluctuations in the solution will no longer disrupt it (Figure 3-9). At this point, additional collisions with other molecules, or with already preassembled or pre-oriented smaller aggregates, are

Figure 3-9 Nucleation energy. To achieve crystallization, nucleation must overcome the kinetic barrier that exists for phase separation (crystallization) from the metastable solution. At its critical size, it is equally likely for the nucleus (symbolized by the aggregate at the peak of the red curve) to fall apart again (left red arrow) or to continue growing into a crystal (right red arrow). Once a nucleus above a critical size defined by the critical free energy of nucleation is formed, additional gain of binding enthalpy overcomes entropic loss during crystal growth, and the system can proceed towards its 2-phase equilibrium state (right side of image). The free energy of nucleation ΔG_n (red curve, scaled up by a factor of 10 for clarity) as a function of critical nucleus radius r is the sum of two competing terms — a volume-dependent ($-vr^3$) term lowering ΔG_n and a surface-dependent (sr^2) term increasing ΔG_n (Sidebar 3-6).

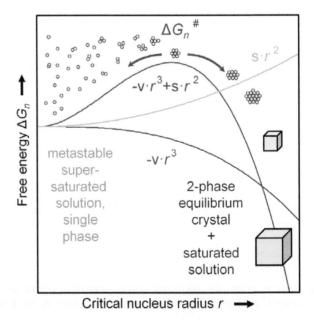

energetically favorable. The molecules will attach to the growing nucleus, and a new protein-rich phase will form. If this phase grows in distinct, periodic long-range order, a crystal will form; if the order is limited, amorphous precipitates, highly concentrated liquid phases, or anything in between may form while the solution relaxes toward equilibrium. The highly concentrated liquid phases are often observed as separated oils in crystallization drops. Even these protein-rich oils can, given a nucleation event, transform into crystals. The artificial induction of nucleation, allowing heterogeneous nucleation, is also the basis behind all seeding techniques, as will be discussed later.

Crystallization diagrams

The conceptualization of events such as nucleation during crystallization becomes somewhat easier if we add the kinetic information to the (thermodynamic) protein solubility diagram. Doing so is somewhat a violation of the pure spirit of a thermodynamic equilibrium phase diagram, but given that the addition of (generally tentative) nucleation information greatly increases our situational awareness during a crystallization experiment, we gladly accept this addition. To distinguish these modified diagrams from the thermodynamic solubility diagrams, we call them simply crystallization diagrams. However, it is important to understand that with the exception of the solubility line, which can be determined experimentally, most of the other information is much less well defined, to the point of being conjectural. These diagrams, nevertheless, greatly aid the conceptual understanding of processes occurring during crystallization.

Figure 3-10 represents a generic crystallization diagram, augmented with additional nucleation information. In the protein solution, we find the molecules in random orientation surrounded by solvent (water and precipitant molecules). Interactions between solvent and protein are stronger than between the protein

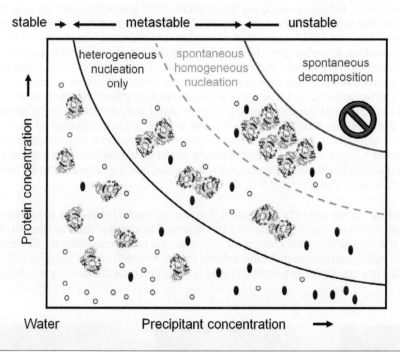

stable → ← metastable → ← unstable

heterogeneous nucleation only

spontaneous homogeneous nucleation

spontaneous decomposition

Protein concentration →

Water   Precipitant concentration →

Figure 3-10 The location of nucleation zones in a protein crystallization diagram. Precipitant molecules are represented as dark blue ovals and water molecules as light blue circles. As a rule of thumb, higher supersaturation is necessary for spontaneous formation of stable crystallization nuclei (homogeneous nucleation), while at low supersaturation nucleation requires external seeds in the form of microcrystals or other particulate matter (heterogeneous nucleation). The zone of homogeneous (spontaneous) nucleation is occasionally referred to as the "labile" zone, but this term should be avoided because it leads to confusion with the different concept of labile or neutral equilibrium used in physical chemistry and thermodynamics.

Sidebar 3-6 The critical radius of nucleation. The flux of particles forming a nucleus and departing from a nucleus can be balanced into a kinetic flow equation[141]. From that equation, the free energy of homogeneous nucleation ΔG_n as a function of the radius r of a spherical nucleus is derived as

$$\Delta G_n = -\frac{4}{3}\pi r^3 kT \ln \beta + 4\pi r^2 \gamma$$

(3-4)

In the Gibbs-Thompson expression 3-4 the advantageous r^3 volume term lowering ΔG_n competes with the r^2 surface term on the right side of the equation. In other words, the bond energy gained within the nucleus volume competes with the loss of degrees of freedom as result of adding more particles to the surface of the nucleus. The specific parameters are the supersaturation coefficient $\beta = C/C_s$, where C_s is the maximum equilibrium concentration and C is the actual concentration of the supersaturated solution; and the adverse increase in free interface energy, denoted as γ. At low supersaturation and/or at a small radius, the surface energy term dominates and the nucleus is unstable, but at high supersaturation and/or larger radius, the volume term dominates and the addition of more particles is favored. For a given supersaturation and surface energy, the critical radius can be found by setting $\delta G/\delta r = 0$ (corresponding to the peak of the curve in Figure 3-9). Dynamic light scattering (DLS) experiments allow determination of the critical nucleus size, which for lysozyme ranges from a few molecules to several 10s of molecules at low β. Similar equations have been established for heterogeneous nucleation.[36]

molecules themselves; the solution is single phase and stable. Once we cross the solubility line, the solution becomes metastable. As described in the section above, collisions between molecules become more frequent and transient nuclei form, but they do not reach critical size. In this region of the supersaturated solution, no spontaneous nucleation occurs, and crystallization needs to be induced by external means such as seeding. We call this region the heterogeneous nucleation zone. Upon further increase of supersaturation, we approach a region where the nuclei can reach critical size, become stable, and spontaneous or homogeneous nucleation occurs. Once nuclei have formed, crystal growth continues at supersaturation well below that of spontaneous nucleation, and slow growth from lower supersaturation generally results in better crystals. Crystal growth finally stops either because equilibrium is reached or for other, kinetic, reasons.

We will use crystallization diagrams extensively when discussing the different crystallization techniques and the pathways the crystallization experiments take through the phase diagram.

The growth of real crystals

The actual growth of protein crystals can be observed by atomic force microscopy, and the work of Chernov and McPherson[37] has greatly contributed to our understanding of protein crystal growth. In an atomic force microscope (AFM), a microscopic needle of silicon nitride (5–40 nm in diameter) mounted on the tip of a cantilever is scanned over a surface with a piezo-mechanism, and the deflections of the tip are measured by a laser. In tapping mode the forces exerted on the scanned substrate are very small, and the method can be used even on protein crystals in their mother liquor without disturbing the growth process.

In the nucleation equation (3-4) the unfavorable term is the surface term, and the effect of newly generated surface is minimized when the new molecules attach themselves to ledges or steps on the crystals. This generally leads to step-like or spiral-like growth patterns in crystals, and protein crystals are no exception (Figure 3-11). The AFM images also allowed investigation of what happens when contamination impedes growth. Because only the proper molecular specimens are stacked onto the crystal, the solution is eventually depleted of these molecules and becomes increasingly rich in contaminants. Eventually, these contaminants block further growth, and a new nucleation island may form on the crystal. As a result of debris and imperfections, that new island may not be perfectly aligned with the previous part of the crystal and a new, slightly misaligned domain forms (Figure 3-12). Therefore, a real single crystal is rarely a true single crystal, but rather a mosaic crystal of many nearly perfectly aligned domains (generally aligned within a few tenths of a degree to no more than a few degrees to be useful for diffraction). The crystal still looks like a perfect single crystal.

Twinning. If a few strongly misaligned domains or nuclei inter-grow as individual crystals, macroscopic twinning can be observed under the microscope. In this case, the resulting crystal clusters or macroscopic twins can often be dissected, and an individual fragment, being a single crystal by itself, can be used for diffraction. If a macroscopically twinned crystal is mounted on a diffractometer, distinct

Figure 3-11 Atomic force microscope images of crystal growth. (Panel A) The atomic force microscope images of the 001 surface of glucose isomerase show the two most common growth patterns observed in crystal growth: step growth starting from 2-dimensional nucleation islands (A, left image) and a double-spiral growth pattern (A, right image). Panel B shows formation of supercritical 2-dimensional nuclei on the 001 surface of cytomegalovirus (CMV), a member of the herpes virus family. As indicated by the arrows, in this case only two virions (B, left image) suffice to generate a critical nucleus from which new step growth commences (B, right image). Images courtesy of Alexander McPherson and Aaron Greenwood, University of California, Irvine.

A

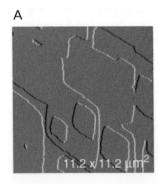

B

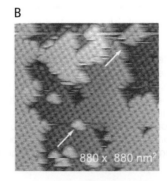
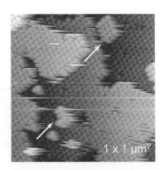

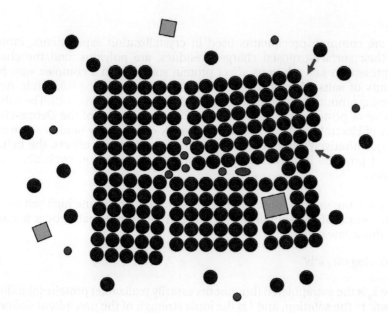

Figure 3-12 Growth of a real mosaic crystal. The schematic drawing shows a crystal growing in a solution of protein molecules (blue spheres). Small impurities (red) and some larger detritus (green squares) are also present in the solution. New molecules attach preferentially to steps and edges (red arrows) and we can recognize a growth defect in the form of a hole; impurities are enclosed at the domain boundaries; and a larger piece of detritus is incorporated at a domain boundary. Individual domains can be substantially misaligned, in this case about 6°; such a highly mosaic crystal would not be useful for diffraction experiments.

diffraction patterns reflecting each of the domain orientations are observed. More insidious and not visually detectable is microscopic merohedral twinning, where the domains alternate in orientation with a fixed relation to a crystallographic axis. Merohedral twins generate diffraction patterns that are abnormal only in their intensity distributions, which is practically impossible to detect by visual inspection of the diffraction pattern. An entire section in Chapter 8 is dedicated to recognition and treatment of merohedral twinning.

> **Box 3-5 Kinetics determine crystallization events.** Once the protein solution has reached thermodynamically metastable supersaturation, nucleation determines how the phase separation into protein-rich phase and saturated protein solution occurs. At high supersaturation, spontaneous homogeneous nucleation of the protein rich phases occurs, while at low supersaturation heterogeneous nucleation must be induced by seeding. Real single crystals are not perfect; they consist of multiple slightly misaligned domains forming a mosaic crystal.

3.6 Crystallization cocktails

Crystallization cocktails are added to the protein stock solution in order to reduce protein solubility to the point where the solution reaches supersaturation. Crystallization reagents are in practice grouped into several classes: the most common ones being *precipitant*, *buffer* (defining the pH of the cocktail), and *additive*, and others such as *detergent* or additional specific reagent classes. A crystallization cocktail generally contains at least some precipitant to be effective. A typical crystallization recipe for readily crystallizing lysozyme is given in Sidebar 3-7. The same reagent can be categorized as a precipitant (usually at high concentration) and also serve at a lower concentration as an additive or as buffer component (for example, Na-acetate). A comprehensive set of cocktails for crystallization screening often contains combinations of components from each class at varying concentration levels. The frequency of each of the reagents in the screen is based on prior experience, and can be further tailored if reliable prior information about the specific protein is available. In the absence of such prior information, it is statistically always better to screen as comprehensively as the resources allow than to focus on unsubstantiated leads, also known as "tips."

Sidebar 3-7 Two recipes for lysozyme crystallization. Prepare the stock solutions A, B, and C listed at the end of the box.

Protein solutions: weigh out 20 mg of lysozyme and place in a 1.5 ml microcentrifugation vial. For the slow growth experiment, add 1 ml of stock buffer A (20 mg ml⁻¹); for the fast growth experiment add only 0.5 ml of stock buffer A (40 mg ml⁻¹). Gently dissolve the protein in the buffer and spin the solution down in a centrifuge to settle foreign matter and denatured protein aggregates.

Slow growth experiment: Fill the reservoir well of a crystallization plate with 0.5 ml of crystallization cocktail B. Combine 2 µl of protein stock (avoid stirring up the spun-down debris from the bottom of the vial with the pipette tip) with 2 µl of crystallization cocktail A from the well on a glass slide. Place the slide over the well and make sure the grease seals tightly. Observe drops under microscope over time; well-formed single crystals should appear after a few hours or overnight.

Fast growth experiment: As above, but use 40 mg ml⁻¹ protein stock and use crystallization cocktail C. Observe drops under microscope immediately; many single crystals nucleate within minutes.

Stock buffer A: 20 mM sodium acetate at pH 4.6.

Crystallization cocktail B: 7–9% w/v NaCl, 250 mM sodium acetate, pH 4.0–4.6. You can screen various protein concentrations and drop size ratios against a variety of crystallization cocktails within the given limits of concentration and pH.

Crystallization cocktail C: 30% w/v PEG-MME 5000, 1.0 M sodium chloride, 0.05 M sodium acetate pH 4.6.

Salts

Salts are common precipitants used in crystallization experiments. Proteins, with their surface-exposed charged residues, are polyions, and the charged ions present in salt solutions affect protein solubility in a complex way. Small amounts of salts often increase protein solubility (salting-in), while further increasing amounts of salt reduce the solubility again (salting-out). The solution behavior of proteins can be described in the framework of the Debye–Hückel theory of electrolytes,[38] but despite the theory qualitatively (and in some cases, even quantitatively) describing salting-in and salting-out effects, the behavior of most proteins in actual multicomponent crystallization cocktails is more complex.

In general, the salting-out section of the solubility line in the high salt concentration region of the solubility diagram can be approximated by a linearized logarithmic law of the form

$$\log (S) = \log (S)_0 + kI \tag{3-5}$$

where S_0 is the extrapolated (but not necessarily realizable) protein solubility for zero salt in the solution, and I is the ionic strength of the precipitant defined as

$$I = \frac{1}{2} \sum_i^{ions} m_i v_i^2 \tag{3-6}$$

with m the molar concentration and v the valence of each ion i in the precipitant solution. The slope of the solubility line is given by a constant k, which is specific for the given solvent–solute system. However, it is important to emphasize that there is no general theory of protein solubility, and proteins do not necessarily have to show a salting-in effect nor a strictly exponential solubility decrease with higher precipitant concentration as suggested by the logarithmic linearization shown in Figure 3-13.

The shape of the schematic solubility curve in Figure 3-13 suggests that in addition to increasing the salt concentration, a reduction of salt (by dialysis of the solution against water, for example) can be used to decrease protein solubility, thereby driving a system into supersaturation. We will discuss in detail the navigation in crystallization phase diagrams in the section about crystallization techniques.

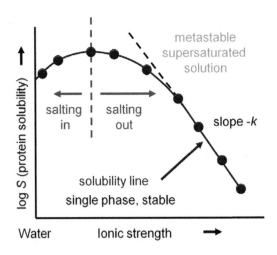

Figure 3-13 Solubility diagram typical for a protein in an ionic solvent. The high ionic strength (concentration) range can be described with a linearized logarithmic law, while the behavior in the low salt or low ionic strength region is more complex. We can distinguish a region where the solubility increases with small salt additions (cation-dominated salting-in) and the anion-dominated salting-out region at higher concentrations.

Salts also form the components of buffer systems, and in the case of bidentate groups such as malonate, tartrate, citrate, or other bicarboxylic acids, they can act as bridging reagents in forming crystal contacts (discussed in the additive section). The classification of a salt as precipitant, buffer, or additive in such cases becomes quite ambiguous, and the interactions with the protein are complex and synergistic. In high concentrations, salts can also act as cryoprotectants,[39] which is an advantage when the crystals are harvested and flash-cooled (described in Chapter 8).

Organic precipitants

The most common precipitants used in crystallization experiments in addition to salts are organic polyalcohols such as polyvinyl alcohol, polyethylene glycol (PEG), and PEG-MME (PEG monomethyl ether) of varying chain lengths between ~200 and ~15000 Da average molecular weight. PEGs ($H–(O–CH_2–CH_2)_n–OH$) were introduced in the 1970s,[40] and PEG-MMEs ($H_3C–(O–CH_2–CH_2)_n–OH$) in the 1980s.[41] Both are popular mild precipitants for proteins, and PEGs as a family are the most successful precipitants for crystal growth. PEGs above 1 kDa are solid white powders, and highly viscous stock solutions of up to 50% w/v of PEG powder in distilled water are freshly prepared for protein crystallization (PEGs are also viable nutrients for bacteria, and shelf life of PEG solutions is thus limited).

Protein solubility S in various PEG solutions can be described by a simple logarithmic law as a function of PEG concentration c_{PEG} similar to the salting-out branch for the protein–salt solubility curve:

$$\log (S) = \log (S_0) + k \cdot c_{PEG} \qquad (3\text{-}7)$$

where S_0 is again the protein concentration (or solubility) extrapolated to zero precipitate concentration (i.e., the ordinate intercept) and k is a constant depending on the particular solvent–solute system and is equal to the slope of the linearized solubility graph (Figure 3-14). Addition of PEGs always decreases protein solubility, and the thermodynamic reason for the linear dependence displayed in the logarithmic solubility plot is a consequence of an excluded volume effect. There is no "pegging-in" effect. Microscopically, the decrease in protein solubility can be explained as a competition of the polyalcohols for the water molecules[42] around the protein, thus forcing the proteins to seek interactions with each other rather than with the polar solvent water. Parts of ordered PEG molecules are occasionally found in the space between protein molecules, where they may participate in crystal contacts, although this is not their main functional role in crystallization (Figure 3-15). One of the most fortunate side effects of PEGs as precipitants is that they are effective cryoprotectants even at relatively low concentrations starting at 5–15%, particularly so for higher MW PEGs.

Low molecular weight alcohols also act as precipitants, and the mechanism of action is the lowering of the dielectric constant (and thus the polarity) of the solvent. Acetone or related compounds act in a similar way, but they tend to be rather harsh and can easily denature sensitive proteins. As a consequence, they are often used at low concentrations, which would group them into the additive category. A drawback of the small molecular weight alcohols such as methanol, ethanol, or isopropanol is their volatility, which makes them hard to control—particularly in very small drops—and necessitates the rapid setup of drops containing such reagents. Reduced volatility made 2-methyl-2,4-pentanediol (MPD) a very popular reagent in the alcohol family,[43] although recent statistics (Figure 3-35) indicate that its effectiveness compared with PEGs is relatively low. MPD does work well for DNA oligomer crystallization,[44] however, and its perceived high success was probably a case of usage bias. A hydrogen-bonded MPD molecule bridging Gln and Asp residues in a protein structure is shown in Figure 12-30.

Additives, detergents, and cofactors

A hotly debated issue (often with little statistical evidence to support the claims) concerns additives. The additives category essentially includes everything that

Sidebar 3-8 **Polyethylene glycols are used in pharmaceutical preparations.** Covalent linking of a PEG to therapeutic proteins or to liposomes containing drugs (pegylation) can increase their solubility, stability, and related absorption-distribution-metabolism-excretion (ADME) properties. For example, pegylated proteins of the interferon-α family are used as an injectable treatment for hepatitis C, and pegylated L-asparaginase (Oncaspar®) is used in the treatment of acute lymphoblastic leukemia. PEGs are also administered as mild laxatives, and PEG 3350 powder is probably cheaply available in pharmacy grade from the local drug store.

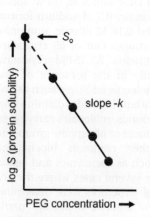

Figure 3-14 Solubility of proteins in PEG. The solubility of proteins in polyethylene glycols decreases exponentially with increasing PEG concentration. A plot of the logarithmic solubility log (S) against PEG concentration is therefore a linear function. S_0 is the extrapolated (but not necessarily achievable) protein solubility at zero PEG concentration.

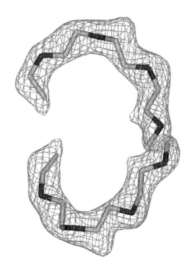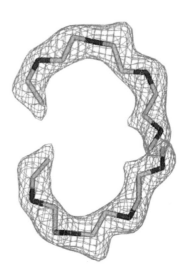

Figure 3-15 Ordered PEG fragments in protein structure. The cross-eyed stereo picture shows a fragment of a PEG molecule built into electron density observed in the solvent neighborhood of a protein molecule (omitted). Note the presence of a 2-fold rotation axis (horizontally across the center of the image). Image courtesy of Jiamu Du, Institute of Biochemistry and Cell Biology, Shanghai Institutes for Biological Sciences.

Sidebar 3-9 Silver bullets for crystallization? An interesting study that emphasizes the importance of additives, largely by mediating favorable intermolecular crystal contacts, has been conducted by Alex McPherson and Bob Cudney.[19] In a controlled experiment they used the same precipitant solutions, either 30% PEG 3350 or 50% tacsimate (a mixture of 1.36 M malonic acid, 0.25 M tribasic ammonium citrate, 0.12 M succinic acid, 0.3 M DL-malic acid, 0.4 M sodium acetate, 0.5 M sodium formate, and 0.16 M dibasic ammonium tartrate). Out of 65 crystallized proteins, 35 (54%) crystallized only in the presence of small molecule additives such as polycarboxylic acids, diamino compounds, molecules carrying sulfonate or phosphate groups, and other common biochemicals such as coenzymes and ligands. In several cases where the crystal structures were determined, the additive molecules participated in crystal contact formation.[47] One practically unavoidable weakness of the study is that many of the proteins used were proteins that have a tendency to crystallize well, and such proteins also tend to crystallize under a wide variety of conditions.[48]

might facilitate or improve crystallization; they may already be present in the initial screening cocktails. Additives are often resorted to in the step of optimization. Common reasons to search for additives to improve crystallization are aggregation, poor morphology, small size, weak diffraction, or erratic reproducibility of crystallization results.

Given the multitude of conceivable additives, it is helpful to group them based on a possible or proposed mechanism of action.[19] However, many additives act in multiple ways, and the effects of the components in a cocktail are synergistic. Additives can act through promotion of intermolecular contacts by divalent metal cations, stabilization of the protein, changing its aggregation state with detergents, changing the solution's dipole moment with small alcohols or highly polar agents such as DMSO, and almost anything else conceivable. In fact, any time a new substance is added to a crystallization mixture and crystallization is observed (for whatever reason), a new additive is born and handled as a hot tip. As with winning lottery numbers, there is seldom a statistical basis to prove overall effectiveness. Given a large enough trial size, even the most unlikely event is eventually going to happen. It is also wise to consider the other constituents of the crystallization cocktail to assure compatibility with additives.

Reagents and substances carried through from the purification into the protein stock can also act as inherent additives. Generally, these are physiologically or biochemically relevant small molecule ligands such as cofactors, substrate analogs, inhibitors, metal ions, or prosthetic groups, added purposefully as stabilizing agents to the purification buffers. Protein–ligand complexes with occupied binding sites are often more stable and conformationally homogeneous than the corresponding apo-proteins, and the importance of adding physiologically relevant binding partners or cofactors to the protein for stability and crystallizability can hardly be overemphasized. Another example for a common additive added to the protein early is the frequently used stabilization reagent β-mercapto ethanol (BME), which prevents oxidation of cysteines during protein preparation. BME can covalently bind to cysteines and is occasionally detected in electron density (Figure 2-27). Heavy metal scavengers such as EDTA and EGTA, and antimicrobials such as azide or phenol are also inherently present in many protein preparations, but their actual role in the crystallization process is probably limited.

A group of additives that has been widely and successfully used are detergents. Detergents are common reagents for resolubilization of proteins from membranes or inclusion bodies and stabilize membrane proteins. However, they are principally useful to increase the solubility of a protein or to act as stabilizing agents, particularly when aggregation is a problem.[45] As an additive, the detergent will be generally mild and used below or around its critical micelle concentration. An example for a detergent frequently used in membrane protein crystallization and as a general additive is octyl β-glucoside. Peptide detergents (peptergents) also have been successfully used to stabilize a variety of proteins.[46] The importance of detergents for membrane protein crystallization will be discussed in Section 3.10.

An example of an additive effect which can be rationally explained on a molecular basis is the formation of intermolecular contacts by intercalated divalent transition metal cations. Cadmium (in sulfate solutions), for example, has long been known to induce the crystallization of horse spleen ferritin and has been rediscovered as a useful agent to promote crystallization (or to increase diffraction quality) in a number of cases.[18] Similar ideas of aiding intermolecular cross-linking have led to a variety of polar or charged cross-linkers[45] such as polyamines and polypeptides as well as polyfunctional carboxylic acids[47] as cross-linking additives.[17] Figure 3-16 shows two examples of common, intermolecular contact mediating crystallization reagents.

Buffers and the effect of pH on protein crystallization

The purpose of the buffer component in a crystallization cocktail is to establish a certain pH (and hence a specific local charge distribution) of the protein independent of the other components and the original protein stock solution's pH, thereby allowing a systematic variation of the pH during crystallization screening. The cocktail buffer thus should have a higher concentration (in the 100 mM range) than the weakly buffered protein stock, so that the crystallization cocktail can drive the pH of the crystallization drop. However, the actual pH in the crystallization drop is rarely experimentally determined, and the final pH value in the drop when crystals have formed is not precisely known. High salt buffers may also act as precipitants, while other precipitant cocktails may not contain any buffer.

Given that protein solubility is at a minimum at the isoelectric point (pI), one might assume that proteins crystallize most frequently at their pI. Evidence demonstrates that this is *not* the case. The first fact to acknowledge is that although at the pI the net charge of the protein is zero, there are still (net compensating) local charges on the protein surface. It is the actual distribution of these local charges that is responsible for the formation of a significant part of the intermolecular interactions giving rise to formation of a crystal, and suitable charge distribution seems to be more important than the level of solubility.

Figure 3-16 **Additives mediating intermolecular contacts.** The left panel shows a Cd^{2+} ion in ferritin bridging three symmetry related molecules. Note that the Cd^{2+} ion is located on a threefold crystallographic axis. The right structure shows a tartrate molecule connecting two molecules of the sweet-tasting protein thaumatin from the African berry *Thaumatococcus daniellii*. Symmetry related molecules are shown in magenta and red. PDB entries 1aew[49] and 1thw.[50]

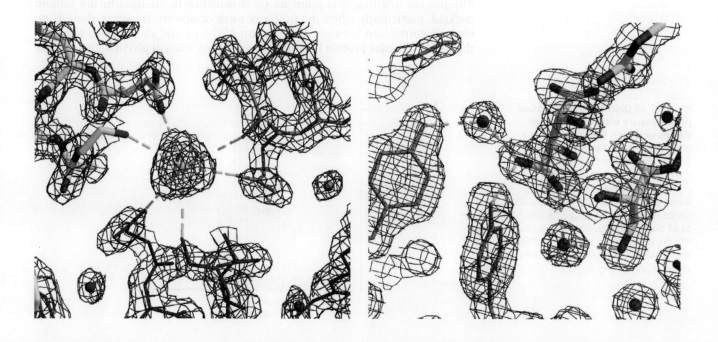

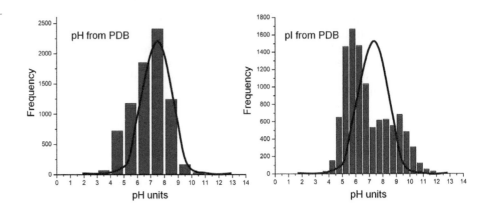

Figure 3-17 Distribution of crystallization pH and distribution of isoelectric point of proteins. The blue curve shows that the peak of the pH distribution around 7.4 falls directly into the gap between the modes of the bimodal pI distribution (right panel). Detailed pair-wise analysis has shown that acidic proteins prefer to crystallize above their pI and basic ones below their pI. The sum of these binned pH distribution gives the resulting overall distribution shown in the left panel.

A quick look at the different shapes of the pI distribution and the crystallization pH distribution in Figure 3-17 shows in agreement with literature[51–53] that no simple direct correlation between pI and crystallization pH exists.[54] Analysis of PDB data[10, 52, 55] has established that the reported pH of crystallization is distributed mono-modally around a physiological pH of ~7.4. This distribution agrees well with the crystallization propensity of pH values established by unbiased random sampling.[56, 57] The pI of each PDB entry can be calculated from SEQRES records, and one can attempt to correlate the reported crystallization pH with the protein's pI.

When the data are binned and analyzed, it becomes obvious that acidic proteins seem to crystallize best about 1–2 units above their pI, and basic proteins prefer to crystallize 1–2 units below their pI. The empirical crystallization pH frequency distribution peaking around 7.4 can thus be interpreted as the sum of distributions that favor crystallization by a few units above or below pI, respectively.[55] Empirical pH distributions for a given pI range have been established, and these binned probabilities generally peak around 7.4, and can be used to select the pH distribution that maximizes crystallization probability (Figure 3-18).

It is interesting to speculate why the highest propensity of proteins to crystallize is generally above or below their pI. All it means is that some net charge seems to increase crystallization propensity, but it is by no means a strict necessity, as seen from Figure 3-18. If the pI is located in the steep part of the titration curve of the protein, small fluctuations in pH could induce large changes in charge and thus destabilize crystallization, nucleus formation, and crystal formation. The protein might at this point be most sensitive to changes in the protein cocktail, particularly when the buffer is weak or absent. However, there is no obvious correlation between the location of the pI and the slope at the pI of the rather complex protein titration curves. In any case, at physiological pH the

Figure 3-18 Discrete crystallization pH frequency distribution for pI values between 8.5 and 10.5. The optimal pH in this range is about 1.5 pH units below the pI. Following the distribution when designing crystallization screens significantly increases the probability of success and hence screening efficiency. The sum of all discrete distributions peaks near the physiological pH for most proteins, thus yielding the known cumulative pH distribution peaking at 7.4.[10]

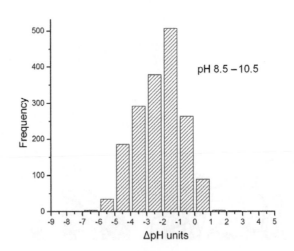

> **Box 3-6 Composition of a crystallization cocktail.** The purpose of a crystallization cocktail is to act as a precipitant reducing the solubility of the protein and to introduce other reagents that are potentially beneficial to crystal formation. Salts and PEGs are the major precipitants, and additives are selected to either facilitate crystal contact formation or otherwise stabilize or improve crystal formation. The pH of the cocktail strongly affects the distribution of charges on the protein surface, and therefore is a major determinant for crystallization. The effects of the reagents in the cocktail are generally synergistic and difficult to predict. At high concentrations, the precipitants (PEGs and salts) can also act as cryoprotectants.

basic amino acids (Arg, Lys, and to varying degree His) are protonated, providing cationic charges, while acidic groups of Asp and Glu can provide the anionic counterparts for dipole interactions and salt bridges (Figure 2-28).

3.7 Crystallization techniques

We have now explored a number of ways (and a nearly unlimited supply of possible reagents) to reduce protein solubility and to move a protein solution into a metastable state: Add or remove salt, add organic precipitants, vary the pH, change the dielectric constant of the solvent using alcohols, use some additives, and also, on occasion, vary the temperature. What remains to be explored is (i) the practical implementation of the experiments and (ii) how to select the most efficient strategies to sample crystallization conditions.

In contrast to what one would expect given the fundamental molecular complexity of the crystallization process as outlined in the previous sections, the actual setup of the common crystallization experiment is deceptively simple (and also easy to automate): The weakly buffered protein solution is combined in ratios of order 1:1 with a crystallization cocktail, placed in a more-or-less closed system, and left alone while approaching equilibrium to the point that kinetics permit. Numerous crystallization techniques have been developed, each with different merits and drawbacks depending on the specific application (for example, initial screening, optimization, harvesting, and ease of automation) and type of research environment. Figure 3-19 provides a quick overview of the most popular methods and some of their advantages and drawbacks, which will be described briefly below. Additional special techniques are described in Alex McPherson's classical textbook on protein crystallization.[10]

The simplest way to obtain crystals is to just mix the protein with crystallization solution to directly obtain a supersaturated solution and let kinetics take its course. Those methods are appropriately called batch crystallization methods.

Figure 3-19 Schematic sketches of popular crystallization techniques. Hanging-drop vapor diffusion is a common method used in small-scale manual setups while sitting-drop vapor diffusion is preferred with robotic setups. The absence of additional sealing requirements and the ease of miniaturization favors automated microbatch screening under oil, although harvesting tends to be more difficult. Use of silicone oils in microbatch wells allows partial exchange of solvent vapor (indicated by the broken arrow). Microdialysis is hard to miniaturize but can be used to grow very large crystals. Miniaturized free-interface diffusion screening chips are gaining popularity, but automation and harvesting issues remain to be resolved. Each method traverses the crystallization phase space in a different path and the same chemical screening conditions can produce widely varying results.

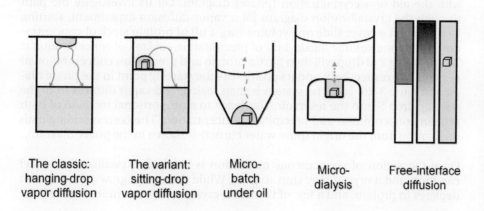

| The classic: hanging-drop vapor diffusion | The variant: sitting-drop vapor diffusion | Micro-batch under oil | Micro-dialysis | Free-interface diffusion |

Many of the first protein crystals crystallized and mentioned in Sidebar 3-4 were actually grown using batch techniques, essentially just adding highly concentrated precipitate to the protein solutions. The most frequent batch implementation employed today is the microbatch method in Terasaki-type microtiter plates or special SBS-style high throughput plates under oil. Alternatively, we can remove a solvent component (usually water), thereby increasing both the precipitant and protein concentration, and drive the system into supersaturation. The vapor-diffusion method, most frequently used and easily implemented as the hanging and sitting drop techniques, makes use of the slow diffusion of water vapor (and other volatile components) from a protein solution drop into a reservoir solution of higher precipitant concentration. It must be noted, however, that vapor-diffusion experiments often start in already supersaturated solutions and are strictly speaking a "vapor-diffusion-assisted batch experiment." A combination of vapor diffusion and batch methods is, furthermore, possible by using water-permeable (silicone) oils mixed with paraffin oils to cover the protein solution drops at the bottom of microbatch wells. Other diffusion-driven techniques and special methods are briefly discussed later.

Vapor-diffusion techniques

By far the most common crystallization technique used in the laboratory is the vapor-diffusion method. As the name indicates, vapor-diffusion techniques rely on the presence of a reservoir of precipitant that absorbs water from the crystallization drop, thus driving the solution into a supersaturated state and enabling crystallization. In the classical format of 24-well Linbro plates covered with siliconized cover slides, hanging-drop vapor diffusion is still used in manual setups, while sitting-drop vapor diffusion is the most prevalent method in automated crystallization plate setups.

Hanging-drop vapor diffusion

In the hanging-drop vapor diffusion technique, a drop of protein solution is placed on a siliconized cover slide and mixed with an approximately equal amount of crystallization solution from a reservoir. The reservoir or well has a greased rim and is sealed with the flipped-over cover slide. In the resulting closed system, water vapor diffuses from the hanging drop into the reservoir, which contains about twice the precipitant concentration of the drop. The method is standard in many small laboratories, and is easy to apply. The drop size is normally limited to no smaller than about 1 + 1 µl, largely because of the limited precision of manual pipetting. Some of the drawbacks of the hanging-drop method are: the relatively high cost of siliconized cover slides; the possibility that drops with low surface tension will spread out or even slip off the slide upon turning it; the necessity of greasing of the rim to seal the wells; and the large, non-standard size of the plates. On the plus side, individual wells are most conveniently opened for crystal manipulation or addition of ligand or heavy-atom solutions to the drop, and the wells can be easily sealed again with the same slide.

Crystallization diagram of a vapor-diffusion experiment

The events during a crystallization experiment can be conveniently visualized with the aid of a crystallization (phase) diagram. Let us investigate the path through the crystallization diagram for a vapor-diffusion experiment, starting by mixing on a cover slide one volume (say, 1 µl) of protein stock of concentration p with one volume (again 1 µl) of precipitation cocktail of concentration c. The resulting 2 µl drop will then have a protein and precipitant concentration of $p/2$ and $c/2$, respectively, and its position as the starting point in the phase diagram (Figure 3-20) is S. The system is then sealed, and vapor diffuses from the starting drop S into the reservoir. This leads to a proportional increase of both protein concentration and precipitant concentration. The corresponding path originating from the origin (pure water corner) is shown in the phase diagram.

Once the region of spontaneous nucleation is reached, crystallization nuclei can form and a crystal may start to grow. While the crystal grows, the solution depletes in protein, and a few of the initial crystals may continue to grow large

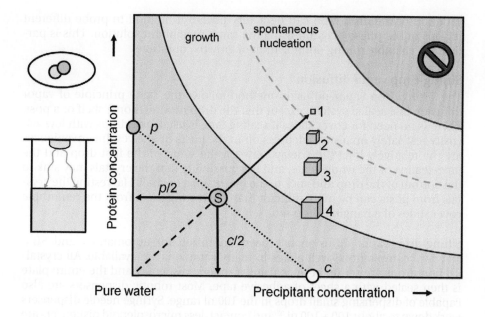

Figure 3-20 Path of a successful vapor-diffusion experiment through the crystallization diagram. The sketch on the left illustrates the principle of the hanging-drop setup. Equal amounts of protein solution (green) and precipitant (blue) are mixed. From the resulting starting drop (S, light green), water vapor diffuses into the precipitation cocktail reservoir and both protein and precipitant concentration in the drop increase. Once the region of spontaneous nucleation is reached, crystallization nuclei can form (1) and a crystal may start to grow. While the crystal grows (2, 3), the solution depletes in protein, and a few of the initial crystals continue to grow large in the growth region of the phase diagram. Once the crystals are in equilibrium with the saturated protein solution (4) they have reached their final size and growth stops. A typical 24-well hanging-drop plate is shown in Figure 3-2.

in the growth region of the phase diagram. Once the crystals are in equilibrium with the saturated protein solution, they have reached their final size and growth terminates.

The situation in Figure 3-20 is greatly idealized. In the depicted scenario, nucleation occurs early in the spontaneous nucleation zone, and only one nucleus has formed, which grows into a formidable crystal. The further into the nucleation zone the experiment proceeds (and the closer it comes to the thermodynamic instability line) the more likely it is that microcrystal showers form. As nearly all protein material is used up while the thousands of tiny crystals are rapidly formed, they cannot grow much larger, and they are rarely suitable for diffraction. Figure 3-21 visualizes such an outcome. In addition, the nucleation events do not necessarily lead to crystals: all kinds of precipitates, liquid protein-rich phases (oils), or additional phases can form. Growth may also stop for kinetic reasons before equilibrium is reached. The phase diagrams also emphasize that

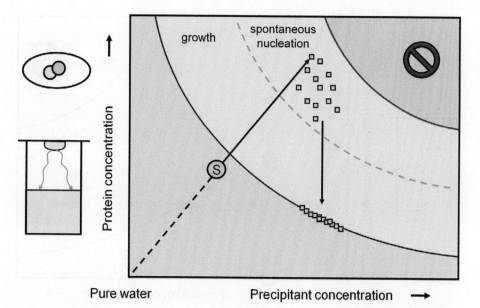

Figure 3-21 Microcrystal formation as the outcome of a vapor diffusion experiment. From the starting drop (S) the system moves deep into supersaturation, and spontaneous formation of many microcrystals occurs. As nearly all material is consumed in the formation of a shower of microcrystals, no significant crystal growth occurs and the tiny crystals containing all protein material are in equilibrium with the saturated protein solution.

variation of the drop ratio can be a very powerful method to probe different regions of the phase diagram using the same precipitant solution. This is particularly valuable during optimization of growth conditions.

Sitting-drop vapor diffusion

The sitting-drop vapor-diffusion method follows the same principle of vapor diffusion in a sealed system, except that the drop now rests on a shelf or a post. There is no need for cover slides if sealing tape is used, and drops with low viscosity rest safely on a small depression in the posts or shelves. Inconveniences are the relatively long time delay between the setup of the first drop and the tape-sealing of the whole plate, and the possibility that the crystals may sink to the bottom of the drop and stick to the plastic post or shelf. Harvesting the crystals from posts can be more difficult than harvesting them from the removable cover slides of a hanging-drop setup.

Sitting-drop vapor diffusion has been optimized for automation, and SBS-standard compliant 96-well plates in many variations are available. All crystallization drops are set up at once using a robotic dispenser and the entire plate is then sealed with a sheet of adhesive tape. Most robotic dispensers are also capable of dispensing small drops in the 100 nl range. Syringe needle dispensers work down to about 100 + 100 nl,[32] and contact-less microsolenoid dispensers are capable of working in the 50 + 50 nl drop range.[58] For the smallest drops, evaporation, diffusion of water into the plate material, surface effects, and rapid kinetics, as well as small achievable crystal size and crystal harvesting create formidable challenges to miniaturization, eventually outweighing the benefit of reduced sample amount requirements. Maximizing comprehensive sampling of crystallization space with the least amount of material plus ease of miniaturization is the major reason favoring the robotic high-throughput setup of nanoliter drops.[59, 60]

Clear drops. A possible outcome of a vapor-diffusion experiment is that all or most of the drops remain clear. Clear drops result if the system does not cross the solubility limit and thus does not reach supersaturation (Figure 3-22). In this case it is thermodynamically impossible to obtain crystals. However, these drops can still be salvaged by increasing the precipitant concentration in the reservoir and resealing the system.[61] Additional vapor diffusion from the clear drop into the reservoir then occurs and will lead to supersaturation and perhaps crystal formation. Clear drops can also be caused by insufficient nucleation, that is, a kinetic barrier toward crystal formation. In this case, crystallization can be induced by purposefully introducing nucleation. Multiple ways exist to artificially promote nucleation: internal nucleation by stirring or vibration (usually

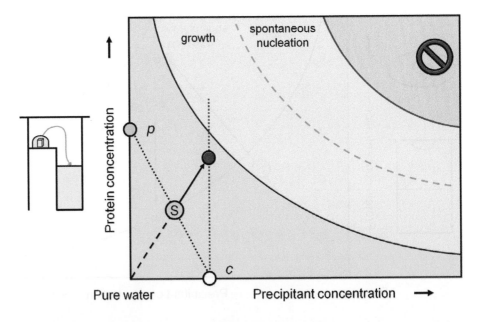

Figure 3-22 A failed sitting-drop vapor-diffusion experiment.
The starting drop (S) cannot reach supersaturation because there is insufficient precipitant in the drop, and the diffusion of vapor into the precipitant reservoir stops (red dot) once the concentration of precipitant in the drop reaches the reservoir concentration (c). The experiment could be resuscitated by increasing the precipitant concentration in the reservoir well and resealing the well again. A typical 96-well sitting-drop crystallization plate is shown in Figure 3-2.

not conducive to formation of well-formed single crystals) or the introduction of external nucleation sites. These so-called seeding techniques are described separately below.

Batch crystallization

As the name indicates, in batch crystallization the protein and precipitant in relatively high concentrations are mixed together to immediately create a supersaturated protein solution, and the system is isolated and kept undisturbed while it approaches equilibrium. In practice, a small drop of weakly buffered protein solution is mixed with an aliquot of the crystallization cocktail and pipetted onto the surface of an oil-covered microtiter plate well. The drop then sinks to the bottom of the well, and is thus isolated from the environment. Variations include placing the individual components under oil, with either the protein drop or crystallization solution first (creating two kinetically different mixing scenarios), and using water-permeable oils to allow water to diffuse into the environment. Water-permeable oils are a mixture of paraffin-based and silicone-based oils. It also should be noted that alcohols, detergents, and lipids can diffuse into the oil (and, to a much smaller degree, into the polymer of the plate material). The microbatch method is well suited for miniaturization and automation, as there is no time delay between the application of the drop and the sealing. The ability to work with minute nanoliter drops makes the microbatch method convenient for initial screening. The most frequently used plates are Terasaki-type microtiter plates and SBS format high-throughput plates with 384- to 1536-well plates.

In batch experiments sealed with water-impermeable oils, concentrations in the drop do not vary, and no movement of the starting point through the phase

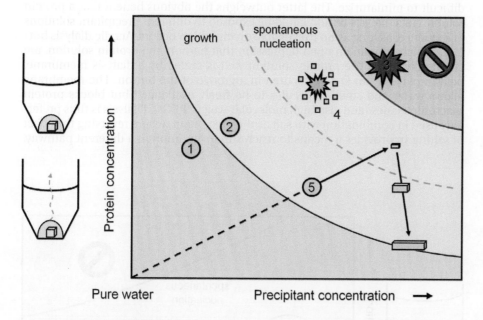

Figure 3-23 Phase diagram visualizing different batch crystallization experiments. Both experiments 1 and 2 remain clear and do not show phase separation, but for different reasons. Drop 1 is not in the supersaturated metastable region and thus it is thermodynamically impossible for phase separation to occur. Drop 2 has been set up with sufficiently high concentrations of protein and precipitant to lie in the supersaturated region, but no phase separation occurs for kinetic reasons because of the absence of spontaneous nucleation. Experiment 3 immediately precipitates upon mixing as it was set up with concentrations beyond the decomposition limit. Experiment 4 was set up in a region of high supersaturation, and a microcrystal shower forms rapidly. Drop number 5 was set up under vapor-permeable oil but with insufficient supersaturation for immediate nucleation. Because of water vapor diffusion out of the drop, it can reach a region of nucleation and usable single crystals can form. Water loss continues during growth, which is why the path of crystal formation is drawn slightly skewed to the right (precipitant rich) side of the diagram.

Sidebar 3-10 Varying the reservoir solutions. In principle the reservoir solution only serves the purpose of absorbing water vapor or other volatile components out of the crystallization drop. Using the same cocktail in both the reservoir and the drop is simply convenient in the setup and ensures that the initial concentration ratios result in vapor moving from drop to reservoir and not in the other direction. This observation opens another avenue to influence kinetics by setting up the same crystallization drops over different reservoirs. The rate at which water vapor transfers out of the drop determines the rate at which supersaturation is reached or increased, and thus can be expected to have a drastic effect on crystallization outcome. When the same drops containing 96 conditions were set up over constant reservoirs of 50 μl of 1.5 M NaCl, 1 M $(NH_4)_2SO_4$, or 50% PEG 3350, respectively, different outcomes were obtained in drops with identical mother liquor.[62] Although these effects are likely compounded with uncertain reproducibility of crystallization experiments (Sidebar 3-12), setup over different reservoirs improves the coverage of the crystallization space.

diagram occurs before phase separation. Should water vapor loss through a permeable oil layer (or through the plate plastic) take place, the pathway can again be depicted in the crystallization phase diagram (Figure 3-23). In batch experiments with impermeable oils, the precipitant concentrations selected are generally higher than in a corresponding vapor-diffusion or diffusion-supported batch experiment.

Batch methods under oil have been successfully used for many years, and they provide a number of unique features. The absence of sealing requirements makes microbatch methods under oil[30, 63, 64] attractive for the small laboratory. Very small drops can be set up without rush because each drop is immediately sealed by the oil, and affordable robots optimized for batch methods are available.[65] The method has been successfully automated, although only for initial screening, to accommodate the 1536-well format.[66] Drawbacks are the non-standard microbatch plate dimensions used in manual setup, which are incompatible with 96-well SBS standard compliant storage and observation systems.

It should be noted that a significant number of the classical vapor-diffusion experiments are actually a combination of batch crystallization and vapor diffusion. This is particularly evident when the initial conditions in a vapor-diffusion experiment are sufficient to precipitate the protein immediately after the drop setup. In addition, the plate polymer materials in both batch and sitting-drop vapor-diffusion setups also allow for limited water diffusion, and as a consequence, in either method, very small drops can dry out within a few days.[67]

Dialysis

Another way to exchange precipitants or to reduce or increase their concentration is dialysis. Dialysis is particularly suitable for growing large crystals, but is difficult to miniaturize. The latter outweighs the obvious benefit that a protein solution in a dialysis button could be exposed to different precipitant solutions in sequence, as long as no irreversible precipitation occurs. Acrylic dialysis buttons, which contain an open depression that harbors the protein solution, are commonly used. The protein solution well is sealed by a dialysis membrane, held in place by an O-ring secured in a groove of the button. The membrane allows water and small molecules to be freely exchanged but blocks protein macromolecules (and also high molecular weight PEGs). Dialysis is thus preferably used in combination with salt as the precipitant, where reversing the effect of salting-in provides a means to reach supersaturation in a different pathway.

Figure 3-24 Pathway of crystallization dialysis experiment. The starting solution (S) in the dialysis button (left insert) can be dialyzed either against distilled water (pathway to the left) or against a solution with higher precipitant concentration. The former uses the reversal of salting-in to achieve supersaturation, whereas the latter employs salting-out as a method to reach supersaturation. Note that the crystals obtained by each method can be entirely different in form, shape, or habit. The shape of the solubility curve is typical for a salt-based precipitate cocktail (compare Figure 3-13); PEGs do not show a salting-in effect.

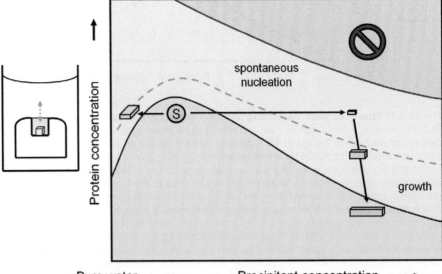

Figure 3-24 shows a sketch of a dialysis button and visualizes the path through a crystallization phase diagram during dialysis against distilled water and against a solution with higher precipitant concentration.

Free-interface diffusion

The basis of free-interface diffusion is to bring the protein solution and precipitant into contact in a narrow vessel without premixing and let the components equilibrate against each other by diffusion only. Free-interface diffusion (FID) methods have been developed for manual setup and for semi-automated setup.

The most significant difference compared with the previously described techniques is that the concentration profiles of protein and precipitant change in different ways across different locations in the system and at different times. Thus, in principle the coverage of phase space in a free-interface diffusion experiment is more complex and more complete[68] than in the spatially homogeneous systems described before. For example, it is quite intuitive that at the initial contact of the protein solution with the precipitant solution, there is a high concentration of both, and the contact region will be in a state of high supersaturation, which is advantageous for nucleation. During the course of the experiment, the diffusion profiles become flattened, and the lower supersaturation is more amenable to controlled growth. Figure 3-25 shows the development of idealized concentration profiles in a FID system over time, and Figure 3-26 provides an interpretation of the spatial and temporal position of the experiment in a crystallization diagram.

Two major variants of FID experiments are common, based on true counter-diffusion in thin-walled glass capillaries or free-interface diffusion in multi-layer soft lithography chips. The counter-diffusion capillary technique[69] is used in combination with gels. In such an experiment, the protein is wicked up by capillary action in the glass capillary, which is then punctured into a gel containing the crystallization cocktail. The gel locks the protein into the capillary, while the precipitant diffuses into the long capillary, efficiently probing the crystallization space (Figure 3-27).

Unlike free-interface diffusion which is in principle a spatially spread out, slow-mixing batch method, the power of capillary counter-diffusion arises from the fact that a very high concentration "wave" of precipitant travels through the

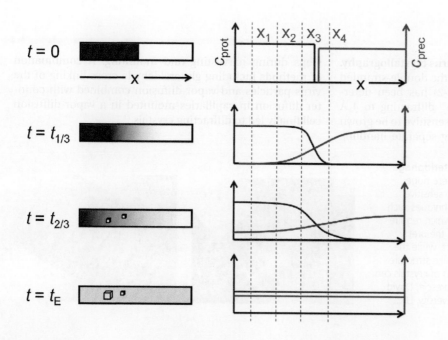

Figure 3-25 The time-development of the concentration profiles in a free-interface diffusion experiment. The left panel depicts the experimental setup and possible outcomes. The right panel shows the concentration profiles for a protein (blue) and a small molecule or salt precipitant (red) at different locations (x_1–x_4). Note the slow diffusion of the protein compared with the small molecules in the precipitation cocktail.

Figure 3-26 Pathway of free-interface diffusion through a crystallization phase diagram. The diagram shows the pathways of each of the four different locations x_1 to x_4 corresponding to the sections of Figure 3-25 and at different times during the process ($t = 0$, start; t_E = end, time when diffusion equilibrium is reached). Note that the pathway at location x_4 starting at pure precipitant never sufficiently enters the nucleation region. In this scenario crystals can only form in the protein-rich part of the diffusion chamber.

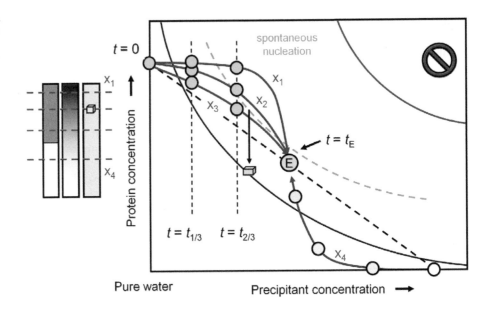

Figure 3-27 Counter-diffusion in thin capillary. The protein is wicked up in the capillary, one end sealed, and the capillary is inserted into a gel saturated with precipitant solution. The gel blocks diffusion of the protein, but allows a wave of precipitant to propagate through the capillary, effectively probing the crystallization space. Image courtesy Juan-Manuel García-Ruiz.

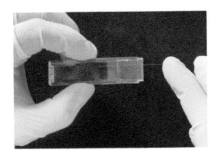

long protein chamber. A drawback is the somewhat larger material requirement (300 nl to 1 μl per experiment), and the method is therefore largely used for growing diffraction quality crystals once initial crystallization conditions are approximately known. Crystals grown in thin capillaries of about 100 μm diameter can be directly exposed to X-rays at room temperature, and useful data can be obtained.

Free-interface diffusion has been semi-automated on a microscale in multilayer soft lithography chips.[72, 73] In these microfluidic chips, very small sample amounts of about 1–2 μl protein stock suffice for a 96-condition screening (at 10 μl reagent each). Chips on an SBS format carrier plate are available, and they can screen several proteins simultaneously against 96 conditions (Figures 3-29 and 3-30). The drawback of the automated systems is the high price of the microfluidic valve control hardware and the rather expensive consumable silicone lithography chips. The rational considerations regarding phase space coverage provided above and preliminary evidence from several laboratories point toward increased screening success rates in free-interface diffusion chips, but no systematic investigations have been made to date. Harvesting of crystals from FID chips is also non-trivial, and they are thus primarily used to rapidly

Sidebar 3-11 Interface diffusion in virus crystallography. The spectacular 66 MDa structure of the double-stranded DNA bacteriophage PRD1 (Figure 2-2) has been determined from capillary-grown crystals[70] diffracting to 4 Å (Figure 3-28). These crystals were too sensitive to be grown by the sitting-drop method requiring separate handling steps during harvesting and mounting. A combination of methods including glutaraldehyde cross-linking of the virus particles and vapor diffusion combined with counter-diffusion in capillaries mounted in a vapor-diffusion cell finally led to diffracting crystals.[71]

Figure 3-28 Crystals of the 66 MDa bacteriophage PRD1 grown in a capillary. Crystals of these huge molecular particles (640 Å diameter) are so sensitive that they cannot be grown using other techniques such as vapor diffusion or batch crystallization which require harvesting and manipulation. Interestingly, the pretty and well-formed crystals (A) did not diffract, while the "fronds" of the crystals growing in a fern-like structure (B) diffracted to 4 Å—a reminder that looks of crystals can be deceptive. Images courtesy of Nicola Abrescia, David Stewart, and Jonathan Grimes, Oxford University, UK.

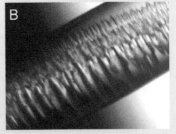

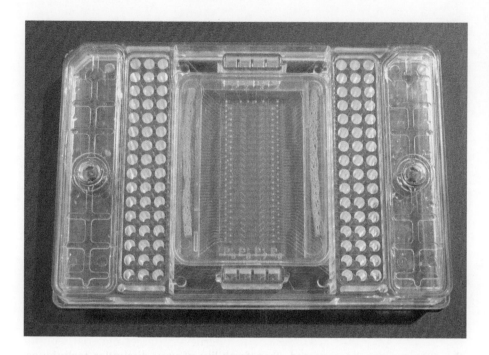

Figure 3-29 A microfluidic free-interface screening chip. The actual silicone microfluidic chip (~20 × 50 mm) is mounted in the center of an SBS footprint compliant plate that houses the sample and precipitant receptacles, and the control valve connections feeding into the chip. This chip model can screen eight proteins (~1 μl each) against 96 reagent conditions (10 μl each), executing a total of 768 simultaneous crystallization screening experiments.

identify promising leads with minimal material during initial crystallization screening. Given the entirely different pathways through the phase diagram, scaling-up or translation from a FID-chip screening experiment to a regular vapor-diffusion setup may require additional effort.

Overlap between crystallization techniques

From the viewpoint of chemical composition, there is no distinction regarding which crystallization technique is chosen. However, the kinetics, as well as the endpoint of the experiment, vary greatly depending on the crystallization setup, and the outcome of the crystallization trials can change drastically. One expects that the more similar in setup the techniques are, the more overlap in

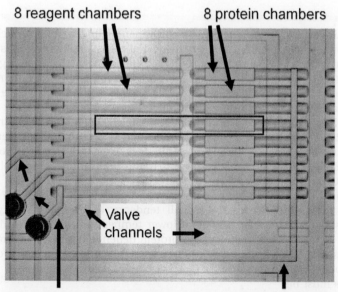

8 reagent chambers   8 protein chambers

Valve channels

8 reagent supply channels   Protein supply channel

Figure 3-30 Free-interface diffusion chambers in a microfluidic 96 × 8 chip. The image shows in strong magnification one of the 96 reaction chamber groups from the chip depicted in Figure 3-29, containing eight free-interface diffusion chambers (one indicated by the blue box) where one protein is screened against eight different reagents. The valve channels control the opening of the "gate" between sample and precipitant, visible in the center, and the sample and precipitant fill valves at the left and right sides of the chamber, respectively.

Figure 3-31 Venn diagrams representing different crystallization scenarios. The circles represent different techniques, and circles of same diameter indicate equal overall success rate for each method. The left panel shows that overlap in outcomes given the same reagents and drop sizes between hanging- and sitting-drop vapor-diffusion (VD) techniques is generally quite large, whereas microbatch experiments may have fewer conditions in common with either of the vapor diffusion techniques.[76] The panel on the right side visualizes a hypothetical scenario representing free-interface diffusion with potentially higher success rates in initial screening (larger circle) but limited overlap with sitting-drop VD technique.

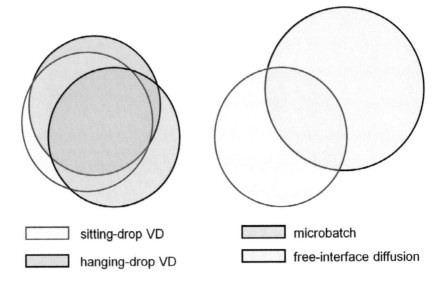

☐ sitting-drop VD

☐ hanging-drop VD

☐ microbatch

☐ free-interface diffusion

the outcome can be expected. Transferability of very dissimilar techniques, for example the scale-up of a microchip FID-screening experiment to vapor-diffusion optimization, is less assured (Figure 3-31).

Controlled experiments have shown that transition from one vapor-diffusion setup to another has little effect on overall success rates,[74] although a few particular conditions are unique to one (sitting-drop) or the other (hanging-drop) setup. This is consistent with the observed variability in the reproducibility of outcomes for exactly the same experimental setup (Sidebar 3-12). The conclusion is that on the one hand, consistent use of the same equipment and technique throughout the whole process assures the best possible reproducibility of results as well as maximum overlap between crystallization screening and optimization.[75] On the other hand, a change of setup technique may be an opportunity to introduce additional variation through the change of instrumental parameters affecting the outcome, particularly if poor results have been obtained with a certain technique for reasons not related to the protein itself.

Sidebar 3-12 How reproducible is protein crystal growth? In an interesting experimental study, Janet Newman and co-workers tried to establish how well protein crystallization experiments can be reproduced using exactly the same setup technique and cocktails.[77] For 96 conditions of the Hampton HT screen, 28 sitting-drop 96-well plates were robotically set up with 200 nl drops each and 28 plates with 100 nl drops. Half of the plates were stored at 277 K, the other half at 293 K. The results showed that although about the same number of crystals were observed for each plate at a given temperature, they were not produced under the same conditions. Only four successful conditions showed crystals in each plate, and only 32% (277 K) and 26% (293 K) of the successful conditions showed crystals in more than one instance.

These results indicate that—at least for lysozyme—reproducibility of experiments is surprisingly poor. Given the same chemical environment, the differences must be attributed to kinetic phenomena, foremost nucleation. Although the statistical analysis was not rigorous in this investigation, nor is it entirely clear to what degree the results can be transferred to other proteins, there are a few lessons to be learned: It may be worthwhile simply replicating experiments and, as mentioned earlier, varying both drop size and protein:cocktail ratio. The observations indicate that reproducibility of crystallization experiments is by no means assured, which is quite troublesome as far as statistical analysis and data mining of crystallization experiments is concerned.

Comparing the experiments at 293 K and 277 K confirmed the overall tendency of lysozyme to crystallize more reliably at lower temperature. However, this effect is probably also protein dependent and no general trend of temperature dependency can be extrapolated for other proteins.

Control of nucleation by seeding

The fact that spontaneous nucleation is very rare at low supersaturation, but frequent at higher levels, poses an interesting dilemma for the crystal grower: In the low nucleation probability region, which is a desirable location for crystal growth, we would have to wait a very long time for few, but probably large, crystals to grow. The long time span necessary for nucleation to occur also increases the risk of protein degradation and other premature losses of the experiment, such as effusion through plate material, leaky well seals, and so forth. In response, the experimenter increases the level of supersaturation in the trials, and promptly, as illustrated in Figure 3-21, many simultaneous nucleation events occur, and showers of tiny microcrystals or other species unsuitable for diffraction experiments form.

Seeding. The most commonly used means to achieve the desired balance between nucleation and growth is to introduce, through seeding, heterogeneous nucleation at low supersaturation, where slow growth is optimal for the formation of diffracting crystals but spontaneous nucleation is improbable.[78] There are two general seeding methods, micro- and macroseeding. Microseeding means that a few tiny fragments of crystalline matter—not necessarily from well-formed crystals, but often from initially obtained anisotropic needles, crystal clusters, or spherulites—are introduced in a crystallization solution of lower supersaturation than is required for spontaneous nucleation. The presence of heterogeneous nucleation seeds in crystallization solutions with low supersaturation, where spontaneous or homogeneous nucleation would not occur, is a plausible mechanism for the success of seeding techniques. Epitaxial growth on existing crystal surfaces has been proposed as another mechanism,[79] but the fact that different materials—and even debris or ground glass—can be used for seeding de-emphasizes the role of epitaxy in microseeding. The fact that denatured protein and other foreign matter can induce excessive nucleation is also the reason for spinning down the protein solution before setting up crystallization trials.

Microseeding. For microseeding, a dilution series of crushed crystal fragments or other particulate matter such as ground glass is prepared and some of the diluted solution is transferred into a new crystallization drop of somewhat reduced supersaturation (less precipitant and/or less protein). A popular variation is streak seeding, where a thin whisker or fiber is swiped across a seed crystal or seed crystal solution and streaked through the new drop. The new crystals are often observed to grow along the direction of the streak containing the nuclei (Figure 3-32). Other high protein containing phases such as protein oils separating from the drop, or even precipitates, can be used as seed sources in streak seeding. As a method of introducing heterogeneous nucleation, microseeding also works between different cocktails and crystal forms. Microseeding can also be automated by adding diluted solutions of crushed crystals to the drops at the time of (robotic) setup. The results have been promising,[80] yielding more crystals than comparable non-seeded control experiments.

Macroseeding. Macroseeding, in contrast to microseeding, implies the transfer of one single, already well-formed, but too small, crystal into a new crystallization solution of identical reagents in an attempt to "fatten" the crystal up to diffraction size. The rationale here is that if crystal growth has ceased because of lack of material, placing the crystal in fresh solution allows growth to proceed. In many cases, however, the reason for the success with seeding appears to be a combination of several seeding effects as mentioned above. Macroseeding requires considerably more skill than streak seeding or microseeding and cannot be readily automated.

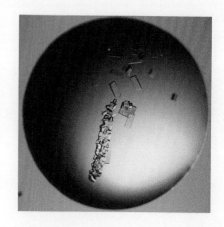

Figure 3-32 Crystals grown along the line of a seed streak. In microseeding, tiny microcrystals, crushed crystals, or other fine particulate matter are used to induce nucleation. The microcrystals or seeds are easily introduced using a whisker that is streaked or swiped through the crystallization drop, giving the name *streak seeding* to this technique. Image courtesy of Andrea Schmidt, EMBL Outstation Hamburg.

Box 3-7 **Crystallization techniques.** The inability to predict *ab initio* the exact conditions favoring protein crystallization means that, in general, several hundred crystallization trials must be set up in a suitable format and design. Crystallization screening experiments are commonly set up manually or robotically in multi-well format crystallization plates. The most common procedure for achieving supersaturation is the vapor-diffusion technique, performed in sitting-drop or hanging-drop format. In vapor-diffusion setups, protein is mixed with a precipitant cocktail, and the system is closed over a reservoir into which water vapor diffuses from the protein solution. During vapor diffusion, both precipitant and protein concentration increase in the crystallization drop and supersaturation is achieved.

Other protein crystallization methods include batch crystallization under oil, dialysis methods, and free-interface diffusion techniques. Microfluidic chips or thin-walled capillaries are used for free-interface diffusion. The advantages of free-interface diffusion methods are a comprehensive coverage of the crystallization phase space, and that very little material is required in the case of microfluidic chip methods. The pathway through crystallization phase space can be visualized with the help of crystallization phase diagrams. Crystallization diagrams combine information about thermodynamically defined phase relations with a tentative assignment of kinetic nucleation regions.

As a rule of thumb, low supersaturation favors controlled crystal growth, while high supersaturation is required for spontaneous nucleation of crystallization nuclei. Seeding is a method to induce heterogeneous nucleation at low supersaturation, which is more conducive to controlled crystal growth.

Robotics in the crystallization laboratory

With the advent of affordable crystallization robotics, automated crystallization setup is now a tempting possibility even for small laboratories.[32, 81] In academia, a primary objective of automation might be to relieve graduate and post-doctoral student talent from intellectually less challenging activities such as high-speed pipetting, cocktail mixing, and plate shuffling. The time saved can be productively used for protein production and unfortunately, as experience has shown, for servicing and programming robots. Raw throughput is generally a minor concern for academic laboratories; even modest robotic crystallization equipment almost always outstrips the capacity of the up-front protein production (most proteins that are difficult to crystallize are also those that are hard to produce, with the relatively high solubility required for crystallization a major concern). The most significant advantage of robotic systems is the ability to miniaturize the crystallization drops. Most of the affordable robots can reliably dispense drops down to 100 + 100 nl, a task practically not achievable with any accuracy through manual pipetting. In addition to miniaturization, reproducibility, and convenience, the easier and more consistent record keeping via automated capturing of the experimental data is probably the most significant benefit of robotic automation in the small laboratory (Sidebar 3-13).

In industrial settings, integrated crystallization platforms have been built that fully automate every step including cocktail preparation, crystallization plate setup, plate sealing, storage, and inspection of crystallization plates. Software can interpret the initial crystallization trial results and design optimization experiments. Recent reviews discuss the relative advantages and difficulties of various automation strategies. Automation presently stops at crystal harvesting, which requires very accurate micromanipulation and supporting real-time machine vision.[82]

Sidebar 3-13 **Automated crystallization setup for the small laboratory.** Based on the assumption of modest throughput requirements, and no necessity for full walk-away automation, two low-budget approaches to automation are conceivable: selection of a single system that can prepare crystallization cocktails (perhaps in a limited fashion) and also set up the crystallization plates;[83] or a dual-station layout using separate cocktail preparation with a generic liquid-handling system followed by a dedicated plate-setup robot.[84] The major reason for separating plate setup from cocktail production is differing requirements for dispensing precision, volume, and speed. Fast, small volume (µl to nl), and very accurate (also in geometric terms) dispensing is mandatory for plate crystallization setup, whereas large volume (ml) handling with modest speed and precision requirements suffices for cocktail production. Another advantage of the separation between the cocktail stage and the plate setup is that simple one-to-one dispensing into reservoir wells and drop aliquots, followed by protein addition with a single needle dispenser, suffices (Figure 3-33) once the cocktails are produced in a 96-well format deep-well block. Deep-well blocks prefilled with crystallization cocktails are also commercially available. In addition, compared with a single-stage setup, failure of one system component does not affect the other. For example, cocktail production can continue while the plate setup robot is inoperative. Figure 3-33 shows a popular robot for 96-well crystallization plate setup.

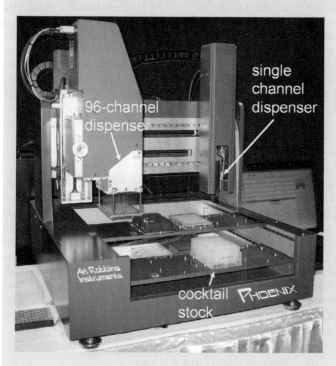

Figure 3-33 A robot for automated crystallization plate setup. The Phoenix robot (Art Robbins Instruments) can set up 96 crystallization trials in about one minute. On the left side, a 96-channel syringe dispenser re-arrays (100 µl each) 96 prefabricated or purchased crystallization cocktails simultaneously from a standard deep-well block into the reservoirs of an SBS-format, 96-well sitting-drop crystallization plate, and places between 1 µl and 100 nl into the drop shelves or wells. From the right side, a contact-less microvalve dispenser nozzle immediately adds the pre-aspirated protein (stock vials in the red block) rapidly and without contact onto each of the precipitant drops. To minimize evaporation, the plate is then immediately sealed with a sheet of pressure-sensitive adhesive. Taking all losses into account, about 12 to 15 µl of protein stock is required for 96 100 + 100 nl drops. The robot design has been based on a prototype developed in an academic laboratory setting.[84]

3.8   Protein crystallization strategies

Having selected an appropriate setup technique for a given laboratory setting, the question remains which reagents and global parameters—such as temperature and pH—to choose for crystallization screening. Prediction of crystallization conditions *a priori* from the protein sequence is not possible because of the specific and local nature of crystal contacts, although statistical analysis does allow some limited predictions about general reagent efficiency and, to some degree, specific crystallization preferences based on empirical classification of proteins.[85] The subject of specific crystallization prediction based on protein properties and their application for protein engineering is discussed separately in Chapter 4. Statistical sampling protocols allow us to develop strategies that optimally and efficiently sample the unlimited combination of screening parameters with a manageable number of experiments. In addition, we can exclude experiments that are likely to fail (such as using sampling at low pH for a protein that becomes unstable under acidic conditions) in order to obtain the most information from the available amount of precious protein material.

The most obvious question for the practitioner is usually which reagents to choose for custom crystallization cocktail preparation, or which one of the bewildering array of prefabricated kits to select. We will see that reagent selection is not as crucial a point as it may appear on first sight. A much more critical question is, how many screening trials are reasonable before hope should be abandoned and a different protein construct pursued? This question can be answered in an unemotional manner using statistical methods of experimental design.

Against all odds

Statistics from several high-throughput structural genomics initiatives reveal that for a given protein construct — of relatively easy to crystallize secreted bacterial proteins — our initial chances of obtaining any crystal are about 30–40%.[58] The probability of that crystal yielding usable diffraction data is again in the range of 30–40%, leaving one with a sobering 10–20% success rate for obtaining a crystal structure in one lucky screening run without further effort. Selecting a proper strategy, a readiness to pursue alternate approaches, and using multiple protein constructs, as well as careful observation and interpretation of the outcome of crystallization trials, are necessary to significantly increase individual odds. Accepting that formation of a perfect protein crystal is an inherently unlikely event, and acknowledging that we have no means of predicting the specific conditions favoring crystal formation for a given protein from first principles, the question arises: How in the absence of a deterministic approach can one beat the odds and still make informed and rational decisions about the path to take during a individual crystallization project?

As a principle, in a multivariate and sparsely explored experimental space, statistical methods can come to the rescue: Interpreting protein crystallization screening as a sampling problem allows one to maximize the odds. Combining sampling statistics with prior knowledge about the specific protein further increases the probability of success. In other words, avoiding all experiments that *a priori* are doomed to fail, while optimally exploring the remaining possibilities with an efficient protocol, maximizes the probability of obtaining diffracting crystals. However, as stated in the opening quote to this chapter, one must realize that, irrespective of how sophisticated our reasoning and statistics are, we can only increase the probability of crystallization success but never guarantee success for any particular protein.[48]

Crystallization as a multivariate sampling problem

Crystallization success analysis can be treated as a sampling problem of an unknown distribution of successful events in a high-dimensional crystallization parameter space, and detailed analysis and reviews regarding the subject are available.[48, 56] Given the huge number of available reagents (about 400 reagents are listed in the biomolecular crystallization database (BMCD) alone)[10, 52] which can be combined at any level, the question arises how to optimally explore this multivariate crystallization space.

One key to increasing efficiency *a priori* in all experimental designs is to select for all probable *factors*, but to eliminate irrelevant nuisance factors and improbable *factor levels*. Removing irrelevant factors from sampling and eliminating deleterious factor levels (such as denaturing pH, excessive precipitant concentrations, or interfering combinations such as high PEG concentrations plus high ionic strength salt solutions,[10] etc.) reduces dimensionality and volume of the multidimensional parameter space that needs to be sampled.

The crystallization data space can be visualized as an *n*-dimensional vortex, whose bases (axes) are extensive parameters such as chemical component and protein concentration, and intensive ones such as temperature, pH, protein properties, or various other defined setup parameters. Crystallization success analysis can then be treated as a sampling problem of an unknown distribution of successes in crystallization parameter space (Figure 3-34). Despite the simplicity of the concept, the high dimensionality, leading to a sparse distribution of data points, and the limited degrees of freedom (parameters) that can be systematically investigated require careful attention to experimental design.

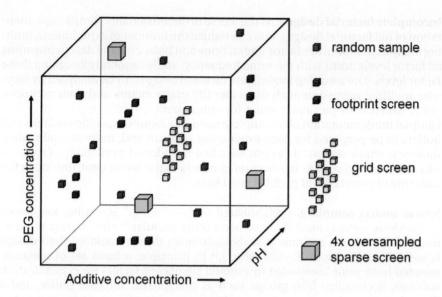

random sample

footprint screen

grid screen

4x oversampled
sparse screen

PEG concentration

Additive concentration →

pH →

Figure 3-34 Visualization of simple 3-dimensional crystallization space. The figure illustrates the varying coverage of the crystallization parameter space by different sampling protocols using 12 trials each. The larger green cube represents 4-fold oversampling in repeated use of the same sparse matrix-type experiment. Grid screening is not normally used to comprehensively screen for initial conditions, but deployed with a rationale to explore systematic variations of the two dimensions considered to be major factors, while keeping other parameters constant.

As an example, consider an exhaustive screening experiment that varies each dimension n (reagents) in k steps (reagent levels). The number of experiments to be set up is then k^n. Using only 10 reagents which we systematically vary at three concentration levels (none, low, high), we would need to set up 59 049 experiments. Even using small protein drops containing only 200 nl of protein solution, we would consume about 12 ml of pure protein preparation, which in all likelihood will exhaust our abilities to prepare material. Clearly, systematic exploration of all parameters is feasible at best only in low dimensions and with low granularity.

Crystallization screen designs

Grid screens. Despite the limitations of systematic exhaustive screening, by selection of specific parameters, either by intuition or prior knowledge, as dominating factors for crystallization, systematic variation designs can rapidly yield valuable information. In their initial 6 × 4 format, these designs are known as grid screen experiments,[51] and commercial kits in 24- and 96-well format are available. Typical examples are pH/PEG screens or pH/ionic strength screens. Without experimenter bias toward established factors, however, repeated 2-dimensional grid screening and its 1-dimensional variant footprint screening[86] become rapidly inefficient.

Factorial designs. In view of the impracticality of exhaustive sampling, the need for a rigorous approach toward efficient crystallization screening designs was recognized early by Carter and Carter,[87] who suggested factorial experimental designs, which allow application of regression and variance analysis, as well as response surface methods for optimization.[88] Factorial designs attempt to balance the occurrence of possible *factors* (reagents, pH, drop size/ratio) and of their combinations during the sampling process. We limit the discussion of designs to the minimum necessary to appreciate its importance for crystallization screening, and refer to Carter[89] and the very readable classical introduction to design of experiments by Box et al.[90]

Full factorial designs. The benefit of the full factorial design is that it provides a complete picture about all possible interactions between factors.[88] For a 2-level, 4-factor full factorial design, 16 experiments are required, and we can interpret our grid screen experiment (pH, PEG) as a simple 2-factor experiment with 6 (pH) and 4 (PEG concentration) levels resulting in 24 combinations. However, as shown above, at high dimensionality such as in crystallization screening, complete designs rapidly become prohibitive.

Incomplete factorial designs. As the "curse of dimensionality" limits implementation of full factorial designs, one can reduce the number of experiments, limiting analysis to first-order factor interactions and balancing the design (meaning all factor levels occur with the same frequency) while randomly assigning those factor levels. The resulting incomplete factorial design can be analyzed by stepwise multiple regression analysis to identify major factors and their contributions, which is of particular value for subsequent optimization experiments. Optimal implementation of incomplete factorials requires specific cocktails and buffers to be prepared for each experiment of a run and, unfortunately, when these systematic statistical designs were first introduced, availability of robotics was not as widespread as it is becoming now (a major factor contributing to the widespread popularity of prefabricated kits).

Sparse matrix sampling. Prefabricated "sparse matrix" screening kits based on previous success analysis have been quite popular,[91] and abundant variations of this first kit are now available (although the rationale for their design is sometimes less than "crystal clear").[92] In principle, a basis set of reagents, selected from prior knowledge (presumed significant factors for crystallization success), is classified into groups such as precipitant, additive, buffer, and a limited number of non-repeating combinations of one (or none) reagent out of each of these classes, usually at varying pH levels, is selected. Although success rates are thus limited to relatively few combinations of a preselected basis set of reagents resulting in incomplete coverage of the sample space, the original formulations have been successfully used in high-throughput screening.[93] In a statistical sense, repeated use of such premixed sparse matrix solutions amounts to oversampling of certain spots in the multidimensional crystallization space (see Figure 3-34).

Random sampling, reagent propensities, and number of experiments

An assessment of grid screen designs,[51] footprint designs,[86] and sparse matrix designs[91] in terms of sampling efficiency,[94] that is, finding crystallization conditions with a minimum number of trials, has demonstrated that—in the absence of any assumptions of prior knowledge—random (stochastic) sampling is most efficient, particularly when success rates are low or the successes are clustered. In principle random sampling means that out of a pool of stock reagents, any random combination of reagents from the four major groups, precipitant, buffer, additive, and detergent, as well as pH levels, are permissible within chemically reasonable limits. The term combinatorial sampling or screening is also used for such types of experimental design.

The benefit of ignorance. The omission of any prior knowledge—with the exception of the reagent basis set selection—and the absence of any assumptions about the protein's properties may at first sight seem a serious limitation of random sampling. However, in addition to reasons related to bias issues affecting the estimate of posterior probabilities in Bayesian models,[95] the adherence to the assumption of ignorance over the years has provided a very valuable, unbiased data set of experiments whose results[48] can now be used to guide our experiments. Figure 3-35 shows a summary of the results, which agree well with the findings from the literature and other high-throughput efforts.[96] The reagent and pH propensities (discussed in Section 3.6) essentially reveal that the highest crystallization propensities are found for PEGs and PEG-MMEs at physiological pH. This result is not new, and the work of Alexander McPherson has been crucial in establishing these facts based on intuition and experience.[51] The unbiased random sampling results prove these findings with quantifiable statistical significance. It is important, however, to appreciate that these propensity predictions reveal nothing but relative probabilities: It is about twice as likely that a given protein crystallizes using some PEG compared with the average reagent, but the protein's overall propensity to crystallize can still be low.

Trial numbers. Another important finding of the random screening analysis was an estimate for the number of trials above which return on investment

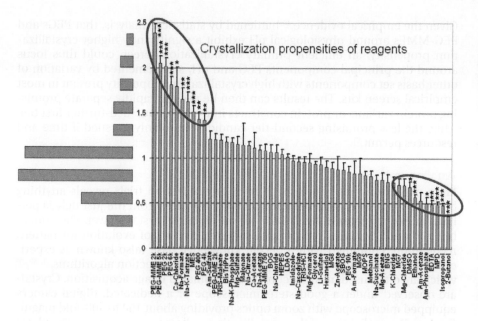

Figure 3-35 Crystallization propensities for 50 common reagents. Propensities are normalized frequencies, where an average crystallizing agent has a propensity of 1.0, and one that crystallizes better than average has a propensity above one. The data, obtained from unbiased random sampling of over 100 proteins, confirm the well-established fact that polyethylene glycols (PEGs) and their methyl ethers as a class are significantly better "crystallizers" than, for example, small molecule alcohols. The trends are statistically significant as indicated by the number of stars, but although over half a million experiments have been analyzed, shortcomings remain: only 50 reagents have been analyzed, synergistic effects are not considered, and the protein set is biased toward highly soluble, bacterially expressed prokaryotic proteins. Redrawn from Rupp and Wang.[48]

(time, supplies, and protein) in further screening diminishes, based on an analysis of cumulative probabilities.[94] The result is that for the average soluble bacterial protein, 288 trials (corresponding to three 96-well plates) should suffice to find crystallization conditions with a high probability. Beyond this point, the return on investment rapidly diminishes, and protein engineering can provide a better alternative than expending resources on continued screening. In practice, it does not matter if this cutoff is set at 288 or maybe even twice as many experiments, but thousands of initial trials are definitely excessive and only serve to harden the evidence, at great expense, that the protein falls into the "poorly crystallizable" category. The initial estimates that ~300 experiments suffice to decide whether a protein construct is a promising candidate for crystallization or not have been further hardened by large-scale experimental data described in the optimization section.

Caveat. The important caveat is to realize that all statistical analysis is invariably biased toward the sampled population (for example globular prokaryotic proteins). Proteins with inherently lower crystallization propensity such as membrane proteins or protein complexes may well require more experiments to establish any crystallization trends. Poor reproducibility (Sidebar 3-12) and batch variation also generate additional variability that can justify repetition of a given set of not-so-promising screening trials. However, even for membrane proteins or hetero-oligomeric protein complexes—once the expression problems are overcome and a stable, conformationally homogeneous soluble protein stock has been prepared—there is little reason to assume that the fundamental principles of crystallization are any different for these classes of more challenging proteins.

Two-tiered approaches

The most important objective of crystallization screening, as already hinted repeatedly in the previous section, is to rapidly determine with a limited initial screen whether a crystallization project will likely succeed and thus justify additional screening and optimization efforts (and thus production of more of the same material). Conversely, if there are already indications that things are not going well, a proactive approach of switching to altered protein constructs or orthologs is more efficient than continued screening. Analyses of the random experiments as well as results from various high-throughput efforts concur with the implementation of a two-tiered approach as the most efficient means of screening.

Given the empirical evidence,[40] hardened by statistical analysis, that PEGs and PEG-MMEs around physiological pH exhibit a significantly higher crystallization propensity, an efficient primary crystallization screen could thus focus around the principal components PEG and pH, supplemented by variation of other basis set components with high crystallization propensity present in most empirical screen kits. The results can then be used to rapidly separate promising protein variants or protein constructs into a to-be-pursued-further first tier, while the less promising second-tier candidates are only pursued if time and resources permit.[97]

Analyzing the outcome of crystallization trials

More often than not, the inspection of crystallization trials reveals anything but well-shaped crystals. To distinguish the hopeless from the just ugly is perhaps one of the most difficult tasks for the aspiring crystal grower. The human brain, optimally hardwired through millions of years of evolution for pattern recognition, combined with a decent knowledge base (also known as experience), at least for now beats computational crystal detection algorithms,[57, 98–100] which are often used in connection with automated image acquisition. Crystals are observed under a good stereomicroscope or a dedicated, digital camera equipped microscope with zoom optics providing about 150 to 400-fold magnification. Proper lighting of the crystallization wells is very important. The plates are usually illuminated from below, but the best arrangement of light sources varies depending on the type of plate used, well geometry, drop size and other experimental parameters. Automated digital observation stations (Figure 3-36) also store the images, providing the additional benefit of automated image analysis[101, 102] and of preserving a documented *timeline* of the development of the crystals.

Frequency of observations. A general rule for timing the trial observations is that the smaller the drops, the more frequent the inspections have to be. Because of the decreasing surface/volume ratio, robotically set up drops in the 100 nl range exchange water vapor rapidly with the reservoir, and water and small solvent molecules also diffuse through the plate plastics and sealing tape. Such microdrop setups as well as FID chips generally have a life time of perhaps a few weeks before the drops are dried out, while properly sealed, manually set up wells with larger, μl-sized drops can last years. A reasonable minimum observation time scale is geometric: immediately after setup, and then roughly

Figure 3-36 A low-cost automated crystallization plate imaging station. The crystallization plate is positioned by an x-y translation stage, and a digital zoom camera takes high-resolution images of the crystallization drops. The images taken in about 2 minutes can then be manually inspected on a computer screen, or processed by automated image recognition software. The depicted instrument is the CrysCam microscope manufactured by Art Robbins Instruments.

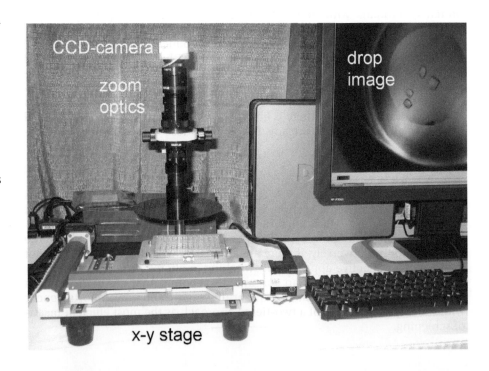

at doubling intervals, selected to give about 10 observations over the life time of the plate. This scheme gives, for example, days 1, 2, 4, 8, 16, 32, 64, 128, and 256, for a life time of the plate of about one year. More frequent inspection seldom hurts, particularly as some crystals appear at some point in time and after a while begin to deteriorate again. Even long abandoned plates from the back shelf of the cold room may occasionally yield pleasant surprises.

Interpretation of outcomes. Crystals are usually classified as "good" crystals when they display well-defined, sharp edges and have similar dimensions (a "blocky" shape). Figure 3-37 shows some well-formed single crystals, which are easy to mount, and often those also diffract well. However, beauty can be deceptive, and pretty crystals sometimes diffract poorly (or not at all), while quite unsightly specimens may diffract well. Next to the blocky crystals, we can discern needles or plates (panels 2 and 3 of Figure 3-37). Needles, if they are not too thin, can still be mounted (sometimes cut in pieces) and used for diffraction. However, the needles depicted in panel 2 are crystals of the precipitate sodium tartrate, and not of the protein (ways to distinguish protein crystals from salt crystals are discussed in the next section). Plate-shaped crystals tend to be more cumbersome to handle than needles, as they are often very fragile and very thin, which poses problems during both mounting and the recording of diffraction data. However, pieces of the thick plates shown in panel 3 were used for data collection to 1.5 Å. As detailed in Chapter 8, given the small beam sizes and brilliance of modern fine focus X-ray sources and synchrotrons, it is often possible to find well-diffracting regions even in otherwise suboptimal crystals of needles and plates.

Figure 3-37 Images of crystallization drops with different experimental outcomes. (1) Perfectly formed octahedral single crystals in hanging drop; (2) a cluster of large needles in microbatch reservoir; (3) thin plates in a hanging drop; (4) microcrystal shower in hanging drop; (5) small crystals together with spherical microcrystal clusters; (6) single crystals growing from a grainy precipitate; (7) irregular dendritic growth of crystals in sitting drop; (8) grainy precipitate in sitting drop; (9) amorphous precipitate and protein "oil" in sitting drop. The black and white pictures have been taken by automated crystal imaging stations. The false colors apparent in images 1 to 6 result from polarization effects in the optically anisotropic (birefringent) protein crystals and the varying depolarization in the injection-molded plastic. The proteins (and their crystals shown in this figure) are actually colorless.

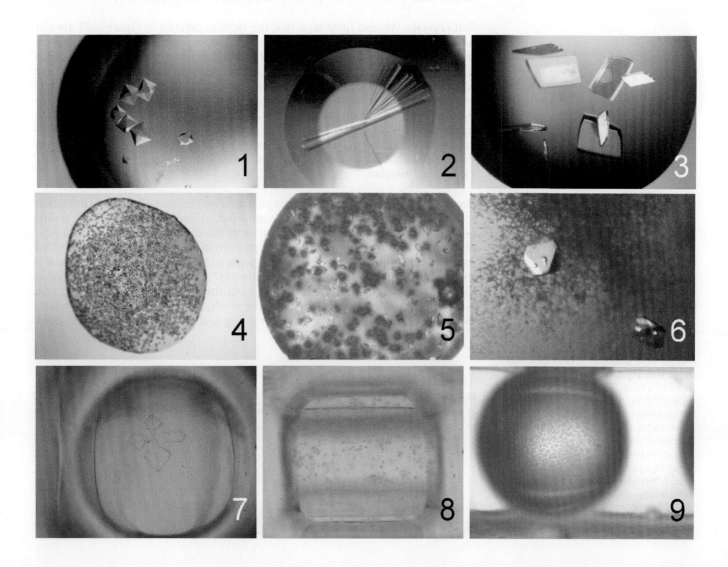

Some crystals are well formed but too small for mounting (panel 4). They will need to be optimized as described in the following chapter or can be used for seeding. The next two images (panels 5 and 6) show small crystals growing together with other frequently observed features in the same drop; fluffy clusters of micro-needles or crystals growing out of a granular, probably microcrystalline precipitate. Such crystals can still be harvested with some skill and give useful data. The last row of black and white images (taken by automated crystal observation stations, because experimenters usually do not photograph bad looking crystals) shows less fortuitous cases, which will need further screening or optimization. Panel 7 shows irregular dendritic growth, panel 8 shows a grainy precipitate, and panel 9 illustrates a slimy or oily, protein-rich precipitate that may still be of some use for seeding.

Polarization and birefringence. An interesting question is why the crystals depicted in Figure 3-37 appear so colorful, despite most proteins being colorless. Microscopes are often equipped with linear polarizer–analyzer assemblies (the polarizer in the base plate below the crystal tray, the analyzer above as a part of the lens assembly). Polarizing filters are useful as they reveal birefringence,[103] which is a good sign indicating anisotropy in the material along the viewing axis, which can help distinguish microcrystalline material from amorphous precipitates. Absence of birefringence in well-formed crystals, however, can result when a trigonal, hexagonal, or tetragonal crystal is viewed along the unique axis, and also in the case of cubic space groups.[104] The effects of birefringence are unfortunately diminished by the depolarizing plastic (mostly polystyrene) of the crystallization trays, and the false colors of the protein crystals actually result largely from rotation of the polarization plane by the protein crystals.

Salt crystals. We also have seen that the tartrate crystals in Figure 3-37 are not easily distinguished from protein by visual inspection alone. Protein crystals, however, can often be distinguished from salt crystals by staining them with a dye such as methylene blue (most food colors and microscopy stains work as well, for obvious reasons). The dye diffuses into the solvent channels, binds to the protein, and thus colors the protein crystal (Figure 3-38). Salt crystals have no such solvent channels and generally stay colorless. In contrast to much harder salt crystals, protein crystals are generally very sensitive and can be easily crushed with a needle or even a stiff fiber, although in rare cases protein crystals can be quite robust as well. In any case, the diffraction pattern will ultimately clarify the situation.

Scoring of outcomes. In order to employ data mining or to design optimization experiments, a quantification of the experimental screening results is necessary. Several scales have been derived to quantify the outcome of crystallization trials, which can range from amorphous precipitates to well-formed crystals. For example, we could assign a quality (Q-) score of 9 ("really good looking crystal") to the crystal shown in panel 1 of Figure 3-37, down to 1 for the rather poor precipitate in panel 9, and a zero for a clear drop. Various Q-scales are in use: 1–10 or 1–7 may be used,[89] or only three levels, clear, precipitate, and crystalline, may be assigned.[93] With proper training and cross-validation, individuals,

Figure 3-38 Soaking of blue dye into lysozyme crystals. The crystals were dyed by adding 0.5 µl of methylene blue solution to a 2 µl drop, and imaged immediately after dye addition, after 1 h, and after 10 h. The solution becomes successively lighter while the crystals absorb nearly all of the dye. It is also important to realize that it takes quite a while even for small molecules to diffuse into a crystal. It is not possible to soak large ligands such as peptides into crystals in seconds and expect to observe any electron density.[27]

as well as automated image recognition and scoring routines, can assign scores in a relatively consistent way within a given laboratory. Unfortunately, what the scores really mean varies, and no strong common scoring metric exists between laboratories. Inconsistent scoring (metric) is a serious problem when mining or comparing crystallization screening results from different sources.[48]

Optimization

Crystallization experiments often go through two procedurally distinguishable phases: an initial screening step is frequently followed by optimization. Screening establishes which conditions produce promising crystals—or at least, some promising leads—and optimization refers to the fine-tuning of those initial conditions in the hope of obtaining well-diffracting crystals for data collection. In fortunate cases, already the initial screening delivers diffraction-quality crystals. For automated processes in particular, this is a very desirable outcome, because automated systems are not as capable of devising an optimization strategy as an experienced experimenter. Common designs for optimization experiments are grid expansion and algorithmic optimization.

Grid expansion. The most popular optimization design is a series of successively finer grids[105] of the major cocktail ingredients and pH, often accompanied by trial of additives such as small amounts of detergents or other stabilizing or otherwise potentially useful additives. Similarly, finer random sampling around the initial conditions including various classes of additives and detergents can be attempted. Depending on the suspected problem, an additive of proven action or biological relevance as described in the additive section may be tried, but often there is very little rationale to guide optimization experiments.

Optimization algorithms. A formal approach toward optimization is to treat it as a special case of crystallization space sampling once major factors have been established. If the optimization experiments are designed in a way that allows formal analysis, a number of statistical methods which predict the best conditions are applicable. Linear regression, variance analysis, and response surface methods for the analysis of incomplete factorials and related designs[107] (factorial factorials, orthogonal arrays, Hardin–Sloane) have been implemented. Neural networks[60], partial least squares, and principal component analysis[106] are also applicable to factorial design optimization. Other multivariate designs used for crystallization optimization are central composite and Box–Behnken design[107] as well as iterative simplex procedures.[108] Each of these methods requires a specific design of experiments, and appropriate analysis software.

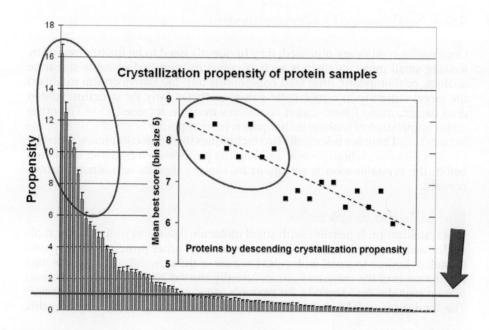

Figure 3-39 Crystallization propensity of proteins. General trends regarding unconditional crystallization probability have been derived by unbiased random sampling.[75] Proteins that crystallize well (quantified as propensity, a measure of how well a specific protein crystallizes compared with the average protein with a propensity of 1.0) also tend to crystallize with higher quality scores (circled in red) as shown in the insert. The compilation agrees well with the empirical estimate that only about one third of proteins crystallize better than average (red line, propensity of 1) in first trials, but those which crystallize do quite well and under multiple conditions. The findings emphasize the rationale that it is more efficient to improve the inherent "crystallizability" of a protein than to engage in excessive further screening that will likely produce only few more, and also in all likelihood poor, crystals. Figure adapted from Rupp and Wang.[48]

Box 3-8 **Crystallization strategies.** Finding suitable conditions for protein crystallization—provided the protein is inherently crystallizable—requires sampling of a nearly unlimited combination of parameters such as reagent combinations, pH, and temperature. Searching for successful crystallization conditions involves sampling of a multidimensional, sparsely populated, and ill-defined sampling space. Efficient sampling requires proper design of experiments.

Random screening experiments with minimum bias have confirmed a number of empirical general rules for crystallization screening and provided in addition a number of important insights affecting screening strategies. Rapid assessment of a protein's crystallization propensity can be gained by using a 2-tiered approach, starting with pH–PEG or index screens and expanding the sample space in the next round.

Random sampling and other large-scale trials have shown that if no promising results are obtained after about 300 trials, it is likely that the protein is a difficult case for crystallization: consider other protein constructs, orthologs, or protein engineering. Accept that the chance of obtaining diffracting crystals of a protein without any additional procedural adjustments or protein modifications is only 10–20%. Whether crystallization will succeed or not is already predetermined by the protein construct itself. If the protein cannot crystallize, it will not, no matter how many crystallization trials are performed. In contrast, proteins that crystallize frequently under multiple conditions also tend to diffract well. This fact lends additional weight to the case for protein engineering rather than expending excessive effort in crystallization of a poorly crystallizing protein that will in all likelihood never yield well-diffracting crystals.

Crystallization propensity of proteins. A very important point, predicted on the basis of efficiency analysis and confirmed by random screen analysis, is the fact that proteins that crystallize reasonably well do so in multiple crystal forms.[48] Figure 3-39 illustrates this fact, suggesting that instead of the frequent beginner's preference of focusing on the first success and trying to force it through a difficult optimization, wider screening in search of more favorable crystal forms is a viable option that makes more efficient use of the available material. As noted above, setting some of the same protein batch that was used for initial screening aside for optimization can avoid problems caused by batch variation.

3.9   Soaking and co-crystallization

Once native crystals are obtained, they frequently need to be further treated by soaking small molecule ligands such as cofactors, non-hydrolyzable substrate analogs, or therapeutic drug lead compounds into the native crystal to obtain the target structure in a molecular complex. Particularly for structure guided drug design, many ligand–target complexes must be screened.[109, 110] The other major application of soaking techniques is to incorporate heavy atoms as phasing markers. There are essentially two techniques to obtain complexes, either by co-crystallization, where protein and binding partner are mixed and incubated before the crystallization is set up, or by soaking ligands into already grown crystals.

Protein–ligand complexes

Co-crystallization is possible with small molecule ligands as well as larger molecules such as peptides, DNA oligomers, and even other proteins. Soaking into crystals is limited to small molecules because of the slow diffusion and the limited diameter of the solvent channels (usually around 20–100 Å wide). Diffusion of long peptides into crystals, for example, does not happen within seconds.[27] The time course of the dye soaking in Figure 3-38 convincingly amplifies this point: diffusion is a slow process, and even small molecule dyes or drugs need several hours to days to travel through the solvent channels into a crystal!

Some ligands that induce conformational changes in proteins and destroy the crystals upon soaking need to be co-crystallized with the protein. Another danger always present upon soaking is the possibility that the crystal disintegrates because of changes in the mother liquor chemistry and pH when the (usually concentrated) small molecule solution is added, or because the ligand itself disrupts crystal contacts. Stabilization of the protein crystals by cross-linking reagents such as glutaraldehyde can reduce these problems. The glutaraldehyde is simply placed in the crystallization reservoir, and vapor diffuses into the crystal. Crystals stabilized by this technique have actually survived soaking in 50% dimethyl sulfoxide (DMSO) containing solutions.[109]

Competitive replacement. A variant of soaking is competitive replacement of an already bound ligand with another molecule. The benefit is that the crystal form obviously is already one in which the protein has assumed a ligand-bound conformation, and the chance that a conformational change disrupts the crystal lattice is reduced. Depending on the dissociation constants, competitive replacement also works when the original ligand is relatively tightly bound, because it can be competed out by a weaker binding molecule if that compound is added in very high concentration.

Ligands of interest as drug leads often have poor solubility in aqueous solution, combined with less tight (meaning large) dissociation constants (K_d) in the early stages of ligand design. This raises two general questions: first, how to get the ligand into solution, and second, what concentration is needed to achieve sufficient occupancy of the ligand site to allow reasonable interpretation of the electron density, which in turn is required to reveal specific ligand–receptor interactions that can be used in lead optimization. It is important that the ligand stock is in reasonably high concentration, so that dilution of the mother liquor drop containing the crystal upon addition of the ligand stock is minimal. The above mentioned stabilization of the crystal with glutaraldehyde may reduce this problem.

Poorly soluble ligands are initially dissolved in pure (neat) DMSO (a powerful polar solvent which even crosses the skin upon contact) and then diluted to the required concentration with water or preferably the reservoir solution. The crystallization cocktail often contains components that, relative to pure water, increase ligand solubility, such as PEGs or other polyalcohols.

Binding constant and site occupation
The final ligand concentration in the crystallization drop necessary to achieve reasonable binding site occupancy should be around 10 times the K_d (or IC_{50}, if K_d is not available), and can be estimated as follows:[109] Assume equilibrium between the receptor–ligand complex [RL] and its dissociated components, receptor [R] and ligand [L]:

$$[R]+[L] \rightleftarrows [RL] \tag{3-8}$$

with the corresponding binding (dissociation) constant K_d defined as

$$K_d = \frac{[R][L]}{[RL]} \tag{3-9}$$

The fraction X_{RL} of occupied receptor sites in equilibrium is given by the amount of receptor in the complex divided by the amount of total receptor present in the solution:

$$X_{RL} = \frac{[RL]}{[R]+[RL]} \tag{3-10}$$

To be useful for estimating the required ligand concentration which has to be present in equilibrium, we would like to have Equation 3-10 expressed as a function of ligand concentration. Substituting [RL] in Equation 3-10 with [R][L]/K_d from (3-9) reduces to the following expression for the fraction of occupied receptor:

$$X_{RL} = \frac{[L]}{K_d + [L]}$$ (3-11)

Equation 3-11 shows two interesting points. First, not unexpectedly, the lower the binding constant K_d (the better the ligand binds), the less ligand we need to add to the solution to achieve the same relative equilibrium occupancy. Second, we can calculate how much ligand we need to add to achieve a desired occupancy of the receptor sites. This also leads to the realization that given very poor binding constants (10 mM and above), quite substantial amounts of ligand must be added to the solution (Figure 3-40). This is a particular challenge in crystallographic fragment screening,[111] (Sidebar 13-4) where libraries of small molecules that do not bind with high affinity are screened against a protein target (the small molecules that indeed bind to the target then serve as building blocks, from which—or from whose analogs—larger molecules with better binding characteristics are designed and synthesized).[112]

As a practical example, let us assume a ligand with a modest binding constant of 20 mM; we want to know how much ligand we need to achieve 80% and 90% binding site occupancy, respectively. We reformulate Equation 3-11 to solve for [L]

$$[L] = \frac{X_{RL} \cdot K_d}{1 - X_{RL}}$$ (3-12)

and the result is that we need 80 mM ligand to achieve 80% binding site occupancy and 180 mM for 90% occupancy. This is 4 and 9.5 times the value of K_d, which gives rise to the rule of thumb that about $10 \times K_d$ is a good estimate for the (in this case quite high) ligand concentration needed to maintain an equilibrium occupation of 90% in the crystallization drop. For high concentrations of ligand, the amount of ligand that is actually consumed by the protein is negligible (the protein usually being below mM concentrations) and the $10 \times K_d$ rule works well. For extremely well-binding ligands, the ligand concentration needs to be kept no lower than either $10 \times K_d$ or a few times the molar concentration of the protein, whichever is higher.

Figure 3-40 **Fraction of occupied receptor sites against ligand equilibrium concentration** for three different binding constants. Note that while at mM and lower K_d range small concentrations of ligand suffice to achieve good binding site occupancy (between 70–90%), quite impractical concentrations of ligand in the crystallization drop are required for poor binders.

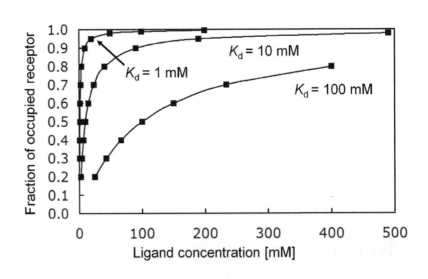

Sidebar 3-14 **Example of a protease inhibitor structure.** Benzamidine provides an example of a strongly binding inhibitor of a serine protease. Benzamidine is a small molecule analog of the basic amino acids arginine and lysine. With a pK_a of 11.6, benzamidine is completely protonated at physiologically relevant pH, just as Arg and Lys are. Benzamidine binds to trypsin with a K_d of about 20 μM, and Figure 3-41 exemplifies how well the electron density of strongly binding ligands can be defined.

Figure 3-41 Small molecule inhibitor bound in active site of a protease. Benzamidine, a 20 μM inhibitor of the serine protease trypsin, is a small molecule arginine-analog which occupies the specificity pocket of trypsin.

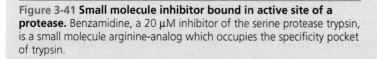

The stock concentrations of ligands often need to be very high. Take our 20 mM binding constant example: if we plan to add 1 μl of ligand stock to a 9 μl drop (dilution ratio of 1:10), we would need, for 80% receptor occupation, a ligand stock concentration of 800 mM! This is quite unlikely to work for a ligand of a MW of several hundred daltons. This example also illustrates why binding constants in the low μM to nM range are desirable for therapeutic drugs to be effective: the lower the drug level required to achieve the desired effect, the lower the potential problems with side effects and toxicity.

Heavy atom soaking

An important application of soaking is the incorporation of heavy atoms into the crystal. As mentioned in Chapter 1 and discussed in further detail in the phasing section (Chapter 10), heavy atoms are the source of isomorphous differences which are exploited for (multiple) isomorphous replacement phasing, and they also act as a source of anomalous signals for anomalous diffraction phasing techniques (Chapter 10). Although the majority of new structures are presently determined via anomalous phasing, often from Se-Met labeled proteins, heavy atoms are still a powerful phasing aid for isolated material and proteins where no suitable labeling systems exist. The reviews by Garman and Murray[113] and Carvin and colleagues[114] are recommended further reading.

As in the case of ligands, heavy atoms can be incorporated by soaking or by co-crystallization. Co-crystallization, where the protein is incubated with a metal ion solution before crystallization, can lead to covalent attachment to sites that change the protein conformation or cause non-isomorphous crystals to grow (isomorphism is a requirement for isomorphous replacement phasing). The more general method is to soak crystals with various reactive heavy atom salt solutions. Compared with drug molecule or ligand soaking, reaching sufficient ligand concentration is relatively easy in the case of metal ions, and the problem is more that the high reactivity of some heavy metal compounds can either destroy the crystals or modify intermolecular contacts to a degree that isomorphism becomes too weak for phasing (Chapter 10).

Typical examples of heavy metal incorporation are the reaction of sulfhydryl groups of cysteine with mercury or gold (Figure 3-42), or tyrosines with iodine. The replacement of a metal ion cofactor such as Ca^{2+} or Zn^{2+} with a heavier atom such as Cd^{2+} is also possible. Typical heavy atom reagents are listed in Table 3-1. Some extreme compounds such as Ta_6Br_{12}-clusters are powerful phasing reagents that can phase many hundred protein residues.[115] Concentrations of heavy metal solutions used for derivatization are typically in the 0.1–10 mM range, with covalently binding mercury compounds being examples at the lower

Name	Formula
Platinum potassium chloride, potassium tetrachloroplatinate(II)	K_2PtCl_4
Aurous potassium cyanide, potassium dicyanoaurate(I)	$KAu(CN)_2$
Mercuric potassium iodide, potassium tetraiodo mercurate(II)	K_2HgI_4
Uranyl acetate, uranium(VI) oxyacetate	$UO_2(C_2H_3O_2)_2$
Mercuric(II) chloride	$HgCl_2$
Potassium uranyl fluoride, potassium uranium(VI) oxyfluoride	$K_3UO_2F_5$
Para-chloromercurobenzosulfonate, PCMBS	$Hg(C_6H_4)SO_4$
Trimethyllead acetate	$(CH_3)_3Pb(CH_3COO)$
Methylmercuric acetate	$CH_3Hg(CH_3COO)$
Ethylmercuric thiosalicylate, thiomersal	$C_2H_5HgSC_6H_4COO^-$
Hexatantalum tetradecabromide	$(Ta_6Br_{12})Br$

Table 3-1 Selected heavy atom reagents. The listed reagents are frequently used for derivatization. The top seven entries are historically the most well used, the alkylated compounds below and the powerful Ta-clusters are more recent and very successful derivatization reagents. Many more are listed in the heavy atom data bank[116] and in the review by M.A. Rould[117]. All these substances are quite toxic when ingested because they bind to proteins and taking corresponding precautions is prudent. The uranium salts are generally prepared from natural uranium (0.7% ^{235}U) or depleted uranium (^{238}U), which both are only a weak α-particle source.

Sidebar 3-15 Xenon derivatization. The noble gas xenon binds to specific sites in a macromolecule, and xenon–protein complexes can therefore serve as heavy atom complexes for isomorphous replacement phasing and as a source of anomalous differences. Derivative crystals are simply produced by pressurizing native crystals in a chamber with xenon gas. An advantage of working with xenon is that it interacts only weakly with the protein, and isomorphism of the derivative with the native crystal is generally high. Xenon often binds to different sites than heavy metals, making it useful when traditional heavy atom soaks fail. Xenon binding is generally reversible so if one has very few crystals, the same crystal could be used again for heavy atom soaks once the Xe has diffused out of the crystal.

That xenon can bind to protein crystals has been known since the early days of protein crystallography, and sperm whale myoglobin was the first protein derivatized with xenon.[120] However, because of the difficulty of pressurizing and flash-cooling at the same time, it took until the mid-1990s until practical pressure cells and protocols were developed.[121, 122]

range, and soaking times range from hours to days. Back-soaking in mother liquor can also be tried, which removes unspecific or weakly bound ions from minor binding sites. Although the back-soaking operation increases the risk of damage, it can improve the background and also generate an intermediate state with different heavy atom sites occupied, which is useful for computing additional isomorphous difference data sets for phasing. Colored heavy atom compounds often color the crystals, but successful derivatization is ultimately confirmed by diffraction analysis (see also Sidebar 3-16).

Quick soaks. Another derivatization method is a quick soak (often less than a minute) of the crystals through heavy halide (iodine or bromine anions)[118] or heavy alkali (cesium and rubidium cations) containing solutions.[119] In contrast to traditional heavy atom derivatives, where specific covalent bonds form over a period of time, these ions are used in high concentrations (0.5–1 M), allowing very short soak times which in turn reduces the risk of damage to the crystals, while at the same time having some cryoprotecting effect. Consequently, these heavy ion sites are located primarily at the surface of the protein, and their occupancy is seldom complete. However, given the strong anomalous signals of these elements, even partially occupied sites are useful for anomalous phasing. Finally, even derivatization with the noble gas xenon is possible (Sidebar 3-15).

Harvesting and mounting of crystals

Once the crystals of a size of about 10 μm to a few 100 μm, preferably in a reasonably isotropic habit (blocky crystals) have been grown, they need to be harvested from the crystallization drop and mounted on the diffractometer. Protein crystals need to remain surrounded by mother liquor during the mounting process and during data collection; otherwise they dry out, disintegrate, and stop diffracting. There are essentially two methods, capillary mounting, which is still occasionally used for room temperature data collection, and cryomounting in small loops for most common data collection at cryogenic temperatures.

Figure 3-42 Heavy atom derivatization of a protein. Shown is the electron density around a gold atom covalently linked to a cysteine residue in the *Clostridium tetani* neurotoxin.[123] A combination of anomalous and isomorphous signals from gold atoms were used to solve the structure of the ganglioside binding domain of the neurotoxin from bacillus *C. tetani*, the causative agent of tetanus infections. PDB entry 1a8d.

Sidebar 3-16 Gel shift assays verify metal binding. Isomorphous incorporation of heavy atoms is ultimately checked by analyzing intensity differences in the diffraction patterns (Chapter 10). A very useful method to check whether heavy atoms have bound to a protein is by native gel shift assay.[124] If a metal binds to a protein, the band on a native, non-denaturing SDS gel will appear distinctly shifted to a higher molecular weight. The benefit of this assay is that denaturing conditions (where the protein does not enter the gel), as well as heavy atoms that do not bind, can be rapidly eliminated even before crystals are grown.

Protein crystals, just like any other organic material, are sensitive to the extremely high X-ray radiation doses they are exposed to during the diffraction experiment. While radical formation through the ionizing X-rays cannot be prevented, the reaction of free radicals which destroy the delicate network of crystal contacts can be kinetically hampered by cooling the crystals to cryogenic temperature. The crystals are commonly flash-cooled in liquid nitrogen (boiling point of 77 K or –196°C). The detrimental formation of ice which destroys the crystals is prevented by cryoprotection. The crystals are either grown in solutions that contain already high concentrations of PEG, salt, or other cryoprotectants, or are briefly swept through an appropriately buffered solution containing a cryoprotectant such as PEG, a cryosalt, sucrose, or glycerol before flash-cooling. The rapid cooling prevents the formation of ice crystals in the surrounding mother liquor, which invariably destroys the crystals.

Harvesting, cryoprotection, and mounting of crystals on the diffractometer are treated in great detail in Chapter 8.

Box 3-9 Ligand and heavy atom soaking. The prevalence of large solvent channels in protein crystals permits small molecules and ions to be readily soaked into crystals. A small drop of concentrated ligand or heavy atom solution is added to the mother liquor of the crystallization drop harboring the crystal. Ligand complexes usually take hours to weeks to form. In small molecule ligand soaking, the limited solubility of the substances in aqueous solutions can be problematic. Specifically bound heavy metal ions are required for isomorphous replacement phasing and are also valuable for anomalous phasing. Native gel shift assays show whether heavy atoms have bound to the protein or not. Successful ligand or heavy atom binding is validated by data collection through analysis of isomorphous difference data.

Sidebar 3-17 **Less than 1% of all deposited protein structures are membrane protein structures.** About a third of all expressed human proteins are presumed to be membrane proteins, and over 60% of all current drug targets are membrane receptors. As discussed in Chapter 2, their primary functions include transport of material and signals across cell membranes as well as motor functions. Despite membrane proteins being a significant class of proteins, it was nearly 30 years, and 195 deposited protein structures, after Kendrew's first myoglobin structure in 1958 that the first integral membrane protein structure, the photosynthetic reaction center isolated from the bacterium *Rhodopseudomonas viridis*, was published in 1985.[125] That research led to a Nobel Prize for crystallographic work being awarded to Johann Deisenhofer, Hartmut Michel, and Robert Huber in 1988. In early 2007, there were 242 coordinate entries of 122 different membrane proteins out of 35 100 total entries in the PDB, still a factor of 1/145 disfavoring the membrane proteins. Clearly, membrane protein crystallization remains a major challenge for crystallography, and only a limited introduction to this rapidly developing field can be given in this text. Many secondary literature sources, review citations, and web resources are provided in introductory reviews by M. Wiener[126] and P. Loll[127] and the general textbook references for this chapter.

3.10 Advanced crystallization problems

So far, we have essentially assumed that our protein molecules are reasonably well soluble globular entities. Proteins in this category are often secreted from the cell or nucleus and perform their function in some cellular compartment or the cytosol. Such proteins can be well handled with the procedures we have already described and, not surprisingly, present the vast majority of all coordinate entries in the PDB. However, there are substantial challenges to crystallization when it comes to more complex assemblies, particularly membrane proteins, protein–DNA complexes, and protein–protein complexes. Many other structural proteins also present their own specific challenges, particularly at the expression and solubility level, which we discuss in Chapter 4. This section is limited to the general crystallization challenges posed by membrane proteins, protein–DNA complexes, and some special techniques such as antibody scaffolding which are of general interest in protein crystallization.

Integral membrane proteins

In membrane proteins, only the exposed cytosolic and extracellular ends or domains reside in the polar solvent environment and in proximity to the polar phosphatidyl head groups of lipid membrane layers, while the transmembrane stems contact the hydrophobic fatty acid tails of the phospholipids inside the membrane (Chapter 2). Consequently, the transmembrane parts of membrane proteins contain a significant number of hydrophobic residues forming a hydrophobic contact area that interacts with the hydrophobic phospholipid tails in membrane layer. The tight membrane association has a number of consequences for production and crystallization of integral membrane proteins, that is, for those which do not have separately expressible ecto- or cytosolic domains.

Removed from the membrane environment, the hydrophobic transmembrane components have no more interaction partners and become unstable in polar, aqueous media, generally forming insoluble aggregates. During overexpression, there is only a small portion of the expressed membrane proteins that can be transported through the cell and properly incorporated into membranes before the cell breaks down. The vast majority of membrane proteins thus form insoluble inclusion bodies during overexpression. Once expression has succeeded (Chapter 4), the membrane proteins are extracted and solubilized with the help of detergents (water-soluble molecules that generally have a hydrophobic tail and a polar group). With their hydrophobic tail, they associate with the transmembrane parts and form a detergent "collar" around the hydrophobic transmembrane stem, as illustrated in Figure 3-43.

For successful crystallization, the primary resolubilization detergent often needs to be exchanged against a milder detergent suitable for crystallization, and small amphiphiles (small polar–apolar molecules similar to detergents) are used to fine-tune the size of the micelle collar. An additional complication for crystallization arises from the fact that many mammalian membrane proteins and receptors are decorated with posttranslational glycosylations (Chapter 4).

Detergent exchange and crystallization

Membrane proteins are solubilized from cell membranes with detergents. In order to keep the detergent collar around the transmembrane part intact, the detergent must be present in a concentration above its critical micelle concentration (CMC). If for example, the protein is dialyzed or buffer exchanged over an ion exchange column against a solution with a detergent below its CMC, the detergent micelle collar will break up and the protein will precipitate. On the other hand, detergent concentrations that are too high above the CMC or the addition of high ionic strength salt solutions can force separation of ionic detergents into a detergent-rich phase, into which the protein generally partitions and denatures. During protein crystallization with vapor diffusion all components in the drop, including the detergent, increase in concentration, and as a consequence, detergent separation in the drops is a common problem.

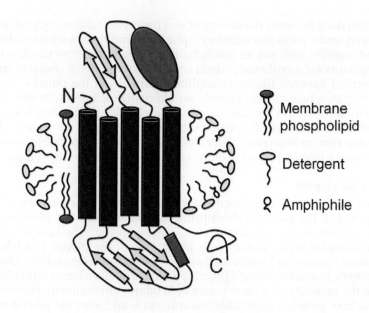

Membrane phospholipid

Detergent

Amphiphile

Figure 3-43 Resolubilized multi-pass, polytopic transmembrane protein with its associated detergent collar. In addition to the detergent collar, membrane fragments are often associated and co-solubilized with the transmembrane stem, as sketched on the left side of the membrane collar. Small amphiphile molecules are often added to fine-tune the size of the membrane collar for subsequent crystallization, as shown at the right side of the membrane collar.

For successful crystallization, the primary resolubilization detergent often needs to be exchanged against a milder detergent suitable for crystallization. Amongst the most popular detergents for (membrane) protein crystallization are: sugar-based detergents of the maltoside or glucopyranoside family (octyl β-glucoside); zwitterionic phosphocholine detergents; diacylglycerols; or polyoxyethylene detergents (hydroxyethyloxytri(ethyloxy)octane (C8E), or n-octyl-tetraoxyethylene (OTE) as visible in Figure 3-44). Many of these detergents also have been successfully used as additives in the crystallization of non-membrane-bound proteins. Detergent collections and resolubilization and crystallization additive kits are available from various manufacturers listed in the web resources, and a useful table of commonly used detergents can be found elsewhere.[128] With the hydrophobic transmembrane section shielded by the detergent, the membrane protein becomes soluble and in principle can be purified and crystallized just like any other soluble, globular protein, but the presence of detergent and the delicate equilibria associated with the detergent balance add another level of complexity to the process.

Figure 3-44 A cross section through the porin structure. The figure illustrates how the protein barrel would be embedded in a bacterial outer membrane. In the left panel circled are detergent molecules visible in electron density next to where the hydrophobic core would contact the fatty acid portion of the membrane phospholipids. The right insert show a magnification of the circled section, with n-octyl-tetraoxyethylene (OTE) molecules modeled into electron density. PDB entry 2por.[129]

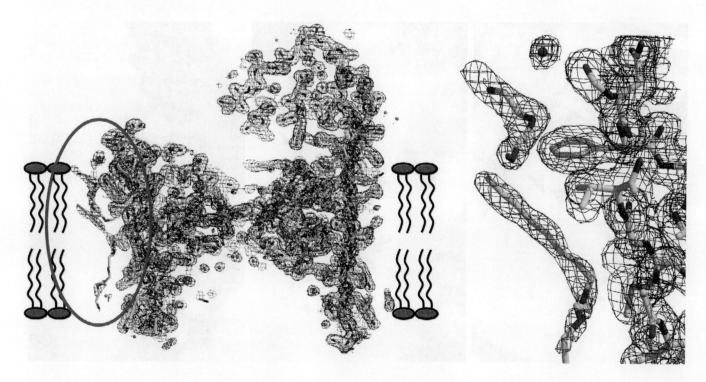

The interactions between the detergent and the crystallization cocktail must be considered, and a particular problem is phase separation when ionic detergents are used together with salt as precipitants. PEGs are therefore the first choice as precipitants for membrane protein crystallization. Crystal contacts are generally formed between the non-membrane domains of the protein. Antibody scaffolding (see below) may provide an alternative if intermolecular contact formation between the membrane protein molecules themselves is impeded. Co-solubilized membrane fragments such as phospholipids associated with the membrane stem or detergent molecules shielding the hydrophobic transmembrane core are often (at least partially) visible in electron density (Figure 3-44).

Lipid cubic phases

A special technique that has been successfully used in a certain class of membrane proteins, the bacterial rhodopsins, halorhodopsins, and photosynthetic reaction centers, is crystallization in lipid cubic phases.[130] The lipid mono-olein forms a complex phase system with water. One of the phases is a bilayered, cubic phase containing 50–80% lipid as well as interconnected solvent channels. The rationale is that membrane proteins would crystallize better either inserted in, or in the presence of, a more native lipid bilayer environment. However, it is not clear how generally applicable this concept is and what the precise mechanism for success is. As Figure 3-45 demonstrates, bacteriorhodopsin can also be crystallized by the conventional sitting-drop vapor-diffusion method.

Antibody scaffolding and F_{ab} structure

The fact that some proteins may not have surface residues in positions suitable for self-assembly into an extended 3-dimensional lattice leads to the following question: Would it be possible to generate some kind of molecular, 3-dimensional scaffolding, in which protein crystals can be "parked" in a regular fashion? Indeed a number of such scaffolds have been proposed, including designed 3-dimensional DNA scaffolds,[131] lectin scaffolds, small fibronectin type-3 domains, and antibodies and antibody fragments. The most generally applicable concept is probably the idea of using F_{ab} antibody fragments[132] at least as molecular assists, if not complete 3-dimensional scaffolds, to crystallize proteins in the form of protein–F_{ab} complexes (Figure 3-46).[133] A recent alternative to F_{ab} antibody fragments (which are more complicated to prepare) is the use of camelid single-domain ($V_{H}H$) antibodies as a crystallization chaperone.[134]

Figure 3-45 Crystals of a membrane protein. Crystals of bacteriorhodopsin D38R mutant (purple) were grown in sitting drops. The transparent crystals visible in the drops (center panel) are from the imidazole buffer present in the crystallization cocktail. Despite the nice looks of the purple crystals, only a sub-percent fraction of the harvested crystals showed good diffraction. Images courtesy of Michael Kolbe, Max Planck Institute for Infection Biology, Berlin.

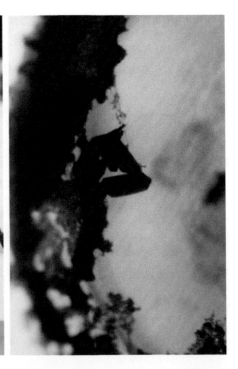

Figure 3-46 Basic anatomy of the F_{ab} antibody fragment. The F_{ab} antibody fragment consists of two chains with two domains each displaying the typical β-immunoglobulin fold. The heavy (blue) and light (green) chain are linked by disulfide bonds. Each of the chains has a variable (V) and constant (C) domain. The antigen-specific regions (six complementarity-determining regions or CDRs) which bind to the target antigen (a KcsA monomer in this case) are located on one end of molecule, with each chain contributing three CDRs. One F_{ab} and the bound KcsA monomer form the asymmetric unit of the KcsA crystal structure. PDB entry 1k4c.[135]

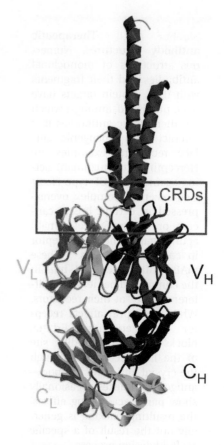

As an example, in Figure 3-47 we show the 2 Å resolution structure of the KcsA potassium channel of *Streptomyces lividans* in complex with a monoclonal F_{ab} antibody fragment.[135] The structure has been determined in Rod McKinnon's laboratory, and beautifully shows the trajectory of multiple K[+] ions passing through the tetrameric transmembrane channel.

Antibodies for scaffolding experiments must be specifically generated for each crystallization target, adding another step of preparatory biochemistry to the structure determination process. Specific, monoclonal IgG antibodies (mABs) raised against a target molecule are commonly isolated from mouse hybridoma cell lines through protein affinity chromatography, and the intact antibody is then proteolytically cleaved using papain.[136] The resulting F_{ab} fragment is incubated with its target, and the complex is gently purified and set up for crystallization. Alternative methods for antibody production are phage display methods.

In the case of the KcsA structure, the antibody fragments form a 3-dimensional skeleton with the tetramer of the potassium channel forming the core. The antibodies partly shield the transmembrane surface of the channel, and form contacts between the tetramers in the plane as well as in the third dimension. The structure was solved by molecular replacement using the antibody fragment structure as search model. Electron density also shows diacylglycerol (a lipid-like detergent) and one long-chain alcohol (nonan-1-ol) molecule per asymmetric unit.

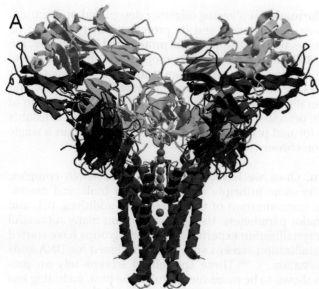

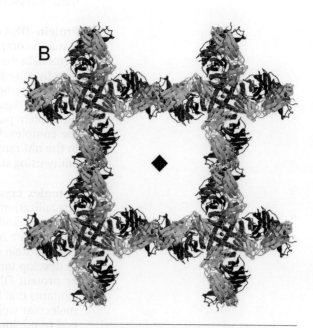

Figure 3-47 Scaffolding of the KcsA potassium channel structure. The left panel shows the tetrameric unit formed from four F_{ab} fragment–KcsA complexes. Note the chain of K[+] ions lined up in the channel. The right panel shows the lateral packing in the paper plane (or cell membrane) with the 4-fold axis (square symbol) of space group *I*4 perpendicular to the paper plane. The F_{ab} fragments form a complete 3-dimensional lattice, scaffolding the tetrameric KcsA channel. PDB entry 1k4c.[135]

Sidebar 3-18 Therapeutic antibody structures. Numerous structures of monoclonal antibodies and their fragments with their protein targets have been published, amongst which are therapeutic antibodies that interact with oncogenic surface receptors. Examples are Herceptin® (trastuzumab) acting against the EGF-type Her2/Neu receptor complex overexpressed in certain breast cancers or Erbitux® (cetuximab) acting against the EGF receptor in colon cancer cells.[137] These mABs (thus the ending -*mab* in the compound name) act at different places at their receptors. While Herceptin affects receptor oligomerization, Erbitux blocks the ligand binding site of the EGF-receptor. Although the crystal lattices of antibody–antigen complexes occasionally show partly scaffolding effects, the resulting packing is generally not the result of a specific scaffold design process.

Protein–DNA complex crystallization

The crystallization of protein–DNA complexes poses a formidable challenge, because a number of critical parameters are added to the already considerable quantity of variables in protein crystallization. Stability, high purity, and conformational homogeneity of the complex are necessary for crystallization. Sequence and the length of the DNA oligonucleotides are additional parameters of significant importance. Although there is no "magic bullet" for protein–DNA co-crystallization, based on the hundreds of co-crystal structures determined so far, a number of important guidelines have been established.

The protein members of a DNA–protein complex are generally overexpressed in *E. coli* and purified using standard chromatographic methods. In the case of multi-domain enzymes, the DNA-binding domain can be expressed and co-crystallized separately. While the biological context may not be entirely complete in these cases, the structures nonetheless provide detailed pictures of the DNA-recognition.

Oligonucleotide design. Single-stranded DNA oligonucleotides are usually synthesized and purified by reverse-phase chromatography, and subsequently annealed to form the required DNA duplex (with the notable exception of the crystal structure of the nucleosome described in Chapter 2). The sequence of synthesized oligomer is either specific for the binding site of the protein partner, or in the case of regulatory proteins, a consensus sequence for the corresponding type of operator. To prevent melting of the oligomer at room temperature, a minimum of about seven base pairs is generally necessary to keep the DNA duplex stable for crystallization. As in the case of pure oligomers described in Chapter 2, fragments of 10 base pairs corresponding to a full turn of DNA are a good starting point for crystallization experiments, because they can stack in the crystal lattice forming an infinite, continuous B-DNA strand. Frequently the length of the DNA duplex needs to be varied, and additional overhanging bases on each end are important variables in the design of a suitable DNA oligomer. Overhanging bases can mediate and stabilize contacts either by standard Watson–Crick base pairing or by forming triplex structures. Overhanging bases may even form unexpected stabilizing interactions with protein side chains. As in the case of proteins alone, none of these interactions can be predicted *a priori* and in general a substantial number of varying DNA oligonucleotides have to be tried in crystallization experiments.

Protein–DNA complex formation. Following oligomer design and purification, a stable complex between the oligomer and its protein partner needs to be formed. In the simplest case, the two components, protein and DNA duplex, are directly mixed in the crystallization drop. In a 1:1 complex, a slight molar excess of the oligomer is often required because DNA annealing is usually not 100% efficient. Oligonucleotides are much more soluble compared with proteins, and the protein partner generally limits the achievable maximum concentration of the complex. Very stable protein–DNA complexes with dissociation constants in the nM range can be formed prior to crystallization and purified as a single entity using size exclusion chromatography.

Complex crystallization. Once we have obtained a protein–DNA complex, crystallization follows the same principle as for any other biological macromolecule, and type and concentration of the precipitant, additives, pH, and temperature form the major parameters. Using the data from many successful DNA–protein complex crystallization experiments, several groups have started to develop targeted crystallization screens specifically designed for DNA and/or protein DNA–crystallization.[138, 139] These specialized screens rely on precipitants that have been shown to be more successful in the past, including low molecular weight alcohols such as glycerol and MPD, and, following the general trend, low molecular weight PEGs. High salt concentration has in general not produced many protein–DNA crystals, presumably because of the potentially destabilizing effect of salts on the mainly polar protein–DNA interface. Compounds that are believed to stabilize DNA such as polyamines (spermine,

Box 3-10 **Special crystallization challenges.** Membrane proteins form the most important class of drug receptors. Isolation, production, and crystallization of membrane proteins for structure determination are difficult because of the hydrophobicity of their transmembrane parts. In general, membrane proteins need to be extracted, purified, and crystallized with a protecting detergent collar around their hydrophobic transmembrane parts. Scaffolding using antibodies may be a generally applicable method for the crystallization of difficult proteins or nucleic acids. A critical parameter in protein–DNA complex crystallization is the length and type of DNA oligomer used in co-crystallization. Single strand overhangs frequently stabilize structures through formation of contiguous DNA strands in the crystal.

N,N′-bis(3-aminopropyl)butane-1,3-diamine; spermidine, N-(3-aminopropyl)-1,3-diaminobutane); Mg^{2+} ions; or cobalt hexaammine, $[Co(NH_3)_6]^{3+}$ are popular additives. A proven way to improve crystal quality from initial crystalline precipitates, microcrystals, or poorly diffracting crystals is to vary the length or type of the DNA oligonucleotide. In many cases the addition or deletion of only one or two bases or the introduction of overhanging bases has a major impact on crystal packing and hence crystal quality. Once a DNA–protein complex suitable for crystallization has been found, crystals are often obtained under many conditions and only fine-tuning of the crystallization conditions may be required to obtain the best diffracting crystals.

The structure of the diphtheria toxin repressor DtxR in complex with a 21 base pair (bp) duplex DNA may serve as an example for successful crystallization of a multi-domain protein–DNA complex.[140] The diphtheria repressor protein is of interest because it tightly regulates expression of a number of essential proteins at the transcriptional level in *Corynebacterium diphtheriae*, and thus may be a relevant drug target. To crystallize the complex, about 40 different oligonucleotides with a core of 19 palindromic consensus sequence base pairs from known DtxR binding sites were synthesized. The duplexes ranged from 15–34 bp in length, and a subset in the range of 20–22bp gave marginally diffracting (8 Å) crystals. After further refinement of the synthesis, a blunt-ended, 21 bp palindromic sequence with a mismatch at bases –4 and 4 (5′-ATTAGGTTAGCCTACCCTAAT-3′) finally yielded crystals that diffracted to 3.2 Å. The crystallization drop contained Tris buffer (pH 7.5) and DDT from the protein stock, $MgCl_2$ from the DNA stock, and MES buffer (pH 6), $CoCl_2$, and polyvinyl alcohol (MW 15 000) as precipitant from the crystallization cocktail. As is frequently the case, binding of the DNA to the repressor leads to a distorted DNA conformation (Figure 3-48).

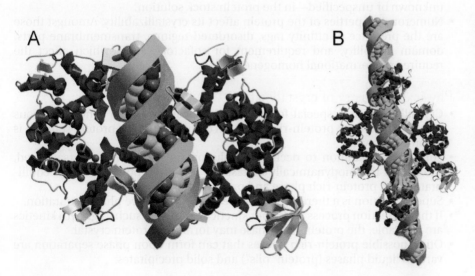

Figure 3-48 The structure of two diphtheria toxin repressor DtxR dimers in complex with a 21 base pair duplex DNA. Note how the helices of the N-terminal winged-helix domain deeply probe the major groove of the slightly distorted DNA. The brown spheres in the protein molecules are Co^{2+} ions. The right panel illustrates how the DNA forms a contiguous stretch of DNA across the crystal. PDB entry 1c0w.[140]

3.11 Key concepts

In Chapter 3 we have developed an understanding of the theory of protein crystallization and the practical aspects of crystal growth. Armed with this knowledge, we can in the next chapter address rational modification of protein properties, which are the most significant determinant whether self-assembly into protein crystals is possible.

Fundamentals of crystallization

- Protein crystals are periodic self-assemblies of large and often flexible macro-molecules, held together by weak intermolecular interactions.
- Protein crystals are generally fragile and sensitive to environmental changes.
- In order to form crystals, the protein solution must become supersaturated.
- In the supersaturated, thermodynamically metastable state, nucleation can occur and crystals may form while the solution equilibrates.
- The most common technique for protein crystal growth is by vapor diffusion, where water vapor equilibrates from a drop containing protein and a precipitant into a larger reservoir with higher precipitant concentration.
- Given the large size and inherent flexibility of most protein molecules combined with the complex nature of their intermolecular interactions, crystal formation is an inherently unlikely process, and many trials may be necessary to obtain well-diffracting crystals.
- The protein molecules in a crystal are connected through a network of few and specific intermolecular interactions.
- Between the packed protein molecules, large voids remain that are filled with solvent.
- The voids between molecules allow the crystal to exchange liquid with the environment, and small molecules such as ligands or drug molecules can be soaked into crystals.
- For experimental phase determination, heavy metal ions can be soaked into the crystal, where they may specifically bind to certain residues and form marker atoms for phase determination.
- Intermolecular packing interactions can change the conformation of surface residues or flexible loops of a molecule, but the core of a protein maintains its solution conformation in the crystalline state, and enzymes generally remain active in the crystalline state.
- Given that a crystal can only form if specific interactions between molecules can occur in an orderly fashion, the inherent properties of the protein itself are the primary factor determining whether crystallization can occur.
- A single residue mutation can make all the difference between successful crystallization and complete failure.
- Important factors related to the protein that influence crystallization are its purity, the homogeneity of its conformational state, the freshness of the protein, and the additional components that are invariably present—but often unknown or unspecified—in the protein stock solution.
- Numerous properties of the protein affect its crystallizability. Amongst those are the presence of affinity tags, disordered regions, transmembrane parts, domain flexibility, and requirement for cofactors, all of which affect the required conformational homogeneity.

Physical chemistry of crystallization

- Crystallization is a special form of phase separation from a homogeneous solution, where the protein-rich phase in equilibrium with protein solution is an ordered crystal.
- For phase separation to occur, the solution must become supersaturated, where it is thermodynamically metastable and will upon nucleation equilibrate into a protein-rich phase and protein solution.
- Supersaturation is a thermodynamic necessity to achieve phase separation.
- If the nucleation process and other kinetic parameters such as growth kinetics are favorable, the protein-rich phase may form as a protein crystal.
- Other possible protein-rich phases that can form upon phase separation are various liquid phases (protein "oils") and solid precipitates.

- The phase relations in a protein solution can be represented in a pseudo-binary phase diagram with protein concentration and the precipitant concentration as parameters.
- The pH of a protein solution has a strong effect on the solubility, but crystallization does not preferentially occur at the isoelectric point, where protein solubility is at its minimum.
- Temperature also affects protein solubility and crystallization, but no general preferences can be predicted.
- Protein crystallization is an entropy-driven process; the release of water molecules across both hydrophobic and polar surface residues during the formation of a crystal contributes to entropy gains of the solvent, exceeding the entropy loss largely caused by the loss of motional degrees of freedom upon crystallization.
- Self-assembly into crystals can be assisted by trial of additives that stabilize the protein, mediate crystal contacts, fine-tune intermolecular interactions, or otherwise modify protein solubility.
- In order to enable molecular self-assembly during crystallization, both thermodynamic and kinetic parameters must favor the formation of a stable, crystalline phase.
- Once the protein solution has reached thermodynamically metastable supersaturation, nucleation determines how the phase separation into protein-rich phase and saturated protein solution occurs.
- At high supersaturation, spontaneous homogeneous nucleation of the protein rich phases occurs, while at low supersaturation, heterogeneous nucleation must be induced by seeding.
- Real single crystals are not perfect; they consist of multiple, slightly misaligned, domains forming a mosaic crystal.

Crystallization techniques

- The inability to predict *ab initio* any conditions favoring protein crystallization requires that in general several hundred crystallization trials must be set up in a suitable format and design.
- Crystallization screening experiments are commonly set up manually or robotically in multi-well format crystallization plates.
- The most common procedure of achieving supersaturation is the vapor-diffusion technique, performed in sitting-drop or hanging-drop format.
- In vapor-diffusion setups, protein is mixed with a precipitant cocktail, and the system is closed over a reservoir into which water vapor diffuses from the protein solution.
- During vapor diffusion, both precipitant and protein concentration increase in the crystallization drop and supersaturation is achieved.
- Other protein crystallization methods include batch crystallization under oil, dialysis methods, and free-interface diffusion techniques.
- Microfluidic chips or thin-walled capillaries are used for free-interface diffusion.
- The benefit of free-interface diffusion methods is a comprehensive coverage of the crystallization phase space, and very little material is required in the case of microfluidic chip methods.
- The pathway through crystallization phase space can be visualized with the help of crystallization phase diagrams.
- Crystallization diagrams combine information about thermodynamically defined phase relations with a tentative assignment of kinetic nucleation regions.
- As a rule of thumb, low supersaturation favors controlled crystal growth, while high supersaturation is required for spontaneous nucleation of crystallization nuclei.
- Seeding is a method to induce heterogeneous nucleation at low supersaturation, which is more conducive to controlled crystal growth.
- After initial screening, optimization of growth conditions around initially established conditions is often necessary to obtain well diffracting crystals.
- Crystals need to be harvested and mounted for data collection. Proper cryoprotection is important for successful cryocooling (quenching or flash-cooling) of protein crystals.
- Data collection at cryogenic temperatures greatly reduces radiation damage to sensitive protein crystals.

- Some extremely sensitive crystals cannot be successfully cryocooled or manipulated and data must be collected at room temperature, despite the increased risk of radiation damage.

Crystallization strategies

- Finding suitable conditions for protein crystallization—provided the protein is inherently crystallizable—requires sampling of an, in principle, unlimited combination of parameters such as reagent combinations, pH, and temperature.
- Searching for successful crystallization conditions requires sampling of a multidimensional, sparsely populated, and ill-defined sampling space.
- Efficient sampling requires proper design of experiments.
- Random screening experiments with minimum bias have confirmed a number of empirical general rules for crystallization screening and provided in addition a number of important insights affecting screening strategies.
- Accept that the chance of obtaining diffracting crystals of a protein in a single screening without any additional procedural adjustments or protein modifications is only 10–20%.
- Whether crystallization will succeed or not is already predetermined by the protein construct itself. If it cannot crystallize, it will not, no matter how many crystallization trials are performed.
- Random sampling and other large-scale trials have shown that if no promising results are obtained after about 300 trials, it is likely that the protein is a difficult case for crystallization: consider other constructs, orthologs, or protein engineering.
- Proteins that crystallize frequently under multiple conditions also tend to diffract well, which lends additional weight to the case for protein engineering rather than expending excessive effort in crystallization of a poorly crystallizing protein that will, in all likelihood, never yield well-diffracting crystals.
- Accept the probabilistic nature of the crystallization game. One can win only by increasing the odds, not by seeking certainty. In other words, do nothing stupid, but sample everything else efficiently.
- Rapid assessment of a protein's crystallization propensity can be gained by using a 2-tiered approach, starting with pH–PEG or index screens and expanding the sample space in the next round of trials.
- Be skeptical of crystallization tips or claims that lack a clear rationale. Causality rules, also for statistically infrequent events.
- The prevalence of large solvent channels in protein crystals permits small molecules and ions to be readily soaked into crystals.
- A small drop of concentrated ligand or heavy atom solution is added to the mother liquor of the crystallization drop harboring the crystal.
- Soaking of ligands into protein crystals can take hours to weeks.
- In small molecule ligand soaking, the limited solubility of the substances in aqueous solutions can be problematic.
- Specifically bound heavy metal ions are required for isomorphous replacement phasing and are also valuable for anomalous phasing.
- Native gel shift assays show whether heavy atoms have bound to the protein or not.
- Successful ligand or heavy atom binding is validated by data collection through analysis of difference data.

Special challenges

- Membrane proteins form the most important class of drug receptors.
- Isolation, production, and crystallization of membrane proteins for structure determination is difficult due to the hydrophobicity of their transmembrane parts.
- In general, membrane proteins need to be extracted, purified, and crystallized with a protecting detergent collar around their hydrophobic transmembrane stems.
- Scaffolding using antibodies may be a generally applicable method for the crystallization of difficult proteins or nucleic acids.
- A critical parameter in protein–DNA complex crystallization is the length and type of DNA oligomer used in co-crystallization. Single strand overhangs frequently stabilize structures through formation of contiguous DNA strands in the crystal.

3.12 Additional reading

1. McPherson A (1999) *Crystallization of Biological Macromolecules*. Cold Spring Harbor, NY: Cold Spring Harbor Laboratory Press.

2. Ducruix A, & Giege R (Eds.) (1999) *Crystallization of Nucleic Acids and Proteins*. Oxford, UK: Oxford University Press.

3. Bergfors T (Ed.) (2009) *Protein Crystallization*. San Diego, CA: International University Line.

4. Chayen N (2007) *Protein Crystallization Strategies for Structural Genomics*. San Diego, CA: International University Line.

5. Han J, & Kamber M (2001) *Data Mining: Concepts and Techniques*. San Francisco: Morgan Kaufmann Publishers.

6. Iwata S (Ed.) (2003) *Methods and Results in Crystallization of Membrane Proteins*. San Diego, CA: International University Line.

7. Doublie S (Ed.) (2007) *Macromolecular Crystallography Protocols, Volume 1: Preparation and Crystallization of Macromolecules*. Totowa, NJ: Humana Press.

3.13 References

1. Coulibaly F, Chiu E, Ikeda K, et al. (2007) The molecular organization of cypovirus polyhedra. *Nature* 446(7131), 97–101.

2. Von Dreele R (2003) Protein crystal structure analysis from high resolution X-ray powder diffraction data. *Methods Enzymol.* 368, 255–267.

3. Margiolaki I, & Wright JP (2008) Powder crystallography on macromolecules. *Acta Crystallogr.* A64(1), 169–180.

4. Dale GE, Oefner C, & D'Arcy A (2003) The protein as a variable in protein crystallization. *J. Struct. Biol.* 142(1), 88–97.

5. Bernal J, & Crowfoot D (1934) X-ray photographs of crystalline pepsin. *Nature* 133, 794–795.

6. Adams MJ, Blundell TL, Dodson EJ, et al. (1969) Structure of rhombohedral 2-zinc insulin crystals. *Nature* 224, 491–495.

7. Glusker JP (1994) Dorothy Crowfoot Hodgkin. *Protein Sci.* 3, 2465–2469.

8. Doye JP, Louis AA, & Vendruscolo M (2004) Inhibition of protein crystallization by evolutionary negative design. *Phys. Biol.* 1(1–2), 9–13.

9. Garfield JL, & Stout CD (1988) Crystallization and preliminary X-ray diffraction studies of a toxic crystal protein from a subspecies of *Bacillus thuringiensis*. *J. Biol. Chem.* 263(24), 11800–11801.

10. McPherson A (1999) *Crystallization of Biological Macromolecules*. Cold Spring Harbor, NY: Cold Spring Harbor Laboratory Press.

11. Sumner JB (1926) The isolation and crystallization of the enzyme urease. *J. Biol. Chem.* 69, 435–441.

12. Kendrew JC, Bodo G, Dintzis HM, et al. (1958) A three-dimensional model of the myoglobin molecule obtained by X-ray analysis. *Nature* 181, 662–666.

13. Alderton G, & Fevold HL (1946) Direct crystallization of lysozyme from egg white and some crystalline salts of lysozyme. *J. Biol. Chem.* 164(1), 1–5.

14. Derewenda ZS (2004) The use of recombinant methods and molecular engineering in protein crystallization. *Methods* 34(3), 354–363.

15. Carugo O, & Argos P (1997) Protein-protein crystal-packing contacts. *Protein Sci.* 6, 2261–2263.

16. Yeates TO, & Padilla JE (2002) Designing supramolecular protein assemblies. *Curr. Opin. Struct. Biol.* 12(4), 464–70.

17. McPherson A (2001) A comparison of salts for the crystallization of macromolecules. *Protein Sci.* 10, 414–422.

18. Trakhanov S, Kreimer DI, Parkin S, et al. (1998) Cadmium induced crystalization of proteins: II. Crystallization of the *Salmonella typhimurium* Histidine binding protein in complex with L-histidine, L-arginine or L-lysine. *Protein Sci.* 7, 600–604.

19. McPherson A, & Cudney B (2006) Searching for silver bullets: an alternative strategy for crystallizing macromolecules. *J. Struct. Biol.* 156(3), 387–406.

20. Segelke B, Knapp M, Kadhkodayan S, et al. (2004) Crystal structure of *C. botulinum* neurotoxin protease in a product bound state: evidence for non-canonical zinc protease activity. *Proc. Natl. Acad. Sci. U.S.A.* 101, 6888–6893.

21. Dale GE, Kostrewa D, Gsell B, et al. (1999) Crystal engineering: deletion mutagenesis of the 24 kDa fragment of the DNA gyrase B subunit from *Staphylococcus aureus*. *Acta Crystallogr.* D55, 1626–1629.

22. Carson M, Johnson DH, McDonald H, et al. (2007) His-tag impact on structure. *Acta Crystallogr.* D63(3), 295–301.

23. Taylor EJ, Goyal A, Guerreiro CIPD, et al. (2005) How family 26 glycoside hydrolases orchestrate catalysis on different polysaccharides: structure and activity of a *Clostridium thermocellum* lichenase. *J. Biol. Chem.* 280(38), 32761–32767.

24. Luecke H, Schobert B, Richter H-T, et al. (1999) Structural changes in bacteriorhodopsin during ion transport at 2 Å resolution. *Science* 286(5438), 255–260.

25. Lamers MH, Perrakis A, Enzlin JH, et al. (2000) The crystal structure of DNA mismatch repair protein MutS binding to a G x T mismatch. *Nature* 407(6805), 711–717.

26. Klaholz BP, & Moras D (2000) Structural role of a detergent molecule in retinoic acid nuclear receptor crystals. *Acta Crystallogr.* D56(Pt 7), 933–935.

27. Rupp B, & Segelke BW (2001) Questions about the structure of the botulinum neurotoxin B light chain in complex with a target peptide. *Nat. Struct. Biol.* 8, 643–664.

28. Pohl E, Brunner N, Wilmanns M, et al. (2002) The crystal structure of the allosteric non-phosphorylating glyceraldehyde-3-phosphate dehydrogenase from the hyperthermophilic archaeum *Thermoproteus tenax*. *J. Biol. Chem.* 277(22), 19938–19945.

29. Krupka HI, Segelke BW, Ulrich RG, et al. (2002) Structural basis for abrogated binding between staphylococcal enterotoxin A superantigen vaccine and MHC-II alpha. *Protein Sci.* 11, 642–651.

30. D'Arcy A (1994) Crystallizing proteins – a rational approach? *Acta Crystallogr.* D50, 469–475.

31. Wilson WW (2003) Light scattering as a diagnostic for protein crystal growth – A practical approach. *J. Struct. Biol.* 142(1), 56–65.

32. Rupp B, Segelke BW, Krupka HI, et al. (2002) The TB structural genomics consortium crystallization facility: towards automation from protein to electron density. *Acta Crystallogr.* D58, 1514–1518.

33. Derewenda ZS, & Vekilov PG (2006) Entropy and surface engineering in protein crystallization. *Acta Crystallogr.* D62, 116–124.

34. Fersht A (1999) *Structure and Mechanism in Protein Science: A Guide to Enzyme Catalysis and Protein Folding*. New York, NY: W.H. Freeman & Company.

35. Vekilov PG (2003) Solvent entropy effects in the formation of protein solid phases. *Methods Enzymol.* 368, 84–105.

36. Veesler S, & Boistelle R (1999) Diagnostic of pre-nucleation and nucleation by spectroscopic methods and background on the physics of crystal growth. In Ducruix A & Giege R (Eds.), *Crystallization of Nucleic Acids and Proteins*. Oxford, UK: Oxford University Press.

37. McPherson A, Kuznetsov YG, Malkin A, et al. (2003) Macromolecular crystal growth as revealed by atomic force microscopy. *J. Struct. Biol.* 142(1), 32–46.

38. Blundell TL, & Johnson LN (1976) *Protein Crystallography*. London, UK: Academic Press.

39. Rubinson KA, Ladner JE, Tordova M, et al. (2000) Cryosalts: suppression of ice formation in macromolecular crystallography. *Acta Crystallogr.* D56(8), 996–1001.

40. McPherson A, Jr (1976) Crystallization of proteins from polyethylene glycol. *J. Biol. Chem.* 251(20), 6300–6303.

41. Brzozowski AM, & Tolley SP (1994) Poly(ethylene) glycol monomethyl ethers – an alternative to poly(ethylene) glycols in protein crystallization. *Acta Crystallogr.* D50(4), 466–468.

42. Tardieu A, Bonnete F, Finet S, et al. (2002) Understanding salt or PEG induced attractive interactions to crystallize biological macromolecules. *Acta Crystallogr.* D58, 1549–1553.

43. Anand K, Pal D, & Hilgenfeld R (2002) An overview on 2-methyl-2,4-pentanediol in crystallization and in crystals of biological macromolecules. *Acta Crystallogr.* D58(Pt 10 Pt 1), 1722–1728.

44. Ducruix A, & Giege R (1999) *Crystallization of Nucleic Acids and Proteins*. Oxford, UK: Oxford University Press.

45. Cudney R, Patel S, Weisgraber K, et al. (1994) Screening and optimization strategies for macromolecular crystal growth. *Acta Crystallogr.* D50, 414–423

46. Yeh JI, Du S, Tortajada A, et al. (2005) Peptergents: peptide detergents that improve stability and functionality of a membrane protein, glycerol-3-phosphate dehydrogenase. *Biochemistry* 44(51), 16912–9.

47. Larson SB, Day JS, Cudney R, et al. (2007) A novel strategy for the crystallization of proteins: X-ray diffraction validation. *Acta Crystallogr.* D63(3), 310–318.

48. Rupp B, & Wang J (2004) Predictive models for protein crystallization. *Methods* 34(3), 390–407.

49. Hempstead PD, Yewdall SJ, Fernie AR, et al. (1997) Comparison of the three-dimensional structures of recombinant human H and horse L ferritins at high resolution. *J. Mol. Biol.* 268(2), 424–448.

50. Ko TP, Day J, Greenwood A, et al. (1994) Structures of three crystal forms of the sweet protein thaumatin. *Acta Crystallogr.* D50(Pt 6), 813–825.

51. McPherson A (1982) *Preparation and Analysis of Protein Crystals*. New York, NY: Wiley.

52. Gilliland GL, Tung M, Blakeslee DM, et al. (1994) The Biological Macromolecule Crystallization Database, Version 3.0. New features, data, and the NASA Archive for Protein Crystal Growth Data. *Acta Crystallogr.* D50, 408–413.

53. Samudzi CT, Fivash M, & Rosenberg JM (1992) Cluster analysis of the biological Macromolecule Crystallization Database. *J. Crystal Growth* 123, 47–58.

54. Page R, Grzechnik SK, Canaves JM, et al. (2003) Shotgun crystallization strategy for structural genomics: an optimized two-tiered crystallization screen against the *Thermotoga maritima* proteome. *Acta Crystallogr.* D59(6), 1028–1037.

55. Kantardjieff K, Jamshidian M, & Rupp B (2004) Distributions of pI vs pH provide strong prior information for the design of crystallization screening experiments. *Bioinformatics* 20(14), 2171–2174.

56. Segelke BW (2001) Efficiency analysis of sampling protocols used in protein crystallization screening. *J. Crystal Growth* 232, 553–562.

57. Rupp B, Segelke BW, Krupka HI, et al. (2002) The TB structural genomics consortium crystallization facility: towards automation from protein to electron density. *Acta Crystallogr.* D58, 1514–1518.

58. Rupp B (2005) A guide to automation and data handling in protein crystallization. In *Crystallization Strategies for Structural Genomics*. San Diego: International University Line.

59. Santarsiero BD, Yegian DT, Lee CC, et al. (2002) An approach to rapid protein crystallization using nanodroplets. *J. Appl. Crystallogr.* 35(2), 278–281.

60. DeLucas LJ, Bray TL, Nagy L, et al. (2003) Efficient protein crystallization. *J. Struct. Biol.* 142(1), 188–206.

61. Dunlop KV, & Hazes B (2003) When less is more: a more efficient vapour-diffusion protocol. *Acta Crystallogr.* D59(Pt 10), 1797–1800.

62. Newman J (2005) Expanding screening space through the use of alternative reservoirs in vapor-diffusion experiments. *Acta Crystallogr.* D61(4), 490–493.

63. Luft JR, Collins RJ, Fehrman NA, et al. (2003) A deliberate approach to screening for initial crystallization conditions of biological macromolecules. *J. Struct. Biol.* 142(1), 170–179.

64. Chayen NE (1998) Comparative studies of protein crystallization by vapour-diffusion and microbatch techniques. *Acta Crystallogr.* D54, 8–15.

65. Stewart PS, & Baldock P (1999) Practical experimental design techniques for automatic and manual protein crystallization. *J. Crystal Growth* 196, 665–673.

66. Luft JR, Wolfley J, Jurisica I, et al. (2001) Macromolecular crystallization in a high throughput laboratory – the search phase. *J. Crystal Growth* 232, 591–595.

67. Bodenstaff ER, Hoedemaker FJ, Kuil EM, et al. (2002) The prospects of nanocrystallography. *Acta Crystallogr.* D59, 1901–1906.

68. Ng JD, Gavira JA, & Garcia-Ruiz JM (2003) Protein crystallization by capillary counterdiffusion for applied crystallographic structure determination. *J. Struct. Biol.* 142(1), 218–231.

69. Ruiz J-M (2003) Counterdiffusion methods for macromolecular crystallization. *Methods Enzymol.* 368, 130–154.

70. Bamford JK, Cockburn JJ, Diprose J, et al. (2002) Diffraction quality crystals of PRD1, a 66-MDa dsDNA virus with an internal membrane. *J. Struct. Biol.* 139(2), 103–112.

71. Cockburn JJ, Bamford JK, Grimes JM, et al. (2003) Crystallization of the membrane-containing bacteriophage PRD1 in quartz capillaries by vapour diffusion. *Acta Crystallogr.* D59(Pt 3), 538–540.

72. Hansen CL, Skordalakes E, Berger JM, et al. (2002) A robust and scalable microfluidic metering method that allows protein crystal growth by free interface diffusion. *Proc. Natl. Acad. Sci. U.S.A.* 99, 16531–16536.

73. van der Woerd M, Ferree D, & Pusey M (2003) The promise of macromolecular crystallization in microfluidic chips. *J. Struct. Biol.* 142(1), 180–187.

74. Schick B, Segelke BW, & Rupp B (2001) Statistical analysis of protein crystallization parameters. *ACA Meeting Series* 28, 166.

75. Rupp B (2003) High throughput crystallography at an affordable cost: The TB Structural Genomics Consortium crystallization facility. *Acc. Chem. Res.* 36, 173–181.

76. Baldock P, Mills V, & Stewart P (1996) A comparison of microbatch and vapor diffusion for initial screening of crystallization conditions. *J. Crystal Growth* 168, 170–174.

77. Newman J, Xu J, & Willis MC (2007) Initial evaluations of the reproducibility of vapor-diffusion crystallization. *Acta Crystallogr.* D63(7), 826–832.

78. Bergfors T (2003) Seeds to crystals. *J. Struct. Biol.* 142(1), 66–76.

79. McPherson A, & Shlichta P (1988) The use of heterogeneous and epitaxial nucleants to promote the growth of protein crystals. *J. Crystal Growth* 90(1–3), 47–50.

80. D'Arcy A, Mac Sweeney A, & Haber A (2003) Using natural seeding material to generate nucleation in protein crystallization experiments. *Acta Crystallogr.* D59(Pt 7), 1343–1346.

81. Segelke B, Schafer J, Coleman M, et al. (2004) Laboratory scale structural genomics. *J. Struct. Funct. Genomics* 5, 147–157.

82. Viola R, Carman P, Walsh J, et al. (2007) Operator-assisted harvesting of protein crystals using a universal micromanipulation robot. *J. Appl. Crystallogr.* 40(4), 539–545.

83. Hazes B, & Price L (2005) A nanovolume crystallization robot that creates its crystallization screens on-the-fly. *Acta Crystallogr.* D61(8), 1165–1171.

84. Krupka HI, Rupp B, Segelke BW, et al. (2002) The high-speed Hydra-Plus-One system for automated high-throughput protein crystallography. *Acta Crystallogr.* D58(10), 1523–1526.

85. Slabinski L, Jaroszewski L, Rychlewski L, et al. (2007) XtalPred: a web server for prediction of protein crystallizability. *Bioinformatics* 23(24), 3403–3405.

86. Stura EA, Nemerow GR, & Wilson IA (1992) Strategies in the crystallization of glycoproteins and protein complexes. *J. Crystal Growth* 122, 273–285.

87. Carter CW, Jr, & Carter CW (1979) Protein crystallization using incomplete factorial experiments. *J. Biol. Chem.* 254(23), 12219–12226.

88. Carter CW (1990) Efficient screening for crystallization conditions. *Methods* 1(1), 12–24.

89. Carter CWJ (1999) Experimental design, quantitative analysis, and the cartography of crystal growth. In Ducruix A & Giege R (Eds.), *Crystallization of Nucleic Acids and Proteins*. New York, NY: Oxford University Press, Inc.

90. Box GEP, Hunter WG, & Hunter JS (1978) *Statistics for Experimenters: An Introduction to Design, Data Analysis, and Model Building*. New York: Wiley and Sons, Inc.

91. Jancarik J, & Kim S-H (1991) Sparse matrix sampling: A screening method for the crystallization of macromolecules. *J. Appl. Crystallogr.* 24, 409–411.

92. Brzozowski AM, & Walton J (2001) Clear strategy screens for macromolecular crystallization. *J. Appl. Crystallogr.* 34(2), 97–101.

93. Kimber MS, Vallee F, Houston S, et al. (2003) Data mining crystallization databases: Knowledge-based approaches to optimize protein crystal screens. *Proteins* 51(4), 562–568.

94. Segelke BW (2001) Efficiency analysis of sampling protocols used in protein crystallization screening. *J. Crystal Growth* 232, 553–562.

95. Rupp B (2003) Maximum-likelihood crystallization. *J. Struct. Biol.* 142(1), 162–169.

96. Page R, & Stevens RC (2004) Crystallization data mining in structural genomics: using positive and negative results to optimize protein crystallization screens. *Methods* 34(3), 373–389.

97. Page R, Grzechnik SK, Canaves JM, et al. (2003) Shotgun crystallization strategy for structural genomics: an optimized two-tiered crystallization screen against the *Thermotoga maritima* proteome. *Acta Crystallogr.* D59(6), 1028–1037.

98. Jurisica I, Rogers P, Glasgow JI, et al. (2001) Intelligent decision support for protein crystal growth. *IBM Systems J.* 402, 248–264.

99. Wilson J (2002) Towards the automated evaluation of crystallization trials. *Acta Crystallogr.* D58(11), 1907–1914.

100. Spraggon G, Lesley SA, Kreusch A, et al. (2002) Computational analysis of crystallization trials. *Acta Crystallogr.* D58(11), 1915–1923.

101. Wilson J (2002) Towards the automated evaluation of crystallization trials. *Acta Crystallogr.* D58(11), 1907–1914.

102. Spraggon G, Lesley SA, Kreusch A, et al. (2002) Computational analysis of crystallization trials. *Acta Crystallogr.* D58(11), 1915–1923.

103. Echalier A, Glazer RL, Fulop V, et al. (2004) Assessing crystallization droplets using birefringence. *Acta Crystallogr.* D60(4), 696–702.

104. Burnett JH, Levine ZH, & Shirley EL (2001) Intrinsic birefringence in calcium fluoride and barium fluoride. *Phys. Rev.* B64(12), 241102(1–4).

105. Cox MJ, & Weber PC (1988) An investigation of protein crystallization parameters using successive automated grid searches (SAGS). *J. Crystal Growth* 90(1–3), 318–324.

106. Zedzik J, & Norinder U (1997) Statistical analysis and modelling of crystallization outcomes. *J. Appl. Crystallogr.* 30, 502–506.

107. Stewart P, & Baldock P (1999) Practical experimental design techniques for automatic and manual protein crystallization. *J. Crystal Growth* 196(665–673).

108. Prater B, Tuller S, & Wilson L (1999) Simplex optimization of protein crystallization conditions. *J. Crystal Growth* 196, 674–684.

109. Danley D (2006) Crystallization to obtain protein-ligand complexes for structure-aided drug design. *Acta Crystallogr.* D62(Pt 6), 569–575.

110. Hassell AM, An G, Bledsoe RK, et al. (2007) Crystallization of protein-ligand complexes. *Acta Crystallogr.* D63(1), 72–79.

111. Rees DC, Congreve M, Murray CW, et al. (2004) Fragment-based lead discovery. *Nat. Rev. Drug Discovery* 3(8), 660–672.

112. Burley S (2004) The FAST and the curios. *Modern Drug Discovery* 7(5), 53–56.

113. Garman EF, & Murray JW (2003) Heavy-atom derivatization. *Acta Crystallogr.* D59, 1903–1913.

114. Carvin D, Islam SA, Sternberg MJE, et al. (2001) The preparation of heavy-atom derivatives of protein crystals for use in multiple isomorphous replacement and anomalous scattering. In *International Tables for Crystallography* F, 247–255.

115. Banumathi S, Dauter M, & Dauter Z (2003) Phasing at high resolution using Ta6Br12 cluster. *Acta Crystallogr.* D59(Pt 3), 492–498.

116. Islam SA, Carvin D, Sternberg MJE, et al. (1998) HAD, a data bank of heavy-atom binding sites in protein crystals: a resource for use in multiple isomorphous replacement and anomalous scattering. *Acta Crystallogr.* D54, 1199–1206.

117. Rould MA (1997) Screening for heavy-atom derivatives. *Methods Enzymol.* 276, 461–472.

118. Dauter Z, Li M, & Wlodawer A (2001) Practical experience with the use of halides for phasing macromolecular structures: a powerful tool for structural genomics. *Acta Crystallogr.* D57(2), 239–249.

119. Nagem RA, Polikarpov I, & Dauter Z (2003) Phasing on rapidly soaked ions. *Methods Enzymol.* 374, 120–37.

120. Schoenborn BP, Watson HC, & Kendrew JC (1965) Binding of xenon to sperm whale myoglobin. *Nature* 207,28–30.

121. Djinovic-Carugo K, Everitt P, & Tucker PA (1998) A cell for producing xenon-derivative crystals for cryocrystallographic analysis. *J. Appl. Crystallogr.* 31(5), 812–814.

122. Soltis SM, Stowell MHB, Wiener MC, et al. (1997) Successful flash-cooling of xenon-derivatized myoglobin crystals. *J. Appl. Crystallogr.* 30(2), 190–194.

123. Knapp M, Segelke BW, & Rupp B (1998) The crystal structure of the ganglioside-binding C-terminal fragment of *Clostridium tetani* at 1.6 Å: solving the phase problem with combined MAD and MIR. *American Crystallographic Association Meeting Series* 25, 90.

124. Boggon TJ, & Shapiro L (2000) Screening for phasing atoms in protein crystallography. *Structure* 7(8), R143–R149.

125. Deisenhofer J, Epp O, Miki K, et al. (1985) Structure of the protein subunits in the photosynthetic reaction centre of *Rhodopseudomonas viridis* at 3 Å resolution. *Nature* 318, 618–624.

126. Wiener MC (2004) A pedestrian guide to membrane protein crystallization. *Methods* 34(3), 364–372.

127. Loll PJ (2003) Membrane protein structural biology: the high throughput challenge. *J. Struct. Biol.* 142(1), 144–153.

128. Fethiere J (2007) Three-dimensional crystallization of membrane proteins. In Doublie S (Ed.), *Macromolecular Crystallography Protocols, Vol. 1*. Totowa, NJ: Humana Press.

129. Weiss MS, & Schulz GE (1992) Structure of porin refined at 1.8 Å resolution. *J. Mol. Biol.* 227(2), 493–509.

130. Landau EM, & Rosenbusch JP (1996) Lipidic cubic phases: a novel concept for the crystallization of membrane proteins. *Proc. Natl. Acad. Sci. U.S.A.* 93(25), 14532–14535.

131. Paukstelis PJ, Nowakowski J, Birktoft JJ, et al. (2004) Crystal structure of a continuous three-dimensional DNA lattice. *Chem. Biol.* 11(8), 1119–1126.

132. Stanfield RL, Zemla A, Wilson IA, et al. (2006) Antibody elbow angles are influenced by their light chain class. *J. Mol. Biol.* 357(5), 1566–1574.

133. Stura EA, Taussig MJ, Sutton BJ, et al. (2002) Scaffolds for protein crystallisation. *Acta Crystallogr.* D58, 1715–1721.

134. Tereshko V, Uysal S, Koide A, et al. (2008) Toward chaperone-assisted crystallography: Protein engineering enhancement of crystal packing and X-ray phasing capabilities of a camelid single-domain antibody (VHH) scaffold. *Protein Sci.* 17(7), 1175–1187.

135. Zhou Y, Morais-Cabral JH, Kaufman A, et al. (2001) Chemistry of ion coordination and hydration revealed by a K^+ channel-F_{ab} complex at 2.0 Å resolution. *Nature* 414(6859), 43–48.

136. Harlow E, & Lane D (1989) *Antibodies: A Laboratory Manual*. Cold Spring Harbor, NY: Cold Spring Harbor Laboratory Press.

137. Weinberg RA (2006) *The Biology of Cancer*. New York, NY: Garland Science.

138. Berger I, Kang CH, Sinha N, et al. (1996) A highly efficient 24-condition matrix for the crystallization of nucleic acid fragments. *Acta Crystallogr.* D52(Pt 3), 465–468.

139. Saida F (2006) Statistical analysis of 15 dimensions in the crystallization space for protein-DNA complexes. *Protein Pept. Lett.* 13(9), 929–939.

140. Pohl E, Holmes RK, & Hol WG (1999) Crystal structure of a cobalt-activated diphtheria toxin repressor-DNA complex reveals a metal-binding SH3-like domain. *J. Mol. Biol.* 292(3), 653–667.

141. Garcia-Ruiz J-M (2003) Nucleation of protein crystals. *J. Struct. Biol.* 142(1), 22–31.

Proteins for crystallography

[In contrast to the fundamental laws of physics] the 'laws' of biology are often only broad generalizations since they describe rather elaborate chemical mechanisms that natural selection has evolved over billions of years.

Francis Crick, in *What Mad Pursuit: A Personal View of Scientific Discovery* (1990), p5

The introductory quote for this chapter, stated by the Nobel Laureate and co-discoverer of the B-DNA structure Francis Crick, serves as a reminder that macromolecular crystallography concerns itself with material that is an integral part of, and partakes in, the complex machinery of cellular life. Even in the laboratory, proteins are with few exceptions produced by living organisms. In contrast to the strict mathematical formalism and principles ruling the computational steps of structure determination, procedures and methods of protein production are thus rather generic recipes that will more often than not require skillful adaption to each individual protein.

The consequences of the individual nature of each protein are twofold: Just as in crystallization, where specific predictions about crystallization success were rather vague if not impossible, protein production generally requires a great deal of experimentation and parallel screening. Designing proper strategies is in turn greatly aided by knowing as much as possible about your specific protein. Cellular environment, complex partners, required cofactors, domain structure, and many more inherent properties help to develop a feel for the particular likes and dislikes of the material. Experimental screening for the best construct, environment, mutants, or domains is frequently required.

Chapter 4 begins with an overview of selected bioinformatics tools that derive information about the protein of interest from its sequence and corresponding data bases. Some of this information can be used to make—within limits—inferences about crystallizability and many of the bioinformatics tools are used in the targeted design of protein engineering strategies. The fact that we need to understand the special requirements for crystallization first, before we can make any rational decision regarding protein design, is the reason the protein engineering chapter is placed after the crystallization chapter, in reverse succession to the actual work flow.

Practical cloning, expression, and purification of proteins require a great deal of experience in biochemical and molecular biology laboratory methods, which cannot be covered in any detail in this chapter. Instead of comprehensive coverage, we focus on outlining which strategies for protein engineering can be applied at either the DNA level or the protein level in order to improve crystallizability of a target protein. We thus provide only a brief introduction to the methods at our disposal to give the less biochemically educated an overview and allow them to discuss their particular needs and protocols with experienced colleagues. Those already familiar with molecular biology techniques will be more interested in the sections emphasizing the special considerations and outlooks for crystallographic structure studies.

After a discussion of protein engineering principles and special aspects of protein production for crystallography, in the final section we review some key biophysical methods that can be used not so much for predicting crystallizability but as postmortem tools to provide some rational clues as to why a particular protein failed to crystallize. Generally these methods provide information about the conformational homogeneity, aggregation state, and stability of the material.

4.1   Overview of protein production

Preparation and purification of proteins for crystallographic studies is a complex and challenging field in its own right, and this chapter can by no means cover all aspects required to become a protein expression and purification expert. To that end, practical laboratory experience is required and discussions with your friendly biochemist or molecular biologist next door, as well as the study of experimental protocol collections specifically written for macromolecular crystallography projects,[1] are quite helpful. The chapter is intended to provide only a limited overview of the methods and techniques with emphasis on the perspective of the practicing structural biologist or protein crystallographer. Modern protein engineering techniques provide extraordinary possibilities (as well as challenges) and new methods and laboratory protocols are constantly developed and extended. The introduction of synthetic DNA, quick PCR-based mutagenesis, sequence-specific proteases, and fusion proteins has provided the practitioner with a near-universally useful set of tools.

Recombinant DNA techniques and heterologous protein expression. The development of recombinant DNA techniques and the polymerase chain reaction (PCR) for gene amplification, combined with enzymatic ligation techniques for gene transfer and manipulation, has enabled heterologous overexpression of protein target genes cloned into expression vectors in a wide variety of cellular hosts, irrespective of often very low natural abundance or cell-cycle dependent expression levels of the target protein. Moreover, the capability to readily add affinity tags or solubility enhancing fusion proteins to the target gene has greatly simplified protein purification, particularly at the few 100 µg to 100 mg scale required for protein crystallization. A wide variety of site-directed mutagenesis tools as well as random mutation techniques allow nearly unlimited variation of the protein target sequence. Combinatorial DNA library designs provide the basis for directed evolution of the protein, a stochastic approach where mutants displaying desirable phenotypes—generally high expression and solubility levels for crystallography—are selected in evolutionary cycles. Finally, it has become affordable to order custom genes (synthetic cDNA) for a price that is comparable to or even beats in-house DNA-manipulation based techniques.

Overview. The steps in a typical protein production cycle comprise several major phases. A very important preparatory step during initial information gathering is extensive on-line literature research about the material to be prepared. Much may be known already about the likes and dislikes of the protein, its cellular context and binding partners, or about any particular and peculiar properties that may have significant impact on preparation as well as on the overall crystallization strategy. In the initial *in silico* phase, the properties of the target protein are analyzed at the peptide sequence level using an array of bioinformatics tools together with the available information about the target protein. Generally included in the *in silico* phase is the design of suitable directional primers[2] for the following cloning stage, where primary PCR amplification of the gene and manipulation of the DNA are performed. Once the target DNA is cloned into a suitable cloning plasmid or expression vector and transfected into a cellular

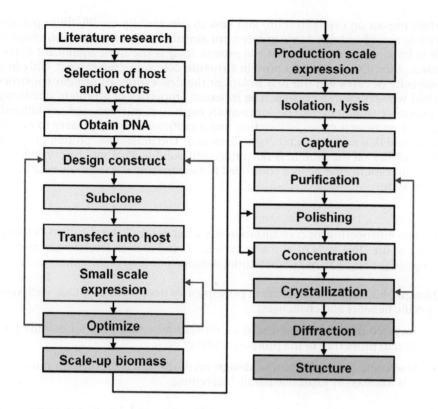

Figure 4-1 Recombinant DNA technologies allow efficient protein expression, purification, and crystallization. One of the most significant advantages of recombinant DNA techniques is the ability to add affinity capture tags which allow rapid separation and concentration of the tagged protein in a single step. In fortunate cases further steps such as purification, polishing, and concentration can be bypassed and the protein directly screened for crystallization. More often than not, further modification to the protocol and ultimately the protein construct will be necessary (red arrows). In summary: • Study all available literature • select host(s) and expression vector(s); • obtain DNA with the open reading frame (ORF) of interest; • design the initial expression constructs; • subclone the cDNA if needed; • introduce the recombinant DNA into the chosen host(s); • perform small-scale expression trials; • analyze the small-scale trials and select optimum expression parameters; • scale up biomass production; • extract the protein of interest from lysed cells or from the expression media; • purify the protein; • condition the protein for crystallization; • perform crystallization and cycle back (red arrows) to purification, expression, or cloning with redesigned construct as needed.

host system, the actual protein production phase begins with overexpression of the target protein in the host cells. Isolation, capture, purification, and "polishing" of the material to concentration and purity sufficient for crystallogenesis conclude the production phase. Expression and solubility screening as well as purification are almost always initially conducted on a small scale. Multiple constructs, expression host strains, and other system parameters are commonly screened in parallel for optimal expression and solubility levels. Modification of the target sequence or vector constructs may become necessary at this stage to obtain sufficient amounts of purified protein for crystallization. The subsequent crystallization trials and diffraction screening will, in the majority of cases, dictate further protein engineering, either at the DNA or the protein level. Key steps are illustrated and outlined in Figure 4-1.

4.2   Engineering proteins that crystallize

A number of crystallization-related parameters of protein production, such as the effect of buffer systems and the role of additives and ancillary molecules following the protein through production, were already addressed in detail in Chapter 3. In this chapter we are focusing on the properties of the protein itself,

Box 4-1 **Protein production for crystallography.** Proteins for crystallographic studies are, with few exceptions, produced by heterologous overexpression in cellular hosts. Heterologous expression requires design and cloning of a recombinant DNA molecule encoding the target protein sequence and its transfer into the expression host. The advantage of recombinant DNA methods is the great flexibility in modifying the protein sequence and the overexpression of proteins with otherwise very low natural abundance.

their impact on crystallization, and how to successfully modify these properties to our advantage. As we pointed out already in the crystallization chapter, it is imperative to realize that *the protein itself* is the most significant determinant for the success of a protein structure determination project. Only in a minority of cases will the first isolate or the first expressed protein construct yield well-diffracting crystals. The inherent properties of a protein decisively determine whether it can be successfully expressed, purified, and crystallized, and often several cycles of iteration and modification are necessary to obtain a protein that expresses and crystallizes well. The most basic protein property is solubility—if the protein is not sufficiently soluble, it cannot be crystallized. Crystallization, after all, is a controlled transition from the solution state to the solid state.

Protein engineering strategies and levels

Protein engineering for crystallization can most generally be classified as taking place at the DNA level or at the protein level, and at each level we can employ targeted (or rational) or combinatorial designs (or even combinations thereof).

Design philosophies. There are in principle two design philosophies, each with specific benefits and challenges:

- **Targeted protein design** based on rational analysis of the (often only predicted) properties of the protein construct.

- **Stochastic or combinatorial design methods** based on random mutagenesis followed by extensive parallel screening.

Analyzing a gene sequence for disordered termini and deciding to remove the likely disordered terminal residues from the sequence is an example of targeted design. Often the targeted design is also combined with a parallel or combinatorial approach such as trying various different extents of terminal truncations. An example of a somewhat more sophisticated combinatorial approach is directed evolution, where, for example, mutant libraries are screened for increased solubility or expression levels as a selection filter and the best constructs are propagated into one or two more rounds of evolution and selection.

Design levels. Targeted as well as combinatorial protein engineering strategies in turn can be applied at two levels in the design process (Figure 4-2):

- at the *DNA level* by modifying the nucleotide sequence (most frequently);

- at the *protein level* by chemical or biochemical modification of the expressed protein. Limited proteolysis of exposed termini and enzymatic removal of oligosaccharides in glycoproteins are examples of engineering at the protein level (see also Sidebar 4-1).

Sequential versus parallel strategies. As a general principle it is almost always more efficient in terms of time and cost to pursue even targeted strategies in parallel instead of a purely sequential approach, in which one modification at a time is examined and in the case of (frequent) expression or crystallization problems, the cycle is repeated from scratch. Each given situation will generally

---

Sidebar 4-1 Common protein modification techniques.

Biochemical/enzymatic:

Proteolysis

Deglycosylation

Phosphorylation/
dephosphorylation

Chemical:

Methylation of primary amines

Reduction/oxidation of disulfides or thiols

Modification of exposed thiols (e.g., with iodoacetamide)

Selective derivatization with heavy atom reagents (e.g., iodination)

Box 4-2 Protein engineering strategies and levels. Protein engineering is possible at the DNA level by modifying the sequence of the protein construct and at the protein level by modifying the expressed protein. In both cases targeted, rational design strategies or random (combinatorial) design strategies can be applied. Frequently, many different protein constructs or expression conditions are screened in parallel for expression and solubility levels. The most fundamental property of a protein suitable for crystallographic studies is sufficient solubility.

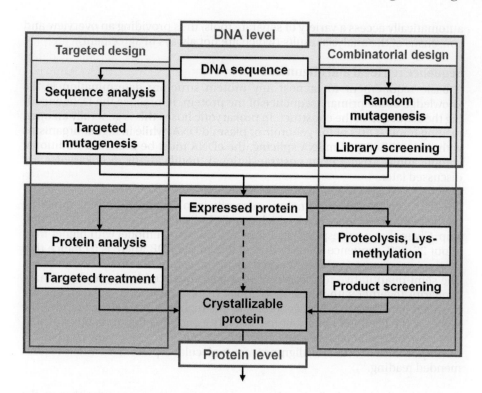

Figure 4-2 Conceptual overview of design strategies toward protein engineering for crystallization. Protein engineering can occur at the DNA level and the protein level. Targeted experimental designs as well as combinatorial designs can be applied to either level. Often several approaches are combined and pursued in a parallel strategy.

require development of a specific hybrid strategy that allows for both parallelism and flexibility.

Salvage strategies. An additional, somewhat arbitrary, term frequently used is salvage strategy, which describes methods that are resorted to after it is established that a protein is not crystallizable. The term (not quite correctly) implies that these strategies are not used as a primary means but rather as a last resort. Chemical lysine methylation at the protein level, for example, could be considered a salvage strategy.

4.3   Targeted design strategies at the DNA level

Each crystallographic study begins with an *in silico* information gathering phase. The more we know about the protein, the easier it is to avoid unnecessary experiments that are *a priori* doomed to fail. Instead, one can focus on the comprehensive exploration and screening of the remaining promising possibilities. A combination of literature search and bioinformatics tools will provide, in addition to specific information, an estimate for the feasibility and the anticipated level of difficulty for a proposed crystallographic study. Such risk assessment is a prudent exercise: Expression and crystallization of a simple secreted bacterial protein may be a good project for a Master's thesis in structural biology, while a membrane protein structure determination probably is not.

Bioinformatics tools

An enormous proliferation of bioinformatics tools exist that can be deployed to examine and analyze your target protein sequence. The database and web server issues of *Nucleic Acids Research* as well as the journal *Bioinformatics* are the primary sources of program and algorithm descriptions. The current web addresses of most bioinformatics tools required for sequence analysis, protein prediction, and general bioinformatics tasks can be found on the *Bioinformatics Links Directory* meta-server.[3] While we give specific references for tools commonly used in the design of target sequences for improved crystallization, it is always worthwhile looking for updated programs and new meta-servers that automatically access a variety of separate tools, thus providing an overview and consensus of the analysis results (which are not always in agreement).

Sequence retrieval and alignment

A basic requirement for almost any protein structure determination is the knowledge of the primary sequence of the protein, in practice the DNA encoding the sequence of the construct. In prokaryotic hosts, this is generally an open reading frame (ORF) of the genomic or plasmid DNA, while in higher organisms with the possibility of mRNA splicing, the cDNA must be available. Additional complications arising from posttranslational modifications in eukaryotes are discussed later.

The first step in sequence analysis generally involves the search for similarity on the peptide sequence level, which usually indicates some structural relationship. Given that a corresponding 3-dimensional search model with higher than about 25–30% sequence identity is available in the PDB, there is a chance that molecular replacement may succeed as the structure determination method. The semi-quantitative relationships between sequence identity and structural similarity, together with the method and caveats of molecular replacement, are provided in Chapter 11. As a general rule, the higher the sequence identity, the higher is the likelihood of structural similarity and thus the probability of success for a molecular replacement structure solution. A recent review emphasizes the importance of accurate alignment for molecular replacement and is recommended reading.[4]

To detect sequence similarity, the query sequence is compared either at the DNA sequence level or at the protein level with a non-redundant (NR) database of all known sequences, but organism-specific databases are available as well. The basic algorithm for comparison of primary biological sequence information is implemented in the Basic Local Alignment Search Tool[5] (*BLAST*), which is available for nucleotide-level search (*BLASTN*) as well as for protein sequence based searches (*BLASTP*). The basic algorithm and its advanced versions *PSI-BLAST* and *PHI-BLAST* are explained in most bioinformatics texts;[6] all versions are available as web tools. The *BLAST* programs return sequences ranked by an expectation score (how likely it would be to obtain each alignment by chance, so the lower the better), list the percentage of identical residues, and flag any sequence that has an associated structure entry. The NCBI web site as well as the EBI and the *PredictProtein* meta-servers provide sequence retrieval, alignment, and other annotation services in addition to the basic alignments.

For sequence searches at the protein level, the advanced *PSI-BLAST*[7] (Position-Specific Iterated *BLAST*) method is commonly employed. In *PSI-BLAST* a profile (a frequency table of each amino acid in each position of the protein sequence) or a position-specific scoring matrix (PSSM, calculated from position-specific scores for each position in the alignment) is automatically generated from multiple pair-wise alignments of the highest scoring hits in an initial basic *BLAST* search. Highly conserved positions then receive high scores and weakly conserved positions receive low scores. The profile obtained is then used to perform a second (or iterative) *BLAST* search and the result of each iteration is used to refine the profile. Additional domain information from a reversed *BLAST* search against the conserved domain database[8] (CDD) and a graphical display of the pair-wise alignments is also provided by the NCBI service.

In the next step, the saved *BLAST* results can be further refined and subjected to multiple sequence alignment, which provides additional information beyond basic pair-wise sequence matching. The multiple alignment may also allow detection of stretches indicative of matching motifs or even previously undetected domains. Such domains may be suitable for separate expression. With some luck, a related structure of a domain may already be known and again provide an avenue for molecular replacement. The multiple sequence alignment program *ClustalW2*[9] is available as a service from the EBI, for example, and provides a variety of representations of the alignment highlighting various physical properties such as hydrophobicity, secondary structure propensity, and others. Hidden Markov Model algorithms such as *HMMER*[10] or the heuristic *DIALIGN-TX*[11] are also popular. If structural information for one or more aligned sequences is available, the inclusion of structural information can further improve the sequence alignments. One example is *3DCoffee*,[12] which also provides alignment output in *Clustal* format from which programs such as *ESPript*[13] can generate publication quality alignment figures. Various other public or commercial programs such as *Molsoft ICM*[14] also provide alignment modules and often produce publication quality graphical output that displays also secondary structure elements.

Secondary structure, transmembrane, and disorder prediction

If the query sequence aligns with a known structure, its secondary structure elements can be readily analyzed using a geometry and hydrogen bond pattern analyzing program such as *DSSP*[15] and the assignment of structure elements to the query sequence is relatively simple. Often structural similarities are not obvious or not present at all if we are dealing with a novel fold, and we want to supplement our sequence alignment with various kinds of predicted structural information. This information ranges from globularity analysis (*GLOBE* and *GlobPlot*)[16] to surface accessibility and secondary structure prediction (*SABLE*,[17] *PHD*,[18] *PSIPRED*,[19] and others), search for disulfide bridges (*DISULFIND*),[20] and detection of disordered regions (*GlobPlot*, *PONDR*,[21] *DisEMBL*,[22] *RONN*, *DISOPRED*[19] and others[23]) to identifying transmembrane helices and signal sequences (*Phobius*,[24] *SignalP*[25]). These programs are generally available, standalone, as web services, or some through meta-servers such as *PredictProtein*.[18] It is important to keep in mind that the (occasionally contradictory) results from these prediction services are not always correct, perhaps in the range of 60–80% of the sequence in secondary structure prediction—*caveat emptor*.

The task now is to combine all the information gathered from different sources in a compact form into a comprehensive property profile to gain an overview of the potential problem areas in the sequence. There is no program that generates one complete output of all the analysis and one can add the results manually to the best suitable sequence alignment or other output. Each situation will differ greatly depending on the specific protein and we therefore will present a few general cases to illustrate how one could develop a rational strategy for optimizing a protein construct.

Analysis of sequence alignments

Let us examine a typical multiple sequence alignment and consider some possible strategies to modify and engineer the target protein at both the DNA and the protein levels. The schematic sequence alignment in Figure 4-3 (which could be augmented by secondary and perhaps even 3-dimensional structural information) shows that we essentially have two regions of reasonable alignment, separated by a linker. Linker regions or interfaces between domains often—but not always—display significantly less sequence conservation and also less defined secondary structure than the conserved structural domains. The domains themselves are flanked by potentially disordered terminal regions. If we find a shorter sequence aligned with only one region of our query, the suspicion grows that we are dealing with a domain structure. This domain structure may also have been detected already by the programs searching domain databases during the alignment process. Note that after domain truncation of the sequence, it might be worthwhile repeating the alignment—a local alignment may then give better results than the global alignment used for the entire multi-domain sequence. If the protein in question has already been isolated, the analysis of its protease sensitivity as outlined in Section 4.6 can provide experimental information on domain boundaries.

The overall picture of the alignment generally suggests a number of possible approaches to protein engineering: The potentially flexible N-terminus could be truncated at the DNA level using various corresponding N-terminal primers. We could also express each domain in different levels of truncation separately, or

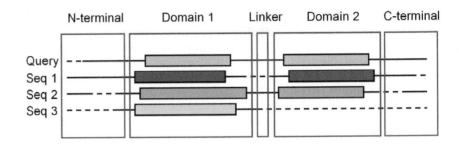

Figure 4-3 **Schematic of alignment**
Figure 4-3 Schematic of alignment suggesting possible strategies for protein engineering. A multiple sequence alignment often reveals the domain organization of a protein of unknown 3-dimensional structure. Query represents the target protein, Seq the similar sequences from the data base. Domains generally show a higher degree of sequence conservation, while loops and termini show more variation in sequence as well as in length. The alignment with a shorter, presumably single domain protein as shown in Seq 3 further supports the proposed domain structure (the dashed lines represent alignment gaps or missing residues). Secondary structure predictions can further delineate the domain boundaries. Expression levels and stability can be affected by as little as a single residue truncation or retention, and multiple constructs are thus generally pursued.

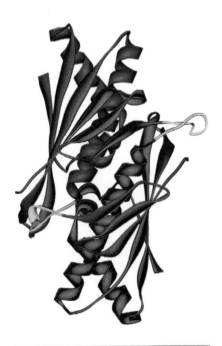

Figure 4-4 Domain swapped dimer. The domain swapped dimer of *Bacillus subtilis* organic hydroperoxide-resistance protein (OhrB) serves as an example where domain truncation would not be effective. Truncating the protein sequence at the disordered loops (green) would likely lead to loss of structural integrity because a part of the β sheet of the domain structure would be missing. These loops, however, remain viable targets for site directed mutagenesis (see Figure 4-6). PDB entry 2bjo.[28]

we could modify the linker region and test the effect on protein crystallization. Another avenue at the expressed protein level might be enzymatic C-terminal truncations with carboxypeptidases A and B, as used, for example, in the preparation of the T-cell receptor.[26] Domain expression and domain truncation are powerful tools for protein crystallography and an additional example is discussed below.

Domain expression and domain truncation

The fact that domains fold independently and are generally stable without their partners (Chapter 2) is of significant relevance for crystallography—it allows us to express and crystallize single domains in isolation. This is particularly helpful if the multi-domain protein in question is either very large, or the domains are flexibly tethered, which generally makes self-assembly into well-ordered crystal lattices difficult. A typical example is the domain structure of the epithelial growth factor receptor (EGF-R, Figure 2-34), a membrane receptor with a single transmembrane helix, where separate domains have been successfully expressed for structure determination.[27]

Domain truncation follows a similar philosophy as expression of single domains, where the expressed sequence is shortened as much as possible toward the presumed domain boundaries, largely to remove flexible N- or C-termini or linker sequences, which likely hamper crystallization. Domain truncation can also be pursued using combinatorial DNA domain truncation libraries as described in Section 4.4. Note that expression levels and stability can be dramatically affected by as little as a single residue truncation or retention. Multiple constructs are thus generally pursued; a reasonable truncation window size is about ±3 to ±5 residues from the selected boundary. Useful information may sometimes be gleaned by studying mRNA splicing sites which can delineate domain boundaries or regions dispensable for stable protein structure.

Caveats. If a domain depends on its surrounding partners for proper folding (as for example seems to be the case for the long, helical translocation domain of botulinum neurotoxin shown in Figure 2-1), the domain-cutting approach will not work. A limited number of domains also include stretches of residues that are distant in the primary sequence, but form integral structural components in the same domain, and here domain cutting also fails. This is the case for example in domain swapped dimers (Figure 4-4) and, most unfortunately, in many integral membrane proteins with multiple transmembrane helices as discussed in Chapter 3. As a general caveat, any design strategy should take full advantage of all available information about the protein and its environment. If it is already known that a binding site is formed at the interface of two domains, a single domain structure will not provide the complete picture regarding the substrate binding. Nonetheless, the single domain structure might well be useful in subsequent studies to phase the complete structure by molecular replacement.

Site predictions

Many proteins contain specific sequences that may indicate a particular binding site and thus give clues to possible binding partners that may stabilize crystallization. However, binding site analysis is a rather sophisticated business and its results need to be interpreted in the specific biological (and structural, if

Box 4-3 **Targeted design strategies at the DNA level.** Design strategies at the DNA level are based on the analysis of the protein sequence and the derived protein properties. Multiple sequence alignments supported by available structural information can identify domain boundaries and terminal regions. Additional tools identifying disordered regions, transmembrane helices, signal peptides, and binding sites augment the analysis and help identify potential problem regions. In eukaryotic sequences mRNA splicing sites may delineate domain boundaries or regions dispensable for stable protein structure.

available) context of the protein. Similarly, searches for glycosylation sites based purely on the quite frequent tripeptide consensus sequence Asn-X-Ser(Thr) are completely pointless if we are dealing with a bacterial protein or the sequence is located within a transmembrane helix, and they are at least suspect for a protein without a signal peptide sequence. The *NetNGlyc* server[29] for example takes such constraints into consideration. For many binding sites, the 3-dimensional arrangement is crucial and advanced servers and meta-servers such as *InterPro*[19] that search multiple domain and site databases take these restrictions into consideration. As always, it may prevent mishaps to be well aware of your protein's biological and functional context.

Estimation of crystallization success

Given the significant amount of sequence-derived information that the bioinformatics tools discussed above can provide us with, one might wonder if we could not combine this information to develop a more quantitative estimate of the crystallization propensity of our particular protein. It would be quite desirable to know beforehand how difficult a structure determination project is likely to be—while still acknowledging that each individual protein requires specific purification and crystallization protocols, and that we have no means of exactly predicting the degree of difficulty from first principles. Only the crystallization experiment will be the ultimate test. However, it still helps to know beforehand how much resources and effort will likely be required for the determination of a difficult but high impact structure.

Predictions based on protein properties

In the absence of first-principle based prediction methods, statistical analysis of outcomes of crystallization experiments given a certain property or predicted structural feature of the protein could—at least in principle—be attempted. In order to correctly quantify these conditional probabilities, negative results are required; we need to know how many trials out of a certain number were successful, given a certain protein property. The importance of balanced data sets including positive and negative results for crystallization prediction has been emphasized,[30, 31] but until recently negative results (i.e., failed experiments) were practically never archived or reported in the literature.

Crystallization data collected and analyzed by the high-throughput facilities sponsored by the NIH Protein Structure Initiative[32] (PSI) have somewhat improved this situation. The particular value of such large-scale, high-throughput experiments is that failures are also systematically recorded and reported, and positive and negative results are available for proper statistical analysis. The drawback is that these dedicated high-throughput centers are not exactly picking the high hanging fruit, that is, the data will be generally biased toward prokaryotic and highly soluble "garden variety" proteins. Unfortunately, one of the most solid empirical rules is that if a project is interesting and promises high impact, it is also probably difficult. Such is frequently true for eukaryotic or mammalian proteins, and nearly always the case with membrane proteins.

Protein properties used in success prediction. While the protein crystallization propensities discussed in Chapter 3 (Figure 3-39) were not conditioned on any specific protein property, data from TargetDB and the Joint Center for Structural Genomics (JCSG) have been analyzed by a logarithmic opinion pool method to combine probability distributions for crystallization success conditional on a number of properties and predictions derived from the protein sequence.[33] The "crystallization feasibility score" is computed by the *XtalPred* server using bioinformatics tools based on: (i) basic sequence-derived information such as sequence length, molecular weight, hydropathy index, instability index, isoelectric point, and content of Cys, Met, Trp, Tyr, and Phe residues; (ii) comparative information such as number of insertions compared with alignments of homologs (if available); and (iii) feature predictions of secondary structure, disordered regions, low complexity regions,[34] coiled-coil predictions, transmembrane regions, and signal peptides.

Homologs. An additional feature of the *XtalPred* server includes the search for close bacterial homologs which are potentially more likely to crystallize, or for general close homologs in the full non-redundant sequence database. The search for (and potential pursuit of all) homologs for crystallization actually goes back to the origins of protein crystallography: The teams of Perutz and Kendrew in Cambridge tried hemoglobin and myoglobin homologs isolated from various species (orthologs), until material isolated from horse heart and sperm whale muscle finally led to success. As far as crystallization is concerned, homologs are basically the result of protein engineering provided for free by nature.

Interpretation. The problem with the results of conditional predictions is that they are probably accurate in the extreme and thus somewhat trivial cases. A secreted bacterial protein with 20 structurally similar homologs already in the PDB is probably a class 1 protein, very likely to succeed. At the other end of the spectrum, an integral transmembrane protein with a few disordered regions will not surprisingly receive a poor class 5 score. The question is what to do with the intermediate results. There is seldom a single showstopper identified that can be engineered away right from the start, so you will have to try in any case. Figure 4-5 represents a case where the prediction is discouraging but no single problem appears that could be readily rectified. Again, the only valid and accurate crystallization probability assessment to this date is to set up crystallization trials and to inspect the results. Numerous other proteins studied by surface entropy reduction[35] also indicate the property-based classification and prediction of crystallization success is a coarse guide at best; successful SER made no difference to the *XtalPred* scores but completely changed the crystallization result. As already mentioned, crystallization is determined by highly specific local interactions that cannot be readily described by global protein properties.

Surface entropy reduction and related methods

Surface entropy reduction (SER) is a site-specific protein engineering method and can be used either right away at the first sequence design stage (generally combined with a parallel approach) or after initial trials in a sequential design to improve poorly diffracting crystals. The theory and motivation behind SER has been detailed in Chapter 3, and in principle aims to identify long-chained

---

Box 4-4 Predicting crystallization success. Inclusion of statistical data relating crystallization success to specific protein properties allows the establishment of conditional probabilities for crystallization success for a given protein. However, these probabilities are largely based on predicted protein properties and cannot pinpoint the specific and local intermolecular interactions that actually determine crystallization. Crystallization predictions are generally more reliable in clear cut cases (very easy or very difficult to crystallize) but provide less guidance in intermediate cases. Crystallization trials remain the only authoritative (and in fact, fast and simple) method of determining crystallization propensity of a specific protein construct.

surface residues with high entropic freedom—foremost Lys, Glu, and Gln (Arg despite its size enjoys less conformational freedom because of its rigid guanidinum group), which can be mutated to short residues such as Ala, Thr, or Val leading to significant improvements in crystallization.[36] One of the risks of course

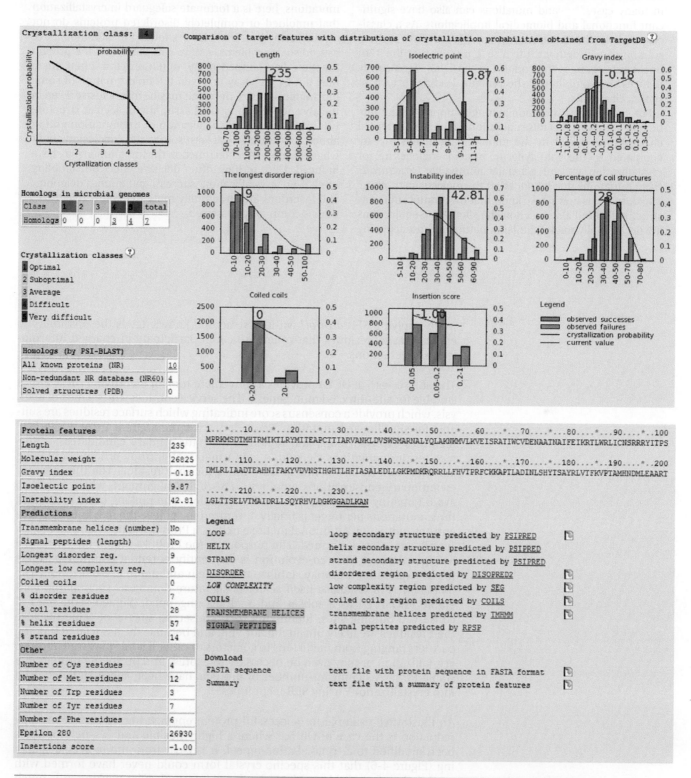

Figure 4-5 Prediction of crystallization success. The panels show the output of the XtalPred server[33] for an unknown (hypothetical) protein from *Pyrobaculum* spherical virus. The source of the poor score is not immediately obvious—the largest perceived risk factor might be the high isolectric point. Also note the contradicting predictions—the predicted N-terminal helical segment is also predicted to be disordered. One would have to conduct the actual experiment to determine with any certainty whether this protein indeed deserves a 4 (or D- grade) in Crystallization 101.

Sidebar 4-2 Consequences of single site mutations. The observation that single residue mutations can radically alter protein properties and thus may also drastically determine solubility[37] and crystallizability has been shown in many cases,[35, 38] and mutations can also have significant functional and biomedical implications. As a classical example, the single substitution Glu6 to Val6 in the β subunit of hemoglobin (the A to T mutation at the 17th nucleotide changes the codon GAG to GTG) significantly reduces the solubility of the protein, leading in turn to sickle cell anemia.

For site-directed mutation aiming to improve solubility or crystallization this raises questions about the potential changes introduced in the mutant molecular structure compared to the wild type. In protein engineering for crystallization we are generally aiming for *structurally silent* mutations that do not affect the molecular structure. The question is how do we know that the modification was structurally (and also functionally) silent? Such objections are occasionally made, similar in nature to those rejecting crystal structures on the grounds that the solid state may not represent the biologically relevant state. We have discussed the limited validity of this criticism in view of the evidence in Chapter 1. In the case of engineered mutations, here is a fortunate safeguard in crystallization that unfolded or completely disordered proteins do not crystallize in the first place, should the mutation have caused such problems. Assays can prove that the protein function is unimpeded by mutation, but often there may not be an assay or, given the finicky nature of some biochemical assays, the result may be inconclusive. Even in the absence of a functional assay, the molecular structure will invariably reveal the location of the mutation and in almost all cases allow at least some valid conclusion about what—if any—effect on function the mutation(s) may have induced. Despite their often decisive effect on solubility levels and crystallization success, mutations, deletions, and insertions are structurally more frequently tolerated[39] than is commonly perceived.

when applying SER *a priori* without structural knowledge is the possibility to engineer away residues that otherwise could participate in charged intermolecular interactions.

The SER server[35] at UCLA provides a convenient tool for identifying surface sites suitable for site-directed mutagenesis. The service conducts three kinds of analysis, which provide a consensus score indicating which surface residues are suitable for site-directed mutation. Secondary structure analysis using *PSIPRED*[19] identifies exposed loops and regions which are primary candidates for surface residue mutations. Second, a conformational entropy profile is obtained and weighted by a solvent exposure index for each residue. The third test for evolutionary conservations employs a *PSI-BLAST* alignment, which serves to avoid targeting conserved residues for mutation. The ranking of no more than three mutations per target is finally optimized to achieve the maximum surface entropy reduction. Additional meta-searches in Prolinks,[40] the PDB, and *Blocks/InterPro* are also performed. The purpose of the Prolinks search for functional linkages derived from co-evolution is to identify potential complex partners for co-expression that may stabilize the protein and aid in crystallization. The *Blocks/InterPro* database itself searches signature databases such as the venerable PROSITE[41] as a subset and analyzes ungapped, aligned sequences of conserved binding sites. Identifying such conserved sites is of value because their residues are likely unsuitable for SER, and because the associated binding partners ranging from metal ions to cofactors or other ligands may well improve crystallization or may even be necessary to express and purify the protein in a stable form. A significant number of proteins have been successfully coerced into crystallization[42] using SER techniques.

An illustrative example for successful protein engineering by surface entropy reduction is the case of OhrB,[28] where a highly flexible and Lys-rich loop has been modified to a triple alanine repeat. It is clear from the molecular packing (Figure 4-6) that this specific crystal form could never have formed with extended surface residues, given the dense intermolecular packing. It is much less evident, however, whether the surface entropy reduction or another concurrent change (of local charge, overall charge, pI, or any other unknown factor) has contributed to the formation of a stable crystal form. As always, it is very difficult to establish causality from simple correlation!

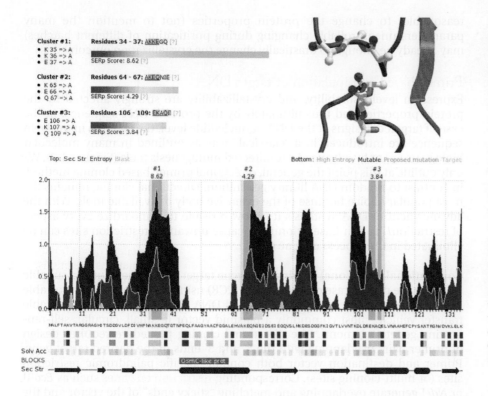

Figure 4-6 **Selection of residues for surface entropy reduction.** The analysis of the wild type (which could not be crystallized) of *Bacillus subtilis* organic hydroperoxide-resistance protein (OhrB) by the SER server at UCLA identifies a stretch of residues 34–37 -A(KKE)- as a cluster of high surface entropy. The inset top right shows the crystal packing in the 2.1 Å structure of the engineered triple mutant, -A(AAA)-. The arrangement of the molecules (blue) and the symmetry related mates (purple) clearly shows that this particular packing and crystal form would not be possible for the native structure given the space requirements of its extended lysine and glutamate residues. While the importance of modifying the protein is evident, it remains open which of the multiple molecular property changes that concur with the SER mutation actually are causal for successful crystallization. The details of the SER analysis and server output can be extracted from the web server instruction pages. PDB entry 2bjo.[28]

Targeted surface mutagenesis is not limited to SER. An illustrative example describing the successful attempt to improve the crystallization behavior of the 2300-residue hetero-tetrameric phenylalanine tRNA synthetase by means of mutating a judicious selection of surface-exposed residues and disordered loops is provided by Evdokimov and colleagues.[43]

Additional targeted, structure-based design strategies

Frequently the initial crystallization trials will yield crystals that do not diffract well but still yield a low resolution structure. In this case, the modification of the protein can be rationalized based on direct structural evidence. The strategies are again very specific, and many examples are listed in a recent compilation.[38] Successful strategies include mutating residues in homologs to those already known to form crystal contacts in homologous structures, eliminating potentially interfering terminal residues, or shortening extensive termini or loops that are absent in the molecular structure. However, Dale and colleagues have also pointed out that the rationale leading to the targeted mutations does not always seem to be causal to the improvement of the crystals, as shown in the case of a DHFR mutant.[38] The important message is that doing something (preferably

---

Box 4-5 **Surface entropy reduction and site specific mutations.** Surface entropy reduction (SER) exemplifies a targeted design strategy at the DNA level. Mutation of high entropy surface residues such as exposed Lys, Glu, and Gln residues to Ala, Thr, or Val can lead to significant improvements in crystallization. However, various other properties such as overall protein charge, local charge distribution, and conformation are also varied at the same time and the precise causality between residue mutations and crystallization success cannot be easily established. A number of targeted protein design studies have shown that the rationale leading to various construct modifications is often not causal to the crystallization success—the readiness to modify the protein is important.

reasonable) to change the protein properties (not to mention the many parameters uncontrollably changing during purification of different batches) may already be enough to drastically change the crystallization outcome.

Primer-based manipulation of target DNA

Expression levels, solubility, and crystallizability are strongly affected by the protein properties and thus ultimately by the protein's DNA sequence. In the case of targeted designs at the DNA or nucleotide level, the changes in the DNA sequence are introduced in a "classical" way as outlined in many molecular biology texts[45] or by advanced site-directed mutagenesis tools (Sidebar 4-3). We only outline at this point the general, directional primer-based cloning method, in contrast to random DNA library generation. Directional cloning remains the most popular choice because of the extensive body of available tools. With the advancement of DNA synthesis it is now possible to standardize every aspect of restriction/ligation-based cloning because unwanted restriction sites can be eliminated in the process of gene design.

Conventional directional cloning. The gene target is amplified out of a suitable cDNA by the polymerase chain reaction (PCR) using high-fidelity, thermostable proof-reading polymerases. The amplified DNA is then ligated into a suitable transfer vehicle, most commonly a plasmid vector (a small, circular, self-replicating genetic element). The directional cloning procedure requires design of primer pairs overlapping with the 5′ and 3′ ends of the desired target gene. Primer and destination vector both contain specific palindromic restriction sites (or multi-cloning sites). Corresponding restriction enzymes such as EcoRI or NdeI generate overlapping and matching "sticky ends" of the vector and the amplified gene insert. Hundreds of specific restriction enzymes are available and it is generally possible to find a restriction enzyme that does not digest the target gene itself. Primer design programs facilitate this task.[2] The fragments are then annealed and joined together by DNA ligase, and the ligation products are transformed into competent E. coli cells. Properly executed restriction and ligation reactions result in highly specific and directional incorporation of the insert DNA into the vector, resulting in the majority of cell colonies harboring the correct construct. An additional degree of fidelity is provided by means of enforced selection (based on antibiotic resistance, killer genes, auxotrophy complementation, etc.). Bacterial (or yeast) colonies can be directly screened for positive

Sidebar 4-3 Single-step PCR mutagenesis (QuikChange™). Mutagenesis of specific amino acid residues is a basic requirement for any targeted protein engineering at the DNA level. Site-directed mutagenesis has become easy by the introduction of single-step PCR mutagenesis (Quik-Change™): The originally described method[44] has been developed into a convenient kit sold and patented by Stratagene. The method relies on two key phenomena: (1) modern high-efficiency strand-displacing DNA polymerases can easily replicate entire plasmids with high fidelity and (2) plasmid DNA originating from most E. coli strains is Dam-methylated at the sequence 5′-GATC-3′ resulting in 5′-Gm6ATC-3′ which is the specific substrate for a DpnI exonuclease; unmethylated DNA is not cleaved by this enzyme. QuikChange mutagenesis is elegantly simple. A small amount of plasmid DNA from almost any common laboratory strain of E. coli is subjected to a limited number of PCR cycles using a high-fidelity proofreading polymerase and two complementary primers, each of which contains the site of desired mutation as well as 15–18-nt sequences flanking the site. Each cycle of the reaction is allowed to run for long enough to have complete replication of the entire plasmid. PCR results in doubly nicked plasmids with the overlap region described by the primer pair. Afterwards the reaction mixture is supplemented with DpnI which rapidly digests the template plasmid DNA but spares the PCR products. Finally, the mixture is transformed into cloning-grade E. coli cells which repair the nicks. Keeping the PCR cycle numbers low and using the high-fidelity enzyme affords highly accurate introduction of specific mutations with overall efficiency over 80% in typical applications. The real power of this mutagenesis technique lies in its ability to introduce mutations with surgical accuracy, and the option for generating arbitrarily large deletions and relatively large insertions. The latter are limited by the length of primer DNA; if "megaprimers" (large DNA fragments generated in a separate PCR step) are used instead of regular synthetic oligonucleotides, the limits on insertion length become solely dependent on the limitations of PCR itself (plasmid sizes of 10 kb or even more are now not an obstacle). The method has been recently expanded to allow for generation of multi-site (typically up to five) variants.

clones by colony PCR (Figure 4-11) with vector-specific primers.[46] The plasmid DNA is extracted from several positive colonies, and the inserts sequenced to assure target gene integrity. Plasmids with confirmed inserts are then frozen and stored for transfer into expression host cells as briefly reviewed in Section 4.4. A good laboratory information and management system (LIMS) will help keep track of the many clones accumulating during a structure determination project.[47] Figure 4-7 shows the principle of directional, primer-based cloning applied in the generation of truncation variants of a target protein.

The traditional restriction enzyme based cloning procedure described above is quite generic but requires several separate steps such as digestion, ligation, selection, and subcloning into different vectors. Time-saving cloning methods and systems have therefore been developed and are described later. Conventional cloning methods combined with site-directed mutagenesis using appropriate primer-based mutation systems and kits[48] (Sidebar 4-3) are rapidly regaining popularity because of the availability of inexpensive synthetic DNA that can be designed to contain all the desired features and none of the detrimental ones. Synthetic DNA enables one to completely standardize conventional cloning.

4.4   Combinatorial designs at the DNA level

The targeted design strategies at the DNA level described above—based on directional, primer based PCR cloning—require that at least some plausible indications exist about where and what to mutate. This is not always the case, and an alternative—particularly valuable for improvement of expression and solubility levels—is combinatorial designs. They generally involve generation of a library of random DNA constructs that are screened in parallel for expression, solubility, and ultimately, crystallization and diffraction. In contrast to targeted approaches using directional cloning, where specific primers as shown in Figure 4-7 are designed to generate one construct at a time, primer-independent DNA library construction requires extensive colony picking and screening.

The challenge of combinatorial designs lies largely in the necessity of efficient colony picking, sequencing, and rapid screening of the surviving colonies for a desirable phenotype marker. As a consequence, robotics such as colony pickers and liquid handlers are helpful equipment.

Random DNA truncation and fragmentation

Random truncation of termini or fragmentation at the DNA level generates a large set of gene variants that can be cloned and expressed in parallel, thereby increasing the chance that one of the constructs ultimately delivers diffracting crystals. Such combinatorial library strategies have been particularly successful in combination with high-throughput pipelines,[49] but in principle they are also suitable for smaller laboratories. Parallel random truncation strategies are particularly

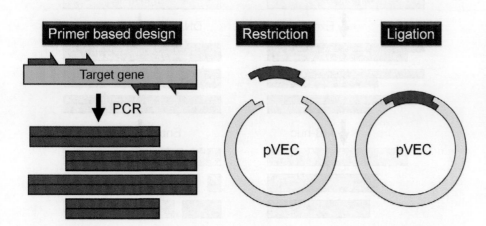

Figure 4-7 **Targeted, primer based generation of truncation mutant library using directional cloning.** Specifically designed 3′ and 5′ primers are used to generate a limited set of truncation mutants, which are ligated into a suitable plasmid vector using standard recombinant DNA protocols. Mutations such as for SER are generated in a similar way using primer based site directed mutagenesis (Sidebar 4-3).

valuable when domain boundaries cannot be rationalized or defined by *in silico* sequence analysis or enzymatic proteolysis.

Truncation libraries. A commonly used method to generate random unidirectional DNA truncation libraries is enzymatic digest of the (PCR amplified) DNA by exonuclease III (ExoIII), followed by removal of the remaining overhang with 5′ nucleases (mung bean or S1 nuclease are common).[49] Commercial kits for unidirectional truncation are available. As a result of the random digestion, 2/3 of the resulting digestion products will be out of frame and all possible truncations will be present in the incubated reaction tube. This requires ligation into a vector that adds the three stop codons for each possible reading frame.[50] In addition, an efficient screening method to separate viable from useless clones is required. After transfection of the library into a cloning strain, the DNA is extracted from viable colonies and an agarose gel reveals which clones contain DNA of expected length. The corresponding bands are excised from the gel and the DNA is isolated and sequenced. Clones containing the proper sequence are then stored for subcloning and transfection into expression hosts.[49] Alternatively, the library can be cloned directly into expression vectors, followed by screening based on expression selectors such as anti-tag antibodies. Positive clones are then sequenced.

Gene fragmentation libraries. In cases when no domain boundaries can be established by comparative sequence analysis or other methods, random fragmentation of the target DNA may lead to stably folding, separately expressible domains (the procedure for establishing domain boundaries is sometimes called domain phasing). The target DNA can be fragmented by sonication and/or enzymatic treatment with DNAse I, followed by polishing of the ends with a proofreading polymerase.[49] The fragments then can be colony picked and screened as described above. The best expressing and soluble fragments can be either directly crystallization screened or further optimized by any of the combinatorial or targeted methods available. Figure 4-8 outlines the generation of unidirectional truncation libraries and fragmentation libraries.

Directed evolution

In vitro directed evolution[51] is a refined combinatorial approach, in which a library of random DNA constructs or mutants is exposed to some selective pressure at the phenotype level (generally an expression or solubility screen, but even crystallization and diffraction limits can be the ultimate selector) and the survivors are again mutated and selected. The number of cycles (generations) and whether gene recombination (DNA shuffling) is applied to further enhance the diversity of the library varies. We will also encounter evolutionary designs in crystallographic search algorithms (Chapter 11).

Figure 4-8 Generation of truncation mutants using directional cloning or gene fragmentation. Unidirectional truncations are obtained by incubating the target DNA for different periods of time with exonuclease III. Mung bean nuclease can be used to remove the overhanging 5′ ends. Gene fragmentation is based on random disruption of the target DNA, followed by end polishing with nucleases. Only few of the fragments will contain expressing and folding domains, and extensive screening is required. The drawing follows a figure design by Hart and Tarendeau.[49]

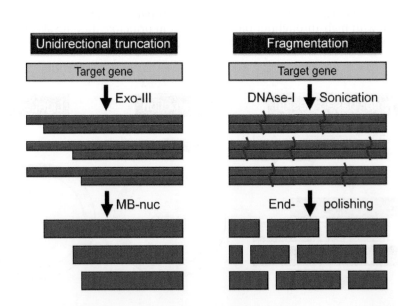

Mutation and selection. Random point mutations are introduced by error-prone PCR (epPCR, using special "mutazyme" polymerases instead of a high-fidelity proofreading polymerase and/or adding Mn^{2+} or Mg^{2+} to the reaction mixture) or by amplification and propagation in mutator strains that are deficient in DNA repair genes. There are a number of caveats that apply as far as various bias issues are concerned,[49] that is, these random mutagenesis libraries are not truly random in a strict statistical sense.[52] In any case, the library of viable clones is screened for a phenotype marker, that is, at the protein expression level. A protein that does not express well is of little interest for crystallography, but even at this point it may be advisable to vary the expression conditions by expanding the screen and examine parameters such as host strain or temperature. The most promising candidates can again be isolated and the library generation and selection cycle can be repeated. Interestingly, repeated recursive random mutation can lead to competition between beneficial mutations which then drive each other to extinction.[53] This problem can be eliminated by DNA shuffling, a form of synthetic, *in vitro* gene recombination.

DNA shuffling. The effectiveness of directed evolution can be further increased by allowing recombination of surviving genes. This technique is frequently used in optimization of industrial enzymes[53] and has been used in the design of improved green fluorescent protein (GFP) variants[54] which themselves are used as fusion reporters for solubility. It is also possible to recombine the mutants with genes of homologs, as used in the determination of the β-propeller structure of paraoxonase.[55] In a general procedure a selected set of mutants generated by epPCR that exhibits a desirable phenotype is subsequently fragmented using, for example, the DNAse procedure described above, and then PCR re-assembled. DNA shuffling (Figure 4-9), often combined with directed evolution (Figure 4-10), requires a great deal of colony picking, sequencing, and screening, which is generally performed using suitable robotics.

4.5   Challenges in protein expression and purification

Once the initial phase of *in silico* analysis is finished and a set of potentially promising constructs have been designed, the actual production of the protein

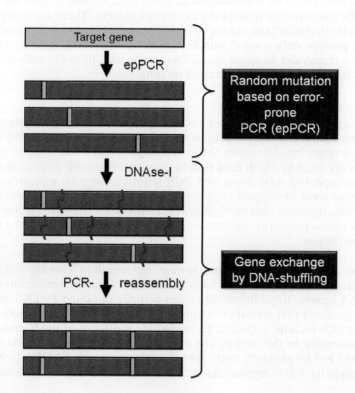

Figure 4-9 DNA shuffling. Because repeated cycles of error-prone PCR can lead to extinction of beneficial mutations, DNA shuffling is a common library generation technique in directed evolution. The original mutants are digested and recombined, leading to some double mutants with improved phenotypes. The drawing follows a figure design by Hart and Tarendeau.[49]

Box 4-6 **Combinatorial designs at the DNA level**. Combinatorial DNA libraries are based on the generation of random mutants of the target gene *in vitro*. Truncation mutant libraries and domain fragment libraries are examples of basic DNA libraries used to generate soluble and crystallizable constructs. Directed evolution is a refined combinatorial technique in which a library of random mutants is exposed to selective pressure such as an expression and solubility screen at the phenotype level. DNA shuffling allows recombining mutants with desirable traits to further improve solubility and ultimately, crystallizability and diffraction quality. Combinatorial DNA library generation requires extensive colony picking, selection, and sequencing and is generally conducted using robotic equipment.

Figure 4-10 **Principle of directed evolution.** From the target gene, a random mutant library is generated by error-prone PCR. The clones undergo phenotype screening (generally high expression levels and solubility for crystallography, indicated by the green dots). The selected and sequence-verified mutants are then exposed to a second round of epPCR or preferably to genetic recombination using DNA shuffling (Figure 4-9). The selection criterion can be a GFP-based fluorescent fusion reporter as shown and explained in Figure 4-14.

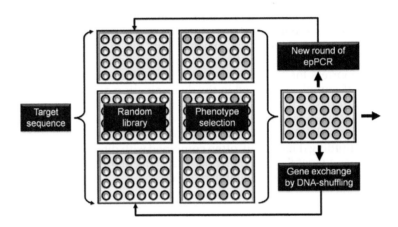

begins. Each expression host requires a specific vector to transfer the target DNA into the host cell. The combination of that vector system combined with the host cells is commonly termed the expression system. There are many options to consider that affect the choice of (sub)cloning, expression, and purification. We can provide only a small selection relevant for crystallography and must leave the remainder to more specialized textbooks, protocol collections, and reviews listed in the general literature section.

Advanced cloning techniques

For straightforward expression of bacterial proteins in bacterial systems, traditional restriction enzyme cloning methods briefly outlined in Section 4.3 often suffice. In many cases however, the interesting protein is mammalian or human and it is not clear in which host system overexpression will ultimately succeed. To avoid repeated subcloning and PCR redesign, creative molecular cloning techniques were developed with the specific goal of construct portability in mind. Note, however, that increasing availability of inexpensive synthetic DNA[56] to some extent preempts the cost- and labor-saving advantages of the methods described below.

Ligation independent cloning. Ligation independent cloning (LIC) allows directional cloning of PCR products without the need for restriction enzymes and DNA ligases. In addition, all clones actually obtained by LIC are recombinant products that contain the appropriate target gene insert. Both primers used for PCR include a specific 12 nucleotide sequence at the 5′ end which is complementary to the vector. The complementary single-stranded ends are generated not by separate restriction digestion but with T4 DNA polymerase, which stops its 3′ to 5′ exonuclease activity once it encounters the same specific deoxynucleotide (dNTP) that has been added as the sole dNTP to the reaction mixture. For example, if dATP was the single dNTP, the so generated single strand overhang extends at the first TA pair encountered. The overhangs are long enough that no separate enzymatic ligation reaction is necessary. Just as for nick sealing in conventional cloning, the host DNA repair and expression machinery joins the strand backbones. The 5′ sense primer can also encode for a protease cleavage site and allow enzymatic removal of whatever tags may be present ahead of the target gene. LIC vectors are available for bacterial and viral expression systems as well as for blue/white cloning strains, and LIC can often be completed within one hour.

Enzyme-free cloning. This cloning technique relies only on PCR as it eliminates all other enzyme-dependent steps.[56] The basic principle of this method is the use of entire genes or gene fragments as very large primers, portions of which anneal to user-designated sequences on the vector. PCR is then performed using strand-displacement proofreading enzymes under conditions that favor elongation of the entire gene-plasmid assembly. The product of such PCR is a doubly nicked circular DNA molecule that is held together by a large overlap (similar to PCR-based mutagenesis) or LIC. This technique is inexpensive, relatively straightforward, and has a cloning efficiency of over 90%.

Topoisomerase cloning. The enzyme DNA topoisomerase I functions both as a restriction enzyme and as a ligase, that is, it can cleave and then rejoin DNA during replication. *Vaccinia* virus topoisomerase I specifically recognizes the pentameric sequence 5′-(C/T)CCTT-3′ and forms a covalent bond with the phosphate group attached to the 3′ thymidine. After it cleaves one DNA strand (allowing the DNA to unwind), the enzyme ligates the ends of the cleaved strand with the insert and releases itself from the DNA. A large variety of commercial TOPO® vectors are available which contain topoisomerase I covalently bound to each 3′ phosphate of the linearized plasmid. This enables the vectors to readily ligate DNA sequences with compatible ends (overhangs or blunt) in about 5 min at room temperature. One drawback is perhaps that old or inappropriately stored reagents can give rise to large numbers of false-positive clones, therefore care should be exercised in following manufacturers' instructions exactly.

Gateway cloning. Another cloning method that elegantly circumvents the typical, multi-step restriction enzyme cloning workflow and allows simple transfer of the DNA inserts from one host system to others is Gateway® recombination cloning. There are two fundamental steps in the Gateway cloning: construction of an entry clone (once) and construction of the expression clone suitable for the desired expression host. The entry clone contains the gene of interest flanked by lambda phage *att*L sequences, and can be produced by conventional cloning or commonly by TOPO cloning into a Gateway entry vector.[57] Site-specific

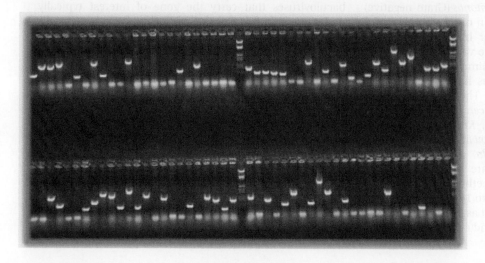

Figure 4-11 Validation of cloning by colony PCR. The image shows an ethidium bromide stained DNA agarose gel of 96 *Y. pestis* clones. The DNA is amplified by colony PCR directly from selected colonies transferred into a PCR tube, where the DNA is subsequently amplified with vector specific primers. Only from colonies that contain the proper full length DNA will a PCR product of expected length be obtained. Lanes 5, 7, 10, and 11 for example show no amplification. Figure reproduced[46] with permission from Springer Verlag.

recombination of *att*L sequences of the entry vector with *att*R sequences of the destination vector by the appropriately termed LR-reaction enzyme mix finally creates the expression clone with the gene of interest. One of the significant advantages of this method is that once the entry vector is sequenced there is no need to sequence any of its derivatives since recombination is highly unlikely to introduce sequence artifacts (with the rare exception of a case where *att*-like sites are present in the ORF of choice). The other significant advantage of this method is its high fidelity (over 99% cloning efficiency) that is enforced by a dual selection screen comprised of an antibiotic resistance switch with the removal of the "killer gene" ccdB, which the Gateway system incorporates to impose additional selective pressure; if the correct product is formed at every recombination step, it will have lost ccdB and therefore become non-toxic to the expression host. An entire system of expression vectors for different hosts ranging from bacteria to insect cells and mammalian cells is commercially available, as are vectors designed for co-expression of multiple genes. All the convenience comes at a price, of course, as the reagents used in this method are proprietary.

Choice of expression system

The most frequent avenue of protein production for crystallization is by expression of the target gene in a suitable cellular system. Heterologous overexpression takes place in foreign host cells; most frequently one of the many specially designed *Escherichia coli* strains, various yeasts, insect cell based expression systems, or a mammalian expression system such as Chinese Hamster Ovary (CHO) cells or Human Embryonic Kidney (HEK) cells is used. In some cases, such as certain membrane proteins, many of the blood-clotting cascade proteins, and some of the immune system components, the material is still isolated from natural sources.

The choice of expression host system depends on a wide variety of parameters; one of the most significant is the absence of posttranslational modifications (PTMs) in the otherwise very robust and simple bacterial (prokaryotic) expression systems. Absence of PTMs in bacterial systems can be an advantage; for

---

Sidebar 4-4 A copious selection of expression hosts.
Escherichia coli. E. coli is the undisputed champion of protein expression. It sports a vast array of cloning and expression tools and is a common gateway to expression in more esoteric hosts. *E. coli* is a cost-effective, robust, and versatile expression platform. Unfortunately many of the more challenging eukaryotic proteins are very difficult (and often impossible) to express in this host.
Other bacteria. Several *Bacillus* and *Lactococcus* (Gram-positive) species, as well as *Pseudomonas* (Gram-negative) bacteria are used for the expression of recombinant proteins. Gram-positive bacteria offer higher likelihood of success with secreted proteins. The variety of expression tools available for these bacteria is limited.
Yeasts. A number of yeast species have been used for recombinant protein production. The most commonly used species are *Saccharomyces cerevisiae*, *Pichia pastoris*, *Schizosaccharomyces pombe*, *Kluyveromyces lactis*, *Hansenula polymorpha*, and *Yarrowia lipolytica*. Methylotrophic yeasts (*Pichia, Hansenula*) offer the incredibly tightly regulated promoters that are triggered upon the yeast's switch to metabolism of methanol. *Hansenula* in particular is known to produce up to 14 g l^{-1} of the desired protein product. Yeasts grow almost as fast as bacteria and are almost as easy to manipulate and culture.

Higher fungi. Besides yeasts, other fungi are also used to express recombinant proteins. Mycelia of *Trichoderma* and *Aspergillus* species are commonly used. One of the significant issues encountered in these systems is the tendency of the host to secrete voracious and persistent proteases.
Insect cells. The second most common expression choice for proteins used in crystallization experiments (after *E. coli*) are cultures of *Trichoplusia ni* or *Spodoptera frugiperda* cells. These cells are infected with recombinant baculoviruses that carry the gene of interest typically under the control of a late viral promoter. Insect cells have been used for the expression of numerous difficult proteins, including membrane proteins and even GPCRs. Unfortunately insect cells are much more delicate to grow than bacteria or yeast and also are considerably slower and more expensive to grow.
Plant cells. Recombinant plants and suspension cell cultures have been used for recombinant protein production. Their use in crystallography is limited to date.
Mammalian cell culture. Mammalian cell lines are used to express a variety of difficult proteins and are considered to be the pinnacle of expression versatility. Unfortunately they are expensive, slow, and somewhat difficult to grow. Hybridoma cells are a special case of mammalian cell culture used to produce massive quantities of recombinant antibodies.

example, the problem of conformationally heterogeneous glycosylation does not happen in non-glycosylating bacterial expression systems. On the other hand, absence of functionally relevant PTMs may render the protein non-functional. Even in this case, if the fold is not compromised, the structure will still provide valuable insight.

Insect cells or higher eukaryotic cell lines usually require expertise in cell culturing and are significantly more challenging to handle than the versatile and robust bacterial or yeast expression hosts. Finally, cell-free systems, containing all the reagents and protein complexes from extracts necessary for a functioning ribosomal translation machinery, are available and have been used to produce proteins for crystallization.[46] Full control and parameterization of expression which is possible in these *in vitro* translation systems must be balanced against a significant cost factor.

Natural sources

There are a number of disadvantages and difficulties associated with classical protein isolation and preparation from tissue. First and foremost, many proteins are very sparsely and selectively expressed in native tissue. The consequence is that significant amounts of tissue material need to be processed in order to obtain a few micrograms of functional protein. Such may suffice for functional studies, but rarely does it satisfy the appetite of a crystallization pipeline. Choosing an organism and tissue where the protein in question is naturally enriched is quite important if a natural source must be exploited.

Another significant complication in the case of isolation is the presence of isoforms in native tissue, which can lead to conformational heterogeneity which in turn is generally problematic for crystallization. Conformational inhomogeneity is also often introduced by varying degrees of inhomogeneous posttranslational amino acid modifications or decorations such as glycosylation—none of those is conducive to formation of perfectly ordered crystals. An additional difficulty in tissue isolation, particularly affecting purification, results from the much more diverse array of other proteins that are initially present during isolation compared with overexpression in a host system. Foremost of concern are endogenous proteases and degradation enzymes which as a result of breakup of compartmentalization during the cell disruption can digest the target protein before it is captured and isolated. Addition of protease inhibitor cocktails is almost always required in the course of natural product isolation. Naturally occurring proteins also do not contain fusions that might assist in solubilization and affinity capture.

Bacterial expression systems

The workhorse of protein production for crystallography (including many complex eukaryotic proteins) is *Escherichia coli* (*E. coli*, EC) to which we will limit this brief discussion. Several other prokaryotic systems derived from *Bacillus*, *Pseudomonas*, or *Lactococcus* species are used as expression hosts in the biotech industry and have been adapted for special uses in crystallography.[62] Bacterial expression systems lack the cellular compartmentalization and the complex translocation machinery of eukaryotes and cannot perform the posttranslational modifications common in eukaryotic cells. The absence of such PTMs in many situations makes protein production more efficient and crystallization easier, but eukaryotic proteins with functionally or structurally relevant PTMs generally require expression of the target gene in corresponding host systems. In certain cases, co-expression with the enzymes required for farnesylation, phosphorylation, or glycosylation may be possible.

Escherichia coli based systems.
The most prevalent and robust heterologous expression systems are based on *E. coli*, and a wide variety of engineered strains with improved properties are available. The selection of expression strains depends on the target gene and can have a drastic effect on expression and solubility levels (often for no obvious reason). Small-scale screening for expression levels in various strains and at different conditions for each construct is thus of advantage. Many of the strains used in protein expression are derived from their

Sidebar 4-5 Concanavalin A: a natural product and a hardy perennial of structural biology. Some of the most celebrated proteins in the history of crystallography came from the humble Jack bean (*Canavalia ensiformis*). The bean gave us urease (the first ever enzyme to be crystallized),[58] canavalin, and the saccharide-binding lectin concanavalin A, a hardy perennial of structural biology.[59] All of them were first isolated by Professor James Sumner[60] who worked on a shoestring budget to overcome numerous scientific challenges, the disbelief of his colleagues, and a physical impediment (his right arm was amputated after a hunting incident). Extraction of these proteins from the beans required grinding, drying, treatment with chloroform, aqueous acetone, and other solvents and, perhaps most notably, recrystallization in a Petri dish on the window sill in cold weather (Sumner's laboratory could not afford a refrigerator at that time). Heroic efforts of Sumner and his colleagues eventually earned him the Nobel Prize in Chemistry (1946).

Concanavalin A is also an example of the extraordinary complexity inherent in the production of certain proteins.[61] The peptide sequence of the mature ConA is circularly permuted with respect to the mRNA and gene sequences. This is not the result of RNA splicing, instead the fully folded protein precursor undergoes as many as seven steps of proteolysis, rearrangement, and re-ligation. Because of incomplete processing of a significant portion of mature ConA in the bean, ConA partly exists as two fragments that are non-covalently held together, which does not impede its crystallization in any way.

generic parent BL21 strain which is deficient in both *lon* and *ompT* proteases, and include for example C41 and C43 (often used for membrane proteins), B834 met⁻ for Se-met expression, and various codon-optimized strains or strains for disulfide bond formation. It is important to note that the effects of strain modifications often extend beyond the original intent of the designers: for example, protein expression in BL21(DE3) pLysS may be greatly improved compared with the same construct expressed in BL21(DE3) even though the protein is entirely non-toxic to the host and therefore the presence of the pLysS plasmid should have no effect. Since such effects are impossible to predict, it is a viable strategy to screen several common *E. coli* strains for expression of their protein. This is a cheap and quick way to improve one's odds for successful expression.

Codon usage and mRNA structure optimization. With the genetic code being degenerate, several nucleotide triplet combinations common in higher organisms are rare in *E. coli*. Codon usage frequency for certain triplets of R, I, P for example is up to ten times lower in *E. coli* than in humans.[45] Codon optimized strains produce transfer RNAs that recognize these rare codons and thus yield higher expression yields (on the DNA level, codon preferences can readily be addressed in total gene synthesis, but not by conventional cloning techniques). Trade names for such codon optimized strains are Rosetta®, Rosetta 2, and others. It is noteworthy that in *E. coli* N-terminal Arg, Lys, Leu, Phe, Tyr, and Trp residues greatly decrease the half-life of the protein, which can be readily addressed by modified primer design during the cloning stage.

The availability of inexpensive synthetic DNA changed the landscape of protein expression experiments—it is now easy to simply eliminate the rare codons during gene synthesis. An equally important advantage is the possibility of optimizing the mRNA sequence with the goal of avoiding mRNA secondary structure that is considered detrimental (through effects such as folding back onto itself) for efficient translation. Codon degeneracy is exploited to direct mRNA toward reduced propensity for secondary structure formation while the message itself is preserved.

Disulfide bonds. In order to form a disulfide bond (Cys-S-S-Cys), chemical oxidation of the two adjacent cysteines (Cys-S-H) must take place. The intracellular environment in bacteria is generally reducing and the *dsb*-based pathway for enzymatic disulfide bond formation is thus easily overwhelmed. As a consequence, complex disulfide bond formation in the cytoplasm often does not occur properly. The formation of proper disulfide bonds also requires that the two participating cysteines are in close proximity, that is, the reduced precursor itself is properly folded. The problem generally increases with the number of disulfide bonds present and often such proteins from higher organisms with multiple disulfide bonds cannot be properly refolded from inclusion bodies when expressed in *E. coli*. The remedy is either to use a yeast or higher expression system, or to express in engineered *E. coli* strains that have mutations in both the thioredoxin reductase (*trxB*) and glutathione reductase (*gor*) genes, which enhances disulfide bond formation in the cytoplasm. Transfection with thioredoxin-fusion clones can further increase the chances for proper disulfide bond formation, since the thioredoxin fusion tag may enhance the formation of disulfide bonds in the periplasm. In cell-free systems and by co-expression (discussed below), disulfide bridge formation can be achieved and controlled by glutathione redox buffers, and/or the addition or co-expression of enzymes that assist disulfide bond formation.

Transformation and induction of expression. Hundreds of plasmid vectors designed with different affinity tags, fusion partners, reporter genes, and any combinations thereof are commercially available for transfection. In the case of bacteria, the plasmid vectors are transfected into competent cells, which means that their cell walls are made partly permeable by electroporation or chemical treatment (competent cells are also commercially available). Bacterial expression systems make use of native and synthetic regulatory elements to allow the experimenter control of induction. The importance of controlled induction becomes clear if we consider the (high) likelihood of the target protein—particularly at high expression levels—being somehow disturbing or even toxic to the host. Therefore, we would want the expression of such genes turned off as completely as possible until sufficient biomass is generated and trigger expression then. For example, the gene of interest can be placed under the control of phage T7 promoter that is normally absent in *E. coli*. The resulting expression construct is harmless to cloning-grade *E. coli* and therefore can be handled without special considerations. For expression the construct is introduced into an *E. coli* strain that harbors T7 polymerase under an *E. coli*-native inducible promoter such as *lac, ara,* and so forth. Induction of the T7 polymerase is induced with IPTG, arabinose, and the like, resulting in high levels of target protein expression.

Temperature dependence of expression levels. As a general rule, protein expression levels decrease with induction temperature (commonly ~10°–37°C). A lower expression level in turn provides more time for the host cells to properly cope with the large amounts of overexpressed material. As a consequence, better controlled folding may take place, more material may be secreted, and less material ends up in inclusion bodies. Screening for optimal expression conditions is generally conducted in parallel experiments on a small scale. For assaying and harvest, the cells are transferred into lysis buffer (containing digestive enzymes such as lysozyme and DNAses), disrupted by detergents, short sonication bursts over ice, or by pressure change in French presses. The supernatant is separated by centrifugation from cell debris and inclusion bodies, and the fractions are analyzed for protein content (see Figures 4-13 and 4-14 for examples of how this can be elegantly accomplished).

Autoinduction. The basis for the elegant and brilliant autoinduction method is the natural frugality of *E. coli* (and other bacteria) and the associated phenomenon of diauxic growth: When bacteria are grown in media containing more than one energy and carbon source (e.g., a mixture of two carbohydrates) they will select the richest source first and ignore the rest. Once the first source is exhausted the cells will begin expressing proteins necessary for transport and metabolism of less efficient fuels. Autoinduction media contain two sugars (glucose and lactose) and glycerol. After inoculation *E. coli* utilizes glucose as the primary source of carbon and energy. During this period other sugar uptake and metabolism systems (e.g., those encoded by the *lac* operon) are silenced. Once the rapidly multiplying cells have consumed their food of choice (glucose) they switch to lactose which requires them to de-repress the *lac* operator and induce the *lac* operon. Concomitantly they activate all other genes under the control of *lac* and *lac*-derived promoters (e.g., the T7 RNA polymerase in BL21(DE3) cells), resulting in the synthesis of the protein of interest. The method was optimized by Bill Studier[63] and requires the production of specific autoinduction media (commercial kits are available). In addition to the convenience of not having to repeatedly sample the cell density and induce (often at inconvenient hours) with IPTG, autoinduced expression typically produces a greater proportion of soluble target protein than standard IPTG induction. Autoinduction is compatible with many host strains such as BL21, Rosetta, and their DE3 lysogen derivatives.

Resolubilization. Formation of inclusion bodies containing highly overexpressing proteins can be an advantage in certain cases, because inclusion body formation is essentially a highly efficient capture method. Inclusion bodies are dense proteinaceous aggregates that form in the cytoplasm in response to certain conditions—most notably when the protein folding machinery does not have the capacity to cope with the influx of nascent polypeptide. Inclusion bodies generally contain misfolded proteins, but there is evidence that partially, or even fully, folded proteins may also be found in them. Inclusion bodies can be sedimented at moderate *g*-force levels and have the visual appearance of very finely ground sand. It is necessary to resolubilize[64] the protein from inclusion bodies with "medium-strength" detergents such as sarcosyl and the protein must be able to refold properly. Some proteins can even withstand treatment with guanidine hydrochloride and can be successfully solubilized, purified, and subsequently refolded by dialysis against a suitable (crystallization) buffer. If inclusion-body-loaded cells are allowed to recover (i.e., if induction is switched off) the inclusion bodies become gradually consumed as the host attempts to refold the proteins. Unfortunately these resolubilized polypeptides are commonly degraded by SOS proteases instead of being successfully refolded.

Expression in yeasts

Yeasts are a single-celled eukaryotic organism with certain advantages for expression of eukaryotic proteins compared with *E. coli*. In contrast to animal cell cultures, yeasts can be grown to high cell densities providing the comparatively high amounts of recombinant material needed for structural studies. Yeasts also divide rapidly (~1 hour for *Pichia* or *Hansenula*) and tend to outgrow or outright kill off any competing organisms that may contaminate the culture. This evolutionary advantage can turn into a drawback because yeast contamination of more delicate cultures (insect or mammalian cells) is very hard to cure. The most commonly used yeast strains are *Saccharomyces cerevisiae*, the methylotrophic yeast *Pichia pastoris*, and *Kluyveromyces lactis*. Proteins expressed in yeasts have access to the eukaryotic protein folding and glycosylation machinery that *E. coli* cells lack, thus providing an important alternative to bacterial expression systems.[65] Another, sometimes neglected, technical drawback of yeasts is their tough chitinaceous cell wall which is hard to penetrate. Lysis of yeasts requires considerably more effort which is why they are less commonly used to produce cytoplasmic proteins.

Secretion. Yeasts are veritable secretion monsters—yields of more than 10 g of protein per liter of culture are not uncommon in industrial protein production. With their compartmentalized cell possessing a dedicated translocation machinery, yeasts are particularly suitable for proteins that are also secreted by their native host (for example glycosidases, serum albumins, cytokines, etc.). To achieve active secretion of proteins in yeast, the gene of interest is cloned into an appropriate vector that contains a signal peptide sequence upstream of the target protein. The choice of the secretion signal is not trivial and is best done empirically; a common starting option is the α-mating factor signal from *Saccharomyces cerevisiae*. Curiously, many genes can be secreted with the aid of their native signal sequences (e.g., human albumin) whereas many others require extensive trials. Recently there have been claims of engineered universal secretion signals[66] which may even cross the boundaries between different expression systems (bacterial, yeast, insect cells, mammalian).

The signal protein enables efficient transport of the fusion product through the yeast secretory pathway. The signal sequence is then either automatically removed by a signal peptidase in the endoplasmic reticulum (ER) and/or a protease in the Golgi apparatus, resulting in the secretion of the native form of the target protein into the growth medium. Alternatively, a protease cleavage site such as for TEV (described below in Section 4.5) can be engineered into the vector for removal of the secretion tag(s) after capture.

Disulfide bond formation and glycosylation. Formation of disulfide bonds requires an oxidative environment. In yeast, this is provided by translocation of the proteins tagged for secretion into the ER, which also contains enzymes that assist in oxidization of disulfide bonds.[67] There is generally a good chance that secreted proteins with multiple disulfide bonds can be overexpressed with proper fold and function in yeasts.

Glycosylation (discussed below in Section 4.6) occurs in yeasts, but follows different patterns than in mammalian systems.[68] *Pichia* may provide some advantage over *S. cerevisiae* because it does not hyperglycosylate. Both *S. cerevisiae* and *P. pastoris* have primarily N-linked, mannose-rich glycosylations, but the length of the oligosaccharide chains added posttranslationally in *Pichia* averages only 8–14 mannose residues per side chain, which is considerably shorter than of those in *S. cerevisiae*, which can attach 50–150 mannose residues.[69] In addition, very little O-linked glycosylation has been observed in *Pichia*. There are additional differences and in general it seems that the glycosylation patterns in *Pichia* resemble more the glycoprotein structure of higher organisms.[70] This is not only relevant for structural studies, but it is believed that the α-1,3 glycan linkages in glycosylated proteins produced from *S. cerevisiae* are primarily responsible for the hyper-antigenic nature of these overexpressed proteins making them unsuitable for therapeutic use. The fission yeast *S. pombe* has also been used to simplify glycosylation patterns of proteins.[71]

Insect and mammalian expression systems

In certain cases the simple bacterial or yeast based systems all fail to express a mammalian target protein. Prime candidates for such trouble are membrane receptors, proteins with complex posttranslational patterns necessary for structural integrity, proteins with complex disulfide bond patterns, and those displaying any combinations thereof. An informal survey of several large expression labs suggests that if *E. coli* does not readily produce the protein of interest, the next choice is most commonly insect cell culture.

Insect cell systems. Baculoviruses exclusively infect insects, and expression systems that utilize insect cells and a baculovirus (BV) have been developed. The most commonly used BV is *Autographa californica* multiple nuclear polyhedrosis virus (AcMNPV), and the host cell line has been derived from the fall armyworm, *Spodoptera frugiperda* (Sf). An *E. coli* transfer vector carries the AcMNPV polyhedrin promoter region and the target gene, flanked by viral DNA. The DNA sequences allow the recombination with the virus DNA, which is co-transfected together with the vector either into Sf insect cells (which requires difficult selection) or using linearized virus DNA into EC for more efficient clone formation and easier clone selection (nearly all virulent BV clones are then positive). Positive clones are then transfected into the insect cells. Although most insect cells can be grown in suspension in serum-free culture, insect cell growth is by no means as straightforward as for bacteria or yeast.[45] Very recently the advent of wavebag technology has made large-scale culture of insect cells relatively easy. The use of frozen infected cells has made it possible to reproducibly infect large-scale cultures without the need of virus titering.[72] Numerous human proteins including membrane receptors[73] have been expressed in insect cells and subsequently crystallized.

Insect cells generally do not add galactose or terminal sialic acid to N-linked glycoproteins, and the glycosylation pattern is therefore different from the glycosylation occurring in mammalian cells. Complete BV/insect cell expression systems are commercially available, including specialized systems providing humanized glycosylation.

CHO and HEK cells. If all systems discussed before fail to produce protein, mammalian cell systems may be the last resort. There are two main options: transient expression and stable cell line creation. Transient expression relies on mass-transfection of cultured mammalian cells with plasmid or linear DNA, or sometimes with recombinant baculovirus. The latter technique has been recently used to produce milligram quantities of GPCRs and glycoproteins.[74] The method is relatively quick and cheap, however, it does not result in continuous production of protein because cells eventually clear out the foreign DNA. Creation of stable cell lines involves multiple rounds of transfection and selection and relies on incorporation of foreign DNA into the nuclear DNA of the host. The product of this process is a stable cell line which can produce the desired protein for as long as it is maintained in culture.

Mammalian cells provide the full array of posttranslational modifications, and lentiviral and adenoviral transfection systems are available. The culture of Chinese Hamster Ovary (CHO) cells or Human Embryonic Kidney (HEK293) cells is non-trivial; they grow slowly only under carefully controlled conditions and are sensitive to contamination. Achieving high cell density and expression yields, and thus large amounts of material, is correspondingly difficult and expensive.[75] As always, a search for engineered cell lines may reveal useful results depending on your specific project. For example, the *Lec*1 CHO mutants provide

Sidebar 4-6 **The far side of expression hosts**. Proteins can be expressed in intact insect larvae using the same viral transfection as used in insect cell culture based expression. The squirming mass of infected larvae explosively liquefies several days post-infection as a result of viral proliferation, thus sparing the experimenter from lysing the larvae (but perhaps at the expense of the experimenters' desire to have lunch, which sometimes is free for crystallographers—Sidebar 10-10).

Other alternative and esoteric expression systems include chicken eggs, goat milk, corn, tobacco, and mouse saliva. Although no reports of crystals grown from recombinant proteins expressed in mouse drool are presently known, the first therapeutic agent (thrombin inhibitor) produced in recombinant goat milk has been recently approved by the FDA.

Link: http://www.c-perl.com/technology/proteinexpression.html

a leuco-phytohemagglutinin resistant cell line unable to synthesize complex and hybrid N-glycans because of the lack of N-acetylglucosaminyltransferase I (GntI) activity,[76] which may provide another avenue to avoid problems with inhomogeneous glycosylation of mammalian proteins.

Cell-free expression systems and chaperones

Cell-free (CF) expression systems[77] (also called *in vitro* translation systems) provide an interesting alternative to host-cell based expression systems. The reaction system is prepared from cell extracts (primarily *E. coli*, but also from wheat germ,[78] insect cells, and rabbit reticulocytes) so that it contains the transcription as well as the translation machinery, effectively forming a complete coupled expression system. The obvious benefit is that all limitations resulting from impairment of the host cell machinery as a result of toxicity,[79] saturation of translocation machinery, cell wall perforation, and so forth, are non-existent. Note, however, that toxicity resulting from direct ribosome interaction can remain a problem.

Molecular chaperones. In addition to the inherent flexibility of CF systems, one can take full control over the transcription–translation process by adding molecular folding chaperones (commonly bacterial heat shock proteins such as DnaK, DnaJ, HtpG, GroEL/GroES, and ClpB), folding catalysts, non-standard transfer RNAs and amino acids, and enzymes necessary for posttranslational modifications. Certain "foldases" can also assist in obtaining the proper protein fold, for example peptidyl prolyl *cis*/*trans* isomerases (PPI's). For oxidative disulfide formation, disulfide oxidoreductase (DsbA) and disulfide isomerase (DsbC) can be added (or co-expressed). Protein disulfide isomerase (PDI) is a eukaryotic protein that catalyzes both protein cysteine oxidation and disulfide bond isomerization, and it also exhibits chaperone activity. Disulfide bridge formation can also be achieved and controlled by glutathione redox buffers or addition of disulfide bond isomerase.

To efficiently express membrane proteins, an *E. coli* based CF system can be used when the reaction is supplied with synthetic lipids. Detergents can also be added to reduce aggregation and increase solubility.

The advantages of the CF system come at a price (in the verbatim sense). The reactions are expensive, and expression levels of mg ml^{-1} cannot always be achieved. Technically the constant supply of nutrients and energy to the cell-free system is quite challenging. Nonetheless, steady progress is made and

Box 4-7 **Selection of expression host systems.** Structural studies by crystallography not only require soluble protein but also conformationally homogeneous material. Solubility as well as the demand for conformational purity requires careful selection of a suitable expression host. Advanced cloning techniques including ligation independent cloning, topoisomerase cloning, and Gateway cloning allow switching of expression systems with relative ease. Expression systems consist of a suitable transfection vector and corresponding host cells, ranging from prokaryotic bacteria to eukaryotic cells such as yeasts, insect cell lines, and mammalian cell lines. *E. coli* based systems are simple and the most common, but are incapable of eukaryotic posttranslational modifications and generally have difficulty expressing properly folded proteins containing multiple disulfide bonds. Yeasts possess cell compartmentalization and the complete eukaryotic secretion and posttranslational modification machinery. Glycosylation patterns are different for yeasts and mammalian cells and some yeasts hyperglycosylate. Glycosylation of mammalian proteins is a major source of conformational inhomogeneity and glycosylation-deficient host cell lines exist. *In vitro* cell-free transcription–translation systems allow full control over the expression conditions, including addition of folding chaperones.

some commercial systems not only eliminate the time-consuming steps of cell-based protein production such as transformation, cell culture maintenance, and expression optimization, but also have a relatively simple tube reaction format eliminating the requirement for specialized instruments (at least in small-scale reactions). Small-scale trials have shown that the *in vitro* CF systems are complementary to standard cell-based expression systems, that is, some proteins that fail in the host cells can still be expressed in CF systems and *vice versa*.[46] In any case it will be worthwhile following the technical development of CF systems, particularly when difficult, high impact proteins justify the investment.

Fusions and tags for expression

Fusion constructs consist of the target protein sequence linked with another sequence. The fusion partner may serve different or multiple purposes. Tags are generally small fusions that serve as affinity capture labels in affinity chromatography. In contrast to regular column (distribution) chromatography such as size exclusion chromatography (SEC), an affinity capture step concentrates the protein. Larger fusion tags are introduced with the (additional) aim of providing increased expression levels and solubility (the two effects are not easily separated). The most commonly used small peptide tags are FLAG-, c-myc-, S-, and Strep II-tag or poly-His and (C-terminal) poly-Arg repeats. Larger fusions with intact proteins or domains such as MBP or GST may also serve as (less efficient) affinity tags but primarily serve to increase the protein solubility. Some, such as the NusA tags, serve only as effective solubility enhancers. Tags at the C-terminus assure that no incomplete translation products (recognizable as a ladder in the SDS-PAGE gels) are captured and they are also used in dual tag systems.

As discussed in Chapter 3, no convincing statistics or predictions are available as to when and whether an expression tag is hampering crystallization. Even quite substantial affinity tags—largely N-terminal—have not prevented crystallization, probably because N-termini often tend to be disordered and not involved in crystal packing. As subcloning and expression testing are relatively cheap and readily performed, pursuit of multiple constructs with varying tags, as well as different lengths and truncations, is a valuable and efficient parallel design strategy to obtain more soluble or better crystallizable protein.[38, 80] Tags generally do not affect the conformation of the rest of the protein, with the exception of a few terminal residues adjacent to the affinity tag. It is common practice to engineer a protease cleavage site into the linker between tag and protein (Figure 4-12), which allows crystallizing the protein with or without the tag.

A recently discovered class of fusion peptide tags which autonomously insert themselves into the membrane are discussed in the membrane protein section.

Small peptide tags

His-tags. The most popular tags for crystallographic studies are poly-His repeats. They come in varying lengths (the His_6-tag is the most common) and with various linkers. The His residues chelate around the Ni or Co ions of the stationary phase nitrilotriacetic acid (Ni-NTA) resin of an immobilized metal affinity chromatography (IMAC) column. The function clearly implies that the His-tag must not be buried and must extend from the protein, and longer His-tags (deca-His-tags) and longer linkers can improve poor affinity. His-tags can be attached either to the N-terminus or to the C-terminus of the construct. C-terminal tags have the advantage of being present only in full-length constructs and antibodies against the tag allow rapid detection of the total expression of the complete construct even in crude cell lysates (Figure 4-13).

Figure 4-12 Generic design of a tagged fusion construct. The protein construct contains an affinity tag for easy affinity capture with a shorter linker to a solubility enhancing fusion protein finally linked to the target sequence. The cleavage site is located right before the N-terminus of the target sequence to leave as few residues as possible behind. The Gateway p-DEST-HisMBP vector is a typical example,[81] with the TEV cleavage site providing for simple IMAC purification when a His-tagged TEV protease is used (Figure 4-13).

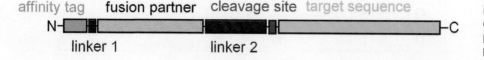

affinity tag   fusion partner   cleavage site   target sequence

N—[   ][ ][   ][   ][   ]—C

linker 1   linker 2

Site-specific viral proteases. An attractive protocol is the use of a His-tag that has a cleavage site for the very specific and efficient Tobacco Etch Virus (TEV) protease designed into the linker. The 27 kDa catalytic domain of TEV protease[82] (which incidentally crystallized with its His_6-tag clearly visible) recognizes a linear epitope of the general form E-X-X-Y-X-Q-G(S) with cleavage occurring between Q and G or Q and S. The most commonly used sequence is ENLYFQG, with only the single G residue of the cleavage epitope left behind. In combination with a His-tagged TEV protease, a simple two-step, one column protocol can provide nearly pure, untagged protein. The His-tag can be combined with an additional, subsequent solubility-enhancing fusion tag such as maltose binding protein[81] (MBP) or NusA[83] to enhance solubility. In a typical protocol the lysate is spun down and the supernatant is loaded on an IMAC-Ni-NTA column. The column is washed and the His-tagged target is eluted with imidazole. The protein is then incubated with His-tagged TEV, which cleaves the tag(s) and linker from the target, and the mixture is again loaded onto the IMAC column. The cleaved His-tag and the tagged TEV protease stay on the column, while the tag-free target protein with the N-terminal G from the cleavage site flows through. Various tagged TEV protease variants (some engineered by directed evolution)[84] and suitable vectors and expression systems are commercially available.

A similar protease from Tobacco Vein Mottling Virus (TVMV) has the commonly used cleavage sequence ETVRFQS and is completely specific with respect to the TEV sequence. A combination of both TEV and TVMV protease sites therefore offers selective removal of two tags.

Thrombin. Thrombin is a less sequence-selective serine protease which is fairly commonly used in protein expression and purification. Its advantage is high processivity (about 1000-fold faster than TEV) but it has the slight potential to cleave proteins at spurious sites other than the intended one. The problem that gave thrombin a somewhat bad reputation generally results from impure thrombin preparations and can be minimized by the use of highly purified protease, available from several commercial suppliers. Active thrombin is (for obvious reasons) very hard to obtain recombinantly and is instead produced and must be highly purified from bovine serum. A typical sequence used for thrombin cleavage is LVPRGS with the cleavage after the arginine residue.

Strep tag. The Strep tag exemplifies a short peptide tag of eight amino acids (WRHPQFGG) originally selected from a genetic random library. The Strep tag binds reversibly and specifically to the same pocket of streptavidin where the natural ligand D-biotin is complexed. It thus provides efficient purification of Strep-tag fusion proteins on affinity columns with immobilized streptavidin. The Strep tag/streptavidin system was systematically optimized over the years

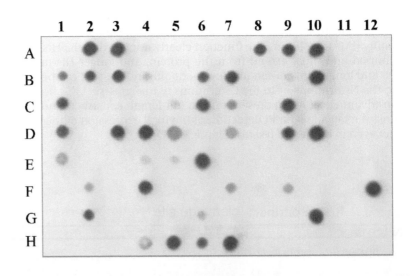

Figure 4-13 **His-tag antibody Western blot of C-terminally His$_6$-tagged proteins.** A stained immuno-blot of lysates from a small scale, parallel expression experiment in 96-well format allows rapid identification of colonies that have overexpressed a large amount of full length constructs (intense dark spots). Figure reproduced[46] with permission from Springer Verlag.

particularly for use in protein production for X-ray crystallography, resulting in the optimized Strep tag II (WSHPQFEK) and the engineering of a variant streptavidin with improved peptide-binding capacity, dubbed Strep-Tactin. The corresponding Strep-tag system provides a useful tool in recombinant protein production.[85]

BAP tag. The biotin acceptor peptide (BAP) is a naturally occurring peptide sequence that undergoes biotinylation on the central lysine residue. In *E. coli* only one protein is biotinylated to any appreciable amount and therefore the BAP tag offers an incredible advantage of achieving a tremendous purification factor in one single affinity step. The two main disadvantages of this tag are its size and the need to co-transform overproducing *E. coli* cells with a biotinyl transferase (BirA) vector because the natural abundance of this enzyme is low.[86]

Larger fusion tags and reporter systems

Fusion tags. Fusion tags consisting of larger peptides or proteins are used—often in combination with affinity capture tags—to improve expression levels and assure proper folding, which prevents aggregation during expression and ultimately increases the solubility of the protein. In many cases the target protein fortunately remains in solution upon enzymatic cleavage of the fusion partner. The maltose binding protein (MBP),[87] thioredoxin (TRX),[88] glutathione S-transferase (GST, which is also available as a monomeric fusion construct in contrast to the native dimer),[89] and NusA[90] are only a few examples. Many of the commercially available fusion tag vectors are combined with a poly-His affinity tag, and some also contain a TEV cleavage site[81] for easy tag removal. Combinations of N-terminal solubility and expression enhancing tags with C-terminal affinity tags for example assure that the entire construct has been expressed.

Reporter tags and systems. Another type of fusion construct is reporter systems that make it easy to track the protein. Particularly for parallel approaches where large numbers of constructs must be screened, colored and/or fluorescent tags that can be detected using plate readers are of advantage. One of the most popular tags is green fluorescent protein (GFP),[91] and certain engineered GFPs may also indicate the folding state of the fused target protein.[92] Unfortunately fusion with GFP can also cause some proteins to aggregate and therefore one must be careful in interpreting the results of such trials—it is particularly suspicious if all samples in a parallel experiment test negative for expression and solubility. Figure 4-14 shows a good example of a laboratory-scale GFP-tagged expression screen.

Intein fusion systems. Inteins are naturally occurring protein splicing elements that have been engineered into small, self-splicing elements so that they can catalyze peptide cleavage at one of its termini and so separate the target gene from any upstream or downstream fusion partners or affinity tags.[93] The cleavage reaction is initiated by addition of thiols, or pH and temperature change. Intein vectors are available that contain a variety of affinity, fusion, and reporter genes such as GFP.

HaloTag. A new fusion tag has been recently proposed on the basis of haloalkane hydrolase. This enzyme has been engineered (catalytic histidine substituted by phenylalanine) to covalently bind haloalkanes and therefore can be used to capture halo-tagged proteins on special resins. The disadvantage of this tag is that the interaction is irreversible and therefore proteins must be released from the substrate by proteolytic cleavage.[94]

Co-expression

In certain cases function and even proper folding, and correspondingly any significant expression levels, depend on the presence of a binding partner in a protein complex. *In vitro* reconstitution of the complex is then generally not possible and the complex partners need to be co-expressed. Molecular chaperones or the disulfide bond forming enzymes discussed above can also be co-expressed with the target protein to achieve proper folding and disulfide bond

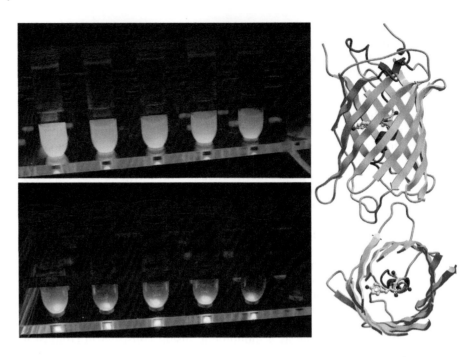

Figure 4-14 Green fluorescent protein (GFP). GFP is a spontaneously fluorescent 238 amino acid protein isolated from the Pacific jellyfish, *Aequorea victoria*.[91] Its biological function is to transduce via energy transfer the blue chemiluminescence of another jellyfish protein, aequorin, into green fluorescent light. The fluorophore originates from the internal S66-Y67-G68 sequence (shown inside the β-barrel in ball-and-stick presentation) which is posttranslationally modified to a 4-(*p*-hydroxybenzylidene)imidazolidin-5-one chromophore structure. The images of the test tubes show both supernatant and cell pellets (bottom rack) containing expressed GFP-fusion protein at various levels. The rightmost tube reveals no expression at all. Note that the leftmost sample contains an engineered cyan fluorescing construct of GFP. Image of fluorescent samples courtesy of Gunter Stier, MPI Heidelberg, Germany.

formation. Other examples of co-expression include rare tRNAs, modifying enzymes (kinases, phosphatases, etc.), and binding partners such as low affinity binding peptides.

Co-expression can be achieved from separate vectors or on the same vector under the same promoter. To achieve the close proximity of the molecules when leaving the ribosomal synthesis machinery, they generally need to be co-located on the same, polycistronic vector. Such vectors and cloning kits are available for various expression systems, and also for the Gateway and Duet® system.

General remarks regarding protein purification

Protein purification follows a general scheme of lysis, capture, purification, polishing, and buffer exchange and concentration (Figure 4-1). Which steps are required and the specifics of each purification step are dictated by the properties of the expressed target protein construct; the protocols and strategies for crystallography generally do not differ from the standard biochemical procedures described in the literature and are thus not discussed here. However, significant points specific for protein production for crystallography have been discussed in the crystallization section—particularly the role of:

- Affinity tags.
- Buffers, salts, and additives.
- Batch variation and contaminants.
- Purity and freshness of the protein.
- Protein concentration.

In view of the last point it is helpful to distinguish between purification steps that concentrate the protein (such as affinity capture steps) and steps based on distribution chromatography, which generally dilute the protein. Protein concentration requirements are generally higher for crystallography than for biochemical studies, simply because crystallization depends on the ability to achieve supersaturation. If the protein concentration is too far from its solubility maximum, supersaturation may be difficult or impossible to reach, and the large majority of drops will remain clear (salvage strategies such as adding more of a hygroscopic precipitant to the reservoir have been discussed in Chapter 3). As a

Box 4-8 **Affinity tags and fusion partners.** Recombinant DNA techniques allow adding or fusing a number of useful sequences to the target sequence. The most common fusion partners are small affinity purification tags and solubility enhancing fusion partners as well as complex partners or chaperones. Reporter genes can be fused to constructs allowing rapid assaying of solubility or folding status as well as cellular location of the target protein. Engineered GFP is one of the most common reporter genes and is frequently used in combination with affinity purification tags.

logical consequence, no fixed number for the protein concentration necessary for crystallization can be given, but a concentration not too far below its solubility limit is required. The absolute values can be as low as 1 mg ml^{-1} and reach up to 100 mg ml^{-1} or more for highly soluble proteins. The necessity for low buffer concentration so that the crystallization cocktail can actually drive the pH has been discussed together with other crystallization related aspects in Chapter 3.

Labeling of proteins, Se-Met

Protein expression in suitable hosts and growth media allows incorporation of atomic labels. In particular Se-Met incorporation for SAD or MAD phasing is a generally applicable and powerful site-specific labeling technique delivering proteins that are highly isomorphous to the native material (Sidebar 4-7). Se-Met labeling is possible in bacteria, yeasts, insect cells, and even in certain mammalian cell lines.

Overexpression in auxotrophs. In bacterial hosts, two general methods exist to incorporate Se-Met: overexpression in a methionine auxotroph (Met−) strain in specially prepared minimal media; or metabolic inhibition in the commonly used regular expression strains. The expression in Met− strains—commonly B834(DE3)—follows a method initially developed by LeMaster for the production of NMR labeled proteins.[95] For the Met− expression method, a methionine-free minimal medium must be prepared (M9 less Met) with glucose and succinate as carbon sources and the usual vitamin, nucleotide, and mineral supplements.[96] The minimal medium is then augmented by 50–60 mg ml^{-1} of Se-Met before inoculation. Note, however, that it is not necessary to use the fully synthetic medium in order to grow the starter culture. Regular LB or TB can be used instead of the Se-Met augmented minimum medium to grow the cells to sufficient density, followed by spinning the cells down and then transferring them then into the synthetic minimal medium. Inevitably, cell growth will slow down and expression yield will be reduced compared with the same process in regular medium. The benefit of the method is that there is in practice no doubt that all methionine residues are in fact Se-Met and partial substitution rarely occurs. Mass-spectroscopic analysis to verify the Se-Met incorporation is generally not necessary.

Metabolic inhibition. The alternate Se-Met protein expression method is based on metabolic starvation or inhibition. The methionine biosynthesis in bacteria can be inhibited by high concentrations of Ile, Lys, and Thr by inhibition of aspartokinase as the first step in Met biosynthesis. The inhibition effect is further enhanced by Leu and Phe. The procedure, therefore, is to grow the cells up to mid-log phase in regular LB and then add the inhibitory amino acids and Se-Met to the medium about 15 min before induction. Cells grow faster to normal density and the yields are better than in Met− strain expression, but there can be a small percentage of partially labeled protein present. Should rich medium be necessary for cell growth and protein production, a special synthetic medium can be made. As long as diffracting crystals grow, partial Se-labeling is not a problem and a mass spectrum can quickly assess the actual degree of Se incorporation. The metabolic inhibition method can also be used in combination with autoinduction[63] as described in the previous section. In common practice this method has essentially replaced use of auxotrophic strains. Note that in principle any other selectively labeled amino acid can be incorporated in a similar fashion.

Phasing power and Se-Met engineering. Each ordered Se atom generally provides enough anomalous signal to phase up to 120 residues (Chapter 6). With an average natural abundance of 1 methionine in 60 residues there is a good chance that enough Se atoms will be present in the target protein. If this is not the case, site directed mutagenesis can be applied to engineer additional methionine residues into the protein. The logical choices for mutation targets are hydrophobic residues of comparable size such as leucine and isoleucine, which is also reflected in the mutation probability (substitution) matrices. The site-directed mutagenesis is generally carried out as described in Sidebar 4-3.

Sidebar 4-7 How isomorphous are Se-Met proteins? A justifiable question is how isomorphous crystal structures of Se-Met labeled proteins are to their native counterparts. As the methionine residues generally do not participate in molecular chemistry, their effect should be largely limited to local variation in stereochemistry. Because of the difference in covalent radii (~1.05(3) Å for S and 1.20(4) Å for Se) the primarily affected stereochemical parameters are the Cγ–Se and Se–Cδ bonds and the enclosed bond angle. The parameters and their standard uncertainty (SU) extracted from *REFMAC* restraint files (Chapter 12) are listed in Table 4-1.

Macromolecular structures, given their flexible and fluid character, can generally accommodate changes on the order of ~0.15 Å well by slight rearrangements and absorbing them in a limited local area. The case of apolipoprotein E3 illustrated in Figure 4-15 shows that even in worst case scenarios isomorphism is maintained in the case of Se-Met labeling. However, this does not necessarily hold for minor changes in discrete solvent structure and potential ligands caused by practically unavoidable variation in the crystallization medium or procedure, as demonstrated in Chapter 12 in the refinement example of the Se-Met structure of the *E. coli* acyl-CoA thioesterase II (tesB). Nonetheless, even a low-resolution Se-Met model is nearly always useful as a (MR) phasing model for the corresponding native high resolution structure.

				dist(Å)	SU(Å)
MET	CG	SD	coval	1.803	0.025
MSE	CG	SE	coval	1.950	0.030
MET	SD	CE	coval	1.791	0.025
MSE	SE	CE	coval	1.950	0.030
				angle(°)	SU(°)
MSE	CG	SE	CE	98.923	2.200
MET	CG	SD	CE	100.900	2.200

Table 4-1 Bond lengths and bond angles for Se-Met and native S-Met.

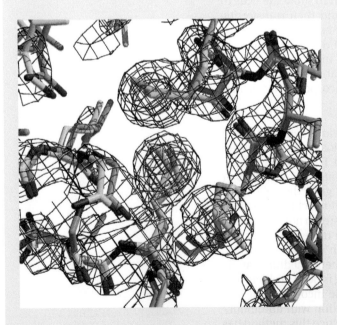

Figure 4-15 **Isomorphism between selenomethionine mutants and native methionine crystal structures.** The figure shows a superposition of a Se-Met structure (yellow sticks, in its electron density) and its native counterpart (orange sticks). Shown is the crystal structure core containing three Met residues of the apolipoprotein E3 (ApoE3) 22 kDa N-terminal domain. The situation represents a worst-case scenario: There are three Met in close proximity; the independently refined Se-Met structure crystallizes in a different crystal form (probably because of absence of 2-β-mercaptoethanol (2BME) in the crystallization cocktail); and the structure itself is a flexible helix bundle. Nonetheless, the local coordinate r.m.s. deviations for the Met residues (0.4 Å) are on the order of coordinate uncertainty of the individual structures (0.28 and 0.44 Å respectively). PDB entries 1bz4 and 1or2.[96]

A method for increasing the anomalous peak signal is oxidation of the Se atoms. Note that increase of the white line peak signal means that the diffraction experiment must be conducted at a precisely determined wavelength, that is, an absorption edge scan is required. The important point seems to be that the oxidation—generally by brief incubation in 0.1% hydrogen peroxide followed by dialysis—is homogeneous and full. As mentioned before, partly oxidized Se-Met samples are presumably locally inhomogeneous or disordered, which in fact weakens the usable anomalous signal. Partial oxidation and conformational (micro)inhomogeneity (together with higher susceptibility to radiation damage) may also be responsible for the generally reduced diffraction limits of Se-Met crystals compared with crystals grown from native material. The practical protocols for various Se-Met labeling procedures and additional references can be found in a review by S. Doublie.[1]

Membrane protein expression and purification

Let us recall from Chapter 2 why membrane proteins pose such difficulty for structural studies. The reason integral membrane proteins—for which the cytosolic domains or ectodomains cannot be separately expressed—are difficult to produce and to crystallize lies foremost in their tight association with the cellular membranes that are required for their stability. Their hydrophobic transmembrane helices contacting the fatty acid tails of the phospholipids inside the membrane contain a significant number of hydrophobic residues, which become unstable in polar, aqueous media and tend to aggregate.

The expression challenge

As a consequence of their tight membrane association, the vast majority of integral eukaryotic membrane proteins cannot be routinely expressed through heterologous overexpression in bacterial host systems, because they rapidly form inclusion bodies during expression. Only in 1998 was the first recombinantly produced membrane protein crystallized, the K+ potassium ion channel.[97] Even in this case, the bacterial KcsA gene from *Streptomyces lividans* was expressed in *E. coli*, itself a bacterial host system. Since then, a number of other bacterial membrane proteins have been expressed in *E. coli*. Slow expression in strains especially suitable for membrane protein expression, the co-expression of chaperones assisting in folding, and low expression temperature reducing the formation of inclusion bodies, have contributed to the success.[98]

The principal limitation to expressing large quantities of membrane proteins is the fact that the volume of the cell membrane available for insertion is orders of magnitude smaller than that of the cytoplasm. In addition, overexpression of membrane proteins generally compromises membrane integrity even in homologous host systems. These and other factors drastically limit even "good" expression yields of membrane proteins to the 100s of $\mu g\ l^{-1}$ range. Eukaryotic expression systems such as insect cells or HEK cells have been used, but cell culture is not trivial and yields are low. Yeast expression systems, in particular *Pichia pastoris* also seem to be suited for eukaryotic membrane protein expression. A further complication for crystallization arises from the fact that many membrane proteins expressed in eukaryotic systems can be decorated with inhomogeneous posttranslational glycosylations, as discussed in Section 4.6.

Box 4-9 **Selenomethionine labeling.** Incorporation of Se-Met instead of regular methionine residues provides heavy atom labels for anomalous phasing while maintaining structural isomorphism to the native protein. One method of incorporating Se-Met is by expression in a methionine auxotroph host strain and supplementing the minimal medium with Se-Met. In the metabolic inhibition technique bacteria are grown in regular medium and Met biosynthesis inhibiting amino acids Ile, Lys, and Thr as well as Se-Met are added to the medium before induction.

Few mammalian membrane proteins have been successfully expressed in bacterial systems and subsequently crystallized. The notable exceptions are certain GPCR receptors, where the neurotensin receptor (NTR) has been expressed as MBP-NTR-thioredoxin double fusion protein in *E. coli*.[99] The recent advent of Mistic fusions (Sidebar 4-8) may provide in addition to antibody scaffolding (Chapter 3) a more general means of mammalian membrane protein expression in bacteria.

Improvements of cell-free *in vitro* expression systems described in the previous section also make them quite suitable for membrane protein expression. In particular, the options of adding lipids as well as chaperones to the mixture and the absence of complications caused by saturation of the host cell's translocation machinery, as well as toxicity and cell lysis issues, highlight *in vitro* expression systems as a potential alternative for membrane protein expression.

Solubilization and purification

Membrane proteins are either resolubilized from inclusion bodies (less common and so far limited to relatively sturdy bacterial outer membrane proteins) or extracted from the expression host membranes with the help of various detergents (Figure 4-17). With their hydrophobic tail or part, detergents associate with the transmembrane parts of the protein and form a detergent collar around the hydrophobic transmembrane stem, which keeps the protein stable during extraction from the membranes.

The presence of detergents necessary for the maintenance of structural stability has a number of significant consequences for membrane protein handling. During purification, care must be taken that the buffers contain enough (often expensive) detergent so that the micelle collar remains intact. In addition, the crystallizer is now faced with an additional parameter that needs to be adjusted

---

Sidebar 4-8 Mystic Mistic. The recent discovery of Mistic (Membrane Integrating Sequence for Translation of Integral membrane protein Constructs) has opened another promising avenue for membrane protein expression. Mistic is an unusual *Bacillus subtilis* dual-topology protein that folds and inserts autonomously into the membrane forming an integral four-helix bundle membrane protein (Figure 4-16).[100] In addition, Mistic lacks any signal sequence and bypasses the cellular translocon complex, and therefore provides a highly suitable fusion partner for expression of eukaryotic membrane proteins.

The auto-inserting ability of Mistic together with the bypassing of the cellular translocon has raised the probability that more structures of crucial membrane proteins of therapeutic interest, such as ion channels and 7-TM G-protein coupled receptors (GPCRs), will become available soon. More than 1000 GPCRs exist, which are involved in regulation of a wide variety of biological functions, and many of them are drug targets. Another avenue to expression of GPCRs via antibody scaffolding[101] was outlined in Chapter 3, and the scaffolding via the insertion of a T4-lysozyme mutant into a flexible loop of the GPCR structure.[102]

Similar to the idea of using Mistic as a transport and integration fusion, members of the bacterial Tol proteins[103] export a wide range of molecules across the periplasmic space and outer membrane of Gram-negative bacteria, and might provide a further avenue for membrane protein expression in bacteria. An engineered version of the anti-apoptotic Bcl-2 family protein Bcl-XL has also been used in a fusion system for membrane proteins.[104]

A good resource for membrane protein production and structure is Steven White's web page: http://blanco.biomol.uci.edu/Membrane_Proteins_xtal.html

Figure 4-16 NMR solution structure of Mistic. The ribbon diagram shows the lowest energy conformer of the irregular four-helix bundle of Mistic. The lipid facing surfaces are unusually polar, and the presence of a concentric ring of solubilizing LDAO detergent molecules was deduced from NMR NOE interactions. PDB entry 1ygm.[100]

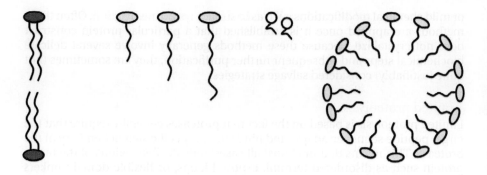

Figure 4-17 Phospholipid, detergents, amphiphiles, and detergent micelle. Phospholipids comprise the majority of the fluid bilayer cellular membrane. A glycerol molecule is linked to two C16 to C18 fatty acids, forming the hydrophobic tail. The third position of the core glycerol is linked to a phosphate group and choline or other polar groups symbolized by the red oval. Detergents similarly have a polar group (blue oval) and shorter hydrophobic tails of varying length. Mild ionic, zwitterionic, and non-ionic detergents are used in membrane protein crystallization. Amphiphiles (white head groups) are small detergent-like molecules that can be further used to adjust the size of the detergent collar around the transmembrane stem of a membrane protein. With increasing concentration, detergents form spherical micelles (right panel). The concentration at which micelles and free detergent are in equilibrium is called the critical micelle concentration or CMC.

during crystallization. Not only must crystal contacts between the hydrophilic, solvent-exposed ectodomain and cytosolic domain be formed, but also the detergent collar has to have a size that allows packing of the molecules. Specific contact-forming interactions between the polar detergent head groups on the outside of the membrane collar are generally absent, largely because the detergent micelle collar, like the cell membrane, is quite fluid. The principal methods of adjusting the micelle collar are detergent exchange and fine-tuning of the micelle collar size with amphiphiles (small polar–apolar molecules similar to detergents). Frequently, detergent molecules (or parts of membrane lipids) are visible in electron density (Figure 4-18)

Box 4-10 **Membrane protein expression.** Because of their intimate association with the lipid molecules of cell membranes, integral membrane proteins are difficult to express and to solubilize. Expression levels of membrane proteins are generally low as a result of overload of the cellular translocation machinery as well as compromised cell membrane stability. The necessity to persistently stabilize the hydrophobic transmembrane part of the protein with lipids and detergents poses additional difficulties for purification. In addition, membrane protein preparations are often unstable—incessant and vigilant quality control is a must.

4.6   Engineering strategies at the protein level

In addition to the DNA-level strategies of genetically engineering the target construct sequence, the already expressed target protein can also be modified in order to render it more amenable to crystallization. The modifications are generally enzymatic digestions as in the case of limited proteolysis or deglycosylation,

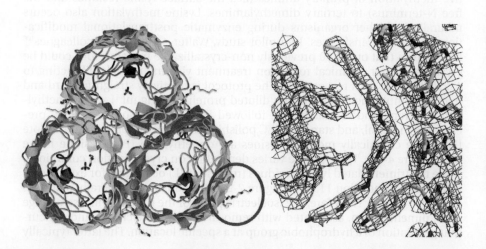

Figure 4-18 The trimeric bacterial outer membrane protein OprP. Published in 2007, the OprP structure revealed a new structural motif in bacterial membrane proteins, a 9-residue arginine ladder. OprP controls the transport of essential phosphate anions into the pathogenic bacterium *Pseudomonas aeruginosa*. The Arg ladder extends from the extracellular surface down to a constriction zone where a phosphate anion is bound (visible in the center of the channels). Note that the crystal structure again contains detergent molecules along the transmembrane barrel, in this case hydroxyethyloxytri(ethyloxy)octane or C8E. The right panel shows the region around the circled detergent molecule in electron density. In this position, the detergent in addition mediates a crystal contact to a symmetry related molecule (magenta). PDB entry 2o4v.[105]

or mild chemical modifications such as in surface lysine methylation. Often these methods are applied once it is established that a particular protein construct does not crystallize. Because these methods generally involve several delicate biochemical steps and subsequent further purification, they are sometimes (not quite justifiably) considered salvage strategies.

Limited proteolysis

Limited proteolysis is based on the fact that proteases generally require that the target peptide sequence adapts and binds to the specific stereochemistry of the protease. This means that in almost all cases, only flexible regions of the target protein such as disordered termini, exposed loops, or flexible domain linkers of the protein are subject to attack by various proteases. This circumstance has been utilized for domain analysis and in biochemical folding studies.[39]

For crystallization, limited proteolysis provides a generic stochastic design approach which can be applied either *a priori*, or as (more commonly) a salvage strategy should the protein fail to crystallize. Removing flexible parts generally generates more compact and conformationally homogeneous molecules, and the same holds for splitting flexibly linked multi-domain molecules into more compact single domains. Human apolipoprotein E for example yielded well-diffracting crystals only after the lipid binding, C-terminal 10 kDa domain and a substantial part of the potentially disordered linker region to the N-terminal 22 kDa LDL receptor binding domain had been removed.[96] The domain structure was initially detected by thrombin cleavage at residues 191–192 in material isolated from plasma. After the initial crystal structure showed no evidence for residues beyond 165, the remaining C-terminal residues were also deleted from the cloning construct, which improved crystallization and led to different, well diffracting crystal forms. On the other hand, even large disordered terminal regions are not necessarily a reason to panic—a construct of the *Clostridium botulinum* neurotoxin A protease, for example, contains a disordered 37 residue N-terminal leader sequence consisting of a His_6-tag, an S-tag (S-peptide fragment of RNase A), and various linkers, and still formed crystals diffracting to 2.0 Å.

For limited truncation experiments for crystallization, small sample amounts are incubated with various proteases and the incubation time (~30 min) and temperature of the trials are varied. SDS-PAGE analysis followed by MS peptide identification will then show whether stable fragments have formed. A special variant of limited proteolysis is C-terminal truncation with carboxypeptidases A and B, which, for example, has been successfully used to digest away the C-terminal 10 to 15 residues of the T-cell receptor[26] and the ligand binding domain of the ionotropic glutamate receptor.[38]

Lysine methylation

A chemical side chain modification at the protein level that has been successfully used both to improve crystallization and as a rescue strategy, is reductive methylation of primary amines (i.e., the surface lysine residues and the free N-terminus) to tertiary dimethylamines. Lysine methylation also occurs naturally in higher organisms during enzymatic posttranslational modification via methyltransferases. In a pilot study, Walter, Grimes, and colleagues[106] have shown that out of 10 previously non-crystallizable proteins, four could be crystallized after chemical reduction treatment yielding crystals diffracting to 4, 2.8, 2.3, and 2.1 Å respectively. The protocol is relatively straightforward and requires simple incubation of the diluted protein (~1 mg ml⁻¹) with dimethyl-amine-borane and formaldehyde, followed by centrifugation to remove precipitated material, and standard SEC polishing followed by concentration. Note that these chemically modified lysines are dimethylated (PDB residue name MLY, Figure 4-19). Their charge varies depending on crystallization conditions, while the trimethylated lysine residues (residue M3L) occurring, for example, in calmodulin at position 115 (see Figure 2-27) or on histones generally maintain a positive charge. Therefore, the isoelectric point of the protein will also change upon dimethylation, associated with removal of a positive charge and concurrent generation of a hydrophobic group at a specific location. The latter typically

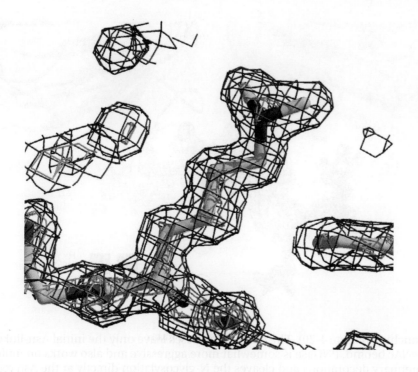

Figure 4-19 Chemically modified lysine residue (dimethyllysine). The electron density clearly shows that this surface residue of the alanine racemase of *Bacillus anthracis* is dimethylated (MLY). See Figure 2-27 for the density of a natively occurring, posttranslationally modified trimethyllysine (M3L) in calmodulin. PDB entry 2vd8.[107]

leads to a ~50% reduction in solubility, and again it is hard to determine which of the multiple change(s) actually represent(s) the dominating or causal factor for improved crystallization.

Glycoproteins

Covalent glycosylation is a frequent posttranslational protein modification of eukaryotic proteins that often causes complications during crystallization. The large and flexible, surface-exposed, concatenated, and branched oligosaccharide moieties can hinder the formation of crystal contacts, and entropic arguments similar to those suggesting surface entropy reduction can be made. In addition, glycosylations can be quite heterogeneous and also differ depending on the host organism, which adds to the complications. Glycosylation generally occurs in proteins that transit through the endoplasmic reticulum in eukaryotic cells.[108] Some mammalian mutant cell lines, however, are incapable of posttranslational glycosylation and thus offer a potential strategy for obtaining functional but unglycosylated proteins.[76]

Two common types of covalent glycosylations exist: N-linked (most frequent) and O-linked. N-linked glycosylation occurs at Asn in the tripeptide consensus sequence Asn-X-Ser(Thr), with X any non-proline residue, while the somewhat less complex O-linked glycosylation occurs at Ser and Thr. In addition, C-terminal glycosylphosphatidylinositol (GPI) anchors exist that bind proteins to the membranes (for example in the prion peptides). The most common monosaccharides forming the decorations are N-acetylglucosamine (N-acetyl-D-glucosamine, GlcNAc, or NAG) which is always the first sugar N-linked to Asn, followed by D-mannose, L-fucose, D-galactose, inositol, and others.

Enzymatic deglycosylation. Oligosaccharide decorations are initially formed by transfer to Asn of a 14-mer oligosaccharide that is then modified by a multitude of glycosyltransferases and glycosidases (which generates enormous variety in the glycoproteome). The enzymatic synthesis in turn allows the process to be reversed and the use of mild enzymatic deglycosylation in order to obtain a conformationally homogeneous glycoprotein (harsh chemical deglycosylation as used in some mass-spectroscopic analyses of glycoproteins is generally not suitable). Among the most commonly used enzymes to truncate N-glycosylations are various endoglycosidases (H and F1-3) which work on oligosaccharides with different mannose contents and have different preferences for the degree of

Figure 4-20 Glycosylated residues stabilizing homodimer. In quercetin 2,3-dioxygenase, a copper-containing enzyme that catalyzes the insertion of molecular oxygen into polyphenolic flavonols, a biantennary (Man-Man)$_2$-Man-GlcNAc-GlcNAc moiety (open circles mannose, full squares N-acetyglucosamine; see additional reading reference 9 for the Oxford glycan symbol notation) is covalently linked to Asn191 and forms a stable dimer contact. Enzymatic deglycosylation with EndoH was necessary to obtain diffraction quality crystals. Although EndoH truncates exposed oligosaccharides after the first GlcNAc, the biantennary sugar at Asn191 remained intact in contrast to the remaining four other oligosaccharide decorations. Image from PDB entry 1juh[112] courtesy of Roberto Steiner, Kings College London.

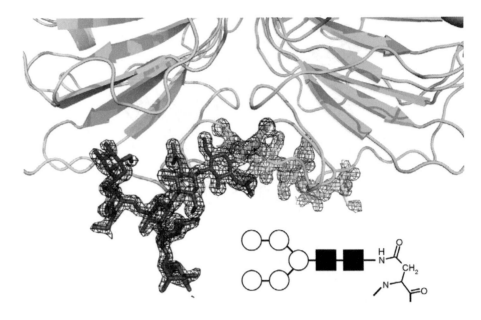

branching (Figure 4-20). The endoglycosidases leave only the initial Asn-linked GlcNAc behind. PNGase is somewhat more aggressive and also works on multi-antennary decorations and cleaves the N-glycosylation directly at the Asn residue, which then becomes oxidized to Asp. A typical example is the enzymatic deglycosylation of the extracellular domain of human neutral endopeptidase (neprilysin) expressed in *Pichia pastoris*.[109] PNGase is also used together with trypsin in protein digests for mass-spectroscopic assessment of glycosylation homogeneity.[110]

Glycosylations are not always detrimental to crystallization, as the case of quercetin dioxygenase in Figure 4-20 demonstrates. Another example is the T-cell receptor structure,[26] where a GlcNAc-(Fuc)-GlcNAc-Man oligosaccharide participates in crystal contact formation.

Site directed mutation and glucosidase inhibition. At the DNA level, the mutation of Asn in the consensus site Asn-X-Ser(Thr) to Asp or Gln removes the N-glycosylation sites entirely. Combinations of various mutations can be rapidly generated with point mutation kits and multiple constructs can be screened for expression, solubility, and crystallization. Another avenue that has yielded a crystallizable variant of angiotensin-1 converting enzyme (ACE) is overexpression in CHO cells in the presence of the glucosidase inhibitor N-butyldeoxynojirimycin (NB-DNJ) which reduced the glycosylation pattern to a single-site oligosaccharide.[111] The material retained activity and yielded crystals diffracting to 2.0 Å. Other alternatives for suppressing glycosylation are the addition of the glycosylation inhibitor tunicamycin to the cell growth medium, or expression in glycosylation-deficient host cell lines.

Box 4-11 **Modifications of the expressed target protein.** Once the target protein is expressed and purified, it can be further modified to improve crystallization. Limited enzymatic proteolysis can be used to remove disordered termini and for domain truncation. Mild chemical surface lysine methylation has been used to improve protein crystallization. Inhomogeneous glycosylation can be treated at the protein level by enzymatic deglycosylation or by expression in glycosylation-deficient strains, or at the DNA level by site-directed mutagenesis of Asn in the glycosylation consensus sequence to Asp or Glu. Modification of cysteine residues with alkylating agents (such as iodoacetamide) is a common protein modification technique and cysteine modification with heavy metals has been occasionally used for successful crystallization.[113]

4.7   Analysis of protein stability and conformational state

We have already discussed in Chapter 3 that stability and purity, in terms of molecular species and also regarding *conformational homogeneity*, are major determinants for successful protein crystallization: It is intuitive that differently shaped building blocks do not lend themselves to arrangement into well-ordered periodic assemblies such as protein crystals. While denaturing SDS gels provide a quick and easy assessment of molecular purity, conformational purity is harder to assess. Moreover, given that conformational purity is not an absolute determinant for crystallization success (about 20–30% of conformationally heterogeneous samples still crystallize)[114] and the analytical experiments usually take more material than a microcrystallization screening experiment; the most efficient experimental design is to screen first for crystallization with as little material as possible, and proceed to *post-mortem* physicochemical analysis once it is established that the material shows little indication of forming any promising crystals. Even if conformational heterogeneity itself is not the cause of failed crystallization, the analysis results at least provide a comforting rationalization for the failure and an incentive to do *something* to the protein or purification protocol, which may already change the outcome irrespective of any causal connection to conformational state.[38]

ThermoFluor stability assays

An elegant method of assessing the effect of buffers, additives, and cofactors on the stability and oligomerization state of a protein is by thermal shift fluorescence assays, or, in short, ThermoFluor (TF) assays.[115] A solvochromatic dye (hydrophobic fluoroprobe) that has a low quantum-yield in an aqueous environment is added to a protein solution. Upon undergoing thermal denaturing and unfolding of the protein, the dye binds preferentially to the now exposed hydrophobic residues, and the increase in fluorescence emission signal is recorded as a function of temperature. The experiments can be set up in 96-well array format in a plate reader combined with a thermocycler. Different salts, buffers, additives, detergents, or ligands are added to the protein, and from the change in the melting temperature those components or class of compounds that increase stability, and thus probably the crystallization success, can be determined.[116] The ThermoFluor method was initially designed for ligand binding studies and Figure 4-21 demonstrates its principle.

Native PAGE and size exclusion chromatography

Native (non-denaturing) polyacrylamide gel electrophoresis (PAGE) as well as size exclusion chromatography (SEC) are well suited for determining the

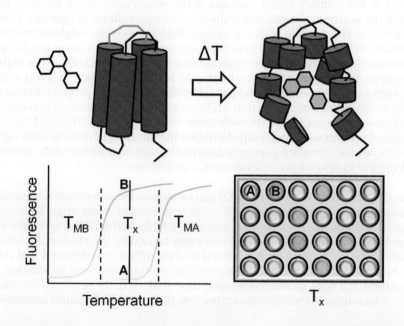

Figure 4-21 Principle of the ThermoFluor stability assay. Fluorescence of a hydrophobic fluoroprobe is quenched in an aqueous environment (indicated by the white molecule) but enhanced when it binds to exposed hydrophobic residues of an unfolded protein (green probe molecule). Increase in melting temperature measured by later onset of fluorescence thus indicates higher stability and additives in the protein solution that increase the protein's stability can be identified. The experiments can be elegantly carried out in 96-well format in an instrument combining a thermocycler and a fluorescence plate reader.

oligomerization state of a protein or the formation of homogeneous protein complexes. In cases of stable oligomers or complexes both methods deliver corresponding single bands or chromatogram traces from which the oligomerization state can be assessed, but the situation becomes more difficult when the oligomers are less stable with buried surface areas in the gray zone below about 700 Å² per molecule. In this case the chemical environment—generally the purification or elution buffer—does strongly affect the equilibrium between monomers and oligomers and it may never be possible to obtain one single oligomeric state. However, as crystallization cocktails generally are slightly chaotropic, a sample that shows an oligomerization equilibrium may still be suitable for crystallization. SEC is often combined with light scattering methods (described below) allowing further assessment of the size distribution and an estimate of the molecular weight for each chromatography fraction.

Light scattering methods

All scattering methods—including X-ray diffraction—are based on the fact that electromagnetic radiation is scattered by inducing polarization in the electrons comprising molecular matter. If no electronic transitions between energy levels occur, the oscillating electrons themselves emit light of the same wavelength and the scattering process is termed elastic. The Mie ratio of the particle radius r to the wavelength λ

$$x_M = 2\pi r / \lambda \tag{4-1}$$

determines the characteristics of the scattering process and we are interested in two limiting situations: In the case of $x_M \ll 1$ (scattering objects small compared to the wavelength) the scattering is isotropic and wavelength and particle size dependent, and is termed Rayleigh scattering (though Lord Rayleigh received the Nobel Prize in Physics in 1904 for the discovery of the element argon). The $x_M \ll 1$ limit is the case for analytical light scattering experiments, where visible light (~500 nm, 0.5 μm, or 5000 Å for green light) is isotropically scattered on protein molecules (with hydrodynamic radius in the 10–100 Å range) in solution (in diffraction experiments, Thomson scattering of short wavelength X-rays of ~1–2 Å takes place on electrons of individual atoms of similar ~1–2 Å size, occurring primarily in a forward direction as discussed in Chapter 6). The total scattering intensity I for the Rayleigh limit is given by

$$I = I_0 \frac{8\pi^4 \alpha^2}{\lambda^4 R^2}(1 + \cos^2 2\theta) \tag{4-2}$$

where I_0 is the primary beam intensity, λ the wavelength, R the detector distance, θ the scattering angle, and α the polarization volume (a measure of the interaction of the particle with the electric field vector). The angle-dependent expression in parentheses is a general polarization term as shown in Chapter 6. We immediately see from the equation that the scattering intensity is highly wavelength dependent (which is why the sky is blue) and given that α is a volume, even more dependent on the particle radius, ultimately proportional to r^6. The immediate consequence is that while we can use light scattering to measure particle size, one must work *extremely* clean—any large dust particles, debris, or aggregated protein will completely dominate the intensity of the scattered light. The samples are therefore generally filtered through nanofilters with minimal dead volume to remove particles above 0.1 or 0.02 μm.

The actual measurement (Figure 4-22) can be conducted in a classical, generally angle-dependent method called static light scattering or multi-angle light scattering (MALS), or by dynamic light scattering, DLS. In the DLS single detector method, fluctuations in the scattered intensity caused by the random Brownian motion of the particles (which is related to their diffusion coefficient and thus their hydrodynamic radius) are analyzed. The method requires, in addition to monochromatic light, an autocorrelation time analysis of the scattered intensity. The DLS technique is therefore also more descriptively called photon correlation spectroscopy (PCS) or quasielastic light scattering (QELS, because a minimal Doppler shift in the wavelength of the scattered light does take place).

Static light scattering and molecular interactions

Static light scattering can be used to derive the actual molecular weight M in extrapolation against zero angle and zero concentration by Zimm plot analysis.[117] It is used in modified form in MALS configuration by the WYATT DAWN instruments, for example. Reduced to single-angle extrapolation, which is possible given particles smaller than $\lambda/20$, we obtain

$$\lim_{c \to 0} \frac{K \cdot c}{R(2\theta)} = \frac{1}{M} + 2B_{22}c \tag{4-3}$$

where K is an optical constant[118] and R is the Rayleigh ratio (usually determined for a scattering angle of $2\theta = 90°$) between incident and scattered light intensity. If we measure R at different concentrations, we can extrapolate the Debye plot to $c \to 0$ and obtain the molecular weight from intercept $1/M$ and B_{22} from the slope of the graph (Figure 4-23).

Expression 4-3 is written in a form equivalent to a thermodynamic virial expansion (essentially a power series) in the particle concentration c that is truncated after the second term. The coefficient B_{22} then represents the second osmotic virial coefficient, which is a direct measure of the thermodynamic excess potential, that is, for (non-ideal) interactions between the molecules.[119] If B_{22} is positive, the excluded volume dominates, resulting in repulsive interaction, while a negative B_{22} indicates attractive interactions between the molecules.

Attractive interactions between molecules are of course necessary to form a crystal, but they should not be too strong, otherwise we will instead obtain a precipitate. We expect therefore that a modest degree of attractive interactions would favor crystallization. This relation ("crystallization window") (Figure 4-23) has indeed been established[120] and it does provide at least an explanation should crystallization have failed. However, keep in mind that as a thermodynamic property, slightly negative B_{22} is a *necessary*, but by no means *sufficient*, condition for crystallization—remember that crystallization is a local phenomenon depending on a multitude of specific and competing local interactions! This fact and the need for measurements at multiple concentrations for each specific cocktail makes static light scattering analysis not practically suitable as a *predictive* technique for crystallization success. The molecular weight estimates from MALS, however, tend to be more accurate than those obtained by DLS.

Dynamic light scattering and size distribution

Conformational heterogeneity frequently prevents protein molecules from self-assembly into the well-ordered crystals required for crystallographic studies. This

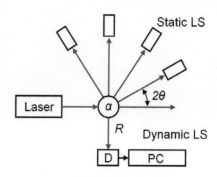

Figure 4-22 Principle of light scattering instruments. The top section of the figure illustrates a static multi-angle light scattering configuration (MALS) while the lower part shows a dynamic light scattering or photon correlation spectroscopy (DLS, PCS) setup. D indicates the detector, and PC the photon correlator (a fast electronic semiconductor circuit chip). Remaining symbols as used in Equation 4-2.

Figure 4-23 Determination of molecular weight and second osmotic virial coefficient from static light scattering. The left panel represents light scattering measurements of a given protein in four different cocktails as a function of the protein concentration and the corresponding extrapolation of Equation 4-3 to $c \to 0$. The red data points are obtained in the case of repulsive interaction (positive B_{22}) and the open symbols in the case of a near ideal system ($B_{22} \sim 0$). None of these conditions can ever lead to crystallization. The filled black symbols indicate strong attractive interactions (quite negative B_{22}); the protein precipitates at low concentrations beyond which no data can be measured. The green points represent a case of modest attractive interactions conducive to (but not necessarily sufficient for) crystallization. The right panel shows experimental B_{22} values for various proteins observed in the "crystallization window." Figures based on graphs in the review by Bill Wilson.[118]

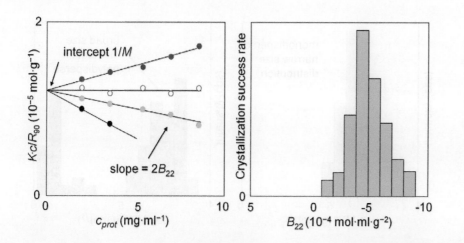

conformational heterogeneity results, to a large degree, from partial folding and unspecific association and aggregation of protein molecules that still remain in solution. A rapid means of accessing the size distribution (mono- and polydispersity), as well as a molecular weight estimate from the hydrodynamic radius, is by dynamic light scattering (DLS). The particles undergoing random thermal motion will move and tumble slower when they are large, which in turn changes the diffusion coefficient D from which, via the Stokes–Einstein relation (4-4), a hydrodynamic radius r_h can be estimated:

$$r_h = kT / 6\pi\eta D \tag{4-4}$$

where η is the viscosity of the medium. Given a shape correction function for non-spherical molecules, the molecular weight can be estimated from hydrodynamic radius r_h, although with a significant uncertainty.

The diffusion coefficient itself is derived from the decay constant Γ of the scattered intensity autocorrelation function (see Chapter 9 for examples of convolution and autocorrelation). Autocorrelation can be qualitatively understood as comparing the scattering function of the solution at consecutive times τ with the starting function at τ_0. The faster the particles diffuse, the more rapidly the scattering function changes and the autocorrelation decays rapidly. To work on the appropriate times cales, the computation has to be extremely fast and thus takes place in a hard wired autocorrelator chip. Once the data points for the experimental autocorrelation function are available, it is fitted with a second order exponential approximation

$$g(\tau) = B + \beta \exp(-2\Gamma\tau) \tag{4-5}$$

where B is the baseline of the function and β just the amplitude. The fit parameter Γ then yields the diffusion coefficient D and an estimate of r_h and M as well as the variance of these values. The monodisperse fit can be extended to polydisperse samples by introducing additional terms into (4-5). Figure 4-24 illustrates typical results from DLS analysis of size distributions in protein samples.

The information about particle size, its variance, and possible polydispersity is actually valuable information for crystallization. Investigations using light scattering methods have shown that there is a strong correlation (~70–80%) between crystallizability and conformational purity.[114] Although the value of these methods in post-mortem investigation of failed crystallization attempts is undisputed—through revealing a probable cause of the failure—they do not provide exclusive predictive guidance.[121] Nevertheless, instruments are reaching the market that can work in high-throughput format with 96 or 384 well plates and sample volumes of 20 µl, and instruments with miniature sample cells can handle sample volumes down to 2 µl.

Figure 4-24 Analysis of size distribution of samples by dynamic light scattering. A desirable outcome of a DLS experiment is shown in the left panel: the size distribution is narrow, and the sample is monodisperse. Notwithstanding other hindrances, the sample is likely to crystallize. The analysis of the scattering data shown in the right panels is less promising: The sample is polydisperse and in addition the two species have a rather large broad size distribution. The probability for successful crystallization is rather low (but crystallization is not impossible). Such size distribution histograms or similar representations are typically provided by DLS instruments.

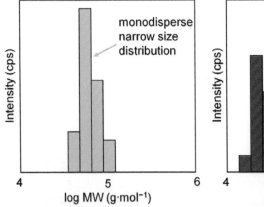

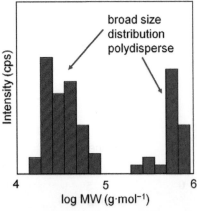

Some dynamic (as well as static) light scattering instruments can also be used in-line after size exclusion columns as additional powerful detectors. This is particularly valuable if oligomer equilibria exist and the elution peaks again contain multiple species with different aggregation states (the same holds for weak protein–protein complexes). SEC dilutes the sample and if the dissociation constant for an oligomer or complex is low, any concentration change will lead to a shift of the equilibrium during the subsequent concentration step. This situation is therefore generally bad news and either different buffer conditions must be found or protein engineering can be attempted to strengthen (or weaken) the oligomer or complex interface. An in-line light scattering experiment will immediately show if an elution peak indeed represents a monomeric species and at the same time provides a molecular weight measurement.

The actual experimental procedure (remember the stringent requirement for cleanliness) and analysis depend on the particular instruments used for the experiment. A useful practical protocol collection based on the DynaPro molecular sizing instrument has been compiled.[122]

Small angle X-ray and neutron scattering

In contrast to solution light scattering, the wavelength of the probing radiation is small in the case of X-rays and (thermal) neutrons relative to the entire scattering object. We thus can expect more detail in the scattering function than for visible light, despite the fact that scattering in solution occurs on randomly oriented molecules, without translational periodicity. The molecular scattering function is thus not discretely sampled in three dimensions, as is the case in X-ray diffraction on single crystals. Instead, the resulting small angle scattering curve can be best described as a spherically averaged molecular envelope or transform (the transform of a molecule is shown in Figure 6-16). Because of the rapid interference of the scattered X-rays with increasing scattering angle, the resulting solution scattering curve decays rapidly with scattering angle and can only be observed in the forward direction, close to the primary beam at very low scattering angles of a few degrees.

Enhanced contrast variation between solvent and molecule using H_2O/D_2O mixtures for neutrons and heavy atom solutions for X-rays can be used to improve the signal-to-noise ratio, but the principal difficulty of extracting 3-dimensional information from a 1-dimensional scattering function remains. The principal means of analysis is the interpretation of the logarithmic scattered intensity I_s as a function of the scattering vector (or momentum transfer) $s = 4\pi\sin(\theta)/\lambda$. From the Guinier approximation

$$I_s = I_0 \exp(-s^2 R_g^2 / 3) \qquad (4\text{-}6)$$

the overall radius of gyration R_g^2 can be determined from the linear part of the log (I_s/I_0) curve, and more sophisticated analysis of the entire scattering curve[123] allows further analysis of the shape function, generally by determining the best fit of various plausible models to the data. The solution conformation of large complexes can be analyzed elegantly when 3-dimensional X-ray or NMR structure models of some members or domains of the complex are available. Such fits with composite models provide a viable avenue to assemble molecular models of isolated parts or domains into a large complex structure in its solution conformation (Figure 4-25).

Circular dichroism spectroscopy and secondary structure

Circular dichroism (CD) spectroscopy or spectropolarimetry provides information about the secondary structure of proteins (and DNA) in solution. Different types of secondary structure, notably helix, sheet, and random coil, have different CD signatures. CD spectroscopy can thus be used for a variety of biophysical studies, including thermal and chemical denaturation/folding studies. It can be of value in troubleshooting protein crystallization to examine whether a soluble protein is unfolded or contains large stretches of random coil conformation, or to obtain a general estimate of the secondary structure composition. CD generally

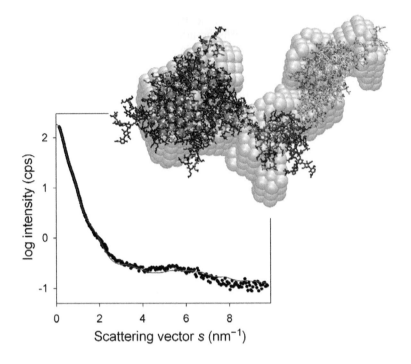

Figure 4-25 Analysis of small angle scattering. The linear part of the logarithmic plot of the relative scattering intensity (Guinier plot) provides an estimate for the overall hydrodynamic radius of gyration, while detailed analysis of the fall-off at higher scattering angles allows fitting of the curve with composite low-resolution models assembled from (partial) crystal structures. The corresponding resolution in Å can be readily calculated from $d = 2\pi/s$, that is, around 6.2 Å at the rightmost edge of the diagram ($s = 10$ nm^{-1}). The example shows the proposed model of the plasminogen-like overall solution structure of the multi-domain hepatocytic growth factor complex.[124] Figure courtesy of Dimitri Svergun, EMBL Outstation Hamburg.

will not provide information about the oligomerization state, and sometimes a protein cannot be represented by a simple linear combination of the three basic secondary structure types.

The most significant drawback of the method is that most of the discriminating information resides in the short UV wavelength region below 200 nm, where most of the organic buffers and detergents we prefer in crystallogenesis absorb strongly. Buffer exchange into weak phosphate buffer or similar may be possible. In addition, as the underlying physical effect of differential absorption of left- and right-hand circularly polarized light is minuscule; the instruments are correspondingly precisely built with UV optics and are relatively expensive. Some synchrotrons also provide extreme far UV CD spectroscopy lines.[125]

The physics of CD and related optical rotatory dispersion (ORD) is well documented[117] and needs only a brief review here. Circular dichroism is observed when optically active matter absorbs left- and right-hand circularly polarized light slightly differently. The result is that the resultant electric field vector becomes offset from the incident vector by a small angle and the resultant polarization becomes slightly elliptic. The differential absorption is expressed as molar dichroism ε, which is finally converted into mean molar ellipticity θ in units of deg · cm^2 dmol^{-1} per residue.

Box 4-12 Assessment of protein stability and conformational state. Protein stability and conformational homogeneity are of paramount importance for protein crystallization. The effect of buffer components and additives on protein stability can be rapidly assessed with ThermoFluor stability assays. Native SDS gels and size exclusion chromatography give a rapid but qualitative assessment of the oligomerization state of a protein. Detailed analysis of particle size distribution and polydispersity is possible using light scattering methods. Small angle X-ray and neutron scattering provide low-resolution shape and size information about the protein.

The spectra are analyzed by fitting them as a linear combination of standard spectra of polypeptides in different secondary structure conformations. The most common reference spectra were derived in the late 1960s for poly-L-lysine in (temperature- and pH-induced) helical, sheet, and random coil conformations.[126] Not every protein can be described as a simple linear combination of these three basic secondary structure types (not to mention additional chromophore interactions in packed proteins) and more sophisticated multi-component and empirical models exist which can provide modest improvement over the basic helix-sheet-coil linear combination.[127] Figure 4-26 shows the three basic reference CD spectra for poly-L-lysine and the result of a fit of myoglobin as a linear combination of these reference spectra.

Nuclear magnetic resonance spectroscopy

Nuclear magnetic resonance (NMR) spectroscopy is a powerful method for the determination of 3-dimensional solution structures of macromolecules.[128] Recent technical advances have made it possible to determine small- and medium-size macromolecular structures (< 30 kDa) and led to the 2002 Nobel Prize in Chemistry for Kurt Wüthrich from the ETH Zürich. The method and its use in routine structure determination is described in recent reviews, and just as in crystallography, it needs well-behaved (i.e. properly folded) proteins to deliver 3-dimensional structures. The fact that NMR can distinguish readily between folded proteins and detect the presence of unstructured regions or aggregates makes it inherently useful as a diagnostic tool for crystallization experiments. Modern instruments can extract within minutes information from small samples (~10 nM) and thus provide some indications about the conformational state of the macromolecules. Production of recombinantly expressed proteins spin-labeled with ^{15}N (and ^{13}C) for multi-dimensional NMR experiments has become relatively cheap.

1-dimensional experiments. A simple 1-dimensional proton (^1H) NMR spectrum showing the signal strength versus chemical shift (measured in ppm of the resonance frequency) already provides information about the conformational state. Because of the inverse relation between spin–spin relaxation time T_2 and peak width (slowly tumbling large molecules have long correlation lifetimes allowing fast spin relaxation and thus yielding broad peaks), large soluble aggregates will not yield an interpretable high resolution NMR spectrum with sharp resonances. In a slightly chaotropic crystallization cocktail such aggregated assemblies, however, may still yield crystals, provided the protein molecules themselves are folded. For non-aggregated molecules that do yield usable 1-dimensional NMR spectra, good discrimination in the backbone amide region below 8.3 ppm downfield, as well as peaks at around ~1 ppm upshift, indicate a folded protein.[129]

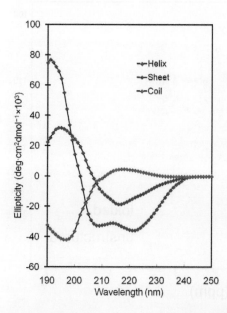

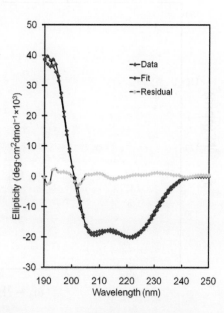

Figure 4-26 Reference CD spectra and fitted CD spectrum of myoglobin. The left panel shows the three basic reference spectra for helix, sheet, and random coil conformation as established for poly-L-lysine. The right panel shows an actual fit of a sperm whale myoglobin sample data against the standards (left) using the program CDFIT, predicting a total helix content of 80% and 20% random coil. The actual value is 77% total helix content. Note that much of the discriminating information is in the short wavelength far UV region sensitive to sample absorption, and that particularly for more complex proteins the fit results may vary depending on the extent of the fitted region.

2-dimensional experiments. The difference between unstructured and structured proteins is even more distinct (and more detail is available in the case of partly disordered structures) in the case of 2-dimensional heteronuclear single-quantum coherence (HSQC) NMR spectra. Such a 2-dimensional spectrum maps the backbone amide groups according to their ^{1}H and ^{15}N resonance frequencies. This method necessitates production of ^{15}N labeled proteins[130] (using ^{15}NH$_4$Cl as the nitrogen source) and requires larger sample amounts than the qualitative 1-dimensional spectra analysis. The benefit is that the effect of environmental conditions such as pH or cofactors on conformation can be readily studied. In any case, because of the non-destructive nature of NMR spectroscopy, the samples can be used for subsequent crystallization experiments and high-throughput structure determination facilities often combine NMR screening with crystallization experiments. Figure 4-27 shows 2-dimensional HSQC spectra for ordered and disordered protein samples.

Stabilization by complex formation. Another set of HSQC spectra comparing an uncomplexed apoprotein with the protein–ligand complex drives across the point made already in Chapter 3 that potential ligands or cofactors are valuable crystallization additives that may dramatically improve crystallization (Figure 4-28).

Mass spectroscopic methods

Mass spectroscopy (MS) is an immensely useful analytical tool with a wide variety of applications supporting crystallographic studies. Any modification that can be traced by a molecular weight change can be readily quantified by MS. In this category fall analytical applications for crystallogenesis such as determining surface lysine methylation, verifying Se-Met or heavy atom incorporation, and many others such as analysis of glycosylation patterns. Digestive protein fragmentation and sequence analysis based on mass spectroscopic fragment matching form the backbone of proteomic expression analysis or proteomics (not discussed here). A mass spectroscopic method that can be used to experimentally determine exposed regions and loops of a protein and thus identify unstructured regions or support the design of domain truncations is deuterium exchange mass spectroscopy (DXMS).

Hydrogen exchange. Exchangeable hydrogen atoms of the peptide backbone NH groups as well as the protons bound to N, O, and S atoms of amino acids have been used to probe dynamics, folding, and intrinsic disorder in proteins,[132]

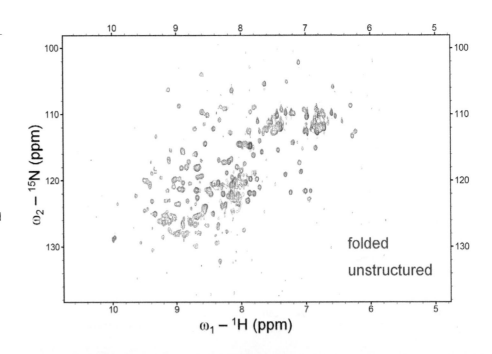

Figure 4-27 HSQC spectrum of folded and unstructured protein.
The 2-dimensional ^{1}H-^{15}N heteronuclear single-quantum coherence (HSQC) NMR spectrum clearly shows the distinct discrimination in the region below 8.3 ppm in ω_1 for the folded 228 residue protein (red, sharp peak contours) compared with the few wide and unresolved peaks for the unstructured protein sample (blue contours). NMR spectra courtesy of Simon Colebrook, Department of Biochemistry, Oxford University, and Joanne Nettleship, Oxford Protein Production facility.

folded

unstructured

$\omega_2 - ^{15}$N (ppm)

$\omega_1 - ^1$H (ppm)

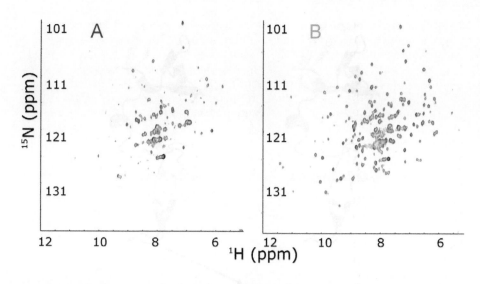

Figure 4-28 HSQC spectra of ligand-free and ligand-bound proteins. The 2-dimensional HSQC spectra of a bacterial methionine aminopeptidase (bMAP, 263 residues) without (A) and with (B) a tightly bound novel inhibitor.[131] Note the drastic increase in number of peaks and the improved discrimination of the spectrum for the bMAP–ligand complex compared with the *apo*-protein. The crystals of the bMAP–ligand complex diffracted to 0.9 Å resolution. Image courtesy of Artem Evdokimov, Procter & Gamble Pharmaceuticals, Mason, OH.

traditionally by NMR techniques[133] and more recently, by deuterium exchange mass spectroscopy (DXMS).[134] The principal idea is that exchangeable hydrogens that are accessible to solvent can be reversibly replaced with deuterium (or tritium, which will not be discussed here). For events involving conformational changes such as folding and denaturation, the time scales of this exchange range from sub-milliseconds to hours.[135]

Deuterium exchange mass spectroscopy. While for small proteins hydrogen exchange can be measured on a residue-by-residue basis using NMR techniques, DXMS[134] provides for large proteins and protein assemblies a means to locate regions that undergo H–D exchange. The key feature of the DXMS technique is coupling of a separation of proteolytic fragments via HPLC with sensitive MS, which allows the determination of which parts of the protein exchange fast, and thus were solvent accessible, and which are less, or not at all, accessible. Depending on the objective of the experiment, a DXMS experiment essentially consists of:

- establishment of a fragmentation map of unexchanged sample;

- exchange-in of D in D_2O buffer;

- quenching of the exchange and denaturation after different incubation times;

- rapid proteolysis over a protease column, HPLC separation, and ESI or Q-ToF MS detection;

- non-trivial analysis of the time-dependent fragmentation map data including corrections for back-exchange.[136]

Box 4-13 Assessment of secondary structure and folding state. It can be of value in troubleshooting of protein crystallization to examine whether a soluble protein is unfolded or contains large stretches of random coil conformation. Circular dichroism spectroscopy allows rapid qualitative assessment of the secondary structure contents of a soluble protein. One-dimensional nuclear magnetic resonance spectra show a distinct pattern in certain regions that are distinctive of folded or unstructured protein. Two-dimensional HSQC NMR spectra of ^{15}N labeled proteins allow a more quantitative and rapid assessment of the folding state of the protein. Deuterium exchange mass spectrometry can pinpoint solvent-exposed loops which may be targets for protein engineering to improve crystallization.

Figure 4-29 Use of DXMS for crystallographic protein engineering. The molecule is a cold-shock domain protein from *Neisseria meningitidis* which forms a domain swapped dimer (note the strand exchange between the red and yellow monomers). The green color indicates surface exposed flexible regions identified by DXMS that are targets for protein engineering. Figure courtesy of Joanne Nettleship, Oxford Protein Production Facility. PDB entry 3cam.[140]

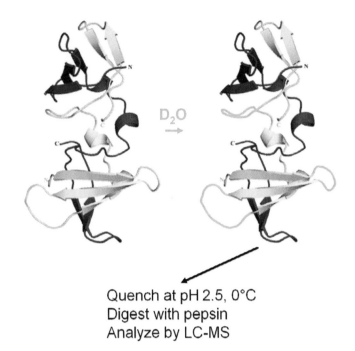

D₂O →

Quench at pH 2.5, 0°C
Digest with pepsin
Analyze by LC-MS

The DXMS method has been used to dissect the sequence of allosteric processes during oxygenation of hemoglobin,[134] to probe the dynamics of cAMP dependent protein kinase,[137] and, of particular interest in the context of protein crystallization, to successfully identify intrinsically unstructured regions in proteins[138] (Figure 4-29). Generation of disorder-depleted constructs via selective C-terminal truncation and loop deletion are typical examples for protein engineering applied as a successful salvage strategy.[139]

4.8   Next steps

At this point of our structure determination project we have successfully expressed and purified our target protein—and hopefully obtained the first crystals worth mounting (and each and every crystal is worth mounting!). We will now leave the laboratory and acquire in the second part of the book the fundamentals of crystallography, which will provide us with the necessary knowledge to confidently collect diffraction data from our precious crystals.

4.9   Key concepts

General protein engineering and design strategies
- Proteins for crystallographic studies are with few exceptions produced by heterologous overexpression in cellular hosts.
- Heterologous expression requires design and cloning of a recombinant DNA molecule encoding the target protein sequence and its transfer into the expression host.
- The advantage of recombinant DNA methods is the great flexibility in modifying the protein sequence and the overexpression of proteins with otherwise very low natural abundance.
- Protein engineering is possible at the DNA level by modifying the sequence of the protein construct and at the protein level by modifying the expressed protein.
- In both cases targeted (rational) design strategies or random (combinatorial) design strategies can be applied.

- Various different protein constructs or expression conditions are commonly screened in parallel.
- The most fundamental property of a protein suitable for crystallographic studies is sufficient solubility.
- Design strategies at the DNA level are based on the analysis of the protein sequence and the derived protein properties.
- Multiple sequence alignments supported by available structural information can identify domain boundaries and terminal regions.
- Additional tools identifying disordered regions, transmembrane helices, signal peptides, and binding sites augment the analysis and help identify potential problem regions.

Predicting protein crystallizability

- Inclusion of statistical data relating crystallization success to specific protein properties allows the establishment of conditional probabilities for crystallization success for a given protein.
- However, the conditional probabilities are largely based on predicted protein properties and cannot pinpoint the specific and local intermolecular interactions that actually determine crystallization.
- Crystallization predictions generally are more reliable in clear cut cases (very easy or very difficult to crystallize) but provide less guidance in intermediate cases.
- Crystallization trials remain the only authoritative (and in fact, fast and simple) method to determine crystallization propensity of a specific protein construct.
- Surface entropy reduction (SER) exemplifies a targeted design strategy at the DNA level. Mutation of high entropy surface residues such as exposed Lys, Glu, and Gln residues to Ala, Thr, or Val can lead to significant improvements in crystallization.
- However, various other properties such as overall protein charge, local charge distribution, and conformation are also varied at the same time and the precise causality between residue mutations and crystallization success cannot be easily established.
- A number of targeted protein design studies have shown that the rationale leading to various construct modifications is often not causal to the crystallization success—the readiness in principle to modify the protein is important.

Combinatorial libraries

- Combinatorial DNA libraries are based on the generation of random mutants of the target gene *in vitro*.
- Truncation mutant libraries and domain fragment libraries are examples of basic DNA libraries used to generate soluble and crystallizable constructs.
- Directed evolution is a refined combinatorial technique in which a library of random mutants is exposed to selective pressure such as an expression and solubility screening at the phenotype level.
- DNA shuffling allows the recombination of mutants with desirable traits to further improve solubility and ultimately, crystallizability and diffraction quality.
- Combinatorial DNA library generation requires extensive colony picking, selection, and sequencing and is generally conducted using robotic equipment.

Protein expression

- Structural studies by crystallography not only require soluble protein but also conformationally homogeneous material.
- Solubility as well as the demand for conformational purity requires careful selection of a suitable expression host.
- Advanced cloning techniques including ligation independent cloning, topoisomerase cloning, and Gateway cloning allow switching of expression systems with relative ease.
- Expression systems consist of a suitable transfection vector and corresponding host cells, ranging from prokaryotic bacteria to eukaryotic cells such as yeasts, insect cell lines, and mammalian cell lines.
- *E. coli* based systems are simple and the most common, but are incapable of eukaryotic posttranslational modifications and generally have difficulty

expressing properly folded proteins containing multiple disulfide bonds.
- Yeasts possess cell compartmentalization and the complete eukaryotic secretion and posttranslational modification machinery.
- Glycosylation patterns are different for yeasts and mammalian cells, and some yeasts hyperglycosylate.
- Glycosylation of mammalian proteins is a major source of conformational inhomogeneity. Glycosylation-deficient host cell lines can reduce this problem.
- *In vitro* cell free transcription–translation systems allow full control over the expression conditions, including addition of folding chaperones
- Recombinant DNA techniques allow adding or fusing a number of useful sequences to the target sequence.
- The most common fusion partners are small affinity purification tags and solubility enhancing fusion partners as well as complex-partners or chaperones.
- Reporter genes can be fused to constructs allowing rapid assaying of solubility or folding status as well as cellular location of the target protein.
- Engineered GFP is one of the most common reporter genes and is frequently used in combination with affinity purification tags.

Protein labeling and protein modifications
- Incorporation of Se-Met instead of regular methionine residues provides heavy atom labels for anomalous phasing methods while maintaining structural isomorphism to the native protein.
- One method of incorporating Se-Met is expression in a methionine auxotroph host strain and supplementing the minimal medium with Se-Met.
- In the metabolic inhibition technique bacteria are grown in regular medium and Met biosynthesis inhibiting amino acids Ile, Lys, and Thr as well as Se-Met are added to the medium before induction.
- Because of their intimate association with the lipid molecules of cell membranes, integral membrane proteins are difficult to express and to solubilize.
- Expression levels of membrane proteins are generally low as a result of overload of the cellular translocation machinery as well as compromised cell membrane stability.
- The necessity to persistently stabilize the hydrophobic transmembrane part of the protein with lipids and detergents poses additional difficulties for purification.
- Once the target protein is expressed and purified, it can be further modified to improve crystallization.
- Limited enzymatic proteolysis can be used to remove disordered termini and for domain truncation.
- Mild chemical surface lysine methylation has been used to improve protein crystallization.
- Inhomogeneous glycosylation can be treated at the protein level by enzymatic deglycosylation or by expression in glycosylation-deficient strains, or at the DNA level by site-directed mutagenesis of Asn in the glycosylation consensus sequence to Asp or Glu.

Stability and conformational analysis
- Protein stability and conformational homogeneity are of paramount importance for protein crystallization.
- The effect of buffer components and additives on protein stability can be rapidly assessed with ThermoFluor stability assays.
- Native SDS gels and size exclusion chromatography give a rapid but qualitative assessment of the oligomerization state of a protein.
- Detailed analysis of particle size distribution and polydispersity is possible using light scattering methods.
- Small angle X-ray and neutron scattering provide low resolution shape and size information about the protein.
- Circular dichroism spectroscopy allows rapid qualitative assessment of the secondary structure contents of a soluble protein.
- One-dimensional nuclear magnetic resonance spectra show distinctive patterns in certain regions that are folded, or unstructured, random coil protein.

- Two-dimensional HSQC NMR spectra of ^{15}N labeled proteins allow a more quantitative and rapid assessment of the folding state of the protein.
- Deuterium exchange mass spectroscopy can pinpoint solvent-exposed loops which may be targets for protein engineering to improve crystallization.

4.10 Additional reading

1. Glick BR, & Pasternak JJ (2003) *Molecular Biotechnology: Principles and Applications of Recombinant DNA*. Washington, DC: American Society for Microbiology.

2. Doublie S (Ed.) (2007) *Macromolecular Crystallography Protocols, Volume 1: Preparation and Crystallization of Macromolecules*. Totowa, NJ: Humana Press.

3. Hughes SH, & Stock AM (2001) Preparing recombinant proteins for X-ray crystallography. *International Tables for Crystallography F*, 65–80.

4. Grisshammer R, & Buchanan SK (Eds.) (2006) *Structural Biology of Membrane Proteins*. Cambridge, UK: Royal Society of Chemistry.

5. Walsh CT (2006) *Posttranslational Modifications*. Greenwood Village, CO: Roberts and Company.

6. Bourne PE, & Weissig H (2003) *Structural Bioinformatics*. Hoboken, NY: Wiley-Liss.

7. Aricescu AR, Assenberg R, Bill RM, et al. (2006) Eukaryotic expression: developments for structural proteomics. *Acta Crystallogr*. D62(10), 1114–1124.

8. Walden H (2010) Selenium incorporation using recombinant techniques. *Acta Crystallogr*. D66(4), 352–357.

9. Marino K, Bones J, Kattla JJ, et al. (2010) A systematic approach to protein glycosylation analysis: a path through the maze. *Nat. Chem. Biol*. 6(10), 713-23.

4.11 References

1. Doublie S (Ed.) (2007) *Macromolecular Crystallography Protocols, Volume 1: Preparation and Crystallization of Macromolecules*. Totowa, NJ: Humana Press.

2. Everett JK, Acton TB, & Montelione GT (2004) Primer Prim'er: a web based server for automated primer design. *J. Struct. Funct. Genomics* 5(1–2), 13–21.

3. Fox JA, Butland SL, McMillan S, et al. (2005) The Bioinformatics Links Directory: a compilation of molecular biology web servers. *Nucleic Acids Res*. 33, W3–W24.

4. Barton GJ (2008) Sequence alignment for molecular replacement. *Acta Crystallogr*. D64, 25–32.

5. Altschul SF, Gish W, Miller W, et al. (1990) Basic local alignment search tool. *J. Mol. Biol*. 215(3), 403–410.

6. Mount DW (2004) *Bioinformatics: Sequence and Genome Analysis*. Cold Spring Harbor, NY: Cold Spring Harbor Laboratory Press.

7. Altschul SF, Madden TL, Schaffer AA, et al. (1997) Gapped BLAST and PSI-BLAST: a new generation of protein database search programs. *Nucleic Acids Res*. 25(17), 3389–3402.

8. Marchler-Bauer A, Anderson JB, Derbyshire MK, et al. (2007) CDD: a conserved domain database for interactive domain family analysis. *Nucleic Acids Res*. 35, D237–D240.

9. Larkin MA, Blackshields G, Brown NP, et al. (2007) Clustal W and Clustal X version 2.0. *Bioinformatics* 23(21), 2947–2948.

10. Krogh A, Brown M, Mian IS, et al. (1994) Hidden Markov models in computational biology. Applications to protein modeling. *J. Mol. Biol*. 235(5), 1501–1531.

11. Subramanian A, Kaufmann M, & Morgenstern B (2008) DIALIGN-TX: greedy and progressive approaches for segment-based multiple sequence alignment. *Algorithms Mol. Biol*. 3, 6-17.

12. O'Sullivan O, Suhre K, Abergel C, et al. (2004) 3DCoffee: combining protein sequences and structures within multiple sequence alignments. *J. Mol. Biol*. 340(2), 385–395.

13. Gouet P, Robert X, & Courcelle E (2003) ESPript/ENDscript: Extracting and rendering sequence and 3D information from atomic structures of proteins. *Nucleic Acids Res*. 31(13), 3320–3323.

14. Abagyan R, Totrov M, & Kuznetsov D (1994) ICM: a new method for protein modeling and design. Applications to docking and structure prediction from the distorted native conformation. *J. Comput. Chem*. 15, 488–506.

15. Kabsch W, & Sander C (1983) Dictionary of protein secondary structure: pattern recognition of hydrogen-bonded and geometrical features. *Biopolymers* 22, 2577–2637.

16. Linding R, Russell RB, Neduva V, et al. (2003) GlobPlot: exploring protein sequences for globularity and disorder. *Nucleic Acids Res*. 31(13), 3701–3708.

17. Adamczak R, Porollo A, & Meller J (2005) Combining prediction of secondary structure and solvent accessibility in proteins. *Proteins* 59(3), 467–475.

18. Rost B, Yachdav G, & Liu J (2004) The PredictProtein server. *Nucleic Acids Res*. 32, W321–W326.

19. Bryson K, McGuffin LJ, Marsden RL, et al. (2005) Protein structure prediction servers at University College London. *Nucleic Acids Res*. 33, W36–W38.

20. Ceroni A, Passerini A, Vullo A, et al. (2006) DISULFIND: a disulfide bonding state and cysteine connectivity prediction server. *Nucleic Acids Res*. 34, W177–W181.

21. Dunker AK, Lawson JD, Brown CJ, et al. (2001) Intrinsically disordered protein. *J. Mol. Graphics Modell*. 19(1), 26–59.

22. Linding R, Jensen LJ, Diella F, et al. (2003) Protein disorder prediction: implications for structural proteomics. *Structure* 11(11), 1453–1459.

23. Ferron F, Longhi S, Canard B, et al. (2006) A practical overview of protein disorder prediction methods. *Proteins* 65(1), 1–14.

24. Kall L, Krogh A, & Sonnhammer EL (2007) Advantages of combined transmembrane topology and signal peptide

prediction – the Phobius web server. *Nucleic Acids Res.* 35, W429–W432.

25. Bendtsen JD, Nielsen H, von Heijne G, et al. (2004) Improved prediction of signal peptides: SignalP 3.0. *J. Mol. Biol.* 340(4), 783–795.

26. Garcia KC, Degano M, Stanfield RL, et al. (1996) An αβ T cell receptor structure at 2.5 \Aring and its orientation in the TCR-MHC complex. *Science* 274, 209–219.

27. Weinberg RA (2006) *The Biology of Cancer*. New York, NY: Garland Science.

28. Cooper DR, Surendranath Y, Devedjiev Y, et al. (2007) Structure of the *Bacillus subtilis* OhrB hydroperoxide-resistance protein in a fully oxidized state. *Acta Crystallogr.* D63(12), 1269–1273.

29. Gupta R, Jung E, & Brunak S (2004) Prediction of N-glycosylation sites in human proteins. http://www.cbs.dtu.dk/services/NetNGlyc/.

30. Rupp B (2003) Maximum-likelihood crystallization. *J. Struct. Biol.* 142(1), 162–169.

31. Rupp B, & Wang J (2004) Predictive models for protein crystallization. *Methods* 34(3), 390–407.

32. Norvell JC, & Zapp-Machalek A (2000) Structural genomics programs at the US National Institute of General Medical Sciences. *Nat.Struct. Biol. Suppl.* 7, 931.

33. Slabinski L, Jaroszewski L, Rychlewski L, et al. (2007) XtalPred: a web server for prediction of protein crystallizability. *Bioinformatics* 23(24), 3403–3405.

34. Gruber M, Soeding J, & Lupas AN (2006) Comparative analysis of coiled-coil prediction methods *J. Struct. Biol.* 155(2), 140–145.

35. Goldschmidt L, Cooper DR, Derewenda ZS, et al. (2007) Toward rational protein crystallization: A Web server for the design of crystallizable protein variants. *Protein Sci.* 16(8), 1569–1576.

36. Derewenda ZS, & Vekilov PG (2006) Entropy and surface engineering in protein crystallization. *Acta Crystallogr.* D62, 116–124.

37. Fox JD, Kapust RB, & Waugh DS (2001) Single amino acid substitutions on the surface of *Escherichia coli* maltose-binding protein can have a profound impact on the solubility of fusion proteins. *Protein Sci.* 10(3), 622–630.

38. Dale GE, Oefner C, & D'Arcy A (2003) The protein as a variable in protein crystallization. *J. Struct. Biol.* 142(1), 88–97.

39. Fontana A, Polverino de Laureto P, Spolaore B, et al. (2004) Probing protein structure by limited proteolysis. *Acta Biochim. Pol.* 51, 299–321.

40. Bowers P, Pellegrini M, Thompson M, et al. (2004) Prolinks: a database of protein functional linkages derived from co-evolution. *Genome Biol.* 5(5), R35,1–5.

41. Hulo N, Bairoch A, Bulliard V, et al. (2008) The 20 years of PROSITE. *Nucleic Acids Res.* 36, D245–D249.

42. Cooper DR, Boczek T, Grelewska K, et al. (2007) Protein crystallization by surface entropy reduction: optimization of the SER strategy. *Acta Crystallogr.* D63(5), 636–645.

43. Evdokimov AG, Mekel M, Hutchings K, et al. (2008) Rational protein engineering in action: the first crystal structure of a phenylalanine tRNA synthetase from *Staphylococcus haemolyticus*. *J. Struct. Biol.* 162(1), 152–169.

44. Griffin AM, & Griffin HG (Eds.) (1995) *Molecular Biology: Current Innovations and Future Trends*. Norwich, UK: Horizon Scientific Press.

45. Glick BR, & Pasternak JJ (2003) *Molecular Biotechnology: Principles and Applications of Recombinant DNA*. Washington, DC: American Society for Microbiology.

46. Segelke B, Schafer J, Coleman M, et al. (2004) Laboratory scale structural genomics. *J. Struct. Funct. Genomics* 5, 147–157.

47. Rupp B (2005) High throughput protein crystallography. In Sundstroem M & Edwards A (Eds.), *Structural Proteomics and High Throughput Structural Biology*. New York. NY: Taylor and Francis.

48. Derewenda ZS (2004) The use of recombinant methods and molecular engineering in protein crystallization. *Methods* 34(3), 354–363.

49. Hart DJ, & Tarendeau F (2006) Combinatorial library approaches for improving soluble protein expression in *Escherichia coli. Acta Crystallogr.* D62(1), 19–26.

50. Longhi S, Ferron F, & Egloff M-P (2007) Protein Engineering. In Doublie S (Ed.), *Macromolecular Crystallography Protocols, Volume 1*. Totowa, NJ: Humana Press.

51. Stemmer WP (1994) Rapid evolution of a protein in vitro by DNA shuffling. *Nature* 370(6488), 389–391.

52. Wong TS, Roccatano D, Zacharias M, et al. (2006) A statistical analysis of random mutagenesis methods used for directed protein evolution. *J. Mol. Biol.* 355(4), 858–871.

53. Bloom JD, Meyer MM, Meinhold P, et al. (2005) Evolving strategies for enzyme engineering. *Curr. Opin. Struct. Biol.* 15(4), 447–452.

54. Crameri A, Whitehorn EA, Tate E, et al. (1996) Improved green fluorescent protein by molecular evolution using DNA shuffling. *Nat. Biotechnol.* 14(3), 315–319.

55. Harel M, Aharoni A, Gaidukov L, et al. (2004) Structure and evolution of the serum paraoxonase family of detoxifying and anti-atherosclerotic enzymes. *Nat. Struct. Mol. Biol.* 11(5), 412–419.

56. Majumder K (1992) Ligation-free gene synthesis by PCR: synthesis and mutagenesis at multiple loci of a chimeric gene encoding OmpA signal peptide and hirudin. *Gene* 110(1), 89–94.

57. Katzen F (2007) Gateway recombinational cloning: a biological operating system. *Expert Opin. Drug Dis.* 2(4), 571–589.

58. Sumner JB (1926) The isolation and crystallization of the enzyme urease. *J. Biol. Chem.* 69, 435–441.

59. Kantardjieff KA, Höchtl P, Segelke BW, et al. (2002) Concanavalin A in a dimeric crystal form: revisiting structural accuracy and molecular flexibility. *Acta Crystallogr.* D58, 735–743.

60. Sumner JB (1919) The globulins of the Jack Bean, *Canavalia ensiformis. J. Biol. Chem.* 37, 137–142.

61. Bowles DJ, Marcus SE, Pappin DJ, et al. (1986) Posttranslational processing of concanavalin A precursors in jackbean cotyledons. *J. Cell Biol.* 102(4), 1284–1297.

62. Moy S, Dieckman L, Schiffer M, et al. (2004) Genome-scale expression of proteins from *Bacillus subtilis. J. Struct. Funct. Genomics* 5(1–2), 103–9.

63. Studier FW (2005) Protein production by auto-induction in high-density shaking cultures. *Protein Expression Purif.* 41(1), 207–234.

64. Buckle AM, Devlin GL, Jodun RA, et al. (2005) The matrix refolded. *Nat. Methods* 2(1), 3.

65. Prinz B, Schultchen J, Rydzewski R, et al. (2004) Establishing a versatile fermentation and purification procedure for human proteins expressed in the yeasts *Saccharomyces cerevisiae* and *Pichia pastoris* for structural genomics. *J. Struct. Funct. Genomics* 5(1–2), 29–44.

66. Tan NS, Ho B, & Ding JL (2002) Engineering a novel secretion signal for cross-host recombinant protein expression. *Protein Eng.* 15(4), 337–345.

67. Woycechowsky KJ, & Raines RT (2000) Native disulfide bond formation in proteins. *Curr. Opin. Chem. Biol.* 4(5), 533.

68. Kukuruzinska MA, Bergh MLE, & Jackson BJ (1987) Protein glycosylation in yeast. *Annu. Rev. Biochem.* 56(1), 915–944.

69. Grinna LS, & Tschopp JF (1989) Size distribution and general structural features of N-linked oligosaccharides from the methylotrophic yeast, *Pichia pastoris. Yeast* 5(2), 107–115.

70. Cregg JM, Vedvick TS, & Raschke WC (1993) Recent advances in the expression of foreign genes in *Pichia pastoris. Biotechnology* 11(8), 905–910.

71. Mobeche I, Ragon M, Moulin G, et al. (2006) N-glycosylation differences between wild-type and recombinant strains affect catalytic properties of two model enzymes: beta-glucosidase and phosphatase. *Microb. Cell Fact.* 5(Suppl 1), 35.

72. Wasilko DJ, Lee SE, Stutzman-Engwall KJ, et al. (2009) The titerless infected-cells preservation and scale-up (TIPS) method for large-scale production of NO-sensitive human soluble guanylate cyclase (sGC) from insect cells infected with recombinant baculovirus. *Protein Expression Purif.*, 65(2), 122–132.

73. Madden DR, & Safferling M (2007) Bacculoviral expression of an integral membrane protein for structural studies In Doublie S (Ed.), *Macromolecular Crystallography Protocols, Vol. 1.* Totowa, N.J.: Humana Press.

74. Dukkipati A, Park HH, Waghray D, et al. (2008) BacMam system for high-level expression of recombinant soluble and membrane glycoproteins for structural studies. *Protein Expr. Purif.* 62(2), 160–170.

75. Hughes SH, & Stock AM (2001) Preparing recombinant proteins for X-ray crystallography. *International Tables for Crystallography F,* 65–80.

76. Puthalakath H, Burke J, & Gleeson PA (1996) Glycosylation defect in Lec1 Chinese hamster ovary mutant is due to a point mutation in N-acetylglucosaminyltransferase I gene. *J. Biol. Chem.* 271(44), 27818–27822.

77. Katzen F, Chang G, & Kudlicki W (2005) The past, present and future of cell-free protein synthesis. *Trends Biotechnol.* 23(3), 150–166.

78. Endo Y, & Sawasaki T (2004) High-throughput, genome-scale protein production method based on the wheat germ cell-free expression system. *J. Struct. Funct. Genomics* 5(1–2), 45–57.

79. Busso D, Kim R, & Kim SH (2004) Using an *Escherichia coli* cell-free extract to screen for soluble expression of recombinant proteins. *J. Struct. Funct. Genomics* 5(1–2), 69–74.

80. Dale GE, Kostrewa D, Gsell B, et al. (1999) Crystal engineering: deletion mutagenesis of the 24 kDa fragment of the DNA gyrase B subunit from *Staphylococcus aureus. Acta Crystallogr.* D55, 1626–1629.

81. Tropea ET, Cherry S, Nallamsetty S, et al. (2007) A generic method for the production of recombinant proteins in *Escherichia coli* using a dual hexahistidine-maltose binding protein affinity tag. In Doublie S (Ed.), *Macromolecular Crystallography Protocols, Vol. 1.* Totowa, NJ: Humana Press.

82. Phan J, Zdanov A, Evdokimov AG, et al. (2002) Structural basis for the substrate specificity of Tobacco Etch Virus protease. *J. Biol. Chem.* 277(52), 50564–50572.

83. de Marco A (2006) Two-step metal affinity purification of double-tagged (NusA-His6) fusion proteins. *Nat. Protoc.* 1(3), 1538–1543.

84. van den Berg S, Lofdahl PA, Hard T, et al. (2006) Improved solubility of TEV protease by directed evolution. *J. Biotechnol.* 121(3), 291–298.

85. Schmidt TGM, & Skerra A (2007) The Strep-tag system for one-step purification and high-affinity detection or capturing of proteins. *Nat. Protoc.* 2(6), 1528–1532.

86. Bucher MH, Evdokimov AG, & Waugh DS (2002) Differential effects of short affinity tags on the crystallization of *Pyrococcus furiosus* maltodextrin-binding protein. *Acta Crystallogr.* 58(3), 392–397.

87. Kapust RB, & Waugh DS (1999) *Escherichia coli* maltose-binding protein is uncommonly effective at promoting the solubility of polypeptides to which it is fused. *Protein Sci.* 8(8), 1668–1674.

88. LaVallie ER, DiBlasio EA, Kovacic S, et al. (1993) A thioredoxin gene fusion expression system that circumvents inclusion body formation in the *E. coli* cytoplasm. *Biotechnology* 11(2), 187–193.

89. Kaplan W, Husler P, Klump H, et al. (1997) Conformational stability of pGEX-expressed *Schistosoma japonicum* glutathione S-transferase: a detoxification enzyme and fusion-protein affinity tag. *Protein Sci.* 6(2), 399–406.

90. de Marco A (2006) Two-step metal affinity purification of double-tagged (NusA-His6) fusion proteins. *Nat. Protoc.* 1(3), 1538–1543.

91. Yang F, Moss LG, & Phillips GN, Jr. (1996) The molecular structure of green fluorescent protein. *Nat. Biotechnol.* 14(10), 1246–1251.

92. Waldo GS, Standish BM, Berendzen J, et al. (1999) Rapid protein-folding assay using green fluorescent protein. *Nat. Biotechnol.* 17, 691–695.

93. Zhang A, Gonzalez SM, Cantor EJ, et al. (2001) Construction of a mini-intein fusion system to allow both direct monitoring of soluble protein expression and rapid purification of target proteins. *Gene* 275(2), 241–252.

94. Los GV, & Wood K (2007) The HaloTag: a novel technology for cell imaging and protein analysis. *Methods Mol. Biol.* 356, 195–208.

95. LeMaster DM, & Richards FM (1985) 1H–15N heteronuclear NMR studies of *Escherichia coli* thioredoxin in samples isotopically labeled by residue type. *Biochemistry* 24, 7263–7268.

96. Segelke BW, Forstner M, Knapp M, et al. (2000) Conformational flexibility in the Apolipoprotein E amino-terminal domain structure determined from three new crystal forms: Implication for lipid binding. *Protein Sci.* 9, 886–897.

97. Doyle DA, Morais Cabral J, Pfuetzner RA, et al. (1998) The structure of the potassium channel: molecular basis of K+ conduction and selectivity. *Science* 280(5360), 69–77.

98. Grisshammer R, & Buchanan SK (Eds.) (2006) *Structural Biology of Membrane Proteins.* Cambridge, UK: Royal Society of Chemistry.

99. Luca S, White JF, Sohal AK, et al. (2003) The conformation of neurotensin bound to its G protein-coupled receptor. *Proc. Natl. Acad. Sci. U.S.A.* 100(19), 10706–10711.

100. Roosild TP, Greenwald J, Vega M, et al. (2005) NMR structure of Mistic, a membrane-integrating protein for membrane protein expression. *Science* 307, 1317–1321.

101. Day PW, Rasmussen SG, Parnot C, et al. (2007) A monoclonal antibody for G protein-coupled receptor crystallography. *Nat. Methods* 4(11), 927–929.

102. Cherezov V, Rosenbaum DM, Hanson MA, et al. (2007) High-resolution crystal structure of an engineered human β_2-adrenergic G protein-coupled receptor. *Science* 318, 1258–1265.

103. Anderluh G, Gokce I, & Lakey JH (2003) Expression of proteins using the third domain of the *Escherichia coli* periplasmic-protein TolA as a fusion partner. *Protein Expression Purif.* 28(1), 173–181.

104. Thai K, Choi J, Franzin CM, et al. (2005) Bcl-XL as a fusion protein for the high-level expression of membrane-associated proteins. *Protein Sci.* 14(4), 948–855.

105. Moraes TF, Bains M, Hancock RE, et al. (2007) An arginine ladder in OprP mediates phosphate-specific transfer across the outer membrane. *Nat. Struct. Mol. Biol.* 14(1), 85–87.

106. Walter TS, Meier C, Assenberg R, et al. (2006) Lysine methylation as a routine rescue strategy for protein crystallization. *Structure* 14(11), 1617–1622.

107. Au K, Ren J, Walter TS, et al. (2008) Structures of an alanine racemase from *Bacillus anthracis* (BA0252) in the presence and absence of (R)-1-aminoethylphosphonic acid (L-Ala-P). *Acta Crystallogr.* F64, 327–233.

108. Walsh CT (2006) *Posttranslational Modifications*. Greenwood Village, CO: Roberts and Company.

109. Dale GE, D'Arcy B, Yuvaniyama C, et al. (2000) Purification and crystallization of the extracellular domain of human neutral endopeptidase (neprilysin) expressed in *Pichia pastoris*. *Acta Crystallogr.* D56, 894–897.

110. Nettleship JE, Aplin R, Radu Aricescu A, et al. (2007) Analysis of variable N-glycosylation site occupancy in glycoproteins by liquid chromatography electrospray ionization mass spectrometry. *Anal. Biochem.* 361(1), 149–151.

111. Gordon K, Redelinghuys P, Schwager SL, et al. (2003) Deglycosylation, processing and crystallization of human testis angiotensin-converting enzyme. *Biochem. J.* 371(2), 437–442.

112. Fusetti F, Schroter KH, Steiner RA, et al. (2002) Crystal structure of the copper-containing quercetin 2,3-dioxygenase from *Aspergillus japonicus*. *Structure* 10(2), 259–268.

113. Evdokimov AG, Anderson DE, Routzahn KM, et al. (2000) Overproduction, purification, crystallization and preliminary X-ray diffraction analysis of YopM, an essential virulence factor extruded by the plague bacterium *Yersinia pestis*. *Acta Crystallogr.* D56, 1676–1679.

114. D'Arcy A (1994) Crystallizing proteins – a rational approach? *Acta Crystallogr.* D50, 469–475.

115. Nettleship JE, Brown J, Groves MR, et al. (2008) Methods for protein characterization by mass spectrometry, thermal shift (ThermoFluor) assay, and multiangle or static light scattering. In Kobe B, Guss M, & Huber T (Eds.), *Structural Proteomics: High-throughput Methods*. Totowa, NJ: Humana Press.

116. Ericsson UB, Hallberg BM, DeTitta GT, et al. (2006) Thermofluor-based high-throughput stability optimization of proteins for structural studies. *Anal. Biochem.* 357(2), 289–298.

117. Cantor CR, & Schimmel PR (1980) *Biophysical Chemistry, Volume II*. New York, NY: Freeman.

118. Wilson WW (2003) Light scattering as a diagnostic for protein crystal growth - A practical approach. *J. Struct. Biol.* 142(1), 56–65.

119. Atkins P (1994) *Physical Chemistry*. New York, NY: Freeman and Company.

120. George A, & Wilson WW (1994) Predicting protein crystallization from a dilute solution property. *Acta Crystallogr.* D50, 361–365.

121. Rupp B, Segelke BW, Krupka HI, et al. (2002) The TB structural genomics consortium crystallization facility: towards automation from protein to electron density. *Acta Crystallogr.* D58, 1514–1518.

122. Borgstahl GEO (2007) How to use dynamic light scattering to improve the likelihood of growing macromolecular crystals. In Doublie S (Ed.), *Macromolecular Crystallography Protocols, Vol. 1*. Totowa, NJ: Humana Press.

123. Konarev PV, Volkov VV, Sokolova AV, et al. (2003) PRIMUS: a Windows PC-based system for small-angle scattering data analysis. *J. Appl. Crystallogr.* 36(5), 1277–1282.

124. Gherardi E, Sandin S, Petoukhov MV, et al. (2006) Structural basis of hepatocyte growth factor/scatter factor and MET signalling. *Proc. Natl. Acad. Sci. U.S.A.* 103(11), 4046–4051.

125. Wallace BA (2000) Conformational changes by synchrotron radiation circular dichroism spectroscopy. *Nat. Struct. Biol.* 7, 708–709.

126. Davidson B, & Fasman GD (1967) The conformational transitions of uncharged poly-L-lysine: α-Helix-Random Coil -β-Structure. *Biochemistry* 6(6), 1616–1629.

127. Johnson WC (1990) Protein secondary structure and circular dichroism: A practical guide. *Proteins* 7, 204–214.

128. Nabuurs SB, Spronk CA, Vuister GW, et al. (2006) Traditional biomolecular structure determination by NMR spectroscopy allows for major errors. *PLoS Comput. Biol.* 2(2), e9.

129. Rehm T, Huber R, & Holak TA (2002) Application of NMR in structural proteomics: screening for proteins amenable to structural analysis. *Structure* 10(12), 1613–1618.

130. Zhao Q, Frederick R, Seder K, et al. (2004) Production in two-liter beverage bottles of proteins for NMR structure determination labeled with either ^{15}N- or ^{13}C-^{15}N. *J. Struct. Funct. Genomics* 5, 87–93.

131. Evdokimov AG, Pokross M, Walter RL, et al. (2007) Serendipitous discovery of novel bacterial methionine aminopeptidase inhibitors. *Proteins* 66(3), 538–546.

132. Englander SW, Sosnick TR, Englander JJ, et al. (1996) Mechanisms and use of hydrogen exchange *Curr. Opin. Struct. Biol.* 6, 18–23.

133. Wüthrich K (1994) NMR assignments as a basis for structural characterization of denatured states of globular proteins. *Curr. Opin. Struct. Biol.* 4, 93–99.

134. Englander JJ, Del Mar C, Li W, et al. (2003) Protein structure change studied by hydrogen-deuterium exchange, functional labelling, and mass spectroscopy Proc. *Natl. Acad. Sci. U.S.A.* 100(12), 7057–7062.

135. Evans PA, & Radford SE (1994) Probing the structure of folding intermediates. *Curr. Opin. Struct. Biol.* 4, 100–106.

136. Weiss DD, Engen JR, & Kaas IJ (2006) Semi-automated data processing of hydrogen exchange mass spectra using HX-Express. *J. Am. Soc. Mass Spectrom.* 17(2), 1700–1703.

137. Hamuro Y, Zawadzki KM, Kim JS, et al. (2003) Dynamics of CAPK type II-β activation revealed by enhanced amide H/D exchange mass spectrometry (DXMS). *J. Mol. Biol.* 327, 1065–1076.

138. Spraggon G, Pantazatos D, Klock HE, et al. (2004) On the use of DXMS to produce more crystallizable proteins: Structures of the *T. maritima* proteins TM0160 and TM1171. *Protein Sci.* 13, 3187–3199.

139. Pantazatos D, Kim JS, Klock HE, et al. (2004) Rapid refinement of crystallographic protein construct definition employing enhanced hydrogen/deuterium exchange MS. *Proc. Natl. Acad. Sci. U.S.A.* 101(3), 751–756.

140. Ren J, Nettleship JE, Sainsbury S, et al. (2008) Structure of the cold-shock domain protein from *Neisseria meningitidis* reveals a strand-exchanged dimer. *Acta Crystallogr.* F64(4), 247–251.

PART II

Fundamentals of Protein Crystallography

Crystal geometry

Crystallography borders, naturally, on pure physics, chemistry, biology, mineralogy, technology and also on mathematics, but is distinguished by being concerned with the methods and results of investigating the arrangement of atoms in matter, particularly when that arrangement has regular features.

Paul Ewald (1948) *Acta Crystallogr.* **1**, 2

Inspection of crystals during crystallization experiments led us to suspect that the regularity of faces and dihedral angles between the crystal faces in some form represents the internal order of that crystal. Such ideas were developed very early on, and the fascinating regularity of crystals strongly influenced the concept of platonic bodies and atomicity during Greek antiquity. Even today, in rather enlightened times (at least in secular aspects), the regularity and beauty of crystals has in some circles given rise to the erroneous superstition that crystals harbor special powers. Notwithstanding such aberrations, the recalcitrance of proteins to crystallize may sometimes invoke perceptions of a not necessarily friendly occult power trying to fool the aspiring crystallographer.

Early attempts at a rational explanation of crystal geometry by Huygens and other scientists followed, and a well-developed descriptive crystallography based on careful measurements of mineral crystals emerged. At the end of the 19th century the idea that the macroscopic appearance of crystals is a manifestation of their microscopic regularity was well established, but the final proof had to wait for the actual X-ray diffraction experiments of Max von Laue and co-workers and by Sir William Henry Bragg and his son Sir William Lawrence Bragg. From that point on, the tremendous impact of crystallography on the understanding of the physics and chemistry of materials has led to the quantum leap in our capability to design and engineer the advanced materials we now use on a daily basis. Since the early 1950s, when protein structure determination became feasible, a similar revolution in our understanding of biological structure and function has taken place, and crystallography plays an indispensable role in modern structure-guided drug discovery.

Chapter 5 develops the fundamental concepts of crystal geometry, first using 2-dimensional examples and then reinforcing the fundamental findings in three dimensions and for real crystals. Many important concepts can be easily introduced and readily understood using simple 2-dimensional examples. As far as possible we will use projections and images of real molecules to examine symmetry and packing in 2- and 3-dimensional lattices. The concepts developed in two dimensions are a large subset of those applying in three dimensions, and they will make it much easier for the student to appreciate the visually more challenging conceptualization in 3-dimensional space.

The introduction of crystal symmetry is limited to operations permitted on asymmetric motifs, while at the same time emphasizing a number of consequences such as solvent channel formation and crystal contact formation, including a brief introduction to non-crystallographic symmetry, which is extensively discussed in Chapter 11. Basic navigation of crystal lattices and the introduction of crystallographic coordinates allow a more formal extension of symmetry in three dimensions. The combination of symmetry operations and lattice translations leads to the 65 chiral space groups. The discussion of space-group symmetry is augmented with only a short formal introduction of group theory (after all, every structure can be solved in *P*1), while the practical interpretation of space-group tables is emphasized.

The application of crystallographic symmetry to molecule coordinates in Cartesian space requires the introduction of orthogonalization and deorthogonalization matrices, tools which in turn allow the proper construction of unit cells and the analysis of molecular packing. The concepts of lattice planes and sampling density finally lead to a preview of the reciprocal lattice as a formal construct, which in this chapter is only intuitively linked to X-ray scattering and diffraction data. X-ray scattering theory and diffraction geometry are treated separately in Chapter 6.

5.1   Basic concepts of crystal assembly

In Chapter 3 crystal contacts and crystal packing were discussed from a molecular viewpoint, focusing on the intermolecular forces which promote the self-assembly of molecules into a protein crystal. We became aware that sparse and anisotropically distributed, weak intermolecular interactions lead to a low contact/volume ratio, and that inherent dynamic flexibility and irregularity of the protein molecules additionally contributes to the fragility and sensitivity of protein crystals.

Until now, however, we have not paid much attention to the symmetry relations between the molecules and their regular, periodic arrangement that is necessary to produce discrete X-ray diffraction. We will therefore examine in this section which, and how many, different arrangements exist for protein molecules to self-assemble into a crystal.

The number of different ways to pack molecules in space is limited. The translational periodicity requirements, combined with internal symmetry, allow only five general ways of arranging 2-dimensional asymmetric objects in the plane (or 2-dimensional space) and only 65 possible types of arrangements of asymmetric objects such as chiral protein molecules in 3-dimensional space. Irrespective of the details of how the molecules self-assemble into a crystal, each protein crystal structure must belong to one of the 65 chiral space groups. The exact construction rules for the space groups will have to wait until later in this chapter, once the necessary crystallographic basics of symmetry and molecular assembly have been covered.

Periodic lattices

We begin with an examination of how to arrange tiles into periodic, 2-dimensional patterns in order to fit a new floor in our X-ray laboratory. We are provided

Box 5-1 **Protein crystals belong to one of 65 space groups.** Only 65 discrete and distinct ways exist to assemble 3-dimensional periodic crystals from asymmetric chiral molecules, through combinations of translational and rotational symmetry. These 65 types of arrangements form 65 chiral space groups, and their symmetry properties and the rules for constructing each crystal structure are described in the *International Tables for Crystallography, Volume A*.[1]

with a collection of arbitrary white tiles with no discernible pattern painted on them (i.e. they harbor no motif) and examine if we can cover the floor with a periodically repeating pattern of these tiles.

Figure 5-1 clearly shows that the attempt to arrange an arbitrarily shaped set of different tiles into a floor-covering pattern is doomed to fail. Even this trivial gedankenexperiment provides a valuable lesson: Our inability to arrange objects of different shapes into a periodic regular pattern is mirrored by the fact that conformationally heterogeneous or impure proteins in general yield no, or only poorly, diffracting crystals.

We can repeat the tiling experiment after sorting the tiles by shape (equivalent to a protein purification step) into different groups of identical tiles. One immediately recognizes that only certain types of tiles can fill the plane. The property common to the oblique, rectangular, and square tiles is that their shape is defined by two sets of intersecting, equidistant parallel edges or lines (Figure 5-2). These lines form the lattice of each tile pattern. As will be demonstrated later (Figure 5-14), even the hexagon can be cut up into three smaller, identical tiles of equal sides enclosing a 120° angle. The new tiles then conform to the definition of a lattice. A square pattern also represents a special case of a *rectangular* plane lattice, which in turn can be reduced to a special case of *oblique* lattice. As discussed in detail in the next section, it requires the concept of internal symmetry to understand what distinguishes the "special" lattices from a "generic" oblique lattice. As an example, a square plane lattice can be rotated by 90° and still fit the same lattice, while a rectangular plane lattice can only be rotated by 180°. A generic oblique lattice possesses no symmetry at all; only the trivial identity operation of rotating it by 360° superimposes it onto itself.

The plane lattice. A plane lattice is an infinitely extendable construct of intersecting parallel and equidistant lines, forming identical unit lattices. The unit lattice of a plane lattice is defined by two lattice vectors, **a** and **b**, of lengths *a* and *b*, which enclose the corresponding angle *γ* between them. The shorthand for a lattice (or basis) spanned by vectors **a** and **b** sharing a common origin is simply [0, **a**, **b**]. Each point of the lattice can be reached by a combination of an integer number of translations along the lattice vectors. Note that the nomenclature we use does clearly distinguish vectors, which have a direction in space, by denoting them in **bold** typeface. In contrast, scalar values are just numbers and are typed in *italics*, as exemplified by the lattice parameters *a* and *b*. We follow the convention of drawing the lattice vector **a** *down* and **b** to the *right* as illustrated in Figure 5-3. This notation is used in the *International Tables for Crystallography* (ITC),[1] an indispensable almanac of crystallography to which we will refer extensively throughout the remainder of this book.

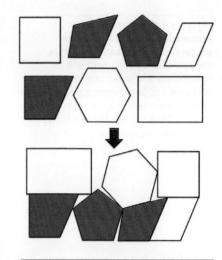

Figure 5-1 Arbitrary 2-dimensional tiles. A set of arbitrary tiles (top) and an unsuccessful attempt to arrange them together into a regular pattern filling the plane (bottom). The inability to assemble different types of tiles into a regular pattern is one of the reasons impure protein preparations tend not to crystallize.

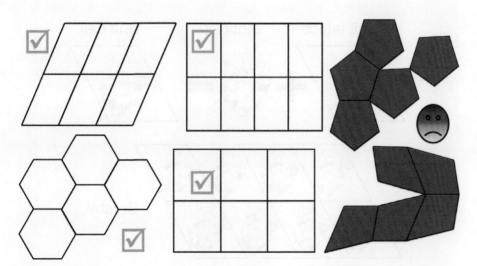

Figure 5-2 Plane filling with tile patterns. Only certain types of tiles taken from the collection shown in Figure 5-1 are useful to tile the floor: Only the objects that allow filling the plane without leaving any uncovered holes can be used. The resulting plane lattices (white) represent oblique, rectangular, hexagonal, and square plane lattices.

Figure 5-3 Assignment and nomenclature of plane lattice vectors. A plane square unit lattice exemplifies the assignment and nomenclature of plane unit lattice vectors **a** and **b**, the corresponding scalar lattice parameters *a*, *b*, and the enclosed angle γ. Table 5-1 lists the remaining plane lattice types.

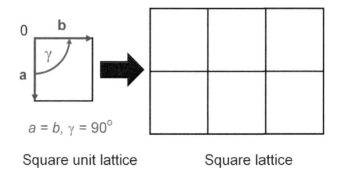

$a = b,\ \gamma = 90°$

Square unit lattice   Square lattice

Appendix A.2 provides a refresher on basic vector algebra and its use in crystallographic computing for those who are not (anymore) familiar with the underlying basic mathematical concepts.

Periodic patterns, motifs, and unit cells

The lattices we described above have nothing in them *per se*; they are just guidelines or a mathematical construct dividing the space into a regular, periodically repeating grid. Without anything in it, the lattice has no material existence—equivalent to our imaginary X-ray laboratory now being successfully floored with white tiles leading to a uniform white appearance. To accentuate our flooring with thematically appropriate decorations, we can print molecules, or motifs, on the tiles.

We begin with the basic oblique lattice tiles and place a molecular painting on them. We select a motif that is a 2-dimensional projection of a triosephosphate isomerase (TIM) α/β- barrel domain structure (Figure 2-37). The molecule is glucocerebrosidase, a recombinant protein administered in enzyme replacement therapy of Gaucher disease.[2] There is good reason to select projections of a real molecule as a motif, because it will allow us to explore a variety of crucial crystallographic concepts in the 2-dimensional tile-arrangement experiments.

Placing the image of our TIM barrel motif onto the oblique tile (the unit lattice) creates an oblique unit cell. In contrast to the lattice, which was an empty construct of space-dividing parallel lines, a unit cell now contains the motif as an object and has substantial existence (Figure 5-4). Once we have created the unit cells, we can stack them into a repeating arrangement that forms the

Figure 5-4 Crystals as translationally periodic arrangements of unit cells. Filling the oblique unit lattice with a motif creates an oblique unit cell. The unit cells can be stacked to form an extended, translationally periodic arrangement of unit cells—the actual crystal.

Unit lattice   Motif   Unit cell

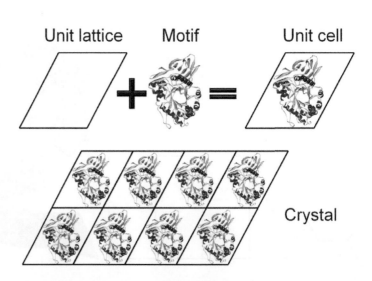

Crystal

actual 2-dimensional crystal corresponding to the tiled floor of the imaginary laboratory. The crystal shown in Figure 5-4 was derived from an oblique lattice and thus belongs to the 2-dimensional *oblique crystal system*. The unit cell does not have any additional, internal symmetry and is thus called primitive (denoted *p*) and it contains only one molecule. Thus we can classify our unit cell (and hence the crystal or crystal structure) as belonging to plane group *p*1 (systematic nomenclature of plane and space groups will be discussed in detail when we progress to 3-dimensional crystals).

The simple *p*1 structure displays a number of important properties of real crystals. Let us first remove the imaginary lattice and revisit the molecular packing of the molecules in Figure 5-5, which we encountered in Chapter 3 and which is repeated here for convenience.

- We recognize various distinct types of crystal contacts, where intermolecular interactions connect the molecules of the crystal. These contacts repeat with the same periodicity as the lattice and form an extended (in real crystals 3-dimensional) network. In our *p*1 example, this contact network is quite isotropic, meaning that this molecular packing probably has uniform properties, and probably forms a reasonably stable crystal.

- We realize that looking, with graphical display software, only at the single unit cell containing a single motif, we would not be able to tell anything about the crystal packing contacts and we could miss the fact that certain loops in this molecule are in conformations that might differ from the solution state because of crystal contacts. This underlines the necessity to also display all translationally or otherwise symmetry-related nearest neighbors (often in neighboring unit cells) for a complete analysis of a crystal structure!

Alternate unit cell origins

In another gedankenexperiment we now place the primitive lattice back over the primitive molecular arrangement illustrated in Figure 5-5. We see in Figure 5-6 that for this depiction of the same crystal a different *lattice* origin has been chosen. In fact, there are unlimited, arbitrary origin choices possible in a *p*1 structure, and there is no necessity that the unit cell is defined in such a way that a single, intact molecule is completely confined within its containment. The different unit cells with different origins still have exactly the same size, and each still contains the complete molecule, although in fragments.

Figure 5-5 Two-dimensional "crystal" in primitive plane group *p*1. Three different types of periodically repeating intermolecular interactions form the crystal contacts. Note that only a few contacts connect our molecules, explaining the general fragility of protein crystals.

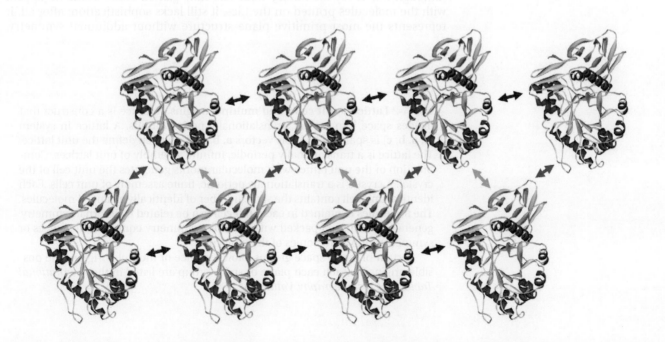

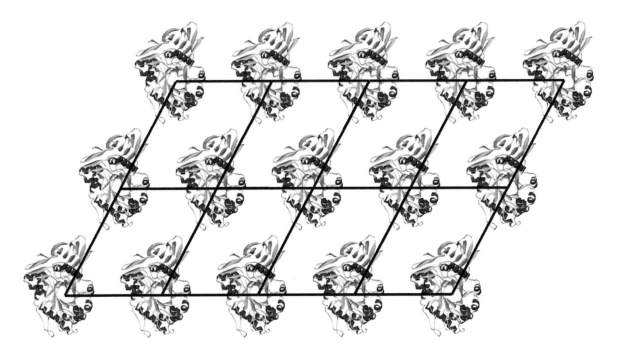

Figure 5-6 Alternate unit cell origin. The same crystal structure as shown in Figure 5-4, but with a different unit cell origin. The structure itself remains unaffected by a different choice of the *p*1 lattice origin in the crystal. Only primitive *p*1 structures and their 3-dimensional relative *P*1 allow an entirely arbitrary choice of origin, but many others allow multiple specific choices of origin.

By selecting a different origin for the unit cell in Figure 5-6, the new unit cell now contains the molecule in four disconnected pieces. The fragmented molecule poses no problem for crystallographic and computational purposes, but it is neither very appealing for display purposes nor can proper connections between atoms be drawn by the display programs. Instead of depositing the coordinates of the fragments contained within one unit cell, the atomic coordinates of the intact molecule, preferably located nearest to the origin of the unit cell, are presented in a PDB file. The molecule then extends beyond the limits of one unit cell, but translation of the parts of the molecule outside the unit cell back into the unit cell generates the equivalent unit cell as depicted in Figure 5-7. For each plane or space group, the ITC states which alternate origin locations exist. Not all plane or space groups have multiple origins.

Symmetry within the unit cell

Although our X-ray laboratory floor has now gained some aesthetic appeal with the molecules printed on the tiles, it still lacks sophistication; after all, it represents the most primitive plane structure without additional symmetry,

Box 5-2 Lattices, unit cells, and multiple origins. A lattice is a construct that divides space into regular, translationally periodic units. A lattice in system [0, **a**, **b**, **c**] is spanned by basis vectors **a**, **b**, and **c**, which define the unit lattice. The lattice is a translationally periodic, infinite assembly of unit lattices. Combination of the unit lattice with molecular motifs generates the unit cell of the crystal. A crystal is a translationally periodic, finite assembly of unit cells. Each identical unit cell contains the same number of identically arranged molecules. The molecules contained in each unit cell can be related by internal symmetry, generating a unit cell packed with multiple, symmetry equivalent instances or symmetry equivalent copies of the molecule.

Certain plane or space groups allow a choice of multiple origins. The possible origins for each each plane or space group are listed in the *International Tables for Crystallography, Volume A.*

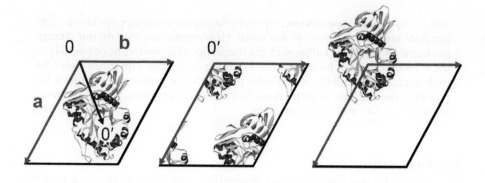

Figure 5-7 Different unit cell origins. Left: the unit cell chosen with the origin so that the whole molecule happens to fit completely within the until cell boundaries, which is rarely possible in reality. Middle: a different choice of origin, with the molecule displayed in fragments within the unit cell boundaries. The origin shift vector 00' is indicated in blue in the left panel. Right: the same origin as in the center panel, but this time the intact molecule is displayed, preferably, but not necessarily, close to the unit cell origin. For crystallographic purposes, the three different representations of the molecule are equivalent and contain the exact same information. For ease of visualization, a representation containing an intact molecule is preferred.

*p*1. One of the most basic ways to create a second copy of the motif on one tile would be to rotate our image of the TIM barrel by 180°. This manipulation is a symmetry operation around a 2-fold rotation axis, or, in short, a *2-fold rotation*, with the axis being perpendicular to the tile or plane. The multiplicity *n* of a plain rotation *axis* and its rotation angle ϕ are related by $\phi = (360/n)°$. A 3-fold symmetry operation around a 3-fold rotation axis or, in short, a *3-fold rotation* thus involves three consecutive rotations by 120°, a 4-fold rotation axis four consecutive rotations by 90°, and a 6-fold rotation six consecutive rotations by 60°, with each rotation successively creating additional copies of the original molecule.

Limitations imposed on symmetry operations. In periodically repeating systems such as crystals, specific limitations exist as to what operations can be used to create additional copies of the motif in a unit cell. The basic rules we will use in the subsequent construction of more complex structure types are:

- The application of a symmetry operation to a motif cannot generate any changes within the motif. This rule excludes any inversions and mirror operations on asymmetric motifs such as protein molecules, because the symmetry-generated, inverted or mirrored molecule would have opposite handedness or chirality. Keeping the object properties invariant upon application of a symmetry operation represents the closure requirement for a mathematical group, as further discussed in Section 5.3.

- As a consequence of the closure requirement and the reasonable limitation that only a finite number of symmetry operations are allowed, a plain rotation can only involve operations by $(360/n)°$. Such a cyclic operation generates, after a certain number of repeated applications, again the original motif in its original position (i.e. at the same point, therefore such rotations are also called point group operations). As an example, in a 4-fold rotation we reach the starting position after four consecutive rotations of 90°, in the process creating three more symmetry related copies of the original motif (called symmetry equivalents of the motif).

- The allowable symmetry operations must be compatible with the translational requirements for the specific lattice. This additional limitation imposed by translational lattice symmetry can be formally expressed as the crystallographic restriction theorem (Appendix A.8). As a consequence, only certain symmetry operations can occur in crystals; most fundamentally, only 2-, 3-, 4-, and 6-fold rotations are allowed. The locations of the rotation axes are indicated in figures and the International Tables for Crystallography by the dyad, triad, tetrad, and hexad symbols ($\blacklozenge$, ▲, ◆, and ●).

- In addition, the rotational symmetry (or point group symmetry) must be compatible with the minimum lattice symmetry. A square plane lattice for example is not compatible with a 3-fold rotation. A number of examples in the following sections will explore basic symmetry operations and the translational constraints.

Box 5-3 **Symmetry operations.** Any crystallographic symmetry operation must generate an identical copy of the motif. Only operations that do not change the motif and are compatible with the translational symmetry or periodicity of the crystal lattice can be used in the assembly of crystals. The requirement for maintenance of handedness excludes any mirror or inversion operations for proteins; the translation restrictions limit all crystallographic rotation operations to 2-, 3-, 4-, and 6-fold rotations.

Two-fold rotations

Armed with the rules for symmetry generation, one can now design a new tile for the virtual X-ray laboratory. We first rotate our molecule counterclockwise by 180° around a point. The 2-fold rotation creates an identical, symmetry equivalent copy of the motif. The position of the 2-fold symmetry axis (denoted by the *dyad* symbol ◗) is selected so that intermolecular packing contacts exist between the two molecules (Figure 5-8).

In the next step of building the 2-dimensional crystal, we translate both molecules related by the 2-fold axis (the complete *unit cell contents*) so that they form a periodic assembly. After the first translation to the right (along our axis **b**), we find that the translational arrangement has generated additional 2-fold axes along the translation vector (Figure 5-9).

Multiple unit-cell choices

The generation of additional symmetry elements also occurs when we translate the new assembly down (along a new **a**) in order to complete the filling of the plane. (Figure 5-10). Upon assignment of a unit cell, we realize that the unit cell origin in the resulting plane group *p*2 is no longer arbitrary, in contrast to the *p*1 cell. The origin must now be located on (any) one of the 2-fold axes. We also can assign a different unit cell, which has the same unit cell volume and again contains the complete unit cell contents (two molecules), but the cell parameters *a* and *γ* differ. Both cells are possible, and the conventions for selecting unit cells are provided in the ITC and listed in Sidebar 8-7.

Inspection of the *p*2 structure reveals that its packing is extremely tight. The solvent channels are narrow, and such a well-packed crystal probably diffracts to high resolution. A relation exists between the solvent content and the resolution,[3] which will be discussed in Section 11.1, introducing the Matthews probability.[4]

Four-fold rotations

Application of a 4-fold rotation to the molecular motif generates a higher level of symmetry. The generation of the unit cell content from the motif is analogous to the previously discussed *p*2 case. We rotate our molecule 90° (360/4)

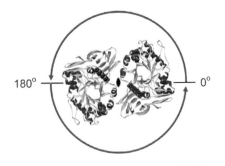

Figure 5-8 Two-fold rotation operation applied to a molecular motif. The 2-fold rotation axis is perpendicular to the paper plane, and its location is depicted by the black dyad symbol (◗). By definition, rotations are applied counterclockwise.

Figure 5-9 Translation of unit cell contents. The translational arrangement of the molecules related by a 2-fold axis perpendicular to the paper plane generates additional 2-fold symmetry axes (depicted by the red dyad symbol ◗).

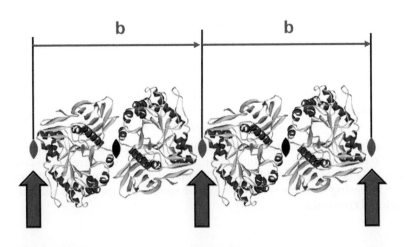

counterclockwise around the 4-fold rotation axis (symbol ◆) to create the first copy, rotate this copy another 90° to create the third molecule, and finally, a third rotation by 90° (equivalent to a 270° rotation from the original motif) generates the third copy (i.e. the fourth molecule). The fourth rotation brings us back to the original motif as required for a crystallographic rotation and is illustrated in Figure 5-11.

Just as in the *p*2 structure, we expect generation of new symmetry elements when completing the structure by translating the unit cells along the unit cell vectors. In

A

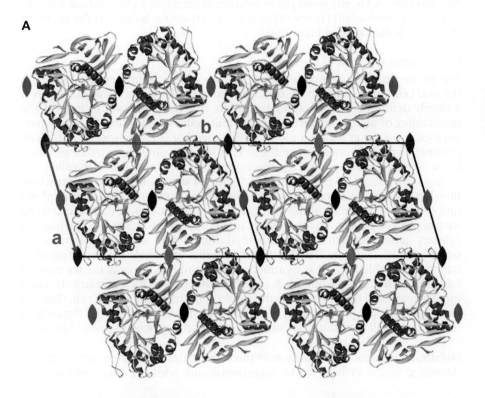

B

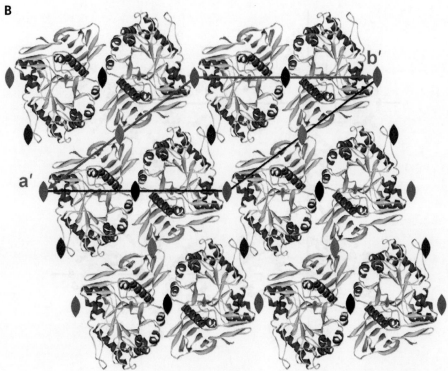

Figure 5-10 Different choices of unit cell and unit cell origins.
A crystal belonging to plane group *p*2 is superimposed with two different unit cells with different origin choices. Note that additional symmetry elements generated by the unit cell translation are located on the cell edges and corners. The crystal packing is very tight, with many intermolecular contacts and narrow solvent channels. Such an unusually well-packed crystal with low solvent content often diffracts well.[3]

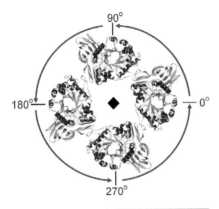

Figure 5-11 Rotation around a 4-fold rotation axis. The tetragonal unit cell is generated by rotation of the molecular motif around a 4-fold axis, depicted by the tetrad symbol (◆). Note that the 4-fold operation has generated a molecular assembly with a distinct solvent channel in the center in the direction of the 4-fold axis.

this case, additional 2-fold axes (dyads) are created in the middle of each unit cell edge, and 4-fold axes (tetrads) are located in the corners and center of the cell. Note that this structure has a large solvent channel in the center of the chosen cell, and a somewhat smaller channel at the corners. This situation is in fact quite typical for crystals with higher symmetry axes, and for protein oligomers of high rotational symmetry (the KcsA structure in Figure 3-47 is a good example).

The 4-fold rotation has now also imposed a distinct minimal symmetry upon the lattice: the lengths of lattice vectors **a** and **b** must be equal, and the angle γ between **a** and **b** is restricted to exactly 90°. The condition $|\mathbf{a}| = |\mathbf{b}| = a$ and $\gamma = 90°$ characterizes a *tetragonal plane lattice* and therefore a plane tetragonal unit cell. One second origin choice exists in the *p*4 structure, located at the center of the unit cell displayed in Figure 5-12.

The asymmetric unit of the unit cell

The *p*4 structure we assembled contains four exactly equivalent molecules in the unit cell, three of which we have generated by the 4-fold rotation. As this is a clearly defined crystallographic operation, it is not necessary to deposit the coordinates of all four molecules in the unit cell. The coordinates of one parent molecule, the generating motif, suffice together with knowledge of the cell parameters and symmetry operations. This smallest object needed to generate the whole unit cell by applying the crystallographic operations is called the asymmetric unit or AU of the unit cell. In our *p*4 structure, the asymmetric unit thus has an area (or a volume in the 3-dimensional case) of one-quarter of the unit cell; but which part of the unit cell shown in Figure 5-12 can be taken? As we can probably guess, an asymmetric unit contains the unit cell origin and the primary generating symmetry element(s). Figure 5-13 shows the asymmetric unit of our *p*4 structure. The left panel contains the whole molecule, again in fragments. In this form, the asymmetric unit would be the only tile we need to fabricate in order to lay out our X-ray laboratory floor, which of course is much more economical than producing the whole, four times larger, unit cell. One can assemble multiple paper copies of either asymmetric unit tile from Figure 5-13 into the unit cell (and thus the entire 2-dimensional crystal) to prove this point.

For similar reasons of economy, only the coordinates of an asymmetric unit are deposited in the PDB files. The asymmetric unit with the reassembled motif

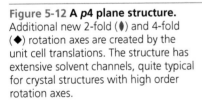

Figure 5-12 A *p*4 plane structure. Additional new 2-fold (◆) and 4-fold (◆) rotation axes are created by the unit cell translations. The structure has extensive solvent channels, quite typical for crystal structures with high order rotation axes.

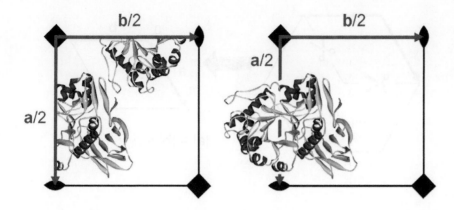

Figure 5-13 Asymmetric unit of the *p*4 structure. The asymmetric unit of *p*4 covers one fourth of the unit cell. The asymmetric unit to the left would be ideal for producing a tile (or in crystallographic computations of the unit cell contents), but the representation to the right is much better suited for displaying the molecule.

as displayed in the right panel of Figure 5-13, together with the fundamental crystallographic information (cell parameters and space group) is provided in the PDB file.

In order to reconstruct the complete unit cell from the AU, in addition to the AU contents we need the unit cell dimensions and angles, as well as instructions on how to assemble the whole unit cell. The assembly information is encoded in the so-called *plane group* (in 2-dimensional space) or space group symbol (in 3-dimensional space). To interpret space group symbols and to properly assemble unit cells, we need to take a closer look at the remaining symmetry operators, which contain the instructions on how to generate the unit cell contents. In addition, some working knowledge of crystallographic coordinates and coordinate systems and knowledge about the interpretation of space group symbols are required. All of the above is provided in great detail later in this chapter. But before we leave the 2-dimensional tile world, let us examine the one remaining plane structure or tile pattern we have not yet generated from the asymmetric unit. In trigonal and hexagonal structures—which both have a hexagonal lattice—the inter-dependence between translational and internal symmetry is especially instructional.

Three-fold and six-fold rotations

We defined as necessary conditions for a   plane lattice that it is spanned by two sets of equidistant parallel lines that intersect at a given angle. This seems to be in contradiction with our initial drawing in Figure 5-2 depicting a hexagonal tile. The hexagon-shaped tile does not fit into a basic grid system spanned by two basis vectors, although it can fill a plane without gaps. The reason this still works is that—given proper 3- or 6-fold internal symmetry—the hexagonal tile can be represented by three equivalent tiles one-third the size of the hexagon. These unit cells now satisfy the translational requirements along a plane lattice (Figure 5-14).

A closer look reveals that in order to be able to assemble the hexagon from the new unit cell tiles we need at least a 3-fold axis (or triad, ▲) in the center of the hexagon, which automatically generates the triads in the unit cell corners and two additional ones on the unit cell diagonal. A structure with such a unit cell is then—because of the internal trigonal symmetry—called *trigonal*, although the lattice they are arranged in is still a hexagonal lattice. The same procedure is used to generate the hexagonal unit cell, where the 6-fold axis (hexad, ●) creates additional 2-fold axes (◆) on the *hexagonal unit cell* edges and in the center of the hexagonal unit cell (Figure 5-15). The internal symmetry has again placed strict requirements on the lattice symmetry. For both trigonal and hexagonal lattices and unit cells the condition is $|\mathbf{a}| = |\mathbf{b}| = a$ and $\gamma = 120°$. Note that the cell parameters a and b are identical but the unit cell vectors are different, that is, $\mathbf{a} \neq \mathbf{b}$.

The trigonal planar structure, plane group *p*3, is generated in the usual way by applying the 3-fold rotation to the TIM barrel motif, and translating the molecules

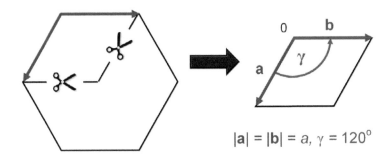

Figure 5-14 The hexagonal plane lattice. A hexagonal tile can be cut up into three equivalent rhombic pieces. Depending on the internal symmetry, the hexagonal lattice can form a hexagonal or trigonal unit cell.

$|\mathbf{a}| = |\mathbf{b}| = a, \gamma = 120°$

along the unit cell axes (Figure 5-16). The asymmetric trigonal unit has one-third of the trigonal unit cell volume. The hexagonal cell can be generated in the same way, except that six consecutive rotations by 60° generate the $p6$ cell. The asymmetric unit then contains only one-sixth of the unit cell—that is, half of the $p3$ AU—because of the additional, newly generated 2-fold symmetry in the unit cell.

Limitations resulting from asymmetry of the motif

Up to this point, we have successfully generated five different plane structures, each representing a specific *plane group*. The structures we have built belong to the five plane groups $p1$, $p2$, $p3$, $p4$, and $p6$ (although we did not explicitly show generation of a $p6$ structure). A quick look at Table 5-1 reveals that 17 plane groups exist. The combination of symmetry elements with lattice translation in two dimensions is called a *plane group*, because it fulfills the requirements defining an algebraic group. We will discuss the implications of this statement in more detail in Section 5.3.

One might wonder why we cannot build the remaining 12 plane structures listed in Table 5-1 from our molecules. The answer lies in the absence of any internal symmetry in our motif, a chiral protein molecule. As any symmetry operation must lead to an identical copy of the original object, the absence of symmetry in the motif places limits the possibilities of their arrangement in two ways:

- *A symmetry operation must generate an identical copy of the motif.* We have already stated this requirement, and the condition clearly eliminates any mirror operations, because the protein is built from L-α-amino acids, and

Figure 5-15 Trigonal and hexagonal plane system. A hexagon can be created from three trigonal or hexagonal unit cells, rotated by 120° and 240°, respectively (the three equivalent unit cells are indicated by the three colors). In a unit cell with hexagonal internal symmetry, additional 2-fold axes are created on the cell edges and in the center of the hexagonal unit cell. These relations are the same in 3-dimensional crystals, where the drawing is identical to a projection down the perpendicular third axis, **c**, pointing upward from the paper plane. The blue outlined unit cells are equivalent in area (volume) and contents to the standard unit cell defined by the vectors **a** and **b**.

> **Box 5-4 The asymmetric unit.** The asymmetric unit of a unit cell contains all the necessary information to generate the complete unit cell of a crystal structure by applying its symmetry operations to the asymmetric unit. The motif (molecule) does not have to be constrained within the boundaries of the asymmetric unit or the unit cell.

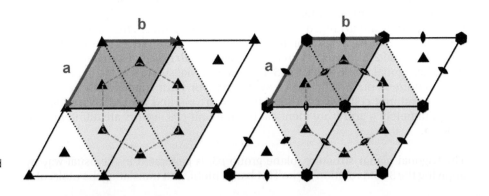

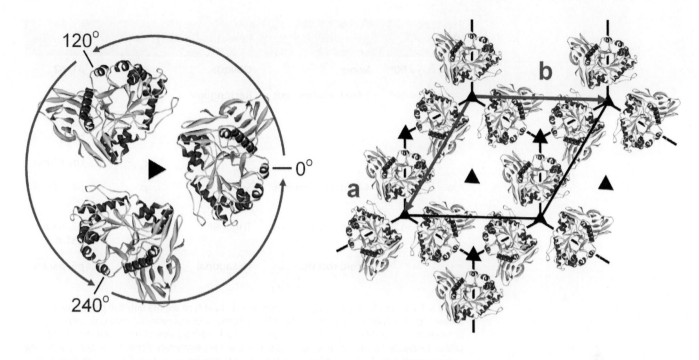

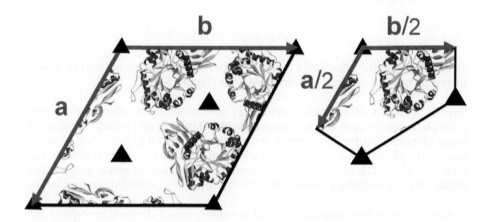

Figure 5-16 The assembly of a trigonal *p*3 structure. After generating the unit cell contents by 3-fold rotation of the motif, lattice translations along **a** and **b** generate the structure. The bottom panel shows the unit cell (left) containing three molecules, and the asymmetric unit (right) containing a single motif, again in fragments. Just as in the *p*4 structure, large solvent channels are present in the trigonal structure. As an exercise, one can cut out the asymmetric unit from paper copies of the figure and assemble the equivalent trigonal unit cell as shown in blue in Figure 5-15.

any copy generated by a symmetry operation must maintain handedness. The rule also eliminates the majority of more complex combined 2- and 3-dimensional symmetry operations that include a mirror operation, such as the glide planes or, in 3-dimensional space, any operation involving inversion (Figure 5-31 provides a 3-dimensional example).

• *An asymmetric motif cannot be located on a non-translational symmetry element.* If we place an asymmetric molecule on a rotation axis, the rotated molecule does not match the original. In contrast, we can turn a plain square by 90°, 180°, 270°, or 360° around an axis through its center, and it is still identical to the original square (in contrast, the combination of a 3-fold axis and a square would not work). We already have seen a manifestation of that rule in the fact that in protein structures the higher plain crystallographic rotation axes must sit in the center of a solvent channel and never pass through a molecule. The exact rule is: a motif located on a symmetry element must at minimum possess the symmetry corresponding to that symmetry operation. In protein structures, only water molecules (which are spherical within the resolution of protein crystallography), metal ions, or small (drug) molecules with a corresponding internal symmetry can therefore be located on symmetry elements or intersections thereof, that is, on so-called special positions.

Lattice properties	Minimum internal symmetry	Two-dimensional crystal system	Cell centering	Plane groups
$a \neq b$, $\gamma \neq 90°$	None	Oblique	p	**p1**, **p2**
$a \neq b$, $\gamma = 90°$	2-fold rotation axis	Rectangular	p	pm, pg, p2mm p2mg, p2gg
			c	cm, c2mm
$a = b$, $\gamma = 90°$	4-fold rotation axis	Square	p	**p4**, p4mm, p4gm
$a = b$, $\gamma = 120°$	3-fold rotation axis	Trigonal	p	**p3**, p3m1, p31m
	6-fold rotation axis	Hexagonal	p	**p6**, p6mm

Table 5-1 The 17 plane groups. The groups listed in bold type are the only five plane groups allowed for asymmetric, chiral motifs. Note that whether a crystal system (hence unit cell) is trigonal or hexagonal depends on the internal symmetry (3-fold or 6-fold axis) of the crystal lattice and cannot be decided from the lattice dimensions or unit cell parameters alone. This is also true for the 3-dimensional case. Note that plane groups that contain mirror operations (*m*) or glide planes (*g*) are not possible for asymmetric motifs, because they change the handedness of the chiral motif. This limitation also holds in the 3-dimensional case.

To demonstrate the effects of a forbidden operation we try to mirror our asymmetric projection of the TIM barrel molecule (Figure 5-17).

Beyond rotations, all remaining symmetry operations in the plane (mirror and glide plane operations) are not allowed for chiral motifs, and we will not discuss those elements further. For completeness, they are listed in Table 5-1. In addition, in the plane groups no structure with translational centering is compatible with an asymmetric motif. This is different in 3-dimensional space, and translational unit cell centering (also called Bravais centering) will be discussed later.

Before we expand our investigations of symmetry into 3-dimensional space and revisit the subject of crystal symmetry in greater formal depth, we take advantage of the easier visualization of projections of our molecule in the plane to introduce additional important aspects of crystal packing.

Figure 5-17 Mirror operations change the handedness of the motif. A mirror operation applied to a chiral structure produces a copy with opposite handedness and is thus not allowed. The change of handedness can be most easily seen by following the sense of direction of the helices pointed out by the colored arrows. The left, original molecule properly has right-handed α-helices, while the mirrored copy has left-handed helices (see also Figure 5-31).

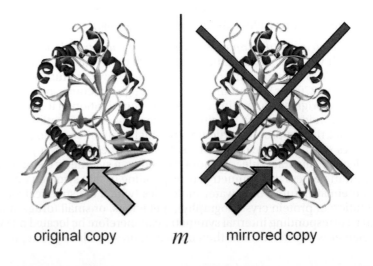

original copy   *m*   mirrored copy

Box 5-5 Asymmetry of the motif places limitations on its position in the unit cell. Asymmetric motifs cannot be located on symmetry elements. A motif located on a symmetry element or a combination thereof (i.e. on a so-called special position) must possess at least the same internal symmetry as that special position.

Non-crystallographic symmetry

Non-crystallographic symmetry (NCS) exists when more than one identical object is present in the asymmetric unit. This is frequently the case; nearly half of all structures in the PDB are dimeric structures or structures with a higher oligomerization state of the molecules. The operations relating the molecules do not have to be exact—two molecules in a dimer can differ in certain parts—and non-crystallographic operations are not limited to crystallographic operations and their combinations. As an example, pentameric molecules are quite common, although no 5-fold crystallographic axes can exist (cholera toxin, porins, and the hetero-pentameric neuroreceptors discussed in Chapter 2 are examples). An advanced treatment of NCS is given in Chapter 11, where additional examples are provided.

Let us examine the frequent case of a dimer formed by nearly perfect 2-fold rotational NCS symmetry. In this case, the two molecules will be related by a so-called pseudo-dyad, or non-crystallographic 2-fold symmetry (NCS) axis, as shown in Figure 5-19.

With the orientation selected for the figure above, our molecules can form an extended antiparallel β-sheet between them. This is a common feature in dimeric structures as shown in Figure 2-15, the concanavalin A dimer interface, or the *rml*C epimerase in Figure 5-21 of Sidebar 5-2. Numerous and strong contacts forming extended secondary structure are in general a good indication that a protein also exists as an obligate dimer in solution. As already discussed

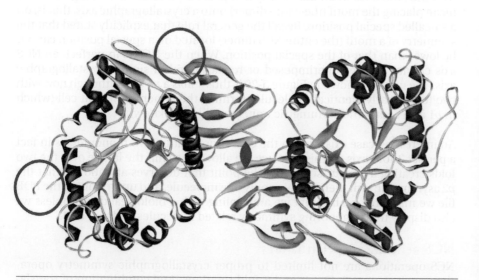

Figure 5-19 A non-crystallographic homo-dimer. The two copies of a molecule related by a 2-fold NCS axis generally are not quite identical: The molecule to the left has lost a part of its N-terminus and the extended loop is not traced (red circles), perhaps because of increased flexibility and an absence of electron density. In general, there are many small differences between the two molecules of NCS-related dimers. To emphasize the differences between the molecules clearly, the color of the helices has been changed. The NCS pseudo-dyad is indicated by the red dyad symbol.

Sidebar 5-1 M. C. Escher's periodic drawings and color symmetry. The periodic, space filling drawings of intricately intertwined creatures by M. C. Escher, the famous Dutch artist, have always had a great attraction for crystallographers and non-crystallographers alike. Escher classified his drawings by symmetry in a system of his own, and in fact rediscovered for himself all 17 plane symmetry groups. In addition, he also discovered many of the extensions resulting from additional *black and white symmetry* (46 plane groups) and *color symmetry*, a concept which plays an important role in the crystallographic study of magnetic materials in the form of spin groups. The extended color or magnetic space groups were derived in the early-mid 20th century largely by theoreticians in the USSR, mostly A. V. Shubnikov and N. V. Belov. The classical text on symmetry aspects of M. C. Escher's drawings is the compilation by C. H. MacGillavry.[5]

Figure 5-18 Color symmetry. Depicted is the view along one of the 3-fold axes of an icosahedron (20 surfaces), decorated with a space-filling Escher pattern. The axis is a plain 3-fold axis only with respect to the shape of the insect motifs; the color symmetry is much more complex—four differently colored motifs are required to decorate the surface with strictly alternating colors. Icosahedral structures are common for viruses.

Figure 5-20 Non-crystallographic dimeric symmetry in a *p*1 structure. In analogy to the *p*2 case depicted in Figure 5-10, additional non-crystallographic symmetry axes (dyad symbols in colors other than red) have been created from the original dimer axis (red) by translational lattice stacking. The single asymmetric unit of the *p*1 unit cell now contains two molecules—one entire NCS dimer forms the single motif.

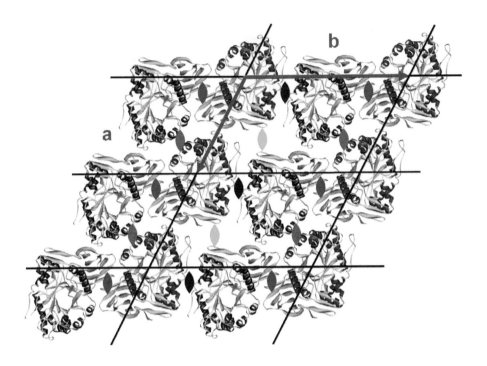

in Chapter 2 in the quaternary assembly section, calculations of the buried surface area between biological dimers yield values from approximately 700 Å² per molecule upwards, reaching several thousand Å² in large and stable dimer interfaces. Intermolecular crystal contact surfaces generally lie in the 100–500 Å² range, with an overlapping gray zone in the mid to high hundreds of Å².

Translation of the entire molecule along the vectors of a *p*1 lattice dimer generates additional NCS axes (Figure 5-20), analogously to what we observed in the *p*2 structure for crystallographic axes. However, because the molecules are not identical, we cannot translate the lattice so that its corners and edges coincide with the NCS axes in an aim to create a pseudo-*p*2 structure. Doing so would mean placing the motif (the entire dimer) onto a crystallographic axis, that is, on a so-called special position. Recall the general rule that explicitly stated that the symmetry of a motif (the entire NCS dimer) located on a special position cannot be lower than that of the special position. When the NCS is imperfect, an NCS axis cannot not be superimposed or replaced with a similar crystallographic axis. The unit cell shown in Figure 5-20 is thus still is a *p*1 cell, although now with two (only nearly identical) molecules forming the dimer in the unit cell (which in *p*1 also equals the asymmetric unit, as we already know).

An interesting case exists when the operation relating two monomers is in fact a perfect 2-fold axis. In this case, we are allowed to shift the lattice over the two folds creating a true *p*2 structure. The point that deserves attention is that the *p*2 asymmetric unit then contains only one molecule, and upon loading the PDB file we might not realize that we are dealing with a biological dimer unless we also display the neighboring symmetry related molecules (Sidebar 5-2).

NCS generated superstructures

NCS operations are not limited to proper crystallographic symmetry operations such as pure 2-, 3-, 4- and 6-fold rotations. Any combination of translation and rotation can relate two or more objects to each other. The last illustration in this section (Figure 5-22) displays an actual case of a *P*1 structure with two molecules in the asymmetric unit (note that in the space group notation of 3-dimensional structures we capitalize the "primitive" lattice symbol *P*). The two calmodulin molecules are nearly identical, but in the second, translated molecule, one domain is just slightly rotated relative to the first molecule.

Sidebar 5-2 Dimers are common and often required for function in protein structures. An example for a perfect 2-fold dimer axis becoming a crystallographic axis in a real crystal structure is provided in Figure 5-21. The structure is the rhamnose epimerase from *Mycobacterium tuberculosis*, a potential drug target because rhamnose, a bacterial sugar, is an essential component of the mycobacterial cell wall, but humans do not synthesize rhamnose.[6] The enzyme crystallized in trigonal space group $P3_221$, and when the asymmetric unit deposited in PDB file 1upi is displayed, only one molecule shows up. However, this molecule is an obligate dimer, and even the binding site is actually formed by residues of both molecules. Displaying the symmetry-related molecules shows that one of the neighboring molecules forms an especially tight interface, with ~1500 Å^2 buried surface area per molecule, which clearly falls into the true oligomer interface range. There is, in addition to the BIOMOL record in the PDB file, another indication that something is not quite right when looking at the single molecule in the asymmetric unit: One can recognize that an isolated β-structure protrudes

suspiciously from the molecule into the empty "solvent" space. As the dimer structure shows, these β-strands are in fact an integral part of the extended β-barrel formed in this dimeric structure. As always, symmetry related molecules and contacts to molecules in neighboring cells must be examined to get a full picture of the molecular structure, and attention must be paid to the BIOMOL remark in the PDB file and the additionally provided protein quaternary structure (PQS) information.

Dimerization also plays an important role in cellular signaling. Several families of membrane receptors exist that transmit cellular signals by dimerization, specifically the epidermal growth factor receptor (EGF-R, Figure 2-34) in humans[7] and some Ser/Thr kinases[8] in *M. tuberculosis*. The isolated kinase domains alone have only low millimolar binding affinity. Dimerization in response to an extracellular signal is required for their activation. In both cases the allosteric activation mechanisms were not understood until the dimeric crystal structures were accidentally obtained and subsequently confirmed by biochemical experiments.

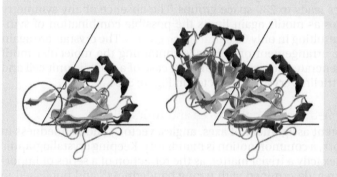

Figure 5-21 Molecular dimer axis coinciding with crystallographic axis. When the contents of the PDB file 1upi—the crystallographic asymmetric unit—is displayed, only one molecule of the obligate dimer shows up because the dimer axis coincides with a crystallographic 2-fold axis. The peculiar protrusion of the β-structure in the depicted monomer indicates that something is amiss, and the completed dimer shows that these β-strands actually forms a continuation of the β-barrel of the dimer structure. Shown in the projection are two of the trigonal unit cell edges. The 2-fold dimer axis coinciding with the crystallographic axis is parallel to unit cell vector *b* but shifted by $z = 1/6$ above the paper plane (space group $P3_221$). PDB entry 1upi.[6]

Such cases of near-perfect translational symmetry always have a distinct manifestation in the diffraction pattern, in this case a so-called translational superstructure pattern. It can be understood as resulting from a nearly perfect translational duplication of a unit cell of half the size with only one molecule in it, and manifests itself through weak layers of reflections along the (reciprocal) axis which was doubled. We will revisit this structure again in Chapter 11, where we learn more about different types of NCS (Figure 11-7).

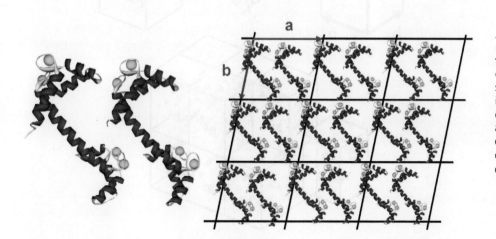

Figure 5-22 Non-crystallographic translational symmetry in a primitive cell. The two calmodulin molecules shown to the left are not exactly identical. The second copy is created by a combination of a translation by ~**a**/2 and a slight rotation of the second domain, easily recognizable by the different orientation of the green Ca^{2+} ions. PDB entry 1up5.[9]

Sidebar 5-3 **Crystal or no crystal?** Figure 5-23 provides an interesting and entertaining puzzle for rest and relaxation. Are we dealing here with a 2-dimensional crystal or not? A crystal would have to be built from identical subunits, and at first perception, the bottom and top row of the rhombic unit cells which are cut in half look distinctly different from the remaining unit cells, the top row appearing much darker than the row at the bottom.

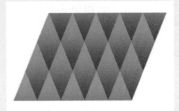

Figure 5-23 Contrast illusion. The effect of apparently different unit cells at the bottom and the top of the "crystal" is an illusion created by a color gradient within the rhomboid-shaped unit cells. They are in fact all the same, including the top and bottom half-cells. The illusion is a variant of the well-known contrast illusion. You can convince yourself by making a copy and cutting out the unit cells. It emphasizes what we said in Chapter 3: looks of (even virtual) crystals can be quite deceptive.

Box 5-6 Non-crystallographic symmetry. In addition to strict crystallographic symmetry between molecules, additional non-crystallographic symmetry can exist locally between molecules in an asymmetric unit. In many cases, this is proper rotational point-group symmetry relating molecular oligomers such as dimers, trimers, tetramers, pentamers, hexamers, and so forth, through n-fold rotation axes. However, any combination of non-crystallographic rotation and translation can relate two or more molecules. Molecular symmetry can also be the combined result of local and crystallographic symmetry, as in virus capsids with up to 60-fold symmetry as a result of combining crystallographic and proper non-crystallographic symmetry.

5.2 Symmetry in three dimensions

The self-assembly of crystals from molecules in three dimensions follows the same principles introduced in the previous section when we were examining 2-dimensional tile patterns. We expect that the added third dimension will add more possibilities of arranging the molecules in real crystals. Indeed, combination of all new symmetry elements and the 14 different 3-dimensional translational lattices leads to 230 space groups. The absence of any symmetry in protein molecules as motifs again limits the possible combination of symmetry operations, resulting in only 65 chiral space groups. The crystals are again formed by periodic arrangement of unit cells containing the molecular motif and its symmetry generated copies. The simplest case of a triclinic unit cell and the corresponding triclinic crystal is depicted in Figure 5-24.

Nomenclature of 3-dimensional lattices and unit cells

To maintain consistent assignment of axes, angles, vectors, and handedness in crystallographic work, a common notion is mandatory. Keeping crystallographic conventions is not exactly a trivial matter, as the retraction of a series of important crystal structures, determined with wrong handedness[10] and published in

Figure 5-24 Assembly of a primitive triclinic 3-dimensional crystal from unit cells. In analogy to the 2-dimensional case, the unit lattice is filled with a motif, and the crystal is built from translationally stacked unit cells. The basis vectors form a right-handed system [0, **a**, **b**, **c**].

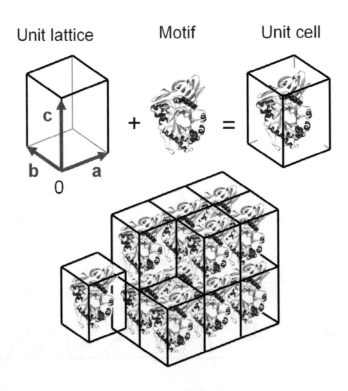

Science and other high profile journals has demonstrated (Sidebar 10-6). The following conventions for designating basis vectors, angles and planes of lattices, unit cells, and crystals in 3-dimensional space apply:

- The three basis vectors of a unit lattice [0, **a**, **b**, **c**] extend from a common origin in a right-handed system; that is, if going counterclockwise from basis vector **a** to basis vector **b**, the third basis vector **c** points upwards (Figure 5-25). The vector product **a** × **b** generates a third vector perpendicular to **a** and **b**, and the vector product **a** × **b** is *positive defined* in a right-handed system. The magnitude of this vector, |**a** × **b**|, is equal to the area spanned by the vectors **a** and **b**. The unit cell volume V_{uc} is given by the triple vector product, $V_{uc} = \mathbf{a} \cdot (\mathbf{b} \times \mathbf{c})$.

- The angle between **a** and **b** is γ, the angle between **b** and **c** is α, and the angle between **a** and **c** is β. Similarly, the plane spanned by **a** and **b** is denoted as C, the plane between **b** and **c** is A, and the plane between **a** and **c** is labeled B.

- The length of a unit cell vector is given by its norm: |**a**| = a, |**b**| = b, and |**c**| = c.

- The cell dimensions and angles are the six cell parameters (or cell constants) a, b, c, α, β, and γ.

Lattice points, lattice vectors, and fractional coordinates

Molecules and their atoms exist in what is called the real space R, in contrast to the reciprocally related reciprocal space R*, which contains the diffraction data and is introduced later. To describe or manipulate a molecular structure, we must be able to unambiguously assign the position of each atom of the molecule. In a rectilinear obverse *crystallographic basis* [0, **a**, **b**, **c**] we assign each atom a position in reference to the crystallographic unit cell. Each point p in a 3-dimensional crystal lattice can be assigned a unique real space lattice vector **r**, extending from the unit cell origin. The components of this real space vector are given in fractions of the unit cell vectors **a**, **b**, and **c**.

For crystallographic purposes, such a normalized representation of atom coordinates and their positional vectors in fractions of the unit cell vectors is very convenient, because the application of crystallographic symmetry operations and related computations in the crystallographic reference system are then entirely independent of cell parameters. The dimensionless crystallographic coordinates x, y, z which are the components of the fractional coordinate vector **x** are called fractional coordinates. In mathematical terms, our real-space

Sidebar 5-4 About coordinate and vector notation. Except when stated otherwise, we follow in this text standard notation commonly used in the crystallographic literature. We distinguish between vectors and matrices on the one hand, denoted by **bold** symbols, and their *norm* and components on the other hand, being *scalars*, denoted in *italics*). For lattice vectors or their component vectors in fractional coordinates we use lower case bold letters such as **r** or **x**. We reserve the capital letters **X**, **Y**, **Z** for Cartesian (orthogonal) world coordinate vectors, while other bold capital letters generally notate matrices, with the notable exception of the complex structure factors **F** and **E** introduced later (Chapter 6). For mathematical treatment involving matrix operations, it is important to note that any lattice vector **x** is in fact a *column vector*, although for convenience in the text sections we write the coordinate triples representing the vector components of a vector **x** as (x y z). We generally do not use commas to separate vector or matrix components, but we do so for explicit lattice points or if clarity requires it. A negative sign in front of a symbol or operator is also often expressed through a *barred* symbol: $-x = \bar{x}$. The generic symbols we use throughout the text are listed in the notation appendix, together with a brief review of vector algebra.

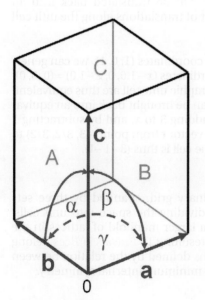

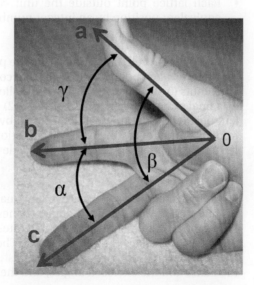

Figure 5-25 Right-handed, 3-dimensional unit lattice. A 3-dimensional unit cell is shown with its unit vectors, angles, and faces assigned in standard crystallographic notation. The angles and faces between two axes are annotated with the remaining complementary letter, for example, vectors **a** and **b** enclose angle γ and span face C, and so forth. In mathematical terms, the unit cell is a parallelepiped: a generic, 3-dimensional body formed by three pairs of parallel planes.

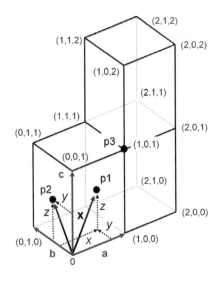

Figure 5-26 Coordinates of unit cell vertices and lattice points. A primitive unit cell is embedded in an infinite crystal lattice, of which three repeats are shown. The unit cell vectors (shown in red) extend from the origin in a right-handed system. The atom located at p1 inside the cell can be reached by real space lattice vector $\mathbf{r}_1 = (\frac{3}{4}\mathbf{a} \; \frac{1}{3}\mathbf{b} \; \frac{1}{2}\mathbf{c})$. Its fractional coordinate vector $\mathbf{x}_1$ is $(\frac{3}{4} \; \frac{1}{3} \; \frac{1}{2})$. Point p2 on unit cell face A is defined by fractional coordinate vector $\mathbf{x}_2 = (0 \; \frac{1}{2} \; \frac{1}{2})$, while lattice point p3 in the corner of the unit cell (1, 0, 1) is equivalent to the unit cell origin (0, 0, 0).

crystallographic lattice vector $\mathbf{r}$ is defined as the scalar product of the basis and the fractional coordinate vector:

$$\mathbf{r} = \begin{pmatrix} \mathbf{a} & \mathbf{b} & \mathbf{c} \end{pmatrix} \cdot \begin{pmatrix} x \\ y \\ z \end{pmatrix} = (x\mathbf{a} + y\mathbf{b} + z\mathbf{c}) = \mathbf{A}^T\mathbf{x}$$

(5-1)

where the components of the result are the translations (in units of Å) along the respective basis vectors (see Appendix A.2 for the matrix notation).

A brief exercise puts the abstract notation into practice. Each of the points in the 3-dimensional lattice depicted in Figure 5-26 can be reached by a number of translations along the lattice or unit cell vectors. To reach a lattice point, the three translations are integers; to reach any point within the unit cell, the translations are fractions of the unit cell vectors.

For example, to reach an atom located at lattice point p3 with fractional coordinates (1, 0, 1) we need to travel one full unit cell to the right along direction $\mathbf{a}$, not at all in direction $\mathbf{b}$, and again one full unit cell length in direction $\mathbf{c}$. Thus, the fractional coordinate vector of an atom located at p3 is $\mathbf{x}_3 = (1 \; 0 \; 1)$, and its crystallographic lattice vector $\mathbf{r}_3$ is $(\mathbf{a} \; 0 \; \mathbf{c})$. We can use the same procedure to reach any point in a lattice, a unit cell, or a crystal.

In order to reach the general point p1 we follow the unit cell vector in direction $\mathbf{a}$ by $\frac{3}{4}$ or $x = 0.75$ of the unit cell, along $\mathbf{b}$ about $\frac{1}{3}$ ($y \approx 0.3333$), and finally up by $\frac{1}{2}$ the length of $\mathbf{c}$. The real space coordinate vector $\mathbf{r}_1$ extending from the origin to p1 is thus $(\frac{3}{4}\mathbf{a} \; \frac{1}{3}\mathbf{b} \; \frac{1}{2}\mathbf{c})$, and the corresponding fractional coordinate vector $\mathbf{x}_1$ is ~$(0.75 \; 0.3333 \; 0.5)$. Similarly, p2 can be reached by translation along $\mathbf{b}$ by one half unit cell length, one half up in $\mathbf{c}$, and no translation along $\mathbf{a}$. Its fractional coordinate vector $\mathbf{x}_2$ thus is $(0 \; \frac{1}{2} \; \frac{1}{2})$.

Lattice translations and equivalent points

Let us take a closer look at the point p3 located at a unit cell corner. We reached p3 from the origin by translation of one unit cell length to the right along $\mathbf{a}$, one unit cell length up along $\mathbf{c}$, and not at all in direction $\mathbf{b}$. The fractional coordinates of p3 are (1, 0, 1). However, point p3 can actually be interpreted as equivalent to the origin of an adjacent unit cell. The following rule results:

- Fractional coordinates of points that lie within a unit cell are limited to the range $0 \leq x, y, z < 1$. Any point with a fractional coordinate component equal to or larger than 1.0 actually lies in an adjacent cell.

- Each lattice point outside the unit cell can be translated back into an equivalent position by an integer number of translations along the unit cell vectors.

Subtracting unit cell translations from the p3 coordinates (1, 0, 1), we can generate the equivalent point at the origin with coordinates $(x - 1.0, 0, z - 1.0) = (0, 0, 0)$. The points (0, 0, 0) and (1, 0, 1) in a crystallographic unit cell are thus equivalent. As a final example, a position (–3, 3/2, 3/2) can be brought back into an equivalent position in the unit cell (0, 0.5, 0.5) by adding 3 to x, and by subtracting 1 from its y and z components. The translation vector $\mathbf{t}$ from point (–3, 3/2, 3/2) to its new equivalent position (0, ½, ½) inside the cell is thus (3 –1 –1).

3-dimensional crystal lattices

A 3-dimensional crystal lattice is an imaginary grid, spanned by three sets of intersecting, equidistant parallel planes dividing the space into unit cells. Internal translational periodicity generates a larger manifold of lattices in the 3-dimensional space than in the plane. The resulting 14 Bravais lattices belong to one of seven 3-dimensional crystal systems defined by the relation between their axes and the enclosed angles, and their minimum internal symmetry.

> **Box 5-7 Crystallographic coordinates.** Each point in a unit cell can be assigned a unique fractional coordinate triple (x, y, z). Its positional real space vector in units of Å is defined as $\mathbf{r} = x\mathbf{a} + y\mathbf{b} + z\mathbf{c}$, and its fractional coordinate vector is $\mathbf{x}$, whose components x, y, z are dimensionless fractional coordinates. The fractional coordinates of points within a unit cell are limited to $0 \leq x, y, z < 1$. Points outside the unit cell can be translated back into equivalent positions inside the unit cell by applying an appropriate translation vector $\mathbf{t}$ to their positional vector $\mathbf{x}$.

Primitive lattices

In 3-dimensional space, we obtain six primitive lattices as special cases of the generic triclinic lattice. The additional primitive lattices are monoclinic, orthorhombic, tetragonal, hexagonal, and cubic (Figure 5-27) Each of these translational lattices is compatible with increasing internal symmetry. As in the 2-dimensional case shown in Figure 5-15 the hexagonal/trigonal lattice splits into a trigonal and hexagonal crystal system depending on the internal symmetry.

Centered lattices

The 3-dimensional space can be filled by stacking multiple copies of lattices into each other as long as the translational vectors of the lattice origins are compatible with the internal symmetry of the corresponding unit cell contents. These non-primitive, *translationally centered* lattices have been described by the French physicist Auguste Bravais (1811–1863), who showed that in addition to the six primitive 3-dimensional lattices, eight more centered 3-dimensional Bravais lattices fulfill the crystallographic symmetry compatibility requirements. Depending on the internal unit cell symmetry, lattices can be centered on one or more *faces* or in the *body center*, and in a special rhombohedral centering arrangement. The corresponding centering symbols (Bravais symbols) we encounter in standard space group settings are: C for face centering on the C plane spanned by $\mathbf{a}$ and $\mathbf{b}$; F for face centering on each of the three different faces of a cell (equivalent to stacking three additional lattices into the original cell); I for body centering; and R for rhombohedral centering.

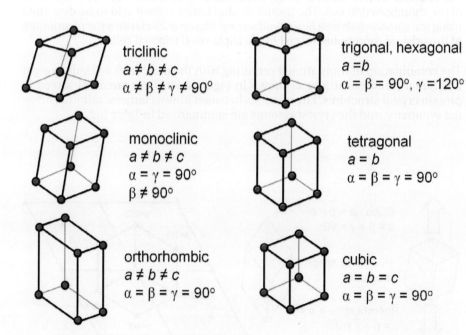

triclinic
$a \neq b \neq c$
$\alpha \neq \beta \neq \gamma \neq 90°$

monoclinic
$a \neq b \neq c$
$\alpha = \gamma = 90°$
$\beta \neq 90°$

orthorhombic
$a \neq b \neq c$
$\alpha = \beta = \gamma = 90°$

trigonal, hexagonal
$a = b$
$\alpha = \beta = 90°, \gamma = 120°$

tetragonal
$a = b$
$\alpha = \beta = \gamma = 90°$

cubic
$a = b = c$
$\alpha = \beta = \gamma = 90°$

Figure 5-27 The six primitive 3-dimensional lattices. The lattices are derived from a general oblique lattice (in which all six cell parameters are different) and are compatible with increasing internal symmetry (Table 5-2). The trigonal/hexagonal lattice splits into two different crystal systems depending on its internal minimum symmetry (3-fold or 6-fold rotation axis along lattice vector $\mathbf{c}$).

Figure 5-28 Centered lattices. A body-centered (*I*) and a C-(face-) centered lattice and the respective Bravais translation vectors. Translation along the Bravais vector (½ ½ ½) leads to the requirement that the octant in the top and rear of the white cube "cut out" by the blue lattice must contain the same arrangement and symmetry of objects as the octant closest to the origin. Similar requirements apply to the single-face-centered lattice to the right, with the translation vector (½ ½ 0) for C-centering.

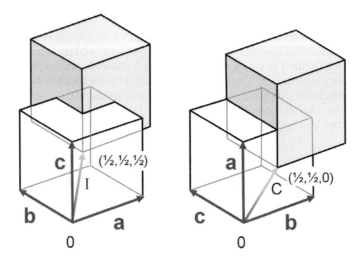

Sidebar 5-5 Rhombohedral centering can cause confusion. In a rather unfortunate decision the PDB has elected to use the symbol *H* for rhombohedral cells and space groups in standard obverse hexagonal setting, that is, for the *R*-centered rhombohedral cell indexed with cell parameters *a = b, c*, $\alpha = \beta = 90°$, $\gamma = 120°$. Some crystallographic programs use space group symbols such as *H*3 or *H*32 internally, but in the standard crystallographic reference (ITC) the centering symbol *H* is reserved for non-standard trigonal or hexagonal *base* centering. These cells can be re-indexed as primitive and you will probably never encounter them. Primitive rhombohedral cells indexed with cell constants *a*, $\alpha \neq 90°$ are discouraged and generally found in older PDB entries or (more commonly) in older inorganic structure literature.

Translational lattice centering is illustrated in Figure 5-28. Translational centering mandates that the overlapping parts of the lattice have the same contents. In the example of *I* centering, the contents of the octant at the top and rear of the first unit cell must be identical to those of the octant close to the origin. Similar restrictions hold for the other octants and apply to other types of centering as well. Translational centering must not to be confused with centrosymmetry, a concept introduced in Section 5.4. The translational centering vectors for the eight centered Bravais lattices shown in Figure 5-30 are (½ ½ 0) for *C*; (½ ½ 0), (½ 0 ½), and (0 ½ ½) for *F*; (½ ½ ½) for *I*; and (⅔ ⅓ ⅓) and (⅓ ⅔ ⅔) for *R*.

The nature of rhombohedral centering is somewhat less obvious. We have already discussed in the introductory section the fact that different choices of unit cells within the same lattice exist. One of these choices is the selection of a primitive rhombohedral cell instead of a centered trigonal cell. The rhombohedral cell can be imagined as derived from the distortion of a cubic cell by pulling along the space diagonals. The rhombohedral unit cell axes still maintain the same length, but all the angles become equally less than 90°. The resulting body is a *rhombus*, and the cell rhombohedral. However, this choice unnecessarily generates a special case of unit cell axis assignment, and it is more consistent and standard practice to select a three times as large, translationally centered trigonal cell instead of the rhombohedral cell. The rhombohedral lattice is then said to be described using the *standard obverse hexagonal setting*. Figure 5-29 clarifies the positioning of the original rhombohedral cell in the triple sized trigonal cell.

The combination of translational centering with the basic lattices leads to eight additional centered lattices as shown in Figure 5-30. They are all observed in protein crystal structures. The relations between Bravais lattices, minimal internal symmetry, and the crystal systems are summarized in Table 5-2.

Figure 5-29 Primitive rhombohedral unit cell. The primitive rhombohedral unit can be derived from a cube by pulling along the space diagonal. Instead of the primitive rhombohedral cell, a rhombohedrically centered trigonal cell is used in preference to describe rhombohedral structures. The right panel shows in projection (**c** perpendicular to paper) the relation between the triple-sized *R*-centered trigonal cell (each one of the red cells) and the corresponding primitive rhombohedral cell (blue).

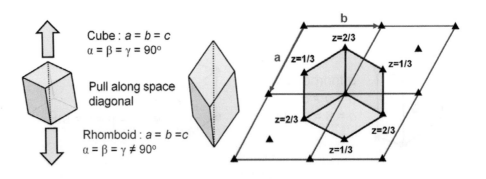

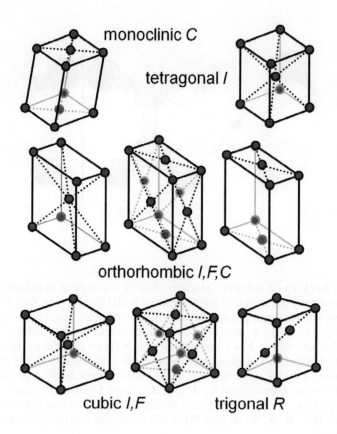

monoclinic C

tetragonal I

orthorhombic I, F, C

cubic I, F   trigonal R

Figure 5-30 Centered 3-dimensional Bravais lattices. In addition to the six primitive 3-dimensional lattices, centered Bravais lattices are derived from the primitive lattices by translational centering. The necessity for additional internal translational symmetry within the unit cells limits the number of combinations to eight. Together, there are thus 14 Bravais lattices. Lattice points located in the rear of the cells are shown with lighter borders for increased clarity.

Box 5-8 The 14 three-dimensional Bravais lattices belong to seven crystal systems. In 3-dimensional space, the combination of internal lattice translations together with the six basic translational lattices leads to 14 Bravais lattices. These lattices fall into seven crystal systems, which are defined by their minimal internal symmetry. Lattice types and crystal systems are listed in Table 5-2.

Symmetry operations in 3-dimensional space

We discussed in Section 5.1 the plain 2-, 3-, 4-, and 6-fold rotation axes as the only symmetry operators available to generate plane structures from chiral motifs. It may not come as a surprise that in 3-dimensional space more possibilities exist to fill the lattices with motifs and additional symmetry operators exist. In the general case we obtain 230 space groups (compared with 17 plane groups) which are formed by combinations of the newly available symmetry operators and the translational Bravais symmetry of the lattices. For asymmetric, chiral motifs, the necessity to maintain the handedness of the motif again eliminates any mirror or inversion operations, and reduces the number of chiral space groups to 65. The only new symmetry operations allowed for protein crystals in 3-dimensional space are screw axes, also termed roto-translations. The symmetry restrictions imposed by the asymmetric, chiral protein motifs exclude mirror planes (m), the related glide planes (d, n, a, b, c), and all inversion axes or improper rotations ($\bar{1}, \bar{2}, \bar{3}, \bar{4}, \bar{6}$). Figure 5-31 illustrates the change of chirality by inversion.

Screw axes

Screw axes are combined 3-dimensional symmetry elements, consisting of an N-fold rotation followed by a successive translation, leading to the alternate name roto-translation used in older literature. While N-fold rotations brought

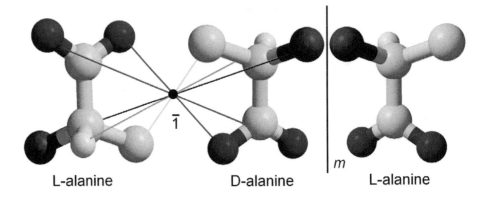

Figure 5-31 Inversion and mirror planes lead to a change in handedness. Just as in the mirror operation, the inversion operation $\bar{1}$ available in 3-dimensional space leads to a change of handedness. Any symmetry operation including inversion such as inversion axes (roto-inversions) are thus not permitted on chiral motifs. The L-alanine molecule to the left is converted into its enantiomorph D-alanine, a stereochemical reaction that in nature is actually performed by alanine racemases in bacteria. A mirror operation converts it back to L-alanine. The CORN rule (Figure 2-5) can be used to determine amino acid handedness.

the motif back to its original position after N successive applications, this is not the case for rotations that are combined with translations; that is, the screw operations are not point group operations. The rotational parts of screw axes are again limited to 2-, 3-, 4-, and 6-fold operations, and the translation vector **t** for an N-fold screw axis N_s has a magnitude of s/N times parallel to the direction **r** of the rotation axis. The index s has values from 1 to $N-1$, and any plain axis is equivalent to a screw axis with $s = 0$. Thus, we have one 2-fold screw axis (2_1,), two 3-fold screw axes (3_1 and 3_2, ▲ and ▲), three 4-fold screw axes (4_1, 4_2, and 4_3, with symbols , , and), and finally five 6-fold screw axes (6_1, 6_2, 6_3, 6_4, and 6_5, symbols , , , , and). A few examples will clarify how operations actually work.

Two-fold screw axes. The simplest screw axis is the 2-fold screw axis 2_1, where we simply rotate the molecule around an axis parallel to a unit cell vector (or cell edge) and then shift the molecule by **t** = (0 ½ 0) equal to half the length of the unit cell vector (Figure 5-32). Space groups containing 2-fold screw axes are the most frequently occurring symmetry operations in protein structures. A later section provides possible explanations for the observed space group preferences.[11]

Screw axes 3_1, 4_1, and 6_1. Packing the unit cell is also straightforward for the case of 3_1, 4_1, and 6_1, where a series of consecutive 120°, 90°, or 60° rotations

Figure 5-32 Two-fold screw axis or roto-translation. In this example, a 2-fold screw axis 2_1 is parallel to the **b** axis of the unit cell. The first part of the combined symmetry operation is to rotate the molecule or motif by 180° (note that the second molecule is not a mirror image; instead we are looking at the "back side" of the molecule), followed by translation of $b/2$ parallel to **b**. The cell is filled with translational copies of the two molecules comprising the unit cell.

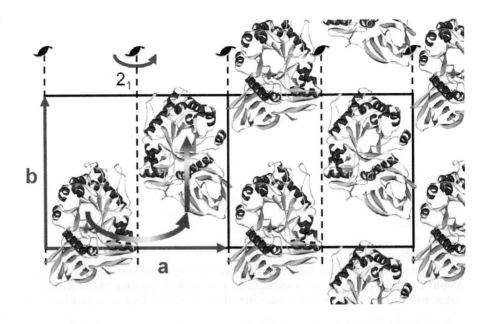

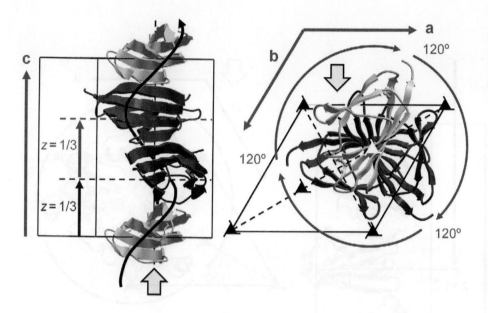

Figure 5-33 Three-fold screw axis 3₁ in a primitive trigonal structure. The figure shows the arrangement of molecules in an actual structure belonging to space group P3₁. In projection down the **c** axis as in the Tables, the original molecule (green) is rotated counter-clockwise by 120°, and shifted up by z = 1/3 along **c** to give the first copy (blue). The second copy (red) is obtained by another rotation of the blue molecule by 120° and shift by another z = 1/3. The green molecule on top is equivalent to the first green molecule and generated by a unit cell translation (0 0 1). The view from the bottom along the **c**-axis (right panel) shows the sequence of the molecular stacking. Looking along the direction of **c** and following the molecules' successive 120° rotations from green to blue to green reveals the *right-handed* helical sequence of the objects. The viewing directions generating the left panel's view are indicated by the arrows. PDB entry 1xak.[12]

plus $1/N$-translations along the corresponding axis stack the molecules above each other. Figure 5-33 shows the stacking in space group $P3_1$ (where the only symmetry element is the 3_1 screw axis) for the small, N-terminal domain of the ancillary protein ORF7A from the coronavirus that causes severe acute respiratory syndrome (SARS). It is believed that this domain of a viral membrane protein plays a role in the pathogenesis of SARS.[12] There are a number of interesting observations one can make: Comparing the projection of the $P3_1$ structure down the **c**-axis with the plain 3-fold rotation in Figure 5-16 shows that, in contrast to the plain rotation axes which must always be located in continuous, often large, solvent channels, the translational shift parallel to the axis allows a tight and continuous stacking, without generation of extended axial solvent channels. Because they require fewer specific contacts between molecules, space groups possessing one single screw axis are much more frequent than those with the same plain rotation axis (Figure 5-45).

We recognize in Figure 5-33 that the molecules in the $P3_1$ unit cell form a continuous β-sheet through crystal contacts, which is not uncommon (we make no apology for re-emphasizing the need to display symmetry-related molecules when analyzing protein crystal structures—it is that important). Looking down the **c**-axis in Figure 5-33, the right-handed helical arrangement of the molecules becomes obvious. The triangle with arms protruding to the left ▲ symbolizes the 3_1 screw axis.

Screw axes N_s with s > 1. Screw axes N_s with $s > 1$ are quite interesting operations. Symmetry related molecules are still formed following the general rule: rotate the generating motif each time $360/N$ and then shift it by a fraction of s times $1/N$ parallel to the rotation axis. Let us examine in Figure 5-34 a $P3_2$ structure ($N = 3$, $s = 2$), which is generated by taking the molecular motif, rotating it counterclockwise by 120°, then shifting it up by $z = 2/3$ parallel to the **c**-axis to generate the first molecule. The next molecule is then rotated another 120°, 240° in total, and shifted by another 2/3 of **c** to $z = 4/3$, which is outside of the unit cell. We need to translate this copy back by $\mathbf{t} = (0\ 0\ \bar{1})$ into the unit cell, which then creates the equivalent second copy inside the unit cell at $z = 1/3$ and 240°, which is also –120°. This negative sign compared with the position of $z = 1/3$ and 120° for the first copy in the 3_1 example leads to the suspicion that the successive stacking of the objects now is in the opposite, clockwise direction, and the resulting screw is left-handed. Figure 5-34 shows a $P3_2$ crystal structure with the molecules arranged in a left-handed helical sequence. The structure is a small protein from *Staphylococcus aureus* with a high structural similarity to its superantigen toxins.[13]

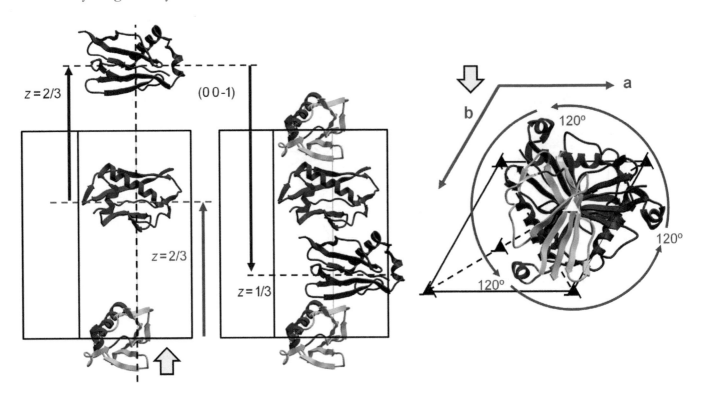

Figure 5-34 Three-fold screw axis 3₂ in a primitive trigonal structure. The figure shows the arrangement of molecules in an actual structure belonging to space group $P3_2$. In projection down the **c** axis as in the Tables, the original molecule (green β-strands) is rotated counter-clockwise by 120°, and shifted up by $z = 2/3$ along **c** to give the second copy (red). The next copy (blue) is obtained by another rotation of the red molecule by 120° and shift by $z = 2/3$. It thus is located outside of the unit cell and needs to be translated back into the unit cell. This procedure again generates a helical stacking of the molecules. The view along the **c**-axis (right panel) shows the sequence of the molecular stacking. Looking in the direction of **c** and following the rotation of the molecules reveals a *left-handed* helical sequence of the objects. The viewing directions generating the left panel's view are indicated by the arrows. PDB entry 1yn4.[13]

The situation for screw axes 4_1, 4_3, 6_1, and 6_5 can be understood following the same explanation we provided above. Both 4_1 and 6_1 (each successive operation 90° rotation plus $z = 1/4$ shift, or 60° rotation and $z = 1/6$, respectively) stack objects in right-handed helical arrangements, 4_3 and 6_5 in left-handed helical arrangements. Screw axes with opposing handedness in their generated helices form enantiomorphic pairs (pairs of opposite handedness) of axes, which is important to remember in practical crystallographic work: Changing the direction of a screw axis during re-indexing or when inverting a marker atom substructure solution requires a change of the screw axis to its enantiomorphic opposite (more explanations follow in Chapters 8 and 10).

Now let us investigate the 4_2 screw axis. The first operation generates a copy at 90° rotation, shifted by 2/4 of along the axis, that is, $z = 1/2$. The second successive rotation generates an object rotated 180° from the first point, and translated by 4/4, that is, by a whole unit cell. Bringing this molecule down places it in the same plane as, but opposite to, the first molecule, which corresponds to a plain 2-fold operation. The third operation generates a molecule at 270° rotation and $z = 1/2$, opposite the second object. Thus, the 4_2 axis is also a plain 2-fold axis, and because the molecules are not as optimally staggered as in the previously discussed cases, $P4_2$ is a very rare space group. In similar fashion we can find that 6_2 and 6_4 are also a 2-fold axis, and a 3_2 or 3_1 screw axis, respectively. Finally, a 6_3 axis is also a 3-fold axis and a 2_1 screw axis.

Box 5-9 Only plain rotations and screw axes are allowed in protein crystal structures. In 3-dimensional space only plain rotations and screw axes are allowed symmetry elements that can act on asymmetric, chiral protein motifs. Higher screw axes form enantiomorphic pairs of opposed handedness, which need to be swapped when changing the handedness of a substructure during structure determination.

With the plain axes and the screw axes, we have covered all possible symmetry operations that are available for the assembly of 3-dimensional crystal structures from asymmetric, chiral motifs. The next task will be to combine the basic symmetry elements with each other and the Bravais translations to generate the 65 chiral space groups. However, this task requires a more formal treatment, and we need to leave our so far rather intuitive approach to the subject and develop the concept of symmetry operators as mathematical instructions.

5.3   Space groups

So far we have examined single symmetry operations. As expected, the combination of up to three independent symmetry elements plus the Bravais translations can generate quite a number of symmetry related objects in the crystallographic unit cell. In the highest symmetry chiral space groups, which are cubic $F432$ and $F4_132$, 96 copies of the asymmetric unit are present in the unit cell. Although these high symmetry space groups are rare for proteins, it is clear that a formal way to construct the unit cell content from the asymmetric unit through the application of symmetry operators is required.

Symmetry operators

Symbolic operators. Let us begin again with a plain 2-fold rotation as depicted in the left panel of Figure 5-35. A 2-fold operation, a rotation of 180° around **c**, is applied to a point P, generating a new copy P′. We recognize that x becomes $-x$, y becomes $-y$, and z remains the same. In the figure next to it, we apply a 2_1 operation to the same point. Now, x becomes again $-x$, y becomes $-y$, and z becomes $z + \frac{1}{2}$ (Figure 5-35). We can formulate these operations in a simple, easy to understand, symbolic form: For the 2-fold axis along **c** we write $\bar{x}, \bar{y}, z$ and for the 2-fold screw axis along **c** we obtain $\bar{x}, \bar{y}, z + \frac{1}{2}$. These representations are the so-called symbolic operators describing each symmetry operation.

Matrix operators. The symbolic notation is quite practical for human interpretation, but not for computational purposes. A simple way to describe these manipulations in mathematical terms is through matrix operators. Each symmetry operation **W** can be expressed as a 3×3 rotation matrix **R** and a translational component **T** (column vector), and is denoted as **W(R,T)**. Let us first investigate the general case of separate rotation and translation terms:

$$\mathbf{x}' = \mathbf{Rx} + \mathbf{T} \tag{5-2}$$

where **x′** is our positional (column) vector for the new (point) position, and **R** is the 3×3 rotation matrix of the operation that acts on the original point position vector **x**, to which the translational (column) vector **T** of the operation is added. Matrix multiplication is non-commutative, that is, the order of elements in the operation is important. We can write the matrix equation (5-2) in explicit terms:

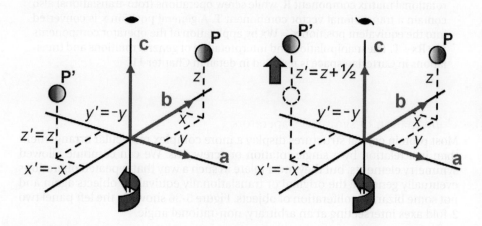

Figure 5-35 **Two-fold rotation and 2-fold screw operations.** The object or point P is transformed into an equivalent object or point P′ by applying a symmetry operator to each fractional coordinate. For the plain rotation (left panel) about **c**, x becomes $-x$, y becomes $-y$, and z remains unchanged. For a 2-fold screw axis (right panel) with its additional translational component parallel to **c**, x becomes $-x$, y becomes $-y$, and z becomes $z + \frac{1}{2}$. The corresponding symbolic operators are $\bar{x}, \bar{y}, z$ and $\bar{x}, \bar{y}, z + \frac{1}{2}$, respectively.

$$\mathbf{x}' = \begin{pmatrix} r_{11} & r_{12} & r_{13} \\ r_{21} & r_{22} & r_{23} \\ r_{31} & r_{32} & r_{33} \end{pmatrix} \cdot \begin{pmatrix} x \\ y \\ z \end{pmatrix} + \begin{pmatrix} t_1 \\ t_2 \\ t_3 \end{pmatrix} = \begin{pmatrix} r_{11}x + r_{12}y + r_{13}z \\ r_{21}x + r_{22}y + r_{23}z \\ r_{31}x + r_{32}y + r_{33}z \end{pmatrix} + \begin{pmatrix} t_1 \\ t_2 \\ t_3 \end{pmatrix} = \begin{pmatrix} r_{11}x + r_{12}y + r_{13}z + t_1 \\ r_{21}x + r_{22}y + r_{23}z + t_2 \\ r_{31}x + r_{32}y + r_{33}z + t_3 \end{pmatrix}$$

For the 2_1 screw axis parallel to **c** discussed above, we obtain in explicit terms:

$$\mathbf{x}' = \begin{pmatrix} -1 & 0 & 0 \\ 0 & -1 & 0 \\ 0 & 0 & 1 \end{pmatrix} \cdot \begin{pmatrix} x \\ y \\ z \end{pmatrix} + \begin{pmatrix} 0 \\ 0 \\ \frac{1}{2} \end{pmatrix} = \begin{pmatrix} -x \\ -y \\ z + \frac{1}{2} \end{pmatrix}$$

which is exactly the form we obtained for our symbolic operator $\bar{x}, \bar{y}, z + \frac{1}{2}$. As an example, this operator would transform a point (0.3, 0.3, 0.3) into (–0.3, –0.3, 0.8).

As a final example, the operators of a 3_2-fold axis along **c** show that non-diagonal terms in the operator matrix lead to "mixing" of vector components. The 3-fold operation generates two more copies in addition to the original at (x, y, z), and the three symbolic operators for space group $P3_2$ (example Figure 5-34, ITC space group No. 145) are:

1. x, y, z

2. $\bar{y}, x - y, z + \frac{2}{3}$

3. $\bar{x} + y, \bar{x}, z + \frac{1}{3}$

which translates into the following three **R** and **T** matrices ($\mathbf{R}_1$ is the unity operation *I* corresponding to the trivial symbolic operator x, y, z):

$$\mathbf{R}_1 = \begin{pmatrix} 1 & 0 & 0 \\ 0 & 1 & 0 \\ 0 & 0 & 1 \end{pmatrix}, \mathbf{T}_1 = \begin{pmatrix} 0 \\ 0 \\ 0 \end{pmatrix} \qquad \mathbf{R}_2 = \begin{pmatrix} 0 & -1 & 0 \\ 1 & -1 & 0 \\ 0 & 0 & 1 \end{pmatrix}, \mathbf{T}_2 = \begin{pmatrix} 0 \\ 0 \\ \frac{2}{3} \end{pmatrix}$$

$$\mathbf{R}_3 = \begin{pmatrix} -1 & 1 & 0 \\ -1 & 0 & 0 \\ 0 & 0 & 1 \end{pmatrix}, \mathbf{T}_3 = \begin{pmatrix} 0 \\ 0 \\ \frac{1}{3} \end{pmatrix}$$

> **Box 5-10 Symmetry operators.** Each crystallographic symmetry operator **W(R,T)** can be described in readily interpretable symbolic form, and in computationally suitable algebraic matrix form. Plain rotations possess only a rotational matrix component **R**, while screw operations (roto-translations) also contain a translational vector component **T**. A general position **x** is converted into the equivalent position $\mathbf{x}' = \mathbf{Wx}$ by application of the operator components $\mathbf{x}' = \mathbf{Rx} + \mathbf{T}$. The manipulation and interpretation of general rotations and translations in Cartesian space is revisited in detail in Chapter 11.

Combination of symmetry operators

Most protein crystal structures display a more complex molecular arrangement than just relation by a single rotation or screw axis. We can combine allowed symmetry elements, but they must relate in such a way that repeated operations eventually generate the original or translationally equivalent objects again and not some bizarre proliferation of objects. Figure 5-36 shows in the left panel two 2-fold axes intersecting at an arbitrary, non-rational angle.

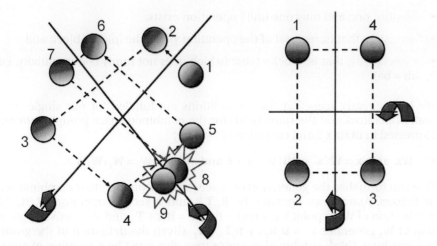

Figure 5-36 Combinations of symmetry elements. The combination of two 2-fold axes in an arbitrary angle leads to uncontrolled proliferation of eventually colliding points (a symmetry cancer, so to speak) and also generates manifolds of 2-fold axes incompatible with any lattice type. In contrast, two 2-folds intersecting at 90° generate only four points, and in 3-dimensional space, a third axis perpendicular to and intersecting both axes is generated. This orthogonal arrangement of 2-fold axes is compatible with the orthorhombic lattice type, and belongs to point group 222.

A pair of two arbitrarily oriented 2-fold axes generates a non-repeating infinite manifold of (eventually colliding) points or molecules, which is incompatible with any discrete translational lattice. A pair of 2-fold axes intersecting at 90° in contrast generates only four points during successive operations. We also see from Figure 5-36 that in three dimensions a third 2-fold axis perpendicular to the paper plane has been created. The arrangement of three perpendicular, intersecting 2-fold axes possesses point group symmetry 222.

Point groups

If single or combined symmetry operations lead after a certain number of operations back to the original molecule, they form in mathematical terms a closed group. The combination of the rotational parts of symmetry operators (i.e. without explicit consideration of the translational contributions in screw axes) forms 32 point groups in general, and for the limited operations allowed on asymmetric, chiral protein molecules (which are plain and screw rotations), 11 enantiomorphic or chiral point groups. These point groups were initially derived from observation of crystal faces, whose normals in a stereographic projection intersect at one single point, hence the name *point groups*. The fact that they were used to classify crystals according to their visual appearance led to the alternate name crystal classes for the 32 crystallographic point groups. The allowed 11 chiral point group symmetries are listed in Table 5-2.

As these point groups were determined from actual crystals, they have to be, by definition, compatible with the underlying crystal system, and what is left is to find their different possible combinations with the 14 Bravais crystal lattices, which leads to the 65 chiral space groups listed in Table 5-2. As the macroscopic, visual inspection of crystals cannot distinguish whether microscopic translations by fractions of unit cells exist, all isometric axes $N_{0,\,\dots,\,N-1}$ (with $N = 2, 3, 4, 6$) belong to the same crystal class. It also cannot be established from visual observation whether the axes *de facto* intersect at one microscopic point; the three 2_1 screw axes in $P2_12_12_1$ do not intersect, but they still belong to crystal class 222. The inability to distinguish for example a 2 from a 2_1 led to the alternate name of *internal symmetry elements* for symmetry operations containing a translational element. Because they do not bring the object back to the same position and do not intersect in a single point, the screw operations are not point group operations although they give rise to macroscopic point group symmetry. The point group symmetry manifests itself in the diffraction pattern symmetry (Chapter 6).

Space group generators

Space groups are in fact mathematical groups of (symmetry) operators. The conditions for forming a group G with elements $g_{1,2,\,\dots\,,j}$ are:

- closure, that is, any combination of operations must again generate a member of the group;

- identity, one and only one unity operation exists;

- inversion, that is, reversal of the operation yields the initial object; and

- associativity, that is, a(bc) = (ab)c (which does not imply commutativity, i.e. ab ≠ ba).

We have already seen that these conditions are fulfilled for our single symmetry operators and the same holds for the combined space group operators. Expressed in matrix form, conditions (b–d) are:

$$\mathbf{x'} = \mathbf{Wx} \text{ and } \mathbf{x} = \mathbf{W^{-1}x'} \text{ with } \mathbf{W^{-1}W} = \mathit{I} \text{ and } (\mathbf{W_1 W_2})\mathbf{W_3} = \mathbf{W_1}(\mathbf{W_2 W_3})$$

One can formalize the generation of space groups from operator combinations as follows. Assume an operator $\mathbf{W_1}(\mathbf{R_1}, \mathbf{T_1})$, and a second operator, $\mathbf{W_2}(\mathbf{R_2}, \mathbf{T_2})$. Application of $\mathbf{W_1}$ to point $\mathbf{x}$ generates first $\mathbf{x'} = \mathbf{R_1 x} + \mathbf{T_1}$, and successive application of $\mathbf{W_2}$ generates $\mathbf{x''} = \mathbf{R_2 R_1 x} + \mathbf{R_2 T_1} + \mathbf{T_2}$. Given the definition of the group, any resulting, third, combined operator then also must be a member of group G: $\mathbf{W_3} = \mathbf{W_2 W_1} \in G$ with rotational component $\mathbf{R_3} = \mathbf{R_2 R_1}$ and translational part $\mathbf{R_2 T_1} + \mathbf{T_2}$. Using the same operator, for example an n-fold rotation, applied k times, can then be computed as $\mathbf{W}_n^k$: the screw operation 4_3 is then simply $\mathbf{W}_4^3 = \mathbf{W_4 W_4 W_4}$. $\mathbf{W_4}$ is termed an operator of *order* 4, regenerating the motif in an identical position after four applications. Groups where all operators can be generated from one single operator, such as P_4, are cyclic groups. $\mathbf{W_4}$ is then the sole generator of the group. The inverse of $\mathbf{W}$, denoted $\mathbf{W^{-1}}$, can also be written in component form, and becomes $\mathbf{W^{-1}} = (\mathbf{R^{-1}}, -\mathbf{R^{-1}T})$. We can see why the more compact 4×4 $\mathbf{W}$ matrix notation is preferred in (and only in) closed group operations in fractional crystallographic coordinate space.

The final result of the application of the group rules is that three generators and the Bravais operator $\mathbf{T_B}$ suffice to form any space group:

$$G \equiv \mathbf{T_B W_1 W_2 W_3} \tag{5-3}$$

which is exactly the form of the Hermann–Mauguin space group symbol whose interpretation is explained in the next section.

For the practice of protein crystallography, it is not necessary (and may come as a relief for many) to expand into a formal treatment of group theory. It suffices to realize that the fact that point and space group operations form mathematical groups is of interest for computational purposes, particularly useful in early days, when much manual calculation and few computers were the rule. Clever use of group symmetry can greatly reduce the effort of rather extensive numerical evaluations in crystallographic or computational chemistry calculations. However, in principle every crystal structure can be solved in space group $P1$ without knowledge of the actual symmetry, which leads to an important observation for data collection: it is always better to collect data under the assumption of lower symmetry (or ignorance of symmetry) than falsely assuming high symmetry and not collecting complete data!

Space group symbols
In crystallographic practice, the space group is—with certain limitations to be discussed later—experimentally determined from the diffraction data. The interpretation of the Hermann–Mauguin space group symbol (after C. Hermann and C.-V. Mauguin), which we have used so far without paying much attention to its form, is basic to understanding diffraction symmetry and necessary to properly generate unit cells and to pack crystal structures.

The space group symbols are actually quite easy to interpret following the general form we derived from basic group theory:

$$G \equiv \mathbf{T_B W_1 W_2 W_3}$$

The first letter $\mathbf{T_B}$ signifies the Bravais lattice type, followed by one to maximally three generating operators $\mathbf{W_i}$. For monoclinic space groups, the operator is a 2

Lattice properties	Minimum internal symmetry	Crystal system	Point group	m	Bravais type	B	Lattice type	Chiral space groups	z, M
$a \neq b \neq c$ $\alpha \neq \beta \neq \gamma \neq 90°$	None	Triclinic	1	1	P	1	aP	$P1$	1
$a \neq b \neq c$ $\alpha = \gamma = 90°$ $\beta \neq 90°$	2-fold rotation axis parallel to unique axis **b**	Monoclinic	2	2	P	1	mP	$P2, P2_1$	2
					C	2	mC	$C2$	4
$a \neq b \neq c$ $\alpha = \beta = \gamma = 90°$	3 perpendicular, non-intersecting 2-fold axes	Orthorhombic	222	4	P	1	oP	$P222, P222_1, P2_12_12, P2_12_12_1$	4
					I	2	oI	$I222, I2_12_12_1$	8
					C	2	oC	$C222_1, C222$	8
					F	4	oF	$F222$	16
$a = b \neq c$ $\alpha = \beta = \gamma = 90°$	4-fold rotation axis parallel to **c**	Tetragonal	4	4	P	1	tP	$P4, P4_1, P4_2, P4_3$	4
					I	2	tI	$I4, I4_1$	8
			422	8	P	1	tP	$P422, P42_12, P4_122,$ $P4_12_12, P4_222, P4_22_12,$ $P4_322, P4_32_12$	8
					I	2	tI	$I422, I4_122$	16
$a = b \neq c$ $\alpha = \beta = 90°$ $\gamma = 120°$	3-fold rotation axis parallel to **c**	Trigonal	3	3	P	1	hP	$P3, P3_1, P3_2$	3
					R	3	hR	$R3$	6
			32	6	P	1	hP	$P312, P321, P3_112,$ $P3_121, P3_212, P3_221$	9
					R	3	hR	$R32$	18
	6-fold rotation axis parallel to **c**	Hexagonal	6	6	P	1	hP	$P6, P6_1, P6_5, P6_2, P6_4, P6_3$	6
			622	12	P	1	hP	$P622, P6_122, P6_522,$ $P6_222, P6_422, P6_322$	12
$a = b = c$ $\alpha = \beta = \gamma = 90°$	Four 3-fold axes along space diagonals	Cubic	23	12	P	1	cP	$P23, P2_13$	12
					I	2	cI	$I23, I2_13$	24
					F	4	cF	$F23$	48
			432	24	P	1	cP	$P432, P4_232, P4_332, P4_132$	24
					I	2	cI	$I432, I4_132$	48
					F	4	cF	$F432, F4_132$	96

Table 5-2 The 65 chiral space groups. Lattice properties, lattice symmetry, the resulting crystal systems, the 11 enantiomorphic point groups and their multiplicity (m), the Bravais lattice translations and their multiplicity (B), lattice type, and the 65 chiral space groups are listed together with the general position multiplicity (M or z; $M = m \cdot B$), for each enantiomorphic space group. The general position multiplicity M is equivalent to the number of asymmetric units that make up the entire unit cell. In the triclinic lattice symbol aP, a stands for anorthic. The symbols follow the Hermann–Mauguin notation. Augmented Table 6-6 includes additional information relevant for data collection.

or 2_1 axis parallel to the unique axis **b** (**b** is perpendicular to the **a**–**c** plane, which encloses the single oblique angle β, thus the name monoclinic). In the orthorhombic systems, the three generating perpendicular rotation or screw axes, in sequence **a**, **b**, **c**, are listed after the Bravais symbol. In the tetragonal, trigonal, and hexagonal systems, the same convention holds, with the first symbol denoting the generating operation along **c**, thus these systems are also called uniaxial. Note that in trigonal space groups sharing point group 32, the location of the 2-fold axis perpendicular to **c** distinctly differs as indicated by the sequence of the symbols. For example, in $P3_221$ equivalent 2-fold axes are parallel to the unit cell edges (as shown in the example of Figure 5-21) at $z = 1/6$, $1/2$, and $5/6$, while in $P3_2 12$ the 2-fold axes and resulting screw axes are located perpendicular to the unit cell edges and diagonals (compare the projection drawings for space groups 153 and 154 in the International Tables). The distinctly different location of the 2-fold axes requires the sampling of different sections of the reciprocal space to obtain a complete data set for each of these space groups (Figure 6-36). This is occasionally overlooked and we will explain the relation of crystal (point group) symmetry and reciprocal space coverage in Chapter 6. In cubic space groups, the distinguishing feature are mandatory 3-fold axes along the four space diagonals of the cube, which are indicated by 3 as the second symbol after the Bravais indicator (see Table 5-2 for examples).

Space group tables

The actions of the symmetry elements upon each other combined with the lattice translations generate many additional symmetry elements in a unit cell, and we need some aid to visualize their locations in the unit cell. A compilation of space group symmetry, origin choices, asymmetric unit extent, generating symmetry operators, and additional useful information for the crystallographer is provided in Volume A of the *International Tables for Crystallography*[1] (ITC) or, in short, Tables. Every crystallographer should be able to extract the most relevant information from these tables. The space group table for space group C2, reproduced here with permission from the International Union for Crystallography (IUCr), may serve as an example.

Header section. In the header section (Figure 5-37) of the space group table we find the short Hermann–Mauguin symbol, followed by the Schönflies symbol (an alternative annotation preferred by chemists), the point group indicator, and the crystal system name. In the second row are the space group number, the full Hermann–Mauguin symbol (which also includes the axes that do not have any associated symmetry elements and are omitted in the short symbol) and the Patterson symmetry (a term we will have to explain later). The space group in the example is C2, No. 5, and its generators are the C-centered Bravais translation (½ ½ 0) and a 2-fold rotation. Note that in crystallographic computing a third form of the space group symbol with explicit origin, the *Hall symbol*,[14] is also occasionally used.

Unit cell projections. The next section of the Tables depicts projections of the unit cell revealing the location of the symmetry elements (Figure 5-38). In the top left corner we see the projection down the unique axis **b**, and all 2-fold and newly generated 2-fold screw axes parallel to **b** and perpendicular to the **ac** plane are shown. We already know the symbols for 2 and 2_1 perpendicular to the projection plane, and the arrows and semi-arrows, respectively, represent these axes parallel to but not in the projection plane, while lines indicate axes in the projection planes. The diagram at the bottom right of Figure 5-38 is the general position diagram, that is, it shows the relative locations of the z general positions. General positions are those not located on any symmetry element.

Origin and asymmetric unit. The text block below the unit cell diagrams lists the origin position(s) for the space group, followed by the limits of the asymmetric unit (Figure 5-39). The "Origin" statement here is actually quite interesting: it states "on 2," which means the origin can be selected anywhere on any of the equivalent 2-fold axes. Such space groups with freely selectable origins

International Tables for Crystallography (2006). Vol. A, Space group 5, pp. 124–131.
Copyright © 2006 International Union of Crystallography

Figure 5-37 Header section of space group C2 in the *International Tables for Crystallography*.

$C2$	C_2^3	2	Monoclinic
No. 5	$C121$		Patterson symmetry $C12/m1$

UNIQUE AXIS b, CELL CHOICE 1

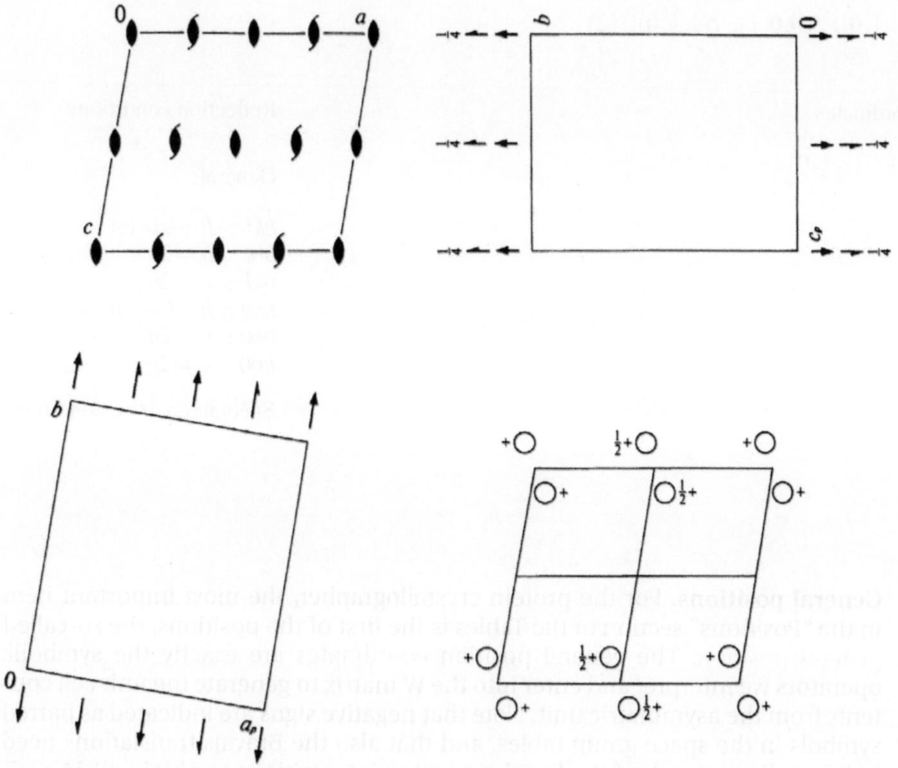

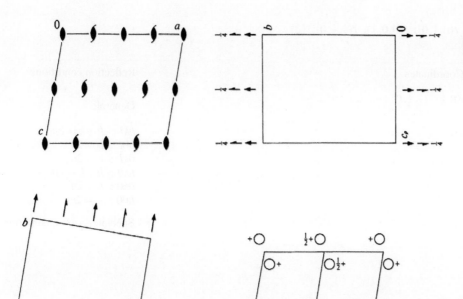

Figure 5-38 **Unit cell projections of space group** *C2* **in the** *International Tables for Crystallography.*

are called polar space groups. The next item contains the limits of the asymmetric unit, whose content suffices to generate the complete unit cell. Under "Symmetry operations" a coded list contains the type, location, and orientation of the symmetry elements used to generate the diagrams. In addition to the identity operation 1, the 2-fold axis along **b** and the Bravais translation generate a 2-fold screw axis parallel to **b** indicated as 2(0, ½, 0) in the **ab** plane, offset 1/4 in the x direction. As an exercise, one can verify that this selection indeed matches the depiction in the diagrams of Figure 5-38.

Generators and point positions. Under "Generators selected" the Tables list first the Bravais generator $\mathbf{T}_B$ and then up to three symmetry operators $\mathbf{W}_{1-3}$ (plus the inversion operator, which is not applicable to the 65 chiral space groups) from the "Symmetry operations" section that generate the equipoint positions listed by number in the following "Positions" section. The header of the "Positions" section contains, under "Coordinates," the Bravais translation vectors, followed by a list of general and special site positions. Each position line begins with the multiplicity M of the position, which for the general position is identical to the number of asymmetric units in the unit cell as provided in the space group in Table 5-2. After the multiplicity, the *Wyckoff* site label followed by the site symmetry is listed. The remaining entries are the m position coordinates in symbolic form (Figure 5-40).

Origin on 2

Asymmetric unit $0 \leq x \leq \frac{1}{2};$ $0 \leq y \leq \frac{1}{2};$ $0 \leq z \leq 1$

Symmetry operations

For $(0,0,0)+$ set

(1) 1   (2) 2 $0,y,0$

For $(\frac{1}{2},\frac{1}{2},0)+$ set

(1) $t(\frac{1}{2},\frac{1}{2},0)$ (2) $2(0,\frac{1}{2},0)$ $\frac{1}{4},y,0$

Figure 5-39 **Origins, asymmetric unit limits, and symmetry operators used to generate the space group diagrams of space group** *C2* **in the** *International Tables for Crystallography.*

Generators selected   (1); $t(1,0,0)$; $t(0,1,0)$; $t(0,0,1)$; $t(\tfrac{1}{2},\tfrac{1}{2},0)$; (2)

Positions

Multiplicity, Wyckoff letter, Site symmetry			Coordinates $(0,0,0)+$ $(\tfrac{1}{2},\tfrac{1}{2},0)+$		Reflection conditions

General:

4	c	1	(1) x,y,z	(2) $\bar{x},y,\bar{z}$

hkl : $h+k=2n$
$h0l$: $h=2n$
$0kl$: $k=2n$
$hk0$: $h+k=2n$
$0k0$: $k=2n$
$h00$: $h=2n$

Special: no extra conditions

2	b	2	$0,y,\tfrac{1}{2}$

2	a	2	$0,y,0$

Figure 5-40 Generators, point positions, and reflection conditions in space group C2 in the *International Tables for Crystallography*.

General positions. For the protein crystallographer, the most important item in the "Positions" section of the Tables is the first of the positions, the so-called general position. The general position coordinates are exactly the symbolic operators we interpret and enter into the **W** matrix to generate the unit cell contents from the asymmetric unit. Note that negative signs are indicated as barred symbols in the space group tables, and that also the Bravais translations need to be applied to each of the listed general point positions to obtain all M positions or operators. The general position multiplicity M or z is then obtained as the product of the multiplicity m of the point group, which can be read from the last of the general position entries (here 2), and the multiplicity of the Bravais translation (2 for C), which yields $z = 4$ for space group C2. In other words, for C2, the two general positions (x, y, z) and $(-x, y, -z)$ get doubled to four general positions by adding the Bravais translation (½, ½, 0) to each of them yielding the following four general positions:

1. x, y, z

2. $\bar{x}, y, \bar{z}$

3. $x + \tfrac{1}{2}, y + \tfrac{1}{2}, z$

4. $\bar{x} + \tfrac{1}{2}, y + \tfrac{1}{2}, \bar{z}$

In the case of centrosymmetric space groups which are verboten for protein structures, we would have to apply in addition the inversion operator $(-x, -y, -z)$ to each of the general point positions so obtained, further doubling z.

Special positions. The remaining lines of the "Positions" section list special positions. Protein molecules, because of the absence of any symmetry, cannot be located on any special position, that is, on or at the intersection of point group symmetry elements. Any motif placed on a special position must possess an internal symmetry at least as high as that of the special position. Spherical atoms can sit on any special position, and indeed heavy atoms or water molecules (which are spherical water "atoms" at protein crystallographic resolution) can be found on special positions (see Figure 3-16 for an example).

Reflection conditions. Of interest for space group determination are the so-called reflection conditions for the general position, which must be fulfilled for a reflection to be present. You may notice that the reflection condition for the general position $h + k = 2n$ (even) includes the other listed conditions. We will cover the extinction rules and their formal derivation in Chapter 6 when discussing the intensities of diffracted X-rays.

The remainder of the Tables provides special projection, subgroup, and super-group symmetries, which are of less interest for routine protein crystallography work. An example in the next section will demonstrate the use of the space group tables for the generation of a packing diagram.

Generation of unit cell contents

Publications of crystal structure papers often show figures of the unit cell packing. Armed with the information from the ITC, we can generate the whole unit cell contents and pack a unit cell using the following simple procedure:

1. Take the asymmetric unit containing the generating motif, and apply the corresponding general symmetry operations according to the general position operator listed in the Tables.

2. Apply the Bravais translations listed at the top of the "Generators selected" section to the generated set of symmetry-equivalent objects.

3. Translate "stray" molecules located outside of the cell back into the cell.

4. Verify positions of symmetry elements against the ITC and complete packing.

5. For a complete view of all possible packing interactions, also generate unit cell translated molecules from neighboring cells that are in contact with a region of interest of the molecule.

We can exercise packing of the unit cell for a structure crystallizing in space group $C2$, a not uncommon space group (Figure 5-44). PDB entry 1jwo[15] is a small SH2 domain of the C-terminal Src kinase (Csk) implicated in the recognition of the HER-2/neu receptor, which is overexpressed in 30% of breast cancers in the Western world and a potential drug target.[16] The space group tables reveal that we expect four molecules in the unit cell, two generated by the 2-fold axis, doubled to four in total by the Bravais C-translation ($\frac{1}{2}$, $\frac{1}{2}$, 0) in the **ab** plane. Figure 5-41 shows the molecule in projection down the unique **b** axis, and in projection looking down the **c** axis, which are the same projections used in the ITC (Figure 5-38). We pick the 2-fold axis in the center of the cell and rotate the molecule by 180°, generating the second copy.

In the second step, shown in Figure 5-42, the Bravais lattice translation of ($\frac{1}{2}$, $\frac{1}{2}$, 0) is applied to both molecules. The operation positions one molecule

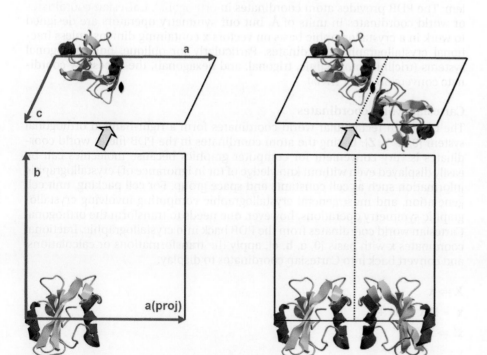

Figure 5-41 **Packing of a $C2$ unit cell: step 1.** The figure shows projections of the unit cell down the unique **b** axis in the top row and the projection looking down **c** in the bottom row. The first molecule is loaded from the PDB file, and then a suitable 2-fold axis is selected and the 2-fold rotation is applied. Any of the 2-fold operations shown in the $C2$ space group tables would work, but the selected one in the center keeps the molecule in the unit cell.

Figure 5-42 Packing of a C2 unit cell: step 2. In the next step of unit cell generation, the original and the molecule generated by the equipoint operator (–x, y, –z) are translated by the Bravais C-centering vector (½ ½ 0). The operation shifts one molecule out of the unit cell, which is brought back into the cell by a translation of (–1 0 0). Several of the unit cells may have to be stacked to examine all possible intermolecular contacts of an individual molecule.

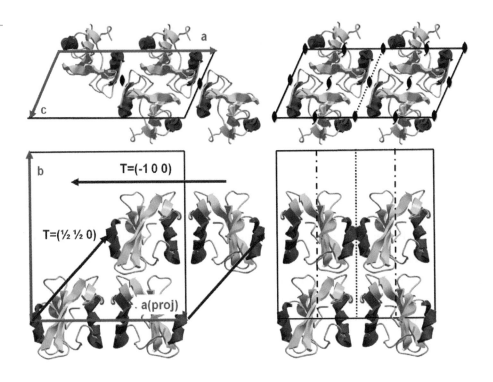

outside the unit cell, but it can be readily translated back into the cell. Inspection of the projections shows that indeed the Bravais translation has generated a set of new 2-fold screw axes in addition to the plain 2-fold axes. The packing diagram reveals that the axes are in the correct positions, in agreement with the projections in the ITC. No collisions occur, and no molecule lies on a plain 2-fold axis (recall that the limitation does not apply to screw axes due to the translational offset of the newly generated molecules).

We can now attempt to write a program and repeat this exercise *in silico* by applying the symmetry operations from the space group table to the coordinates of each atom in the corresponding PDB entry. However, we encounter a problem: The PDB provides atom coordinates in orthogonal, Cartesian coordinates or world coordinates in units of Å, but our symmetry operators are designed to work in a crystallographic basis on vectors **x** containing dimensionless fractional crystallographic coordinates. Particularly for oblique, non-orthogonal systems (triclinic, monoclinic, trigonal, and hexagonal), the necessary coordinate conversion is not trivial.

Cartesian world coordinates

The Cartesian rectangular world coordinates form a right-handed orthogonal system [0, **X**, **Y**, **Z**]. Listing the atom coordinates in the PDB file in world coordinates is very convenient for computer graphics because molecules can be easily displayed even without knowledge of (or in ignorance of) crystallographic information such as cell constants and space group. For cell packing, unit cell generation, and most general crystallographic computing involving crystallographic symmetry operations, however, one needs to transform the orthogonal Cartesian world coordinates from the PDB back into crystallographic, fractional coordinates **x** with basis [0, **a**, **b**, **c**], apply the transformations or calculations, and convert back into Cartesian coordinates to display:

$$\mathbf{X} \Rightarrow \mathbf{x}$$
$$\mathbf{x}' = \mathbf{W}\mathbf{x} = \mathbf{R}\mathbf{x} + \mathbf{T}$$
$$\mathbf{x}' \Rightarrow \mathbf{X}'$$

There are two connected issues to address when converting between fractional and Cartesian coordinates: First, the fractional coordinates are normalized in the interval $0 \leq x, y, z < 1$ irrespective of cell dimensions, and second, the angles in crystallographic systems are not necessarily orthogonal. Starting from orthogonal fractional systems, the coordinate transformation reduces to a simple multiplication of the coordinate values by the corresponding cell constants, that is, $X = ax$ or inversely, $x = X/a$. As an example, in a tetragonal cell with $a = b = 40$ Å, and $c = 60$ Å, an atom with fractional coordinates (½, ½, ½) would be at located at Cartesian world coordinates (20, 20, 30). However, for oblique systems, the transformation is not so simple. In the hexagonal system depicted in Figure 5-43, point P at (0, 1, 0) is projected on the Cartesian axes to find the Cartesian coordinates, and the X coordinate of P becomes $b \cdot \cos \gamma$ (not a!) and the Y coordinate becomes $b \cdot \sin \gamma$.

The orthogonalization matrix. The general procedure for transformation from fractional crystallographic coordinates to orthogonal Cartesian world coordinates and vice versa is to apply a transformation matrix to the positional vector. The orthogonalization matrix **O** and its inverse **O**$^{-1}$, the deorthogonalization matrix, serve this purpose. The orthogonalization matrix performs two tasks: (i) the orthogonalization of the general anorthic crystallographic basis, and (ii) the transformation from dimensionless fractional coordinates to Cartesian coordinates in Å. The matrix elements of the orthogonalization matrix thus have the units of Å.

Any point with fractional coordinate vector **x** can be converted into the Cartesian coordinate vector **X** by multiplication with the orthogonalization matrix **O**:

$$\mathbf{X} = \mathbf{O}\mathbf{x} \text{ and } \textit{vice versa } \mathbf{x} = \mathbf{O}^{-1}\mathbf{X} \tag{5-4}$$

Establishing this matrix depends on the convention used for the axis projection. The following convention shown in Figure 5-43 is followed by the PDB and practically all macromolecular crystallographic programs:

- Cartesian axis **X** is collinear with crystallographic axis **a**.
- **Y** is collinear with $(\mathbf{a} \times \mathbf{b}) \times \mathbf{X}$.
- **Z** is collinear with $(\mathbf{a} \times \mathbf{b})$ (or with the reciprocal axis **c***, see Chapter 6).

Using the above convention, the calculation yields the explicit general form of the orthogonalization matrix **O**:

$$\mathbf{O} = \begin{pmatrix} a & b\cos\gamma & c\cos\beta \\ 0 & b\sin\gamma & \dfrac{c(\cos\alpha - \cos\beta\cos\gamma)}{\sin\gamma} \\ 0 & 0 & \dfrac{V}{ab\sin\gamma} \end{pmatrix} \tag{5-5}$$

with

$$V = abc(1 - \cos^2\alpha - \cos^2\beta - \cos^2\gamma + 2\cos\alpha\cos\beta\cos\gamma)^{1/2} \tag{5-6}$$

and $V = \det(\mathbf{O})$. The derivation of **O** from the metric tensor **G** is provided in Appendix A.3.

The inverse procedure, transforming Cartesian coordinates as provided by the PDB files into fractional coordinates, is performed by using the deorthogonalization matrix, **O**$^{-1}$ (with elements in units of Å^{-1}) which symbolically expands into

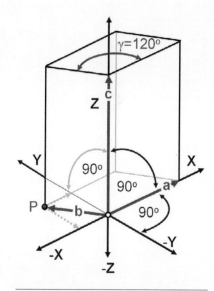

Figure 5-43 Conversion of fractional into Cartesian coordinates. The orthogonal Cartesian world coordinate system [0, **X**, **Y**, **Z**] shares the origin with the crystallographic coordinate system [0, **a**, **b**, **c**]. In the depicted case of a uniaxial hexagonal crystallographic system **a** and **X**, and **c** and **Z** coincide, and the coordinates of point P (0,1,0) become $(b \cdot \cos \gamma, b \cdot \sin \gamma, 0) = b \cdot (-0.5, 0.866, 0)$.

$$
\mathbf{O}^{-1} = \begin{pmatrix}
\dfrac{1}{a} & -\dfrac{\cos\gamma}{a\sin\gamma} & \left(\dfrac{b\cos\gamma c(\cos\alpha-\cos\beta\cos\gamma)}{\sin\gamma}-bc\cos\beta\sin\gamma\right)\dfrac{1}{V} \\[3ex]
0 & \dfrac{1}{b\sin\gamma} & -\dfrac{ac(\cos\alpha-\cos\beta\cos\gamma)}{V\sin\gamma} \\[3ex]
0 & 0 & \dfrac{ab\sin\gamma}{V}
\end{pmatrix}
\tag{5-7}
$$

As an exercise we take the CRYST1 record from PDB file 1upi

```
CRYST1   64.905   64.905   87.203   90.00   90.00 120.00 P 32 2 1
```

and calculate the deorthogonalization matrix $\mathbf{O}^{-1}$. For angles α and $\beta = 90°$, all sine terms in matrix 5-7 are multiplied by 1 and all cosine-containing terms evaluate to 0, which greatly simplifies the calculation and reduces Expression (5-6) to $V = abc(1-\cos^2\gamma)^{1/2}$, which computes to $318\,140$ Å^3. Using this enumeration and the remaining terms of $\mathbf{O}^{-1}$

$$
\mathbf{O}^{-1} = \begin{pmatrix}
\dfrac{1}{a} & -\dfrac{\cos\gamma}{a\sin\gamma} & 0 \\[3ex]
0 & \dfrac{1}{b\sin\gamma} & 0 \\[3ex]
0 & 0 & \dfrac{ab\sin\gamma}{V}
\end{pmatrix}
$$

we obtain exactly the matrix part of the SCALE records in the PDB file 1upi:

```
SCALE1      0.015407  0.008895  0.000000        0.00000
SCALE2      0.000000  0.017791  0.000000        0.00000
SCALE3      0.000000  0.000000  0.011467        0.00000
```

Given the deorthogonalization matrix $\mathbf{O}^{-1}$, we can convert the PDB file now into fractional coordinates, apply the symmetry operations to each atom, and use the inverse of $\mathbf{O}^{-1}$, the orthogonalization matrix $\mathbf{O}$, to transform the new coordinates back into Cartesian space and then display the molecule. The combined transformation is then

$$\mathbf{X}' = \mathbf{O}(\mathbf{R}(\mathbf{O}^{-1}\mathbf{X})+\mathbf{T})$$

That is exactly the operation the molecular graphics program (*Molsoft ICM Browser Pro*[17]) performed when generating the packing figures illustrating this chapter. As an exercise, enumerating orthogonalization matrix $\mathbf{O}$, prove that the coordinate transformation given in Figure 5-43 for transforming $(0, 1, 0)$ into $(b \cdot \cos\gamma,\, b \cdot \sin\gamma,\, 0)$ is correct.

With this exercise in applied crystallographic computing, we leave the theory of space groups and investigate whether any general preferences for certain space groups can be extracted from the experimental data deposited at the PDB.

Space group preferences
We already know that due to their chiral nature, proteins can crystallize only in one of the allowed 65 chiral space groups (the rare exception are racemic mixtures such as those used in the determination of the structure of rubredoxin[18]

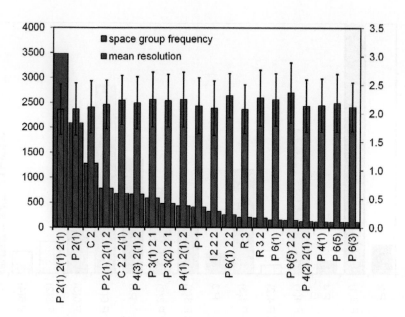

Figure 5-44 Space group preferences.
The distribution of space groups (red bars) and the mean reported resolution and its variance (blue bars, right-hand axis) of the corresponding structures are plotted for the 20 most frequent space groups from a non-redundant data set including both monomeric and oligomeric proteins.[3] Nearly 90% of all structures crystallize in one of these 20 space groups. Note that no correlation exists between the observed maximum resolution and the frequency of a space group. Screw axis subscripts are printed in parentheses.

in $P\bar{1}$ or centrosymmetric bilayers in engineered structures[19]). Analysis of the occurrence of space groups for the structures reported in the PDB shows that the distribution of space groups is by no means even. Although all allowed 65 chiral space groups have been observed, their occurrence ranks from several thousand for the most frequent space groups $P2_12_12_1$, $P2_1$, and $C2$ to a few instances for the space groups at the extreme tail of the distribution (outside of the range shown in Figure 5-44). The question arises why this strong preference for certain space groups such as orthorhombic $P2_12_12_1$ or monoclinic $P2_1$ occurs.

The first explanation that comes to mind could be that some space groups just pack much better. But packing cannot be the sole reason for their preference, because there is no correlation between the observed highest resolution and the frequency of a space group, which one would expect if packing quality alone dictated space group preference.

An observation noted already was that screw axes allow translationally staggered packing with fewer required specific contacts when compared with plain rotation axes. This is most distinctly visible in the case of higher rotation axes, along which large solvent channels form (Figures 3-47, 5-12, and 5-16). Analysis of the space group distribution confirms that the freedom introduced by translational elements has an effect: all the space groups with single screw axes are more frequent than the corresponding groups with plain axes, except those screws which themselves generate plain rotation axes ($P4_2$, $P6_2$, and $P6_4$; Figure 5-45). The preference for translational symmetry elements is also evident from the overall distribution of space groups shown in Figure 5-44: by far the most prevalent space groups are the ones that contain translational elements, either in the form of screw axes or resulting from translational Bravais cell centering, with the only exception being triclinic $P1$ without any symmetry.

Wukovitz and Yeates have investigated the possible degrees of freedom for the packing in the chiral space groups in more detail and arrived at an entropic model that qualitatively explains the space group preference for space groups of monomeric proteins.[11] In their model the total number of degrees of freedom D in which a molecule can be arranged in a certain space group is given by

$$D = S + L - C \tag{5-8}$$

where S is the number of degrees of freedom (DoF) for placing the first molecule into the unit cell, with values of 3 for $P1$ (translation only), 5 for uniaxial space

Figure 5-45 Plain rotation axes versus screw axes. Space groups with single plain screw axes (red) and those which generate plain axes (cyan) are less frequent than the corresponding space group with a single screw axis. Space group $P2_1$ is off the scale (about 100 times more frequent than $P2$).

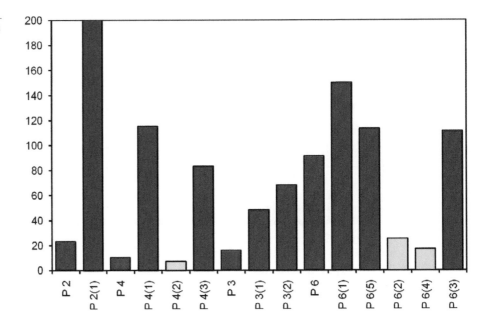

groups, and 6 for dihedral. L is the number of freely selectable cell parameters which is 6 for $P1$ (all axes and angles freely selectable) down to 1 for cubic space groups (where only one axis can be selected and the angles are fixed at 90°). The parameter C is the minimal number of unique contacts that must exist to establish a 3-dimensional network between the molecules. This number ranges from 5 in $P222$ to 2 in $P2_12_12_1$. The values of the total degrees of freedom D qualitatively explain the trend in the observed space group distribution (7 for $P2_12_12_1$, 6 for the next following groups, and 5 and 4 for the less frequent and rare groups).

With more numerous recent data, Yeates and colleagues have extended the analysis to space groups with oligomeric motifs, which show a similar trend in distributions.[20] The interesting result is that symmetrization seems to be inherently favorable for crystal formation, which is potentially useful for protein engineering by synthetic symmetrization (such as forming linked dimers of recalcitrantly crystallizing molecules).

With this final exploration of symmetry and space groups, we leave the space group section and turn our attention to a key concept in crystallography, the reciprocal lattice.

Box 5-11 **Space groups.** The combination of plain rotation axes and screw axes with the 14 Bravais lattices gives rise to the 65 chiral space groups. To generate the packed unit cell of the crystal structure, the Cartesian world coordinates of the model need to be transformed into fractional crystallographic coordinates before the symmetry operations are applied. The conversion of world coordinates into crystallographic coordinates is given by the deorthogonalization matrix (given as the SCALE record in the PDB file), and the reverse operation is given by its inverse, the orthogonalization matrix.

Space groups with the highest degrees of freedom for connecting the molecules in a 3-dimensional network occur preferentially, and by far the most common space groups are $P2_12_12_1$, $P2_1$, and $C2$ (numbers 19, 4, and 5). Single screw axes are preferred over the corresponding plain rotation axes.

5.4   The reciprocal lattice

An enormously useful concept in crystallography is that of the reciprocal lattice. The reciprocal lattice is a mathematical construct that greatly simplifies metric calculations as well as the description of diffraction events and the conditions necessary for diffraction to occur. The first task will be to navigate lattices and assign lattice planes, followed by a formal construction of the reciprocal lattice and exploration of the relations linking the reciprocal and the direct (crystal) lattice.

Lattice planes

An important task crucial to crystallographic work is the assignment and exact annotation of lattice planes. The rules for the annotation of planes in a lattice are quite similar to the annotation of lattice points. Lattice planes are formal constructs that simply slice the 3-dimensional lattice in a distinct and periodic way. Let us briefly return to an oblique 2-dimensional lattice (which can be interpreted as a projection of a monoclinic 3-dimensional lattice down the unique axis) and define a number of important properties of lattice planes (which are actually lattice "lines" in two dimensions).

Two-dimensional lattices

Figure 5-46 shows a 2-dimensional oblique lattice with basis [0, **a**, **b**] together with three sets of lattice planes. From red to blue to green, each set of lattice planes "slices" the lattice into progressively thinner slabs. This is a relevant observation: the closer together the lattice planes lie, as indicated by the progressively shorter interplanar distance vectors **d**, the finer the sampling of the lattice.

From Figure 5-46 we can derive a scheme to classify or index the lattice planes. We can observe where any one plane of a set intersects through two lattice points and count how many unit cells in each direction we have to travel from the common origin. For the red planes, we intercept lattice points at (1**a**, 1**b**); for the blue set at (1**a**, −2**b**); and for the green set at (1**a**, 3**b**). This direct indexing scheme yields the Weiss indices (*u v*) of each set of lattice planes. One drawback of the real space indexing scheme is that the planes parallel to a lattice vector (Figure 5-47) intersect the corresponding lattice vector in the infinite, which is at least mathematically a nuisance. There are other important reasons to use the *reciprocal* Miller indices to uniquely identify sets of lattice planes.

Miller indices. An indexing scheme that will turn out to be immensely useful and at the same time avoids the infinity issue of the direct Weiss indices is reciprocal Miller indexing. Taking the reciprocal of the lattice point intercepts or Weiss indices and normalizing them to integers by multiplying with the smallest common

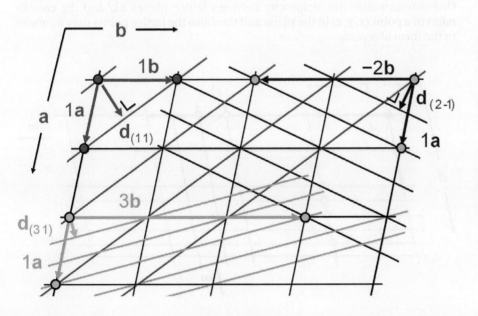

Figure 5-46 Sets of 2-dimensional lattice "planes". Three sets of lattice planes (actually lattice lines in two dimensions) with increasingly tighter spacing from red to blue to green are shown. The reciprocal Miller indices of a set of lattice planes are obtained from the lattice point intercepts (indicated by the correspondingly colored dots) by taking the reciprocal intercept values and multiplying them by their smallest common denominator. The Miller indices (*h k*) thus are (1 1) for the red set, (2 −1) for the blue set, and (3 1) for the green set. The corresponding, increasingly shorter interplanar distance vectors $\mathbf{d}_{hk}$ are shown as correspondingly colored vectors.

denominator yields the Miller indices (h k) of a lattice plane. This scheme works for any set of planes through any lattice. Following that scheme we obtain, for the set of lattice planes from Figure 5-46, the following indices: for the red lattice planes with Weiss indices (1 1) the inverse (1/1 1/1) remains (1 1); for the blue set with lattice intercepts (1 –2) the inverse (1/1 1/–2) becomes normalized by multiplication with 2 which results in Miller indices (2 –1). The green set of lattice planes in Figure 5-46 finally has the Miller indices (3 1).

An alternative procedure for obtaining reciprocal Miller indices is to count how many times a set of lattice planes intercepts the corresponding unit cell vector. The blue set for example intercepts **a** two times, but –**b** only once. The Miller indices of this set are thus (2 –1). This procedure automatically inverts the direct indices, and also works well for planes parallel to lattice vectors, as shown in Figure 5-47. For the red set with Weiss indices (½, ∞) we obtain by either procedure Miller indices (2 0); for the blue set, (0 3); and finally for the tightly spaced green lattice planes, the Miller indices (5 0).

Three-dimensional lattices and Miller indices

The reciprocal Miller indexing provides the Miller indices h, k, and l of the lattice plane. The concept of *reciprocity* is fundamental in crystallography, and it will greatly simplify the treatment of diffraction geometry. Reciprocal relations between real space (or direct space), in which we treat lattices, crystals, and all other real objects, and reciprocal space, in which we describe scattering processes and the resulting diffraction patterns, will remain a constant theme during the subsequent chapters.

One qualitative observation becomes immediately obvious from inspection of Figures 5-46 and 5-47: the higher the Miller indices, the closer the planes and the smaller the interplanar spacing between them, as indicated by the interplanar distance vectors, which are normal vectors to the planes and represent the shortest distance between the planes. Once we investigate the concept of reciprocity in more detail, we will derive important quantitative relationships between the Miller indices of a set of lattice planes (hkl), their interplanar spacing d_{hkl}, and the direction of diffracted X-rays.

Indexing of 3-dimensional lattice planes (Figure 5-48) follows the same procedure as described for the 2-dimensional case. Note that we are always describing a set of planes: there is no such thing as, for example, one single (0 2 0) plane, there are two of them in one unit cell. The fact that the (0 1 0) plane is also a member of the set of (0 2 0) planes simply means that their normal vectors or distance vectors d_h point in the same direction along direction [010].

One can formalize the reciprocity between lattice planes hkl and the coordinates of a point (x, y, z) in the plane and thus also the lattice points they intersect in the form of a plane equation:

Figure 5-47 Lattice "planes" parallel to unit cell vectors. The Miller indices of planes parallel to unit cell edges can be derived by counting how many times the set of planes crosses a unit cell axis. The planes thus are (2 0), red; (0 3), blue; and (5 0), green. Higher indices imply closer interplanar spacing. The light blue arrows again indicate the corresponding interplanar distance vectors, $\mathbf{d}_{hk}$.

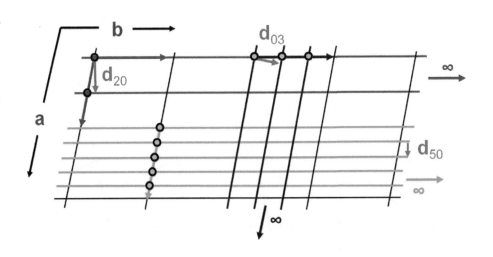

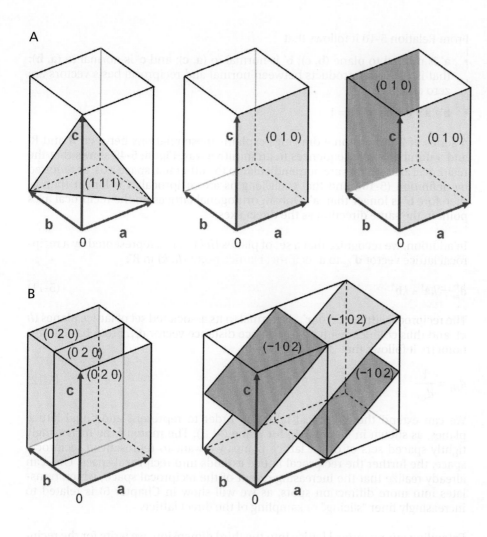

A

B

Figure 5-48 **Lattice planes in 3-dimensional space.** Note that, for example, the (0 1 0) plane is also a member of the set of (0 2 0) planes; they belong to the same family or zone of planes. The distance vector $\mathbf{d}_{hkl}$ is perpendicular to both planes (0 1 0) and (0 2 0) but only half as long for (0 2 0) as for (0 1 0).

$$hx + ky + lz = n \text{ or in vector notation, } \mathbf{h} \cdot \mathbf{x} = n. \tag{5-9}$$

Equation 5-9 is nothing more than a concise mathematical description of the procedure used to derive the Miller indices h, k, l via the lattice points x, y, z which the lattice planes $(h\,k\,l)$ intersect.

Construction of the reciprocal lattice

The reciprocal lattice is a formal construct, developed by P. Ewald in 1921, and its usefulness and convenience in the description of diffraction events makes it an indispensable tool for the crystallographer. We will frequently make use of the relations between real and reciprocal lattices, and it is important to work through the next section until it is well understood.

A reciprocal lattice $[0, \mathbf{a}^*, \mathbf{b}^*, \mathbf{c}^*]$ in reciprocal space R* with reciprocal lattice vectors $\mathbf{r}^*$ is spanned by reciprocal axes $\mathbf{a}^*$, $\mathbf{b}^*$, and $\mathbf{c}^*$ and shares a common origin with the real or direct crystal lattice spanned by $[0, \mathbf{a}, \mathbf{b}, \mathbf{c}]$ in real space R with lattice vectors $\mathbf{r}$. Any point in real or direct space R can be described by a position vector $\mathbf{r}$, and any point in reciprocal space can be described by a general reciprocal space vector $\mathbf{r}^*$. The axes of the direct lattice and the reciprocal lattice are related through the following condition (a dyad product, Appendix A.3):

$$\begin{pmatrix} \mathbf{a}^* \\ \mathbf{b}^* \\ \mathbf{c}^* \end{pmatrix} (\mathbf{a} \quad \mathbf{b} \quad \mathbf{c}) = \begin{pmatrix} \mathbf{a}\mathbf{a}^* & \mathbf{b}\mathbf{a}^* & \mathbf{c}\mathbf{a}^* \\ \mathbf{a}\mathbf{b}^* & \mathbf{b}\mathbf{b}^* & \mathbf{c}\mathbf{b}^* \\ \mathbf{a}\mathbf{c}^* & \mathbf{b}\mathbf{c}^* & \mathbf{c}\mathbf{c}^* \end{pmatrix} = \mathbf{I} \tag{5-10}$$

From Relation 5-10 it follows that

- $\mathbf{a}^*$ is normal to plane $(\mathbf{b}, \mathbf{c})$; $\mathbf{b}^*$ is normal to $(\mathbf{a}, \mathbf{c})$; and $\mathbf{c}^*$ is normal to $(\mathbf{a}, \mathbf{b})$; that is, all scalar products between normal and reciprocal basis vectors are zero except

- $\mathbf{a}^* \cdot \mathbf{a} = \mathbf{b}^* \cdot \mathbf{b} = \mathbf{c}^* \cdot \mathbf{c} = 1$

A series of 2-dimensional drawings explains these relations between R and R* and a number of consequences in an intuitive way. Figure 5-49 shows that the reciprocal axes vectors are perpendicular to the other real space axis, as required by definition (5-10), and that their lengths are reciprocal: $\mathbf{a}$ is longer than $\mathbf{b}$, therefore $\mathbf{b}^*$ is longer than $\mathbf{a}^*$. Only in orthogonal lattices do the reciprocal axes point in the same direction as the direct axes.

In addition, we recognize that a set of planes $(h\ k)$ in R is represented by a reciprocal lattice vector $\mathbf{d}^*_{hk}$ to a reciprocal lattice point (h, k) in R*:

$$\mathbf{d}^*_{hk} = h\mathbf{a}^* + k\mathbf{b}^* \tag{5-11}$$

The reciprocal lattice vector $\mathbf{r}^*_{hk}$ is normal to its associated set of lattice planes $(h\ k)$, and thus collinear with the real space distance vector $\mathbf{d}_{hk}$. From basic trigonometry it follows that

$$d^*_{hkl} = \frac{1}{d_{hkl}} \tag{5-12}$$

We can extend the reciprocal lattice in order to represent additional lattice planes, as shown in Figure 5-50 for plane $(-1\ 1)$. The more of the increasingly tightly spaced sets of direct lattice planes we want to represent in reciprocal space, the further the reciprocal lattice extends into reciprocal space. One can already realize that the increasing extent of the reciprocal space (which translates into more diffraction spots, as we will show in Chapter 6) is related to increasingly finer "slicing" or sampling of the direct lattice.

Extending our reciprocal lattice into the third dimension, we write for the reciprocal lattice vector $\mathbf{d}^*$ or $\mathbf{r}^*$ in analogy to the crystallographic real space vector $\mathbf{r}$:

$$\mathbf{d}^*_{hkl} = h\mathbf{a}^* + k\mathbf{b}^* + l\mathbf{c}^* = (\mathbf{a}^* \quad \mathbf{b}^* \quad \mathbf{c}^*)\begin{pmatrix} h \\ k \\ l \end{pmatrix} = (\mathbf{A}^*)^{\mathrm{T}}\mathbf{h} \tag{5-13}$$

Figure 5-49 Construction of a 2-dimensional reciprocal unit lattice. A direct lattice (red) and its corresponding reciprocal lattice (blue) are shown. The direct and reciprocal lattices share the common origin O. The reciprocal axis $\mathbf{a}^*$ is perpendicular to real axis $\mathbf{b}$, and reciprocal axis $\mathbf{b}^*$ is perpendicular to real axis $\mathbf{a}$. The lattice vector $\mathbf{d}^*_{11}$, pointing to reciprocal lattice point RLP $(1,1)$ is perpendicular to the set of lattice planes with Miller indices hk of $(1\ 1)$ and collinear with the interplanar distance vector $\mathbf{d}_{11}$. The interplanar spacing for the set of planes $(1\ 1)$ is given by $|\mathbf{d}_{11}| = d_{11}$. Note that in the case of an orthogonal real lattice the direction of each reciprocal axis coincides with the direction of its corresponding real axis.

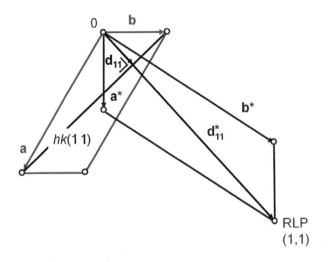

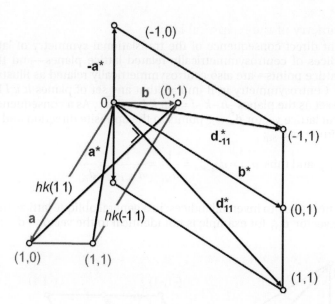

Figure 5-50 **Extending the reciprocal lattice.** Following the construction rules we can extend the reciprocal lattice to fill the reciprocal space. Each and every lattice vector r_{hk} corresponds to a distinct set of lattice planes hk. The tighter the lattice plane spacing, the larger the extent of the reciprocal space becomes.

where **h** is the corresponding index (column) vector, and $(\mathbf{A}^*)^T$ is the matrix containing the reciprocal cell vectors in its columns (i.e. the reciprocal axis magnitudes a^*, b^*, and c^* in the diagonal elements).

While the relation between the unit cell volume V and the reciprocal lattice unit volume $V^* = 1/V$ is quite intuitive and generally valid, the metric relations between direct and reciprocal axes are quite complicated for oblique lattices. The general expression in terms of reciprocal axes is given by

$$d^*_{hkl} = \frac{1}{d_{hkl}} = \left(h^2 a^{*2} + k^2 b^{*2} + l^2 c^{*2} + 2hka^*b^*\cos\gamma^* + 2hla^*c^*\cos\beta^* + 2klb^*c^*\cos\alpha^*\right)^{1/2}$$

$$(5\text{-}14)$$

As a consequence, the general algebraic expressions of d_{hkl} as a function of the real space lattice constants are quite complicated. They can be derived by symbolic evaluation of a convenient general expression often used in crystallographic computing that relates the lattice index vector **h** directly to the lattice spacing $1/d$ using the vector product with the transpose of the real space deorthogonalization matrix (proof in Appendix A.3.):

$$d^*_{\mathbf{h}} = \frac{1}{d_{\mathbf{h}}} = \left|(\mathbf{O}^{-1})^T \mathbf{h}\right|$$

$$(5\text{-}15)$$

The metric relations between direct axes and lattice plane spacing are simple only for orthogonal systems, where they assume the general form

$$d_{hkl} = \left(\frac{a^2}{h^2} + \frac{b^2}{k^2} + \frac{c^2}{l^2}\right)^{1/2}$$

$$(5\text{-}16)$$

The relations between reciprocal and direct lattice for an orthogonal lattice simplify to

$$a^* = 1/a, \ b^* = 1/b, \ c^* = 1/c, \ V^* = 1/V = a^*b^*c^*, \text{ and } \alpha, \beta, \gamma, \alpha^*, \beta^*, \gamma^* = 90°$$

$$(5\text{-}17)$$

The general metric and reciprocal relations are listed in Appendix A.3.

Centrosymmetry of the reciprocal lattice

An important direct consequence of the translational symmetry of lattices is that the indices of centrosymmetrically related lattice planes—and thus the reciprocal lattice points—are also centrosymmetrically related as illustrated in Figure 5-51. Centrosymmetry in R implies that any set of planes $h\,k\,l$ belongs to the same set as the planes $-h\,-k\,-l = -(h\,k\,l) = \overline{hkl}$. As a consequence, in R^* the reciprocal lattice vector $\mathbf{d}^*_{-h-k-l}$ points in the opposite direction and has the same magnitude as $\mathbf{d}^*_{hkl}$:

$$\mathbf{d}^*_{hkl} = -\mathbf{d}^*_{-h-k-l} \text{ and thus } d^*_{hkl} = d^*_{-h-k-l} = \frac{1}{d_{hkl}} = \frac{1}{d_{-h-k-l}} = \frac{1}{d_{\overline{hkl}}} \qquad (5\text{-}18)$$

Note that is necessary to invert all indices: in a general oblique lattice, the reciprocal lattice vector $\mathbf{d}^*_{111}$ for example is not identical to the vector $-\mathbf{d}^*_{\overline{1}11}$; only in

Figure 5-51 Centrosymmetry of reciprocal space. The centrosymmetrically related lattice planes *hk* and *–h–k* generate reciprocal lattice vectors pointing in opposite directions and of equal magnitude.

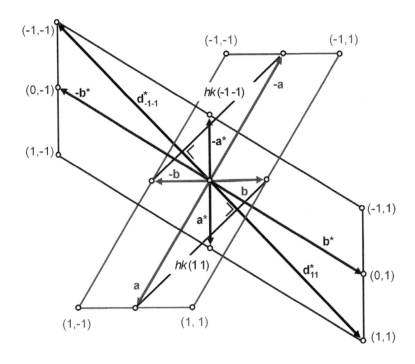

Box 5-12 Relations in the reciprocal lattice. The axes of the crystal lattice in real space R and its reciprocal lattice in R^* are related through the following conditions: $\mathbf{a}^*$ is normal to plane $(\mathbf{b}, \mathbf{c})$, $\mathbf{b}^*$ is normal to $(\mathbf{a}, \mathbf{c})$, and $\mathbf{c}^*$ is normal to $(\mathbf{a}, \mathbf{b})$, which means that all scalar products between normal and reciprocal basis vectors are zero except $\mathbf{a}^* \cdot \mathbf{a} = \mathbf{b}^* \cdot \mathbf{b} = \mathbf{c}^* \cdot \mathbf{c} = 1$.

Sets of parallel and equidistant lattice planes are defined by their Miller indices *hkl*. The Miller indices are integer numbers indicating the number of intercepts of a set of lattice planes with each of the unit cell axes. A set of planes *hkl* in R is represented in R^* by a reciprocal lattice vector $\mathbf{d}^*_{hkl}$ extending from the reciprocal lattice origin to a reciprocal lattice point (h, k, l). The reciprocal lattice vector $\mathbf{d}^*_{hkl}$ is normal to the set of lattice planes *hkl*, and is collinear with the interplanar distance vector $\mathbf{d}_{hkl}$. The magnitude of $\mathbf{d}^*$ is reciprocal to the magnitude of $\mathbf{d}_{hkl}$. The reciprocal lattice is centrosymmetric, and each reciprocal lattice vector $\mathbf{d}^*_{hkl}$ is opposite and of the same magnitude as $\mathbf{d}^*_{-h-k-l}$.

The closer the interplanar spacing and thus the larger the indices *hkl*, the more tightly the planes *hkl* "slice" or sample the cell. Tight sampling means more information, thus high *hkl* (equivalent to reciprocal lattice points far from the lattice origin) provide more detail about the sampled structure.

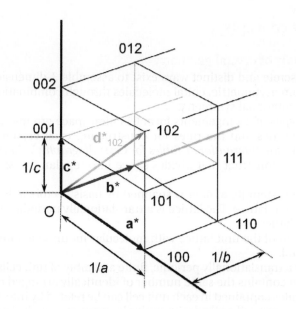

Figure 5-52 Orthorhombic reciprocal lattice. A small section of a 3-dimensional, orthorhombic reciprocal lattice in the positive octant (all indices *h*, *k*, *l* positive) around the origin is shown together with the reciprocal unit cell vectors and the corresponding direct cell lengths. Many more figures illustrating the reciprocal lattice and its use in the exploration of diffraction space and geometry are included in Chapters 6 and 8.

the special case of tetragonal or cubic lattice symmetry, would this example for relation between sets of lattice planes (1 1 1) and (–1 1 1) be true.

The same construction rules as used in Figures 5-49 to 5-51 can finally be employed to construct a 3-dimensional reciprocal lattice. For simplicity we just show a part of the positive octant (*h*, *k*, *l* running from 0 to infinite) around the origin of an orthorhombic reciprocal lattice in Figure 5-52.

5.5   Next steps

We have at this point obtained a basic understanding of crystal geometry and basic symmetry operations, and can navigate in a crystal structure. Together with the construction rules of the reciprocal lattice, we have acquired the fundamentals necessary to begin exploration of the diffraction process.

Sidebar 5-6 The concept of sampling density. Sampling is an important concept in information theory: the tighter the discrete sampling of any object, the more information we obtain about it. Sampling theory also holds for crystallography, and intuitively we have already assumed that the more tightly we sample the lattice, the more we will know about the molecules packed inside, and we can correctly guess that there has to be a correlation between the extent to which a crystal diffracts and the resulting sampling density. In subsequent chapters we will explore this correlation in detail. Slicing the direct cell with tightly spaced lattice planes is the real space equivalent to sampling the reciprocal space by diffraction. We intuitively realize that a set of planes in a cell, or a reciprocal lattice point, is probably directly related to a point or reflection in the diffraction pattern. This is an important, qualitative assumption, which we will prove in the next chapter.

It is quite interesting to contemplate the meaning of the dimensions of the reciprocal lattice constants. As the dimensions of cell constants are in Å, the reciprocal units must be reciprocal Å or Å⁻¹. For example, if the cell dimension *a* is 10 Å, a^* is 1/(10 Å) or 0.1 Å⁻¹ or 0.1 "per Å". For a reciprocal lattice point (5, 0, 0) corresponding to real lattice plane *hkl* indexed (5 0 0) the value is $5 \times 0.1 = 0.5$ per Å. The question is *what* per Å do these numbers imply? Here again the analogy to sampling becomes evident: We can interpret these numbers as the sampling density for the corresponding set of lattice planes: While planes (1 0 0) sample a 10 Å cell only at 0.1 samplings per Å, the (5 0 0) planes slice the lattice much finer, at 0.5 samplings per Å. The latter is equivalent to one sampling every 2 Å, and at that slicing, we may not be able to resolve individual atoms, but we will gain a fairly realistic idea about the location of the atoms in the cell. From many of these reciprocal samplings in different directions, we can assemble the molecule, and the finer the sampling, the more detailed the molecular reconstruction.

5.6   Key concepts

Fundamentals of crystal geometry

- Only 65 discrete and distinct ways exist to assemble 3-dimensional, periodic crystals from asymmetric, chiral molecules through combinations of translational and rotational symmetry.
- These 65 types of arrangements form 65 chiral space groups, and their symmetry properties and the rules for constructing each crystal structure are described in the *International Tables for Crystallography, Volume A*.
- A lattice is a construct that divides space into regular, translationally periodic units.
- A lattice in system [0, **a**, **b**, **c**] is spanned by basis vectors **a**, **b**, and **c**, which define the unit lattice. The lattice is a translationally periodic, infinite assembly of unit lattices.
- Combination of the unit lattice with molecular motifs generates the unit cell of the crystal.
- A crystal is a translationally periodic, finite assembly of unit cells. Each identical unit cell contains the same number of identically arranged molecules.
- The molecules contained in each unit cell can be related by internal symmetry, generating a unit cell packed with multiple, symmetry equivalent instances or symmetry equivalent copies of the molecule.
- Certain plane or space groups allow choice of multiple origins.
- The possible origins for each unit type of unit cell (i.e. each plane or space group) are listed in the *International Tables for Crystallography, Volume A*.
- The asymmetric unit of a unit cell contains all the necessary information to generate the complete unit cell of a crystal structure by applying its symmetry operations to the asymmetric unit.
- The motif (molecule) does not have to be constrained within the boundaries of the asymmetric unit or the unit cell.
- Each point in a unit cell can be assigned a unique fractional coordinate triple (x, y, z).
- A positional real space vector in units of Å is defined as $\mathbf{r} = x\mathbf{a} + y\mathbf{b} + z\mathbf{c}$, while $\mathbf{x}$ is the corresponding fractional coordinate vector whose components x, y, and z are dimensionless fractional coordinates.
- The fractional coordinates of points within a unit cell are limited to $0 \leq x, y, z < 1$.
- Points outside the unit cell can be translated back into equivalent positions inside the unit cell by applying an appropriate translation vector $\mathbf{t}$ to their positional vector $\mathbf{x}$.

Basic crystal symmetry

- Any crystallographic symmetry operation must generate an identical copy of the motif.
- Symmetry operations must not change the molecular motif upon application. This limits available symmetry operations for chiral, asymmetric protein molecules to plain rotation axes and screw axes.
- Mirror operation, inversion, and any combination of symmetry operations including mirror operations or inversion are not allowed for asymmetric, chiral molecules because they change the handedness of the molecule.
- Required compatibility with translational periodicity of the lattice limits rotations and screw axes to 2-, 3-, 4-, and 6-fold operations.
- Allowable combinations of translational lattice symmetry with internal unit cell symmetry define seven crystal systems, dividing into 14 Bravais lattice types, 11 chiral (32 general) point groups, and ultimately 65 chiral (230 general) space groups. A protein crystal structure must belong to one of the 65 chiral space groups.
- Asymmetric motifs cannot be located on symmetry elements.
- A motif on a symmetry element or a combination thereof (i.e. on a so-called special position) must possess at least the same internal symmetry as that special position.
- Only the generating motif and not all symmetry generated copies are deposited in a PDB coordinate file.

Diffraction basics

It is the job of the crystallographer to measure the diffraction pattern, and to use it to figure out the structure of the objects doing the scattering. The task is not trivial.
Richard E. Dickerson (2005) *Present at the Flood*, p 140

While optical diffraction was already described and quantified by renaissance scientists, the discovery of X-rays and subsequently of X-ray diffraction are events of the late 19th and early 20th century, closely linked with the concurrent development of quantum mechanics. Diffraction is in principle the interaction of electromagnetic radiation with periodically arranged matter, which in microscopic, quantum-mechanical detail is a complex process. As in the previous chapters, we will allow qualitative and intuitive explanation whenever possible, and revert to more formal rigor when necessary.

Chapter 6 begins with a qualitative description of the physical nature of the scattering of X-rays. Progressing from scattering of X-rays by free electrons to atoms and finally to assemblies and periodic arrangements of atoms, we will derive the fundamental conditions for the occurrence of discrete diffraction.

Once the basics of diffraction are established, it becomes necessary to quantify the intensity of diffracted X-rays as a function of the actual atomic or molecular contents of the diffracting crystals, which leads to the introduction of the structure factors and our first calculation of complex structure factors. The absence of the phase angles for the experimentally determined structure factor amplitudes leads to anticipation of the phase problem in the reconstruction of the molecular structure.

The concept of the reciprocal lattice, where each set of equidistant lattice planes is represented as a discrete reciprocal lattice point, will lend itself ideally to the interpretation of diffraction as Bragg reflection on sets of crystal lattice planes. The representation of diffraction conditions in reciprocal space with the assistance of the Ewald construction will provide the fundamentals of diffraction geometry and a first introduction of how X-ray diffraction data are collected.

Of particular interest in view of anomalous diffraction phasing techniques is the phenomenon of absorption and anomalous scattering of X-rays. A qualitative explanation of the underlying physics is followed by a demonstration of the practical consequences of the loss of centrosymmetry in the diffraction data. This section is technically demanding, but studying it thoroughly will be well rewarded with the understanding of how to develop successful strategies for anomalous phasing experiments.

At the end of the chapter, we revisit the symmetry of reciprocal space, bringing together the fundamentals necessary for the development of diffraction data collection strategies and explaining how reciprocal space symmetry affects structure factor amplitudes and X-ray diffraction data. The concepts of centric reflections, their phase restrictions, as well as reflection multiplicity and ε-factors are necessary for the subsequent discussion of intensity statistics in Chapter 7 as well as for the development of data collection strategies and the practice of data collection discussed in Chapter 8.

6.1　Scattering of X-rays

Before the discovery of X-rays in the late 19th century, the life of a crystallographer was quite different than it is today. Crystal structure determination essentially was limited to careful measurement of the angles between well-developed crystal faces and plotting normal vectors to the planes in stereographic projections, thereby establishing the crystal class or point group of the crystal. Other physical properties, such as measurements of the piezoelectric effect or double refraction, gave additional clues to the internal symmetry of the crystals, but no direct experimental evidence regarding the location of atoms in crystalline matter was available. Theory was ahead of the experiments, as space group theory was already well developed through the work of Fedorov, Schönflies, and others. The purely descriptive nature of crystallography was rapidly changed, beginning in 1895, when Wilhelm Conrad Röntgen, a physicist in Würzburg, Germany, described a new type of invisible and penetrating radiation emanating from a cathode-ray (electron) tube (Sidebar 6-1). In the following decades,

---

Sidebar 6-1 The discovery of X-rays. Wilhelm Conrad Röntgen, a physicist in Würzburg, Germany, described a new type of invisible radiation emanating from a cathode-ray (electron) tube in 1895. The emitted invisible radiation could be detected using a fluorescent screen. He coined the name *X-rays* still used in English today (but named "Röntgenstrahlen" after their inventor in German). With participation of body parts of a horrified Frau Röntgen, he established the penetrating nature of his newly discovered radiation. Frau Röntgen was convinced that the skeletonized image of her hand was an omen of death and never returned to his laboratory, but still (or perhaps because of that) lived to the then biblical age of 78. Nevertheless,

Herr Professor Röntgen quickly found new victims, as the image in Figure 6-1 demonstrates. Various X-ray machines soon appeared as gaudy contraptions in vaudeville shows, but the lethal effects of exposure to the ionizing X-rays became rapidly evident. Nonetheless, within a year after publication of their discovery, X-rays were already being used for medical imaging in surgery. In 1901 Röntgen was awarded the first Nobel Prize in Physics, signifying the commencement of a long series of Nobel Prizes awarded to crystallographers and for crystallographic studies up to the present day—and probably continuing into the future, provided you study this textbook carefully.

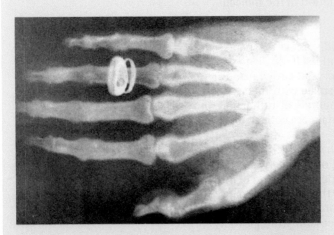

Figure 6-1 An X-ray transmission image taken in 1896 by Wilhelm Conrad Röntgen. The hand belongs to Röntgen's faculty colleague Rudolph Albert von Kölliker. The image is of a film negative, blackened where X-rays excited optical fluorescence in a fluorescence screen placed facing the film located below the hand. We learn from the exposed areas that soft tissue, largely composed of light atoms such as C, O, and N, (number of electrons $z = 6, 7, 8$), only weakly absorbs hard X-rays. The calcium phosphate ($z = 20, 16, 8$) in the bone absorbs considerably more, and the presumably gold rings ($z = 79$) absorb very strongly, while the gemstone set in the larger ring must consist of a lighter material than gold, as it absorbs less than the gold ring. However, it is likely not a diamond ($z = 6$), because it absorbs significantly more strongly than the soft tissue (average $z = 6.7$) of the hand. We can only hope that Prof. von Kölliker did not purchase a diamond (C) and received a cheap cubic zirconia (ZrO_2) crystal ($z = 40$ for Zr) instead. Source: Public Domain.

Max von Laue and his assistants in Munich and the father–son team of Sir W. Henry Bragg and Sir W. Lawrence Bragg in Leeds and Cambridge established that X-rays behave like electromagnetic waves and as a consequence, get diffracted by the periodically arranged atoms in crystals.

X-ray diffraction is the result of the interaction of electromagnetic radiation with the electrons of the atoms in the crystals. We will therefore have to examine in some detail the nature of X-rays and the mechanism of their fundamental interaction with matter, the scattering of X-rays by electrons.

The nature of X-rays and their interaction with matter

X-rays are high energy electromagnetic radiation, and thus part of the electromagnetic spectrum. Electromagnetic radiation interacts with matter primarily through its oscillating electric field vector and, to a much lesser extent, the magnetic field vector. The reason is that the interaction of the electric field vector due to dielectric polarizability is several orders of magnitude higher than the diamagnetic susceptibility. As a consequence, we will focus on the electronic interactions (and understand why magnetic miracle cures are generally ineffective). The polarizability α is simply the interaction constant relating the induced dipole moment **p** to the electric field vector **E**:

$$\mathbf{p} = \alpha\mathbf{E} \tag{6-1}$$

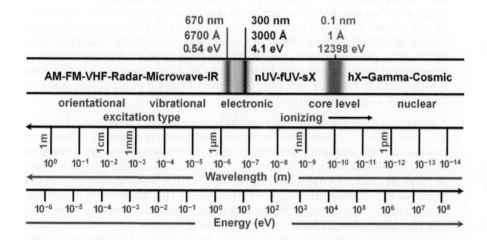

Figure 6-2 The electromagnetic spectrum. Electromagnetic radiation extends from low energy radio frequency via microwaves and visible light all the way to X-rays and cosmic radiation. Beginning with ultraviolet light (UV), electromagnetic radiation becomes ionizing, meaning that it can separate electrons from atomic cores and induce chemical changes and radiation damage. AM: amplitude modulation, FM: frequency modulation, VHF: very high frequency band, IR: infrared, nUV: near ultraviolet, fUV: far ultraviolet, sX: soft X-rays, hX: hard X-rays, Gamma: γ-radiation.

Box 6-1 Important relations between energy, wavelength, and mass.

$E = h\nu$	energy E as a function of frequency ν; h = Planck's constant $h = 6.6261 \cdot 10^{-34}$ Js
$\nu = c/\lambda$	relation between frequency ν, wavelength λ, and speed of light c $c = 299792.5 \cdot 10^4$ cm s^{-1}
$1/\lambda = n$	wavenumber
1 eV = 8065.5 cm^{-1}	energy expressed as wavenumber
λ (Å) = 12397.639/E (eV)	conversion of energy (in eV) to wavelength (in Å), which allows using the terms *energy* and *wavelength* interchangeably when discussing absorption, scattering, and diffraction of X-rays
$\lambda = \dfrac{h}{p} = \dfrac{h}{mc}$	de Broglie wavelength, for particles of mass m moving at speed of light c

At low frequency, electromagnetic radiation interacts with permanent electric dipoles, which for small molecules can reorient freely up to the high picosecond (ps) range (large molecules cannot undergo such rapid reorientation, leading to slow rotational relaxation of large proteins which limits solution NMR). Once the frequency exceeds the microwave region (10^{11} Hz), small molecules also cannot reorient fast enough, the interaction drops off rapidly, and we leave the range of orientation polarization and enter the distortion polarization range, dominated by vibrational modes up to infrared (IR) frequencies (where vibrational IR and Raman spectroscopy operate). Once past IR frequency, only the electrons of an atom themselves are polarizable enough to interact with the electric field vector and we enter the visible light range, and electron polarization remains the basis for interaction with electromagnetic radiation up to the highest X-ray energy ranges.

The dielectric polarizability is directly correlated with the refractive index of a material through the Clausius–Mossotti equation. The change of the refractive index with frequency causes optical dispersion (i.e. blue light gets refracted more strongly than red light; this dispersive refraction can be observed in rainbows or glass prisms). At very high frequencies, even electronic polarizability becomes very small and asymptotically approaches practically identical (vacuum) values for all matter.

From the qualitative statements above we can draw a number of conclusions for X-rays, which are very high energy electromagnetic radiation, with energies about 10^4 times higher than visible light (see Figure 6-2).

- **Absence of lenses for X-rays:** As the polarizability of materials in the X-ray range and thus also the refractive index is practically identical for all materials, we cannot build any refractive lenses for X-rays. Lenses depend on the change of refractive index when light, or any electromagnetic radiation, transitions from one medium to the other, and there is practically no difference in refractive index between media for X-rays. As a consequence, no refractive lenses exist for X-rays, precluding any direct imaging of an object, quite in contrast to a light microscope.

- **Anomalous dispersion for macromolecular phasing:** The effects of dispersion and absorption also exist for X-rays. Dispersion is generally a change of a property with frequency (or energy), and the dispersion we are interested in as crystallographers is anomalous X-ray dispersion, that is, the energy-dependent change of the scattering factor of X-rays when absorption occurs at the so-called absorption edge of an element. This anomalous dispersion or anomalous scattering of X-rays is a most important property of matter, and provides the basis for a variety of powerful anomalous diffraction phasing techniques allowing us to overcome the principal inability to directly image molecules because of the lack of X-ray lenses.

- **Radiation damage:** X-ray absorption involves high-energy electronic transitions in core levels, and X-rays are ionizing radiation. Ionizing radiation means that high energy electrons can be separated from inner atomic levels, and the free electrons can directly react with other atoms, break bonds, or generate free radicals, all of which cause severe radiation damage to organic (and even inorganic) matter. Protein crystals are organic matter, and radiation damage when exposed to highly brilliant X-ray beams during diffraction experiments is therefore a common occurrence.

On a microscopic level, we can view the interaction of the oscillating electric field vector (Figure 6-3) with electrons as an induction of oscillations of the electrons, which then, as accelerated charges, themselves emit electromagnetic waves of the same frequency. The phase difference between these emitted partial, scattered waves upon superposition gives rise to interference and diffraction phenomena. This purely wave-based picture is quite useful but suffers from one crucial weakness: The incoming X-rays are generally not coherent, that is, there is no fixed phase relationship between them. Illustrations of scattering or

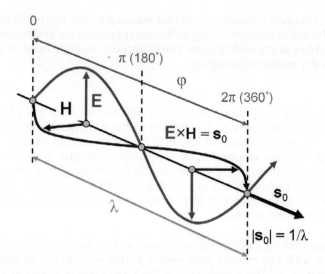

Figure 6-3 Anatomy of an electromagnetic wave. The electric field vector **E** is perpendicular to the magnetic field vector **H**, and both oscillate perpendicular to the propagation vector s_0 of the wave of wavelength λ. The vector s_0 in the direction of wave propagation has a magnitude of $1/\lambda = n$. In physics, the term *wavenumber* is common for n and s_0 is the *wave vector*, occasionally labeled **k**. The position of any point of the wave in reference to the periodic origin node can be expressed in terms of phase shift $\Delta\varphi$ ranging from 0 to 2π (in units of radians or rad, 1 rad = $180/\pi$ degrees). The interaction of the electric field vector **E** with dielectric matter is about six orders of magnitude larger than the interaction of the magnetic field vector **H**, thus we generally neglect the effects of **H** in the discussion of scattering and diffraction.

diffraction phenomena showing two or more X-rays arriving in phase are thus physically somewhat misleading.

The particle–wave picture. In the description of scattering processes a very useful concept is that of single X-ray photons traveling as wave packets or particles with the internal oscillation frequency defined by their energy or wavelength. When photons travel through a crystal, either of two things can happen: (i) nothing, which happens over 99% of the time; (ii) the electric field vector induces oscillations in all the electrons coherently within the photon's coherence length ranging from a few 1000 Å for X-ray emission lines to several microns for modern synchrotron sources. At this point, the photon ceases to exist, and we can imagine that the electrons themselves emanate virtual waves, which constructively overlap in certain directions, and interfere destructively in others. The scattered photon then appears again in some direction, with the probability of that appearance proportional to the amplitude of the combined, resultant scattered wave in that particular direction. This description follows the analogy to the basic process of a quantum mechanical change of state, a virtual annihilation–generation process. The sum of all scattering events of independent, single photons then generates the diffraction pattern. This is a quite reasonable picture, as evidenced by the fact that we can discretely count each single X-ray photon scattered in different directions one by one on a photon counting (area) detector, until a distinct diffraction pattern emerges.

Box 6-2 X-rays are scattered by the electrons in matter. The underlying physical process of X-ray scattering is the interaction of the electric field vector **E** of a propagating electromagnetic wave packet representing each X-ray photon with the electrons in matter. The electric field vector **E** is perpendicular to the propagation direction of the wave or wave vector, and oscillates with a frequency corresponding to the energy and inverse of the wavelength of the X-rays. Energy and wavelength are related through the conversion formula λ (in Å) = 12397.639/E (in eV). No correlation in time and space, that is, no coherence, between individual photons is required for X-ray scattering. However, the scattering process itself is coherent within the coherence length of each photon; that is, a fixed phase relation exists between all partial waves emanating from all the electrons excited by an individual photon.

As the interference between the partial waves emitted by the electrons of the atoms is the key to understanding diffraction phenomena, and consequently for a large number of crystallographic computations, we now need to address this somewhat dry matter thoroughly.

Superposition of electromagnetic waves

In the particle–wave picture of a scattering process, an X-ray photon excites simultaneously vibrations in all electrons within its coherence length. The resonating electrons emit waves of identical frequency and in fixed phase relation to each other which recombine, or superimpose, to give a resulting scattered wave, with the probability of the photon to be scattered in a certain direction proportional to the amplitude (squared, as we will see) of the recombined wave in that direction. Constructive interference increases this probability; destructive interference decreases it.

We can describe an electromagnetic wave as two perpendicular, propagating sine waves with the electric field vector perpendicular to the magnetic field vector (Figure 6-3). The wavelength of the X-ray wave is λ, and its propagation vector $\mathbf{s}_0$, the wave vector, is perpendicular to both $\mathbf{E}$ and $\mathbf{H}$ and defined by their vector product. The magnitude of the wave vector is defined as $1/\lambda$, a very important feature for later work:

$$\mathbf{s}_0 = \mathbf{E} \times \mathbf{H} \quad \text{with} \quad |\mathbf{s}_0| = \frac{1}{\lambda} \tag{6-2}$$

The phase angle (or simply, the *phase*) of the electric wave vector in relation to the wave origin is defined in the range 0–2π rad(ians) or 0–$360°$ over one period of the wave.

As the interaction between electrons and the electric field vector $\mathbf{E}$ alone is the source of the scattering process (interactions of $\mathbf{H}$ with dielectric media are many orders of magnitude lower and do not contribute to coherent X-ray scattering), we can describe electromagnetic radiation as a classical, plane, transversal sine wave with its amplitude given by the magnitude of the electric field vector oscillating with a given frequency v defined by the photon energy $E = hv$. Using the plane wave description, superposition of waves can be easily understood following the graphical amplitude addition procedure shown in Figure 6-4: we add the electric field vector amplitudes of the partial wave at each point along the period of 2π, and plot the result for the combined wave. In the depicted case, the amplitudes and frequency of both waves are identical and the waves are phase shifted by $180°$, expressed as $\Delta\varphi = \pi$. As seen from the drawing, the two waves cancel out completely: the resultant amplitude is zero. In contrast, a phase difference of $0°$ or $360°$ results in perfect constructive interference, where the new wave has twice the amplitude of the two partial waves.

Figure 6-4 Graphical superposition of waves. To obtain the resulting wave (Σ), the amplitudes are summed at discrete points over the period of 2π. In this case, two waves of the same amplitude are phase shifted by $\Delta\varphi = \pi$, and the waves cancel out completely. The peak value for the amplitude is the magnitude $|\mathbf{F}|$, or in short F, of the wave described as a complex number $\mathbf{F}$ (cf. Figure 6-5).

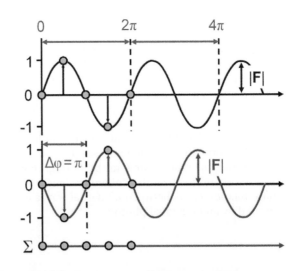

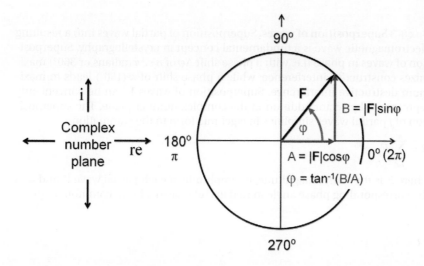

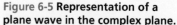

Figure 6-5 Representation of a plane wave in the complex plane. We can represent any plane wave and its phase relation to a fixed origin as a vector **F** in the complex plane. The real part of the complex number is plotted horizontally along the abscissa, and the imaginary component is perpendicular on the ordinate. The magnitude of the vector is given by the amplitude of the wave |**F**|, and its direction by the phase angle φ relative to a fixed origin. The convention is to measure the phase angle in counterclockwise direction. In the illustrated case, the vector **F** has a phase angle of about +50° relative to the origin. The presentation of the wave as a vector in the complex number plane allows a number of useful algebraic assignments, which are annotated in the figure. This representation of complex numbers in the complex plane is also called an Argand diagram.

The point-wise graphical addition procedure is intuitive, but quite cumbersome for any practical purposes and particularly so for computation. A more elegant description exists by splitting the wave into amplitude and phase parts. One then can represent the scattered wave in the complex number plane (Figure 6-5), and perform vector addition to simplify the partial wave summations.

A plane wave of amplitude F and phase angle φ relative to a fixed origin 0 or reference point can be represented in the complex number plane, spanned by the real part axis **re** and the imaginary part axis **i**. The axes thus form a coordinate basis system [0, **re**, **i**], with i defined as the square root of -1, $(-1)^{1/2}$, from which it follows that $i \cdot i = -1$. We draw a circle with radius of amplitude F and draw a vector **F** at a phase angle of φ, measured counterclockwise from the real axis. Basic trigonometry leads to a number of important and useful relations which are shown in Figure 6-5.

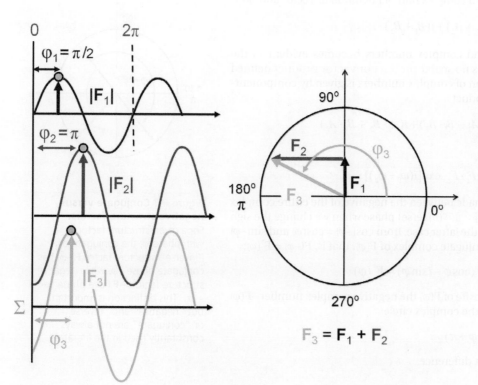

Figure 6-6 Addition of plane waves. Superposition of waves becomes very convenient using the vector presentation of the waves in the complex plane, where it is reduced to the addition of the corresponding vectors. The resulting vector and its phase angle can be directly visualized from the diagram, and readily algebraically summarized. We can see from the example diagram that the resulting wave represented by F_3 has a larger amplitude than the two seperate components and has a phase angle of about 160°.

Box 6-3 **Superposition of waves.** Superposition of partial waves into a resulting electromagnetic wave is a fundamental concept in crystallography. Superposition of waves in phase (i.e. with a phase shift $\Delta\varphi$ of $n \cdot 2\pi$ radians or 360°) maximizes constructive interference, while a phase shift of π (180°) leads to maximum destructive interference. Superposition of waves $\mathbf{F}$ can be conveniently represented by vector addition in the complex number plane. The superposition of j partial waves simplifies in algebraic form to the summation

$$\mathbf{F} = \sum_{j=1}^{n} F_j \cdot e^{i\varphi_j}$$

where F_j is the norm, magnitude, or amplitude of each partial wave $\mathbf{F}_j$ and φ_j is the corresponding phase angle in radians relative to a fixed, common origin.

For a plane wave represented as a complex number $\mathbf{F}$, the real component along **re** is then the real part, $A = |\mathbf{F}| \cos\varphi = F \cos\varphi$, and the imaginary part in the direction of **i** is $B = |\mathbf{F}| \sin\varphi = F \sin\varphi$, with φ the phase angle. Thus,

$$F = |\mathbf{F}| = \left(A^2 + B^2\right)^{1/2} = \left((A+iB)(A-iB)\right)^{1/2} = \left(\mathbf{F}\mathbf{F}^*\right)^{1/2} \tag{6-3}$$

with $\mathbf{F}^*$ the complex conjugate of $\mathbf{F}$, and F or $|\mathbf{F}|$ the norm or magnitude of $\mathbf{F}$.

The presentation of $\mathbf{F}$ in the complex plane reduces electromagnetic wave superposition to simple vector addition, as Figure 6-6 illustrates.

Sidebar 6-2 **Complex numbers and their representation.** Note that complex numbers are not really vectors. The rule for component-wise addition is just the same, which makes it convenient to think of complex number addition as vector addition:

$$\mathbf{F}_1 + \mathbf{F}_2 = (A_1 + iB_1) + (A_2 + iB_2) = (A_1 + A_2) + i(B_1 + B_2)$$

The difference between vectors and complex numbers becomes evident in the case of multiplication rules: There is no scalar product or vector product defined for complex numbers. Multiplication of complex numbers is given by component-wise algebraic evaluation of the product

$$\mathbf{F}_1 \cdot \mathbf{F}_2 = (A_1 + iB_1) \cdot (A_2 + iB_2) = (A_1 \cdot A_2 - B_1 \cdot B_2) + i(A_1 \cdot B_2 + B_1 \cdot A_2)$$

which in exponential form yields

$$\mathbf{F}_1 \cdot \mathbf{F}_2 = F_1 \cdot \exp(i\varphi_1) \cdot F_2 \cdot \exp(i\varphi_2) = F_1 \cdot F_2 \cdot \exp[i(\varphi_1 + \varphi_2)]$$

An important distinction must be made between the negative of the entire complex number $-\mathbf{F}$ and the negative or conjugate (inverse) phase when we change the sign of the phase term φ from φ to $-\varphi$. In the latter case, from $\cos(-\varphi) = \cos(\varphi)$ and $\sin(-\varphi) = -\sin(\varphi)$ we obtain for $\mathbf{F}(-\varphi)$ the conjugate complex of $\mathbf{F}(\varphi)$, that is, $\mathbf{F}(-\varphi) = \mathbf{F}^*(\varphi)$:

$$\mathbf{F}(-\varphi) = F(\cos(-\varphi) + i\sin(-\varphi)) = F(\cos\varphi - i\sin\varphi) = \mathbf{F}^*(\varphi)$$

In contrast, for the negative or opposite of $\mathbf{F}(\varphi)$ the negative complex number $-\mathbf{F}(\varphi)$ points in the opposite direction in the complex circle:

$$-\mathbf{F}(\varphi) = F(-\cos(\varphi) - i\sin(\varphi)) = \mathbf{F}(\varphi + \pi)$$

Figure 6-7 illustrates this important difference.

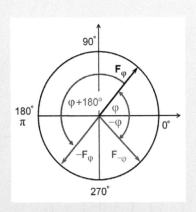

Figure 6-7 Conjugate versus negative structure factors. For a given structure factor **F** with phase φ, the conjugate or "inverse" structure factor **F**$(-\varphi)$ has conjugate phase $-\varphi$, while negative structure factors **−F** have phase ($\varphi + \pi$). This distinction is important, but "negative" and "inverse" or "conjugate" are not always consistently used in the literature.

With the tedium of graphically adding the waves now reduced to a simple complex number addition, we can use a number of algebraic relations to our advantage. From the form of the wave vector $\mathbf{F}$

$$\mathbf{F} = A + iB = |\mathbf{F}|(\cos\varphi + i\sin\varphi) = F(\cos\varphi + i\sin\varphi) \qquad (6\text{-}4)$$

and Euler's formula for the complex exponential function, which is periodic in the interval 0 to 2π

$$e^{i\varphi} = \cos\varphi + i\sin\varphi \qquad (6\text{-}5)$$

it follows that

$$\mathbf{F} = |\mathbf{F}| \cdot e^{i\varphi} = F \cdot e^{i\varphi} \qquad (6\text{-}6)$$

Using Relation 6-6 the addition of n waves $\mathbf{F} = \mathbf{F}_1 + \mathbf{F}_2 + \mathbf{F}_3 \dots \mathbf{F}_n$ becomes a very practical and simple exponential summation:

$$\mathbf{F} = \sum_{j=1}^{n} |\mathbf{F}_j| \cdot e^{i\varphi_j} = \sum_{j=1}^{n} F_j \cdot e^{i\varphi_j} = \sum_{j=1}^{n} F_j \cdot \exp(i\varphi_j) \qquad (6\text{-}7)$$

Summation 6-7 is so fundamental for crystallography that the proof of Euler's formula (6-5) through expansion into a McLaurin series is given in Appendix A.5. In the following chapters, we will make extensive use of the complex notation of scattered photons as waves. As far as notation goes, we will adhere to our already stated convention of vector notation and use *italics* for the norm or magnitude of a **bold** vector or complex number and use the norm signs $|\mathbf{F}|$ interchangeably with F in cases where emphasis is needed.

Scattering diagrams

One of the fundamental tools in crystallography is the geometric description of the scattering events taking place when photons get scattered by electrons of

---

Sidebar 6-3 Euler's number, e. The name of the *irrational* number e was introduced by the Swiss mathematician and polymath Leonard Euler (1707–1783), who spent much of his career in St. Petersburg at the Russian Academy of Sciences. Also a *transcendental* number, e cannot be represented through a discrete algebraic equation, and is obtained as the limiting value of

$$\lim_{n\to\infty}(1 + 1/n)^n = e = 2.718281828\dots$$

Historically this definition of e was first derived from compound interest computations: if we place a certain amount of money in the bank, and get 100% interest once at the end of the year, we have doubled the money ($n = 1$), but if we compound it monthly, we obtain 2.6 times the money ($n = 12$), already close to the limiting value of 2.718… for $n \to \infty$.

The great importance of e for physics in general rests in the fact that the exponential function e^x plays a role in the many processes where the change of a quantity is directly related to its magnitude, that is, in processes described by differential equations. The fact that the exponential function e^x is the only function where the value of the function equals its derivative makes it eminently useful in differential calculus applications. The inverse function of e^x is the natural logarithm, that is, $\ln(e^x) = x$.

The importance of e for crystallography results from the fact that the complex exponential function e^{ix} is periodic, and thus can be used to describe periodic or oscillating physical processes such as the transversal waves representing X-rays. The fundamental relation between the trigonometric and the exponential representation of a complex number is given by Euler's formula:

$$e^{i\varphi} = \cos\varphi + i\sin\varphi$$

The relation above can be use to prove (see Appendix A.5) one of the most elegant equations in mathematics, the Euler Identity

$$e^{i\pi} + 1 = 0$$

containing five algebraic fundamentals. Contrary to common belief, this formula is not inscribed on Leonard Euler's tombstone. Such an honor belongs to Ludwig Boltzmann, whose tomb at the Zentralfriedhof in Vienna (Gruppe 14 C, Grab No. 10) bears the inscription $S = k \ln W$, which is of fundamental importance in statistical thermodynamics as well as in maximum entropy methods in crystallography.

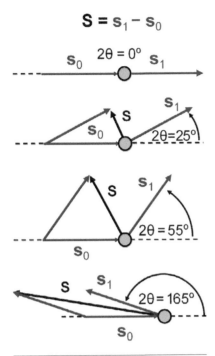

Figure 6-8 Scattering diagrams.
Scattering processes can be visualized using scattering vectors. The scattering vector is the vector difference between scattered and incoming wave vector: **S** = **s₁**−**s₀** In elastic processes no energy transfer takes place and incoming and emanating wave vectors have the same magnitude of $1/\lambda$. Less than 1% of all incoming photons are scattered; in > 99% of all cases no scattering takes place. The magnitude of the scattering vector **S** increases with scattering angle and reaches $2/\lambda$ for 180° backscattering, that is, when the momentum transfer is maximal. The reason for assigning the scattering angle units of 2θ becomes obvious once the diffraction of X-rays is interpreted as reflection on lattice planes (Figure 6-15).

the atoms in a crystal lattice. The description of scattering events in terms of the wave vectors is a generally applicable and very useful concept. The wave vector is the vector in the propagation direction of the wave with the norm or length of $1/\lambda$. In the description of a scattering event, the scattering vector **S** or momentum transfer is given by the difference between incoming wave vector **s₀** and wave vector **s₁** of the scattered wave:

$$\mathbf{S} = \mathbf{s}_1 - \mathbf{s}_0 \text{ with } |\mathbf{S}| = |\mathbf{s}_1 - \mathbf{s}_0| \tag{6-8}$$

If the scattering process is elastic without transfer of energy between electron and photon, the wave vectors **s₀** and **s₁** maintain the same magnitude of $1/\lambda$, and all excited electrons emit in phase. This is the case in the coherent scattering relevant for diffraction. In physics and small angle scattering, |**S**| is sometimes replaced with q (momentum transfer) in dimensions of $2\pi/\lambda$, that is, $q = 2\pi S$. Figure 6-8 visualizes selected scattering events through scattering diagrams.

Scattering of X-rays by free electrons

Let us now analyze the process of scattering of X-rays by electrons. The simplest case is scattering of an X-ray photon by a free electron. There are essentially two possibilities. Either the scattering is elastic—without any energy loss and the emitted photon has the same frequency as the incoming photon—or the electron accepts some momentum from the incoming photon, and the emitted photon has less energy. The second process is called inelastic Compton scattering, and does not interest us as far as diffraction is concerned; however, it is of some interest in the context of advanced techniques of X-ray production (Compton X-ray sources).[1] Compton processes are also incoherent, that is, they do not maintain phase integrity. Therefore, they do not contribute to discrete diffraction, but generate diffuse background scattering. Moreover, Compton scattering is highest in the backward direction, which is converse to elastic, coherent Thomson scattering. We will limit the discussion to elastic scattering, originally formulated by Joseph John Thomson, who identified the electron as a subatomic particle and received the unshared Nobel Prize in Physics in 1906 for his work on electron properties.

We imagine a single electron that is induced to oscillate by the electric field vector **E** of a photon. The oscillating electron itself then emits radiation of the same frequency, and a scattered photon will appear in a discrete direction with a certain probability depending on the scattering probability or scattering function for X-rays on the free electron. We intuitively expect that the scattering function will not be isotropic in all directions, because the electric field vector oscillates in a distinct plane. Treating unpolarized X-rays as a beam of photons polarized in random orientation, the classical Thomson formula for coherently scattered X-ray intensity I_c on a charged particle evaluates to

$$I_c = I_0 \frac{e^4}{m^2 r^2 c^4} \left(\frac{1 + \cos^2 2\theta}{2} \right) \tag{6-9}$$

with e the electronic charge, m the electron mass, r the diameter of the electron (the Thomson equation (6-9) derived from classical electrodynamics holds for any spherical charged object) and c the light speed. I_0 is the incoming intensity of the photon beam. The interesting quality in Equation 6-9 is the cosine-dependent polarization factor in the brackets. As shown in Figure 6-9, scattering is strongest in both forward and backward directions, and weakest perpendicular to the incoming beam. We rarely observe backward scattering in protein single crystal diffraction, but the effect contributes to, and is distinctly visible in, diffraction patterns of well-diffracting material. When computing normalized scattering intensities from reflection data, the polarization factor is corrected for by instrument specific software, as its exact form depends on whether, and how, the incoming beam is already pre-polarized by a monochromator or other X-ray optics.

As an added bonus, integration of the Thomson equation (6-9) over all scattering angles yields the total scattering power P, from which the scattering cross

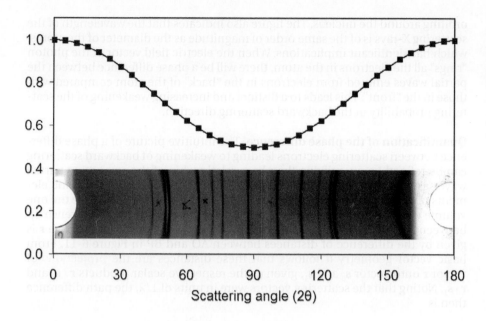

Figure 6-9 The polarization factor for Thomson X-ray scattering on electrons. The scattered intensity is largest in forward and backward directions, and weakest perpendicular to the direction of the incoming photons (scattering angle 90°). The powder X-ray diffraction film of a strongly scattering metal alloy (insert) also shows backward reflections, a phenomenon generally not observable for protein crystals, given their relatively weak, forward-only scattering.

section (P/I_0) can be computed. The values obtained for the cross section show that the fraction of X-ray photons scattered on electrons is indeed very low, in the range of 0.1–1% of the incident beam intensity.

Scattering of X-rays by electrons of a single atom

The negatively charged electrons of atoms are not free, but move around the positively charged nucleus in stable, defined orbitals represented by probability distributions directly related to the square of each electron's complex wave function, ψ. The amplitude of a scattered wave emanating from an atom will thus depend on the distribution of the electrons, that is, on how many electrons scatter from any given position $\mathbf{r}$ in the atom, which is described by the electron density $\rho(\mathbf{r})$.

In a first approximation we can treat the electron distribution or electron density $\rho(\mathbf{r})$ as spherical (which, as we know from basic quantum chemistry, is not exactly correct for higher orbitals). Nevertheless, the spherical approximation suffices for the calculation of atomic scattering factors (except in ultra-high resolution, charge-density crystallography work) and under the assumption of spherical electron density, the atomic scattering function f will have spherical symmetry as well.

We can qualitatively derive the general shape of the atomic scattering function f based on the simple picture in Figure 6-10 of an atom consisting of electrons

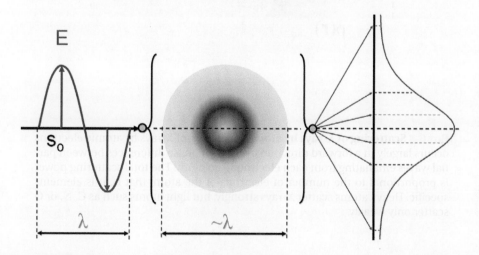

Figure 6-10 Scattering of X-rays by a single atom. If the wavelength of the scattered radiation and the size of the scattering object are of comparable dimensions, scattering occurs mostly in a forward direction. Note that the electric field vector is not drawn to scale; the field vector's range of interaction is actually much larger than the vector amplitude drawn in this figure. The rightmost panel shows the atomic scattering function which is proportional to the square root of the intensity of the scattered X-ray photons recorded on a detector.

orbiting around the nucleus. The figure also indicates that the wavelength of the scattering X-rays is of the same order of magnitude as the diameter of the atoms, which has significant implications. When the electric field vector of the photon "rings" all the electrons in the atom, there will be a phase difference between the partial waves emitted from electrons in the "back" of the atom compared with those in the "front". This leads to a distinct and increasing weakening of the scattering probability in the backward scattering direction.

Quantification of the phase difference. The intuitive picture of a phase difference between scattering electrons leading to weakening of backward scattering can be refined by quantifying the phase relationship of the emanating partial waves as a function of their distance. We consider two scattering volume elements of the electron density $\rho(\mathbf{r})$ of the atom. For simplicity, we place the one volume element at the origin, the other one at point P. The path difference Δp between a wave emanating from point O at the origin and P at a distance $\mathbf{r}$ is given by the difference of distances between AO and BP in Figure 6-11. From basic vector geometry it follows that these distances are the projections of vector $\mathbf{r}$ onto vector $\mathbf{s}_0$ and $\mathbf{s}_1$, given as the respective scalar products $\mathbf{r} \cdot \mathbf{s}_0$ and $\mathbf{r} \cdot \mathbf{s}_1$. Noting that the scattering vectors were in units of $1/\lambda$, the path difference then is

$$\Delta p = (\lambda \mathbf{r} \bullet \mathbf{s}_1 - \lambda \mathbf{r} \bullet \mathbf{s}_0) = (\mathbf{s}_1 - \mathbf{s}_0) \bullet \mathbf{r}\lambda = \mathbf{S} \bullet \mathbf{r}\lambda \qquad (6\text{-}10)$$

We arrive from path difference Δp to phase difference $\Delta \varphi$ expressed in radians by multiplying by $2\pi/\lambda$, resulting in

$$\Delta \varphi = 2\pi(\mathbf{s}_1 - \mathbf{s}_0) \cdot \mathbf{r} \qquad (6\text{-}11)$$

which by immediately substituting $\mathbf{S} = \mathbf{s}_1 - \mathbf{s}_0$, gives the crucially important general expression for the phase difference or relative phase angle of the scattered wave in reference to a fixed origin,

$$\Delta \varphi = 2\pi \mathbf{S} \cdot \mathbf{r} \qquad (6\text{-}12)$$

Figure 6-11 Scattering of X-rays by electrons of an atom. The diagram shows the scattering vector representation of partial waves emanating from two scattering volume elements of the electron density $\rho(\mathbf{r})$ of an atom. The spherically symmetric electron density is symbolized by the shading. For simplicity, we place one volume element at the origin, the other one at point P. The path difference Δp between the shorter upper path through P and the lower path though O is given by the difference between the scalar vector products $\mathbf{r} \cdot \mathbf{s}_0$ and $\mathbf{r} \cdot \mathbf{s}_1$. Given that the scattering vectors were in units of $1/\lambda$, the path difference is then $\Delta p = (\lambda \mathbf{r} \bullet \mathbf{s}_1 - \lambda \mathbf{r} \bullet \mathbf{s}_0) = (\mathbf{s}_1 - \mathbf{s}_0) \bullet \mathbf{r}\lambda = \mathbf{S}_1 \bullet \mathbf{r}\lambda$ We arrive from path difference Δp to phase difference $\Delta \varphi$ between the partial waves emanating from O and P expressed in radians by multiplying by $2\pi/\lambda$, resulting in $\Delta \varphi = 2\pi(\mathbf{s}_1 - \mathbf{s}_0) \cdot \mathbf{r} = 2\pi \mathbf{S} \cdot \mathbf{r}$. Note that there is only one incoming electric field vector perpendicular to the wave propagation, not two incoming waves or X-rays in phase.

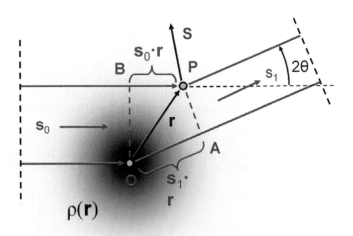

Box 6-4 Scattering of X-rays by atoms. Scattering of X-rays by atomic electrons occurs largely in a forward direction because of phase differences between partial waves emanating from each electron in an atom. The total scattering power is proportional to the number of electrons of the atom, that is, it is element-specific. Heavy atoms scatter X-rays strongly, but light atoms such as C, N, or O scatter only weakly.

Atomic scattering factor. Each partial wave emanating from a volume element of $\rho(\mathbf{r})$ is described in exponential vector form as a wave $\exp(i\varphi)$ with relative phase $\varphi = 2\pi\mathbf{Sr}$ and a magnitude corresponding to the electron density in that volume element. The total resulting wave is finally obtained by integration over the entire atom volume. As the electron density is in first approximation spherical and thus centrosymmetric, the imaginary sine terms in the trigonometric form of the scattering exponent cancel out, and the integral over the function is real:

$$f_{\mathbf{S}} = \int_{\mathbf{r}}^{V(atom)} \rho(\mathbf{r}) \exp(2\pi i \mathbf{Sr})\, d\mathbf{r} \qquad (6\text{-}13)$$

The magnitude of $f_{\mathbf{S}}$ plotted against the scattering angle indeed looks like the red graph in Figure 6-10. The shape of the scattering function is Gaussian, centered on $\mathbf{S} = 0$. From analysis of Equation 6-13 we can deduce a number of important properties:

- The function $f_{\mathbf{S}}$ describes the scattering of the entire atom as a function of the scattering direction and thus of the scattering angle. $f_{\mathbf{S}}$ is therefore called the atomic scattering factor or *atom form factor*.

- The atomic scattering factor $f_{\mathbf{S}}$ falls off rapidly with scattering angle, and setting $\mathbf{S} = 0$, the exponent becomes $\exp(0) = 1$ and the integration is over all electron density in the atom volume, which is exactly the number of electrons z. Thus, $f_{\mathbf{S}=0} = z$. This is simply a manifestation of the fact that the scattering probability increases proportionally with the number of available scattering objects, that is, the number of electrons.

- The scattering function is symmetric, because the electron density is in first approximation also centrosymmetric.

- Because we integrated a density of electrons with dimension $[e^-/\text{Å}^3]$ over the entire atomic volume, the dimension of the scattering factor is $[e^-]$.

- As we prove later, the atomic scattering factor integral over the electron density has the form of a Fourier integral, and therefore $f_{\mathbf{S}}$ is the Fourier transform (FT) of the electron density $\rho(\mathbf{r})$ of the atom:

$$f_{\mathbf{S}} = \int_{\mathbf{r}}^{V(atom)} \rho(\mathbf{r}) \exp(2\pi i \mathbf{Sr})\, d\mathbf{r} = \mathrm{FT}[\rho(\mathbf{r})] \qquad (6\text{-}14)$$

Atomic scattering factors

The actual atomic scattering functions look like the red Gaussian graph depicted in Figure 6-10 and their functions can be derived from first principles by calculating the scattering contribution of each electron from its electronic wave functions (which are the solutions of the Schrödinger equation). In practice, an empirical approximation for the atomic scattering functions, calculated from Hartree–Fock wave functions, was established by Don Cromer and Joseph Mann[2] at Los Alamos National Laboratory. Their nine-parameter Gaussian summation is used in practically all protein crystallography programs to compute the wavelength-independent atomic scattering factor or atomic form factor $f_{\mathbf{S}}^0$ of an atom as a function of scattering angle θ:

$$f_{\mathbf{S}}^0 = \sum_{i=1}^{4} a_i \cdot \exp\left(-b_i |\mathbf{S}|^2 / 4\right) + c = \sum_{i=1}^{4} a_i \cdot \exp\left(-b_i (\sin\theta/\lambda)^2\right) + c \qquad (6\text{-}15)$$

The nine parameters a_i, b_i, and c are the Cromer–Mann coefficients for each atom, and the resulting scattering curve is a superposition of four Gaussian exponentials of the form $\exp(-x^2)$ centered on $\mathbf{S} = 0$ (or $\sin\theta = 0$) plus a constant term c. Figure 6-12 shows actual atomic scattering functions calculated from the Cromer–Mann formula, and illustrates the shape of the scattering factor function $f_{\mathbf{S}}^0$ as a function of $\sin\theta/\lambda$. The Cromer–Mann coefficients are listed in the International Tables for Crystallography.[3]

We annotate the atomic scattering factor $f_{\mathbf{S}}$ sometimes explicitly as $f_{\mathbf{S}}^0$ to indicate that this is the wavelength-independent atomic scattering factor, in contrast

Sidebar 6-4 Exponents must be dimensionless. Note that throughout all equations derived, the exponents do not have any dimensions. This often provides a useful consistency check. For example, we can express a scattering term as $\exp(2\pi i \mathbf{Sr})$, where the reciprocal space scattering vector $\mathbf{S}$ has the dimensions of $1/\text{Å}$ while the direct space position vector $\mathbf{r}$ has the dimensions of Å, and the exponent becomes dimensionless. We can also express the scattering term as $\exp(2\pi i \mathbf{hx})$ in already dimensionless quantities such as $\mathbf{h}$ (the index triple of a reciprocal lattice point) and $\mathbf{x}$ (dimensionless fractional crystallographic coordinates). Mixing $\mathbf{S}$ and $\mathbf{x}$, however, does not work. The same holds for exponents containing the B-factor in dimensions of Å^2: The term $(\sin\theta/\lambda)^2$ has the dimensions Å^{-2} and the temperature factor term $\exp[-B(\sin\theta/\lambda)^2]$ properly becomes dimensionless.

Figure 6-12 Scattering factors for selected atoms calculated from Cromer–Mann coefficients. The scattering factor curves for H, C, O, S, and Fe atoms at rest are calculated from their respective nine Cromer–Mann coefficients (see, for example, Table 6-21). For $\sin\theta/\lambda = 0$, the scattering factor is identical to the number of electrons z. The scattering occurs largely in a forward direction (i.e. falls off at higher scattering angles), and is further attenuated by vibrational or displacive disorder of the scattering atoms (Figure 6-13). Note that hydrogen atoms with their single electron contribute very little to the overall scattering of a protein and they are therefore normally not visible in the electron density, except at very high resolution (beginning at approximately 1.2 Å).

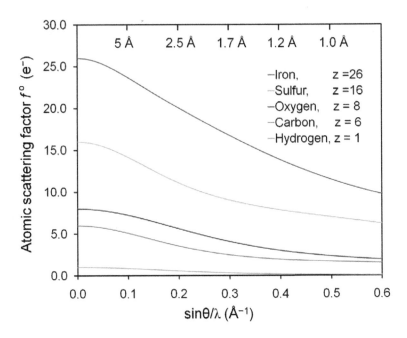

to the wavelength-dependent, anomalous scattering factor contributions, introduced later in this chapter. The abscissa of Figure 6-12 is plotted in units of $\sin\theta/\lambda = S/2 = |\mathbf{S}|/2 = 1/2d$ providing a common, wavelength-independent normalization of functions of the scattering angle (the term $\sin\theta/\lambda$ is inversely related to the interplanar spacing d of the diffracting lattice planes and equals $1/2d$, as we will derive shortly from the Bragg equation).

The form of the scattering factor curves in Figure 6-12 has a significant implication for X-ray diffraction. As a principle, the intensity of the scattered X-rays will always decrease with increasing scattering angle, no matter how well a crystal diffracts, even when the atoms are at rest.

Scattering factors of ions. One may wonder what happens to the atomic scattering factor curve when we remove or add valence electrons and form ions. Metals, for example, are incorporated into proteins as cations and not as free atoms, and the number of electrons for a Mg^{2+} ion for example is 10 and not 12 as for atomic Mg, a significant difference. However, the contribution of valence electrons to the scattering function dominates only the lowest $\sin\theta/\lambda$ values, while the remaining core electron contribution remains practically unchanged. The scattering curves for ions thus are "flattened" for very low $\sin\theta/\lambda$ and correctly reflect the reduced electron count at $\sin\theta/\lambda = 0$, but they are well approximated by the atomic core contributions for the remainder of the atomic scattering factor curve. Nonetheless, it is good practice to set the ionization state to

Cromer–Mann coefficients for C, $z = 6$				
i	1	2	3	4
a_i (e^-)	2.310	1.020	1.589	0.865
b_i ($Å^2$)	20.844	10.208	0.569	51.651
c (e^-)	0.216	—	—	—

Table 6-1 Cromer–Mann coefficients for carbon. The Cromer–Mann coefficients a_i, b_i, and c are the coefficients of a four-Gaussian (plus constant) superposition approximating the atomic scattering functions or atom form factors. As an exercise, calculate the scattering factor f^0 for $\sin\theta/\lambda = 0$ using the Cromer–Mann formula given above.

Box 6-5 **Atomic scattering factors.** The electron density around an atom is in first approximation treated as spherical, and the atomic scattering factor curves for each element are Gaussian functions, continuously decreasing with scattering angle. The atomic scattering curves (or atom form factors) are computed from the 9-parameter Cromer–Mann approximation for each scattering angle. The atomic scattering factor is measured in units of electrons, e^-, and evaluates for zero scattering angle to the number of electrons of the scattering atom, z. The total scattering power of a molecule is thus proportional to the scattering from all its atoms, i.e. proportional to the sum of all z.

a reasonably probable value, and protein crystallography refinement programs recognize ions in form of "MG+2" or "FE+3" and use the appropriate scattering factor curves.

Debye–Waller factor, atomic displacement, and B-factor

In the discussion of scattering of X-rays by atoms we qualitatively explained the increasing attenuation of the scattering with diffraction angle as a result of the phase difference between the partial waves emanating from electrons at different positions across the atom. In addition, atoms in a crystal lattice or in a molecule are not fixed in absolutely rigid positions, but they will vibrate around a mean position. The atoms can also be displaced in the lattice because of disorder, that is, in each unit cell the atoms tend to be at a slightly different (or in unfortunate cases, quite different) position. Higher temperature implies higher vibrational energy of an atom, and therefore larger displacement from its mean equilibrium position. This thermal vibration effect is generally indistinguishable from static or dynamic displacive disorder given a single diffraction data set.

In our qualitative picture, additional phase difference will result from the displacement of atoms against each other, causing increased interference of partial waves emanating from the electrons. The additional attenuation of the atomic scattering function will therefore be an additional Gaussian, wavelength-dependent term, called the Debye–Waller factor T:

Sidebar 6-5 **Neutron scattering.** When the wavelength of the X-ray photons (or any exciting radiation) is comparable to the size of the scattering object, scattering occurs largely into the forward direction, as shown in Figure 6-10. In terms of classical scattering theory, this is not the case when the size R of the scattering object is very small compared with the scattering wavelength λ, that is, the Mie ratio $2\pi R/\lambda \ll 1$ (compare the "Light scattering methods" section in Chapter 4). Thermal neutrons are uncharged particles of a de Broglie wavelength comparable to that of X-rays, and they are also used for diffraction experiments. Neutrons do not interact with the charged electrons, but they are scattered by the more than three orders of magnitude smaller atomic nucleus. As a consequence, their scattering cross section is practically independent of the scattering angle. For the discovery of the relation $\lambda = h/p$, Prince Louis-Victor Pierre Raymond de Broglie received (for his Ph.D. thesis) the Nobel Prize in Physics in 1929, another (in this case truly "noble") Nobel Prize relevant to crystallography.

Neutron scattering factors for light atoms, notably for hydrogen, are of similar magnitude as for heavier atoms.

This explains why neutron scattering is sometimes used for small-angle scattering for enhanced solvent contrast experiments (Chapter 4), and for high-resolution diffraction studies when the location of hydrogen atoms is of special interest. Neutron crystallography requires a neutron source (a nuclear reactor or a spallation neutron source). Exposures are relatively long even with strong neutron sources, and the resulting necessity for relatively large crystals provides an additional challenge. The advantage of using neutrons over X-rays in diffraction studies is the ability to directly determine exact hydrogen bond directionality or to determine the protonation state of enzymes. Soaking of the crystals in D_2O which is usually conducted to suppress inelastic background scattering from ^1H also leads to hydrogen–deuterium exchange of dynamically exchangeable hydrogen positions, which can be clearly located due to the sign difference in scattering factors of ^1H and ^2D. Remarkable results have been obtained through neutron diffraction,[4] and numerous figures in Niimura and Bau's 2008 review illustrate recent achievements.[5]

$$T_\mathbf{S} = \exp\left(-B_{iso}\left|\mathbf{S}\right|^2 / 4\right) = \exp\left(-B_{iso}(\sin\theta / \lambda)^2\right) \tag{6-16}$$

where B_{iso} is the so-called isotropic displacement parameter or B-factor which is directly related to the mean square isotropic displacement $\langle u_{iso}^2 \rangle$ of the atom from its equilibrium or mean position:

$$B_{iso} = 8\pi^2 \langle u_{iso}^2 \rangle \tag{6-17}$$

Using Equation (6-17) we can calculate that a B-factor of $8\pi^2 = 79$ Å² corresponds to a root mean square (r.m.s.) displacement $\langle u_{iso}^2 \rangle^{1/2}$ of 1 Å, a relatively large value compared with the atomic radius of ~0.77 Å for a carbon atom. The attenuation of the scattering function as a function of the scattering angle (Figure 6-13) can be expressed in terms of $\sin\theta/\lambda$:

$$f_\mathbf{S}^B = f_\mathbf{S}^0 \cdot T_\mathbf{S} = f_\mathbf{S}^0 e^{-B_{iso}(\sin\theta/\lambda)^2} \tag{6-18}$$

For proteins, the B-factors of individual atoms vary depending on the position of the atom in the protein and the overall dynamic behaviour of the molecule. A residue in a tightly packed protein core may have B-factors of around 5 Å², but surface-exposed residue side chains may have much higher B-factors of maybe 50–70 Å². For poorly diffracting low-resolution structures, B-factors can become very large, and the absolute values become uncertain because of scaling difficulties. Note from Figure 6-13 that for a carbon atom with a B-factor of 100 Å² the scattering contribution is essentially zero beyond 2.5 Å. This means that molecules or parts thereof with very "floppy" atoms—the equivalent of a very large positional displacement—*cannot* and *will not* diffract well, and we have already seen the results in the form of weak or missing electron density in the illustrations of disordered loops (Figure 2-18). The relation of the extent of diffraction and the average B-factor of the atoms in a structure will further examined in Chapter 9.

Effect of partial occupancy. Quite frequently a binding site in a protein structure or a discrete solvent molecule position may not be fully occupied. In this situation, the scattering factor curve is just proportionally attenuated. Note that in principle the shape of the scattering curve of a partially occupied atom looks different than that of a fully occupied atom with a high B-factor. Although in theory this also manifests itself in different shapes of the electron density

Figure 6-13 Scattering factor curves for carbon atoms with increasing *B*-factor. The scattering factor curve for carbon ($z = 6$) is calculated according to Equation 6-18 for the atom at rest ($B = 0$) and $B = 5$, 25, and 100 Å². The isotropic displacement factor B in the exponent of Equation 6-18 causes additional attenuation with increasing scattering angles. Attenuation is significant for higher vibrational amplitudes or large displacement of atoms from their mean position, so that atoms with high B-factors contribute little to the diffraction pattern and are thus not or only poorly visible in electron density.

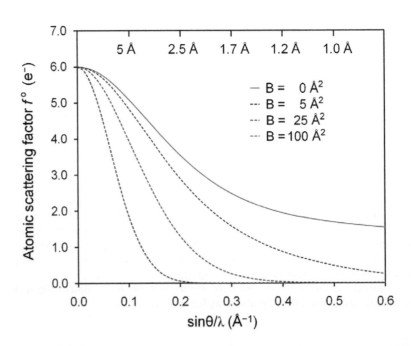

(Figure 9-6), in protein electron density maps it is generally not possible to distinguish a fully occupied atom with a high B-factor from a partially occupied atom with a low B-factor. This can make the assignment of unknown atom types for density observed at discrete solvent positions uncertain, and chemical plausibility needs to augment the assignment of atom type (Chapter 12).

Displacement anisotropy. Given the irregular and flexible shape of a protein molecule, the displacement of atoms in a protein molecule is likely not perfectly isotropic but generally anisotropic, that is, of different displacement magnitude along the three principal axes of an anisotropic displacement ellipsoid. Accounting for individual anisotropic B-factors in protein model refinement is only possible if data of roughly 1.2 Å or better are available. We leave a detailed discussion of anisotropic B-factors, their displacement tensors, and the description of correlated movements of entire regions of the protein through TLS (Translation-Libration-Screw) motions for refinement in Chapter 12.

Scattering from adjacent atoms

So far our discussion has been limited to the scattering of X-rays from single atoms, with the scattering functions of atoms continuously decreasing with scattering angle. In the following we will examine the effects of scattering from two atoms in the vicinity of each other, and expand the conclusions to the discrete scattering of X-rays on atoms arranged in a periodic crystal lattice. This discrete scattering is also termed diffraction of X-rays.

Let us examine the scattering of X-rays on two atoms next to each other. The electric field vector of the photon excites all electrons in both atoms, and interference between partial waves emanating from the electrons of both atoms results. The scattered intensity for a single atom was described by its atomic scattering or form factor, representing the Fourier transform of the electron distribution of the atom. For two atoms, one again expects to observe an atomic transform, but this time modulated by additional interference.

Figure 6-14 depicts the theoretical scattering curve of X-rays from two adjacent identical atoms, recorded on a linear detector. The suspicion of additional interference is confirmed, and we recognize that the scattering is now twice as intense compared with a single atom because we have twice the electrons available to scatter the photons, and the modulation of the scattering curve is obvious. We further suspect that the position of these modulations is somehow related to the distance of the scattering atoms. This assumption is correct, and we need to derive a formal relation between the distance of the scattering objects d and the position of the peaks in the scattering function.

The relation between spacing of scattering objects and distance of the modulations from the origin can be derived from the scattering diagram on the left side of Figure 6-14. The path difference can be derived as before from Figure 6-11, and as shown in the case of discrete volume elements, the corresponding phase difference is

$$\Delta\varphi = 2\pi(\mathbf{s}_1 - \mathbf{s}_0)\cdot\mathbf{r} = 2\pi\mathbf{S}\cdot\mathbf{r} \qquad (6\text{-}19)$$

Sidebar 6-6 Debye–Waller factor, temperature factor, and B-factor. The theory of thermal motions in crystal lattices was originally developed by Albert Einstein, and refined by Peter Debye and Ivar Waller in the 1920s. Peter Debye received the Nobel Prize in Chemistry in 1936 for this work and other studies in physical chemistry; another example of a Nobel honor awarded for work related to crystallography. In tribute to their work, the entire exponential term T

$$T_s = \exp\left(-B_{iso}\,|\mathbf{S}|^2/4\right)$$
$$= \exp\left(-B_{iso}(\sin\theta/\lambda)^2\right)$$

is called the Debye–Waller factor. The parameter B_{iso} in the exponent is the isotropic B-factor. Because of its original derivation the B-factor is sometimes also referred to as the temperature factor. Given that displacive disorder can have the same effect, the historic term *temperature factor* for B should be avoided in crystallographic literature;[6] B is in fact an isotropic displacement parameter.

Box 6-6 Attenuation of atomic scattering and the B-factor. The scattering factors for atoms in a crystal structure are additionally attenuated by displacement of the atoms from their mean position. Positional displacement results from thermal vibration around the resting position as well as from displacive disorder in a crystal lattice. All these effects compound and are accounted for through the exponential B-factor attenuation in a Debye-Waller factor T_s. A molecule whose atoms have high B-factors or exhibit a high degree of disorder thus cannot diffract well.

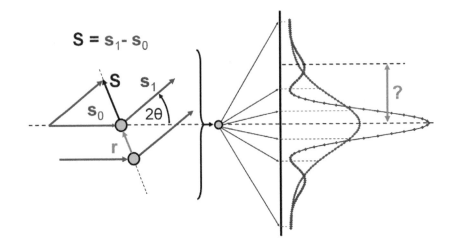

Figure 6-14 Scattering of X-rays by two adjacent atoms. The left panel of the figure shows a scattering diagram (Figure 6-11) for two adjacent atoms; the right panel shows the corresponding scattering function of two atoms in blue and a single scattering factor curve in red. The scattering curve in the right panel suggests that some relation exists between the scattering angle at which diffraction maxima appear and the distance between the scattering objects. The total scattering by two atoms is twice as intense as from a single atom, and the scattering curve of the atoms is modulated by further interference from scattering contributions originating from the two different atoms. In practice, the straight-ahead scattering at $2\theta = 0°$ cannot be separated from the unscattered direct (incoming or primary) beam.

The maximum in the scattering function will occur when there is maximal constructive interference between the partial waves. The phase difference $\Delta\varphi = 2\pi$ or integer multiples $n \cdot 2\pi$ will generate maximum constructive interference, while at phase shifts of π we expect minima in the scattering function. Therefore, for maximal constructive interference, the difference between the scattering vectors has to be $\Delta\varphi_{max} = 2\pi\mathbf{S}\cdot\mathbf{r} = n\cdot 2\pi$ and therefore, $\mathbf{S}\cdot\mathbf{r} = n$. In a 1-dimensional crystal lattice, the distance vector $\mathbf{r}$ between two identical atoms would correspond to the unit cell basis vector $\mathbf{a}$. For a 3-dimensional structure, the condition for maximum constructive interference would have to be fulfilled in all three dimensions, and we obtain three independent equations, one for each unit cell vector:

$$\mathbf{S}\cdot\mathbf{a} = n_1, \quad \mathbf{S}\cdot\mathbf{b} = n_2, \text{ and } \mathbf{S}\cdot\mathbf{c} = n_3 \qquad (6\text{-}20)$$

The three equations (6-20) determine the position of the diffraction maxima (i.e. the peaks in the diffraction pattern) and are known as the Laue equations. We will show in the subsequent sections that the integers in the Laue equations can be interpreted as the Miller indices (h, k, l) of the lattice planes of the unit cell containing the scattering objects. As a consequence, we can establish a direct relation of the scattering direction defined by $\mathbf{S}$ to the reciprocal lattice vector $\mathbf{d}^*_{hkl}$, as quantified in the next section.

The Bragg equation

A trick that drastically simplifies the interpretation of X-ray diffraction introduced by Sir William Lawrence Bragg (Sidebar 6-7) consists of turning the scattering diagram in such a way that the scattering vector $\mathbf{S}$ points straight up. This small adjustment has far reaching consequences: one realizes now that the scattering diagram can be interpreted as the reflection of X-rays on a set of planes in the crystal, to which the scattering vector $\mathbf{S}$ is normal.

Figure 6-15 Interpretation of X-ray diffraction as reflection on a lattice plane. The graphical interpretation of the Bragg equation allows treatment of X-ray diffraction as reflection on a set of planes in the crystal. The total path difference between the two excited partial waves is now $2d \cdot \sin\theta$, which must equal a multiple of $n\lambda$ for maximum constructive interference. The Bragg equation states this fundamental diffraction condition: $n\lambda = 2d \cdot \sin\theta$.

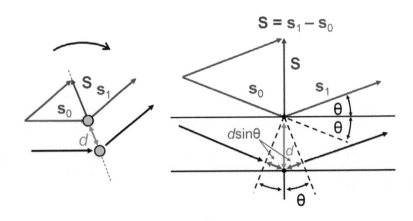

Sidebar 6-7 From Laue diffraction to Bragg equation. The mathematical formulation of the reflection conditions in a 3-dimensional crystal in the form of (6-20) was formulated first by Max von Laue, who received the Nobel Prize in Physics in 1914 for the discovery of diffraction of X-rays by crystals. His theoretical derivations were tested in 1912 (when he was Privatdozent in Munich) by Arnold Sommerfeld's assistants W. Friedrich and P. Knipping, who conducted the first diffraction experiments with X-rays. These experiments nearly did not happen, because Sommerfeld was occupied at this time with the study of the directional dependence of the Bremsstrahlung, and thought little of testing von Laue's "Raumgitterhypothese." While the experiments unambiguously established both the electromagnetic wave nature of X-rays and that matter consists of discrete atoms, the three separate diffraction conditions expressed as Laue equations for each direction in the crystal lattice are not overly convenient and not easy to visualize in practical use. Laue's diffraction pictures were actually obtained using "white" (polychromatic) X-ray radiation, and were finally interpreted by Sir William Lawrence Bragg.

Born in Australia, W. Lawrence Bragg was exposed to X-rays at the early age of 5 years, when he crashed his tricycle and broke his arm. His father W. Henry had recently read about Röntgen's experiments and used the newly discovered X-rays to examine his own son's broken arm. The affinity to X-rays appears to have stuck with W. Lawrence, and he continued to work in Cambridge on X-ray diffraction in collaboration with his father, who ran an X-ray laboratory in Leeds. Sir W. Henry built the first "X-ray spectrometer," a prototype of the modern diffractometer that used monochromatic X-rays, which greatly simplifies the interpretation of the diffraction patterns. W. Lawrence Bragg finally came up with the brilliant idea to interpret X-ray diffraction as *reflection* on discrete lattice planes (*hkl*), leading to the famous and indispensable Bragg equation that relates the diffraction angle θ to the lattice spacing d_{hkl}:

$$n\lambda = 2d_{hkl}\sin\theta \qquad (6\text{-}21)$$

When the Braggs received the Nobel Prize in Physics jointly in 1915, W. Lawrence was the youngest Nobel recipient ever, at an age of 25 years. An interesting reminiscence of his early work with W. L. Bragg in Cambridge was compiled by Max Perutz,[7] who worked for over three decades on the structure of hemoglobin[8] and received the 1962 Nobel Prize in Chemistry together with John Kendrew for their studies of the structures of globular proteins.

The interpretation of scattering as reflection on lattice planes allows the establishment of two extremely important relations:

- We can quantify the relation of the scattering angle θ to the interplanar distance d_{hkl} for the set of reflecting planes *hkl*. This follows immediately from the path difference $2d_{hkl}\sin\theta$ which can be directly read off the drawing in Figure 6-15 and which must be an integer multiple of the wavelength for maximum diffraction. This yields the famous Bragg equation,

$$n\lambda = 2d_{hkl}\sin\theta \quad \text{or} \quad \frac{1}{d_{hkl}} = \frac{2\sin\theta}{n\lambda} = d_{hkl}^* . \qquad (6\text{-}22)$$

- The Bragg equation (6-22) is scalar, that is, it relates only the magnitude of **S** to the scattering angle. In Chapter 5 we derived the basic relation between d_{hkl} and the corresponding reciprocal lattice vector $\mathbf{d}_{hkl}^*$, which in turn allows the establishment of the relation between the direction of the reciprocal scattering vector **S** and the corresponding set of reflecting planes *hkl* in the crystal lattice. Both the scattering vector **S** and the reciprocal lattice vector **d*** are perpendicular to the set of reflecting planes, and we expect that we will find a specific relation between the two vectors when diffraction occurs. We will derive and visualize this relationship shortly with the help of the Ewald sphere.

One point is paramount to keep in mind: There are not two incoming X-rays involved—it is the electromagnetic field vector of one and the same photon exciting all the atoms within its coherence length. A defined phase relation between the waves scattered by objects is required for constructive interference, but there is no phase relation between different photons from an X-ray source. Common X-ray sources such as X-ray generators and synchrotrons in normal operating mode are incoherent light sources, in contrast to lasers. We use lasers in classroom demonstrations of optical diffraction (Sidebar 6-10) because they are monochromatic and brilliant photon sources, not because of any coherence between photons.

Box 6-7 **Diffraction and the Bragg equation.** Interference between the partial waves scattered by neighboring atoms leads to a modulation of the scattering function. For a single molecule, the combined molecular scattering function is an irregularly modulated, decaying envelope function. The scattering of X-rays by a single molecule is too small to be measured, but by arrangement of many molecules in a periodic lattice, amplification and discrete sampling of the molecular scattering function occurs. The process described is the diffraction of X-rays on a crystal. Diffraction can be interpreted as reflection on sets of equidistant lattice planes, and the Bragg equation establishes a quantitative relation between the lattice spacing and the diffraction angle of discrete reflections:

$$n\lambda = 2d_{hkl}\sin\theta.$$

Scattering from a molecule

As a virtual experiment we now examine the scattering of X-rays by a single, whole molecule. The scattering function is, as always, a superposition of all atomic scattering factors in the molecule, some heavy ones scattering more, most light ones less. In a protein molecule, the scattering atoms are located at irregular distances from each other, and the modulation of scattering by the interference of neighboring atoms will not show any systematic periodicity. As a consequence, the scattering function (or scattering envelope) of a molecule is in essence a continuous, decaying complex function with an irregularly structured modulation to it, as shown in Figure 6-16.

The scattering function for the entire molecule is given by the superposition of all partial waves from all atoms in the molecule. The contribution of each atom j to the scattered intensity is given by the atomic scattering factor $f_{s,j}^{0}$. From the exponential summation formula for wave superposition (6-7) and the relative phase of each contribution given as $\varphi = 2\pi \mathbf{S}\mathbf{r}_j$ we obtain for the partial wave summation

$$\mathbf{F}_\mathbf{S} = \sum_{j=1}^{atoms} f_{s,j}^{0} \cdot \exp(2\pi i \mathbf{S}\mathbf{r}_j) \tag{6-23}$$

$\mathbf{F}_S$ describes the scattering function or molecular diffraction envelope of the entire molecule as shown in Figure 6-16. Note that the molecular envelope obtained by scattering is centrosymmetric, despite the molecule not being at all symmetric. This is a general property of reciprocal space, that is, the diffraction space. As the molecule itself is not centrosymmetric in contrast to a single atom, the molecular envelope function is complex (recall the imaginary number i in the exponent) but the recorded diffraction pattern is only a scalar function of the magnitude of the scattering function. The phase information is lost.

Figure 6-16 Molecular scattering function of a single molecule. The right panel shows the molecular scattering function of the 2-dimensional projection of the TIM-barrel molecule on the left. The molecular scattering function is a computed Fourier transform of the image on the left, and it is the equivalent of a single molecule diffraction pattern, here essentially a modulated, 2-dimensional Gaussian function rapidly decaying with scattering angle, just as for the atomic 1-dimensional scattering factors. Color indicates scattered intensity. The scattering of a single molecule is in practice too weak to be useful for structure determination with present X-ray diffraction methods, but X-ray lasers may provide a future avenue for exploiting single molecule diffraction (Sidebar 6-8).

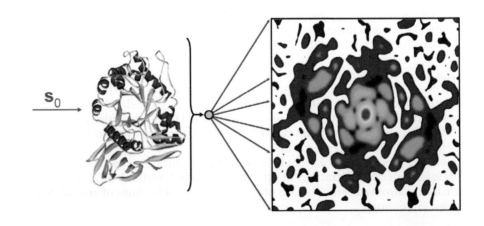

In practice, the X-ray scattering off a single molecule is miniscule and practically impossible to detect. We thus need more molecules to observe stronger total scattering intensity, and we need them assembled in a periodic lattice, so that systematic amplification of the reflections can occur. The above requirements are met by a crystal, and the amplification factor is proportional to the number of repeats. For an isotropic 100 μm crystal containing an array of 100 Å large molecules, we gain a factor of 10^4 in scattering power in each dimension, totaling a 10^{12} times stronger diffraction signal than from a single molecule. In addition, the discrete sampling at the peaks of a diffraction pattern greatly increases the signal-to-noise ratio in practical measurements of the diffraction intensities.

Diffraction from a molecular crystal

Given that a crystal is a periodic arrangement of atoms and molecules, each point or atom has multiple exact translational equivalents in space along the three crystallographic lattice directions. The scattering function therefore will eventually exactly repeat in certain directions in the crystal. For example, along the **c** unit cell axis, the same repeat is already in the next cell, translated by (0, 0, 1c). We can generalize this as a translation by u in **a**, v in **b**, and w in **c**. The total scattering from the unit cell is then a summation of all molecular unit cell scattering contributions $\mathbf{F}_s^{cell}$ in the crystal. In directions where these repeats u, v, and w, are exact integers, the vectors $\mathbf{F}_s^{cell}$ will add up in exactly the same direction, because they have the same phase. We already know from the Laue equations that for constructive interference, the repeats must occur in integer multiples along the unit cell vectors. The total scattering then becomes proportionally stronger with increasing number of repeating unit cells. The maxima in the scattering function also become sharper the more unit cells are aligned, because any small phase error for directions where unit cells do not scatter exactly in phase will also amplify rapidly. As a consequence, no more scattering occurs in directions that do not correspond exactly to integer hkl values. The qualitative scenario of summing unit cell contributions is visualized in Figure 6-17.

Quantification of the situation depicted in Figure 6-17 leads to a fundamental result. The total scattering factor of a crystal consisting of $n(a) \times n(b) \times n(c) = u \times v \times w = N$ unit cells is given by summing up of all molecular transforms in the unit cell $\mathbf{F}_s^{cell}$ that make up the crystal. In unit cell direction **a**, the phase difference between unit cells is given by $2\pi\mathbf{Sa}$, and the scattering function in direction **a** is

$$\mathbf{F}_s^a = \sum_{u=0}^{u-1} \mathbf{F}_s^{cell} \cdot \exp[2\pi i(u\mathbf{Sa})] \tag{6-24}$$

Factoring the scattering contribution of each cell out of the sum (6-24) and extending the summation over all directions in the crystal, the total scattering by the crystal becomes a product of the three summations:

$$\mathbf{F}_s^{cryst} = \mathbf{F}_s^{cell} \cdot \sum_{u=0}^{u-1} \exp[2\pi i(u\mathbf{Sa})] \cdot \sum_{v=0}^{v-1} \exp[2\pi i(v\mathbf{Sb})] \cdot \sum_{w=0}^{w-1} \exp[2\pi i(w\mathbf{Sc})] \tag{6-25}$$

Sidebar 6-8 Single molecule diffraction? If phase information could be conserved, a reconstruction of the scattering molecule is, in principle, possible from the molecular transform. A free electron X-ray laser (FEL), providing an extremely brilliant and coherent source of photons, could in theory be used to produce coherent scattering from a single molecule before it is disintegrated into plasma (where the electrons separate from nuclei) because of the enormous energy density of the X-ray laser beam. The phases can be retrieved from a coherent 2-dimensional scattering image by oversampling procedures.[9] Combined with yet-to-be developed, highly sensitive detection techniques and sampling of multiple molecules, the structure could in principle be reconstructed from single-shot exposures of multiple molecules in different orientations. The recent proof of principle was partly successful,[10] but many formidable technical challenges must be addressed if the method is ever to become generally applicable[11] for structural biology.

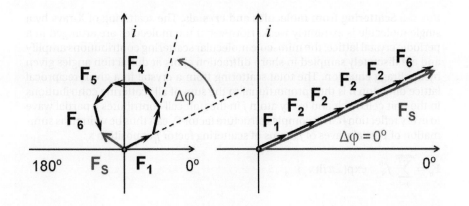

Figure 6-17 Summation of molecular scattering functions. If n repeated molecules in the crystal scatter exactly in phase, after a certain number of integer translations along the cell constants, the scattering functions $\mathbf{F}_n$ add up and the resulting scattering vector, called the structure factor, is strong (right panel). If there is a phase difference due to non-integer repeat distances, the relative phase differences add up and the resultant structure factor is much weaker (left panel). The more lattice repeats, the more sensitive the vector addition becomes to phase differences—for 100 structure factor repeats, a misalignment of as little as 3.6° will lead to full extinction. As a consequence, the diffraction pattern becomes sharper with an increasing number of aligned unit cells.

As visualized in Figure 6-17, the summation (6-25) is largest when $F_\mathbf{S}^{cryst} = N \cdot \mathbf{F}_\mathbf{S}^{cell}$. This is the case only when each of the $u \times v \times w = N$ exponential terms in the sum enumerates to 1. This again requires that the exponent is $n \cdot 2\pi i$ because $\exp(n2\pi i) = 1$ with integer n. Therefore also $\mathbf{Sa}$, $\mathbf{Sb}$, and $\mathbf{Sc}$ must be integers, which is equivalent to the conditions spelled out in the Laue equations.

The total scattering function from a crystal is thus proportional to the scattering from the N unit cell contents $F_\mathbf{S}^{cryst} = N \cdot \mathbf{F}_\mathbf{S}^{cell}$. We can therefore use the general expression for molecular scattering, except that the summation now goes over all atoms in the unit cell:

$$\mathbf{F}_\mathbf{S} = \sum_{j=1}^{atoms} f_{\mathbf{s},j}^0 \cdot \exp(2\pi i \mathbf{S} \mathbf{r}_j) \qquad (6\text{-}26)$$

The structure factor equation (6-26) is one of the most fundamental equations of crystallography and will be used constantly in the following sections of this text.

The structure factor

The scattering function equation (6-26) for the entire unit cell contents depends explicitly on the scattering vector $\mathbf{S}$ and the coordinate vectors $\mathbf{r}_j$ of all atoms in the crystal unit cell. The equation can be made entirely general by expressing the same situation in terms of the reciprocal lattice, where the scattering vector is expressed as the reciprocal Miller indices hkl of the reflecting planes, and the positional vector $\mathbf{r}$ is expressed in fractional coordinates $\mathbf{x}$. This eliminates both the explicit wavelength ($\mathbf{S}$ has the dimension 1/Å) and the cell dimensions (in Å) out of the exponent. With the definition of the real space lattice vector $\mathbf{r}_j$ for each atom j (see Chapter 5)

$$\mathbf{r}_j = \mathbf{A}^{\mathrm{T}}\mathbf{x}_j = (\mathbf{a}x_j + \mathbf{b}y_j + \mathbf{c}z_j) \qquad (6\text{-}27)$$

we obtain for the dot product in the exponent of Equation 6-26

$$\mathbf{S}\mathbf{r}_j = \mathbf{S}\mathbf{a}x_j + \mathbf{S}\mathbf{b}y_j + \mathbf{S}\mathbf{c}z_j \qquad (6\text{-}28)$$

Applying the Laue equations (6-20) and substituting using Miller indices (equivalent to reciprocal lattice point indices) for the integers, it follows that

$$\mathbf{S}\mathbf{r}_j = \mathbf{S}\mathbf{a}x_j + \mathbf{S}\mathbf{b}y_j + \mathbf{S}\mathbf{c}z_j = hx_j + ky_j + yz_j = \mathbf{h}\mathbf{x}_j \qquad (6\text{-}29)$$

Substituting (6-29) back into the scattering function $\mathbf{F}_\mathbf{S}$ (6-26) finally gives, for the structure factor expressed as a function of index vector $\mathbf{h}$,

$$\mathbf{F}_\mathbf{h} = \sum_{j=1}^{atoms} f_{\mathbf{s},j}^0 \cdot \exp(2\pi i \mathbf{h}\mathbf{x}_j) \qquad (6\text{-}30)$$

The scattering function of the unit cell expressed in fractional coordinates $\mathbf{x}_j$ and reciprocal lattice indices $\mathbf{h}$ is the structure factor $\mathbf{F}_\mathbf{h}$ of reflection $\mathbf{h}$. We can

---

Box 6-8 **Scattering from molecules and crystals.** The scattering of X-rays by a single molecule is extremely weak. However, if the molecules are arranged in a periodic crystal lattice, the miniscule molecular scattering contributions amplify and are discretely sampled in sharp diffraction spots at diffraction angles given by the Bragg equation. The total scattering from a crystal in a given reciprocal lattice direction $\mathbf{h}$ is thus proportional to the sum of all scattering contributions in the unit cell. Each and every atom j in the unit cell contributes a partial wave to every reflection $\mathbf{h}$. The complex structure factor $\mathbf{F}_\mathbf{h}$ can thus be written as summation of partial waves of j atoms of scattering factor f_j at position $\mathbf{x}_j$:

$$\mathbf{F}_\mathbf{h} = \sum_{j=1}^{atoms} f_{\mathbf{s},j}^0 \cdot \exp(2\pi i \mathbf{h}\mathbf{x}_j)$$

therefore compute exactly the scattering function for each diffracted X-ray, given the reciprocal Miller index triple of the lattice planes that reflect the X-rays. We have not yet established, however, in which *direction* in space the X-rays diffract

Sidebar 6-9 Diffraction and Fourier transforms. Diffraction can be formally described as the transformation of the electron density from real space R into reciprocal space R˚. The reciprocal space diffraction pattern is then given by the amplitudes of the Fourier transform (FT) of the crystal structure. Fourier transforms are a very important concept in crystallography, and Chapter 9 is dedicated to the discussion of Fourier transforms and their crucial role in the reconstruction of the molecular structure from diffraction data.

Figure 6-18 provides a preview of how diffraction improves when X-rays are scattered by an array of regularly arranged objects: the continuous molecular scattering turns into discrete diffraction from a crystal. The images in the right column are computed Fourier transforms of the objects in the left column representing molecules arranged in a periodic lattice with increasing number of repeats.

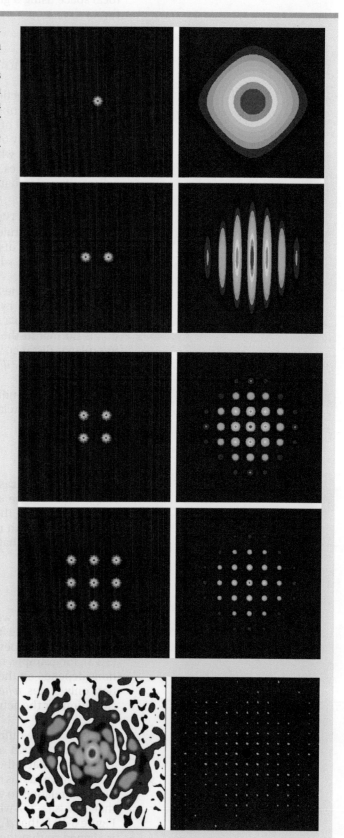

Figure 6-18 Scattering from a single molecule turns into discrete diffraction when the molecules are periodically arranged in a crystal lattice. Each pair of panels shows the scattering object (left side) and the resulting diffraction pattern (right side). The scattering object is a cross-shaped, nearly globular "molecule," and its molecular scattering function has a somewhat square shape. With a growing number of periodically arranged objects, the molecular envelope becomes sampled in increasingly sharp, discrete spots, and molecular scattering turns into discrete diffraction from a crystal. In the first four rows, the diffraction patterns in the right column are calculated Fourier transforms of the diffracting objects depicted on the left side. In row two, note the vertical streaking of the diffraction pattern in the direction lacking periodic order. The last row provides a preview of the concept of Fourier convolution: The diffraction pattern results from periodic sampling of the molecular envelope. Shown is a molecular envelope (left side), sampled with a periodic lattice function, which generates the diffraction pattern of the molecular structure. The black dot symbolizes the beam stop blocking the strong **F**(000) reflection coinciding with the primary X-ray beam direction.

given the indices h, k, l of the planes in a crystal. Because the scattering vector $\mathbf{S}_{hkl}$ and the corresponding reciprocal lattice vector $\mathbf{d}^*_{hkl}$ are both perpendicular to a reflecting lattice plane, the geometric relations between crystal orientation and fulfillment of the diffraction condition can be readily understood in reciprocal space using a graphic representation of the diffraction process called the Ewald sphere.

6.2   Diffraction geometry

In the following section we will derive a convenient way of visualizing the geometric conditions under which diffraction occurs. The reciprocal lattice, which represents each set of equidistant lattice planes as one discrete reciprocal lattice point, will naturally be very suitable for the interpretation of the Laue equations and of Bragg's law, which established the diffraction conditions as reflection on a set of lattice planes. As an added bonus, we will gain a first insight into how to collect diffraction data.

The Ewald construction

The construction rules for the reciprocal lattice, established in Chapter 5 during the discussion of basic crystal and lattice geometry, lead us to the following important finding, which we review briefly.

A set of lattice planes in real space R with Miller indices (hkl) is represented in reciprocal space R* by a reciprocal lattice vector $\mathbf{d}^*_{hkl}$ extending from the reciprocal lattice origin to a reciprocal lattice point h, k, l. The reciprocal lattice vector $\mathbf{d}^*_{hkl}$ is normal to the set of lattice planes (hkl), and is thus collinear with the interplanar distance vector $\mathbf{d}_{hkl}$. The magnitude of $\mathbf{d}^*$ is reciprocal to the magnitude of $\mathbf{d}_{hkl}$, that is, $d^*_{hkl} = 1/d_{hkl}$.

The diffraction conditions from the Bragg equation $n\lambda = 2d_{hkl}\sin\theta$ can be rewritten so that its relation to the reciprocal lattice becomes distinct as

$$d^*_{hkl} = \frac{1}{d_{hkl}} = \frac{2\sin\theta}{n\lambda}$$

(6-31)

One detail that occasionally confuses is the quiet disappearance of n from the Bragg equation. We already know (and can easily demonstrate for the general case by evaluating the equation for the interplanar distance d_{hkl}) that the spacing for a plane with multiple indices $(n \cdot h, n \cdot k, n \cdot l) = n(hkl) = n\mathbf{h}$ is n-times smaller than d_{hkl}, that is,

$$\frac{1}{n \cdot d_{hkl}} = \frac{1}{d_{n(hkl)}}$$

(6-32)

which means that we can interpret the nth-order diffraction condition as diffraction originating from an n-times higher order set of planes belonging to the same zone. This does *not* imply that if a reflection (1 1 1) is observed, reflection (3 3 3), for example, must also be automatically observed; the *intensity* of reflections depends on the actual content of the unit cell, and the diffraction condition only states *when* diffraction is possible—a necessary but not sufficient condition for the actual occurrence of a reflection with finite intensity!

The form of the diffraction condition $d^*_{hkl} = 2\sin\theta/\lambda$ can be very conveniently visualized through a geometrical construction introduced by Paul Ewald in 1921. If we place the scattering diagram for the Bragg equation into a Ewald sphere with a radius of $1/\lambda$, the significance of Bragg's form of the reflection condition in relation to the reciprocal lattice immediately becomes obvious. Recall that $|\mathbf{s}_0| = |\mathbf{s}_1| = 1/\lambda$ when studying Figure 6-19. The key conclusion derived from Figure 6-19 is that $d^*_{hkl} = |\mathbf{S}| = 2\sin\theta/\lambda$, and it follows that if the reciprocal lattice is aligned so that the vectors $\mathbf{S}_{hkl}$ and $\mathbf{d}^*_{hkl}$ coincide, the reflection condition for

271

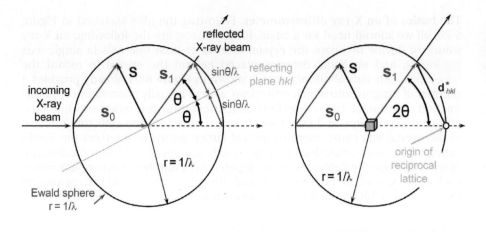

Figure 6-19 Ewald construction and reflection condition. Placing the sketch for reflection on a set of lattice planes *hkl* as shown in Figure 6-15 into the Ewald sphere of radius $1/\lambda$ allows the interpretation of diffraction in relation to the reciprocal lattice vectors *hkl*. As the left-hand panel shows, the two magenta sections each have a magnitude of $\sin\theta/\lambda$ following basic trigonometry. According to the Bragg equation, the distance of $2(\sin\theta/\lambda) = |\mathbf{S}|$ also equals $1/d_{hkl}$, which is also the length of the reciprocal lattice vector $\mathbf{d}^*_{hkl}$. Thus, the geometric reflection condition for the set of lattice planes *hkl* is necessarily fulfilled, when its reciprocal lattice point *hkl* lies on the Ewald sphere.

a set of lattice planes *hkl* in a crystal is necessarily fulfilled. This is equivalent to the reciprocal lattice point *hkl* lying on the Ewald sphere when the origin of the reciprocal lattice is placed at the intersection of the Ewald sphere with the incoming beam direction.

In the diffraction case $\mathbf{S} = \mathbf{d}^*$, and from the definition of the reciprocal lattice vector $\mathbf{S} = \mathbf{d}^* = (h\mathbf{a}^* + k\mathbf{b}^* + l\mathbf{c}^*)$ and substituting the Laue equations $\mathbf{S}\cdot\mathbf{a} = h$, $\mathbf{S}\cdot\mathbf{b} = k$, and $\mathbf{S}\cdot\mathbf{c} = l$ into it, we obtain exactly the reciprocity conditions we defined in Chapter 5 when introducing the reciprocal lattice:

$$\mathbf{Sa} = h = (h\mathbf{aa}^* + k\mathbf{ab}^* + l\mathbf{ac}^*) \Rightarrow \mathbf{aa}^* = 1 \quad \mathbf{ab}^* = 0 \quad \mathbf{ac}^* = 0$$
$$\mathbf{Sb} = k = (h\mathbf{ba}^* + k\mathbf{bb}^* + l\mathbf{bc}^*) \Rightarrow \mathbf{ba}^* = 0 \quad \mathbf{bb}^* = 1 \quad \mathbf{bc}^* = 0 \quad (6\text{-}33)$$
$$\mathbf{Sc} = l = (h\mathbf{ca}^* + k\mathbf{cb}^* + l\mathbf{cc}^*) \Rightarrow \mathbf{ca}^* = 0 \quad \mathbf{cb}^* = 0 \quad \mathbf{cc}^* = 1$$

Ewald sphere and reciprocal lattice

If we superimpose the reciprocal lattice with the Ewald sphere as indicated in Figure 6-19, any reciprocal lattice point that lies on the Ewald sphere will automatically fulfill the diffraction condition $d^*_{hkl} = 2\sin\theta/\lambda$ and $\mathbf{S} = \mathbf{d}^*$. Figure 6-20 shows this situation again together with a section of the reciprocal lattice. However, in practice only few reciprocal points will lie on the Ewald sphere right away, and we will observe few diffraction spots at any given orientation of the crystal. The other reciprocal lattice points can be brought to intersection with the Ewald sphere by rotating the crystal, which automatically rotates the crystal's reciprocal lattice.

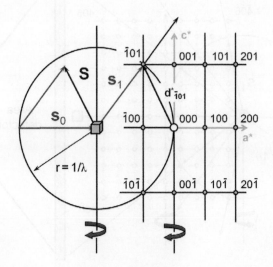

Figure 6-20 Reciprocal lattice and Ewald sphere. As few reciprocal lattice vectors will fulfill the diffraction condition of lying on the Ewald sphere in each specific crystal orientation, rotation of the crystal—and thus concurrent rotation of its reciprocal lattice—will bring more reciprocal lattice points to intersect the Ewald sphere whereupon the diffraction conditions are fulfilled.

The basics of an X-ray diffractometer. Following the idea sketched in Figure 6-20, all we should need for a basic diffractometer are the following: an X-ray source; a device to rotate the crystal around at least one axis (a single axis goniostat); and an X-ray detector placed behind the crystal to record the reflections. Our virtual diffractometer shown in Figure 6-21 already provides a number of insights into which reflections we can actually observe during a diffraction experiment. In the selected initial orientation, few reflections fulfill the diffraction condition, and only two of them actually hit the detector. Rotation of the crystal will cause more reciprocal lattice points to intersect the Ewald sphere. How many reflections we can record, and how complete the collection of reflections on the detector will be, depends on a number of other parameters such as the size of the Ewald sphere, and the diffraction limit of the crystal and its symmetry, which we will now examine further.

Diffraction limits and excluded regions

Inspection of the Ewald sphere in Figure 6-21 shows that only the part of the reciprocal lattice intersecting the $1/\lambda$ sphere can actually be sampled. The number of reflections sampled will thus depend on the size of the $1/\lambda$ sphere: the larger the $1/\lambda$ sphere, the more reflections will be sampled, in other words, the shorter the wavelength, the more reflections can be sampled. This fact is used to advantage in small molecule crystallography to collect as many reflections as possible from well-diffracting materials using the characteristic radiation of molybdenum (Mo) with a short wavelength of 0.7107 Å.

The $1/\lambda$ sphere in principle samples a donut-like torus of the reciprocal space when it is rotated though 360° around a single axis. We recognize that some reflections located on or close to the reciprocal axis of rotation will never intersect the Ewald sphere and thus will not be recorded (indicated by the blue shading in Figure 6-22). However, this is not as critical as it may appear at first sight. Protein crystals generally do not diffract very well, which we can represent by drawing a resolution sphere of $1/d_{min}$ extending from the center of the reciprocal space. This sphere contains all the reciprocal lattice points that actually carry any real diffraction intensity; beyond the diffraction limit d_{min}, no reflections are observed. We see that in the chosen example of a 2.0 Å diffraction limit, diffraction is limited to forward reflections up to an angle θ_{max}, which we can compute from the Bragg equation: $d_{min} = \lambda / 2 \sin \theta_{max}$.

The equation above evaluates for Cu Kα radiation of 1.5418 Å and $d_{min} = 2$ Å to 45.3° for reflection angle 2θ, in agreement with our drawing in Figure 6-22. From the same equation we also can calculate that the smallest observable d-spacing

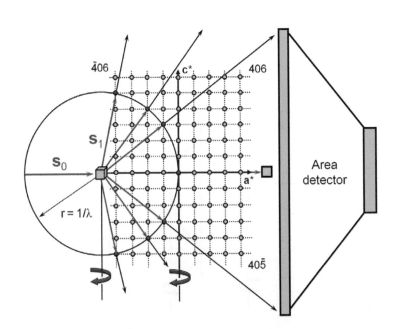

Figure 6-21 Geometry of a basic single-axis diffractometer. To collect the X-ray reflections resulting from diffraction by a crystal, we need to place an X-ray detector behind the crystal. When we rotate the crystal the reciprocal lattice also rotates and more reciprocal lattice points intersect the Ewald sphere and fulfill the reflection condition. In practice, the crystal is rotated by small increments of about 1° or less while a section or *frame* of the data is collected during each rotation increment. How many reflections need to be recorded depends on symmetry and diffraction limits of the crystal. The small blue block in front of the detector center symbolizes the primary beam stop.

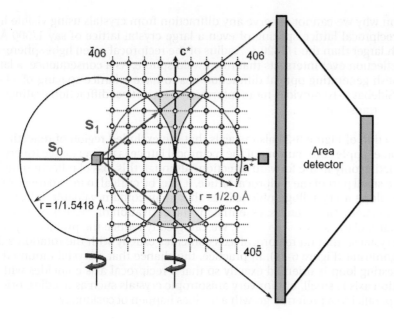

Figure 6-22 **Diffraction and resolution limits.** The maximal extent to which a particular crystal diffracts can be visualized by its resolution sphere (blue sphere). This crystal diffracts to 2 Å, and only the reciprocal lattice points or reflections within this resolution sphere carry any actual intensity; the ones outside produce no reflections. For the given 2 Å resolution sphere, the maximum diffraction angle is ~45°, and the detector can indeed collect all possible reflections from this crystal. However, the reflections contained in the blue "apple core" zone (cusp) or "blind region" can never intersect the reciprocal sphere and thus will not be recorded. This can be alleviated by tilting the crystal (Figure 6-24).

for $2\theta = 180°$ (backward scattering) for a given wavelength is $\lambda/2$, which is 0.77 Å for Cu Kα radiation, far beyond the diffraction limit of most protein crystals. However, we can also see from Figure 6-22 that for well-diffracting material we need to either have a very large area detector to collect all the data in one scan, or place the detector closer to the crystal, or offset the detector and perhaps collect complete data in several consecutive scans. Alternatively, we could select a shorter wavelength to collect the data; however, we do pay a price of reduced total scattering intensity in this case.

The tiny size of the limiting sphere for visible light (the wavelength for a green laser is long compared with X-rays: 532 nm, 5320 Å, or about 1/2 µm) is also the

Sidebar 6-10 **Optical transforms and diffraction.** Diffraction phenomena can be readily demonstrated through optical transforms. When a diffraction grating or periodic objects representing a 2-dimensional crystal printed on a transparent slide are illuminated with a brilliant, monochromatic beam of light from a laser pointer, a nice diffraction pattern can be projected onto a wall taking the role of the detector. The limiting sphere for visible light (532 nm or about 1/2 µm for a green laser) requires a lattice or mesh with a lattice spacing of ~10–100 µm. Diffraction slides with different 2-dimensional unit cell gratings and plane space groups are available from the Institute for Chemical Education (ICE) at the University of Wisconsin—Madison.

They make an excellent teaching aid for exploration of the reciprocal space, extinction rules, and the effect of different wavelength (try a red, green, or blue laser). Figure 6-23 shows a slide with eight different 2-dimensional lattices (diffraction gratings) and a corresponding diffraction pattern from an ICE kit (http://ice.chem.wisc.edu/Catalog/SciKits.html). Optical transforms were also used as a means of trial-and-error structure determination in the very early days of crystallography[12] (see also Sidebar 9-6).

Figure 6-23 **Optical diffraction grids and diffraction pattern.** The Institute for Chemical Education at the University of Wisconsin in Madison has generated a set of optical diffraction slides that make excellent teaching aids. The left panel shows one of the slides, and the right panel a corresponding diffraction pattern from a square plane lattice.

reason why we cannot observe any diffraction from crystals using visible light: the reciprocal lattice spacing of even a large crystal lattice of say $1/500$ $Å^{-1}$ is much larger than the $1/5320$ $Å^{-1}$ radius of the reciprocal green light sphere, and no reflection ever intersects that tiny Ewald sphere. As a consequence, a lattice or mesh generating optical diffraction has to have a lattice spacing of ~1–100 µm. Sidebar 6-10 provides more information on optical diffraction gratings and optical transforms.

Inspection of Figure 6-22 also reveals that the excluded region of reflections in the blue "apple core" cut-out of the reciprocal lattice is not very large for crystals diffracting only in the forward direction. Although this loss of reflections affects only a small part of the reciprocal lattice, it may be critical to record these reflections along a crystallographic axis as they may contain valuable information about systematic absences, essential for the derivation of space group symmetry, particularly so for screw axes. We can greatly reduce the problem by tilting the crystal so that no reciprocal axis coincides exactly with the rotation axis of the goniostat (Figure 6-24). In practice, the chance that a crystal mounted in a harvesting loop is oriented exactly so that a reciprocal axis coincides with the rotation axis is small, but for very anisotropic crystals such as needles, orientation parallel to a preferred growth axis does happen occasionally.

A further reason why it is not as difficult as it may appear to record complete diffraction data is that reciprocal space symmetry reflects the real space crystal point group symmetry. This makes it considerably easier to bring all necessary reciprocal lattice points into diffraction condition and to obtain complete and redundant diffraction data. However, before we can examine the relations of crystal symmetry and reciprocal space symmetry in detail (Section 6.5), we need to derive a formal quantification that relates intensities of a reflection to the contents of the unit cell.

Figure 6-24 Offset of reciprocal axes and symmetry reduce the number of unrecorded reflections. For the case that a unique axis of the crystal in monoclinic, trigonal, hexagonal, and tetragonal systems coincides with the rotation axis, tilting of the crystal so that the axis is offset from the rotation axis reduces the number of ultimately missed reflections (red points). The symmetry of the reciprocal space additionally means that only a smaller, unique part of the reciprocal space must be sampled. In the depicted example, the green and purple regions are symmetry related, and they each possess a centrosymmetrically related wedge that contains—in the absence of anomalous signal—the same intensities. If anomalous data are to be collected, centrosymmetrically opposed wedges of reciprocal space (or their symmetry equivalents) must also be sampled.

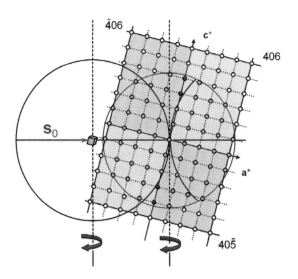

Box 6-9 Diffraction conditions in reciprocal space. The vector representation of the Bragg equation in the Ewald sphere of radius $1/\lambda$ shows that diffraction only occurs when the scattering vector **S** is collinear and of equal magnitude with a reciprocal lattice vector $\mathbf{d}_h^*$. This is the case when a reciprocal lattice point (h, k, l) lies on the Ewald sphere. Rotating the crystal, which corresponds to rotation of the reciprocal lattice, brings reciprocal lattice points successively into diffraction condition. During each small increment of rotation, we record a section of the diffraction pattern until a complete set of reflections is sampled.

6.3   The intensity of diffracted X-rays

At this point in our discourse we understand from the Laue and Bragg equations that the direction of diffracted X-rays relates to the direction of the corresponding reciprocal lattice vector. Based on the Ewald construction and given the crystal orientation, we know, for each reflection *hkl*, *where* it will be collected on the detector. But so far, our reciprocal lattice is a point lattice that does not carry any real intensity. The intensity of each reflection (as illustrated for a reciprocal lattice in Figure 6-25) depends on the presence of molecular motifs in the real space crystal lattice, and is determined by the contents of the unit cell of the crystal.

The intensities of all unique reflection intensities are collected in a diffraction experiment to reconstruct the electron density of the molecular object which diffracted the X-rays. Determination of cell parameters and space group alone does not suffice, because entirely different molecules could pack with the same symmetry in identical unit cells that possess the same reciprocal lattice. In the following we will therefore derive the relation between the unit cell contents and the intensity of diffracted X-rays in each reciprocal lattice direction *hkl*.

Structure factor and structure factor amplitude

The scattering function in a given direction **h** emanating from a crystal irradiated with X-rays is given by the complex structure factor $\mathbf{F_h}$, a vector representing the diffracted X-rays. The structure factor itself is a summation of the scattering contributions of each and every atom in the unit cell:

$$\mathbf{F_h} = \sum_{j=1}^{atoms} f_{s,j}^0 \exp(2\pi i \mathbf{h}\mathbf{x}_j) \qquad (6\text{-}34)$$

The term in the exponent of Equation 6-34 contains the relative phase angle for the partial wave emanating from each atom, which depends solely on the *direction* of scattering, **h**, and the *position* of the atoms *j* in relation to the origin, given by the fractional coordinate vector $\mathbf{x}_j$. Remember that $\mathbf{h}\mathbf{x}_j$ is a dimensionless scalar product (dot product). The magnitude of each partial atomic contribution is given by the atomic scattering factor, which is closely approximated by a Gaussian function centered on $\sin\theta = 0$ and decaying rapidly with diffraction angle (Figure 6-12). The atomic scattering factor represents the Fourier transform of the atom's real space electron density, $\rho(\mathbf{r}_j)$. The summation (6-34) clearly states the very important fact that each and every atom in the unit cell (*unit cell*, not just *asymmetric unit*) contributes to each and every reflection. This is the source of non-linearity complicating crystallographic reciprocal space refinement (Chapter 12).

We can visualize the structure factor summation in a complex Argand diagram. Figure 6-26 shows the graphical structure factor summation for a structure containing six light atoms (C, N, and O) and one slightly heavier atom, say S.

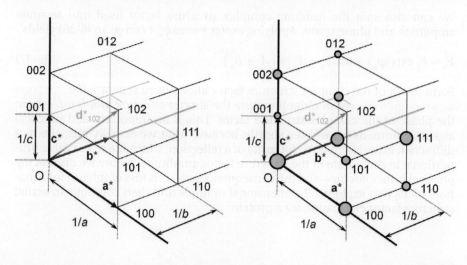

Figure 6-25 The reciprocal lattice and reflection intensities. The reciprocal lattice in itself is a mathematical point lattice devoid of any actual contents (left panel). The "contents" of the reciprocal lattice in terms of observed diffraction intensities result from scattering by all the electrons of the atoms in the unit cell, sampled in discrete lattice directions at reciprocal lattice points *hkl*.

Figure 6-26 Structure factor F$_h$ as summation of individual scattering contributions. Six light atoms (C, N, and O, z = 6, 7, 8) and one heavier S atom (z = 16, long vector) contribute to the total scattering. The length of the individual vectors is given by the atomic scattering factor of the atoms, proportional to z, the number of electrons. The resulting structure factor F$_h$ (green) has a phase angle φ of ~315° (or −45°).

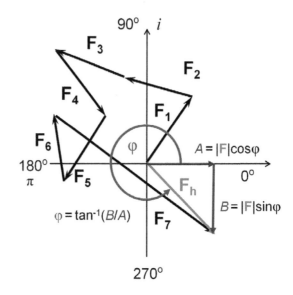

For a practical calculation, we set

$$2\pi \mathbf{h}\mathbf{x}_j = \varphi_j$$

and apply Euler's formula

$$\exp(i\varphi) = \cos\varphi + i\sin\varphi$$

which allows us to group together all real cosine terms and all imaginary sine terms of the complex structure factor, and sum them into separate partial sums A and B for each hkl:

$$\mathbf{F_h} = \sum_{j=1}^{atoms} f_j(\cos\varphi_j + i\sin\varphi_j) = \sum_{j=1}^{atoms} f_j(\cos\varphi_j) + \sum_{j=1}^{atoms} f_j \cdot i(\sin\varphi_j) = A_\mathbf{h} + i \cdot B_\mathbf{h} \tag{6-35}$$

In the general case of a non-centrosymmetric protein structure, computation will yield complex structure factors

$$\mathbf{F_h} = A_\mathbf{h} + i \cdot B_\mathbf{h} = |\mathbf{F_h}|(\cos\varphi_\mathbf{h} + i\sin\varphi_\mathbf{h}) \text{ with } \varphi_\mathbf{h} = \tan^{-1}(B_\mathbf{h} / A_\mathbf{h}) \tag{6-36}$$

where the sine terms have non-zero values, and the presence of an imaginary component will indicate a phase angle different from 0° or 180°. Structure factors with a phase angle of 0° or 180° are a special case of centric or phase restricted reflections. Because the imaginary sine terms of these special centric reflections are zero, their structure factors are real and positive for φ = 0° and have a negative sign for φ = 180°.

We can also split the resulting complex structure factor itself into separate amplitude and phase terms. Applying $\exp(\varphi_\mathbf{h}) = \cos\varphi_\mathbf{h} + i\sin\varphi_\mathbf{h}$ to (6-36) yields

$$\mathbf{F_h} = F_\mathbf{h}\exp(i\varphi_\mathbf{h}) \text{ with } F_\mathbf{h} = |\mathbf{F_h}| = \left(A_\mathbf{h}^2 + B_\mathbf{h}^2\right)^{1/2} \tag{6-37}$$

Form (6-37) of the complex structure factor allows us to separate the structure factor amplitude—or in older literature the shorter *structure amplitude*—from the phase of the complex structure factor. This is an important and practical algebraic form, as we will see shortly, because what we actually measure in a diffraction experiment is the intensity of a reflection, which is a scalar value proportional to the square of the structure factor amplitude. Before we discuss the physical principles for—and the consequences of—this loss of (phase) information, let us first explore a 2-dimensional example and then compute an actual structure factor calculation for a protein.

Occupancy and B-factor revisited

A few details must be added to make our structure factor formula (6-34) useful for actual structure factor calculations. In Section 6.1 we examined the fact that atoms which are displaced from a mean (equilibrium) position exhibit attenuated scattering. This attenuation diminishes the diffracted intensities and is quantified by the isotropic displacement parameter or, in short, B-factor in the exponent of the Debye–Waller factor T. The most general description of T is given as

$$T_j(\mathbf{h}) = \exp(-2\pi\langle(\mathbf{S}u_j)^2\rangle) \tag{6-38}$$

where $\mathbf{u}_j$ is the displacement vector in Å of atom j and the corresponding scattering vector $\mathbf{S} = (\mathbf{A}^*)^T\mathbf{h} = (h\mathbf{a}^* + k\mathbf{b}^* + l\mathbf{c}^*)$. Equation 6-38 is quite general and includes anisotropic displacement, which is discussed in Chapter 12. In the isotropic case the exponent becomes scalar because the displacement vector $\mathbf{u}$ is replaced with the isotropic displacement u and we obtain

$$T_j(\mathbf{h}) = \exp(-2\pi\langle(2(\sin\theta/\lambda)u_j)^2\rangle) = \exp(-8\pi^2\langle u_j^2\rangle(\sin\theta/\lambda)^2) \tag{6-39}$$

With our definition of the isotropic B-factor in terms of the mean square isotropic displacement $B_{iso} = 8\pi^2\langle u_{iso}^2\rangle$ we obtain the Gaussian exponential for the isotropic B-factor attenuation of the atomic scattering factor:

$$T_j(\mathbf{h}) = \exp(-B_j|\mathbf{S}|^2/4) = \exp(-B_j(\sin\theta/\lambda)^2) \tag{6-40}$$

which is readily included in the structure factor formula

$$\mathbf{F_h} = \sum_{j=1}^{atoms} T_j(\mathbf{h})\cdot f_{\mathbf{S},j}^0 \cdot \exp(2\pi i\mathbf{h}\mathbf{x}_j) = \sum_{j=1}^{atoms} f_{\mathbf{S},j}^0 \cdot \exp(-B_j(\sin\theta/\lambda)^2)\exp(2\pi i\mathbf{h}\mathbf{x}_j) \tag{6-41}$$

For numeric purposes, grouping the B-factor exponential and the scattering factor exponential together makes sense: they both depend on the scattering vector and thus the diffraction angle θ or resolution d according to $2\sin\theta/\lambda = 1/d = |\mathbf{S}|$.

We must also be able to account for partial occupancy of a particular atomic site. While the continuity of the peptide chain in proteins mandates that the total occupancy is 1.0 for each residue in a chain, side chains may split into multiple distinct conformations. In this case, the sum of occupancy factors n_j must be constrained to 1.0. For ligands such as drug molecules, metal ions, or discrete components of the solvent, the actual occupancy is often less than 1.0 and is rarely known exactly. We extend the structure factor formula with the occupancy n_j for each atom

$$\mathbf{F_h} = \sum_{j=1}^{atoms} n_j f_{\mathbf{S},j}^0 \exp(-B_j(\sin\theta/\lambda)^2)\exp(2\pi i\mathbf{h}\mathbf{x}_j) \tag{6-42}$$

and we realize that the occupancy n_j and the B-factor are correlated; that is, the effect of a higher B-factor leading to attenuated scattering can also be achieved by lower occupation, which is also decreasing the scattering contribution from the atom at position $\mathbf{x}_j$. Because of this correlation, it is generally difficult to determine or refine both occupancy and B-factor independently with accuracy.

The first structure factor calculation

To become familiar with the practical use of the structure factor equation (6-42) we will calculate the structure factors of a simple 2-dimensional structure containing two identical atoms in a rectangular unit cell (Figure 6-27). To make the calculation easier, we set the temperature factor B_j to zero for each atom, that is,

$$\exp(-B_j\sin^2\theta/\lambda^2) = \exp(0) = 1$$

Figure 6-27 **A simple 2-dimensional structure.** Two identical atoms at rest ($B = 0$) at full occupancy ($n = 1$) are located at the origin and at the center of a rectangular unit cell. The task is to complete the structure factor table to the right.

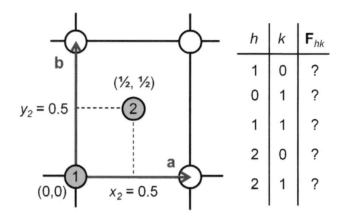

h	k	F_{hk}
1	0	?
0	1	?
1	1	?
2	0	?
2	1	?

and assume full occupancy $n_j = 1.0$. Furthermore, the atoms shall be of the same element, so that we can also eliminate f_j as a proportionality (still scattering vector dependent) from the explicit summation (6-42) and all we need is to evaluate

$$\mathbf{F_h} = f_s^0 \sum_{j=1}^{atoms} \exp(2\pi i \mathbf{hx}_j)$$

One can make the calculation even simpler by assuming for now that the atomic scattering remains constant with scattering angle and just use the number of electrons, z, as is the case for point atoms (we are effectively computing normalized structure factors $\mathbf{E_h}$, further explained in Chapter 7). Applying Euler's formula $\exp(i\varphi) = \cos\varphi + i\sin\varphi$, our structure factor equation then becomes

$$\mathbf{F_h} = z \sum_{j=1}^{2} \cos 2\pi(\mathbf{hx}_j) + i \cdot \sin 2\pi(\mathbf{hx}_j)$$

In explicit notation for the vector components we obtain

$$\mathbf{F}_{hk} = z \sum_{j=1}^{2} \cos 2\pi(hx_j + ky_j) + i \cdot \sin 2\pi(hx_j + ky_j)$$

which further expands into

$$\mathbf{F}_{hk} = z(\cos 2\pi(hx_1 + ky_1) + i\sin 2\pi(hx_1 + ky_1) + \cos 2\pi(hx_2 + ky_2) + i\sin 2\pi(hx_2 + ky_2))$$

We are ready now to insert explicit hk pairs for each reflection and explicit positions with $\mathbf{x}_1 = (0, 0)$ and $\mathbf{x}_2 = (\frac{1}{2}, \frac{1}{2})$. We obtain for $hk = 1,0$

$$\mathbf{F}_{1,0} = z(\cos 2\pi(1 \cdot 0 + 0 \cdot 0) + i\sin 2\pi(1 \cdot 0 + 0 \cdot 0) + \cos 2\pi(1 \cdot \frac{1}{2} + 0 \cdot \frac{1}{2}) + i\sin 2\pi(1 \cdot \frac{1}{2} + 0 \cdot \frac{1}{2})) =$$
$$= z(\cos 2\pi(0) + i\sin 2\pi(0) + \cos 2\pi(\frac{1}{2}) + i\sin 2\pi(\frac{1}{2})) =$$
$$= z(\cos 0 + i\sin 0 + \cos \pi + i\sin \pi) =$$
$$= z(1 + 0 - 1 + 0) = 0$$

One can enumerate the remaining reflections in the same way and produce the structure factor listing in Table 6-2 for the first few pairs of indices h, k.

Examination of Table 6-2 provides a number of important lessons we can learn already from this simple structure factor calculation:

- For large structures, with many atoms and many reflections, the direct summation which we employed in the manual calculation will become very slow. This is a practical issue in crystallographic computing, where Fast Fourier Transform (FFT) methods are implemented that replace the slow direct summation (Sidebar 9-1).

h	k	F_{hk}
1	0	0
0	1	0
1	1	$2z$
2	0	$2z$
0	2	$2z$
2	1	0
1	2	0
...	...	...

Table 6-2 Two-dimensional structure factor listing. The structure factors $\mathbf{F}_{hk}$ for each reflection hk are tabulated for the 2-dimensional structure illustrated in Figure 6-27.

- Because of the presence of only zero-valued sine terms, all structure factors were purely real; there was no imaginary part. This is a result of the centrosymmetry of our 2-dimensional structure: for each atom j, $\mathbf{x}_j = -\mathbf{x}_j$. In centrosymmetric 3-dimensional structures, which are not possible for proteins, all phases are either $0°$ or $180°$. However, even in chiral protein structures, some reflections are centric and have restricted phase values depending on the crystal symmetry (Section 6.5). Atoms in lattice planes that correspond to centric zones (directions of reciprocal space) are centrosymmetrically related, just as in our 2-dimensional structure.

- There is a distinct pattern in these structure factor values: Only reflections with $h + k = 2n$ (i.e. even sum) have non-zero values. This is a manifestation of the translational centering of the cell. We mentioned in Chapter 5 that certain extinction rules or "conditions limiting reflections" exist for space groups possessing translational Bravais symmetry or space group symmetry operators with translational components, such as screw axes. The extinction rules are typical for a space group (but not unique for all space groups) and are a direct manifestation of the effect of translational symmetry on the phases of the structure factor calculation.

Structure factor calculation for a protein

We shall now calculate the structure factors for a protein structure from the atomic coordinates deposited in the Protein Data Bank (PDB). We perform the structure factor calculation by slow but accurate direct summation, and therefore we pick a small protein, bovine pancreatic trypsin inhibitor,[13] a hardy perennial of high-resolution protein crystallography.

From the CRYST1 record of entry 1bpi we extract the cell parameters and the space group symbol:

```
CRYST1   75.390   22.581   28.600   90.00   90.00   90.00 P 21 21 21   4
```

The cell parameters allow us to set up the metric calculations and to compute the deorthogonalization matrix $\mathbf{O}^{-1}$ which we need to convert the Cartesian world coordinates $\mathbf{X}_j$ listed in the PDB file for each atom j back into fractional crystallographic coordinates $\mathbf{x}_j$. The deorthogonalization matrix given in the SCALE records of the PDB file is generally not accurate enough because of format limits (and there is no guarantee that it actually matches the deposited cell parameters). Decoding of the space group symbol provides us with the symmetry operators $\mathbf{W}_G$ which we need to generate the entire unit cell contents from the deposited asymmetric unit contents. For space group $P2_12_12_1$ we have to apply four operators (ITC-A, space group No. 19) including the identity operator. We also generate a list of unique index triples $\mathbf{h}$ up to the desired maximum resolution (say, 2 Å).

From a library file we pick up the necessary Cromer–Mann coefficients to compute the atomic scattering factors for O, C, N, S, and P (an ordered phosphate molecule from the solvent is present in the crystal structure):

$$f_{s,j}^0 = \sum_{i=1}^{4} a_{i,j} \cdot \exp[-b_{i,j}(\sin\theta / \lambda)^2] + c_j \tag{6-43}$$

Because we need the scattering-vector-dependent terms $|(\mathbf{O}^{-1})^T\mathbf{h}|^2/4 = (\sin\theta/\lambda)^2$ repeatedly, we pre-compute those and the scattering factors for each atom type and for each $\mathbf{h}$ in the desired resolution range. The computation of structure factors in practice becomes more efficient if we first evaluate the contributions of all atoms in the asymmetric unit and then expand the unit cell. This saves the re-computation of terms that are invariant under symmetry operation. With G the number of crystallographic symmetry operators $\mathbf{W}_G$ and summation over all atoms j in the asymmetric unit we obtain

$$\mathbf{F_h} = \sum_{G=1}^{sym} \sum_{j=1}^{atoms} f_{s,j}^B \exp[2\pi i\mathbf{h}(\mathbf{W}_G\mathbf{x}_j)] = \sum_{G=1}^{sym} \sum_{j=1}^{atoms} f_{s,j}^B \exp[2\pi i\mathbf{h}(\mathbf{R}_G\mathbf{x}_j + \mathbf{T}_G)] \tag{6-44}$$

We make the atomic scattering factors in the inner summation and the B-factor correction now explicit as a function of $\mathbf{h}$:

$$\mathbf{F_h} = \sum_{G}^{sym} \sum_{j}^{atoms} n_j f_j^0(\mathbf{h}) \exp(-B_j |(\mathbf{O}^{-1})^{\mathrm{T}} \mathbf{h}|^2/4) \exp[2\pi i \mathbf{h}(\mathbf{R}_G \mathbf{X}_j + \mathbf{T}_G)] \qquad (6\text{-}45)$$

Now we need to consider that the coordinates in the PDB file are Cartesian coordinates $\mathbf{X}_j$ and not fractional crystallographic coordinates, so we need to deorthogonalize them and we are finally ready to go:

$$\mathbf{F_h} = \sum_{G}^{sym} \sum_{j}^{atoms} n_j f_j^0(\mathbf{h}) \exp(-B_j |(\mathbf{O}^{-1})^{\mathrm{T}} \mathbf{h}|^2/4) \exp[2\pi i \mathbf{h}(\mathbf{R}_G (\mathbf{O}^{-1} \mathbf{x}_j) + \mathbf{T}_G)] \qquad (6\text{-}46)$$

Table 6-3 provides a listing of the structure factors computed from Equation 6-46 for each index triple hkl sorted by decreasing interplanar spacing d up to a resolution of 2 Å. The data items of Table 6-3 are the indices h, k, l, structure factor amplitude F, phase angle φ, lattice plane spacing d, and $(\sin\theta/\lambda)^2$ for each reflection hkl. Also included is a flag that indicates whether the reflections are centric (explained in the table caption and in Section 6.5) and their epsilon-factors for the correction of expectation values in intensity statistics (Section 6.5).

The listing in Table 6-3 teaches us a number of very interesting properties of our structure factors:

- For this small protein diffracting to an average resolution of 2.0 Å, we have already calculated 7316 reflections. This allows us to project that for large proteins and large unit cells, we will have tens of thousands or even hundreds of thousands of reflections to collect and to (repeatedly) compute during cycles of reciprocal space structure refinement.

- The reflections are listed in pairs: reflection $h\,k\,l$ and its centrosymmetrically related mate $-h\,-k\,-l$. We learned in Chapter 5 that the reciprocal space is centrosymmetric and, indeed, each of the pairs has exactly the same intensity but conjugate phase, $\varphi(-h\,-k\,-l) = -\varphi(h\,k\,l)$. In the absence of anomalous scattering contributions this is always true and is known as Friedel's law; these pairs of centrosymmetrically related reflections are so-called Friedel pairs. Friedel pairs are a special case of Bijvoet pairs, as explained in Section 6.5.

- While no general limitations exist regarding the phase values and amplitudes for a structure factor in protein structures, certain structure factors have special properties. Centric structure factors are centrosymmetrically related reflections that are additionally related by the point group symmetry of the crystal. They are phase restricted and are invariably of the same intensity, irrespective of anomalous contributions. General centric reflections play a useful role in computing intensity statistics without anomalous signal contributions. Some of those centric reflections have structure factors that are limited to a phase of 0° or 180°. As a consequence, the structure factor of this subgroup of centric reflections is real and has no imaginary part. They were therefore targeted in initial phasing attempts of the first protein structures (Sidebar 6-14). In contrast to the purely real centric structure factors, general centric structure factors are complex and do have an imaginary part. For example, depending on the presence of translational symmetry elements, some centric structure factors are phase restricted to 90° or 270° in orthorhombic space groups (Section 6.5).

- A distinct feature of certain structure factors is that they compute to exactly zero—that is, their reflections are totally extinct. Systematic extinction or absence is a manifestation of the presence of translational symmetry within the unit cell; in the example of $P2_12_12_1$ it is caused by presence of the translational component of the 2-fold screw axis along each direction $\mathbf{a}$, $\mathbf{b}$, and $\mathbf{c}$. The reflection conditions for each of the reciprocal axes are $h00 = 2n$, $0k0 = 2n$, and $00l = 2n$, respectively. Indeed, only even indices for these serial reflections show non-zero structure factor amplitudes in our calculation.

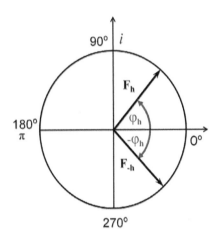

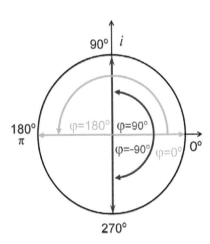

Figure 6-28 Special structure factors. A general Friedel pair (top panel); pairs of special centric structure factors with $\varphi = 0°$ and $\varphi = 180°$ (green), and an example of general phase restricted centric reflections, $\varphi = 90°$ and 270°, blue (bottom panel).

To gain some practice with the vector representation we represent common special structure factors in Argand diagrams (Figure 6-28).

h	k	l	F(hkl)	φ(hkl)	d(hkl)	(sin θ/λ)²	centric	ε
1	0	0	0.00	---	75.39	0.000044	1	2
-1	0	0	0.00	---	75.39	0.000044	1	2
2	0	0	228.04	180.00	37.69	0.000175	1	2
-2	0	0	228.04	180.00	37.69	0.000175	1	2
0	0	1	0.00	---	28.60	0.000305	1	2
0	0	-1	0.00	---	28.60	0.000305	1	2
1	0	1	10.42	90.00	26.74	0.000349	1	1
-1	0	-1	10.42	270.00	26.74	0.000349	1	1
3	0	0	0.00	---	25.13	0.000395	1	2
-3	0	0	0.00	---	25.13	0.000395	1	2
2	0	1	401.84	270.00	22.78	0.000481	1	1
-2	0	-1	401.84	90.00	22.78	0.000481	1	1
0	1	0	0.00	---	22.58	0.000490	1	2
0	-1	0	0.00	---	22.58	0.000490	1	2
1	1	0	469.25	270.00	21.63	0.000534	1	1
-1	-1	0	469.25	90.00	21.63	0.000534	1	1
2	1	0	439.40	0.00	19.37	0.000666	1	1
-2	-1	0	439.40	0.00	19.37	0.000666	1	1
3	0	1	398.01	270.00	18.88	0.000701	1	1
-3	0	-1	398.01	90.00	18.88	0.000701	1	1
4	0	0	715.45	180.00	18.85	0.000703	1	2
-4	0	0	715.45	180.00	18.85	0.000703	1	2
0	1	1	389.38	270.00	17.72	0.000795	1	1
0	-1	-1	389.38	90.00	17.72	0.000795	1	1
1	1	1	366.98	332.08	17.25	0.000839	0	1
-1	-1	-1	366.98	27.92	17.25	0.000839	0	1

---- 7290 more reflections ----

h	k	l	F(hkl)	φ(hkl)	d(hkl)	(sin θ/λ)²	centric	ε
8	11	1	97.89	255.07	2.00	0.062445	0	1
-8	-11	-1	97.89	104.93	2.00	0.062445	0	1
18	6	10	111.95	45.79	2.00	0.062465	0	1
-18	-6	-10	111.95	314.21	2.00	0.062465	0	1
25	3	10	32.91	139.90	2.00	0.062467	0	1
-25	-3	-10	32.91	220.10	2.00	0.06246	0	1
12	8	9	72.12	11.90	2.00	0.062469	0	1
-12	-8	-9	72.12	348.10	2.00	0.062469	0	1
3	11	3	92.18	300.09	2.00	0.062471	0	1
-3	-11	-3	92.18	59.91	2.00	0.062471	0	1

---- end of list ----

Table 6-3 Calculated structure factors for PDB entry 1bpi. The structure factor amplitudes are grouped in centrosymmetrically related pairs (i.e. **h** = –**h**), while the shading color indicates their properties: magenta and white, general acentric structure factor (or reflection); yellow, extinct reflections; cyan, general centric, phase restricted structure factor; green, centric structure factors with real components only. Centric structure factors are centrosymmetrically related in reciprocal space and also related by point group symmetry operations. They are phase restricted to certain values, and the centrosymmetrically related mates (Friedel opposites) of centric reflections are invariably of the same intensity, irrespective of anomalous scattering contributions (explained in Sections 6.4 and 6.5).

Structure factor amplitude and intensity of diffraction

When a diffracted photon hits a detector of any kind, a physical interaction process takes place that generally involves some electronic excitation or ionization of the detector material. While we explain the function of the respective detectors in Chapter 8, we will focus here on the principles of the conversion of diffracted photons into the measured reflection intensities.

Let us consider first the basic process: As the scattered X-ray photons arrive one by one, without any fixed phase relation between them (X-ray sources used for diffraction are non-coherent sources of radiation), the intensity recorded by a

Box 6-10 **Special structure factors.** Certain classes of structure factors have special properties. Centric structure factors are structure factors that are related by point group symmetry and by centrosymmetry in reciprocal space. As a consequence their Friedel mates invariably have the same amplitude and are restricted in their phase values. Centric structure factors can assume only certain fixed phase values such as 0°, 90°, 180°, 270°, 120°, 240°, and so forth. A special subgroup of centric structure factors has phase values of 0° or 180° and are real, with a sign corresponding to + for 0° and − for 180°. Some centric structure factors compute to exactly zero. Such systematic extinctions or absences due to translational elements in the space group symmetry assist in the space group determination. Centric structure factors have different intensity distributions than general, acentric, reflections (Chapter 7). In the presence of anomalous scattering contributions, general acentric Friedel pairs $\mathbf{F(h)}$ and $\mathbf{F(-h)}$ have different amplitudes $\mathbf{F(h)} \neq \mathbf{F(-h)}$ (detailed in Section 6.4). For centric Friedel pairs, $\mathbf{F(h)} = \mathbf{F(-h)}$ irrespective of any anomalous scattering contributions.

detector will simply be a signal proportional to the number of photons counted at that position. How many photons arrive at any position in a given time depends on the scattering probability $P(\mathbf{h})$ in that particular direction, which we have already derived by computing the complex structure factor for each reflection hkl. Given that we measure for long enough, the accumulated photon count will give quite an accurate measure of some proportionality to the magnitude or amplitude of the complex structure factor. The photon count simply represents the (squared, as we will show shortly) structure factor amplitude. The fact that we measure only the amplitude of a reflection and lose information about its phase angle during the detection process leads us to anticipate that there will be a problem in reconstructing the molecular structure from the diffraction data. Indeed, that difficulty is properly termed the phase problem in crystallography, and a considerable effort in the design of a crystallographic study involves the development of a strategy for solving the phase problem. Later chapters will thus cover the principles of structure reconstruction from diffraction data (Chapter 9) and the invariably necessary solution of the phase problem (Chapters 10 and 11).

To derive a quantitative relation between structure factor amplitude and intensity, we need to consider two factors: the fundamental proportionality between the structure factor amplitude and the detected signal resulting from detection principles, and the effects that modify that signal depending on experimental parameters such as diffraction geometry, X-ray wavelength, sample size and absorption, primary beam polarization, and others.

A general principle derived from quantum mechanics states that the observable of any Hermitian complex function or operator is a scalar value. For example, we know that the probability $P(\mathbf{x})$ (a real, scalar value) of finding an electron of an atom at a certain position $\mathbf{x}$ in space is proportional to the product of its complex wave function ψ and its conjugate complex ψ^* at this position:

$$P(\mathbf{x}) = \psi_{(\mathbf{x})}\psi_{(\mathbf{x})}^* \quad \text{with normalization} \int_{\mathbf{x}} \psi\psi^* \, d\mathbf{x} = 1 \qquad (6\text{-}47)$$

More generally, the expectation value of an observable E of a process defined by an operator $\hat{E}$ is defined by the integral

$$\langle E \rangle = \int_{\mathbf{x}} \psi\hat{E}\psi^* \, d\mathbf{x} \qquad (6\text{-}48)$$

In a loose analogy we can set our "absorption operator" to 1 (absorption of the photon happens), and the complex wave function in our case is the complex structure factor $\mathbf{F}$. The fundamental, principal proportionality between the observed intensity I and structure (factor) amplitude F then can be expressed as

$$1 \propto \mathbf{F}\mathbf{F}^* = (A + iB) \cdot (A - iB) = A^2 + B^2 = |\mathbf{F}|^2 = F^2 \tag{6-49}$$

Having derived that the diffraction intensity is proportional to the square of the structure factor amplitude, we can now investigate the remaining, experiment-dependent parameters.

Given unpolarized X-rays, the Thomson formula (6-9) revealed that a fundamental, angle-dependent polarization factor P for scattering from each electron applies

$$P = (1 + \cos^2 2\theta) / 2 \tag{6-50}$$

which takes on more complicated, instrument-specific forms for monochromated or pre-polarized radiation.[14]

The remaining factors depend further on the particulars of the diffraction experiment. The first factor to consider is that upon rotation of the reciprocal lattice through the Ewald sphere with a constant angular speed ω, reciprocal lattice points far from the origin will pass though the Ewald sphere with far greater velocity (and thus spend less time diffracting) than those lattice points close to the origin. Which reflections move at what speed through the Ewald sphere will of course depend on the particulars of the diffraction geometry,[14] which is generally accounted for by the raw data processing software provided with the diffractometer. As an example, for a single rotation axis intersecting the incoming beam in the frequently used rotation method, the Lorentz factor L becomes[15, 16]

$$L_{\mathbf{h}} = 1 / |\mathbf{h} \cdot (\mathbf{e}_\mathrm{R} \times \mathbf{s}_0)| \tag{6-51}$$

where $\mathbf{e}_\mathrm{R}$ is the rotation axis unit vector, $\mathbf{h}$ is the reciprocal index vector for the reflection hkl, and $\mathbf{s}_0$ is our incoming X-ray beam wave vector.

The Lorentz factor can become large for reflections skimming tangentially through the Ewald sphere. A section of reflections located in a small region along the rotation axis (the Lorentz exclusion region) is thus affected by such a huge correction (and dispersed over so many frames) that their correction becomes unreliable and they are generally excluded from data integration.

The properties of the crystal naturally affect the total scattered intensity. Crystal size, a combination of primary and secondary extinction (accounted for largely as an empirical correction together with the absorption A), and choice of wavelength λ also affect the actually observed reflection intensity. A full quantification of the integrated intensity of a reflection hkl is given[14] by

$$I_{hkl} = I_0 k \frac{N}{U} \lambda^3 LPAF_{\mathbf{h}}^2 \tag{6-52}$$

where I_0 is the incoming X-ray intensity and k includes the remaining elementary constants from the Thomson formula $k = e^4 / m^2 c^4$.

Let us take a brief look at the additional terms in Equation 6-52. Quite obvious is the direct dependency of I on the intensity of incoming X-rays I_0—the more photons we send into the crystal, the more will be diffracted. N (the number of unit cells) divided by the unit cell volume U accounts for the basic fact that more material in the beam diffracts more. Quite revealing is the fact that, given the λ^3 term in Equation 6-52, X-rays of longer wavelength diffract significantly better than those of shorter wavelength. However, the practical use of long wavelengths is limited by the smaller Ewald spheres (limiting how many reflections can be recorded), and most dramatically, by strong absorption of X-rays by the atoms of the protein and even by air (Sidebar 6-11). However, as we will see, longer wavelengths can also be of advantage for anomalous phasing from light elements such as sulfur.

Box 6-11 **The intensity of scattered X-rays.** During a diffraction experiment, only the magnitude of each structure factor, $F_\mathbf{h}$, termed the structure factor amplitude, is accessible through an intensity measurement of reflection $I_\mathbf{h}$, which is proportional to F_h^2. The absence of phases implies a loss of critical information necessary to reconstruct the electron density, giving rise to the crystallographic phase problem.

A number of experimental factors such as crystal size, primary beam intensity, diffraction geometry, and wavelength affect the conversion of raw intensities, obtained as photon count, into structure factor amplitudes. The corrections are performed during raw data integration and scaling. The general proportionality is given by

$$F_\mathbf{h} = k \cdot k_\mathbf{h}' \cdot I_\mathbf{h}^{1/2}$$

where k contains all angle independent terms and k' contains all conversions that are dependent on the scattering angle. The structure factor amplitudes are thus proportional to the square root of the reflection intensity.

In practice, correction factors to the measured raw intensities fall into two major groups: instrument corrections, which are accounted for in the raw data processing step, depending on the diffractometer geometry, through the instrument software; and the remaining experimental corrections, which are applied in the scaling of the experimental observations. In the most general form, the observed structure factor amplitude $F_\mathbf{h}(obs)$ can thus be derived from the intensity as

$$F_\mathbf{h}(obs) = k \cdot k_\mathbf{h}' \cdot I_\mathbf{h}^{1/2} \tag{6-53}$$

where k contains all angle-independent corrections, and k' contains the angle-dependent corrections for each value of $\mathbf{h}$. Finally, for the subsequent computations and comparisons against calculated (model) structure factors $F_\mathbf{h}(calc)$, the observed structure factor amplitudes $F_\mathbf{h}(obs)$ must be brought onto an absolute scale by means of Wilson scaling, which also accounts for the average (or overall) B-factor attenuation specific for the respective protein structure (discussed in detail in Chapter 7).

6.4   X-ray absorption

One phenomenon that is central to understanding the success of modern protein crystallography is X-ray absorption and the associated anomalous dispersion (leading to anomalous scattering and thus anomalous diffraction) in the vicinity of an X-ray absorption edge. Anomalous phasing techniques, which contribute to the vast majority of all *de novo* (i.e. in the absence of an already available phasing model) protein structure determinations, are an indispensable tool in modern protein crystallography. At least a qualitative working knowledge of X-ray absorption and anomalous dispersion is necessary to understand the anomalous phasing techniques discussed in Chapter 10.

Variation of absorption with photon energy

All materials absorb X-rays, and as a general rule, the absorption decreases rapidly with increasing energy (or shorter wavelength). At a given wavelength, the mass absorption coefficient μ is higher for heavier elements than for light elements. The latter is clearly evidenced in the X-ray transmission image of Prof. von Kölliker's hand shown in Figure 6-1. The fact that harder, high energy X-rays are more penetrating and are less absorbed than soft, low energy X-rays explains why medical X-ray exposures taken at 70 keV radiation from tungsten tubes are relatively harmless compared with involuntary exposures with highly

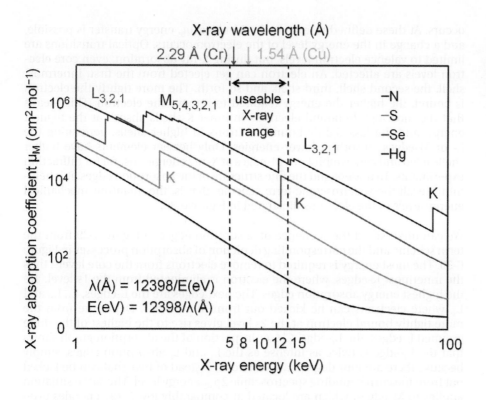

Figure 6-29 X-ray absorption of selected elements. Absorption curves for this figure were calculated using the program FPRIME[17] from relativistic photoelectric cross sections.[18] Note the sharp increase in absorption at specific energies for each element. These absorption edges are labeled (K, L_i, and M_i) according to the atomic core level transitions involved (Figure 6-30).

focused copper laboratory X-ray sources around 8 keV. Figure 6-29 shows X-ray absorption as a function of energy for a number of selected elements. In addition to confirming our qualitative statements made above, Figure 6-29 contains a wealth of additional information.

We recognize that the plot in Figure 6-29 is double logarithmic; that is, it covers a huge range particularly in the magnitude of the X-ray absorption coefficient. From the X-ray absorption coefficients we can calculate the amount of X-rays absorbed given the dimensions of a crystal through the Lambert–Beer absorption law:

$$I_T = I_0 e^{-t\rho\mu} \tag{6-54}$$

where I_T is the transmitted radiation, I_0 the incident beam intensity, t the sample thickness in cm, ρ the material density in g cm^{-3}, and μ the material's mass absorption coefficient in cm^2 g^{-1}.

For a 100 µm diameter protein crystal containing only light atoms (C, O, and N) we obtain an acceptable 30% radiation absorption at the lower limit of usable X-ray energy range of ~5 keV. However, when soaked with heavy atoms, such a crystal will—despite the low relative abundance of the heavy atom in the crystal—absorb heavily and because of increased absorption will likely suffer serious radiation damage. Radiation damage is thus generally an even higher risk for heavy-atom derivative (or Se-labeled) crystals compared with native, light-atom-only crystals. The upper limit of usable wavelengths is determined by various factors, such as the X-ray source characteristics (Chapter 8) and the general decay of the scattering power with decreasing wavelength ($I \approx \lambda^3$).

X-ray absorption edges

A most striking feature of Figure 6-29 is that the absorption curves for heavy elements are not smooth curves, but exhibit distinct, step-like features, suitably termed X-ray absorption edges. Given that the electric field vector in X-rays excites electrons to oscillate, we expect—just as in the case of absorption and dispersion of visible light—that at certain frequencies electronic resonance

Sidebar 6-11 Absorption of X-rays by air. Even light elements (such as C, N, and O) absorb X-rays. Absorption and scattering by air is therefore appreciable, and the intensity of characteristic Cu radiation (~1.54 Å) drops about 1% per cm in air. For Cr radiation (~2.29 Å), air absorption is even stronger. The appreciable air absorption and scattering is the reason why early multi-wire proportional detectors, which had modest pixel resolution in the mm range and required large crystal–detector distances, had helium-flushed collimators and He-flushed boxes between the detector and the crystal (Figure 8-7). With modern equipment and short beam paths, air absorption and scattering is less of a concern, and such contraptions are generally omitted, although they can be an advantage when soft wavelengths are used, for example for sulfur single-wavelength anomalous diffraction (S-SAD) phasing.

occurs. At these defined resonance frequencies, ω_0, energy transfer is possible, and a change in the energy level of the electron occurs. Optical transitions are limited to valence electron levels, but during X-ray absorption, even core electron levels are affected. An electron can get ejected from the first, innermost shell, the second shell, third shell, and so forth. The more tightly the electron is bound, the higher the energy required to remove the electron. This means that the most tightly bound electrons in inner K-shells absorb at the highest energy, and the less tightly bound electrons in higher shells, generating the L- or M-edges, absorb at lower energies. Only heavier elements have higher shells filled and can exhibit L- or M-edges at X-ray energies useful for diffraction experiments. To understand the fine structure of the absorption edges, it is helpful to recall the spectroscopic term scheme, that is, the quantum-mechanical energy levels of the electrons illustrated in Figure 6-30.

We can understand the sequence of absorption edges in Figure 6-29 from the term scheme and the corresponding depiction of absorption processes in Figure 6-30. The most energy is required to remove electrons from the core levels; thus the innermost K-edges, where the electron is eliminated from the 1s level, are the highest energy absorption edges. The next edges are the L-edges, L_1, L_2, and L_3, where electrons can be kicked out from three levels. Departure from the more tightly bound electron at the 2s level gives rise to the highest of the three grouped L-edges, the L_1-edge. Closer inspection of the absorption graph shows that the L_3-edge is twice as intense as the L_1 and L_2 absorption edges, simply because there are four degenerate electrons, instead of two, that can be kicked out from the corresponding spectroscopic $2p_{3/2}$ energy level. The same situation applies to M-edges, which are located at comparably low X-ray energies even for heavy elements.

Absorption edge fine structure (XAS). Close inspection of an experimentally recorded L_3-edge as illustrated in Figure 6-31 shows that absorption edges display a distinct absorption edge fine structure, overlaying the (in reality sigmoid) theoretical edge jump. At the low energy side of the edge in the X-ray Absorption Near Edge Structure (XANES) region, pre-edge features result from transition of the photoelectrons into "forbidden" unoccupied states, followed by a distinct "white line." The name white line derives from the fact that when materials absorb strongly at this energy, a weakly exposed, bright line appears on an X-ray film negative during the recording of an X-ray absorption spectrum in transmission geometry. The most pronounced white lines result from transitions of $2p_{3/2}$ core level electrons into unoccupied nd-electron levels ($n > 3$) of the absorbing atom in L_3- and L_2-edges, but weaker white line transitions can sometimes be

Figure 6-30 Spectroscopic term scheme and X-ray absorption process. The left panel shows the spectroscopic term scheme nomenclature with the possible quantum numbers given for each electronic energy level. The right panel shows the energy levels occupied with electrons corresponding to the degeneracy of the atomic level. During X-ray absorption, high energy photons (red wiggles) eject electrons from the atom. The most energy is required to remove electrons from tightly bound core levels (at the bottom of the right panel). The departing electrons (red) leave behind an unoccupied hole (white) in an excited, ionized state. The hole can be filled by an electron from a higher shell, which in turn produces X-ray emissions characteristic for the particular element.

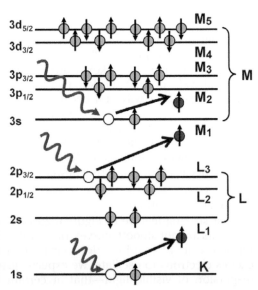

Shell	n	l	$j=l+m_s$	$m_j = (-j...\,j)$
		2	5/2	±1/2 ±3/2 ±5/2
			3/2	±1/2 ±3/2
M	3	1	3/2	±1/2 ±3/2
			1/2	±1/2
		0	1/2	±1/2
		1	3/2	±1/2 ±3/2
L	2		1/2	±1/2
		0	1/2	±1/2
K	1	0	1/2	±1/2

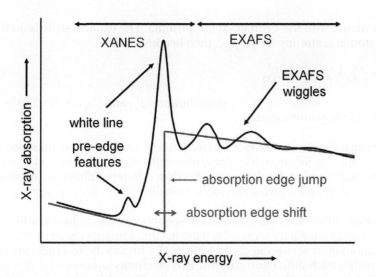

Figure 6-31 The fine structure of an X-ray absorption edge. The figure shows the fine structure of an experimental absorption edge overlaying the theoretically computed edge jump (red step-curve). The features are discussed in detail in the text.

observed in L_1- and K-edges (for example, Se) as well. The peak shape of the white line follows a Lorentz–Cauchy distribution—similar only in shape to a Gaussian normal distribution (Chapter 7) but "peakier" and with wider wings—typical for electronic transition processes.

With a further increase of energy, the electrons have absorbed enough energy to depart the atom, but on their way out they interact with ("bounce off") neighboring atoms. If these atoms are heavy, a distinct, decaying wiggle-pattern is observed, from which the nearest neighbor distances can be determined through Fourier transform analysis[19] of the Extended X-ray Absorption Fine Structure (EXAFS) spectrum.

The excited atomic states resulting from the "holes" left by the departed electrons can relax either in one step through dropping of an inner shell electron, giving rise to characteristic X-ray emission (this specific relaxation is used in the generation of X-rays in laboratory X-ray generators and in elemental analysis), or through more complex, multi-step, energy transfer processes. The resulting X-ray fluorescence reveals the same structure as the absorption spectrum and is actually what is recorded in practice during an X-ray excitation scan in the course of anomalous diffraction data collection. The fact that the actual energy levels of the scattering atom depend on the chemical state of the atom is reflected in a small chemical shift of the absorption edge. This chemical shift requires (among other, instrumentation-related reasons) that for precise determination of the absorption edge maxima, an actual edge scan is performed.

Anomalous scattering factors

The effect of resonance at the edge energy E_0 or frequency ω_0 of a bound electron leads to a singularity in the free electron Thomson formula as a result of the term $1/(\omega^2 - \omega_0^2)$ becoming infinite at $\omega = \omega_0$. The quantum-mechanical derivation of anomalous scattering factors is quite involved, and for the practical understanding a qualitative classical derivation treating a bound electron as a dampened oscillator will suffice. The general dispersion term for the atomic scattering function then assumes a form

$$f = \frac{\omega^2}{\omega^2 - \omega_0^2 - ik\omega} \tag{6-55}$$

with the complex dampening term $-ik\omega$ preventing the discontinuity at the edge or resonance frequency where $\omega = \omega_0$. The formal expansion of the resonant atomic scattering factor equation leads to two important consequences for practical crystallography: First, the atomic scattering factor now contains wavelength-dependent contributions $f'_{(\lambda)}$ and $f''_{(\lambda)}$ in addition to the wavelength-independent (but scattering-angle-dependent) atomic scattering factor f_s^0 which are approximated with the Cromer–Mann formula. The explicit expression for the entire atomic scattering factor $f_{(s,\lambda)}$ then becomes

$$f_{(s,\lambda)} = f^0_{(s)} + f'_{(\lambda)} + i \cdot f''_{(\lambda)} \tag{6-56}$$

The anomalous scattering factor contributions $f'_{(\lambda)}$ and $f''_{(\lambda)}$ are practically independent of the scattering angle.

Dispersive differences. The fact that we observe additional contributions to the atomic scattering factor at different wavelengths means that around absorption edges we will measure, for the same reflection, different diffraction intensities at different wavelengths, giving rise to the so-called dispersive differences.

Anomalous differences. As a second consequence of anomalous scattering, the presence of an imaginary term (which means that a phase change of +90° occurs in the anomalous scattering contributions $f''_{(\lambda)}$) breaks the internal centrosymmetry within each diffraction pattern. This generates anomalous differences in magnitude between the two structure factors $\mathbf{F_h}$ and its centrosymmetrically related mate $\mathbf{F_{-h}}$ of a Friedel pair (the same is true for $\mathbf{F_h}$ and a symmetry equivalent of its Friedel mate $\mathbf{F_{-h}}$, forming a Bijvoet pair, Section 6.5). The Friedel pairs of equal magnitude then become pairs of different magnitude because of the presence of anomalous scattering contributions. Both the dispersive and the anomalous (or Bijvoet) differences can be exploited for experimental *de novo* phasing of crystal structures, an indispensable step in any determination of a new protein structure, as discussed in the electron density reconstruction and phasing chapters.

The numeric values of anomalous X-ray scattering factors depend on the electronic state, that is, the chemical environment of the dispersive atom, and thus cannot be precisely calculated without prior structural knowledge (although in XANES and EXAFS spectroscopy,[20] inferences about chemical state and nearest neighbors are possible by fitting the observed absorption spectra). For free atoms, the dispersive f' and anomalous f'' scattering component (and thus the mass absorption coefficient) can be computed from photoelectric cross sections derived from relativistic Slater–Dirac wave functions.[18] The program FPRIME by Don Cromer[17] allows the computation of the theoretical values at arbitrary wavelengths, and the program has been used to compute the absorption curves in Figure 6-29 and for the anomalous structure factor calculations presented later in this chapter.

X-ray absoption edge scans

To optimize the anomalous and dispersive signal for anomalous phasing, we generally need to actually measure the amount of anomalous contributions through an X-ray absorption edge scan (generally in fluorescence mode in the form of an X-ray excitation spectrum). Chemical (and monochromator) shifts affect the precise edge position, and white lines resulting from XANES transitions can significantly increase the available signal. Recording an X-ray edge scan first requires an energy-tunable X-ray photon source, generally a synchrotron, whose operating principles are explained in Chapter 8. The technical acquisition of the X-ray edge scan is straightforward: An X-ray fluorescence detector is placed next to the crystal, about perpendicular to the incident beam, while the energy of the X-rays is stepped in small increments of about 0.2 to 1 eV across the absorption edge of the anomalously scattering element. The intensity of the total X-ray fluorescence signal varies proportionally to the absorption, and from the recorded X-ray excitation spectrum, the normalized atomic absorption coefficient $\mu_{(\lambda)}$ is obtained.

The imaginary anomalous scattering contribution, f'', is proportional to the atomic absorption coefficient μ at that wavelength

$$f''_{(\lambda)} = \frac{mc}{4e^2 \hbar} E_{(\lambda)} \mu_{(\lambda)} \tag{6-57}$$

and f'' is thus directly available from the experiment. In addition, the real part f' of the anomalous scattering contribution can be indirectly obtained from the absorption scan though enumeration of a Kramers–Kronig transform (which is just a general formalism that relates the real part of any analytic complex function to its imaginary part):

$$f'_{(\lambda)} = \frac{2}{\pi} \int_0^\infty E' \cdot \frac{f''_{(E')}}{(E^2 - E'^2)} dE' \qquad (6\text{-}58)$$

In the absence of a dedicated program, the f' component can also be obtained through simple numeric differentiation of the measured f'' spectrum. Remember that energy, frequency, and wavelength are equivalent, and can be converted into each other using the formulas in Box 6-1.

Figure 6-32 shows a normalized X-ray edge scan and its resulting Kramers–Kronig transform. Note that the real part f' of the anomalous scattering contribution is *negative* and the absolute values, given in units of electrons, are quite small. In the given example for the red curve, the edge jump reflecting f'', measures about +3.5 electrons, and the f' contribution about –7 electrons. For the example scan showing the strong white line peak, we can gain nearly twice the anomalous signal by selecting exactly the peak energy.[21] However, the anomalous contribution from the few dispersive atoms is, in general, still weak compared with the scattering power of the whole molecule, which is proportional to the sum of all the electrons in the unit cell.

Selection of wavelengths

Given that the measured signal is very small, it is an advantage to choose the wavelengths carefully so that we obtain a maximal signal for both the dispersive and the anomalous differences. In a multi-wavelegth anomalous diffraction (MAD) phasing experiment, at least two sets of anomalous data sets of Friedel or Bijvoet pairs, measured at different wavelengths, are necessary to uniquely determine the protein phases (Chapter 10). As protein crystals suffer radiation damage (particularly at or above heavy atom absorption edges), it is a good strategy to collect first those data that carry the most useful signal.

As we can see from Figure 6-32, the maximal anomalous f'' signal can be obtained from data at the absorption peak wavelength. As the largest anomalous differences between Bijvoet pairs are measured at the peak wavelength, the peak data set (or at least a data set *above* the absorption edge) is generally

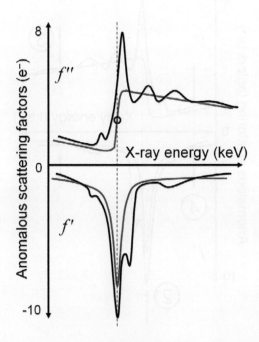

Figure 6-32 Derivation of f' from the normalized X-ray edge scan.
The f' curve is derived by Kramers–Kronig transform of the normalized X-ray edge scan. The red graph depicts the situation ignoring any X-ray edge fine structure, emphasizing that the negative peak of the f' function coincides with the steepest slope at the inflection point (blue circle) of the edge step in the f'' scan.

the first data set measured. In case the crystal decays rapidly in the beam, the structure may still be phased from strong anomalous differences alone by the single-wavelength anomalous diffraction (SAD) method (Chapter 10). In view of the possibility of radiation damage induced diffraction decay, it can be of advantage to collect anomalous data in small, temporally close centrosymmetrically opposed wedges of reciprocal space to minimize integration and scaling and errors, a subject we will discuss in detail in Chapter 8.

The next best wavelength to choose in a MAD experiment would be one that optimizes the dispersive differences between wavelengths, and the optimal dispersive f' signal is obtained at the negative peak of the f' curve (which coincides with the inflection point of the f'' scan). This inflection data set will still contain about half of the anomalous f'' peak signal, and have significant dispersive differences from any data set recorded either below or above the edge. If a data set recorded using a Cu-anode laboratory source at 1.5418 Å is available, it may be already used as a low energy remote data set against which strong dispersive differences can be obtained. The choice of the third data set depends to a degree on the state of the crystal. If radiation decay is already progressing, collecting another data set at low energy (higher absorption of light atoms) may rapidly destroy the crystal. The high energy remote data set above the edge is a common choice, because it carries both the anomalous internal signal, as well as the dispersive signal against the inflection data set. However, data collection with higher redundancy at fewer wavelengths instead of adding a high remote data set may have advantages as well. More formal strategies for optimizing the signal have been developed[22] and will be discussed in Chapter 8. The locations of the typical wavelengths selected in a MAD experiment are illustrated in reference to the anomalous scattering factors in Figure 6-33.

The fact that relative intensity differences exist between data recorded at different wavelengths is easy to understand: If the atomic scattering factors are different because of anomalous contributions, then the resulting structure factors are different, and as a final consequence, the measured intensities at each wavelength will be different. That anomalous differences exist *within* a data set recorded at one and the same wavelength is, however, not so obvious and requires some explanation, which is provided in the following section.

Figure 6-33 Choice of wavelengths for anomalous data collection. The peak data set (1) contains maximal anomalous Bijvoet differences. Dispersive differences between data sets are maximal for low-energy remote and high-energy remote data sets against the inflection data set (2). High-energy remote data (3) carry, in addition to the dispersive signal against inflection data, internal anomalous differences; the low remote data (4) produce only dispersive differences against inflection data. Remote data sets are usually recorded a few hundred eV below or above the absorption edge.

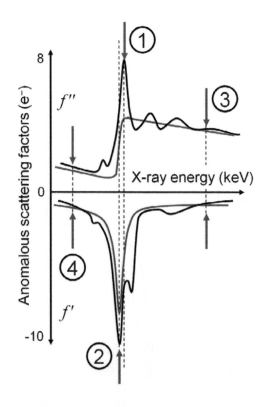

Friedel pairs

A relation that followed directly from the centrosymmetry of the reciprocal lattice was that in the absence of anomalous differences, centrosymmetrically opposed (conjugate) pairs of structure factors $\mathbf{F_h}$ and $\mathbf{F_{-h}}$, have exactly the same magnitude ($F_h = F_{-h}$) and conjugate phase angle, that is, $\varphi_h = -(\varphi_{-h})$. Therefore, in the absence of an anomalous signal, the reflection intensities I_h and I_{-h} are equal, and this law is called Friedel's law, and corresponding reflections are called Friedel pairs. We can algebraically show that Friedel's law is true by changing the sign of $\mathbf{h}$ in the general complex structure factor formula written with separate real and imaginary terms:

$$\mathbf{F_h} = F_h(\cos\varphi_h + i\sin\varphi_h) = A_h + iB_h$$
$$\mathbf{F_{-h}} = F_h(\cos(-\varphi_h) + i\sin(-\varphi_h)) = F_h(\cos\varphi_h - i\sin\varphi_h) = A_h - iB_h \qquad (6\text{-}59)$$

Note that a distinct difference exists between the pairs $\mathbf{F_h}$ and $\mathbf{F_{-h}}$ of centrosymmetrically related reflections on the one hand and the negative structure factors $-\mathbf{F_h}$ and $\mathbf{F_h}$ on the other hand: $-\mathbf{F_h} \neq \mathbf{F_{-h}}$. The phase of $-\mathbf{F_h}$ is $\varphi_h + 180°$, while the phase of $\mathbf{F_{-h}}$ is $-\varphi_h$ (revisit Sidebar 6-2).

The amplitude of the reflections in a Friedel pair is as usual obtained by the product of $\mathbf{F}$ and its conjugate complex $\mathbf{F^*}$ which computes as

$$I_h \approx F_h^2 = (A_h + iB_h)(A_h - iB_h) = A_h^2 + B_h^2$$
$$I_{-h} \approx F_{-h}^2 = (A_h - iB_h)(A_h + iB_h) = A_h^2 + B_h^2 \qquad (6\text{-}60)$$

From the derivation it follows that the intensities I_h and I_{-h} of a Friedel pair are equal with $\varphi_{-h} = -\varphi_h$, but only as long as there are no contributions that change in any way either coefficients A or B. The question is, what happens if the structure contains a dispersive atom, that is, an anomalous scatterer?

The breakdown of Friedel's law

We represent the scenario of a single dispersive atom present in a protein structure in the Argand Figure 6-34. $\mathbf{F_P}$ represents the sum of all the normally scattering atoms of the protein, and $\mathbf{F_A}$ represents an anomalously scattering atom A in the protein. In the absence of anomalous scattering contributions, Friedel's law holds, and the structure factor amplitudes are symmetric about the real axis as visualized in the leftmost panel of Figure 6-34. If an atom A scatters anomalously, its total scattering factor becomes

$$f_{(\mathbf{s},\lambda)} = f_{(\mathbf{s})}^0 + f_{(\lambda)}' + i \cdot f_{(\lambda)}'' \qquad (6\text{-}61)$$

The additional real dispersive contribution f' to the normal scattering factor $f_{(\mathbf{s})}^0$ must have the same phase as the normal contribution, because the atom position does not change, and thus the phase angle remains the same. As f' is

---

Sidebar 6-12 Polarization anisotropy of anomalous scattering. The breakdown of Friedel's law, which gives rise to the anomalous differences between Bijvoet pairs (Section 6.5), is not the only symmetry breakdown that occurs in reciprocal space as a result of anomalous diffraction. Anomalous diffraction is also polarization dependent and, given the strong polarization of the X-ray beam in the storage ring plane on synchrotrons, the (minuscule) anisotropy effect can actually be observed in specific crystal orientations. The polarization anisotropy of anomalous scattering (AAS) leads to a breakdown of symmetry in diffraction intensity also between normally symmetry related reflections. Early work of David and Lieselotte Templeton indicated[23] that this weak additional anisotropy effect might be exploited to improve anomalous macromolecular phasing, but to date it is not routinely used (i.e. it is ignored) in anomalous phasing. In fact, it was feared that the anisotropy might actually "cripple" the MAD phasing technique.[24] However, it has been shown that, given careful measurements and crystal alignment, the effect can in fact be beneficially exploited to improve anomalous phasing from Br atoms.[25] The previously scalar anomalous scattering factor contributions f' and f'' then become orientation dependent and are expressed as anisotropic scattering tensors $\mathbf{f'}$ and $\mathbf{f''}$.

negative, it reduces the total scattering factor $f^0_{(S)}$ of atom A by a small amount. However, the anomalous contribution f'' lags the phase of the dispersive contribution by 90°. Thus, its phase relative to the dispersive phase is always –270°, or +90°, and the algebraic value for the anomalous contribution is $i \cdot f''$. It is this complex anomalous phase contribution that breaks the centrosymmetry, and both magnitude and phase of the reflections forming a Friedel or Bijvoet pair are now distinctly different (right-hand panel of Figure 6-34).

We can prove algebraically that the anomalous and imaginary contribution $i \cdot f''$ to the atomic scattering factor leads to a mixing of cosine and sine terms in the real and imaginary parts of the structure factor contribution $\mathbf{F}_A$, and thus to a breakdown in symmetry (symmetry breakdown through mixed trigonometric terms in complex functions is a general rule). For the complex structure factor $\mathbf{F}^A_\mathbf{h}$ of an anomalous scattering atom A,

$$\mathbf{F}^A_\mathbf{h} = (f_0 + f' + if'')(\cos\varphi + i\sin\varphi) = (f_r + if'')(\cos\varphi + i\sin\varphi) =$$
$$= f_r\cos\varphi + if''\cos\varphi + if_r\sin\varphi - f''\sin\varphi = f_r\cos\varphi - f''\sin\varphi + i(f''\cos\varphi + f_r\sin\varphi)$$
$$(6\text{-}62)$$

and

$$\mathbf{F}^A_{-\mathbf{h}} = (f_0 + f' + if'')(\cos\varphi - i\sin\varphi) = (f_r + if'')(\cos\varphi - i\sin\varphi) =$$
$$= f_r\cos\varphi + if''\cos\varphi - if_r\sin\varphi + f''\sin\varphi = f_r\cos\varphi + f''\sin\varphi + i(f''\cos\varphi - f_r\sin\varphi)$$
$$(6\text{-}63)$$

In the equations above, the symmetry between the pairs of reflections is broken, equivalent to the scenario in the vector diagrams in the right panels in Figure 6-34. A Friedel pair with identical amplitudes thus becomes a general Bijvoet pair with different intensities F^+ and F^- for the related reflections. The fact that anomalous scattering contributions break the centrosymmetry of the reciprocal space is of crucial importance for phase determination methods based on intensity differences resulting from anomalous scattering.

Selection of dispersive atoms

Another reason for the widespread use of anomalous phasing techniques is rooted in the fact that quite a large variety of heavy (and not so heavy) atoms can be used as sources of anomalous signal (Table 6-4). A detailed view of the useable X-ray energy range, which on most synchrotrons extends from ~5 keV to 15 keV, shows that numerous elements have K- or L-edges (and actinides even M-edges), that fall into this region (Figure 6-35). Only a small section of the periodic system,

Figure 6-34 Breakdown of Friedel's law through anomalous scattering contributions. The vector $\mathbf{F}_P$ represents the partial sum of all of the non-anomalous atom contributions, and the remaining vectors are the contributions from the anomalously scattering atom, $\mathbf{F}_A$. The real and imaginary components of $\mathbf{F}_A$ are indicated in green ($\mathbf{f^0 + f'}$) and magenta ($\mathbf{f''}$), respectively. When the indices hkl are inverted to $-h-k-l$ as indicated by the (·) superscripts, normal scattering contributions are mirrored across the real axis, while the anomalous component becomes mirrored in relation to the imaginary axis. The result is that in the presence of anomalous scattering contributions, Friedel's law breaks down and the structure factor amplitude, and thus the intensity of Bijvoet mates (F^+ and F^-), becomes different.

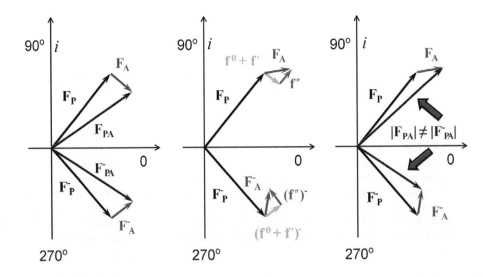

roughly the 5th period between Rb and Te, is not accessible for classical MAD measurements around an absorption edge. However, for anomalous data collection of Bijvoet differences it is, in principle, only necessary that the selected wavelength is above an absorption edge. This is certainly not as optimal as sitting on a white line peak right at the edge, but it will still yield anomalous signal if the energy is not too much (up to a few 1000 eV) above the edge. Many metalloenzymes, oxidases, reductases, or electron transfer enzymes contain naturally occurring transition metals such as Fe, Zn, Cu, or others. Very heavy atoms such as Hg which form useful derivatives also have strong L-edges in the energy range accessible for diffraction experiments at most synchrotrons. The vast majority of new structures are determined by anomalous phasing methods, often via MAD from Se-labeled proteins whose preparation we have described in Chapter 4, or by single-wavelength anomalous diffraction (SAD) phasing utilizing density modification (Chapter 10) to break the phase angle ambiguity inherent in SAD phasing.

An increasingly popular single-wavelength phasing technique, sulfur single-wavelength anomalous diffraction (S-SAD) phasing, is based on the small but useable signal from native S observable even far above the K-edge energy at 2472 eV of cysteine and methionine residues.[26] Redundant data collection at wavelengths as long (low energy) as practicable (~2 to 2.3 Å) increases the anomalous sulfur signal. However, softer X-rays are also strongly absorbed, and the resulting radiation damage limits the usable energy to > 5 keV. Characteristic Cr-radiation from in-house X-ray sources at 2.29 Å serves that purpose well.[27] A most peculiar but successful exploitation of radiation damage as a source of difference signal is radiation-damage-induced RIP phasing,[28] where the "burn-induced" intensity differences between data before and after heavy radiation exposure are used for phasing.

Estimating the anomalous signal

Many thousands of electrons are contributing to each structure factor and thus to each reflection. In a derivative crystal containing an anomalously scattering heavy atom, or Se in Se-methionine labeled crystals, there are in general few anomalously scattering atoms, maybe one for a hundred or more residues. This means that essentially we have to measure, on average, a miniscule anomalous or dispersive difference signal of several electrons in an orders-of-magnitude larger background of "normal" scattering contributions. This is not trivial, and emphasizes the need for accurate data collection in phasing experiments. Crick and Magdoff[29] have derived a basic formula to estimate the relative intensity

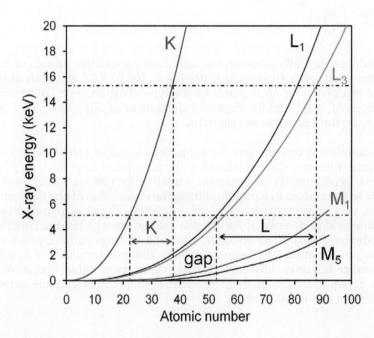

Figure 6-35 **Absorption edges in the range useful for anomalous diffraction data collection.** Most elements, with the exception of a gap approximately extending over the elements of the 5th period, have K- or L-edges in the range of X-ray energies (~5 to 15 keV) available at most synchrotrons for anomalous MAD diffraction data collection at an absorption edge. Note, however, that any energy above the absorption edge of an element will provide anomalous differences, which allows SAD or SIRAS phasing.

E	z	K(eV)	K(Å)	L₁(eV)	L₁(Å)	L₂(eV)	L₂(Å)	L₃(eV)	L₃(Å)	M₁(eV)	M₁(Å)
C	6	283.8	43.6844	–	–	–	–	–	–	–	–
N	7	401.6	30.8706	–	–	–	–	–	–	–	–
O	8	532.0	23.3038	–	–	–	–	–	–	–	–
S	16	2472.0	5.01523	229.2	54.0909	164.8	75.2284	164.8	75.2284	–	–
Ca	20	4038.1	3.07017	437.8	28.3180	350.0	35.4218	346.4	35.7900	44.3	–
Mn	25	6539.0	1.89595	769.0	16.1218	651.4	19.0323	640.3	19.3622	82.3	–
Fe	26	7112.0	1.74320	846.1	14.6527	721.1	17.1927	708.1	17.5083	91.3	–
Co	27	7708.9	1.60822	925.6	13.3942	793.8	15.6181	778.6	15.9230	101.0	–
Ni	28	8332.8	1.48781	1008.1	12.2980	871.9	14.2191	854.7	14.5053	110.8	–
Cu	29	8978.9	1.38075	1096.1	11.3107	951.0	13.0364	931.1	13.3150	122.5	–
Zn	30	9658.6	1.28359	1193.6	10.3868	1042.8	11.8888	1019.7	12.1581	139.8	88.681
Se	34	12657.8	0.97945	1653.9	7.49600	1476.2	8.39835	1435.8	8.6347	229.6	53.997
Br	35	13473.7	0.92014	1782.0	6.95715	1596.0	7.76794	1549.9	7.9990	257.0	48.240
Kr	36	14325.6	0.86542	1921.0	6.45374	1727.2	7.17788	1674.9	7.4020	292.8	42.342
Pd	46	24350.3	0.50914	3604.3	3.43968	3330.3	3.72268	3173.3	3.9069	671.6	18.460
Ag	47	25514.0	0.48592	3805.8	3.25756	3523.7	3.51836	3351.1	3.6996	719.0	17.243
Cd	48	26711.2	0.46414	4018.0	3.08552	3727.0	3.32644	3537.5	3.5046	772.0	16.059
I	53	33169.4	0.37377	5188.1	2.38963	4852.1	2.55511	4557.1	2.7205	1072.0	11.565
Xe	54	34561.4	0.35871	5452.8	2.27363	5103.7	2.42915	4782.2	2.5925	1148.7	10.793
Eu	63	48519.0	0.25552	8052.0	1.53970	7617.1	1.62761	6976.9	1.7770	1800.0	6.8876
Ir	77	76111.0	0.16289	13418.5	0.92392	12824.1	0.96675	11215.2	1.1054	3174.0	3.9060
Pt	78	78394.8	0.15814	13879.9	0.89321	13272.6	0.93408	11563.7	1.0721	3296.0	3.7614
Au	79	80724.9	0.15358	14352.8	0.86378	13733.6	0.90272	11918.7	1.0402	3425.0	3.6197
Hg	80	83102.3	0.14919	14839.3	0.83546	14208.7	0.87254	12283.9	1.0093	3562.0	3.4805
Pb	82	88004.5	0.14088	15860.8	0.78165	15200.0	0.81563	13035.2	0.9511	3851.0	3.2193
U	92	115606.1	0.10724	21757.4	0.56981	20947.6	0.59184	17166.3	0.7222	5548.0	2.2346

Table 6-4 X-ray absorption edge energies for selected elements. The theoretical atomic absorption edge energies in eV and Å are listed for common elements used in anomalous phasing. The light elements (C, N, O) provide no appreciable anomalous contributions, but beginning with S, a usable anomalous signal can be obtained. Highlighted boxes emphasize elements that can be used for edge-specific, complete MAD experiments in the 5–15 keV window accessible by most synchrotron X-ray sources.

changes between isomorphous heavy atom derivative data and native data (discussed in Chapter 10) that has been adapted[30] for anomalous diffraction ratios:

$$\frac{\Delta F}{F} = \frac{\Delta f_A}{f_P} \cdot \left(\frac{n_A}{2n_P} \right)^{1/2}$$

(6-64)

In this adapted formula we assume n_A anomalous atoms in a protein of n_P non-hydrogen atoms with an average scattering factor f_P of 6.7 electrons at $\sin \theta / \lambda$ = 0. The expression for Δf_A depends on the type of experiment. For anomalous differences, $\Delta f_A = 2f''$ and for dispersive differences $\Delta f_A = |f'(\lambda_1) - f'(\lambda_2)|$, where λ_1 and λ_2 are the respective wavelengths.

As an example one can estimate the anomalous signal of a single Se atom from anomalous difference data measured at the absorption edge, with an f'' of 3.8 electrons. Depending on the organism, a methionine residue occurs on average every 27 to 60 residues in a protein. Taking the lower value of one Se atom in 60 residues, we can estimate the expected anomalous signal as follows. Based on the amino acid preference of *Escherichia coli*, an average residue contains 8.1 non-hydrogen atoms, so we obtain n_P = 486 non-hydrogen atoms, which results according to Equation 6-64 in an anomalous diffraction ratio $\Delta F / F$ of 0.036 or 7.2% average intensity difference. This is readily measurable, and in practice useable anomalous difference can be obtained for proteins with one ordered Se atom in up to about 120 residues.

One physical principle works to our advantage in anomalous data collection: The electronic levels involved in X-ray absorption are core levels, and as we discussed for the atomic scattering factor, the scattering from the core levels decays less rapidly with scattering angle than the scattering from outer electronic levels. The anomalous scattering factors are thus practically independent of the scattering angle, and contribute relatively more at higher diffraction angles. This positive effect is partly compensated by the increasing relative error in the increasingly weaker intensity measurements at higher resolution.

Anomalous structure factor calculation for a protein

We can now explore what actual effect anomalous scattering has on the structure factors and scattered intensities of a protein containing a dispersive atom. We use our previous example, BPTI, from Section 6.3, except that we substitute the single methionine residue Met 52 with selenomethionine, where the S-δ is replaced with Se (recall Sidebar 4-7). Table 6-5 compares three data sets: one fictitious set without any consideration of anomalous signal; one with anomalous contributions computed at a low energy remote wavelength ~200 eV below the Se absorption K-edge; and the peak data set computed at the maximum anomalous contribution f'' at the theoretical Se absorption edge energy of 12658 eV, corresponding to 0.97645 Å.

The first pair of F/φ columns, computed without any anomalous contribution, follows Friedel's law and the columns provide the theoretical phases for the real-valued centrics (0°, 180°) and general phase restricted centric reflections (90°, 270°) for space group $P2_12_12_1$. In the next columns, we find amplitudes F at a low remote wavelength ~200 eV below the Se edge. The anomalous scattering contributions f' and f'' are listed above the low remote in the header, and we see that there are some dispersive and a small anomalous contribution present. This presence of anomalous contributions manifests itself in a structure factor amplitude difference between reflections without phase restrictions, while all centric reflection pairs still have exactly the same structure factor amplitudes.

Low remote data. Despite the fact that the low remote wavelength is well below the Se edge, there are still small anomalous contributions present in the low remote data. They originate mostly from six additional sulfur atoms in Cys–Cys disulfide links in the structure. The calculated data for our favorable case show that the highest amplitude differences in the low remote column are below 2% for the given data collection wavelength, which would very likely be insufficient to obtain usable difference data. Nevertheless, with redundant data collection and proper (longer) wavelength selection, even the sulfur anomalous data can provide usable anomalous signal. We will discuss this special S-SAD (sulfur single-wavelength anomalous diffraction) phasing in Chapter 10.

Peak data. Some reflections in the peak data show strong structure factor amplitude differences up to ~8%. This translates into maximal intensity differences of ~16%, and such differences are well measurable in this favorable case of one Se atom per 58 residues. The average amplitude difference for the entire data set computes to ~4%, in agreement with our estimate of 3.6% from Formula 6-64 in the previous section. This is again in agreement with the practical estimate that approximately one ordered Se atom per 120 residues suffices to provide usable anomalous signal. Note also that the phase restricted centric reflection pairs, which are unaffected by anomalous contributions and thus should have identical intensities (subject to measurement errors), provide a baseline for the overall data quality (discussed in detail in Chapter 8).

Dispersive difference data. In addition to the anomalous Bijvoet differences in the peak data set, we can also measure quite reasonable dispersive differences between the low remote data set and the peak data set. Although the peak data set is not optimal for dispersive differences (we would select the inflection point wavelength for maximum dispersive f' contribution), we find 2–3% differences on F, which probably could be measured.

Table 6-5 Computed anomalous structure factors for a Se-Met labeled protein. A representative sample of structure factor amplitudes and phases are grouped in centrosymmetrically related pairs. The shading color indicates their phase properties: centric reflections with real structure factors (yellow); general centric reflections (magenta); Friedel pairs with no phase restrictions (alternating in white and green). Dispersive differences between the remote and peak wavelength are present for all reflections, while anomalous differences between Friedel pairs exist only for non-centric reflections with unrestricted phases. The anomalous differences are on the order of a few percent of the structure factor amplitude.

					no signal	below edge	maximum f''			
Energy eV					–	12458	12658			
Wavelength Å					–	0.99518	0.97645			
f' (Se)					0.0	–3.6	–5.2			
f'' (Se)					0.0	0.5	3.8			
h	k	l	$F(hkl)$	$\varphi(hkl)$	$F(hkl)$	$F(hkl)$	centric	$d(hkl)$	$(\sin\theta/\lambda)^2$	
2	0	0	82.6	180.0	79.0	77.7	1	37.7	0.000176	
-2	0	0	82.6	180.0	79.0	77.7	1	37.7	0.000176	
2	0	1	112.2	270.0	113.0	113.5	1	22.8	0.000482	
-2	0	-1	112.2	90.0	113.0	113.5	1	22.8	0.000482	
2	1	0	114.5	0.0	116.6	117.9	1	19.4	0.000666	
-2	-1	0	114.5	0.0	116.6	117.9	1	19.4	0.000666	
2	0	1	129.3	270.0	126.8	125.7	1	18.9	0.000702	
-2	0	-1	129.3	90.0	126.8	125.7	1	18.9	0.000702	
4	0	0	197.0	180.0	198.9	200.0	1	18.9	0.000704	
-4	0	0	197.0	180.0	198.9	200.0	1	18.9	0.000704	
1	1	1	115.2	323.8	113.9	116.0	0	17.3	0.000840	
-1	-1	-1	115.2	36.2	112.9	109.3	0	17.3	0.000840	
2	1	1	34.4	354.2	35.4	36.4	0	16.0	0.000972	
-2	-1	-1	34.4	5.9	34.6	33.9	0	16.0	0.000972	
4	0	1	49.4	90.0	50.9	51.4	1	15.7	0.001009	
-4	0	-1	49.4	270.0	50.9	51.4	1	15.7	0.001009	
3	1	1	72.7	249.3	75.6	77.6	0	14.5	0.001192	
-3	-1	-1	72.7	110.7	75.0	75.4	0	14.5	0.001192	
1	1	2	69.7	118.7	69.9	69.4	0	11.9	0.001757	
-1	-1	-2	69.7	241.3	69.7	70.4	0	11.9	0.001757	
2	1	2	45.4	252.1	45.5	44.2	0	11.5	0.001889	
-2	-1	-2	45.4	107.9	46.7	48.7	0	11.5	0.001889	
1	2	1	59.0	356.6	59.0	60.2	0	10.4	0.002311	
-1	-2	-1	59.0	3.5	59.3	58.4	0	10.4	0.002311	
6	1	1	104.7	305.0	105.9	106.3	0	10.3	0.002379	
-6	-1	-1	104.7	55.0	105.1	105.4	0	10.3	0.002379	
1	1	3	111.5	7.4	111.0	108.7	0	8.7	0.003285	
-1	-1	-3	111.5	352.6	111.7	114.3	0	8.7	0.003285	
2	2	2	99.0	104.6	97.4	94.9	0	8.6	0.003360	
-2	-2	-2	99.0	255.5	99.3	101.8	0	8.6	0.003360	
5	2	1	27.0	78.3	31.4	34.0	0	8.6	0.003366	
-5	-2	-1	27.0	281.7	30.4	31.7	0	8.6	0.003366	
12	5	1	151.2	144.4	151.4	150.8	0	3.6	0.018897	
-12	-5	-1	151.2	215.6	152.3	153.4	0	3.6	0.018897	
2	6	2	99.6	307.6	101.3	103.7	0	3.6	0.019049	
-2	-6	-2	99.6	52.4	100.9	99.7	0	3.6	0.019049	
16	4	8	56.8	211.3	56.5	55.9	0	2.5	0.038666	
-16	-4	-8	56.8	148.7	56.8	57.2	0	2.5	0.038666	
23	3	6	42.3	111.5	41.9	40.5	0	2.5	0.038684	
-23	-3	-6	42.3	248.5	42.4	43.5	0	2.5	0.038684	

A crucial fact favoring anomalous phasing techniques is that we can collect all the anomalous and dispersive difference data from *one and the same* crystal. In theory this implies that no merging and non-isomorphism errors, which often plague classical isomorphous difference data from heavy atom derivatives (Chapter 10), should be present in the anomalous data. Radiation damage diminishes this advantage to a certain degree, but there is generally a good chance that one can collect multiple and redundant difference data of very high accuracy from one single crystal. This fact is one of the fundamental strengths of the anomalous phasing techniques, contributing to their widespread success in protein crystallography.

> **Box 6-12 X-ray absorption and anomalous scattering.** All matter absorbs X-rays. Heavy elements absorb X-rays more strongly than light elements and absorption decreases for harder X-rays. Above characteristic absorption edge energies, absorption occurs and electrons are removed from core levels by X-ray photons. As a consequence, the atomic X-ray scattering factor contains significant imaginary components above the edge energy. These wavelength-dependent, anomalous components to the X-ray scattering factor lead to a break in the centrosymmetry of reciprocal space, leading to measurable differences between related Bijvoet pairs of reflections.
>
> Dispersive differences exist between data sets recorded at different wavelengths above and below the absorption edge. The perfect isomorphism of anomalous and dispersive difference data sets collected from one single crystal is a major reason for the success of multi-wavelength anomalous diffraction (MAD) phasing techniques. Single-wavelength anomalous diffraction (SAD) techniques in combination with solvent modification are powerful and increasingly dominating alternatives to multi-wavelength phasing experiments.

6.5   Reciprocal space symmetry revisited

The previous section clearly emphasized the necessity for accurate and adequate data collection. The first task during data collection is usually to determine how much data must be acquired to obtain a complete data set containing the set of so-called unique reflections representing the asymmetric unit of the reciprocal space. The question thus is: What is the extent of the reciprocal asymmetric unit and how does it relate to the real space crystal symmetry?

During the examination of the reciprocal lattice properties in Chapter 5 we noticed that the reciprocal lattice is centrosymmetric. For a certain set of planes *hkl*, described by the lattice vector **h**, the centrosymmetric reciprocal lattice vector –**h** is identical in magnitude and points in opposite direction with indices –*h*–*k*–*l*. This leads us to assume that elements of crystal symmetry will also manifest themselves as symmetry in reciprocal space, that is, as symmetry in the diffraction pattern. If this is true, then we do not need to collect absolutely every data point in the entire reciprocal sphere, but only (or at least) the part containing all the unique reflections necessary to reconstruct the molecular structure. This part is the asymmetric unit of the reciprocal space, in analogy to the asymmetric unit of the unit cell of the crystal in real space. Just as the asymmetric unit in a crystal structure suffices to reconstruct the contents of the entire crystallographic unit cell, the unique data set suffices to fill the entire reciprocal space. In the case of anomalous data collection, the reflections in the centrosymmetrically related wedge (or their symmetry related mates) need to be collected in addition to the unique data wedge.

Laue symmetry

Let us examine what effect the same symmetry operations that act on molecular motifs in the crystal have on structure factors and thus on reciprocal space symmetry. We already know from Chapter 5 that we can formulate any of the m general point group operators listed in the International Tables as $\mathbf{W}_m(\mathbf{R}_m,\mathbf{T}_m)$, with $\mathbf{R}_m$ the 3×3 rotational (point group operation) matrix and $\mathbf{T}_m$ the operator's translational component. Let them both act on a complex structure factor. The component in the structure factor formula that is affected by $\mathbf{W}_m$ is the fractional coordinate vector **x** of each atom. For a structure factor $\mathbf{F_h}$ that is subject to **W** we thus can write for each of the m operators

$$\mathbf{F_h} = \sum_{j=1}^{atoms} f_j \exp[2\pi i h(\mathbf{Wx})] = \sum_{j=1}^{atoms} f_j \exp[2\pi i h(\mathbf{Rx} + \mathbf{T})]$$

$$= \sum_{j=1}^{atoms} f_j \exp[2\pi i h(\mathbf{Rx})] \cdot \exp(2\pi i h\mathbf{T}) \tag{6-65}$$

Remembering that the index vector **h** preceding the matrix is a row vector (short for $\mathbf{h}^T$) and the matrix multiplication associativity $\mathbf{h}(\mathbf{Rx}_j) = (\mathbf{hR})\mathbf{x}_j$ is maintained (Appendix A.2), we can reformulate

$$\mathbf{F_h} = \sum_{j=1}^{atoms} f_j \exp[2\pi i (\mathbf{hR})\mathbf{x})] \cdot \exp(2\pi i h \mathbf{T}) = \mathbf{F_{hR}} \exp(2\pi i h \mathbf{T}) \qquad (6\text{-}66)$$

This expression can be rearranged to $\mathbf{F_{hR}} = \mathbf{F_h}\exp(-2\pi i h \mathbf{T})$ and split into amplitude and phase term, from which we obtain following relations:

$$\left|\mathbf{F_{hR}}\right| = \left|\mathbf{F_h}\right| \text{ and thus } I_{hR} = I_h \text{ and } \varphi_{hR} = \varphi_h - 2\pi \mathbf{hT} \qquad (6\text{-}67)$$

From Equation 6-67 it immediately follows that (i) intensities of symmetry related reflections are equal, and (ii) if the space group operator includes a translational element, then a defined relation exists between the phases of the symmetry related reflections. We also realize that the translational component **T** (and also any additional Bravais translation) *only affects phases* and not amplitudes. Thus, related space groups with plain rotations and roto-translations (screw axes) such as $P222$ and $P2_12_12_1$ or even $I222$ have the same reciprocal space symmetry; that is, reciprocal space symmetry can only reflect the point group symmetry m. Reflections which are related through point group symmetry must invariably have the same intensities, even in the presence of anomalous scattering contributions (the only rarely exploited exception is in the presence of anisotropic anomalous scattering; Sidebar 6-12).

In addition to the restriction to point group symmetry, in the absence of anomalous contributions the reciprocal lattice, and thus the reciprocal space, is generally centrosymmetric, irrespective of the absence or presence of an actual center of inversion in the space group. Combination of inversion with the point group symmetry generates additional mirror symmetry, and thus every rotational element of a point group also becomes a mirror element, expressed as the symbol n/m, where n is the multiplicity of the corresponding axis. For example, point group 222 in real space then becomes the so-called Laue group $\frac{2}{m}\frac{2}{m}\frac{2}{m}$ or, in short, mmm in reciprocal space. In this example, we would have to collect only the positive octant of the reciprocal space, limited to the range from zero to the highest available indices in each direction. Figure 6-36 illustrates this example for orthorhombic space groups and a similar scenario for trigonal Laue groups $\bar{3}$ and $\bar{3}m1$.

The augmented Table 6-6 of space groups includes the Laue symbol describing the symmetry of the reciprocal space, and the extent in indices h, k, l, of the reciprocal space that must be sampled[31] to record a complete diffraction data set containing all unique reflections. Note that Table 6-6 contains a subdivision of Laue group $\bar{3}$ into $\bar{3}1m$ and $\bar{3}m1$, because a significant difference exists between the trigonal space groups in terms of reciprocal space coverage, depending on the location of the 2-fold axis (cf. Figure 6-36).

Figure 6-36 The asymmetric unit of reciprocal space for orthorhombic and trigonal Laue groups. The asymmetric units of reciprocal space are shown as the wedge cut out of the reciprocal sphere. The corresponding limits for indices hkl that are necessary to collect the asymmetric unit of reciprocal space are listed in Table 6-6. The presence of the mirror in the hk plane reduces the amount of space that needs to be sampled from 1/6 in $\bar{3}$ (center) to 1/12 in $\bar{3}m1$ (right panel). As the location of the 2-fold axis in real space is different in trigonal space groups with point group 32, the segments of reciprocal space in Laue groups $\bar{3}1m$ and $\bar{3}m1$ that need to be collected are different, but cover the same amount of 1/12 of the reciprocal space volume (Table 6-6). At least these index triples within the asymmetric unit of reciprocal space (or their symmetry-related mates) need to be collected to obtain a complete data set, without consideration of anomalous contributions. If anomalous data are to be collected for phasing, at least the wedge of centrosymmetrically related Friedel pairs –**h** of reflection **h** (or their symmetry mates, the Bijvoet mates) must be collected to obtain a complete (anomalous) data set.

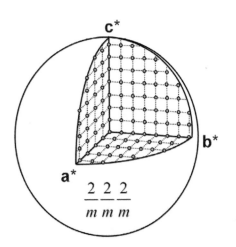

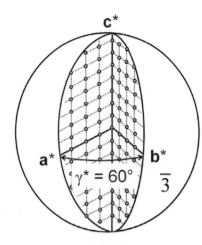

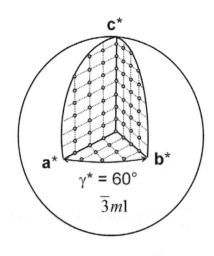

The Laue group is established early during data collection by examining the symmetry of the reciprocal space. The Laue symmetry confirms the first estimate of the crystal system or lattice type obtained from the cell parameters, which is needed to develop a data collection strategy that samples the reciprocal space properly so that a complete data set containing as much of the unique data as possible is obtained. In the case of anomalous data collection, Friedel or Bijvoet mates of the reflections in the unique wedge also need to be collected. If we are not interested in anomalous contributions, the centrosymmetrically related part of the reciprocal space does not need to be sampled, and any observed Friedel pair members can be averaged. As a principle, it seldom hurts to collect data in anomalous mode and to keep the anomalous pairs separate. If there is indeed no usable anomalous signal, reflections can be averaged and used as such. However, if, for example, a heavy ion from the solvent binds specifically to the molecules in the crystal, their anomalous signal may (serendipitously) suffice to provide anomalous differences for phasing. The statistical treatment of data averaging is discussed in Chapter 7, and the practice of collecting, integrating, merging, and scaling of diffraction data will occupy an important part of the discussion of data collection in Chapter 8.

Symmetry relations and Friedel pairs of the asymmetric reciprocal unit

It is a very enlightening exercise to generate the symmetry-related wedges and the centrosymmetrically related Friedel wedge from the wedge of unique data. This is because understanding the coverage and symmetry of reciprocal space is important for the development of more complex data collection strategies (in practice, most data collection programs in fact generate reasonable standard data collection strategies).

Take as an example space group $P2_1$, the second most frequent space group for protein structures (Figure 5-44). Its point group is 2, with a multiplicity m of 2, the same as for chiral space groups $P2$ or $C2$—translational symmetry elements do not affect the point group symmetry of the reciprocal space. According to Table 6-6, the standard asymmetric unit of reciprocal space extends from $-h$ to h, 0 to k, and 0 to l, with the highest resolution d_{min} determining the diameter of the resolution sphere and thus the longest vector $\mathbf{h}$. Reflections h, k, 0 are excluded for $h < 0$ to avoid counting them twice when generating the symmetry related reflections.

Generating the symmetry related wedges of reciprocal space. We apply the point group part $\mathbf{R}$ of the general position operator of space group $P2_1$ to a general reflection h, k, l. In the absence of a Bravais translation, the general position multiplicity M equals point group multiplicity $m = 2$, and we obtain from the general positions

(1) $x, \quad y, z$

(2) $-x, \frac{1}{2} + y, -z$

by ignoring $\mathbf{T}$ the two operator matrices $\mathbf{R}_1$ and $\mathbf{R}_2$ ($\mathbf{R}_1$ always being the identity operator)

$$\mathbf{R}_1 = \begin{pmatrix} 1 & 0 & 0 \\ 0 & 1 & 0 \\ 0 & 0 & 1 \end{pmatrix} \text{ and } \mathbf{R}_2 = \begin{pmatrix} -1 & 0 & 0 \\ 0 & 1 & 0 \\ 0 & 0 & -1 \end{pmatrix}$$

We apply $\mathbf{R}_2$ to the general reflection h, k, l and obtain for $\mathbf{h}_2 = \mathbf{h}\mathbf{R}_2$ (or $\mathbf{R}_2^T\mathbf{h}$ if you prefer to work with column vectors) the symmetry related reflection $-h$, k, $-l$. This reflection lies in the symmetry related wedge of reciprocal space extending from $-h$ to h, 0 to k, and 0 to $-l$ (with h, k, 0 excluded for $h > 0$), as illustrated in Figure 6-37.

Generating the Friedel wedge of the asymmetric unit of reciprocal space. In an identical fashion as we generated the symmetry mate of h,k,l we can now generate the Friedel mate by applying the inversion operator $\bar{1}$ to $\mathbf{h}$. We obtain the Friedel mate $-\mathbf{h}$ of $\mathbf{h}$ of general reflection h, k, l as $-h$, $-k$, $-l$ and the Friedel wedge

Lattice properties	Minimum internal symmetry	Crystal system	Point group	PG mult m	Laue group	Asymmetric reciprocal unit	Coverage of R*	Bravais type	Bravais mult.	Lattice type	Chiral space group (No.)	Patterson symmetry	M, z
$a \neq b \neq c$ $\alpha \neq \beta \neq \gamma \neq 90°$	none	triclinic	1	1	$\bar{1}$	h: $-\infty$ to ∞ k: $-\infty$ to ∞ l: 0 to ∞ Exclude $0kl$ if $l < 0$ and $0k0$ if $k < 0$	1/2	P	1	aP	P1 (1)	$P\bar{1}$	1
$a \neq b \neq c$ $\alpha = \gamma = 90°$ $\beta \neq 90°$	2-fold rotation axis parallel to unique axis **b**	monoclinic	2	2	$2/m$	h: $-\infty$ to ∞ k: 0 to ∞ l: 0 to ∞ Exclude $hk0$ if $h < 0$	1/4	P C	1 2	mP mC	P2 (3), P2₁ (4) C2 (5)	P2/m C2/m	2 4
$a \neq b \neq c$ $\alpha = \beta = \gamma = 90°$	3 perpendicular, non-intersecting 2-fold axes	orthorhombic	222	4	$2/mmm$	h: 0 to ∞ k: 0 to ∞ l: 0 to ∞	1/8	P I C F	1 2 2 4	oP oI oC oF	P222 (16), P222₁ (17), P2₁2₁2 (18), P2₁2₁2₁ (19) I222 (23), I2₁2₁2₁ (24) C222₁ (20), C222 (21) F222 (22)	Pmmm Immm Cmmm Fmmm	4 8 8 16
$a = b \neq c$ $\alpha = \beta = \gamma = 90°$	4-fold rotation axis parallel to **c**	tetragonal	4	4	$4/m$	h: 0 to ∞ k: 0 to ∞ l: 0 to ∞	1/8	P I	1 2	tP tI	P4 (75), P4₁ (76), P4₂ (77), P4₃ (78) I4 (79), I4₁ (80)	P4/m I4/m	4 8
			422	8	$4/mmm$	h: 0 to ∞ k: 0 to ∞ l: 0 to ∞ Exclude $0kl$ if $k > 0$	1/16	P I	1 2	tP tI	P422 (89), P42₁2 (90), P4₁22 (91), P4₁2₁2 (92), P4₂22 (93), P4₂2₁2 (94), P4₃22 (95), P4₃2₁2 (96) I422 (97), I4₁22 (98)	P4/mmm I4/mmm	8 16
$a = b \neq c$ $\alpha = \beta = 90°$ $\gamma = 120°$	3-fold rotation axis parallel to **c**	trigonal	3	3	$\bar{3}$	h: 0 to ∞ k: 0 to ∞ l: $-\infty$ to ∞ Exclude $h0l$ if $l < 0$ and $0kl$ if $l \leq 0$	1/6	P R	1 3	hP hR	P3 (143), P3₁ (144), P3₂ (145) R3 (146)	$P\bar{3}$ $R\bar{3}$	3 9
			32	6	$\bar{3}1m$	h: 0 to ∞ k: h to ∞ l: $-\infty$ to ∞ Exclude $0kl$ if $l < 0$	1/12	P	1	hP	P312 (149), P3₁12 (151), P3₂12 (153)	$P\bar{3}1m$	6
					$\bar{3}m1$	h: 0 to ∞ k: 0 to ∞ l: $-\infty$ to ∞ Exclude $hk0$ if $h < k$	1/12	P R	1 3	hP hR	P321 (150), P3₁21 (152), P3₂21 (154) R32 (155)	$P\bar{3}m1$ $R\bar{3}m$	6 18
	6-fold rotation axis parallel to **c**	hexagonal	6	6	$6/m$	h: 0 to ∞ k: 0 to ∞ l: 0 to ∞	1/12	P	1	hP	P6 (168), P6₁ (169), P6₅ (170), P6₂ (171), P6₄ (172), P6₃ (173)	P6/m	6
			622	12	$6/mmm$	h: 0 to ∞ k: 0 to ∞ l: 0 to ∞ Exclude $0kl$ if $k > 0$	1/24	P	1	hP	P622 (177), P6₁22 (178), P6₅22 (179), P6₂22 (180), P6₄22 (181), P6₃22 (182)	P6/mmm	12
$a = b = c$ $\alpha = \beta = \gamma = 90°$	Four 3-fold axes along space diagonal	cubic	23	12	$m\bar{3}$	h: 0 to ∞ k: h to ∞ l: h to ∞	1/24	P I F	1 2 4	cP cI cF	P23 (195), P2₁3 (198) I23 (197), I2₁3 (199) F23 (196)	$Pm\bar{3}$ $Im\bar{3}$ $Fm\bar{3}$	12 24 48
			432	24	$m\bar{3}m$	h: 0 to ∞ k: h to ∞ l: h to ∞ Exclude hkh if $h < k$	1/48	P I F	1 3 4	cP cI cF	P432 (207), P4₂32 (208), P4₃32 (212), P4₁32 (213) I432 (211), I4₁32 (214) F432 (209), F4₁32 (210)	$Pm\bar{3}m$ $Im\bar{3}m$ $Fm\bar{3}m$	24 48 96

Table 6-6 The 65 chiral space groups and their reciprocal space symmetry. The table columns include lattice properties, minimal lattice symmetry requirements, the resulting crystal system, the 11 enantiomorphic point groups and their multiplicity m, Laue symmetry, extent of the reciprocal space unit cell, reciprocal cell space coverage, Bravais type and its multiplicity, lattice type, and the 65 chiral space groups (ITC-A number in parenthesis), Patterson symmetry, and the general position multiplicity M or Z, which is the number of symmetry related, equivalent copies of the asymmetric unit and its contents each space group generates. The symbols follow the Hermann–Mauguin notation. In the triclinic primitive lattice symbol aP, a stands for *anorthic*. The Patterson symmetry is generated from the space group symmetry plus inversion. Patterson symmetry plays an important role in the determination of the heavy atom substructure, and is discussed in Chapter 9. Polar space groups with a floating origin along the unique axis are shown in red.

extends from $-h$ to h, 0 to $-k$, and 0 to $-l$ (with h, $-k$, 0 excluded for $h > 0$), again shown in Figure 6-37. Note and verify that the handedness of the right-handed reciprocal coordinate system has changed upon inversion, while it did not do so when generating a symmetry equivalent wedge.

Bijvoet pairs and absolute configuration

Just as the asymmetric unit of reciprocal space containing the set of unique reflections **h** has m symmetry related wedges $\mathbf{h}_m = \mathbf{hR}_m$ within the Ewald sphere, the centrosymmetrically related wedge will have m symmetry related wedges, too. Each of the m symmetry related mates $\mathbf{hR}_m$ of reflection **h** (or F^+) will also have a Friedel mate, $(-\mathbf{h})\mathbf{R}_m$ (or F^-). These reflections F^+ and F^- form Bijvoet pairs. Similarly, starting from the Friedel mate $-\mathbf{h}$ (or F^-) of a refection **h** and then applying the m point group followed by inversion generates $-(-\mathbf{h})\mathbf{R}_m$ reflections (or F^+), and the pairs F^+ and F^- are again Bijvoet mates (the next section and Figure 6-38 illustrate this further). Some of these Bijvoet pairs will be symmetry mates of **h** as well, and carry no anomalous difference intensity signal, but most will not be symmetry mates of **h** and therefore intensity differences between the two Bijvoet mates will exist. The fact that many Bijvoet pairs exist in higher symmetry space groups allows a great deal of flexibility in the data collection, and many anomalous differences can be observed between different wedges of reciprocal space.

Generating Bijvoet mates. We take again our general reflection h, k, l, generate its Friedel mate $-h, -k, -l$, and then apply $\mathbf{R}_2$ to $-h, -k, -l$ and obtain for $-\mathbf{h}_2 = -\mathbf{hR}_2$ the Bijvoet mate $h, -k, l$ of h, k, l. This reflection now lies in a symmetry related wedge of the Friedel wedge that extends from $-h$ to h, 0 to $-k$, and 0 to l (with $h, k, 0$ excluded for $h < 0$), as illustrated in Figure 6-38. This Bijvoet mate of hkl is of course also a Friedel mate of symmetry related reflection $-h, k, -l$. We will observe anomalous differences between each of the Friedel pairs (h, k, l and $-h, -k, -l$) and ($h, -k, l$ and $-h, k, -l$), as well as between the Bijvoet pairs ($h\,k\,l$ and $h, -k, l$) and ($-h, k, -l$ and $-h, -k, -l$).

Figure 6-37 **Symmetry related wedge and Friedel wedge of asymmetric unit of reciprocal space.** The figure shows the unique wedge of reciprocal space for point group 2. The unique reciprocal axis is k, and the angle β^* between h and l is generally different from 90°. The asymmetric unit extends from $-h$ to h, 0 to k, and 0 to l; reflections $h, k, 0$ located in the shaded area are excluded for $h < 0$ to avoid including them twice when generating the symmetry related reflections. Applying the 2-fold symmetry operation generates the magenta wedge extending from $-h$ to h, 0 to k, and 0 to $-l$ (with $h, k, 0$ excluded for $h > 0$). The Friedel wedge (blue) of the unique wedge is generated by inversion of **h** to $-\mathbf{h}$; it extends now from $-h$ to h, 0 to $-k$, and 0 to $-l$ (with $h, -k, 0$ excluded for $h > 0$).

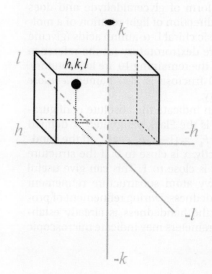

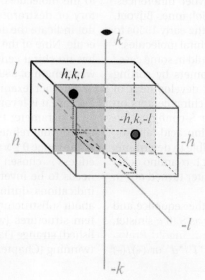

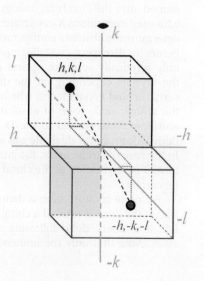

Figure 6-38 Bijvoet pairs, Friedel pairs, and symmetry related pairs. The Bijvoet pairs are generated by applying a point group symmetry operation **R** to the Friedel opposite −**h** of reflection **h**. The reflection $h, -k, l$ is thus located in a symmetry related wedge (red) of the Friedel wedge (blue) of the asymmetric unit of reciprocal space. The right panel shows the pair-wise relations between the four reflections generated by point group symmetry 2 and inversion $\bar{1}$. FR, Friedel pairs; BV, Bijvoet pairs; and S, symmetry related pairs of reflections.

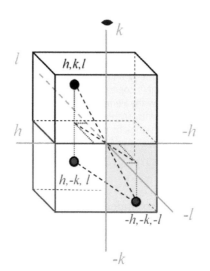
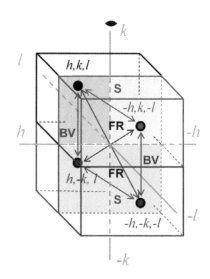

Absolute configuration. The fact that the imaginary anomalous scattering component is always offset by +90° from the real part of the anomalous scattering means that the anomalous differences will be different depending on the absolute configuration of the anomalous scatterer (i.e. the relative position of the dispersive atom in reference to the chiral centers). If we call the structure factors in the direct, right-handed wedge (and its symmetry related wedges) **F⁺** and those in its inverse (and its symmetry related wedges) **F⁻**, we can compute the Bijvoet differences $\Delta F = F^+ - F^-$ and resulting anomalous intensity differences between each pair $\Delta I = I^+ - I^-$ and compare the sign of the observed differences with the sign computed for the two possible enantiomer models. The fit between observed and calculated data will be better for the correct handedness. The magnitude of the intensity difference signal is proportional to $4f''$, indicating that it makes good sense to pick a wavelength with the highest anomalous signal (preferably the peak of a white line) for optimal difference signal. For proteins we know the absolute configuration of the molecule, because proteins are synthesized from L-amino acids. As long as we keep F^+ associated with a right-handed coordinate system, the protein structure will show the correct handedness, that is, right-handed α-helices and L-amino acids. The handedness ambiguity in substructure determination (Chapter 10) is simply resolved by trying both substructure enantiomorphs. Only the correct

Sidebar 6-13 Absolute configuration. The intensity differences between centrosymmetrically related pairs of reflections are anomalous differences or Bijvoet differences, named after the Dutch crystallographer Johannes Bijvoet, who used anomalous X-ray scattering in the early 1950s to determine the absolute configuration of chiral molecules.[33] Before his discovery, crystallographers could in some cases only physically separate optical enantiomers by sorting the crystals depending on the different development of certain chiral crystal faces; whether the chiral centers on the atomic scale were of absolute R- or S-configuration and how these related to the D-, L-amino acid labeling derived from glyceraldehyde was impossible to determine. In the case of amino acids, the guess that L-α-amino acids have S-configuration at the chiral Cα center turned out to be correct.

Absolute conformation is defined by the sequence and size of the ligands around a chiral center as S (S = sinister, left; R = rectus, right) following the Cahn-Ingold-Prelog rules. Note that only the nomenclature "*l*"/"*d*" or (+)/(−)

refers to the rotation of light (l = l(a)evus, to the left; d = dexter, to the right), while the D- or L- stereo-descriptor relates to the molecule's conformation in reference to levo-rotatory or dextrorotatory form of glyceraldehyde and does not indicate the actual direction of light rotation of a molecule. Nine of the 19 basic chiral L-α-amino acids (glycine has no chiral center) are dextrorotatory (at the reference wavelength of 589 nm); the remaining 10 are levorotatory. As another example, D-fructose is also named *levulose* because it is levorotatory.

A parameter that can indicate the absolute configuration during refinement is the Flack parameter[34] x defined in the range $0 \leq x \leq 1$ as $F_{\mathbf{h},x}^2 = (1-x)F_{\mathbf{h}}^2 + xF_{-\mathbf{h}}^2$. If the handedness is chosen correctly, x is close to 0; if the structure needs to be inverted, x is close to 1. This can give useful indications during heavy atom substructure refinement about substructure handedness. During refinement of protein structures (where the handedness is already established) strange Flack parameters may indicate microscopic twinning (Chapter 8).

substructure handedness will give an interpretable protein map. Determination of absolute conformation is, however, important in small molecule crystallography, see Sidebar 6-13 and Glusker and Trueblood[32] for details.

Centric and phase restricted reflections

In a previous section we showed that pairs of symmetry related, phase restricted structure factors F_{hR} and F_h which are related through point group symmetry operations must invariably have the same reflection intensities or structure factor amplitudes. If any one of these point group symmetry related structure factors F_{hR} also happens to be centrosymmetrically related, that is, fulfills the condition

$$hR = -h \tag{6-68}$$

then this refection is a centric reflection (or, in short, *centric*). Bijvoet pairs of centric reflections cannot carry any anomalous difference signal even in the presence of anomalous scattering contributions despite their centrosymmetric relation, because they also fulfill a symmetry relation of the corresponding space (point) group. Not all space groups have centric reflections; for example in $P3$, none of the three point group operators R applied to h generates a reflection hR that is also a Friedel mate $-h$ of itself.

Use of centrics in anomalous intensity statistics. Because the intensity of the Friedel or Bijvoet mates of centric reflections has to be the same irrespective of (isotropic) anomalous contributions, they are useful in the calculation of intensity statistics by comparing centric pairs that cannot carry any anomalous signal with unrestricted Bijvoet pairs that are allowed to carry anomalous contributions. Intensities of phase restricted, centric pairs will merge with significantly better statistics than non-centric Bijvoet pairs—if those Bijvoet pairs in fact do carry significant anomalous difference signal. If the differences in measured intensities between these groups or reflections are statistically significant, usable anomalous signal is likely present and its strength can be estimated. Examples of how to use merging statistics to estimate anomalous signal are given in Chapter 8.

The relation $hR = -h$ implies that the reflection must lie (and can only lie) in a plane of reciprocal space where a symmetry related wedge and a centrosymmetrically related wedge make contact. Figure 6-39 shows this relation in our example of $P2_1$ for a reflection $h, 0, l$ where application of the operator R_2 leads to $-h, 0, -l$ which obviously is also the Friedel mate of $h, 0, l$. Note that in real space atoms that give rise in reciprocal space to centric reflections are located centrosymmetrically opposed respective to the corresponding zone axis. This in turn leads to collinear atomic scattering contributions to the centric structure factor (cf. Figure 6-40).

Phase restrictions. In the structure factor Table 6-5 we have already observed that all phase restricted reflections form anomalous pairs with exactly the same intensities. The question we have not answered as of yet is what additional phase restrictions apply to centric reflections. We have shown in the subsection on Laue symmetry that for symmetry related reflections F_h and F_{hR} the following phase relation exists:

$$\varphi_{hR} = \varphi_h - 2\pi hT \tag{6-69}$$

From the above relation, it follows that a set of reflections that satisfies a centric relation for its indices so that $hR = -h$ places additional restrictions on its phase. The phase of a centric structure factor can be either

$$\varphi h = \pi hT \text{ or } \pi(hT + 1) \tag{6-70}$$

that is, one of the two possible values (depending on the translational symmetry elements) that are separated by π. Let us apply an operator of our favorite space

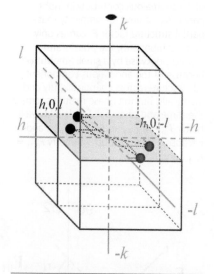

Figure 6-39 Centric reflections and centric zones in reciprocal space. The centric reflections obeying the relation $hR = -h$ are located in zones where symmetry related and centrosymmetrically related wedges make contact (purple plane). For reciprocal space symmetry 2, the centric zone is $h, 0, l$, as shown for the two non-equivalent reflections 3, 0, 1 and 1, 0, 3 and their related Friedel pairs. Pairs of centric reflections carry no anomalous difference signal even in the presence of anomalous scattering contributions.

group $P2_12_12_1$, number 19, to a number of structure factors. We pick non-trivial operator $\mathbf{W}_2$ from the ITC-A (International Tables for Crystallography, Volume A), in symbolic form written as $\bar{x}+1/2, \bar{y}, z+1/2$. We can write the operator components in matrix form

$$\mathbf{R}_2 = \begin{pmatrix} -1 & 0 & 0 \\ 0 & -1 & 0 \\ 0 & 0 & 1 \end{pmatrix} \text{ and } \mathbf{T}_2 = \begin{pmatrix} 1/2 \\ 0 \\ 1/2 \end{pmatrix}$$

where $\mathbf{R}_2$ describes an operation around a 2-fold axis along $\mathbf{c}$. Let us apply the operators to $hkl = 200$. The index $\mathbf{h}$ becomes $\mathbf{hR}$ which simply gives -200. The centric phase of for 200 is given by $\varphi_c = \pi \mathbf{hT} = \pi(2 \cdot \frac{1}{2} + 0 \cdot 0 + 0 \cdot \frac{1}{2}) = \pi(180°)$ or by $\pi(\mathbf{hT} + 1) = \pi(2 \cdot \frac{1}{2} + 0 \cdot 0 + 0 \cdot \frac{1}{2} + 1) = 2\pi$ (0°), respectively. For $\mathbf{F}_{201}$ we obtain $\varphi_c = 0.5\pi$ or 1.5π for possible phase values (90° or 270°), again in agreement with the values in Table 6-5. For the centrosymmetrically related index pair (111) and $(-1-1-1)$ we do not find any operator in space group $P2_12_12_1$ that relates them. Therefore, these structure factors have no restrictions on their phase and will show anomalous differences. Depending on the combination of translational (including Bravais) symmetry elements, additional phase restrictions for centric reflections of 30°, 60°, 120°, 240°, 300°, and 330° are possible for example in trigonal and hexagonal space groups; or values of 45°, 90°, 270°, 225°, and so forth in tetragonal and cubic space groups. A frequent special case is the centric structure factors with phases of either 0° or 180°; they do not have any imaginary component, they are real, and they can be simply assigned a sign, + for 0° and – for 180° phase angle. The often-made general statement[35] that centric structure factors must have 0° or 180° phase angle is incorrect.

The fact that a centric structure factor assumes one of the two possible centric phase values (depending on the translational elements present) that are separated by π does not imply that the Friedel mate of a centric structure factor with $\varphi(\mathbf{h})$ has a phase of $\varphi + 180°$—it still has the conjugate phase of $\varphi(-\mathbf{h}) = -\varphi(\mathbf{h})$ and $F(\mathbf{h}) = F(-\mathbf{h})$. Even in the presence of anomalous signal, the amplitude $F(\mathbf{h}) = F(-\mathbf{h})$ of centric Friedel pairs invariably remains the same, but the phases will become $\varphi(\mathbf{h}) + \delta\varphi$ (or $-\delta\varphi$) and $\varphi(-\mathbf{h}) + \delta\varphi$ (or $-\delta\varphi$), respectively, depending on the direction and magnitude of the f'' contribution which is always phase shifted 90° counterclockwise from the real f' part of the anomalous scatterer(s). The scattering contributions of the centrosymmetrically related atoms in a centric projection have conjugate phase $+\alpha$ and $-\alpha$ around one of the possible phases φ_c or $\varphi_c + 180°$ which the centric structure factors of the corresponding centric zone can take, and their contributions are necessarily collinear. The

Figure 6-40 Centric reflections and their phase relations. As a somewhat more complicated general example for a centric structure factor that has phase values different from 0° or 180°, $\mathbf{F}_{011}$ in space group $I4_122$ and its symmetry related structure factors and Friedel mates are shown. The structure factors are the result of pair-wise centric atom contributions which have conjugate phase centered around the resultant phase, which is phase restricted, in this particular case 45° (shown in detail for $\mathbf{F}_{011}$). Its symmetry related structure factor $\mathbf{F}_{01-1}$ has a phase offset by π, that is, $\varphi_{01-1} = \varphi_{011} + 180°$ and also has a Friedel mate; it is $\mathbf{F}_{0-11}$ which itself is a symmetry mate (again a Bijvoet mate) of $\mathbf{F}_{011}$'s Friedel mate $\mathbf{F}_{0-1-1}$. In the case of an anomalous contribution (right panel; we assume for simplicity that partial structure factor $\mathbf{F}_3$ carries only an f'' component), the phases are shifted clockwise by a small amount φ''. Because of the anomalous component all structure factors change slightly and identically, but the phase relations remain the same and of course Friedel's law still holds: $F_{011} = F_{0-1-1}$ and $F_{01-1} = F_{0-11}$, and because of point group symmetry $F_{0-11} = F_{0-1-1}$ and $F_{011} = F_{01-1}$.

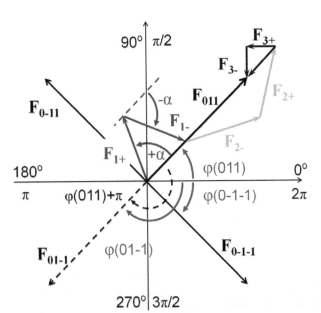

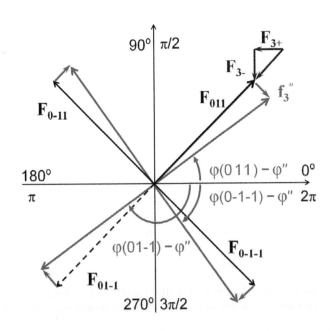

Sidebar 6-14 Centric reflections in early phasing. The fact that certain centric structure factors have phase values limited to precisely 0° or 180° and are thus real, with a sign of either + or – corresponding to the cosine of 0° or 180°, made them the first target for phase determination in the earliest protein crystallographic studies:[36] If electron density maps from centric reflections $h0l$ in monoclinic structures are calculated, the electron density reconstruction calculated from their amplitudes and the correct phases corresponds to a projection of the protein structure along Y (down the unique axis **b**) into the X-Z plane. Unfortunately, as protein unit cells are large, the Fourier projections extend over a great depth of material (~32 Å for monoclinic $C2$ hemoglobin) and are thus not interpretable in terms of molecular structure details. The 2 Å electron density projection generated by Fourier synthesis using centric structure factor amplitudes $h0l$ and their correct phases is shown in Figue 6-41. There is not much more discernible than the shape (projection) of the molecule.

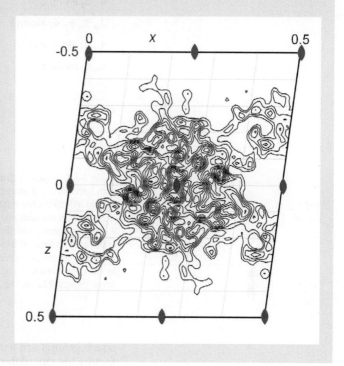

Figure 6-41 Centric Fourier projection. The projection is computed from of sections $h0l$ along zone axis [010], that is, the unique axis **b**. The projection is for monoclinic hemoglobin crystallizing in space group $C2$. Compare with the projection of $C2$ shown in Chapter 5 and note that the 2-fold screw axes become normal 2-fold rotation axes in the projection.

anomalous scattering contribution can also point in the opposite direction from the resultant; therefore the f'' contribution can point either clockwise or counterclockwise for a given centric structure factor and its Friedel pair (Figure 6-40).

Intensity distribution of centric reflections. Centric reflections have a different intensity distribution than general acentric reflections: centric reflections are more often either very strong or very weak than general acentric reflections. We derive the corresponding, underlying structure factor distributions in Chapter 7. The difference in these distributions also allows distinguishing centrosymmetric space groups from (acentric) chiral space groups.

Systematic absences

Any symmetry operation that includes translational elements leads to systematic absences of certain reflections in the diffraction pattern. The reason is additional destructive phase difference between atoms which are translated along a fraction of the unit cell axis and thus lie exactly "in between" a set of certain lattice planes. If the translation is along a screw axis, the extinction will only affect reflections along the corresponding reciprocal axis, and these extinctions are called serial extinctions, as only a series of reflections on the reciprocal axis are affected. For example, a 2_1 screw axis along **b** allows reflections only for $0k0$ with $k = 2n$. In other words, only reflections 020, 040, and so forth, are possible, and every second reflection (010, 030, etc.) must be systematically absent.

If translational elements affect two directions in the lattice such as in Bravais centering of a plane, the extinctions affect the corresponding reciprocal plane, and these extinctions are termed zonal extinctions. For example in space group $C2$, only reflections hkl with $h + k = 2n$ (even) are allowed. These are exactly the extinctions we have computed by hand in our 2-dimensional structure factor calculation, which was essentially a projection of a simple $C2$ case. For the remaining Bravais translations, integral extinctions affect the entire reciprocal lattice. The reflection conditions for the symmetry elements observed in chiral

Symmetry element	Reflection condition
2_1 along a, b, or c	either $h00$, $0k0$, or $00l$: $2n$
4_2, 6_3 along c	$00l$: $2n$
3_1, 3_2, 6_2, 6_4 along c	$00l$: $3n$
4_1, 4_3 along a or c	$h00$ or $00l$: $4n$
6_1, 6_5 along c	$00l$: $6n$
C	$h + k$: $2n$
I	$h + k + l$: $2n$
F	h, k, l all even or all odd

Table 6-7 Conditions limiting reflections observed in chiral space groups. In space groups with symmetry operators possessing translational components and in centered Bravais lattices certain reflections are systematically absent. The reflection conditions are listed for each symmetry element with translational components occurring in chiral space groups. Space groups with several limiting operations will show corresponding combinations of systematic absences. Absences that affect a series (line) of lattice points are serial extinctions, those which affect zones such as C (planes of reciprocal lattice points) are zonal extinctions, and those affecting the entire lattice such as I or F are integral extinctions. Serial extinctions can be masked by integral extinctions; chiral space groups $I222$ and $I2_12_12_1$ for example show the same absences.

space groups are listed in Table 6-7. They automatically result when structure factors are calculated for the entire unit cell.

In the majority of cases, knowledge of the Laue group combined with systematic absences allows unambiguous determination of the space group. In a few cases, however, a unique determination of the space group from data symmetry and extinctions alone is not possible. For example, in the case of $I222$ versus $I2_12_12_1$ the serial extinctions caused by the screw axes coincide with the integral Bravais extinctions of I. In this case, the structure solution will not succeed in the wrong space group, which normally becomes obvious in the early stages of substructure and phase determination.

General position multiplicity and ε-factors

The multiplicity of a general point position (the number of copies generated by the symmetry operators) is given by the product of the multiplicity of all symmetry operations that are applied to it. The multiplicity M of a general position x, y, z is thus given as $M = m \cdot B \cdot I$, where m is the point group multiplicity (Table 6-6), B is the number of Bravais translations, and I is 1 for an acentric space group and 2 for a centric space group. Special positions located on symmetry elements or on the intersection of symmetry elements have higher symmetry. The atoms at their position are $n > 1$ times mapped onto themselves depending on the order of the symmetry element(s) they are sitting on, and a special position thus has a lower multiplicity $m = M/n$. The same is true for certain classes of reflections: their indices $\mathbf{h}$ map onto themselves ε times upon application of the point group symmetry operations, and such a reflection therefore has ε times the intensity of a general reflection.[37] The correction factor $\varepsilon(\mathbf{h})$ is the epsilon-factor of reflection $\mathbf{h}$. In the example of space group $P2_12_12_1$ (PG 222) provided in Table 6-3, each reflection $h00$, $0k0$, and $00l$ maps twice onto itself and therefore $\varepsilon = 2$ for these reflection series. The multiplicity of the reflection $\mathbf{h}$ is defined in analogy to the real space case as

$$m(\mathbf{h}) = M/\varepsilon(\mathbf{h}) \tag{6-71}$$

In intensity statistics and for computation of normalized structure factors, the fact that these non-general reflections have higher expectation values for their

Box 6-13 **Summary of reciprocal space symmetry.** Reciprocal space reflects the symmetry of the crystal in real space. The symmetry of the reciprocal space is determined by the point group of the crystal structure. In the absence of anomalous contributions, real space point group symmetry combined with the centrosymmetry of the reciprocal lattice results in the Laue symmetry of reciprocal space. The asymmetric unit cell of reciprocal space is determined by one of the 12 Laue group and contains the unique data necessary to reconstruct the crystal structure.

Pairs of reflections that are centrically related by the point group part of each general position symmetry operator must invariably have identical intensities irrespective of anomalous contributions.

For anomalous data collection, the centrosymmetrically related wedge of the reciprocal space asymmetric unit containing Friedel mates $-\mathbf{h}$ of reflection $\mathbf{h}$ also needs to be collected. Bijvoet pairs are formed between a reflection $\mathbf{h}$ and each symmetry equivalent of its Friedel mate $-\mathbf{h}$.

Centric structure factors are structure factors related by point group symmetry and by centrosymmetry in reciprocal space: $\mathbf{h} = -\mathbf{hR}$. As a consequence, Bijvoet and Friedel mates of centric reflections invariably have the same amplitude and are restricted in their phase values. Centric reflections are therefore useful for estimating the data quality in absence of anomalous contributions, while the anomalous signal is estimated from intensity differences in acentric pairs of reflections. Centric reflections are located in zones where point group symmetry related wedges and centrosymmetrically related wedges of reciprocal space overlap.

In most cases systematic absences of reflections related to specific symmetry operators with translational components allow the unique determination of the space group.

The indices of certain classes of reflections map onto themselves upon application of point group symmetry. These reflections therefore have higher expectation values for their intensities, which is accounted for in intensity statistics by the epsilon-factor $\varepsilon(\mathbf{h})$.

intensities than general reflections needs to be adjusted with the ε-factors tabulated in the ITC-B or in Iwasaki and Ito.[37]

Figure 6-42 shows examples for reflections 0, k, 0 which have $\varepsilon = 2$ and are absent for $k \neq 2n$ (i.e. odd k) in space group $P2_1$ because of the presence of the translational element $\mathbf{T}_2 = (0, \frac{1}{2}, 0)$ of the second general position symmetry operator. The epsilon zones or series are located on symmetry elements in reciprocal space, in this case the 2-fold axis along the unique reciprocal axis k.

6.6   Key concepts

Scattering of X-rays

- X-rays are high energy photons in the range of ~100 to 100 000 eV, with a usable range of 5000 to about 15 000 eV for diffraction experiments.
- Photons can be described as packets of traveling electromagnetic waves, whose electric field vector interacts with the charged electrons of matter. The scattering of X-rays by electrons is weak. Only a small fraction of incident radiation gets scattered.
- The electromagnetic field component of X-rays excites oscillations of the electrons of atoms which themselves emit partial waves that recombine into a resulting wave or photon.
- Superposition of waves in phase (i.e. with a phase shift $\Delta \varphi$ of $n \cdot 2\pi$ radians or 360°) maximizes constructive interference, while phase shifts of π (180°) leads to maximum destructive interference.

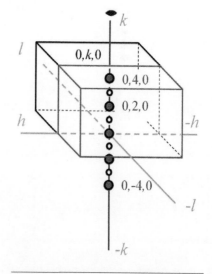

Figure 6-42 Location of epsilon zones and extinctions in reciprocal space. In the example of space group $P2_1$, the reflections along the zone axis 0, k, 0 have an ε-factor of 2, because their multiplicity is only 1 instead of the general reflection multiplicity of 2. In addition, because of the translational vector $\mathbf{T}_2 = (0, \frac{1}{2}, 0)$ a generated equivalent atom is exactly π out of phase for reflections 0, k, 0 with odd k, and because of this serial extinction such reflections are systematically absent.

- The probability of a photon being observed in a specific direction is proportional to the amplitude of the resulting wave in this direction.
- No correlation in time and space, that is, no coherence, between individual photons is required for X-ray scattering. However, the scattering process itself is coherent within the coherence length of each photon; that is, a fixed phase relation exists between all partial waves emanating from electrons excited by an individual photon.
- Waves can be described in the complex plane as complex numbers in a vector-like representation whose length reflects the amplitude of the wave and whose direction determines the phase angle relative to a fixed origin.
- Summation of partial waves can be performed algebraically by complex number (vector) addition.

Structure factors

- The electron density of an atom is to a first approximation treated as spherical, and the atomic scattering factor curves for each element are Gaussian functions, continuously decreasing with scattering angle.
- The atomic scattering curves (or atom form factors) are computed from the 9-parameter Cromer–Mann exponential approximation as a function of scattering angle.
- The atomic scattering factor is measured in units of electrons, e^-, and evaluates for zero scattering angle to the number of electrons of the scattering atom, z.
- The total scattering power of a molecule is proportional to the scattering from all its atoms and thus proportional to the sum of all z.
- The scattering factors for atoms in a crystal structure are additionally attenuated by displacement of the atoms from their mean position.
- Positional displacement results from thermal vibration around the resting position as well as from disorder in a crystal lattice.
- All displacive effects compound and are accounted for through the exponential B-factor attenuation in a Debye–Waller factor T_S. A molecule or parts thereof which have atoms with high B-factors or exhibit a high degree of disorder cannot diffract well.
- Interference between the partial waves scattered by neighboring atoms leads to a modulation of the scattering function.
- For a protein molecule, the combined molecular scattering function is an irregularly modulated, decaying envelope function.
- The scattering of X-rays by a single molecule is too small to be measured, but by arrangement of many molecules in a periodic lattice, amplification and discrete sampling of the molecular scattering function occurs. The process described is diffraction of X-rays by a crystal.
- The total scattering from a crystal in a given reciprocal lattice direction $\mathbf{h}$ is thus proportional to the sum of all scattering contributions in the unit cell.
- Each and every atom j in the unit cell contributes a partial wave to every reflection $\mathbf{h}$. The complex structure factor $\mathbf{F_h}$ is thus a summation of partial waves of j atoms of scattering factor f_j at position $\mathbf{x}_j$:

$$\mathbf{F_h} = \sum_{j=1}^{atoms} f^0_{S,j} \cdot \exp(2\pi i \mathbf{h}\mathbf{x}_j)$$

- Centric structure factors are structure factors that are related by point group symmetry *and* by centrosymmetry.
- As a consequence, their Friedel mates have invariably the same amplitude and are restricted in their phase values.
- Centric structure factors can assume only certain fixed phase values such as 0°, 90°, 180°, 270°, 120°, 240°, and so forth.
- A special group of centric structure factors have phase angles of 0° or 180° and are real, with a sign corresponding to + for 0° and – for 180°.
- Some structure factors compute to exactly zero. Such systematic extinctions or absences due to translational elements in the space group symmetry assist in the space group determination.
- Centric structure factors have different intensity distributions compared with general, acentric reflections.

- In the presence of anomalous scattering contributions, general acentric Friedel pairs $\mathbf{F}(\mathbf{h})$ and $\mathbf{F}(-\mathbf{h})$ become different with $F(\mathbf{h}) \neq F(-\mathbf{h})$. For centric Friedel pairs, $F(\mathbf{h}) = F(-\mathbf{h})$ irrespective of any anomalous contributions.
- Diffraction can be formally described as the transformation of the electron density from real space R into reciprocal space R^*.
- The reciprocal space diffraction pattern is given by the amplitudes of the Fourier transform (FT) of the crystal structure.
- Fourier transforms are a very important concept in crystallography and are crucial in the reconstruction of the molecular structure from diffraction data.

Diffraction of X-rays

- Diffraction on a periodic array of molecules in a crystal can be interpreted as reflection of X-rays on sets of lattice planes hkl. The relation between the diffraction angle and the lattice plane spacing d_{hkl} is given by Bragg's law $n\lambda = 2d\sin\theta$.
- The direction of the scattering vector is normal to the set of reflecting planes hkl, and reflection occurs when the scattering vector is collinear and equal in magnitude with the reciprocal lattice vector $\mathbf{d}^*_{hkl}$.
- The Ewald construction shows that diffraction occurs when a reciprocal lattice point hkl intersects the Ewald sphere of radius $1/\lambda$. The reciprocal space can thus be sampled while rotating the crystal (and thus its reciprocal lattice) in increments and recording a diffraction pattern during each rotation increment.
- The scattering from a crystal in direction hkl is determined by the complex structure factor $\mathbf{F}_{hkl}$, obtained by summation of all partial waves emanating from each atom in the unit cell.
- The intensity of scattered X-rays in direction $\mathbf{h}$, $I_{\mathbf{h}}$, is proportional to the squared amplitude $F_{\mathbf{h}}^2$ of the complex scattering factor.

Absorption of X-rays

- Absorption of X-rays occurs at any wavelength and in any matter. Light elements absorb X-ray less than heavy elements, and for a given element, high energy X-rays are less absorbed than soft X-rays.
- At characteristic absorption edge energies, resonance absorption occurs and the atomic scattering factor contains anomalous, wavelength-dependent contributions.
- The imaginary component of the anomalous scattering factor leads to the breakdown of reciprocal lattice centrosymmetry, and significant anomalous differences exist within data sets recorded above the absorption edge of a dispersive atom, that is, an anomalous scatterer.
- Dispersive differences exist between data recorded at different wavelengths.
- The collection of multiple, in principle perfectly isomorphous anomalous and dispersive difference data from one single crystal contributes to the widespread success of multi-wavelength anomalous diffraction (MAD) phasing and related anomalous phasing techniques.
- Single-wavelength anomalous diffraction (SAD) in combination with phase modification and phase extension techniques is becoming an increasingly dominating phasing method.

Reciprocal space symmetry

- Symmetry of the crystal in real space manifests itself in specific symmetry of the reciprocal space.
- Reciprocal space symmetry is determined by point group symmetry of the crystallographic space group. In the absence of anomalous scattering contributions, the centrosymmetry of the reciprocal lattice combined with the point group symmetry leads to the 12 Laue groups describing the symmetry of the reciprocal space.
- The amount of unique data that needs to be collected is determined by the extent of the asymmetric unit of the reciprocal lattice.
- Reflections related by point group symmetry operators of the crystal space group invariably have the same intensity, irrespective of anomalous scattering contributions.

- For anomalous data collection, the centrosymmetrically related wedge of the reciprocal space asymmetric unit containing Friedel mates –**h** of reflection **h** also needs to be collected.
- Bijvoet pairs are formed between a reflection **h** and each symmetry equivalent of its Friedel mate –**h** and *vice versa*.
- Centric structure factors are related by point group symmetry and by centro-symmetry in reciprocal space: **h** = –**hR**.
- Bijvoet and Friedel mates of centric reflections invariably have the same amplitude and are restricted in their phase values. Centric reflections are therefore useful for estimating the data quality in the absence of anomalous contributions, while the anomalous signal is estimated from intensity differences in acentric pairs of reflections.
- Centric reflections are located in zones where point group symmetry related wedges and centrosymmetrically related wedges of reciprocal space overlap.
- The indices of certain classes of reflections map onto themselves upon application of point group symmetry. These reflection therefore have higher expectation values for their intensities, which is accounted for in intensity statistics by the epsilon-factor $\varepsilon(\mathbf{h})$.
- Translational space group symmetry including Bravais symmetry in real space leads to systematic absences in reciprocal space, which in most, but not all, cases can be used to determine the space group.
- The reciprocal space symmetry and not the unit cell dimensions and angles determine the crystal system.

6.7   Additional reading

1. Blow D (2002) *Outline of Crystallography for Biologists*. Oxford, UK: Oxford University Press.

2. Drenth J (2007) *Principles of Protein X-ray Crystallography*. New York, NY: Springer.

3. Giacovazzo C, Monaco HL, Viterbo D, et al. (2002) *Fundamentals of Crystallography*. Oxford, UK: Oxford Science Publications.

4. Hahn T (Ed.) (2002) *International Tables for Crystallography, Volume A: Space-Group Symmetry*. Dordrecht, The Netherlands: Kluwer Academic Publishers.

5. Shmueli U (Ed.) (2001) *International Tables for Crystallography, Volume B: Reciprocal Space*. Dordrecht, The Netherlands: Kluwer Academic Publishers.

6. Rossmann MG, & Arnold E (Eds.) (2001) *International Tables for Crystallography, Volume F: Crystallography of Biological Macromolecules*. Dordrecht, The Netherlands: Kluwer Academic Publishers.

7. Sands DE (1995) *Vectors and Tensors in Crystallography*. New York, NY: Dover Publications, Inc.

6.8   References

1. Hartemann VF, Baldis HA, Kerman AK, et al. (2000) Three-dimensional theory of emittance in Compton scattering and X-ray protein crystallography. *Phys. Rev.* E64, 16501-1-16501–26.

2. Cromer DT, & Mann JB (1968) X-ray scattering factors computed from numerical Hartree-Fock wave functions. *Acta Crystallogr.* A24, 321–324.

3. Wilson AJC, & Prince E (Eds.) (1999) *International Tables for Crystallography, Volume C: Mathematical, Physical and Chemical Tables*. Dordrecht, The Netherlands: Kluwer Academic Publishers.

4. Hazemann I, Dauvergne MT, Blakeley MP, et al. (2005) High-resolution neutron protein crystallography with radically small crystal volumes: application of perdeuteration to human aldose reductase. *Acta Crystallogr.* 61(10), 1413–1417.

5. Niimura N, & Bau R (2008) Neutron protein crystallography: beyond the folding structure of biological macromolecules. *Acta Crystallogr.* A64(1), 12–22.

6. Trueblood KN, Burgi HB, Burzlaff H, et al. (1996) Atomic displacement parameter nomenclature. Report of a subcommittee on atomic displacement parameter nomenclature. *Acta Crystallogr.* A52(5), 770–781.

7. Perutz MF (2002) *I Wish I'd Made You Angry Earlier: Essays on Science, Science, Scientists, and Humanity*. Oxford, UK: Oxford University Press.

8. Cullis AF, Muirhead H, Perutz MF, et al. (1962) The structure of haemoglobin. IX. A three-dimensional Fourier synthesis at 5.5 Å resolution: description of the structure. *Proc. R. Soc. London* A265, 161–187.

9. Miao J, Kirtz J, & Sayre D (2000) The oversampling phasing method. *Acta Crystallogr.* D56, 1312–1315.

10. Chapman HN, Barty A, Bogan MJ, et al. (2007) Femtosecond diffractive imaging with a soft-X-ray free-electron laser. *Nat. Phys.* 2(12), 839.

11. Henderson R (2002) Excitement over X-ray lasers is excessive. *Nature* 415, 833.

12. Taylor CA, & Lipson H (1964) *Optical Transforms: Their Preparation And Application To X-Ray Diffraction Problems.* UK: G. Bell & Sons.

13. Parkin S, Rupp B, & Hope H (1996) Structure of bovine pancreatic trypsin inhibitor at 125 K: Definition of carboxyl-terminal residues Gly57 and Ala58. *Acta Crystallogr.* D52(1), 18–29.

14. Giacovazzo C, Monaco HL, Viterbo D, et al. (2002) *Fundamentals of Crystallography.* Oxford, UK: Oxford Science Publications.

15. Milch JR, & Minor TC (1974) The indexing of single-crystal X-ray rotation photographs. *J. Appl. Crystallogr.* 7, 502–505.

16. Kabsch W (1988) Evaluation of single-crystal X-ray diffraction data from a position-sensitive detector. *J. Appl. Crystallogr.* 21, 916–924.

17. Cromer DT (1983) Calculation of anomalous scattering factors at arbitrary wavelengths. *J. Appl. Crystallogr.* 16, 437.

18. Cromer DT, & Liberman D (1970) Relativistic calculation of anomalous scattering factors for X-rays. *J. Chem. Phys.* 33(5), 1891–1898.

19. Agrawal BK (1991) *X-Ray Spectroscopy.* New York, NY: Springer Verlag.

20. Rehr JJ, & Albers RC (2000) Theoretical approaches to x-ray absorption fine structure. *Rev. Mod. Phys.* 72(3), 621–654.

21. Phillips JC, & Hodgson KO (1980) The use of anomalous scattering effects to phase diffraction patterns from macromolecules. *Acta Crystallogr.* A36(6), 856–864.

22. Burla MC, Carrozzini B, Cascarano GL, et al. (2004) MAD phasing: choosing the most informative wavelength combination. *Acta Crystallogr.* D60(9), 1683–1686.

23. Templeton DH, & Templeton LK (1982) X-ray dichroism and polarized anomalous scattering of the uranyl ion. *Acta Crystallogr.* A38(1), 62–67.

24. Fanchon E, & Hendrickson WA (1990) Effect of the anisotropy of anomalous scattering on the MAD phasing method. *Acta Crystallogr.* 46(10), 809–820.

25. Sanishvili R, Besnard C, Camus F, et al. (2007) Polarization-dependence of anomalous scattering in brominated DNA and RNA molecules, and importance of crystal orientation in single- and multiple-wavelength anomalous diffraction phasing. *J. Appl. Crystallogr.* 40(3), 552–558.

26. Dauter Z, Dauter M, & Dodson ED (2002) Jolly SAD. *Acta Crystallogr.* D58, 496–508.

27. Yang C, Pflugrath JW, Courville DA, et al. (2003) Away from the edge: SAD phasing from the sulfur anomalous signal measured in-house with chromium radiation. *Acta Crystallogr.* D59(11), 1943–1957.

28. Nanao MH, Sheldrick GM, & Ravelli RB (2005) Improving radiation-damage substructures for RIP. *Acta Crystallogr.* D61(Pt 9), 1227–1237.

29. Crick FH, & Magdoff BS (1956) The theory of the method of isomorphous replacement for protein crystals. I. *Acta Crystallogr.* 9, 901–908.

30. Hendrickson WA, & Ogata CM (1997) Phase determination from multiwavelength anomalous diffraction measurements. *Methods Enzymol.* 276, 494–516.

31. Suh I-H, Kim K-J, Choo G-H, et al. (1993) The asymmetric unit of X-ray intensity data in the seven crystal systems. *Acta Crystallogr.* A49, 369–371.

32. Glusker JP, & Trueblood KN (1990) *Crystal Structure Analysis. A Primer.* New York, NY: Oxford University Press.

33. Bijvoet JM (1954) Structure of Optically Active Compounds in the Solid State. *Nature* 173, 888–891.

34. Flack H (1983) On enantiomorph-polarity estimation. *Acta Crystallogr.* A39(6), 876–881.

35. Lattman EE, & Loll PJ (2008) *Protein Crystallography: A Concise Guide.* Baltimore, MD: The John Hopkins University Press.

36. Bragg WL, & Perutz MF (1954) The structure of haemoglobin. VI. Fourier projections on the 010 plane. *Proc. R. Soc. London* A225, 315–329.

37. Iwasaki H, & Ito T (1977) Values of ε for obtaining normalized structure factors. *Acta Crystallogr.* A33(1), 227–229.

Statistics and probability in crystallography

> There comes a time in the life of a scientist when he must convince himself either that his subject is so robust from a statistical point of view that the finer points of statistical inference are irrelevant—or that the precise mode of inference he adopts is satisfactory.
>
> A. F. W. Edwards (1992) in *Likelihood—An account of the statistical concept of likelihood and its application to scientific inference*, p. xv

It may come as a bit of surprise to find a separate chapter dedicated to probability theory and statistics in a biomolecular crystallography textbook. However, in view of the leading quote to this chapter, crystallography definitely falls into the second category of science. *Maximum likelihood* and the related concept of *Bayesian inference* are fundamental to practically all modern methods of crystallographic structure determination and affect all stages from data collection to initial phase determination by heavy atom or molecular replacement methods, extending to electron density reconstruction and model refinement.

Contemporary methods are heavily based on a probabilistic approach to the subject of statistics, and future developments are likely to be even more so. Experience shows that the likelihood-based view of statistics is treated either at a suboptimal time or context, or not at all in bioscience curricula. The practitioner of crystallography may also benefit from a brief review of the statistical key concepts presented here in a crystallographic context. Not all the derivations must be followed in a successive way; it may suffice to focus on the results in a first reading, and to turn back to this chapter for reference and the detailed derivations when they are later used in the advanced sections of the text. You can trust that whatever is derived here will be needed and used at a later stage. The section about intensity statistics is a good example; it will be needed again to understand scaling, twinning, phasing, and the nature of crystallographic data in general.

After a brief introduction emphasizing the importance and relevance of statistical methods in modern protein crystallography, a short review of basic probability and probability distributions follows. Deriving the Poisson distribution from basic X-ray photon counting statistics, we obtain the ubiquitous Gaussian normal distribution as a limiting case for large numbers of events. Based on the normal distribution and the central limit theorem, we apply the rules of error propagation to data reduction, merging, and weighted averaging, and provide related definitions for *R*-values and correlation coefficients.

Armed with the fundamentals of probability and statistics, we are ready to introduce for the first time the concepts of Bayesian inference and maximum likelihood in a general crystallographic context. A brief general derivation of likelihood maximization yields the classical least squares method as a special case of likelihood optimization. A simple example demonstrating where Bayesian inference overcomes a serious problem is the treatment of negative intensities, encountered in the data processing stage. A final (semi-serious) example emphasizes the danger of relying on poor data without inclusion of adequate prior information.

The derivation of basic, unconditional structure factor probabilities in the form of the Wilson distribution follows, establishing the means to bring structure factors from different sources onto a common scale. We then develop a general approach to conditional structure factor probability distributions and derive their likelihood functions for incomplete partial models and models with errors. This leads directly to Sim weights, the Luzzati factor, and finally the sigma-A coefficients accounting for both incompleteness and errors in the conditional structure factor probabilities. As an added bonus, the Wilson distributions are obtained as unconditional limiting cases of the Rice and Woolfson distributions.

This chapter may provide some difficulty for students exposed to advanced statistics for the first time, and in a condensed course for bioscience students, it may not be necessary to follow each derivation in detail. Nevertheless, at least a qualitative review of the general topics of probabilities, Bayesian inference, and maximum likelihood is advisable, not least in order to be able to refer back to these from later chapters.

7.1   Crystallography: applied probability theory

We have already seen in Chapter 3 that deductive and analytical methods are not always available to solve a practical problem of inference. Crystallization propensities were a typical example, where we attempted to infer or improve the probability of crystallization of a protein given some global prior information about the historic performance of reagents and—as much as possible—some specific information about the protein. In fact, quite a number of problems in crystallography are based on probabilistic inference and the probability distributions of key properties.

Examples include photon counting statistics, which determine the quality of our experimental diffraction data, structure factor probability distributions that form the basis for empirical intensity scaling, extensive use of phase angle probability distributions in modern phasing, and phase extension techniques. During refinement, prior probabilities in the form of the likelihood of a protein structure model based on chemical plausibility (the geometric restraints) play an indispensable role. In all areas, from crystallization to structure validation, the maximum likelihood principle and the related Bayesian inference models, relying on incorporation of independent prior knowledge, form the foundation of modern biomolecular crystallography.

Box 7-1 **Probability and likelihood in crystallography.** Numerous key steps in contemporary crystallographic structure determination are based on probabilistic methods of inference. Maximum likelihood models and related Bayesian inference models that incorporate prior knowledge are implemented in most crystallographic programs, ranging from data collection to phasing and refinement.

Moreover, given the non-deterministic process of protein structure "determination," the evaluation of our structure model is again based on its probability of being correct, expressed as how well it complies with the experimental data and how compatible it is with our prior knowledge of what a healthy protein structure should look like. Prior knowledge helps to keep weak experimental evidence in check against our well-established expectations. However, given very strong data (evidence), even ingrained expectations of how a protein structure should look can, must, and will be corrected and expanded.

7.2 Probability distributions

Many physical processes such as X-ray scattering or radioactive decay are not determined or may not even occur in a given observation interval. The measurements of these properties may have different values each time and fluctuate around an expectation value. Each of these observations or events occurs with a certain probability, for which a probability distribution function (PDF) can be established. Examples of probability distributions we encounter in crystallographic work are the Poisson distribution in photon counting statistics, the Gaussian normal distribution in error models, the Lorentz–Cauchy distribution for atomic emission line widths, and the Rice[1] or Sim distribution[2] and the Wilson[3] and Woolfson[4] distributions in intensity statistics and likelihood-based error models.

A distribution can be *discrete* (as our experimental diffraction intensity data in crystallography always are) or can be *continuous*. Continuous distributions are often referred to as probability density functions. In many common situations this distinction is obvious from the context and of no further consequence, but in certain contexts (for example integration or differentiability) this distinction may be necessary. Figure 7-1 shows a basic, 1-dimensional or univariate and unimodal (one peak) probability distribution function and defines some of its general properties in the figure caption.

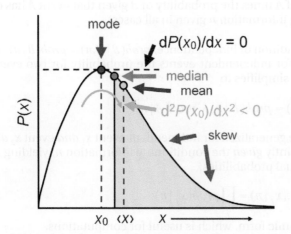

Figure 7-1 Properties of a general univariate and unimodal probability distribution. The mode x_0 of the probability distribution *prob(x)* or *P(x)* is defined by its peak value, which is the most probable value. The first derivative *dP(x)/dx* (the *slope* of the probability distribution graph) is zero at the most probable value x_0. As the most probable value represents a maximum of the function, the second derivative of the function $d^2P(x)/dx^2$ (the curvature) at x_0 must be negative (i.e. the slope gets continuously smaller and smaller) as we pass the maximum. The median of the distribution is located at the value of *x* that splits the area under the curve into equal areas (white, blue). A distribution is skewed when it is asymmetric around the mode (i.e. mean and mode are different). Positive skewness indicates a shift toward higher values of *x*. For distributions symmetric about x_0, the mode, the median, and the mean are identical and all given by the arithmetic mean $\bar{x}$ which is a special case of the expectation value $\langle x \rangle$ defined as the weighted average or centroid value of all values of *x*. Other commonly used descriptive statistics of distributions are the *n*th central moments (or moments about the mean) $\mu_n = E(x - x_0)^n$ with the expectation value *E* defined as the integral over the distribution. The first central moment is zero, the second moment about x_0 is the variance σ_x^2 (indicating the width of the distribution), the third is the skewness, and the fourth is the kurtosis, a measure how "squat" or "peaky" the distribution is. The moments about the mean must not be confused with moments of a distribution about the origin or zero (the raw moment), also used in intensity statistics and defined as $\langle x^k \rangle$ for the *k*th raw moment.

Box 7-2 Probability distribution functions. Numeric values of a physical property ranging from photon count to structure factors and stereochemical geometry restraints follow certain distributions in their actually observed or theoretically expected values. The function that describes the distribution of these values is the probability distribution function (PDF) of that property or parameter.

7.3   Basic probability algebra

In the following we describe or define a few rules and terms of probability algebra that will be frequently used in our discussion of probabilities and distributions.

Conditional probabilities. The probability of an event is conditioned on the particulars of that event. Take for example a toss of an n-faced die. If the die is fair, the probability of obtaining any one of the faces x in the range 1 to n is

$$prob(x\,|\,n) = 1/n \tag{7-1}$$

Put into words this reads: the conditional probability of a face showing value x, given that the die has n sides, is $1/n$. The bar symbol | stands for the conditioning statement *given* (sometimes a semicolon is also used as the conditioning symbol). *Given* here means that we assume that the conditioning terms used to describe the probability are known for the case, or for a certain model or hypothesis. For an unbiased die, the probability of observing a certain face value x, conditioned on the fact that the die has six faces, is $1/6$.

Independence of probabilities. If two events A and B (such as the throw of two dice) are independent,

$$prob(B\,|\,A,n) = prob(B\,|\,n) \tag{7-2}$$

This just means that event B is independent (unconditional) of the outcome of event A.

Joint probabilities and product rule. The general joint probability for two events A and B to occur jointly given conditioning information n is given by the product rule

$$prob(A,B\,|\,n) = prob(A\,|\,n) \times prob(B\,|\,A,n) \tag{7-3}$$

The comma in the above relations annotates the *joint symbol* or logical *and*. In other words: The joint probability that events A and B occur is defined by the probability of A times the probability of B given that event A has occurred (with conditioning information n given in all cases).

From the condition of independence, $prob(B\,|\,A,n) = prob(B\,|\,n)$, it immediately follows that for independent events the probability for two events A and B to occur jointly simplifies to

$$prob(A,B\,|\,n) = prob(B\,|\,n) \times prob(A\,|\,n) \tag{7-4}$$

which can be generalized for independent event x_1 *and* event x_2 *and* up to event x_j to occur jointly *given* the conditioning information n, yielding the product of their individual probabilities:

$$prob(x_1,x_2,\ldots,x_j\,|\,n) = \prod_j prob(x_j\,|\,n) \tag{7-5}$$

or in logarithmic form, which is useful for computations,

$$\ln\!\left(prob(x_1,x_2,\ldots,x_j\,|\,n)\right) = \sum_j \ln(prob(x_j\,|\,n)) \tag{7-6}$$

Example: The joint probability that we score a 4 *and* a 6 in two independent throws of a fair 6-sided die is simply the product of the two independent probabilities

$$prob(4,6\,|\,6) = prob(4\,|\,6) \times prob(6\,|\,6) = 1/6 \times 1/6 = 1/36$$

Dependent (or correlated) probabilities. On the other hand, if individual probabilities are dependent on each other, that is, are correlated, the generality of the simple product rule suffers. Assume that the evil die decides to make dependent, or to correlate, the throws by making each subsequent throw equal or larger than the preceding one: if we throw first a 4 (event *A*), the subsequent throw (event *B*) can only be a 4, a 5, or a 6. The probability of throwing a 6 after a 4 is now $1/6 \times 1/3 = 1/18$, and the conditional probability for each joint event changes depending on the first throw. Moreover, the joint probability also changes depending on the sequence of events. Consider the following scenario: if we first throw a 6, the probability of event *B* (throwing a 4), if event *A* is a 6, is zero, and the joint probability is zero as well. We therefore note that for dependent or correlated events the combined outcome $prob(A,B \mid n)$ depends on our knowledge of the first trial—or in other words, knowledge of one data point affects our ability to predict the value of another data point:

$$prob(B \mid A,n) \neq prob(B \mid n) \tag{7-7}$$

Independence is thus a requirement if we plan to use the product of individual probabilities to form a joint probability such as shown in Equation 7-5.

Bias. A biased die would still deliver independent throws, but we would notice after a certain number of throws that one face appears more frequently. The preference for that face will bias our probability distribution toward the number on the face (generally some kind of model in crystallography). As a consequence of the phases dominating electron density reconstruction (Chapter 9), model or phase bias is a common nuisance in crystallographic model building and refinement, and measures to minimize bias are implemented in these situations.

Normalization and sum rule. Mutually exclusive and exhaustive probabilities are summed, providing a means for normalization of the probability function: in our example of the roll of a die, at least one face must show up, and all others not. As one event must occur, the sum of all individual probabilities can be normalized:

$$\sum_{j=1}^{n} prob(x_j \mid n) = 1 \tag{7-8}$$

Just as the logical *and* statement indicated the *product* of probabilities, the summation corresponds to the logical *exclusive or* statement *xor*: we can roll either a 1 *or* a 2 *or* a 3 *or* a 4 *or* a 5 *or* a 6, but only *one* of them will show up with certainty. In the continuous case, we can integrate over the entire probability density function:

$$\int_{-\infty}^{+\infty} prob(x \mid n)dx = 1 \tag{7-9}$$

Confidence interval. The probability that an event *x* with a magnitude between x_1 and x_2 occurs within a given confidence interval is obtained by integrating the PDF over that chosen interval:

$$prob(x_1 \leq x < x_2 \mid n) = \int_{x_1}^{x_2} prob(x \mid n)dx \tag{7-10}$$

Mean or expectation value. The probability weighted average (the *mean* or *expectation value* $\langle x \rangle$ or $E(x)$) of a probability density function is the integral over the quantity weighted by its probability distribution:

$$\langle x \rangle = E(x) = \int_{-\infty}^{\infty} x\, P(x)\, dx \tag{7-11}$$

Marginalization and nuisance variables. Our property *x* can be a function of another variable α whose value we do not explicitly know (generally the phase angle when we are establishing probability distributions for structure factor amplitudes). The probability distribution for our property then becomes $prob(x,\alpha \mid n)$. This variable α, called a nuisance variable, can be eliminated by expressing the desired probability $prob(x \mid n)$ as a summation over all individual probabilities which the discrete variable α can assume in the interval *a, b*:

$$prob(x \mid n) = \sum_{i=a}^{b} prob(x,\alpha_i \mid n) \qquad (7\text{-}12)$$

In practice, the nuisance variable α is often continuous, which allows the transition from the discrete summation to the *integral* (thus the name "integrating out the variable" for the procedure). Assuming for example an interval of $0 \le \alpha \le 2\pi$ (i.e. the whole phase circle), we obtain

$$prob(x \mid n) = \int_{\alpha=0}^{2\pi} prob(x,\alpha \mid n)d\alpha \qquad (7\text{-}13)$$

Integrating out variables will prove itself of great value in the establishment of likelihood functions. A typical example is the derivation of conditional probability distributions for structure factor amplitudes F, which are derived from the distributions of complex structure factors $\mathbf{F}$. We observe (and compare with a phased model) only the experimental structure factor amplitudes F, requiring us to eliminate the explicit phase angle from the probability distribution (Section 7.11).

Bayes' theorem. Rearranging the product rule for probabilities yields an interesting and very useful result. From the trivial statement that if (*A and B*) are true also (*B and A*) must be true

$$prob(A,B \mid n) = prob(B,A \mid n) \qquad (7\text{-}14)$$

follows that

$$prob(A,B \mid n) = prob(A \mid B,n) \times prob(B \mid n) = prob(B \mid A,n) \times prob(A \mid n) \qquad (7\text{-}15)$$

and hence

$$prob(A \mid B,n) = \frac{prob(B \mid A,n) \times prob(A \mid n)}{prob(B \mid n)} \qquad (7\text{-}16)$$

Equation 7-16 is Bayes' theorem, which is of fundamental importance for the logic of inference, and thus also for macromolecular crystallography, as we will briefly introduce here (and cover in detail later): let us just replace A with "*model*", B with "*data*", and let I represent any available conditioning information:

$$prob(model \mid data,I) = \frac{prob(data \mid model,I) \times prob(model \mid I)}{prob(data \mid I)} \qquad (7\text{-}17)$$

In words: The posterior probability $prob(model \mid data,I)$ of our model given the data is the product of the data likelihood function $prob(data \mid model,I)$, or sampling probability indicating the probability that the measured data would be observed given our model, times the prior probability $prob(model \mid I)$ of our model without considering the data. Both *prior* and *posterior* are not meant in a temporal sense here, but refer to our knowledge before and after having examined the data. The denominator $prob(data \mid I)$ is, in the context of parameter estimation (or fitting), only a proportionality constant. In this situation, Bayes' theorem reduces to a basic proportionality:

$$prob(model \mid data,I) \propto prob(data \mid model,I) \times prob(model \mid I) \qquad (7\text{-}18)$$

The posterior $prob(model \mid data,I)$ then becomes the model likelihood (often just called the *likelihood* without qualifier) which is sometimes (for obvious reasons) also called *inverse probability*.[5] We shall return to likelihood and Bayes' theorem and their ubiquitous use in crystallography after examination of a few commonly encountered probability distributions and their descriptive statistics. It will turn out that these apparently disconnected items are in fact derived directly from the maximum likelihood principle.

> Box 7-3 **Probability basics.** Fundamental terms describing probabilities as used in crystallography are: conditioning, joint probability, independence, bias, normalization, and marginalization. Conditioning implies that a property or event is conditioned (dependent on) another parameter. Joint probabilities of independent events are obtained by multiplication of individual probabilities. Mutually exclusive probabilities are added, thus providing a means for normalization. **Marginalization** refers to the removal, or integrating out, of a parameter so that it does not explicitly appear in the probability distribution function. Bayes' theorem results from basic algebra of logic and states that the posterior probability is proportional to the product of the data likelihood function and the prior probability.

For those enjoying a good game of dice, a series of further examples and explanations of definitions are given in the lucid introduction by A. McCoy[1] and in the introductory text by D. S. Sivia.[6]

7.4   Photon counting statistics and Poisson distribution

Already in the very first step of the actual structure determination, during data collection, we are confronted with a statistical process and its probability distribution. The incoming photons from an X-ray source do not arrive in a perfectly timed stream, nor is it determined whether an individual photon gets scattered or not.

We can measure the number of photons arriving at an X-ray detector for say 1 s and record the number of counts n. For a very weak reflection, the probability of counting nothing in 1 s is still large, but for a strong reflection, the probability of measuring zero counts in 1 s is vanishingly low. Each time period, a varying number of counts n will be recorded in each measurement interval. The number of measured counts in each period will deviate from the true or expectation value $\langle n \rangle$ following a certain probability distribution for measuring n counts, given that the mean or expectation value is $\langle n \rangle$:

$$prob(n \,|\, \langle n \rangle) \tag{7-19}$$

The actual form of the probability distribution function (7-19) depends on the number of events and the physical process underlying the events. For a random process of uncorrelated events such as the counting of photons scattered by atoms (with a low individual probability p for each photon, but a large number of scattering events) the probability function $prob(n|\langle n \rangle)$ is a Poisson distribution $P(n)$:[7]

$$prob(n \,|\, \langle n \rangle) = \frac{1}{n!}\langle n \rangle^n \exp(-\langle n \rangle) = P(n) \tag{7-20}$$

Note that the Poisson distribution $P(n)$ depends explicitly on the *a priori* unknown expectation value, $\langle n \rangle$, and we suspect that the shape of the distribution may change depending on the magnitude of that true value $\langle n \rangle$. Note also that this probability distribution is discrete and positive defined, as the number of counts n is always an integer and we can never actually count fewer than 0 counts. In practice we could establish the probability function $P(n)$ as follows: each time period we measure a certain number of counts, we add one into a corresponding *bin* containing the count value. The result of our photon counting and binning experiment is shown in Figure 7-2.

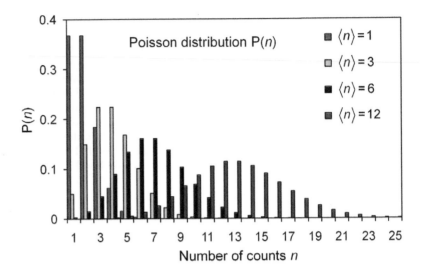

Figure 7-2 The Poisson distribution.
The Poisson distribution is a discrete probability distribution function, applicable (amongst other discrete sampling processes) to photon counting statistics. With increasing number of counts, it approximates in shape the continuous Gaussian normal distribution (Figure 7-3), which can be recognized in the red histogram for $\langle n \rangle = 12$. The ordinate $P(n)$ is normalized so that the sum of the probabilities in all bins equals unity.

The binned histogram of experimental counts shows that the Poisson probability distribution $P(n)$ indeed changes its form with increasing true value of the photon counts. For a larger number of counts, above $n \approx 10$, the Poisson distribution can be approximated by the well-known symmetric Gaussian normal probability distribution familiar from the description of random measurement errors. Some experimental distributions such as the distribution of the B-factors of all atoms in a protein structure model are decidedly not normal; many other probability distributions which we will use later on, such as the Rice and Sim distributions, are also distinctly not normal, but may approach a normal limit under certain conditions.

Absolute and relative error

During an X-ray diffraction experiment we are not collecting each individual reflection the many times needed to establish the actual probability distribution function. Although several instances of the same reflection may be measured independently, in general we measure the counts for an individual reflection only once. Therefore, a most basic question in data collection that we need to address is: given that we measure a certain number of counts n in a given time interval, what is our confidence that a value we measured once does in fact represent the true value $\langle n \rangle$?

The variance σ_n^2 of any probability distribution is defined via the maximum likelihood principle (Section 7.9) as the (arithmetic) *mean squared deviation* of each measurement from its expectation value $\langle n \rangle$:

$$\sigma_n^2 = \langle (n - \langle n \rangle)^2 \rangle = \langle n^2 \rangle - \langle n \rangle^2 \tag{7-21}$$

In the case of the Poisson distribution, $\langle n^2 \rangle = \langle n \rangle^2 + \langle n \rangle$ and thus the variance is $\sigma_n^2 = \langle n \rangle$ and the standard uncertainty, s.u., or standard deviation, SD, defined as the square root of the variance, becomes $\sigma_n = \sqrt{\langle n \rangle}$. Note that here the number under the root is the unknown true value, but one can show[6] that the best estimate for the true $\langle n \rangle$ is indeed our observed value n, and we finally obtain

$$\sigma_n = \sqrt{n} \tag{7-22}$$

For an intensity measurement of 900 counts per second (cps) the absolute error will thus be ±30 cps. From relation 7-22 it immediately follows that the larger the count rate, the smaller the relative error

$$\sigma_{rel}(\%) = 100 \cdot \sqrt{n} / n \tag{7-23}$$

Box 7-4 The Poisson distribution and photon counting error. The Poisson distributionis $P(n|\langle n \rangle)$ a single-parameter probability distribution that only depends on the number of event counts n. With increasing number of counts ($n > 10$), the Poisson distribution approximates the Gaussian normal distribution. The variance σ^2 of the Poisson function is n, and the estimated standard error σ of the Poisson distribution is the square root of n. The Poisson distribution occurs in intensity (photon counting) statistics and also in crystallization probability statistics. The higher a photon count is, the lower its relative measurement error is.

expressed as % of the measured value will be. Inversely, the relative counting error for very weak reflection intensities will be quite high.

The photon counting error obtained via the σ-estimate is only the first step in error estimation. There will be background to subtract and multiple reflections to be averaged or merged. Numerous other experimental random errors as well as systematic errors will contribute to the total measurement error. Reflections are thus generally measured multiple times, that is, with high redundancy, and weighted average intensities and their combined standard deviation are calculated (Section 7.8).

7.5   The normal distribution

We have qualitatively observed in Figure 7-2 that the discrete Poisson probability distribution $P(n)$ becomes for a large number of events n the Gaussian normal probability distribution $G(x)$ of the well-known form

$$G(x) = prob(x\,|\,\sigma_x, \langle x \rangle) = \frac{1}{\sigma_x\sqrt{2\pi}}\, \exp\left(-\frac{\left(x - \langle x \rangle\right)^2}{2\sigma_x^2}\right) = \frac{1}{\sigma_x\sqrt{2\pi}}\, \exp\left(-\frac{1}{2}\,Z^2\right)$$

(7-24)

In the case of the Poisson function, the shape of the distribution $prob(n|\langle n \rangle)$ changed as a function of the sole integer parameter $\langle n \rangle$ describing it. In the case of the normal probability distribution function $G(x|\sigma_x, \langle x \rangle)$ the overall shape of the distribution remains the same independent of the true value $\langle x \rangle$. The width of the distribution, however, depends on its variance defined through the second parameter, the standard uncertainty or standard deviation (SD) σ_x. In order to compute the normal distribution we must know the numeric values for both the mean $\langle x \rangle$ and standard deviation σ_x.

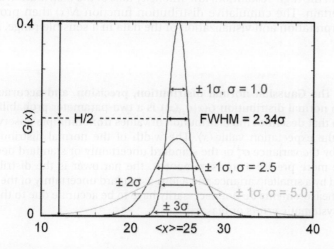

Figure 7-3 The normal distribution. The Gaussian normal distribution is shown for a mean value $\langle x \rangle = 25$ with three different standard deviations σ_x (1, 3, 5 units). The Gaussian normal distribution is symmetric; therefore its mean, mode, and median are the same. For the red 1σ distribution, about 68% of all values will fall into the $\pm 1\sigma$ region, 95% into the $\pm 2\sigma$ region, 99.7% into the $\pm 3\sigma$ region, 99.99% into $\pm 4\sigma$, and 99.9999% fall within $\pm 5\sigma$. In other words, observing a value outside of the $\pm 5\sigma$ confidence interval occurs only once in $\sim 10^6$ cases; that is, it is very unlikely that such a value will be observed. The full width at half maximum (FWHM) covers $\pm 1.17\sigma$ corresponding to a 75% probability that a value will fall in its range (the FWHM is often used to describe functions that do not have a defined variance, such as the Cauchy-Lorentz function in spectroscopy). The percentage of values that fall into a given confidence interval of $\pm n\sigma$ can be computed from the error function or normal error integral, erf(n).

Figure 7-4 Accuracy versus precision.
The left panel shows an imprecise, but accurate measurement, with its red mean value right on the true value indicated by the coordinate intersection in the center. This accurate measurement has a high random error. In contrast, the right panel depicts a quite precise but inaccurate measurement. We expect a large systematic error despite the precision of the measurement.

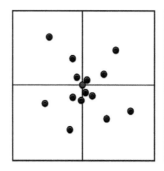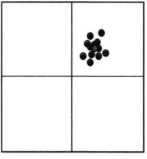

Accuracy versus precision

The interpretation of Figure 7-3 is quite straightforward, demonstrating that the larger the variance σ^2 (or the standard deviation σ), the wider the distribution. *Precise* measurements will thus have narrow distribution of values around the arithmetic mean (Figure 7-4). This does not necessary imply that the mean value is also *accurate*, that is, representing the true mean. Precision is limited by random errors, while accuracy requires the absence of systematic errors, and we need to keep a clear distinction between accurate and precise values in statistical discussions. I can devise an experiment that measures the sum of $2 + 2$ quite precisely as 5.002 ± 0.003. Unfortunately, this precise experiment is highly inaccurate and incompatible with prior knowledge. Note that without (some) prior knowledge of the answer (or target location) there is no way of detecting a systematic deviation in the sampled data.

From Figure 7-3 we also deduce that an experimental observation more than $\pm 3\sigma$ from the mean is quite improbable (i.e., 99.7% of measurements of a random variable will fall into this range), and the probability for a value lying outside of a $\pm 3\sigma$ confidence interval is only 0.3%, that is, one in 333 measurements will lie outside. Note that such a value is not impossible, it is just improbable. Observed values that lie outside of expected error limits (sometimes expressed as a Z-value, giving the number of standard deviations the value deviates from the expected mean) cannot be simply discarded, but should be investigated. Only if either a physical reason for their abnormality is found (stepped on power cord, dropped wrench onto goniostat, residue not in electron density, etc.) or a statistical *outlier test*[8] (or at least some agreed upon criterion) indicates so can, and should, they be removed.

Cumulative probability distributions

A commonly used representation of a probability distribution is by its cumulative distribution function (CDF). This presentation is particularly handy when only a part of the distribution is known. For example, in crystallographic data we generally have a good idea how the strong and well-measured data are distributed, but the high-resolution, low-intensity tails of the distribution are much more uncertain. The cumulative distribution function $N(x)$ then provides an easier interpretation and visualization of the data in a sensible range. A typical

Box 7-5 **The Gaussian normal distribution, precision, and accuracy.** The Gaussian normal distribution $G(x \,|\, \sigma_x, \langle x \rangle)$ is a two-parameter probability distribution that describes the distribution of values from multiple observations around the expectation value $\langle x \rangle$. The width of the normal distribution is defined by the variance σ_x^2 or the standard uncertainty or standard deviation σ_x. The more precise a measurement is, the narrower is the distribution, indicated by a smaller variance and a lower standard uncertainty of the mean. Precise measurements do not necessarily have to be accurate due to the presence of systematic errors or bias.

case where such cumulative distributions are used is the recognition of crystal twinning[9, 10] (Chapter 8). Cumulative distribution functions are commonly annotated as $N(x)$ in order to distinguish them from ordinary probability distributions, written as $prob(x)$ or $P(x)$. Cumulative distributions are obtained by integration of the probability function. $N(x)$ then is a measure for what fraction of data points $\leq x$ are covered by a corresponding probability distribution $P(x)$, or in other words, $N(x)$ is the probability that a data point or random variable will be smaller than a given value of x. For the Gaussian normal distribution $G(x)$, its CDF $N(x)$ is obtained by integration of $G(x)$ from $-\infty$ to x which can be expressed in terms of the error function (erf):

$$N(x) = \int_{-\infty}^{x} G(x)\,dx = \frac{1}{\sigma_x \sqrt{2\pi}} \int_{-\infty}^{x} \exp\left(-\frac{(x - \langle x \rangle)^2}{2\sigma_x^2}\right) = \frac{1}{2}\left[1 + \mathrm{erf}\left(\frac{x - \langle x \rangle}{\sigma_x \sqrt{2}}\right)\right] \qquad (7\text{-}25)$$

The error function gives $N(x)$ the typical sigmoid shape shown in Figure 7-5. The error function $\mathrm{erf}(n)$ itself provides the probability that the error of a single measurement of x lies in the range between $x - n\sigma$ and $x + n\sigma$.

We will encounter cumulative distributions in intensity statistics and use them in Chapter 8 when examining diffraction data for the presence of merohedral twinning.

Sampling and the central limit theorem

If data are normally distributed, their distribution is also symmetric about the mean, and we can readily accept that—given enough random samplings—the sample mean or average $\overline{x}$ will eventually approach the true distribution mean $\langle x \rangle$. This result is not immediately plausible if we sample from a probability distribution that is actually different in shape from a normal distribution. Most real

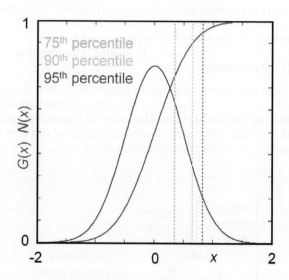

75th percentile
90th percentile
95th percentile

Figure 7-5 Cumulative probability distribution and percentiles. The cumulative probability distribution $N(x)$ (red graph) is shown for a Gaussian distribution $G(x)$ with $x_0 = 0$ and $\sigma = 0.5$ together with percentile ranges, which indicate what percentage of all observable values fall into the range from $-\infty$ up to the corresponding x.

Box 7-6 Cumulative distribution functions. Another way of representing an ordinary PDF is by its cumulative distribution function (CDF), written as $N(x)$ in most crystallographic texts. $N(x)$ is obtained by integration of the PDF, and gives the probability that a data point or random variable following a probability distribution will be smaller than a given value of x. Cumulative distributions are useful for analyzing intensity distributions of diffraction data.

PDFs underlying our observations in crystallography in fact are asymmetric and different from a normal distribution—just take a preview of the acentric Wilson distribution for structure factor amplitudes in Figure 7-10. How can we justify in such cases accepting the sample mean and its variance as an expectation value for the true mean?

The central limit theorem (CLT)—whose propositions can be derived as a special case of the more general, likelihood inference model introduced later—states the somewhat counterintuitive fact that when random samples are drawn from almost any, even decidedly non-normal distributions, the sample mean will in fact approach the true mean of the underlying distribution in the limit of a large number of samples N (Figure 7-6). The variance of the underlying distribution is σ^2_{PDF} related to the variance of the sample mean $\sigma^2_{\bar{x}}$ as $\sigma^2_{\bar{x}} = \sigma^2_{PDF} / N$. The sample size N does not have to be really large to achieve a reasonable estimate for the expectation value $\langle x \rangle$. Similar to the Poisson distribution, the sampling distribution rapidly approaches a normal distribution for about $N \geq 10$.

The practical implication of the CLT is that in many cases when we need to combine probability distributions in crystallography, it suffices to combine the sample variances and mean values instead of convoluting the entire underlying PDFs. There are, however, a few additional caveats that affect the applicability of the CLT, particularly in crystallography:

- The sampling of the distribution must be random. If the samples become dependent or correlated (see the dice example in Section 7.3) the sample mean will not reflect the actual mean of the underlying PDF. In crystallography, measurement errors in certain instances are correlated; for example, if in an anomalous pair the F^+ is affected by some type of error, its Bijvoet mate F^- will likely be affected by the same error. Similar considerations hold for native and isomorphous structures in heavy atom phasing: they are necessarily correlated.

- The distribution must be unimodal. If a distribution has two peaks for example, and these peaks are far apart, the resulting mean and its associated huge variance are very poor descriptors of this distribution. This is the reason why in the case of macromolecular phasing—where we generally find multiple possible solutions for the sought-after phase angle—we need to consider the entire phase probability distribution functions and combine those, not just average their mean values.

- The distribution cannot be otherwise pathological, such as non-integrable (with the Lorentz–Cauchy distribution as a rare example), or with a few huge values dominating the distribution. The latter for example requires detection and removal (or more elegantly, down-weighting) of outliers in measured intensities in difference data for substructure solution.

Figure 7-6 The central limit theorem. The central limit theorem states that the sample mean drawn at random from almost any parent PDF will be normally distributed, given enough trials. The more samplings, the narrower and more normal the distribution of the sample mean becomes, irrespective of the shape and variance of the actual sampled PDF. The green sample mean distribution is not quite normal yet, but the red distribution is near normal.

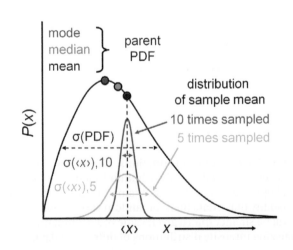

The general variance equation (7-21) is derived on the assumption that the true value $\langle x \rangle$ is known. This is not exactly the case; we only know the best estimate for x in the form of the computed sample mean value $\bar{x}$ as the average of N independent measurements. In this case, the best estimate of the variance of the mean $\sigma_{\bar{x}}^2$ and its square root, the standard uncertainty $\sigma_{\bar{x}}$, changes, explicitly written, to

$$\sigma_{\bar{x}}^2 = \frac{N}{N-1}\sigma_x^2 \quad \text{or} \quad \sigma_{\bar{x}} = \sqrt{\frac{1}{N-1}\sum_{j=1}^{N}(x_j - \bar{x})^2} \tag{7-26}$$

For large N, $N-1$ approximates N and (7-26) becomes identical to (7-21).

7.6   Descriptive statistics used in crystallography

Given the ubiquitous applicability of the CLT, it is not surprising that a large number of statistical descriptors are actually derived from or related to the normal distribution. The following summarizes these descriptors and points out some of their applications in crystallography. Be sure to distinguish between the descriptors of the underlying PDF (or population) that we sample from, and the descriptors of the distribution of the sample mean. These are distinctly different terms although their definitions may look the same!

Arithmetic mean. The expectation value of almost any probability distribution function that is randomly sampled experimentally N times is readily computed as the arithmetic mean:

$$\langle x \rangle = \bar{x} = \frac{1}{N}\sum_{j=1}^{N}x_j = \sum_{j=1}^{N}\langle x_j \rangle \quad \text{with} \quad \langle x_j \rangle = x_j/N \tag{7-27}$$

Residual. The difference between any expectation value $\langle x \rangle$ and an individual measured value x_j is the residual or *deviate*:

$$r_j = x_j - \langle x \rangle \tag{7-28}$$

Variance of any distribution. The variance σ_{PDF}^2 of any probability distribution function is defined as the mean squared deviation of each value from its expectation value:

$$\sigma_{PDF}^2 = \langle (x - \langle x \rangle)^2 \rangle = \langle x^2 \rangle - \langle x \rangle^2 = \langle r_j^2 \rangle = \frac{1}{N}\sum_{j=1}^{N}(x_j - \langle x \rangle)^2 \tag{7-29}$$

Variance of the sample mean. The variance $\sigma_{\bar{x}}^2$ of the mean of the samples we drew from the parent PDF is the maximum likelihood estimate[11] defined as the mean squared deviation of each measurement from the sample mean $\bar{x}$. For a population sample, one single measurement allows no conclusion at all about the variance. Hence, the denominator becomes $N-1$ instead of N (compared to Equation 7-29) and we obtain for the variance $\sigma_{\bar{x}}^2$ of the sample:

$$\sigma_{\bar{x}}^2 = \frac{1}{N-1}\sum_{j=1}^{N}(x_j - \bar{x})^2 \tag{7-30}$$

The variance of the sample mean $\sigma_{\bar{x}}^2$ relates to the variance of the sampled parent distribution (see Figure 7-6) as

$$\sigma_{\bar{x}}^2 = \frac{1}{N}\sigma_{PDF}^2 \tag{7-31}$$

Standard deviation. The standard deviation (SD) σ or standard uncertainty (s.u.) is the square root of the variance. We obtain for standard deviation $\sigma_{\bar{x}}$ of the sample

$$\sigma_{\bar{x}} = \sqrt{\frac{1}{N-1}\sum_{j=1}^{N}(x_j - \bar{x})^2} \qquad (7\text{-}32)$$

RMS deviation. The root mean square deviation or RMSD of sampled values from the mean of a distribution is equivalent to the square root of the variance:

$$RMSD = \langle r_j^2 \rangle^{1/2} \qquad (7\text{-}33)$$

The RMSD is also computed for other situations where the mean or expectation value used in the computation of the residuals $\langle r_j^2 \rangle$ is derived from a target distribution which is not necessarily the mean of the sample distribution. This is the case when the RMSD of bond lengths and angles of a structure model from restraint targets is calculated. The sample RMSD is then generally smaller than or equal to the standard deviation of the target distribution. For the superposition of two structures, only two molecules make up the sample, and the residual is simply the difference between the coordinates of each corresponding atom pair. The interpretation of the RMSD as a standard deviation is then meaningless.

Sum of residuals squared. The sum of residuals squared (SRS)

$$SRS = \sum_{j=1}^{N}(x_j - \langle x \rangle)^2 \qquad (7\text{-}34)$$

plays an important role in the context of minimization and crystallographic refinement. A process that minimizes the variance or sum of (normalized) residuals squared between measured data and a parameterized model will provide the best fit between model and data. This is the basis for parameter estimation or parameter refinement. The SRS is directly related to Chi-squared.

Chi-squared. The χ^2 measure or the sum of normalized residuals squared is frequently used in parameter estimation (crystallographic refinement):

$$\chi^2 = \sum_{j=1}^{N}\left(\frac{x_j - \langle x \rangle}{\sigma_x}\right)^2 = \sum_{j=1}^{N} Z_j^2 \qquad (7\text{-}35)$$

For normally distributed data, χ^2 is of the order of ~N. It can therefore also be used to test statistical models, generally in the form of the reduced Chi-squared measure.

Reduced Chi-squared. The χ_R^2 statistic

$$\chi_R^2 = \frac{\chi^2}{N-P} = \frac{1}{N-P}\sum_{j=1}^{N}\left(\frac{x_j - \langle x \rangle}{\sigma_x}\right)^2 \qquad (7\text{-}36)$$

is employed in statistical tests deciding whether a distribution is a normal distribution. Here P is the number of parameters needed to describe the model distribution, that is, $N - P = D$, the number of degrees of freedom. For large numbers N and a normal distribution (two parameters), $\chi_R^2 \simeq 1$ if the experimental errors indeed are normally distributed. If $\chi_R^2 \gg 1$, the mean values are questionable; if $\chi_R^2 \ll 1$, the variances are suspect. The reduced Chi-squared is thus used in data integration and scaling to fine-tune the error model.

Z-scores. The normalized residuals Z are a statistical standard score and represent the deviation of the measured value from the distribution mean in units of the standard deviation. Z is negative when the observed value is below the mean and positive when it is above:

$$Z_j = \frac{x_j - \langle x \rangle}{\sigma_x}$$

The probability that a value lies outside a given range from $-Z$ to Z is obtained from the error function. For $Z \pm 1$: 32%; ± 2: 5%; ± 3: 2%; ± 4: 0.01%; ± 5: 0.0001%. This means that only one in 10 000 observations will lie outside $Z = \pm 4$ or outside of 4σ.

RMSZ scores. When many Z-values have been calculated, as is the case for deviations from target values, they themselves follow a distribution. The RMSZ score

$$RMSZ = \sqrt{\frac{1}{N}\sum_{j=1}^{N} Z_j^2} \tag{7-37}$$

is a normalized standard score and thus equals 1 when the target and sample distribution are identical. Sample distributions narrower (tighter) than the target distribution lead to RMSZ < 1, while wider (looser) distributions lead to RMSZ > 1. Note the similarity to the χ^2-test described above.

Mean absolute error. The mean absolute error

$$MAE(x) = \frac{1}{N}\sum_{j=1}^{N} |x_j - \langle x \rangle| = \langle |x - \langle x \rangle| \rangle \tag{7-38}$$

of normally distributed data and its relation to the standard deviation

$$\frac{MAE(x)}{\sigma_x} = \sqrt{\frac{2}{\pi}} = 0.79788 \quad \text{or} \quad MAE(x) = \sqrt{\frac{2}{\pi}} \cdot \sigma_x \tag{7-39}$$

allows us in Section 8.3 to derive that for the linear merging R-value R_{merge} defined in Section 7.8

$$R_{merge} \simeq 0.8 / \langle |I| / \sigma(I) \rangle \tag{7-40}$$

and as a consequence,

$$R_{merge}(noise) \quad 0.8 / 0.8 = 100\%$$

Variance of a normalized structure factor distribution. The variance of a structure factor distribution is annotated with the symbol Σ (capital sigma) and is defined by the sum of all atomic scattering factor contributions squared. It relates to the expectation value of the intensity and is corrected by the ε-factor (Chapter 6) when Σ is derived from experimental intensities:

$$\Sigma_N = \sum_{j=1}^{N} f_j^2 = \frac{1}{\varepsilon_h}\langle F^2 \rangle = \frac{1}{\varepsilon_h}\langle I \rangle \tag{7-41}$$

7.7 Integration of probability distributions

The proof[12] of the relation $MAE(x) = (2/\pi)^{1/2}\sigma_x$ may serve as an exercise how to obtain expectation values from PDFs. To obtain the expectation value of x we need to integrate over the variable weighted by its corresponding probability distribution (7-11):

$$\langle x \rangle = \int_{-\infty}^{\infty} x\, P(x)\, dx \tag{7-42}$$

In our case we need to integrate over the Gaussian normal distribution

$$G(x) = \frac{1}{\sigma_x \sqrt{2\pi}} \exp\left(-\frac{(x - \langle x \rangle)^2}{2\sigma_x^2}\right)$$

The integration then takes the form

$$\langle |x - \langle x \rangle| \rangle = \int_{-\infty}^{\infty} |x - \langle x \rangle| \, G(x) \, dx = \frac{1}{\sigma \sqrt{2\pi}} \int_{-\infty}^{\infty} |x - \langle x \rangle| \exp\left(\frac{-(x - \langle x \rangle)^2}{2\sigma_x^2} \right) dx$$

The first step in solving this type of integral is by substitution of the normalized residual

$$t = \frac{(x - \langle x \rangle)}{\sigma_x} \quad \text{and} \quad dx = \sigma_x \, dt$$

which leads to

$$\langle |x - \langle x \rangle| \rangle = \frac{\sigma_x}{\sqrt{2\pi}} \int_{-\infty}^{\infty} |t| \exp\left(\frac{-t^2}{2} \right) dt$$

Given the symmetry about the mean for $G(x)$ one can simplify the integration range

$$\langle |x - \langle x \rangle| \rangle = \frac{2\sigma_x}{\sqrt{2\pi}} \int_{0}^{\infty} t \exp\left(\frac{-t^2}{2} \right) dt$$

and by application of the chain rule we obtain for the defined integral

$$\langle |x - \langle x \rangle| \rangle = \sigma_x \sqrt{\frac{2}{\pi}} \left[-\exp\left(\frac{-t^2}{2} \right) \right]_{0}^{\infty}$$

Evaluating the integration boundaries leads to the result

$$\langle |x - \langle x \rangle| \rangle = \sigma_x \sqrt{\frac{2}{\pi}} (0 - (-1)) = \sigma_x \sqrt{\frac{2}{\pi}}$$

from which, using definition (7-38) of the *MAE*, immediately follows

$$\frac{MAE(x)}{\sigma_x} = 0.79788 \tag{7-43}$$

7.8   Covariance and basic error propagation

Given the propositions of the central limit theorem, the general case for the uncertainty $\sigma(f)$ for any function of n variables $f(\mathbf{x})$ where $\mathbf{x} = (x_1, x_2, \ldots x_n)$ with individual uncertainties of $\sigma(x_i)$ is given by

$$\sigma(f) = \left(\sum_{i=1}^{n} \sum_{j=1}^{n} \left(\frac{\partial f}{\partial x_i} \cdot \frac{\partial f}{\partial x_j} \cdot C_{i,j} \right) \right)^{1/2} \tag{7-44}$$

where $C_{i,j}$ is the covariance of each pair of variables. Covariance is also an important concept in crystallographic refinement and essentially describes, as the name says, how one variable changes with variation of the other. The general error propagation formula (7-44) with its partial derivatives looks frightening, but a few examples will put the mind at ease.

Take as an example a measurement x_1 of 900 counts s^{-1} with an estimated random counting error of ±30 cps or 3.3% (square root of 900). This peak may sit on top of a separately measured background x_2 of 100 cps, with a random error of ±10 cps. If the errors are limited to random photon counting statistics, the two measurements x_1 and x_2 contributing to the total intensity are independent, and thus mathematically uncorrelated; the summation in (7-44) requires only the variances $\sigma^2(x_i) = C_{i,i}$ or their square root, the standard deviation $\sigma(x_i)$, for each measurement:

$$\sigma(f) = \left(\sum_{i=1}^{n} \left(\frac{\partial f}{\partial x_i} \cdot \sigma(x_i) \right)^2 \right)^{1/2} \tag{7-45}$$

The error propagation formula (7-45) for uncorrelated variables greatly simplifies our case. The partial derivatives of f are easy to determine, because for adding measurements, $f = x_1 + x_2$, and both $\partial f / \partial x_1$ and $\partial f / \partial x_2$ are equal to 1. For the difference function $x_1 - x_2$ we obtain the derivatives 1 and –1, respectively. As the partial derivatives are squared in (7-45), the error propagation for either the sum or the difference of measurements, $x_1 \pm x_2$, is thus given as

$$\sigma_{x_1 \pm x_2} = \left(\sigma_{x_1}^2 + \sigma_{x_2}^2 \right)^{1/2} \tag{7-46}$$

For our example of a 900 cps peak on a 100 cps background, the background-corrected peak intensity is 800 ± 32 counts s^{-1}, which increased the relative error to 4%. It is also quite possible that for weak reflections the background estimate (measured left and right of a weak peak in high background) can be higher than the peak and we obtain negative intensities. For example, a peak of 4 ± 2 and a background of 9 ± 3 gives an intensity of -5 ± 3.5 cps. Our suspicion that the negative intensities I will cause trouble when converting to structure factor amplitudes F by taking the square root $F = I^{1/2}$ is justified, and we will cover the treatment of negative intensities in connection with Bayes' theorem in Section 7.9. Note, however, that refining against intensities (i.e. F^2)—already available in few programs such as *SHELXL*—avoids this complication.

Converting intensities to structure factor amplitudes

During data processing we need to convert intensities I into structure factor amplitudes F. Given that I and its standard uncertainty $\sigma(I)$ are known, we can readily compute F as the square root of I, but what is the estimated error $\sigma(F)$ for the structure factor amplitude? We again use the error propagation formula (7-45) and set $f(x) = I = F^2$:

$$\sigma(F \cdot F) = \left(\left(\frac{\partial F^2}{\partial F} \cdot \sigma(F) \right)^2 \right)^{1/2} = \left((2F \cdot \sigma(F))^2 \right)^{1/2} = \left(4F^2 \cdot \sigma^2(F) \right)^{1/2} \tag{7-47}$$

Expanding with $1/(F \cdot F) = 1/I$ leads to

$$\frac{\sigma(I)}{I} = \left(\frac{4F^2 \cdot \sigma^2(F)}{F^4} \right)^{1/2} = \frac{2\sigma(F)}{F} \tag{7-48}$$

which further simplifies to

$$\sigma(F) = \frac{1}{2} \cdot \frac{\sigma(I) \cdot F}{I} = \frac{1}{2} \cdot \frac{\sigma(I)}{\sqrt{I}} \tag{7-49}$$

stating that the relative error in the structure factor amplitudes F is only half as big as the relative error in the measured intensities I. We can now convert from intensities to structure factor amplitudes simply by rewriting (7-49) as follows:

$$F = \sqrt{I} \pm \frac{1}{2} \cdot \frac{\sigma(I)}{\sqrt{I}} \tag{7-50}$$

Using Equation 7-50, our background-corrected reflection with $I = 800$ cps, $\sigma(I) = 32$ cps, and a relative error of 4% will give a structure factor amplitude F of 28.3 with $\sigma(F)$ of 0.6, and a relative error of 2%.

The I-to-F conversion procedure given by (7-50) of course fails the moment we encounter zero or negative intensities. Neither the mean value nor the error estimate can be computed. In addition, Equation 7-50 also gives unrealistically

Sidebar 7-1 A first treatment for the small or negative intensity problem. A first empirical fix for the problem caused by negative intensities is to use the approximation

$$(F + \sigma(F))^2 = I + \sigma(I)$$

from which follows the quadratic equation

$$(\sigma(F))^2 + 2F\sigma(F) - \sigma(I) = 0$$

which solved for $\sigma(F)$ finally yields

$$\sigma(F) = (I + \sigma(I))^{1/2} - F$$

This expression works for zero intensity and for negative intensities with high error, while giving a less inflated variance $\sigma(F)$ for small values of I. It is implemented in the CCP4 program *TRUNCATE* when the more advanced French and Wilson treatment described later is not selected.

high values for $\sigma(F)$ as soon as I becomes very small. Unfortunately, negative intensities for very weak or absent reflections are not uncommon in protein work as a result of background subtraction. What are we supposed to do when we encounter negative reflection intensities? Just ignoring such reflections only throws away information, because we know that this reflection has very likely at least no large value. Ignoring the negative sign and taking the square root as structure factor amplitude is improper as well, because it lends increasing significance to increasingly improbable values. As we will understand shortly, the problem is caused by the inappropriateness of the central limit theorem assumptions in this case, and requires more sophisticated Bayesian error estimates[13] when converting from intensities $I_{\mathbf{h}}$ to structure factor amplitudes $F_{\mathbf{h}}$.

R-values and correlation coefficients

During the course of a crystallographic structure determination we frequently encounter the need to compute a measure for the fit of structure factor amplitudes calculated from models to the experimental observations. A general measure that is frequently used in connection with intensities and structure factors is a linear residual, called the *R*-value, where *R* stands for *reliability* or better *residual index*. *R*-values for a quantity or function *F* have the general form

$$R_F = \frac{\sum_{i=1}^{n} |F(i)_{\text{data}} - F(i)_{\text{model}}|}{\sum_{i=1}^{n} F(i)_{\text{data}}} \tag{7-51}$$

where $F(i)_{\text{data}}$ and $F(i)_{\text{model}}$ stand for observed and calculated data on a common scale (see Section 7-10). For a perfect fit with no difference between calculated and observed values, the *R*-value is zero, while with increasing difference between the compared quantities, *R* increases. For example, the *R*-value measuring the fit of structure factor amplitudes of a refined protein structure model against the observed amplitudes might be 0.20, or expressed as percentage, 20%. For a non-centrosymmetric random atom structure, we expect, based on the structure factor probabilities described in Section 7.10, an R_F-value of ~59%, and ~83% for a centrosymmetric random atom structure.[9] Note that these *R*-values depend on a proper scale factor *k* to be meaningful. For *R*-values of quantities of the same data set (such as anomalous Bijvoet pairs F^+ or F^-) or already scaled data, *k* is often 1. However, in cases such as comparison of initial and partial trial structures with experimental maps or data, we often lack a reliable scale factor.

Box 7-7 Error propagation and basic conversion of intensity to structure factor amplitudes. Error propagation rules lead to two important results for the manipulation of intensities. If an intensity value is combined from two independently measured variables such as peak and background counts, the errors propagate via the sum of their individual variances:

$$\sigma_{x_1 \pm x_2} = \left(\sigma_{x_1}^2 + \sigma_{x_2}^2 \right)^{1/2}$$

The relative error in structure factor amplitudes when converted from intensities is only $1/2$ of the relative error in intensities:

$$F = \sqrt{I} \pm \frac{1}{2} \cdot \frac{\sigma(I)}{\sqrt{I}}$$

The basic error propagation formula is, however, insufficient for cases of very small, zero, or negative intensities. Bayesian error models allow a more accurate estimate of structure factor amplitudes from small or negative intensities.

In cases where we cannot properly establish a common scale for model data and observed data, the (linear) correlation coefficient, CC, which is scale-independent, comes to the rescue. The correlation coefficient CC is defined in the range of $-1 \leq CC \leq 1$, with -1.0 indicating a perfect anticorrelation between the data, and 1.0 perfect correlation. The derivation is straightforward from error propagation, and substituting the explicit variances into the definition

$$CC = \sigma_{xy} / \sigma_x \cdot \sigma_y \qquad (7\text{-}52)$$

we obtain

$$CC = \frac{\sum_{i=1}^{n}(x_i - \bar{x}) \cdot (y_i - \bar{y})}{\left(\sum_{i=1}^{n}(x_i - \bar{x})^2 \cdot \sum_{i=1}^{n}(y_i - \bar{y})^2\right)^{1/2}} \qquad (7\text{-}53)$$

where the bar symbol indicates the arithmetic mean values.

The summations in (7-53) extend over each value of the functions to be correlated, for example, the electron density values at each grid point $\mathbf{r}$ in the asymmetric unit of an electron density map. The real space correlation coefficient (RSCC) between an "experimental" electron density map $\rho(\mathbf{r})_{obs}$ (computed from figure-of-merit weighted observed amplitudes and experimental or model phases) and the calculated model density map $\rho(\mathbf{r})_{cal}$ (computed from the model phases and calculated amplitudes) is a very useful measure during model building and model validation.[14] The real space correlation coefficient is according to (7-53) defined as

$$RSCC = \frac{\sum_{\mathbf{r}}(\rho(\mathbf{r})_{obs} - \overline{\rho(\mathbf{r})_{obs}}) \cdot (\rho(\mathbf{r})_{cal} - \overline{\rho(\mathbf{r})_{cal}})}{\left(\sum_{\mathbf{r}}(\rho(\mathbf{r})_{obs} - \overline{\rho(\mathbf{r})_{obs}})^2 \cdot \sum_{\mathbf{r}}(\rho(\mathbf{r})_{cal} - \overline{\rho(\mathbf{r})_{cal}})^2\right)^{1/2}} \qquad (7\text{-}54)$$

A good model typically has overall real space correlation coefficients greater than 0.90, but the RSCC varies for each residue, depending on how well each residue of the model fits the actual electron density map. A residue-by-residue plot of the real space correlation coefficient is thus a very powerful and quick measure to judge the local quality of a structure model,[15] as shown in Chapters 12 and 13.

One final remark regarding the R-value: The linear residual is actually an odd statistic; you will not find anything like an R-value in a "real" statistics textbook. The insufficiency of this basic linear residual becomes painfully obvious in the merging R-values for intensities, where by introduction of variance-like terms (the variance is a *bona fide* likelihood measure) somewhat more meaningful definitions have been derived (Chapter 8).

Box 7-8 **R-values and correlation coefficients.** Measures frequently used in crystallography for relative errors are the R-value and the correlation coefficient. While meaningful R-values require a common scale between the compared or merged data, the correlation coefficient is scale-independent and can be used when the absolute scale of the data to be correlated is unknown. A wide variety of different R-values and correlation coefficients are used in crystallography, depending on the type of data to be merged or compared.

Weighted averages and merging of observations

In crystallographic diffraction intensity data sets, the same reflection or its symmetry mates are generally measured multiple times. The redundant measurements increase precision, and if the measurements occur in different diffraction geometry, also the accuracy of the measurement is likely improved because of compensation of systematic errors.

Assume we measured the same reflection three times, as 900 ± 30 cps (the notation placing the estimated uncertainty in parentheses is equivalent), 1000(32), and 1100(50) cps, resulting in an arithmetic mean of 1000 cps. This is different from the sampling case, where we had no individual error estimate for any given single observation. Now the simple arithmetic mean is not the best estimate, because the relative error of our highest measurement is larger than for the other two measurements. The question is what is the best estimate and its error for combined independent measurements of the same quantity?

The best estimate for n independent measurements is the weighted average, where the individual weights w_i are given by the inverse of the variance: $w_i = 1/\sigma_i^2$:

$$\langle x \rangle = x_{best} = \sum_{i=1}^{n} w_i x_i \Big/ \sum_{i=1}^{n} w_i \tag{7-55}$$

with the uncertainty of our best value obtained via error propagation from the individual weights as

$$\sigma_{best} = \left(\sum_{i=1}^{n} w_i \right)^{-1/2} \tag{7-56}$$

The weighted average as an estimate for the best value is a fundamental result in statistics and derives directly from the fact that the derivative of the weighted sum of residuals at the peak of the probability distribution is zero; that is, this value has the maximum likelihood[7] to occur (Figure 7-8). For our example (900(30), 1000(32), and 1100(50) cps) we obtain 971(20) for the best value and its associated uncertainty, showing that the relative error of the best value is reduced to ~2% compared to the individual measurement errors of ~3% to ~5%. Note that in this particular case the difference between each of the measurements is larger than their individual standard deviation. This indicates that some source of systematic error in the measurement could be present.

A commonly used quality indicator for diffraction data which we will frequently encounter in future sections is the linear merging R-value, R_{merge}, for combining N observations of reflection $\mathbf{h}$:

$$R_{merge} = \frac{\sum_{i=1}^{N} \left| I_{(\mathbf{h})i} - \langle I_{(\mathbf{h})} \rangle \right|}{\sum_{i=1}^{N} I_{(\mathbf{h})i}} \tag{7-57}$$

For our example of merging three symmetry related reflections and using the best value of 971 for $\langle I_{(\mathbf{h})} \rangle$, we obtain for the linear $R_{merge} = 0.076$ or 7.6%.

Merging R-values are generally computed for the entire data set or in resolution shells. The summation then extends over all reflections $\mathbf{h}$ that lie within the respective resolution range, and the linear merging R-value becomes

$$R_{merge} = \frac{\sum_{\mathbf{h}} \sum_{i=1}^{N} \left| I_{(\mathbf{h})i} - \bar{I}_{(\mathbf{h})} \right|}{\sum_{\mathbf{h}} \sum_{i=1}^{N} I_{(\mathbf{h})i}} \tag{7-58}$$

> **Box 7-9 Best estimates for independent measurements.** The best estimate x_{best} for a value obtained or expected from n independent multiple observations is the weighted average $\langle x \rangle$:
>
> $$\langle x \rangle = x_{best} = \sum_{i=1}^{n} w_i x_i \Big/ \sum_{i=1}^{n} w_i \quad \text{with} \quad \sigma_{best} = \left(\sum_{i=1}^{n} w_i \right)^{-1/2}$$
>
> Weighted averages play an important role in crystallography in data merging as well as in establishing phase probability distributions and best phases.

However, this basic merging R-value (7-58) is statistically on shaky grounds, and more sophisticated merging statistics such as the multiplicity-weighted R-value R_{meas}[16] and the precision indicating merging R-value R_{pim}[17] have been developed and are reported by data processing programs. We will discuss the use and meaning of various merging R-values in the "Fundamentals of data collection" section in Chapter 8.

7.9 Likelihood and Bayesian inference

In several previous chapters we have already made use of statistics and concerned ourselves with the probability of certain events occurring. We were asking questions such as "What is the probability that a certain count of photons represents the true value?" The answer to this question was a well-defined probability distribution function $prob(n|\langle n \rangle)$ known as the Poisson distribution. We asked similar questions for the probability of crystallization $prob(C|D,I)$ of a certain protein given perhaps some initial screening data D and specific other prior information such as hydrophobicity index, isoelectric point, or solubility, and our quest was for a conditional probability of crystallization given some general conditioning information. In this ill-defined multivariate case, the answer to our quest unfortunately was not a neat function, but a number of hand-waving arguments and guesses.

In crystallization prediction as well as in the treatment of negative intensities we found that classical frequency-based statistics do not cope well with—and generally ignore—the inclusion of independent prior knowledge in the estimate of probabilities. Crystallography, however, because it concerns itself with real matter, can draw from a wealth of prior information based on the molecular nature of its objects. An elegant approach to introducing prior knowledge is based on Bayesian inference, a general branch of statistics where prior information is explicitly used in the formulation of probability distribution functions and maximum likelihood target functions. In view of its ubiquitous use in modern crystallography, we provide here a basic introduction focusing on specific crystallographic applications. For deeper study, refer to the easily readable tutorial by D. S. Sivia,[6] the more comprehensive text by E. T. Jaynes,[18] and the particularly lucid introduction to likelihood by A. W. F. Edwards.[19] Implementation of Bayesian frameworks to protein crystallography was pioneered, beginning in the late 1970s and 1980s, by S. French,[20] G. Bricogne,[21] R. Read,[22] and others.

Deductive versus inductive reasoning

An example demonstrating the difference between deductive logic and inductive inference is the following:

1. Heme-containing crystals are red.

2. This crystal is a heme-containing crystal.

We can clearly state from the premises, via deduction, that this particular heme-containing crystal must be red. In other words, given that the crystal contains heme, it must be red. Contra factum argumentum non facit.

However, in everyday inductive inference, the problem we are faced with is usually different, in the form of the probability of a red crystal also being a heme-containing crystal, *prob(heme|red)*, and formulated inductively:

1. Heme-containing crystals are red.

2. This crystal is red.

The importance of prior information for answering whether the red crystal is also a heme-containing crystal immediately becomes quite relevant. Short of analyzing that particular crystal, the single experiment does not inform us in any way about the probability that the red crystal indeed contains heme. There is, however, already some important information in the experiment, namely from statement (1) we know that if the crystal was not red, it could not be a heme-containing crystal. The probability *prob(red|heme)* that a heme-containing crystal is red is in fact 1.

To solve the problem, we need to determine the posterior probability *prob(heme|red)* of a red crystal indeed containing a heme group, given prior information—independent of this particular measurement—which we have about red protein crystals in fact being heme crystals. This prior probability of a red crystal containing heme—independent or unconditional of our present experiment—*prob(heme)*, perhaps acquired through independent analysis of a large data set of random crystals, shall be 0.8, that is, 8 out of 10 red crystals are in fact heme-containing crystals.

The posterior probability or likelihood *prob(heme|red)* that our crystal is a heme-containing crystal thus depends on *prob(red|heme)* that a heme-containing crystal is red and the independent prior probability *prob(heme)* that a red crystal is in fact a heme-containing crystal. As these probabilities are independent, we can form according to (7-18) the basic joint probability product

$$prob(heme \mid red) = prob(red \mid heme) \times prob(heme) = 1.0 \times 0.8 = 0.8$$

The posterior probability that our red crystal is a heme-containing crystal is 80%. We can also immediately understand why *prob(heme|red)* is a posterior probability: we know only after conducting the experiment that the crystal is in fact red. Assume the case that the crystal is green: then the experiment clearly shows, because the probability of observing a green crystal containing heme is 0, that

$$prob(heme \mid green) = prob(green \mid heme) \times prob(heme) = 0.0 \times 0.8 = 0.0$$

We just solved a common everyday problem of crystallographic inference by simple Bayesian reasoning. We shall now explore the concepts of likelihood and Bayesian inference—both indispensable parts of modern macromolecular crystallography—in more detail. It is unfortunately often the case that in more complicated situations the exact probability distribution functions are hard to establish. Nonetheless, we commonly base our everyday estimates of the outcome of our actions heavily and quite reliably on prior knowledge. We just accept that an open flame is very likely hot; we do not need to get burned every time to find out.

Bayes' theorem

The formal solution to the fundamental problem of how to implement (in a mathematically tractable form) deductive logic in the context of inductive infer-

Box 7-10 **Bayesian inference.** Bayesian inference or reasoning is a generally applicable procedure that allows the treatment of inductive logic in a formal deductive framework based on fundamental rules of probability algebra.

ence was provided in 1763 by Reverend Thomas Bayes in England.[23] Although Bayes' theorem was largely forgotten in the following two centuries, and for rather sophisticated technical reasons denounced in the mid-1920s, it was revived in 1946 through Richard Cox's work, which demonstrated that Bayes' theorem can be derived straightforwardly from the sum and product rules for probabilities (as derived in Section 7.3) and Boolean logic.[6] Formulated in a general context of evaluating a probability for occurrence of an event H (or the probability of a hypothesis H or a *model*) in view of our actual data D and the conditioning information I we possess, Bayes' theorem reads as

$$prob(H \mid D, I) = \frac{prob(D \mid H, I) \times prob(H \mid I)}{prob(D \mid I)} \tag{7-59}$$

Bayes' theorem (7-59) essentially states that the posterior probability $prob(H|D,I)$ of our hypothesis or model being true, given our data D and conditioning information I, depends directly on the key part in the formula, the data likelihood function $prob(D|H,I)$ and the independent prior probability $prob(H|I)$ of our hypothesis without having analyzed the data. While it is readily understood what prior probability means (for example, the statement "80% of all red crystals are heme crystals" qualifies), we need to clarify what the data likelihood function implies. The likelihood function $prob(D|H,I)$ is a measure of how likely it is that the data are actually observed, given our current hypothesis (or model in parameter estimation). As we can see from the product in the numerator of (7-59), $prob(D|H,I) \times prob(H|I)$, the data likelihood function can also be interpreted as a modifier for our prior belief for the hypothesis. Weak data will not overcome a strong prior probability or knowledge, but a strong experiment can either modify or affirm our prior beliefs.

The product of the data likelihood function and prior probability is normalized by the marginalization constant $prob(D|I)$ which takes on different forms depending on the context in which we use Bayes' formula. If we apply Bayes' theorem to the case of hypothesis testing, the marginalization term is the sum of the probabilities of all alternate hypotheses

$$prob(D \mid I) = \sum_{i=1}^{hypotheses} prob(D \mid H_i, I) \times prob(H_i \mid I) \tag{7-60}$$

In the case of parameter estimation, the marginalization term $prob(D \mid I)$ is essentially a normalization constant that can be figured out *a posteriori* (if needed at all) and thus is included as a proportionality factor in the posterior probability (because the peak of the probability function $P(x)$ remains at the same value of x, regardless of any multiplicative terms $nP(x)$). Bayes' formula then simplifies to

$$prob(H \mid D, I) \propto prob(D \mid H, I) \times prob(H \mid I) \tag{7-61}$$

This is the form of Bayes' equation we are most frequently dealing with in crystallography. The posterior probability—which we can generally compute via the data likelihood function together with the prior—is then a measure for our sought-after model likelihood $L(H \mid D, I)$:

$$L(H \mid D, I) \propto prob(D \mid H, I) \times prob(H \mid I) \tag{7-62}$$

Absence of prior knowledge. It is also possible that we do not have any or only very limited clues about the prior probability of our hypothesis or model. Then the prior probability function is uniform (essentially a flat line) and is *de facto* only a proportionality constant. In addition, if there are variables that need to be evaluated but do not contribute to the variation of the likelihood function, so-called nuisance variables, they can be integrated out and also absorbed in the proportionality constant. In this case Bayes' equation reduces to

$$L(H \mid D, I) \propto prob(D \mid H, I) \tag{7-63}$$

Sidebar 7-2 Prior knowledge keeps weak data in check; and strong data modify beliefs. One of the most powerful aspects of Bayesian models is to keep weak data in reasonable bounds using prior probabilities or knowledge. Weak or poor crystallographic data will not overcome established prior expectations, but strong data (evidence) can force us to revise and expand our prior expectations about molecular structure and function, and even lead to a scientific revolution (you may note the sociological implication of Bayesian reasoning in Thomas S. Kuhn's theory of scientific paradigm change).[24]

A historical example of a small scientific revolution resulting from strong crystallographic evidence overcoming ingrained but incorrect prior expectations is the discovery of the four-membered β-lactam ring, evident for the first time in the molecular structure of penicillin determined by Dorothy Crowfoot Hodgkin and colleagues in 1949. Contemporary organic chemists in Oxford forcefully resisted her notion that a four-membered ring of three carbon atoms and a nitrogen atom could be quite stable. It is reported that one famous chemist threatened "to give up science and grow mushrooms" if that proposed chemical formula for the β-lactam was ever shown to be correct. To the detriment of mushroom farming, he did not keep this promise.[25]

Sidebar 7-3 Terminology of likelihood functions and probabilities. Be aware that the statistical literature as well as crystallographic texts sometimes use the term *likelihood* without qualifier for both the data likelihood function $prob(D|H,I)$ which is a function of the *data*, and the (posterior) model likelihood $L(H|D,I)$, which is a function of the *model* (and what we are actually interested in). Note also that the proportionality in Equation 7-61 implies that the model likelihood is not necessarily normalized, in contrast to a true probability. When we form likelihood ratios for purposes of hypothesis (model) testing, this subtle distinction becomes irrelevant—as long as the underlying data and thus $prob(D|I)$ is the same (compare Sidebar 12-10). The same holds for optimization: the most probable parameter values will remain the same, irrespective of any proportionality connecting likelihood and probability.

Parameter optimization. In crystallography we often wonder which set of parameters of a parameterized model provides the best model, and we are interested in parameter estimation (or optimization, refinement, or fitting of a structure model to the data). Given the basic proportionality in likelihood equations (7-61) and (7-63), maximizing $prob(D|H,I)$ by adjusting the model parameters also maximizes our model likelihood $L(H|D,I)$, that is, provides the best model. This is the basis of maximum likelihood methods in crystallography, and we will expand on the topic quite extensively in the chapters on phasing (Chapter 10), molecular replacement (Chapter 11), and refinement (Chapter 12). We shall now examine model optimization based on the maximum likelihood principle in more detail.

Note that the quest for the optimal set of parameters for a given parameterization of a model is not the same as the quest for the best physical model (which may be differently parameterized). The former is a question of minimization, and the second is one of hypothesis testing, generally determined through the likelihood ratio of the competing hypotheses (explained in more detail in Sidebar 12-10).

Maximum likelihood

We will concern ourselves now with the implication of Bayes' theorem for parameter estimation (or model fitting), beginning with the posterior probability for a single parameter. Bayes' theorem connects the posterior probability distribution (or in short the *posterior*), the (data) likelihood function, and the prior probability (or the *prior*) through the following proportionality:

$$prob(model \mid data, I) \propto prob(data \mid model, I) \times prob(model \mid I) \qquad (7\text{-}64)$$

where *model* now stands for some parameterized (structure) model of interest, whose fit to a set of (diffraction) *data* we are interested in. In general, we can calculate the right-hand terms for specific models, which then provide us with the sought-after (posterior) likelihood of the model $L(model|data, I)$.

Maximum posterior versus maximum likelihood

Bayes' formula joins two fundamental probabilities defining the posterior probability: the data likelihood function and the independent prior probability. In many cases, for example substructure heavy atom refinement, we have no significant prior knowledge that would *a priori* restrict our model of the heavy atom sites (beyond perhaps a minimum distance between atoms). In such cases of no prior assumptions, the prior $prob(model|I)$ is only a constant and thus can be ignored or absorbed in the proportionality constant (see Sidebar 7-4 for a caveat). Irrespective of the presence of an independent prior, our data likelihood function $prob(data|model, I)$ will be conditioned on certain assumptions I about our model, such as incompleteness or errors in the model. In the future, we will omit explicit notation of this conditioning information for simplicity.

An actual protein structure model is significantly limited in its parameter freedom as it needs to reasonably comply with all (all!) known laws of nature such as

Box 7-11 Bayes' theorem. Bayes' theorem states that the posterior probability of a hypothesis or model is given by our prior information or knowledge about it, modified by the data likelihood function which is a measure how likely the experimental data are reproduced given our hypothesis:

$$prob(H \mid D, I) = \frac{prob(D \mid H, I) \times prob(H \mid I)}{prob(D \mid I)}$$

stereochemical restraints and chemical plausibility. In the case of model refinement, we thus do have significant prior information that allows us to judge the model independently of looking at the data. The prior $prob(model|I)$ thus will be a very important part (reality check) in the determination of our posterior probability or model likelihood. A completely nonsensical model with correspondingly low prior probability that fits the data well still has a low likelihood of being correct.

Because of the fundamental proportionality between model likelihood $L(model|data)$ and the data likelihood function $prob(data|model)$, any procedures that maximize the data likelihood function are called *maximum likelihood* methods. Strictly speaking, methods that maximize the joint probability $prob(data|model) \times prob(model)$ including the prior are termed *regularized maximum posterior methods*, but the term *maximum likelihood* methods is also used if the prior term—generally as a prior likelihood term in the sense of a regularization term—is included in the joint probability. Technically this distinction does not require different treatment.

Crystallographic models and conditioning information

Let us first examine what a model or hypothesis, the data, and the conditioning information I actually are in a crystallographic context. The experiment delivers data that consist of structure factor amplitudes, F_{obs}, and the mathematical model, which is generally represented as a set of calculated structure factors, $\mathbf{F}_{calc}$, that are functions of the adjustable parameters such as coordinates and B-factors of the underlying atomic structure model. The first step in establishing the data likelihood function as a product of the individual structure factor probabilities is developing an appropriate error model for conditional probabilities of one structure factor ($\mathbf{F}_{obs}$) conditioned on another ($\mathbf{F}_{calc}$). This is most easily understood in the basic conditional likelihood case: given a calculated structure factor computed from the the model parameters, what is the probability of observing an experimental measurement (the data) of that value?

The corresponding basic data likelihood function for an experimental structure factor $\mathbf{F}_{obs}$ given a model defined by $\mathbf{F}_{calc}$ can then simply be written as

$$prob(data\,|\,model, I) = prob(\mathbf{F}_{obs}\,|\,\mathbf{F}_{calc}) \tag{7-65}$$

To actually develop the (often quite complicated) conditional probability distribution for a specific scenario, two major points must be considered:

- The model structure factors can be computed because we know the model coordinates, which directly provide the phases. These model structure factors are complex. In refinement, however, we maximize the fit between the scalar properties F_{obs} and F_{calc}, and the phase must be marginalized to obtain the practically useful likelihood function $prob(F_{obs}|F_{calc})$. The marginalization or elimination of the phase from the likelihood function can be readily done by integrating out the phase angle, and we will exercise this explicitly in the following sections.

- The conditioning information I, which we noted explicitly as a reminder in the data likelihood function, is accounted for in the complex (also in the

Sidebar 7-4 The problem of ignorance in the Bayesian method. Some statisticians are critical of Bayes' approach on sophisticated grounds that center around the fact that complete ignorance in principle does not imply a flat prior, which actually is a somewhat stronger statement of equal ignorance over all values of $P(H|I)$. While this argument and more intricate details which can be found in A. W. F. Edwards' text[19] are under certain circumstances relevant, it generally does not affect the application of Bayesian methods in crystallography.

Box 7-12 Maximum likelihood. The highest posterior probability indicating the best model can be found by maximizing the data likelihood function in maximum likelihood cases where no prior probability term is present, or by maximizing the joint probability of the data likelihood function and the prior probability in maximum posterior models and Bayesian models.

sense of complex number) error model for the structure factors, largely as consideration of measurement errors, positional errors, and incompleteness of the model. This is a key point: in maximum likelihood models we can explicitly account for incompleteness or errors in our model or hypothesis, an option not (as) readily available in the normal-distribution-based error model underlying basic least squares minimization models.

In addition, through the prior probability term, we have the possibility of including independently obtained prior information about the probability of the model itself, which modifies our resulting posterior probability. The separate prior term provides an additional powerful means to keep the posterior in reasonable, physically determined bounds. Both the acceptance of errors and incompleteness in the model and the inclusion of prior information—largely as a prior likelihood term in the form of stereochemical plausibility of the model—in the calculation of the posterior probability are of significant importance in crystallography.

Another example where the prior probability provides a means to impose physically reasonable restraints will be encountered in the computation of structure factor amplitudes from negative intensities, which cannot be treated with the classical error propagation rule. Structure factor amplitudes always have to be positive; in the worst case, the probability for a scattering event is zero, but it can never be negative.

Log-likelihood

Irrespective of how a specific probability distribution function $prob(x)$ looks, as long as it is a reasonable and continuous function, any proportional function $n \cdot prob(x)$ will have its maximum at the same value of x. The same holds for the logarithm of $prob(x)$, and from our basic likelihood equation trivially follows for the log-likelihood LL

$$LL = \ln(prob(model \,|\, data)) \propto \ln(prob(data \,|\, model)) \tag{7-66}$$

The natural $\log_e$ (ln) is selected instead of $\log_{10}$ because many probability density functions are related to the exponential Gaussian normal distribution and $\ln[\exp(x)] = x$. In this context we still use the term log-likelihood, not ln-likelihood.

Figure 7-7 of an actual crystallographic likelihood function (the Wilson distribution for acentric structure factors on an arbitrary scale) and the corresponding log-likelihood function emphasize this point.

The point of taking the logarithm is, first, that individual probability values can become very small, and they are numerically more easily handled when using the log-likelihood, or LL. Second, multiplication of joint probabilities reduces to simple logarithmic addition. Following the product rule for independent probabilities, the total likelihood function L, which we seek to maximize in crystallographic refinement to obtain the best fit between data and model, is the joint probability for all individual structure factor probabilities:

Box 7-13 Maximum likelihood and error models. Maximum likelihood (ML) methods play a key role in protein crystallography, because in many instances the assumption of a normal distribution of random errors in data and models is not fulfilled, and central-limit-based methods such as least squares do not provide optimal solutions. Bayesian and maximum likelihood methods account for correlation (dependence) in data, as well as non-normal errors and incompleteness in models, and are thus implemented in a wide range of crystallographic programs used in phasing, data processing, map reconstruction, and macromolecular refinement.

$$L \propto \prod_{\mathbf{h}} prob(F_{\mathbf{h}(obs)} \mid F_{\mathbf{h}(calc)}) \tag{7-67}$$

or in logarithmic form,

$$LL \propto \ln \prod_{\mathbf{h}} prob(F_{\mathbf{h}(obs)} \mid F_{\mathbf{h}(calc)}) = \sum_{\mathbf{h}} \ln prob(F_{\mathbf{h}(obs)} \mid F_{\mathbf{h}(calc)}) \tag{7-68}$$

For parameter estimation or fitting, many useful minimization algorithms exist, and instead of maximizing LL, we can just minimize the negative log-likelihood to find the most probable value of our posterior:

$$\max\left[L(model \mid data)\right] \propto \min\left[-LL\right] = \min\left[-\sum_{\mathbf{h}} \ln prob(F_{\mathbf{h}(obs)} \mid F_{\mathbf{h}(calc)})\right] \tag{7-69}$$

We will encounter the value of $-LL$ and the log-likelihood gain as a statistic in a number of crystallographic maximum likelihood-based programs such as *SHARP*,[26] *BUSTER/TNT*,[27] *PHASER*,[28] *CNS*,[29] or *REFMAC*.[30] The fundamental equation stating that the log-likelihood can be represented as the sum of the individual log-probabilities and that their maximum will thus also represent a maximum of the entire likelihood is the so-called *absolute criterion* formulated and justified by R. A. Fisher in 1922.[11]

As we noted, the reason why we frequently employ maximum likelihood targets in crystallographic work is quite simple: The classical least squares minimizations rely on the central-limit-based assumption of normal distribution of independent random errors. While this is quite defendable and valid for a complete and correct substructure solution or for a nearly perfect protein structure model that is just being polished, this is certainly not the case for bad or incomplete initial models. The deviation from the assumption of random error and model completeness affects heavy atom and phase refinement (Chapter 10) as well as molecular replacement searches (Chapter 11), electron density reconstruction (Chapter 9), and model refinement protocols (Chapter 12).

Approximation of log-likelihood functions

To generalize and simplify the mathematical treatment of the—generally complicated—likelihood function (formulating a proper likelihood function is often the most difficult part when implementing ML methods), we follow the standard procedure of expanding the log-likelihood LL into a Taylor series. For simplicity, we replace for now the multiparametric model with a simple variable x and extend to the multivariate case later. In the single parameter case, the Taylor expansion of $f(x)$ around x_0 is

$$f(x) = \sum_{n=0}^{\infty} \frac{1}{n!} \cdot \frac{\partial^n f(x_0)}{\partial x^n} \cdot (x - x_0)^n \tag{7-70}$$

We apply the Taylor expansion to a log-likelihood function around its maximum, $LL(x_0)$, and terminate the series at the second order term:

$$LL \cong LL(x_0) + \frac{\partial LL(x_0)}{\partial x}(x - x_0) + \frac{1}{2} \cdot \frac{\partial^2 LL(x_0)}{\partial x^2}(x - x_0)^2 \tag{7-71}$$

The first summand $LL(x_0)$ is just a constant value for the maximum which we do not know yet; it is therefore just a proportionality constant. In the second term, the first derivative (the slope of the likelihood function) is of course zero around its maximum x_0, and the whole linear term disappears. The critical remaining term is thus the second order term, which must be negative for a maximum. The second derivative or curvature is the change of the slope, and if we travel over a positive peak, the slope constantly becomes less, and the curvature $\partial^2 f(x)/\partial x^2$ is negative (Figure 7-1).

Let us now simplify this even more and think what we have actually accomplished by the Taylor series termination. Given that the curvature term at the

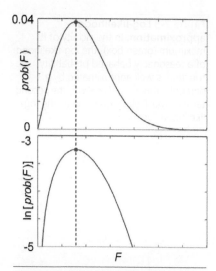

Figure 7-7 A likelihood function and its logarithm. The acentric Wilson distribution (an unconditional probability distribution or likelihood function for structure factors) and its natural logarithm. Note that the logarithmic representation—in addition to simplifying multiplication of probabilities to addition and increasing numeric stability—also focuses on the "business end" of the function, namely the region around maximum probability or maximum likelihood (red dot). The maximum occurs at the same value of F for both the linear and logarithmic representation.

Figure 7-8 Log-likelihood approximation. In the vicinity of its maximum (green box), the log-likelihood of a reasonably behaved probability function is well approximated by a parabolic function $-k(x-x_0)^2$, the second term of the Taylor expansion of the log-likelihood function.

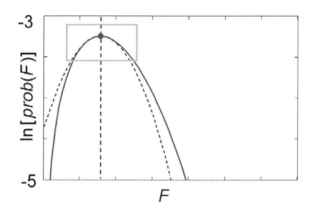

maximum is just another (negative) constant in the univariate case, we can reduce the log-likelihood function (7-71) further to

$$LL \cong -\frac{1}{2}k(x-x_0)^2 \qquad\qquad (7\text{-}72)$$

The graph of this function is nothing but an inverted parabola centered on x_0; we have just replaced the complicated (log-) likelihood function with a simple parabolic approximation, which is actually quite good as long as we are reasonably close to x_0, as Figure 7-8 illustrates.

Let us now examine the same situation in terms of a non-logarithmic representation. With the first derivative zero and $LL(x_0)$ a constant, our likelihood function (7-72) can now be written as proportional to the exponent of the second order Taylor term

$$L \cong \exp\left(-\frac{1}{2}k\cdot(x-x_0)^2\right)=\exp\left(\frac{1}{2}\cdot\frac{\partial^2 LL(x_0)}{\partial x^2}(x-x_0)^2\right) \qquad (7\text{-}73)$$

This function has a somewhat familiar feel: If we set

$$\sigma_x=\left(-\frac{\partial^2 LL(x_0)}{\partial x^2}\right)^{-1/2} \quad\text{then}\quad L \cong c\cdot\exp\left(-\frac{(x-x_0)^2}{2\sigma_x^2}\right) \qquad (7\text{-}74)$$

Now let

$$c=\frac{1}{\sigma_x(2\pi)^{1/2}} \qquad\qquad (7\text{-}75)$$

and we obtain the familiar Gaussian normal distribution

$$L \cong \frac{1}{\sigma_x(2\pi)^{1/2}}\cdot\exp\left(-\frac{(x-x_0)^2}{2\sigma_x^2}\right) \qquad\qquad (7\text{-}76)$$

Relation 7-76 is a very nice result as it shows that the normal distribution is nothing but a simplified approximation of a general likelihood function: we have successfully approximated our likelihood function L in the vicinity of x_0 with a normal distribution (Figure 7-9). It also allows us to describe the variance for *any* parameter x as $x = x_0 \pm \sigma$ just as in the special case of the normal distribution.

Maximum likelihood principle, least squares, and central limit theorem
In addition, we immediately grasp the concept of maximum likelihood (or least

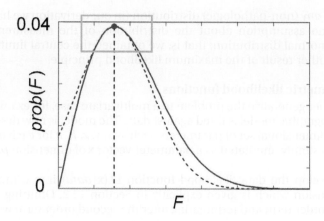

Figure 7-9 **Principle of maximum likelihood and least squares.** In the vicinity of the maximum, a probability distribution can be approximated by a normal distribution. The probability function is largest when the squared residual in the exponent is minimal, hence the name *least squares*. In the general maximum likelihood case, the functions in the exponent of the second Taylor term can be quite complicated, but the same principle applies.

squares) refinement from Relation 7-76. The data likelihood function as well as the posterior probability or model likelihood will have the maximum (most probable) value if the exponent of the likelihood function

$$\exp\left(-\frac{(x - x_0)^2}{2\sigma_x^2}\right) \tag{7-77}$$

is largest, that is, exactly when the squared residual (or for more than one data item the sum of squared residuals) is minimal.

The case of a single data point can be readily expanded. Assume we have measured N data points x_k; the combined likelihood function L will then—always provided the individual probabilities (and thus the measurements and errors) are independent—be the product of the N individual data likelihood functions $L_k(x_k|x_0)$. We bring any terms leading the exponent into a constant c

$$L_k = c_k \cdot \exp\left(-\frac{(x_k - x_0)^2}{2\sigma_{x(k)}^2}\right) \tag{7-78}$$

and multiply the individual likelihoods. The product of exponential terms then becomes an exponential sum and we obtain the final result for the complete model likelihood L:

$$L \propto \prod_k L_k = K \exp\left(-\sum_{k=1}^{N}\frac{(x_k - x_0)^2}{2\sigma_{x(k)}^2}\right) \tag{7-79}$$

The product constant K in (7-79) can also be absorbed in the proportionality, and we see that the posterior probability or model likelihood L will be at its maximum when the data likelihood function is at its maximum. This is the case when the exponent is largest, or the normalized sum of residuals squared (*SRS* or Q) is minimal. The least squares minimization is thus a special case of a maximum likelihood method. The key difference is that in least squares we have explicitly assumed a normal error model; we have assumed that the model value x_0 is invariably correct, and only the observations x_k were weighted according to their experimental variance $\sigma_{x(k)}^2$. In the general maximum likelihood case, we have the option to develop any suitable error model for the model parameters (i.e. our $\mathbf{F}_{calc}$). Most of the time developing the proper error models for the likelihood functions is not trivial, but it is crucial for the success of ML methods.

The central limit theorem. By setting x_0 in the likelihood equation (7-79) to a mean of multiple measurements, we have also derived that independent sampling of *any* (non-pathologic) distribution (in our derivation we have made absolutely no assumption about the distribution of the measurements x_k) follows the normal distribution; that is, we obtained the central limit theorem (CLT) as another result of the maximum likelihood principle.

Multi-parametric likelihood functions

We can further generalize the problem to a multivariate case, for example when our crystallographic model is fitted against data. The model is now described by a multi-(p) dimensional set of parameters such as x, y, z, B, n for each atom, and the model is simply annotated as a parameter vector **x** of dimension p.

We now develop the data likelihood function $LL(\mathbf{x}|data)$ into a multivariate Taylor expansion which is given explicitly in Section 12.2. Omitting the zero-valued first order term and terminating after the second order term we obtain

$$LL \cong \frac{1}{2} \cdot \sum_{i=1}^{p} \sum_{j=1}^{p} \frac{\partial^2 LL(\mathbf{x}_0)}{\partial x_i \partial x_j}(x_i - x_{0,i})(x_j - x_{0,j}) \tag{7-80}$$

Using short matrix notation for the second derivative matrix $\nabla^2 LL(\mathbf{x}_0)$ we obtain an exponential quadratic form:

$$LL \cong \frac{1}{2}\left((\mathbf{x} - \mathbf{x}_0)^\mathrm{T} \cdot \nabla^2 LL(\mathbf{x}_0) \cdot (\mathbf{x} - \mathbf{x}_0)\right) \tag{7-81}$$

The likelihood function (7-81) now represents a multidimensional probability vertex, and the parameter uncertainty is given by the $N \times N$-dimensional covariance matrix $\boldsymbol{\sigma}_{ij}^2$ defined as

$$\boldsymbol{\sigma}_{i,j}^2 = -\left(\nabla^2 LL(\mathbf{x}_0)\right)^{-1} \tag{7-82}$$

The point of this general derivation of the multivariate posterior probabilities is to have them available in the practical part for the minimization of multivariate problems in crystallography, that is, the refinement of model coordinates and parameters. All we need to do in principle is to adjust the parameters, so that the condition $\nabla LL(\mathbf{x}_0) = 0$ is met; that is, the partial derivatives of the log-likelihood function are zero, and therefore LL as well as L is at the peak of its likelihood, at maximum likelihood. Various different algorithms exist, such as the classical least squares, steepest gradients, or stochastic optimizations such as evolutionary algorithms or Monte Carlo and simulated annealing algorithms, to achieve maximum likelihood by minimization of the negative log-likelihood, $-LL$.

Optimization of crystallographic log-likelihood

We turn now to the case of multivariate optimization of a crystallographic likelihood function. For simplicity, we assume a simple least squares function, but the principle is the same for any more complicated likelihood function. Our data items are now a set of N observed structure factor amplitudes $F_\mathbf{h}(obs)$ with corresponding variance $\sigma_{F(\mathbf{h})}^2$. The expectation value is the calculated structure factor

---

Box 7-14 Maximum likelihood by minimization of negative log-likelihood. Minimizing the negative log-likelihood instead of maximizing the likelihood function has a number of advantages. The very small individual structure factor probabilities making up the joint probability product of the entire likelihood function are numerically easier to handle by logarithmic addition. In addition, various proven minimization algorithms such as conjugate gradient, steepest descent, or stochastic algorithms such as simulated annealing already exist that can be used to carry out the actual computation. A smaller residual implies a larger exponential term in the likelihood function or a log-likelihood gain in the logarithmic picture.

amplitude $F_h(calc)$, itself a function of p parameters (generally the coordinates and B-factors of each atom) that make up the parameter vector $\mathbf{x}$: $F_h(calc) = f(\mathbf{x}, \mathbf{h})$. The likelihood function for each structure factor then is

$$L_h = k \cdot \exp\left(-\frac{\left(F_h(obs) - F_h(calc)\right)^2}{2\sigma^2_{F_h(obs)}}\right) = k \cdot \exp\left(-\frac{1}{2}\left(\frac{F_h(obs) - F_h(calc)}{\sigma_{F_h(obs)}}\right)^2\right) \quad (7\text{-}83)$$

For N observed structure factors, the joint likelihood becomes

$$L = K \cdot \exp\left(-\frac{1}{2}\sum_h^N\left(\frac{F_h(obs) - F_h(calc)}{\sigma_{F_h(obs)}}\right)^2\right) \quad (7\text{-}84)$$

Substituting the χ^2-residual

$$\chi^2 = \sum_h^N\left(\frac{F_h(obs) - F_h(calc)}{\sigma_{F_h(obs)}}\right)^2 \quad (7\text{-}85)$$

into (7-84) we obtain for the likelihood

$$L = K \cdot \exp\left(-\frac{\chi^2}{2}\right) \quad (7\text{-}86)$$

and for the log-likelihood

$$LL = k' - \frac{\chi^2}{2} \text{ or } -LL = -k' + \frac{\chi^2}{2} \quad (7\text{-}87)$$

In the logarithmic picture, an increase in the exponential term (smaller residual, higher likelihood) then corresponds to a log-likelihood gain or *LLG*.

The derivation above reinforces that minimizing the normalized sum of residuals squared, χ^2, will maximize the log-likelihood (or minimize $-LL$). Numerically this can be achieved—as soon as the number of observations N at least equals the number of parameters p—by calculating the analytical expressions for all the first derivatives with respect to the parameters that determine our model, setting them to zero, and setting up a corresponding system of linear normal equations (the so-called design matrix). The linear system can then be solved by a variety of system solvers.[31, 32] Optimization algorithms for refinement are explained in detail in Chapter 12. For now, there are two points to emphasize:

- Recall that basic least squares is not always the appropriate crystallographic minimum function because we have not allowed or accounted for any non-normal errors or incompleteness in the model. We will develop these more general likelihood functions accounting for incompleteness and complex errors in models in the next sections.

- Although a linear system of equations becomes numerically stable for $N = p$, this does not mean that the solution makes much sense in terms of fitting the model to the data when the system is not reasonably overdetermined, that is, we have multiple times more data than adjustable model parameters. The importance of the data-to-parameter ratio N/P is discussed in Chapter 12 in more detail.

Bayesian treatment of negative intensities

Let us examine a practical application of Bayesian inference for the case of negative intensities. We have already realized that we run into a serious problem in converting from intensities to structure factor amplitudes when they become zero or negative:

$$F = \sqrt{I} \pm \frac{1}{2} \cdot \frac{\sigma(I)}{\sqrt{I}} \qquad (7\text{-}88)$$

We also agreed that deleting the negative reflections discards valuable information, while taking the square root of the absolute intensity lends increasing significance to increasingly improbable values. A more reasonable approach is to set intensities to zero, because we know that they cannot be negative. In fact, this is already an application of some very naive Bayesian reasoning. Unfortunately, application of the prior of positivity alone does not solve the problem that we still have no reliable error estimate $\sigma(F)$ for the zero structure factor amplitudes.

Why are accurate error estimates so important? We already know that the data likelihood function explicitly depends on the error estimates in the form of the conditioning information I for the data: $prob(data\,|\,\mathbf{x}, I)$, where $\mathbf{x}$ is again a multivariate model or proposition whose likelihood we are interested in. It is therefore imperative for the structure determination procedures such as searches or refinement that reasonable error estimates are available. Depending on the detector and integrating and scaling software, up to about 2–5% of intensities in a data set can become negative and would be lost if not handled properly.

The treatment of negative intensities was one of the first areas in protein crystallography where Bayesian models were applied.[13] The data-processing program *TRUNCATE* in the CCP4[33] program suite, which we will use in practical examples, is an implementation of Bayesian inference in protein crystallography. The approach that French and Wilson[13] have taken is to base the prior probability distribution of the structure factor amplitudes on the statistical expectation values derived by Wilson (Section 7.10), which automatically imposes positivity on structure factor amplitudes and thus on intensities, and provides also adequate error estimates.

While the formulas derived by French and Wilson for the posteriors of I, F, and their standard errors are somewhat elaborate, we can examine a simpler model without a distinction between centric and acentric cases under the basic assumption of positivity and a normal error distribution proposed by D. S. Sivia.[6] The posterior probabilities of F and σ_F are derived as the first and the second derivative of the likelihood function, respectively, as we exercised for the general case in the previous section. Using the results[34]

$$F = \frac{1}{2}\left(2I + \left(4I^2 + 8\sigma_I^2\right)^{1/2}\right)^{1/2} \quad \text{and} \quad \sigma_F = \left(\frac{1}{F^2} + \frac{2(3F^2 - I)}{\sigma_I^2}\right)^{-1/2} \qquad (7\text{-}89)$$

we can compute Table 7-1 and see how the Bayesian posteriors compare with the standard square-root formula (7-88) we derived from the classical error propagation.

Comparison of the results computed for the Bayesian posteriors (7-89) with the classical expression (7-88) shows that for intensities where $I \gg \sigma(I)$, the Bayesian posterior is practically identical with the classical estimate given by the standard square-root formula (7-88). In contrast to the square-root formula, the Bayesian posterior still gives small positive intensities for negative values, albeit with an increasingly higher error estimate. The same holds for the structure factors, and we obtain positive structure factor amplitudes, again with a significant error. Nevertheless, this is a more reasonable estimate than negative, zero, or simply omitted reflections, and provides us with error estimates for weak and unreliable reflections.

There are of course details that even the more elaborate original French and Wilson derivation cannot account for. If the assumptions (prior beliefs) that went into the Wilson statistics (Section 7.10) are not fulfilled, as is the case for heavy atoms in special positions or high pseudo-symmetry (due to special NCS)

I(obs)	σI(obs)	I(Bayes)	σI(Bayes)	F(obs)	σF(obs)	F(Bayes)	σF(Bayes)
−16.0	8.0	1.8	3.0	—	—	1.34	0.90
−9.0	6.0	1.7	2.3	—	—	1.30	0.85
−4.0	4.0	1.5	1.8	—	—	1.21	0.76
−2.0	2.8	1.2	1.5	—	—	1.10	0.66
−1.0	2.0	1.0	1.2	—	—	1.00	0.58
0.0	2.0	1.4	1.4	0.00	—	1.19	0.59
1.0	2.0	2.0	1.6	1.00	1.00	1.41	0.58
2.0	2.8	3.2	2.4	1.41	1.00	1.80	0.66
4.0	4.0	5.5	3.6	2.00	1.00	2.34	0.76
9.0	6.0	10.7	5.2	3.00	1.00	3.27	0.84
16.0	8.0	17.8	7.6	4.00	1.00	4.22	0.90
25.0	10.0	26.9	9.7	5.00	1.00	5.18	0.93
49.0	14.0	50.9	13.7	7.00	1.00	7.14	0.96
100.0	20.0	102.0	19.8	10.00	1.00	10.10	0.99
1000.0	63.0	1002.0	63.3	31.62	1.00	31.65	1.00

Table 7-1 Posterior probabilities for structure factors. Given are the I(obs) data and their standard error in the left two columns. F(obs) and σF(obs) are derived from the classical square-root formula 7-88, which fails to give results for negative intensities (red). For values with $I \gg \sigma(I)$, the Bayesian posteriors give practically the same results as the standard square-root formula (green). For values where the experimental error approaches the measured value (yellow), the Bayesian posterior begins to deviate but maintains small but positive values for I and F, with appropriately increasing error estimates (or decreasing degree of belief). The intensity values I(obs) and σI(obs) have been selected so that the standard error for the structure factors computes to 1.0 using the classical square-root formula.

in the data, the posterior estimates will also not be entirely correct. In general, however, the Bayesian posteriors are in almost any case superior to just tossing out negative or weak intensities, and they also provide the much-needed error estimates for subsequent, likelihood-based procedures in electron density reconstruction and structure refinement.

Are you crystallography-positive?

We finish this strenuous section with an entertaining example of using Bayes' theorem in the context of hypothesis testing. The example is borrowed from population biology and adapted to crystallography, and it strikingly demonstrates the effect (and caveats) of introducing prior knowledge into the inference process.

Assume a genetic marker exists for a terrible disease, namely the irresistible urge to carry out crystallographic studies. This is arguably a most debilitating

Box 7-15 Bayesian treatment of negative intensities. Negative intensities can result during background subtraction from very small intensity values. Applying statistical expectation values based on Wilson structure factor amplitude statistics as a Bayesian prior imposes the necessary positivity on intensities, and sensible structure factor amplitudes and standard uncertainties can be computed from negative intensities, thus keeping valuable information about those otherwise unmanageable reflections.

anomaly, and you want to be tested for that condition (maybe a cure can be found through structure-based drug discovery). The test is 99.9% double accurate, meaning that only 1 in 1000 cases will be either a false positive or a false negative. After two weeks of anxious waiting and sleepless nights, the shocking result is received that you are crystallography-positive (C^+), apparently with 99.9% certainty.

A few days after this social death sentence has been delivered, you begin to wonder what, truly, is the probability of a solitary life in front of a dimly lit synchrotron hutch, given the test result. You recall that prior probabilities may modify the posterior probability of your hypothesis that you are C^+. You decide to evaluate the probability of the hypothesis H that you are C^+, given the likelihood of the data D (the positive test) and the prior probability of being affected. Research of the American Crystallographic Association member directory shows that out of the population of 55 million people in your eligible demographic group, only 2200 in fact became registered crystallographers. The prior probability for ending up C^+ in the absence of the test is thus 1/25000 or 0.00004 or 0.004%. To obtain the posterior probability for our hypothesis of actually being C^+ after being tested positive (hypothesis H_+) we need to evaluate the following Bayesian posterior for hypothesis testing:

$$prob(H_i \mid D, I) = \frac{prob(D \mid H_i, I) \times prob(H_i \mid I)}{\sum\limits_{i=1}^{hypotheses} prob(D \mid H_i, I) \times prob(H_i \mid I)} \qquad (7\text{-}90)$$

Let us begin to evaluate the terms of Equation 7-90 from right to left. The prior probability of being C^+ without having been tested, $prob(H_+ \mid I)$, is easily determined; we know it is 0.00004. The likelihood $prob(D \mid H_+, I)$, that is, that our data (the test result) is correct for the hypothesis that we really are C^+ given (based on) the conditioning information of the test reliability is 0.999.

The denominator of (7-90) is essentially a normalization given by the sum of the probability of all alternate hypotheses. We already have $prob(D \mid H_+, I) \times prob(H_+ \mid I)$, so we need to formulate the same expression for the alternate hypothesis $prob(D \mid H_-, I) \times prob(H_- \mid I)$. The alternate hypothesis to our present one of being C^+ based on the test is that you are C^- given (despite) the test. The likelihood for $prob(D \mid H_-, I)$, that is, that you tested positive despite being a C^- genotype, is given by the false positive rate, which is 0.001. The prior probability of being C^-, $prob(H_- \mid I)$, is $1 - prob(H_+ \mid I)$ and thus 0.99996.

Plugging all the numbers into Bayes' formula, we obtain

$$prob(H_+ \mid D, I) = \frac{0.999 \times 0.00004}{0.999 \times 0.00004 + 0.001 \times 0.99996} = 0.0384$$

The result proves that, considering prior knowledge, your probability of being crystallography-positive has been reduced from a shocking 99.9% to a mere 4%. This means that despite the positive test, there is a good chance that you will get away with a scare and will not end up spending countless nights at synchrotrons or setting up millions of futile crystallization trials. After a second positive test, however, the probability climbs to 98% (which is still less than the 99.9% for the single test accuracy).

Caveat. The observant reader may have already discovered that there is a small problem with our reasoning—reading this book may introduce a risk factor that requires an adjustment to the prior probability, significantly increasing your chances of a random walk in reciprocal space. We therefore should adjust our prior probability of ultimately becoming a crystallographer perhaps by a factor of 10. In this case, the probability to be C^+ after being tested positive jumps to 28%, still less than the test would make one believe, and retesting would be required to confirm the diagnosis. So we clearly see that including all available and correct prior information—to the degree it is known to us—is necessary to obtain valid Bayesian estimates. This fact has given cause to the criticism of subjectivity in Bayesian inference, but it is evident that while the initial low estimate may have conveyed a false sense of security, even the risk-adjusted result is much less alarming than the unconditional belief in the single test result.

As a general rule, Bayesian models recover much faster from uniform priors reflecting a state of ignorance than they recover from plausible but incorrect priors.[6] The example in this section should have driven across the point of how important it is to include all available and reliable prior information in inference-based reasoning and hypothesis testing. Bayesian inference provides the instrument to include useful prior knowledge in the likelihood estimates for many of our crystallographic computations in data processing, phasing, and refinement. Keep in mind, however, that the universal, all-overriding GIGO principle holds no matter how sophisticated our statistical analysis: Garbage In, Garbage Out.

7.10 Unconditional structure factor probability distributions

Distributions of structure factor amplitudes F and their expectation values $\langle F \rangle$ play an important role in crystallographic computing. On many occasions we need to bring data sets containing I_{obs} and the corresponding structure factor amplitudes F_{obs} and/or calculated structure factor amplitudes F_{calc} onto a common scale. For example, data can be sourced from different crystals, from native and derivative data sets, and experimental and calculated structure factors need to be scaled for refinement. In order to accomplish proper scaling of the data, we need to establish a standard measure, or an absolute scale. The reference to absolute scale can be established through comparison of the experimental data with the theoretical structure factor distributions (centric and acentric) calculated under the assumption of randomly distributed atoms. These structure factor (amplitude) distributions have been derived by A. J. C. Wilson[3] in 1949 and are not conditioned on, that is, they are independent of, any model structure factors.

The classical distributions are readily derived as a limiting, unconditional case of a more general, conditional, and likelihood-based treatment of structure factor probabilities, which is the avenue we will pursue in a subsequent section. These conditional distributions form the basis of all likelihood functions applied in modern crystallography and are of corresponding importance. While it is not absolutely necessary to fully follow the derivation of the classical distribution in each detail, having some idea about the unconditional structure factor distributions may help the reader to appreciate the elegance of the likelihood-based derivations.

The Wilson distribution for a random atom model

In the following we briefly sketch a "classical" derivation for the centric and acentric Wilson distribution for the structure factor amplitudes, $prob(F)$. The likelihood-based derivation as a special unconditional case of a Rice distribution is provided in the next section.

Random atom model. Our interest in expected structure factor amplitude distributions is to define an absolute scale that is independent of whatever the actual contents of the unit cell may be, that is, the distribution of "random" structure factor amplitudes. For this purpose it does not suffice to just fill the reciprocal lattice points arbitrarily with randomly assigned amplitude (or intensity) values: Any crystal structure consisting of real atoms will have to obey the overall intensity attenuation due to the scattering angle dependency of the atomic scattering factors and the interaction of all scattering contributions in the unit cell to a resulting complex structure factor.

To obtain "random" structure factors and their probability distribution $prob(\mathbf{F})$ we need to examine a random atom structure. In a random structure, the random variables are the atomic coordinates $\mathbf{x}_j$ of all j atoms in the unit cell, which in turn define the individual structure factor contributions $\mathbf{f}_{j,h}$ and their sum again yields the entire complex structure factor $\mathbf{F}_h$ for each reflection $\mathbf{h}$. For the complex structure factors and their variance (considering the effects of centrosymmetry) we derive their probability distribution, and finally obtain the probability distributions for their amplitudes $prob(F)$ and (generally in normalized form) for the intensities. We consider first a primitive random structure without symmetry in the unit cell.

Random walk and the CLT. Given a structure consisting of a sufficiently large number of about equal atoms with random coordinates, the resulting structure factor will be a random walk-like sum of all atomic contributions in the complex structure factor plane (shown for the general case in Figure 7-17). The central limit theorem states that given these conditions—irrespective of the actual form of the underlying function $f(x)$—the probability distribution for obtaining any value x follows the Gaussian normal probability distribution, $G(x)$. Following the standard form of $G(x)$ given by Equation 7-24 we can write down the structure factor distribution $prob(\mathbf{F}\,|\,\sigma_F,\langle\mathbf{F}\rangle)$ as a function of their expectation value $\langle\mathbf{F}\rangle$ with a variance σ_F^2 as

$$prob(\mathbf{F}\,|\,\sigma_F,\langle\mathbf{F}\rangle) = \frac{1}{\sigma_F\sqrt{2\pi}}\ \exp\left(-\frac{\left(\mathbf{F}-\langle\mathbf{F}\rangle\right)^2}{2\sigma_F^2}\right) \tag{7-91}$$

In order to evaluate the probability distribution function (7-91) we need to determine the expressions for the two parameters in the equation: the expectation value $\langle\mathbf{F}\rangle$, and the variance of the structure factor distribution σ_F^2 as a function of the random coordinate variable $\mathbf{x}_j$. Both are defined as average values over all N atomic contributions:

$$\langle\mathbf{F}\rangle = \sum_{j=1}^{N(atoms)} \langle\mathbf{F}_j\rangle \tag{7-92}$$

and according to the sum rule for variances

$$\sigma_F^2 = \sum_{j=1}^{N(atoms)} \sigma_{F,j}^2 \tag{7-93}$$

Note that according to the CLT these relations (7-92) and (7-93) also hold when $\langle\mathbf{F}_j\rangle$ itself, as well as the variance, is a function of any another random variable, as is the case in our random-atom (coordinate) situation.

Centric structure factors. In the case of centric structure factors, the derivation of the structure factor amplitude distribution is relatively straightforward. Recall from Chapter 6 that atoms that give rise to centric reflections in centric zones are located centrosymmetrically opposed and their contributions to the structure factor are collinear (Figure 6-40). The position vector of such an atom $\mathbf{x}_j$ becomes $-\mathbf{x}_j$, hence the contribution of a pair of identical centrosymmetrically opposed atoms to the structure factor becomes

$$\mathbf{f}_h = f_j\left(\cos(2\pi\,\mathbf{hx}_j) + \cos(-2\pi\,\mathbf{hx}_j) + i\sin(2\pi\,\mathbf{hx}_j) + i\sin(-2\pi\,\mathbf{hx}_j)\right) = 2f_j\cos(2\pi\,\mathbf{hx}_j)$$

Thus for all the atom pairs contributing to a centric reflection, the summation over one half of the atoms suffices. We obtain for the resultant signed structure factor

$$\mathbf{F}_h = \sum_{j=1}^{N/2} 2f_j\cos 2\pi\mathbf{hx}_j \tag{7-94}$$

The random walk of pair-wise contributions with conjugate phase is then limited and consists of collinear segments as shown in Figure 6-40. In a non-primitive cell, the presence of translational symmetry can change the resultant phase to special values different from 0° and 180°, but the distribution of atom contributions along the resultant phase remains the same.

Expectation value. The structure factor $\mathbf{F}_h$ now needs to be expressed as a function of our random variable $\mathbf{x}_j$ which is in turn required to derive the mean or expectation value of $\langle \mathbf{F} \rangle$ and its variance σ_F^2. Both are needed to establish the desired distribution $prob(\mathbf{F} \mid \sigma_F, \langle \mathbf{F} \rangle)$.

Let the mean value of the individual summands in (7-94), expressed as a function of random variable $\mathbf{x}_j$, be another random variable, say v_j:

$$\langle v_j \rangle = 2 f_j \langle \cos 2\pi \mathbf{h}\mathbf{x}_j \rangle \qquad (7\text{-}95)$$

With the assumption of random coordinates, all possible $\mathbf{x}_j$ including their negatives $-\mathbf{x}_j$ are equally probable. On average, the cosine terms will also cancel out and thus $\langle v_j \rangle = 0$. The mean of all centric structure factors $\langle \mathbf{F} \rangle$ will therefore also be equal to zero:

$$\langle \mathbf{F} \rangle = \sum_{j=1}^{N/2} \langle v_j \rangle = 0 \qquad (7\text{-}96)$$

Variance. Next we need to evaluate the variance σ_v^2 of our random variable v_j from which we can obtain σ_F^2. Using the definition of the variance

$$\sigma_{v,j}^2 = \langle v_j^2 \rangle - \langle v_j \rangle^2 \qquad (7\text{-}97)$$

we obtain $\sigma_{v,j}^2 = \langle v_j^2 \rangle$ because $\langle v_j \rangle = 0$. We can make the right term of (7-97) explicit by back-substitution and obtain

$$\langle \sigma_{v,j}^2 \rangle = 2^2 f_j^2 \langle \cos^2 2\pi \mathbf{h}\mathbf{x}_j \rangle \qquad (7\text{-}98)$$

Now we need to evaluate the cosine term of (7-98): Random coordinates imply random phase angles, and for randomly distributed angles α $[-\pi \leq \alpha \leq \pi]$ the term $(\cos^2 \alpha)$ is limited to the interval $[0 \ldots 1]$. Its mean therefore is ½, and we obtain

$$\sigma_{v,j}^2 = 2 f_j^2 \qquad (7\text{-}99)$$

We obtain the variance of centric structure factors σ_F^2 finally by summing over all the individual variances $\sigma_{v,j}^2$ considering that the summation extends only over half of the reflections for the centrosymmetric arrangement of atoms. The variance of the centric structure factor distribution of a random atom model with N atoms is then

$$\sigma_F^2 = \frac{1}{2} \sum_{j=1}^{N} \sigma_{v,j}^2 = \sum_{j=1}^{N} f_j^2 = \Sigma_N \qquad (7\text{-}100)$$

Remember the short notation Σ_N for the variance of structure factors; it is general and holds also for acentric distributions. It will appear frequently in derivations of structure factor distributions. The important gain is that because of (7-100) we can now express the probability distribution on an absolute scale without the need of a separate parameter for the variance.

Centric Wilson distribution. Given the results of (7-96) and (7-100) we can express the probability function on an absolute scale without the need of a separate second parameter for the variance: Substituting the values for the two parameters $\langle \mathbf{F} \rangle = 0$ and $\sigma_F^2 = \Sigma_N$ into the Gaussian structure factor probability distribution (7-91) yields for the probability distribution function $prob(\mathbf{F})$ for centric structure factors

$$prob(\mathbf{F}) = \frac{1}{\sqrt{2\pi\Sigma_N}} \exp\left(-\frac{\mathbf{F}^2}{2\Sigma_N}\right) \quad \text{(centric)} \qquad (7\text{-}101)$$

What is left now is to convert Expression 7-101 to amplitudes, which generally requires integrating over all phases. In the case of a (primitive) centric distribution, we only need to consider the sign of the structure factor and can directly convert to amplitudes. Because for a centric structure factor the probability of having phase φ or $\varphi + 180°$ is equal and the amplitudes are the same for both vectors (even in random cells with crystallographic symmetry), we just need to double the value to obtain the expression for the structure factor amplitudes:

$$prob(F) = \left(\frac{2}{\pi\Sigma_N}\right)^{1/2} \exp\left(-\frac{F^2}{2\Sigma_N}\right) \quad \text{(centric)} \qquad (7\text{-}102)$$

which is just a 1-dimensional Gaussian centered about zero (Figure 7-10) and called the Wilson distribution[35] for centric structure factors. Note that for the unsigned amplitudes $\langle F \rangle > 0$ while for the signed (equivalent to complex) structure factors $\langle \mathbf{F} \rangle = 0$.

Acentric Wilson distribution. Without providing the exact proof,[36] which is similar except that we need to use the general form of the structure factors $F = (A^2 + B^2)^{1/2}$ and integrate over a 2-dimensional Gaussian around the complex structure factor $\mathbf{F}$, we obtain the Wilson distribution $prob(F)$ for acentric structure factor amplitude as

$$prob(F) = \frac{2F}{\Sigma_N} \exp\left(-\frac{F^2}{\Sigma_N}\right) \quad \text{(acentric)} \qquad (7\text{-}103)$$

which is a special case of the more general, conditional Rice or Sim distribution that plays an important role in maximum likelihood statistics and will be derived later in this chapter. The functions are shown in their general form in Figure 7-10 and normalized with additional statistics in Figure 7-12.

Epsilon-factor. So far we have not considered possible symmetry in the unit cell except for a centrosymmetric arrangement of atoms. The variance symbol Σ_N indicates the sum of squares of all atomic scattering factors, but a correction applies when these variances are derived from intensities because of the individual symmetry of certain classes of reflections in reciprocal space. These reflections fall into so-called epsilon-zones, ε, which we will encounter again in data processing. The value for ε is a statistical weight that measures how often a reflection $\mathbf{h}$ is superimposed onto the same spot in reciprocal space due to crystal point group symmetry, and the ε-factors for sets of reflections are tabulated[37] as the ratio of general position multiplicity divided by the point group multiplicity for that reflection (see Chapter 6). The correction for the variance Σ_N is then

$$\Sigma_N = \sum_{j=1}^{N} f_j^2 = \frac{1}{\varepsilon_\mathbf{h}}\langle F^2 \rangle = \frac{1}{\varepsilon_\mathbf{h}}\langle I \rangle \qquad (7\text{-}104)$$

Figure 7-10 Centric and acentric Wilson distributions. The Wilson distribution is an unconditional probability distribution for structure factor amplitudes and can be calculated from a model as well as from observed intensities. They thus provide a means of scaling experimental observations and calculated data onto a common (and absolute) scale. Figure 7-12 for normalized structure factor amplitude distributions contains additional very useful statistical descriptors.

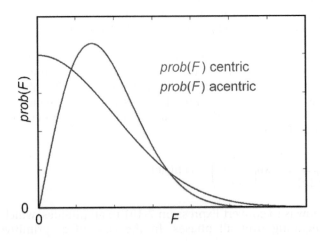

$prob(F)$ centric
$prob(F)$ acentric

Variance: its relation to intensities. With the general definition of the variance as $\sigma_F^2 = \langle F^2 \rangle - \langle F \rangle^2$ and $\langle F \rangle = 0$ it also follows that $\sigma_F^2 = \langle F^2 \rangle$, and by comparing with (7-100) we obtain a direct relation to the intensity of measured reflections:

$$\sigma_F^2 = \sum_{j=1}^{N} f_j^2 = \Sigma_N = \frac{1}{\varepsilon_\mathbf{h}} \langle F^2 \rangle = \frac{1}{\varepsilon_\mathbf{h}} \langle I \rangle \tag{7-105}$$

There are two consequences of the result that the mean intensity of the random atom structure is given by the sum of the squared atomic scattering factors: First, it essentially means that it does not matter where the atoms are in the periodic unit cells, it just matters how much is in there to scatter. The variance and the shape of the distributions will thus be universal, irrespective of molecular details. Second, as

$$\sigma_F^2 = \sum_{j=1}^{N} f_j^2 = \frac{1}{\varepsilon_\mathbf{h}} \langle F^2 \rangle \tag{7-106}$$

we can now express the probability distribution on an absolute scale without the need of a separate parameter for the variance. Equation 7-106 provides the basis for bringing experimental observations and calculated structure factors onto a common, absolute scale by Wilson scaling.

Differences between centric and acentric distributions. The distributions for centric and acentric structure factors are distinctly different. Centric reflections are more often either very strong or very weak than general acentric reflections. This can be qualitatively understood as follows: First, because centric reflections result from centrosymmetrically arranged atoms, their collinear contributions to the structure factor on average cancel out and the centric distribution is centered around zero. Second, if the crystal has symmetry, even if the atomic positions are randomly distributed in the asymmetric unit cell, they will not be completely random when viewed in projection along a symmetry axis, which manifests itself in enhanced intensities (i.e. a longer tail of the centric distribution) of the centric reflections located in corresponding reciprocal centric zones. None of these limitations apply to acentric reflections.

The above results, expanded in the next section when we introduce normalized intensity distributions $P(E^2)$, provide two important insights:

- From the intensity statistics we can make inferences whether a structure is centric or not. In proteins, we know that the structure cannot be centric. Deviations from expected intensity statistics can give valuable indications whether the data are either poorly collected or merged or that twinning is present.

- In order to use the experimental data in structure determination and refinement, the observations first need to be brought onto a common, absolute scale. This is possible by means of Wilson scaling described in a later section.

Normalized structure factors

In many cases when we need to compare or analyze intensity distributions, normalized intensities that are on an absolute scale and corrected for the specifics of the experiment such as overall B-factor attenuation are used. Let us first investigate the properties of a "point atom," where all electrons are presumed to

---

Box 7-16 Intensity statistics. Structure factor distribution statistics for a random structure lead to different intensity distributions for centric and acentric reflections. Deviations from expected intensity distributions can indicate anomalies such as pseudo-symmetry or twinning.

be concentrated in one point. If an atom had no dimensions, then there could also be no phase difference between its scattering electrons. Because it was this phase difference that gave rise to the angle-dependent decay of the atomic scattering factor, an atom with all its electrons concentrated in one point scatters equally in all directions, with a scattering factor of f_j^0 (or z electrons), as shown in Figure 7-11.

In order to scale up the actually measured, double attenuated scattering amplitude of a real atom $F_h(obs)$ and turn it into scattering amplitude of a point atom P_h, we need to

- invert the B-factor attenuation, and

- correct for the scattering factor attenuation.

Unitary structure factors. The first correction we achieve by simply inverting the B-factor attenuation, that is, dropping the negative sign in the exponent. The attenuation due to the scattering factor decay we correct by the factor z_j/f_j, the number of electrons of atom j divided by the atomic scattering factor at the respective d-spacing. We obtain for a single atom

$$P_{h,j} = \frac{z_j \cdot \exp(B_j S^2/4)}{f_j} \cdot F_{h,j}(obs) \qquad (7\text{-}107)$$

To obtain the total normalized scattering amplitude we sum over all atoms under the assumption of the same overall B-factor as experimentally determined from a Wilson plot. We obtain the unitary structure factors

$$U_h = \frac{\exp(BS^2/4)}{\displaystyle\sum_{j=1}^{atoms} f_j} \cdot F_h(obs) = \frac{P_h}{\displaystyle\sum_{j=1}^{atoms} f_j} \qquad (7\text{-}108)$$

These unitary structure factors now represent scattering from point atoms, are limited to the range $0 \le U_h \le 1$, and are independent of scale.

Normalized structure factors. While unitary structure factors are useful for a number of probability-based calculations, they still suffer from the predicament that the multiplicity of ε-zones has not been accounted for. This can be overcome by defining normalized structure factors, E_h, introduced by Jerome Karle and Herbert Hauptman[38] (Sidebar 10-2):

Figure 7-11 Scattering curve for point atom and real atoms. The scattering of a point atom is independent of the scattering angle and equivalent to z electrons over the whole range of scattering vectors. Compare the constant point atom scattering line with the scattering curves showing attenuation by phase difference between electrons within the atom, quantified by the atomic scattering factors, and the additional attenuation by the B-factor, that is, by displacement from the equilibrium (or mean) position.

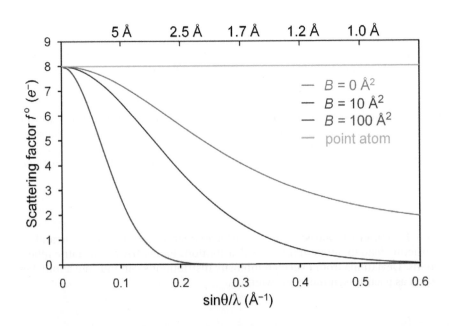

$$E_{\mathbf{h}} = \frac{\exp(BS^2/4)}{\left(\varepsilon \sum\limits_{j=1}^{atoms} f_j\right)^{1/2}} \cdot F_{\mathbf{h}}(obs) = \frac{F_{\mathbf{h}}}{(\varepsilon\Sigma_N)^{1/2}} \qquad (7\text{-}109)$$

The following useful relations (Chapter 10) hold for normalized structure factors:

$$E_{\mathbf{h}} = \frac{F_{\mathbf{h}}}{\langle F_{\mathbf{h}}^2\rangle^{1/2}} = \frac{F_{\mathbf{h}}}{\langle I_{\mathbf{h}}\rangle^{1/2}} = \frac{F_{\mathbf{h}}}{(\varepsilon\Sigma_N)^{1/2}} \quad \text{and thus} \quad E^2 = \frac{F^2}{\varepsilon\Sigma_N} \qquad (7\text{-}110)$$

The normalized structure factor amplitudes $E_{\mathbf{h}}$ can be readily obtained from the Wilson-scaled and binned mean intensities $\langle I_{\mathbf{h}}\rangle$ or mean squared amplitudes $\langle F_{\mathbf{h}}^2\rangle$, and it follows that $\langle E_{\mathbf{h}}^2\rangle = 1$. The normalized structure factors represent sharp point atom scatterers, and are convenient for certain sharpened electron density map constructions, probability distributions, and in the discussion of phase relations and inequalities in direct methods as discussed in Chapter 10.

Normalized structure factor distributions. We can now substitute the relation $E^2 = F^2/\varepsilon\Sigma_N$ directly into the equations for the Wilson distributions (7-101) and (7-103) where ε was not explicitly included, and immediately obtain for the unconditional normalized structure factor distributions $P(E)$

$$P(E) = \left(\frac{2}{\pi}\right)^{1/2} \exp\left(-\frac{E^2}{2}\right) \quad \text{(centric)} \qquad (7\text{-}111)$$

$$P(E) = 2E \cdot \exp\left(-E^2\right) \quad \text{(acentric)} \qquad (7\text{-}112)$$

which are plotted in Figure 7-12. The values for a variety of mean normalized amplitude E values $\langle E\rangle$ as well as $\langle|E^2-1|\rangle$ used as a diagnostic for centrosymmetry of a structure are also shown in Figure 7-12.

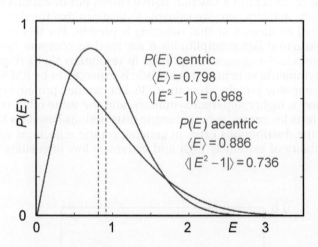

P(E) centric
$\langle E\rangle = 0.798$
$\langle|E^2-1|\rangle = 0.968$

P(E) acentric
$\langle E\rangle = 0.886$
$\langle|E^2-1|\rangle = 0.736$

Figure 7-12 Normalized Wilson distribution. The Wilson distribution for normalized structure factor amplitude probabilities is distinctly different for centric (blue graph) and acentric reflections (red graph). The graphs demonstrate that centric reflections show a higher probability of extremely low and extremely high (longer tail of distribution) intensities, while for acentric reflections, the maximum probability is distinctly shifted toward more moderate intensities. Both distributions are special cases of general conditional probability functions of ubiquitous importance in likelihood based methods. The acentric Wilson distribution is a special case of the Rice distribution, and the centric Wilson distribution is a special case of the Woolfson distribution, derived together in the next section from general conditional likelihood functions. Also shown are the mean E values $\langle E\rangle$ (dashed lines) and the numbers for $\langle|E^2-1|\rangle$, a diagnostic for testing whether a structure is centrosymmetric or non-centrosymmetric (as protein structures normally are). Note that $\langle\mathbf{E}\rangle$ is zero for the centric distribution.

Box 7-17 Normalized structure factors and their probability distributions. Under the assumption of point atoms where all electrons are located at the atom center, no angle-dependent attenuation of the atomic scattering factors occurs and the normalized structure factors $E_{\mathbf{h}}$ can be computed. They are used in analysis of intensity distributions, sharpened map constructions, and in testing phase relations in direct methods. The cumulative probability distributions of E^2 (normalized intensities) are particularly useful for statistical intensity distribution analysis of experimental data.

Cumulative normalized intensity distributions

The normalized structure factor distributions assume a particularly simple form when expressed in terms of $Z = E^2$. Z can be readily obtained in binned shells from the intensities on an absolute scale as $Z = I_{abs} / \langle I_{abs} \rangle$. The distributions $P(Z)$ have the following form:

$$P(Z) = \left(\frac{1}{2\pi Z} \right)^{1/2} \exp\left(-\frac{Z}{2} \right) \quad \text{(centric)} \tag{7-113}$$

$$P(Z) = \exp(-Z) \quad \text{(acentric)} \tag{7-114}$$

For statistical analysis, the normalized structure factor distribution is plotted as a cumulative probability distribution $N(Z)$. The plots of $N(Z)$ (and of higher raw moments of E as a function of resolution) are used for detection of microscopic (merohedral) twinning and other statistics; we will analyze them in the twinning section in Chapter 8. They are again different for centric (containing the error function erf) and acentric reflections:

$$N(Z) = \text{erf}\left(\frac{Z}{2} \right)^{1/2} \quad \text{(centric)} \tag{7-115}$$

$$N(Z) = 1 - \exp(-Z) \quad \text{(acentric)} \tag{7-116}$$

The cumulative probability distributions for centric and acentric reflections are graphed in Figure 7-13.

Mean $|E^2-1|$ value and expected random R-value

The structure factor distributions derived above were originally derived by A. J. C. Wilson[35] in 1942 and provide a number of important insights. First, from the intensity statistics we can make inferences about whether a structure is centrosymmetric or not. In proteins, we know that the structure cannot be centrosymmetric (except in a few non-native cases), but deviations from non-centric intensity statistics can give valuable indications that the data are either poorly collected or merged or that twinning is present. For this purpose, the normalized structure factor amplitudes E are used to compute $\langle |E^2-1| \rangle$, the mean E-squared minus one, often plotted in resolution shells (Figure 7-14). For a centrosymmetric structure, this statistic is expected to be 0.968 and for a non-centrosymmetric protein structure 0.736. For twinned protein crystals, the intensities have a higher apparent symmetry, and the value often is less than 0.736. Other tests for centric versus acentric distributions based on higher raw moments of the distributions exist. In general, centric reflections will show a higher probability of extremely high and extremely low intensities, while for

Figure 7-13 Cumulative probability distribution of normalized intensities. The cumulative probability distributions $N(Z) = N(E^2)$ for centric and acentric reflections. Deviations from these expected distributions are often indicative of crystal twinning (Chapter 8).

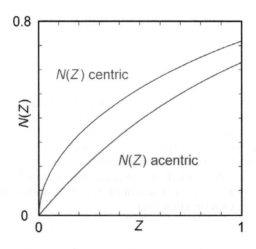

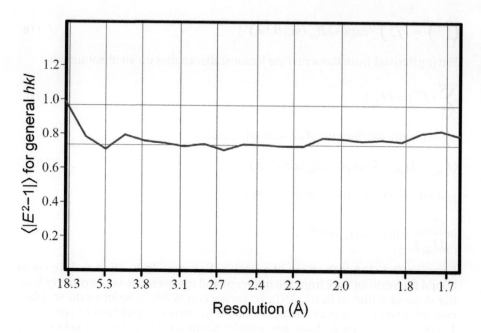

Figure 7-14 |E²–1|-plot for an acentric structure. The plot of $\langle|E^2-1|\rangle$ versus resolution bin shows that the intensity distributions for this protein structure are representative of a non-centrosymmetric (chiral) structure. The small peak at low resolution results from the fact that, depending on symmetry, in very low resolution shells there are proportionally more centric than acentric reflections compared with high resolution shells. Data are plotted by *SADABS* (G. Sheldrick) for the vta1 example used in the data collection in Chapter 8.

acentric reflections, the maximum probability is distinctly biased toward more moderate intensities (Figure 7-12). Comparison of the entire experimental and theoretical distributions can again give clues of aberrant distributions due to twinning or other problems with crystals or the data collection.

Upper limits for *R*-values. Another interesting result derived by Wilson[9] is the upper limiting value for the linear residual or R_F-value for a random atom model. For a non-centrosymmetric structure such as a protein structure we expect based on the intensity statistics a worst case (random) R_F-value of ~59%, and ~83% for a centrosymmetric random structure. Even worse R_F-values than 0.59 can sometimes be obtained for initial protein structure models, but these are generally (i) bad news, and (ii) may result from poor initial scaling between measured and calculated structure factor amplitudes.

Wilson plots and initial scaling

The first important step during statistical analysis and for further use of the experimental data is to bring the experimental data onto a common scale with theoretical model data or absolute data, which requires an overall linear scale factor together with a correction for the overall *B*-factor attenuation specific for each data set and crystal. The first approximate scaling is performed through isotropic overall *B*-factor scaling, or Wilson scaling.[35]

The count rate of intensities we measure depends on a number of proportionality factors such as measurement time, incoming beam intensity, exposed crystal volume, and so forth, which we discussed in Chapter 6. The measured or "observed" average (Lorentz- and polarization-corrected) intensity $\langle I_{obs}\rangle$ in each resolution shell is thus only proportional to the absolute intenity $\langle I_{abs}\rangle$, that is, $\langle I_{obs}\rangle \propto \langle I_{abs}\rangle$.

In a real protein structure, each atom will have individual displacement and additional displacive lattice imperfections will further increase the positional displacements so that an overall attenuation of atomic scattering will reduce the measured intensities (Chapter 6). We can factor out an average or overall isotropic *B*-factor B_{iso} from the atomic scattering factors as follows:

$$f_j^B = f_j^0 \cdot \exp(-B_{iso}(\sin\theta/\lambda)^2) \qquad (7\text{-}117)$$

Squaring the expression containing the atomic scattering factors brings it into useful form for the intended intensity comparison in the next step:

Sidebar 7-5 Binning and discretization. Binning is a common procedure of rendering continuous functions discrete, that is, to approximate continuous functions by discrete measurements. The general form of the substitution of a continuous integration with a discrete summation in n bins in the range A to B with the mean bin value $\langle x_j \rangle$ is:

$$\int_{x=A}^{B} f(x)\,dx = \sum_{A,j=1}^{B,j=n} f(\langle x_j \rangle) \qquad (7\text{-}122)$$

For the validity of (7-122) it is important that the bin width is selected so that the data samples within each bin vary only modestly. Improper bin selection is a popular means of "adjusting" histogram statistics. A typical warning sign occasionally overheard is a statement similar to this one: "The measurements did not make sense, so we had to use statistics".[6]

In crystallography, the experimental data are commonly binned in resolution shells of equal width in $(\sin\theta/\lambda)^2$. Binning in shells of $1/d^3$ is also possible, with the advantage that each bin contains an equal number of reflections which somewhat reduces the noise in low resolution bins due to the lower number of reflections they contain otherwise.

$$\left(f_j^B\right)^2 = \left(f_j^0\right)^2 \cdot \exp\!\left(-2B_{iso}(\sin\theta/\lambda)^2\right) \qquad (7\text{-}118)$$

Having derived from the structure factor statistics that on an absolute scale

$$\sum_{j=1}^{atoms} \left(f_j^0\right) = \langle I_{abs} \rangle \qquad (7\text{-}119)$$

we can write for the sum of all scattering atoms

$$\langle I_{obs} \rangle = \langle I_{abs} \rangle \cdot k \cdot \exp\!\left(-2B_{iso}(\sin\theta/\lambda)^2\right) \qquad (7\text{-}120)$$

and take the logarithm, finally yielding

$$\ln\frac{\langle I_{obs} \rangle}{\langle I_{abs} \rangle} = \ln k - 2B_{iso}(\sin\theta/\lambda)^2 \qquad (7\text{-}121)$$

Equation (7-121) has the form of a straight line $y(x) = \ln k - 2B_{iso}x$ and we could obtain the scale factor k from the intercept and the overall B-factor directly from the slope of a plot of $\ln\left(\langle I_{obs}\rangle/\langle I_{abs}\rangle\right)$ versus $(\sin\theta/\lambda)^2$. In order to do so, however, we need to compute the mean values in narrow, equal bins of $(\sin\theta/\lambda)^2 = 1/(2d)^2 = S^2/4$. These bins are actually shells of the reciprocal sphere and contain an increasing number of reflections in each bin. Sometimes the bins are plotted in units of $4(\sin\theta/\lambda)^2$, thus grouping them into shells of width $1/d^2$. For the observed intensities $\langle I_{obs}\rangle$ the binning is straightforward, because all we have to do is sort the reflections according to their $(\sin\theta/\lambda)^2$ and compute the bin average $\langle I_{obs}\rangle_{bin}$.

To establish a first guess for the absolute intensities in bins $\langle I_{abs}\rangle_{bin}$ (we generally know neither the atomic positions nor any individual B-factors at this point), the values are estimated from the unit cell contents. We provide the number of residues in the asymmetric unit, and based on the average number of light atoms (C, N, O, and H) we can compute the absolute scattering $\langle f_{abs}^2\rangle_{bin} = \langle I_{abs}\rangle_{bin}$ without B-contribution using the Cromer–Mann approximation (6-15). Average atom numbers per residue are 5 (C), 1.35 (N), 1.5 (O), and 8 (H), with 15.9 average atoms/residue. These estimates are generally sufficient for the initial scaling and inspection of the Wilson plot. Both the scale factor k and overall B are later refined as parameters during the actual structure refinement procedure. The overall B-factor refinement is generally anisotropic,[39] where the scalar B_{iso} is replaced with the (overall) anisotropic displacement tensor $\mathbf{U}$ (Chapter 12).

Let us examine an isotropic Wilson plot for the example of bovine trypsin shown in Figure 7-15. The molecule has about 250 residues, and we compute a Wilson plot and a straight line fit for 1.5 Å intensity data using the CCP4[33] program *TRUNCATE*.[13]

Inspection of the plot shows that significant deviations from linearity exist, because the distances in a real protein are not random, but obey a common stereochemistry. For example, the Cα–Cα distance is very ubiquitous at around 3.8 Å, and other "preferred" distances exist primarily at low resolution. The resulting deviations are called Debye effects, and corresponding corrections for scaling have been suggested.[40, 41] We generally cannot use Wilson plots reliably for scaling of low resolution data, which becomes thus very ambiguous and often leads to absurd values for (prematurely) refined individual B-factors (examples are provided in Chapter 13). In low resolution cases, maximum likelihood based methods can still provide a more reliable estimate[42] of the overall B-factor, and these B-factor estimates can in turn be used to sharpen low resolution map coefficients.[43]

In the case of the high resolution data analyzed in Figure 7-15, the values for $\ln k$ and for B are readily obtained from simple linear regression in the range between highest resolution and about 3.2 Å. We read from the plot –3.2 for $\ln k$, and the factor to convert the experimental data to absolute scale is $1/k$, which

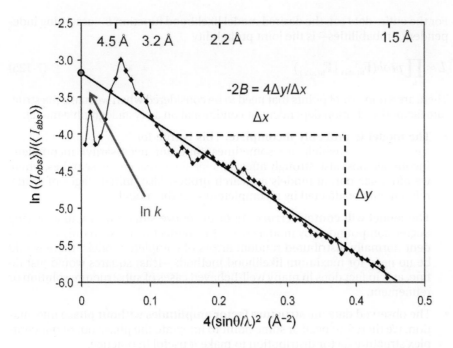

Figure 7-15 Wilson plot for a small protein. Shown is the Wilson plot for bovine trypsin (~250 residues, PDB entry 2j9n) plotted from 1.5 Å data using the CCP4 program *TRUNCATE*. The intercept of a linear regression through the high-resolution branch of the data delivers the scale factor, while the slope determines the overall *B*-factor needed to bring experimental data back up onto an absolute scale. Note the significant deviations from linearity for resolution bins above 3.5 Å. Part of the low-resolution clutter is also because with $1/d^2$ binning the low resolution bins or shells contain fewer reflections.

evaluates to 24.5. We can manually estimate $-2B$ from the slope $\Delta y/(\Delta x/4)$, with $\Delta y = -1.5$ and $\Delta x \approx 0.25$ Å^2, as -24, thus $B \sim 12$ Å^2. The actual least squares fit result of 12.7 ± 0.3 Å^2 is within one standard uncertainty of the final mean overall B-factor which refined to 12.5 Å^2 according to PDB entry 2j9n. The final equation to bring the intensities back onto an absolute scale is essentially the "inverse" of Equation 7-118

$$f_{abs}^2 = f_{obs}^2 \frac{1}{k} \exp(2B\sin^2\theta / \lambda^2) = f_{obs}^2 \frac{1}{k} \exp(BS^2/2) \qquad (7\text{-}123)$$

Note the absence of the minus sign in the exponent, which means we are scaling back up to the absolute scale that we lost in the experiment through the overall B-factor attenuation.

7.11   Conditional structure factor probability distributions

Conditional structure factor distributions form the basis of maximum likelihood methods. Any physical model—which can be just a few substructure atoms, a partial structure model used as a search probe, or a structure model whose parameters are adjusted during refinement—gives rise to calculated structure factors $\mathbf{F}_{calc}$, generally computed from some parameterized mathematical model representing the physical structure model. We are then asking the basic question: Given our model defining $\mathbf{F}_{calc}$, what is the probability that we make a certain observation $\mathbf{F}_{obs}$? This conditional probability distribution for each individual structure factor given as

$$prob(\mathbf{F}_{obs} \mid \mathbf{F}_{calc}) \qquad (7\text{-}124)$$

Box 7-18 Wilson scaling. One of the first steps in data preparation is to bring the experimentally observed data of an unknown structure onto an absolute scale so that they can be further processed or refined. Initial scaling is done by overall B-factor scaling, where the scale factor $1/k$ and the initial isotropic overall B-factor B are determined from a linear fit to a Wilson plot. In later refinement stages, anisotropic overall B-factor scaling is employed.

For the entire data set, the desired model likelihood function L—assuming independent probabilities—is the joint probability

$$L \propto \prod_{\mathbf{h}} prob(\mathbf{F}_{\mathbf{h}(obs)} \mid \mathbf{F}_{\mathbf{h}(calc)}) \tag{7-125}$$

There are a number of points that need to be considered when we make the structure factor distribution dependent or conditional on a crystallographic model:

- **The model is probably incomplete.** This is true for heavy atom substructure solutions, which are sometimes (but not necessarily) incomplete. Incompleteness also strongly affects molecular replacement searches, where we often use partial models as search probes. Also, initial stages of model refinement are affected by incompleteness of the model.

- **The model will contain errors.** All of the above situations are to a varying degree compounded by model errors. In a perfect world with only independent normally distributed random errors of complete models, there would be no need for maximum likelihood methods—least squares would just do fine, as it in fact does in many well-behaved cases of substructure solution or refinement.

- The observed data are **structure factor amplitudes without phase** information. We therefore need at some point to integrate the phase out of our complex structure factor distribution to make it useful in practice.

We need to address all of the above problems in the derivation of structure factor probabilities. Additional special problems such as correlation of errors introducing dependence in the joint probability (7-125) are addressed in the corresponding chapters when the respective likelihood functions are derived.

Accounting for incomplete models: Sim weights

One of the first problems in crystallography, establishing correct phase probabilities, was recognized during the early days of isomorphous replacement as a problem requiring likelihood treatment, and it has been reformulated nearly two decades later in the context of maximum likelihood by French,[20] Bricogne,[44] Read,[22] and others. Leaving aside for a moment the issue of model correctness, model completeness is by no means assured; there are, for example, often multiple low-occupancy sites in a heavy atom substructure that have been missed in the substructure solution. Similar considerations hold for the partial models used in molecular replacement searches and incomplete starting models during initial refinement stages.

Sim distribution. In the late 1950s, G. A. Sim derived a structure factor probability distribution[45] based on the proposition that a model used to compute phase probabilities is incomplete. The total structure factor contributions of N atoms (of which we observe the diffraction pattern) can be formally split up into a contribution of P known atoms (our incomplete or partial model, for which we can calculate the structure factors) and that of Q atoms, which we do not know anything about (but about which we can make certain statistical assumptions):

$$\mathbf{F}_N = \mathbf{F}_P + \mathbf{F}_Q = \sum_{j=1}^{P} \mathbf{F}_j + \sum_{j=P+1}^{N} \mathbf{F}_j \tag{7-126}$$

The corresponding probability distribution derived under the assumption of incompleteness is the Sim distribution, which has the form of a Rice distribution[46] known in statistics (related to the non-central χ-distribution). We will not discuss Sim's derivation of his distribution[47] here, because we will obtain the same distribution in the more general framework of likelihood functions[48] in the next section, but will briefly discuss the ubiquitous weighting function for the probability distributions that Sim derived.[45]

Sim weights. The Sim weights for general acentric and for centric reflections, m_a and m_c, respectively, are

$$m_a = \frac{I_1(X)}{I_0(X)} \text{ with } \lim_{x \to \infty} \frac{I_1(X)}{I_0(X)} = 1 \text{ and } m_c = \tanh\left(\frac{X}{2}\right) \qquad (7\text{-}127)$$

where I_n represents the nth order modified Bessel functions of the first kind (Figure 7-16). The modified or hyperbolic Bessel functions are common in crystallography, because they generally are the result of integrating out the phase as a nuisance variable in maximum likelihood structure factor probability distributions. The modified Bessel functions of the first kind are essentially functions of X that exponentially increase with X, but the fraction $I_1(X)/I_0(X)$ asymptotically approaches 1 for $X \to \infty$. The argument X of the weighting function is given by

$$X = \frac{2F_N \cdot F_P}{\Sigma_Q} = \frac{2F_{obs} \cdot F_{calc}}{\sum_{j=P+1}^{N} f_j^2} \qquad (7\text{-}128)$$

The structure factor amplitude $F_{calc} = F_P$ is computed from the known atoms of the model, and the atomic scattering factor summation in the denominator of (7-128) goes over all the missing Q atoms. The difficulty is to determine the variance Σ_Q of the distribution of the missing atoms, but it can be estimated from the discrepancy between observed and calculated structure factors. Bricogne[44] and Read[22] have derived the following estimates ($n = 1$ for centrics and $n = 2$ for non-centric reflections)

$$\sum_{i=1}^{m} f_i^2 = \langle | F_{obs}^2 - F_{calc}^2 | \rangle \qquad (7\text{-}129)$$

and

$$n(F_{obs} - F_{calc})^2 \qquad (7\text{-}130)$$

respectively.

The Read estimate (7-130) reportedly seems to give somewhat better results,[49] and in either case, the Sim weights revert to the standard crystallographic figure of merit $m = \langle \cos(\Delta\varphi) \rangle$ (shown in Chapter 10) when all atoms are assumed to be known. Sim weights are used in the computation of the best phases from heavy atom positions and in phase combination the same way the standard Crick and Blow weights m are used. Such Sim-weighted electron density maps are constructed as described in Chapter 9 with Fourier coefficients $mF_{obs}\exp(i\alpha_{BEST})$ using the centroid phases α_{BEST} (Chapter 10) and Sim weights m derived above.

Accounting for errors in the model: anticipating σ_A

So far, our assumption was that our $\mathbf{F}_{calc}$ were derived from atom positions that had no positional errors $\Delta\mathbf{r}_j$. The positional error can be accounted for in

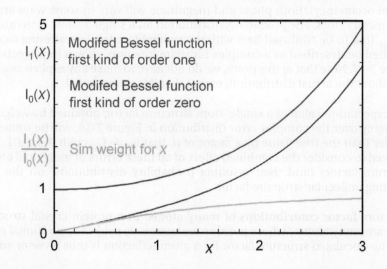

Figure 7-16 Modified Bessel functions and Sim weights. The Bessel functions of the first kind $I_n(X)$ for orders $n = 0$ and 1, and the Sim weight $I_1(X)/I_0(X)$ are plotted against argument X. For large values of X, the Sim weight asymptotically approaches unity, that is, if the number of unknown atoms equals zero, $X \to \infty$ and $w \to 1$.

the structure factor probability distribution by a variance term, called sigma-A or for short, σ_A. The variance term σ_A was introduced by Srinivasan[50] and Ramachandran[51] in the mid-1960s, and improved σ_A estimates were derived by Main[52] and Read.[22] We will encounter the Sim weights m, σ_A, and its relative, the Luzzati D-factor, during the derivation of the general maximum likelihood distributions for observed structure factors $\mathbf{F}_{obs}$ conditional on their calculated counterparts $\mathbf{F}_{calc}$, and we will see that effects of coordinate error and of missing parts are in fact equivalent in terms of normalized structure factor probabilities. For the purpose of phase combination in heavy atom phasing, the Sim weights can be adjusted for errors in the partial model by introduction of the σ_A term and their argument X becomes

$$X = \frac{2\sigma_A E_{obs} \cdot E_{calc}}{1 - \sigma_A^2} \tag{7-131}$$

In map reconstruction with partial model phases, σ_A is in practice determined by minimizing a residual function based on the normalized cross-validation structure factors from the partial model and the complete (= observed) structure factor amplitudes.[22] The map reconstruction is then performed in the usual way by a Fourier synthesis with map coefficients

$$\mathbf{F}_{BEST} = (2mF_{obs} - DF_{calc}) \exp(i\varphi_{calc}) \tag{7-132}$$

where F_{calc} and φ_{calc} are computed from the partial model and D is the Luzzati factor, which we will revisit in the framework of maximum likelihood in the next section and detail in the refinement and model building discussion in Chapter 12.

Improving the structure factor error model

The first step in establishing the likelihood function as a product of the individual structure factor probabilities

$$L \propto \prod_{\mathbf{h}} prob(\mathbf{F}_{\mathbf{h}(obs)} \mid \mathbf{F}_{\mathbf{h}(calc)}) \tag{7-133}$$

is developing an appropriate error model for the conditional structure factor probabilities. This is most easily understood in the basic conditional likelihood case: given a calculated value (from the model) of a structure factor, what is the probability of observing an experimental measurement (the data) of that value? In other words, we need to derive as a first step the conditional probability distribution for $\mathbf{F}_{obs}$ given $\mathbf{F}_{calc}$: $prob(\mathbf{F}_{obs} \mid \mathbf{F}_{calc})$.

Structure factor contribution of a single atom. Let us step back and look at the structure factor of a single atom. There are essentially two errors that for a single atom can exist independently of each other: phase error (the structure factor "vector" points in a different direction) as a result of positional error $\Delta\mathbf{r}$; and error in magnitude (length of the vector) as a result of B-factor variation (or partial occupancy). Both phase and magnitude will vary in some ways around their mean values. The probability distribution for a single atomic structure factor $\mathbf{f}_{calc}$ (not to be confused here with the symbol for atomic scattering factor f) can then be described as a complex bivariate function, shown in projection in Figure 7-17. Note that at this point, we do not have to make any explicit assumption about the actual distribution, except that it is monomodal.

The expectation value of a single-atom structure factor, obtained by weighted-averaging over the complex error distribution in Figure 7-16, will be somewhat smaller than the true value by a factor of d, that is, $d \cdot \mathbf{f}_{calc}$ with $0 \le d \le 1$. Next we need to consider the combined effect of all these errors of individual atomic structure factors (and their resulting probability distributions) on the total resulting molecular structure factor.

Structure factor contributions of many atoms. In a protein crystal structure, there are many atoms j whose position we know only subject to positional errors $\Delta\mathbf{r}_j$. The calculated structure factor for a given reflection is thus a vector sum of

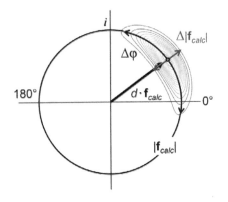

Figure 7-17 Probability distribution for a single atom structure factor. Phase error $\Delta\varphi$ (the structure factor "vector" points in different directions) results from positional error of the atom, and the error in magnitude $\Delta|\mathbf{f}_{calc}|$ (length of the vector) originates from variation in B-factor or occupancy. The resulting complex distribution is shown in a contour plot peaked at the true value $\mathbf{f}_{calc}$ (green dot at the contour peak). By vector averaging over the whole distribution, the expected mean value (black vector) will be somewhat shorter than $\mathbf{f}_{calc}$ by a factor of d. Note that with no positional information, the phase of an atomic structure factor could be anywhere between $0 \le \alpha \le 2\pi$, but the intensity would still have an expectation value determined by the Wilson distribution. We will use this argument in Chapter 9 when examining the effects of phase error versus intensity errors on map construction.

the corresponding atomic structure factors. The errors will accumulate, and given the large number of atoms, the resulting probability function for observing $\mathbf{F}_{obs}$ given $\mathbf{F}_{calc}$ will be a Gaussian normal distribution—a consequence of the central limit theorem—irrespective of the specific form of the underlying probability distribution for the individual atomic scattering factors $d \cdot \mathbf{f}_{calc}$. The likelihood function will then be a 2-dimensional Gaussian (the coordinate axes being real and imaginary axes of the complex plane) with a complex variance σ_Δ^2, centered on $D \cdot \mathbf{F}_{calc}$, because the sum vector will again fall somewhat short (Figure 7-18).

The Luzzati factor D. The expression for D was derived by Luzzati[53] in the 1950s straightforwardly from the structure factor distributions:

$$D_\mathbf{h} = \langle \cos 2\pi(\Delta\mathbf{r}_j \mathbf{s}) \rangle \tag{7-134}$$

If there are no errors in the known positions (coordinate deviation $\Delta\mathbf{r}_j = 0$), D becomes 1. Note that the individual variances σ_Δ^2 are a function of $\mathbf{h}$, and we have in practice to account for the expected intensity factor ε for each reflection. For simplicity, we assume in the following derivations that the epsilon-factor as tabulated[37] is included in the variance (the symbol Σ is generally used in the literature to indicate that the epsilon-correction is included in a variance term employed in the description of structure factor probabilities).

The resulting conditional probability distribution for a general, acentric reflection[54] is then described by a 2-dimensional Gaussian in the complex plane

$$prob(\mathbf{F}_{obs} \mid \mathbf{F}_{calc}) = \frac{1}{\pi\sigma_\Delta^2} \exp\left(-\frac{\left|\mathbf{F}_{obs} - D\mathbf{F}_{calc}\right|^2}{\sigma_\Delta^2}\right) \quad \text{(acentric)} \tag{7-135}$$

with its corresponding 2-dimensional variance σ_Δ^2 in the complex plane as illustrated in Figure 7-18. The interpretation of the figure is straightforward: if an observed structure factor amplitude circle is barely intersecting the Gaussian peak, then the probability of making this observation is low and the model does not seem to describe reality very well. We will quantify this qualitative observation shortly.

Equation 7-135 in fact represents the general probability distribution for a structure factor conditioned on another one (or in other words, its expectation value given the model), and serves as a starting point in the derivation of many crystallographic maximum likelihood functions. For centric reflections with phases restricted to φ and $\varphi + 180°$, the probability distribution becomes again a Gaussian of the form

$$prob(\mathbf{F}_{obs} \mid \mathbf{F}_{calc}) = \frac{1}{(2\pi\sigma_\Delta^2)^{1/2}} \exp\left(-\frac{\left|\mathbf{F}_{obs} - D\mathbf{F}_{calc}\right|^2}{2\sigma_\Delta^2}\right) \quad \text{(centric)} \tag{7-136}$$

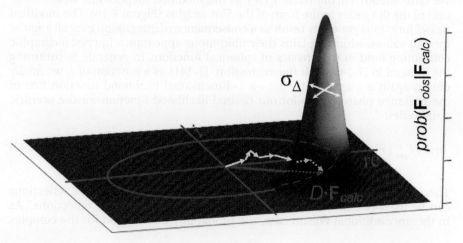

Figure 7-18 Complex conditional structure factor probability distribution. The probability distribution for an acentric structure factor resulting from vector summation of many atomic structure factors $d \cdot \mathbf{f}$ (short pink arrows with individual phase and amplitude errors as shown in Figure 7-17) is a 2-dimensional Gaussian centered on $D \cdot \mathbf{F}_{calc}$.

Marginalizing the phase and the Rice distribution

The probability distribution functions we derived above have a small problem for practical use: they are conditional probabilities of a complex structure factor $\mathbf{F}_{obs}$ given the value of complex structure factor $\mathbf{F}_{calc}$. However, we measure only the structure factor amplitude F_{obs}, lacking phase information in the observations, and are actually seeking $prob(F_{obs} \mid F_{calc})$. The nuisance variable phase is eliminated by integration of the phase angle in the likelihood function and the procedure is called integrating out or marginalization of the nuisance variable α (nuisance is not to be taken literally in this context). The resulting probability distribution is the Rice or Sim distribution, which plays an important and ubiquitous role in crystallography.[1]

The residual complex difference in the exponent of the acentric probability distribution (7-135) can be transformed to an explicit phase term using the cosine rule (Appendix A.4)

$$\left| \mathbf{F}_{obs} - D\mathbf{F}_{calc} \right| = \left(F_{obs}^2 + D^2 F_{calc}^2 + 2 F_{obs} D F_{calc} \cos\alpha \right)^{1/2} \tag{7-137}$$

For the purpose of integrating out the phase angle α, the probability function needs to be transformed via Jacobian transformation[1] from the Cartesian basis (with the separate real and imaginary terms as used to plot the function in Figure 7-17) into polar coordinates as a function of the phase angle α:

$$prob(F_{obs}, \alpha \mid F_{calc}) = F_{obs} \times prob(\mathbf{F}_{obs} \mid \mathbf{F}_{calc}) \tag{7-138}$$

Our desired probability $prob(F_{obs} \mid F_{calc})$ can then be obtained by integrating over the entire phase circle:

$$prob(F_{obs} \mid F_{calc}) = \int_0^{2\pi} prob(F_{obs}, \alpha \mid F_{calc}) \, d\alpha \tag{7-139}$$

Marginalization of phase. Substituting the expressions (7-137) to (7-139) into acentric distribution (7-135) we derive[55] for our desired probability distribution in the general acentric case

$$prob(F_{obs} \mid F_{calc}) = \frac{F_{obs}}{\pi \sigma_\Delta^2} \cdot \exp\left(-\frac{F_{obs}^2 + D^2 F_{calc}^2}{\sigma_\Delta^2} \right) \cdot \int_0^{2\pi} \exp\left(\frac{2 F_{obs} D F_{calc}}{\sigma_\Delta^2} \cos\alpha \right) d\alpha \tag{7-140}$$

Often such exponential integrals as in the above equation do not have analytical solutions and need to be solved numerically, but in this case an analytical solution for the integral exists:

$$\int_0^{2\pi} \exp(X \cdot \cos\alpha) \, d\alpha = 2\pi I_0(X) \tag{7-141}$$

We have already encountered $I_0(X)$ as the modified (hyperbolic) Bessel function of the 0th order in the form of the Sim weights (Figure 7-16). The modified Bessel functions generally result as a consequence of integration over all angular (phase) values, which explains their ubiquitous appearance in crystallographic computing (and in the physics of spherical functions in general). Substituting the integral in (7-140) with the expression (7-141) as a function of I_0 we finally obtain again a Rice distribution—a 1-dimensional likelihood function free of the nuisance phase term—for our desired likelihood function for the acentric amplitudes:

$$prob(F_{obs} \mid F_{calc}) = \frac{2 F_{obs}}{\sigma_\Delta^2} \cdot \exp\left(-\frac{F_{obs}^2 + D^2 F_{calc}^2}{\sigma_\Delta^2} \right) \cdot I_0\left(\frac{2 F_{obs} D F_{calc}}{\sigma_\Delta^2} \right) \tag{7-142}$$

The same considerations as applied above to the general acentric reflections can be used to derive the probability distributions for centric reflections.[1] As in the unconditional Wilson case, the probability distribution for the complex structure factor is a 1-dimensional Gaussian, because for centric reflections the phase angle is limited to φ or $\varphi + 180°$. The integration for centric reflections becomes a summation over the two possible centric phase angles

$$prob(F_{obs} \mid F_{calc}) = \left(\frac{1}{\pi\sigma_\Delta^2}\right)^{1/2} \exp\left(-\frac{F_{obs}^2 + D^2 F_{calc}^2}{2\sigma_\Delta^2}\right) \cdot \sum_{\varphi,\varphi+\pi}^{i} \exp\left(\frac{F_{obs}DF_{calc}}{\sigma_\Delta^2}\cos\alpha_{i,calc}\right)$$

$$(7\text{-}143)$$

and with $e^x + e^{-x} = 2\cosh x$ we finally obtain the Woolfson distribution:[4]

$$prob(F_{obs} \mid F_{calc}) = \left(\frac{2}{\pi\sigma_\Delta^2}\right)^{1/2} \exp\left(-\frac{F_{obs}^2 + D^2 F_{calc}^2}{2\sigma_\Delta^2}\right) \cdot \cosh\left(\frac{F_{obs}DF_{calc}}{\sigma_\Delta^2}\right) \qquad (7\text{-}144)$$

We will now normalize these functions and analyze them in detail, where we will make a number of crucial observations of great relevance for crystallographic work.

Normalizing and analyzing the Rice and Woolfson distributions

The likelihood-based Rice and Woolfson distributions can both be further simplified by normalization and the introduction of the single σ_A^2 parameter substituting the anticorrelated Luzzati factor D and the probability distribution variance σ_Δ^2. The Luzzati D and the variance are anticorrelated: if the positional error $\Delta\mathbf{r}$ is large (i.e. the model is bad) and thus the variance σ_Δ^2 is large, then D will be small. When the structure factor probability is expressed using normalized structure factors

$$E_{calc} = F_{calc} / (\varepsilon\Sigma_P)^{1/2} \text{ and } E_{obs} = F_{obs} / (\varepsilon\Sigma_N)^{1/2} \qquad (7\text{-}145)$$

the complex variance σ_Δ^2 and the anti-correlated D can be combined into the single σ_A^2 parameter.[51] The combined variance then becomes

$$\sigma_A^2 = D^2 \Sigma_P / \Sigma_N \text{ and } \sigma_\Delta^2 = 1 - \sigma_A^2 \qquad (7\text{-}146)$$

Sigma-A extends from 0 to 1, where a value approaching 1 means that the model is essentially correct. Expressing the conditional probability distributions in normalized structure factors E and utilizing the relations for σ_A^2 we obtain the probability distributions in a normalized form that is frequently used in likelihood functions of structure factor distributions:

$$prob(\mathbf{E}_{obs} \mid \mathbf{E}_{calc}) = \begin{cases} \dfrac{1}{\pi(1-\sigma_A^2)} \exp\left(-\dfrac{\left|\mathbf{E}_{obs} - \sigma_A\mathbf{E}_{calc}\right|^2}{1-\sigma_A^2}\right) & \text{(acentric)} \\[4mm] \dfrac{1}{\left(2\pi(1-\sigma_A^2)\right)^{1/2}} \exp\left(-\dfrac{\left|\mathbf{E}_{obs} - \sigma_A\mathbf{E}_{calc}\right|^2}{2(1-\sigma_A^2)}\right) & \text{(centric)} \end{cases}$$

$$(7\text{-}147)$$

and after integrating out the phase as above for the probabilities on $\mathbf{F}$ we obtain for the normalized Rice and Woolfson distributions

$$prob(E_{obs} \mid E_{calc}) = \begin{cases} \dfrac{2E_{obs}}{1-\sigma_A^2} \cdot \exp\left(-\dfrac{E_{obs}^2 + \sigma_A^2 E_{calc}^2}{1-\sigma_A^2}\right) \cdot I_0\left(\dfrac{2\sigma_A E_{obs}E_{calc}}{1-\sigma_A^2}\right) & \text{(acentric)} \\[4mm] \left(\dfrac{2}{\pi(1-\sigma_A^2)}\right)^{1/2} \exp\left(-\dfrac{E_{obs}^2 + \sigma_A^2 E_{calc}^2}{2(1-\sigma_A^2)}\right) \cdot \cosh\left(\dfrac{\sigma_A E_{obs}E_{calc}}{1-\sigma_A^2}\right) & \text{(centric)} \end{cases}$$

$$(7\text{-}148)$$

Figure 7-19 The Rice distribution for acentric structure factors.
The Rice distribution, ubiquitous in maximum likelihood structure factor statistics, is the conditional probability or likelihood function of the expectation value for amplitude E_{obs} given a model amplitude E_{calc} and a specific value of σ_A indicating our belief in our model value: $prob(E_{obs}|E_{calc},\sigma_A)$. For the unconditional case, $\sigma_A = 0$ and $E_{calc} = 0$, the Rice distribution becomes the acentric Wilson distribution. The red graph in panel A with $E_{calc} = 0$ and $\sigma_A = 0.2$ very closely resembles that situation. For increasingly larger values of E and also of σ_A the Rice function rapidly approximates the Gaussian normal distribution (panels B, C). For a near perfect model, the probability distributions all become increasingly narrow Gaussians centered on the respective conditioning value, representing the least squares limit (panel D). The vertical light blue line represents an observed E_{obs} value of 2.4. The likelihood of observing that value given a model E_{calc} is obtained by the magnitude at the intersection of the E_{obs} line and the corresponding E_{calc} peak. Note that the likelihood of observing a given value changes dramatically with σ_A, emphasizing the importance of accurate σ_A estimates, generally obtained from cross-validation data in order to avoid bias.

Box 7-19 Conditional probability distributions and model errors. Conditional structure factor probability distributions $prob(\mathbf{F}_{obs}|\mathbf{F}_{calc})$ allow the inclusion of model errors when estimating the expectation value of an observation in the form of $\mathbf{F}_{calc}$ from the structure model. The effect of incompleteness on the expectation value can be estimated by Sim weights m derived from likelihood distributions under the assumption of an incomplete model, while positional errors of model atoms are accounted for by the Luzzati D-factor. Incorporation of bivariate errors (phase and amplitude) as a complex Gaussian function and transformation into normalized structure factors shows that the Luzzati D-factor and the complex variance σ_Δ^2 are anticorrelated, and can be represented by the single parameter σ_A.

Interpretation of the Rice function and the unconditional Wilson limit. Let us now examine the function $prob(E_{obs}|E_{calc},\sigma_A)$ conditioned on various values of $E(calc)$ and σ_A^2. We can immediately see from the equations that for the limiting case of $\sigma_A^2 = 0$ and $E(calc) = 0$ and thus with Bessel function $I_0(0) = 1$ or $\cosh(0) = 1$ we obtain exactly the non-conditional, normalized Wilson distributions as derived in the classical way in the preceding sections:

$$P(E) = \left(\frac{2}{\pi}\right)^{1/2} \exp\left(-\frac{E^2}{2}\right) \quad \text{(centric)} \tag{7-149}$$

$$P(E) = 2E \cdot \exp\left(-E^2\right) \quad \text{(acentric)} \tag{7-150}$$

Figure 7-19 illustrates the graphs for the acentric Rice distribution for varying values of E and σ_A (the centric distributions behave the same in principle). With

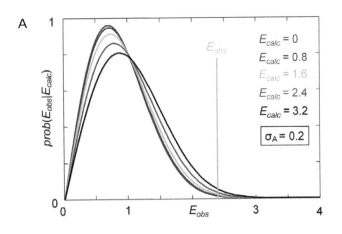
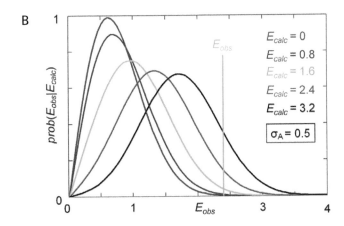
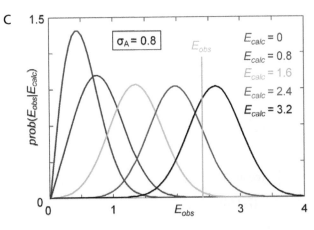
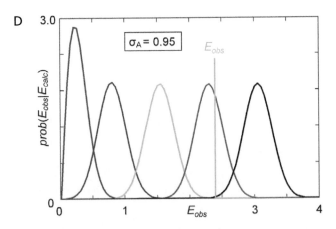

> **Box 7-20 Likelihood treatment of probability distributions for structure factors.** Integrating out the phase angle from the complex structure factor probability distribution yields 1-dimensional probability distributions $prob(F_{obs}|F_{calc})$ for the structure factor amplitudes. The distribution for general acentric reflections is the Rice distribution, also known in crystallography as the Sim distribution, and for centric reflections the Woolfson distribution. In the limiting case of no available conditioning information ($F_{calc} = 0$, $\sigma_A = 0$) the Rice distribution reverts to the acentric Wilson distribution, while the Woolfson distribution reverts to the centric Wilson distribution.

increasing E_{calc} and increasing σ_A^2 the probability distributions become more and more like Gaussians, until in the limiting case of a correct model they are (infinitely) sharp Gaussians at exactly the conditioning value, and $E_{calc} = E_{obs}$. It is also quite obvious that the value of σ_A^2 affects both the amplitude and variance of the function, thus an accurate estimate of σ_A^2 is crucial—a subject we address in the refinement part of Chapter 12.

7.12   Key concepts

Probabilities and probability distributions

- Numerous key steps in contemporary crystallographic structure determination are based on probabilistic methods of inference. Bayesian inference models that incorporate prior knowledge and the related maximum likelihood models are implemented in most crystallographic programs, ranging from data collection to phasing and refinement.
- Numeric values of a physical property ranging from photon counts to structure factors and geometric restraints follow certain distributions in their actually observed values. A function that describes the distribution of the expected values is called the probability distribution function of that property.
- Fundamental terms describing probabilities as used in crystallography are conditioning, joint probability, independence, bias, normalization, and marginalization.
- Conditioning implies that a property or event outcome is dependent on a certain given parameter.
- Joint probabilities of independent events are obtained by multiplication of individual probabilities.
- Mutually exclusive probabilities are added, thus providing a means for normalization.
- Marginalization refers to the removal or integrating out of a parameter so that it does not explicitly appear in the probability distribution function.

Common probability distributions and their properties

- The Poisson distribution $\mathbf{P}(n|\langle n \rangle)$ is a single-parameter probability distribution that only depends on the number of event counts n. With increasing number of counts ($n > 10$), the Poisson distribution approximates the Gaussian normal distribution.
- The variance σ^2 of the Poisson function is n, and the estimated standard uncertainty σ of the Poisson distribution is the square root of n. The Poisson distribution occurs in intensity (photon counting) statistics and also in crystallization probability statistics. The higher a photon count is, the lower is its relative error.
- The Gaussian normal distribution $G(x|\sigma_x, \langle x \rangle)$ is a two-parameter probability distribution that describes the distribution of values from multiple observations around the mean value $\langle x \rangle$.
- The width of the normal distribution is defined by the variance σ_x^2 or the standard uncertainty or standard deviation σ_x. The more precise a measurement

is, the narrower is the distribution, indicated by a smaller variance and a lower estimated standard uncertainty of the mean.

- Another way of representing an ordinary PDF is by its cumulative distribution function (CDF), written as $N(x)$ in crystallographic texts. $N(x)$ is obtained by integration of the ordinary PDF, and gives the probability that a data point or random variable following a probability distribution will be smaller than a given value of x.
- Cumulative distributions are useful for analyzing intensity distributions of diffraction data.
- Precise measurements are not necessarily accurate, because of the presence of systematic errors or bias.
- A fundamental theorem of frequency-based statistics is the central limit theorem. It states that irrespective of the actual form of the function describing the distribution of a variable x, the best estimate for the most probable value is the mean sample value.
- A process that minimizes the variance or sum of (normalized) residuals squared between measured data and a parameterized model will provide the best fit between model and data. This is the basis for parameter estimation or refinement (colloquially called *fitting*).

Error propagation

- Error propagation rules lead to two important results for the manipulation of intensities.
- If an intensity value is combined from two independent measured variables such as peak and background, the errors propagate via the sum of their individual variances.
- The relative error in structure factor amplitudes when converted from intensities is only $1/2$ of the relative error in intensities.
- Measures frequently used in crystallography for relative errors are the R-value and the correlation coefficient.
- While R-values require a common scale between the compared or merged data, the correlation coefficient is scale-independent and can be used when the absolute scale of the data to be correlated is unknown.
- The best estimate x_{best} for a value obtained or expected from n independent multiple observations is the weighted average.
- Weighted averages play an important role in crystallography in data merging as well as in establishing phase probability distributions and best phases.

Intensity statistics and Wilson scaling

- Structure factor distribution statistics for a random structure lead to different intensity distributions for centric and acentric reflections.
- Deviations from expected intensity distributions can indicate anomalies such as pseudo-symmetry or twinning.
- Experimentally observed intensities and the derived structure factor amplitudes are brought onto an absolute scale for further data processing and refinement.
- An initial scale factor and an overall isotropic B-factor are obtained from a straight line fit to the high-resolution region of a Wilson plot.
- In later stages of refinement, anisotropic B-factor scaling is generally applied.
- Under the assumption of point atoms, no angle-dependent attenuation of the atomic scattering factors occurs, and normalized structure factors $E_{\mathbf{h}}$ can be computed.
- Normalized structure factors are used in analysis of intensity distributions, various map constructions, and in testing phase relations in direct methods.
- The cumulative probability distributions of E^2 (normalized intensities) are particularly useful for statistical intensity distribution analysis of experimental data.
- Conditional structure factor probability distributions in the form of likelihood functions $prob(\mathbf{F}_{obs}|\mathbf{F}_{calc})$ allow the inclusion of non-normal model errors when estimating the expectation value of an observation in the form of $\mathbf{F}_{calc}$ from the structure model.

- The effect of incompleteness on the expectation value can be estimated by Sim weights m derived from likelihood distributions under the assumption of an incomplete model, while positional errors of model atoms are accounted for by the Luzzati D-factor.
- Incorporation of bivariate errors (phase and amplitude) as a complex 2-dimensional Gaussian function and expression of the resulting probability distribution in the form of normalized structure factors shows that the Luzzati D-factor and the complex variance are anticorrelated and can be substituted by the single parameter, σ_A.
- Integrating out the phase angle from the complex structure factor probability distribution yields 1-dimensional probability distributions $prob(\mathbf{F}_{obs}|\mathbf{F}_{calc})$ for the structure factor amplitudes.
- The distribution for general acentric reflections is the Rice distribution, also known in crystallography as the Sim distribution; and for centric reflections, the Woolfson distribution.
- In the limiting case of no available conditioning information ($F_{calc} = 0$, $\sigma_A = 0$) the Rice distribution reverts to the acentric Wilson distribution, and the Woolfson distribution reverts to the centric Wilson distribution.

Bayesian inference and maximum likelihood

- Bayesian inference or reasoning is a generally applicable procedure that allows the treatment of inductive logic in a formal deductive framework based on fundamental rules of probability algebra.
- Bayes' theorem states that the posterior probability of a hypothesis or model is given by our prior information or knowledge about it, modified by the likelihood function which is based on our experimental data.
- Bayesian models allow obtaining useful posterior probabilities even in case of negative intensities.
- The highest posterior probability indicating the best model can be found by maximizing the likelihood function in maximum likelihood cases where no prior probability term is present, or by maximizing the joint probability of the likelihood function and the prior probability in maximum posterior models.
- Maximum likelihood methods play a key role in protein crystallography, because in many instances the assumption of normal distribution of errors in data models is not fulfilled, and central-limit-based methods such as least squares do not provide optimal solutions.
- Bayesian and maximum likelihood methods are implemented in a wide range of crystallographic programs used in data processing, map reconstruction, and macromolecular refinement.
- Minimizing the negative log-likelihood instead of maximizing the likelihood function has a number of advantages. The very small individual structure factor probabilities making up the joint probability product of the entire likelihood function are numerically easier to handle by logarithmic addition.
- In addition, various existing minimization algorithms, such as conjugate gradient, steepest descent, or even basic linear system solvers, can be used to carry out the actual computation.
- A smaller residual implies a larger exponential term in the likelihood function or a log-likelihood gain in the logarithmic picture.
- Negative intensities can result during background subtraction from very small intensity values.
- Applying statistical expectation values based on Wilson structure factor amplitude statistics as a Bayesian prior imposes the necessary positivity on intensities, and sensible structure factor amplitudes and standard uncertainties can be computed from otherwise intractable negative intensities.
- One of the most powerful aspects of Bayesian models is to keep weak data in reasonable bounds using prior probabilities or knowledge.
- Weak or poor crystallographic data will not overcome established prior expectations, but strong data (evidence) can force us to revise and expand our prior expectations about molecular structure and function.

7.13 Additional Reading

1. Taylor JR (1982) *An Introduction to Error Analysis*. Mill Valley, CA: University Science Books.

2. Sivia DS (1996) *Data Analysis – A Bayesian Tutorial*. Oxford, UK: Oxford University Press.

3. Edwards AWF (1992) *Likelihood – An Account of the Statistical Concept of Likelihood and Its Application to Scientific Inference*. Baltimore, MD: The Johns Hopkins University Press.

4. Jaynes ET (2003) *Probability Theory: The Logic of Science*. Cambridge, UK: Cambridge University Press.

7.14 References

1. McCoy A (2004) Liking likelihood. *Acta Crystallogr.* D60(12), 2169–2183.

2. Sim G (1959) The distribution of phase angles for structures containing heavy atoms. II. A modification of the normal heavy-atom method for non-centrosymmetrical structures. *Acta Crystallogr.* 12(10), 813–815.

3. Wilson AJC (1949) The probability distribution of X-ray intensities. *Acta Crystallogr.* 2, 318–321.

4. Woolfson M (1956) An improvement of the "heavy-atom" method of solving crystal structures. *Acta Crystallogr.* 9(10), 804–810.

5. Eddy SR (2004) What is Bayesian statistics? *Nat. Biotechnol.* 22, 1177–1178.

6. Sivia DS (1996) *Data Analysis – A Bayesian Tutorial*. Oxford, UK: Oxford University Press.

7. Taylor JR (1982) *An Introduction to Error Analysis*. Mill Valley, CA: University Science Books.

8. Read RJ (1999) Detecting outliers in non-redundant diffraction data. *Acta Crystallogr.* D55, 1759–1764.

9. Wilson AJC (1950) Largest likely value for the reliability index. *Acta Crystallogr.* 3, 397–398.

10. Yeates TO (1997) Detecting and overcoming crystal twinning. *Methods Enzymol.* 276, 344–58.

11. Aldrich J (1997) R.A. Fisher and the making of maximum likelihood 1912–1922. *Statist. Sci.* 12(3), 162–176.

12. Tickle IJ (2008) Personal communication.

13. French S, & Wilson K (1978) On the treatment of negative intensity observations. *Acta Crystallogr.* A34(4), 517–525.

14. Brändén CI, & Jones TA (1990) Between objectivity and subjectivity. *Nature* 343, 687–689.

15. Rupp B (2006) Real-space solution to the problem of full disclosure. *Nature* 444(7121), 817.

16. Diederichs K, & Karplus PA (1997) Improved R-factors for diffraction data analysis in macromolecular crystallography. *Nat. Struct. Biol.* 4(4), 269–275.

17. Weiss M (2001) Global indicators of X-ray data quality. *J. Appl. Crystallogr.* 34(2), 130–135.

18. Jaynes ET (2003) *Probability Theory: The Logic of Science*. Cambridge, UK: Cambridge University Press.

19. Edwards AWF (1992) *Likelihood – An Account of the Statistical Concept of Likelihood and Its Application to Scientific Inference*. Baltimore, MD: The Johns Hopkins University Press.

20. French S (1978) A Bayesian three-stage model in crystallography. *Acta Crystallogr.* A34, 728–738.

21. Bricogne G (1997) Bayesian statistical viewpoint on structure determination: Basic concepts and examples. *Methods Enzymol.* 276, 361–423.

22. Read RJ (1986) Improved Fourier coefficients for maps using phases from partial structures with errors. *Acta Crystallogr.* A42, 140–149.

23. Bayes T (1763) An essay towards solving a problem in the doctrine of chances. *Philos. Trans. R. Soc.* 53, 370–418.

24. Kuhn TS (1970) *The Structure of Scientific Revolutions*. Chicago: University of Chicago Press.

25. Glusker JP (1994) Dorothy Crowfoot Hodgkin. *Protein Sci.* 3, 2465–2469.

26. Bricogne G, Vonrhein C, Flensburg C, et al. (2003) Generation, representation and flow of phase information in structure determination: recent developments in and around SHARP 2.0. *Acta Crystallogr.* D59(11), 2023–2030.

27. Blanc E, Roversi P, Vonrhein C, et al. (2004) Refinement of severely incomplete structures with maximum likelihood in BUSTER-TNT. *Acta Crystallogr.* D60(12), 2210–2221.

28. Read RJ (2001) Pushing the boundaries of molecular replacement with maximum likelihood. *Acta Crystallogr.* D57, 1373–1382.

29. Brünger AT, Adams PD, Clore GM, et al. (1998) Crystallography & NMR system: A new software suite for macromolecular structure determination. *Acta Crystallogr.* D54, 905–921.

30. Murshudov GN, Vagin AA, & Dodson ED (1997) Refinement of macromolecular structures by the maximum-likelihood method. *Acta Crystallogr.* D53, 240–255.

31. Press WH, Flannery BP, Teulkolsky SA, et al. (1986) *Numerical Recipes*. Cambridge, UK: Cambridge University Press.

32. Tronrud D (2004) Introduction to macromolecular refinement. *Acta Crystallogr.* D60(12), 2156–2168.

33. CCP4 (1994) The CCP4 suite: Programs for protein crystallography. *Acta Crystallogr.* D50, 760–763.

34. Sivia DS, & David WIF (1994) A Bayesian approach to extracting structure-factor amplitudes from powder diffraction data. *Acta Crystallogr.* A50, 703–714.

35. Wilson AJC (1942) Determination of absolute from relative X-ray intensity data. *Nature* 150, 152.

36. Giacovazzo C, Monaco HL, Viterbo D, et al. (2002) *Fundamentals of Crystallography*. Oxford, UK: Oxford Science Publications.

37. Iwasaki H, & Ito T (1977) Values of ε for obtaining normalized structure factors. *Acta Crystallogr.* A33(1), 227–229.

38. Hauptman H, & Karle J (1955) A relationship among the structure-factor magnitudes for P1. *Acta Crystallogr.* 8(6), 355.

39. Trueblood KN, Burgi HB, Burzlaff H, et al. (1996) Atomic displacement parameter nomenclature. Report of a subcommittee on atomic displacement parameter nomenclature. *Acta Crystallogr.* A52(5), 770–781.

40. Zwart PH, & Lamzin VS (2004) The influence of positional errors on the Debye effects. *Acta Crystallogr.* D60(2), 220–226.

41. Morris RJ, Blanc E, & Bricogne G (2004) On the interpretation and use of $<|E|^2>(d*)$ profiles. *Acta Crystallogr.* D60(2), 227–240.

42. Popov AN, & Bourenkov GP (2003) Choice of data-collection parameters based on statistic modelling. *Acta Crystallogr.* D59(7), 1145–1153.

43. Brunger AT, DeLaBarre B, Davies JM, et al. (2009) X-ray structure determination at low resolution. *Acta Crystallogr.* D65(2), 128–133.

44. Bricogne G (1976) Methods and programs for direct-space exploitation of geometric redundancies. *Acta Crystallogr.* A32(5), 832–847.

45. Sim G (1960) A note on the heavy-atom method. *Acta Crystallogr.* 13(6), 511–512.

46. Rice SO (1954) Mathematical analysis of random noise. In Wax N (Ed.), *Selected Papers on Noise and Statistics*. New York, NY: Dover Publications.

47. Sim G (1960) The cumulative distribution of structure amplitudes. *Acta Crystallogr.* 13(1), 58–59.

48. Read RJ (2003) New ways of looking at experimental phasing. *Acta Crystallogr.* D59, 1891–1902.

49. Drenth J (2007) *Principles of Protein X-ray Crystallography*. New York, NY: Springer.

50. Srinivasan R (1966) Weighting functions for use in the early stage of structure analysis when a part of the structure is known. *Acta Crystallogr.* 20(1), 143–144.

51. Srinivasan R, & Ramachandran GN (1966) Probability distribution connected with structure amplitudes of two related crystals. VI. On the significance of the parameter σA. *Acta Crystallogr.* 20(4), 570–571.

52. Main P (1979) A theoretical comparison of the α, γ and $2F_o$-F_c syntheses. *Acta Crystallogr.* 35(5), 779–785.

53. Luzzati V (1952) Traitement statistique des erreurs dans la determination des structures cristallines. *Acta Crystallogr.* 5(6), 802–810.

54. Read RJ (1990) Structure-factor probabilities for related structures. *Acta Crystallogr.* A46, 900–912.

55. Read RJ (1997) Model phases: probabilities and bias. *Methods Enzymol.* 277, 110–128.

PART III

From Crystal to Data

Instrumentation and data collection

The photographing, indexing, measuring, correcting and correlating of some 7000 reflexions was a task whose length and tediousness it will be better not to describe.

Max F. Perutz (1949) *Proc. R. Soc. London*, **A195**, 474–499

The instrumentation and data collection chapter focuses on practical aspects of data collection. As data quality ultimately and mercilessly determines the quality and information content of the resulting structure, the last actual physical experiment—before the *in silico* phase of structure determination begins—deserves careful preparation and execution.

The instrumentation subchapter begins with a section describing the main components of a diffractometer and its ancillary devices for data collection. The discussion is necessarily compact and not overly technical, and focuses on modern instrumentation in practical use at the time of writing, with a few forward-looking sidebars on new developments. Although there is not much the casual user can do to improve the performance of a diffractometer or a synchrotron beam line, the instrument parameters significantly determine the usefulness of a particular setup for a specific purpose. Knowledge or at least awareness of the instrument parameters and their effect on data collection therefore helps to make an informed decision which resource to select for data collection.

In preparation for practical data collection a discussion of crystal harvesting and mounting, with special emphasis on cryotechniques and their role in suppressing radiation damage is provided. The section is followed by a review of the properties of real crystals, discussing the origin of mosaicity and the possibility of twinning as a consequence of the domain structure of real crystals. An in-depth discussion of microscopic merohedral twinning and its detection concludes this section.

The actual data collection section focuses on the rotation method commonly used in macromolecular crystallography. A review of the Ewald construction and its use in the description of diffraction geometry is followed by the analysis of the anatomy of a rotation image. In the following we discuss data collection as a branched decision process, where from initial inspection of the image to indexing and strategy development the suitability of a crystal for the task at hand is thoroughly evaluated. A section dealing in more depth with aspects of data statistics, merging *R*-values, signal levels, space group determination, and indexing of cells concludes the theory part.

The principal steps of data processing, which were first described with emphasis on fundamental aspects, are now reinforced through practical examples of a high-resolution synchrotron data collection and the challenge of collecting high redundancy data for native sulfur single-wavelength anomalous diffraction (S-SAD) phasing. We follow the steps of data processing using the program *MOSFLM* for a high resolution synchrotron data set. We then focus on strategy generation for high redundancy data and only briefly describe the processing of the in-house data with the *PROTEUM* software.

The chapter concludes with the preparation of the experimental data section of "Table 1," a mandatory item in any crystallographic structure-determination paper.

8.1   X-ray instrumentation

The instrument used for X-ray data collection is a diffractometer, measuring the position (angular distribution) and intensity of the diffracted X-rays. The basic components of a diffractometer are:

- An intense source of hard X-rays in the 5–25 keV range.

- Suitable optics to select monochromatic X-rays and to focus or collimate them into a brilliant beam of X-rays.

- A mechanical device, the goniostat, to orient the crystal in the primary X-ray beam.

- A detector for the diffracted X-rays, generally a 2-dimensional area detector.

Additional components that support and enhance the function of a diffractometer suitable for protein crystallography include:

- A cryocooler to keep the protein crystals continuously within a stream of cold nitrogen gas to reduce radiation damage.

- A microscope focused onto the diffractometer center to align and center the crystal in the X-ray beam.

- For anomalous scattering experiments, a fluorescence detector to record the X-ray excitation spectrum for optimal choice of wavelengths.

- An optional sample changing robot for high-throughput or remote operation.

- Data collection software controlling goniostat and performing primary processing of raw data from the detector.

Several images of in-house laboratory instruments (Figures 8-1 and 8-7) and of synchrotron beam lines (Figure 8-11) identify these components.

X-ray sources

The reliable production of X-rays is a prerequisite for any X-ray diffraction experiment. There are three general types of devices that are used for generation of X-rays: laboratory X-ray sources, which rely on the emission of characteristic radiation from materials when electrons from outer shells fall back into core level holes generated by electron bombardment; synchrotron sources, which exploit the emission of X-rays by electrons moving in an orbit at high energy; and Compton sources, an area under development, where high repetition rate lasers are used to excite oscillation in electron beams and thus generate X-rays. Radioactive elements also emit hard X-rays or γ-rays, which are sometimes used in the X-ray laboratory for flood field calibration of detectors.

X-ray emission
Bremsstrahlung. The emission of bremsstrahlung (loosely translated from German as "breaking radiation") is a general phenomenon predicted by classical

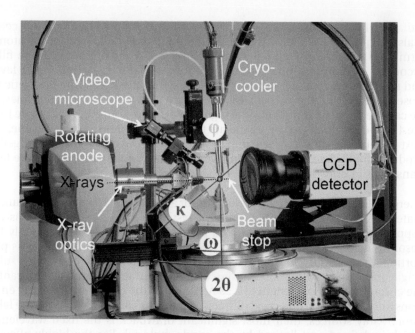

Figure 8-1 A contemporary laboratory X-ray diffractometer for macromolecular crystallography. A rotating anode X-ray source is closely coupled with integrated focusing optics delivering high photon flux at low operating power. In the center of the diffractometer is a full 4-circle κ-goniostat for orienting and rotating the crystal in multiple positions in the X-ray beam, thus enabling redundant data collection and in-house S-SAD phasing experiments. The CCD area detector is located to the right, and the diffractometer is also equipped with a cryocooler and a video microscope. The 2θ- and the ω-axis are collinear, with 2θ the detector offset angle. Image courtesy Matt Benning, Bruker AXS.

electrodynamics when a force F acts on a charged particle. Any force acting on a particle of mass m leads after Newton's $F = m \cdot a$ to acceleration (or deceleration, if the sign of the force is negative). When a high energy electron "brakes" or decelerates in the target material, it thus emits radiation. If 100% of the electron's kinetic energy is released as bremsstrahlung, the emitted X-rays have the maximum energy, equivalent to the acceleration voltage between the electron source and the target anode material (Figure 8-2).

Characteristic radiation. A high energy electron can also transfer its energy to a core level electron of the target material so that the electron departs the atom and a core hole is generated. This high energy state can relax by filling this hole with an electron from a higher shell. When this electron transitions down into the tighter bound core level, it emits the difference in energy as an X-ray photon with a frequency characteristic of this material, hence the name characteristic

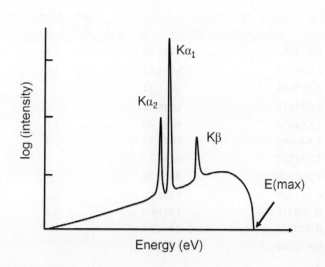

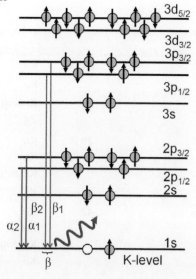

Figure 8-2 X-ray emission spectrum. When an anode material is bombarded with high energy electrons, characteristic X-ray fluorescence emission is observed on top of a continuous bremsstrahlung spectrum up to the maximum acceleration energy (or kinetic energy) of the electrons. Note the logarithmic scale in the left panel: the characteristic fluorescence radiation is ~10^3 times as intense as the background of bremsstrahlung. Right panel: The characteristic radiation emanates when holes in core shells generated by the high energy electron bombardment are filled by electrons from upper shells, and the energy difference is emitted as X-ray radiation. The panel shows transitions from higher levels to a K-level hole state. As a consequence, corresponding K-lines characteristic for the anode material are emitted. Kβ-line multiplets are generally not resolved.

radiation. In analytical X-ray spectroscopy, the term characteristic fluorescence is also frequently used for this type of emitted radiation. Looking at the atomic term level scheme in Figure 8-2 we can see that a 1s core hole can be filled according to quantum-mechanical selection rule $\Delta l = \pm 1$ only from 2p levels and not from the 2s level. The two transitions into the K-shell are energetically close together, and the two emission lines are called the $K\alpha_1$ and $K\alpha_2$ lines. As we can see from the term table, the higher energy $K\alpha_1$ transitions originate from the 4-fold degenerate $2p_{3/2}$ level and are thus twice as intense as the $K\alpha_2$ line. The next possible transition comes from 3p levels, which lie even closer and together produce the $K\beta$ line. Cu has no 4p levels, so there are no additional core level transitions or emission lines.

Energies and corresponding wavelengths for a number of frequently used anode materials are listed in Table 8-1. Protein crystallography uses nearly exclusively copper radiation (1.54 Å), which is a good compromise with reasonably low absorption and still good diffraction (recall that the diffracted intensity is proportional to λ^3). Chromium anodes are occasionally used for in-house sulfur single anomalous diffraction phasing (S-SAD), but absorption is significant at the chromium $K\alpha$ wavelength of 2.29 Å. Although the characteristic X-ray emissions are quite strong and good diffraction patterns can be recorded using laboratory X-ray sources, their most significant drawback is the fixed wavelength which is defined and limited by the anode material. For the highly effective multi-wavelength anomalous diffraction (MAD) phasing we need to be able to tune the wavelength. Although it would be in principle possible to use a monochromator and select any wavelength from the bremsstrahlung background, the approximately three orders of magnitude lower intensity of the bremsstrahlung makes this impractical for single crystal diffraction experiments.

Laboratory X-ray generators

The simplest laboratory X-ray generators use X-ray tubes, based on the same principle as the cathode ray tubes that Röntgen used in 1895 when he incidentally discovered X-rays (Chapter 6). It is important to realize that any time high energy electrons decelerate in a target material, X-rays are emitted, and any device generating a high energy electron (or charged particle) beam is thus in principle capable of producing X-rays, be it planned or by accident. The operating principle of the X-ray tubes is quite simple and is shown in Figure 8-3. A glowing tungsten filament in an evacuated tube emits electrons. A high voltage

Anode element	z	Emission line	Wavelength (Å)	Energy (eV)	Line width (eV)
Cr	24	$K\alpha_2$	2.293652	5405.20	2.4
		$K\alpha_1$	2.289755	5414.42	2.0
		$K\alpha$(avg.)	2.291048	5411.34	
		$K\beta$	2.084912	5946.36	
Cu	29	$K\alpha_2$	1.544414	8027.40	3.0
		$K\alpha_1$	1.540593	8047.32	2.4
		$K\alpha$(avg.)	1.541867	8040.67	
		$K\beta$	1.392246	8904.78	
Mo	42	$K\alpha_2$	0.713607	17373.2	6.7
		$K\alpha_1$	0.709317	17478.3	6.4
		$K\alpha$(avg.)	0.710747	17443.1	
		$K\beta_1$	0.632303	19607.1	

Table 8-1 Selected characteristic X-ray wavelengths. X-rays with longer wavelengths (lower photon energy) provide stronger diffraction (Equation 6-52) but experience higher absorption (Figure 6-29 and Sidebar 6-11). Cr anodes can be of advantage if S-SAD data are to be collected, because the characteristic Cr radiation (5.4 keV) is closest to the S absorption edge (2.5 keV). The highest observable resolution d_{min} is given by $\lambda/2$. Cu ($d_{min} = 0.77$ Å) is most commonly used for routine protein work, while short wavelength Mo radiation ($d_{min} = 0.355$ Å) has the advantage that very high resolution data from well-diffracting crystals can also be collected on an area detector without repositioning the detector to cover higher diffraction angles. Mo is thus the standard anode material for small molecule crystallography. The "average" $K\alpha$ wavelength is the weighted average (2:1) of $K\alpha_1$ and $K\alpha_2$ radiation. Energies are obtained from wavelength by conversion: E (eV) $= 12397.639/\lambda$ (Å). Values are taken from International Tables for Crystallography, Volume C, Table 4.2.2.1.

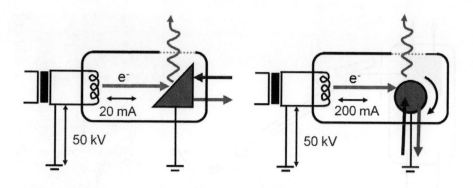

Figure 8-3 Schematic of laboratory X-ray sources. X-ray sources for the home laboratory are generally sealed tubes (left panel) or rotating anodes (right panel). Both rely on the emission of characteristic radiation specific for the anode material (red targets). A rotating water-cooled target can dissipate more heat than a stationary tube target, and small rotating anodes with integrated focusing optics (Figure 8-1) achieve photon fluxes rivaling 2nd generation synchrotron beam lines. The X-rays depart the evacuated tube or anode housing through a thin Be window (dashed green line).

of several tens of kV is applied between the electron emitting cathode (thus the name "cathode ray tubes" as a general name for such devices including old-style TV monitors) and the positively charged anode that the electrons accelerate toward. When the electrons impact the target material, they either generate heat and bremsstrahlung, or they remove an electron from the inner shell of the target (anode) material, where upon filling of the hole with a higher shell electron, characteristic X-ray radiation is emitted. More than 99.8% of the total energy is transferred into heat, thus the target anode must be well cooled, or it is sputtered off or even melted away by the electron beam. The power that can be dissipated by a stationary target in a water cooled sealed X-ray tube is limited to ~2–3 kW.

The need to dissipate substantial amounts of heat generated by the intense electron beam hitting the target prompted improvement of laboratory generators by introducing a rotating target. The anode targets rotate at up to 10 000 rpm, and as they are necessarily encased in a vacuum housing, they need to have effective vacuum seals and water feed-throughs into the (hollow) water-flushed rotating target. Energies up to several tens of kW can be dissipated on conventional rotating targets (although for fine electron beam focus, as used in protein single crystal diffraction experiments, generally only lower electron beam currents giving a maximal power of a few kW can be used). State-of-the-art rotating anodes with integrated focusing X-ray optics (Figure 8-1) deliver a X-ray photon flux of up to ~ 6 · 10^{11} X-rays mm^{-2} s^{-1} rivaling that of 2nd generation synchrotrons (although without the benefit of tunability). Such intense home laboratory sources allow the collection of single-wavelength anomalous diffraction data originating from native sulfur atoms which can effectively be used for protein phasing. Such data need to be collected with very high accuracy and precision because of the miniscule anomalous signal originating from the sulfur atoms,[1] and we will use an in-house collected S-SAD data set in the data processing and S-SAD phasing example.

X-ray optics

The X-rays emanating from the target area of the tube are radiated over a large solid angle and not directly useful as such. Only a small fraction of the generated X-rays is accepted through the entrance slits onto a filter, monochromator, or focusing optics which select the useful part of the spectrum around the Kα lines.

Filters are simple metal foils made of an element that absorbs X-rays above the characteristic radiation of the anode, and thus suppresses the characteristic Kβ line. The next element below the anode material in the periodic table is suitable as a filter material, for example Ni for copper anodes. Filters do not separate the two Kα lines.

Crystal monochromators are more efficient in the separation of the intense Kα lines from the polychromatic bremsstrahlung and the Kβ radiation. The incoming radiation is diffracted at a selected crystal plane (often (002) of pyrolytic graphite or silicon (111) for Cu home sources). The crystal disperses the X-rays by diffraction angle according to Bragg's law, and the desired wavelength is selected through narrow exit slits into a collimator. It is generally not necessary to separate the two Kα lines for protein crystallography, although focusing

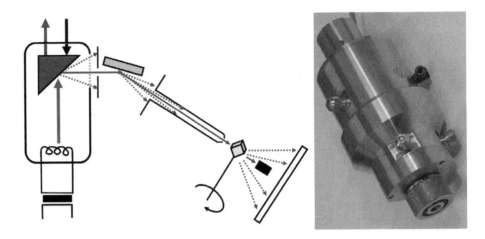

Figure 8-4 X-ray optics. The left panel shows a schematic of a basic diffractometer with a monochromator attached to a sealed tube. X-rays emanating from the anode (red) pass though entrance slits and are diffracted according to wavelength at the monochromator crystal (green) from which the narrow bands of $K\alpha_1$ and $K\alpha_2$ are selected by exit slits. A collimator in front of the crystal limits the X-rays further to a narrowly focused parallel bundle of suitable diameter, approximating the crystal size. The small black block in front of the detector plate symbolizes the beam stop. The right panel shows a compact Helios optic (image courtesy of Bruker AXS) that includes crossed focusing multilayer mirrors, which are adjusted by the thumbscrews extending from the body.

monochromators or multilayer mirror optics capable of separating them are available. Once the beam exits a non-focusing monochromator, it is still divergent and is further cut down in intensity by a collimator, a narrow tube (often helium-filled to reduce absorption by air) with small entrance and exit holes that allow only nearly parallel X-rays with low divergence to exit and diffract off the sample crystal (Figure 8-4). We clearly see that only a tiny fraction of generated X-rays ever reaches the crystal; we need an intense X-ray beam with low divergence (a brilliant beam) for diffraction experiments.

Separation and focusing of X-rays at the same time can be achieved by a sequence of mirrors, either total reflection mirrors or designed multilayer mirrors. Such modern and compact focusing X-ray optics provide significant gains in intensity and brilliance over the monochromator/collimator setup, and modern laboratory rotating generators can reach photon flux densities that rival those of 2nd generation synchrotrons, approaching 10^{11} X-rays $\cdot$ mm^{-2} s^{-1}.

Synchrotrons

Synchrotrons are an indispensable tool for modern protein crystallography. The ability to cut a precisely defined, narrow band of monochromatic X-rays out of the "white" or polychromatic primary synchrotron X-ray beam is the key to the MAD phasing techniques in protein crystallography. In addition, the high brilliance of the X-ray beam allows the collection of usable data from crystals down to 10 μm size, which generally do not show much diffraction in a reasonable time on a laboratory source.

Classical electrodynamics predicts that when electrons move in a circular orbit around atoms, they experience a centripetal force and thus acceleration, and as a consequence the electrons must emit radiation. This phenomenon actually implies the consequence that an electron circling an atomic nucleus should spiral into it while emitting light, which puzzled early atomic theorists such as Bohr tremendously—they had to postulate that the atomic orbits are stable by some magic. The resolution of the contradiction surfaced when orbital stability

Box 8-1 The diffractometer. The instrument used to record the X-rays diffracted by a crystal is the X-ray diffractometer. The basic components of a diffractometer are an X-ray source, suitable X-ray optics, a goniostat to orient and rotate the crystal, and an X-ray (area) detector. Practically all diffractometers used in macromolecular crystallography also include a cryocooler to keep the crystals at cryogenic (near liquid nitrogen) temperatures during X-ray exposure to minimize radiation damage. As a reminder, the X-rays used in crystallography are extremely intense and focused beams of ionizing radiation; proper training and precautions are mandatory when operating a diffractometer.

resulted directly from the solutions of the Schrödinger equation. Nonetheless, the effect of radiation loss when electrons circle in an orbit on macroscopic scale exists as predicted, and the electrons need to be supplied with energy to maintain a stable circular path. The first circular electron accelerator, a cyclotron, was invented and built in 1929 by Ernest O. Lawrence, the 1939 Nobel Laureate in Physics, at the University of California, Berkeley. The first cyclotron was a simple contraption of only ~10 cm in diameter. The main driver for the development of circular particle accelerators was to serve as compact high energy particle sources for nuclear physics experiments, and the radiation emitted when operating them was largely considered a nuisance.

The perception of synchrotron radiation drastically changed once electrons could be accelerated to relativistic speeds, where the "nuisance" radiation gains some interesting properties: It becomes restricted to the plane of the electron orbit, from which it emanates tangentially in a tightly focused cone in the forward direction. This makes the radiation itself useful, and the machines specifically built to produce radiation were synchrotrons, so named because they keep the electrons in a closed, synchronous path (in contrast to cyclotrons, where electrons or particles were just spiraling out).

In a modern synchrotron, electrons are produced in an electron gun, pre-accelerated in a linear injector, boosted further in energy, and fed into the storage ring in bunches (Figure 8-5). In order to keep the electrons in a stable orbit, energy lost by conversion into radiation needs to be resupplied; a task for the radio frequency generators, which add a small compensating energy "kick" to the electrons on each pass. The actual deflection of the electrons into a closed path happens in bending magnets, and numerous multi-pole steering electromagnets keep the electron beam orbit stable. The electrons are contained in a flat ultra high vacuum beam tube to minimize collision losses. Some synchrotrons are filled just once at periodic intervals with electrons, and then the electron beam current slowly decays through electron losses until the next injection, while modern, 3rd generation sources such as the Advanced Photon Source (APS) are continuously fed with electrons.[2]

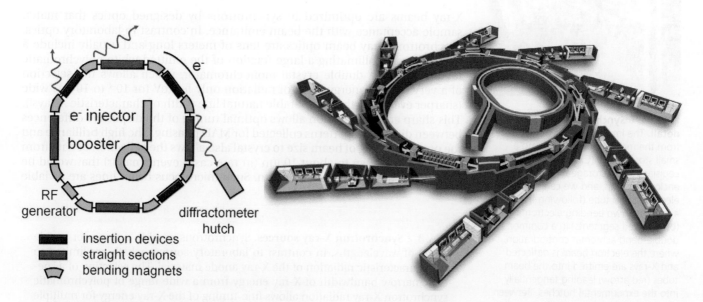

Figure 8-5 Schematics of a synchrotron. The left panel shows the location of the basic components in a synchrotron storage "ring." Electrons produced in a small electron gun or injector are pre-accelerated in the booster ring and then injected into the ring. A radio frequency generator or "kicker" compensates energy losses of the circling electrons and keeps them at constant energy. Bending magnets and insertion devices located in straight sections produce a tangential beam of white (polychromatic) X-rays, which are monochromated to a narrow band width in the beam line optics and finally reach the experimental hutch with the diffraction equipment. The right panel shows the layout of the Soleil synchrotron in France, reproduced courtesy of EPSIM 3D/JF Santarelli, Synchrotron Soleil.

In order to achieve high intensity, focusing, and sufficient energy of the emitted synchrotron radiation, the electrons must be accelerated to relativistic speeds and acquire energy in the GeV range. As the capability to deflect electrons is essentially limited by achievable magnetic fields, high energy synchrotrons become very large, while hard UV or soft X-ray synchrotrons, used for lithographic mask etching in electronic chip production, can fit into a room. The higher the electron beam energy, the higher the critical energy or wavelength λ_c of the synchrotron ring:

$$\lambda_c = 5.59 \frac{r}{E^3} \tag{8-1}$$

where r is the diameter in meters and E is the energy in GeV. This is of interest to crystallography because the intensity of the emitted X-rays starts to decay rapidly even before the critical energy, and limits the highest useful energy (or shortest wavelength) at which diffraction experiments can be conducted in a reasonable time frame.

The bending magnets (Figure 8-6) are needed for electron beam deflection and they also act as sources of radiation, but even more powerful insertion devices can be placed in straight sections. These insertion devices are undulators and wigglers, so called because they "wiggle" the electrons through a high, alternating magnetic field produced by a series of permanent magnet pairs with alternating polarity. This again exerts accelerating force on the electrons, and very intense and focused radiation emanates.

Brute intensity or flux density of photons alone are not helpful, because high divergence, as we have seen in Figure 8-4 for laboratory optics, leads to losses of most of the radiation in the X-ray optics path. High primary beam divergence translates into unacceptably high spread of the reflected X-rays, that is, broad and thus overlapping diffraction spots. What we need is high brilliance at a given wavelength, defined as the number of photons per second that pass through an area of 1 mm² with a given divergence measured in millirad:

$$B_\lambda = \text{photons} \cdot s^{-1} \cdot mm^{-2} \cdot mrad^{-2}$$

X-ray beams are optimized in synchrotrons by designed optics that match sample acceptance with the beam emittance. In contrast to laboratory optics, synchrotron X-ray beam optics are tens of meters long and usually include a set of mirrors collimating a large fraction of the emitted white polychromatic radiation onto a double crystal monochromator, which allows the selection of a very narrow energy band of radiation only few eV (or 10^{-4} to 10^{-5} Å) wide (sharper even than the comparable natural line width of characteristic X-rays!). This sharp energy selection allows optimal tuning of the dispersive differences between diffraction patterns collected for MAD phasing. The high brilliance and the optimal tuning of beam size to crystal also allows the collection of data from microcrystals down to about 10 µm (in rare cases even smaller) that would be too tiny for in-house data collection. Such microfocus beam lines are a viable

Figure 8-6 Synchrotron storage ring detail. The image shows a view down from the maintenance bridge into a small storage ring (CAMD Baton Rouge, Louisiana). This storage ring is not enclosed on top, and we can see the electron beam tube (following the green arrow) and two bending electromagnets (blue curved segments) in a common double-bend achromat configuration, where the electron beam is deflected and X-rays are emitted into the beam tubes (red arrow) leading tangentially into the experimental hutches. Between the two bending magnets, four small red steering quadrupole magnets and a sextupole (larger red steering magnet in the center) are visible, together with an ultra high vacuum pump to maintain the storage ring tube vacuum. Image courtesy of Henry Bellamy, CAMD, U. Louisiana, Baton Rouge.

Box 8-2 **Synchrotron X-ray sources.** Synchrotrons provide X-rays over a wide range of wavelengths, in contrast to laboratory sources which are limited to the characteristic radiation of the X-ray anode material. The selection of a specific, narrow bandwidth of X-ray energy from a wide range of polychromatic synchrotron X-ray radiation allows fine-tuning of the X-ray energy for multiple anomalous diffraction (MAD) experiments. The X-rays emanating from a synchrotron ring are generated by the action of a magnetic field in bending magnets or insertion devices on the electrons circling in the synchrotron storage ring. The high brilliance of synchrotron X-rays allows the collection of data from very small crystals (10 µm and sometimes below), and dedicated microfocus beam lines exist on most synchrotrons.

resource when only the smallest crystals can be successfully grown,[3] and are often combined with advanced mounting robotics.[4]

The high intensity of X-rays generated by modern 3rd generation synchrotron sources causes severe radiation damage, and thus has prompted the near-exclusive use of cryocrystallography techniques, in which crystals are kept at cryogenic (near liquid nitrogen) temperatures to minimize radiation damage. Synchrotron radiation has additional features that make it attractive for advanced applications. A wider bandwidth of polychromatic X-rays can be used in so-called Laue-diffraction experiments, where many more lattice planes in the crystal fulfill reflection conditions at the same time. This enables time-resolved X-ray diffraction studies of biologically relevant reactions. Because synchrotron radiation is pulsed as a result of the bunched nature of the electron beam, it can be exploited for examining rapid time-dependent phenomena, and because it is highly polarized, it can be used to investigate polarization-dependent and angle-dependent effects.[5]

X-ray area detectors

Irrespective of the radiation source, once the X-rays are diffracted by a sample crystal, they need to be reliably and efficiently detected. Practically all detectors for contemporary macromolecular crystallography experiments are area detectors which cover a large solid angle of diffracted radiation.

Film and imaging plates

Area detectors in the form of sheets of X-ray sensitive film encased in light-shielding cartridges were used early on in protein crystallography. The exposed films had to be developed and processed off-line in a dark room and then read by a densitometer. As film has a high sensitivity but limited dynamic range, stacks of film on top of each other were packed into each cassette and each one had to be read separately (giving rise to the introductory quote for this chapter by Max Perutz).

The first generation of area detectors replacing the X-ray film packages were imaging plates (IPs), which store the X-ray photon energy in a phosphorescent material. A layer of small crystalline grains consisting of BaFBr crystals doped with lanthanide ions (commonly Eu^{2+}) and organic binders is sandwiched between a support layer and protective layer on a flexible backing forming the imaging plate. X-ray photons excite photoluminescence in the crystals, which remain in excited states for many hours. Initially theses image plates were again read off-line outside of the beam line hutch using a large He–Ne laser scanner exciting the luminescence centers to release the energy as 390 nm photons, which are amplified with a photomultiplier and stored as a digital image. After readout, imaging plates are erased for reuse with high intensity light. As a consequence of miniaturization of lasers and electronic components, this type of detectors has been refined by combining the plate, readout, and erasure electronics into compact and reliable detectors allowing on-line electronic readout of the diffraction spot intensities. Modern IPs have high quantum efficiency, a wide dynamic range of about 10^5–10^6, good linearity of response, a high spatial resolution, a large active area, and are the least expensive detectors.[7] Their drawback in synchrotron use is the slow readout, which is generally not an issue for less intense home-lab X-ray sources.

Multiwire array detectors

The first area detectors that detected X-ray photons directly were photon counting multiwire array detectors. They are position-sensitive proportional counters and consist of a large beryllium entrance window acting as cathode of a proportional counter chamber (Be with $z = 4$ has a very low X-ray absorption coefficient but it is quite expensive to manufacture in large sheets and also toxic). The chamber contains an array of tungsten wires with a high voltage bias against the cathode window. When X-ray photons are absorbed and ionize the counting gas, electron cascades generate an electronic pulse that can be recorded. Multiwire detectors for crystallography were pioneered by Ron Hamlin and Xuong Nguyen-

Sidebar 8-1 Potential X-ray sources of the future: Compton Sources and FEL. There are two other X-ray sources that may become relevant for biological crystallography, the X-ray Free Electron Laser (FEL) and the Compton Sources (CS). In a CS, electrons in a small, room-sized storage ring (MeV range) are undulated by a high intensity laser focused into the oncoming electron beam. In the collision zone the electrons inelastically backscatter high energy X-ray photons (Compton effect). The resulting X-rays are emitted in a narrow energy bandwidth range of a few 100 eV and can be tightly focused and mono-chromated, but currently the repetition rate of commercial high intensity (terawatt) lasers is not high enough to produce continuous X-rays sufficiently intense for routine use in diffraction experiments.[6]

The FEL requires a high energy electron beam that is sent through a very long wiggler. The alternating magnetic field vectors in the wiggler impose alternating accelerating forces on the electrons, which emit X-rays in phase. The resulting coherent X-ray flash is extremely brilliant, and could be potentially used for determining molecular envelopes by single-molecule scattering (discussed in Sidebar 6-8).

Natural X-ray sources in the form of radioactive material (^{55}Fe) emitting photons with energy of 5.9 KeV with a half-life of 2.7 years are occasionally used for detector calibration.

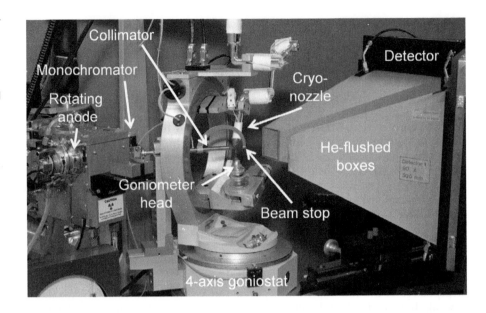

Figure 8-7 Diffractometer with dual multiwire array detectors. These fully electronic detectors, developed at the University of California at San Diego under Ron Hamlin, Chris Nielsen, and Xuong Nguyen-Huu,[8] have revolutionized X-ray data collection and were in use from the late 1970s to early 1990s. Note the large source–crystal–detector distances that require a He-flushed beam tunnel,[9] and He-boxes to reduce air absorption and background scatter. The helium-supplying plastic tubes leading to the collimator and the boxes are distinctly visible. Helium boxes are now again becoming fashionable at synchrotrons when long wavelengths are used for sulfur single-wavelength anomalous diffraction (S-SAD) phasing experiments. The goniostat of the depicted diffractometer is a closed 4-circle Eulerian cradle, now only rarely used in macromolecular crystallography because of the increased solid angle coverage achievable with modern area detectors.

Huu at the University of California[8] and introduced to protein crystallography in 1983. They revolutionized the field and remained in use until the current state of the art, the charge-coupled device (CCD) detectors, emerged. Despite high sensitivity and excellent counting statistics, the large pixel size determined by the wire spacing in the counter chamber (which could not be made much smaller than 1 or 2 mm because of arcing between the electric counter wires) requires large crystal–detector distances with helium boxes (Figure 8-7) to prevent absorption, and the diffraction pattern could be collected only in a relatively small solid angle during each exposure.

CCD detectors

The predominantly used detectors today are CCD detectors. At the core of a CCD detector are 2-dimensional charge-coupled device (CCD) semiconductor array chips (essentially highly specialized video camera chips) with up to $4k \times 4k$ pixels which directly deliver a digital image of the diffraction pattern. In contrast to imaging plates, which need to store the photon energy for delayed readout, on a CCD detector the X-ray photons fall onto a thin fluorescent screen (where they are directly converted to visible light) which is bonded to an optical glass fiber taper leading to a photon-sensitive CCD chip bonded to the other end of the taper. Gd_2O_2S (terbium doped) is a commonly used efficient "phosphor" material for photon conversion screens, which are also highly absorbing and must be kept thin.

The photons generated by the X-rays absorbed in the fluorescent screen pass through the taper onto the chip and generate free electrons in the semiconductor

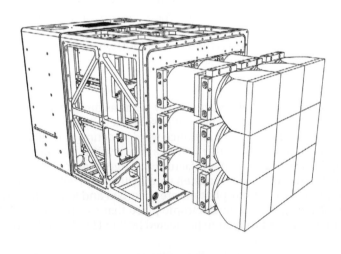

Figure 8-8 Schematic of a 3×3 module CCD area detector. Because of the limited resolution of CCD chips and the difficulty and expense of producing large glass fiber tapers, several modules are combined into a single detector covering a large solid angle. Electronics in the back of the housing perform readout, and software compensates for gaps between modules and for geometric distortion. Schematic of the Quantum 315 detector courtesy of Chris Nielson, ADSC, San Diego.

of the CCD in proportion to the number of photons. Depending on the specific design of the detector, fast readout electronics generate a raw electronic image of the diffraction pattern, which is further processed by the data collection computer. To reduce their dark current (noise), the sensitive complementary metal oxide semiconductor (CMOS) chips are generally cooled. CCD detectors exhibit high sensitivity, high quantum efficiency, low noise, and excellent linearity of response. However, CCD detectors can be saturated by very intense X-rays (the dynamic range generally covers 16 bits, i.e., $2^{16} = 65\,536$ counts), and multiple data collection passes with very short exposures for intense reflections may be necessary to capture data from crystals that diffract strongly. Exposure times for single images can be as fast as seconds on synchrotrons, and a whole data set can sometimes be collected within minutes. Because of the limited capability of producing large optical fiber tapers or CCD chips, four or nine detector elements are often combined into large 2×2 or 3×3 module CCD detectors (Figures 8-8 and 8-9).

An interesting phenomenon affecting CCD detectors is the generation of *zingers* which are sharp spots (often single pixel size) in the diffraction image. They result from photons generated in the glass taper by high energy cosmic radiation passing through the detector (and the building) or by traces of radioactive thorium in the taper.

Cosmic zingers originate from muons (μ^-), which are negatively charged elementary particles generated by cosmic ray protons in the upper layer of the atmosphere. They are extremely penetrating but can be absorbed into, and ionize, the silicon of the CCD. They can be distinguished from reflections by their sharpness (high counts only in a few discrete pixels with the adjacent pixels showing only background). A μ^- striking the CCD across the layer of the chip shows up as a stripe. Muon zingers affect all CCD-type detectors. Some zingers are the result of particles (He^{2+}) originating from the decay of radioactive contaminants in the taper glass. If they materialize very close to either the fluorescent screen or the CCD chip (α particles are not very penetrating), α-zingers look like weak reflections when interacting with the fluorescent screen or sharp saturating spots when they hit the CCD chip. Zingers can be accounted for by comparing two images taken for each exposure, as zingers will appear at different locations on each of the images.

Sidebar 8-2 Already arriving at your beamline: pixel array detectors. A Pixel Array Detector (PAD) consists of a pixelated silicon diode layer bonded to a CMOS chip. The X-rays are directly absorbed at the silicon pixel layer (no converting layer and taper needed) and converted into an electrical signal that is read out individually from the underlying CMOS chip. The advantages are very large dynamic range, low dead time, low noise, and extremely fast signal readout. The possibility of full-frame readout in mixed analog and digital readout mode[10] in the millisecond range makes these hybrid detectors particularly valuable for shutterless synchrotron data collection and time resolved X-ray diffraction experiments.[11] The first PADs designed and built by major detector manufacturers are already installed on several synchrotron beam lines, having significant impact on the practice and strategies of data collection: In addition to much faster accumulation of routine data, the high quality of (anomalous) diffraction data pushes the limits for S-SAD and related phasing techniques.

Figure 8-9 View into a synchrotron hutch equipped with CCD detector. Beam line ID23 at the ESRF in Grenoble is equipped with a 3×3 module CCD detector (white box with nine cable sets leading to the modules). Note the table that allows the detector to be lifted vertically if a larger diffraction angle needs to be covered. We are looking at the back of the detector; the beam tunnel is coming from the front of the hutch, out of the paper plane. Image courtesy of Chris Nielson, ADSC, San Diego.

Figure 8-10 Basic two-axis goniostat with fixed χ offset angle. This type of goniostat is very common for macromolecular diffractometers. It has the benefit of being simple while still allowing two different scan types (ω-scan or φ-scan) providing more degrees of freedom for more flexible data collection strategies than a basic single axis φ-goniostat. The two axes and the X-ray beam coming from the right intersect in the diffractometer center or eucentric point, where the crystal is also located. Image of goniostat courtesy of Matt Benning, Bruker AXS.

Goniostats

In order to collect a complete set of diffraction data, we need to sample an entire asymmetric unit of reciprocal space. This requires being able to rotate the crystal around at least one axis. The instrument that allows this is the goniostat or goniometer. Goniostats come in various designs; the most common instrument for protein work is the single-axis goniostat. The single (or primary) rotation axis is commonly called the phi (φ) axis. Given the limited, mostly forward, diffraction by protein crystals and a sufficiently large area detector, even in rare cases of a unique crystal axis aligned parallel with the φ rotation axis, a single rotation covers practically the entire diffracting reciprocal space, as discussed in Chapter 6. Limited additional adjustments are possible on the goniometer head (screwed with a standard thread to the goniostat), which allows for translational movement in x, y, and z to center the crystal in the beam. Some (eucentric) goniometer heads allow additional limited offset in two perpendicular arcs. Figure 8-11 shows a single-axis goniostat in a synchrotron hutch.

More complex goniostats exist which allow a higher degree of freedom for the crystal orientation. A common design for protein crystallography is a 2-axis goniostat in fixed κ (kappa) geometry. Here the rotation axis ω is perpendicular to the beam, and the φ axis is offset by a fixed χ-angle, often for simplicity of calculations the magic angle of 54.74°. The offset arm (Figure 8-10) usually passes under the optics, and thus allows the full range of motion without danger of equipment collisions. The 4-axis κ-goniostat additionally provides rotation around the offset κ-axis, with the 4th axis being the θ-axis of the detector arm co-aligned with ω. Finally, the Eulerian cradle or circle is a goniostat that also allows full coverage of all crystal orientations, but because of its large size and propensity to collide with other parts of the diffractometer it has found limited use in routine protein crystallography work. Figure 8-7 shows such a 4-cycle goniostat together with labeling of major components of the diffractometer. It is mandatory that all goniostat axes precisely intersect with each other and with the X-ray beam in the diffractometer center or eucentric point. A user normally cannot affect goniostat and diffractometer alignment, but bumping into the goniostat or otherwise mishandling it will require that the beam and goniostat alignment is checked and readjusted if necessary.

To maintain exposure to the X-rays during rotation, the crystal must be brought precisely into the unique diffractometer center. Manual centering is typically accomplished by observing the crystal from two angles 90° apart in rotation, which are selected so that the goniometer head translations are perpendicular to the line of sight of the microscope. At each observation angle the crystal is centered on the crosshairs of a microscope aligned with the diffractometer center. After a few iterations the crystal is centered on the rotation axis. This x-y adjustment is preceded or followed by translating the crystal along the rotation axis (in the z-direction) into the beam center. In addition to measuring offsets from the center, automated centering software implemented on some synchrotron beam lines also uses focal point optimization to center crystals.

Cryocooling equipment

The susceptibility of protein crystals to radiation damage, particularly at the intense synchrotron sources, requires that crystals are kept at cryogenic

---

Box 8-3 X-ray area detectors. The large number (up to many 100 000s) of X-ray reflections that need to be recorded because of the large unit cells of macromolecular crystals requires the near-exclusive use of area detectors for data collection. The most common detectors are cost-effective imaging plate (IP) detectors and charge-coupled device (CCD) detectors, with the latter far more common on synchrotrons because of their faster read-out times. These detectors generate a raw digital diffraction image file that is corrected and processed by detector-specific raw image processing software.

Sidebar 8-3 **Cryoequipment.** Cryotechniques were introduced to protein crystallography as early as the late 1960s. The necessity to rapidly quench or flash-cool the crystals, its effect on mosaicity, and the need for cryoprotection were realized early on, as summarized in a recent review.[12] The technique was popularized by Greg Petsko[13] and Håkon Hope,[14] and refined by many others. Cryocooling of protein crystals is now routinely applied in protein crystallography.

Early cryocoolers were simple boil-off devices, with a heater in a LN_2-tank that regulated the evaporation of liquid nitrogen. The cold gas was conducted through evacuated and mirrored glass tubes connected by flexible joints to a heated nozzle close to the crystal. The glass tubes of such a boil-off device can be seen in Figure 8-7. Their major problem was the icing up of the joints and the inherent fragility of the glass parts. Because of warming of the boiled-off nitrogen gas during the long travel distance from LN_2 container to nozzle, low temperatures together with a low flow were hard to achieve (~125 K at best).

An informative video describing cryotechniques has been produced by Stephen Ealick and co-workers at the Cornell High Energy Synchrotron Source (CHESS). It can be downloaded from the web: search for cryocryst_shortvideo.wmv.

temperatures (generally slightly above liquid nitrogen (LN_2) boiling temperature of 77 K or –196°C) during data collection (Section 8-2). This is achieved by a cryocooler gently blowing a stream of cold nitrogen gas over the crystal.

A common cryocooler design evaporates LN_2 in a coil in the cooler head, and extracts the heat of evaporation from the same gas again in a second coil in contact and counter-flowing to the LN_2 coil. Thus the cooling gas is brought to near liquid nitrogen temperature close to the nozzle, and very low temperatures close to the boiling temperature of LN_2 can be achieved. A common nuisance particularly in high humidity environments can be the formation of frost through condensation of water vapor on the sample, goniometer head, or nozzle exit. Frosting is counteracted by nozzle heaters, heated baffles, or a secondary warm and dry nitrogen (or air) stream concentrically shielding the cryogenic nitrogen stream. Frost formation on the sample is generally just a nuisance; it does not destroy the crystal and can sometimes be blown off or knocked off. Frost forming on the crystal, as well as ice crystals in the mother liquor, produce distinct diffraction rings (Figure 8-25).

Mounting robotics

In principle every protein crystal deserves X-ray screening (recall that looks are deceptive), and fast and reliable storage and mounting procedures are needed to safely and efficiently process a large number of samples. In addition to saving valuable synchrotron beam time (lost largely during repeated opening, closing, and securing of the diffractometer hutch), mounting robots (also called automounters) allow the safe removal and re-storage of a sample, should the initial analysis of diffraction snapshots cast doubt on the crystal quality. Potentially better crystals can be mounted and examined without the risk of losing the best one found so far.

Automated mounting of the cryo-pins carrying the harvested and cryoprotected crystals on the diffractometer increases reliability and enhances utilization of valuable synchrotron beam time. Major synchrotron facilities and larger biotech companies have developed mounting robots for their HTPX beam lines,[15, 16] and commercial systems, which are also suitable for laboratory sources, are available (Mar Research, Rigaku Americas, Bruker AXS). The robotic mounting system developed at the Advanced Light Source (ALS) Macromolecular Crystallization Facility[17] may serve as an example of a practical development (Figure 8-11).

Sample transport and storage systems generally consist of a storage puck or storage cylinder that harbors the cryo-pins and is normally submerged in liquid nitrogen. The so-called Hampton-style and the SPINE cryo-pins have been adopted as a *de facto* standard and fit various storage and mounting systems, but they are not entirely compatible with each other.

Figure 8-11 View inside the experimental hutch of beam line 5.0.3 at the Advanced Light Source (ALS) in Berkeley. Several key components of the instrumentation can be distinguished: A 4-module CCD detector in the left foreground, the single-axis goniostat (blue base) equipped with a centering goniometer head, and the video camera (black box) used for crystal centering above the beam tunnel. A cryocooler (cylindrical vessel with nozzle pointing toward the crystal) delivers cryogenic nitrogen gas to the crystal. Opposite to the goniometer axis, the gripper of the mounting robot is visible. The mounting robot can randomly access any sample with a cooled, robotic gripper which transfers the sample from the storage Dewar (large white vessel with red plug) to the diffractometer within seconds, maintaining the crystal temperature below 110 K. The Dewar contains pucks loaded at the crystallization laboratory with the crystals mounted on cryo-pins, and several pucks at a time are transferred into the robot-hutch liquid nitrogen vessel. Crystals are centered automatically on the computer-controlled goniometer head. Image courtesy of G. Snell, ALS Berkeley.

8.2   Crystal harvesting and mounting

Successful data collection starts with proper harvesting, cryoprotection, and mounting of the delicate protein crystals. The quality of the crystal and the data will ultimately determine the achievable quality of the final structure, and careful mounting without insult to the precious crystals is the first step to assure successful data collection. Because of the prevention or minimization of radiation damage caused by the ionizing X-rays, cryomounting in cryoloops is the most prevalent mounting technique.

Harvesting and mounting of crystals

Once our crystals have grown to a size of about 10 μm to a few 100 μm, preferable in a reasonably isotropic fashion ("blocky" crystals being the most desirable), they need to be harvested from the crystallization solution and mounted on the diffractometer. Protein crystals need to remain in the mother liquor during the mounting process and the data collection; otherwise they dry out, disintegrate, and stop diffracting[18] (Sidebar 3-2). There are essentially two methods: capillary mounting, which is still used for room temperature data collection, and cryomounting, for data collection at cryogenic temperatures. Some specialized techniques have been developed for growing the crystals directly in loops, which then are cryocooled without further manipulation.[19] Proper flash-cooling and cryomounting does require some practice, and a tray of lysozyme crystals may come in handy for that purpose.

Capillary mounting

In capillary mounting, the crystal, viewed under the microscope, is sucked up together with mother liquor from its growth drop into a thin-walled quartz-glass capillary of suitable diameter. The liquid is wicked out of the capillary, leaving just enough solvent around the crystal to maintain humidity and to provide some surface tension to adhere the crystal to the inner wall of the capillary. The capillary is then sealed with wax or some inert glue and centered on the goniostat for data collection. Although no longer used for routine data collection, capillary mounting is still important if problems occur during cryocooling, such as crystal damage or unacceptable increase in mosaicity leading to poor diffraction. A practical alternative to capillary mounting is thin, protective Mylar sleeves that can be slipped over crystals harvested in cryoloops, thereby maintaining a constant humidity environment. Crystals grown during free-interface diffusion experiments in capillaries or microfluidic chips are also directly mounted on the diffractometer for room temperature screening or data collection.

Cryomounting and cryoprotection

The vast majority of diffraction data are routinely collected from cryoquenched (flash-cooled) crystals. The crystals are harvested while viewed under the microscope (a good base with support for the lower arms and hands is important) and a single crystal is scooped out of the mother liquor with a nylon loop or microfabricated Kapton mount[20] attached to a steel pin with a magnetic base (Figure 8-12). The Hampton-style bases and pins have evolved as a *de facto* standard and can be used and handled by most laboratory robotics and synchrotron

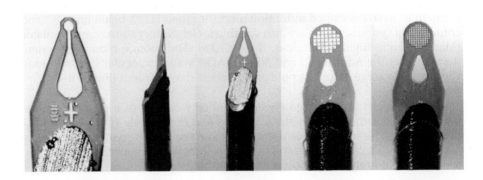

Figure 8-12 Cryoloops. Cryomounting loops and meshes micro-fabricated from laser-etched Kapton[20] are reproducibly manufactured, sturdy but flexible, and a significant improvement over the floppy nylon or hair loops used previously.

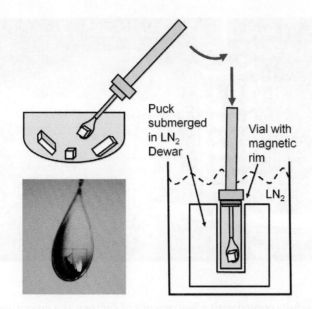

Puck submerged in LN₂ Dewar

Vial with magnetic rim

LN₂

Figure 8-13 Harvesting and flash-cooling process. The crystal is scooped up from the drop in a mounting loop on a pin, and dipped quickly into liquid nitrogen. A vial with a magnetic rim placed in a submerged storage puck protects the mounting pin. The insert shows a harvested crystal in an older style nylon loop that is quite large for the crystal. Pre-fabricated loops are now available in many sizes, allowing to select a harvesting loop that matches crystal dimensions. The cross-hairs indicate the crystal location determined by image recognition software. Insert image courtesy of Andrea Schmidt, EMBL Hamburg.

beam lines, and the very similar SPINE pins have been developed by European high-throughput laboratories. It is advisable to verify that the equipment you plan to use in-house or at the user facility is compatible with your favorite mounting pin bases.

The captured crystal may be directly dipped into liquid nitrogen (LN₂) if there is sufficient cryoprotectant already in the crystallization cocktail (Figure 8-13). If this is not the case, the crystal is briefly swished through a cryoprotectant and then quickly dipped into liquid nitrogen. The cryo-pin with the crystal is then transferred into the nitrogen cold-gas stream of a diffractometer with pre-cooled protective tongs or LN₂-containing vials to maintain cryogenic temperature. Alternatively, the cryo-pin with the crystal can be mounted directly on the goniostat with the cryostream deflected with a card or a shutter. The card is then rapidly removed and the crystal is cooled by the cryogenic gas stream. This cooling technique is simple but yields significantly slower quench rates. A cryoprotectant acts as an anti-freeze, preventing the mother liquor water from crystallizing during the rapid cooling (also called quenching) and the mother liquor solidifies as an amorphous phase (vitrifies). The formation of crystalline ice outside of a crystal in the surrounding mother liquor almost invariably destroys the diffraction quality of the crystal largely because of the change of density during the phase transition from water to ice. Inside the crystal, ice formation is generally hampered as the protein itself acts as a cryoprotectant. Common cryoprotectants are glycerol, ethylene glycol, MPD, sucrose, and low molecular weight PEGs, all in a buffer compatible with and at the same pH as the crystallization cocktail. Crystallization cocktail spiked with additional cryoprotectant can also be used. Proper selection of loop size, as small as possible while still large enough to loop the crystal out of the growth drop, minimizes the amount of external mother liquor.

Unfortunately, it is not possible to predict which cryoprotectant works for a specific protein, and some screening is generally necessary. Some crystals cannot be successfully cooled at all (see Sidebar 3-11, the PRD1 bacteriophage head, for an example). The microscopic reasons are likely growth defects caused by impurities[21] or general problems with the domain structure of the crystals[22] (Chapter 3). Some crystallization cocktails already contain enough salt[23] (~0.5 M and above) or higher concentrations (~10% and above) of a cryoprotectant such as glycerol, a sugar, a PEG, or PEG-MME, so that the crystals can be directly quenched in liquid nitrogen. In cases where a potential cryoprotectant is present but not in sufficient concentration, the reservoir cocktail spiked with additional amounts of the cryoprotectant is a good start. If no cryocomponent at all is contained in the cocktail, PEGs or glycerol at increasing concentrations and at the same pH as the growth cocktail are an alternative.[24] However, this simple strategy may not always be sufficient to find a suitable cryoprotectant. Hen egg white lysozyme, for example,

Figure 8-14 Various cryotools. The picture shows a variety of common tools used during cryomounting and cryostorage of protein crystals. The blue vessel is an unbreakable foam Dewar made by SpearLabs from closed cell foam. On top of it rests a magnetic wand used to hold and manipulate the mounting pins. A Hampton base with a copper pin is shown; stainless steel pins are more frequently used. The storage cylinder to the left of the foam vessel is an SSRL design. It fits the bore of a shipping Dewar and the pins can be auto-mounted by the SSRL beam line robotics. Below the cylinder is a simple aluminum cane that holds a series of cryo-vials with a magnetic rim securing and protecting the mounting pins. Several of these canes can be stored in a shipping or storage Dewar. At the bottom of the image is a cryo-tong that is pre-cooled under LN_2 and used to transfer pins from storage or quenching vessels onto the diffractometer. Tools and samples are kept at cryogenic temperatures at all times during mounting.

can be successfully cooled with a few percent of sucrose as a cryoprotectant or by removing the mother liquor surrounding the crystal by swiping the loops through Paratone-N, a high viscosity mineral oil. It is also possible to swish the crystals through low-viscosity perfluoroethers to remove excess mother liquor.

The actual quenching or flash-cooling of the crystal is done by quickly dipping the mounting pin with the looped crystal into liquid nitrogen, where a storage puck is submerged. The base of the pin is held safely in place by a magnetic rim on the puck (Figure 8-14) or a vial located in the puck. The loaded puck can be transferred into a shipping or storage container filled with liquid nitrogen. Once cryocooled, the crystals are essentially permanently conserved. Different designs of storage pucks exist, but the magnetic bases and vials are standardized. The actual mounting of the cryo-pins on the diffractometer is generally automated on synchrotrons, as described above.

Advantages of cryocrystallography

The benefits of cooling protein crystals to cryogenic temperatures are multiple. First and foremost, radiation damage due to radicals formed by the ionizing X-ray photons is practically eliminated on in-house X-ray sources, and is greatly reduced even at powerful synchrotron beam lines. This in turn means that complete data can be collected from a single crystal, and no merging or non-isomorphism issues between partial data sets are introduced. Because data collection at ~100 K usually prolongs the crystal lifetime by a factor of ~70,[25] the problem of radiation damage had largely been considered resolved until the advent of third generation synchrotrons in the late 1990s. Radiation damage still does occur at extremely high X-ray doses on synchrotrons (Figure 8-15). Given that free radical reactions are kinetically hampered at cryogenic temperatures below about 130 K and infrared imaging has shown that local heating of protein crystals does not seem to be significant,[26] other, likely solid-state, quasi-particle transfer mechanisms are probably responsible for high-dose radiation damage still occurring even at cryogenic temperatures.[27]

Second, data collection at cryogenic temperatures close to the liquid nitrogen boiling point of 77 K (practically, in the range of 90–120 K) reduces thermal vibrations according to physical principles. In addition, flexible parts of the protein can snap into specific conformations and become distinct in electron density. Both effects lead to increased resolution, which means more data and thus a more detailed structure. However, the effect of resolution improvement is most pronounced at higher resolution.[14] In practice, a crystal with 3–4 Å resolution limit will not show dramatically improved diffraction upon cryocooling, but the improvement from 1.7 Å at room temperature to 1.2 Å at cryogenic temperature would be very dramatic. This again shows that careful screening for the best diffracting crystal form will pay big dividends later in the structure determination!

Symptoms of radiation damage to protein crystals detectable during data collection include decrease in diffraction intensity and resolution; increase in unit cell volume; and color changes.[28] In refined models, site-specific damage, beginning with breakage of disulfide bonds, demethylation of methionine, and sulfoxidation,[29] followed by decarboxylation of aspartates, glutamates, and the C-terminus, and then loss of the hydroxyl group from tyrosines[30–32] can be observed. Even a phasing method, radiation-damage-induced phasing (RIP), exists that exploits the difference between data sets obtained from native and severely radiation-damaged crystals.[33] Note that the presence of heavy atoms or other strongly absorbing constituents (such as selenomethionine) in the crystal or the mother liquor can dramatically increase the damage to crystal. James Holton has compiled a list of these effects in addition to expected lifetimes on various synchrotron beam lines[27]. Ongoing systematic studies on different aspects of radiation damage[34] are slowly increasing our understanding of the underlying physics and chemistry of the possible mechanisms.

Annealing, hyperquenching and slow cooling

Flash-cooling in liquid nitrogen sometimes fails, and the crystals show either limited diffraction or large mosaicity, which both render the crystal useless for data collection. Short of resorting to room-temperature data collection, with its drawbacks of more difficult mounting and much more risk of radiation damage, there are two relatively simple techniques that can be used to improve the results: annealing and hyperquenching.

Annealing. In protein crystallography, annealing refers to a cyclic thermal treatment of cryocooled protein crystals, which are (sometimes repeatedly) warmed up and cooled again. Not all crystals survive such a treatment. The exact process taking place during protein crystal annealing is not clear, but a likely explanation is that during annealing of a crystal, the strain accumulated during crystal growth or during quenching is relaxed and the alignment of domains in the crystal improves. Either case may lead to drastic reduction in mosaicity[35] with a substantially increased resolution. Strain in crystals is also induced through a mismatch between the thermal expansion of solvent and bulk protein.[36]

Hyperquenching. Hyperquenching[37] is a recently introduced cryotechnique that delivers very high quenching rates. Quick cooling is necessary to prevent formation of ice crystals, which invariably leads to destruction of the crystals. Upon very rapid cooling, water (or the water-based crystallization cocktail) remains in a glass-like (vitreous) state, and the formation of crystalline ice (largely in the mother liquor surrounding the crystal) is prevented. The cooling rate necessary for pure water to remain vitreous is too high (~10^6 K s^{-1}), which is why cryoprotectants must be used. At any rate, dipping crystals directly into liquid nitrogen would seem to be the optimal and fastest procedure to quench crystals. However, a cushion of cold gas overlays the liquid nitrogen and the crystal is slowly pre-cooled while it moves through this cold gas cushion. The cooling in the gas cushion is much slower because of its lower heat transfer

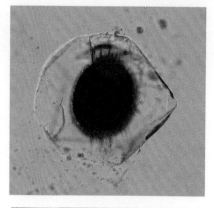

Figure 8-15 Radiation damage.
The image shows massive radiation damage (black area) in a crystal of protein Atu1728 from *Agrobacterium tumefaciens* caused by intense X-ray exposure. The crystal, 350 × 350 μm, was exposed 2 min per frame for total of 180 exposures (6 h) on a Rigaku FR-E X-ray generator (one of the most powerful rotating anode laboratory X-ray generators available) equipped with VariMax HR optics and a 0.1 mm collimator. Image kindly provided by Aiping Dong and Xiaohui Xu, University of Toronto.

Sidebar 8-4 Low temperature versus room temperature structure. A frequently asked question is whether cryoquenching causes any changes to the room temperature structure. Because of the rapid cooling we are actually not looking at the protein structure equilibrated at cryogenic temperature, but at a quenched room-temperature (or somewhat below) state. Nonetheless, during the quenching small displacements of groups of atoms against each other can happen. Minor movements of exposed side chains are common, and often distinct, multiple conformations of flexible side chains are observed, which were not found in the room temperature structure.[30] Similarly, even the core of proteins can display minor deviations exceeding coordinate uncertainty.[29] As a general rule, the solvent structure is better defined at lower temperature (i.e. more water atoms can be built in corresponding electron density). The latter is only a concern when specific biological activity is assigned to a water molecule (for example, acting as a nucleophile in an enzymatic reaction mechanism), but such waters are usually conserved. The overall structure and features of room-temperature and corresponding low-temperature structures generally agree very well.

Box 8-4 **Harvesting and cryocrystallography.** Protein crystals are invariably susceptible to radiation damage caused by radicals and ionized species generated by the ionizing high-energy X-ray photons. Flash-cooling to cryogenic temperatures can hamper or slow down—to a large degree, but not entirely at extreme dose rates—the reactions leading to radiation damage. Foremost, cryoprotectants prevent the formation of ice upon flash-cooling in the mother liquor surrounding the crystal. Crystallization cocktails may already contain sufficient amounts of cryoprotectants such as high concentrations of PEGs, salts, glycerol, or sucrose in the crystallization mother liquor. If not, the crystals are briefly swished through a compatible cryosolution such as mother liquor spiked with a compatible cryoprotectant. Upon rapid quenching or flash-cooling, the mother liquor in and around the crystals remains in a glass-like, vitreous state. Crystalline ice diffracts and ice formation is indicated by distinct ice diffraction rings (Figure 8-25).

rate, and the instant cooling effect needed for high quenching rates is never achieved. However, when the gas cushion is blown away with a dry nitrogen stream,[37] super-fast quenching rates can indeed be achieved (Figure 8-16). This in turn means that lower cryoprotectant concentrations are needed, which are often already present in the cryo-cocktail. Thus the extra (and loss-leading) step of swishing or soaking the crystals in additional cryoprotectant can be omitted.

Slow cooling. The complete removal of water from the surface of the crystal either mechanically or by swiping through Paratone-N or perfluoroethers in certain cases allows slow cooling of the crystals without harm. Although interesting phase transitions could be observed in some cases, it is presently unknown whether slow cooling is a generally applicable cryocooling method.

Real crystals: mosaicity and twinning

The molecules in protein crystals do not arrange themselves flawlessly in each direction across the entire length of the single crystal in a perfect 3-dimensional network. There are always impurities in the crystallization mother liquor, mostly in the form of other species, foreign detritus, and aggregates of denatured material. As shown by atomic force microscopy (AFM) studies of protein crystals[21, 39] (Figure 3-11), these foreign constituents cause interruptions in the growth patterns, and a single crystal is thus generally not one single perfect entity but consists of multiple, similarly aligned domains. Such a real crystal is a mosaic crystal, and the misalignment of the individual domains is described by the mosaicity of the crystal.

Each domain will diffract at a slightly different orientation, and the reflections of each domain will fall at slightly different, overlapping positions on the detector. Protein single crystals can exhibit quite large mosaicity (up to several degrees), and environmental effects such as stress from cryocooling can increase mosaicity, while procedures such as annealing (repeated warming and cooling to relieve the stress) may improve mosaicity. Large mosaicity causes X-ray reflections to extend over many successive rotation images, thus becoming problematic for data integration.

A most insidious form of perfect microscopic twinning, merohedral twinning, occurs when domains are related through a twinning operation with a defined relation to a crystallographic axis so that their diffraction patterns exactly overlap. As this type of twinning cannot be detected by visual inspection of the diffraction pattern alone, the entire data set must be tested for deviations from the expected intensity distributions.

Macroscopic and non-merohedral twinning

When distinct, separate crystals grow from different seeds or nucleation sites, the so-called macroscopic twinning is generally visible and recognizable under

Figure 8-16 Robotic cryoquenching of a mounted crystal. The robot arm carries the crystal on a cryo-pin with a magnetic base and is about to dip it into the storage puck submerged in liquid nitrogen Note how the liquid nitrogen is boiling, and that the crystal will travel through a cushion of cold nitrogen gas first before it is quenched into the liquid nitrogen itself. Blowing the gas cushion away with dry nitrogen drastically increases the quenching rates.[37] This robot is also a prototype capable of semi-automatically harvesting crystals, and autonomous crystal harvesting is a remaining frontier in full automation of high-throughput crystallography.[38]

the microscope, and the crystal cluster can often be dissected into separate single crystals. If major misalignment of two or more discrete parts or domains of the crystal is not detected under the microscope, the diffraction pattern will reveal a superposition of discrete diffraction patterns in intensity ratios proportional to the amount of exposed material of each domain orientation. In favorable circumstances, one of these patterns dominates and can be assigned a unit cell (indexed) and be processed by itself without significant contribution from the remaining domain(s), which are simply ignored. This scenario is illustrated in the top row of Figure 8-17.

In the most common case of mosaic crystals, the individual domain orientation will not be visible under the microscope and the crystal will look perfectly single. However, the degree of mosaicity will reveal itself in the diffraction pattern. A useable crystal will consist of many domains with a given distribution of domain misalignment of less than 1° or slightly more. At the fixed Bragg angle of a still image (or small rotation increment), only the parts of domains fulfilling the Bragg conditions will generate reflections. These reflections are still relatively sharp, and the entire extent of mosaicity can only be detected by checking how long the same reflections appear in neighboring crystal orientations. Reflections that remain over many successive oscillation frames (and are not located in the Lorentz exclusion region, Chapter 6) are a telltale sign of large mosaicity. The center panel of Figure 8-17 illustrates a crystal with large mosaicity.

In some situations, the cell dimensions of a crystal are such that they either match or are reasonably close in two dimensions when domains are specifically oriented to each other. In this case, the crystals also look unsuspicious, but the diffraction pattern in 3 dimensions will reveal two different interpenetrating lattices. If the spots are reasonably resolved and one orientation dominates, the major component can often be indexed and integrated separately. This situation is depicted in the bottom panel of Figure 8-17, and is called epitaxial or non-merohedral twinning. At present, only the program *SHELXL* provides the option to refine structures from non-merohedrally twinned crystals where the data cannot be entirely separated.

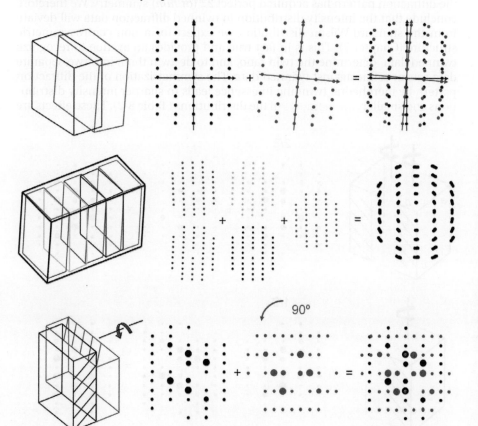

Figure 8-17 Macroscopic twinning, mosaicity, and non-merohedral twinning. The top panel shows macroscopic twinning of a crystal. The two growth domains are distinctly different as are their diffraction patterns. It may be possible to dissect this crystal, to expose only one domain, or (sometimes) to index and integrate one of the patterns if it is dominant. The center panel shows a mosaic crystal, with a relatively large (6°) mosaic spread. The diffraction pattern shows a corresponding mosaic spread of the spots. The bottom panel shows non-merohedral twinning, where two orientations are epitaxially related. The diffraction pattern is an interpenetration of the two patterns which can, in some cases but not always, be separately indexed.

Box 8-5 **Mosaicity and non-merohedral twinning.** Real protein single crystals are not perfect translationally periodic single-domain crystals, because growth defects lead to a number of imperfections. Practically all protein crystals are mosaic crystals, where multiple growth domains are slightly misaligned against each other. The resulting mosaic spread is the dominant contribution to finite diffraction peak width. Crystals can be macroscopically intertwinned, where differently oriented crystals generate different, superimposed diffraction patterns. Epitaxial or non-merohedral twinning occurs when two or more discrete domain orientations are prevalent and cell dimensions match or nearly match in these domain orientations.

Merohedral twinning

In addition to the normally occurring slight misalignment of domains in a mosaic crystal, domains can also grow in different orientations that can be precisely described by a symmetry operation relative to the crystal. In this case, reflections from the distinctly oriented domains will perfectly superimpose and the diffraction pattern will look unsuspicious and normal. In Figure 8-18 cases are shown where two types of domains grow in a specific, perfect orientation to each other, related by a symmetry operation. This symmetry operator is called the twin operator, in the depicted case a 2-fold axis along the **c** axis.

Merohedral and hemihedral twinning. Take a careful look at what happens to the combined diffraction pattern in Figure 8-18. In the top panel, the combined pattern consists of 2/3 of domains in one orientation and 1/3 in the related twin orientation. The resulting diffraction pattern is much more symmetric than the original pattern; in fact, the symmetry is close to 22 in this projection. In this case, the lattice symmetry exceeds the crystal symmetry, and the twinning is termed merohedral twinning. If the twin fraction α of a single twin operator is 0.5 (a perfect hemihedral twin with exactly equal amounts of each domain orientation as shown in the lower panel of Figure 8-18), the depicted projection of the diffraction pattern has acquired perfect 22 (or *mm*) symmetry. We therefore conclude that the intensity distribution in twinned diffraction data will deviate from the standard Wilson distribution we expect for a non-centrosymmetric structure (Chapter 7). This is in fact true and provides an avenue to recognize twinned data, determine the twin ratio, and to de-twin them into two separate data sets or refine against twinned data. The symmetrization of the diffraction pattern by merohedral twinning in essence leads to sharper intensity distributions and smaller raw moments of the distributions (Table 8-2). These effects are

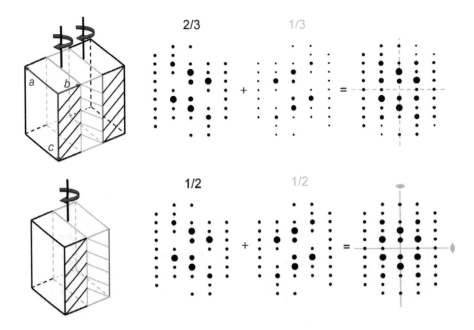

Figure 8-18 Hemihedral twinning. The superposition of two twin-related diffraction patterns in partial hemihedral twinning (merohedral twinning with a single twin operator) with a twin fraction $\alpha \neq 0.5$ generates a diffraction pattern that is more symmetric than the original pattern (top row). As a consequence, intensity statistics differ from expected values for Wilson statistics for untwinned structures and the distributions become sharper. For perfect hemihedry with $\alpha = 0.5$, the apparent symmetry is perfect and confers higher space group symmetry (lower row).

used to detect and analyze merohedral twinning. In the most common case of only one single twin operator, this simplest case of merohedral twinning is called hemihedral twinning.

The vast majority of actual twinning cases encountered in macromolecular crystallography belong into the category of hemihedral twinning, with partial hemihedry ($\alpha < 0.5$) most frequent. In order for this type of twinning to occur, the point group (PG) of the crystal's space group must be lower than the symmetry of the lattice (known as its holohedry). This is possible in the case of trigonal, tetragonal, and hexagonal systems for space groups with point group symmetry 3, 4, 6, and 32 and in cubic (tetrahedral) space groups based on PG 23. As an exercise, look at the right-side projection of space group $P3$ in the ITC-A: the projection diagram can be rotated by 180°, which swaps the axes and the unit cell content, but the lattice still looks exactly the same. This twin operator for example would be a 2-fold axis parallel to **c**, or written as operator $-h, -k, l$. Perfect hemihedry, as illustrated in Figure 8-18, generates apparent higher PG symmetry; we then observe 32, 6, 422, 622, or 432 as apparent point groups, respectively. Table 8-2 provides a list of most common hemihedral twinning operators in the chiral space groups.

Table 8-2 Hemihedral twin operators of the chiral space groups. The table lists whether partial hemihedral twinning is possible for a given observed chiral space group, and provides the possible twin operator(s). For perfect twinning, the twin operator will generate a higher space group (Laue) symmetry; for these cases a warning regarding the possibility of lower point group (PG) symmetry is listed for the apparent observed space group of higher symmetry. A more extensive table listing each pair of apparent and true space groups together with the corresponding operator explicitly is given in the 1999 CCP4 study weekend proceedings.[41]

Crystal system	Point group	Laue group	Bravais type	Observed chiral space groups	Twinning type, twin operator
Triclinic	1	$\bar{1}$	P	$P1$	None
Monoclinic	2	$2/m$	P	$P2$, $P2_1$,	Pseudo-hemihedry only in $P2_1$ and $C2$ with $a' = a + nc$ and $\cos(\beta^*) = nc/2a$, or, if $a > c$, $\cos(\beta^*) = na/2c$
			C	$C2$	
Orthorhombic	222	$2/mmm$	P	$P222$, $P222_1$, $P2_12_12$, $P2_12_12_1$	Pseudo-hemihedry only with two axes (nearly) identical
			I	$I222$, $I2_12_12_1$	
			C	$C222_1$, $C222$	
			F	$F222$	
Tetragonal	4	$4/m$	P	$P4$, $P4_1$, $P4_2$, $P4_3$	Partial ($k\,h\,{-}l$)
			I	$I4$, $I4_1$	
	422	$4/mmm$	P	$P422$, $P42_12$, $P4_122$, $P4_12_12$, $P4_222$, $P4_22_12$, $P4_322$, $P4_32_12$	None, perfect ($k\,h\,{-}l$) twin in PG 4 may appear 422, e.g. apparent $I422$ is true $I4$
			I	$I422$, $I4_122$	
Trigonal	3	$\bar{3}$	P	$P3$, $P3_1$, $P3_2$	Partial, either ($-h\,{-}k\,l$), ($k\,h\,{-}l$), or ($-k\,{-}h\,{-}l$)
			R	$R3$	Partial ($k\,h\,{-}l$)
	312	$\bar{3}1m$	P	$P312$, $P3_112$, $P3_212$	Partial ($-h\,{-}k\,l$), perfect twin ($k\,h\,{-}l$) with true PG 3 may appear 312
	321	$\bar{3}m1$	P	$P321$, $P3_112$, $P3_221$	Partial ($-h\,{-}k\,l$), perfect twin ($-k\,{-}h\,{-}l$) with true PG 3 may appear 321
			R	$R32$	None, perfect ($k\,h\,{-}l$) twin with true PG 3 may appear 32
Hexagonal	6	$6/m$	P	$P6$, $P6_1$, $P6_5$, $P6_2$, $P6_4$, $P6_3$	Partial ($k\,h\,{-}l$), perfect twin $-h\,{-}k\,l$) with true PG 3 may appear 6
	622	$6/mmm$	P	$P622$, $P6_122$, $P6_522$, $P6_222$, $P6_422$, $P6_322$	None, perfect twin in lower PG 6 may appear 622
Cubic	23	$m\bar{3}$	P	$P23$, $P2_13$	Partial ($k\,h\,{-}l$)
			I	$I23$, $I2_13$	
			F	$F23$	
	432	$m\bar{3}m$	P	$P432$, $P4_232$, $P4_332$, $P4_132$	None, perfect twin in lower PG 23 may appear 432
			I	$I432$, $I4_132$	
			F	$F432$, $F4_132$	

Pseudo-merohedral twinning. In addition to the cases of true merohedry described above, specific axis ratios can allow twinning, for example in addition in orthorhombic crystals with $a = b$ (appearing tetragonal) or in monoclinic crystals with $\beta \sim 90.0°$ (appearing orthorhombic). In such cases of pseudo-merohedral twinning the apparent higher symmetry implies that the unit cells will appear too small (too many symmetry related copies) to harbor a monomeric motif. When multiple molecules make up the motif, the distinction based on the Matthews probabilities (Chapter 11) is not so clear cut, because it might alternatively be possible that an oligomer axis coincides with a crystallographic axis.

Pseudo-symmetry. As a general rule, twinning analysis becomes more complicated when twinning coincides with translational NCS[40] or pseudo-symmetry. In the case of pseudo-translational symmetry, the intensity distribution tends toward bimodal because one subset of reflections becomes systematically enhanced and the other systematically weakened. The trend toward bimodal shape in turn makes the whole intensity distribution broader. This is essentially the opposite effect of merohedral twinning and manifests itself in the twinning analysis (Table 8-3) as *negative twinning*. The presence of a combination of compensating twinning and translational pseudo-symmetry can lead to near-normal intensity distribution statistics and obscure the detection of twinning.

Intensity statistics for twinned crystals

We can define a statistic H that is a function of the twinning fraction α and which relates the measured diffraction intensities (or the square of the structure amplitudes) as follows:

$$H = \frac{\left|I_{obs}(\mathbf{h}_1) - I_{obs}(\mathbf{h}_2)\right|}{I_{obs}(\mathbf{h}_1) + I_{obs}(\mathbf{h}_2)} \tag{8-2}$$

where $I_{obs}(\mathbf{h}_1)$ and $I_{obs}(\mathbf{h}_2)$ are the observed intensities resulting from a mixture of the true twin related reflections $I(\mathbf{h}_1)$ and $I(\mathbf{h}_2)$ in a hemihedral twinning case. With the corresponding fractions written as

$$I_{obs}(\mathbf{h}_1) = (1-\alpha)I(\mathbf{h}_1) + \alpha I(\mathbf{h}_2) \text{ and } I_{obs}(\mathbf{h}_2) = (\alpha)I(\mathbf{h}_1) + (1-\alpha)I(\mathbf{h}_2) \tag{8-3}$$

we can solve for the true intensities $I(\mathbf{h}_1)$ and $I(\mathbf{h}_2)$ the following linear system:

$$I(\mathbf{h}_1) = \frac{(1-\alpha)I_{obs}(\mathbf{h}_1) - \alpha I_{obs}(\mathbf{h}_2)}{1-2\alpha} \text{ and } I(\mathbf{h}_2) = \frac{-\alpha I_{obs}(\mathbf{h}_1) + (1+\alpha)I_{obs}(\mathbf{h}_2)}{1-2\alpha} \tag{8-4}$$

Equations 8-4 allow the recovery of the true intensities from observed twin related pairs provided: (i) we know the twinning operator (otherwise we cannot form the pairs of related reflections); (ii) we know the twinning fraction α; and (iii) that the twinning fraction is not exactly 0.5 (where Equations 8-4 become singular) which is the case for perfect merohedral twinning.

The twinning fraction can be recovered from analysis of the cumulative probability distribution $N(H)$ for the statistic H from the pairs of twin related intensities.[42] $N(H)$ takes the following algebraic forms:

$$N(H) = \cos^{-1}\left[H/(2\alpha - 1)\right]/\pi \quad \text{(centric)} \tag{8-5}$$

$$N(H) = \left[1 + H/(1-2\alpha)\right]/2 \quad \text{(acentric)} \tag{8-6}$$

The centric function has the sigmoid shape of the arccosine and the acentric function is linear in H. Both functions can be computed and plotted for a number of discrete values for the twinning fraction α for centric and acentric reflections and compared with the actually observed data. In general there are not many centric reflections, so the acentric plots are more meaningful (Figure 8-19). The theoretical experimental distribution that is followed most closely by the experimental data (normalized in resolution shells) provides an estimate of the twinning fraction. This was the first useful statistical procedure for analyzing

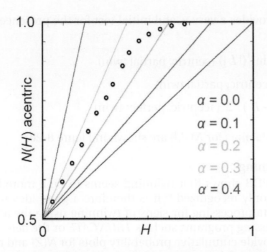

Figure 8-19 **Cumulative probability distribution N(H) for acentric reflections.** The plot shows the theoretical curves for acentric data and the experimental data points. A twinning ratio α of about 0.26 can be interpolated from the graph.

merohedral twinning in macromolecular data, derived by Todd Yeates[43] extending earlier work. Once α is known, the data can be de-twinned (the fractions computed and combined) and treated as normal.

For perfect merohedral twinning, where the expressions (8-5 and 8-6) derived above for H become singular, a number of additional tests based on the moments of E (normalized structure factors) or Z (normalized intensities, $Z = I / \langle I \rangle$) can be inspected. The CCP4 program *TRUNCATE*, for example, provides such plots, where the moments are plotted in resolution shells. Their expectation values are given in Table 8-3.

The drawback of the above method is that to determine which pairs of reflections are related, one still needs to know the twin operator, that is, try the possible ones, given the apparent space group. The expressions for the cumulative probability distributions of H again become singular for $\alpha \to 0.5$, that is, for perfect hemihedral twinning.

Padilla and Yeates[44] have thus modified the above statistic by considering pairs of locally related reflections instead of twin related reflections. The statistic L is more robust in the presence of anisotropic diffraction and pseudo-centering than the cumulative $N(Z)$ plots ($Z = I / \langle I \rangle$) derived in Chapter 7. Importantly, L can be evaluated without prior knowledge of the twin law:

$$L = \frac{I_{obs}(\mathbf{h}_1) - I_{obs}(\mathbf{h}_2)}{I_{obs}(\mathbf{h}_1) + I_{obs}(\mathbf{h}_2)} \qquad (8\text{-}7)$$

One can again plot a cumulative distribution function $N(|L|)$ and check for deviations from the expected values for untwinned data. The practical benefit is that Function 8-7 also delivers a defined curve for perfectly twinned data,

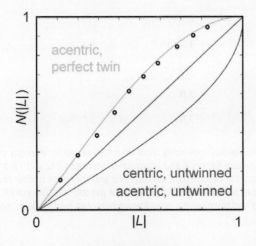

Figure 8-20 **Yeates–Padilla plot for the cumulative probability distribution N(|L|).** The graph shows the expected cumulative distribution curves for acentric and centric untwinned data and acentric experimental data (open circles) for a perfect twin. For partial twins, deviations from untwinned data will result in experimental data points located between the calculated curves.

and therefore provides a very useful initial test for the presence of any kind of twinning:

$$N(|L|) = (2/\pi)\sin^{-1}(|L|) \text{ (centric, partial twin)} \tag{8-8}$$

$$N(|L|) = |L| \text{ (acentric, partial twin)} \tag{8-9}$$

$$N(|L|) = |L|(3 - L^2)/2 \text{ (acentric, perfect twin)} \tag{8-10}$$

The Yeates–Padilla plots for $N(|L|)$ are shown in Figure 8-20.

Testing for twinning

Analysis of the PDB shows that twinning seems to be a more frequent problem than commonly recognized.[40] It is therefore a good idea to always check during initial data processing for signs of twinning using the above-described tests. Data processing programs such as *TRUNCATE* or the *detect_twinning.inp* script in *CNS* provide cumulative probability plots for $N(Z)$ and resolution shell dependent graphing of the moment statistics listed in Table 8-3, and a web service plotting the $N(|L|)$ statistics and their cumulative distributions is available. It is imperative to try various reflection error (sigma) cutoff values if the tests are not conclusive, because the intensity difference-based statistics H and L are necessarily sensitive to inaccurate measurements.

Practical implications of twinning. Twinning affects all aspects of structure determination, although with different severity. Difference-data-based experimental phasing is particularly sensitive to wrong intensities caused by twinning, and only in rare cases is experimental phasing possible with twinned data. Molecular replacement solutions are in general less sensitive to twinning, and in these cases the problem usually becomes evident in the refinement stage, when the refinement stalls at high R-values with no obvious improvement possible in the electron density maps. At present the programs *SHELXL*, *REFMAC*, and *PHENIX/Refine* allow refinement against twinned data, with the benefit that the twin fraction can be accurately refined, in contrast to de-twinning the data based on a twin fraction estimated from the $N(H)$ plots.

Statistic	Untwinned centric	Perfect twin centric	Untwinned acentric	Perfect twin acentric	Partial centric	Partial acentric
$\langle|H|\rangle$					$2(1-2\alpha)/\pi$	$(1-2\alpha)/2$
$\langle|H^2|\rangle$					$(1-2\alpha)^2/2$	$(1-2\alpha)^2/3$
$\langle|L|\rangle$	$2/\pi$	1/2	1/2	3/8	See P&Y[44]	
$\langle|L^2|\rangle$	1/2	1/3	1/3	1/5	See P&Y[44]	
$\langle|E^2-1|\rangle$	0.968	too low	0.736	too low	too low	too low
$\langle E\rangle$	0.798	0.886	0.886	0.940		
$\langle E^3\rangle$	1.596	1.329	1.329	1.175		
$\langle E^4\rangle, \langle Z^2\rangle$	3.0	2.0	2.0	1.5	in between	
$\langle Z^3\rangle$	15.0	6.0	6.0	3.0		
$\langle Z^4\rangle$	105.0	24.0	24.0	7.5		

Table 8-3 Statistical descriptors used for twin detection. The table lists various commonly reported mean values for intensity probability distributions, including the twin statistics H and L as well as normalized structure factors E and normalized intensities Z ($Z = I/\langle I\rangle$) and their higher kth raw moments. Symmetrization of the diffraction pattern by twinning causes sharper intensity distributions and therefore smaller raw moments for twinned data. Deviations in the opposite direction ("negative twinning") can be caused by the presence of pseudo-translational NCS symmetry. P&Y, Padilla and Yeates.[44] Presence of a combination of compensating twinning and translational pseudo-symmetry can lead to near "normal" intensity distribution statistics and obscure the detection of twinning.

Box 8-6 Hemihedral twinning. The most common form of microscopic twinning in protein crystals is merohedral twinning. If two (hemihedral twinning) or more domain orientations occur with a specific twinning operation in special orientations to a crystallographic axis, the diffraction patterns overlap perfectly in reciprocal space. The diffraction patterns from a merohedral twin appear normal and unsuspicious, but exhibit unusual intensity distributions. The symmetrization of the diffraction pattern by merohedral twinning in essence leads to sharper intensity distributions and smaller raw moments of the intensity distributions (Table 8-3). The shape of the cumulative intensity distributions and raw moment analysis of these distributions are used to detect and analyze merohedral twinning. Deviations of intensity distributions indicating broader than normal distributions and larger raw moments than untwinned data can indicate pseudo-translational NCS. In the most common case of only a single twin operator, this simplest case of merohedral twinning is hemihedral twinning. Hemihedral twinning is a relatively common and probably underreported phenomenon in macromolecular data sets; each data set should be carefully analyzed for indications of hemihedral twinning.

8.3   Fundamentals of data collection

Collection of diffraction data is the last physical experiment that is conducted before the *in silico* phase of structure determination. As a reminder, data quality mercilessly decides the quality and informative content of the final structure model, and data collection thus deserves much attention and care. Moreover, losing a crystal or collecting poor data is an expensive late-stage failure in the experimental part of structure determination process, which is better avoided. Details matter: for example, verify that all mounting and adjustment screws on the goniometer head as well as the mounting pins are properly secured, and that the cryostream is optimally adjusted.

The actual routine data collection is automated to a high degree. Practically all present data collection software includes a strategy planner that provides a reasonable data collection strategy based on analysis of the initial diffraction images. Particularly on synchrotron beam lines, user support is generally quite efficient. Various standard strategies are implemented, assuring that a complete data set will be obtained. However, in suboptimal cases a number of decisions must be made by the user whether to proceed with data collection or to abandon the crystal (or dismount it and store it while testing additional crystals). It is therefore necessary to be able to judge the diffraction quality of crystals and to decide whether they can actually lead to a useful data set.

Before we review data collection from a strategic perspective, we shall briefly expand on diffraction geometry and how the reflections are actually recorded in the most common configuration used in macromolecular data collection, the rotation (or oscillation) method. Figure 8-21 provides as a reminder an overview of the structure determination process.

Anatomy of a rotation image
The most common data collection method in macromolecular crystallography is the rotation method. As the name indicates, it involves a simple rotation of the crystal around a single axis in small increments—generally 0.1° to about 1.5°—while the crystal is exposed to X-rays and the reflections are recorded on an area detector. The diffraction image recorded during each of these small slices of rotation is often called a frame. Given the large coverage of solid angle of present area detectors, rotation around a single axis (often perpendicular to the X-ray beam and termed φ) in most cases suffices to obtain complete data. The range of rotation necessary to sample the unique asymmetric unit of reciprocal space depends primarily on the Laue group (crystal point group plus inversion), the actual orientation of the crystal, and whether anomalous data need to be collected, in which case the centrosymmetrically opposed Friedel

Sidebar 8-5 Low *and* high resolution parts of the diffraction data are important. What exactly constitutes a useful data set largely depends on the purpose of the data collection—phasing and refinement have different requirements. The low-resolution part of the diffraction data is important for phasing, and high-resolution data help model building and refinement. As a consequence, there are no unimportant parts in the data that can be neglected during data collection. The same holds for completeness of data: In Chapter 9 we will see that random omission of data or of small parts—often the case in high-resolution data—is generally unproblematic. In contrast, systematically omitting sections (such as missing wedges because of insufficient φ-rotation range, Figure 9-12) or systematically missing only a small percentage of high intensity data—often in the low-resolution part of the data and giving rise to the dominating I^2 or E^2 terms in phasing (Figure 9-11)—is invariably grievous.

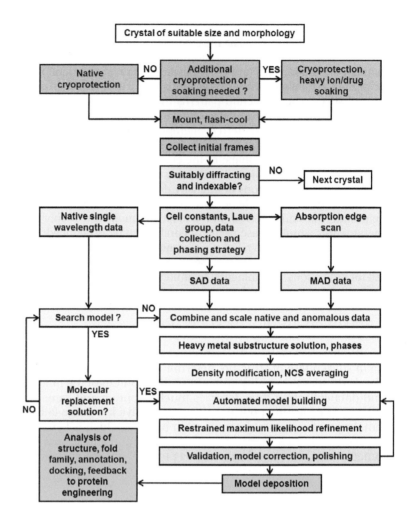

Figure 8-21 Overview of data collection strategies and structure determination. The choice of data collection strategy depends on the intended use of the data set. For molecular replacement and refinement, single wavelength data sets without special consideration of anomalous data suffice. Anomalous phasing techniques require collection of at least one Bijvoet pair of each reflection, effectively doubling the amount of data that needs to be collected. Multi-wavelength anomalous diffraction (MAD) data collection in addition requires that data sets are collected at various wavelengths, whose optimal values are determined by an experimental X-ray absorption edge scan.

mates or corresponding Bijvoet mates of each reflection also need to be collected. The unit cells of reciprocal space have been derived in Chapter 6 and are listed for each Laue group in Table 6-6.

As the description of diffraction geometry centers around the concepts of reciprocal space and the visualization of diffraction by means of the Ewald sphere, now may be the time to repent and return to Chapter 6 in case you skipped the diffraction geometry sections. Sidebar 8-6 briefly reviews the basic concepts needed to follow this section.

When a crystal is exposed to X-rays, only a limited number of reciprocal lattice points (RLPs) actually lie on the Ewald sphere and diffract. Because the reciprocal lattice is densely packed, given the large unit cells of protein molecules, in multiple points in each lattice plane will likely fulfill diffraction conditions. If the RL planes are parallel to the rotation axis and normal to the incident X-ray beam, the RLPs will generate reflections that lie on a cone centered on the primary or incident beam axis. If this cone intersects a flat detector, the reflections then lie on concentric circles. In the general case, the diffraction cones will be offset from the incident beam direction, and the intersection curves of the cone with the flat detector are ellipses. The special situation described above is shown in the upper left panel of Figure 8-23. If the detector or film is curved, the intersection curve is a more complicated second-order intersection curve, not an ellipse.

Diffraction images and lunes. If we now rotate the crystal by a small increment, other RLPs of the same planes will intersect the Ewald sphere, and another cone of reflections, somewhat offset by the progressive rotation, will be diffracted onto the detector. If we rotate more, the next set of reflections will be recorded. This rotation procedure works fine until the lunes and their reflections start to overlap. The rotation increment (width of the "slice") for a single frame or image is thus limited by the overlap of the lunes. The larger the cell constants, the closer the lunes and thus the smaller the rotation increment needs to be. The lunes are most distinct when the reciprocal lattice planes are parallel or near parallel to the rotation axis, less so in arbitrary directions (see the actual diffraction images later in the chapter). If a lattice plane touches the Ewald sphere so that the entire diffraction circle is filled with spots, a (nearly) undistorted, small section of the diffracting reciprocal lattice plane is projected into the detector and the cell constants can be quickly estimated from the spot distance, generally with some measuring tool of the image-display software.

Partial and full reflections. As a result of crystal mosaicity and beam divergence, the diffraction spots have finite width, and while some RLPs will be recorded on one single rotation image (full reflections), some will be only partly recorded on one image, and continue on the next one. Combining the partial reflections from different images is done by the integration module of the data collection software.

Sidebar 8-6 Review of diffraction geometry and Ewald construction. The Ewald construction (Chapter 6) shows the diffraction geometry in reciprocal space presentation. If we place the scattering diagram for the Bragg equation into a Ewald sphere with a radius of $1/\lambda$, the significance of Bragg's form of the reflection condition in relation to the reciprocal lattice immediately becomes obvious. If the reciprocal lattice is aligned so that the vectors $\mathbf{S}_{hkl}$ and $\mathbf{d}^*_{hkl}$ coincide, the reflection condition for a set of lattice planes hkl in a crystal is necessarily fulfilled. This is the case when and only when the reciprocal

lattice point hkl lies on the Ewald sphere as illustrated in Figures 8-22 and 8-39.

Offset of the crystal axis from the rotation axis alleviates the problem of the blind region, but be aware that if systematic serial absences fall into the missing region (blue shade) as is the case when screw axes are aligned with the rotation axis, a unique assignment of the space group may not be possible. In this case, all space groups that are compatible with the respective point group must be tried in the subsequent phasing stage.

As only a few reflections at any given orientation lie on the Ewald sphere,

the crystal is rotated by small increments (~0.1° to ~1.5°) so that additional reciprocal lattice points intersect the Ewald sphere and diffract. If the rotation range is too large, the diffraction lunes and reflections begin to overlap (illustrated in Figure 8-23).

The maximum resolution of reflections that are recorded is determined for a given wavelength λ by the crystal–detector distance D and the width (radius) of the detector W. For a circular detector, the maximum 2θ angle is given by

$$2\theta = \tan^{-1}(W/D) \qquad (8\text{-}11)$$

and the highest recorded resolution follows from the Bragg equation $d(min) = \lambda/2\sin\theta$.

On a square or rectangular detector, the maximal observable resolution at the corners of the detector is correspondingly larger.

Figure 8-22 Ewald sphere and diffraction limits. The figure shows the Ewald sphere, centered at the crystal, its reciprocal lattice, and reciprocal lattice points which fulfill the diffraction condition (red dots). Diffraction can only be observed for reciprocal lattice points that lie inside the resolution sphere (blue circle). If a reciprocal axis is aligned with the rotation axis, a small number of reflections along the rotation axis (blue zone) are never recorded. The fraction of reflections lost in this blind region is generally low (less than 5% at 2 Å with 1.54 Å Cu radiation), smaller at shorter wavelengths, and only significant if crystals diffract to high resolution (i.e. large θ-angles).[45]

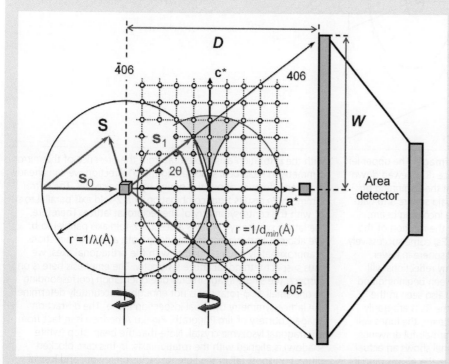

Fine slicing. A finer slicing of the rotation range in frames of about 0.1°–0.3° has certain benefits,[47] and an increasing number of data collection programs will probably implement this option in the near future.[48] With fine slicing (where each rotation increment of ~0.1–0.3° is substantially smaller than the mosaicity) a full 3-dimensional spot profile can be fitted, allowing very accurate background subtraction for low intensity reflections and larger mosaicity. Wide slicing in contrast allows only the 2-dimensional integration over the spot size on each image.

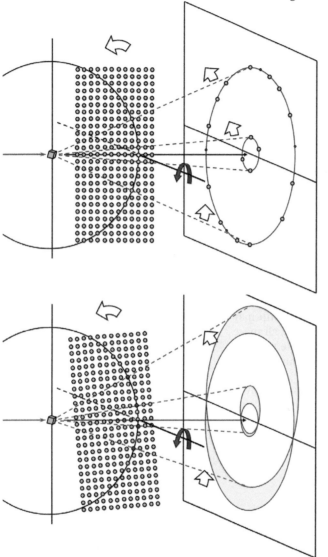

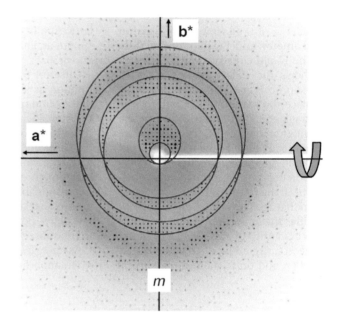

Figure 8-23 Diffraction lunes in a rotation image. The upper left panel shows a projection of the reciprocal lattice (RL) viewed down the rotation axis (perpendicular to the plane) at the beginning of the rotation. A few spots already lie on the Ewald sphere and with this particular RL oriented perpendicular to the incoming beam; their reflections are located on a circle. During the rotation of the crystal neighboring reciprocal lattice points (RLPs) come successively into diffraction condition (red points on Ewald sphere in lower left panel). With a dense reciprocal lattice, many reflections will fill the (cyan) area—the diffraction lune—between beginning and endpoint of the rotation (lower left panel, and also seen in the actual diffraction image in the right panel). If the RL is arbitrarily aligned relative to the rotation axis and X-ray beam, the lunes will not appear as distinct as in this drawing. Similar helpful drawings are shown in Z. Dauter's review.[46] The right panel shows an actual diffraction pattern for the same scenario, where the **a***-axis (and in the case of this orthogonal lattice also the crystal **a** axis) is aligned

with the φ rotation axis. Note that from the presence of the mirror symmetry in reciprocal space along the **b*** direction we can already make an inference about the symmetry of the crystal. The mirror symmetry *m* results from combination of a 2-fold axis parallel to **b*** with the centrosymmetry of the reciprocal lattice. Therefore, the lattice symmetry contains at least a 2-fold axis parallel to **b**. We also see that the lattice spacing in both **a*** and **b*** directions is approximately equal. Together with the orthogonal axes, we thus suspect tetragonal lattice symmetry. The emphasis here is on *suspect*; the single image in only one orientation (corresponding to one small 1° φ-rotation) is not enough to accurately determine the lattice symmetry by visual inspection alone. The diffraction image (courtesy of Jim Pflugrath, Rigaku Americas) is in fact from a tetragonal lysozyme crystal. Note that the beam stop (white shadow) is aligned with the rotation axis: in this case blocked diffraction spots fall in the Lorentz exclusion region, and loss of reflections is minimized.

Box 8-7 **The rotation method of data collection.** The most commonly used data collection technique in macromolecular crystallography is the rotation method. The crystal is rotated around a single axis by small successive increments (~1°) while exposed to the X-ray beam, and during each rotation increment a diffraction image or frame is recorded. The spot intensities in individual images are then integrated and the reflection intensities are determined. Fine slicing using small rotation increments (~0.3° or less) allows complete 3-dimensional peak profiles to be fitted to the reflections, which can give more accurate integration, particularly for crystals with large mosaicity.

Instrumental peak broadening. The Ewald sphere itself has a finite width, which on synchrotrons with a bandwidth of 1 eV is about 10^{-4} or smaller, and is practically negligible. The natural line width of laboratory sources is also small, about 3×10^{-4}, but some monochromators do not separate the $K\alpha_1$ and $K\alpha_2$ lines, and at high diffraction angles, the reflections split. The shorter focusing pathway (source–crystal distance) of home sources generally leads to somewhat higher beam divergence than on synchrotrons, but in both cases, mirror optics focus the X-rays onto the crystal or the detector plane. Synchrotron beam lines commonly allow matching the beam size to the crystal, which minimizes diffuse scattering from excessive surrounding mother liquor and minimizes air scattering. On a synchrotron, proper beam alignment is normally provided by the beam line staff, but it is good practice to double-check and write down all available beam parameters and monochromator (wavelength) settings. By far the largest source of finite spot width is the mosaicity of the crystal, which is discussed in the next section.

Data collection as a multi-level decision process

Data collection can be viewed as a multi-level decision process: At several points, strategic decisions need to be made whether to accept a given level of data quality (and hence, a certain probability of success or failure in the structure determination), or to proceed to the next crystal. Clearly, reliable mounting robotics provide the advantage of safely un-mounting and storing an acceptable, but not optimal, crystal for later use, and to proceed to evaluate hopefully better crystals. Problems and irregularities can occur at several stages during data collection, but often reveal themselves and become critical only later. At present, the data collection expert systems are not fully developed to completely handle the entire decision process. Such an expert system requires tight interfacing of the indexing, integration, scaling, and data reduction programs with the beam line hardware and robotics control software (reference 49 exemplifies the substantial complexities involved).

The decision process begins immediately after the first X-ray exposure of the crystal and the inspection of the resulting snapshot (generally two rotation images offset by 90°) of the diffraction pattern. Whether a crystal is worth pursuing generally depends on the purpose of the study and the stage in structure determination. Figure 8-24 provides an overview of important inspection and decision-making steps in the actual data collection experiment.

Initial inspection of the diffraction image

Once the crystals are centered on the goniostat, a first diffraction image is recorded, generally with a ~1° rotation range. The crystal is exposed for a few seconds on a powerful synchrotron, and up to several minutes on laboratory X-ray sources. The raw electronic signal from the detector is then corrected by the detector software for detector-specific non-uniformity, defects, distortion, and so forth, provided in a unique setup file for the individual detector. The corrected raw image (several MB in size) includes a header section with relevant experimental parameters such as detector distance, beam center position, wavelength, and so on, and is displayed in a graphical display window. The image is then visually inspected for a number of qualitative key parameters of the crystal. Good diffraction implies single, resolved, and strong spots, extending far out in the diffraction angle to high resolution. The first diffraction snapshot can also

Figure 8-24 Crystal inspection and decision making during data collection. The first inspections relying primarily on visual appearance of the pattern are shaded in pink. If deemed satisfactory, the first images are then indexed and the mosaicity is estimated. Once the indexing succeeds, a strategy is developed based on the initial diffraction symmetry and data covering at least the asymmetric unit of the reciprocal space (plus the Friedel Bijvoet opposites in the case of anomalous data) are collected. A final risk factor is radiation damage of the crystal; it will manifest itself in diminishing diffraction over time.

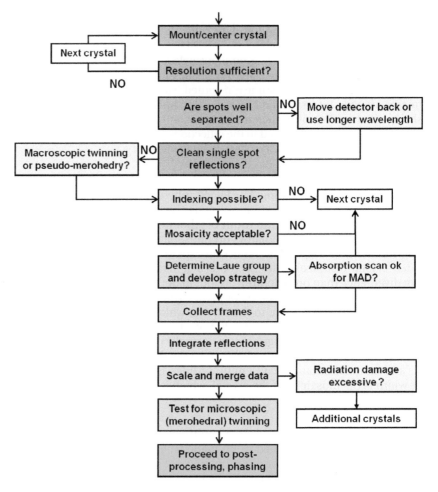

reveal the presence of *ice rings*[50] (Figure 8-25). Although diffraction from ice or frost affects the reflections in proximity to the ice ring, frames with not too excessive ice rings can be processed with little difficulty[51, 52] if the crystal still diffracts. Salt crystals (if not yet detected by lack of dye absorption and hardness) will generally show only few, but sharp, spots at resolutions higher than about 10 Å because of their much smaller cell dimensions.

Resolution. The first diffraction image immediately reveals the extent to which the crystal diffracts. The decision whether to leave a crystal on the diffractometer is based on whether the crystal diffracts sufficiently or not. As a general rule,

Figure 8-25 Ice rings. Diffraction image (1° rotation) typical for a 3 × 3 module detector. The horizontal and vertical white lines separate the nine modules of the detector as shown in Figure 8-8, but the image is zoomed and the peripheral areas are not completely shown. When crystalline ice forms in the mother liquor or through frost deposition on the crystal, the ice crystallites are generally randomly orientated and generate concentric diffraction rings typical for a powder diffraction pattern. The lattice spacing (Å) for the ice rings is provided in the corresponding labels. The regions affected by the ice rings can be excluded from data processing with not too much detriment. Ice rings can also be used to verify the location of the primary beam position. Image and ice rings simulated using MLFSOM[53] by James Holton, UCSF.

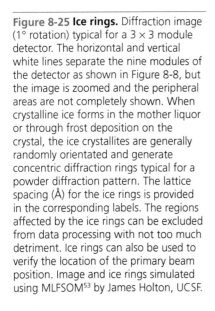
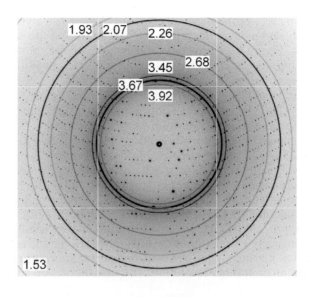

low-resolution data (>4.0 Å) can still be useful for phasing experiments, but often are insufficient to refine a quality model. For the purpose of model refinement, the general rule is that higher resolution always pays off in later stages as far as model quality and detail goes, and also in terms of ease of (and confidence in) your model building. The decision whether to pursue a low resolution structure will be influenced by whether a structure serves for example the purpose of fold determination, or must satisfy the more stringent quality criteria for a drug target structure. Both manual and automated model building, as well as refinement, become considerably more reliable once the resolution reaches or exceeds about 2.5–2.8 Å. In view of the increased difficulty in accurately building and refining low-resolution models (see Sidebar 10-6 for a cautionary tale), and considering the reduced information content in low-resolution models, trying to find the best diffracting crystal will pay off.

Exposure time. The first snapshots will also determine which exposure time is optimal. Too short exposure times mean noisy images with poor resolution, and too long exposure times finally lead to saturated high intensity spots (Figure 8-26). One temptation in particular must be resisted: Overexposing the crystal in an attempt to record high-resolution data at the expense of saturating the important low-resolution data invariably leads to problems with phasing and to poor electron density maps (Figure 9-11). Always check for excessive low resolution overflows in the diffraction image—the dynamic range of CCD detectors is limited (often 16 bit, ~65 000 counts). If necessary, collect a low resolution data set with shorter exposure time and merge it later to fill in any missing high-intensity reflections. The low- and high-resolution data sets must have a reasonable overlap in resolution shells so that they can be merged together properly. A salvage scaling procedure allowing the rescue of non-overlapping data sets after initial refinement has been described.[54] Exposure to X-rays can be selected either by fixed time or, preferably, in dose mode, where the exposure is adjusted so that the crystal is exposed to the same amount of photons. The dose mode has the advantage of compensating for primary beam fluctuations. In dose mode, a decay of diffraction intensity can then generally be attributed to crystal decay or anisotropy and not to decreasing beam intensity.

Anisotropy. Diffraction can be highly anisotropic, that is, the resolution in certain reciprocal lattice directions is sometimes significantly different. It is therefore good practice to include at least a second, 45° to 90° offset image in the initial inspection to avoid later disappointment and wasted data collection time. A special data truncation and anisotropic scaling method for severely anisotropic diffraction data is described in Chapter 12. Also keep in mind that the unaided eye is a poor judge for ultimate resolution; the actually useful resolution after

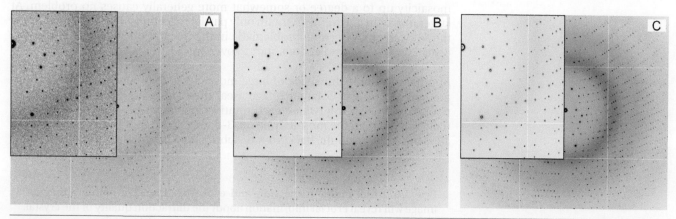

Figure 8-26 Effects of exposure time on diffraction image quality. The same image, exposed for successive times (30 ms, 1 s, 64 s) giving about eight times improvement in S/N between images. For each panel, the insert shows a magnification of the lower right quadrant. The first image (A) is too noisy, and we are not fully exploiting the diffraction limit of this well diffracting crystal. The second image (B) is just right; we obtain good resolution and only very few reflections are slightly saturated (yellow spot centers in insert). The third image (C) is heavily overexposed; practically all low resolution reflections are saturated (yellow spots in insert), while we are not gaining much more in terms of ultimate resolution. Images simulated using MLFSOM.[53]

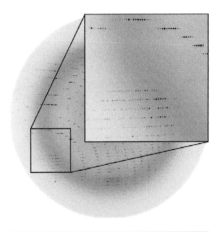

Figure 8-27 Diffraction image of crystal with 300 Å unit cell. The crystals of Rv1347c, a putative antibiotic resistance protein[55] from *M. tuberculosis*, belong to orthorhombic space group $P2_12_12_1$ with cell parameters a = 75.6 Å, b = 77.4 Å, and c = 297.6 Å, with eight molecules of 210 residues each per asymmetric unit. The data were recorded on a MAR345 image plate detector which was positioned as close as possible while still capturing all reflections up to the resolution limit of 2.2 Å. The insert shows that this is about the closest crystal–detector distance where the closely spaced reflections along **c*** are still resolved. Image files courtesy Clyde Smith and Graham Card, U. of Auckland, NZ, and SSRL, Stanford, CA.

data processing is often better than it appears from visual inspection of the images. Make sure to leave some extra space extending beyond the last visible spots when adjusting the detector distance.

Large unit cells. When the cell constants are very large (~300–400 Å and above), the RLPs lie very densely and the reflections begin to spatially overlap on the detector. In this case, the crystal–detector distance can be increased (or a longer wavelength can be selected) to resolve the diffraction spots. In addition, the individual diffraction lunes lie now very close together and even after small rotation increments begin to overlap. Therefore, both the crystal–detector distance must be increased and the rotation increment reduced. However, moving the detector back comes at the price of decreasing coverage in solid angle (Sidebar 8-6) and thus the high resolution data may no longer fall onto the detector. A large detector is obviously of benefit here, as is a well-collimated, narrow beam with small divergence, both of which are advantages of synchrotron data collection. Offsetting the detector and/or a second, high-resolution data collection run at the offset detector position may be necessary to cover the high-resolution data. Fortunately (or rather, unfortunately), very large unit cells and very high resolution seldom occur together. Figure 8-27 shows a diffraction image of crystal with a 300 Å unit cell.

Split reflections and multiple diffraction patterns. When reflections visibly split into distinct diffraction patterns or streak, it is commonly an indication of multiple discrete domain orientations in the crystal. If one of these patterns dominates and can be indexed separately, it may be possible to obtain usable diffraction data. Epitaxial, non-merohedral twinning can also lead to separate but interpenetrating lattices, as illustrated in Figure 8-17. In certain cases, if even one of these the patterns can be indexed independently the data may be usable. This is not always the case if subsets of reflections from the two (or more) sub-lattices overlap. Dismounting the crystal and checking other crystals will reveal if this is an isolated problem or typical for the particular crystal form.

Mosaicity. By far the largest source of finite spot width is the mosaicity of the crystal. As described above, in a real mosaic crystal a large number of individual growth domains are slightly misaligned against each other. The spread in orientation causes the corresponding reciprocal lattice points to remain in reflection condition over a longer rotation increment (see the center panel of Figure 8-17). In the Ewald picture, the RLPs of a mosaic crystal appear elliptically elongated perpendicular to the Ewald sphere. Therefore, the reflections do not necessarily look suspicious on a single exposure (although the spots may appear more fuzzy), but in a rotation image the lunes widen in the direction of the rotation axis (Figure 8-28). As long as the mosaicity is smaller than the rotation angle, mosaicity up to a degree or somewhat more generally causes no problem. At larger mosaicity the frames contain only partial reflections which, depending on the software, may become problematic in the integration stage and render the crystal useless once reflections start to overlap.

Another inevitable drawback of large mosaicity is that because the reflections are spread out over a larger image area, their signal-to-noise ratio becomes poor. With fine slicing (where each rotation increment of ~0.1–0.3° is substantially smaller than the mosaicity,) a full 3-dimensional spot profile can be fitted, allowing very accurate background subtraction for low-intensity reflections

Box 8-8 Initial image inspection and exposure time. The initial diffraction image will reveal crucial information about the diffraction quality of the crystal. The diffraction limit (highest resolution), presence of ice rings, spot separation, sharpness of spots, and mosaicity are primary concerns when inspecting a diffraction image. The exposure time must be selected so that no significant overflow of intense reflections occurs. Overflows affecting low-resolution, high-intensity reflections will invariably lead to problems during phasing.

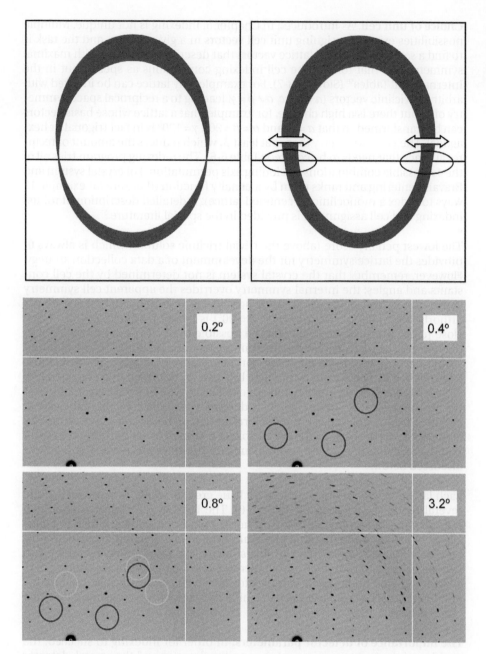

Figure 8-28 Manifestation of mosaicity. Estimating mosaicity from a single diffraction pattern is not trivial. In cases where the lunes are approximately centered around the rotation axis (top panel), the lunes widen along the axis direction. In a general orientation (bottom panels, same 1° rotation image with increasing mosaicity) the lunes are not as distinct and we only see an increasing number of partial reflections (extending over more than one image, red and green circles) but no distinct or tell-tale elongation of the spot shapes. Once the mosaicity becomes excessive and exceeds the rotation range, we observe the streaking as sketched in Figure 8-17. Images simulated using MLFSOM[53] courtesy of James Holton, UCSF.

and—within reasonable limits—larger mosaicity. With excessive mosaicity the spots can be spatially overlapped in all three dimensions and thus cannot be resolved or separated even by fine slicing.

Indexing

Indexing means the assignment of a consistent set of three reciprocal basis vectors $\mathbf{a}^*$, $\mathbf{b}^*$, and $\mathbf{c}^*$, which span the reciprocal lattice represented by the diffraction spots. The corresponding direct vectors $\mathbf{a}$, $\mathbf{b}$, and $\mathbf{c}$ (or scalar unit cell dimensions a, b, and c and angles α, β, and γ between them) define the crystal unit cell. Depending on the quality of the data and when using Fourier indexing algorithms, it is generally possible to index the diffraction pattern based on a single frame or snapshot[52] (Sidebar 8-8). In practice, more than one frame is used for indexing, or to verify the indexing, for several reasons: Crystals may not diffract isotropically, and snapshots in different orientations assure that diffraction anisotropy does not cause unacceptably low resolution in certain directions or orientations of a crystal. A single frame also may not contain enough strong reflections (~50–100) to unambiguously assign a unit cell.

Sidebar 8-7 **Rules for the choice of unit cells.** Given the multiple possibilities of assigning a unit cell to a given reciprocal lattice, the question arises, which is the most appropriate choice of unit cell? The following rules have been established:

1. The system [0, **a**, **b**, **c**] shall be right-handed.
2. The basis vectors should coincide with directions of highest symmetry.
3. The cell shall be the smallest one that satisfies rule 2.
4. Lattice vector **a** shall be the shortest lattice vector.
5. None of the remaining vectors not along **a** shall be shorter than **b**.
6. None of the vectors in the *ab* plane is shorter than **c**.
7. The angles should be either all <90° (acute) or all ≥90° (obtuse).

Centered cells arise because rule 2 supersedes rule 3. The presence of centering can be deduced from the resulting zonal (*C*) or integral (*F*, *I*) extinctions.

Choice of unit cell. As introduced in Chapter 5, indexing is not unique. Multiple possibilities exist for assigning unit cell vectors in a given lattice, and the task is to find a set of transformed lattice vectors that describes the lattice with maximal symmetry and that follows the cell indexing conventions as spelled out in the International Tables[56] (Sidebar 8-7). For example, any lattice can be indexed with arbitrary triclinic vectors ($a \neq b \neq c$, $\alpha \neq \beta \neq \gamma$) leading to a reciprocal space symmetry of $\bar{1}$, but there is a high chance, for example, that a lattice whose basis vectors can be transformed so that $a = b$, and $\alpha = \beta = 90°$, $\gamma = 120°$ is in fact trigonal or hexagonal. The Laue symmetry then is at least $\bar{3}$, which reduces the amount of reciprocal space that needs to be covered (Table 6-6). The indexing program tries all of the 44 possible combinations (including axis permutations) of crystal system and Bravais centering and ranks them by a penalty function (there are for example 12 ways to index a monoclinic, *C*-centered lattice). A detailed description of robust indexing and cell assignment is provided in the special literature.[57]

The lowest penalty score (above the trivial triclinic solution which is always 0) provides the lattice symmetry for the development of a data collection strategy. However, remember that the crystal system is not determined by the cell constants and angles; the internal symmetry overrides the apparent cell symmetry deduced from the cell constants. It is possible, for example, that an apparently orthorhombic cell ($a \neq b \neq c$, $\alpha = \beta = \gamma = 90°$) is in fact monoclinic, with β very close to 90° (in this case it may not be possible to fit the entire molecule into the unit cell, as discussed in the preliminary analysis section). In case of doubt always pick a lower lattice symmetry for the purpose of strategy determination. Redundant data can always be merged, but missing data cannot be regenerated. As several space groups may belong to one lattice type or even Laue group, it is not always possible to unambiguously determine the space group of the crystal at this early indexing stage. Proper determination of the crystal lattice symmetry is necessary to develop a strategy to collect a complete set of diffraction data. The data collection strategy also depends on the selected phasing strategy, as discussed below.

Each lattice type contains one or more Laue groups, which in turn can include several space groups (consult Table 6-6). The Laue group can be determined once the internal symmetry of the diffraction pattern is known, which is generally established in the merging phase after the data collection. If the data processing software is integrated with the strategy generator, it may be possible to deduce the Laue symmetry from smaller wedges of reciprocal space during data collection,[58] and optimize the strategy for the given reciprocal space symmetry. Systematic absences are used to determine the exact space group after the Laue group is fixed.

The importance of detector parameters. In order for indexing to succeed, the position of the incident primary beam on the detector and the crystal–detector distance needs to be known to a precision of less than half of the RL spacing. In cases of indexing failure of an otherwise unsuspicious diffraction pattern, a check of these instrument parameters is the first remedy that should come to mind. Defining correct starting values for the detector parameters is most important for quality indexing, and the image header does not always contain proper experimental setup data.

Sometimes it is not possible to find a consistent unit cell for the crystal. Large mosaic spread, large spot size, streaking and overlap, crystal slippage, multiple or satellite spots due to macroscopic twinning, and excessive ice rings can cause problems. Changing spot-finding parameters, adding additional frames, or trying a different indexing program can solve the problem. Once the detector parameters and the crystal orientation matrix are refined, spot predictions are checked, and when they fit, the remaining data are collected following one of the strategies outlined below.

Real crystals: modulated structures

Mosaicity and twinning can be considered as one type of deviations from perfect periodicity in a crystal. It is also possible that the deviations from strict

Sidebar 8-8 About indexing algorithms. While visual inspection of the distance between spots within a lune in a single diffraction frame gives valuable first clues about the approximate length of (reciprocal) unit cell axes, assigning a consistent and unique set of reciprocal space vectors by just looking at a diffraction pattern is seldom possible, largely because the crystal and thus the reciprocal lattice are generally in an arbitrary orientation. In addition, because of angular distortion, the distance between the spots is not the true (reciprocal) lattice spacing. With 4-circle diffractometers providing full control over the crystal orientation, it is possible to record a pseudo-precession image that provides an undistorted view of the RL.[59] The term *pseudo-precession* refers to the venerable instrument of the precession camera, an ingenious mechanical contraption perfected by Martin Julian Buerger[60] that positions a movable film cassette so that an undistorted section of the reciprocal lattice is recorded. From precession photographs, the reciprocal cell constants can be directly measured, and given the wavelength and crystal–detector distance, the real space cell constants can be computed.

Once the raw electronic image is corrected, the indexing is preceded by a spot-finding routine based on image-recognition algorithms. While early auto-indexing algorithms used in macromolecular crystallography were based on finding a consistent pattern of distances in a series of successive rotation images,[61] modern indexing algorithms are generally based on Fourier methods and can generally index a lattice from one single image frame.[62]

Orientation matrix. Three preliminary parameters that are later refined must be known: The origin of the image, given by the position of the incident primary beam on the detector (to a precision of less than half of the RL spacing), the crystal–detector distance, and the wavelength. Knowing these parameters, the indexing program can acquire the position vectors **x** of a number of RLPs whose indices **h** and relation to the crystal orientation are given by a matrix equation

$$\mathbf{x} = \mathbf{\Phi Ah} \tag{8-12}$$

where Φ is the 3×3 rotation matrix for the rotation around the spindle axis φ of the indexed image and **A**

is the crystal orientation matrix in reciprocal dimensions.[57] Given the indices and position, the orientation and rotation matrix are refined, new reflection positions are calculated, and the predictions are tested against subsequent frames.

Fourier methods in indexing. In order to determine the orientation matrix and to assign values for the indices in the above equation, a set of basis vectors must first be found. Most programs use a Fast Fourier Transform (FFT) algorithm to find a consistent set of lattice directions. The *DPS* algorithm[63] for example uses a 1-dimensional FFT analysis of the projections of the RL vectors onto a given direction vector **t**. For each of about 7300 discrete, evenly distributed directions **t**, the distribution of the Fourier coefficients is analyzed. Thirty directions with distinct periodicity (largest Fourier coefficients) are saved, and from them the three linearly independent basis vectors are chosen, which via Relation 8-12 in turn provides the sought-after crystal orientation matrix **A**.

The *DPS* (Data Processing Suite) indexing routine or Rossmann algorithm[62] described above is used in popular data collection programs such as *MOSFLM* by Harry Powell,[52] which is associated with the CCP4 project. Other data processing and indexing programs are *Xengen* by Andy Howard,[64] *XDS* by Wolfgang Kabsch,[61] *HKL2000/3000* by Wladek Minor and Zbyszek Otwinowski,[65] and their custom versions for various detector manufacturers. Most detector manufacturers also provide packaged data collection software integrated with their detector control software (for example Bruker's *SAINT/PROTEUM*; Rigaku Americas' *d*TREK*[47] by Jim Pflugrath; MAR Research with a combined public/proprietary *automar* package; and ADSC with programs written by Chris Nielson and Andy Howard).[66] Each of the program packages has their particular strength, and if one program fails to index, it is a viable option to try another one before abandoning the crystal. Most beam lines have the common indexing programs installed. Non-proprietary programs can read most detector image formats.

Figure 8-29 illustrates a 2-dimensional example of the principle of Fourier indexing and the relations between the reciprocal lattice and the direct unit cell. Chapter 9 discusses Fourier techniques in detail.

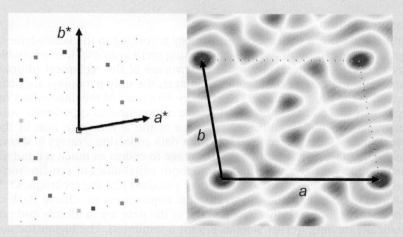

Figure 8-29 Basic 2-dimensional Fourier indexing. The left panel shows a reciprocal lattice with a limited number of reflections as they would appear when located in a diffraction image lune. The right image shows the amplitude-based Fourier transform of that lattice. Note the clear appearance of (distance vector) peaks at the unit cell corners in the real space transform. The images were generated using Kevin Cowtan's interactive structure factor tutorial: Select the structure factor applet, remove all but a few reflections representing a diffraction lune, and set their phase to zero to generate the cosine transform. http://www.ysbl.york.ac.uk/~cowtan/ (also shown in the practical guide by Elspeth Garman and Bob Sweet).[67]

crystal periodicity occur in a periodic fashion. These modulations can occur with a period commensurate with the lattice spacing or incommensurate with the lattice periodicity, and in one, two, or three dimensions. In Chapter 5 we have already encountered a 1-dimensionally commensurately modulated structure in the example of the largely translational NCS in $P1$ calmodulin (PDB entry 1up5).[68] In this case, the modulation occurred with double the period of the basic lattice and only in direction **a**. Consequently, one observes a superstructure (Figure 11-7) giving rise to a typical diffraction pattern with alternating strong and weak layers of reflections along **a*** (Figure 11-8).

A structure modulation generates additional reflections **H** whose position in reciprocal space depends on the modulation vector **q** relative to the primary reciprocal unit cell. For a single modulation vector,

$$\mathbf{H} = h\mathbf{a}^* + k\mathbf{b}^* + l\mathbf{c}^* + m\mathbf{q} \tag{8-13}$$

Equation 8-13 defines a 4-dimensional (reciprocal) superspace vector. For the calmodulin example, the modulation is solely in unit cell direction **a** with exact doubling of the unit cell, and the modulation vector **q** in reciprocal space thus is 0.5**a***, as illustrated schematically in the top panel of Figure 8-30. Crystal structure modulation can also occur with a period that is not commensurate with the basic lattice periodicity, that is, the modulation period and the basic lattice period are never exactly in phase, regardless of how many unit cells are repeated (Figure 8-30, second row).

Incommensurately modulated protein crystals have indeed been reported, although rarely.[69] The incommensurate or aperiodic modulation in protein crystals generally manifests itself in the form of weak satellite reflections close to the primary Bragg reflections. It is not exactly clear what causes structure modulation in proteins, but it is likely that modulated structures form in the vicinity of (displacive) phase transitions, as documented for small-molecule crystals. Some support for this hypothesis stems from the observation that modulations in protein crystals are sensitive to thermal treatments such as cryocooling and annealing, and modulation is difficult to control. While commensurately modulated superstructures can be refined normally, presently no standard refinement programs for incommensurately modulated protein structures exist. It is, however, worthwhile (as always!) inspecting the diffraction images for indications of additional peaks or satellites that could indicate the presence of a modulated structure. It is possible that modulation is present more frequently than reported, because the patterns can generally be indexed normally by ignoring the modulation peaks. Whether effects such as unusually large B-factors or related abnormalities are indicative of ignored structure modulation remains to be seen. If only the main reflections are used in refinement, the modulated areas will show weak or no electron density, depending on the degree of modulation. Again, keeping the images for processing with software capable of processing incommensurate diffraction data and refining modulated structures at a later point in time might be wise.

Strategy generation

After initial indexing, a data collection strategy according to the lattice type or Laue group needs to be devised. Each phasing technique requires a suitable strategy to obtain the necessary minimal coverage of the reciprocal (diffraction) space. In the case of basic rotation images, the only parameter that can be selected is the rotation range. The strategy generator of the data collection programs can make suggestions, and it also allows one to experiment with rotation ranges to obtain sufficient coverage of the asymmetric unit of the reciprocal space (Table 6-6; an example for a data collection strategy is shown later in Figure 8-36). It is seldom a disadvantage to collect as much redundant data as possible, and given the high-throughput capabilities of synchrotrons, a few general strategies suffice to cover most standard phasing techniques.[45, 70] As explained in Chapter 9, there is no absolute requirement to achieve 100% of data coverage, as long as only a few percent of the data are randomly and not systematically missing (revisit Sidebar 8-5 for a reminder). The higher the lattice symmetry and Laue symmetry, the easier it is to obtain complete data, and

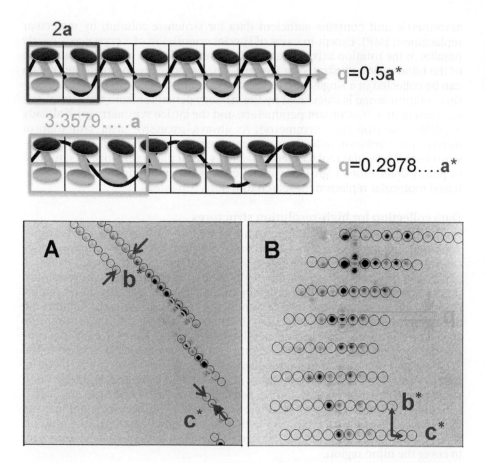

Figure 8-30 Modulated protein structures. The top part of the figure illustrates schematically a commensurate superstructure with doubling of the principal unit cell leading to a superstructure (Figures 11-7 and 11-8 provide the corresponding example of calmodulin, PDB entry 1up5). The second schematic drawing illustrates incommensurate, aperiodic modulation of a structure reflecting the situation in the profilin:β-actin complex described below. Soaking crystals in acidic substitute mother liquor causes profilin to dissociate from actin and an incommensurately modulated actin-superstructure forms in the **b** direction. Both images show the satellite reflections in the **b*** direction. Predicted locations of main reflections are circled in dark blue; satellite reflections are circled in light blue. Note that in the region of reciprocal space shown in image B, the satellite reflections are in some cases more intense than the main reflections. The satellite reflections have a modulation vector **q** = 0.00**a*** + 0.29**b*** + 0.00**c***. Figure of unpublished data courtesy of G. Borgstahl and J. Porta, Eppley Institute for Cancer Research, University of Nebraska Medical Center, Omaha, NE.

completeness in the 90% or better range is in practice often readily achievable with a roughly 90° oscillation range or even less. Exceptions are low-symmetry space groups, which depending on crystal orientation may require full 360° rotations to achieve high completeness. In addition, for anomalous data collection, the centrosymmetrically opposed part (i.e. 180° mirrored wedge or its corresponding Bijvoet symmetry mates) of the unique reciprocal space unit must be collected.

Recall from the discussion of counting statistics and error propagation in Chapter 7 that in principle—if only true random counting errors were present—collecting the same data in one single rotation scan of say 9 s exposures is equivalent to recording them in three scans of 3 s exposures. The benefit of recording in three consecutive scans (sometimes colloquially referred to as "sweeps") is that if the crystal suffers radiation damage, at least one—perhaps weak but still fairly complete—data set may be recorded, while a data set obtained by longer exposure with only 33% completeness is definitely useless or requires that another isomorphous crystal be found. An additional advantage of the multi-scan strategy may provide is the cancelling-out of some systematic errors. A minor drawback of the multi-scan method is more overhead in readout and during data processing. Remember from Poisson counting statistics that to obtain twice ($n = 2$) the signal-to-noise (S/N) ratio, four times (n^2) as much data collection time is needed.

Data collection strategy for molecular replacement phasing

The simplest case of data collection is a single-wavelength data set without consideration of the anomalous signal (although anomalous contributions from all atoms in the crystal are present at varying degrees at all wavelengths, they are miniscule for the light elements (H, C, N, and O) comprising most of the scattering matter in native proteins, but special data collection techniques described in Section 8.4 can be employed to extract the minute anomalous signal of sulfur for S-SAD phasing). A single-wavelength data set covering the reciprocal space asymmetric unit contains sufficient data for structure solution by molecular replacement (MR). Except in cases of special orientation of a crystal axis nearly parallel to the rotation axis, and at very high resolution (as a result of the shape of the blind region illustrated in Figure 8-22), a practically complete set of data can be collected in a single scan of successive frames.[45] The extent of the necessary rotation range is calculated by the strategy generator from the crystal orientation matrix, instrument parameters, and the lattice symmetry (or if already available, the true Laue symmetry). As always, excessive pixel saturation of intense low-resolution reflections may require a second scan with shorter exposure times. Molecular replacement phasing is particularly sensitive to saturated or poorly determined high-intensity reflections which dominate the Patterson-based molecular replacement search functions (Chapter 11).

Data collection for high-resolution structures

In fortunate cases, crystals diffract to high resolution. The term is used loosely, and shall indicate in the context of data collection crystals diffracting better than about 1.5 Å; diffraction in excess of 1.2 Å is denoted as the onset of atomic resolution, or ultra-high resolution. As illustrated in Chapter 1, high resolution permits the discernment of fine details in the structure, which can be understood as a manifestation of tighter sampling intervals or slices throughout the crystal. As the number of reflections increases with the volume of sampling space, even a numerically less impressive increase in resolution leads to a substantial increase in recorded data (Figure 12-11), thus drastically improving the accuracy and precision of the subsequent structure refinement. Given the (rare) case of very strong data with resolution of at least 1.2 Å (Sheldrick's Rule)[71, 72] and small protein size, structures can be determined *ab initio* via direct methods[73] (Chapter 10). Recall that the excluded blind region can become significant at high resolution and with reciprocal lattice axes oriented parallel to the rotation axis. Reorientation of the crystal using a κ-offset (Figure 8-38) may be necessary to cover the blind region.

Anomalous diffraction data

In Chapter 6 we extensively discussed that the presence of anomalous scatterers breaks the centrosymmetry of the reciprocal space and centrically opposed Friedel mates F_{-h} of each reflection F_h (or generally at least one of the Bijvoet pairs of $\mathbf{h}$) must be recorded. Of those, only the non-centric pairs can show intensity differences in the presence of anomalous scattering contributions. These anomalous differences, as well as dispersive differences between data recorded at different wavelengths, form the basis for the most powerful phasing techniques available to date. Anomalous data collection therefore requires that in addition to the unique wedge of data covering the asymmetric reciprocal unit, the centrosymmetrically opposed wedge ($\varphi + 180°$) or corresponding Bijvoet pairs in symmetry related wedges must be collected.

The sensitivity of the crystals to radiation damage—particularly when they contain strongly absorbing heavy atoms—suggests the strategy of collecting anomalous data in so-called inverse beam geometry (or related strategies). After each rotation increment, the φ-axis is rotated by 180°, and the corresponding inverse image containing the Friedel or Bijvoet mates is collected. This reduces potential difficulty in the later scaling stage, when the reflections from temporally distant exposures need to be combined and merged into a single data set. Splitting the data set into small inverse blocks has the additional benefit that, after recording the first obverse and inverse images, the significance of the anomalous difference signal can be determined, and the data collection expert (system) should adjust data collection times accordingly. Alternatively, if no usable anomalous signal can be expected to be collected in a reasonable time, it may be more efficient to abandon the data collection and to proceed to another crystal.

In orthogonal symmetries with intersecting axes, it is often possible to record both reflections of a Friedel or Bijvoet pair in the same image and get complete coverage of anomalous data in one scan. However, for the general use of this technique in low-symmetry cases, a multi-axis diffractometer is required.

Single wavelength anomalous data

Anomalous data collection for single-wavelength experiments does not require measurement of an X-ray absorption edge scan. The experiment must simply be conducted at or somewhat above the absorption edge of the selected marker element, but the exact determination of the absorption edge spectrum (which is required to optimize dispersive ratios in MAD experiments) is not necessary. In cases where "white lines" in the absorption spectrum are present, experimentally determining the exact absorption maximum is of advantage in obtaining maximal anomalous differences. The most pronounced white lines result from transitions of $2p_{3/2}$ core level electrons into unoccupied $3d$-electron levels of the absorbing atom in L_3- and L_2-edges, but weaker white line transitions can also sometimes be observed in L_1- and K-edges (for example Se).

Special considerations are required when using native sulfur of the Met and Cys residues as anomalous markers. Even the longest usable X-ray wavelengths (around 2.5–2 Å) are far above the K absorption edge of sulfur, and the anomalous difference signal is often as low as ~0.5% of the total signal.[74] Geometric and X-ray optical constraints of the tunable beam line components, as well as rapidly increasing absorption at longer wavelengths (lower X-ray energies) set a limit to how long a wavelength can be used experimentally. However, given sufficiently redundant data collection via integration of multiple scans covering the reciprocal space unit many times (720° and more of rotation with varying crystal orientations, not just rotating the crystal around the same axis multiple times), data with S/N ratios sufficiently high to extract anomalous intensity differences can be collected. In combination with powerful density modification techniques, the S-SAD phasing method proposed more than 20 years ago by B. C. Wang[75] has been shown to be quite successful in its modern implementations.[74, 76] Given that no special marker atoms need to be introduced into the protein, the method is gaining rapid acceptance in protein crystallography. We will process an anomalous S-SAD data set in the example section of this chapter, and use the data for S-SAD phasing in Chapter 10.

Multiple anomalous diffraction data

MAD phasing[77, 78] exploits two effects resulting from anomalous contributions by using not only Bijvoet differences within each data set, but also the dispersive differences between data sets recorded at different wavelengths. To optimize these differences, an accurate experimental edge scan for the phasing element needs to be recorded, and between two and four MAD wavelengths are selected as discussed in detail in Chapter 6. In summary, the anomalous differences are largest at the absorption edge maximum (maximum of f''), and the dispersive differences are largest between the remote data sets and the data set recorded at the inflection point of the edge jump, corresponding to a minimum in f'. The high-energy remote data are usually recorded a few hundred eV above the edge and still contain substantial internal anomalous signal. The least anomalous difference signal, but still with dispersive contributions, is expected from the optional low-energy remote data set, collected several hundred eV below the absorption edge. In view of signal loss through radiation damage, the

---

Box 8-9 **Indexing and strategy generation.** Indexing is the assignment of a consistent set of unit cell vectors matching the diffraction pattern. Most indexing methods are based on Fourier analysis of a subset of spot positions of one or more diffraction images. Once a unit cell has been selected, its lattice symmetry determines the minimal symmetry of the reciprocal space. Based on this symmetry, a data collection strategy is determined. The goal of the strategy is to collect at least enough reflections to cover the entire asymmetric unit of the reciprocal space and obtain complete data. Accurate data require higher redundancy, achieved by multiple independent measurements of equivalent reflections. Anomalous data for phasing purposes also require collection of the centrosymmetrically related reflections. For MAD phasing, data sets collected at different wavelengths provide additional dispersive difference data.

most common strategy is to obtain peak wavelength data first (still enabling SAD phasing with a reasonable chance), followed by inflection point data (this second data set providing mostly dispersive differences), and the high-energy remote data set (with redundant internal anomalous differences and large dispersive differences against inflection data). All other basic data collection strategy considerations regarding completeness and S/N ratio discussed in the previous sections apply to MAD data as well. Note that radiation damage can render temporally remote dispersive data sets practically non-isomorphous. The dispersive data sets may then become more or less non-isomorphous SAD data sets.[27] MAD data collection as the phasing method of choice in protein crystallography[78] is increasingly being rivaled by SAD phasing because of improved density modification methods that allow the breaking of the inherent phase ambiguity in the SAD phasing equations (Chapter 10).

Raw data correction, integration, and scaling

Once all image frames are collected, the data collection program must convert the raw pixel intensity of the measured and indexed diffraction spots into a list of properly scaled intensities and their estimated standard errors. In each frame, a number of corrections must be applied to the raw pixel data of the frame to obtain the reflection intensities. In addition to detector-specific spatial corrections and calibrations, the Lorentz and polarization (LP) corrections are applied for the given beam geometry and crystal orientation (Chapter 6). The raw pixel intensity of the diffraction spots must be integrated,[48] and partial reflections dispersed over multiple frames are combined onto single reflections. The integration of the pixel area is generally a 2-dimensional integration for wide slicing, and a 3-dimensional spot profile in fine slicing. The counts outside the reflection profile provide the background area, and based on photon-counting statistics (Chapter 7), we obtain a value for the intensity and the estimated error for each observation of a reflection. The fact that noise and background subtraction can yield negative intensities, and how to treat them with Bayesian methods, was also discussed in Chapter 7.

Error model statistics. The absence of abnormalities in the series of integrated frames can be checked by analyzing the merging R-values, which should be the same for each frame (around 0.04–0.07 for most data sets, cf. Figure 8-32). The same holds for the error model, which can be checked for consistency in each frame through the reduced chi-squared χ^2. If the errors in the measurements are normally distributed and no systematic errors arising from radiation damage or other effects are present, $\chi^2 \simeq 1$. If there are outliers in either the merging statistics or the χ^2 test, the affected frames should be examined for experimental errors and removed if no explanation can be found. Removing a few outlier frames is

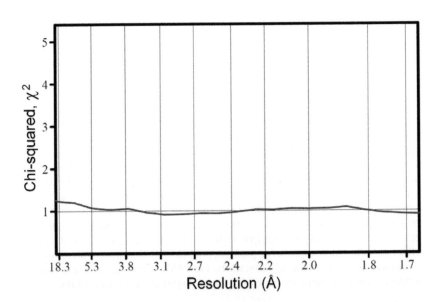

Figure 8-31 Error model statistics. The plot of reduced χ^2 binned in resolution shells will be approximately unity over the entire range of the variance if the data are normally distributed, that is, representative of random errors. Plot was generated by George Sheldrick's *SADABS* program for the high redundancy S-SAD data of VTA1.[79]

not critical, particularly in high redundancy data, but it improves the overall accuracy of the data. If the frame-by-frame or resolution-dependent χ^2 values deviate significantly from 1.0, some parameters in the integration error model might be wrong. In the VTA1 synchrotron data integration example described below, the detector gain was initially set incorrectly in the image header which gave systematically higher χ^2 values. A properly adjusted error model will show the resolution (in)dependence graphed in Figure 8-31.

Up to this point, multiple incidences of the same reflections are treated independently. However, in addition to the spatial integration, we also must account for the fact that the X-ray reflections were temporally separated during recording. Therefore, in the next step the integrated reflections must be properly scaled and combined to account for effects of fluctuating beam intensity, if data are not collected in dose mode. The decay of primary intensity is commonly a sawtooth function in those synchrotrons where the electrons are injected at once and the beam current slowly decays until the next filling. In addition, loss of diffraction resulting from progressive radiation damage needs to be accounted for by proper scaling (if possible at all for extreme cases). In cases of severe radiation damage, even the cell constants may shift during exposure, and it may be necessary to manually repeat the indexing in small groups of frames in case the automatic indexing fails between frames. In the section about processing high redundancy data we will take a closer look at some instructive plots of integrating and scaling statistics.

Data reduction and quality analysis

The integrated raw data are essentially a long list of indices, the reflection intensity of each individual reflection, its standard error, and additional batch information for each frame. Save at least this integrated but unmerged multi-batch data file before further processing the data (it is of course better to also archive the image frames; disk space is cheap and reprocessing or proof of data provenance may later be required). Short of the lattice type selected after indexing, we have not made yet any assumptions (although we may have good guesses) about the actual space group; only the basic lattice type and the corresponding lowest symmetry space group are determined. For example, this could be space group $P4$ for lattice type tP (tetragonal primitive).

The initial reduction of data begins with the merging and scaling of multiple measurements of identical reflections according to the initial indexing. The result is a reduced data set that contains all reflections, but not yet the further merged symmetry equivalents. Friedel pairs should in any case always be kept separate until the very last merging steps before refinement and map reconstruction. Some refinement programs, notably *REFMAC*, already include refinement against anomalous data sets, and it is expected that refining against anomalous data will become practice.

At this point, it is a good idea to save also a scaled but unmerged data file before further processing the data. The program used on the synchrotron or laboratory source may not be the ideal program to determine exact merging statistics, the outlier detection may need to be repeated, and reindexing and determination of the space group may be necessary with a different or external program. For example, the reflection data processing program *XPREP* works best when it has scaled but unmerged data at its disposal; this is particularly important for the enumeration of proper statistics for anomalous difference data. *XPREP* also has a more powerful space group search program and other analysis routines that are not always available in combined data collection suites.

Merging statistics and redundancy
The intensity of the reflections becomes weaker with increasing resolution as a consequence of the falloff in atomic scattering factors and of atomic and lattice displacement. Data collection statistics are thus provided as overall numbers for the entire data set as well as being binned in (about 20–40) resolution shells with an equal number of reflections. The consequence of increasing relative error in

Sidebar 8-9 Good reasons to keep unmerged data. In case of the slightest doubt about the crystal or Laue symmetry, generate the strategy and collect the data in the lower symmetry; redundant data can always be merged, but missing data can rarely be divined. Keep at least the raw integrated, scaled, but unmerged data (it is best to archive the primary image data). You might need unmerged intensity data later for testing different space groups, higher or lower symmetry, or might find later some unexpected anomalous signal that may provide some crucial (serendipitous) phasing information.

Situations where NCS axes are parallel to crystallographic axes may turn out to be cases of higher space group symmetry where the NCS axis is actually a crystallographic axis (an example is given in Chapter 11), and you need to merge the data and generate new statistics in the correct space group. In the opposite case of lower than initially proposed symmetry, you must again have the unmerged data at hand to generate the unique data for a space group with reduced symmetry. Another reason to keep the unmerged data is that the commonly reported simple linear merging R-value is suboptimal for important practical applications as discussed in the next section. Other merging R-values that consider symmetry-related redundancy may be available only from post-processing programs.

the measurement of the weak reflections is a diminishing signal-to-noise ratio (SNR) with increasing resolution. It is commonly (but not necessarily) expressed as average $\langle |I|/\sigma(I) \rangle$ summed over all N reflections in a resolution shell:

$$\langle |I|/\sigma(I) \rangle = \frac{1}{N} \sum_{\mathbf{h}}^{N} \frac{|I_{(\mathbf{h})}|}{\sigma(I_{(\mathbf{h})})} \tag{8-14}$$

During data collection we will have made multiple, redundant observations of identical or symmetry-related reflections, which we need to ultimately merge into the unique data set representing the asymmetric unit of the reciprocal space as defined in Table 6-6. For molecular replacement phasing and refinement, a single data set containing the unique asymmetric wedge of the reciprocal space will suffice, while for phasing using anomalous data, we also need to collect and keep separate the inverse wedge of reciprocal space. The redundancy is given by how many independent observations of reflections we made, and is explicitly listed in the infamous "Table 1" containing data collection and refinement statistics.

Linear merging R-value. The most commonly used quality indicator when merging N reflections $\mathbf{h}$ within a resolution range is the linear merging R-value, R_{merge}:

$$R_{merge} = \frac{\displaystyle\sum_{\mathbf{h}} \sum_{i=1}^{N} \left| I_{(\mathbf{h})i} - \overline{I}_{(\mathbf{h})} \right|}{\displaystyle\sum_{\mathbf{h}} \sum_{i=1}^{N} I_{(\mathbf{h})i}} \tag{8-15}$$

where the inner summation extends over all N redundant observations for a given reflection $\mathbf{h}$ and $\overline{I}_{(\mathbf{h})}$ is the averaged intensity of each reflection. The outer summation extends over the desired resolution range. Depending on what kind of reflections are actually merged, this merging R-value can be termed R_{int} indicating general merging on identical (but otherwise unmerged) intensities, R_{sym} for merging of symmetry-related reflections, R_{cryst} for merging between different crystals, R_{anom} for merging Bijvoet pairs carrying anomalous signal, and so on. Merging statistics are computed as overall quality indicators for entire data sets as well as in resolution shells for more detailed analysis of intensity data. As a consequence of increasing relative error in the measurement of the weak reflections the merging R-values increase rapidly with higher resolution (Figure 8-32).

Redundancy-independent merging R-value. Unfortunately, the basic linear merging R-value is statistically on shaky ground, because it does not account for the redundancy of the data. Lower redundancy will always give lower linear R_{merge} because the contribution of each reflection $\mathbf{h}$ to the numerator of Equation 8-15 is a function of N. Thus, even with only random errors present, the linear merging R will rise with increasing redundancy.[80] The value it asymptotically approaches is called R_{meas} or R_{rim} (redundancy-independent merging R-value):[81]

$$R_{rim} = \frac{\displaystyle\sum_{\mathbf{h}} \left(\frac{N}{N-1}\right)^{1/2} \sum_{i=1}^{N} \left| I_{(\mathbf{h})} - \overline{I}_{(\mathbf{h})} \right|}{\displaystyle\sum_{\mathbf{h}} \sum_{i=1}^{N} I_{(\mathbf{h})}} = R_{meas} \tag{8-16}$$

Being an asymptotic value, the redundancy-independent merging R-value R_{rim} or R_{meas} will always be higher than the conventional linear merging R-value (which may be a reason for its reluctant acceptance as a data item in the macromolecular crystallography "Table 1").

Precision-indicating merging R-value. In contrast to what the linear R-merge would suggest, the intensities are actually becoming more precise as we are merging more and more observations. We can account for this improvement in precision in a similar way as in the computation of the variance of a distribution (which *is* a measure for precision) by introducing a term $1/(N-1)^{1/2}$ in the linear merging R-value, yielding R_{pim}, the precision-indicating merging R-value:[82]

$$R_{pim} = \frac{\sum\limits_{h} \left(\dfrac{1}{N-1}\right)^{1/2} \sum\limits_{i=1}^{N} \left| I_{(ih)} - \bar{I}_{(h)} \right|}{\sum\limits_{h} \sum\limits_{i=1}^{N} I_{(ih)}} \tag{8-17}$$

The precision-indicating merging R-value R_{pim} decreases with redundancy N. R_{pim} seems to be a particularly useful statistic for the estimation of data quality for anomalous diffraction data sets.

Anomalous merging R-value. Empirical tests[81] have shown that when the ratio R_{anom}/R_{pim} exceeds about 1.5 (i.e. the precision-indicating R-value becomes ~1.5 times smaller than the anomalous merging R), substructure solution based on anomalous difference data seems to be likely to succeed. The basic assumption here is that R_{anom} (based on anomalous differences) represents the signal and R_{pim} (for all reflections) the noise. This is not quite correct as the anomalous data contain a noise contribution as well, but the guideline seems to work reasonably well.[83] The anomalous merging R-value R_{anom} is computed as

$$R_{anom} = \frac{\sum\limits_{h} \left| I_{(-h)} - I_{(+h)} \right|}{\sum\limits_{h} \bar{I}_{(h)}} = 2\frac{\sum\limits_{h} \left| I_{(-h)} - I_{(+h)} \right|}{\sum\limits_{h} \left(I_{(-h)} + I_{(+h)} \right)} \tag{8-18}$$

However, there is another point to consider beyond simply increasing precision through redundancy. Every intensity measurement will, in general, contain some other error besides the fundamental uncertainty from counting statistics alone. Systematic and other random errors exist whose qualities and quantities depend on the particular experimental conditions. Crystal orientation, absorption in the crystal, absorption from the mounting loops or poor crystal centering, instrumental parameters such as shutter jitter and beam flicker: all may be present in each individual measurement. In average values merged from many individual measurements, there will, in all likelihood, be some compensation of systematic errors as well, and the merged average will present a more accurate estimate of the true value than, say, one single measurement of the same total intensity. This is why redundant data for S-SAD, or other difference data phasing methods that depend on accurate and precise data, are preferrably collected from multiple scans in different crystal orientations.

Some crystallographic data processing programs also report the squared weighted merging R-value R_w[84] with the weights the inverse variance:

$$R_w = \frac{\sum\limits_{h} \sum\limits_{i=1}^{n} w_{(h)i} \left(I_{(h)i} - \langle I_{(h)} \rangle \right)^2}{\sum\limits_{h} \sum\limits_{i=1}^{n} w_{(h)i} I_{(h)i}^2} \tag{8-19}$$

Another statistic occasionally encountered, the pooled coefficient of variation (PCV), is related to R_{rim} by $\text{PCV} = (\pi/2)^{1/2} R_{rim}$.

Once the data are reduced to the unique asymmetric unit of reciprocal space in the correct space group (we need at least the Bravais part of the space group here, not just Laue symmetry, because integral extinctions affect the number of reflections), the final completeness of the data is computed by resolution bins.

Signal-to-noise ratio, completeness, and merging *R* value

The crystallographic S/N ratio $\langle |I|/\sigma(I)\rangle$ defined in (8-14) provides an estimate for the usefulness of the data as a function of resolution; usefulness here is not a defined statistical measure but depends on the objective. For refinement and map reconstruction, even a high-resolution shell with a relatively low $\langle |I|/\sigma(I)\rangle$ will in principle still contain a few valuable reflections, and most modern likelihood-based refinement programs will apply proper statistical weights and do not need artificial SNR-based high-resolution cutoffs (and low-resolution cutoffs are also unnecessary since bulk solvent correction has been introduced). The importance of proper error estimates, however, for maximum-likelihood-based programs cannot be overstated (a point in favor of fine slicing with 3-dimensional peak profiling).

Difference-data-based phasing methods on the other hand require data with high SNR and thus do justify larger $\langle |I|/\sigma(I)\rangle$ cutoff values: Because we create small differences of large values, they are quite sensitive to noise in the data (Section 10.8). This is amplified by the fact that large squared difference terms used in the form of ΔI^2 or ΔE^2 dominate the phasing equations. Excessive inclusion of weak high-resolution data shells (where really only a few spots are visible on the image) during image processing can lead to poor overall integration statistics. In this case it is advisable to re-process the data with a higher $\langle |I|/\sigma(I)\rangle$ cutoff and thus lower resolution based on completeness and $\langle |I|/\sigma(I)\rangle$ obtained after the initial merging of the data. In practice (and because of the potental of additional systematic errors), a SNR of about 1.5 to 2.0 seems to be an acceptable compromise for a low-resolution cutoff during data processing. Data at high resolution also tend to be less complete because of reflections falling beyond detector edges, in blind regions, or resulting from other geometric diffractometer limitations.

Completeness and effective resolution. Randomly missing data will generate noise in map reconstruction, while systematically missing data will cause more perturbing aberrations, such as failed phasing, or lead to streaking in maps (Chapter 9). The level of detail in a map will not be representative of the nominal last shell resolution if only 50% of the data are recorded. If these data are strong (and not systematically missing), there is no reason to truncate the data, but it must be expected and accepted that the map (and therefore the final model) will not be as good as if complete data to this resolution were available. The relation between diffraction limit and optical resolution of an electron density map is discussed in Sidebar 9-4, and an effective resolution as a function of completeness *C* can be estimated[81] as $d_{eff} = d_{min} \cdot C^{-1/3}$.

Relation of merging *R* and SNR. Consider the following: The relation between the mean absolute error (*MAE*) of a normally distributed variable x and its standard deviation σ is given (Chapter 7) by

$$\frac{MAE(x)}{\sigma(x)} = \sqrt{\frac{2}{\pi}} = 0.79788 \tag{8-20}$$

We expand both sides of (8-20) with the expectation value of $\langle x \rangle$ and obtain

$$\frac{MAE(x)}{\langle x \rangle} = \frac{\langle |x - \langle x \rangle| \rangle}{\langle x \rangle} = \frac{\frac{1}{N}\sum_{j=1}^{N}|x_j - \langle x \rangle|}{\langle x \rangle} = \sqrt{\frac{2}{\pi}} \cdot \sigma(x)/\langle x \rangle \tag{8-21}$$

The relation to R_{merge} becomes immediately clear if we substitute in Equation 8-21 the explicit expressions for the *MAE* of intensities $\Delta I_{(\mathbf{h})}$ and the mean intensity $\overline{I}_{(\mathbf{h})}$

$$\overline{\Delta I}_{(\mathbf{h})} = \frac{1}{N}\sum_{i=1}^{N}\left|I_{(\mathbf{h})i} - \overline{I}_{(\mathbf{h})}\right| \text{ and } \overline{I}_{(\mathbf{h})} = \frac{1}{N}\sum_{i=1}^{N}I_{(\mathbf{h})i} \tag{8-22}$$

and obtain for a single reflection $\mathbf{h}$

$$R = \frac{\sum_{i=1}^{N}\left|I_{(\mathbf{h})i} - \overline{I}_{(\mathbf{h})}\right|}{\sum_{i=1}^{N} I_{(\mathbf{h})i}} = \sqrt{\frac{2}{\pi}} \cdot \sigma(I_{(\mathbf{h})}) / \overline{I}_{(\mathbf{h})}$$

(8-23)

which averaged over a shell provides the relation of the linear merging R-value with the signal-to-noise ratio $\langle |I| / \sigma(I) \rangle$:

$$R_{merge} \approx 0.8 / \langle |I| / \sigma(I) \rangle$$

(8-24)

As a consequence, for a resolution shell with $\langle |I| / \sigma(I) \rangle \approx 0.8)$ the merging R-value becomes ~0.8/0.8 or 100%. Note that the meaning of the merging R-value for a shell with random noise data is entirely different from the maximum expected R-value on F for a random structure ($R_F = 0.59$).

Equation 8-24 is of course a very useful and interesting relation. First, it allows us an internal consistency check of the data: if the merging R-value is much less than expected from the $\langle |I| / \sigma(I) \rangle$, something like outlier downscaling or outlier removal may have been excessive; if it is too large, there may be systematic errors, perhaps from detector calibration or other physical problems, or it may be that data items are taken from inconsistent sources such as later and earlier stages of data processing.

In addition, Relation 8-24 allows a crude *a posteriori* empirical estimate for R_{merge} in each resolution shell when we have only merged data (or table entries in other papers) available. At $\langle |I| / \sigma(I) \rangle$ of 4 we would expect an R_{merge} of about 0.2 (20%), and for the lowest resolution shell with $\langle |I| / \sigma(I) \rangle$ of say 2 we estimate ~40% for the merging R-value.

R-sigma. The estimated standard errors for intensities $\sigma(I)$ contain contributions from the individual (integrated but unmerged) intensities as well as the contributions from merging the intensities. As a variance measure, we expect $\sigma(I)$ to decrease during the addition of more measurements. Therefore, R_{sigma}[85]

$$R_\sigma = \frac{\sum_{\mathbf{h}} \sigma_{(\mathbf{h})i}}{\sum_{\mathbf{h}} I_{(\mathbf{h})i}}$$

(8-25)

in general tends to be lower than the corresponding linear R_{merge}. If this is not the case it probably indicates that either the data used to determine R_{merge} were not the original, identical-reflections-only merged data, or the integration error model was poor (i.e. the normalized χ^2 becomes significantly larger than 1.0) or not on an absolute scale. A typical plot of the resolution dependency of the linear merging value R_{int} compared with R_{sigma} is shown in Figure 8-32.

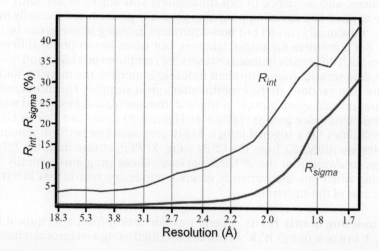

Figure 8-32 Merging R-values versus resolution. The plot shows a typical behavior of merging R-values R_{int} and R_{sigma} with resolution. R_{sigma} is generally lower than the corresponding R_{int}. Data were plotted by *SADABS* for the VTA1 in-house data processed in Section 8.4.

> **Box 8-10 Descriptors of data quality.** Common indicators for data quality are merging R-values, number of independent observations (redundancy), and the signal-to-noise ratio, computed for the entire data set as well as in resolution bins. The linear merging R-value R_{merge} always increases with redundancy, while the redundancy-independent merging R-value R_{rim} or R_{meas} provides an independent upper limit. The precision-indicating merging R-value R_{pim} accounts for increasing precision with increasing redundancy. A ratio of $R_{anom}/R_{pim} > 1.5$ is an empirical indicator for successful use of the data in SAD phasing. The linear merging R-value and the signal-to-noise ratio are related by $R_{merge} \sim 0.8 / \langle |I|/\sigma(I) \rangle$.

Space group determination

The reduced data are now analyzed for determination of the Laue symmetry and the space group. In many cases, the Bravais symmetry and chiral space group can be determined from systematic absences. The 11 pairs of enantiomorphic space groups which display the same serial extinctions ($l = 3n$ only for $P3_1$ and $P3_2$, for example) cannot be distinguished. In the cases of tetragonal $I222$ and $I2_12_12_1$ and of cubic $I23$ and $I2_13$, the integral absences caused by the Bravais centering coincide with the serial extinctions from the three screw axes. In all these ambiguous cases, the structure solution will only succeed, or will have better statistics, in the correct space group, thereby revealing which of the possible space groups is correct.

Most programs guess the correct space group (or the correct enantiomorphic pair) by considering extinctions, space group frequency, and the highest possible internal symmetry by merging the potential symmetry equivalents and computing the average merging R-value. If the merging R-values are unusually high, the apparent space group symmetry may be too high. Remember that the internal symmetry of the diffraction pattern (Laue symmetry) determines what the highest possible space group symmetry is, not the apparent lattice type. In images from special orientations, the diffraction symmetry may also be detectable by visual inspection. For example, in a crystal orientation with $00l$ in the detector plane, Laue symmetry $4/mmm$ will show a clear mirror symmetry that would be absent in $4/m$. The cell contents (number of molecules in the proposed asymmetric unit cell) will also provide clues to the possible space group assignment.

Table 6-7 lists all the systematic extinctions for symmetry elements of chiral space groups, and we will demonstrate a practical example for space group determination in the experimental section of this chapter.

Reindexing of unit cells

We have already realized in Chapter 5 that unit cells can be indexed in multiple ways. Even when all the rules regarding highest Bravais symmetry, orthogonal preferences, and sequence of cell dimensions and angles as set forth in the ITC are applied (the indexing programs generally perform satisfactorily in this regard), individually correct but non-equivalent indexing schemes can be used. This is not a problem for single data sets, but when isomorphous differences between such differently indexed data sets are computed or a high- and low-resolution data set with non-equivalent indexing is merged, the merging statistics often are near random, to the experimenter's great surprise. The most common cases are uniaxial space groups in trigonal, tetragonal, and hexagonal systems and a few cubic space groups (Table 8-4). Figure 8-33 shows two different indexing possibilities for a trigonal lattice. Many programs can perform reindexing; examples are *REINDEX* from the CCP4 suite, *XPREP*, *xtriage* from the *PHENIX* package, or *dtcell* from the *d*TREK* package. Those programs typically issue a warning when one inadvertently enters a reindexing matrix that inverts the handedness of the structure.

The reindexing matrix P. The general transformation from a reciprocal basis $[0, \mathbf{h}, \mathbf{k}, \mathbf{l}]$ to a new one $[0, \mathbf{h}', \mathbf{k}', \mathbf{l}']$ can be described using a reciprocal reindexing

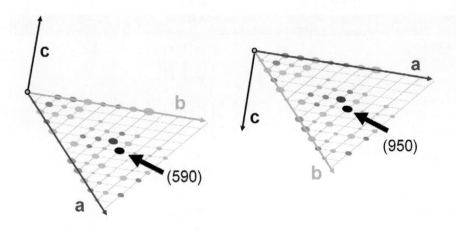

Figure 8-33 Multiple indexing possibilities. The figure shows a trigonal cell that is reindexed with the transformation *hkl* → *kh–l*. Consistent indexing between data sets is important when merging data or generating difference data sets. Incompatible indexing can often be recognized by near random merging statistics between otherwise good data sets.

matrix **P** with elements p_{ij} that is formed by expressing the new system axes in terms of the old reciprocal axes:

$$\mathbf{h'} = p_{11}\mathbf{h} + p_{12}\mathbf{k} + p_{13}\mathbf{l}$$
$$\mathbf{k'} = p_{21}\mathbf{h} + p_{22}\mathbf{k} + p_{23}\mathbf{l}$$
$$\mathbf{l'} = p_{31}\mathbf{h} + p_{32}\mathbf{k} + p_{33}\mathbf{l} \tag{8-26}$$

or in compact matrix notation

$$\mathbf{H'} = \mathbf{PH} \text{ and the inverse } \mathbf{H} = \mathbf{P^{-1}H'} \tag{8-27}$$

where **H** is the index column vector. The reindexing matrices listed in Table 8-4 can be confirmed using the expressions provided above.

The determinant of **P** has a number of interesting properties:

- if det(**P**) < 0, the system changed handedness (generally undesirable);

- if det(**P**) = 0, the chosen new basis vectors are not linearly independent;

- the magnitude $|\det(\mathbf{P})|$ provides the ratio of the (reciprocal) unit cell volumes, V'/V.

8.4   Processing of S-SAD data: a practical example

A phasing technique that has gained increasing importance for *de novo* experimental structure determination is single-wavelength anomalous diffraction (SAD) phasing utilizing the weak anomalous signal of natively present sulfur atoms in proteins (S-SAD). The theory of anomalous diffraction is explained in Chapter 6. Because the X-ray absorption K-edge of sulfur is located at very low energies (2472 eV or ~5 Å), in-house X-ray sources with copper (1.54 Å) or chromium (2.29 Å) anodes are suitable for data collection above the absorption

Box 8-11 Space group determination and reindexing. The space group can be determined from a combination of merging statistics, systematic absences, and space group preferences with the exception of the remaining handedness ambiguity resulting from screw axes in the 11 pairs of enantiomorphic space groups and the pairs *I*222 and *I*$2_1 2_1 2_1$ and *I*23 and *I*$2_1$3. In ambiguous cases, all possible space groups must be tested during the phasing stage. For 27 chiral space groups, non-equivalent indexing possibilities exist. While the choice of setting is free for individual data sets, the reflections must be consistently indexed for the generation of difference data sets.

Space groups	Symbolic reindexing operator	Reindexing matrix
P4, P4₁, P4₃, P4₂, I4, I 4₁, *P6, P6₁, P6₅, P6₃, P6₂, P6₄*	$hkl \rightarrow kh\text{-}l$	$\begin{pmatrix} 0 & 1 & 0 \\ 1 & 0 & 0 \\ 0 & 0 & \bar{1} \end{pmatrix}$
P3, P3₁, P3₂	$hkl \rightarrow kh\text{-}l$, $hkl \rightarrow \text{-}h\text{-}kl$, $hkl \rightarrow \text{-}k\text{-}h\text{-}l$	$\begin{pmatrix} 0 & 1 & 0 \\ 1 & 0 & 0 \\ 0 & 0 & \bar{1} \end{pmatrix}, \begin{pmatrix} \bar{1} & 0 & 0 \\ 0 & \bar{1} & 0 \\ 0 & 0 & 1 \end{pmatrix}, \begin{pmatrix} 0 & \bar{1} & 0 \\ \bar{1} & 0 & 0 \\ 0 & 0 & \bar{1} \end{pmatrix}$
R3	$hkl \rightarrow khl$	$\begin{pmatrix} 0 & 1 & 0 \\ 1 & 0 & 0 \\ 0 & 0 & 1 \end{pmatrix}$
P321, P3₁21, P3₂21, P312, P3₁12, P3₂12	$hkl \rightarrow \text{-}h\text{-}kl$	$\begin{pmatrix} \bar{1} & 0 & 0 \\ 0 & \bar{1} & 0 \\ 0 & 0 & 1 \end{pmatrix}$
P23, P2₁3, I23, I2₁3, F23	$hkl \rightarrow k\text{-}hl$	$\begin{pmatrix} 0 & 1 & 0 \\ \bar{1} & 0 & 0 \\ 0 & 0 & 1 \end{pmatrix}$

Table 8-4 Non-equivalent indexing and reindexing operators. Chiral space groups allowing non-equivalent indexing and the possible reindexing operators are given. Note that these space groups can also exhibit hemihedral twinning, as explained in the twinning section. Enantiomorphic pairs of space groups are shown in red. Table rearranged from Dauter.[45]

edge. However, the anomalous signal contribution from the sulfur atoms is miniscule, and data need to be collected very carefully and with high redundancy (Sidebar 8-11). The need for accurate data collection is easily understood based on the fact that small intensity differences between anomalous Bijvoet pairs (i.e. centrosymmetrically related, non-centric reflections) are used to obtain the difference data from which the heavy atom (sulfur) positions are determined. In order to solve the structure, two inherent ambiguities must be overcome: the handedness ambiguity present in the heavy atom solution, and the phase ambiguity always present in single derivative or single anomalous phasing equations (Chapter 10). The S-SAD method is thus always used in combination with density modification techniques which resolve the ambiguity in the phasing equations. Both enantiomorphs of the heavy atom solution are tested, and only the correct handedness will give an interpretable initial map after density modification and phase extension (Table 10-2).

The diffraction image data used in the following processing exercise have been kindly provided by Aritra Pal and George Sheldrick.[79] The material is viscotoxin A1 (VTA1, PDB entry 3c8p), one of the cytotoxins with possible anti-tumor properties isolated from European mistletoe (*Viscum album*). It is a small, 46-residue, 5 kDa protein containing three disulfide bonds, and thus an ideal candidate for S-SAD phasing. In reverse order of history of the actual structure determination, we begin the exercise under the assumption of full ignorance with processing of a native synchrotron data set.

High resolution synchrotron data

In order to collect diffraction data on a synchrotron, the crystals must be brought to the facility. This can be done in two ways, either by bringing the trays with the crystal to the beam line and mounting the crystals there, or flash-cooling the crystals in-house and shipping them to the synchrotron in a dry-shipping Dewar. These Dewars contain an absorbent that takes up liquid nitrogen, thereby preventing spillage, and they can be shipped via air freight. For short distances, transport of crystallization plates is feasible. With plates well packed and cushioned in a Styrofoam container to keep the temperature constant, most crystals survive car or train trips. Rules for carry-on items in airplanes change daily and irrationally; one must thus check with the corresponding airline.

Sidebar 8-10 Sample images and data sets for practice.
Data processing is perhaps the most opaque black-box part of structure determination for the beginning macromolecular crystallographer. It is best learned by practicing with a few sample data sets using all the locally available data processing software (if one program fails, another one may still be able to generate a usable data set in marginal cases). Example data sets with varying degrees of real-life problems are available from the *Autostruct* web site http://www.ccp4.ac.uk/autostruct/testdata/. Another avenue is to use the excellent *MLFSOM* program (the "reverse" anagram of *MOSFLM*) by James Holton[53] to generate simulated test images with certain defects and to examine the effects of changing experimental parameters. The images demonstrating icing and exposure effects, for example, have been created by *MLFSOM*.

It is also advisable to keep the image files. Sooner or later the situation will occur where you have prematurely merged data or have gotten better at data processing through practice and wish to reprocess the data. There are efforts underway to publicly archive the image data files, and a common image archiving format, *imgCIF*, is being developed (so far, each detector manufacturer has their own image format, and the data processing software needs a module specifically written for reading each proprietary format). Keeping primary data is just good scientific practice; try not to leave a beam line without images written to DVD or backup disk. As a minimum, save the unmerged multi-batch reflection file for possible later reprocessing.

We will collect the VTA1 diffraction images on beam line BW7B at the EMBL outstation Hamburg on a large MAR-345 imaging plate detector. With a short wavelength of 0.84299 Å (14.7 keV), the diffraction pattern is "compressed" in the forward direction and data up to 1.1 Å can be recorded without the need for a detector offset or a second scan at a detector distance of 174 mm. However, we would not be able to collect reasonable anomalous S-SAD data at this short wavelength because its energy is too high above the sulfur K absorption edge of 2472 eV and therefore the anomalous contribution would be too small. The experimental setup data such as detector distance, beam position, and so on are provided in the image headers.

Indexing and strategy generation

We assume that we have successfully cryoprotected, mounted, and centered a VTA1 crystal on the goniostat. We close the synchrotron hutch and load the first snapshot (a 1° rotation image) into the interactive version of *MOSFLM*, a program of the CCP4 suite written by Harry Powell and Andrew Leslie.[86] To our great contentment we find plenty of good looking, sharp diffraction spots to about 1.3 Å, but only a few spots are visible at the detector edge of 1.1 Å (in contrast to the square CCD detectors, the active area of the MAR345 image plate detector is circular). We recognize distinct diffraction lunes, and no ice rings obscure the image. We take a second exposure at a 90° offset angle, which reveals no anisotropy in the extent of the diffraction pattern. We confirm that 1.25 Å is a reasonable cutoff for data integration. Excessive inclusion of higher resolution shells with mostly empty spots will only generate noise, which results in worse overall integration statistics.

In preparation for indexing, we adjust the beam xy-position read from the header to fit the center mark which this particular beam stop provides, and blank out the area around the beam stop shadow from spot finding and processing. We select both images for indexing, and the spot-finding routine finds plenty of strong spots (Figure 8-34). The Fourier-based DPS indexing algorithm[52, 63] now generates the list of possible lattices, ranked by a penalty function (Figure 8-35). The program selects the tP (tetragonal primitive) lattice as the most likely choice. We concur because it has the highest symmetry of the remaining possible lattices with the same low penalty. The orthorhombic lattices are essentially a pseudo-tetragonal indexing of the same lattice, and the pseudo-monoclinic indexing ($\beta = 90°$) is just another orthorhombic axis assignment with lower internal symmetry. The initial space group with lowest tetragonal lattice symmetry is thus $P4$, Laue group $4/m$. The initial mosaicity estimate of 0.45° is also reasonable for a 1° rotation increment, and we check the spot predictions on both images; and indeed the predictions fit well. Figure 8-34 shows the respective screenshots.

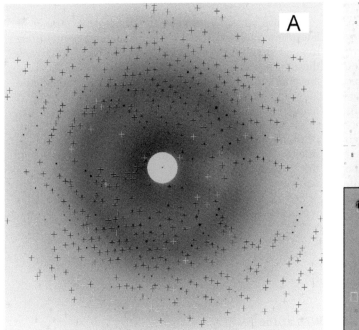

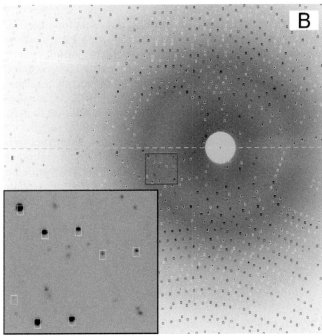

Figure 8-34 Spot finding and indexing of images. (A) The center section of the initial 1° diffraction pattern of the VTA1 crystal loaded into *iMOSLFM* with detected spots indicated by cross marks and the blanked-out beam stop region. We can distinguish distinct lunes of diffraction spots. (B) The same frame after indexing with a tetragonal primitive lattice (*tP*, *a* = 65.8, *c* = 47.2 unrefined, lowest symmetry space group *P*4). The colored boxes mark the spot predictions according to the initial indexing. The spot marks fit the reflections well. Blue spot prediction boxes indicate full reflections, partials (reflections that extend over more than one image) are yellow, and along the φ-rotation axis (green dashed line) the green boxes indicate reflections located in the Lorentz exclusion region.

These reflections traverse the Ewald sphere tangentially and remain in diffraction condition too long, that is, over too wide a rotation range, to allow reasonable Lorentz factor correction and integration. We also see a few predicted overlaps in this region (red prediction boxes). The insert shows a magnified region (large red box) where some weak spots appear that do not belong to the primary diffraction pattern. These may be from tiny fragments of secondary crystal pieces or other diffracting foreign matter not belonging to the primary crystal. They do not pose a problem as long as they are minor in contribution and do not compromise the indexing process. As these foreign spots do not fall into predicted spot positions, they are simply ignored during intensity integration.

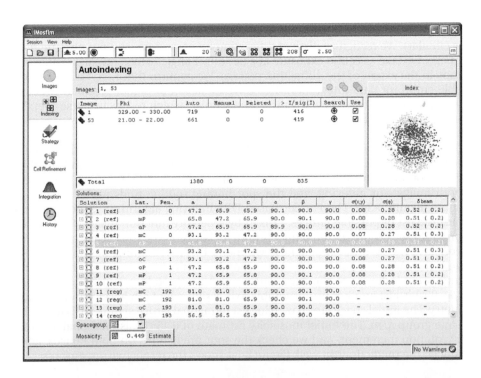

Figure 8-35 Indexing panel of *iMOSFLM*. Two images with different rotation angles (offset by 53° in this case) have been loaded and spots detected. The Fourier indexing algorithm lists the possible lattice types ranked by a penalty function. The tetragonal primitive lattice *tP* (gray highlight) is the best solution with highest lattice symmetry. All tetragonal space groups are possible choices. In the absence of further information, we select the tetragonal space group with the lowest symmetry, *P*4 with Laue symmetry 4/*m*, to design a data collection strategy.

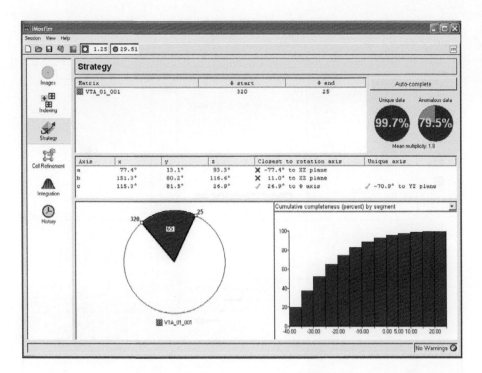

Figure 8-36 Strategy panel of
iMOSFLM. Assuming reciprocal space (Laue) symmetry 4/*m*, a 65° rotation range gives nearly 100% complete data. Inspecting the cumulative completeness panel (bottom right), we realize that also a smaller rotation range of about 50° provides nearly the same coverage. Note that for anomalous data collection, we would need to revise this strategy and select a larger rotation range to also cover the Friedel pairs. As the true space group can only have higher symmetry, the actual data completeness and redundancy will be higher.

Now is the time to develop the data-collection strategy based on the tetragonal lattice symmetry. We employ the *MOSFLM* strategy generator (Figure 8-36), which with default settings suggests a rotation range of 90° covering 1/8 of the reciprocal sphere (Table 6-6). As we are not (yet) interested in complete anomalous data, we can save some data collection time by selecting a somewhat smaller φ-range of 65° starting from our first image. The cumulative completeness diagram shows us that we could still achieve nearly 100% coverage with an even smaller rotation range, which is reassuring in case the crystal suffers radiation damage or we run out of beam time. We program and start the data collection software and collect 53 frames from 329° to 382° (or 22°) in φ before the beam unexpectedly shuts down. The missing frames have a negligible effect on completeness, and because we likely have higher true space group symmetry than *P*4, both completeness and redundancy for the true space group will be higher. We can play with the strategy generator and try a space group in the higher tetragonal Laue group 4/*mmm* and recognize that we still have complete coverage, now with a higher redundancy of 2.6 (because the reciprocal asymmetric unit with 4/*mmm* symmetry is smaller, and the same rotation range covers more symmetry related reflections).

Cell refinement, integration, and scaling

Having all 53 frames collected, we load them into *iMOSFLM* and refine the cell constants and experimental parameters using about 5–10 frames. With this refined cell constant fixed, we can now integrate the spots. The program provides a number of useful diagnostics, some of which are explained in the caption of Figure 8-37. As we encountered no significant problems, we write out an integrated, but unmerged, multi-batch file in CCP4 *mtz* format to be processed with the scaling program *SCALA*[87] in combination with *TRUNCATE*, which provides a number of useful statistics and analysis—including basic twinning tests—of our data set. *SCALA* also provides absorption correction through a secondary beam absorption correction[88] (default) or a 3-dimensional detector model.[61]

We start the graphical *CCP4i* interface to the CCP4 suite, and from the "Data Reduction" menu we select "Scale and Merge Intensities" which opens the input panel for *SCALA* and *TRUNCATE*. We select our multi-batch *mtz* file written by *iMOSFLM* as input and proceed with defaults, but we need to enter the approximate unit cell content (92 residues) for Wilson scaling in *TRUNCATE*. The result is a multi-column data file that includes the unique scaled and merged data for

Figure 8-37 Integration panel of iMOSFLM. The selected default panels provided by the integration module (there are more options available for further analysis) show a number of useful statistics in frame-by-frame graphs, that is, over the time course of the data collection. We see that the beam center remains stable (top center), but the crystal slips slightly (below). The S/N (bottom center) also did not decrease appreciably during the course of the data collection, so we do not expect much radiation damage. The column of panels to the right includes analysis of the spot profiles, which are well defined and cover the spots. The reflection profiles and integration statistics for the weak reflections beyond 1.25 Å were poor, and we therefore limit the data integration to a maximum resolution of 1.25 Å.

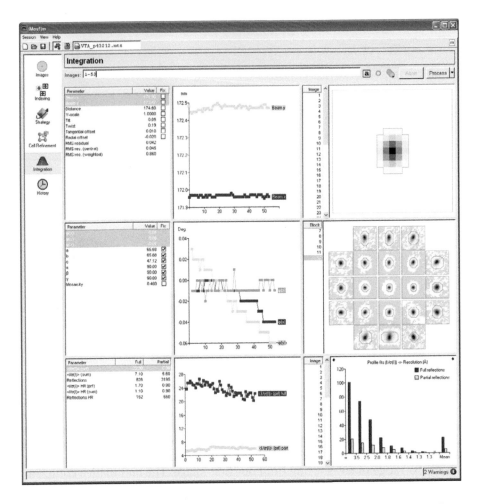

initial space group *P4*. The columns generally contain at a minimum the reflection indices, their intensity and corresponding structure factor amplitude, and their estimated standard deviation, all kept separate for centrosymmetrically related reflections to preserve the anomalous data if needed.

Laue symmetry and space group determination

Before we further analyze the data, we determine the actual space group using the program *XPREP* written by George Sheldrick and available from Bruker AXS. Using the *CCP4i* GUI "Reflection Data Utilities" we pick "Convert from MTZ"

Systematic absence exceptions:

	41/43	42	n--	-b-	-c-	-n-	-21-	--c
N	4	3	958	728	715	723	23	506
N I>3s	0	0	509	448	459	431	0	330
<I>	2.4	3.0	145.7	137.1	155.3	146.3	1.5	124.2
<I/s>	0.9	1.1	8.0	10.7	10.9	10.3	0.4	11.0

Identical indices and Friedel opposites combined before calculating R(sym)

Option	Space Group	No.	Type	Axes	CSD	R(sym)	N(eq)	Syst.	Abs.	CFOM
[A]	P42(1)2	# 90	chiral	1	4	0.035	24802	0.4 /	8.0	21.29
[B]	P4(2)2(1)2	# 94	chiral	1	20	0.035	24802	0.4 /	8.0	6.06
[C]	P4(1)2(1)2	# 92	chiral	1	245	0.035	24802	0.4 /	8.0	1.70
[D]	P4(3)2(1)2	# 96	chiral	1	245	0.035	24802	0.4 /	8.0	1.70

Table 8-5 Space group determination using XPREP. Output from *XPREP* during space group determination of VTA1. The data items are discussed in the text.

```
INTENSITY STATISTICS FOR DATASET # 1    vta_01_001_scala2.sca
```

Resolution	#Data	#Theory	%Complete	Redundancy	Mean I	Mean I/s	R(int)	Rsigma
Inf - 3.41	1436	1606	89.4	1.49	688.0	36.77	0.0184	0.0288
3.41 - 2.71	1448	1490	97.2	1.75	444.1	33.83	0.0203	0.0296
2.71 - 2.36	1488	1506	98.8	1.83	234.1	30.46	0.0201	0.0312
2.36 - 2.14	1484	1505	98.6	1.85	169.4	27.56	0.0231	0.0330
2.14 - 1.99	1429	1442	99.1	1.87	129.3	24.67	0.0269	0.0362
1.99 - 1.87	1494	1506	99.2	1.88	80.8	20.31	0.0365	0.0440
1.87 - 1.77	1561	1567	99.6	1.89	50.3	16.22	0.0435	0.0553
1.77 - 1.69	1526	1530	99.7	1.90	39.2	14.06	0.0541	0.0647
1.69 - 1.62	1608	1614	99.6	1.89	29.1	11.62	0.0673	0.0812
1.62 - 1.56	1611	1615	99.8	1.89	24.3	9.74	0.0826	0.0993
1.56 - 1.51	1552	1553	99.9	1.89	20.5	7.98	0.0982	0.1207
1.51 - 1.46	1746	1747	99.9`	1.89	18.2	7.04	0.1177	0.1389
1.46 - 1.42	1604	1604	100.0	1.88	14.8	5.81	0.1600	0.1756
1.42 - 1.38	1782	1783	99.9	1.87	11.6	4.64	0.2089	0.2211
1.38 - 1.35	1474	1475	99.9	1.88	9.1	3.68	0.2296	0.2713
1.35 - 1.32	1607	1607	100.0	1.88	7.7	3.08	0.2642	0.3215
1.32 - 1.29	1768	1769	99.9	1.88	6.4	2.57	0.3096	0.3871
1.29 - 1.25	1924	1926	99.9	1.87	5.4	2.13	0.3551	0.4626
1.35 - 1.25	5805	5808	99.9	1.88	6.6	2.65	0.2998	0.3753
Inf - 1.25	28542	28845	98.9	1.85	102.2	13.88	0.0344	0.0440

```
Merged [A],   lowest resolution = 16.96 Angstroms,        693 outliers downweighted
```

Table 8-6 Merging statistics for VTA1 provided by *XPREP*. Output from XPREP during merging of reflections of VTA1, grouped in resolution shells with a width proportional to $1/d^3$ containing approximately equal numbers of reflections. Merging statistics for the highest resolution shell and overall merging statistics for the entire data set are highlighted. The remaining statistics are discussed in the body text.

to convert the intensities from the unmerged *mtz* file into a text file in *SHELX* format and read it into *XPREP*, select (chiral) space group determination, and obtain a listing of possible space groups (Table 8-5).

The top of Table 8-5 lists the systematic absence violations for all symmetry elements compatible with the selected lattice type (all glide plane elements such as *b*, *c*, and *n* can be safely excluded for chiral space groups). The number of absence violations detected and their significance are provided in each column (relevant extinctions are highlighted). We see that the $4_1/4_3$ absences (reflection condition 00*l* with *l* = 4*n* only, along the unique axis *c*) and 2_1 absences (perpendicular to *c*, *h*00 with *h* = 2*n* only and 0*k*0 with *k* = 2*n*) show the lowest number of violations, but 4_2 cannot yet be excluded based on extinctions alone. However, the combined figure of merit (CFOM), which also includes space group preferences, strongly suggests the true space group is either $P4_12_12$ or $P4_32_12$. Note that the overall merging *R*-sym for symmetry related reflections upon merging from lower Laue symmetry 4/*m* into higher 4/*mmm* is of course the same for all suggested space groups, and it is reasonably low (3.5%). We thus concur with *XPREP* and choose $P4_32_12$, being aware of a 50% chance the assignment may have to change later to $P4_12_12$ during substructure solution.

We now check the merging statistics from 4/*m* to 4/*mmm* in a shell-by-shell binned listing while merging the Friedel pairs (Table 8-6). The statistical analysis listed in Table 8-6 is encouraging, with normal behavior of increasing merging *R*(*int*) with all reflections including Friedel pairs merged. The values for *R*(*int*) also are consistent with the derived estimate R~ $0.8/\langle I/\sigma(I)\rangle$, indicating that the reduced χ^2 is around unity and the errors are random. Note that the redundancy and merging statistics here are only for the merging from 4/*m* to 4/*mmm* symmetry, not the true redundancy which we need to extract from a reprocessing

Table 8-7 Merging statistics summary for VTA1 provided by *SCALA*. Summary listing of full merging statistics from SCALA during merging of reflections of VTA1. Note that the precision indicating merging *R*-value R_{pim} is lower than the standard, linear merging *R*-value.

Summary data for Project: vta1 Crystal: native Dataset: sync	Overall	OuterShell
Low resolution limit	29.37	1.32
High resolution limit	1.25	1.25
Rmerge	0.050	0.456
Rmeas (all I+ & I-)	0.068	0.741
Rpim (all I+ & I-)	0.032	0.259
Total number of observations	120777	17236
Total number unique	28867	4179
Mean((I)/sd(I))	22.9	2.1
Completeness	99.3	100.0
Multiplicity	4.2	4.1

run of *SCALA* in the correct space group (the observant reader might notice that this is also indicated by R_{sigma} not being smaller than R_{int}). Nevertheless we can write out this merged data set for later use as a native data set in phase extension and refinement with the *SHELX* programs. If we plan to use other refinement programs that cannot (yet) refine against intensities, we will use the final data set written by *SCALA/TRUNCATE* because *TRUNCATE* has a more sophisticated way of treating negative intensities with Bayesian estimates when converting from intensities to structure amplitudes (see Chapter 7).

Reprocessing in correct Laue symmetry

What remains to be done is to quickly rescale and merge the data in space group $P4_32_12$ (i.e. the correct Laue symmetry $4/mmm$) in *SCALA/TRUNCATE* and with merged Friedel pairs so that we obtain the proper statistics for the data collection part of the infamous "Table 1" (data collection statistics) in our future publication. The summary output of *SCALA* is listed in Table 8-7, and in Section 8.5 we will also analyze several graphs of intensity statistics provided by the program before we produce our "Table 1." We could also have used *XPREP* instead of *SCALA* to merge from entirely unmerged $4/m$ intensities to $4/mmm$; the result would be the same.

Collecting highly redundant anomalous data

For collection of the anomalous data for S-SAD phasing we select a home-laboratory diffractometer with a Cu Kα rotating anode and a Bruker Smart 6000 CCD detector. The wavelength of Cu is 1.54187 Å (~8 keV) which is much closer to the X-ray absorption edge of sulfur (2.5 keV) than the high energy photons (14.7 keV) of beam line BW7B that we used to collect the high resolution "native" data. We therefore expect, given careful and redundant data collection, a weak but measurable anomalous difference between Bijvoet pairs. Initial indexing is conducted as described above using the *PROTEUM* software, which provides a 3-dimensional reciprocal space explorer (Figure 8-39).

Strategy for highly redundant data collection

For the purpose of phasing, we do not need to record data to the highest possible resolution; good quality data (high redundancy, low final merging *R*-values, high S/N ratio) is the objective. The first improvement in data collection is to use fine slicing so that 3-dimensional spot profiles can be fitted. We will therefore use a rotation increment of only 0.2° compared with the 1° rotation range selected for the native synchrotron data. For each 180° of rotation we thus have to collect 900 images! Knowing the reciprocal space (Laue) symmetry of $4/mmm$, we can now devise a strategy to record the data with high redundancy in multiple crystal orientations. For $4/mmm$ symmetry, 45° φ-scans (1/16 of the reciprocal space) suffice to cover the unique wedge of reciprocal space (Table 6-6). Because the detector is positioned with the primary beam in the center, we will achieve up to ~4 times the coverage (the redundancy of 4.1 computed by *SCALA* for our native VTA1 synchrotron data set with 53° scan range confirms this estimate).

Sidebar 8-11 S-SAD data and redundancy. Given the miniscule anomalous signal originating from sulfur atoms ($Z = 16$), highly accurate and precise data are required for S-SAD phasing. Therefore, the reflections must be measured with high redundancy. Note that the emphasis here is on accuracy, not just precision; the purpose of redundant data collection is to achieve high precision (reducing random errors) while at the same time minimizing systematic errors. This in turn implies that the individual measurements should be independent.

The need for independent measurements means that just rotating the crystal for several full 360° turns around the φ-axis, for example, provides only repeated measurements of the same reflections in identical diffraction geometry, which is—assuming no radiation damage or other time-dependent effects—the same as just measuring the reflections for longer. The data then become more precise (recall that the counting error is proportional to $N^{1/2}$, i.e. four times the data collection time will provide half the relative measurement error) but any systematic error affecting the measurements will remain the same. Typical systematic errors are introduced by absorption and anisotropy of the crystal itself, absorption by the mounting hardware and mother liquor, poor centering, slipping crystals, and other experimental setup issues.

Systematic errors tend to compensate when different diffraction geometries are selected to record multiple independent measurements of the same reflections. The general procedure is to use a goniostat where the pin axis (φ-axis) can be offset, commonly in κ- or χ-geometry, or alternatively to use a simple goniometer head arc that allows manual selection of different κ-angles for each rotation range around the ω-axis (Figure 8-38). The crystal is then rotated around the ω-axis during 180° or 360° scans while the crystal is kept at a fixed κ-offset. In addition, the pin axis (φ-axis) can also be manually offset by turning the pin base (which generally requires recentering).

Alternatively, on a true 3-axis or 4-axis goniostat, the crystal can also be rotated around each new φ-axis and around ω, which again provides different diffraction geometries during each scan. In addition, if the crystal diffracts to high resolution (high 2θ-angles), some scans may need to be repeated at offset detector positions. However, for phasing purposes, high quality low and medium resolution data are more important than very high resolution data. The latter

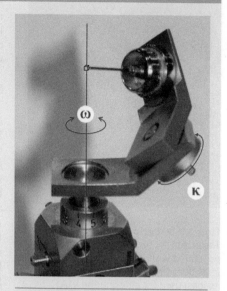

Figure 8-38 Poor crystallographer's κ-goniostat. A small attachment to the goniometer head (originally developed for simplified cryomounting at Yale) is also useful for anomalous data collection to achieve variety in diffraction geometry during anomalous data collection. Both the κ-axis and the pin rotation axis can be adjusted manually for each ω-scan. The magnetic pin and the κ-arc are made by Hampton Research. Image of arc courtesy Katherine Kantardjieff, CSU Fullerton.

are important for phase extension and refinement, but do not need the high redundancy strategy required for anomalous data collection.

Because we also need to collect the anomalous wedge, a 180° scan will provide an anomalous redundancy of 8 (16 for all reflections merged). As explained in Sidebar 8-11, just rotating φ for $4 \times 360°$ is not providing any further independent measurements, and would increase precision but not accuracy. On our 3-circle diffractometer the φ-axis is offset by $\chi = 54.74°$ (the magic angle) and for additional variety in diffraction geometry we can rotate either around the ω-axis or the φ-axis. Finally, for good measure (literally), we can also add a scan with an offset detector θ-angle to record high resolution data which are useful later for phase extension and can be merged with the high resolution synchrotron data. We arrive at the strategy listed in Table 8-8.

Scan	Scan type	ω	φ	2θ
1	180° ω	0°–180°	0°	0°
2	180° ω	30°–210°	0°	30°
3	180° ω	0°–180°	60°	0°
4	360° φ	0°	0°–360°	0°
5	180° ω	0°–180°	120°	0°
Anomalous redundancy calculated by strategy generator			~30	
Total exposure time (60 s per image)			90 h	

Table 8-8 Strategy for highly redundant data collection. The table entries contain the scan ranges and diffractometer angle settings for a highly redundant data collection strategy for the collection of anomalous data required for native sulfur single-wavelength anomalous diffraction (S-SAD) phasing for VTA1 (Laue symmetry 4/mmm).[79]

Figure 8-39 Reciprocal space view.
The image shows the reciprocal space populated with two thin shells of reflections (from two sets of 90° offset images) used for indexing the VTA1 fine-sliced high redundancy data. The reflections are located on two corresponding Ewald spheres whose shape is distinctly recognizable (reciprocal space viewer of the Bruker *PROTEUM* software). The remaining images would completely populate the 3-dimensional reciprocal space within the resolution sphere (Sidebar 8-6).

For 6 sulfur atoms per 46 residues in VTA1, the expected anomalous diffraction ratio $\Delta F/F$ is about 1.5% (use http://www.ruppweb.org/Xray/MAD_phasing. html to confirm). We thus need to measure the intensity data with a precision of 3% or better. Empirical evidence indicates that we can achieve this accuracy with a redundancy of about 20–30 (on anomalous data).

Scaling and merging statistics

The data processing with the graphical Bruker *PROTEUM* interface to *SAINT* follows the same principal steps we used with *iMOSFLM*. Because we have 5400 frames to process, we batch process the data directly in *SAINT* and read the raw data into the absorption correction program *SADABS* (written by George Sheldrick for Bruker). What remains then is to inspect the scaled, but as yet unmerged, file with *XPREP*, determine a resolution cutoff for the anomalous difference data to be used in SAD phasing, and write out the anomalous difference data file. We will analyze the difference data in more detail once we have conquered the fundamentals of phasing, which will help to understand why and how we select resolution cutoffs for phasing. For now it suffices to say that we achieved an anomalous redundancy of 29 and a ratio of $R_{anom}/R_{pim.}$ of 0.0260/0.0096 = 2.70 which is much larger than the empirical lower limit of 1.5 and thus a quite favorable predictor for the outcome of our planned S-SAD phasing experiment.[83]

A number of graphs provided in one form or another by the data processing programs are worth examining to prevent later disappointment. Amongst those are the raw integrating and scaling statistics R-values per frame (Figure 8-40), and the quality of the error model in the form of a χ^2 plot (shown already in Figure 8-31), and finally the resolution-binned plots of various merging statistics and mean $I/\sigma(I)$ (for example Figure 8-32).

8.5　First analysis and descriptive statistics of data sets

We will now conduct a few more statistical tests and collect the descriptive data for our VTA1 synchrotron data set. At this stage we have obtained a merged, unique data set whose extent in resolution, redundancy, and completeness together with the corresponding statistics is known. It is generally a good idea at this point to assemble the information for "Table 1" or at least keep clear records where the data statistics are to be found. They will be dearly wanted later—often much later—when the structure is finally ready to be published. It is rather embarrassing to spend untold hours during PDB submission trying to find raw data statistics or unmerged data distributed over undocumented network resources with non-descriptive directory names. Keeping at least a rudimentary database such as provided by the *CCP4i* GUI for data harvesting[89] or a LIMS system such as *Xtrack*[90] is important for orderly bookkeeping.

Figure 8-40 Overall scale factor and merging statistics for multiple data collection scans. The normalized overall scale factor (top panel) is plotted against the frame number of each scan (see Table 8-8 for the scan setups). The scale factor brings the individual frames onto a common scale. The individual variations are caused by changing diffraction geometry, absorption, crystal illumination, centering errors, beam fluctuation, and so on, during the progression of each scan. The variation of the scale factor is relatively smooth (with the exception of a few irregularities in scan 4) and there are no individual frames or scan regions that stick out as particularly bad. Scan 5 shows a spike in merging R-value that is consistently higher than in the other scans. Note that significant radiation damage would manifest itself in progressively increasing scale factors and consistently increasing merging R-values because of rapidly weakening intensities. The deviations in scan 5 therefore must have a different explanation.

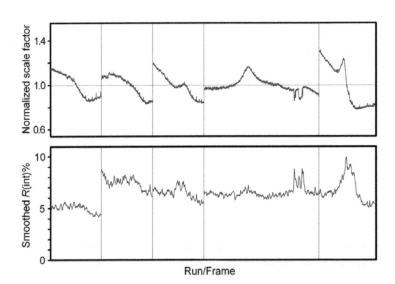

Wilson plots and checks for anisotropy

At some point in the final merging and scaling of the data, either the data processing program or a post-processing program such as *TRUNCATE* will perform diagnostics based on the intensity distributions of the experimental data. This may or may not be transparent to the user, and it is generally a good idea to check whether the programs provide any graphical presentation of the analysis, usually binned in some resolution-dependent shells. The most basic plot is the Wilson plot (Figure 8-41) establishing an overall *B*-factor and bringing the data onto a absolute scale, which is needed for the subsequent structure solution and refinement steps. The visual inspection of the Wilson plot (Chapter 7) also reveals whether any anomalies are present, and whether the linear fit occurred over a reasonable data range. Particularly low-resolution data are susceptible to poor Wilson scaling, and a proper adjustment of the low-resolution and high-resolution boundary for the plot may have to be done manually, so that the interpolation spans a reasonably linear portion of the Wilson plot. Examples for poor overall *B*-factor scaling are provided in Chapter 13.

Most scaling programs also provide a plot of the diffraction anisotropy (Figure 8-41), and it is wise to have a look at these plots. Although most scaling programs can deal with a limited amount of anisotropy by anisotropic *B*-factor scaling, extreme diffraction anisotropy requires a more sophisticated treatment of the raw data, including truncation of useless parts of the reciprocal space that contain no information. A procedure for treatment of severe anisotropy is discussed in Section 12.2.

Testing for twinning

Microscopic, merohedral twinning is by far the most insidious complication that can afflict diffraction data and, if undetected, will invariably lead to some (if not terminal) complications during phasing and refinement. Always test for twinning.

Inspection of Table 8-2 shows that for space group $P4_32_12$ no partial merohedral twinning is possible. The remaining possibility for perfect twinning would imply that the true space group symmetry would be lower, an unlikely proposition because VTA1 is not known to form any higher oligomers (we would need at least a tetramer to fill the *P*4 cell which has twice the asymmetric unit volume; you can convince yourself using the Matthews probabilities calculator). Nevertheless, we check the corresponding cumulative probability graphs and raw moment plots provided by *TRUNCATE* as a habit-forming exercise. As expected, the result (Figure 8-42) shows no indication of twinning.

Figure 8-41 Wilson plot and anisotropy fall-off. The Wilson plot (left panel) provides the basis for bringing experimental and calculated data onto a common absolute scale (Chapter 7) during refinement and should be inspected for anomalies in the linear region beyond 5–4 Å. If necessary, the region for the linear fit must be adjusted later during the refinement stage. Limited anisotropy detected in resolution fall-off plots can be dealt with in refinement through anisotropic *B*-factor scaling, but excessive anisotropy requires additional data truncation and normalization procedures (Chapter 12). The resolution fall-off plot for VTA1 (right panel) shows no anisotropy of the diffraction data. The graphs are screenshots from *TRUNCATE* log-files displayed with *LOGGRAPH* implemented in the CCP4i GUI.

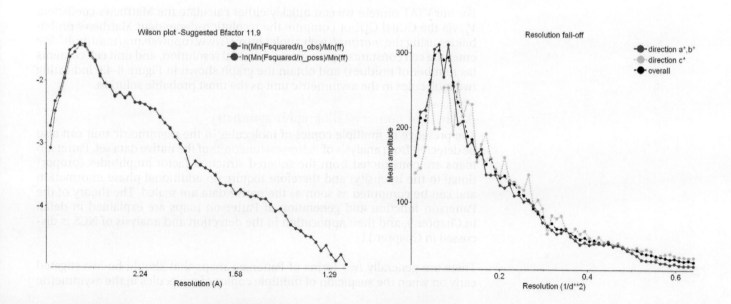

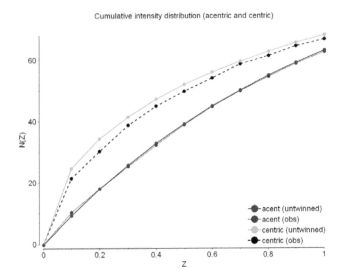

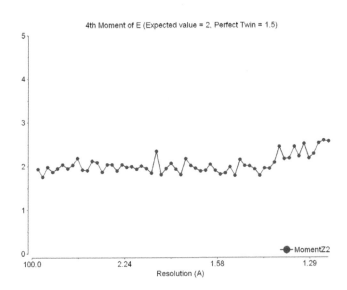

Figure 8-42 Testing for partial and perfect twinning. The plot of the cumulative probability distribution function (CDF, left panel) and the 4th raw moment of the normalized acentric structure factor distribution (right panel) show that there are no indications of either partial or perfect twinning for VTA1 (space group $P4_32_12$ in fact does not allow for partial merohedral twinning). The small deviation from the theoretically expected CDF curve for centric reflections is largely because there are only few centric reflections compared with acentric reflections. The experimental CDFs for centric reflections are therefore generally not as reliable as the acentric CDFs. The CDFs are explained in Chapter 7, and for raw moment values consult Table 8-3. The plots are screenshots from *TRUNCATE* log-files displayed with *LOGGRAPH* implemented in the CCP4i GUI.

Determining the unit cell contents

In Chapter 5 we have already encountered cases of non-crystallographic symmetry (NCS) where more than one molecule occupies the asymmetric unit of the crystal structure. As soon as we have obtained useful diffraction data we can estimate how many copies of the protein we crystallized are likely in the asymmetric unit. To determine this, we do not even need to know the exact space group: a look at the rightmost column (z, the general position multiplicity) of space group Table 6-6 confirms that the point group or Laue symmetry plus the Bravais centering suffice to determine the number of asymmetric units that comprise the unit cell. Dividing the unit cell volume by the general position multiplicity z provides us thus with the volume of the asymmetric unit V_a, that is, the space available to accommodate the protein molecule or molecules.

The computation of the solvent content (analyzed first by Matthews in 1968[91] and 1976)[92] and the corresponding, resolution-dependent Matthews probabilities[93] are described in Chapter 11. While the Matthews coefficient calculation works well in clear cut cases with few molecules in the asymmetric unit, multiple solutions become possible for higher assembly numbers. Additional biological evidence from size exclusion chromatography, light scattering, ultracentrifugation, or native polyacrylamide gels may confirm the proper copy number. In addition, certain types of Patterson maps (discussed in detail in Chapter 11) can also give valuable clues about the nature of the NCS present in a molecular structure.

For our VTA1 protein we can quickly either calculate the Matthews coefficient V_M via the CCP4i GUI or compute the resolution-dependent Matthews probabilities using the *mattprob* web applet http://www.ruppweb.org/mattprob/. We enter the cell constants, space group, nominal resolution, and unit cell contents (as number of residues) and obtain the graph shown in Figure 8-43, indicating two molecules in the asymmetric unit as the most probable solution.

Analysis of non-crystallographic symmetry

The presence of multiple copies of molecules in the asymmetric unit can also be detected from analysis of Patterson functions of the native data set. Patterson maps are constructed from the squared structure factor amplitudes (proportional to the intensity) and therefore require no additional phase information and can be computed as soon as the native data are scaled. The theory of the Patterson function and generation of Patterson maps are explained in detail in Chapter 9, and their application in the detection and analysis of NCS is discussed in Chapter 11.

There are generally two forms of Patterson maps that should be investigated early on when the suspicion of multiple copies of molecules in the asymmetric

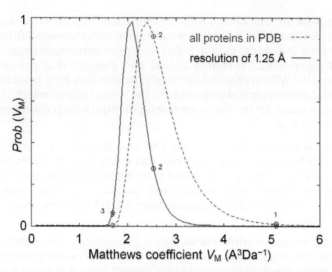

Figure 8-43 Estimate of unit cell contents. Calculation of the Matthews coefficient V_M and the corresponding solvent content V_s provides an estimate of the number of molecules in the asymmetric unit cell. Using the resolution-dependent Matthews probability calculator we obtain the highest probability for the presence of two VTA1 molecules in the asymmetric unit with a V_M of 2.54 Å³ Da⁻¹ corresponding to a solvent content of 52%.

unit is raised by the Matthews probability analysis. In cases of purely rotational, proper NCS symmetry, the Patterson self-rotation functions will reveal orientation and multiplicity of the NCS rotation axis; native Patterson maps may reveal translational NCS as well as NCS rotation axes that are parallel, or very nearly parallel, to crystallographic symmetry axes (which are otherwise obscured by the crystallographic symmetry axis peaks in self-rotation maps). The thorough analysis of these Patterson maps requires a deeper understanding of NCS operators and representation of rotations in spherical coordinates and stereographic projections, and is discussed in detail in Chapter 11.

Assembling "Table 1"

Publications describing crystal structures generally lead the experimental structure determination section with the infamous "Table 1" containing descriptive data collection statistics. There are no specifications for mandatory data items, but some good practice guidelines have been agreed upon in the crystallographic community.[94] Although "Table 1" is increasingly relegated to the supplemental online section of high impact journals, it contains valuable information regarding raw data quality as well as the consistency of the data processing. In Chapter 13 we will discover a questionable structure in a high impact magazine journal just by examining "Table 1." Sometimes "Table 1" also contains the phasing and refinement statistics, which we will treat in the corresponding chapters.

At minimum, the experimental data collection section of "Table 1" should contain information about

- **Space group, unit cell parameters, wavelength**. Without those, no data processing is possible and they are logically the first items in the table. Ambiguities in the case of enantiomorphic space group pairs will be resolved and can be updated later. Note that cell parameters are notoriously inaccurate[95] (despite precise refinement during data processing) in macromolecular data sets. Empirical analysis against refined bond lengths[96] shows that the accuracy of unit cell parameters is at best on the order of 10^{-2} to 10^{-3} (1–0.1 %). Listing them with 6-digit precision is more a sign of poor judgment than of crystallographic prowess.

- **Unit cell contents.** The Matthews coefficient and solvent content provide clues about the number of molecules in the asymmetric unit cell. It may not be known yet what the exact number of molecules is, but it will become determined later and the information included in "Table 1."

- **Resolution limits.** Given that the overall achievable information content (detail) of a structure depends primarily on resolution, this information is crucial. Moreover, the remaining statistics are listed separately for the highest resolution shell, so we need to provide its resolution range. Beyond the basic resolution limits, the effective resolution $d_{eff} = d_{min} \cdot C^{-1/3}$ could be listed (C is the completeness).

- **Wilson *B*-factor.** The overall *B*-factor estimated from the linear region of the Wilson plot should be listed and should agree with the overall *B*-factor determined from individual atomic *B*-factors by the refinement program. Listing overall experimental *B*-factors should also indicate that you have in fact inspected the Wilson plot and detected no anomalies. Low-resolution data in particular require careful inspection of the Wilson plot to select a reasonable resolution range for the linear extrapolation from which the overall *B*-factor is derived (Chapter 7).

- **Merging *R*-values.** We discussed the meaning of the various merging *R*-values in Section 8.3. It is important that the merging *R*-values provided are those from the primary scaling and merging program (*SCALA*, *SCALEPACK*, *SADABS*, or others), and not arbitrarily picked from a later merging stage. Only the primary scaling programs have full information about all experimental observations.

- **Redundancy and completeness.** Again, redundancy (the number of independent observations) must come from the primary scaling program. If necessary, rerun the data processing with correct Laue symmetry and/or space group to obtain correct final numbers overall and for the highest resolution shell.

- **Signal strength.** The ratio $\langle I/\sigma(I)\rangle$ is the primary indicator for data quality and is particularly informative for the last resolution shell. It should be consistent with the linear merging *R*-value, R_{merge}.

Any additional information that might be considered useful can also be included in "Table 1." In particular, data items that are needed later as minimal information during PDB deposition are worth including, especially when no automated data harvesting is possible or as a safeguard should the harvesting files get lost for some reason. Following the above guidelines and using the information we obtained for our VTA1 synchrotron data, we assemble the following "Table 1" (Table 8-9).

Table 8-9 Data collection statistics. The first table in the experimental section of a publication describing a crystal structure determination contains data collection statistics. The data items for the VTA1 high resolution synchrotron data set are taken from the *SCALA* summary and the *TRUNCATE* Wilson plot. Despite the data being processed by different individuals with entirely different software (*iMOSFLM/SCALA/TRUNCATE*) than in the original publication[79] (*HKL2000/SADABS/XPREP*), the diffraction data statistics agree very well.

Data collection statistics, viscotoxin A1	
X-ray source	EMBL/DESY BW7B
X-ray detector	MAR345 imaging plate
Wavelength (Å)	0.84299
Space group	$P4_32_12$
a, b (Å)	65.68
c (Å)	47.12
Matthews coefficient (Å^3 Da^{-1})	2.54
Solvent content (%)	52
Molecules in asymmetric unit cell	2
Resolution range, overall (Å)	29.37–1.25
Resolution range, highest shell (Å)	1.32–1.25
Overall *B*-factor, Wilson plot (Å^2)	11.9
Number of observations[*]	120777 (17236)
Unique reflections[*]	28867 (4179)
Completeness (%)[*]	99.3 (100.0)
Redundancy[*]	4.2 (4.1)
Mean $I/\sigma(I)$[*]	22.9 (2.1)
R_{int}	0.050 (0.456)
R_{rim}	0.068 (0.741)
R_{pim}	0.032 (0.259)

[*]Values in parentheses are for the highest resolution shell.

Box 8-12 **Initial analysis of diffraction data and contents of "Table 1".** The following initial analysis of the scaled and merged unique data is commonly carried out: inspection of a Wilson plot for approximate linearity in the high resolution part beyond about 5–4 Å; anisotropy fall-off analysis to detect severe diffraction anisotropy; twinning tests for the detection of partial and perfect merohedral twinning; and estimation of the unit cell contents through calculation of Matthews probabilities. Tests for the presence and location of non-crystallographic symmetry (NCS) elements using Patterson methods are also conducted at this stage.

Data items listed in the experimental data collection section of "Table 1" should include: wavelength; space group; unit cell parameters; Matthews coefficient, solvent content, and unit cell contents; overall Wilson B-factor; resolution range; completeness; signal-to-noise ratio; and merging R-values (both overall statistics and for the highest resolution shell).

8.6   Next steps

At this point in our studies we understand how to collect and process diffraction data. Before we can progress to the next step of experimental phase determination, we need examine how to actually reconstruct the electron density from diffraction data and phases. We will also review in detail how quality and defects in diffraction data impact the final model quality.

8.7   Key concepts

A brief review of key concepts is provided below. Questions derived from these key concepts will aid in rapid assessment of student progress.

Instrumentation

- The instrument used to record the X-rays diffracted by a crystal is the X-ray diffractometer.
- The basic components of a diffractometer are an X-ray source, suitable X-ray optics, a goniostat to orient and rotate the crystal, and an X-ray (area) detector.
- Practically all diffractometers used in macromolecular crystallography also include a cryocooler that keeps the crystals at cryogenic (near liquid nitrogen) temperature during X-ray exposure to minimize radiation damage.
- The X-rays used in crystallography are extremely intense and focused beams of ionizing radiation; proper training and precautions are mandatory when operating a diffractometer.
- Synchrotrons provide X-rays over a wide range of wavelengths, in contrast to laboratory sources which are limited to the characteristic radiation of the X-ray anode material.
- The selection of a specific, narrow bandwidth of X-ray energy from a wide range of polychromatic synchrotron X-ray radiation allows fine-tuning of the X-ray energy for multiple-wavelength anomalous diffraction (MAD) experiments.
- The X-rays emanating from a synchrotron ring are generated by the action of a magnetic field in bending magnets or insertion devices on the electrons circling in the synchrotron storage ring.
- The large number (up to many 100 000s) of X-ray reflections that need to be recorded because of the large unit cells of macromolecular crystals requires the near-exclusive use of area detectors for data collection.
- The most common detectors are cost-effective Imaging Plate (IP) detectors and Charge-Coupled Device (CCD) detectors, with the latter nearly exclusively used on synchrotrons because of their faster read-out times.
- Area detectors generate a raw digital diffraction image file that is processed and corrected by detector-specific image processing software.

Cryotechniques and real crystals

- Protein crystals are invariably susceptible to radiation damage caused by radicals and ionized species generated by the ionizing high energy X-ray photons.
- Flash-cooling to cryogenic temperatures can hamper or slow down—to a large degree, but not entirely at extreme dose rates—the reactions leading to radiation damage.
- Cryoprotectants prevent foremost the formation of ice upon flash-cooling in the mother liquor surrounding the crystal.
- Crystallization cocktails may already contain sufficient amounts of cryoprotectants such as high concentrations of PEGs, salts, glycerol, or sucrose in the crystallization mother liquor. If not, the crystals are briefly swished through a compatible cryosolution such as mother liquor spiked with a compatible cryoprotectant.
- Upon rapid quenching or flash-cooling, the mother liquor in and around the crystals remains in a glass-like, vitreous state.
- Crystalline ice diffracts and ice formation is indicated by distinct ice diffraction rings.
- Real protein single crystals are not perfect translationally periodic single-domain crystals, because growth defects lead to a number of imperfections.
- Practically all protein crystals are mosaic crystals, where multiple growth domains are slightly misaligned against each other.
- The resulting mosaic spread is the dominant contribution to finite diffraction peak width.
- Crystals can be macroscopically intertwinned, where differently oriented intergrown crystals generate different, superimposed diffraction patterns.
- Epitaxial or non-merohedral twinning occurs when two or more discrete domain orientations are prevalent and cell dimensions match or nearly match in these domain orientations.
- The most common form of microscopic twinning in protein crystals is merohedral twinning.
- If two (hemihedral twinning) or more domain orientations occur with a specific twinning operation in special orientations to a crystallographic axis, the diffraction patterns overlap perfectly in reciprocal space.
- The diffraction patterns from a merohedral twin look normal and unsuspicious, but exhibit unusual intensity distributions.
- The symmetrization of the diffraction pattern by merohedral twinning in essence leads to sharper intensity distributions and smaller raw moments of the intensity distributions.
- The shape of the cumulative intensity distributions and raw moment analysis of these distributions are used to detect and analyze merohedral twinning.
- Deviations of intensity distributions indicating broader than normal distributions and larger raw moments than untwinned data can indicate pseudo-translational NCS.
- The most common case of a single twin operator is the simplest case of merohedral twinning, namely hemihedral twinning.
- Hemihedral twinning is a relatively common and probably underreported phenomenon in macromolecular data sets; each data set should be carefully analyzed for indications of hemihedral twinning.

Data collection

- The most commonly used data collection technique in macromolecular crystallography is the rotation method.
- The crystal is rotated around a single axis by small successive increments (~1°) while exposed to the X-ray beam, and during each rotation increment a diffraction image or frame is recorded.
- The spot intensities of individual images are then integrated and the actual reflection intensities are determined.
- Fine slicing using small rotation increments allows complete 3-dimensional peak profiles to be fitted to the reflections, which can give more accurate integration, particularly for crystals with large mosaicity.
- The initial diffraction image will reveal crucial information about the diffraction quality of the crystal.

- The diffraction limit (highest resolution), presence of ice rings, spot separation, sharpness of spots, and mosaicity are primary concerns when inspecting a diffraction image.
- The exposure time must be selected so that no significant overflow of intense reflections occurs. Overflows affecting low resolution, high intensity reflections will invariably lead to problems during phasing.
- Indexing is the assignment of a consistent set of unit cell vectors matching the diffraction pattern.
- Once a unit cell has been selected, its lattice symmetry determines the minimal symmetry of the reciprocal space.
- Based on the lattice symmetry, a data collection strategy is determined.
- The goal of the strategy is to collect at least enough reflections to cover the entire asymmetric unit of the reciprocal space in order to obtain complete data.
- Accurate data require higher redundancy, achieved by multiple independent measurements of equivalent reflections.
- Anomalous data for phasing purposes also require the collection of centrosymmetrically related reflections or their symmetry equivalents.
- For MAD phasing, data sets collected at different wavelengths provide additional dispersive difference data.

Data quality and analysis

- Common indicators for data quality are merging R-values, number of independent observations (redundancy), and the signal-to-noise ratio, computed for the entire data set as well as in resolution bins.
- The linear merging R-value R_{merge} always increases with redundancy, while the redundancy-independent merging R-value R_{rim} or R_{meas} provides an independent upper limit.
- The precision-indicating merging R-value R_{pim} accounts for increasing precision with increasing redundancy. A ratio of $R_{anom}/R_{pim} > 1.5$ is an empirical indicator for successful use of the data in SAD phasing.
- The linear merging R-value and the signal-to-noise ratio are related by $R_{merge} \sim 0.8 / \langle |I| / \sigma(I) \rangle$.
- The space group can be determined from a combination of merging statistics, systematic absences, and space group preferences with the exception of the remaining (screw axis) handedness ambiguity in the 11 pairs of enantiomorphic space groups and the pairs $I222$ and $I2_12_12_1$ and $I23$ and $I2_13$.
- In ambiguous cases, all possible space groups must be tested during the phasing stage.
- For 27 chiral space groups non-equivalent indexing possibilities exist. While the choice of setting is free for individual data sets, the reflections must be consistently indexed for the generation of difference data sets.
- Protein structures generally have a solvent content of ~25–80%, with proteins diffracting to high resolution tending to have lower solvent content.
- Based on the distribution of the solvent content, the probability of the presence of multiple copies of a molecule in the asymmetric unit can be estimated from the Matthews probabilities.
- The following initial analysis of the scaled and merged unique data is commonly carried out: inspection of the Wilson plot for approximate linearity in the high resolution part beyond about 5–4 Å; anisotropy falloff analysis to detect severe diffraction anisotropy; twinning tests for the detection of partial and perfect merohedral twinning; and estimation of the unit cell contents through calculation of Matthews probabilities.
- Tests for the presence and location of non-crystallographic symmetry (NCS) elements using Patterson methods are also conducted at this stage (described in Chapter 11).
- Data items listed in the experimental data collection section of "Table 1" should include wavelength; space group; unit cell parameters; Matthews coefficient, solvent content, and unit cell contents; overall Wilson B-factor; resolution range; completeness; signal-to-noise ratio; and merging R-values (both overall statistics and for the highest resolution shell).

8.8   Additional reading

1. Evans G, & Walsh M (2006) Data collection and analysis – Proceedings of the CCP4 study weekend. *Acta Crystallogr.* D62(1), 1–124.

2. Turkenburg J, & Brady L (1999) Date collection and processing – Proceedings of the 1999 CCP4 study weekend. *Acta Crystallogr.* D55(10), 1631–1772.

3. Dauter Z (1999) Data-collection strategies. *Acta Crystallogr.* D55, 1703–1717.

4. Holton JE (2009) A beginner's guide to radiation damage. *J. Synchrotron Radiat.* 16(2), 133–142.

5. Garman EF, & Sweet R (2007) X-ray data collection from macromolecular crystals. In Doublie S (Ed.), *Macromolecular Crystallography Protocols, Volume 2.* Totowa, NJ: Humana Press.

6. Pflugrath JW (2004) Macromolecular cryocrystallography-methods for cooling and mounting protein crystals at cryogenic temperatures. *Methods* 34(3), 415–423.

7. Rossmann MG, & Arnold E (Eds.) (2001) *International Tables for Crystallography, Volume F: Crystallography of Biological Macromolecules.* Dordrecht, The Netherlands: Kluwer Academic Publishers.

8. Wilson AJC, & Prince E (Eds.) (1999) *International Tables for Crystallography, Volume C: Mathematical, Physical and Chemical Tables.* Dordrecht, The Netherlands: Kluwer Academic Publishers.

9. Garman EF, Pearson AR, & Vonrhein C (2010) Experimental Phasing and Radiation Damage - Proceedings of the CCP4 study weekend. *Acta Crystallogr.* D66(4), 325–501.

10. Garman E (2010) Radiation damage in macromolecular crystallography: what is it and why should we care? *Acta Crystallogr.* D66(4), 339–351.

8.9   References

1. Debreczeni JE, Bunkoczi G, Girmann B, et al. (2003) In-house phase determination of the lima bean trypsin inhibitor: a low-resolution sulfur-SAD case. *Acta Crystallogr.* D59(2), 393–395.

2. Arndt UW (2001) Radiation sources and optics. In *International Tables for Crystallography* F, 125–142.

3. Lindley PF (1999) Macromolecular crystallography with a third-generation synchrotron source. *Acta Crystallogr.* D55(10), 1654–1662.

4. Rupp B (2005) High throughput protein crystallography. In Sundstroem M & Edwards A (Eds.), *Structural Proteomics and High Throughput Structural Biology.* New York. NY: Taylor and Francis.

5. Helliwell JR (2001) Synchrotron-radiation instrumentation, methods and scientific utilization. In *International Tables for Crystallography* F, 155–166.

6. Hartemann VF, Baldis HA, Kerman AK, et al. (2000) Three-dimensional theory of emittance in Compton scattering and X-ray protein crystallography. *Phys. Rev.* E64, 16501–16526.

7. Gruner SM, Eikenberry EF, & Tate MW (2001) Comparison of X-ray detectors. In *International Tables for Crystallography* F, 143–153.

8. Hamlin R, Cork C, Howard A, et al. (1981) Characteristics of a flat multiwire area detector for protein crystallography. *J. Appl. Crystallogr.* 14(2), 85–93.

9. Parkin S, & Rupp B (1995) A helium flushed beam tunnel. *J. Appl. Crystallogr.* 28, 243.

10. Angello SG, Augustine F, Ercan A, et al. (2004) Development of a mixed-mode pixel array detector for macromolecular crystallography. *Nucl. Sci. Symp. Conf. Rec. 2004 IEEE* 7, 4667–4671

11. Ercan A, Tate MW, & Gruner SM (2006) Analog pixel array detectors. *J. Synchrotron Radiat.* 13(2), 110–119.

12. Rodgers DW (2001) Cryocrystallography techniques and devices. In *International Tables for Crystallography* F, 202–208.

13. Petsko GA (1975) Protein crystallography at sub-zero temperatures: cryoprotective mother liquors for protein crystals. *J. Mol. Biol.* 96, 382–392.

14. Hope H (2001) Introduction to cryocrystallography. In *International Tables for Crystallography* F, 197–201.

15. Karain WI, Bourenkov GP, Blume H, et al. (2002) Automated mounting, centering and screening of crystals for high-throughput protein crystallography. *Acta Crystallogr.* D58, 1519–1522.

16. Muchmore SW, Olson J, Jones R, et al. (2000) Automated crystal mounting and data collection for protein crystallography. *Structure* 8(12), 243–246.

17. Snell G, Cork C, Nordmeyer R, et al. (2004) Automatic sample mounting and alignment system for biological crystallography at a synchrotron source. *Structure* 12, 1–12.

18. Perutz MF (1985) Early days of protein crystallography. *Methods Enzymol.* 114, 3–18.

19. McPherson A (2000) *In situ* X-ray crystallography *J. Appl. Crystallogr.* 33, 397–400.

20. Thorne RE, Stum Z, Kmetko J, et al. (2003) Microfabricated mounts for high-throughput macromolecular cryocrystallography. *J. Appl. Crystallogr.* 36(6), 1455–1460.

21. McPherson A, Malkin AJ, Kuznetsov YG, et al. (2001) Atomic force microscopy applications in macromolecular crystallography. *Acta Crystallogr.* D57, 1053–1060.

22. Malkin AJ, & Thorne RE (2004) Growth and disorder of macromolecular crystals: insights from atomic force microscopy and X-ray diffraction studies. *Methods* 34(3), 273–299.

23. Rubinson KA, Ladner JE, Tordova M, et al. (2000) Cryosalts: suppression of ice formation in macromolecular crystallography. *Acta Crystallogr.* D56(8), 996–1001.

24. Garman EF, & Doublie S (2003) Cryocooling of macromolecular crystals: Optimization methods. *Methods Enzymol.* 368, 188–216.

25. Nave C, & Garman E (2005) Towards an understanding of radiation damage in cryocooled macromolecular crystals. *J. Synchrotron Radiat.* 12, 257–260.

26. Snell EH, Bellamy HD, Rosenbaum G, et al. (2007) Non-invasive measurement of X-ray beam heating on a surrogate crystal sample. *J. Synchrotron Radiat.* 14, 109–115.

27. Holton JM (2007) XANES measurements of the rate of radiation damage to selenomethionine side chains. *J. Synchrotron Radiat.* 14, 51–72.

28. Garman E, & Nave C (2002) Radiation damage to crystalline biological molecules: current view. *J. Synchrotron Radiat.* 9, 327–328.

29. Dunlop KV, Irvin RT, & Hazes B (2005) Pros and cons of cryocrystallography: Should we also collect a room-

temperature data set? *Acta Crystallogr.* D61(Pt 1), 80–87.

30. Burmeister WP (2000) Structural changes in cryo-cooled protein crystals owing to radiation damage. *Acta Crystallogr.* D56, 328–341.

31. Ravelli RBG, & McSweeney SM (2000) The "fingerprint" that X-rays can leave on structures. *Struct. Fold. Des.* 8, 315–328.

32. Weik M, Ravelli RBG, Kryger G, et al. (2000) Specific chemical and structural damage to proteins produced by synchrotron radiation. *Proc. Natl. Acad. Sci. U.S.A.* 97, 623–628.

33. Nanao MH, Sheldrick GM, & Ravelli RB (2005) Improving radiation-damage substructures for RIP. *Acta Crystallogr.* D61(Pt 9), 1227–1237.

34. Holton JE (2009) A beginner's guide to radiation damage. *J. Synchrotron Radiat.* 16(2), 133–142.

35. Kriminski S, Caylor CL, Nonato MC, et al. (2002) Flash-cooling and annealing of protein crystals. *Acta Crystallogr.* D58, 459–471.

36. Juers DH, & Matthews BW (2004) The role of solvent transport in cryo-annealing of macromolecular crystals. *Acta Crystallogr.* D60(3), 412–421.

37. Warkentin M, Berejnov V, Husseini NS, et al. (2006) Hyperquenching for protein cryocrystallography. *J. Appl. Crystallogr.* 39, 805–811.

38. Viola R, Carman P, Walsh J, et al. (2007) Automated robotic harvesting of protein crystals—addressing a critical bottleneck or instrumentation overkill? *J. Struct. Funct. Genomics* 8(4), 145–152.

39. McPherson A, Kuznetsov YG, Malkin A, et al. (2003) Macromolecular crystal growth as revealed by atomic force microscopy. *J. Struct. Biol.* 142(1), 32–46.

40. Lebedev AA, Vagin AA, & Murshudov GN (2006) Intensity statistics in twinned crystals with examples from the PDB. *Acta Crystallogr.* D62(1), 83–95.

41. Chandra N, Acharya KR, & Moody PCE (1999) Analysis and characterization of data from twinned crystals. *Acta Crystallogr.* D55(10), 1750–1758.

42. Yeates TO (1997) Detecting and overcoming crystal twinning. *Methods Enzymol.* 276, 344–58.

43. Yeates TO (1988) Simple statistics for intensity data from twinned specimens. *Acta Crystallogr.* A44(Pt 2), 142–144.

44. Padilla JE, & Yeates TO (2003) A statistic for local intensity differences: robustness to anisotropy and pseudo-centering and utility for detecting twinning. *Acta Crystallogr.* D59(7), 1124–1130.

45. Dauter Z (1999) Data-collection strategies. *Acta Crystallogr.* D55, 1703–1717.

46. Dauter Z, & Wilson KS (2001) Principles of monochromatic data collection. In *International Tables For Crystallography* F, 177–195.

47. Pflugrath JW (1999) The finer things in X-ray diffraction data collection. *Acta Crystallogr.* D55, 1718–1725.

48. Leslie AGW (1999) Integration of macromolecular diffraction data. *Acta Crystallogr.* D55, 1969–1702.

49. McPhillips TM, McPhillips SE, Chiu H-J, et al. (2002) Blu-Ice and the Distributed Control System: software for data acquisition and instrument control at macromolecular crystallography beamlines. *J. Synchrotron Radiat.* 9, 401–406.

50. Sweet RM (1998) The technology that enables synchrotron structural biology. *Nature Struct. Biol. Suppl.* 5, 654–656.

51. Garman E (1999) Cool data: quantity AND quality. *Acta Crystallogr.* D55, 1641–1653.

52. Powell HR (1999) The Rossmann Fourier autoindexing algorithm in MOSFLM. *Acta Crystallogr.* D55, 10690–10695.

53. Holton JM (2008) MLFSOM – a program to simulate diffraction images. *J. Synchrotron Radiat.* to be published.

54. Kantardjieff KA, Höchtl P, Segelke BW, et al. (2002) Concanavalin A in a dimeric crystal form: revisiting structural accuracy and molecular flexibility. *Acta Crystallogr.* D58, 735–743.

55. Card GL, Peterson NA, Smith CA, et al. (2005) The crystal structure of Rv1347c, a putative antibiotic resistance protein from *Mycobacterium tuberculosis*, reveals a GCN5-related fold and suggests an alternative function in siderophore biosynthesis. *J. Biol. Chem.* 280, 13978–13986.

56. Burzlaff H, Gruber B, Wolff PM, et al. (2002) Crystal lattices. In Hahn T (Ed.), *International Tables for Crystallography* Volume A. Dordrecht, The Netherlands: Kluwer Academic Publishers.

57. Sauter NK, Grosse-Kunstleve RW, & Adams PD (2004) Robust indexing for automatic data collection. *J. Appl. Crystallogr.* 37(3), 399–409.

58. Sauter NK, Grosse-Kunstleve RW, & Adams PD (2006) Improved statistics for determining the Patterson symmetry from unmerged diffraction intensities. *J. Appl. Crystallogr.* 39(2), 158–168.

59. Xuong NH (1985) Strategy for data collection from protein crystals using a multiwire counter area detector diffractometer *J. Appl. Crystallogr.* 18, 342–350.

60. Buerger MJ (1964) *The Precession Method in X-Ray Crystallography.* New York: Wiley.

61. Kabsch W (1988) Evaluation of single-crystal X-ray diffraction data from a position-sensitive detector. *J. Appl. Crystallogr.* 21, 916–924.

62. Rossmann MG (2001) Automatic indexing of oscillation images. In Rossmann MG & Arnold E (Eds.), *International Tables for Crystallography* F, Dordecht, The Netherlands: Kluwer Academic Publishers

63. Steller I, Bolotovsky R, & Rossmann MG (1997) An algorithm for automatic indexing of oscillation images using Fourier analysis. *J. Appl. Crystallogr.* 30, 1036–1040.

64. Howard AJ, Gilliland GL, Finzel BC, et al. (1987) The use of an imaging proportional counter in macromolecular crystallography. *J. Appl. Crystallogr.* 20(5), 383–387.

65. Otwinowski Z, & Minor W (1997) Processing of X-ray diffraction data collected in oscillation mode. *Methods Enzymol.* 267, 307–326.

66. Howard AJ, Nielsen C, & Xuong NH (1985) Software for a diffractometer with multiwire area detector. *Methods Enzymol.* 114, 452–472.

67. Garman EF, & Sweet R (2007) X-ray data collection from macromolecular crystals. In Doublie S (Ed.), *Macromolecular Crystallography Protocols*, Volume 2. Totowa, NJ: Humana Press.

68. Rupp B, Marshak DR, & Parkin S (1996) Crystallization and preliminary X-ray analysis of two new crystal forms of calmodulin. *Acta Crystallogr.* D52(2), 411–413.

69. Lovelace JJ, Murphy CR, Daniels L, et al. (2008) Protein crystals can be incommensurately modulated. *J. Appl.Crystallogr.* 41, 600–605.

70. Minor W, Tomchick D, & Otwinowski Z (2000) Strategies for macromolecular synchrotron crystallography. *Struct. Fold. Des.* 8(5), 105–110.

71. Sheldrick GM (1990) Phase annealing in SHELX-90: direct methods for larger structures. *Acta Crystallogr.* A46, 467–473.

72. Morris RJ, & Bricogne G (2003) Sheldrick's 1.2 Å rule and beyond. *Acta Crystallogr.* D59, 615–617.

73. Sheldrick GM, Hauptman HA, Weeks CM, et al. (2001) Ab initio phasing. In *International Tables for Crystallography* F, 333–354.

74. Ramagopal UA, Dauter M, & Dauter Z (2003) Phasing on anomalous signal of sulfurs: what is the limit? *Acta Crystallogr.* D59, 1020–1027.

75. Wang BC (1985) Resolution of Phase Ambiguity in Macromolecular Crystallography. *Methods Enzymol.* 115, 90–112.

76. Dauter Z, Dauter M, & Dodson ED (2002) Jolly SAD. *Acta Crystallogr.* D58, 496–508.

77. Hendrickson WA (1991) Determination of macromolecular structures from anomalous diffraction of synchrotron radiation. *Science* 254, 51–58.

78. Hendrickson WA (2000) Synchrotron crystallography. *Trends Biochem. Sci.* 25, 637–643.

79. Pal A, Debreczeni JÉ, Sevvana M, et al. (2008) Crystal structures of viscotoxins A1 and B2 from the European mistletoe. *Acta Crystallogr.* D64, 985–992.

80. Diederichs K, & Karplus PA (1997) Improved R-factors for diffraction data analysis in macromolecular crystallography. *Nat. Struct. Biol.* 4(4), 269–275.

81. Weiss M (2001) Global indicators of X-ray data quality. *J. Appl. Crystallogr.* 34(2), 130–135.

82. Weiss MS, & Hilgenfeld R (1997) On the use of the merging R factor as a quality indicator for X-ray data. *J. Appl. Crystallogr.* 30, 203–205.

83. Weiss MS, Sicker T, & Hilgenfeld R (2001) Soft X-rays, high redundancy and proper scaling: A new procedure for automated protein structure determination via SAS. *Structure* 9, 771–777.

84. van Beek CG, Bolotovski R, & Rossmann M (2003) The use of partially recorded reflections for post refinement, scaling and averaging of X-ray data. In *International Tables for Crystallography* F, 236–242.

85. Schneider TR, & Sheldrick GM (2002) Substructure solution with SHELXD. *Acta Crystallogr.* D58, 1772–1779.

86. Leslie A, Powell H, Winter G, et al. (2002) Automation of the collection and processing of X-ray diffraction data – a generic approach. *Acta Crystallogr.* D58, 1924–1928.

87. Evans PR (1997) Scala. *Joint CCP4 and ESF-EACBM Newsl.* 33, 22–24.

88. Blessing R (1995) An empirical correction for absorption anisotropy. *Acta Crystallogr.* A51(1), 33–38.

89. Potterton E, Briggs PJ, Turkenberg M, et al. (2003) A graphical user interface to the CCP4 program suite. *Acta Crystallogr.* D59, 1131–1137.

90. Harris M, & Jones TA (2002) Xtrack – a web-based crystallographic notebook. *Acta Crystallogr.* D58(10), 1889–1891.

91. Matthews BW (1968) Solvent content of protein crystals. *J. Mol. Biol.* 33, 491–497.

92. Matthews BW (1976) X-ray crystallographic studies of proteins. *Annu. Rev. Phys. Chem.* 27, 493–523.

93. Kantardjieff KA, & Rupp B (2003) Matthews coefficient probabilities: Improved estimates for unit cell contents of proteins, DNA, and protein-nucleic acid complex crystals. *Protein Sci.* 12, 1865–1871.

94. Kleywegt GJ (2000) Validation of protein crystal structures. *Acta Crystallogr.* D56, 249–265.

95. Taylor R, & Kennard O (1986) Accuracy of crystal structure error estimates. *Acta Crystallogr.* B42(1), 112–120.

96. EU 3-D Validation Network (1998) Who checks the checkers? Four validation tools applied to eight atomic resolution structures. *J. Mol. Biol.* 276, 417–436.

PART IV

Determining Your Structure

Reconstruction of electron density and the phase problem

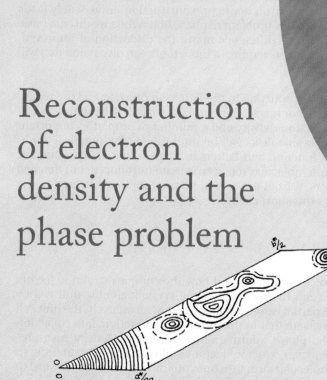

h k 0 projection of triclinic α-globlglobin. According to Dr. A.L. Patterson, this should properly be called a Fourier projection. Mr. Joseph Fourier insists with equal firmness that it be referred to as a Patterson projection.

Figure and caption by R. E. Dickerson (2005) *Present at the Flood*, p 217

The reconstruction of electron density from complex structure factors and its inversion, the computation of complex structure factors from electron density, are amongst the most fundamental and frequent tasks in the course of crystallographic structure determination. In practice, the experimental structure factor amplitudes and separately supplied phases from a phasing experiment are needed—a consequence of phase information being lost in the physical detection of the diffracted photons, fittingly termed the *phase problem* in crystallography.

We will lay out in this short but important chapter the mathematical principles of *Fourier transforms* as far as they are needed to derive the equations used in practical crystallography. The *Fourier convolution theorem* and a basic discussion of *convolutions* relevant for crystallography in general are also discussed in this section. While it may not be necessary to follow the derivations in all detail in a short course, the results will be universally used in subsequent chapters.

The theory section is followed by an investigation of the effects that a number of characteristics of the experimentally measured data have on the electron density reconstruction. We explore the principles first with the help of simple, 1-dimensional model structures and then discuss the features in numerous examples based on the reconstruction of 3-dimensional electron density maps. Effects of resolution, truncation, incompleteness, and lattice disorder and their practical consequences are demonstrated.

After a brief refresher of the phase problem, we appreciate that the Fourier coefficients used in map reconstruction—the measured structure factor amplitudes and the separate phases—originate from different sources and their associated errors exert different effects on the electron density reconstruction. The dominance of the phases on the electron density reconstruction immediately leads to the appreciation of the serious problem of phase bias. While we discuss basic map types such as $2F_o - F_c$ and difference maps, the discussion of improved, likelihood-based map coefficients requires a more thorough discussion that will be provided in Chapter 12.

The chapter ends with an introduction to the *Patterson function* and the resulting interatomic distance vector maps. The Patterson maps will be derived as an auto-correlation of the electron density and a number of properties important for a variety of crystallographic tasks will be established. The practical applications of the Patterson function and Patterson maps in initial data analysis, substructure solution, and molecular replacement are introduced, but detailed in the respective chapters. A brief preview of phasing methods treated in the following chapters closes this short chapter.

9.1   The reconstruction of electron density

At this point in our studies, we understand how the complex structure factors **F(h)** can be computed from the contents of the crystallographic unit cell by summation of scattered partial waves emanating from all atoms in the unit cell (Chapter 6). We are also aware that in the process of diffraction data collection (Chapter 8), we lose the phase information: From each intensity measurement we obtain only the scalar structure factor amplitude or $F_{obs}(\mathbf{h})$. In order to reconstruct the actual molecular structure from our diffraction data, we need to address two separate problems.

1. We need to provide the missing phases $\alpha(\mathbf{h})$ for each structure factor amplitude $F(\mathbf{h})$. The lack of phases generates the so-called phase problem in crystallography and we will spend considerable effort on discussing the methods to provide the missing phases; experimental phasing is a core discipline of crystallography. Experimental phasing and molecular replacement phasing will be treated in Chapters 10 and 11 respectively.

2. We need a formalism to somehow invert the diffraction process, that is, to reconstruct the molecular structure from the complex structure factors, which we know for each reciprocal index vector **h** in the form of the experimental phases $\alpha(\mathbf{h})$ and the associated measured structure factor amplitudes $F_{obs}(\mathbf{h})$. The basics of the electron density reconstruction procedure by means of Fourier transformation will concern us in this chapter.

We assume for now that we can actually obtain phases and address first the problem of how to reconstruct the molecular structure from the diffraction data.

The process of diffraction transforms information from the real (or direct) space domain R into the reciprocal space domain R^*. As the physical source of diffraction was the scattering of X-ray photons by the electrons distributed around the atoms of the molecules, we expect that a mathematical inversion of the diffraction process will again deliver an electron density distribution, $\rho(\mathbf{r})$, of the diffracting molecules.

Fourier transformations

We can formalize the switching from one domain representation to its reciprocal representation through the mathematical formalism of Fourier transformation. The Fourier transformation is named in honor of Jean Baptiste Joseph Fourier (1768–1830), a French mathematician and physicist (although the final proof for the theorem named after Fourier was actually given by Jacques Charles François Sturm in 1829).

The Fourier transformation relates functions in mutually reciprocal domains (this can be real space and reciprocal space as in crystallographic applications, or frequency and time as used in spectroscopy) in invertible and unique form. A function $G(x)$ can be Fourier-transformed ("FTed") into $F(x^*)$ and back into $G(x)$ with complete conservation of information (i.e., the FT is a bijective function):

$$FT(G(x)) = F(x^*) \quad \text{and} \quad FT^{-1}(F(x^*)) = G(x) \qquad (9\text{-}1)$$

where the * indicates the reciprocal domain. The Fourier transform FT and its inverse FT^{-1} are defined as complex Fourier integrals (which are derived as eigenfunctions that satisfy the general Sturm–Liouville eigenvalue problem with periodic boundary conditions):

$$F(x^*) = \int_R G(x) \cdot \exp(2\pi i x^* x)\,dx \quad \text{and} \quad G(x) = \int_{R^*} F(x^*) \cdot \exp(-2\pi i x^* x)\,dx^* \quad (9\text{-}2)$$

We have written the Fourier integrals in a form so that their application to our task of back-transforming the complex structure factors $\mathbf{F(h)}$ into their real space representation, the electron density $\rho(\mathbf{r})$, will be readily recognizable: In its most general form, we can write the scattering function as a continuous integral (in contrast to the discrete summation we use in practice) of the scattering function for a general position vector $\mathbf{r}$ in direct space and its related (scattering) vector $\mathbf{r}^*$ in reciprocal space as

$$\mathbf{F}(\mathbf{r}^*) = \int_R \rho(\mathbf{r}) \cdot \exp(2\pi i \mathbf{r}^* \mathbf{r})\,d\mathbf{r} \qquad (9\text{-}3)$$

where $\rho(\mathbf{r})$ stands for the electron density at any given point $\mathbf{r}$ in real space R. For the resulting function $\mathbf{F}(\mathbf{r}^*)$ in R^*, an inverse Fourier transform exists that lets us recover or back-transform $\rho(\mathbf{r})$ from $\mathbf{F}(\mathbf{r}^*)$:

$$\rho(\mathbf{r}) = \int_{R^*} \mathbf{F}(\mathbf{r}^*) \cdot \exp(-2\pi i \mathbf{r}^* \mathbf{r})\,d\mathbf{r}^* \qquad (9\text{-}4)$$

In the above Fourier integrals, the integrand(s) preceding the exponent are called the Fourier coefficient(s). The Fourier transform FT is completely invertible, that is, given that

$$\mathbf{F}(\mathbf{r}^*) = FT[\rho(\mathbf{r})] \quad \text{and} \quad \rho(\mathbf{r}) = FT^{-1}[\mathbf{F}(\mathbf{r}^*)] \qquad (9\text{-}5)$$

it follows that

$$FT^{-1}\left[FT\big(\rho(\mathbf{r})\big)\right] = \rho(\mathbf{r}) \qquad (9\text{-}6)$$

Equation 9-4 provides a fundamental insight, because the fact that $\rho(\mathbf{r})$ is just the inverse or back-transform of $\mathbf{F}(\mathbf{r}^*)$ provides a convenient formalism to reconstruct the electron density from the complex structure factors (or structure factor amplitudes plus phases), and on the other hand, to take electron density and convert it back into a set of complex structure factors, a procedure called map inversion. We will make extensive use of this free exchange of direct and reciprocal space information in many crystallographic applications such as density modification, real space refinement, fast structure factor calculations, and map averaging.

Computing the Fourier integrals and summations

We can immediately recognize the complete analogy of the general Fourier transform equation (9-3) and the integration of the electron density over the entire atom volume, which yielded in Chapter 6 the atomic scattering factors as a reciprocal space representation of the electron density:

$$f(\mathbf{S}) = \int_{\mathbf{r}}^{V(atom)} \rho_{at}(\mathbf{r})\exp(2\pi i \mathbf{S}\mathbf{r})d\mathbf{r} = FT[\rho_{at}(\mathbf{r})] \qquad (9\text{-}7)$$

The same correspondence exists between the electron density of the entire unit cell and its scattering function, the complex structure factor. Here the integration extends over the entire unit cell volume V:

$$F(S) = \int_r^{V(cell)} \rho(\mathbf{r}) \cdot \exp(2\pi i S\mathbf{r}) d\mathbf{r} = FT[\rho(\mathbf{r})] \tag{9-8}$$

To make the structure factor equation explicit in $\mathbf{h}$, we use the reciprocal and direct space position vector definitions

$$S = h\mathbf{a}^* + k\mathbf{b}^* + l\mathbf{c}^*$$

$$\mathbf{r} = x\mathbf{a} + y\mathbf{b} + z\mathbf{c} \tag{9-9}$$

leading to $\mathbf{Sr} = \mathbf{hx}$. We therefore need to integrate in real space over the entire unit cell volume in fractional coordinates, which implies the substitution $d\mathbf{r} = abc\, dx\, dy\, dz = V\, dx\, dy\, dz = V\, d\mathbf{x}$ and we obtain the triple integral

$$F(\mathbf{h}) = V \int_{x=0}^{1} \int_{y=0}^{1} \int_{z=0}^{1} \rho(xyz) \cdot \exp(2\pi i\mathbf{hx}) dx\, dy\, dz = V \int_{x=0}^{1} \rho(\mathbf{x}) \cdot \exp(2\pi i\mathbf{hx}) d\mathbf{x} \tag{9-10}$$

We have now expressed $F(\mathbf{h})$ in the form of a Fourier integral, and can immediately perform the back-transformation of (9-10) from reciprocal space R^* to real space R. We apply the Fourier transform from reciprocal space (which extends over all reflections $\mathbf{h}$) into real space:

$$FT[F(\mathbf{h})] = FT\left[V \int_{x=0}^{1} \rho(\mathbf{x}) \cdot \exp(2\pi i\mathbf{hx}) d\mathbf{x} \right]$$

$$= \rho(\mathbf{x}) = \frac{1}{V} \int_{h=-\infty}^{+\infty} F(\mathbf{h}) \cdot \exp(-2\pi i\mathbf{hx}) d\mathbf{h} \tag{9-11}$$

Because R^* is discrete, we can replace the continuous Fourier integration over R^* in (9-11) with a discrete triple summation extending over all reciprocal lattice index vectors $\mathbf{h}$. The Fourier coefficients are then the complex structure factors $F(\mathbf{h})$ in the range $-\infty \le \mathbf{h} \le \infty$.

$$\rho(\mathbf{x}) = \frac{1}{V} \sum_{h=-\infty}^{+\infty} F(\mathbf{h}) \cdot \exp(-2\pi i\mathbf{hx}) \tag{9-12}$$

In practice we generally evaluate $\rho(\mathbf{x})$ on a grid of points with fractional grid position coordinates (x, y, z) over the asymmetric unit cell. With the unit cell given in dimensions of Å, the normalization factor of $1/V$ preceding the discrete Fourier summation provides the correct units of $e^-/\text{Å}^3$ for the electron density $\rho(xyz)$:

$$\rho(x, y, z) = \frac{1}{V} \sum_{h=-\infty}^{+\infty} \sum_{k=-\infty}^{+\infty} \sum_{l=-\infty}^{+\infty} F(hkl) \cdot \exp[-2\pi i(hx + ky + lz)] \tag{9-13}$$

We can now use this explicitly written summation for the electron density and express the back-transformation into structure factors in terms of discrete grid point coordinates:

$$F(hkl) = V \sum_{x=0}^{1} \sum_{y=0}^{1} \sum_{z=0}^{1} \rho(x,y,z) \cdot \exp(2\pi i(hx + ky + lz)) \tag{9-14}$$

Summation 9-14 above is used to convert real space electron density maps back into reciprocal space as a set of structure factors, a procedure called map inversion. With the pair of Fourier summations (9-13 and 9-14) we possess now a convenient formalism to reconstruct the electron density from the complex structure factors, and on the other hand, to take electron density and convert it back into a set of complex structure factors. We can visualize the transform and back-transform concept as shown in Figure 9-1.

The Fourier convolution theorem

In the introductory chapters we have already shown qualitatively that the crystal structure is a convolution of a periodic lattice function with a molecular motif:

h	k	l	F(hkl)	φ(hkl)
2	0	0	228.0	180.0
1	0	1	10.4	90.0
2	0	1	901.8	270.0
1	1	1	367.0	332.1
1	2	3	149.3	37.8
8	9	1	97.9	255.1
7	7	2	111.5	139.7

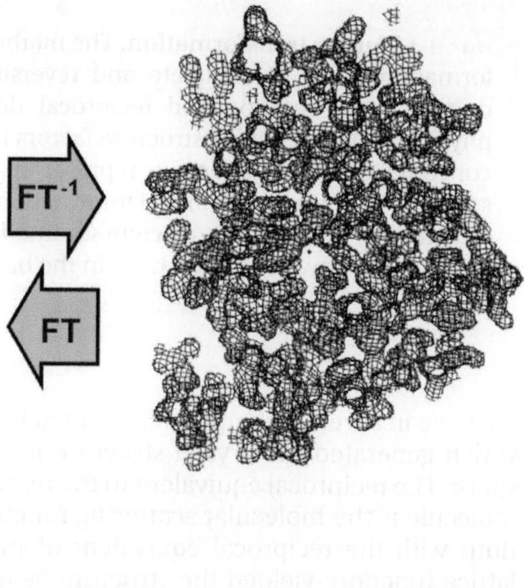

FT⁻¹

FT

Figure 9-1 Back-transformation of complex structure factors into electron density. The back-transformation of complex structure factors (provided as a list of structure factor amplitudes plus their phases) from the reciprocal into the real space domain by discrete Fourier summation produces the electron density of the scattering molecule. Using the same formalism, any electron density can be transformed into complex structure factors (map inversion). The Fourier transformation from/to the reciprocal space domain (complex structure factors) to/from the real space domain (electron density) is completely reversible without loss of information.

Sidebar 9-1 The Fast Fourier Transform (FFT) method and structure factor calculations. The nested summations in the discrete Fourier transform of both the electron density calculation (9-13) and the structure factor equation (9-14) are very slow to compute, and the same holds for the direct structure factor summation from an atomic model. In crystallographic computations the cycling from structure factors to electron density and back is a fundamental and recurrent task, and therefore a more efficient approach to executing Fourier transforms than the direct summations is needed. There are two avenues of attack on the problem: the general approach of selecting sampling intervals (grid points) so that fewer computations are required in the transform itself. Clever grid point selection in Fast Fourier Transform (FFT) implementations reduces the problem from an order N^2 to an order $N \cdot \log_2(N)$ problem—an enormous saving in computations for large N, as you may convince yourself. In addition, space-group specific symmetries can be exploited to reduce identical computations in crystallographic applications.

The FFT algorithm. Let us assume a 1-dimensional FT over N points and write it in the general form

$$F_k = \sum_{n=0}^{N-1} f_n e^{2\pi i n k / N} = \sum_{n=0}^{N-1} f_n W^{nk} \quad \text{with} \quad e^{2\pi i / N} = W$$

It can be shown (the Danielson–Lawson theorem)[1] that the above Fourier summation can be split in two summations with length $N/2$ each over even and odd indices, respectively:

$$F_k = F_k^{even} + W^k F_k^{odd}$$

This procedure can be applied recursively, that is, we can break up each sub-summation again in two separate even–odd summations. If the original N is a power of 2, we can repeat that separation until we end up with a FT of length one belonging to a certain pattern of even–odd combinations such as $F_k^{eeoooeoeoeoeoee}$, whose value is just one of the original Fourier coefficients f_n. The bookkeeping for which coefficient belongs to which sub-summation is accomplished by a bit-reversal algorithm. Once we know the sequence of coefficients, we just reverse the process by combining adjacent one-point transforms into two-point transforms, evaluate them, combine all the two-point transforms into four-point FTs and so on, until we have obtained our complete FT of order N. This is the principal structure of the Fast Fourier Transform (FFT) algorithm, which has reduced the number of matrix operations in the evaluation of W^{nk} from order N^2 to order $N \cdot \log_2(N)$. The generic FFT algorithm has been further extended by Lynn Ten Eyck[2, 3] to exploit space group symmetry in crystallographic applications. His papers also summarize many useful relations and properties of Fourier transforms.

Structure factor calculation using FFT. We can also use the FFT algorithm to compute the structure factors from an atomic model. Instead of executing the explicit structure factor summation extending over all atoms and all positions in the unit cell, we first calculate the electron density map of the model in the asymmetric unit, and obtain the structure factors from the complete unit cell density by applying space group-specific FFT routines. The shape of the atomic electron density is approximated by a Gaussian function with an increasing number of terms appropriate to the resolution up to which the structure factors are calculated.[4] The Fourier back-transform of reciprocal space atomic scattering function exponentials (as used for example in the Cromer–Mann approximation, Chapter 6) to a real space Gaussian describing the shape of atom j with occupation n_j and B-factor B_j looks as follows:[5]

$$\rho(r) = n_j \sum_{i=1}^{k} a_i \left(\frac{4\pi}{B_j + b_i} \right)^{3/2} \exp\left(-\frac{4\pi^2}{B_j + b_i} r^2 \right)$$

where the summation extends over as many coefficients as were used in the description of the atom form factor.

Despite the route through first constructing a map and then applying the FFT to transform it into reciprocal space appearing circuitous, it is still substantially faster than direct summation, and the speed gain increases with the size of the molecule. An even faster, molecular transform-interpolation based method exists, which is described in Chapter 11.

> **Box 9-1 Fourier transformation.** The mathematical formalism of Fourier transformation allows a complete and reversible transformation between direct domain representation and reciprocal domain representation of the same physical reality. Complex structure factors representing reciprocal space can be converted into electron density representing direct space and back into complex structure factors without loss of information. In practice, the continuous Fourier integrals are replaced by a discrete summation over all structure factor indices h, k, l or all density grid points x, y, z in the back-transform, respectively.

We have in several examples filled or packed an empty unit cell with molecules, which generated the crystal structure in the direct domain representation of space. The reciprocal equivalent to the real space electron density $\rho(\mathbf{r})$ of a single molecule is the molecular scattering function, and its sampling (or multiplication) with the reciprocal equivalent of the real crystal lattice, the reciprocal lattice function, yielded the structure factors for each reciprocal lattice point. Given that the inverse transformation exists, we can imagine the diffraction process (and its inversion, structure determination) in terms of convolutions as shown in Figure 9-2.

The Fourier convolution theorem for two general functions $G(\mathbf{x})$ and $F(\mathbf{x})$ can be formally written as

$$\mathrm{FT}\big[G(x) \otimes F(x)\big] = \mathrm{FT}\big[G(x)\big] \cdot \mathrm{FT}\big[F(x)\big] \tag{9-15}$$

where the operator symbol $\otimes$ implies the convolution of the two functions. The convolution $Conv(\mathbf{u})$ between $F(\mathbf{x})$ and $G(\mathbf{x})$ itself is mathematically defined through a convolution integral

$$Conv(\mathbf{u}) = F(\mathbf{x}) \otimes G(\mathbf{x}) = \int_x F(\mathbf{x})G(\mathbf{u} - \mathbf{x})\,d\mathbf{x} \tag{9-16}$$

Despite the unpleasant looks of the integral (9-16), the convolution is easy to visualize. For a lattice function, which is a periodic Dirac δ-function (1 at the lattice point, 0 elsewhere), the convolution with the molecular function simply

Figure 9-2 Convolutions in real and reciprocal space. We can visualize the crystal structure as a convolution (indicated by the operator ⊗) of the real space crystal lattice with the molecular contents, and the diffraction pattern as a sampling (or multiplication) of the molecular transform with the reciprocal lattice. The two domains of space, the direct and reciprocal domains, can be transformed into each other, and back again, by the application of Fourier transformations, or FTs. The figure is essentially a visualization of the Fourier convolution theorem: $FT\big[\rho(\mathbf{r}) \otimes L(\mathbf{r})\big] = FT\big[\rho(\mathbf{r})\big] \cdot FT\big[L(\mathbf{r})\big]$, with $\rho(\mathbf{r})$ the molecular electron density and $L(\mathbf{r})$ the lattice function.

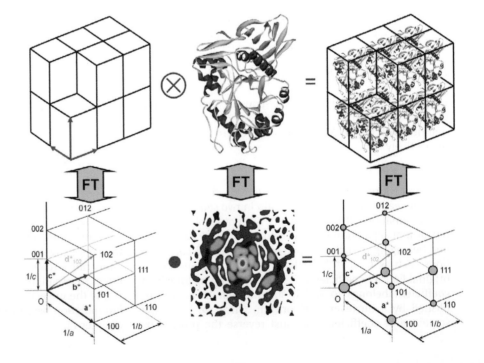

fills the lattice with the molecular objects, as illustrated in the top panel of Figure 9-2. For their transforms, the sampling (or multiplication) of the molecular transform function (envelope) by the reciprocal lattice function yields the discrete diffraction pattern. In contrast to convoluting a function $F(\mathbf{x})$ (say, a molecule) with a sharp δ-function, we can probe it along the sampling distance $\mathbf{u}$ with a point spread function, such as a Gaussian function $G(\mathbf{u})$. Such a convolution results in a blurred image of the molecule, somewhat equivalent to the application of a B-factor that smears out the individual atomic positions of a static structure model (Figure 9-3).

A common case in crystallography is the convolution of two Gaussian functions (a situation encountered in error propagation or in the convolution of electron density peaks discussed later in Section 9.4). If $G_A(\sigma_A, x_A)$ and $G_B(\sigma_B, x_B)$ are both Gaussian functions defined by their standard deviation and mean value (Figure 7-3) their convolution $G_A \otimes G_B$ will be another Gaussian, which can be derived explicitly from error propagation as shown in Chapter 7:

$$G_A \otimes G_B = G_C((\sigma_A^2 + \sigma_B^2)^{1/2}, x_A + x_B) \qquad (9\text{-}17)$$

The convolution of electron density functions as well as the Fourier convolution theorem will be quite useful in the derivation of the Patterson function, direct methods, and related general electron density manipulations.

Box 9-2 Convolution of functions. The convolution of two functions F and G is defined by a convolution integral. Numerically the convolution is obtained by multiplication of the functions over a moving window, imaginable as "probing" of one function with the other. Typical functions appearing in convolutions are Gaussian point spread functions, lattice functions, molecular envelopes, and electron density. The Fourier convolution theorem allows us to separately transform two functions and multiply the resulting individual Fourier transforms. An example is the sampling of a molecular scattering envelope (the transform of the molecule) by the reciprocal lattice function (the transform of the direct lattice function) yielding the structure factors and thus the diffraction pattern.

Calculation of electron density

In practical crystallographic work, the structure factor amplitudes and phases in general originate from different sources: $F_\mathbf{h}$ is proportional to the square root of the measured intensity data, while the phase angle $\alpha_\mathbf{h}$ is provided from experimental phases or computed phases from a structure model. We therefore separate the complex structure factor $\mathbf{F_h}$ into an explicit amplitude term $F_\mathbf{h}$ and a phase angle $\alpha_\mathbf{h}$ term using the Euler relation: $\mathbf{F_h} = F_\mathbf{h} \cdot \exp(i\alpha_\mathbf{h})$. Substituting this

Figure 9-3 Convolution of a molecular image with a Gaussian function. The image on the left (our TIM barrel) conveys unrealistic precision because it is generated using the mean position of the atoms without any consideration of atomic B-factors. A more realistic picture may be the slightly blurred image, conveying a sense of overall dynamic motion in the molecule. The rightmost panel shows a high degree of positional uncertainty, and the loss of detail is representative of our actual state of knowledge in the case of a low resolution structure. Note that a molecular graphics display program would deliver the same apparently precise image shown on the left side irrespective of B-factors and resolution (and thus ignoring any coordinate uncertainty) of the atomic model.

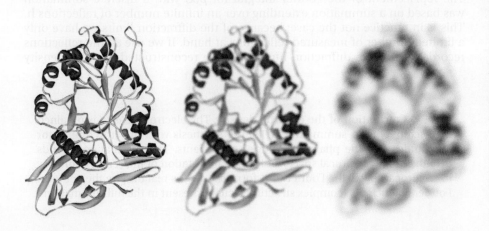

expression into the electron density summation (9-12) we obtain for the electron density at grid point $\mathbf{x}$

$$\rho(\mathbf{x}) = \frac{1}{V} \sum_{\mathbf{h}=-\infty}^{+\infty} F_{\mathbf{h}} \cdot \exp(i\alpha_{\mathbf{h}}) \cdot \exp(-2\pi i(\mathbf{hx})) \tag{9-18}$$

In this final form of the density summation, the Fourier coefficients $F_{\mathbf{h}} \cdot \exp(i\alpha_{\mathbf{h}})$ now explicitly include both the structure factor amplitude $F_{\mathbf{h}}$ and its associated phase, $\alpha_{\mathbf{h}}$. Expression 9-18 is in principle directly useful for practical calculation of electron density.

An interesting point the observant reader may have picked up is why, despite its Fourier coefficients—the structure factors—being *complex*, the Fourier summation for the electron density yields a real physical observable, namely the actual electron density at a point x, y, z in real space. This is a direct and general consequence of the centrosymmetry of reciprocal space, as we have shown in Chapter 5, and not of direct space as erroneously stated.[6] Given that the reciprocal space is centrosymmetric, and the FT of a symmetric function is real, $\rho(\mathbf{x})$ must be real. This is evident from the basic Argand diagram, where we see that for each structure factor $\mathbf{F}_{\mathbf{h}} = A + iB$ the centrosymmetrically related conjugate structure factor is $\mathbf{F}_{-\mathbf{h}} = A - iB$. Thus, if we sum the contributions from one wedge of reciprocal space with those from the opposite centrosymmetric wedge, the imaginary B-terms containing the sine terms of the phase angle cancel out and the summation over the remaining cosine terms is real. The Fourier summation can then be written as two times the summation over half of the reciprocal space, that is, twice the summation over the cosine terms. The $F(000)$ term, which is the sum of all electrons in the unit cell, then acts just as an additive constant and can be omitted from the summation for the purposes of electron density calculation:

$$\rho(\mathbf{x}) = \frac{2}{V} \sum_{h=0}^{+\infty} \sum_{k=-\infty}^{+\infty} \sum_{l=-\infty}^{+\infty} F_{\mathbf{h}} \cos(\varphi_{\mathbf{x}} - \alpha_{\mathbf{h}}) \tag{9-19}$$

The blue highlighted terms again indicate the experimentally supplied Fourier coefficients as separate structure factor amplitudes and the corresponding experimental or model phase angles. Make sure to clearly distinguish between $\varphi_{\mathbf{x}}$ and $\alpha_{\mathbf{h}}$: The FT phase term $\varphi_{\mathbf{x}} = 2\pi\mathbf{hx}$ here contains the fractional coordinate vector $\mathbf{x}$ of each *grid point* in the density reconstruction $\rho(\mathbf{x}) = \rho(xyz)$ in real space, while $\alpha_{\mathbf{h}}$ is the phase angle of each reflection $\mathbf{h}$ and derived from the fractional *atom coordinate* vectors $\mathbf{x}_j$ for each atom j. The explicit proof for (9-19) is provided in Appendix A.6 as an exercise in practical computing with complex numbers and trigonometric functions.

Given this practically useful formula, we can explore the properties of the electron density reconstruction by generating some electron density maps.

Fourier transforms, truncations, and grids

The replacement of the Fourier integral for $\rho(\mathbf{x})$ with a discrete summation was based on a summation extending over an infinite number of reflections $\mathbf{h}$. This is in practice not the case; because of the diffraction limits we have only a limited number of measured reflections at hand. If we use all the reflections recorded up to the diffraction limit, then the reconstructed electron density

---

Box 9-3 Calculation of the electron density. The electron density is obtained by discrete Fourier summation or Fourier synthesis with the structure factor amplitudes and the phases as Fourier coefficients. The summation extends over the entire reciprocal space, and as a manifestation of the centrosymmetry in reciprocal space, the Fourier transform $\rho(x, y, z)$ is real despite the complex Fourier coefficients (complex structure factors) present in the full summation.

Sidebar 9-2 **Calculation of electron density from anomalous data.** The observant reader may have recognized that our argument that the electron density summation is real was based on the observation that—given the centrosymmetric relation between $\mathbf{F_h}$ and $\mathbf{F_{-h}}$—the imaginary sine terms cancel out (i.e. Friedel's law applies). In the presence of anomalous scattering contributions, this is obviously not the case for general acentric reflections, where the Friedel or Bijvoet pairs, and F^+ and F^-, have different amplitudes. In practice in macromolecular refinement the anomalous differences are generally small enough so that one ignores the theoretical argument and simply merges the anomalous pairs using their weighted averages as amplitudes when computing the electron density. However, when the anomalous signal becomes large enough, the maps need to be constructed from (and structures refined against) a more accurately computed amplitude average. The problem was addressed by A. L. Patterson[7] in 1963 and we give the result for the simple case of one dispersive atom. Let $\mathbf{F_\pm}$ be the complex structure factor of a reflection $\mathbf{h}$ and $-\mathbf{h}$, respectively, with A and B the non-anomalous contributions and A_d, and B_d the anomalous (dispersive) contributions. Then

$$F_\pm^2 = A^2 + B^2 + (\delta_1^2 + \delta_2^2)(A_d^2 + B_d^2) + 2\delta_1(AA_d + BB_d) \pm 2\delta_2(AB_d + BA_d)$$

with

$$\delta_1 = f_d' / f_d \text{ and } \delta_2 = f_d'' / f_d$$

where f_d is the non-anomalous part of the scattering contributions of the dispersive atom and f_d' and f_d'' are its anomalous scattering factors (Chapter 6). One could now either use separate contributions to compute the maps or refine against, or compute, the proper average to obtain the "experimental" value $F(obs)$ and use it in the normal way. With

$$S(obs) = [F_+^2(obs) + F_-^2(obs)] / 2$$

we obtain the final result for the corrected "true" $F(obs)$:

$$F(obs) = \left(S(obs) - 2\delta_1(AA_d + BB_d) - (\delta_1^2 + \delta_2^2)(A_d^2 + B_d^2)\right)^{1/2}$$
(9-20)

The formula above can be extended to more than one type of dispersive atom.[7] Some refinement programs such as *REFMAC* and *SHELXL* allow refinement against anomalous data.

(given a proper sampling interval) will provide no more and no less than the information sampled and contained in the diffraction pattern. The available information about our structure is thus limited only by the extent of diffraction, which is affected by individual B-factors and/or lattice disorder. Because the B-factor is a measure for the mean displacement of an atom from its equilibrium position, one would expect a somehow "smeared out" or broadened peak of electron density at the equilibrium position compared with an atom at rest. We shall verify this with a few examples of Fourier transforms.

We can use Equation 9-19 to compute some actual Fourier transforms of a simple 1-dimensional structure containing two atoms, one carbon atom at fractional coordinate $x = 0.1$ and one oxygen atom at $x = 0.7$. From the calculated complex structure factors, we compute the Fourier summation as a function of available data (that is, of resolution). The FT generates the familiar electron density peaks at the position of the atoms, with relative intensities proportional to the electron count in the scattering atoms. With a lower resolution limit (less data) as a consequence of higher B-factors, fewer structure factor terms are available for summation in the density reconstruction, and the electron density peaks broaden significantly (Figure 9-4).

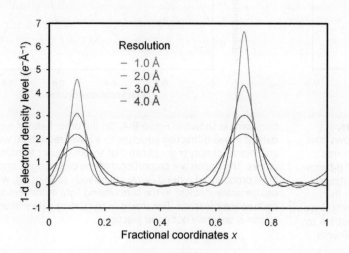

Figure 9-4 Reconstruction of 1-dimensional electron density via Fourier transformation. The graphs are 1-dimensional Fourier transforms with 1-dimensional complex structure factors computed to different limiting resolutions of 1, 2, 3, and 4 Å providing the Fourier coefficients. In the 1-dimensional structure, one carbon atom ($z = 6$) is located at fractional coordinate $x = 0.1$ and one oxygen atom ($z = 8$) at $x = 0.7$. The peak positions correctly reflect the positions and the relative peak heights correspond to the scattering power of the atoms. With decreasing resolution, fewer terms are available for Fourier summation of the density reconstruction and the peaks broaden, but their positions remain accurate, although less precise.

Data truncation. The situation is somewhat different if we have more data available than we actually use for the Fourier reconstruction. In this case, we truncate information at a point where significant signal is still present and the transform will be incomplete. The consequence is the occurrence of truncation ripples, which are most distinct around heavy atoms where they are, in fact, occasionally observed (see Figure 10-10). In mathematical terms, the truncation in reciprocal space is equivalent to multiplication of the data with a Heaviside unit-step function (which is 1 up to the selected resolution cutoff, and 0 beyond). According to the Fourier convolution theorem, the real space equivalent is then a convolution of the true electron density with the Fourier transform of the Heaviside function which is of the form $\sin(x)/x$. The effect is a broadening of the reconstructed electron density and the appearance of ripples (whose distance from the peak center depends on the resolution where the truncation took place). Figure 9-5 shows the relation between scattering factors, B-factor, and the consequence of resolution truncation in electron density reconstruction.

Grid spacing. In Figure 9-4 we also realized that a Fourier reconstruction grid of 100 points over the unit cell, which was appropriate and necessary to obtain a smooth high resolution data transform, becomes quite excessive for the 4 Å data. Computational effort can be reduced without information loss by selecting an appropriate Fourier grid spacing. From theoretical considerations, in a bandwidth-limited situation (which diffraction with a limited resolution represents) the Nyquist theorem indicates that the sampling interval (grid spacing) needs to be at least $d_{min}/2$ to preserve the information in the transform; anything finer than that, in principle, just increases the visual esthetics of the map (which can be helpful for manual model building or illustration). As a rule of thumb, the grid spacing in Å is selected at least two to about four times finer than the maximum resolution, that is, for a 2 Å structure we could use a 0.5 Å grid and obtain a (visually at least) good looking map.

Effect of atom occupancy and B-factor on electron density

Another interesting question we can answer by inspecting our simple Fourier transforms is what effect high B-factors have compared with partial occupancy of an atom. Both effects obviously will lead to an overall signal attenuation. We

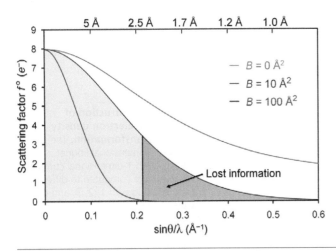

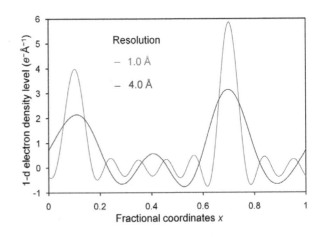

Figure 9-5 Relation between atomic scattering factors, B-factors, and resolution truncation. The left panel shows how atomic scattering factors become attenuated by B-factor terms. The graphs indicate clearly that a structure with an overall B-factor of 100 will—even in the best noise-free case, solely by limitation of the atomic scattering power—never diffract any further than 2.5 Å, beyond which there is just no more scattered intensity. If we use at $B = 10$ Å² all the data to the resolution limit of about 1 Å to reconstruct the electron density, we will obtain a smooth Fourier transform as shown in Figure 9-4. On the other hand, if we truncate data of a well-diffracting structure to a lower resolution, we obtain an electron density (right panel) that shows distinct truncation ripples. The ripples are proportional to the peak height, and thus most pronounced around heavy atoms such as Se or Hg, where they can generate "holes" in the surrounding light atom density and obscure their features. Discarding or truncating data without good reason is generally not good practice.

have already mentioned the correlation of B-factor B_j and occupancy n_j and the resulting difficulty in discerning them by electron density appearance alone.

The example in Figure 9-6 depicts the reconstructed electron density (from 1.8 Å structure factors) of a structure containing an oxygen atom (perhaps representing an ordered water molecule) and a discrete Mg ion, also potentially present in ordered solvent. In one case, both temperature factors were set to 25 Å², but occupation of the Mg was reduced to 0.6, and in the second case, the Mg B-factor was increased to 70 Å² while keeping its occupation full at 1.0. The following conclusions result from inspection of Figure 9-6: without additional chemical or anomalous scattering information, it is practically impossible to tell the low occupancy heavy ion (Mg) from a normally occupied light atom (O, or a water molecule, for example). In a real case scenario, bond distances and coordination geometry of the atom in question may help to distinguish a partially occupied solvent ion from a water molecule. Mg²⁺ would likely show octahedral coordination and shorter (2.1 Å) distances to coordinated water atoms (Appendix A.11).

Comparison of the two electron density reconstructions in Figure 9-6 also demonstrates that the distinction between an under-occupied atom with standard B-factor and the same fully occupied atom type with high B-factor is difficult. Although in the calculated Fourier transforms the slightly different peak shape can be recognized (broader at high B-factor, as we have shown), in practice with noisy protein electron density maps this distinction is generally not straightforward, and again prior knowledge about the chemical environment may give additional clues.

3-dimensional electron density maps

So far we have used a simple 1-dimensional representation of the Fourier transform of a 1-dimensional structure in a linear graph, where the value of the electron density was plotted as the y-value at every point x along the crystal. In analogy, a 3-dimensional electron density summation computes the value of the electron density at each grid point xyz in a 3-dimensional unit cell box. Each grid point is then assigned a certain bin level, and points of the same level are

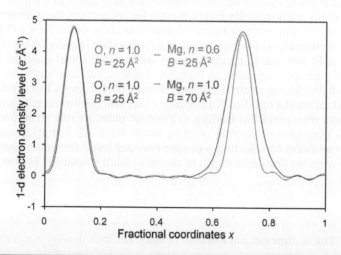

Figure 9-6 Effect of partial occupancy versus high B-factor on electron density. Comparison of the two electron density reconstructions (blue, red) of a structure containing an oxygen atom (z = 8) and a magnesium atom (z = 12) shows that from electron density alone: i) a partially occupied heavier atom cannot in practice be distinguished from a lighter atom with the same B-factor; ii) distinction between an atom with low occupancy and "normal" B-factor from a fully occupied atom with high B-factor is, in practice and for protein maps, not easily possible, but differences in peak shape (as a result of different shape in fall-off of the respective scattering function) do exist. Drawing the respective atomic scattering factor curves for the depicted scenario as an illustrative exercise is recommended.

Sidebar 9-3 Fourier transforms of special functions important in crystallography. A number of Fourier transforms of special functions play an important role in crystallography. The first FT we have already encountered, the FT of the electron density:

- The FT of a *Gaussian function*. We approximated the electron density with a Gaussian distribution, and the transform in the form of the scattering factor was again a Gaussian, but with inverse variance: a broad electron density peak describing an atom with high B-factor will have a narrow scattering function, that is, the scattering will fall off rapidly with diffraction angle.

- The FT of a *constant function* is zero everywhere except at the origin. This FT is of interest for density modification and bulk solvent correction, where a constant value added to a map does not affect the structure factors $F(\mathbf{h})$ except $F(000)$.

- The FT of a *delta-function* is again a delta-function, but with inverse spacing; the transform of the lattice function $L(\mathbf{r})$ is the reciprocal lattice, $RL(\mathbf{r}^*)$. If the real lattice or its sampling is finite, the reciprocal transform will have peaks with a width reciprocal to the sampling density at the lattice (see Figure 9-4).

- The FT of a *step function* (or Heaviside function) is of the form $\sin(x)/(x)$, which leads to truncation ripples when data are truncated at resolutions lower than the diffraction limit.

- Finally, the FT of a sphere or *spherical function* is the reciprocal space G-function, also known as the reciprocal space interference function, which plays an important role in estimating the range of influence that modified structure factors have on each other. This will be relevant in density modification and molecular replacement, where the extent of the molecule is to a first approximation considered a sphere.

Box 9-4 **Effect of data resolution, truncation, and *B*-factors on electron density.** The electron density reconstruction or Fourier synthesis generates electron density peaks at the position of the atoms, with the peak height proportional to the number of electrons in the case of resting atoms. When all available data are used, the electron density peaks will broaden with decreasing resolution, but the transform remains smooth. When data are truncated, truncation ripples appear, particularly around heavy atoms, that can obscure neighboring atoms if the truncation is severe. Increasing *B*-factor also broadens the electron density peaks, corresponding to convolution of the peak with a Gaussian point spread function. Distinguishing between a partially occupied atom and an atom with high *B*-factor by peak shape alone is generally not possible in protein electron density maps.

connected with contour lines. Two-dimensional projections of maps are shown later in Section 9.4, and the 3-dimensional grid or mesh representing surfaces of equal electron density we have repeatedly seen in our electron density images. The electron density needs to be contoured at a level that suits the intended purpose of showing the features of interest as clearly as possible. Maps will generally have a wide range of electron density levels between a minimum and maximum level.

In practice, maps are contoured at levels expressed in number of standard deviations of the mean electron density, defined as the square root of the density variance

$$\sigma(\rho_r) = \left(\frac{1}{g} \sum_{i=1}^{g} (\rho(\mathbf{r}_i) - \langle\rho\rangle)^2 \right)^{1/2}$$

(9-21)

where the summation extends over all grid points g. A 5 sigma (5σ) level thus means that the features shown have a density level 5σ above the mean or average electron density level, and this high contour level will reveal only strong features. A lower level, say 1σ, will show all features above 1σ, which might already include some noise even in an otherwise useful map. Note that sigma levels are relative and for each map the levels chosen for inspection must be suitable for the purpose. Searching for dominant features such as metal ions will be easier at higher electron density levels generating cleaner but less detailed maps, while modeling split side chain conformations will require lower contour levels, at the risk of perhaps beginning to interpret noise at too low levels. In practice, levels of ~0.8–1.0σ are pretty much the lowest useful levels in non-difference maps used for model building. In Chapter 13 we will encounter cautionary tales of risky map interpretation leading to poor, or plain wrong, structures. Figure 9-7 shows examples of varying contour levels for a 1.5 Å map of quite good quality. In addition to selecting a proper contour level, the zoom level (size of displayed map section) and depth of the map (slab thickness) is also adjusted

Box 9-5 **Three-dimensional electron density.** Electron density $\rho(x, y, z)$ is represented in 3-dimensional space as a mesh or grid of density values. Points of equal density level are connected by lines forming 3-dimensional contours or a 3-dimensional electron density map. The maps are contoured on a relative scale in levels of standard deviations σ of the mean density. It is important to select contour levels and depth (slab thickness) appropriate to the task when displaying an electron density map. Very low contour levels reveal too much noise, and very high contours lack detail, but can be useful to visually detect strong peaks.

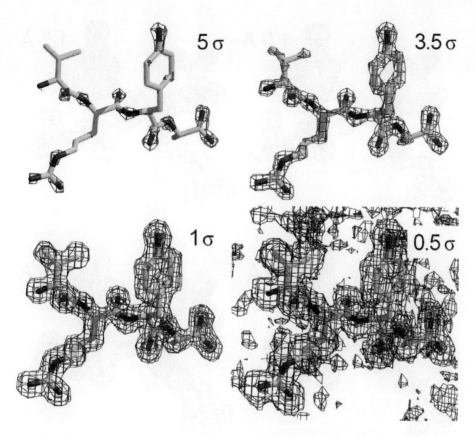

Figure 9-7 Selecting map contour levels. At very high 5σ map contour levels (top left), only the highest electron density peaks are visible, in this case mostly around the oxygen atoms and some nitrogen atoms. This high contour level might still be useful for looking for heavy solvent ions bound to the protein. In the 3.5σ level, the contours of all atoms become visible except of those with a higher *B*-factor (hydrophobic Val), while 1σ is a useful level during detailed model building to produce a map of decent appearance. The final panel (right bottom) demonstrates that at the 0.5σ level, noise starts to obscure the electron density of the peptide, and the selected level is too low to be useful for model building or figure preparation.

in the display programs, depending on the purpose of the map inspection. For a detailed view of single residues, 5–7 Å is a practical slab range; to trace molecule contours, 10–15 Å or larger depth is often suitable.

Resolution and information content of electron density maps

Similar to the 1-dimensional map examples shown at different resolution above, we can compute 3-dimensional electron density reconstructions from diffraction data extending to varying resolution. Recall that the resolution is commonly defined as the maximum interplanar *d*-spacing to which X-ray reflections were observed (see Sidebar 9-4 for a more advanced interpretation). As we need structure factor amplitudes and phases for the reconstruction, for this exercise we employ computed structure factors, $\mathbf{F}_{calc}$ or $\mathbf{F}_c$. The resulting electron densities shown in Figure 9-8 thus represent best case situations without any experimental error. They are nevertheless very useful to train the eye and to develop a feeling for what level of detail can be expected in a structure given the resolution of the data. The example we use is a peptide of the sequence Val-Arg-Tyr-Ala.

Resolution and information level. In agreement with the argument made in Chapter 6 that tighter sampling of the crystal lattice by diffraction of X-rays from increasingly more tightly-spaced sets of lattice planes provides more information, the detail we can discern at 1 Å resolution is quite impressive, and discrete atoms become distinctly visible. Therefore, we place the onset of atomic resolution in the crystallographic realm around 1.2 Å. While we can still confidently build an atomic model of a structure at lower resolution than 1.2 Å, we no longer see distinctly separated atoms. In the range of 1.2–2.0 Å we can still clearly discern residue types and structures in this resolution range are commonly termed high resolution structures. While good data at 2.5 Å still can give good structures, the detail is diminished, and at progressively worse resolution in the 3–4 Å range the previously distinct electron density turns into sausage-like tubes. In the 4 Å example it is not possible to discern any residue types without sequence

Figure 9-8 Electron density reconstruction at decreasing resolution. The electron density shape progressively changes from distinct atomic spheres discernable at 1.0 Å to a sausage- or tube-like electron density without distinguishable side chain definition at 4.0 Å. The electron densities are reconstructed from error-free, *B*-factor attenuated $\mathbf{F}_{calc}$ data and thus represent noise-free, best case scenarios.

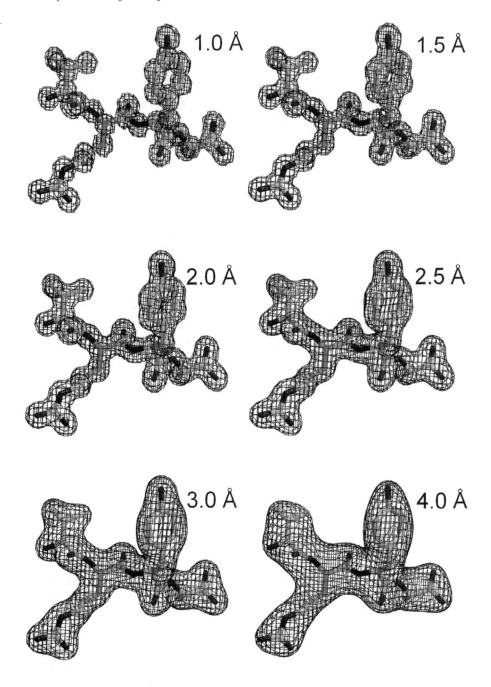

information, and in this low resolution range we depend very heavily on prior knowledge of sequence and molecular stereochemistry. At resolutions lower than 2.5 Å, an unambiguous assignment of bound ligands or drug molecules from electron denisty alone is seldom possible.

The practical consequences of the primary dependence of information content on resolution are clear: If we are interested in atomic details of an enzymatic mechanism or aim for structure based drug design or lead optimization, a 2.5 Å structure will very likely not provide enough detail, while for a huge virus capsid structure 4.0 Å resolution may present a major achievement. However, the nominal resolution alone does not provide any information about the actual structure quality, which depends on many more factors, including overall data quality, completeness of the data, phase accuracy, and model building and refinement practice. Above all, structure quality is a *local property*, which will be discussed in great detail in Chapters 12 and 13.

Sidebar 9-4 Nominal resolution and peak separation in electron density. The *resolution* of a structure quoted in crystallographic tables is generally the *d*-spacing corresponding to the 2θ diffraction angle limit of the highest resolution shell, d_{min}. It is commonly assumed that d_{min} represents the limit of resolution (LoR), the smallest distance or detail that can be resolved in the electron density map. This is, however, only a conservative approximation.

For optical lenses and visible light, the Rayleigh criterion leads to a LoR of $0.61 \cdot \lambda$, which by simple analogy for a 2-dimensional X-ray case translates to $0.61 \cdot d_{min}$. R. W. James estimated in 1948 in the inaugural volume of *Acta Crystallographica*[8] that in the 3-dimensional X-ray case the LoR becomes $0.715 \cdot d_{min}$. Exact analysis by R. E. Stenkamp and L. H. Jensen[9] showed that for point atoms and amplitude-based reconstruction (which is the fact for Fourier reconstruction in the X-ray case, in contrast to intensity-based optical functions) the LoR becomes $0.917 \cdot d_{min}$. This limit works reasonable well even for real atoms with *B*-factors up to 10 Å². You can convince yourself that this is true by using the small FT calculator on http://www.

ruppweb.org/. Calculate structure factors with following settings: Cell 20 Å, two fully occupied oxygen atoms with $B = 10$ Å², and resolution $d_{min} = 2$ Å. Vary the distance between the atoms from $\Delta x = 1.0$ (2 Å) to 0.9 (1.8 Å), 0.8 (1.6 Å), and finally 0.7 (1.4 Å), and compute the Fourier transform for each structure (Figure 9-9).

As expected, the result for the 1-dimensional X-ray case is only slightly more optimistic than in the 3-dimensional case: while $0.9 \cdot d_{min}$ still shows recognizable peak separation, the $0.8 \cdot d_{min}$ case is marginal and at $0.7 \cdot d_{min}$ there is definitely no more peak separation. You can repeat the experiment with higher *B*-factors and different atoms.

In addition to the atomic *B*-factors, the limit of resolution depends in practice on the fractional completeness of data *C*, and for (randomly) incomplete data the effective resolution[10] becomes $d_{eff} = d_{min} \cdot C^{-1/3}$. An even more elaborate criterion for the optical resolution of an electron density map based on the variance of the Patterson origin peak and a spherical interference function is reported by the data validation program *SFCHECK*.[11]

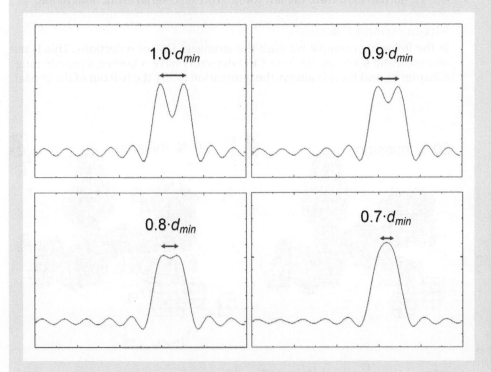

Figure 9-9 Electron density peak resolution. Given data to 2.0 Å resolution, the electron density peaks of two atoms with increasingly close spacing can be recognizably separated at distances slightly smaller than the nominal resolution of the data used in the reconstruction. The $0.9 \cdot d_{min}$ reconstruction still shows recognizable peak separation, the $0.8 \cdot d_{min}$ case is marginal, and at $0.7 \cdot d_{min}$ peak separation can no longer be observed.

9.2   Effects of incomplete data on electron density reconstruction

So far we have discussed only the effects of truncation, which can be easily avoided in density reconstruction by using all available structure factors up to the resolution limit. However, on occasion data are not collected completely for a variety of reasons, ranging from instrument limitations to incorrect (too low) Laue symmetry. Lattice imperfections also contribute to limited resolution, and diffraction data can be highly anisotropic.

Randomly missing reflections

As a general rule, randomly missing data do not create much of a problem. In this category fall thin shells of data that need to be removed because of ice rings, or little pieces of reciprocal space missing at the detector boundaries at high resolution (see for example the data in Figure 11-8), or data removed at random for cross-validation (discussed in Chapter 12). Figure 9-10 demonstrates that in fact quite a large percentage of data can be randomly omitted, and despite increasing noise, the remaining terms still recover the electron density. This should bring peace to the mind of those who fear that setting aside a small percentage of data for cross-validation will give a noticeably worse structure refinement than one against all data.

Omitting a high percentage of data randomly is not something that happens in real data collection. The situation does become more critical when data are systematically missing, such as complete regions or wedges of reciprocal space that are absent because of poor data collection strategy, overloads, radiation damage, premature crystal death, or wrong Laue group (again a reminder: collect your data in lower Laue symmetry if there are any doubts about the true symmetry). Another example we have already discussed is the case when a rotation axis is nearly exactly aligned with a unique axis (Figure 6-22). In this case, none of the reflections along the corresponding reciprocal axis are collected, which may preclude the determination of a space group operator translation (screw axis vs. normal axis) from the few (or no) recorded serial extinctions alone.

Missing strong reflections

In the following example we omit the strongest 10% of reflections. This is not an unrealistic scenario, because CCD detectors have a limited dynamic range (Chapter 8), and there is always the temptation to "fry the hell out of the crystal"

Figure 9-10 Effect of randomly omitted data. The panels show the effect of an increasing amount of randomly missing data. In the top left panel with 20% of the data randomly deleted, there is barely a difference noticeable compared with the maps generated from complete data shown in Figure 9-8. Even when the reconstruction misses 50% and 80% of data, the molecule is still traceable despite the increase in noise. About 800 out of 4000 reflections are all that is left in the reconstruction of the bottom electron density. The density in the bottom right panel is recontoured 80% missing, but at a higher σ-level, and the molecule is still traceable. Comparing the bottom left and bottom right panels emphasizes the importance of selecting a suitable density level for model density visualization and model building.

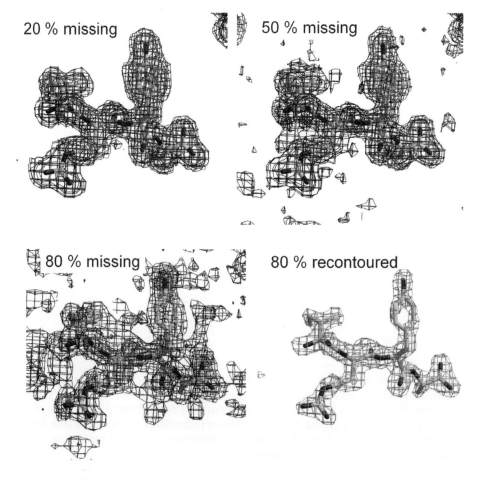

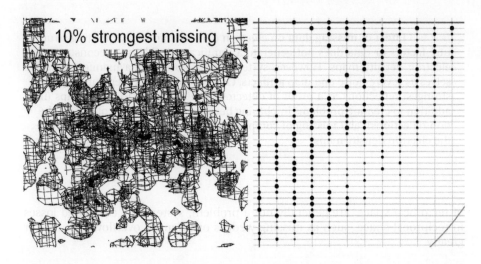

Figure 9-11 Effect of intensity overflows. In contrast to randomly missing reflections, eliminating only the strongest 10% of reflections resulting from detector overflows leads to significant noise levels in the map. The (*h*, *k*, *0*) section of the reciprocal lattice shows that the 90° rotation scan covering the positive reciprocal space octant for Laue group 222 is complete, but holes in the data exist because of deleted high-intensity overflows, particularly at low resolution.

in a misguided attempt to aim for the highest obtainable resolution. This is fatal: as we can see from Figure 9-11, the map with the strongest reflections missing as a result of detector saturation becomes really noisy and poorly interpretable. In addition, experimental phasing methods, as well as molecular replacement, depend heavily on the most intense (generally low resolution) reflections. The remedy is to collect two scans of the data, one with short exposure times to prevent overloads, and a second scan with longer exposure times to obtain a good signal-to-noise ratio up to the diffraction limit of the crystal. The two data sets can then be merged and scaled together into a complete data set with a large dynamic range.

Missing data wedges

Equally unforgiving is failing to cover the entire rotation range necessary to obtain complete unique data during data collection. This can be the result of either poor strategy, incorrect space group/indexing, or premature crystal death because of progressive radiation damage. The effect is again of concern. Scanning less than 70–80% of the complete reciprocal unit cell creates disturbing streaks perpendicular to the systematically missing reciprocal zones. This is particularly troublesome for structures containing β-sheets, where the density begins to connect across the β-strands, which makes it painful, or impossible, to properly trace the protein chain (Figure 9-12).

The conclusion of our observations is clear: completeness of data is important. While randomly missing parts, such as ice rings or one or the other limited part

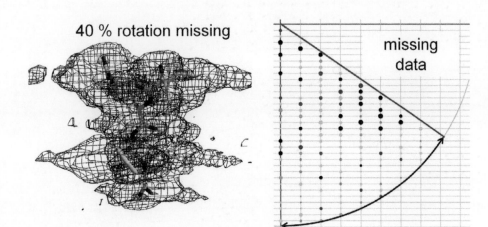

Figure 9-12 Effect of incomplete data collection range. Collecting incomplete data that miss entire rotation ranges leads to streaking of the map, which is particularly distracting when building sheet structures. Note that even the high density contours (red) are not contiguous anymore and are also streaked. The right panel shows the incomplete data scan around the perpendicular reciprocal axis *l*.

Box 9-6 Effect of missing data on electron density. Randomly omitting structure factors or data does not generate a significant problem in reconstructed electron density maps. However, systematically absent data manifest themselves as severe noise and map artifacts. Omission of a small percentage of the strongest reflections as a result of detector overflows leads to noisy maps (in addition to creating problems in phasing because of the absence of low resolution reflections), and missing of complete wedges of data leads to streaking in the electron density maps.

of reciprocal space, are generally not a problem, systematic, correlated losses in reciprocal space, such as missing wedges or intensity overflows, do matter greatly.

Manifestation of lattice disorder and anisotropy

According to the Fourier convolution theorem, we can also apply the reverse FT formalism to compute the reciprocal lattice function of a periodic crystal lattice. This allows us to investigate the effect of B-factors or, more generally, any effects of displacive disorder in the crystal upon the diffraction pattern itself.

Lattice disorder. To model disorder, we generate a model of a periodic lattice with successively increasing random displacement from the perfect lattice positions and examine the effect on its Fourier transform. Figure 9-13 shows the lattice function and its Fourier transform in the form of a diffraction pattern. To make the pattern look more realistic, the lattice points have some dimension, and the Fourier transformation then leads to peaks that also have a corresponding spread. The perhaps counterintuitive result is that while resolution decreases, the diffraction peaks—in contrast to the electron density peaks—*do not broaden* with increasing disorder or B-factors. We also see that despite considerable disorder in the lattice depicted in panel C, the high correlation that remains between the lattice points still generates some low resolution diffraction, and the diffraction spots remain as sharp as they were before. However, diffraction is completely lost when the lattice points are truly randomly distributed and have no remaining translational (periodic) correlation between them.

Figure 9-13 The effect of lattice disorder on diffraction. Increasing random displacement (disorder) in the lattice points or molecules (left in each panel) manifests itself in the Fourier transforms (diffraction patterns) in the right image in each panel as a decrease in the extent to which distinct diffraction occurs (panels A–C), equivalent to decreasing resolution. Despite increasing disorder, the diffraction spots do not broaden. Complete random lattice points generate only a noise pattern (panel D). Note that despite displaying only noise, the diffraction pattern still remains centrosymmetric.

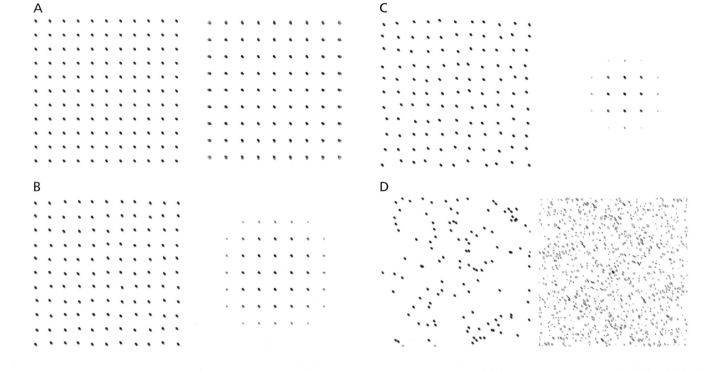

> **Box 9-7 Effect of lattice disorder on diffraction pattern.** The Fourier transform of the lattice function can be used to investigate the effects of disorder on the diffraction pattern itself. While reduced periodic order leads to decreased resolution, the peak width itself does not increase with increasing disorder or B-factor. Anisotropic disorder results in anisotropic resolution limits of the diffraction pattern.

The positional displacement exerts the same attenuating effect on the diffraction pattern as increased atomic B-factors, and crystal disorder is also accounted for in the overall B-factor. As discussed in Chapter 7, the determination of the overall B-factor is in practice the first step in scaling experimental data to calculated structure factors.

Diffraction anisotropy. As the B-factor is a general measure of scattering attenuation, it absorbs all types of displacements, be they real thermal motion, disorder, or any other displacement effects. An anisotropic distribution of these properties will necessarily lead to anisotropic attenuation. Disorder in a protein crystal is frequently anisotropic because adequate intermolecular interactions may exist in two dimensions or in layers (often the case in membrane protein structures), while contacts in the third dimension may be weak. The scenario of anisotropic lattice disorder and its effect on the diffraction pattern can again be visualized using our lattice-function Fourier transforms. Figure 9-14 shows such a scenario and its effect on the diffraction pattern. The attenuation is increased and the diffraction pattern is weakened in the direction of the lattice disturbance. Anisotropic scaling methods discussed in Chapters 10 and 12 can account for diffraction anisotropy.

9.3 The phase problem and phases in electron reconstruction

So far, the Fourier back-transformation used to reconstruct the electron density from complex structure factors has seemed like a straightforward procedure, with the exception perhaps that the direct summation becomes quite computationally intense for many reflections and large cells. However, the serious problem that we have already introduced is that the phase angle of the complex structure factor is completely missing in our intensity (amplitude) measurements. This is the source of the crystallographic phase problem (Figure 9-15). Because of the loss of information in the physical process of measuring the

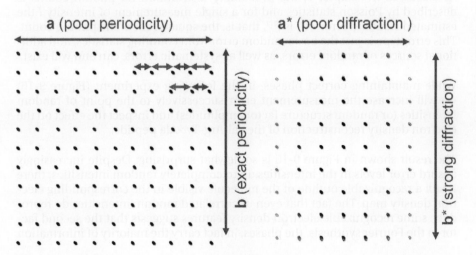

a (poor periodicity)   a* (poor diffraction)

b (exact periodicity)   b* (strong diffraction)

Figure 9-14 Effect of anisotropic disorder on diffraction pattern. The diffraction pattern (right panel) is weakened along **a*** as a result of poor translational periodicity (small random displacement) in **a**, while the exact line spacing in direction **b** is unaffected and diffraction remains strong.

Figure 9-15 The crystallographic phase problem. The measurable component of the Fourier transform of the crystal is only the scalar structure factor amplitude $F(\mathbf{h})$ proportional to the square root of $I(\mathbf{h})$. The missing phases $\varphi(\mathbf{h})$ must be supplied by additional phasing experiments or in the form of model phases via molecular replacement. The two necessary Fourier coefficients in the back-transform formula are emphasized in blue.

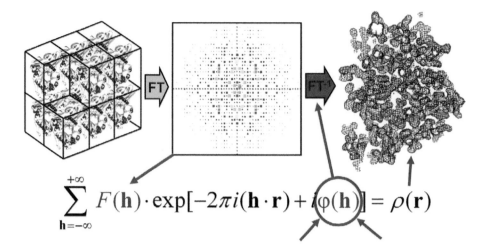

$$\sum_{\mathbf{h}=-\infty}^{+\infty} F(\mathbf{h}) \cdot \exp[-2\pi i(\mathbf{h}\cdot\mathbf{r}) + i\varphi(\mathbf{h})] = \rho(\mathbf{r})$$

intensities, no general formal solution of this predicament exists. However, we can still perform the Fourier back-transformation, if we can supply the phases for each reflection—or at least as many as are necessary to obtain an interpretable electron density map for initial model building—from some other source, generally through additional phasing experiments or by starting phases from a correctly positioned structurally similar model via molecular replacement. The theory and techniques to obtain the phases will be subject of subsequent chapters; for now we just assume that we have the phases $\varphi(\mathbf{h})$ for each measured reflection $\mathbf{h}$ available.

Before we discuss the principal methods at our disposal to generate the missing phases, it is very instructive to investigate the effect that the amount and the quality of the data, as well as phases, have in practice on the electron density reconstruction. In addition, we will find quite striking and consequential differences concerning the influence of the Fourier amplitudes $F(\mathbf{h})$ and of their associated phase $\varphi(\mathbf{h})$ on the electron density reconstruction.

Effect of intensity errors on electron density reconstruction

In crystallographic practice, electron density is reconstructed from structure factor amplitudes derived from measured intensities and experimentally determined phases. Both will not be perfectly accurate and will carry errors. The electron density examples shown previously in Figure 9-8 represented an ideal, best-case scenario. In reality, intensity and phase errors, as well as incomplete data, will lead to noise and other, sometimes systematic, errors in maps. Let us investigate first the effect of random errors during the intensity measurement. As discussed in Chapter 7, the probability distribution for photon counts is described by Poisson statistics and for a single measurement of intensity I the estimated standard error $\sigma(I)$ is $I^{1/2}$, that is, the square root of the photon count. This error type is just the basic random error from counting statistics, and additional sources of random errors, as well as systematic errors, can and will exist.

While maintaining correct phases, in the following experiment (Figure 9-16) we will increase the measurement error successively to the point of random intensities (or random structure factor amplitudes) and inspect the effect on the electron density reconstruction of the Val-Arg-Tyr-Ala peptide.

The result shown in Figure 9-16 is somewhat surprising: Despite increasingly absurd error levels in the intensities up to completely random intensities, there is still a recognizable outline of the molecule visible in the corresponding electron density map. The fact that even uncorrelated random intensities do reproduce some recognizable electron density features suggests that the second factor in the Fourier synthesis, the phases, in fact carry the majority of information

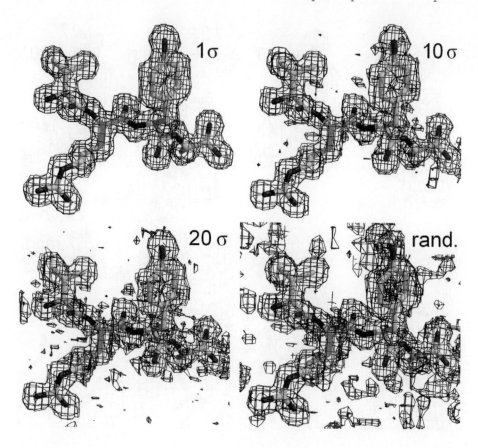

Figure 9-16 Effect of intensity measurement noise on electron density reconstruction. The electron density is calculated for data with increasing random measurement (amplitude) error while keeping correct phases. With increasing error, the maps become progressively noisy, but even at absurd error levels and even with random intensities (obeying only the Wilson distribution), the outline of the molecule remains surprisingly distinct. The mean measurement error is given in number of standard deviations $\sigma(I)$.

needed for electron density reconstruction. This is quite a shocking discovery, as we know that the phase information is actually missing from diffraction data and must be supplied externally through additional phasing experiments.

Effect of phase errors on electron density reconstruction

To emphasize the importance of correct phases, we repeat the above electron density reconstruction experiment, now keeping the intensities (or structure factor amplitudes) correct while introducing increasing phase error (Figure 9-17).

The results of the phase perturbation illustrated in Figure 9-17 confirm our suspicion that the electron density reconstruction is much more sensitive to mean phase error than to errors in measured intensities. Beyond about 70° phase error, even with perfect intensity data the map becomes very noisy and is barely interpretable anymore, and random phases indeed generate a random map. From the complex vector diagram, the much more severe effect of poor phases can be readily understood: A complex structure factor **F** with wrong intensity but correct phase still points in the correct direction, but with wrong phases, the complex "vector" **F** can point in all imaginable directions in its magnitude circle. Sidebar 9-5 provides additional insight regarding the dominance of phases on the Fourier reconstruction of electron density.

The fact that phases determine the density reconstruction much more than intensities means double trouble. First, we do not have measured phases to begin with, and thus need to obtain high quality phases from phasing experiments, and second, if we have wrong or very poor phases, accurate intensity data collection alone will not overcome the phase errors. The inability of correct intensities to overcome poor phases gives rise to the phenomenon of phase or model bias. Phase bias[12] is a clear and present danger in molecular replacement phasing[13] and we will spend considerable time in the molecular replacement and refinement chapters exploring methods to minimize phase bias.

Figure 9-17 Influence of phase error on electron density map reconstruction. Even in the case of perfectly correct intensities or structure factor amplitudes, past about ~70° of mean phase error, the maps become uninterpretable. Random phases finally produce a random map, in contrast to random amplitudes (Figure 9-16). Poor data combined with poor phases obviously creates a particularly ghastly situation for electron density construction.

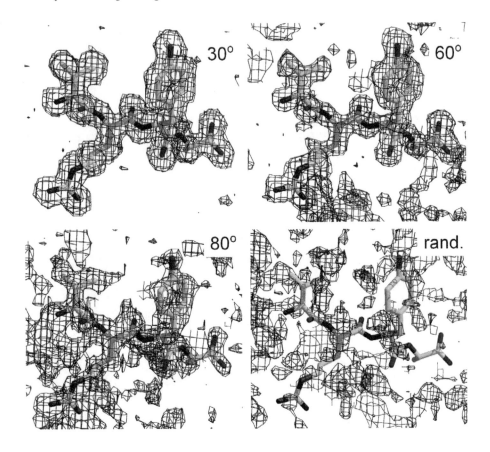

Phase or model bias

To demonstrate the effect of model bias, we can reconstruct the electron density from structure factor amplitudes of one molecule (the actual structure to be determined) with the phases calculated from another, similar structure (say an available molecular replacement starting model). The Fourier coefficients for the map reconstruction are then

$$F_{obs} \cdot \exp(i\varphi_{calc}) \tag{9-22}$$

We will show the effect on a mutant of our Val-Arg-Tyr-Ala peptide, that is, we change the sequence to Phe-Tyr-Lys-Ala. In a second example, we additionally rotate the mutant model so that the mutant peptide chain now runs in the opposite direction. Then we generate maps from intensity data of the "unknown" new structures (mutant and rotated mutant) given the phases from the old starting model. The resulting maps are shown in Figure 9-18.

The result of applying model phases to the experimental data from the new model is, not unexpectedly, quite concerning. Both electron density reconstructions shown in the lower panel of Figure 9-18 are still dominated by the model

Box 9-8 The role of phases in electron density reconstruction. While the map reconstruction by Fourier synthesis is relatively robust against noisy or poorly measured structure factor amplitudes, the phase error strongly affects the electron density reconstruction. As phases dominate the Fourier reconstruction, model phases will seriously bias the resulting map towards the model, that is, introduce model or phase bias. Model bias can be minimized through special map coefficients and weighting schemes, and actively through a variety of bias minimization methods.

Sidebar 9-5 Errors and the best map coefficients. The fact that phases have much more profound impact on electron density reconstruction than amplitudes can be derived from Parseval's Theorem, which can also be used to show that the "best" map (with the least residual error) is in fact the one constructed from "best" Fourier coefficients (which are, depending on the problem, generally centroid, Sim, or σ_A-based, maximum likelihood coefficients). The reviews by Randy Read[12, 14] are highly recommended additional reading.

Parseval's Theorem states in words that the mean square of a complex function $F(x)$ and the mean square of its Fourier transform are proportional:

$$\langle (FT[\mathbf{F}(\mathbf{x})])^2 \rangle \propto \langle \mathbf{F}(\mathbf{x})^2 \rangle$$

We can compute this expression via the Fourier integrals as before for both $\mathbf{F}$ and ρ and obtain

$$\langle \rho(x)^2 \rangle = (1/V^2) \sum_{\mathbf{h}} \mathbf{F}_{\mathbf{h}}^2 = (1/V^2) \sum_{\mathbf{h}} F_{\mathbf{h}}^2$$

Because the FT is additive, it also applies to sums or differences, that is,

$$\langle (\rho(true) - \rho(calc))^2 \rangle = (1/V^2) \sum_{\mathbf{h}} (F_{\mathbf{h}}(true) - F_{\mathbf{h}}(calc))^2$$

The residual density or map error is thus defined by the sum of the squared residuals (SRS) between the true value for F (our measured $F(obs)$) and our best calculated (model) value $F(calc)$. The best map with minimum residual density will therefore always be the map calculated when the SRS in F is minimal, which is exactly what we are trying to achieve during reciprocal space refinement. The quality of this map ultimately depends on how well we estimate that residual, which is generally in the form of a maximum likelihood residual.

From geometric considerations using the Argand diagram one can now derive[12, 14] that the SRS (expressed as r.m.s. deviation) from the true value for randomized amplitudes is much smaller than the deviation from the true value for random phases. The factor is

$$\frac{r.m.s.(F, phs)}{r.m.s.(F, amp)} = \left(\frac{2}{(4-\pi)/2} \right)^{1/2} = 1.41/0.66 = 2.16$$

In other words, there are no limitations on the phase angle, but the magnitude of random amplitudes is still defined and limited by the underlying structure factor distribution for random atoms, that is, the Wilson distribution as derived in Chapter 7. As a consequence, amplitudes can never be as wrong as phases and the phases dominate the electron density map reconstruction.

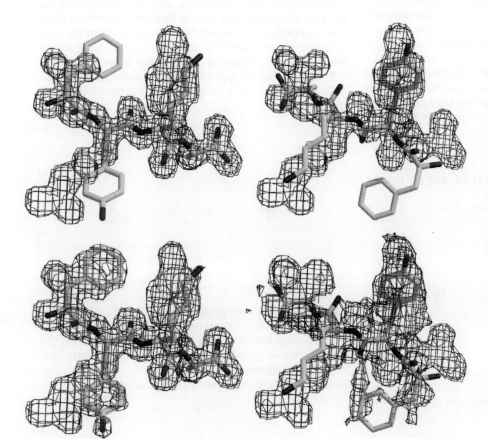

Figure 9-18 Phase bias in electron density maps. The upper panels show a mutant peptide Phe-Tyr-Lys-Ala (left) and the same peptide rotated (leading to reverse chain direction) simply superimposed on the electron density of the original Val-Arg-Tyr-Ala peptide. The lower panels show the electron density reconstructed using the diffraction data from the new models above, but using the old starting phases from the original peptide. The result is quite sobering: the shape of the electron density is still dominated by the starting model, and only weak outlines of the correct molecule density are visible. In the lower right panel, not even the direction of the peptide could be assigned for the reversed peptide with any certainty.

phases and the map reconstructions show the density of the old molecule, with more or less ghost-like outlines of the correct, new structure. It would not be possible to build a correct model into this type of electron density map. We are again reminded that phase bias minimization techniques will be very important particularly for molecular replacement phasing.

Basic electron density map types and map coefficients

In the previous sections we have generated a variety of basic maps, but have not yet formally classified the maps in terms of the Fourier coefficients that were used in the Fourier synthesis to generate them. We are aware that there are two distinct parts in the Fourier synthesis that are needed for reconstruction: structure factor amplitudes, derived from observed intensities, and the additionally supplied phases. Each of these Fourier coefficients can originate from different sources, which gives the map type its name. For example, the electron density

Sidebar 9-6 Animal husbandry and phase bias. A variety of creatures from the animal kingdom have historically been employed as motifs in the description of crystallographic symmetry operations, or have served in demonstrations of diffraction phenomena through optical transforms (Sidebar 6-8). Amongst the bugs and sea urchins,[15] piglets and hands,[16] crystallographers' faces,[14] and others, the most popular creature in the macromolecular crystallographer's faunal arsenal seems to be the duck. It makes its debut in plate 44 in Taylor and Lipson's early work on the demonstration of diffraction by optical transforms,[17] and has since been referred to as the Fourier Duck, *Anas fouriensis*. The drawing prefacing this chapter, sketched by Richard Dickerson to illustrate a parody by Jerry Donohue,[18] shows *A. fouriensis* taking off toward the sun and escaping the confinements of a low resolution electron density map—a situation many graduate students and post-docs may be able to relate to.

The breakthrough in the career of *A. fouriensis* occurred when it starred together with *Felis fouriensis* in Kevin Cowtan's (web-) *Book of Fourier*. With his kind permission, the section describing phase bias is reproduced below. It needs no comment owing to its clarity.

Animal magic:
Here is our old friend; the Fourier Duck, and his Fourier transform:

And here is a new friend; the Fourier Cat and *its* Fourier transform:

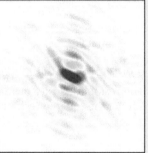

Now we will mix them up. Let us combine the magnitudes from the Duck transform with the phases from the Cat transform (you can see the brightness from the duck and the colors from the cat). If we then transform the mixture (the "mixed" Fourier coefficients for this image reconstruction are $F_{duck} \cdot \exp(i\varphi_{cat})$) we get the following:

The Cat image which contributed the phases is still visible, whereas the Duck image which contributed the magnitudes has gone!

Crystallographic interpretation:
In X-ray diffraction experiments, we collect only the diffraction magnitudes and not the phases. Unfortunately the phases contain the bulk of the structural information. That is why crystallography is difficult.
Kevin Cowtan, http://www.ysbl.york.ac.uk/~cowtan/fourier/fourier.html

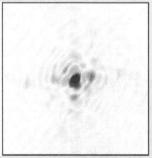

maps we generated for Figure 9-8 were reconstructed from calculated structure factor amplitudes (F_c or F_{calc}) and calculated phases (φ_c or φ_{calc}). The map was thus computed from Fourier coefficients $F_c \cdot \exp(i\varphi_c)$ or simply annotated as an F_c/φ_c map. In the next series of electron density maps (Figure 9-10), we used simulated noisy intensity data, equivalent to observed structure factor amplitudes F_{obs} of varying quality, and calculated phases, producing a $F_o \cdot \exp(i\varphi_c)$ or F_o/φ_c map. The maps shown in the phase-error series (Figure 9-17) were reconstructed using "observed" structure factor amplitudes F_o and "experimental" phases φ_{exp} of varying quality derived from simulated phasing experiments.

The maps in Figure 9-18 demonstrating the effect of phase bias were also F_o/φ_c maps, where the intensities originated from the measured data for the new structure, while starting phases came from the old model. Two questions arise:

1. Can we compute other map types from the same coefficients that may make the interpretation of electron density easier?

2. Are there ways and means to obtain modified map coefficients to reconstruct better electron density maps?

Types of map coefficients. Indeed, both ways to improve maps exist—the first approach simply implies using different combinations of Fourier coefficients, which presents the available information in more interpretable form. Basic difference maps $(F_o - F_c) \cdot \exp(i\varphi_c)$ and combined $(2F_o - F_c) \cdot \exp(i\varphi_c)$ maps, for example, fall into this first category. Further improvements require modification of the Fourier coefficients themselves, and examples for such maps will be encountered when we discuss phasing and phase improvement methods in subsequent chapters. Amongst those are maps with coefficients based on improved phase probabilities such as obtained by Sim weights,[19] omit maps,[20] sigma-A (σ_A) maps[21] reconstructred from maximum likelihood coefficients[22] $(2mF_o - DF_c) \cdot \exp(i\varphi_{wt})$ with model phases φ_{wt}, maps with figure-of-merit-corrected observed structure factor amplitudes $F_o \cdot$ f.o.m. with weighted and/or averaged phases,[13] or improved experimental phases obtained by a variety of density modification methods[23–25] discussed in Chapter 10. A detailed discussion of modified maximum likelihood map coefficients accounting for incomplete or incorrect models (introduced in Chapter 7) follows in Chapter 12.

Difference maps

Figure 9-19 demonstrates that by simply selecting a proper combination of map coefficients, we greatly improve the interpretability of the electron density. The top panel shows the difference density map $(F_o - F_c) \cdot \exp(i\varphi_c)$, where F_o represents the observed structure factor amplitudes, and both F_c and phases φ_c come from the starting (phasing) model. The red contours represent negative density values, that is, regions where the model *has* density but where there *should not be* density. Indeed, these regions coincide with the starting model and clearly indicate that these regions are not the correct place for the residues in our mutant structure. The purple density shows positive regions, where we miss density in the maps. As expected, these regions coincide with regions where our mutant's side chains are located. The regions where the difference map does not show any contours represent regions where model density and actual density agree; this is the backbone region in the case of our mutant structure.

The absence of regions that are actually correct makes the sensitive difference maps[26] very well suited for detecting model errors or for finding missing density, but not optimal for initial model building. A compromise is the $(2F_o - F_c) \cdot \exp(i\varphi_c)$ map, sometimes imagined as a superposition of the difference map and model-phased $F_o \cdot \exp(i\varphi_c)$. However, the actual theoretical reason for using maps of the $(2F_o - F_c)$ type is that in incomplete models, atoms missing from the model are only reconstructed at half the density or less. Using $2F_o - F_c$ as Fourier coefficients in the map reconstruction weights the missing atoms up to normal density levels.[27] The lower panels of Figure 9-19 show a $2F_o - F_c$ map, and we recognize that it would be quite possible to correct this mutant structure, particularly in view of the fact that we know the mutant sequence.

Figure 9-19 $F_o - F_c$ difference maps and $2F_o - F_c$ maps. The $F_o - F_c$ difference maps in the top panel show negative difference density (where there should not be any density) in red and positive difference density (where there should be density) in purple. Both regions correctly reveal the difference between the starting model (yellow sticks) and the correct model (orange sticks, shown in the right panels). The sensitive difference maps are thus particularly valuable for detailed model correction. The $2F_o - F_c$ maps in the bottom panel can be interpreted as a combination of the difference map $(F_o - F_c)\cdot\exp(i\varphi_c)$ and a $(F_o)\cdot\exp(i\varphi_c)$ map. The $2F_o - F_c$ map is contoured to amplify positive density and is well suited to early model building stages. Another common color scheme is red for negative difference density and green for positive.

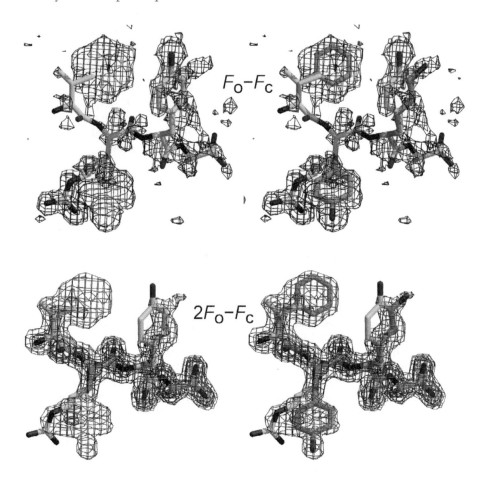

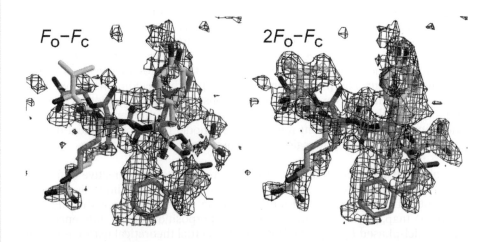

Sidebar 9-7 Good phases and good intensity data are important. There is an important lesson to be learned from the failure of the electron density reconstruction from poor starting phases as demonstrated in Figure 9-20: only **good phases = good structure**. However, the presumptuous conclusion that one might get away with sloppy intensity data collection would be quite misleading, because another rule comes into effect: **poor data = poor phases**. Very accurate data collection is necessary to produce data good enough to extract the experimental phases. The reason is that the data employed in experimental phase determination are intensity *difference* data, where small differences of large $I(hkl)$ values are used to extract phase information. Clearly, in such a scenario even small errors of each intensity measurement can have huge effects on the resulting intensity difference.

Caveat. While these plain maps with unmodified map coefficients and model phase values do allow for correction of not-too-wrong models such as our mutant peptide—with identical backbone and Cβ-atoms compared with the original peptide—they become much less interpretable in cases where large differences between starting phases (model) and actual structure exist. Even with good intensity data of reasonable resolution it is generally impossible to recover the true structure from substantially incorrect model phases, as illustrated in Figure 9-20 for the reverse traced mutant peptide model; it is even more likely

Figure 9-20 Poor starting phases give poor electron density maps. In the case of the reverse traced mutant peptide (orange sticks) as the true structure providing the intensities, no basic map type is able—despite good 1.5 Å data—to give sufficient clues as how to correct the starting model (yellow sticks) that provided the phases. The difference map informs us in some parts about what is wrong, but none of the maps has sufficient reconstructive power to produce an outline of the correct orange molecule.

> **Box 9-9 Selecting proper map coefficients.** The suitability for the task, as well as the quality of the map, can be improved by selecting proper map coefficients. Fourier coefficients can be combined to give basic difference maps $(F_o - F_c) \cdot \exp(i\varphi_c)$ and combined $(2F_o - F_c) \cdot \exp(i\varphi_c)$ maps. The latter are preferred for model building, the sensitive difference maps for model correction. Further improvement is possible through modified map coefficients based on improved phase probabilities such as σ_A-coefficients or related $(2mF_o - DF_c) \cdot \exp(i\varphi_{wt})$ maximum likelihood map coefficients.

to be impossible when data are poor and of low resolution. Given good data, density reconstruction with suitably modified Fourier coefficients may come to the rescue. Without providing details, which must wait until we have covered the refinement and bias minimization in later chapters, we have already noted that various map types exist with suitably improved coefficients that often can—within reasonable limits—recover maps that allow a correct model to be built from relatively poor initial phases. For example, an iterative brute-force approach[13] was able to eventually reconstruct the density for our mutant in over 900 refinement–rebuilding cycles. However, there is no way to recover a correct structure from poor data and incorrect phases, particularly at low resolution (equaling few X-ray reflection data).

9.4   Patterson function and Patterson maps

An important type of map reconstruction that can be performed directly from intensities, that is, the squared structure factor amplitudes—without knowledge of any phases—are Patterson maps. Patterson maps contain interatomic distance information, and the fact that they can be constructed without phases makes them very valuable as a tool in the early stages of a structure determination. Typical applications are locating heavy metal atoms and anomalous scatterers, internal symmetry search, or finding the orientation of a molecule in molecular replacement phasing. Patterson maps are based on the Patterson function, introduced by A. L. Patterson[28] in 1934.

The Patterson function $P(\mathbf{u})$ is defined at any point $\mathbf{u} = (u, v, w)$ by a convolution integral of the electron density over the whole unit cell in real space R:

$$P(\mathbf{u}) = \int_R \rho(\mathbf{r})\rho(\mathbf{r} + \mathbf{u})\, d\mathbf{r} \tag{9-23}$$

Let us analyze what Equation 9-23 implies. We realize that the integrand is the product of the real space electron density at a point $\mathbf{r}$ with the electron density at a point $\mathbf{r} + \mathbf{u}$, that is, translated from the other point by the distance vector $\mathbf{u}$. This observation leads to the conclusion that the Patterson function will have large values at a point $\mathbf{u}$ when both electron densities $\rho(\mathbf{r})$ and $\rho(\mathbf{r} + \mathbf{u})$ have high values (which is at atom positions). The function value thus will be largest whenever a distance vector $\mathbf{u}$ corresponds to an interatomic distance, and a map constructed from the Patterson function will contain peaks at all interatomic distances (or more accurately, at interatomic distance vector tips in 3-dimensional space). As the map contains peaks for all interatomic distances, we realize that this will be a very crowded map, with $N(N-1)$ peaks for N atoms, excluding the "self-peaks" at the origin (zero distance). As a consequence, the Patterson map will not be directly interpretable for structures with many atoms, such as proteins. However, given that the Patterson peak height is the product of the corresponding electron density peak heights, for few heavy atoms in a light atom structure these peaks will be very distinct. We can clearly see the potential usefulness of this map type in the search for heavy atom positions in the intensity difference data between isomorphous derivative crystals and native crystals or between anomalous differences or dispersive intensity differences.

Autocorrelation of the electron density

The task is now to generate the interatomic distance vector maps based on the Patterson function in the absence of phases. Fortunately, a Patterson map can be constructed from a Fourier synthesis using the intensities, or F_h^2, as Fourier coefficients. The proof is provided as follows. The convolution of two functions is defined by a convolution integral of the general form

$$Conv(\mathbf{u}) = f(\mathbf{r}) \otimes g(\mathbf{r}) = \int_R f(\mathbf{r})g(\mathbf{u}-\mathbf{r})d\mathbf{r} \tag{9-24}$$

We recognize that this convolution integral (9-24) is equivalent to the Patterson function (9-23) if we set $f(\mathbf{r}) = \rho(\mathbf{r})$ and $g(\mathbf{r}) = \rho(-\mathbf{r})$. The convolution then becomes an autocorrelation, and we can write the Patterson function as a basic autocorrelation of the electron density

$$P(\mathbf{u}) = \rho(\mathbf{r}) \otimes \rho(-\mathbf{r}) \tag{9-25}$$

From the Fourier convolution theorem

$$FT\left[f(\mathbf{r}) \otimes g(\mathbf{r})\right] = FT\left[f(\mathbf{r})\right] \cdot FT\left[g(\mathbf{r})\right] \tag{9-26}$$

follows that we can construct a Patterson map from the product of the FTs of the electron density

$$FT\left[\rho(\mathbf{r}) \otimes \rho(-\mathbf{r})\right] = FT\left[\rho(\mathbf{r})\right] \cdot FT\left[\rho(-\mathbf{r})\right] \tag{9-27}$$

Fortunately, we already know that the Fourier transform of the electron density at a point $\mathbf{r}$ is the complex structure factor $FT\left[\rho(\mathbf{r})\right] = \mathbf{F}(\mathbf{r}^*)$. For discrete summation, we replace $\mathbf{r}^*$ with $\mathbf{h}$ and using the compact notation $\mathbf{F}(\mathbf{h}) = \mathbf{F_h}$ the FT of the convolution becomes

$$FT\left[P(\mathbf{u})\right] = \mathbf{F_h} \cdot \mathbf{F_{-h}} = \mathbf{F_h} \cdot \mathbf{F_h^*} = F_h^2 \tag{9-28}$$

Inverting the FT we finally obtain

$$P(\mathbf{u}) = FT^{-1}(F_h^2) = \sum_{\mathbf{h}=-\infty}^{+\infty} F_h^2 \exp(-2\pi i \mathbf{h}\mathbf{u}) \tag{9-29}$$

As in the case for the electron density reconstruction $\rho(xyz)$, we can apply Euler's formula and use the relation $F_h = F_{-h}$ to derive the explicit form useful for numeric evaluation:

$$P(uvw) = \frac{2}{V}\sum_{h=0}^{+\infty}\sum_{k=-\infty}^{+\infty}\sum_{l=-\infty}^{+\infty} F_h^2 \cos 2\pi(hu + kv + lw) \tag{9-30}$$

Generating the Patterson function. The autocorrelation of the electron density of the unit cell defined in Equation 9-25 can be readily visualized by shifting a window containing the unit cell electron density over itself, and plotting a measure for the degree of overlap for each shifting distance $\mathbf{u}$. The value of the Patterson function $P(\mathbf{u})$ can then be plotted, and we obtain the Patterson map. For the real space unit cell we use the same 1-dimensional structure we have used to generate the 1-dimensional Fourier examples in Figures 9-4 and 9-6, with the carbon atom located at fractional coordinate $x = 0.1$ and the oxygen atom at $x = 0.7$. Figure 9-21 illustrates a sequence of superpositions for the values of $u = 0.2, 0.35, 0.4$, and 0.6. For $u = 0$, the superposition is maximal, and just yields the origin peak of the 1-dimensional Patterson function $P(u)$.

The fact that the Patterson function results from the autocorrelation of the molecular structure implies another important consequence for the interpretation of Patterson maps: We can place a trial structure over a Patterson map generated from measured F_h^2 by positioning each of the trial structure atoms at the origin and then examining the Patterson map for peaks at the resulting interatomic distances. If the trial structure (two atoms suffice as initial trial

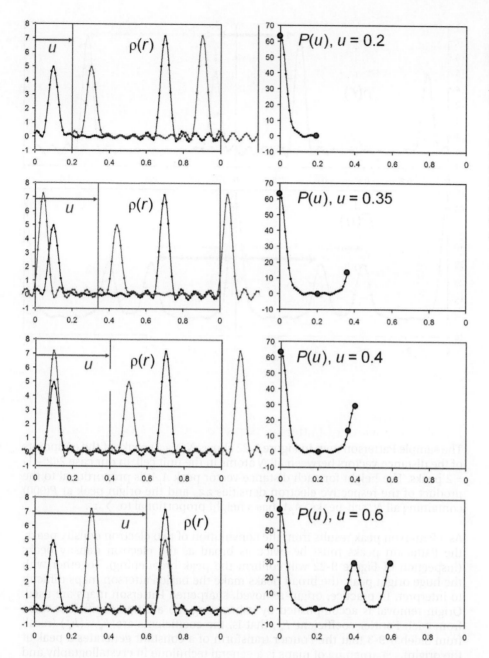

Figure 9-21 The Patterson function as autocorrelation of the molecular structure. Positioning the electron density of the molecular structure (C, $x = 0.1$; O, $x = 0.7$) over itself at each distance u ($0 \leq u \leq 1$) and computing the convolution yields the value of $P(u)$ for each u. The numeric value of the convolution can be readily calculated from the square of the structure factors (i.e., experimentally determined intensities) using the summation (9-30). Note that the structure is not centrosymmetric, but the Patterson function is centrosymmetric (illustrated in additional detail in Figure 9-22).

model) is correct, then we will observe a Patterson peak at the corresponding distance. If the first vector distance fits, the trial structure can be expanded and the procedure can be repeated until the structure is completed. This procedure forms the basis for Patterson vector search methods used to determine small molecule structures and heavy atom substructures in protein crystallography.

Centrosymmetry of Patterson space

Inspection of (9-30), which depends only on cosine terms, reveals that the Patterson function must be inherently centrosymmetric (even for non-centrosymmetric electron density, i.e. non-centrosymmetric structures), and the Patterson symmetry is thus a combination of space group symmetry plus inversion (as listed in Table 6-6). The relation between real space electron density and Patterson space centrosymmetry can be readily understood because any interatomic distance vector $\overrightarrow{AB}$ necessarily has to have a mate $\overrightarrow{BA}$ pointing in the opposite direction. Considering the centrosymmetry, the Patterson map has $N(N-1)/2$ *unique* peaks.

Figure 9-22 The Patterson function and its relation to electron density.
Electron density (top panels) and corresponding Patterson function (bottom panels) of a structure with a carbon atom at $x = 0.1$ and an oxygen atom at $x = 0.7$. Two unit cells of each map are shown. The Patterson map contains $N(N-1)$ peaks at the interatomic distance vectors between all N atoms in the unit cell. The peak height of Patterson peaks is proportional to the product of the density value at the peak electron density positions. Thus, because both peaks represent C–O or O–C distance vectors, they are equally high, but twice as broad as the electron density peaks as a result of the convolution. The zero- or self-peak $P(000)$ is proportional to the sum of all atomic z^2, and is suppressed in practical maps. Note that the Patterson map is always centrosymmetric, even if the corresponding structure is not. The y-axis for $P(u)$ is provided in units of $e^2\text{Å}^{-1}$. Note that the Patterson peaks can be generated by placing each atom successively at the origin. As the Patterson peaks represent the interatomic distance vectors, searching for correct distance vectors provides a means for determining a molecular structure by Patterson vector search methods.

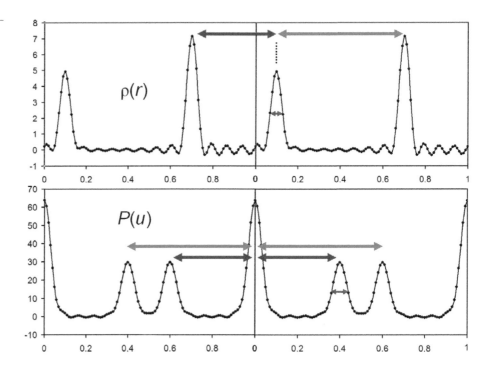

Number and width of Patterson peaks

The sample Patterson map in Figure 9-22 shows $N(N-1)$ peaks at the endpoints of the distance vectors between all N atoms in the unit cell, in our case $2\cdot(2-1)$ = 2 peaks. The height for each distance vector peak $\overrightarrow{A_iA_j}$ is proportional to the product of the respective electron densities z_iz_j, and the origin peak at $P(000)$ containing all N "self-peaks" $\overrightarrow{A_jA_j}$ has a height proportional to $\sum_{atoms} z_j^2$.

As a Patterson peak results from the convolution of two electron density peaks, the Patterson peaks must be twice as broad as the electron density peaks (inspection of Figure 9-22 will confirm the peak broadening). Together with the huge origin peak, the broad peaks make the basic Patterson maps difficult to interpret. In practice, origin-removed, sharpened Patterson maps are used. Origin removal is accomplished by subtracting the average coefficient value from each Fourier coefficient $F_\mathbf{h}^2$, that is, the coefficients are $F_\mathbf{h}^2 - \langle F_\mathbf{h}^2 \rangle$ (recall from Sidebar 9-3 that the Fourier transform of a constant generates a peak at the origin). Sharpening of maps is a general technique in crystallography and implies elimination of the fall-off (attenuation) of the structure factor magnitude with increasing scattering angle $\sin\theta/\lambda$. We have introduced normalized structure factors in Chapter 7, where the assumption of point atoms resulted in normalized structure factor amplitudes $E_\mathbf{h}$ defined as $E_\mathbf{h} = F_\mathbf{h}/\langle F_\mathbf{h}^2 \rangle^{1/2}$. For practical use of Patterson maps, somewhat less extreme sharpening is desirable (sharpening also leads to increased truncation ripples) and "mixed" structure factor amplitudes of the form $(E_\mathbf{h}^3 \cdot F_\mathbf{h})^{1/2}$ are commonly used.[29]

Applications of the Patterson function

The Patterson function and the interpretation of Patterson maps play a significant role in the following practical applications:

- Determination of marker atom positions (marker atom substructure) from isomorphous difference data and/or determination of anomalously scattering atoms in anomalous/dispersive difference data. This requires the deduction of atom positions from the Patterson maps, and will be discussed in Chapter 10.

- Determination of the rotational component, that is, the search model orientation, during a molecular replacement phasing experiment and detection of

> **Box 9-10 The Patterson function and Patterson maps.** The Patterson function is based on the autocorrelation of the electron density. In contrast to the electron density maps which require both the phases and the structure factor amplitudes as Fourier coefficients, a Patterson map can be computed directly from intensities without additional phases. The Patterson map contains peaks at the interatomic distance vector tips and is centrosymmetric irrespective of the structure symmetry. It is used in determining heavy atom positions, in NCS searches and map averaging, and in molecular replacement.

proper NCS axes orientation using self-rotation Patterson maps (discussed in detail in Chapter 11).

- Detecting translational NCS parallel to unit cells and of NCS axes parallel to crystallographic axes from native Patterson maps (discussed in detail in Chapter 11).

9.5   Next steps

At this point of our studies we know how to construct basic electron density maps using different types of Fourier coefficients, but we have not yet found a general method to obtain the experimental phases we need to reconstruct the electron density of an as yet unknown protein structure. It is now time to progress to the fascinating and crucial subject of phasing at the core of protein crystallography. Several approaches to solving the phase problem exist, and the choice of phasing strategy depends on our state of prior knowledge and technical or biochemical resources. As already mentioned in the introduction, the phasing methods can be roughly grouped into about four general areas:

- **Experimental substructure phasing**, which is generally applicable and does not depend on any prior structural information about the protein molecule.

- **Molecular replacement**, which requires the availability of a structurally similar model as a molecular search probe.

- **Density modification** techniques are used following the substructure phasing methods in practically all *de novo* structure determinations, and they are powerful phase improvement and phase extension methods.

- **Direct methods**, which exploit the fact that phase relations and inequalities exist between certain sets of structure factors. Although limited to small proteins and atomic resolution data (1.2 Å or better) for *ab initio* structure determination, they are very successful in heavy atom substructure determination required for experimental substructure phasing.

The details of phase determination will concern us for the next two chapters. Phasing is a mathematically and conceptually challenging part of structure determination, and it is a most gratifying experience when the first phased map of a protein appears and a new protein structure becomes visible for the first time ever.

9.6   Key concepts

A brief review of the key concepts is provided below. Questions formed from these key concepts will aid in rapid assessment of student progress.

Fourier transforms and maps

- The mathematical formalism of Fourier transformation allows a complete and reversible transformation between direct domain representation and reciprocal domain representation of the same physical reality.

- Complex structure factors representing reciprocal space can be converted into electron density representing direct space, and back into complex structure factors without loss of information.
- In practice, the continuous Fourier integrals are replaced by a discrete summation over all structure factor indices h, k, l or all density grid points x, y, z in the back-transform, respectively.
- The electron density is obtained by a discrete Fourier summation or Fourier synthesis with the structure factor amplitudes and the phases as Fourier coefficients. The summation extends over the entire reciprocal space, and as a manifestation of the centrosymmetry in reciprocal space, the Fourier transform $\rho(x, y, z)$ is real despite the complex Fourier coefficients (complex structure factors) used in the full summation.
- The convolution of two functions F and G is defined by a convolution integral.
- The Fourier convolution theorem allows us to separately transform two functions and multiply the resulting individual Fourier transforms. An example is the sampling of a molecular envelope (the transform of the molecule) by the reciprocal lattice function (the transform of the lattice function) yielding the complex structure factors or the diffraction pattern.
- The electron density reconstruction or Fourier synthesis generates electron density peaks at the position of the atoms, with the peak height proportional to the number of electrons.
- When all available data are used, the electron density peaks will broaden with decreasing resolution, but the transform remains smooth. When data are truncated, truncation ripples appear, particularly around heavy atoms, that can obscure neighboring atoms if the truncation is severe.
- Increasing B-factor also broadens the electron density peaks, corresponding to convolution of the peak with a Gaussian point spread function. Distinguishing between a partially occupied atom and an atom with high B-factor by peak shape alone is generally not possible in protein electron density maps.
- Electron density $\rho(x, y, z)$ is represented in 3-dimensional space as mesh or grid of density levels. Points of equal density level are connected by lines forming 3-dimensional contours or a 3-dimensional electron density map.
- Electron density maps are contoured on a relative scale in levels of standard deviations σ of the mean density. It is important to select contour levels and depth (slab thickness) appropriate to the task when displaying an electron density map.
- Randomly omitting structure factors or data does not generate a significant problem in reconstructed electron density maps. However, systematically absent data manifest themselves as severe noise and map artifacts.
- Omission of a small percentage of the strongest reflections resulting from detector overflows leads to noisy maps (in addition to creating problems in phasing) and missing of complete wedges of data leads to streaking in the electron density maps.
- While reduced periodic lattice order leads to decreased resolution, the peak width itself does not increase with increasing disorder or B-factor. Anisotropic disorder generates anisotropic resolution limits of the diffraction pattern.

Phase problem and map reconstruction

- While the map reconstruction by Fourier synthesis is relatively robust against noisy or poorly measured structure factor amplitudes, the phase error strongly affects the electron density reconstruction.
- As phases dominate the Fourier reconstruction, model phases will strongly bias the resulting map toward the model, that is, introduce model or phase bias. Model bias can be minimized through special map coefficients and weighting schemes, and through a variety of additional bias minimization methods.
- The suitability for the task as well as the quality of the map can be improved by selecting proper map coefficients.
- Fourier coefficients can be combined to give basic difference maps $(F_o - F_c) \cdot \exp(i\varphi_c)$ and combined $(2F_o - F_c) \cdot \exp(i\varphi_c)$ maps. The latter are preferred for model building, the sensitive difference maps for model correction.

- Further improvement is possible through modified map coefficients based on improved phase probabilities obtained using Sim-weights, omit-maps, σ_A-coefficients or the related $(2mF_o - DF_c) \cdot \exp(i\varphi_{wt})$ maximum likelihood map coefficients.

Patterson function and Patterson maps

- The Patterson function is based on the autocorrelation of the electron density. In contrast to the electron density maps which require the phases and structure factor amplitudes as Fourier coefficients, a Patterson map can be computed directly from intensities without phases.
- The Patterson map contains peaks at the interatomic distance vector tips and is centrosymmetric irrespective of the structure symmetry. It is used in the determining heavy atom positions, in NCS searches and map averaging, and in molecular replacement.

9.7 Additional reading

1. Read RJ (2002) Model Phases: Probabilities, bias and maps. In *International Tables for Crystallography F*, 325–331.

2. Giacovazzo C, Monaco HL, Viterbo D, et al. (2002) *Fundamentals of Crystallography*. Oxford, UK: Oxford Science Publications.

3. Rossmann MG, & Arnold E (Eds.) (2001) *International Tables for Crystallography, Volume F: Crystallography of Biological Macromolecules*. Dordrecht, The Netherlands: Kluwer Academic Publishers.

4. Carter CWJ, & Sweet RM (1997) Macromolecular Crystallography. *Methods Enzymol.* 276, 277. London, UK: Academic Press.

5. Carter CW, & Sweet R (2003) Macromolecular Crystallography. *Methods Enzymol.* 368, 374. London, UK: Academic Press.

9.8 References

1. Press WH, Flannery BP, Teulkolsky SA, et al. (1986) *Numerical Recipes*. Cambridge, UK: Cambridge University Press.

2. Ten Eyck L (1977) Efficient structure-factor calculation for large molecules by the fast Fourier transform. *Acta Crystallogr.* A33(3), 486–492.

3. Ten Eyck L (1973) Crystallographic fast Fourier transforms. *Acta Crystallogr.* A29(2), 183–191.

4. Grosse-Kunstleve RW, Sauter NK, Moriarty NW, et al. (2002) The Computational Crystallography Toolbox: crystallographic algorithms in a reusable software framework. *J. Appl. Crystallogr.* 35(1), 126–136.

5. Agarwal R (1978) A new least-squares refinement technique based on the fast Fourier transform algorithm. *Acta Crystallogr.* A34(5), 791–809.

6. Julian MM (2008) *Foundations of Crystallography with Computer Applications*. Boca Raton, FL: CRC Press.

7. Patterson AL (1963) Treatment of anomalous dispersion in X-ray diffraction data. *Acta Crystallogr.* 16(12), 1255–1256.

8. James RW (1948) False detail in three-dimensional Fourier representations of crystal structures. *Acta Crystallogr.* 1(3), 132–134.

9. Stenkamp RE, & Jensen LH (1984) Resolution revisited: limit of detail in electron density maps. *Acta Crystallogr.* A40(3), 251–254.

10. Weiss M (2001) Global indicators of X-ray data quality. *J. Appl. Crystallogr.* 34(2), 130–135.

11. Vaguine AA, Richelle J, & Wodak SJ (1999) SFCHECK: a unified set of procedures for evaluating the quality of macromolecular structure-factor data and their agreement with the atomic model. *Acta Crystallogr.* D55, 191–205.

12. Read RJ (2002) Model Phases: Probabilities, bias and maps. In *International Tables for Crystallography F*, 325–331.

13. Reddy V, Swanson S, Sacchettini JC, et al. (2003) Effective electron density map improvement and structure validation on a Linux multi-CPU web cluster: The TB Structural Genomics Consortium Bias Removal Web Service. *Acta Crystallogr.* D59, 2200–2210.

14. Read RJ (1997) Model phases: probabilities and bias. *Methods Enzymol.* 277, 110–128.

15. Rossmann MG (2001) Molecular replacement – historical background. *Acta Crystallogr.* D57, 1360–1366.

16. Drenth J (2007) *Principles of Protein X-ray Crystallography*. New York, NY: Springer.

17. Taylor CA, & Lipson H (1964) *Optical Transforms: Their Preparation And Application To X-Ray Diffraction Problems*. UK: G. Bell & Sons.

18. Dickerson RE (2005) *Present at the Flood*. Sunderland, MA: Sinauer Associates.

19. Sim G (1959) The distribution of phase angles for structures containing heavy atoms. II. A modification of the normal heavy-atom method for non-centrosymmetrical structures. *Acta Crystallogr.* 12(10), 813–815.

20. Bhat TN, & Cohen GH (1994) OMITMAP: An electron density map suitable for the examination of errors in a macromolecular model. *J. Appl. Crystallogr.* 17, 244–248.

21. Read RJ (1986) Improved Fourier coefficients for maps using

phases from partial structures with errors. *Acta Crystallogr.* A42, 140–149.

22. Murshudov GN, Vagin AA, & Dodson ED (1997) Refinement of macromolecular structures by the maximum-likelihood method. *Acta Crystallogr.* D53, 240–255.

23. Cowtan KD, & Main P (1996) Phase combination and cross validation in iterated density-modification calculations. *Acta Crystallogr.* D52, 43–48.

24. Terwilliger TC (2000) Maximum likelihood density modification. *Acta Crystallogr.* D56, 965–972.

25. Kleywegt GJ, & Read RJ (1997) Not your average density. *Structure* 5, 1557–1569.

26. Rould MA, & Carter CW (2003) Isomorphous difference methods. *Methods Enzymol.* 374, 145–163.

27. Luzzati V (1953) Resolution d'une structure cristalline lorsque les positions d'une partie des atomes sont connues: Traitement statistique. *Acta Crystallogr.* 6, 142–152.

28. Patterson AL (1934) A Fourier Series Method for the Determination of the Components of Interatomic Distances in Crystals. *Phys. Rev.* 46(5), 372–376.

29. Schneider TR, & Sheldrick GM (2002) Substructure solution with SHELXD. *Acta Crystallogr.* D58, 1772–1779.

Experimental phasing

Alice looked in amazement at the magic sphere which the Cheshire cat had handed her. It consisted of thousands of tiny sparkling beads of various sizes neatly arranged in a regular lattice. "What is it?" asked Alice. "It is the molecule of divine wisdom. On top of it, it attracts eternal grant funding. But it's in secret code." A strange gleam appeared in Alice's eyes: "How do I crack the code?" The grin of the Cheshire cat completed a counterclockwise rotation by 90° until it appeared like a crescent moon. "You'll need the phases" said the Cheshire cat and picked its teeth with a feather of the poor Fourier duck.

B. Rupp, *Born to Phase*

Experimental phasing is a core task in protein crystallography. Unlike small-molecule crystallography, where thanks to atomic resolution data—combined with a comparably small number of atoms—structure determination is practically assured, macromolecular crystallography requires experimentally determined phases in addition to the diffraction data. Despite the ever increasing power of newly developed computer programs, automation is more difficult to achieve because of the complexity of the problem and the generally poorer data in macromolecular crystallography. Devising a proper phasing strategy is essential and Chapter 10 provides the necessary foundation.

We begin the chapter with the conceptual introduction of *marker atom substructure methods*, which are based on the idea of solving a smaller more tractable problem first, and bootstrapping the phase determination from the simplified substructure case. Heavy atoms and anomalous scatterers can act as marker atoms and give rise to difference data that in turn are used to determine the position of the marker atoms. The marker atom substructure is subsequently used to generate initial protein phases.

After the conceptual introduction we move on to the subject of actual substructure solution. *Patterson methods* and *direct methods* form the backbone of substructure determination. Initial data preparation is discussed and a simple single atom derivative example is exercised throughout the section.

The fourth section of this chapter expands on the principles of protein phase determination from substructure positions by solving the phasing equations. Intuitive graphical presentations in the form of Harker diagrams and more rigorous formal derivations of the phasing equations are provided for different phasing scenarios. We cover isomorphous as well as anomalous cases, and encounter for the first time the power of density modification techniques to break the phase ambiguity inherent in the phasing equations.

To combine the experimental phases into a consistent set of best protein phases, we employ phase probability distributions. This is a most important subject, and we will progress from building foundations from basic Blow and Crick phasing to consideration of partial phasing models and errors in the model.

Heavy atom refinement and phasing are intimately connected. Based on the introduction of likelihood functions for structure factor probabilities in Chapter 7 we introduced error models in the phasing and refinement of heavy atom positions. Many previously disjointed concepts including the various phase probabilities as well as error models will fall together in the greater framework of maximum likelihood, which is generally applicable and will stay with us through the rest of this text and our crystallographic career.

Substantial improvement of the initial protein phases is routinely achieved through the use of density modification techniques. They are based on the fact that protein crystals contain on average 50% solvent, which has distinctly different properties than the protein. Solvent flattening, solvent flipping, sphere of influence, histogram matching, and other methods are discussed. We defer the extension to powerful NCS averaging until the next chapter because of the direct relation to molecular replacement.

The final section discusses two practical examples of the most common phasing techniques, SAD and MAD phasing. We use the popular *SHELXC/D/E* set of programs to quickly progress from anomalous data to high quality protein maps. Emphasis here is on the interpretation of diagnostics to assess data quality, the quality of the substructure solution, and map quality during density modification and phase extension.

10.1   Solving the phase problem

At this point of our curriculum we know how to construct electron density maps by Fourier synthesis using different types of Fourier coefficients, but we have not yet found a general method for obtaining the experimental phases we need to reconstruct the electron density of an as yet unknown protein structure. The task is to obtain a phase value for *each* measured reflection. The available phasing methods can be roughly grouped into four general areas.

- **Marker atom substructure methods** do not depend on prior structural information about the protein molecule and are therefore generally applicable. Substructure methods involve the determination of a marker atom substructure from difference data, generally by Patterson search methods and/or direct methods. From the marker substructure, starting phases for the electron density reconstruction of the protein can be calculated. We refer to methods that do not need prior structural information as *de novo* or experimental phasing methods.

- **Density modification** techniques are used after the substructure phasing methods in practically all *de novo* structure determinations and are powerful phase improvement methods. In cases of very high local symmetry (noncrystallographic symmetry, NCS), such as the 532 symmetry in virus capsids, *de novo* phasing using only density modification and density averaging is possible. Density modification techniques are indispensable in the resolution of phase ambiguity in the case of phases derived from a single set of difference data, where two possible phase values result. Density modification does not require an atomic model, and can be considered as an extension of experimental phasing.

- **Molecular replacement** methods require availability of a structurally similar model as a molecular search probe. *Replacement* is thus to be understood as "positioning" of the search probe in the crystal structure, not as "substitution." The phases of the correctly placed model are used as starting phases for map reconstruction. Although in many cases molecular replacement is a quick way to a protein structure, model phase bias can be substantial because the phases dominate the electron density reconstruction. Because of the similarities between map averaging used in density modification and molecular replacement search techniques, we treat them together in Chapter 11.

- **Direct methods** exploit the fact that phase relations exist between certain sets of structure factors. They require high resolution (1.2 Å or better) data for *ab initio* determination of a protein structure and have so far been limited to relatively small proteins. They are, however, very important and successful in the marker atom substructure determination required for the *de novo* substructure phasing techniques and will be discussed in this chapter.

In the following section we will provide a short introduction to substructure methods and follow the outline of the methods with more detailed discussions in individual sections about substructure solution, initial phasing, and basic density modification.

10.2 Marker atom substructure methods and electronic differences

The most generally applicable *de novo* phasing methods are based on the determination of a marker atom substructure. A marker atom provides the source of an electronic difference relative to an isomorphous reference structure. The electronic difference in turn implies different atomic scattering factors, and thus different structure factor amplitudes and different intensities relative to the reference structure. The difference data are then used to determine the location of the source of the electronic difference, the marker atom. By creating difference intensities between data sets with and without the contributions from the marker atoms, the initial problem is reduced to solving a substructure of a few up to a few hundred atoms versus many thousands to tens of thousands of atoms in the entire protein structure.

The difference principle is generally applicable, regardless of what underlying electronic difference introduced by the marker atom gives rise to the differences in atomic scattering factors. The differences can result from an additional heavy atom soaked into the crystal, a different kind of atom replacing an atom, or a marker atom such as Se in selenomethionine introduced by protein engineering. Intensity differences also arise as a consequence of dispersive and anomalous scattering factor contributions from anomalously scattering atoms, native or introduced, as discussed in Chapter 6.

The concept of isomorphous difference data is readily understood by investigating a fictitious subtraction of a native protein crystal from a crystal containing a marker atom, say a heavy atom derivative crystal, as shown in the top row of Figure 10-1.

The consequence of the scenario shown in Figure 10-1 is easy to understand: The difference data represent a very much simplified situation, as we need to search only for one (or a few) marker atoms in the difference data instead of many thousands in the case of the whole protein. Although the difference intensity pattern in all likelihood will contain useful information about the marker atom, it is important to realize that it is not really a true diffraction pattern of the heavy atom "difference" crystal: Each complex structure factor of the derivative $\mathbf{F}_{PA}$ is the complex (vector) sum of the partial structure factor contributions from the marker atom $\mathbf{F}_A$, and from the protein $\mathbf{F}_P$

$$\mathbf{F}_{PA} = \mathbf{F}_P + \mathbf{F}_A \text{ and thus } \mathbf{F}_A = \mathbf{F}_{PA} - \mathbf{F}_P \tag{10-1}$$

but their amplitudes are generally *not* additive (Figure 10-2):

$$F_A \neq F_{PA} - F_P \tag{10-2}$$

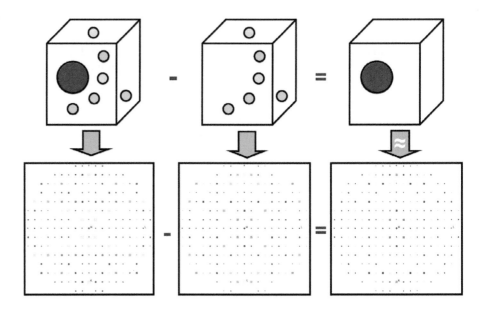

Figure 10-1 The concept of isomorphous difference data. The top line shows the gedankenexperiment in real space of subtracting a native protein crystal from an exactly isomorphous derivative crystal. The light atoms "cancel" out, and only the heavy marker atom remains in the difference crystal. While we cannot produce a real difference crystal, we can very well obtain a "difference diffraction pattern" from the differences between experimental data of the derivative and the native protein. The difference diffraction pattern has the same reciprocal dimensions and thus the same number of reflections, but represents the much simpler scenario of the "difference crystal."

Despite the inequality (10-2), difference data are still representative of the marker atom substructure and thus useful in the substructure determination, as we will discuss in the next section. For now, we shall concern ourselves with the sources of the electronic differences and the fundamental requirement of isomorphism for difference data. The question arises which sources of electronic differences are available for practical use in difference data generation. Multiple possibilities exist to introduce or utilize electronic differences in a crystal structure and the marker atoms providing the source of the electronic differences can be either natively present or artificially introduced.

Isomorphous differences

The traditional markers providing the source of electronic differences in protein crystals are heavy metal atoms. They can be soaked into a crystal as described in Chapter 3, and thus provide a derivative crystal. Intensity differences between a native crystal and a derivative crystal can be exploited for substructure solution. The methods are generally called isomorphous replacement methods, and, depending on how many derivatives are used, they are single isomorphous replacement (SIR) or multiple isomorphous replacement (MIR) methods. A

Figure 10-2 The complex structure factors for native protein and derivative. The complex structure factors for the derivative $\mathbf{F}_{PA}$ is the vector sum of native structure factor for the protein alone $\mathbf{F}_P$ and heavy atom structure factor $\mathbf{F}_A$. The structure factor amplitudes, indicated here by the double-pointed arrows, however, are not additive: $F_A \neq F_{PA} - F_P$. The radius of the circles is given by the amplitude of the corresponding structure factor. A complex structure factor with unknown phase can point to anywhere on its magnitude circle, the source of the phase problem.

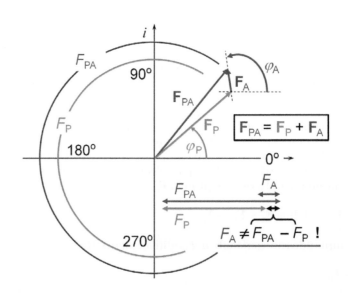

stringent requirement for all difference methods is isomorphism between the derivatives and native crystals. Isomorphous replacement methods were the first successfully applied phasing methods in protein crystallography and theoretically developed from the mid-1930s onward and put to practice in the mid-1950s in the first protein structure determinations (Sidebar 10-1).

Anomalous and dispersive differences

Any heavy atom marker—which includes even the natively present sulfur atoms in proteins and other native heavier atoms—can also act as a source of anomalous signal. Anomalous phasing techniques have largely supplanted the traditional multiple isomorphous replacement phasing techniques. As discussed in the anomalous scattering section in Chapter 6, the anomalous scattering contributions to the heavy atom structure factor provide the source for anomalous Bijvoet differences between centrosymmetrically related wedges (or corresponding symmetry equivalents) of reciprocal space within one data set, or for dispersive differences between data sets collected at different X-ray energies. The methods are generally called anomalous diffraction (or anomalous dispersion, AD) techniques and the heavy atom substructure can be derived from single-wavelength anomalous diffraction (SAD) data or from multi-wavelength anomalous diffraction (MAD) data. While MAD methods require exactly tunable X-ray wavelengths and thus generally a synchrotron source, SAD phasing techniques can be used as long as anomalous data can be recorded above an X-ray absorption edge. The use of anomalous data in conjunction with isomorphous replacement data as a general tool to break the handedness ambiguity in the phasing equations was introduced by Brian Matthews.[10] A significant advantage of anomalous phasing techniques is (subject to systematic and radiation damage induced errors) the absence of non-isomorphism between anomalous or dispersive diffraction data from one single crystal.

Non-isomorphism and estimating isomorphous signal change

Despite the general applicability of difference methods, one stringent necessity exists for the validity and success of difference-data-based phasing methods, namely that of isomorphism. We can understand the need for isomorphism from the gedankenexperiment shown in the top panel in Figure 10-1: If we subtract crystals from each other that do not have the atoms in exactly the same (or very nearly so) position, we will have "partial" or even whole atoms left over at many positions, which does not produce a clean and simple difference crystal.

Box 10-1 Principles of marker atom substructure phasing. The basis of experimental marker atom substructure phasing is difference intensity data, resulting from electronic differences caused by heavy or anomalous marker atoms. Sources of scattering differences with respect to a native reference crystal are introduced by heavy metal atoms. Anomalous scatterers provide anomalous differences within a data set and dispersive differences between data sets of the same crystal recorded at different wavelengths. Structural isomorphism is a fundamental requirement for difference methods. Isomorphism can be difficult to achieve for heavy atom derivative crystals, but is inherent in anomalous or dispersive data from the same crystal.

The difference data represent only the marker atom substructure of a few to a few hundred atoms, representing a greatly simplified situation compared with the thousands and tens of thousands of atoms in a protein structure. The heavy or anomalous marker atom positions can be determined from difference data by a variety of substructure solution methods based on Patterson methods and/or direct methods.

Once the marker atom positions are known, phasing equations for the protein phases can be solved, the initial protein phases can be determined, and the electron density maps can be reconstructed by Fourier synthesis.

Sidebar 10-1 Early history of isomorphous replacement. Isomorphous replacement was the phasing method that made possible the determination of the first three macromolecular structures, myoglobin,[1,2] hemoglobin,[3] and the first enzyme, lysozyme.[4] The method was used first by Green, Ingram, and Perutz at the MRC Laboratory in Cambridge in 1954 to phase the centric reflections of horse hemoglobin.[5] However, the Perutz team had difficulties obtaining usable derivatives, and the first complete protein structure of sperm whale myoglobin, determined by John Kendrew's team was reported in 1958[1] and followed in 1961 by a model experimentally phased to a then incredible (and still respectable!) 2.0 Å.[2] Shortly thereafter Michael Rossmann developed the dual-derivative Patterson synthesis[6] that solved the common origin problem and the hemoglobin structure was finally reported at 5.5 Å by the Perutz group[3] in 1962. In the same year, the group of David Phillips published the structure of hen egg white lysozyme at 6 Å,[4] and in 1965 the resolution was extended to 2.0 Å.[7] Personal reminiscences and accounts of some of the pioneering crystallographers are provided in the introductory chapters of the *International Tables for Crystallography, Volume F*; *Methods of Enzymology, Volume 368* (both sources are listed in the Additional reading section); and the inaugural volume of *Protein Science*.[8] The personalized historic account of Richard Dickerson[9] includes commented reprints of seminal key publications from this exciting early period of protein crystallography.

As a consequence, the difference pattern will contain much noise in addition to any marker atom information. The, in principle, perfect isomorphism of anomalous difference data obtained from one single crystal is a major factor contributing to the power of anomalous phasing techniques.

There are generally two sources of non-isomorphism from a practical point of view: those which change the cell constants (which is readily recognizable from refined cell constants upon initial indexing of the data), and those where the relative orientation of the molecules changes, which does not necessarily lead to a cell change. Often both effects coincide and can give rise to substantial non-isomorphism, which renders the derivatives useless. Changing the cell constants essentially means that we sample the derivative crystal on a reciprocal grid slightly out of phase with the native grid, and that the grid-mismatch effect becomes relatively larger at higher resolution (provided the derivative diffracts well in the first place).

Francis Crick and Beatrice Magdoff[11] estimated in the mid-1950s that small shifts of the molecule in a cell of 0.1 Å, or relative cell changes of 0.5%, may generate as much as 20–30% intensity differences, and increased radiation damage in derivative crystals compounds the problem of non-isomorphism. They derived an estimate for the expected change in reflection intensities as a function of the heavy atom derivatization:

$$\frac{\Delta I}{I} = k \frac{f_A}{f_P} \cdot \left(\frac{n_A}{n_P}\right)^{1/2} = 2\frac{\Delta F}{F} \tag{10-3}$$

where f represents the respective scattering factors at $\sin\theta/\lambda = 0$ (with the average f_P of 6.7 e⁻) and n is the respective number of atoms. For centric and acentric reflections, k is 2 and $\sqrt{2}$, respectively. The same formula can also be used to estimate the corresponding effect of anomalous and dispersive scattering contributions (Section 6.4 provides an example).

The result of the estimate is that—given high isomorphism and careful data collection—one single heavy metal such as Hg can produce useful intensity differences for a protein of several hundred kDa molecular weight. In practice, derivative data that merge with the native data with an overall R_{merge} value on F of approximately 15–30% are potentially useful, but the ultimate criterion remains whether heavy atom positions are found or not during the substructure solution. Some modern programs such as SHARP[12] or Phaser[13] treat non-isomorphism in a rather sophisticated way, while other programs do not attempt such adjustments.

Obtaining initial protein phases

In the following, we will investigate how we can, in principle, obtain phases through difference-based methods given the marker atom substructure. Knowing that we can do this will motivate us to develop the means to actually solve the marker atom substructure. Given the intensity data for the native crystal and for the derivative crystal, let us assume that we have a means of determining the heavy atom position(s) from difference data.

The knowledge of the marker atom *position* automatically implies knowledge of marker atom *phases*, as we can readily compute the complex structure factor from the standard structure factor summation derived in Chapter 6. We can now investigate the scenario given only the structure factor amplitudes F_P and F_{PA} from the native crystal and the derivative crystal, respectively, and the complex structure factors $\mathbf{F}_A$ for the heavy marker atom. The graphical representation of the phasing equations in the complex plane is called a Harker diagram after David Harker, who made substantial contributions to the interpretation of Patterson maps and the phase inequalities employed in direct methods.

A structure factor $\mathbf{F}$ of known magnitude F but unknown phase φ can lie anywhere on a circle of radius F in the complex plane. We can thus draw a circle with amplitude F_P for the native structure factor amplitude. We also know from the heavy atom substructure solution both the magnitude and the phase, and thus the direction of, the heavy atom structure factor $\mathbf{F}_A$. We can therefore graphically solve the phasing equation by drawing a second circle with radius F_{PA} and its origin offset by vector $\mathbf{F}_A$, as shown in Figure 10-3.

It becomes immediately obvious from Figure 10-3 that the solution for the phase angle φ_P is ambiguous. The 2-fold degeneracy in the SIR or SAD case can be resolved:

- by adding more derivatives (MIR);

- by using additional anomalous signal (SIRAS, MIRAS);

- by using dispersive and anomalous signal together (MAD);

- with the help of density modification procedures;

- or with dedicated, direct methods based programs.[14]

Each of these methods will be discussed in the main phasing section. Table 10-1 provides a quick overview of the most frequently used phasing methods and the data they require.

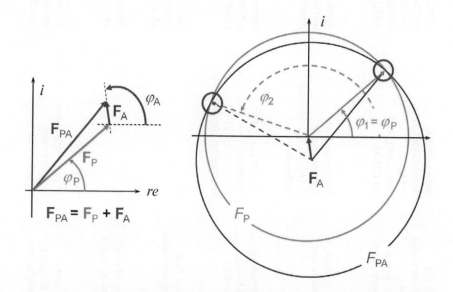

Figure 10-3 Graphical solution of the phasing equations. The left panel shows how the complex structure factors for protein, derivative, and heavy atom are related for a generic, acentric reflection *hkl*. Intensity measurements only reveal the magnitude of the structure factors for the derivative F_{PA} and the native F_P, but from the difference data, the position(s) of the heavy atom(s) can be determined. Therefore, the entire complex structure factor for the heavy atom $\mathbf{F}_A$ is known, and the two possible solutions for the protein phase angle for each reflection can be determined by the graphical procedure shown in the left panel. However, only one phase angle (φ_P) will be correct; the other one (φ_2) will be incorrect. In practice we do not know (yet) which one of the solutions is the correct one.

Box 10-2 Methods of experimental phase determination. Depending on the difference signal source and the data available, various methods to determine the protein phases can be distinguished and are often combined. Single isomorphous replacement (SIR), generic single anomalous diffraction (SAD), and native sulfur-based single anomalous diffraction (S-SAD) require the support of density modification or direct methods to break the inherent 2-fold phase ambiguity of the phasing equations.

The phase ambiguity can also be broken by adding orthogonal anomalous signal (AS) from the same crystal (SIRAS) or providing phases from at least a second derivative in multiple isomorphous replacement (MIR), or by combining both possibilities (MIRAS). Multiple anomalous diffraction (MAD) provides orthogonal dispersive and anomalous data from the same crystal and the phase ambiguity can be resolved directly from the phasing equations.

In optimal cases, one heavy atom can provide useful difference data to phase a protein structure of several hundred residues; heavy metal clusters can phase proportionally more. As a result of the weaker signal, anomalous phasing requires slightly more marker atoms per residue, but even sulfur atoms can provide a source of usable anomalous signal.

Phasing method	Phasing marker	Remarks	Available data	Marker	Difference data	Data to be phased
SAD via sulfur atoms (S-SAD)	S in Met, Cys residues	Highly redundant data collection, must be combined with density modification	Anomalous pairs F_{PS}^+, F_{PS}^- also serve as native data	Sulfur positions, $\mathbf{F}_S$	ΔF_{ano} from $|F_{PS}^+ - F_{PS}^-|$	Merged F_{PS}
SAD via naturally bound metals	Naturally bound metal ion, cofactor	Requires density modification for resolution of phase ambiguity	Anomalous pairs F_{PA}^+, F_{PA}^- also serve as native F_P	Anomalous scatterer positions, $\mathbf{F}_A$	ΔF_{ano} from $|F_{PA}^+ - F_{PA}^-|$	Merged F_{PA}
SIR(AS) via isomorphous metals	Heavy atom ion, specifically bound anions Br⁻, I⁻, I³⁻, also Xe	Isomorphous phasing power proportional to $z_{(H)}$, anomalous signal or density modification needed to break phase ambiguity	Native data F_P, isomorphous data F_{PA} in pairs F_{PA}^+, F_{PA}^- for SIRAS	Isomorphous/anomalous scatterer positions, $\mathbf{F}_A$	ΔF_{iso} from $|F_{PA} - F_P|$ and ΔF_{ano} from $|F_{PA}^+ - F_{PA}^-|$	Native F_P
MIR(AS) via isomorphous metals	Heavy atom ions, clusters, specifically bound anions Br⁻, I⁻, I³⁻	As above, except multiple derivatives or anomalous signal break phase ambiguity. Hg, Pt, Au, etc. phase several hundred residues, heavy atom clusters more.	Native data F_P, isomorphous data $n \cdot (F_{PA})$ in pairs $n \cdot (F_{PA}^+, F_{PA}^-)$ for MIRAS	Isomorphous/anomalous scatterer positions, $n \cdot (\mathbf{F}_A)$	$n \cdot \Delta F_{iso}$ from $n \cdot |F_{PA} - F_P|$, and ΔF_{ano} from $n \cdot |F_{PA}^+ - F_{PA}^-|$ pairs	Native F_P
MAD via Se	Se in Se-Met residues	1 Se phases 100–200 residues, introduced by expression host	Bijvoet pairs at n wavelengths, $n \cdot (F_{PA}^+, F_{PA}^-)_{\lambda n}$, optional native data	Anomalous scatterer positions $\mathbf{F}_{Se}$	ΔF_{ano} from $|F_{PA}^+ - F_{PA}^-|_\lambda$ pairs; ΔF_A from $|F_{\lambda i} - F_{\lambda j}|$ pairs	Best merged data F_{PA}, optional native F_P
MAD via isomorphous metals	Heavy atom, specifically bound	Strong signal on XAS "white lines," particularly at L-edges, can phase several hundred residues	Bijvoet pairs at n wavelengths $n \cdot (F_{PA}^+, F_{PA}^-)_{\lambda n}$, native data F_P (not needed for phasing)	Anomalous scatterer positions $\mathbf{F}_A$	ΔF_{ano} from $|F_{PA}^+ - F_{PA}^-|_\lambda$ pairs; ΔF_λ from $|F_{\lambda i} - F_{\lambda j}|$ pairs	Native F_P, optional best merged F_{PA}
Direct methods	None	Near atomic resolution data (1.2 Å or better), relatively small proteins	F_P	All non-H atom positions ab initio	N/A	Native F_P
Density modification	None	Needs multiple copies of motif in asymmetric unit for ab initio phasing	F_P	Multiple copies of a subunit	N/A	Native F_P
MR via model structure	Positioned search model	Needs search model with structural similarity, subject to model bias, particularly at low resolution	Native data F_P and search model structure factors F_C	Entire model serves as search probe	N/A	Native F_P

Table 10-1 Overview of common phasing methods. SAD: single-wavelength anomalous diffraction; MAD: multi-wavelength anomalous diffraction; SIR: single isomorphous replacement; MIR: multiple isomorphous replacement; (AS): with anomalous scattering; MR: molecular replacement.

10.3   Marker atom substructure solution

By creating difference intensities between data sets with and without the scattering contributions originating from the marker atoms, the initial structure solution problem is reduced to solving a substructure of a few up to a few hundred atoms, versus many thousands or tens of thousands of atoms in the whole protein structure. Once the marker atom substructure is obtained, we can proceed to solving the phasing equations for all reflections of the protein as outlined in the following sections.

After obtaining the initial marker atom position from a substructure solution, it can be refined against the available derivative data. Because the heavy atom or anomalous marker atom refinement and the initial protein phase calculations are intimately connected, most modern programs combine both steps. The modern crystallographic programs for heavy atom substructure solution such as *SHELXD*,[15] *SnB*,[16] *SOLVE*,[17] and *CNS*[18] (*SHARP*[19] does not find heavy atoms but can extend heavy atom substructure) are very powerful and complete structure determination program packages such as *CNS*, *SOLVE/RESOLVE*, and *Phaser*,[13] or *autoSHARP*[20] and the *PHENIX*[21] suite are highly automated. Given reasonable difference data it is very likely that a heavy atom substructure solution is going to succeed. Figure 10-4 provides a quick overview of the key steps of substructure solution and experimental phasing.

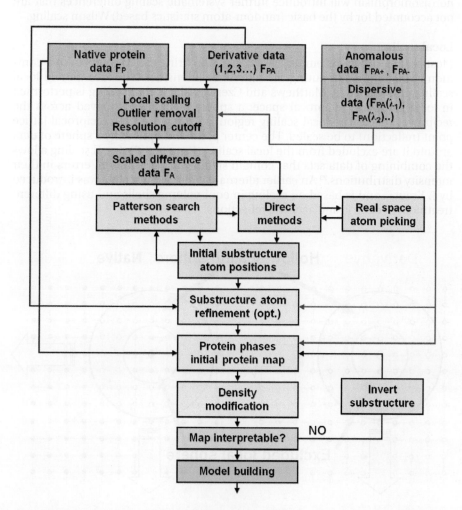

Figure 10-4 Overview of marker atom substructure-based experimental phasing. The most general experimental phasing methods are marker atom substructure-based methods (blue shading) that allow *de novo* phase determination from difference data (purple shading). Following proper scaling, the difference data enter a Patterson-based or a direct methods-based initial substructure site search. Once initial marker atom sites are found, they may be refined against the appropriate data, and phases for the native protein data set (or in the case of native anomalous scatters, the anomalous data) are calculated. If subsequent density modification does not deliver an interpretable map, the substructure is inverted and new phases are generated using the other substructure enantiomorph.

We will limit the discussion of the "classical" method of solving a difference Patterson map by hand to one rather simple example and then rapidly progress to automated Patterson vector searches and direct methods procedures, maximum likelihood heavy atom refinement, and associated phase probability computation methods. Given the phase bias issues, obtaining the best possible map before any model building is attempted will not only make the subsequent model building much easier and much more accurate, but as a consequence will also minimize model bias in the final structure. It is thus always worthwhile putting effort into producing the best possible substructure solution and best possible initial phases. Also noteworthy is that, given good difference data, dedicated macromolecular substructure solution programs such as *SHELXD* often deliver *complete* and *accurate* substructure solutions, effectively obsoleting the step of heavy atom position refinement and allowing us to directly proceed to density modification.

If only a few heavy atoms are present as substructure markers, there is a good chance that a Patterson map of the difference data, which is an interatomic vector map (Chapter 9), will in fact show a limited number of heavy atom–heavy atom peaks and be interpretable. Patterson methods were the classical substructure solution technique in the determination of the first three macromolecular structures, myoglobin,[1, 2] hemoglobin,[3] and the first enzyme, lysozyme[4] (Sidebar 10-1).

Preparing the difference data

Irrespective of the method used to solve the substructure, the first point we have to address is to ensure that the two data sets containing the native data $F_P{}^2$ and the derivative data $F_{PA}{}^2$ are on a common scale. This can be accomplished in principle by the basic Wilson scaling procedure, as discussed in Chapter 7. However, heavy atom derivatives tend to absorb strongly and probably anisotropically, and non-isomorphism will introduce further systematic scaling differences that are not accounted for by the basic (random-atom statistics based) Wilson scaling.

Local scaling

Most scaling programs employ local scaling to bring the data sets onto a common scale before computing the difference intensities. The idea behind local scaling, introduced by Matthews and Czerwinski,[22] is that scaling is performed in local groups of reciprocal space: a sphere of indices is moved across the reciprocal space. The local scaling region is centered on the reciprocal lattice point (reflection) to be scaled. The center point and optionally a sphere of data around it are excluded from the local scaling (Figure 10-5). Local scaling allows the combining of data sets that contain significant non-random errors in their intensity distributions.[23] An earlier alternative to Wilson scaling was introduced by J. Kraut,[24] and is based on scaling in equi-volume shells and using different treatment of centric and acentric reflections.

Figure 10-5 Local scaling. Derivative and native data are scaled in local regions (cyan hollow sphere) around each reflection to be scaled (the larger drawn reciprocal lattice point in the center), which is excluded from the scaling process. A local sphere of reflections can also be excluded to avoid scaling away the signal in the case of correlated errors. The process is repeated for each reflection. The figure is adapted from C. W. Carter.[25]

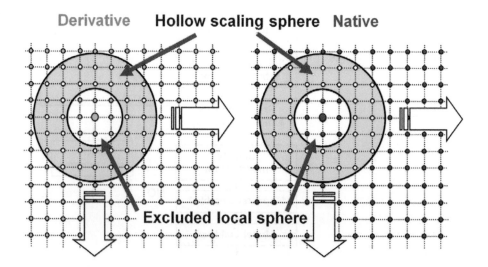

Anisotropic scaling

Many crystals do not diffract isotropically, that is, their diffraction limits vary significantly in different directions in reciprocal space. As we have shown in Chapter 9 in the case of the Fourier transform of an anisotropically perturbed lattice function (Figure 9-14), anisotropic attenuation of diffraction is largely a result of intrinsic displacive disorder in the lattice packing of the crystal, often introduced through stress during the heavy atom soaking. The situation is not uncommon, and can be accounted for by anisotropic scaling. Likelihood functions are particularly susceptible to systematic (and thus correlated) errors in structure factors because the use of joint probabilities in the structure factor distributions explicitly requires the independence of individual probabilities.

Most scaling programs account for diffraction anisotropy (Figure 10-6) by fitting anisotropic B-factors[26] so that the resolution of the scaled diffraction intensities becomes constant in all directions. This works fine as long as the anisotropy is not too severe; the anisotropic scaling essentially weighs up the reflections in the poor direction and attenuates the good ones. In Section 12.2 anisotropic displacement and anisotropic diffraction corrections are discussed in detail.

Most scaling programs provide additional statistical analysis that can be useful in cases when a derivative data set fails to provide a substructure solution, suggesting a change in scaling strategy. As long as heavy atom positions can be found, the choice of initial scaling method is not that relevant, because the initial scaling is in any case further adjusted in the heavy atom refinement process.

Outlier detection and resolution truncation

Most scaling programs (including difference data scaling), or automated packages that include a scaling step, also perform the important task of outlier detection in addition to resolution truncation.

High intensity differences. Because they are used in substructure determination as either squared intensity or squared normalized structure factor amplitudes, extremely high differences will dominate the Patterson function or triplet relations used in direct methods to determine the substructure. To suppress abnormally high difference values a basic outlier cutoff, based on root mean square deviations (r.m.s.d.) $\langle \Delta F^2 \rangle^{1/2}$ from the average difference in the corresponding resolution shell, can be applied and differences that exceed about four times the r.m.s.d. are generally rejected or down-weighted. More sophisticated tests for outliers in non-redundant data sets, based on statistical intensity expectation values,[27] have also been suggested. The outliers are preferably scaled down rather than just deleted. Scaling down outliers instead of deleting them has the advantage that no sharp and arbitrary resolution cutoff has to be selected by the user.

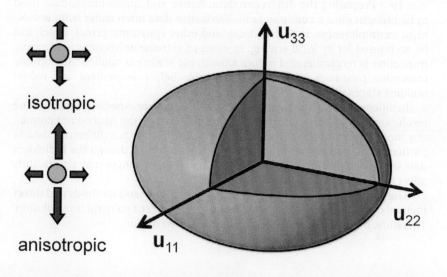

Figure 10-6 Anisotropic scattering ellipsoid. Anisotropic properties are described by a second rank tensor, which can be visualized as a property ellipsoid with principal axes $\mathbf{u}_{ij}$. If the displacement disorder in a crystal structure expressed as anisotropic displacement parameters vary in different directions of reciprocal space, the diffraction or resolution limits will be anisotropic as well.

Noise removal. High-resolution difference data that contain largely noise, either because of weak anomalous signal or low isomorphism, are truncated because they add only noise to Patterson maps or give poor E-values which in turn affect direct methods. They are removed by applying a so-called sigma-cutoff, that is, reflections (generally in shells) with a signal-to-noise ratio $\langle I/\sigma(I)\rangle$ (Chapter 8) of less than ~1–3 are omitted. Proper resolution truncation of anomalous difference data can drastically affect the chance of substructure solution and if no, or only poor, solutions are obtained a more aggressive cutoff, based on the internal correlation of anomalous data, may improve the situation[15] as discussed in detail in the practical examples in Section 10.8. While there is a good physical reason for resolution truncation of data during substructure solution, removing low $\langle I/\sigma(I)\rangle$ reflections is not proper practice for later stages of structure refinement against single data sets using likelihood-based refinement programs. In this case, even weak, high sigma reflections contain more information than no reflections at all!

Scaling programs instructed to generate difference data will generally provide a list of various merging R-values by resolution shell. A general difference merging R-value, R_Δ, between data pairs of type A and type B has the form

$$
R_\Delta = \frac{\sum_{\mathbf{h}}\left|I_{A(\mathbf{h})} - I_{B(\mathbf{h})}\right|}{\sum_{\mathbf{h}} \overline{I}_{(\mathbf{h})}} = 2\frac{\sum_{\mathbf{h}}\left|I_{A(\mathbf{h})} - I_{B(\mathbf{h})}\right|}{\sum_{\mathbf{h}}\left(I_{A(\mathbf{h})} + I_{B(\mathbf{h})}\right)}
\tag{10-4}
$$

The data pairs (A,B) can be native and isomorphous data, anomalous data $(I_{(+\mathbf{h})}$ and $I_{(-\mathbf{h})})$, or dispersive differences, and the same formula holds for amplitudes $F_{\mathbf{h}}$. As a first approximation, R_Δ of F will be in the range of ~10–30% for useful isomorphous difference data, and will as usual increase with resolution. At higher R-values, the data probably contain mainly noise due to non-isomorphism and are omitted from the difference data set. Anomalous difference data are particularly sensitive to noise, and we will discuss special procedures to treat them in the VTA1 example (Section 10.8) for which we have processed high redundancy S-SAD data in Chapter 8.

High resolution limits of around 4–3.5 Å are often sufficient to determine the substructure solution, while density modification techniques (Section 10.7) are capable of extending the protein phases to higher resolution. If upon merging a near-random R_F-value around ~60% over the entire data range results, there is reason to suspect inconsistent indexing between the data sets. This can happen in certain hexagonal, trigonal, tetragonal, and cubic space groups. In this

---

Box 10-3 Preparing the difference data. Native and isomorphous data need to be brought onto a common scale. Derivative data often suffer from anisotropy, incompleteness, radiation decay, and other systematic errors, which can be accounted for by local scaling. In cases of extreme anisotropy, anisotropic truncation is recommended before anisotropic scaling is applied. Anisotropic truncation treatment is particularly valuable before refinement and model building stages.[28]

In difference data, the weak high resolution data are especially poor and the resolution needs to be truncated from the point where the data do not contain any more information and thus add only noise. Anomalous difference data in particular can become noisy at modest resolution, even though the individual data sets may diffract to high resolution. Different data ranges may significantly affect the outcome of a substructure solution.

The highest difference intensities affect both Patterson methods and direct methods, and a few bad strong reflections can prevent substructure solution. Therefore, intensity outliers are removed or scaled down.

case, the data need to be indexed with a consistent cell, and the indices of one data set may need to be transformed. The 27 space groups in which multiple non-equivalent indexing possibilities exist and their re-indexing operators and transformation matrices are listed in Table 8-4.

Difference Patterson maps

Let us briefly recall from Section 9.4 what a Patterson map shows. The Patterson function is an autocorrelation of the electron density of the structure and thus represents a map of interatomic distance vectors. If there are N strong peaks in the electron density, then the Patterson map

$$P(\mathbf{u}) = FT^{-1}(F_h^2) = \sum_h F_h^2 \exp(-2\pi i \mathbf{hu}) \tag{10-5}$$

will show $N(N-1)$ strong interatomic distance values. The Patterson reconstruction requires only intensities (squared amplitudes) and no phase. Therefore, a Patterson map generated from difference data that reflect the marker atom substructure alone should contain only peaks between the marker atoms. The questions are:

- Which intensity differences should be used? One could either subtract the intensities directly and generate the Patterson map from coefficients ($F_{PA}^2 - F_P^2$), or convert first to Fs, and take their squared difference ($F_{PA} - F_P)^2$ as Fourier coefficients.

- Given that the differences in the structure factor amplitudes do not reflect the true vector differences (Figure 10-2 and Equation 10-2), to what degree do the difference data actually provide a reasonable representation of our "difference crystal"?

Addressing the first question, we can substitute $\mathbf{F}_{PA} = \mathbf{F}_P + \mathbf{F}_A$ and using $F^2 = \mathbf{F} \cdot \mathbf{F}^*$ we expand as follows

$$F_{PA}^2 - F_P^2 = (\mathbf{F}_P + \mathbf{F}_A)(\mathbf{F}_P^* + \mathbf{F}_A^*) - \mathbf{F}_P\mathbf{F}_P^* = F_A^2 + \mathbf{F}_A\mathbf{F}_P^* + \mathbf{F}_P\mathbf{F}_A^* \tag{10-6}$$

The result is that this map from ($F_{PA}^2 - F_P^2$) coefficients will contain the correct heavy atom positions only in the F_A^2 term, but also substantial noise contributions from the many protein–heavy atom distances.

Isomorphous difference Patterson map

A more useful map is the isomorphous difference Patterson map generated from ($F_{PA} - F_P)^2$ coefficients, which combines the contributions of centric and acentric reflections. This is of great value because centric projections in a Patterson map are directly interpretable in terms of interatomic distances, a not insignificant advantage for our quest.

Centric case. Let us first consider the centric case. Recall that general centric reflections are phase-restricted reflections where the phase angles are limited to 0° or 180° or pairs of special angles such as 90°/270° (Chapter 6). We consult the vector diagram and realize that in these cases the difference $F_{PA} - F_P$ is actually an accurate estimate for F_A as long as F_A is small, compared to F_{PA} and F_P. This is generally the case, and only if the observed values are very small can the rare situation of a "crossover" occur. We obtain the centric coefficients F_A either as

$$F_A = F_{PA} - F_P \tag{10-7}$$

if F_{PA} is larger than F_P and

$$F_A = F_P - F_{PA} \tag{10-8}$$

if F_{PA} is larger than F_A. In both cases,

$$F_A^2 = \left(F_P - F_{PA}\right)^2 = \Delta F_{ISO}^2 \tag{10-9}$$

and the centric projection of the isomorphous difference Patterson map contains the exact representation of the heavy atom positions. We will therefore shortly examine centric projections or Harker sections of Patterson maps.

Acentric case. For the general, acentric case, using the cosine rule we can derive[6] from Figure 10-2 the following relation:

$$F_A^2 = F_P^2 + F_{PA}^2 - 2F_P F_{PA} \cos(\varphi_P - \varphi_{PA}) \tag{10-10}$$

As long as both F_{PA} and F_P are substantially larger than F_A, the difference in the phase angles will be nearly 0 and therefore the cosine term ~1. Thus again,

$$F_A^2 \approx F_P^2 + F_{PA}^2 - 2F_P F_{PA} = \left(F_{PA} - F_P\right)^2 = \Delta F_{ISO}^2 \tag{10-11}$$

For small derivative and protein structure factors crossovers are rare, as in the centric case, and in addition it can be shown that in the acentric case the Patterson map contains the heavy atom positions and again a noise term whose magnitude depends on the phase angle difference between the derivative and the heavy atom structure factor:[29]

$$\Delta F_{ISO}^2 = F_A^2 \cos^2(\varphi_{PA} - \varphi_A) \tag{10-12}$$

Anomalous difference Patterson map

Another type of Patterson map that is frequently used is the anomalous difference Patterson map, with coefficients

$$\Delta F_{ANO}^2 = \left(F_{PA}^+ - F_{PA}^-\right)^2 \tag{10-13}$$

The derivation of the contents of the coefficients in terms of F_A is similar to the isomorphous case, and we expect from the complementarity between isomorphous and anomalous differences (Section 10.4) that the analogous formula for the anomalous case will contain a sine term instead of the cosine term:

$$\Delta F_{ANO}^2 = F_A^2 \sin^2(\varphi_{PA} - \varphi_A) \tag{10-14}$$

Similarly to the combination of anomalous contribution and isomorphous differences allowing us to break the two-fold phase ambiguity in the SIRAS case, in a combined anomalous and isomorphous difference Patterson map the angle-dependent noise terms will cancel out:

$$\Delta F_{ANO}^2 + \Delta F_{ISO}^2 = F_A^2 \left(\sin^2(\varphi_{PA} - \varphi_A) + \cos^2(\varphi_{PA} - \varphi_A)\right) = F_A^2 \tag{10-15}$$

The combined isomorphous and anomalous Patterson map with coefficients F_A^2 (MAD case) or $\Delta F_{ANO}^2 + \Delta F_{ISO}^2$ (SIRAS case) thus will give—with the provision of F_A being small compared to F_{PA} and F_P—for the acentric case a map of comparable quality to a centric projection.

Having determined that Patterson maps computed from difference data indeed provide a reasonably accurate representation of the marker atom arrangement, we are ready now to examine the maps and find a way to determine the heavy atom positions from distance vectors, that is, "solve" the Patterson map.

Harker sections and distance vectors

Using scaled difference data, we can compute an isomorphous difference Patterson map generated from $(F_{PA} - F_P)^2$ coefficients. This map will contain peaks at the interatomic distance vectors. Now the question is how to find these peaks and how to translate the corresponding set of interatomic distances into a useful atomic position.

The general interatomic position vectors can be computed from the difference between the general position generators listed in the space group tables. Doing so, we realize the following:

1. For a single atom, there will be a set of $m^2 - m = m(m - 1)$ distance vectors, where m is the multiplicity of the general position. For example, if $m = 3$, we will find six peaks, 1–2, 1–3, 2–3, 2–1, 3–1, and 3–2 in the Patterson unit cell. If we have n atoms in the asymmetric unit cell, the cell will contain $m \cdot n = N$ atoms, and correspondingly $N^2 - N$ peaks. Although this number increases geometrically with increasing m or n, we need to search only the asymmetric unit of the Patterson space for a set of consistent peaks according to the symmetry of the Patterson space (space group plus inversion, Table 6-6).

2. The possible difference vectors between symmetry related equivalent atoms fall into certain sections, that is, one coordinate assumes a fixed value such as 0, $1/2$, $1/3$ or $1/4$. These sections (centric projections) are the Harker sections. If we have only one atom in the asymmetric unit, the Harker sections contain all Patterson peaks between symmetry related atoms, termed the self-vector peaks. If more than one atom in the asymmetric unit is present, there will be cross-vector peaks in non-Harker sections. These cross-peaks will automatically be accounted for if a correct and consistent solution is found, and they are used in manual solution as well as by several heavy atom search programs to test and score possible solutions. The Harker sections, general and special distance vectors, and cross-vectors can be found in the Patterson peak compilation (Patterson Atlas) by D. Ward.[30]

3. As the Patterson map is centrosymmetric, both the correct enantiomorph of the substructure and the inverted substructure will correctly reproduce the Patterson map. Therefore, both enantiomorphs of the substructure are tried in the subsequent protein phase determination. The wrong handedness of the substructure will not give a correct map. Table 10-2 lists the possible combinations of enantiomorph ambiguity with the 2-fold phase ambiguity.

Manual solution of a difference Patterson map

In almost all practical cases, the heavy atom positions are determined by automated programs. They are exceptionally powerful, largely because they combine different approaches—some weak difference data often yield solutions only by one method and not by another—and the programs can rapidly test many possible solutions for the best one. It is nevertheless very instructive to execute one example of a difference data solution by manual Patterson map interpretation.

Let us assume we have properly scaled diffraction data from a native crystal and from an isomorphous derivative. We employ as an example the ~450 residue *C. tetani* neurotoxin C-fragment and a single gold ($z = 79$) derivative, prepared by soaking the native crystals with a $KAu(CN)_2$ complex salt solution. The gold atom was found to be covalently bound to one of the cysteine residues (Figure 3-42). The space group is orthorhombic $P2_12_12_1$, the most frequent space group for protein crystals (Figure 5-44), with cell constants of $a = 71.2$, $b = 79.4$, and $c = 93.8$ Å, and contains one toxin molecule in the asymmetric unit. We expect from the Crick and Magdoff estimate (10-3) an overall relative intensity difference of about 26%, which—with reasonable isomorphism and good data collection—should provide useful difference data even in the case of partial occupancy of the heavy atom. The difference data are usable to about 3.0 Å.

Location of Harker sections and Patterson vectors

We can derive the Harker section location and the Patterson vectors directly from the space group operators. From the symbolic operators we can generate a set of conditions that all distance vectors must fulfill. Applying space group symmetry, a heavy atom in $P2_12_12_1$ occupies the four general positions

(1) $x,$ $y,$ z

(2) $1/2 + x,$ $1/2 - y,$ $- z$

(3) $- x,$ $1/2 + y,$ $1/2 - z$

(4) $1/2 - x,$ $- y,$ $1/2 + z$

From the difference of these positions, we obtain 12 Patterson vectors (u, v, w) which reduce to a unique set of three general non-zero distance vectors with coordinates

Diff.	u,	v,	w
$(1-2)$	½,	½ + 2y,	2z
$(1-3)$	2x,	½,	½ + 2z
$(1-4)$	½ + 2x,	2y,	½

The (self) distance vectors (u, v, w) between a set of equivalent atoms in the general position x, y, z are thus located in the centric projections or Harker sections $(u, v, ½)$, $(u, ½, w)$, and $(½, v, w)$ for space group $P2_12_12_1$, and from the Harker vectors above we can derive the heavy atom coordinates.

For space group $P2_12_12_1$ with general position multiplicity of four and only one heavy Au atom in the asymmetric unit, we expect four heavy atoms in the unit cell. According to our peak formula, the Patterson map then should contain $4 \cdot (4-1) = 12$ non-origin Patterson peaks in the whole Patterson vector space $0 \leq u, v, w \leq 1$. However, we need to search only the asymmetric unit of the Patterson space, and for Patterson symmetry $Pmmm$ (obtained from space group symmetry $P2_12_12_1$ plus inversion) this is one octant $0 \leq u, v, w \leq 0.5$ containing three peaks. The centric projections or Harker sections for space group $P2_12_12_1$ containing the peaks are $(u, v, ½)$, $(u, ½, w)$, and $(½, v, w)$, that is, the faces of the $0 \leq u, v, w \leq 0.5$ cube as shown in Figure 10-7.

Visual inspection of the Harker sections provides us with three approximate Patterson peak positions, (½, 0.22, 0.45), (0.32, ½, 0.04), and (0.18, 0.28, ½). The visual estimate is consistent with the peak search results from the CCP4 Patterson peak search program *PEAKMAX*:

Figure 10-7 Harker sections for space group $P2_12_12_1$. The Harker sections of a Patterson map for space group $P2_12_12_1$ $(u, v, ½)$, $(u, ½, w)$, and $(½, v, w)$ contain the interatomic distance vectors between each set of equivalent atoms. Other sections contain the cross-peaks between different atom sets. The red map contours represent negative values. In the depicted case, the Harker peaks from the single Au heavy atom in the 3 Å Patterson map are strong, and clearly identifiable. We can read the distance vector components u, v, w directly from the map sections: (½, 0.22, 0.45), (0.32, ½, 0.04), and (0.18, 0.28, ½).

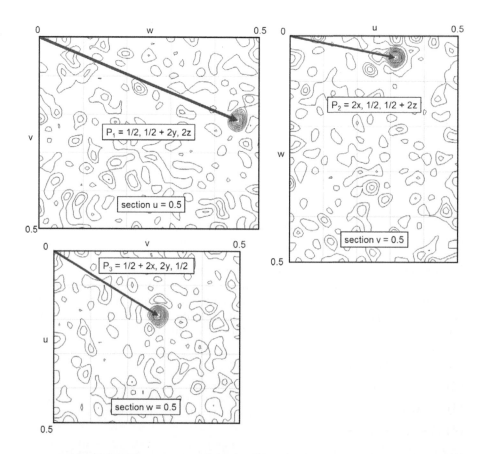

```
--------------- u ----- v ----- w ------ height --
ATOM1   Ano   0.0000   0.0000   0.0000      138.46
ATOM2   Ano   0.3150   0.5000   0.0472       12.76
ATOM3   Ano   0.1878   0.2826   0.5000       11.84
ATOM4   Ano   0.5000   0.2167   0.4518       10.24
ATOM5   Ano   0.0837   0.5000   0.0000        4.46
ATOM6   Ano   0.0000   0.5000   0.0509        3.63
ATOM7   Ano   0.3761   0.5000   0.0000        3.48
... etc
```

Note the high origin peak at 0, 0, 0 and the sharp drop-off in peak height after the first three correct and expected solutions. A clear drop-off in ranking (such as height of peaks, Patterson correlations in atom searches, occupancy, or combined figures of merit) that distinguishes correct solutions from spurious ones is generally a good sign.

From distance vectors to atom position. Having found the Patterson peaks, in the next step we need to extract a consistent set of atom coordinates from the Harker vectors. Using the first difference vector (1–2)

Diff.	u,	v,	w
(1–2)	½,	½ + 2y,	2z

and the values for u, v, w from the first Patterson peak from our manual peak search (½, 0.22, 0.46) we obtain ½ + 2y = 0.22 and thus y = –0.14; and 2z = 0.46 and thus z = 0.23. We can now take for example (1–4) as the second vector to solve for the missing x:

Diff.	u,	v,	w
(1–4)	½ + 2x,	2y,	½

We obtain ½ + 2x = 0.18 and thus x = –0.16. However, if we solve for y from this vector, we obtain y = +0.14 instead of –0.14. Is this a problem? Not really, because we know that our Patterson vector space symmetry $Pmmm$ implies that for any vector u, v, w there is also a vector u, –v, w and the assignment of y = –0.14 corresponds to this symmetry related Patterson peak. What is important, however, is that the obtained solution is self-consistent for the resulting set of coordinates (Figure 10-8). Our solution for the heavy atom position of (–0.16, –0.14, 0.23) is consistent and correct, while (0.16, –0.14, 0.23) belongs to the enantiomorphic substructure.

Note that in triclinic $P1$, the first atom can be placed anywhere, and then determines the origin. Similarly, in polar space groups with a single unique rotation or screw axis, the coordinate along the polar axis is floating. In $P2_1$, for example, the first atom can be placed anywhere along the y axis and then fixes the origin.

Multiple origins and Cheshire groups

Let us now examine if these coordinates (–0.16, –0.14, 0.23) correspond to the gold atom position in the deposited PDB file:

```
CRYST 71.180 79.380 93.810 90.00 90.00 90.00
HETATM 3659 AU AU 501 95.590 28.474 21.186 1.00 50.87
```

To obtain fractional coordinates from the (in this case) orthorhombic, orthogonal Cartesian coordinates, we just divide them by the cell constants a = 71.2, b = 79.4, and c = 93.8 Å, and we obtain (1.34, 0.36, 0.23). Translation along **a** by one unit cell with **T** = (–1, 0, 0) yields (0.34, 0.36, 0.23). This is not the same result that we obtained from the Patterson map interpretation, and we detect that the coordinates are (½ + x, ½ + y, z), that is, obviously shifted by (½, ½, 0). Inspection of

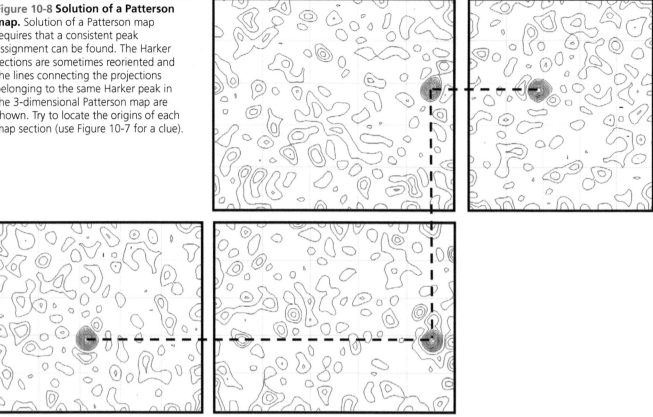

Figure 10-8 Solution of a Patterson map. Solution of a Patterson map requires that a consistent peak assignment can be found. The Harker sections are sometimes reoriented and the lines connecting the projections belonging to the same Harker peak in the 3-dimensional Patterson map are shown. Try to locate the origins of each map section (use Figure 10-7 for a clue).

the *Origin* statement in the $P2_12_12_1$ space group table (Figure 10-9) reveals that the origin can be located at the midpoint of (any) three non-intersecting pairs of parallel 2_1 axes. These origins can be constructed by permuting shifts by ½ along the cell axes as shown in the projections in Figure 10-9, giving eight possible origin choices (000, ½00, 0½0, 00½, ½½0, 0½½, ½0½, ½½½).

The different possible origin and enantiomorph choices (1 to 16 depending on the space group) can be combined with the symmetry operators, generating the complete set of operators (in our $P2_12_12_1$ case 16) of the so-called Cheshire group. Cheshire groups can be understood as the symmetry groups of the symmetry operators, and generally play a role in crystallographic search procedures (including heavy atom searches and translation searches in molecular replacement) and are formally derived as the *Euclidean normalizers* of the space groups.[31] The limits of the Cheshire cell that needs to be searched have been tabulated by Hirshfeld[32] and can be found in the CCP4 documentation. The question of possible origin shifts also arises in direct methods, where reflections related by origin shifts are called structure semi-invariants, because the amplitudes but not the phases of the structure factors are invariant upon origin shift.

Because complex structure factors for the same unit cell with different origins have invariant amplitudes but different phases, care must be taken to assure the *same origin* for the marker atom substructure has been chosen when combining structure factors from different phasing programs.

Heavy atom cross-Fourier maps
The choice of origin does not affect the final structure, and it would not matter which origin was chosen for final deposition of a PDB file (although one generally strives to have the majority of the molecule within the unit cell). Our solution for the heavy atom position is equivalent to the one found in the actual PDB

International Tables for Crystallography (2006). Vol. A, Space group 19, pp. 206–207.
Copyright © 2006 International Union of Crystallography

$P2_12_12_1$ D_2^4 222   Orthorhombic

No. 19 $P2_12_12_1$ Patterson symmetry $Pmmm$

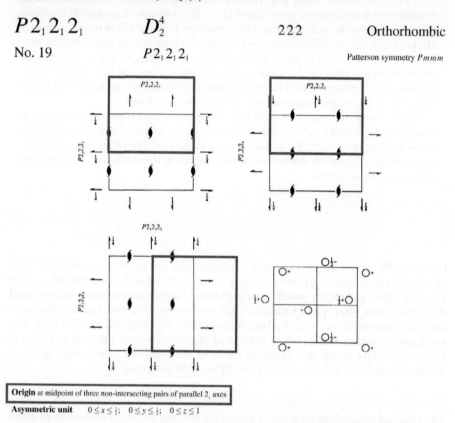

Origin at midpoint of three non-intersecting pairs of parallel 2_1 axes

Asymmetric unit $0 \le x \le \frac{1}{2}$; $0 \le y \le \frac{1}{2}$; $0 \le z \le 1$

Figure 10-9 Multiple origins in space groups. The three red unit cells superimposed on the projections in space group $P2_12_12_1$ have alternate origins, each generated by a shift of ½ in along a unit cell axis. There are eight possible origin choices in space group $P2_12_12_1$ which can be obtained by permutation of the shift vectors: 000, ½00, 0½0, 00½, ½½0, 0½½, ½0½, and ½½½. International Tables section reproduced with permission of the IUCr.

coordinate file, and would give a correct and equivalent structure, just shifted by (½, ½, 0). However, if we plan on using a second derivative, the origin choice needs to be consistent for all derivatives. To bring them on a common origin, we can construct a so-called cross-Fourier map. This is a difference Fourier map with coefficients

$$\Delta\mathbf{F} = m\left|F_{PA2} - F_P\right|\exp(i\varphi_{SIR1}) \tag{10-16}$$

using the SIR centroid phases φ_{SIR1} and their figure of merit m obtained from the first derivative and the isomorphous differences for the second derivative PA2. The coordinates of the second and all further derivative atoms can then be directly picked up from the difference Fourier map. The initially determined heavy atom substructure(s) can be further refined during the phasing process.

For two derivatives, a derivative Patterson map can be constructed with coefficients $(F_{PA1} - F_{PA2})^2$ which, in analogy to the derivation for the difference maps, contains positive peaks at the positions between atoms of one derivative, and negative peaks in the mixed peaks. The benefit is that no absolute scaling in necessary, but even for few atoms, this map gets very crowded leading to problems resolving the peaks. This map type was successfully used to determine the Hg positions in the first structure of horse hemoglobin.[5]

Once major sites have been identified, low occupancy sites that were missed in initial Patterson peak searches can be picked up from a derivative-difference Fourier map, that is, an electron density map computed from the Fourier coefficients

$$\Delta\mathbf{F} = m\left|F_{PA}(obs) - F_{PA}(calc)\right|\cdot\exp\left(i\varphi_{PA}(calc)\right) \tag{10-17}$$

Box 10-4 Patterson maps and Harker sections. Difference Patterson maps constructed from Fourier coefficients $(F_{PA} - F_p)^2$ contain interatomic distance vectors between the heavy atoms. The centric projections of the maps in special Harker sections provide a direct measure for the interatomic self-distance vectors and can be solved to yield the heavy atom positions. In space groups with multiple origins only the Cheshire cell needs to be searched for the first atom. In polar space groups with a floating origin, the first atom can be placed freely along the unique axis direction and the origin is then fixed.

Once the first derivative positions have been determined, the required common origin can be assured by constructing an isomorphous or anomalous cross-Fourier map from SIR or SAD phases of the first derivative and searching the map directly for the difference peaks indicating the marker atom positions of the second derivative.

Complex Patterson maps of metal substructures containing more than a few atoms are rarely solved by hand, largely because of the tedium involved and the fact that most computer programs combine different approaches—some weak difference data often yield solutions only by one method and not by another—and the programs can rapidly test many possible solutions for the best one. For historic interest, a detailed review of stepwise (semi) manual solution of Patterson maps using the classical method we outlined above was given in McRee's handbook accompanying his *XTALVIEW* program.[33]

Automated Patterson search methods

The manual procedure described above for solving Patterson maps in practice becomes intractable for complex substructures, hence a variety of computer methods have been developed to solve them. Patterson methods can also be used to provide strong restraints for the starting values of trial phases in direct methods such as in the program *SHELXD* which we will describe later. The question arises, how can we efficiently implement search methods that can handle a substructure that may contain a significant number of marker atoms?

Just as we did in the manual exercise, programs that search or interpret Patterson maps start with a peak list containing the Harker peaks and cross-peaks. Once this peak list is established, the further search for heavy atom sites can proceed in different ways. We will briefly describe basic real space grid searches and Patterson vector based methods.

Real space grid searches

The simplest way to find heavy atom sites is to place a first, single atom successively at each grid point within the asymmetric unit, expanding the cell, and to compute a linear correlation coefficient (7-53) for each resulting calculated Patterson map with the Patterson map constructed from the experimental difference data. The best correlation will then likely indicate a correct heavy atom site. If more significant Harker peaks remain, another atom can be added and the procedure repeated. In this case the correlation with the cross-peaks will also be scored to test whether the coordinates correspond to a correct second site.

Such basic real space grid searches as implemented in *RSPS* from the CCP4 package or *Xhercules* from the *XTALVIEW* suite[33] are quite robust and still useful for finding a limited number of first positions, but they do not have the power of the more complex vector search programs described later. As a practical example we examine the search results of the *RSPS* program applied to our Patterson map of the gold derivative.

```
REMARK File written by RSPS on 22/06/07
CRYST 71.180 79.380 93.810 90.00 90.00 90.00
REMARK TYPE  SITE   X       Y       Z      SCORE BFAC
ATOM  1 AU MTL 1 60.013 28.451 21.216 11.56 20.00
ATOM  2 AU MTL 1 60.013 50.929 21.216 11.56 20.00
ATOM  3 AU MTL 1 46.757 68.141 21.216 11.56 20.00
ATOM  4 AU MTL 1 60.013 11.239 21.216 11.56 20.00
ATOM  5 AU MTL 1 11.167 28.451 21.216 11.56 20.00
ATOM  6 AU MTL 1 11.167 11.239 21.216 11.56 20.00
ATOM  7 AU MTL 1 60.013 68.141 21.216 11.56 20.00
ATOM  8 AU MTL 1 11.167 50.929 21.216 11.56 20.00
ATOM  9 AU MTL 1 24.423 68.141 21.216 11.56 20.00
ATOM 10 AU MTL 1 11.167 68.141 21.216 11.56 20.00
ATOM 11 AU MTL 1 46.757 11.239 21.216 11.56 20.00
ATOM 12 AU MTL 1 24.423 11.239 21.216 11.56 20.00
ATOM 13 AU MTL 1 24.423 28.451 21.216 11.56 20.00
ATOM 14 AU MTL 1 24.423 50.929 21.216 11.56 20.00
ATOM 15 AU MTL 1 46.757 28.451 21.216 11.56 20.00
ATOM 16 AU MTL 1 46.757 50.929 21.216 11.56 20.00
ATOM 17 AU MTL 2 24.384 28.429  0.000  5.27 20.00
ATOM 18 AU MTL 2 24.384 11.261  0.000  5.27 20.00
```

We see that 16 equivalent peaks were found, as we expect from the multiplicity of the Cheshire group for $P2_12_12_1$. Again, the score drops drastically after the correct 16 peaks. Our position from the PDB file

```
HETATM 3659 AU AU 501 95.590 28.474 21.186 1.00 50.87
```

corresponds to atom 13 in the list (95.59 − 71.18 = 24.41). Note that it was by pure chance that so far the solutions we obtained independently by real space peak search and by manual interpretation had (i) the same and (ii) correct handedness!

Patterson vector methods

While the real space grid searches rapidly become inefficient when more than a few atoms have to be found—which is frequently the case for Se-Met labeled structures of large proteins—so-called Patterson vector methods are used for more complex heavy atom searches. They are well suited for automation and are also used for validation of substructures found independently by direct methods against Patterson maps.

The fact that the Patterson function results from the autocorrelation of the molecular structure has an important consequence for the interpretation of Patterson maps: We can place a trial structure over the Patterson map by positioning each of its atoms successively at the origin and checking the Patterson map for peaks at the resulting interatomic distances. In more than one dimension, the procedure will include examination of different orientations (rotations) of the trial structure. If the trial structure (two atoms suffice as initial model) is correct, then we will observe a Patterson peak at the corresponding distance (Figure 9-21). If the first vector distance fits, the trial structure can be expanded and the search procedure can be repeated until the structure is completed. This procedure is the basis for Patterson superposition methods used to determine small molecule structures and marker atom substructures in protein crystallography.

Patterson superposition methods were initially developed for small molecule crystallography, where they are routinely used because they do not require the presence of a heavy atom. A substructure of many identical heavy atoms such as in a Se-Met anomalous difference case indeed resembles a small molecule structure consisting only of similar light atoms such as C, N, and O. Patterson vector methods are therefore also useful in macromolecular crystallography. They are implemented for example in the *SHELX* suite of programs, which has been extended to include the powerful macromolecular substructure solution and phasing modules *SHELXD* and *SHELXE* (Sidebar 10-4).

A variety of measures exist to determine the fit of our trial structure to the Patterson vector map, each with certain advantages. They represent general image-seeking functions (that is, they try to find the structure fragment in the cluttered Patterson map), and are the *Patterson product function*, the *Patterson summation function*, and the Patterson minimum function (PMF). The minimum function is generally the most robust of the scoring functions as far as background noise is concerned. The Patterson-seeded direct methods program *SHELXD* for example uses the PMF as a measure of the probability of an initial dual-atom trial structure being correct.

Reciprocal space searches, Fourier methods, and combined techniques

Given that the Patterson function can be directly derived from the intensities, trial structures could in principle also be scored on their match to observed intensities. However, brute-force summations such as used in the general structure factor summation or Patterson summations are too slow to allow efficient searches over a large search space. Nonetheless, fast translation search functions (initially developed for molecular replacement searches,[34] Chapter 11) have been implemented in the *CNS* program package[18] and are used for substructure determination.[35] Fragment translation searches have the advantage of being exhaustive.

If all Patterson maps were as near-perfect, simple, and clear as our example map, iterative search-and-pick procedures as well as Patterson superposition searches should work well until all atom sites are found and a unique solution is obtained. Unfortunately, the situation is not as trivial. First, a unit cell that contains say $N = 20$ marker atom sites will contain 380—nearly N times as many—peaks, each twice as broad as the electron density peaks as a result of the convolution in a Patterson unit cell of the same dimensions. There will be considerable crowding and peak overlap. Second, particularly in the anomalous cases with small scattering differences, the difference data will be very noisy. This means that initially high scoring starting positions may be wrong, the error propagates into the next round of atom search, and a large number of possible solutions of rather similar merit will result. Therefore, automated substructure solution and phasing programs deploy multiple strategies to identify a correct— or at least the most probable—substructure solution.

As discussed already, isomorphous or anomalous difference Fourier maps can be used to find additional sites once one or more initial atom positions have been determined by Patterson vector searches. The program *SOLVE*,[36] for example, is a complete, automated structure solution package that essentially determines the initial positions by Patterson peak search, refines the positions by a modified origin-removed Patterson map correlation,[37] and uses difference methods to find additional heavy atom sites. To identify the best solution, *SOLVE* applies a number of scoring methods that range from simple figures of merit for the resulting protein phases to map contrast and difference Fourier validation. A final, combined Z-score then allows quite reliable ranking of the substructure solutions.

Marker atom substructure solution by direct methods

An entirely different method to determine the marker atom substructure is through direct methods. As already pointed out in Chapter 7, the underlying assumption of Wilson statistics that the set of random variables giving rise to the structure factor distributions—the atomic coordinates—is independent does not hold for a real molecular structure. Direct methods exploit the resulting *implicit relations* between structure factor amplitudes to solve the phase problem without any additional assumptions.

Box 10-5 **Solving the substructure from Patterson maps.** Heavy atom positions are determined from isomorphous or anomalous difference Patterson maps by automatic methods. Brute force grid searches systematically place atoms at grid positions and compute a correlation score. The best positions are saved and the procedure repeated. Patterson-vector based methods exploit the fact that the Patterson map is the autocorrelation of the underlying electron density and search for fit of fragments. Various scoring functions exist, foremost Patterson minimum functions. Once an initial two-atom (vector) fragment is found, additional atoms are added by Patterson superposition and the PMF search function is evaluated.

Patterson maps are imperfect, noisy, and crowded if a significant number of atoms are present (197 atoms has so far been the maximum solved), and automated programs generate multiple possible heavy atom substructure models that are ranked according to a combined scoring function. Automated programs often employ a combination of Patterson and Fourier methods to test possible solutions for consistency.

Because of the centrosymmetry of the Patterson space, the handedness of the substructure solution is not determined. The solution can either be the correct substructure enantiomorph or have the wrong handedness, with a 50% chance for either case.

Ab-initio structure determination versus substructure solution

Like the Patterson-based techniques, direct methods were originally developed for the determination of small molecule structures. A fundamental necessity for direct methods is that the interatomic distance between the atoms to be found is larger than the d-spacing d_{min} corresponding to the resolution of the available diffraction data. While sub-atomic resolution is routinely achievable in small molecule crystallography, atomic resolution data are rarely available for proteins. For the direct determination of a complete protein structure, direct methods need atomic resolution data of at least ~1.2 Å or better, a situation achievable only for small and/or sturdy proteins.[38] This requirement for atomic separation is known as Sheldrick's rule[39] and coincides with a strong maximum in the distribution of the squared normalized structure factor amplitudes $\langle E_h^2 \rangle$ for proteins at around ~1.1 Å.[40] A detailed discussion of *ab initio* determination of (small) protein structures is provided in ITC volume F.[41]

Despite the limitation of atomic resolution for *ab initio* determination of a complete protein structure, direct methods play a critical role in marker atom substructure solution. In the case of a substructure solution, it is unlikely that the heavy atoms or the anomalously scattering atoms such as Se are any closer than ~3.5 Å because of stereochemical limitations and repulsive interaction between the atoms, and available isomorphous or anomalous difference data are often useful up to this resolution or even better. In addition, combined methods that cycle between reciprocal space and real space atom searches are very powerful, and substructures consisting of no less than 160[42] and now 197 Se atoms (PDB entry 2pnk) have been solved using the Patterson-seeded, dual-space direct methods program *SHELXD*.

A multi-solution hybrid phasing method based on phase extension from molecular fragment searches using the maximum likelihood program *Phaser* followed by density modification and fragment rebuilding with *SHELXE* is described in Sidebar 10-10. As it does not start from a specific probe molecule, it might also be considered an *ab initio* method, but it is not related to direct methods.

Triplet relations and tangent formula

In addition to the fundamental assumption of separate atoms (atomicity), the electron density reconstruction of a correct structure should only be positive, which leads to a second property of the electron density function critical for assignment of phase relations. The original formulation of phase probabilities

Sidebar 10-2 A Nobel Prize in Chemistry for crystallographic methods. Because their work was fundamental for the development of direct methods of phase determination,[47] which allow automated determination of small molecule structures and powerful determination of heavy metal substructures for protein phasing, Jerome Karle and Herbert Hauptman shared the Nobel Prize in Chemistry in 1985. Again, a major contribution toward the understanding of molecular structure and chemistry came from the fundamental theory of crystallography.

between structure factor amplitudes through inequalities have been formulated by Harker and Kasper, and Karle and Hauptman[43] (Sidebar 10-2), and many more extensions and powerful probabilistic methods such as maximum entropy methods[44] have been developed. While the details of phase relations and their probabilities are quite complicated and exceed what we need to understand substructure solutions, we will provide an outline of the most important principal equations based on the squaring method introduced in 1953 by David Sayre[45] and outlined in the general references.[46]

We have already interpreted the Patterson function as a convolution of the direct space electron density

$$P(\mathbf{u}) = \rho(\mathbf{r}) \otimes \rho(-\mathbf{r}) \tag{10-18}$$

which we have shown to be equivalent to squaring in reciprocal space:

$$P(\mathbf{u}) = \mathrm{FT}^{-1}(\mathbf{F}_\mathbf{h}^2) \tag{10-19}$$

Sayre's equation. Now let us, in analogy, convolute structure factors and examine the result. Using again the general convolution integral

$$Conv(\mathbf{u}) = f(\mathbf{r}) \otimes g(\mathbf{r}) = \int_R f(\mathbf{r})g(\mathbf{u}-\mathbf{r})\,d\mathbf{r} \tag{10-20}$$

we obtain, by replacing the integral with a discrete summation, a convolution of structure factors in reciprocal space, known as the Sayre equation:

$$Conv(\mathbf{h}) = \sum_k \mathbf{F}_\mathbf{k} \cdot \mathbf{F}_{\mathbf{h-k}} \propto \mathbf{F}_\mathbf{h} \tag{10-21}$$

The convolution of structure factors in reciprocal space will be equivalent to squaring of the electron density in direct space,

$$Conv(\mathbf{h}) = \mathrm{FT}(\rho_\mathbf{r}^2) \tag{10-22}$$

and thus

$$\mathbf{F}_\mathbf{h} \propto \sum_k \mathbf{F}_\mathbf{k} \cdot \mathbf{F}_{\mathbf{h-k}} = \mathrm{FT}(\rho_\mathbf{r}^2) \tag{10-23}$$

or, expressed using normalized structure factors,

$$\mathbf{E}_\mathbf{h} \propto \sum_k \mathbf{E}_\mathbf{k} \cdot \mathbf{E}_{\mathbf{h-k}} = \mathrm{FT}(\rho_\mathbf{r}^2) \tag{10-24}$$

Triplet relations.
Both Sayre's equation (10-21), and the related equation (10-23) are extremely interesting. Expanding both sides of Sayre's equation by $\mathbf{F}_{-\mathbf{h}}$ and applying Euler's equation we obtain

$$F_\mathbf{h}^2 \propto \sum_k F_\mathbf{h} \cdot F_\mathbf{k} \cdot F_{\mathbf{h-k}} \cdot \exp\left(i(\varphi_{-\mathbf{h}} + \varphi_\mathbf{k} + \varphi_{\mathbf{h-k}})\right) \tag{10-25}$$

Equation 10-25 has interesting implications: if $F_\mathbf{h}^2$ is very large and (necessarily) real, then the product in the sum should also be large and (nearly) real. If in addition the values for $F_\mathbf{k}$ and $F_{\mathbf{h-k}}$ are also large, then the exponent has to be close to zero. In this case, the phase triplet relations are

$$\varphi_{-\mathbf{h}} + \varphi_\mathbf{k} + \varphi_{\mathbf{h-k}} \simeq 0 \tag{10-26}$$

Structure invariants. Depending on the relation of the index triples involved, the generic triplet phase probability (10-26) above can have certain properties. Structure invariants are independent of the origin of the structure, and semi-invariants are invariant in relation to positions in the cell with the same point group (i.e., the structure factors have the same amplitude but different phase). The triplet relations can thus be used to find an initial trial phase set for sets of reflections.

To determine which structure factors are genuinely large on an absolute scale, we need to remove both the *B*-factor attenuation and the scattering factor attenuation. We discussed in Chapter 7 that the fall-off of the structure factor amplitudes with resolution can be eliminated by introducing normalized structure factors $E_\mathbf{h}$ defined as

$$E_\mathbf{h} = \frac{(\varepsilon I_\mathbf{h})^{1/2}}{\langle \varepsilon F_h^2 \rangle^{1/2}} = \frac{\varepsilon^{1/2} F_\mathbf{h}}{(\varepsilon \Sigma)^{1/2}} \tag{10-27}$$

The normalized structure factors represent sharp point atom scatterers at rest, and an electron density map based on *E* value will be highly sharpened.

Only the strongest *E* values are assigned starting phases and the trial phases are checked for consistency with the triplet relations. An alternative explanation for the triplet relations is that when selecting high *E* values, the atoms have to scatter largely in phase (otherwise *E* would not be so high) and therefore are likely located on the corresponding set of lattice planes *hkl*. Under this assumption, geometric relations exist between atoms located on certain related sets of lattice planes and the phase relations for the reflections representing these lattice planes can be derived.[29]

Tangent formula. Picking the strongest reflections will generate multiple sets of probable trial phase sets. They generally suffice to fix the origin or to determine the absolute configuration, as used in some programs to determine the correct handedness of the substructure solution.[14] However, the triplet relations of strong reflections *alone* do not suffice to complete the structure. The question is how to determine the remaining phases.

The left side of Equation 10-24 is a simplified form of the Hauptman–Karle tangent formula,[43] which establishes phase relations between sets of reflections:

$$\tan(\varphi_\mathbf{h}) = \frac{\sum_k \left| E_\mathbf{k} \cdot E_\mathbf{h-k} \right| \sin(\varphi_\mathbf{k} + \varphi_\mathbf{h-k})}{\sum_k \left| E_\mathbf{k} \cdot E_\mathbf{h-k} \right| \cos(\varphi_\mathbf{k} + \varphi_\mathbf{h-k})} \tag{10-28}$$

The famous tangent formula can be used to find additional phases by recycling with the squared density (phase extension) and also to refine the phases of reflections that are presumed to satisfy the phase relations (tangent refinement). Equation 10-24 essentially implies that a Fourier transform of the squared electron density is directly proportional to the normalized structure factors. Thus, if an *E* value is particularly high, there is a high probability that the atoms are located at a peak of ρ^2, which provides a phase value. A trial structure can thus be tested (or recycled) for consistency with the requirement expressed in the tangent formula (10-28).

Reciprocal space extension and refinement can also be performed using a minimal function, as implemented in the program Shake and Bake (*SnB*).[16] Although further extensions to purely reciprocal space based phase extension and refinement have been implemented, the strongest improvement in speed and convergence of direct methods can be achieved by cycling reciprocal space methods with real-space peak picking from the resulting initial Fourier maps. These methods are called dual-space direct methods.

Dual-space direct methods

The dual-space direct methods programs *SnB*,[16] *SHELXD*,[15] and the more Patterson-oriented *HySS*[48] in the *PHENIX* package have been specifically written or adapted for macromolecular substructure solution. *SnB* and *SHELXD* follow a similar strategy that differs in detail. In addition, *SHELXD* selects the initial trial atoms not randomly, but picks two-atom fragments (vectors) from a Patterson map (unsharpened for substructure solution). Translational searches for the strongest Patterson superposition are scored with a modified PMF as described in the Patterson section, and the best initial solution is expanded up to the expected number of marker atoms by subsequent Patterson vector superposition. This procedure leads to multiple starting sets of atoms that are all consistent with the Patterson map. Each of these starting sets is then subjected to dual space cycling and refinement, and a final figure of merit is based on the fit to the Patterson map. The procedure is then repeated with the next starting set of atoms and the solution with the best final overall correlation score CC(all) between E(obs) and E(calc) is always kept. The Patterson seeding improves the speed of the overall program execution by about an order of magnitude.

Given these first, non-random starting atoms, a phase set is calculated and extended in reciprocal space by direct methods through the tangent formula (10-28). However, the purely reciprocal-space phase refinement does not consider any chemical constraints, and given the less than perfect difference data, the solution can diverge to stereochemically unreasonable situations; the most spectacular one in non-translational space groups is the "uranium-atom solution," where all phases are zero (fulfilling the tangent formula) and the atoms merge into one huge Fourier peak. Figure 10-10 shows the map resulting from such a failed uranium solution.

Minimum distances. Prior knowledge applied in direct space can constrain and correct the purely reciprocal space-based initial direct methods solution: A peak search in a Fourier map computed from trial phases yields atom positions to which a minimum distance criterion is applied, thus eliminating unreasonably close atoms. An additional constraint is that only the expected number of marker atoms (plus a few for good measure as explained below) should be present. Furthermore, the random elimination of about 30% of the Fourier peaks from the peak list when back-calculating the new phase set for reciprocal space refinement is likely to eliminate bad peaks in the next cycle while correct ones will likely be regenerated. This real-space random-elimination method has been shown to also be extremely effective in model phase correction and improvement.[49] The solution landscape for a multi-trial or multi-solution method generally consists of many non-solutions and a few possible solutions including the best one (Figure 10-11).

Occupancy refinement. After the peak search in the final cycle, *SHELXD* refines occupancies of the substructure atoms. This is useful to identify partial sites, and

Figure 10-10 Uranium-atom solution and Fourier truncation ripples. The "uranium solution" results when all atoms generated during a failed direct methods substructure solution collapse into one huge peak. The monstrous peak generates correspondingly strong Fourier truncation ripples, which in less dramatic incarnation can occasionally be observed in protein electron density maps around heavy atoms.

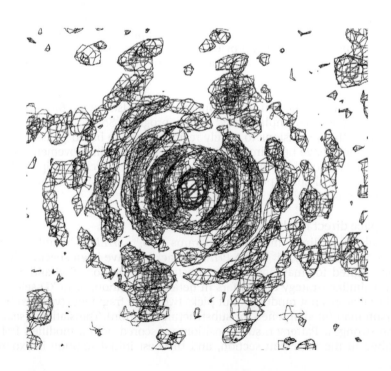

the occupancy of the few additional sites found should generally drop sharply after the number of expected or correct sites. A sharp drop in occupancy nearly always indicates a correct solution (see Figure 10-30 in the practical section of this chapter). In addition to a standard atom list in PDB format, the solution is also presented in a "crossword table." The crossword table includes overall diagnostics as R-values, correlation coefficients CC between E(obs) and E(calc) and the overall PATFOM, and a table where both the distance between each atom pair and the PMF of their interatomic vectors are listed. As the distances and interatomic vectors take the space-group symmetry into account, this table can be examined for the presence of non-crystallographic symmetry in the heavy atom solution. While this is an excellent exercise as demonstrated in the paper describing SHELXD,[15] automated NCS search with other programs or by the subsequent heavy atom refinement and protein phasing program is the norm.

Let us have a look at the heavy atom positions delivered by SHELXD for our gold derivative. The data were locally scaled, F_H difference data were generated, and an instruction file was prepared with XPREP. Here is the SHELXD heavy atom output in PDB file format:

```
CRYST1  71.180  79.380  93.810  90.00  90.00  90.00
SCALE1   0.014049  0.000000  0.00000
SCALE2   0.000000  0.012598  0.00000
SCALE3   0.000000  0.000000  0.010660
HETATM 1AU  HAT 1  46.742  11.227  68.098  1.000  20.00
HETATM 2AU  HAT 2  72.389  11.192  67.663  0.196  20.00
HETATM 3AU  HAT 3  24.315   6.305  73.973  0.135  20.00
HETATM 4AU  HAT 4  24.408  16.283  72.637  0.034  20.00
END
```

In the atom position rows of the atom list we find again that the occupancy (in the column highlighted) drops off significantly after the correctly found single gold atom:

```
HETATM 1AU  HAT 1  46.742  11.227  68.098  1.000  20.00
```

Subtracting $c/2 = 46.91$ from the z coordinate we find that with the new $z = 21.19$ our solution corresponds to ATOM 11 in the table of equivalent Cheshire positions generated by RSPS.

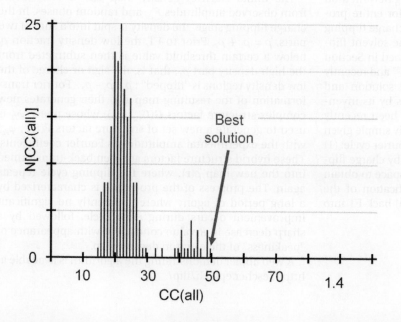

Figure 10-11 Multi-solution landscape. The figure shows the overall correlation coefficient between E(obs) and E(calc) for 200 trials of a Patterson-seeded search in SHELXD. Typical for all multi-solution methods like SHELXD or SnB is the distinct separation into a bulk of non-solutions and a few solutions, including the best one. If the solution landscape is indiscriminate, it is unlikely that a correct solution has been found. The pattern of "near misses" shown in the solution distribution is typical for the disulfide splitting, where some nearly correct solutions may miss a sulfur atom. Although these near complete solutions would also suffice to solve the structure, it is best to run as many trials as reasonable to assure that the best solution is indeed found. Histogram generated by HKL2MAP.[52]

Box 10-6 Direct methods. Direct methods exploit the deviations from random-atom (Wilson) structure factor distributions as a result of the non-random distribution of atoms in a true molecular structure. The non-randomness leads to implicit relations between structure factor amplitudes and to phase relations. Triplet phase relations are used to find trial structures, which are expanded using the tangent formula.

Dual space direct methods recycle and modify trial structures by peak search in the electron density trial map, and exploit chemical restraints and plausibility such as minimum distances to remove insensible atoms and place new ones. They converge much faster than reciprocal space-only methods. Particularly efficient is Patterson seeding, where initial positions for the dual-space direct methods are picked from a list of two-atom vectors compatible with the Patterson map, and the atom-pairs are then employed as a search fragment in a translational search based on the full symmetry Patterson minimum function.

Although special direct methods programs can solve the phase ambiguity in SAD data, the substructure handedness is generally not determined during direct methods substructure solution.

As the power of real space cycling combined with direct methods is quite convincing, it can be expected that more program packages will likely implement such substructure solution strategies in the future. It should be noted that additional techniques and procedures exist for improving phase determination and refinement in direct methods. While most are extensions of tangent-formula-based technique,[41] maximum entropy methods take a radically different approach.[50] A recent method that seems to perform quite well is charge flipping[51] (Sidebar 10-3), further extending the toolbox of different methods available to solve difficult and large heavy atom substructures.

10.4   Solving the phasing equations

Given the intensity data for the native crystal and for the derivative crystal, let us assume that we have correctly determined the initial heavy atom position(s)

Sidebar 10-3 Charge flipping. An interesting new method for substructure solution (and potentially for entire protein structures given sufficient data) is the charge flipping (CF) method. CF is somewhat related to the solvent flipping density modification technique described in Section 10.7. CF was conceptually described in 2004[53] and recently applied to macromolecular (sub)structure solution and phase extension.[54] A review of the methods by its inventors Gábor Oszlányi and András Sütö has been recently published.[51] The concept of CF is surprisingly simple given its power. At its core is the basic four-step Fourier cycle: (1) real space modification of electron density by charge flipping; (2) Fourier transform into reciprocal space to obtain corresponding structure factors; (3) modification of the structure factors in reciprocal space; and (4) back-FT into real space.

$$\rho(\mathbf{r}) \rightarrow \quad FLIP \quad \rightarrow \quad g(\mathbf{r})$$
$$\uparrow FT^{-1} \qquad\qquad\qquad FT \downarrow$$
$$\mathbf{F}(F_{obs}, \varphi_{FL}) \leftarrow \mathbf{G}(G_{FL}, \varphi_{FL})$$

The initial electron density map $\rho(\mathbf{r})$ is constructed from observed amplitudes F_{obs} and random phases. In the charge flipping stage, the density is split into a sum of two parts, $\rho = \rho_1 + \rho_2$. Prior to FT, the low density fraction ρ_2 below a certain threshold value is then subtracted from the high density part ρ_1, that is, the sign or charge of the low density regions is "flipped" : $\rho = \rho_1 - \rho_2$. Fourier transformation of the resulting map $g(\mathbf{r})$ then generates new complex structure factors $\mathbf{G}(G_{FL}, \varphi_{FL})$, whose phase φ_{FL} is used to generate a new set of structure factors $\mathbf{F}(F_{obs}, \varphi_{FL})$ with the experimental amplitudes as Fourier coefficients. These hybrid structure factors are then back-transformed into the new map $\rho(\mathbf{r})$, where the flipping cycle repeats again. The progress of the procedure is characterized by a long period of agony where apparently no significant improvement occurs during each cycle, followed by a sharp decrease in R-value concurrent with appearance of "peakiness" of the electron density map.

A web applet demonstrating the algorithm is available at http://escher.epfl.ch/flip/

from the difference data. The knowledge of the marker atom *position* automatically implies knowledge of marker atom *phases*, as we can readily compute the complex structure factor for the heavy atoms from the standard structure factor summation derived in Chapter 6. We can now investigate in Harker diagrams the fundamental experimental phasing scenario given the structure factor amplitudes F_P and F_{PA} from the measured data and the structure factor $\mathbf{F}_A$ for that heavy marker atom.

SIR phasing and SIR maps

A structure factor $\mathbf{F}$ of known magnitude F but unknown phase φ can lie anywhere on a circle of radius F in the complex plane. Thus we can draw a circle with amplitude F_P for each protein structure factor amplitude. We know from the heavy atom substructure solution both the magnitude and the phase and thus the direction of the heavy atom structure factor $\mathbf{F}_A$. We graphically solve the phasing equation by drawing a second circle with radius F_{PA} offset by $-\mathbf{F}_A$, as shown in Figure 10-12.

The two possible solutions are equivalent to the two numeric values we derive for the protein phase angle φ_P from the algebraic expression

$$\varphi_P = \varphi_A + \cos^{-1}\left(\frac{F_{PA}^2 - F_P^2 - F_A^2}{2 F_P F_A}\right) = \varphi_A \pm \alpha \tag{10-29}$$

Because of the two possible values for the cosine term ($\cos\varphi = \cos-\varphi$), the solutions are symmetric along the direction of the vector $\mathbf{F}_A$, with phase angles $\varphi_A \pm \alpha$ and $\varphi_{SIR} = \varphi_A$.

Given the two solutions, we can still try to construct a so-called single isomorphous replacement (SIR) map, but the question is what such a map actually shows. According to Figure 10-12, the Fourier coefficients $\mathbf{F}_{SIR}$, with their phase of φ_{SIR}, are given by the weighted mean (or the centroid) of the correct protein structure factor $\mathbf{F}_P$ and the wrong vector $\mathbf{F}_2$

$$\mathbf{F}_{SIR} = \left(\mathbf{F}_P + \mathbf{F}_2\right)/2 = \left(F_P \exp(i\varphi_P) + F_2 \exp(i\varphi_2)\right)/2 \tag{10-30}$$

The heavy atom structure factor $\mathbf{F}_A$ is collinear with the vector $\mathbf{F}_{SIR}$ and thus

$$\varphi_A = \varphi_{SIR} = \left(\varphi_P + \varphi_2\right)/2 \tag{10-31}$$

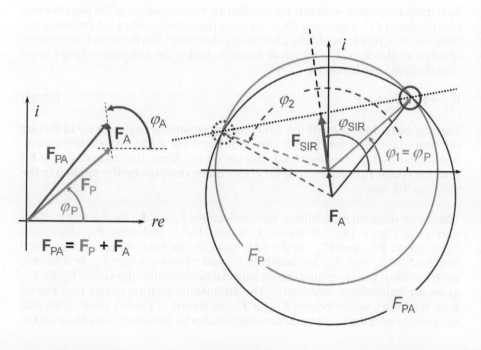

Figure 10-12 **Graphical solution of the SIR phasing equations.** The left panel shows how the complex structure factors for protein, derivative, and heavy atom are related for a generic, acentric reflection *hkl*. Intensity measurements only reveal the magnitude of the structure factors for the derivative F_{PA} and the native F_P, but from the difference data, the position(s) of the heavy atom(s) can be determined. Therefore, the entire complex structure factor for the heavy atom $\mathbf{F}_A$ is known, and the two possible solutions for the protein phase angle then can be determined by the graphical procedure shown in the right panel. Only one phase angle (φ_P) will be correct; the other one (φ_2) will be incorrect. In practice we do not know (yet) which one of the solutions is the correct one. Note that the solutions are symmetrically distributed around the SIR centroid phase φ_{SIR}, complementary to the SAD case shown in Figure 10-13.

which we substitute together with the fact that correct and wrong structure factor amplitudes F_P and F_2 are identical, into (10-30) and obtain

$$2\mathbf{F}_{SIR} = F_P \exp(i\varphi_P) + F_P \exp(2i\varphi_A)\exp(-i\varphi_P). \tag{10-32}$$

Equation (10-32) essentially states that if $\varphi_A = \varphi_P$, that is, the vectors are collinear, then the SIR phases are correct. Unfortunately, this is not the case in general, and the phase values φ_A and φ_P are not correlated. Thus, the SIR map is in fact reconstructed 50% from the correct coefficients in the left part of the sum in (10-32), and 50% from the right part containing uncorrelated phases of φ_A and φ_P, that is, noise terms. In other words, the SIR map constructed from the centroid phases φ_{SIR} is a superposition of the true map and a noise map. Given that also the "true" phase values will be affected by experimental errors, this does not bode well for direct map interpretation and model building. However, as we will see shortly, the SIR maps can be considerably improved by density modification techniques.

SAD phasing and SAD maps

Anomalous scattering of X-rays by heavy marker atoms can also provide a source of electronic differences. The result is intensity differences between two reflections of a Bijvoet pair, and the corresponding set of SAD data collected from a single crystal at a wavelength corresponding to a X-ray energy above an absorption edge will contain *pairs* of reflections with different intensities for non-phase-restricted reflections (compare the anomalous structure factor table computed for BPTI in Chapter 6).

Let us briefly refresh our knowledge about anomalous scattering contributions. The atomic scattering factor always contains additional, wavelength-dependent contributions, which become significant in the vicinity of, or above, the X-ray absorption edge of the dispersive marker atom:

$$f_{(\theta,\lambda)} = f^0_{(\theta)} + f'_{(\lambda)} + i \cdot f''_{(\lambda)} \tag{10-33}$$

We know from Chapter 6 that the anomalous scattering contribution f'' is phase retarded by 90° relative to the real part and thus is always perpendicular by +90° to the vector of the real scattering component. This leads to a phase difference and to a measurable difference in the intensities of the two related reflections in a Bijvoet or anomalous pair. Recall that as $\mathbf{F}_P$ and $\mathbf{F}_A$ are collinear for phase restricted reflections, these reflections are thus not affected and will remain pairs with identical amplitudes.

To remain consistent with our nomenclature, we are grouping the negative real contribution of f' together with f^0. As both real contributions are collinear, the complex structure factor of the anomalous atom then can be written as the sum of two contributions. The structure factor including the anomalous heavy atom contribution then becomes

$$\mathbf{F}_{PA} = \mathbf{F}_P + \mathbf{F}_A + i \cdot \mathbf{F}''_A \tag{10-34}$$

The graphical derivation of the SAD phasing equations in Figure 10-13 follows the procedure used for the SIR case, but there are a few distinct differences. Remember that in the basic SAD case we do not know the "native" F_P nor $\mathbf{F}_P$; thus the vector $\mathbf{F}_P$ is drawn only for clarity and to emphasize the relation to the previous SIR case.

The vector diagram for a Bijvoet pair (reflections $\mathbf{F}_+$ and $\mathbf{F}_-$) for the same reflection as in Figure 10-12 is shown in Figure 10-13, including the anomalous components $\mathbf{F}''_{A+}$ and $\mathbf{F}''_{A-}$. In the SAD situation for each reflection the paired amplitudes F_{PA+} and F_{PA-} are available instead of one average F_{PA}. In addition, we know from the anomalous atom substructure solution the vector $\mathbf{F}_{A+}$ (or $\mathbf{F}_{A-}$, given the handedness ambiguity). The anomalous phasing circles now extend from respective vector origins $\mathbf{F}''_{A+}$ or $\mathbf{F}''_{A-}$ as shown in the left panel. Note that because we selected $\mathbf{F}_{A+}$ as the enantiomer choice for the anomalous atom vector,

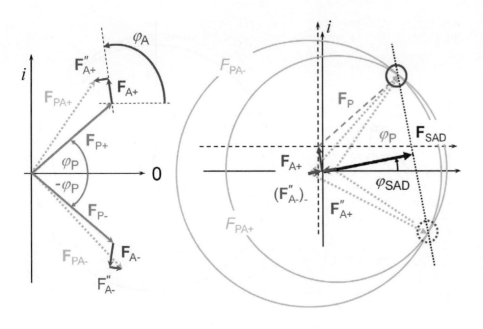

Figure 10-13 SAD phasing equations.
The right panel shows complex real and imaginary contributions for the dispersive contributions to F_{PA+} and its Bijvoet mate F_{PA-}. The structure factor magnitudes for F_{PA+} and the Bijvoet opposite F_{PA-} have been measured. In analogy to the isomorphous difference data, the position of the anomalous scatterer and thus $\mathbf{F}_{A+}$ (or $\mathbf{F}_{A-}$ in the case of the enantiomer substructure) is known. The two possible solutions for the protein phase angle then can be determined by drawing circles of radius F_{PA+} and F_{PA-} from the corresponding vector origins. Because we selected the $\mathbf{F}_{A+}$ enantiomer, we need to invert the anomalous contribution $\mathbf{F''}_{A-}$ for drawing the phasing circles. As in the SIR case, only one phase angle (φ_P) is correct; the other one will be wrong. However, in contrast to the SIR case, the solutions are now symmetrically distributed around $\varphi_A - 90°$, complementary to the SIR case, where the solutions were symmetric around the heavy atom phase. As we do not know F_P nor $\mathbf{F}_P$ (which is drawn for clarity here only), we cannot yet determine which one of the solutions is the correct one from SAD data alone.

the anomalous contribution $\mathbf{F''}_{A-}$ needs to be inverted to obtain the proper origin of the F_{PA-} phasing circle.

The geometric solution of the phase equation again delivers both a correct and a wrong phase value. The interesting part is that compared with the SIR case, where the solutions were symmetric around the heavy atom phase φ_A, the SAD phases are symmetric around $\varphi_A - 90°$. Thus the situation shown here is complementary to the SIR case, and we can make use of this fact in the combination of SIR and SAD information from the same heavy atom to break the phase ambiguity. The SAD map constructed from the centroid phases φ_{SAD} and using the average of F_{PA+} and F_{PA-} as a substitute for F_P in the Fourier coefficients will again be a superposition of a true map and a noise map.

Anomalous signal resolves phase ambiguity

Given the orthogonality of the SAD and SIR phase solutions for the same marker atom, a logical way of breaking the phase ambiguity is by combining available SIR and SAD information into the so-called SIRAS phasing procedure.[10] Most heavy atoms used for isomorphous replacement will also provide a source of anomalous signal and collecting the centrosymmetrically related reflections in addition to the unique cell of the reciprocal space provides anomalous differences for the derivative. From either isomorphous difference data or anomalous difference data (or even better, using both data sets simultaneously) we can determine the position of the dispersive atom. Given the heavy atom position, we can then solve the phase equations simply by following the same procedure as in the separate SIR and SAD cases (Figures 10-12 and 10-13). Because the pairs of SIR and SAD centroid phases are 90° offset from each other, they share only one common intersection at the true protein phase value and we can break the phase ambiguity, as illustrated in Figure 10-14.

The derivation of the algebraic form of the SIRAS equations follows the same procedure outlined for the SIR case which was derived by Matthews[10] as provided in detail in the general references.[46, 55]

Another method of breaking the phase ambiguity is by density modification, which can be directly applied to SIR or SAD centroid maps (Section 10.7).

Resolving SIR/SAD phase ambiguity by density modification

Irrespective of the difficulty in directly interpreting a SIR or SAD centroid map, as long as the outline of the molecule is distinguishable from the background noise, subsequent and iterative application of density modification techniques

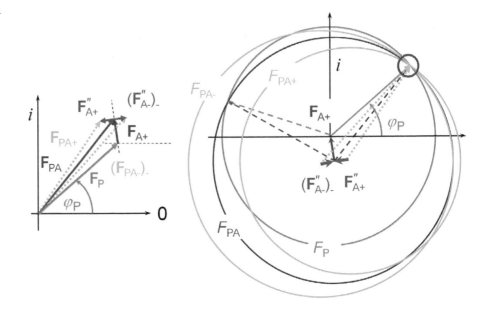

Figure 10-14 Graphical solution of phase equations for SIR with anomalous scattering contributions (SIRAS). The procedure of solving the phasing equations for SIR with anomalous contributions is a combination of the SIR and SAD cases. In the associated SIR Harker diagram, the circles of magnitude F_{PA+} and F_{PA-} are centered at the corresponding vector origins of $\mathbf{F}_{A+}$ and $(\mathbf{F}_{A-})_{-}$, and both intersect at the correct phase angle. The phase ambiguity has thus been resolved.

such as solvent flattening, solvent flipping, or histogram matching can improve the SIR phases to the point that a model can be built. These methods have been introduced by B. C. Wang and co-workers[56] to break phase ambiguity and have become an indispensable tool in the arsenal of phasing techniques. As a consequence, they are implemented in many programs. The same principles that allow resolving phase ambiguity in the SIR and SAD case are also useful for improvement of any starting phases and we will extensively use and discuss density modification methods in Section 10.7. The basic procedure of solvent flattening is to define a molecular envelope and set the density outside this region (or within the solvent mask) to a constant value. The back-transform of the modified map then yields improved structure factors and a new map is computed from improved, combined phases φ_{comb} and the original structure factor amplitudes F_{obs}. The procedure is then iterated until convergence, when no more improvement takes place.

Sidebar 10-4 A fast track from data to electron density map using the *SHELX* suite of programs. The programs in the *SHELX* suite have been developed by George Sheldrick in Göttingen.[57] Originally designed for automated structure solution in small molecule crystallography, its powerful substructure solution programs were quickly adopted by the macromolecular crystallographic community. Recognizing this trend and realizing the need for fast structure solution in high-throughput protein crystallography, Sheldrick has extended the program suite to include fast anomalous data preparation and phase extension programs. The programs provide a number of valuable diagnostics and are ideally suited to obtaining a quick result and to evaluating whether the data are good enough to proceed to more sophisticated phase improvement methods. It turned out that in many cases, Sheldrick's unique "sphere of influence" phase extension method is powerful enough to produce high quality maps that can be directly used for automated or manual model building. A version of *SHELXE* now also includes an integrated auto-tracing and model building module.[119] Together with the Free Lunch[58] phase extension method (Sidebar 10-11), the package provides a complete fast-track from data processing to initial models ready for polishing and refinement.

In our SAD example shown in Figure 10-15 and the VTA1 example in the practical section we processed the anomalous data with *SHELXC*, an automated public version of the more extensive Bruker AXS program *XPREP*. The single anomalous difference data were then fed into the substructure solution program *SHELXD*[15] and the sulfur substructure was determined. Subsequently, using the substructure positions from *SHELXD* and the full SAD data, the 2-fold ambiguity in the phasing equations was initially resolved and then improved using an automated iterative density modification applied to the SAD map using the phasing program *SHELXE*.[59] The publicly available Sheldrick programs *SHELXC,D,E* have been provided with a graphical interface wrapper named *HKL2MAP* by Thomas Schneider,[52] and CCP4i also provides a basic graphical (input) interface to the *SHELXC/D/E* programs.

In our first S-SAD phasing example we take anomalous 1.5 Å sulfur SAD data measured using a Cu home source at 1.54178 Å (which is about 3600 eV above the sulfur edge) of our small BPTI molecule containing three disulfide bridges and one methionine residue. The intensity measurements were converted into single anomalous difference data, the metal substructure was determined in space group $P2_12_12_1$, and the phase ambiguity was resolved using the programs of the *SHELX* suite (Sidebar 10-4).

We will not at this point discuss the details of the density modification procedure and the statistics used to evaluate data and map quality (Section 10.7) but instead focus directly on the appearance of the maps. The initial SAD map calculated from the SAD centroid phases and the anomalous data is shown in the left panel of Figure 10-15. The result is surprising: although we had determined the correct sulfur substructure, we obtain a map that is complete rubbish, showing no discernible features, just noise. We expected, for an unmodified SAD map based on a good heavy atom solution, at least some discernible outline of the molecule in a noisy background. What went wrong here?

The heavy atom enantiomorph problem

We recall from the previous section that because of the inherent centrosymmetry of the Patterson space, the marker atom substructure does not necessarily have to have the correct handedness. In the case of a non-centrosymmetric substructure (which is practically always the case with anomalous S-SAD or Se-Met data) we have in general a 50% chance of obtaining the correct enantiomorph of the substructure and a 50% chance of obtaining the inverted solution. In the case of anomalous data, the wrong substructure handedness gives complete noise maps (left panel of Figure 10-15) and the attempt to apply any density modification will not improve matters. We therefore invert the substructure enantiomorph by changing all fractional coordinate vectors from x to $-x$ and change also the space group enantiomorph in the case of an enantiomorphic pair (such as $P3_1$ or $P3_2$) and repeat the SAD map construction. The 11 enantiomorphic pairs of space groups are are $P3_1$-$P3_2$, $P3_121$-$P3_221$, $P3_112$-$P3_212$, $P4_1$-$P4_3$, $P4_122$-$P4_322$, $P4_12_12$-$P4_32_12$, $P6_1$-$P6_5$, $P6_2$-$P6_4$, $P6_122$-$P6_522$, $P6_222$-$P6_422$, and cubic $P4_132$-$P4_332$.

Exception from inversion about the origin. There are three exceptions to the inversion about the origin: In chiral space groups $I4_1$, $I4_122$, and $F4_132$, the origin is not located on an enantiomorph axis, and the center of inversion does not coincide with the origin. Their inversion operators are $(-x, \frac{1}{2}-y, -z)$, $(-x, \frac{1}{2}-y, \frac{1}{4}-z)$, and $(\frac{1}{4}-x, \frac{1}{4}-y, \frac{1}{4}-z)$, respectively.[60]

After inverting the substructure coordinate handedness (no enantiomorph change for space group $P2_12_12_1$ is necessary) the result shown in the right panel of Figure 10-15 looks much better, and we can discern a molecular shape with

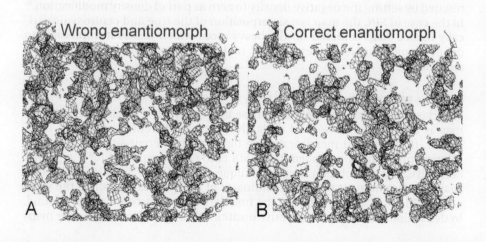

A Wrong enantiomorph

B Correct enantiomorph

Figure 10-15 Initial SAD maps based on inverted and correct substructure enantiomorph. Panel A shows an initial S-SAD map constructed from 1.5 Å data and phases obtained from the sulfur substructure in the wrong handedness. The map has poor contrast, low connectivity, no discernable features, and is essentially noise. Panel B shows the S-SAD map this time constructed from the same data but with phases from the correct enantiomorph of the sulfur substructure. Although there is still plenty of noise in the map, solvent channels between the molecular envelopes are discernible, and the connectivity of the ghost-like molecule hidden behind the noise is better than in the complete noise map in the left panel. The qualitative differences between the two maps can be hard to detect by eye in the beginning of a crystallographic career, and the statistics provided by the phasing programs provide support and validation of the visual inspection.

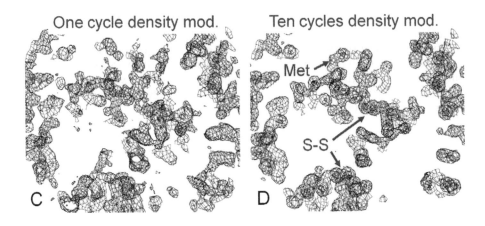

Figure 10-16 SAD map improved by density modification. The left panel shows a SAD map section computed from the same centroid phases as in the right panel of Figure 10-15, but after a first, crude cycle of density modification. The map becomes more contiguous, with increasing contrast between solvent and protein. The right panel shows the map after 10 cycles of solvent flattening, and the molecular parts are now clearly traceable, with the highest density at the disulfides and a clearly visible methionine residue. As an exercise, try to identify other residues in the electron density map. The improvement in map quality through density modification recognizable from panel B (Figure 10-15) to panels C and D is quite dramatic.

Sidebar 10-5 The future is SAD. In 1981 Wayne Hendrickson and Martha Teeter demonstrated that even the small anomalous signal of native sulfur atoms alone can be used to determine the structure of a protein.[65] Their test case was a small and well diffracting plant seed protein, crambin, which contains six sulfur atoms in disulfide bridges among its 46 residues and is structurally similar to our VTA1 S-SAD phasing test case. Although their originally applied method of resolved anomalous phasing is not in use anymore, their paper was fundamental and proved the predictions made earlier by Bijvoet that solution of structures from SAD data alone should in principle be possible. Since this seminal publication, instrumentation and computational methods have evolved to a point where S-SAD and general SAD phasing in combination with density modification becomes an increasingly prominent phasing technique, rivaling the powerful MAD method,[66] which was popularized through contributions from Hendrickson's laboratory based on earlier work of Karle,[67] Phillips and Hodgson,[68] Templeton, and others. See Sidebar 10-7 for more about when to use either SAD or MAD phasing.

more distinct connectivity, still overlaid by noise as expected. Taking the initial SAD map based on the correct substructure enantiomorph into 10 cycles of density modification yields a clearly interpretable map that shows the strongest intensities at the sulfur peaks (Figure 10-16).

Two ambiguities must be resolved. The important lesson here is that in SAD and SIR we have in principle *two* ambiguities to resolve: the substructure handedness and the 2-fold phase ambiguity. This is occasionally mixed up and has caused serious problems as discussed in Sidebar 10-6. While maps using anomalous signal to break phase ambiguity do not yield interpretable maps given wrong substructure handedness, MIR and MAD structures can give maps which show inverted protein structures when at the same time the substructure is inverted and the Bijvoet pairs are swapped. This is easy to detect in good maps as all helix density appears left-handed, but in low resolution structures completely wrong models can be built into poor density if other warning signs are ignored (see also Sidebar 12-8). We will revisit these issues again in the substructure solution section and Table 10-2 lists the various combinations of inversion and ambiguity that can exist.[61]

Centrosymmetric substructure. Irrespective of the inherent chirality of the protein part of a crystal structure, the heavy atom substructure can be centrosymmetric. Examples are one or two sites in a $P1$ structure or a single atom in space group $P2_1$. Centrosymmetry can also arise when heavy atoms lie on a special position, such as on the 3-fold axis in the case of R_3 Zn-insulin.[62] A centrosymmetric heavy atom solution is of course unambiguous and does not need to be inverted, but the effect the heavy atom centrosymmetry has on the phasing situation is different, depending on whether we have a SAD, SIR, MAD, or MIR case. The SAD map will be a double image of positive density for the correct image and negative density of the inverted image. This case can still be rescued by setting the negative density to zero as part of density modification.[63] In the case of SIR, the map is a superposition of the true and centrosymmetrically related map centered on the heavy atom which makes it impossible to define a unique molecular envelope.[64] Table 10-2 summarizes the handedness ambiguities.

Solving phase ambiguity by MIR

The classical way of resolving the phase ambiguity for a single derivative SIR map is by preparing a second derivative from a different heavy metal compound. If the second metal derivative binds at a different site, the graphical solution procedure we applied in the SIR case can be repeated for the second derivative, and the solution becomes unique, as shown in Figure 10-17. There is again the requirement that the coordinates of both derivatives share the same origin. This can be ensured once the first derivative structure has been found, by determining the second derivative positions through difference Fourier maps

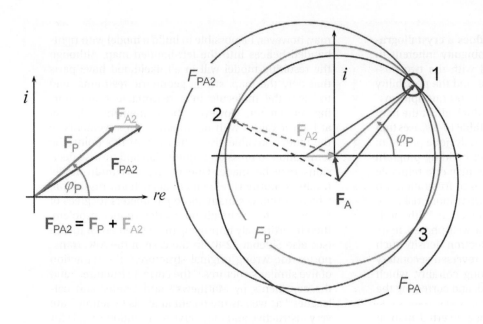

Figure 10-17 The classical MIR case of breaking the phase ambiguity. In addition to the first derivative from the SIR case (Figure 10-12), a second derivative with a different metal binding site is prepared, and the graphical solution procedure for the phase equations is repeated, determining unambiguously (subject to measurement and heavy atom position errors to be discussed later) the correct phase angle. The phasing circle intersections labeled 1, 2, and 3 correspond to the probability distribution peaks in Figure 10-19.

calculated from coefficients $(F_{PA2} - F_P) \cdot \exp(\varphi_p)$ with the phases φ_p of the protein from the first derivative SIR or SIR(AS) case.

Consideration of experimental errors. The procedure of multiple isomorphous replacement, or MIR, in general involves the preparation of further derivatives in addition to the principally necessary second derivative. The need for more derivatives results from experimental errors in observed native and derivative structure factor amplitudes and errors in the heavy atom positions. As a consequence, the phasing circles will not coincide as perfectly as we have drawn them so far in our figures. For the treatment of these errors we need to resort to a probabilistic treatment of the phase angle distribution (Section 10.5). The goal of obtaining the best protein phases will naturally also influence the refinement of the heavy atom positions.

Phasing technique	Substructure properties	Resulting map properties
SAD, single anomalous signal	Correct handedness	Correct + noise
	Inverted handedness	Noise only
	Centrosymmetric	Correct + inverted
SIR alone (without anomalous HA signal measured)	Correct handedness	Correct
	Inverted handedness	Inverted
	Centrosymmetric	Double image with inversion center
IR + AS (with anomalous HA signal measured), or MAD	Correct handedness	Correct
	Inverted handedness	Noise
	Centrosymmetric	Correct

Table 10-2 Handedness ambiguity in phasing techniques. There are two ambiguities in experimental phasing that need to be resolved: the principal 2-fold degeneracy of the phasing equations in the case of single isomorphous or isomorphous replacement, and the enantiomorph ambiguity of the marker atom substructure. For green entries the phase ambiguity can be resolved by density modification. Red entries invariably cannot be unraveled. Table extended from Brian Matthews[61] and George Sheldrick's review.[60] IR, isomorphous replacement techniques; AS, anomalous scattering techniques; HA, heavy atom. Note that the SIRAS case and the MAD case can revert to the "IR alone" case when the anomalous signal is just noise but the isomorphous or dispersive signal, respectively, is still good.[63]

Sidebar 10-6 How many hands does a crystallographer need? The handedness ambiguity inherent in the phase equations—combined with the two possible substructure enantiomorphs and the possibility of flipping the Bijvoet pairs—generates multiple possible handedness combinations which affect the map reconstruction as summarized in Table 10-2. Cases that produce true noise maps are generally not a problem; no density modification or model building program will ever produce an interpretable map or a refinable structure and short of a great deal of frustration, no permanent harm is done. Somewhat more risky are situations when the map itself is inverted. With modest to high resolution data and a well-phased map, there will be clear signs in the electron density such as left-handed helices and other reversed secondary structure features, including wrong chirality, which all clearly indicate what happened, and correcting the swapped indices fixes the problem.

With poor phases giving a noisy inverted map at low resolution (~4 Å or worse) with poor detail, it may, however, be possible to build a model with right-handed helices into the left-handed map. Although the resulting model will in all likelihood have parts that defy the map, strong geometric restraints tend to keep the molecule in a reasonable shape during refinement. However, the refinement will stall at high R-values with no apparent improvements to the model possible given the map. At this point, the temptation to press on is quite high and unwise decisions may be made. One of those is multi-model (multi-conformer) refinement at such low resolution, which introduces many more parameters (degrees of freedom) into an already poorly determined problem, thereby artificially improving the refinement statistics (see also Sidebar 12-8). In the case of the ABC transporter, the wrong original structure,[69] the retraction of five similar structures,[70] the correct structure,[71] and the comments by Matthews[61] and Dauter and colleagues[64] as well as the recent analysis by Jeffrey[72] are very instructive and may serve as a cautionary tale for the aspiring crystallographer.

MAD phasing equations

The multi-wavlength anomalous diffraction phasing method (MAD; the term anomalous dispersion is sometimes used interchangeably with anomalous diffraction in biological crystallography) breaks the 2-fold phase ambiguity by including a second, dispersive data set at a different wavelength in the phasing equations. Similarly to the SIRAS case, we now have dispersive signal between wavelengths orthogonal to the anomalous signal between the Bijvoet pairs and the intersection on the phasing circle is uniquely defined. In analytical form, the phasing equations derived by Karle[67] and Hendrickson, Smith, and Sheriff[73] relating the structure factor amplitudes of the anomalous marker atom F_A and the total amplitude of the protein F_T (including the f_0 of the marker but not its anomalous contributions f' and f''), to each Bijvoet pair F_+ and F_- are more complicated than for the previous SIR and SAD cases:

$$F_+^2 = F_T^2 + aF_A^2 + bF_T F_A \cos\alpha + cF_T F_A \sin\alpha$$
$$F_-^2 = F_T^2 + aF_A^2 + bF_T F_A \cos\alpha - cF_T F_A \sin\alpha$$

(10-35)

The phase angle α is not the protein phase angle, but is defined as the difference (or phase shift) between the true protein phase (structure without anomalous contribution) φ_T and the (still unknown) phase of the anomalous scatting contribution φ_A:

$$\alpha = \varphi_T - \varphi_A$$

(10-36)

The constants a, b, and c in phasing equations (10-35) depend on the anomalous scattering contributions of the marker atom:

$$a = \frac{f'^2 + f''^2}{f_0^2}, \quad b = \frac{2f'}{f_0}, \text{ and } c = \frac{2f''}{f_0}$$

(10-37)

As defined in Chapter 6, f_0 is the normal scattering contribution, f' is the dispersive component, and f'' is the anomalous component of the scattering factors of the marker atom.

With more than one wavelength, the equation system (10-35) becomes determined and the variables F_A, F_T, and α can be computed for each reflection $F_\pm$. From the F_A values we can determine the dispersive atom positions using one of the substructure solution methods described in Section 10.3, and from the resulting positions we compute structure factors $\mathbf{F}_A$ and thus φ_A which in turn gives φ_P, the sought-after protein phase angle. Thus, we know amplitude F_T and phase φ_T for the protein and can calculate a Fourier synthesis that gives the initial protein map. Although the phase equations are now determined uniquely, the handedness ambiguity of the substructure solution remains and both substructure enantiomorphs are tried in the phase calculation.[59] Because in the general case $\alpha - \varphi_A \neq \alpha + \varphi_A$, the wrong substructure handedness will give an uninterpretable noise map (Table 10-2).

Although the original MAD equations allow a refined solution for all available data simultaneously,[74] the MAD case is often alternatively treated as a combination of independent SIR and SAD phases. As we will see in the section about maximum likelihood phase refinement, neither is necessarily the best treatment of phase combination, as the obtained protein phase angles are not independent. In order to properly treat phase combination, we need to first take a closer look at the probability distributions for the protein phases.

10.5   Basic probability distributions for protein phases

Intersections of circles with similar radius and with a small origin offset always occur at a shallow angle, a suboptimal situation as far as precise location of the intersection on the phase circles is concerned. In addition, the error of the intensities turns the phasing circles into "rings" of 2-dimensional distributions themselves. Combined with the error in the offset of the circles resulting from errors in the complex heavy atom structure factor, we thus have a rather "fuzzy" state of knowledge of the protein phases. This fuzziness can ultimately be properly accounted for through maximum likelihood error models and probability distributions, but the fundamental principles can be well explained using the classical Blow and Crick[75] derivations.

What we are looking for is a general formalism to express the probability of a certain protein phase angle, beginning with the ideal phasing circles as a model. We can again start from the SIR case shown in Figure 10-12, and define a closure error $\varepsilon(\varphi)$ for the deviation of the experimentally possible phase angles from the true phase angle of the $\mathbf{F}_P$ vector. For either of the correct solutions, in the absence of experimental errors the vector triangle $\mathbf{F}_{PA} = \mathbf{F}_A + \mathbf{F}_P$ perfectly closes and the closure error is zero. The further a phase angle deviates from the true value, the bigger the gap $\varepsilon(\varphi)$ becomes, as shown in Figure 10-18.

Sidebar 10-7 When to get MAD instead of SAD. The choice of when to use SAD or MAD phasing depends on a number of factors and can sometimes even be decided during data collection (provided the X-ray source is tunable). As a general rule, SAD phasing—requiring density modification to break the phase ambiguity—works best when the solvent content of the crystal is high. In addition, high quality anomalous difference data usable (not just measured) to a resolution of 2.5 Å or better should be available and high resolution native data for phase improvement and extension by density modification are of advantage.

MAD data, given their redundancy and often stronger signal because of precise wavelength selection, are often stronger than SAD data and also work well for low resolution phasing. MAD phasing does not depend on density modification to break the phase ambiguity and also works with low solvent content.

In both techniques different resolution cutoffs, many (thousands of) trial cycles, and trying different programs may help to obtain or improve substructure solutions even from recalcitrant difference data.

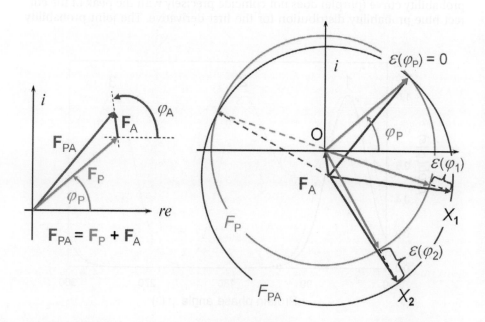

Figure 10-18 Closure error in phasing circles. The closure error $\varepsilon(\varphi)$ for the phasing circles is zero for the two possible solutions of the phase equations, and increases the further away from a solution (φ_P or the other possible value) a phase angle φ lies. The distances OX can be computed for each phase angle from the trigonometric form given in the text. We can thus interpret the closure error $\varepsilon(\varphi)$ as a residual between the distances OX and the measured protein amplitude F_P. Given the closure error as a residual, a probability distribution $prob(\varphi|\varepsilon)$ analogous to the normal (Gaussian) error distribution can be formulated. When the closure error $\varepsilon(\varphi)$ is solely attributed to errors in F_{PH}, it can be treated as residual $F_{PH} - F_P$ as in the original deriviation by Blow and Crick[75] and the residual can be minimized by refining the heavy atom positions.

The closure error $\varepsilon(\varphi)$ can be derived from the known components in Figure 10-18 by the cosine rule, from which we can enumerate the distances OX for each angle φ as

$$\overline{OX}_\varphi = \left(F_{\text{P}}^2 + F_{\text{A}}^2 + 2F_{\text{P}}F_{\text{A}}\cos(\varphi_{\text{A}} - \varphi) \right)^{1/2} \tag{10-38}$$

The closure error is then given as

$$\varepsilon(\varphi) = \overline{OX}_\varphi - F_{\text{P}} \tag{10-39}$$

and we can compute the values for $\varepsilon(\varphi)$ every few degrees. For the true protein phase angle φ_{P}, the distance OX is exactly F_{P}, and the closure error is zero. We can thus interpret the difference $OX_\varphi - F_{\text{P}}$ as a residual, and under assumption of normal distribution of the underlying errors formulate a probability distribution for the protein phases $P(\varphi)$ in analogy to the Gaussian normal distribution (Chapter 7) as

$$P(\varphi) = prob(\varphi\,|\,\varepsilon) = N\cdot\exp\!\left(-\frac{\varepsilon^2(\varphi)}{2\langle\varepsilon^2\rangle} \right) \tag{10-40}$$

with N a normalization factor. The mean square closure error $\langle\varepsilon^2\rangle$ in the denominator serves as a measure for the variance of the phase probability distribution.

Phase combination and best phases

We can use the actual F_{P}, F_{PA}, and $\mathbf{F}_{\text{A}}$ values from our MIR phasing diagram example in Figure 10-17 to compute the closure errors and the resulting probability function for each derivative separately and then express the combined joint probability for the experimentally determined protein phase angle as the product of the independent individual probabilities:

$$prob(\varphi_{\text{P}}\,|\,\varepsilon_1,\varepsilon_2,\ldots,n) = \prod_{j=1}^{n} prob_j(\varphi_j\,|\,\varepsilon_j) = N\cdot\exp\!\left(-\sum_{j=1}^{n}\frac{\varepsilon_j^2(\varphi_j)}{2\langle\varepsilon_j^2\rangle} \right) \tag{10-41}$$

Monomodal joint probability. The individual probability distributions (10-40) as well as the joint probability (10-41) for our two derivatives used in the MIR case depicted in Figure 10-17 are graphed in Figure 10-19. The blue curve is computed from exact values for the first derivative, and the maximum probability is indeed obtained for phase angles of ~40° and ~160°, just as expected from the Harker diagram (Figure 10-17). For the second derivative, we have introduced a small error in F_{PH}, and thus the correct peak of its associated phase probability curve (purple) does not coincide precisely with the peak of the correct blue probability distribution for the first derivative. The joint probability

Figure 10-19 Phase probabilities for isomorphous replacement. The graphs show the phase probabilities for a MIR case derived from the previous examples. The blue probability distribution is based on the exact values from the MIR case in Figure 10-17. For derivative 1, the phase angles of ~40° and ~160° are equally probable. The probabilities for the second derivative have been calculated with a small amplitude error on F_{PH}, which leads to slightly wrong values for the most probable phases in the purple graph compared to the perfect situation depicted in Figure 10-17. The numbers above the individual probability distributions correspond to the numbers in Figure 10-17 indicating the phasing circle intersections. The joint probability curve (red) peaks in the middle between the two most probable values. In this example, the most probable phase and the best phase are the same.

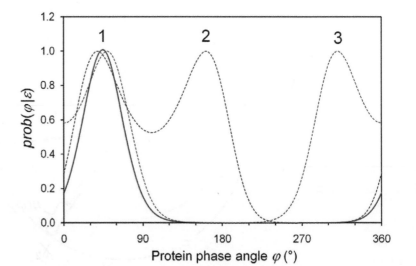

for the protein phase, computed via (10-41), is graphed in the red curve. As expected, the combined probability peak lies in the middle of the two closely spaced individual peaks and the solution is unique (i.e., the joint probability distribution is monomodal). In this case, it is simple to decide which phase to use to compute the best protein map. The Fourier coefficients for the best map are the most probable phase given by the peak of the red joint probability distribution and the measured protein amplitude:

$$\mathbf{F}_{\text{BEST}} = F_{\text{P}} \cdot \exp\left(i\varphi_{\text{PEAK}}\right) \tag{10-42}$$

Multimodal phase probability. In the less perfect, but common, multimodal case the simple approach of picking the maximum value of the probability distribution—the most probable value—is not reasonable at all and we will have to find a procedure to calculate the best map coefficients. The example in Figure 10-20 shows a bimodal phase probability distribution where the most probable phase value indicated by the highest peak in the red joint PDF is not the best phase value. We guess that the best estimate will be somewhere between the two red peaks, probably shifted a bit toward the higher peak. As Bayesians in training, we also have a feeling that if the two peaks in the joint PDF are far apart, our belief in the correctness of the estimated phase angle and thus its usefulness for the map reconstruction will be much lower compared with the unique solution shown in Figure 10-19. Clearly, some weighted averaging of the probabilities over the phase circle will be necessary and, as shown first by Blow and Crick in 1959, the best phase[75] is found by weighted averaging of the possible phase values, that is, by computing of the centroid phases.

Best phases. The probability-weighted phase average or expectation value for the best phases or centroid phases is obtained by the centroid integral, that is, the normalized integration of the probability distribution (Chapter 7) over all phase angles:

$$\varphi_{\text{BEST}} = \langle\varphi\rangle = \frac{\int\limits_{2\pi} \varphi P(\varphi)\,d\varphi}{\int\limits_{2\pi} P(\varphi)\,d\varphi} \tag{10-43}$$

Recall from Chapter 7 that in analogy the best estimate for n discrete measurements is the weighted average

$$\langle x \rangle = x_{best} = \sum_{i=1}^{n} w_i x_i \Big/ \sum_{i=1}^{n} w_i \tag{10-44}$$

The integration over the probability function (10-43) can then be replaced by a discrete summation over all possible protein structure factors $\mathbf{F}(\varphi) = F_{\text{P}} \cdot \exp(i\varphi)$ with the weights being the phase probabilities $p(\varphi)$ over the whole phase range

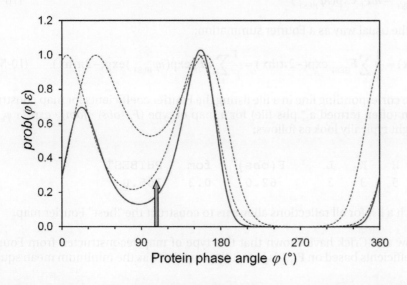

Figure 10-20 **Phase probabilities for multimodal distribution.** In this example of a phase combination, the joint probability is bimodal, and neither peak position in the joint probability function (red graph) represents the best phase angle. The best phase angle is indicated by the position of the green arrow representing the centroid phase obtained from weighted averaging of the phase probabilities as described in the text. The magnitude of the green arrow (0.3) represents its figure of merit (*fom*), a measure of our degree of belief in its accuracy.

$0 \leq \alpha < 2\pi$. The centroid vector $\mathbf{F}_{\text{BEST}}$ or $\langle \mathbf{F} \rangle$ can be calculated by summation every, say, 5° of

$$\mathbf{F}_{\text{BEST}} = \frac{\sum\limits_{\varphi=0}^{2\pi} P(\varphi)\mathbf{F}(\varphi)}{\sum\limits_{\varphi=0}^{2\pi} P(\varphi)} = \frac{F_{\text{P}}\left(\sum\limits_{\varphi=0}^{2\pi} P(\varphi)\cos\varphi + i\sum\limits_{\varphi=0}^{2\pi} P(\varphi)\sin\varphi\right)}{\sum\limits_{\varphi=0}^{2\pi} P(\varphi)} = \frac{F_{\text{P}}\left(\sum\limits_{\varphi=0}^{2\pi} A + i\sum\limits_{\varphi=0}^{2\pi} B\right)}{\sum\limits_{\varphi=0}^{2\pi} P(\varphi)}$$

(10-45)

The best phase angle φ_{BEST} immediately results from $\mathbf{F}_{\text{BEST}}$ as

$$\varphi_{\text{BEST}} = \tan^{-1}\left(B_{\text{BEST}} / A_{\text{BEST}}\right)$$

(10-46)

and the corresponding amplitude F_{BEST} as

$$F_{\text{BEST}} = \frac{(A_{\text{BEST}}^2 + B_{\text{BEST}}^2)^{1/2}}{\sum P(\varphi)}$$

(10-47)

Finally, we can express our degree of belief through a weight or figure of merit (m, w, or *fom* are frequently used abbreviations) in the range $0 \leq m \leq 1.0$, defined as the relative length of the resulting centroid vector $\mathbf{F}_{\text{BEST}}$:

$$m = \frac{F_{\text{BEST}}}{F_{\text{P}}} \text{ or } \mathbf{F}_{\text{BEST}} = mF_{\text{P}} \cdot \exp(i\varphi_{\text{BEST}})$$

(10-48)

By substituting (10-48) back into the probability summation (10-45) we also realize that the figure of merit m is a measure for the mean phase error $\Delta\varphi$:

$$m = \langle\cos(\Delta\varphi)\rangle$$

(10-49)

In our bimodal example, we computed a best phase φ_{BEST} of 105° and an amplitude for F_{BEST} of 19 based on an $F_{\text{P}}(obs)$ of 62. This results in a figure of merit of $m = 0.3$ with a corresponding phase error of 73°. The green arrow in Figure 10-20 is located at the best phase angle φ_{BEST} of 105°, and its length of 0.3 corresponds to the figure of merit. This is certainly a poor phase estimate for this particular reflection and the low figure of merit will weight it down correspondingly in the Fourier synthesis of the initial experimentally phased protein electron density map.

Constructing the initial protein electron density map

Once the best phases and their figure of merit are computed, the initial protein map is reconstructed by taking the best structure factors $\mathbf{F}_{\text{BEST}}$ as Fourier coefficients

$$\mathbf{F}_{\text{BEST}} = mF_{\text{P}} \exp(i\varphi_{\text{BEST}})$$

(10-50)

in the usual way as a Fourier summation:

$$\rho(\mathbf{x}) = \frac{1}{V}\sum_{\mathbf{h}} \mathbf{F}_{\text{BEST}} \exp(-2\pi i\mathbf{h}\mathbf{x}) = \frac{1}{V}\sum_{\mathbf{h}} mF_{\text{P}} \exp(i\varphi_{\text{BEST}})\exp(-2\pi i\mathbf{h}\mathbf{x})$$

(10-51)

The corresponding line in a file listing the Fourier coefficients for map construction (often termed a *.phs-file) for a map of type $(F_{\text{P}} (obs) \cdot fom) \cdot \exp(i \cdot \varphi_{\text{BEST}})$ might typically look as follows:

H	K	L	F(obs)	fom	PHIBEST
5	3	3	62.0	0.3	105.0

Such a list for all reflections allows us to construct the "best" Fourier map.

Blow and Crick have shown that this type of map reconstructed from Fourier coefficients based on $\mathbf{F}_{\text{BEST}}$ is in fact the map that has the minimum mean square

Box 10-7 Probability distributions for phases. After a marker atom substructure has been determined, the phase angle for the protein structure factor can be computed from the phasing equations. A phase angle will have a high probability if it is close to the correct value and a low probability if it is far off the target value. The phase probability distribution can be expressed as a function of a closure error, which in the Blow and Crick approximation is attributed to a sole error in the heavy atom structure factor amplitude.

The best phase angle is computed by probability weighted vector averaging of all possible phase values. From the best phase angle obtained, a best protein structure factor is calculated. The protein map in turn is reconstructed from the best protein structure factors as Fourier coefficients. The protein structure factors are weighted by their figure of merit, equivalent to the cosine of their estimated phase error and indicative of their reliability. A map constructed from best phase angles has the minimum residual error in the electron density construction.

residual error to the true map, that is, the smallest residual error. The proof that minimization of the sum of residuals squared of the form

$$SRS = \sum_{h} \left(\mathbf{F}_{BEST} - \mathbf{F}_{OBS}\right)^2 \tag{10-52}$$

also leads to a minimum in the mean squared residual of the electron density $\langle (\rho_{BEST} - \rho_{TRUE})^2 \rangle$, that is, the best map, has been derived (Sidebar 9-5) as a direct consequence of Parseval's theorem stating that the mean square value of a Fourier-transformed variable is also the mean square of its transform.[76]

10.6   Heavy atom refinement and phasing

At this point, we need to review what information we have at hand to phase our protein structure. We have determined initial heavy atom coordinates and thus the complex structure factors $\mathbf{F}_A$ representing our substructure solution. However, as a result of subtracting large values that only differ modestly, the difference data (ΔF_{iso}, ΔF_{ano}, or ΔF_{disp}) were probably much more noisy and therefore of significantly lower usable resolution than the corresponding measured data pairs they were derived from (F_P, F_{PA}), (F_{PA+}, F_{PA-}), or ($F_{\lambda1}$, $F_{\lambda2}$). This implies that for poor cases one generally needs to refine the atom positions from their initial interpolated grid position against the full derivative or anomalous data. From the refined positions we compute the initial protein phase probabilities and the best phases as outlined above. Once we have determined initial protein phases, we will extend and improve the protein phases to native resolution by density modification techniques. Often these steps are iterated and combined; heavy atom refinement programs nearly always include protein phase calculation and refinement, followed by density modification to improve protein phases even further.

Classical heavy atom refinement
The derivation of protein phase probabilities as outlined in Section 10.5 followed a general treatment of the closure error $\varepsilon(\varphi)$ between structure factor amplitudes as a function of the phase angle φ. Blow and Crick[75] attribute the entire closure error in the vector triangle $\mathbf{F}_{PH} = \mathbf{F}_P + \mathbf{F}_H$ (Figure 10-21) to the amplitude of the heavy atom derivative, F_{PH}. The closure residual can be written as the difference between $F_{PH}(obs)$ and $F_{PH}(calc)$:

$$\varepsilon(\varphi) = F_{PH}(obs) - F_{PH}(calc) \tag{10-53}$$

Following the cosine rule, $F_{PH}(calc)$ relates to the measured property F_P, the initial experimental phase angle φ_P, and the refinement parameters heavy atom position, B-factor, and occupancy (which provide F_H and φ_H in the form of $\mathbf{F}_H$) as

$$F_{PH}(calc) = \left(F_P^2 + F_H^2 + 2F_P F_H \cos(\varphi_H - \varphi_P)\right)^{1/2} \tag{10-54}$$

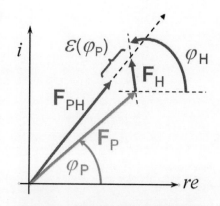

Figure 10-21 Closure error attributed to errors in derivative structure. When the closure error $\varepsilon(\varphi)$ is solely attributed to errors in F_{PH}, it can be treated as residual $F_{PH}(obs) - F_{PH}(calc)$ as in the original deriviation by Blow and Crick.[75] The closure residual can then be minimized by refining the heavy atom positions.

The total sum of residuals squared for each derivative i, SRS_i, and all structure factor amplitude residuals for all reflections $\mathbf{h}$ becomes

$$SRS_i = \sum_{\mathbf{h}} m(\varphi)_{\mathbf{h}} \left(k_i F_{PH}(obs)_{i,\mathbf{h}} - F_{PH}(calc)_{i,\mathbf{h}} \right)^2 \tag{10-55}$$

with m the figure of merit for the phase angle of each reflection $\mathbf{h}$ defined by $m = \langle \cos(\Delta\varphi) \rangle$.

The SRS can be minimized using the basic least squares method introduced in the statistics chapter. Once the heavy atom parameters for each heavy atom derivative are determined (treated as an SIR case), a new protein phase angle can be computed from combined Sim-weighted phase angle probabilities from all derivatives, yielding a best protein phase angle as explained before, and the refinement cycle can be repeated for each derivative.

The progress of the classical heavy atom refinement is generally tracked by R-Kraut and R-Cullis (Josef Kraut and Ann Cullis being pioneers of early protein crystallography) for each derivative, and both R-values can be found as quality indicators in older papers. R-Cullis is defined for centric reflections as

$$R_{Cullis} = \frac{\sum_{\mathbf{h}} \left| \left| F_{PH}(obs) \pm F_P(obs) \right| - F_H(calc) \right|}{\sum_{\mathbf{h}} \left| F_{PH}(obs) \pm F_P(obs) \right|} \tag{10-56}$$

where the $\pm$ sign is negative if the phases have the same sign, otherwise positive. R-values below 0.6 are acceptable, but only a limited number of centric reflections are included. R-Kraut is enumerated as the sum of closure residuals divided by the sum of the derivative amplitudes:

$$R_{Kraut} = \frac{\sum_{\mathbf{h}} \left| F_{PH}(obs) - F_{PH}(calc) \right|}{\sum_{\mathbf{h}} F_{PH}(obs)} \tag{10-57}$$

As R-Kraut is a function of the actually refined quantity F_{PH}, it is useful to track the progress of the refinement, but its absolute values are not as meaningful because a high degree of heavy metal substitution gives quite high R-Krauts. Another measure occasionally found is the phasing power, defined for each derivative as the sum of heavy atom contributions divided by the sum of closure residuals:

$$P_{iso} = \frac{\sum_{\mathbf{h}} F_H(calc)}{\sum_{\mathbf{h}} \left| F_{PH}(obs) - F_{PH}(calc) \right|} \tag{10-58}$$

Maximum likelihood heavy atom refinement

The Blow and Crick phasing procedure (not quite correctly called the phase refinement method) works relatively well for highly isomorphous derivatives and good data; the procedure was successfully applied by Dickerson et al. in the high resolution refinement and phasing of the myoglobin structure.[2] Nevertheless, there are a couple of problems with the basic Blow and Crick method, and we need to investigate improved methods to derive the protein phases. Phasing programs such as $SHARP$[20]—which can handle even cases with high non-isomorphism or poor and incomplete heavy atom solutions—employ probabilistic maximum likelihood phasing models. As phase improvement and heavy atom refinement are closely related, we will use the likelihood functions derived for the conditional structure factor probabilities and extend the basic σ_A-based error model further.

Assumptions in early heavy atom refinement and phasing

The assumptions that went into the basic Blow and Crick heavy atom and phase refinement and their consequences can be summarized as follows:

- The underlying assumption of the closure error triangulation—that all experimental errors can be attributed to errors in F_{PH}, and that F_P and $\mathbf{F}_H$ are correct—is certainly not an adequate description of the underlying physical reality, because errors in all measured intensities, as well as in the derived heavy atom positions, will contribute complex errors.

- Heavy atom substructure models are often incomplete. Some atoms may be absent from the heavy atom substructure and multiple minor sites may have easily been missed. Application of probabilistic weights that account for incompleteness, known as Sim weights, alleviates this problem.

- The marker atom structure will also contain errors, in coordinate positions as well as B-factors and occupancies. For single atoms, coordinate position errors lead to *phase* errors, while B-factor and occupancy errors lead to errors in the *amplitude* of their individual structure factors. Coordinate errors are accounted for in structure factor distributions through the Luzzati D-factor, which can be combined with the incompleteness error into the single sigma-A (σ_A) coefficient accounting for errors in a partial model.

- The presence of both phase error and amplitude error requires derivation of the conditional structure factor probabilities for complex structure factors. So far, we have developed probability distributions for the phase angles, which are not directly accessible quantities; the structure factor amplitudes are what we measure. Maximum likelihood statistics provide a means of integrating the phases out of the complex structure factor probability distributions if needed and obtaining likelihood functions for the observable quantity, the structure factor amplitudes.

- The application of a least squares target assumes that the errors in Fs for each derivative are *independent*. This is certainly—given isomorphism between the derivatives—not a correct assumption. Indeed, during refinement of their chymotrypsin structure Blow and Matthews[77] noted that in the computation of the best heavy-atom coordinates for phase determination, the SIR protein phase angles for the heavy-atom derivative to be refined should be left out of the refinement to minimize bias.

- The practice of treating each derivative as a SIR case against one and the same native and then combining the SIR probabilities lends proportionally too much weight to the native observation.

- The phase angle used in basic heavy atom refinement is the single best value, ignoring the actual phase angle probability distribution. Considering the full phase probabilities, rather than only the single best value, for each SIR case

---

Box 10-8 Improving heavy atom refinement. The classical heavy atom refinement procedure following the Blow and Crick approximation necessarily has imperfections. Attribution of all errors to the heavy atom derivative structure factors, the assumption of independence of errors in structure factors, and lending too much weight to the single native in multiple derivative situations are a few.

Most significantly, the phase angle, which is adjusted in each refinement cycle, is not an actually measured variable, which violates least squares principles. Maximum likelihood statistics allow integrating out of the phase angle from complex structure factor probability distributions, which provides probability distributions and likelihood functions for the actual observable, the structure factor amplitudes.

Improvements to the error model include Sim weights, which take into account the incompleteness of the heavy atom model. Further improvement is achieved by the introduction of σ_A coefficients, a variance term in the probability distribution of each structure factor, which also accounts for errors in the partial substructure model.

Sidebar 10-8 Probabilistic phasing models and when they are needed. It is important to recognize that the early protein crystallographers were quite aware of the limitations of the Blow and Crick based phasing techniques. They realized that a probabilistic model based on maximum likelihood targets which takes accurate probability distributions into consideration would be desirable. The initial publication of Blow and Crick[75] contains the proper description (and projections similar to the 3-dimensional probability distributions in Figure 10-22) of phase probabilities as complex, bivariate distributions. It is also quite interesting that very late in his career, David Blow expressed to Brian Matthews[80] that when he wrote their pioneering paper in 1959 he thought of it essentially as an academic exercise, not as a technique that would be routinely used for protein structure determination. When the early phasing methods were developed (1950 to early 1970s) electronic computing was in its infancy and a full probabilistic or maximum

likelihood treatment was just not feasible. Other early error models have been implemented, for example the HLE and HUE estimates which include anomalous contributions in the error model[81] and with associated *R*-values as quality indicators. We will not discuss these earlier models but focus instead on modern maximum likelihood methods of phase probability determination.

Note however that the more *complete* and *accurate* the heavy atom substructure solution is (including minor sites), the less need there is for likelihood methods for phase refinement and phase extension. In anomalous difference data, non-isomorphism problems are also largely absent. This is the reason anomalous heavy atom positions determined by *SHELXD* are very accurate and the subsequent *SHELXE* phasing program does not attempt to refine the heavy atom positions, although it uses σ_A-based weights to compute the map coefficients during phasing and phase extension.

when computing the joint phase probability was, in fact, the first application of a basic maximum likelihood probability function in experimental phasing, implemented in the CCP4 program *MLPHARE*[78] (*M*aximum *L*ikelihood *PHA*se *RE*finement).

- Most painful for the Bayesian mind, as pointed out already by Bricogne,[12] is that keeping the protein phase values constant while minimizing the closure error by refining the heavy atom parameters, and then *a posteriori* adjusting the protein phases and repeating the process until convergence, is *de facto* not proper least squares practice: the parameter sought (protein phase) is not a parameter in the refinement model. From a Bayesian point of view, this is similar to plugging the posterior probability back into the prior for re-analysis of the data. This practice introduces bias and artificially narrow error estimates.[79] Maximum likelihood statistics allow us to integrate the phase angles out of the complex structure factor probabilities and obtain likelihood functions for the actual, measurable quantity, the structure factor amplitudes.

Accounting for incomplete model knowledge
So far we have assumed in our phase probability combination that the heavy atom solution expressed as the complex heavy atom structure factors $\mathbf{F}_A$ was essentially correct and complete. In Chapter 7 we have shown that the errors in model structure factors—incompleteness and positional error—can be accounted for in likelihood functions by the combined variance parameter σ_A.

Model incompleteness. A brief refresher condensed from Chapter 7 might be useful at this point. In the late 1950s G. A. Sim derived[82] a structure factor probability distribution based on the proposition that the heavy atom model giving rise to the joint phase probabilities is incomplete. The total structure factor contribution of N atoms (of which we observe the diffraction pattern) can be formally split up into a contribution of P known atoms (our incomplete model, for which we can calculate the structure factors) and that of Q atoms, which we do not know anything about (but about which we can make some statistical assumptions):

$$\mathbf{F}_N = \mathbf{F}_P + \mathbf{F}_Q = \sum_{j=1}^{P} \mathbf{F}_j + \sum_{j=P+1}^{N} \mathbf{F}_j \qquad (10\text{-}59)$$

The corresponding probability distribution derived under the assumption of incompleteness is the Sim distribution in crystallography, known in statistics as the Rice distribution. The best Sim weights for general acentric and for centric reflections, m_a and m_c, respectively, are

$$m_a = \frac{I_1(X)}{I_0(X)} \text{ with } \lim_{x \to \infty} \frac{I_1(X)}{I_0(X)} = 1 \text{ and } m_c = \tanh\left(\frac{X}{2}\right) \qquad (10\text{-}60)$$

where I_n represents the modified Bessel function of the first kind for order n (Figure 7-16). The argument of the weighting function, X, is defined as

$$X = \frac{2F_N \cdot F_P}{\Sigma_Q} = \frac{2F_{obs} \cdot F_{calc}}{\sum_{j=P+1}^{N} f_j^2} \qquad (10\text{-}61)$$

The structure factor amplitude F_{calc} (F_P) is computed from the known atoms in the model and the atomic scattering factor summation in the denominator of (10-61) goes over all the missing Q atoms. The difficulty is to determine the variance for the number of missing atoms, but it can be estimated from the discrepancy between observed and calculated structure factors. Bricogne[83] and Read[84] have derived the following estimates ($n = 1$ for centric and $n = 2$ for noncentric reflections):

$$\sum_{i=1}^{m} f_i^2 = \langle | F_{obs}^2 - F_{calc}^2 | \rangle \text{ or } n(F_{obs} - F_{calc})^2, \text{ respectively} \qquad (10\text{-}62)$$

Sim weights are used in the computation of the best phases from heavy atom positions and in phase combination the same way the standard Crick and Blow weights are used (to which they revert for a complete model). The initial electron density maps are constructed as described above with Fourier coefficients $mF_{obs} \cdot \exp(i\varphi_{BEST})$ using the centroid phases and Sim weights for m.

Positional errors in the model. So far, our assumption was that our $\mathbf{F}_A$ or $\mathbf{F}_{calc}$ were derived from heavy atom positions $\mathbf{r}_i$ that had no positional errors $\Delta\mathbf{r}_i$. The positional error can be accounted for in the structure factor probability distribution by a variance term, called sigma-A, or in short σ_A. The variance σ_A was introduced by Srinivasan[85] and Ramachandran[86] in the mid-1960s and improved σ_A estimates were derived by Main[87] and Read.[84] In essence, the effects of coordinate error and missing parts are in fact equivalent in terms of normalized structure factor probabilities. For the purpose of phase combination in heavy atom phasing the Sim weights can be adjusted for errors in the partial model through the additional σ_A variance term and the argument X in the equations for the Sim weights becomes

$$X = \frac{2\sigma_A E_{obs} \cdot E_{calc}}{1 - \sigma_A^2} \qquad (10\text{-}63)$$

In map reconstruction with partial model phases, σ_A is in practice determined by minimizing a residual function based on the normalized structure factors from the partial model and the total (observed) structure factors.[84] The map reconstruction is then performed in the usual way by a Fourier synthesis with map coefficients

$$\mathbf{F}_{BEST} = (2mF_{obs} - DF_{calc})\exp(i\varphi_{calc}) \qquad (10\text{-}64)$$

where F_{calc} and φ_{calc} are computed from the partial model and D is the Luzzati factor as derived in Chapter 7 and revisited again in the framework of maximum likelihood map coefficients and refinement in Chapter 12.

Likelihood functions and phase probabilities

While Chapter 7 dealt with maximum likelihood principles on a general basis, we will now discuss their application to a specific crystallographic situation. We can formulate the posterior probability for our crystallographic model in question using Bayes' formula introduced in Chapter 7:

$$prob(model \,|\, data) \propto prob(data \,|\, model) \times prob(model) \qquad (10\text{-}65)$$

The posterior probability of our model (or, for short, model likelihood) given the data is the product of the data likelihood function *prob(data|model)* of observing the data given a particular model and the prior probability of that model. In situations such as refinement of protein structure models, we may have independent prior probabilities regarding our model without the measured data based on the stereochemical or potential energy restraints.

In the case of heavy atom refinement, we generally entertain no prior assumption about any stereochemistry of the heavy atom model and the model likelihood for each reflection is just proportional to its (data) likelihood function:

$$prob(\mathbf{F}_{calc} \mid F_{obs}) \propto prob(F_{obs} \mid \mathbf{F}_{calc}) \qquad (10\text{-}66)$$

However, we recognize that our experimentally available data points are the structure factor amplitudes, F_{obs}, but the model we seek to optimize is defined by the complete structure factors $\mathbf{F}_{calc}$ including the phase term α. We are not only adjusting the amplitudes during refinement, but are also directly adjusting the model phases α by refining positional parameters, B-factors, occupancy, and probably other parameters ranging from solvent models to other physical variables included in our model description. This situation can be dealt with by marginalization, or integrating-out, of the so-called nuisance variable α as demonstrated in Chapter 7.

Following the product rule for independent probabilities, the total likelihood function L for our substructure atom model (or any structure model, in fact) becomes the joint probability for all reflections $\mathbf{h}$:

$$L \propto \prod_{\mathbf{h}} prob(\mathbf{F}_{\mathbf{h}(obs)} \mid \mathbf{F}_{\mathbf{h}(calc)}) \qquad (10\text{-}67)$$

We maximize the likelihood for our heavy atom model *heavy* in the usual way by minimizing the negative log-likelihood *LL*:

$$\max\left[L(heavy \mid data)\right] \propto \min[-LL(data \mid heavy)] = \min\left[-\sum_{\mathbf{h}} \ln prob(\mathbf{F}_{\mathbf{h}(obs)} \mid \mathbf{F}_{\mathbf{h}(calc)})\right]$$

$$(10\text{-}68)$$

The basic idea behind maximum likelihood (heavy atom) refinement is straightforward: the likelihood function is maximized by varying the (heavy atom) model parameters until the model that represents the data best is obtained. Note that by multiplying the probabilities to obtain the joint likelihood we have made an implicit assumption about independence between the data (structure factors) and their conditioning information, including their errors. This condition is well fulfilled in final refinement of a crystal structure, but not so in heavy atom refinement, where a correlation must necessarily exist between the structure factors of a native crystal and its isomorphous derivative. Similarly, dispersive data, as well as anomalous pairs, are correlated and maximum likelihood probabilities and full Bayesian models can account for that.[88]

The key task in applying ML methods is to formulate the appropriate likelihood function that describes the corresponding phasing situation and its error (or probability) model, which is different depending on whether we have a SIR, MIR, SAD, or MAD experiment. The first applications of maximum likelihood heavy atom refinement have been independently formulated by Otwinowski and Read (implemented in the CCP4 program *MLPHARE*[78]) and in a considerably more rigorous framework by Bricogne (implemented in *SHARP*[19]). The program package *SOLVE*[89] by Terwilliger and Berendsen also employs ML targets in a variety of places, including its density modification and model-building component *RESOLVE*. The same holds for the *CNS* program package[18] and the programs of the evolving *PHENIX* project.[90]

The MIR likelihood function and non-isomorphism
The first step in establishing the likelihood function as a product of the individual

structure factor probabilities is developing an appropriate error model for conditional probabilities of one structure factor on another. This is most easily understood in the basic conditional likelihood case: given a calculated value (the model) of a structure factor, what is the probability of observing an experimental measurement (the data) of that value?

We begin with the basic data likelihood function (which is ultimately proportional to the model likelihood or posterior)

$$prob(data \mid model) = \prod_{\mathbf{h}} prob(\mathbf{F}_{obs} \mid \mathbf{F}_{calc}) \qquad (10\text{-}69)$$

which requires us to develop a conditional probability distribution for $\mathbf{F}_{obs}$ given $\mathbf{F}_{calc}$. While writing relation 10-69 down is quite simple, establishing the proper likelihood function is generally the trickiest part of the process,[91] because the function needs to accurately describe the phase probabilities and their error model and should have reasonably well-behaved numerical behavior, especially as far as the necessary integration over the probability distributions (and phase angle) is concerned.

MIR likelihood. Our desired total MIR likelihood function would be the joint probability of all structure factors $\mathbf{F}_i$ of all N crystals, conditioned on the derivative heavy atom contribution $\mathbf{H}_i$:

$$L_{MIR} = \prod_{\mathbf{h}} prob(\mathbf{F}_0, \mathbf{F}_1, \mathbf{F}_2 .. \mathbf{F}_N \mid \mathbf{H}_0, \mathbf{H}_1, \mathbf{H}_2 .. \mathbf{H}_N) \qquad (10\text{-}70)$$

where we treat the native ($N = 0$) just as a derivative without a heavy atom contribution ($\mathbf{H}_0 = 0$). As usual, we later need to integrate the nuisance variables α_i out to obtain the probability for the structure factor amplitudes:

$$L_{MIR} = \prod_{\mathbf{h}} \int_{\alpha_0}^{2\pi} \int_{\alpha_1}^{2\pi} \int_{\alpha_2}^{2\pi} .. \int_{\alpha_N}^{2\pi} prob(F_0, \alpha_0, F_1, \alpha_1, F_2, \alpha_2 .. F_N, \alpha_N \mid \mathbf{H}_0, \mathbf{H}_1, \mathbf{H}_2 .. \mathbf{H}_N) d\alpha_0 \, d\alpha_1 \, d\alpha_2 .. d\alpha_N$$
$$(10\text{-}71)$$

This "grand" likelihood function (10-71) allows us to treat all derivatives involved in refinement simultaneously and avoid the circuitous classical route through multiple SIR situations (which we pretend are independent). However, to establish a proper MIR likelihood function, we need to address the issue of correlation between structure factors from the derivative(s) and the native structure factors before forming the joint probability for all structure factors. Because of the complexity involved, we can only outline the derivation of the MIR likelihood function in the major conceptual steps, following the complete analysis as provided lucidly by Read,[92] Bricogne,[12] and McCoy.[93]

Independence of probabilities. The trick to making the structure factor probabilities independent is to act as if we know the true structure factor $\mathbf{F}_T$ for the native protein. A derivative structure factor for derivative j then is given by $D\mathbf{F}_T + \mathbf{H}_i$ (the argument for introducing the Luzzati factor D remaining the same as in the derivation of the general conditional structure factor probability in Chapter 7). The protein part of the structure factor $\mathbf{F}_T$ is then correlated, but the remaining parameters and errors in the heavy atom part are independent. If we integrate the true structure factor out as a nuisance variable (as we generally do for the phase angle) then we have obtained independence in the remaining individual probabilities and the joint probability. Under the provision of independence, the joint likelihood function L for all structure factors can be computed in the usual way by multiplication of the individual probabilities (or summation of the individual log-likelihoods).

To integrate the nuisance variable $\mathbf{F}_T$ out, we need to form the integral over the whole complex plane (with x in the integral representing the real part and y the imaginary part of $\mathbf{F}_T$). For each reflection, the joint MIR probability for all derivatives then becomes

$$prob(\mathbf{F}_0,\mathbf{F}_1..\mathbf{F}_N \mid \mathbf{H}_0,\mathbf{H}_1..\mathbf{H}_N)$$

$$= \int_x \int_y prob(\mathbf{F}_T) prob(\mathbf{F}_0,\mathbf{F}_1..\mathbf{F}_N \mid \mathbf{H}_0,\mathbf{H}_1..\mathbf{H}_N) d\mathbf{F}_T \tag{10-72}$$

$$= \int_x \int_y prob(\mathbf{F}_T) \prod_{i=0}^{N} prob(\mathbf{F}_i \mid \mathbf{F}_T,\mathbf{H}_i) d\mathbf{F}_T.$$

Accounting for non-isomorphism. Given (10-72), we can account for the errors in heavy atom structure factor $\mathbf{F}_i$ resulting from non-isomorphism or measurement error before we eliminate the nuisance phase by integration over 0 to 2π to evaluate the sought-after likelihood $prob(\mathbf{F}_i|\mathbf{F}_T,\mathbf{H}_i)$. The method to account for the error in $\mathbf{F}_i$ resulting from non-isomorphism is inflating the variance, meaning that we add an additional variance term for each derivative structure factor to the general conditional probability $prob(\mathbf{F}_1|\mathbf{F}_2)$ as derived above. Inflating the variance can be visualized as broadening the Harker "craters" of the 2-dimensional probability distributions shown in Figure 10-22A, B, and E. Starting from the basic acentric 2-dimensional probability distribution (shown in Figure 7-18)

$$prob(\mathbf{F}_1 \mid \mathbf{F}_2) = \frac{1}{\pi\sigma_\Delta^2} \exp\left(-\frac{|\mathbf{F}_1 - D\mathbf{F}_2|^2}{\sigma_\Delta^2} \right) \tag{10-73}$$

inflating the variance and considering the split of the heavy atom component into $D\mathbf{F}_T + \mathbf{H}_i$ we obtain

$$prob(\mathbf{F}_i \mid \mathbf{F}_T,\mathbf{H}_i) = \frac{1}{\pi(\sigma_\Delta^2 + \sigma_i^2)} \exp\left(-\frac{|\mathbf{F}_i - (D_i\mathbf{F}_T + \mathbf{H}_i)|^2}{\sigma_\Delta^2 + \sigma_i^2} \right) \tag{10-74}$$

Integrating the phase out to get the probability for the derivative structure factor amplitude leads again to a Rice distribution:

$$prob(F_i \mid \mathbf{F}_T,\mathbf{H}_i) = \frac{2F_i}{\sigma_\Delta^2 + \sigma_i^2} \exp\left(-\frac{F_i + |D_i\mathbf{F}_T + \mathbf{H}_i|^2}{\sigma_\Delta^2 + \sigma_i^2} \right) \cdot I_0\left(\frac{2F_i|D_i\mathbf{F}_T + \mathbf{H}_i|}{\sigma_\Delta^2 + \sigma_i^2} \right) \tag{10-75}$$

The equation above allows us to explicitly calculate the probabilities which we substitute into (10-72) to obtain the product over all reflections **h** which finally gives the complete MIR likelihood function:

$$L_{MIR} = \prod_{\mathbf{h}} \int_x \int_y prob(\mathbf{F}_T) \prod_{i=0}^{N} prob(F_i \mid \mathbf{F}_T,\mathbf{H}_i) d\mathbf{F}_T \tag{10-76}$$

or written as the –log-likelihood to be minimized with an appropriate numeric algorithm:

$$-LL_{MIR} = -\sum_{\mathbf{h}} \ln\left(\int_x \int_y prob(\mathbf{F}_T) \prod_{i=0}^{N} prob(F_i \mid \mathbf{F}_T,\mathbf{H}_i) d\mathbf{F}_T \right) \tag{10-77}$$

Interpretation. This mother of a likelihood function (10-77) is actually nothing more than an expression of our uncertainty about our state of knowledge in the phasing equations. This is readily visualized in terms of the Harker construction, except that now we do not operate with sharp circles representing an exact magnitude of a structure factor, but a "probability crater" representing the complex Gaussian distribution of the measurement errors as well as the substructure model errors. The following Figures 10-22 and 10-23 are calculated in a *Mathcad* spreadsheet from the functions derived above and follow a similar presentation by Randy Read.[92] Note, however, that Blow and Crick had already realized, and visualized in contour plots[75], that these complex probability distributions should be used in the calculation of the best phases.

SAD, SIRAS, and MAD likelihood functions

SAD likelihood function. Following similar considerations that led to the derivation of MIR likelihood, the likelihood functions for other experimental phasing situations can be derived. Improved SAD likelihoods are particularly

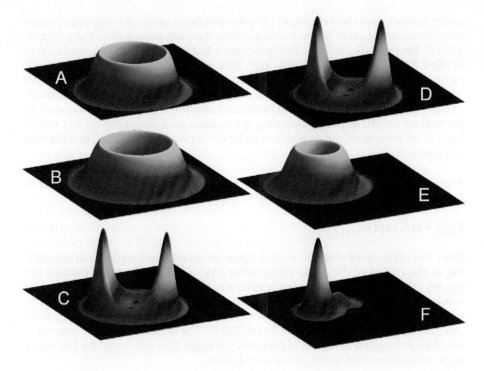

Figure 10-22 MIR likelihood function.
The figure shows the structure factor probability distributions for three derivatives (A, B, E) and their respective joint MIR likelihood functions visualized in 3-dimensions as a Harker construction in the complex plane. The left column shows the native structure factor probability distribution (A) and the first derivative structure factor probability distributions (B) in the form of "Harker craters" which are combined into a joint probability function corresponding to a symmetric SIR situation with two overlapping Harker circles (C, D). In the right column, the joint SIR probability function (C, D) is combined with a second derivative (E) delivering a nearly monomodal joint probability distribution (F) that clearly resolves the phase ambiguity. The heavy atom parameters are then adjusted (which moves the center positions of their probability craters) to achieve the maximal MIR likelihood function (which is given by the volume under the joint probability function in panel F).

relevant, given the increasing use of single-wavelength anomalous diffraction experiments. The SAD situation is predicated by the clear relation that must exist between the pairs of anomalous structure factors, F_{obs}^+ and F_{obs}^-, and the SAD likelihood function must also consider that the dispersive marker atom itself is part of the model of normal scattering atoms. SAD likelihood functions have been derived by Pannu and Read,[94] Bricogne,[19] and Terwilliger and Berendsen.[89] To give a feel for the complexity involved, the SAD likelihood function as derived by McCoy[95] (quote: "simple algorithm") and implemented in *Phaser*[13] is given below:

$$prob(F_o^-, F_o^+ \mid \mathbf{F}_H^-, \mathbf{F}_H^+) = \frac{F_o^-}{\pi \Sigma^-} \int_{\alpha=0}^{2\pi} \exp\left(-\frac{\left|\mathbf{F}_o^- - \mathbf{F}_H^-\right|^2}{\Sigma^-}\right) \cdot \Re(F_o^+, F_c^+, \Sigma^-) \, d\alpha^- \qquad (10\text{-}78)$$

where

$$F_c^+ = \left|\mathbf{F}_H^+ + \mathbf{D}(\mathbf{F}_o^- - \mathbf{F}_H^-)\right| \qquad (10\text{-}79)$$

and

$$\Re(F_o^+, F_c^+, \Sigma^-) = \frac{2F_o^+}{\Sigma^+} \exp\left(-\frac{(F_o^+)^2 + (F_c^+)^2}{\Sigma^+}\right) I_0\left(\frac{2F_o^+ F_c^+}{\Sigma^+}\right) \qquad (10\text{-}80)$$

SIRAS likelihood function. In the SIRAS or MIRAS case, the derivative signal is also split into a set of anomalous pairs as in the SAD case and, therefore, the heavy atom contribution also has pairs $\mathbf{H}^+$ and $\mathbf{H}^-$. Therefore, the full likelihood function would have to account for anomalous as well as isomorphous correlations, which is presently not feasible. The situation can be remedied by refining in the anomalous part against the difference ΔF_{ano} instead of the individual pair members. The correlation between the difference of pairs ΔF_{ano} and the magnitude of their mean value $\langle F_{obs}^+ + F_{obs}^- \rangle$ is much less pronounced than the correlation between each anomalous pair, and the anomalous contribution to the probability distribution can be accounted for in the form of a least squares term instead of a full likelihood term.[93]

MAD likelihood functions. The MAD case can be treated in different ways. Originally, non-likelihood MAD equations were formulated by Hendrickson[73] (discussed in Section 10.4), but most programs handle the MAD case as a combination of independent SIR and SAD cases where the dispersive data are treated

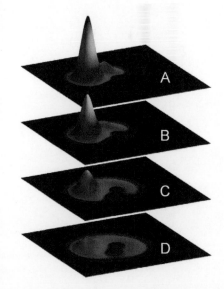

Figure 10-23 Effect of lack of isomorphism on likelihood function.
Lack of isomorphism can be expressed as increasing the variance of the corresponding derivative's probability distribution function. Successive panels A–D show the effect of increasing non-isomorphism (increasing variance) of the first derivative on the joint probability distribution. The initially sharp probability function degrades to a pancake of a rather shallow (low probability) and more ambiguous (tending towards bimodal) probability distribution.

as derivatives, and then combine the probabilities—under the not-always-sustainable assumption of independence—into the joint probability. The program *SOLVE*, for example, follows this procedure but in the final refinement cycle applies Bayesian correlated phasing[88] that considers that the measurement errors in dispersive MAD data are necessarily correlated (because they are all from the same crystal). The approach that *SHARP* takes is quite sophisticated[12] and the problem is that in the full likelihood treatment of the dispersive data as derivatives all phase pairs (including anomalous and dispersive terms) would have to be integrated out, which has to be done numerically for all but the first integral.

Protein maps. Each of the heavy atom refinement programs finally produces Fourier coefficients for a protein map of the general form

$$\mathbf{F}_{\text{BEST}} = mF_{\text{P}}\exp(i\varphi_{\text{BEST}}) \tag{10-81}$$

from the phase probabilities with a figure of merit $m = \langle\cos(\Delta\varphi)\rangle$ corresponding to the cosine of the mean phase error $\langle\Delta\varphi\rangle = \cos^{-1}m$. However, initially the phases can be generated only to the extent of the highest usable derivative data resolution. Phase improvement and phase extension to the full resolution of the native data is achieved through density modification, either in separate programs such as *DM*,[96] *SQUASH*,[97] *SOLOMON*,[98] or *RESOLVE*.[17] These programs are often integrated in the same program packages that were used for the heavy atom refinement and initial phase calculation.

Combining phases from different sources

In the process of phasing, phase improvement, and phase extension we frequently have to combine phases from different sources. This is greatly aided by using a consistent parameterization of the phase probabilities, because having a different formulation for the phase probabilities in each phasing situation is not conducive to free exchange and combination of phase information in crystallographic computations. The problem of general phase combination was addressed by Hendrickson and Lattman.[99] The four parameters used in their phase probability model, the Hendrickson–Lattman coefficients *A*, *B*, *C*, and *D*, allow simple phase combination between phases obtained from different phasing or phase-improvement techniques. The phase probability as a function of four terms $\cos\varphi$, $\sin\varphi$, $\cos 2\varphi$, and $\sin 2\varphi$, can be expressed in the form

$$prob(\varphi \mid A,B,C,D) = K\exp(A\cos\varphi + B\sin\varphi + C\cos 2\varphi + D\sin 2\varphi) \tag{10-82}$$

Box 10-9 Maximum likelihood functions for MIR, SIR, SAD, and MAD. The log-likelihood function for MIR is obtained from the sum of probability distributions formulated to treat all derivatives at once, in contrast to combined SIR cases. The problem of dependence or correlation of errors between derivatives and native is accounted for by assuming that the protein part of the derivative is the same and known in both derivative and protein, and integrating it out of the probability distribution. The error resulting from non-isomorphism is addressed by inflating the variance (adding additional variance) of the probability distribution.

After integrating out the phase, a final –log-likelihood function is formulated by summation of all (log) structure factor probabilities, which can be minimized with a suitable numeric algorithm. The equations for SIR, SAD, and MAD are derived in similar fashion.

The likelihood function represents our state of knowledge or uncertainty of the system and can be visualized by plotting the complex distributions and forming their joint probabilities in analogy to the Harker circles.

Phases from different sources can readily be combined by parameterizing the separate phase probabilities with Hendrickson–Lattman coefficients.

To obtain the Hendrickson–Lattman (HL) coefficients, each source of phase probability information is expressed in terms of $\cos\varphi$, $\sin\varphi$, $\cos 2\varphi$, and $\sin 2\varphi$, and the specific form of the coefficients A, B, C, and D and a normalization constant K depend on the particular phase probability distribution. For example, we may want to combine phase information from isomorphous replacement and from complementary anomalous data. The combined joint probability is then defined as

$$prob(\varphi)_{comb} = \prod_{j=1}^{n} prob_j(\varphi)_{iso} \cdot \prod_{i=1}^{m} prob_i(\varphi)_{ano} \qquad (10\text{-}83)$$

Given that the individual probabilities were formulated using HL coefficients, we can accomplish the multiplication of probabilities simply by adding the HL coefficients in the exponent:

$$prob(\varphi \mid A,B,C,D)_{comb}$$
$$= \exp\left(\sum_k K_k + \sum_k A_k \cos\varphi + \sum_k B_k \sin\varphi + \sum_k C_k \cos 2\varphi + \sum_k D_k \sin 2\varphi\right) \qquad (10\text{-}84)$$

The sum over the Ks can be taken out of the exponent and absorbed in a common normalization constant N, which cancels out completely in the computation of best phases (10-45). The HL coefficients thus allow the combining of phase information from varying sources such as density modification, non-crystallographic symmetry averaging, or partial models. The option of phase combination via HL coefficients is implemented in most crystallographic programs.

10.7 Improving the initial electron density map

Once we have obtained an initial protein map, the map quality and resolution can generally be improved—and almost always quite drastically—by density-modification techniques. Density modification is an encompassing term that applies to a number of methods—largely real space based—that incorporate prior knowledge or certain assumptions we have about the physical properties of a protein structure (and its electron density map) to improve the protein phases. Amongst the methods are solvent flattening (and its relative, solvent flipping), histogram matching, or density averaging in cases where NCS is present in the structure or maps from multiple crystal forms are available. Reciprocal space maximum likelihood solvent flattening and map-likelihood phasing[100] (called "prime and switch") is implemented in the *RESOLVE* program. The basis for all methods is that the physical properties (either in real space or in their reciprocal space transform) of a protein molecule are distinct from those of the disordered solvent in the intermolecular channels of a crystal structure. Iterative density modification is extraordinarily powerful, and can lead from poor and practically uninterpretable initial protein maps to high quality electron maps with low mean phase error, which can be directly used for manual or automatic model building.

Solvent modification and phase extension

Solvent flattening/flipping exploits that fact that in the disordered solvent region the electron density $\rho(\mathbf{x})$ is essentially constant (flat) at about 0.33 e⁻/Å³ for pure water (but it can be higher, depending on the crystallization cocktail) while protein has an average electron density of about 0.44 e⁻/Å³ (Sidebar 11-6). A fundamental necessity for real space based solvent modification is that the initial protein map is at least of a quality that allows contiguous solvent regions to be distinguished from the molecular envelope and a solvent mask to be generated.

Solvent masks and solvent flattening

The first task for a procedure implementing solvent modification is to define the solvent regions. The various solvent mask definition procedures implemented in solvent density modification programs are based on the method pioneered by

Sidebar 10-9 Properties of a good electron density map. Density modification programs exploit the fact that a correct electron density map has distinctly different properties in the disordered solvent region compared with the discrete density in the protein region. They essentially try to make a poor (noisy) map look more like a "real" protein structure map. Large solvent regions provide better discrimination between protein and solvent than more densely packed structures, and the methods generally work better with higher solvent content and poorly below 30% (where the solvent content essentially approaches that of close packed spheres, Chapter 11). Some commonly employed criteria are used to distinguish a poor map from a good one:

Connectivity: In the protein regions, the density should display contiguous stretches representing secondary structure, while the solvent should contain only short, disjointed stretches of noise density.[101]

Contrast: The contrast between the flat solvent region and the protein molecule should be high, that is, a clear delineation between protein and solvent should exist.

Density histogram: a good map has a positive skew in a sharp electron density distribution, while noise maps have a broader random (Gaussian-like) distribution.

Standard deviation of the local r.m.s. density: The protein and solvent region have distinctly different density variance, while a noise map has uniformly the same density variation.[102]

B. C. Wang[56] in 1985. The principle is readily visualized with the help of Figure 10-24 and resembles the local scaling procedure. A 3-dimensional grid, somewhat coarser than the electron density grid, is placed over the low resolution map. At each grid point (x, y, z) the electron density $\rho(\mathbf{x})$ within a sphere of a given radius of ~10 Å is computed. If the density is above a certain cutoff value, the grid point is deemed to be in protein density (circles with a blue center in Figure 10-24); if the value is below the cutoff, the grid point is assigned to the solvent region. If the map is of redeemable quality, contiguous solvent and protein regions will be found.

Within a solvent region, the electron density $\rho(\mathbf{x})$ is set to the solvent average $\langle \rho_s \rangle$ (0.33 e$^-$/Å^3 on absolute scale) and new structure factors are computed by FT (map inversion) of both the protein part of the map and the now flat substructure of the solvent part of the molecule. The new phases are then combined with the original phases, which will yield an improved map, and the procedure can be repeated with adjusted parameters (solvent content and integration radius) while adding (unphased) high resolution data until convergence is achieved and no more improvement takes place. The procedure is generally most effective when the solvent content is large, and a correct initial solvent content estimate from Matthews probabilities is also a key parameter for proper mask definition. Given too large a solvent content the mask will tend to nibble into protein, and too small a solvent content estimate renders the procedure ineffective. The key steps of solvent modification and phase extension are illustrated in Figure 10-25.

The fact that we are combining phase probabilities to generate the new map immediately raises a flag, namely the question of necessary independence of the individual probabilities which we combine into the new joint probability. The new information (the flattened solvent region) generally has a correlation to the old information (the protein region in the map), which introduces bias of the new phase angles toward the old map. This can be largely overcome by a modified procedure, called solvent flipping.[98]

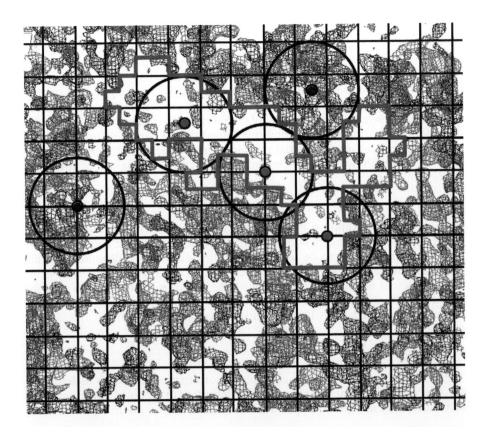

Figure 10-24 Principle of mask generation. A 3-dimensional grid somewhat coarser (exaggerated in this drawing) than the map grid is placed over the initial protein map. At each grid point, the average electron density within a sphere of a given radius (shown as circles) is computed. If the density is above a certain cutoff value (circles with a blue center), the grid point is deemed to be in protein and assigned the corresponding average density ρ_{ave}. If the value is below the cutoff, the grid point is assigned to the solvent region (red-centered circles). If the map is of redeemable quality, contiguous solvent and protein regions will be found.

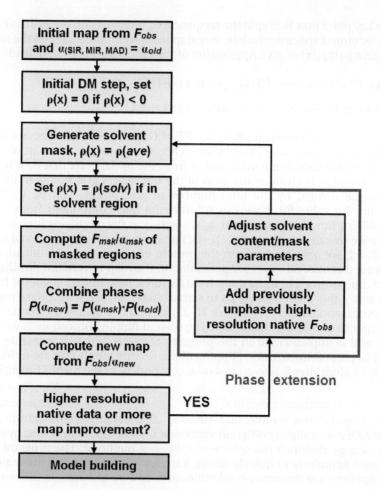

From a masked representation of protein and solvent, structure factors and phases are computed. The masked-map phases are combined with the initial phases (from a SIR, SAD, MIR, or MAD experiment) and improved protein phases are obtained by phase (probability) combination. Initially unphased, high resolution native F_{obs} for which no experimental substructure phases were available can be successively added and phases computed from the mask extended to a higher resolution. The process of phase extension and mask improvement is continued until convergence is achieved.

Flowchart:

Initial map from F_{obs} and $\alpha_{(SIR, MIR, MAD)} = \alpha_{old}$

↓

Initial DM step, set $\rho(x) = 0$ if $\rho(x) < 0$

↓

Generate solvent mask, $\rho(x) = \rho(ave)$

↓

Set $\rho(x) = \rho(solv)$ if in solvent region

↓

Compute F_{msk}/α_{msk} of masked regions

↓

Combine phases $P(\alpha_{new}) = P(\alpha_{msk}) \cdot P(\alpha_{old})$

↓

Compute new map from F_{obs}/α_{new}

↓

Higher resolution native data or more map improvement? — YES → Add previously unphased high-resolution native F_{obs} → Adjust solvent content/mask parameters → (back to Generate solvent mask)

Phase extension

↓

Model building

Solvent flipping

Abrahams and Leslie implemented solvent flipping and first demonstrated its application in the phasing of the bovine mitochondrial F1 ATPase,[98] the peripheral part of a transmembrane protein found in bacterial plasma membranes. Solvent flipping does not set the density values in a solvent region to a constant low value, but inverts the grid point (or density voxel) values in the solvent region. Flipping the solvent density does introduce independence of the partial maps and thus allows proper phase probability combination.

For solvent flipping, we define a solvent mask as described above, but set the average density $\langle \rho_s \rangle$ of the solvent to zero. This can be accomplished by subtracting the constant value $\langle \rho_s \rangle$ in real space from each density grid point, which maintains the map contrast and translates in reciprocal space into a change of F_{000} but leaves the other structure factors unchanged (the FT of a constant is zero everywhere except at the origin, Sidebar 9-3). On the other hand, a function in reciprocal space that is zero everywhere except at the reciprocal origin (F_{000}) becomes a constant value in real space. This is simply a result of the Fourier integration, and the proof is straightforward.[103]

Having set the mean solvent density $\langle \rho_s \rangle$ to zero, we can apply a modifying mask function $g(\mathbf{x})$ to the entire map density $\rho(\mathbf{x})$. The density $\rho(\mathbf{x})$ at any grid point in real space is multiplied by $g(\mathbf{x})$ which is 1 for protein density values and 0 for solvent regions. The reciprocal space equivalent of $g(\mathbf{x})$ is a function $\mathbf{G_h}$ applied to $\mathbf{F_h}$, where $\mathbf{G_h}$ is the FT of $g(\mathbf{x})$. From the FT convolution theorem (Chapter 7) it follows that

$$FT(\mathbf{F_h} \otimes \mathbf{G_h}) = FT(\mathbf{F_h}) \cdot FT(\mathbf{G_h}) = \rho(\mathbf{x}) \cdot g(\mathbf{x}) \tag{10-85}$$

Sidebar 10-10 A hybrid method for *ab initio* structure determination at 2 Å or higher resolution. A novel hybrid phasing method (*ARCIMBOLDO*) that is based on multi-solution fragment search with the maximum likelihood based program *Phaser*[13], followed by density modification and further fragment building and auto-tracing with *SHELXE*[119] has recently been described.[120] Several structures that could not be determined otherwise have been solved using this brute-force *de novo* method, which seems to work at resolutions of 2 Å or better. The maximum likelihood based program *Phaser* is used to search for several helical poly-alanine (or other general secondary structure) fragments, and the multiple results (none of which is sufficiently discriminate to be called a true solution) are the submitted in parallel to the powerful combination of density modification and auto-tracing with *SHELXE*. True solutions can then be distinguished based on improved map correlation coefficients and the number of residues built. The method is being further developed and may provide a new and general avenue to the solution of the phase problem in protein structures. Although termed an *ab initio* method, it is not related to the (true) *ab initio* direct methods based on probabilistic phase relations.

The key point now is to split the reciprocal $\mathbf{G_h}$ function into two parts, $\mathbf{G_{h=0}} + \mathbf{G_{h\neq0}}$, the reciprocal space equivalent to real space solvent part $g_S(\mathbf{x})$ and the real space protein part $g_P(\mathbf{x})$ of $g(\mathbf{x})$. Application of the convolution theorem leads to

$$\rho(\mathbf{x}) \cdot \text{FT}(\mathbf{G_h}) = \rho(\mathbf{x}) \cdot \text{FT}(\mathbf{G_{h\neq0}}) + \rho(\mathbf{x}) \cdot \text{FT}(\mathbf{G_{h=0}})$$
$$= \rho(\mathbf{x}) \cdot g_S(\mathbf{x}) + \rho(\mathbf{x}) \cdot g_P(\mathbf{x}) = \rho(\mathbf{x}) \cdot g_S(\mathbf{x}) + \rho(\mathbf{x}) \cdot \gamma \qquad (10\text{-}86)$$

Given that $g_S(\mathbf{x})$ corresponds to the FT of $\mathbf{G_{h\neq0}}$, the FT of the second part $\mathbf{G_{h=0}}$, denoted γ, is a constant in real space. Therefore, the rightmost part of (10-86) is the part that carries the information from the initial map, that is, it represents bias. Its value is given by the ratio of the number of grid points in the protein envelope divided by the total number of grid points, that is, V_P/V. Having obtained a value for γ, we can subtract the bias component from the initial modifying function $g(\mathbf{x})$ and use $g_S(\mathbf{x}) = g(\mathbf{x}) - \gamma$ as the modifying function for $\rho(\mathbf{x})$ after we rescale it to the range $(1, -1)$. For a solvent content of 0.5 and the initial $g(\mathbf{x}) = 1$, we obtain 0.5 within the protein envelope and -0.5 outside. We can rescale this interval to $1, -1$. Application to the map thus maintains the protein part, but inverts (i.e., flips) the solvent features. By subtracting out the bias component γ, this procedure helps to reduce bias, and the joint phase probability recombination (step 6 in Figure 10-25) in practice fulfills the requirement of independent phase probabilities. The solvent flipping procedure is very powerful, and is implemented in the program *SOLOMON* that is available in CCP4 and as part of *AutoSHARP*. Flipping of the density voxels that are unlikely to be protein atom sites is also an essential component of the *SHELXE* algorithm.

Sphere of influence method

The popular macromolecular phasing and density modification program *SHELXE* uses a slightly different version of solvent flipping, termed by its inventor George Sheldrick the sphere of influence method.[59] The program *SHELXE* is used frequently to quickly derive a useful initial electron density map in all except the most desperate borderline cases and we will use it in the next section to demonstrate the power of density modification.

In contrast to the binary methods that essentially integrate over the volume of a Wang sphere, Sheldrick calculates the variance of the density on the surface of a 2.42 Å sphere. The reason for this choice of radius is to maintain some plausible chemical information and 2.42 Å is in fact the dominant 1–3 atom distance in macromolecular structures (a case of likelihood by intuition).[63] Thus, when the sphere of influence is located at a protein atom position, the density variance at the shell surface will be high. Being also an accomplished small molecule crystallographer, George Sheldrick approximates the 2.42 Å sphere as a C_{60} fullerene, and adds the 32 points at the face centers of the polyhedron to form a 92-vertices sphere.

The density modification procedure starts with generation of a sharpened density map with coefficients $\left(F_T \cdot E_T\right)^{1/2}$ from native protein (not including any anomalous contributions) and centroid starting phases derived either from SIR or SAD data, or from initial phases obtained via the MAD equations (10-35) as explained in detail in the example in Section 10.8. To ensure that high-z atoms at the solvent periphery such as sulfur or phosphorus (or Se in Se-Met) are not discarded because they do not comply with the 1,3 distance criterion of the 2.42 Å sphere, the 5% highest density values are simply treated as protein. The covered density region is then split into solvent, protein, and a crossover region. The variable (fuzzy) crossover region (inflating the variance by intuition) prevents the solvent mask becoming locked in because of a wrong initial solvent content estimate. Negative density in the protein regions is again set to zero and positive protein density is sharpened. The solvent correction then has a γ-correction applied as explained above in the solvent flipping section. The fuzzy region allows a smooth transition from solvent to protein and the final phase combination uses σ_A weighted coefficients, which accounts for the partial bias from the parts of the map assigned as protein. The procedure is again repeated until convergence.

Histogram matching

Histogram matching is a technique originally used for image enhancement in image processing programs. It exploits the fact that the quality of an image can be improved by matching the distribution of its voxel intensity values to that of a high quality image of a similar object. Histogram matching was adapted to crystallography in the late 1980s by Lunin and co-workers[104] as a complementary approach to density modification. Electron density histogram matching is based on the fact that the distribution of the electron density values $\rho(\mathbf{x})$ in a map is practically the same at a given resolution for each protein structure, irrespective of individual structural details.

In practice we plot the contiguous probability distribution of the electron density in i discrete bins, and compare the density histogram of the initial map with that of a theoretical map at the same resolution. Figure 10-26 shows that the density distribution of a poor initial map is much broader, more Gaussian indicating randomness, and shifted to higher mean values than the theoretical distribution. We thus adjust the bin positions by a shift b_i so that the histograms match and then multiply the value of each density grid point value in a bin so that it matches the theoretical mean bin value:

$$\rho(new) = a_i\rho(old) + b_i \qquad (10\text{-}87)$$

The histogram-matching procedure works even better when sharpened (B-factor corrected) structure factors are used to generate the maps.[105] Further improvements are implemented by also using the derivatives (gradient) of the electron density probability function leading to 2-dimensional histogram matching. Histogram matching is implemented in a number of programs such as DM[96] and $SQUASH$.[97] The latter also makes use of Sayre's equation[106] (10-21) derived earlier and which additionally restrains the electron density in positivity. Figure 10-26 also shows how density modification by the sphere of influence method as implemented in $SHELXE$ (which makes no prior assumptions about any density histogram) drastically improves the actual density histograms for our MAD structure determination example exercised in Section 10.8.

A variant of histogram matching that involves local density-template matching[108] is implemented in $RESOLVE$. In addition, the procedure computes the properties of each density point through a Patterson function solely by using the information from the surrounding points. Therefore, the new and improved map points $\rho_i(new)$ are independent of the old density value $\rho_i(old)$ at the same location and the phase combination can be done in the usual way through joint probabilities.

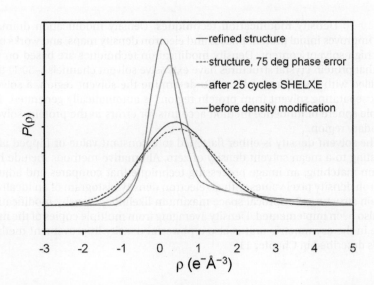

Figure 10-26 Electron density histograms. The electron density probability distribution is plotted for our example case (PDB entry 1a8d) at 1.9 Å. The density distribution from calculated values shown in the blue graph is relatively sharp and has a distinct positive skew. Compare it with the density distribution of the same structure with a near-random mean phase error of 75° (much broader and symmetric magenta distribution). Also shown is the dramatic improvement of the density distribution by solvent modification using Sheldrick's sphere of influence method[59] (which does not employ any histogram matching). The red distribution belongs to a poor initial MAD map close to a near random distribution, which improved dramatically after 25 cycles of $SHELXE$ density modification (green graph) to yield a well-interpretable map (example in the next section). Peter Zwart, LBL, kindly computed the histograms using the Crystallographic Computing Toolbox (cctbx).[107]

Maximum likelihood reciprocal space methods

Instead of adjusting the density values in a map based on the distinction of solvent versus protein region, we can alternatively ask ourselves what the likelihood function of a good protein map should look like, and then maximize its likelihood. That approach is the basis for map-likelihood phasing,[100, 109] and has been implemented in *RESOLVE*. The joint likelihood considers experimental structure factor probabilities, as well as prior information about the structure factor distribution and expectations for the properties of the protein map in the form of the density histogram, and is thus best described as a Bayesian reciprocal space maximum posterior method. The corresponding grand likelihood function is then the sum of several log-likelihoods:

$$LL(\mathbf{F_h}) = LL_0(\mathbf{F_h}) + LL_{obs}(\mathbf{F_h}) + LL_{MAP}(\mathbf{F_h}) \qquad (10\text{-}88)$$

The first term on the right of (10-88) includes the prior expectation based on intensity statistics alone, such as Wilson statistics; the second term LL_{obs} includes the probability of the structure factors (and phases) based on the experimental map. The key part is the map likelihood function LL_{MAP} for our expectations of a good map, defined as

$$LL_{MAP}(\mathbf{F_h}) \simeq \frac{N_{ref}}{V} \int_V LL[\rho(\mathbf{x}, \mathbf{F_h})]\, dx\, dy\, dz \qquad (10\text{-}89)$$

The map-likelihood function is then split again into a density probability for protein and solvent region, and LL_{MAP} is optimized by varying the parameters defining $\rho(\mathbf{x})$. The method is again used iteratively until an optimal map is obtained and it can also be useful in bias removal.[100]

Density averaging, dummy atom placing, and Free Lunch

Non-crystallographic symmetry implies that multiple copies of the same molecule occupy the asymmetric unit. As a consequence, there will be redundancies both in real space and in reciprocal space. These redundancies can be quite powerfully exploited for further map improvement. As map averaging and molecular replacement share the search for (multiple) copies of molecules, we will treat both techniques in Chapter 11.

Iterative dummy atom placement and refinement becomes possible as soon as discrete atoms are discernible in the electron density maps, including discrete water atoms in the nearest coordination shells. The atomicity limit is fairly relaxed in dummy atom placement, and can be applied as soon as the resolution of the data exceeds about 2.5 Å. In addition to serving as a real space density modification technique, iterative dummy atom refinement has also been implemented

---

Box 10-10 Density modification techniques. Density modification dramatically improves initial protein phases and electron density maps, and works best with high solvent content. Density modification techniques are based on the fact that protein crystal structures have extensive solvent channels (~50%) that are filled with disordered solvent. To determine the solvent region, a solvent mask separating solvent from protein regions is automatically generated. The flexible sphere of influence method accounts for errors in the protein–solvent boundary region.

The solvent density is either flattened to a constant value or flipped after adjusting to a mean solvent density of zero. Alternative methods include histogram matching, an image processing technique that compares and adjusts electron density pixel values with an electron density histogram of an idealized protein structure. Reciprocal space maximum likelihood density modification has also been implemented. Density averaging from multiple copies of the molecule in the asymmetric unit is also a powerful density improvement method and is described in Chapter 11.

for model rebuilding and is one of the most powerful model bias minimization techniques when combined with map averaging. We will therefore discuss the implementation of dummy atom methods in the programs *ARP/wARP*[110] and in *Shake&wARP*[49] in sections about model bias minimization in Chapter 11.

The Free Lunch phase extension method as implemented in *SHELXE* is described in Sidebar 10-11. It is presently being developed further to incorporate partial model building which will also improve the phase extension procedure (similar in effect to the dummy atom phase extension). Future releases of *SHELXE* will include auto-tracing (and perhaps free dinner in addition to Free Lunch).[119]

10.8   Fast track from data to high quality protein map

Most modern program packages perform the steps of heavy atom search, initial phase generation, and density modification in one step, and deliver (hopefully) a protein map that is interpretable, either visually or by an automated program. To develop a feel for the processes taking place during automated substructure solution and to demonstrate the impressive effects of solvent modification we will solve two protein structures in separate steps using the *SHELX* programs. The first structure will be the VTA1 S-SAD structure[112] from the in-house anomalous data collected on a Cu-anode and processed in Chapter 8. For this example, we will use *SHELXC/SHELXD/SHELXE* and the *HKL2MAP* GUI.[52] The second example will be a Se-MAD structure of modest quality which we will use for low resolution model building in Chapter 12. A very lucid introduction to *autoSHARP*, which is a powerful alternative for complex MIR cases and when the fast-track phasing fails, has been provided by Gérard Bricogne and co-workers.[20]

The big advantage of the *SHELX* fast phasing suite is that the programs can give a very fast (few minutes) indication if a heavy atom solution and a traceable map can be produced from the data, even while the crystal is still mounted (Sidebar 10-4). One can for example collect more data on the same crystal if higher signal-to-noise is likely to be sufficient for solving the structure, or if the situation seems entirely hopeless, try another crystal without wasting more beam time. In addition, the *SHELXE* maps are of excellent quality and a recent auto-tracing module augments the suite [119] so that, in most routine cases, you will be able to leave the synchrotron (or your home laboratory) with an initial model, ready for refinement and polishing!

Caveat. One will not always have the good fortune of having 1.25 Å near-atomic resolution native data and high quality anomalous phases to 2.5 Å available, as Figure 10-27 reveals. Model building in particular—manual and automated alike—becomes much more reliable with higher resolution data. In isomorphous difference phasing, non-isomorphism between difference data sets and resulting incomplete or weak heavy atom solutions create additional difficulties, and likelihood-based phase refinement methods (*SHARP*, *SOLVE*, *CNS*, *PHENIX*) accounting for non-isomorphism may in such cases provide an advantage over the *SHELX* programs.

S-SAD phasing with highly redundant in-house data

The data for the following S-SAD phasing example have been processed in Chapter 8 from 5400 frames of highly redundant anomalous S-SAD data collection scans. The basic strategy for SAD phasing is as follows:

- examine the data quality;
- extract the anomalous difference data set;
- solve the heavy atom substructure;
- break the phase ambiguity;
- combine with native high-resolution data (if present); and
- extend the phases to high resolution (not forgetting to get some Free Lunch on the way).

Sidebar 10-11 Free Lunch for protein crystallographers. An interesting and unique approach to improved phase extension, termed the Free Lunch algorithm,[58] is implemented[111] in the current version of *SHELXE*. Free Lunch is so-called because it requires no additional information over what is commonly provided for phase extension; it is not entirely clear yet where the evident improvement originates from. The basic idea is just to add more high resolution reflections with extrapolated amplitudes and cycle the phase extension procedure.

The phases for the unmeasured, high resolution reflections that are added are computed by *SHELXE* as described for the sphere of influence method above. To get their corresponding amplitudes on a correct scale, the map is inverted, and the amplitudes are set to the mean amplitudes obtained from the experimental Wilson plot extrapolated to the resolution limit of the added, unmeasured data. The phase and amplitude extension to unmeasured reflections can also be supported by including partial models if available, similar to the *ARP/wARP* phase extension.

The Free Lunch method, skillfully combined with partial model tracing, has brought maps with a mean phase error of ~80° (decidedly not interpretable, as consultation of Figure 9-17 will remind you) to a final mean phase error of only 17°. As Free Lunch (and *SHELXE*) comes at no cost, it is worth trying it in any case. It generally works when native data of at least 2 Å are available.

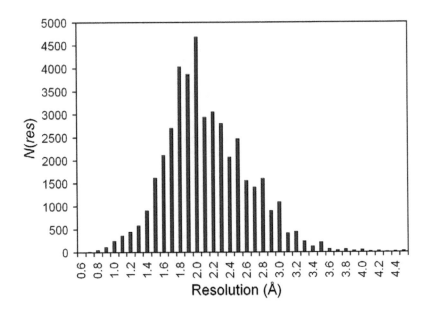

Figure 10-27 Reported resolution of PDB structures. The most frequent reported resolution is in the 1.95–2.05 Å bin, with the mode of the distribution at 2.1 Å. No attempt was made to exclude redundant entries (multiple determinations of the same protein in the same crystal form).

The process flow becomes evident in the form of the *HKL2MAP* GUI[52] (Figure 10-28).

Data analysis and preparation: *SHELXC*

We can roughly estimate whether the data we have collected will likely lead to a successful S-SAD phasing experiment. The first indication comes from an estimate of the ratio R_{anom}/R_{pim} (Chapter 8) which, based on empirical evidence,[111] should be above 1.5. We first read the integrated but otherwise unmerged data into *XPREP* and compute the precision-indicating merging R-value R_{pim} by merging all intensities except Friedel pairs. From this symmetry-merged data set, we compute the linear anomalous merging R-value R_{anom} by merging its Friedel pairs. The same values could be extracted with a few more steps from *SCALA* using the CCP4 GUI or other data processing software. Given the anomalous redundancy of ~30 of our data, we are relatively confident that the ratio R_{anom}/R_{pim} will turn out in our favor.

The result from the redundancy estimate is quite encouraging and for the anomalous data up to 1.7 Å we obtain a ratio of R_{anom}/R_{pim} of 0.0246/0.0114 = 2.16, definitely larger than the empirical lower limit of 1.5. This estimate tends to be somewhat biased in our favor because of the relative increase in anomalous signal at high resolution. We therefore check the ratio also for 2.5 Å, which yields 0.0124/0.075 = 1.65, which is still a favorable predictor for the outcome of our planned S-SAD phasing experiment.[113]

High resolution cutoff. An important step in the data processing is proper selection of a high resolution cutoff for the phasing step (the reasons root foremost in the nature of the *E*-value peak search as explained in the next section).

Figure 10-28 Flow of S-SAD phasing. The basic steps of SAD phasing are neatly arranged in the *HKL2MAP* GUI: Analysis and preparation of anomalous difference data, search for the anomalous heavy atoms, and determination and extension of phases.

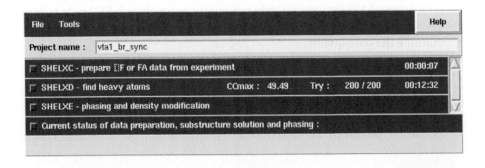

An empirical criterion is to truncate the data at a value where the signal-to-noise ratio for the difference signal

$$\langle \Delta F / \sigma(\Delta F)\rangle = \langle | F_{\mathbf{h}}^{+} - F_{\mathbf{h}}^{-} | / \sigma(F_{\mathbf{h}}^{+} - F_{\mathbf{h}}^{-})\rangle \tag{10-90}$$

falls below ~1.3. We can see from Figure 10-29 that in our example $\langle \Delta F / \sigma(\Delta F)\rangle$ falls off below 1.3 around 2.5 Å. The default cutoff criterion of $d_{min} + 0.5$ (2.2 Å) is too optimistic in this S-SAD case. Other criteria based on the correlation coefficient between signed anomalous differences also exist.[15, 60]

The practical processing of the data is straightforward: read the unmerged anomalous file and (optionally) a merged high resolution data set for phase extension into *SHELXC* (free for academics and essentially a stripped-down command line version of the Bruker *XPREP* program although the algorithmic details are different in both programs) and examine the output tables, or use the graphical interface program *HKL2MAP*.

Searching for heavy atom locations: *SHELXD*

The Patterson-seeded, dual-space direct methods program *SHELXD* picks a set of Patterson vectors from a Patterson map, and then proceeds to generate trial solutions, which it extends with direct methods cycled with real space peak picking. The solutions are evaluated using the correlation coefficient CC(all) between *E(obs)* and *E(calc)*. The relation between anomalous or difference amplitudes provided as input and the resulting anomalous (or isomorphous in the case of SIR) difference Patterson maps is straightforward as explained in Section 10.3. It is quite interesting to consider the actual nature of the *E* data, which are used in the direct methods part for picking the largest *E*-values. Starting from the MAD equations (10-35) one can derive[60] that the difference amplitudes relate to the experimental heavy atom amplitude via

$$\Delta F_{ANO} = |F^{+}| - |F^{-}| = c | F_{A} | \sin\alpha$$

$$\Delta F_{ISO} = |F_{deriv}| - |F_{nat}| = b | F_{A} | \cos\alpha \tag{10-91}$$

for the SAD and SIR cases, respectively (note again the orthogonality between the two cases). Here α is the phase shift between the true phase and the heavy atom phase, that is, $\varphi_{T} = \varphi_{A} - \alpha$ and the constants include the scattering factors as defined in formulas (10-37). Upon normalizing the difference amplitudes to normalized structure factor amplitudes E, the constants cancel out, but the E values will still be wrong, depending on the cosine or sine terms in (10-91), respectively. How can the program still find the sites so effectively? As only the largest E values are selected for the peak search, those with $\sin\alpha = \pm 1$ will

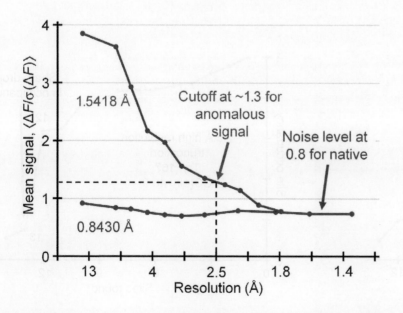

Figure 10-29 Empirical resolution cutoff for SAD phasing. The plot shows the signal level of the anomalous amplitude differences for the high redundancy anomalous data collected at the Cu wavelength. In the case of the high resolution data collected at the high energy wavelength of 0.8340 Å (red graph) practically no more S anomalous signal is measureable. An empirically well determined high resolution cutoff is around the point where the anomalous signal-to-noise ratio drops below 1.3.

likely be the largest ones. These large E-values are in turn the most likely correct ones and therefore the peak search works better than expected.[60] It also explains why the method needs a sensible resolution cutoff: large but incorrect differences primarily occur between noisy high resolution intensities.

Varying resolution cutoffs. Because substructure solution using the relatively weak sulfur anomalous signal is quite sensitive to data quality, it is a good idea to try a few high resolution cutoffs and monitor both the discrimination of the solution landscape (either in the CC(all) histogram (Figure 10-11) or in the CC(all) vs. PATFOM/CC(weak) scatter plots) and the falloff in heavy atom occupancies. There should be a distinct separation between solutions and non-solutions as well as a distinct fall-off around the expected number of correctly found atoms (Figure 10-30). The solutions that meet the criteria then can all be submitted to the phasing and phase extension stage with *SHELXE*, and the best map will be finally selected for model building. Note also, from the trial number of 167 required to obtain the best solution in a distinctly advantageous situation, that in less optimal cases many more trial runs (as many as the resources allow) should be run to assure the best possible solution is actually found.

Disulfide links. The heavy marker atoms we are searching for in our case are the relatively light ($z = 16$) dispersive S atoms. Our VTA1 protein has three cysteine bridges per molecule and, because the S–S bond distance is ~2.03 Å, we cannot expect that the peak search can resolve individual S atoms given the anomalous difference data resolution of about 2.5 Å. The substructure solution program thus needs to be instructed to search for six disulfide pairs, not for 12 sulfurs.[60] A default parameter setting the closest distance between heavy atoms can also be set. The default of 3.5 suffices for most situations. Note that if an atom is missing in the initial peak list, it will likely be found later in the final peak list with updated heavy atom positions that *SHELXE* provides after the last cycle. Note that the "super-sulfur" search concept[114] is in principle also applicable to heavy atom clusters such as Fe_4S_4 or $(Ta_6Br_{12})^{2+}$ and for tri-substituted "magic triangle" iodobenzene derivatives.[115]

We can again use the *HKL2MAP* interface or the simple command line scripts provided at the end of this section to submit our ΔF data to *SHELXD* with a resolution cutoff of 2.5 Å set in the GUI (or in the instruction `shel 99 2.5` in the input file). Above ~2.8 Å there were also no more correct solutions. The resulting best solution is shown with a non-solution in Figure 10-31.

Phase determination and phase extension: *SHELXE*

Once we have selected the best heavy atom solution, the heavy atom list, the difference data file, and (optional) native high resolution data are read by *SHELXE*.

Figure 10-30 Effect of resolution truncation on heavy atom solution. Anomalous difference data are sensitive to noise, and need to be truncated at an optimal high resolution cutoff. A clear solution (right panel) is distinguished from a poor or wrong solution (left panel) by a discriminate solution landscape indicated by a clear drop in occupancy after a number of correct high occupancy sites were found. Note that a high number of trials may be necessary to find all sulfur sites with *SHELXD*.

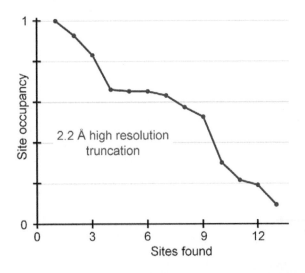

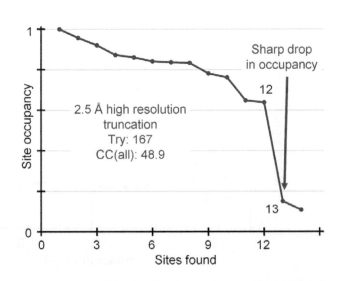

The program determines the phases from the SAD phasing equations given above, breaks the phasing equation ambiguity and extends the phases by the sphere of influence algorithm as detailed in Section 10.7.

Resolving phase ambiguity. The first step in resolving the phase equation ambiguity is done with a low density elimination algorithm,[60] using only the highest electron density peaks. In this first step, the phase improvement is small, but enough to determine the correct solution for a subset of reflections. If the heavy atoms are also included in the native data (which is always the case for S-SAD and often for Se-SAD or -MAD cases) their calculated phases φ_A are also a weak estimate for the protein phases and can be used to further break the phasing ambiguity. Note that at this point we do not know about the correct handedness of the substructure solution; we need to submit both substructure enantiomorphs and examine which map provides the better statistics (Figure 10-31).

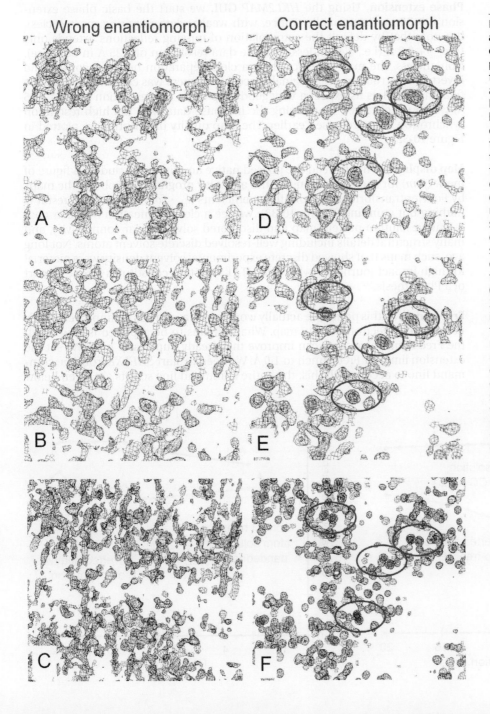

Wrong enantiomorph Correct enantiomorph

Figure 10-31 Stages of phase ambiguity resolution and enantiomorph selection. While the phase ambiguity is resolved internally in the first step by a low density elimination algorithm, the metal substructure handedness is determined by trying both enantiomorphs (inverted, left column; correct, right column). The top row shows the maps at 2.5 Å after the initial ambiguity resolution (but no further density modification) for both enantiomorphs. Note the map from the correct enantiomorph (D) appears less noisy and shows the unresolved "super-sulfur" density blobs (red outline) in correct positions. The center row (B, E) show further improvement by sphere of influence solvent modification for the correct substructure enantiomorph (E), but still no phase extension beyond 2.5 Å has been attempted. The contrast in the correct enantiomorph map has further increased, and the outline of the molecule becomes distinct. The bottom row (C, F) shows the maps after phase extension to 1.25 Å in 20 cycles of phase extension. We clearly can distinguish side chain shapes and atoms begin to visibly resolve at this resolution (F). As it may not be trivial for the beginner to distinguish correct from poor maps with the unaided eye, attention to the map statistics (Figure 10-32) is vitally important.

Map quality assessment. During iterative sphere-of-influence phase extension a cycle-by-cycle listing containing the key parameters map contrast and map connectivity is generated. The contrast in the map is the difference between what is believed to be solvent region and what represents protein. Clearly, a high contrast is a good sign. Connectivity refers to the connectivity of the electron density within the map, which also should be high, but seems to be a less sensitive parameter compared with map contrast. Note, however, that as the resolution increases toward atomic, the map connectivity tends to decrease because the atoms are less connected. The contrast can also decrease as more side chains are resolved with increasing resolution. The final listing reports a resolution-binned correlation coefficient for the map as a key indicator; values above ~0.65 are considered acceptable. Once the pseudo-free correlation coefficient (based on the comparison of E_{obs} and E_{calc} for 10% of the data left out at random in the calculation of a density modified and Fourier back-transformed map) approaches 0.8, an interpretable map is likely.

Phase extension. Using the *HKL2MAP* GUI, we start the basic phase extension (20–30 cycles normally suffice, with weaker data requiring more cycles), but note that by default the extrapolation of missing reflections extends only to d_{min} + 0.2 Å (i.e., 1.45 Å, less than the data resolution of 1.25 Å in our case). Nonetheless, the map statistics show a clear separation between correct and inverted substructure (the original space group guess was correct) and also the final map CC lets us expect a very good map up to the resolution limit of 1.25 Å, which we obtained by merging the in-house data with the high resolution synchrotron data processed earlier. The map quality diagnostics are plotted in Figure 10-32.

Map display. We finally read the phased native amplitudes including a figure of merit (*fom*) for each reflection file into a display program, and obtain the maps by Fourier transformation from the coefficients $F_{obs} \cdot fom \exp(i\varphi)$. The result is nothing but stunning (Figure 10-33): we see a clear delineation between near atomic resolution protein map and disordered solvent, clear connectivity, and many structural details including well-resolved discrete solvent atoms. Not long ago, such maps that showed discrete experimental solvent density were material for high impact journals[116] while our VTA1 structure constitutes a routine part of a Ph.D. thesis.[112]

Free Lunch. At this point one actually wonders what can still be improved in this near-perfect electron density map. What remains to be explored is how much the Free Lunch algorithm can improve the already impressive map. We set the extension limit for Free Lunch to 1.0 Å. We need to start *SHELXE* from the command line to accomplish this; the entire command line scripts are given below

Figure 10-32 Map contrast and final map correlation coefficient. The map contrast is a key indicator in the discrimination between correct and inverted heavy atom substructure enantiomorph (left panel). For the final map, the map correlation coefficient is an indicator for the resulting map quality. Given that the map correlation coefficient is approaching 0.95 over large parts of the map including the high resolution shells, we expect an exceptional experimental electron density map.

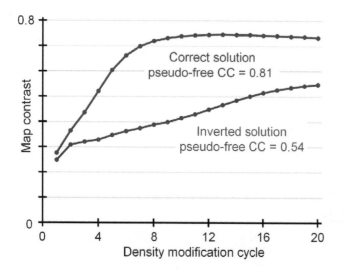

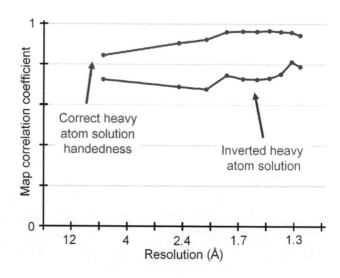

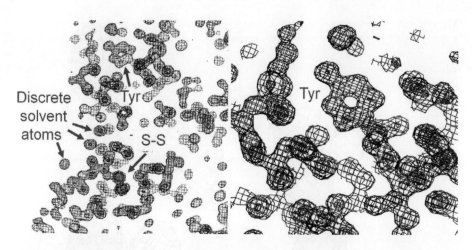

Figure 10-33 S-SAD experimental electron density of VTA1 at 1.25 Å resolution. As expected from the map correlation coefficient, which was above 0.9 for the high resolution reflections, we obtain an experimental map of outstanding quality. The delineation of protein and solvent is clear, and even discrete solvent molecules are visible. In the overview panel we also recognize high density of the S atoms in an S–S bridge, and from a helix viewed in cross section (compare Figure 12-28) protrudes a tyrosine residue with a nice hole in the aromatic ring. Because protein crystallographers tend to present holes in phenyl rings of experimentally phased maps with great pride, we share the detail view in the right panel.

and are self-explanatory (common *SHELXE* flags are: **–m**, phase extension cycles; **–s**, solvent fraction; **–b**, print updated heavy atoms; **–h**, set when native data do include the heavy atom – *important*; and optional **–i**, invert the heavy atom substructure). The switch **–a***n* used in the example below instructs *SHELXE* to conduct *n* cycles of autotracing, with **–n***m* specifying the number of NCS related copies in the asymmetric unit. Both switches are available in the most recent *SHELXE* release.

```
shelxc vta1 < vta1.inp
shelxd vta1_ fa
shelxe vta1 vta1_fa —s0.5 —m20 —e1.0 —h —b -a3 -n2
```

with the input to *SHELXC* (also including instructions used by *SHELXD*)

```
cell 1.54178 65.73 65.73 47.16 90 90 90
spag P43212
nat allmerged
sad unmerged17
sfac S
find 6
dsul 6
shel 99 2.5
mind -3 1.5
ntry 200
```

Practice makes perfect (structures). As we started from an already near-perfect map, the additional improvement is subtle but distinct. Additional detail can be recognized largely in the solvent-exposed regions of the protein (Figure 10-34). In summary, Free Lunch as implemented in *SHELXE* is certainly worth pursuing. Given the simplicity of the few input lines and rapid execution, experimentation with the *SHELXC/D/E* is certainly one of the fastest ways toward learning and exploring anomalous phasing (SAD, MAD) or anomalous signal supported isomorphous SIRAS phasing.

Submission to model building program

As we have obtained such a stunning map from *SHELXE*, we shall submit it right away together with the sequence information to the *ARP/wARP* model building server[117] at the EMBL Hamburg. We will explain manual and automated model building in great detail in Chapter 12; for now it suffices that we take the **vta1. phs** file containing structure factor amplitudes, standard deviations, figure of merit, and phase, and convert it into *mtz* format using the CCP4 GUI. The model returned by the *ARP/wARP* server (Figure 10-35) is complete for both copies and lacks only the N-terminal and C-terminal Lys residues.

Figure 10-34 Electron density map improvement after Free Lunch extrapolation to 1.0 Å resolution.
Shown is a superposition of the original 1.25 Å experimental map obtained with standard sphere of influence phase extension (blue map) with the 1.0 Å map extrapolated by the Free Lunch algorithm contoured at the same level (green map). While the maps are practically identical in the interior parts of the protein, small but distinct improvements can be detected primarily in the solvent-exposed regions. Shown is the recovery of atoms in a main chain break of a flexible solvent-exposed region. Note how the dynamic nature of this region also manifests itself in the elliptical shape of the electron density, indicative of anisotropic atom displacement. The model therefore must be refined with anisotropic displacement parameters.

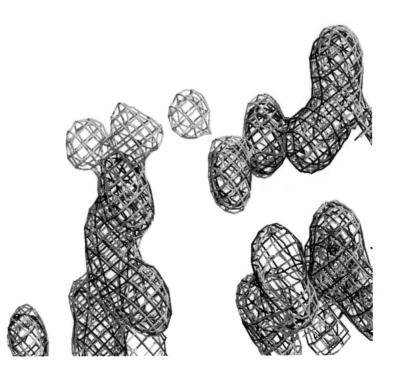

MAD phasing with Se substructure

Solution of the MAD phasing equations is slightly different from the SAD case, and it is worthwhile briefly repeating key steps in the fast phasing exercise with MAD data. In addition, we also want to build a low resolution structure in the model building section and, therefore, data with lower resolution will be used. The molecule is the *E. coli* medium chain length acyl-CoA thioesterase II (*tes*B). It is of interest because it is a protein that interacts with the product of the HIV-1 *Nef* gene.[118] The MAD data have been kindly provided by Z. Dauter and colleagues and can be downloaded from George Sheldrick's web site http://shelx.uni-ac.gwdg.de/~trs/mad/mad.html.

To make our exercise slightly more interesting, we will not use the high resolution S-Met native structure that extends to 1.9 Å, but use only the 2.5 Å Se-Met low energy remote data (0.9801 Å, 12.649 keV) as the native data set. We have additional inflection point data (0.9793 Å, 12.660 keV), peak data (0.9787 Å,

Figure 10-35 Ribbon model of VTA1.
The phase extended 1.25 Å experimental map was submitted to the *ARP/wARP* web service http://cluster.embl-hamburg.de/ARPwARP/remote-http.html and a near complete model of the VTA1 dimer, lacking only the N-terminal and C-terminal lysine residues, was built. View down the 2-fold NCS axis.

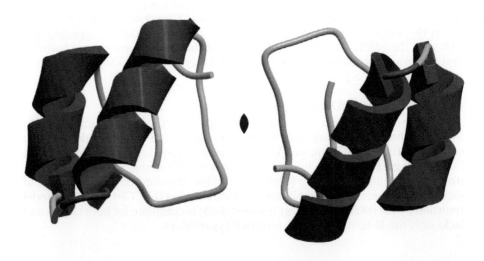

12.667 keV), and high energy remote data (0.9747 Å, 12 .719 keV) available. The crystal contains two molecules of 286 residues with four Met residues each, which should, according to the anomalous Crick formula (Chapter 6), allow a decent substructure solution given reasonable data. In addition, given a solvent content of about 68%, we expect solvent modification to work quite well. Figure 10-36 reviews the key steps in MAD phasing with the *SHELX* programs.

Data preparation and substructure solution

The first step includes the generation of a combined anomalous difference data set based on solution of the over-determined MAD phasing equations. The same statistics as in the SAD case inform us about the anomalous signal levels in each data set. The anomalous signal level measured as $\langle \Delta F / \sigma(\Delta F) \rangle$ follows the f'' values as expected from theory: practically absent in the low remote energy data because it is below the absorption edge, followed by inflection data, largest at the absorption peak wavelength and lower again at the high remote energy (which in this case was not measured very high above the peak and thus contains substantial anomalous signal). Because we have multiple wavelengths available in the MAD case, we can also check for the correlation of the signal between the wavelengths. The highest correlation will be between the data with strong f''. A good resolution cutoff value for MAD experiments is the point where the anomalous CC falls below ~30%. Figure 10-37 provides an overview of the diagnostics for our MAD data, and we see that the three data sets with anomalous contributions correlate well above 0.3 up to the data collection limit of 2.5 Å.

Because the MAD equations are over-determined, if we have more than two MAD data sets available (Section 10.4), we could actually refine the f' and f'' values, but only *XPREP* allows us to manually and stepwise refine the initially entered anomalous scattering factors.

Once the scattering factors are determined, the MAD equations (10-35) are solved for F_A, F_T, and α; and an "F_A file" containing F_A, $\sigma(F_A)$, and α is written by *SHELXC* up to the useful resolution range. Note that α is not yet the protein phase, but is defined as the phase shift between true protein phase and the heavy atom phase: $\alpha = \varphi_T - \varphi_A$. The input file for the subsequent *SHELXD* program

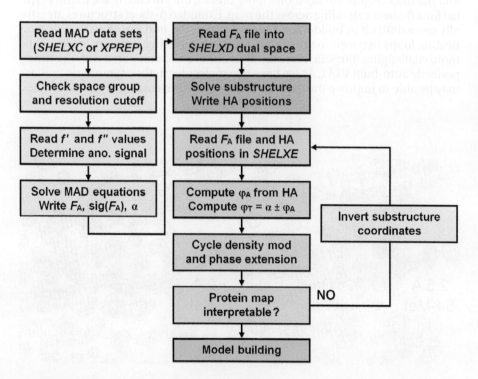

Figure 10-36 **Substructure determination and phase extension using *SHELXC/D/E*.** The flow diagram represents the key steps in the MAD substructure solutions described in Section 10.4. After data preparation in *XPREP* or *SHELXC*, the substructure is determined by the Patterson-seeded dual space direct methods program *SHELXD*, and the subsequent phasing and phase extension is executed by *SHELXE*.

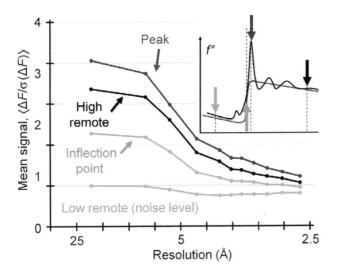

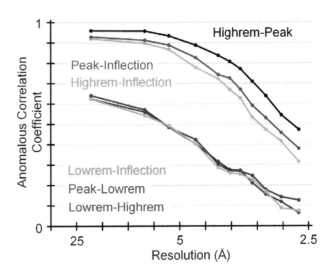

Figure 10-37 **Anomalous signal level and anomalous correlation coefficient.** The anomalous signal level follows the *f″* values for the corresponding wavelengths (insert left). A strong correlation between the anomalous signal in the data sets recorded at or above the absorption edge is an indication of usable data sets. An empirical cutoff for MAD data is where the anomalous correlation coefficient CC(ano) between the data sets containing anomalous signal (top curves right) drops below 0.3 or 30%. In this example, all data to the resolution limit of 2.5 Å are usable to generate the combined F_A difference data file for the heavy atom search.

needs to contain at least the eight expected anomalous Se scatterers; for good measure we let the program look for 12 atoms. The solution is again rapidly found and unambiguous.

Resolving handedness ambiguity and phase extension

After substructure solution, we employ the program *SHELXE* to solve the phasing equations and to check for the correct substructure enantiomorph. The program reads the heavy atom positions (from which it can compute φ_A), and it also reads the F_A file, from which it takes the phase value α and computes an initial protein phase angle $\varphi_T = \alpha + \varphi_A$, which can either be correct, if we selected the correct handedness, or, according to Table 10-2, in the MAD case invariably contains noise if we picked the wrong handedness (given correct F^+ and F^- assignment). Therefore the discrimination between the enantiomorphs is generally sharper in the MAD case than in comparable SAD situations.

Given the lower resolution of our MAD data, the map this time will not be as detailed as in the VTA1 high resolution S-SAD example. The connectivity is lower and the map (Figure 10-38) shows more breaks, but we clearly see density typical for a β-sheet extending across the map. Extended β-sheet structures are usually more difficult to build than α-helical structures, and the often flexible connecting loops between the β-strands tend to be missing. This molecule will be more challenging (but still not really difficult) to build compared with the nearly perfectly auto-built VTA1. As we have two molecules in the asymmetric unit, we may be able to improve the map further by density averaging of the two copies.

Figure 10-38 **Experimental electron density maps phased from Se-MAD substructure.** The left map is generated using the low remote Se-Met data set as native data in density modification and phase extension from the 2.5 Å Se-substructure phases. Note the β-strand density extending across the map. As the low-remote energy "native" also contains the Se atom, it is distinctly visible. The missing connection between the Se atom and the rest of the Met residue is common; it is a result of negative truncation ripples around the big Se peak extinguishing the connecting density. The map on the right side is phase extended to 1.9 Å using native S-Met data. The residue shapes now become quite distinct, and the connectivity has improved.

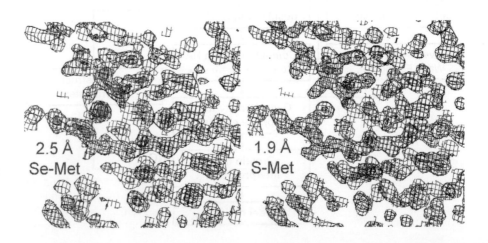

This procedure will be described in Chapter 11. Finally, if we are able to collect a native data set to higher resolution, we can extend the phases and improve the map drastically (Figure 10-38, right panel). If the data set exceeds 2 Å in resolution, the Free Lunch algorithm may provide additional improvements. A native test data set to 1.9 Å is available for our example from George Sheldrick's web site. Try it yourself and experience what a difference high resolution data can make; it is always worthwhile aiming for a native high resolution data set, given the power of density modification and phase extension procedures.

10.9 Next steps

At this point we are able to obtain a quality protein map through experimental phasing. The examples and data analyzed during phasing should enable you to use a variety of phasing programs to your advantage and we could proceed directly to the next step, to model building and refinement. In many instances, as in our example cases, the crystals contain more than one copy of the molecule in the asymmetric unit cell, and the redundancies caused by this local or non-crystallographic symmetry (NCS) can be exploited to further—and often drastically, in the case of higher subunit numbers—improve the initial protein map. Because the map averaging methods are closely related to molecular replacement (MR) where a search probe of a similar molecule must be placed in the unit cell, we will discuss both related subjects NCS and MR in Chapter 11 before we proceed to model building and refinement in Chapter 12.

10.10 Key concepts

Phasing methods

- Phase determination methods fall into four general groups: substructure methods, density modification techniques, molecular replacement methods, and direct methods.

Substructure methods

- The basis of experimental substructure phasing is difference intensity data, originating from electronic differences resulting from heavy or anomalous marker atoms.
- Sources of scattering differences with respect to a native reference crystal are introduced by heavy metal atoms, while anomalous scatterers provide anomalous differences within a data set and dispersive differences between data sets of the same crystal recorded at different wavelengths.
- Structural isomorphism is a fundamental requirement for difference methods, which can be difficult to achieve for heavy atom derivative crystals, but is inherent in anomalous or dispersive data from the same crystal.
- The difference data represent only the metal substructure of a few to a few hundred atoms, which is a greatly simplified situation compared with the thousands and tens of thousands of atoms in a protein structure.
- The heavy or anomalous marker atom positions can be determined from difference data by a variety of substructure solution methods based on Patterson methods and/or direct methods.
- Once the marker atom positions are known, phasing equations for the protein phases can be solved, the initial protein phases can be determined, and the electron density maps can be reconstructed by Fourier synthesis.
- Depending on the difference signal source and the data available, various methods to determine the protein phases can be distinguished and may also be combined.
- Single isomorphous replacement (SIR), generic single-wavelength anomalous diffraction (SAD), and native sulfur-based single-wavelength anomalous diffraction (S-SAD) require density modification or direct methods support to break the inherent 2-fold phase ambiguity of the phasing equations. The phase ambiguity can also be broken by adding orthogonal anomalous signal

(AS) from the same crystal (SIRAS) or providing phases from at least a second derivative in multiple isomorphous replacement (MIR), or combining both possibilities (MIRAS).

- Multi-wavelength anomalous diffraction (MAD) provides orthogonal dispersive and anomalous differences from the same crystal and the phase ambiguity can be resolved directly from the phasing equations.

Heavy atom substructure solution

- Native and isomorphous data must generally be brought onto a common scale. Derivative data often suffer from anisotropy, incompleteness, radiation decay, and other systematic errors, which can be accounted for by local scaling and anisotropic scaling.
- In difference data, the weak high resolution data are especially poor and the resolution needs to be truncated from the point where the high resolution data do not contain any more information and thus add only noise.
- Different data ranges may significantly affect the outcome of a substructure solution.
- The highest difference values affect both Patterson methods and direct methods, and a few bad strong reflections can prevent substructure solution. Therefore, intensity outliers are removed or scaled down.
- Difference Patterson maps constructed from Fourier coefficients $(F_{PA} - F_P)^2$ contain interatomic distance vectors between the heavy atoms.
- The centric projections of the maps in special Harker sections provide a direct measure of the interatomic self-distance vectors, which can be solved to yield the heavy atom positions.
- In space groups with multiple origins only the Cheshire cell needs to be searched for the first positions. In polar space groups with floating origin, the first atom can be placed freely along the unique axis direction which then fixes the origin.
- Brute force grid searches systematically place atoms at grid positions and compute a correlation score. The best positions are saved and the procedure is repeated.
- Patterson-vector based methods exploit the fact that the Patterson map is the autocorrelation of the underlying electron density and search for fit of fragments. Various scoring functions exist; foremost are Patterson minimum functions. Once an initial two-atom (vector) fragment is found, additional atoms are added and the full symmetry PMF is evaluated.
- Patterson maps are imperfect, noisy, and crowded if a significant number of atoms are present (197 atoms so far has been the maximum), and automated programs generate multiple possible heavy atom substructure models that are ranked according to a combined scoring function. Automated programs often employ a combination of Patterson and Fourier methods to test possible solutions for consistency.
- Because of the centrosymmetry of the Patterson space, the handedness of the substructure solution is not determined. The solution can be either the correct substructure enantiomorph or have inverted handedness, with a 50% chance of either.
- Direct methods exploit the non-random distribution of atoms in a molecular structure, leading to implicit relations between structure factor amplitudes and to phase relations. Triplet phase invariants and semi-invariants are used to find trial structures, which are expanded using the tangent formula.
- Dual space direct methods recycle and modify trial structures by peak search in the electron density trial map, and exploit chemical restraints and plausibility such as minimum distances to remove insensible atoms and place new ones. They converge much faster than reciprocal-space-only methods.
- A particularly efficient direct method is Patterson seeding, where initial positions for the dual-space direct methods are picked from a list of two-atom vectors compatible with the Patterson map.
- Although special direct methods programs can solve the phase ambiguity in SAD data, the substructure handedness is generally not determined during direct methods substructure solution.

Phasing equations and phase probabilities

- After a marker atom substructure has been determined, the phase angle for the protein structure factor can be computed from the phasing equations.
- The best phase angle is computed by probability-weighted vector averaging of all possible phase values.
- From the best phase angle obtained, a best protein structure factor is calculated. The protein map in turn is reconstructed from the best protein structure factors as Fourier coefficients. The protein structure factors are weighted by their figure of merit, equivalent to the cosine of their estimated phase error and indicative of their reliability.
- A map constructed from best phase angles has the minimum residual error in the electron density construction.
- Improvements to the error model include Sim weights, which take into account the incompleteness of the heavy atom model.
- Further improvement is achieved by the introduction of σ_A coefficients, a variance term in the probability distribution of each structure factor, which accounts for errors in a partial model.

Heavy atom refinement and phasing

- Maximum likelihood statistics allow integrating out of the phase angle from complex structure factor probability distributions, which provides probability distributions and likelihood functions for the actual observable, the structure factor amplitudes.
- The expression of structure factor probabilities as maximum likelihood functions allows the formulation of conditional probabilities for complex structure factors.
- Incorporation of bivariate errors (phase and amplitude) as a complex Gaussian function and expression as normalized structure factors shows that the Luzzati D-factor and the complex variance are anticorrelated can be substituted by the single parameter, σ_A.
- Integrating out the phase angle from the complex structure factor probability distribution in the general acentric case yields a 1-dimensional probability distribution for the structure factor amplitudes, a Rice function, also known in crystallography as the Sim distribution.
- The log-likelihood function for MIR is obtained from the sum of probability distributions formulated to treat all derivatives at once, in contrast to classical combined SIR cases.
- The problem of dependence or correlation of errors between derivatives and native is accounted for by assuming that the protein part of the derivative is the same and known in both derivative and protein, and integrating it out of the probability distribution.
- The error resulting from non-isomorphism is addressed by inflating the variance (adding additional variance) of the probability distribution.
- After integrating out the phase, a final –log-likelihood function is formulated by summation of all structure factor probabilities, which can be minimized with a suitable numeric algorithm. The equations for SIR, SAD, and MAD are derived in similar fashion for each specific phasing situation.
- The likelihood function represents our state of knowledge or uncertainty of the system and can be visualized by plotting the complex distributions and forming their joint probabilities in analogy to the Harker circles.
- Phases from different sources can be readily combined by parameterizing the separate phase probabilities with Hendrickson–Lattman coefficients.

Density modification techniques

- Density modification dramatically improves initial protein phases and electron density maps.
- Density modification techniques are based on the fact that protein crystal structures have extensive solvent channels (~50%) that are filled with disordered solvent.
- To determine the solvent region, a solvent mask separating solvent from protein regions is automatically generated.

- The flexible sphere of influence method automatically accounts for errors in the protein–solvent boundary region.
- The solvent density is either flattened to a constant value or flipped after adjusting to a mean solvent density of zero.
- Alternative methods include histogram matching, an image processing technique that compares and adjusts electron density voxel values with an electron density histogram for an idealized protein structure.
- Reciprocal space maximum likelihood density modification has also been implemented.
- Density averaging from multiple copies of the molecule in the asymmetric unit can be effectively used to improve map quality.

10.11 Additional reading

1. Blundell TL, & Johnson LN (1976) *Protein Crystallography*. London, UK: Academic Press

2. Stout GH, & Jensen LH (1989) *X-Ray Structure Determination*. New York, NY: Wiley Interscience.

3. Drenth J (2007) *Principles of Protein X-Ray Crystallography*. New York, NY: Springer.

4. Giacovazzo C, Monaco HL, Viterbo D, et al. (2002) *Fundamentals of Crystallography*. Oxford, UK: Oxford Science Publications.

5. Rossmann MG, & Arnold E (Eds.) (2001) *International Tables for Crystallography, Volume F: Crystallography of Biological Macromolecules*. Dordrecht, The Netherlands: Kluwer Acdemic Publishers.

6. Carter CWJ, & Sweet RM (1997) Macromolecular crystallography. *Methods Enzymol.* 276, 277. London, UK: Academic Press.

7. Carter CW, & Sweet R (2003) Macromolecular crystallography. *Methods Enzymol.* 368, 374. London, UK: Academic Press.

8. Baker EN, & Dauter M (2003) Experimental phasing – Proceedings of the CCP4 study weekend. *Acta Crystallogr.* D59(11), 1881–2050.

9. Garman EF, Pearson AR, & Vonrhein C (2010) Experimental Phasing and Radiation Damage—Proceedings of the CCP4 study weekend. *Acta Crystallogr.* D66(4), 325–501.

10. McCoy AJ, & Read RJ (2010) Experimental phasing: best practice and pitfalls. *Acta Crystallogr.* D66(4), 458–469.

10.12 References

1. Kendrew JC, Bodo G, Dintzis HM, et al. (1958) A three-dimensional model of the myoglobin molecule obtained by X-ray analysis. *Nature* 181, 662–666.

2. Dickerson RE, Kendrew JC, & Strandberg BE (1961) The crystal structure of myoglobin: Phase determination to a resolution of 2 Å by the method of isomorphous replacement. *Acta Crystallogr.* 14(11), 1188–1195.

3. Cullis AF, Muirhead H, Perutz MF, et al. (1962) The structure of haemoglobin. IX. A three-dimensional Fourier Synthesis at 5.5 Å resolution: description of the structure. *Proc. R. Soc. London* A265, 161–187.

4. Blake CC, Fenn RH, North AC, et al. (1962) Structure of lysozyme. A Fourier map of the electron density at 6 angstrom resolution obtained by x-ray diffraction. *Nature* 196, 1173–1176.

5. Green DW, Ingram VM, & Perutz MF (1954) The structure of haemoglobin. IV. Sign determination by the isomorphous replacement method. *Proc. R. Soc. London* A225, 287–295.

6. Rossmann M (1960) The accurate determination of the position and shape of heavy-atom replacement groups in proteins. *Acta Crystallogr.* 13(3), 221–226.

7. Blake CC, Koenig DF, Mair GA, et al. (1965) Structure of hen egg-white lysozyme. A three-dimensional Fourier synthesis at 2 Angstrom resolution. *Nature* 206(986), 757–761.

8. Dickerson RE (1992) A little ancient history. *Protein Sci.* 1(1), 182–186.

9. Dickerson RE (2005) *Present at the Flood*. Sunderland, MA: Sinauer Associates.

10. Matthews B (1966) The extension of the isomorphous replacement method to include anomalous scattering measurements. *Acta Crystallogr.* 20(1), 82–86.

11. Crick FH, & Magdoff BS (1956) The theory of the method of isomorphous replacement for protein crystals. I. *Acta Crystallogr.* 9, 901–908.

12. de La Fortelle E, & Bricogne G (1997) Maximum-likelihood heavy atom-parameter refinement for multiple isomorphous replacement and multiwavelength anomalous diffraction methods. *Methods Enzymol.* 276, 472–494.

13. McCoy AJ, Grosse-Kunstleve RW, Adams PD, et al. (2007) Phaser crystallographic software. *J. Appl. Crystallogr.* 40(4), 658–674.

14. Hao Q, Gu YX, Zheng CD, et al. (2000) OASIS: a computer program for breaking phase ambiguity in one-wavelength anomalous scattering or single isomorphous substitution (replacement) data. *J. Appl. Crystallogr.* 33, 980–981.

15. Schneider TR, & Sheldrick GM (2002) Substructure solution with SHELXD. *Acta Crystallogr.* D58, 1772–1779.

16. Weeks CM, & Miller R (1999) The design and implementation of SnB v2-0. *J. Appl. Crystallogr.* 32, 120–124.

17. Terwilliger TC (2004) SOLVE and RESOLVE: Automated structure solution, density modification and model building. *J. Synchrotron Radiat.* 11(1), 49–52.

18. Brünger AT, Adams PD, Clore GM, et al. (1998) Crystallography & NMR system: A new software suite for macromolecular structure determination. *Acta Crystallogr.* D54, 905–921.

19. Bricogne G, Vonrhein C, Flensburg C, et al. (2003) Generation, representation and flow of phase information in structure determination: Recent developments in and around SHARP 2.0. *Acta Crystallogr.* D59(11), 2023–2030.

20. Vonrhein C, Blanc E, Roversi P, et al. (2007) Automated structure solution with autoSHARP. *Methods Mol. Biol.* 364(2), 215–253.

21. Adams PD, Afonine PV, Bunkoczi G, et al. (2010) PHENIX: a comprehensive Python-based system for macromolecular structure solution. *Acta Crystallogr.* D66(2), 213–221.

22. Matthews BW, & Czerwinski EW (1975) Local scaling: A method to reduce systematic errors in isomorphous replacement and anomalous scattering measurements. *Acta Crystallogr.* A31(4), 480–487.

23. Rould MA (1997) Screening for heavy-atom derivatives. *Methods Enzymol.* 276, 461–472.

24. Kraut J, L.C. S, High DF, et al. (1962) Chymotrypsinogen: A three-dimensional Fourier synthesis at 5 angstrom resolution. *Proc. Natl. Acad. Sci. U.S.A.* 48, 1417–1424.

25. Rould MA, & Carter CW (2003) Isomorphous difference methods. *Methods Enzymol.* 374, 145–163.

26. Trueblood KN, Burgi HB, Burzlaff H, et al. (1996) Atomic displacement parameter nomenclature. Report of a subcommittee on atomic displacement parameter nomenclature. *Acta Crystallogr.* A52(5), 770–781.

27. Read RJ (1999) Detecting outliers in non-redundant diffraction data. *Acta Crystallogr.* D55, 1759–1764.

28. Strong M, Sawaya MR, Wang S, et al. (2006) Toward the structural genomics of complexes: Crystal structure of a PE/PPE protein complex from *Mycobacterium tuberculosis. Proc. Natl. Acad. Sci. U.S.A.* 103(21), 8060–8065.

29. Drenth J (2007) *Principles of Protein X-ray Crystallography.* New York, NY: Springer.

30. Ward DL (1998) *Patterson Peaks.* Dayton, OH: Polycrystal Book Service.

31. Koch E, Fischer W, & Müller U (2002) Normalizers of space groups and their use in crystallography. *International Tables for Crystallography A*, 877–905.

32. Hirshfeld F (1968) Symmetry in the generation of trial structures. *Acta Crystallogr.* A24(2), 301–311.

33. McRee DE (1993) *Practical Protein Crystallography.* San Diego, CA: Academic Press.

34. Navaza J, & Vernoslova E (1995) On the fast translation function for molecular replacement. *Acta Crystallogr.* A51, 445–449.

35. Grosse-Kunstleve RW, & Brünger AT (1999) A highly automated heavy-atom search procedure for macromolecular structures. *Acta Crystallogr.* D55, 1568–1577.

36. Terwilliger TC, & Berendzen J (1999) Automated MAD and MIR structure solution. *Acta Crystallogr.* D55, 849–861.

37. Terwilliger TC, & Eisenberg D (1987) Isomorphous replacement: effects of errors on the phase probability distribution. *Acta Crystallogr.* A43(1), 6–13.

38. Uson I, & Sheldrick GM (1999) Advances in direct methods for protein crystallography. *Curr. Opin. Struct. Biol.* 9, 642–648.

39. Sheldrick GM (1990) Phase annealing in SHELX-90: Direct methods for larger structures. *Acta Crystallogr.* A46, 467–473.

40. Morris RJ, & Bricogne G (2003) Sheldrick's 1.2 Å rule and beyond. *Acta Crystallogr.* D59, 615–617.

41. Sheldrick GM, Hauptman HA, Weeks CM, et al. (2001) Ab initio phasing. *International Tables for Crystallography F*, 333–354.

42. vonDelft F, & Blundell TL (2002) The 160 selenium atom substructure of KMPHT. *Acta Crystallogr.* A58, C239.

43. Hauptman H, & Karle J (1955) A relationship among the structure-factor magnitudes for P1. *Acta Crystallogr.* 8(6), 355.

44. Morris RJ, Blanc E, & Bricogne G (2004) On the interpretation and use of <|E|2>(d*) profiles. *Acta Crystallogr.* D60(2), 227–240.

45. Sayre D (1952) The squaring method: a new method for phase determination. *Acta Crystallogr.* 5, 60–65.

46. Giacovazzo C, Monaco HL, Viterbo D, et al. (2002) *Fundamentals of Crystallography.* Oxford, UK: Oxford Science Publications.

47. Bricogne G (1984) Maximum entropy and the foundation of direct methods. *Acta Crystallogr.* A40, 410–445.

48. Grosse-Kunstleve RW, & Adams PD (2003) Substructure search procedures for macromolecular direct methods. *Acta Crystallogr.* D59, 1966–1973.

49. Reddy V, Swanson S, Sacchettini JC, et al. (2003) Effective electron density map improvement and structure validation on a Linux multi-CPU web cluster: The TB Structural Genomics Consortium Bias Removal Web Service. *Acta Crystallogr.* D59, 2200–2210.

50. Bricogne G (2001) The maximum entropy method. *International Tables for Crystallography F*, 346–348.

51. Oszlanyi G, & Suto A (2008) The charge flipping algorithm. *Acta Crystallogr.* A64, 123–134.

52. Pape T, & Schneider TR (2004) HKL2MAP: a graphical user interface for macromolecular phasing with SHELX programs. *J. Appl. Crystallogr.* 37(5), 843–844.

53. Oszlanyi G, & Suto A (2004) Ab initio structure solution by charge flipping. *Acta Crystallogr.* A60, 134–141.

54. Dumas C, & van der Lee A (2008) Macromolecular structure solution by charge flipping. *Acta Crystallogr.* D64, 864–873.

55. Matthews BW (2001) Heavy atom location and phase determination with single wavelength diffraction data. *International Tables for Crystallography F*, 293–298.

56. Wang BC (1985) Resolution of Phase Ambiguity in Macromolecular Crystallography. *Methods Enzymol.* 115, 90–112.

57. Sheldrick G (2008) A short history of SHELX. *Acta Crystallogr.* A64(1), 112–122.

58. Caliandro R, Carrozzini B, Cascarano GL, et al. (2007) Advances in the free lunch method. *J. Appl. Crystallogr.* 40(5), 931–937.

59. Sheldrick GM (2002) Macromolecular phasing with SHELXE. *Z. Kristallogr.* 217, 644–650.

60. Sheldrick GM (2007) SAD phasing: Basic concepts and high throughput. In Read RJ & Sussman JL (Eds.), *Evolving Methods for Macromolecular Crystallography.* The Netherlands: Springer.

61. Matthews BW (2007) Five retracted structure reports: Inverted or incorrect? *Protein Sci.* 16, 1013–1016.

62. Dauter Z (2002) One-and-a-half wavelength approach. *Acta Crystallogr.* D58, 1958–1967.

63. Sheldrick GM (2008) personal communication.

64. Wang J, Wlodawer A, & Dauter Z (2007) What happens when the sign of anomalous differences or the handedness of substructure are inverted? *Acta Crystallogr.* D63, 751–758.

65. Hendrickson WA, & Teeter MM (1981) Structure of the hydrophobic protein crambin determined directly from the anomalous scattering of sulphur. *Nature* 290, 107–113.

66. Hendrickson WA (1991) Determination of macromolecular structures from anomalous diffraction of synchrotron radiation. *Science* 254, 51–58.

67. Karle J (1980) Some developments in anomalous dispersion for the structural investigation of macromolecular systems in biology. *Int. J. Quantum Chem.* 7, 357–367.

68. Phillips JC, & Hodgson KO (1980) The use of anomalous scattering effects to phase diffraction patterns from macromolecules. *Acta Crystallogr.* A36(6), 856–864.

69. Chang G, & Roth CB (2001) Structure of MsbA from *E. coli*: A homolog of the multidrug resistance ATP binding cassette (ABC) transporters. *Science* 293(5536), 1793–1800.

70. Chang G, Roth CB, Reyes CL, et al. (2006) Retraction. *Science* 314(5807), 1875.

71. Dawson RJP, & Locher KP (2006) Structure of a bacterial multidrug ABC transporter. *Nature* 443(7108), 180–185.

72. Jeffrey P (2009) Analysis of errors in the structure determination of MsbA. *Acta Crystallogr.* D65(2), 193–199.

73. Hendrickson WA, Smith JL, & Sheriff S (1985) Direct phase determination based on anomalous scattering. *Methods Enzymol.* 115, 41–55.

74. Hendrickson WA, & Ogata CM (1997) Phase determination from multiwavelength anomalous diffraction measurements. *Methods Enzymol.* 276, 494–516.

75. Blow DW, & Crick FHC (1959) The treatment of errors in the isomorphous replacement method. *Acta Crystallogr.* 12, 794–802.

76. Read RJ (1997) Model phases: probabilities and bias. *Methods Enzymol.* 277, 110–128.

77. Blow DM, & Matthews BW (1973) Parameter refinement in the multiple isomorphous-replacement method. *Acta Crystallogr.* A29(1), 56–62.

78. Otwinowski Z (1991) Maximum likelihood refinement of heavy atom parameters. In Wolf W, Evans PR, & Leslie AGW (Eds.), *Proceedings of the 1991 CCP4 Study Weekend*. CLRC Daresbury Laboratory, Warrington, UK.

79. Sivia DS (1996) *Data Analysis – A Bayesian Tutorial*. Oxford, UK: Oxford University Press.

80. Matthews B (2004) David Mervyn Blow: A scholar and a gentleman (1931–2004). *Acta Crystallogr.* D60(10), 1695–1697.

81. Blundell TL, & Johnson LN (1976) *Protein Crystallography*. London, UK: Academic Press.

82. Sim G (1960) The cumulative distribution of structure amplitudes. *Acta Crystallogr.* 13(1), 58–59.

83. Bricogne G (1976) Methods and programs for direct-space exploitation of geometric redundancies. *Acta Crystallogr.* A32(5), 832–847.

84. Read RJ (1986) Improved Fourier coefficients for maps using phases from partial structures with errors. *Acta Crystallogr.* A42, 140–149.

85. Srinivasan R (1966) Weighting functions for use in the early stage of structure analysis when a part of the structure is known. *Acta Crystallogr.* 20(1), 143–144.

86. Srinivasan R, & Ramachandran GN (1966) Probability distribution connected with structure amplitudes of two related crystals. VI. On the significance of the parameter σA. *Acta Crystallogr.* 20(4), 570–571.

87. Main P (1979) A theoretical comparison of the α, γ and $2F_o–F_c$ syntheses. *Acta Crystallogr.* 35(5), 779–785.

88. Terwilliger TC, & Berendzen J (1997) Bayesian correlated MAD phasing. *Acta Crystallogr.* D53(5), 571–579.

89. Terwilliger TC, & Berendsen J (2001) Automated MAD and MIR structure solution. *International Tables for Crystallography F*, 303–309.

90. Adams PD, Grosse-Kunstleve RW, Hung LW, et al. (2002) PHENIX: building new software for automated crystallographic structure determination. *Acta Crystallogr.* D58, 1948–1954.

91. Pannu N, & Read R (1996) Improved structure refinement through maximum likelihood. *Acta Crystallogr.* A52, 659–668.

92. Read RJ (2003) New ways of looking at experimental phasing. *Acta Crystallogr.* D59, 1891–1902.

93. McCoy A (2004) Liking likelihood. *Acta Crystallogr.* D60(12), 2169–2183.

94. Pannu NS, & Read RJ (2004) The application of multivariate statistical techniques improves single-wavelength anomalous diffraction phasing. *Acta Crystallogr.* D60(1), 22–27.

95. McCoy AJ, Storoni LC, & Read RJ (2004) Simple algorithm for a maximum-likelihood SAD function. *Acta Crystallogr.* D60(7), 1220–1228.

96. Zhang KYJ, Cowtan KD, & Main P (2001) Phase improvement by iterative density modification. *International Tables for Crystallography F*, 311–324.

97. Zhang KYJ (1993) SQUASH – combining constraints for macromolecular phase refinement and extension. *Acta Crystallogr.* D49, 213–222.

98. Abrahams JL, & Leslie AGW (1996) Methods used in the structure determination of the bovine mitochondrial F1 ATPase. *Acta Crystallogr.* D52, 30–42.

99. Hendrickson WA, & Lattman EE (1970) Representation of phase probability distributions for simplified combination of independent phase information. *Acta Crystallogr.* B26, 136–143.

100. Terwilliger T (2001) Map-likelihood phasing. *Acta Crystallogr.* D57(12), 1763–1775.

101. Baker D, Krukowski AE, & Agard DA (1993) Uniqueness and the ab initio phase problem in macromolecular crystallography. *Acta Crystallogr.* D49(1), 186–192.

102. Terwilliger TC, & Berendzen J (1999) Evaluation of macromolecular electron-density map quality using the correlation of local r.m.s. density. *Acta Crystallogr.* D55(11), 1872–1877.

103. Kleywegt GJ, & Read RJ (1997) Not your average density. *Structure* 5, 1557–1569.

104. Lunin VY (1988) Use of the information on electron density distribution in macromolecules. *Acta Crystallogr.* A44, 144–150.

105. Zhang K (1990) Histogram matching as a new density modification technique for phase refinement and extension of macromolecules. *Acta Crystallogr.* A46, 41–46.

106. Zhang KYJ (1990) The use of Sayre's equation with solvent flattening and histogram matching for phase extension and refinement of protein structures. *Acta Crystallogr.* A46, 377–381.

107. Grosse-Kunstleve RW, Sauter NK, Moriarty NW, et al. (2002) The Computational Crystallography Toolbox: crystallographic algorithms in a reusable software framework. *J. Appl. Crystallogr.* 35(1), 126–136.

108. Terwilliger TC (2003) Statistical density modification using local pattern matching. *Acta Crystallogr.* D59(1688–1701).

109. Terwilliger TC (2000) Maximum likelihood density modification. *Acta Crystallogr.* D56, 965–972.

110. Perrakis A, Sixma TK, Wilson KS, et al. (1997) wARP: Improvement and extension of crystallographic phases by weighted averaging of multiple-refined dummy atomic models. *Acta Crystallogr.* D53, 448–455.

111. Uson I, Stevenson CEM, Lawson DM, et al. (2007) Structure determination of the O-methyltransferase NovP using the "free lunch algorithm" as implemented in SHELXE. *Acta Crystallogr.* D63(10), 1069–1074.

112. Pal A, Debreczeni JÉ, Sevvana M, et al. (2008) Crystal structures of viscotoxins A1 and B2 from the European mistletoe. *Acta Crystallogr.* D64, 985–992.

113. Weiss MS, Sicker T, & Hilgenfeld R (2001) Soft X-rays, high redundancy and proper scaling: A new procedure for automated protein structure determination via SAS. *Structure* 9, 771–777.

114. Debreczeni JE, Bunkoczi G, Girmann B, et al. (2003) In-house phase determination of the lima bean trypsin inhibitor: a low-resolution sulfur-SAD case. *Acta Crystallogr.* D59(2), 393–395.

115. Beck T, Krasauskas A, Gruene T, et al. (2008) A magic triangle for experimental phasing of macromolecules. *Acta Crystallogr.* D64, 1179–1182.

116. Burling FT, Weis WI, Flaherty KM, et al. (1996) Direct observation of protein solvation and discrete disorder with experimental crystallographic phases. *Science* 271, 72–77.

117. Cohen SX, Ben Jelloul M, Long F, et al. (2008) ARP/wARP and molecular replacement: the next generation. *Acta Crystallogr.* D64(1), 49–60.

118. Li J, Derewenda U, Dauter Z, et al. (2000) Crystal structure of the *Escherichia coli* thioesterase II, a homolog of the human Nef binding enzyme. *Nat. Struct. Biol.* 7(7), 555–559.

119. Sheldrick GM (2010) Experimental phasing with SHELXC/D/E: combining chain tracing with density modification. *Acta Crystallogr.* D66, 479–485.

120. Rodríguez DD, Grosse C, Himmel S, et al. (2009) Crystallographic Ab Initio protein solution far below atomic resolution. *Nature Methods* 6(9), 651–653.

Non-crystallographic symmetry and molecular replacement

It occurred to me that both the ability of proteins to crystallize in different space groups and their frequent property of being made up of identically folded polypeptide chains might form a basis for an alternative process of solving the phase problem.

Michael G. Rossmann, around 1960, in Cambridge, UK.
In *Acta Crystallogr.* **D57**, 1360–1366

Non-crystallographic symmetry (NCS) implies the presence of more than one copy of a molecule in the asymmetric unit of the unit cell. This local symmetry can be exploited to further improve the initial experimentally phased electron density maps. The corresponding method of NCS map averaging involves rotations and translations of objects, procedures it shares with the phasing technique of molecular replacement, a very popular and increasingly sophisticated phasing method applicable when a search model structurally similar to the molecules in the target structure is available.

We begin the section with a formal discussion of general transformations in Cartesian space. The interpretation of rotation matrices in particular often causes confusion and a number of examples will help the reader to become comfortable with the computation of Euler axis and principal Euler angle for any transformation, in world coordinates and in the fractional (crystallographic) reference system.

Armed with the understanding of transformations, we can then discuss the various types of NCS, how they manifest themselves and how they can be detected in experimental data. Once we know how to find and interpret the NCS operators, we can exploit NCS for map averaging. A formal treatment that provides some important insights into the reciprocal space interference function needed later in the molecular replacement section is followed by a practical example that convincingly shows the improvement that can be achieved by combined map averaging and density modification.

The molecular replacement section starts with an overview of the various methods available to locate a structurally similar search probe in an unknown crystal structure, given only diffraction data. We begin with the conceptually more intuitive modern 6-dimensional evolutionary search algorithms, where we provide an instructive example and insights into the general features of solution landscapes. A section about model selection and model pruning is included.

We then step back to the classical rotation–translation search based methods and discuss the theory, scoring functions, and general strategies for combined rotation–translation searches in modern molecular replacement programs. We finish the section with a discussion of maximum likelihood rotation and translation functions and the unique features of the modern, automated maximum likelihood molecular replacement program *Phaser*.

11.1   Non-crystallographic symmetry

When non-crystallographic, local symmetry (NCS) is present in a molecular structure, the asymmetric unit harbors multiple copies of the molecule, and the diffraction pattern contains redundant information.[1] The presence of oligomers is very common; nearly half of all structures in the PDB are dimeric structures (80% of oligomers) or structures with a higher oligomerization state of the molecules.[2] Obligate oligomers, with tight intermolecular contacts and buried surface areas exceeding ~700–800 Å^2 per monomer, often form the motif of the asymmetric unit. In all cases that involve multiple identical subunits in the asymmetric unit, the initial experimentally phased electron density maps will show the electron density of two or more molecules, which can be averaged to improve the map quality. The expected improvement in the signal to noise ratio of the maps is proportional to $N^{1/2}$, where N is the number of independent molecular copies in the asymmetric unit. The strong dependence of the achievable improvement on the copy number explains why multi-copy map averaging is so powerful for highly symmetric virus structures, to the point that it allows phasing virus data from a hollow sphere as an initial phasing model.

The concept of map averaging can be extended to maps from different crystals—either in different crystal forms, or different maps independently determined from the same crystal form. In all cases it is necessary to find a transformation that superimposes one map onto the other. To accomplish this, the NCS operator superimposing the maps can be determined from the heavy metal substructure symmetry, or if this fails, by self-rotation of the map in Patterson space and subsequent phased translation functions.[3]

Directly related with detecting symmetry between subunits in the asymmetric unit is the search for a known structure model similar to the target molecule in another crystal structure. This method is molecular replacement and the associated rotation and translation searches in Patterson space were initially developed for the detection of subunit symmetry. In both techniques, NCS averaging and molecular replacement, a fundamental task is rotation and translation of molecules or maps. We therefore need to investigate the analysis of general transformations and transformation operators in Cartesian space. In addition, as the strategies for experimental phase improvement and molecular replacement phasing vary depending on the presence of subunits, we need to concern ourselves with how to determine from the diffraction data what the probability is that a certain number of subunits of our crystallized molecule could be present in the asymmetric unit.

Determining the unit cell content

One of the first questions to answer as soon as we have collected diffraction data is how many copies of the protein crystallized in the asymmetric unit. To determine this, we do not need to know the exact space group: a look at the rightmost column (z or M, the general position multiplicity) of space group Table 6-6 confirms that the Laue symmetry plus the Bravais centering suffice to determine the number of asymmetric units that comprise the unit cell. Dividing the unit cell volume by the general position multiplicity z provides us with the volume of the asymmetric unit V_a, that is, the space that is available to accommodate our protein molecule, or molecules.

For molecules to pack and form stable crystals, a 3-dimensional network of intermolecular contacts must exist (Figure 5-5). The need to form crystal contacts limits how far the molecules can be apart while still, depending on shape and surface residue distribution, forming a stable network. This upper empirical limit[4] for loose packing is around 80–85% solvent content. At the other extreme, assuming a spherical shape for a globular protein, we can provide a lower estimate for the solvent content based on a simple packing calculation for rigid spheres in a closed packed lattice. In a closed packed lattice 26% of the volume is occupied by solvent and although a tighter packing of "brick-shaped" or interlocking molecules could be conceived, only in rare cases have solvent contents below ~25% been observed.

Matthews coefficient

Solvent content data were first analyzed by Matthews in 1968[5] and 1976.[6] For all the 226 then known different crystal forms of globular proteins, mainly in the molecular weight range <70 kDa, Matthews plotted the volume of the asymmetric crystallographic unit against the molecular weight of the corresponding protein. The ratio of these two values Matthews defined as V_M, now known as the Matthews coefficient or Matthews volume. V_M has the dimensions of $Å^3 Da^{-1}$, and generally lies between 1.5 and 6 $Å^3 Da^{-1}$, with a mean around 2.5 $Å^3 Da^{-1}$.

V_M bears a straightforward relationship to the fractional volume of solvent in the crystal. The fraction of the asymmetric volume occupied by a protein molecule $x(p)$ is given as

$$x(p) = \frac{1.66 \cdot \bar{v}}{V_M} = \frac{1.23}{V_M} \qquad (11\text{-}1)$$

where $\bar{v}$ denotes the partial specific volume of the protein. The partial specific volume (the inverse of the specific density) can be computed from the sum of all the partial specific volumes of the residues, and $\bar{v}$ is practically constant at 0.74 $cm^3 g^{-1}$ for proteins.[7] From the definition of $x(p)$ as the fraction of the asymmetric volume occupied by protein immediately follows that the fraction of solvent $x(s) = 1 - x(p)$. Matthews found that the percentage of the crystal volume occupied by solvent, the solvent content V_S, ranged from 27% to 78%, with the most common value being about 43%.

The computation of the Matthews coefficient is straightforward. For the example of an orthorhombic $P2_12_12_1$ unit cell of dimensions of $a = 71.18Å$, $b = 79.38Å$, and $c = 93.81$ Å, and $z = 4$, we obtain $V_a = 132\,513$ $Å^3$. Given a molecular weight of 26 kDa for the basic motif (either a single molecule or a known oligomer) V_M computes from V_a/MW as 5.1 $Å^3 Da^{-1}$. The corresponding solvent content V_S according to (11-1) is 76%. As this value is at the upper extreme of the observed solvent content values, there are likely to be two motifs of 26 kDa in the asymmetric unit, which gives a more probable value of 2.55 for V_M. We conclude that the asymmetric unit most likely contains two of our motifs. The solvent fraction V_S of the dimer computes as 0.517, or about 52%.

Box 11-1 **Matthews coefficient and Matthews probabilities.** Protein structures in general have a solvent content of ~25–80%, with proteins diffracting to high resolution tending to have lower solvent content. Based on the distribution of the solvent content, the probability of the presence of multiple copies of a molecule in the asymmetric unit can be estimated indirectly from the Matthews coefficient and directly from the associated Matthews probabilities. The more molecules can be accommodated in the asymmetric unit, the less discriminating the multiple, physically plausible copy numbers become. Biological evidence and further crystallographic evidence from Patterson self-rotation maps or native Patterson maps must then complement the estimate of Matthews probabilities.

While the basic Matthews coefficient calculation above works well in clear-cut cases, it does not allow quantification of the probability to find a certain number of molecules in the unit cell. Multiple solutions are possible, particularly for higher assembly numbers. Additional biological evidence from size exclusion chromatography, light scattering, ultracentrifugation, or native polyacrylamide gels helps to select the proper oligomer number. In cases of proper non-crystallographic symmetry, the Patterson self-rotation functions will also reveal the multiplicity of the NCS rotation.

As an additional consideration based on the packing argument emphasizing the need for specific intermolecular contacts, we would expect that well-diffracting crystals might also pack more tightly, and poor diffraction may be indicative of—or result from—loose packing and higher solvent content. Again, a probabilistic approach that establishes conditional probability distributions incorporating the available prior knowledge provides the answers.

Matthews probabilities

Given the much larger number of structures available in the PDB by 2003, Kantardjieff and Rupp[4] established conditional probability distributions for the Matthews coefficient given the resolution limit d_{min} to which the crystals diffract. The probability distributions confirm a significant and plausible correlation between solvent content and diffraction limit: crystals with low solvent content tend to diffract better. Further analysis also showed that, in contrast, molecular weight is not a significant determinant for the diffraction limit.

The probability distribution for V_M is described in several resolution bins by an empirical five-parameter-double exponential function (a modified extreme value distribution) $prob(V_M|d_{min})$ suitable for the description of highly asymmetric distributions:

$$prob(V_M \mid d_{min}) = p_0 + a \cdot e^{(-e^{(-z)} - z \cdot s + 1)} \text{ with } z = \frac{(V_M - \bar{V}_M)}{w} \qquad (11\text{-}2)$$

The adjustable parameters are a function of bin resolution d_{min}. Function (11-2) is implemented in a web program *MATTPROB* that provides normalized Matthews probabilities for the occurrence of a certain number of copies $prob(N)$ given the observed diffraction limit. Table 11-1 and Figure 11-1 illustrate a case where the probability of a dimer versus a trimer in the asymmetric actually reverses compared with the classical estimate, once the diffraction limit is included as conditioning knowledge. Matthews probabilities are implemented in the automated maximum likelihood molecular replacement program *Phaser*[8] and in a related estimate in the *PHENIX*[9] crystallographic program suite.

```
+-------------------------------------------------------------------------------+
|    a          b          c        alpha      beta      gamma      volume      |
|  71.1800    79.3800    93.8100    90.000     90.000    90.000     530051.6    |
+-------------------------------------------------------------------------------+
| Space group P 21 21 21 , space group number   19, Laue class mmm  , z =4      |
+-------------------------------------------------------------------------------+
|  N(mol)      Prob(N)     Prob(N)      Vm         Vs          Mw               |
|           for resolution overall   A**3/Da   % solvent      Da               |
+-------------------------------------------------------------------------------+
|    1         0.00        0.01        5.64       78.19      23500.00           |
|    2         0.41        0.85        2.82       56.37      47000.00           |
|    3         0.58        0.13        1.88       34.56      70500.00           |
|    4         0.00        0.01        1.41       12.75      94000.00           |
+-------------------------------------------------------------------------------+
```

Table 11-1 Matthews probabilities. The Matthews probability provides a normalized measure for the probability that a certain number of copies of a molecule are present in the asymmetric unit, given the diffraction limit. In the selected case of 1.6 Å resolution, the probability for a trimer is actually higher than for a dimer once the better-than-average resolution is taken into account.

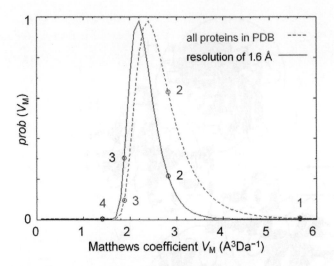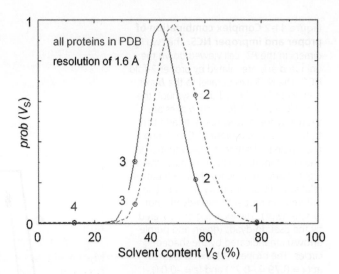

NCS transformations

In Chapter 5 we introduced non-crystallographic symmetry (NCS) through the general definition of more than one identical object being contained in the asymmetric unit. NCS operations relating the molecules act on a *local* volume in each asymmetric unit of the crystal, while crystallographic operators apply to the *entire* crystal lattice. NCS operations can be exact point group operators (proper NCS rotation axes), or they can be any combination of general rotations and translations (improper NCS). Note that the meaning of *proper* here differs from that in the case of crystallographic axes, where improper axes refers to roto-inversion. Non-crystallographic operations are not limited to any specific combination of rotation and translation operations, and sometimes can be largely translational (Figure 11-7). The NCS can also be limited or be different between different domains of a multimer that itself contains oligomeric subunits. An illustrative example is the antibody F_{ab} fragments, where the structurally highly similar domain pairs L_V and H_V, and L_C and H_C, are themselves related by different local NCS operators leading to a wide variation in elbow angles.[10] In addition to a proper, 2-fold NCS axis relating the entire antibody F_{ab} fragment, two differently oriented NCS pseudo-dyads relate their L_V and H_V and L_C and H_C domains (Figure 11-2).

Note that the term "local operator" is used here to indicate what we recognize as the "intra-dimer" or "intra-oligomer" operator acting on the local volume occupied by the selected, generally biologically sensible oligomer. For example, consider an NCS operator relating molecules 1 and 2 of a dimer in unit cell A. There are many more NCS operators that relate monomer 1 in unit cell A to monomer 2 in unit cell B and so on. These related "inter-oligomer" NCS operators resulting from crystal symmetry include translational elements and cannot be proper (point group) NCS operators. As all these operators are a result of crystal symmetry, and any preferred intra-oligomer NCS operator suffices to describe the NCS relations in the crystal.

Figure 11-1 Probability distribution of Matthews coefficient and solvent content. The probability distributions for Matthews coefficient (left panel) and solvent content (right panel) are graphed for a given resolution of 1.6 Å compared with the overall distribution of all samples in the PDB. The relative probabilities of V_M and V_S for 1 to 4 copies in the asymmetric unit are labeled. A monomer and tetramer are extremely unlikely, but both a trimer and a dimer are possible. Note the reversal of the probabilities once the resolution limit is taken into account.

Box 11-2 **Non-crystallographic symmetry.** When multiple copies of the same molecule are present in the asymmetric unit, they are related by local non-crystallographic symmetry (NCS). NCS can be classified according to the operations that relate the monomers. If the molecules are related though closed point group rotations, such as precise 2-, 3-, 4-, 5-, or 6-fold rotations, the NCS is termed proper. Any other NCS operation combining general rotations and translations is called improper.

Figure 11-2 Complex combination of proper and improper NCS. The two F_{ab} dimers in the $P2_1$ cell viewed down the unique **b** axis are related by proper 2-fold NCS. The NCS axis parallel to the **b** axis is indicated by the red dyad symbol; the crystallographic 2-fold screw axes are shown by the blue symbols. In addition to the local 2-fold proper NCS relating the whole F_{ab} fragments, the variable (V) and constant (C) domains of each chain (light chain L red, heavy chain H blue) are related by local symmetry through two different, improper pseudo-dyads (not exactly 180°). The local regions covered by the pseudo-dyads (green and purple arrows) are indicated by the shaded circles. The proper NCS axis is located at ($x = 0.79 = -0.21$) and ($z = -0.01$). Figure 11-10 shows how the location of the proper 2-fold NCS axis parallel to the 2-fold crystallographic screw axis can be found with the help of native Patterson maps. PDB entry 12e8.[11]

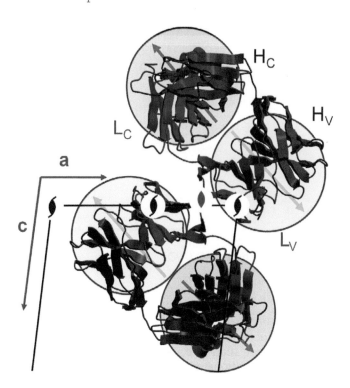

NCS operators

Because by NCS operations we are mapping objects in Cartesian space through general rotations and translations that are not limited to the crystallographic group operations with their strict relation to crystallographic axes and translations, they cannot be represented by a single augmented 4×4 **W**-matrix (Chapter 5), and we need to examine the anatomy of a general rotation in Cartesian space. Figure 11-3 illustrates the mapping of one molecule onto another one, visualizing the concept of describing any arbitrary transformation $\mathbf{W}(\mathbf{R},\mathbf{T})$ by rotational and translational components. The rotation $\mathbf{R}$ takes place around the center of mass of the molecule, followed by a translation $\mathbf{T}$. Any general transformation $\mathbf{W}(\mathbf{R},\mathbf{T})$ relating two objects A and B defined in location by their coordinate vectors $\mathbf{x}_A$ and $\mathbf{x}_B$ can thus be written as a combination of a rotation and a translation:

$$\mathbf{x}_B = \mathbf{W}_{AB}(\mathbf{R},\mathbf{T})\mathbf{x}_A = \mathbf{R}_{AB}\mathbf{x}_A + \mathbf{T}_{AB}$$

where the rotation matrix $\mathbf{R}_{AB}$ is the 3×3 direction cosine matrix describing the rotation and $\mathbf{T}_{AB}$ is the translational vector.

Rotations in a Cartesian system

Any rotation around a point in space can be defined by a single axis $\mathbf{E}$ (the Euler axis) and the corresponding rotation angle κ, the principal Euler angle (Euler's rotation theorem). The corresponding rotation matrix $\mathbf{R}$ itself can be interpreted as a succession of three basic rotations $\mathbf{R}_1$, $\mathbf{R}_2$, $\mathbf{R}_3$, of the reference frame around three independent axes. The combined transformation $\mathbf{R}_{123}$ is given by the product of the individual operators, that is, $\mathbf{R}_{123} = \mathbf{R}_3\mathbf{R}_2\mathbf{R}_1$. Note that the first operation is applied first to the object, then the second, and then the third:

$$\mathbf{R}_{123}\mathbf{x} = \mathbf{R}_3(\mathbf{R}_2(\mathbf{R}_1\mathbf{x})) \tag{11-3}$$

Multiple and sometimes confusing conventions and descriptions are possible as far as the choice of axes, angles, and the sequence of the rotations is concerned. Convention refers to the sequence in which the rotations about reference axes $\mathbf{X}_0$, $\mathbf{Y}_0$, $\mathbf{Z}_0$ are executed to obtain the new coordinate frame $\mathbf{X}'$, $\mathbf{Y}'$, $\mathbf{Z}'$. The most common conventions defining the sequential rotations with reference to the fixed original axes are: the *XYZ* or roll-pitch-yaw convention, used

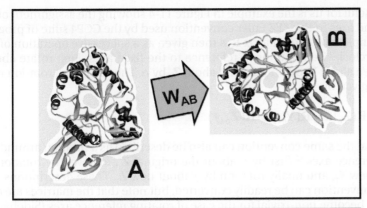

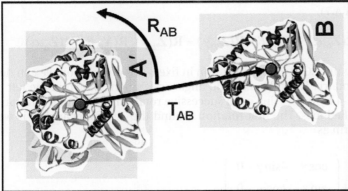

Figure 11-3 Mapping of objects through rotation and translation.
The mapping or transformation $\mathbf{W}_{AB}$ of one object (A) onto another one (B) is performed stepwise by first determining the proper orientation (rotation matrix $\mathbf{R}$) around its center of geometry or mass (red dot) and then translating the object by $\mathbf{T}$ to its new position.

in technical programs and rotating about $\mathbf{X}_0$, $\mathbf{Y}_0$, $\mathbf{Z}_0$; the CCP4 *ZYZ* convention that defines the rotations about $\mathbf{Z}_0$, $\mathbf{Y}_0$, $\mathbf{Z}_0$; and the *XPLOR/CNS ZXZ* convention, rotating about $\mathbf{Z}_0$, $\mathbf{X}_0$, $\mathbf{Z}_0$. The rotation matrices $\mathbf{R}$ for each of the conventions are different (Sidebar 11-1).

Sidebar 11-1 Conventions and description of rotations. Convention refers to the sequence in which the rotations about reference axes $\mathbf{X}_0$, $\mathbf{Y}_0$, $\mathbf{Z}_0$ are executed to obtain the new coordinate frame $\mathbf{X}'$, $\mathbf{Y}'$, $\mathbf{Z}'$. Assuming one consistent direction of rotations—either clockwise or counterclockwise—alone provides 12 possible ways to combine the three separate rotation axes. The crystallographic programs in the CCP4 suite and *XPLOR/CNS* use counterclockwise rotations, given that the rotation axis is viewed toward the origin.

In addition to the different conventions, the rotations can be described by successive rotations about fixed reference axes, or about the "new" rotated axes. In reference to the fixed original axes the most common conventions are the *XYZ* or roll-pitch-yaw convention, used in technical programs rotating about $\mathbf{X}_0$, $\mathbf{Y}_0$, $\mathbf{Z}_0$; the CCP4 *ZYZ* convention that defines the rotations about $\mathbf{Z}_0$, $\mathbf{Y}_0$, $\mathbf{Z}_0$; and the *XPLOR/CNS ZXZ* convention, rotating about $\mathbf{Z}_0$, $\mathbf{X}_0$, $\mathbf{Z}_0$.

Each of the conventions in one of the descriptions yields a 3×3 rotation matrix $\mathbf{R}$ that contains different combinations of trigonometric terms in its matrix elements depending on the conventions and descriptions used. When enumerated, the matrix elements are the direction cosines

of the angles between the basis vectors of the reference system and those of the rotated coordinate system, leading to the name direction cosine matrix (DCM) for $\mathbf{R}$.

The direction cosine matrix (DCM) is a real square matrix whose transpose $\mathbf{R}^T$ equals its inverse $\mathbf{R}^{-1}$. The column vectors are the rotated coordinate system basis vectors spanning the rotated basis [0, $\mathbf{X}'$, $\mathbf{Y}'$, $\mathbf{Z}'$]. As the transpose of $\mathbf{R}$ is also its inverse, the row vectors must contain the original system basis vectors [0, $\mathbf{X}_0$, $\mathbf{Y}_0$, $\mathbf{Z}_0$]. A DCM has a trace of $1 + 2\cos\kappa$, from which the principal Euler angle κ can be determined. For a perfect 180° rotation $\mathbf{R}$ is symmetric with a trace of -1, which is often useful in recognizing 2-fold rotations directly from the rotation matrix. Finally, the eigenvector of $\mathbf{R}$ belonging to the real eigenvalue is the Euler axis $\mathbf{E}$.

Conversion programs such as *ROTMAT* in the CCP4 suite and *CONVROT*[12] can perform conversions between the various conventions and settings used by different programs. The program description of *CONVROT* also contains a listing of the conventions used by most current crystallographic programs. A concise description of the formal properties and conventions used in the description of rotation matrices[13] can be found in Goldstein's *Classical Mechanics*.[14]

Most useful for us is the example in Figure 11-4 showing the assignment of rotations and axes in the *ZYZ* Euler convention used by the CCP4 suite of programs. The complete rotation matrix **R** is then given as a successive operation of three counterclockwise rotations in reference to the fixed axes. First rotate about $\mathbf{Z}_0$ by γ, then about $\mathbf{Y}_0$ by β, and again about $\mathbf{Z}_0$ by α. The combined rotation matrix then is given by

$$\mathbf{R}_{ZYZ} = \mathbf{R}(\mathbf{Z}_0, \alpha) \cdot \mathbf{R}(\mathbf{Y}_0, \beta) \cdot \mathbf{R}(\mathbf{Z}_0, \gamma) \qquad (11\text{-}4)$$

Note that the same convention can also be described as a rotation around moving reference axes,[15] first by α about the original $\mathbf{Z}_0$, followed by rotation by β about new $\mathbf{Y}_1$, and finally rotation by γ about new $\mathbf{Z}_2$. These descriptions of the same convention can be readily converted, but note that the matrices are different and become non-trivial for the case of rotating reference axes (Sidebar 11-1, Appendix A.7).

$$\mathbf{R}_{ZYZ} = \mathbf{R}(\mathbf{Z}_0, \alpha) \cdot \mathbf{R}(\mathbf{Y}_0, \beta) \cdot \mathbf{R}(\mathbf{Z}_0, \gamma) = \mathbf{R}(\mathbf{Z}_2, \gamma) \cdot \mathbf{R}(\mathbf{Y}_1, \beta) \cdot \mathbf{R}(\mathbf{Z}_0, \alpha) \qquad (11\text{-}5)$$

Following the CCP4 *ZYZ* convention in fixed-axis description we can derive the rotation matrix **R** for the axis and angle assignment shown in the left panel of Figure 11-4 by expressing each successive rotation as a simple rotation around the specified axis. The first rotation around the $\mathbf{Z}_0$ axis by angle γ is written in matrix form as

$$\mathbf{R}(\mathbf{Z}_0, \gamma) = \begin{pmatrix} \cos\gamma & -\sin\gamma & 0 \\ \sin\gamma & \cos\gamma & 0 \\ 0 & 0 & 1 \end{pmatrix} \qquad (11\text{-}6)$$

We then combine all three rotation matrices $\mathbf{R}(\mathbf{Z}_0, \gamma)$, $\mathbf{R}(\mathbf{Y}_0, \beta)$, and $\mathbf{R}(\mathbf{Z}_0, \alpha)$, one for each rotation, corresponding to the operations in the left panel of Figure 11-4, and obtain for the complete rotation $\mathbf{R}_{ZYZ}$

$$\mathbf{R}_{ZYZ} = \begin{pmatrix} \cos\alpha & -\sin\alpha & 0 \\ \sin\alpha & \cos\alpha & 0 \\ 0 & 0 & 1 \end{pmatrix} \cdot \begin{pmatrix} \cos\beta & 0 & \sin\beta \\ 0 & 1 & 0 \\ -\sin\beta & 0 & \cos\beta \end{pmatrix} \cdot \begin{pmatrix} \cos\gamma & -\sin\gamma & 0 \\ \sin\gamma & \cos\gamma & 0 \\ 0 & 0 & 1 \end{pmatrix}$$

$$(11\text{-}7)$$

Figure 11-4 Rotation in Cartesian space in Euler *ZYZ* convention. The original coordinate system is defined as [0, $\mathbf{X}_0$, $\mathbf{Y}_0$, $\mathbf{Z}_0$] (blue axes), and the final orientation is described by the system [0, **X'**, **Y'**, **Z'**] (red axes). The left panel visualizes the process using fixed reference axes, where we first rotate the object about $\mathbf{Z}_0$ by angle γ, yielding the system [0, $\mathbf{X}_1$, $\mathbf{Y}_1$, $\mathbf{Z}_1$], then rotate that system around $\mathbf{Y}_0$ by angle β, yielding [0, $\mathbf{X}_2$, $\mathbf{Y}_2$, $\mathbf{Z}_2$], and finally again by α about $\mathbf{Z}_0$, yielding the new rotated system [0, **X'**, **Y'**, **Z'**]. The same operation can be described (albeit with different matrices) as rotation about the newly generated axes, as shown in the right panel. We first rotate again about $\mathbf{Z}_0$ but this time by α, providing $\mathbf{X}_{1'}$ and $\mathbf{Y}_{1'}$ as the node axis (green vector **N**). In the next step, we rotate about the new $\mathbf{Y}_{1'} = \mathbf{Y}_{2'}$ by β, providing us with the new $\mathbf{Z}_{2'}$ and $\mathbf{X}_{2'}$. Then we rotate about this new $\mathbf{Z}_{2'} = \mathbf{Z'}$ by γ and obtain the remaining two final **X'** and **Y'**. The complete rotation **R** is then the product of the individual rotations (note the order of application in matrix notation) $\mathbf{R}_{ZYZ} = \mathbf{R}(\mathbf{Z}_0, \alpha) \cdot \mathbf{R}(\mathbf{Y}_0, \beta) \cdot \mathbf{R}(\mathbf{Z}_0, \gamma) = \mathbf{R}(\mathbf{Z}_2, \gamma) \cdot \mathbf{R}(\mathbf{Y}_1, \beta) \cdot \mathbf{R}(\mathbf{Z}_0, \alpha)$. Note that the starting and ending coordinate frames are the same irrespective of the rotation reference, but the intermediate axes are not.

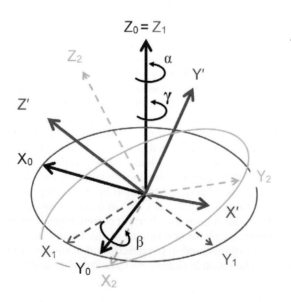

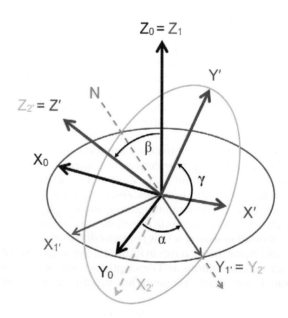

The combined rotation matrix $\mathbf{R}_{ZYZ}$ is also known as the direction cosine matrix (DCM). We symbolically evaluate the DCM by successive matrix multiplication as

$$\mathbf{R}_{ZYZ} = \begin{pmatrix} \cos\alpha\cos\beta\cos\gamma - \sin\alpha\sin\gamma & -\cos\alpha\cos\beta\sin\gamma - \sin\alpha\cos\gamma & \cos\beta\sin\beta \\ \sin\alpha\cos\beta\cos\gamma + \cos\alpha\sin\gamma & -\sin\alpha\cos\beta\sin\gamma + \cos\alpha\cos\gamma & \sin\alpha\sin\beta \\ -\sin\beta\cos\gamma & \sin\beta\sin\gamma & \cos\beta \end{pmatrix}$$

$$(11\text{-}8)$$

The DCM (11-8) has a number of interesting properties. It is a real square matrix whose transpose $\mathbf{R}^T$ equals its inverse $\mathbf{R}^{-1}$ (from which it follows that $\mathbf{R}^T\mathbf{R} = I$) and whose determinant $\det\mathbf{R}$ equals +1. The column vectors are the resultant coordinate system vectors spanning the rotated basis $[0, \mathbf{X}', \mathbf{Y}', \mathbf{Z}']$. As the transpose of $\mathbf{R}$ is also its inverse, the row vectors must contain the original system basis vectors $[0, \mathbf{X}_0, \mathbf{Y}_0, \mathbf{Z}_0]$. The norm of each of these vectors must be 1. As a further consequence, the determinant $\det\mathbf{R}$ must be +1; it maintains handedness and dimensions of the object upon application. Another useful property of $\mathbf{R}$ is that for a 180° rotation the matrix is symmetric: In this case, $\mathbf{R} = \mathbf{R}^{-1}$, because a rotation by +180° and −180° brings one to the same position. With the definition that for any rotation matrix $\mathbf{R}^{-1} = \mathbf{R}^T$, it immediately follows that $\mathbf{R} = \mathbf{R}^T$, that is, $\mathbf{R}$ is a symmetric matrix for a 180° rotation—which is often useful in recognizing 2-fold rotations directly from the rotation matrix. A DCM furthermore has a trace of $1 + 2\cos\kappa$, from which the principal Euler angle κ can be determined.

When superimposing molecules, we are often confronted with having to find the rotation angles and axes given the DCM that a program returns. We provide them here for the roll-pitch-yaw convention XYZ (θ_3, θ_2, θ_1) which is equivalent to the ZYX (θ_1, θ_2, θ_3) description and frequently used in superposition programs. The rotations about XYZ are defined as

$$\mathbf{R}_{XYZ} = \begin{pmatrix} \cos\theta_1 & -\sin\theta_1 & 0 \\ \sin\theta_1 & \cos\theta_1 & 0 \\ 0 & 0 & 1 \end{pmatrix} \begin{pmatrix} \cos\theta_2 & 0 & \sin\theta_2 \\ 0 & 1 & 0 \\ -\sin\theta_2 & 0 & \cos\theta_2 \end{pmatrix} \begin{pmatrix} 1 & 0 & 0 \\ 0 & \cos\theta_3 & -\sin\theta_3 \\ 0 & \sin\theta_3 & \cos\theta_3 \end{pmatrix} \quad (11\text{-}9)$$

and yield the following DCM:

$$\mathbf{R}_{XYZ} = \begin{pmatrix} \cos\theta_1\cos\theta_2 & \cos\theta_1\sin\theta_2\sin\theta_3 - \sin\theta_1\cos\theta_3 & \cos\theta_1\sin\theta_2\cos\theta_3 + \sin\theta_1\sin\theta_3 \\ \sin\theta_1\cos\theta_2 & \sin\theta_1\sin\theta_2\sin\theta_3 + \cos\theta_1\cos\theta_3 & \sin\theta_1\sin\theta_2\cos\theta_3 - \cos\theta_1\sin\theta_3 \\ -\sin\theta_2 & \cos\theta_2\sin\theta_3 & \cos\theta_2\cos\theta_3 \end{pmatrix}$$

$$(11\text{-}10)$$

The values for the angles can be readily expressed through the elements of the DCM:

$$\theta_1 = \tan^{-1}(R_{12}/R_{11})$$
$$\theta_2 = -\sin^{-1}(R_{13})$$
$$\theta_3 = \tan^{-1}(R_{23}/R_{33})$$

$$(11\text{-}11)$$

Note that the inverse tangent requires a quadrant check, as implemented in the *FORTRAN atan2* function. The angles themselves are defined in the ranges $-\pi \le \theta_1, \theta_3 < \pi$, and $0 \le \theta_2 < \pi$. Two possible sets of angle combinations yielding identical rotations are possible: $\theta_1' = \theta_1 - \pi$, $\theta_2' = \pi - \theta_2$, and $\theta_2' = \theta_2 - \pi$ [or in other words, $(\pi + \theta_1, \pi - \theta_2, \pi + \theta_3)$ is an identity operator in Eulerian space]. The above identity operator is for the XYZ convention; in CCP4 ZYZ convention it is $(\pi + \alpha, -\beta, \pi + \gamma)$.

Eigenvalues and eigenvector of the direction cosine matrix

The matrix equation represented by $\mathbf{R}$ can be solved and yields the three eigenvalues $\{1, e^{+ik}, e^{-ik}\}$. The eigenvector $\mathbf{E} = (l, m, n)$ belonging to the real eigenvalue 1 is the Euler axis, with the angle κ the single rotation angle (principal Euler angle) around the Euler axis $\mathbf{E}$.

The Euler angle κ can be readily computed from the trace $1 + 2\cos\kappa$ of the DCM:

$$\kappa = \cos^{-1}\left(\frac{trace\mathbf{R} - 1}{2}\right) = \cos^{-1}\left(\frac{R_{11} + R_{22} + R_{33} - 1}{2}\right) \tag{11-12}$$

and using the definition of the skew matrix $\mathbf{A}^\times$

$$\mathbf{A}^\times = \frac{1}{2\sin\kappa}(\mathbf{R}^\mathrm{T} - \mathbf{R}) = \begin{pmatrix} 0 & n & -m \\ -n & 0 & l \\ m & -l & 0 \end{pmatrix} \tag{11-13}$$

we obtain (noting the singularity at κ exactly 0° or 180°, where the axis can be inferred from the DCM directly) the elements of the Euler axis:

$$\mathbf{E} = \frac{1}{2\sin\kappa}\begin{pmatrix} R_{32} - R_{23} \\ R_{13} - R_{31} \\ R_{21} - R_{12} \end{pmatrix} \tag{11-14}$$

Computing the Euler axis $\mathbf{E}$ and Euler angle κ is often useful to detect the type of rotation that relates two objects. The actual rotation type is not always easy to recognize from the full rotation matrix, particularly when the off-diagonal elements become dominant. For example, consider the following general rotation matrix $\mathbf{R}$:

```
-------------- R --------------
-0.870616 -0.342573 -0.353087
 0.359975 -0.932799  0.017422
-0.335328 -0.111935  0.935428
```

Although we suspect—given that the diagonal elements are not too far off the values of (–1.0, –1.0, 1.0) and the matrix is nearly symmetric—that the rotation is close to a 2-fold rotation around $\mathbf{Z}$, it is hard to say how close the axis is to a true dyad. With $\mathbf{E}$ computed as (–0.181, –0.025, 0.983) and $\kappa = 159.0°$, we see that the axis orientation is indeed still largely along $\mathbf{Z}$, but the rotation angle of the pseudo-dyad deviates by 21° from the ideal 2-fold rotation of 180°.

Spherical coordinates

The direct interpretation of the DCM in terms of the rotation angles is, as we have seen, not easy and a common alternative representation of molecular rotations that is perhaps more intuitive (but algebraically less elegant) uses spherical coordinates. Spherical coordinates describe a rotation in a spherical coordinate system in terms of the azimuthal angle φ (horizontal rotation), the lateral angle ψ (up/down rotation), and the rotation angle κ around the new axis $\mathbf{E}$ defined by the rotations about φ and ψ as shown in Figure 11-5. The spherical coordinate representation is particularly useful when visualizing Patterson space searches for self-rotations indicative of near-perfect NCS axes (Figure 10-12 shows an example).

The rotation angles ranges necessary for coverage of all possible rotations are $0 \leq \varphi < 2\pi, 0 \leq \psi \leq \pi/2, -\pi < \kappa \leq \pi$. As a rotation about (φ, ψ, κ) is the same as a rotation about $(\varphi + \pi, \pi - \psi, -\kappa)$, the alternate angle range that must be covered in a superposition search is $(0 \leq \varphi < 2\pi, 0 \leq \psi \leq \pi, 0 < \kappa \leq \pi)$. The DCM expressed in polar angles is quite bulky and given in Appendix A.7. The computation of the Euler axis (which in polar coordinate representation is the axis that is rotated about κ as you may have noticed already) and the Euler angle κ follows the same procedure as described above, with

$$\kappa = \cos^{-1}\left[(R_{11} + R_{22} + R_{33} - 1)/2\right] \tag{11-15}$$

The elements of the skew matrix again define the directional components of the Euler axis as provided by Equation 11-14. Given the components of the Euler

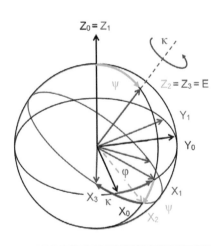

Figure 11-5 Definition of rotation angles in spherical coordinates. The coordinate system is first rotated about the azimuthal angle φ, then rotated up or down the lateral angle ψ, and finally rotated by κ about the axis $\mathbf{Z}_3$, which is the Euler axis $\mathbf{E}$, and κ is the principal Euler angle.

> **Box 11-3 Transformations in Cartesian world coordinates.** A transformation or mapping $\mathbf{W}(\mathbf{R},\mathbf{T})$ of an object onto another can be performed through a combination of rotation and translation. The translation is determined by the translation vector $\mathbf{T}$. The rotation is described through the direction cosine matrix $\mathbf{R}$ which is a real square matrix whose transpose $\mathbf{R}^T$ equals its inverse $\mathbf{R}^{-1}$ (from which follows that $\mathbf{R}^T\mathbf{R} = \mathit{I}$) and whose determinant $\det \mathbf{R}$ equals $+1$. Any rotation can be described as a rotation round the single eigenvector $\mathbf{E}$ of the rotation matrix, the Euler axis, and the associated principal Euler angle. The rotation matrix can be expressed in various conventions in orthogonal coordinates through three Euler angles (α, β, γ) or in sperical coordinates with azimuthal, lateral, and axial rotations through the angles (ϕ, ψ, κ). The last rotation axis is the Euler axis, and κ is the associated principal Euler angle.

axis (l,m,n) we obtain by comparison of the DCMs

$$\mathbf{E} = \frac{1}{2\sin\kappa}\begin{pmatrix} R_{32} - R_{23} \\ R_{13} - R_{31} \\ R_{21} - R_{12} \end{pmatrix} = \begin{pmatrix} \sin\psi\cos\varphi \\ \sin\psi\sin\varphi \\ \cos\psi \end{pmatrix} \tag{11-16}$$

from which we can compute the Euler axis $\mathbf{E}$.

Crystallographic and Cartesian transformations

Crystallographic transformations. Symmetry operations between oligomeric molecules can in some cases coincide with crystallographic transformations, and the NCS transformation relating to the molecule then becomes a true crystallographic transformation. Crystallographic transformations in the crystallographic reference system $[0, \mathbf{x}, \mathbf{y}, \mathbf{z}]$ are not limited to a local volume but by definition apply to the whole unit cell and the entire crystal lattice. Crystallographic rotations are consequently limited to pure point group rotations $\mathbf{R}^n$ with $n = 1$, 2, 3, 4, or 6, and the translational components resulting from screw-axes are limited to $\mathbf{T} = m/n$ with $1 \le m < n$. Rotation matrices $\mathbf{R}$ that correspond to crystallographic operations are easy to recognize: As they occur in fixed relation to crystallographic axes, they can only contain -1, 0, or 1 in their matrix elements r_{ij} once the operator $\mathbf{W}(\mathbf{R},\mathbf{T})$ is deorthogonalized, and $\mathbf{T}$ will contain the corresponding fractions of unit cell translations. As a reminder, for the example of a rotation around a 2_1-fold axis parallel to $\mathbf{z}$, the Euler angle $\kappa = 180°$ and $\mathbf{R}$ becomes the familiar operator for 2-fold rotation around $\mathbf{z}$ (Figure 5-35) plus the corresponding translation by $\mathbf{z}/2$:

$$\mathbf{W}(\mathbf{R}^{(2)}_{001},\mathbf{T}) = \begin{pmatrix} -1 & 0 & 0 \\ 0 & -1 & 0 \\ 0 & 0 & 1 \end{pmatrix} + \begin{pmatrix} 0 \\ 0 \\ \frac{1}{2} \end{pmatrix} \quad \text{or in symbolic notation,} \quad \overline{x}, \overline{y}, z+\tfrac{1}{2}$$

Let us examine two practical examples: In the case of orthogonal axes, a symmetry operator that relates crystallographic copies is immediately recognizable, because there are no other elements than 0 and ± 1 in the DCM. We take a $P4_1$ crystal structure and compute the superposition between two symmetry related copies:

```
CRYST1 48.200 48.200 63.000 90.00 90.00 90.00 P 41

-------------- R --------------       --- T ---
  0.0000      1.0000      0.0000         0.0000
 -1.0000      0.0000      0.0000        48.2000
  0.0000      0.0000      1.0000       -15.7500
```

The rotation matrix translates into symbolic operation $y, \overline{x}, z$, and with the translations exactly b (equivalent to no translation) and $-c/4$ equivalent to $c3/4$, we obtain for the entire symbolic operator $y, \overline{x}, z + \frac{3}{4}$, which is indeed the third operator listed in the International Tables Volume A for space group $P4_1$.

Cartesian transformations. In oblique crystal systems, NCS operators are not as readily recognizable as special. We need to transform the NCS or superposition operation $\mathbf{W}(\mathbf{R,T})$ that took place in rectilinear Cartesian world coordinates back into fractional crystallographic coordinates. We use the subscript W for the Cartesian world operator and C for the crystallographic equivalent. Then

$$\mathbf{R}_C = \mathbf{O}^{-1}\mathbf{R}_W\mathbf{O} \text{ and } \mathbf{T}_C = \mathbf{O}^{-1}\mathbf{T}_W \tag{11-17}$$

where $\mathbf{O}$ is the orthogonalization matrix and $\mathbf{O}^{-1}$ is the deorthogonalization matrix.

Let us apply this procedure to a perfect NCS rotation in a special orientation. Such NCS operations can indicate possible higher symmetry space groups and we want to examine whether the NCS operator in world coordinates possibly has a distinct relation to the crystallographic system. As an example we show the following matrix obtained from NCS superposition of two monomers forming a dimer, the asymmetric unit of a trigonal crystal structure in polar space group $P3_2$ from the PDB:

```
CRYST1 66.066 66.066 87.620 90.00 90.00 120.00 P 32 6
SCALE1      0.015136  0.008739  0.000000       0.00000
SCALE2      0.000000  0.017478  0.000000       0.00000
SCALE3      0.000000  0.000000  0.011413       0.00000
```

The SCALE records contain the real space deorthogonalization matrix $\mathbf{O}^{-1}$ from which we can obtain the real space orthogonalization matrix $\mathbf{O}$ simply by inversion

$$\mathbf{O}^{-1} = \begin{pmatrix} 0.015136 & 0.008739 & 0 \\ 0 & 0.017478 & 0 \\ 0 & 0 & 0.011413 \end{pmatrix} \text{ and } \mathbf{O} = \begin{pmatrix} 66.066 & -33.033 & 0 \\ 0 & 57.215 & 0 \\ 0 & 0 & 87.620 \end{pmatrix}$$

Applying the deorthogonalization (11-17) to the real world NCS operator $\mathbf{W}_W(\mathbf{R}_W, \mathbf{T}_W)$ relating the two molecules of the dimer in the asymmetric unit

```
-------------- Rw -------------     --- Tw ---
-0.500569  -0.865697  -0.000028     33.023445
-0.865697   0.500569   0.000522     19.067408
-0.000438   0.000286  -1.000000      0.159851
```

yields the operator $\mathbf{W}_C(\mathbf{R}_C, \mathbf{T}_C)$ in crystallographic coordinates:

```
-------------- Rc -------------     --- Tc ---
-1.000377   0.000753   0.000363      0.650759
-0.999615   1.000377   0.000799      0.333258
-0.000330   0.000352  -1.000000      0.001824
```

The computation of the Euler axis and principal Euler angle from the real world NCS operator $\mathbf{R}_W$ yields within numerical accuracy a perfect 2-fold orientation axis ($\kappa = 180.03°$, which we also can realize by visual inspection of $\mathbf{R}_W$ from the fact that the matrix is nearly perfectly symmetric and has a trace of -1. The program *CONVROT*[12] informs us that the Euler axis $\mathbf{E}_W$ in Cartesian space is (-0.500, 0.866, 0.000) and thus lies in the *ab*-plane. Applying the deorthogonalization matrix yields the direction of the Euler axis $\mathbf{E}_C$ in reference to the crystallographic system:

$$\mathbf{E}_C = \mathbf{O}^{-1}\mathbf{E}_W = \begin{pmatrix} 0.00000 \\ 0.015136 \\ 0.00000 \end{pmatrix} \tag{11-18}$$

What can we learn from these results? First, we see from the above transformation of the Euler axis into crystallographic coordinates that it only points in the y direction equivalent to $(0, 1, 0)$, so that the NCS operation is for all practical purposes a perfect 2-fold rotation parallel to the crystallographic y axis. This alone is not uncommon, although the perfection of angle and orientation raises some suspicion.

A further inspection of the deorthogonalized NCS operator $\mathbf{W_C}(\mathbf{R_C}, \mathbf{T_C})$ shows that in the crystallographic system $\mathbf{R_C}$ corresponds to a crystallographic 3-fold rotation coinciding with z ($\bar{x}, \bar{x} + y, \bar{z}$ in symbolic notation) plus a fractional translation of ($2/3, 1/3, 0$). Both are compatible with the trigonal system, and we can try to apply the local operation to the entire crystal. The 3-fold rotation by 120° swaps the 2-fold axis from parallel $\boldsymbol{b}$ to parallel $\boldsymbol{a}$, and then we apply the purported NCS translation or its equivalent of $-1/3, 1/3, 0$ to the entire unit cell and investigate the result (Figure 11-6).

The cell shift has now brought the perfect 2-fold NCS axis into the B-face of the unit cell, parallel to the $\boldsymbol{a}$-axis. If we fix the floating origin from original space group $P3_2$ so that the position of the 2-fold NCS axis is fixed at $z = 1/6$, we have generated space group $P3_221$—with one molecule in the asymmetric unit. Inspection of the International Tables A, space group No. 154, will confirm our finding. Compare the new cell in Figure 11-6 with Figure 5-21: the unit cells are identical and indeed the structure can be refined without problem in space group $P3_221$.

Types of NCS transformations

We can distinguish three common cases of NCS transformations mapping molecules onto each other, each with specific properties of $\mathbf{R}$, $\mathbf{T}$, and the corresponding Euler axis and principal Euler angle.

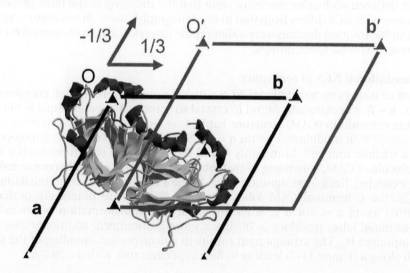

Figure 11-6 **Dimer axis coinciding with a crystallographic axis in higher symmetry space group.** By transforming a suspect NCS operator from Cartesian world coordinates into crystallographic coordinates and applying it to the entire unit cell, a higher symmetry space group is formed. The perfect 2-fold NCS in the $P3_2$ structure is in fact the 2-fold crystallographic axis in $P3_221$. Compare with Figure 5-21.

Box 11-4 Transformations to the crystallographic reference frame. To interpret a NCS operation in the crystallographic reference system, its operators need to be converted into a crystallographic basis with crystallographic coordinates. A transformation in Cartesian world coordinates $\mathbf{W}(\mathbf{R_W}, \mathbf{T_W})$ is transformed into the equivalent operation $\mathbf{W}(\mathbf{R_C}, \mathbf{T_C})$ in the crystallographic reference system with

$$\mathbf{R_C} = \mathbf{O}^{-1}\mathbf{R_W}\mathbf{O} \quad \text{and} \quad \mathbf{T_C} = \mathbf{O}^{-1}\mathbf{T_W}$$

where $\mathbf{O}$ is the real space orthogonalization matrix.

Perfect (proper) NCS rotation. In this case, the rotation matrix **R** of the NCS operation is either exactly (or very close to) a closed point group operation, but not limited to the 2-, 3-, 4-, and 6-fold rotations allowed for crystallographic axes. Five-fold and higher axes are possible. Near-perfect rotations are frequent in NCS;[16] many dimers, trimers, and higher assemblies are related by perfect or near-perfect pseudo-rotations such as pseudo-dyads, pseudo-triads, and so forth. As the rotation axis is not necessarily in any special orientation relative to a crystallographic axis, the **R** matrix will contain significant non-zero off-diagonal elements, but the similarity to a single-axis rotation is often still distinctly recognizable in the elements of the DCM, and calculating the Euler axis **E** and Euler angle κ from the DCM helps in more complex cases.

Near-closed point group rotations manifest themselves in two ways in reciprocal space: always in Patterson self-rotation maps (Section 11.3) and in native Patterson maps only when a 2-fold axis is parallel or very near parallel to another 2-, 4-, or 6-fold crystallographic axis or screw axis, as discussed in the next section. If the intersection of multiple proper NCS axes is located at the origin of the unit cell, which is often the case in high symmetry oligomers or virus structures, their NCS operators will not have any translational component.

Near-perfect NCS translation. In this case, the NCS operation does not contain significant rotational terms, **R** is close to *I*, and the translation vector **T** dominates the NCS. Such translations are also occasionally found, and we provide an example showing the manifestation of real-space translational symmetry in reciprocal space in the next section.

General (improper) NCS transformations. In this case, both **R** and **T** contribute to the NCS, and neither has any special relation to pure rotations (proper NCS) or special transitions. All rotation angles can be arbitrary. The local symmetry relating the non-identical variable and constant domains in F_{ab} antibody fragments shown in Figure 11-2 is an example of a general, improper 2-fold relation between molecular domains. Note that the meaning of the term *proper* in context with NCS differs from that in crystallographic axes: an *improper* crystallographic rotation denotes a crystallographic inversion axis (roto-inversion) not allowed in chiral space groups.

Translational NCS in real space

Pure or near-pure translational NCS without significant rotational component (i.e., **R** ≈ *I*) is occasionally found in crystal structures. A good example is the triclinic calmodulin (CAM) structure 1up5 (a commensurately modulated superstructure with modulation vector $\mathbf{q} = 2\mathbf{a}^*$, Section 8.3) that contains 2 molecules in a triclinic unit cell. Commonly found *P*1 calmodulin structures harbor one molecule of CAM, consisting of two lobes of two EF hands each, connected by an extended, flexible recognition helix. In the *P*1 crystal form with translational NCS, the C-terminal CAM lobes (residues 86–147) are practically perfectly shifted along a vector **a**/2, while the local NCS transformation between the N-terminal lobes (residues 5–76) has a more pronounced, additional rotation component $\mathbf{R}_N$. The arrangement results in a near-perfect doubling of the unit cell along **a** (Figure 11-7) leading to new superstructure with $a = 59.7$ Å.

Inspection of the two separate NCS matrices for the N-terminal and C-terminal lobes obtained from the molecular superposition program *LGA*[17] confirms the largely translational character of both NCS operations. For the N-terminal NCS transformation we obtain

$$\mathbf{R}_N = \begin{pmatrix} 0.992 & 0.127 & 0.014 \\ 0.127 & 0.992 & -0.027 \\ -0.010 & 0.029 & 0.999 \end{pmatrix} \text{ and } \mathbf{T}_N = \begin{pmatrix} -24.7 \\ 0.7 \\ 0.7 \end{pmatrix}$$

and for the C-terminal near-perfect translational NCS with $\mathbf{R}_C$ close to the identity operator:

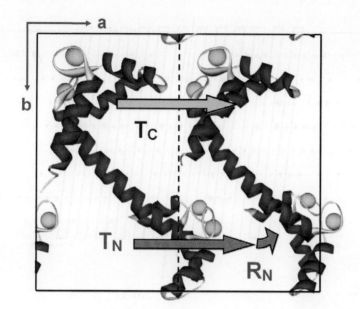

Figure 11-7 Near-perfect translational NCS. The two CAM molecules in PDB entry 1up5[18] are related by translational NCS. Each of the two lobes contains two Ca-binding EF motifs (Chapter 2), connected by a flexible linker helix. The NCS transformation between the C-terminal lobes (green arrow) is nearly perfectly translational ($R_c \sim I$, $T_c = a/2$), while the second NCS transformation relating the N-terminal domains contains a more significant rotational component R_N. The triclinic unit cell can be described as commensurately modulated superstructure of a smaller P1 cell with one molecule in the asymmetric unit and with $a/2$ as its cell parameter.

$$\mathbf{R}_C = \begin{pmatrix} 0.997 & 0.071 & -0.024 \\ 0.071 & 0.997 & 0.008 \\ -0.010 & -0.006 & 0.999 \end{pmatrix} \simeq \begin{pmatrix} 1 & 0 & 0 \\ 0 & 1 & 0 \\ 0 & 0 & 1 \end{pmatrix} = \mathbf{I} \text{ and } \mathbf{T}_C = \begin{pmatrix} 29.1 \\ 2.11 \\ -1.32 \end{pmatrix} \simeq \begin{pmatrix} a/2 \\ 0 \\ 0 \end{pmatrix}$$

Manifestation of translational NCS in reciprocal space

Translational NCS in real space is also manifested in reciprocal space according to the general averaging equation (11-24) introduced in Section 11.2. In those cases, $\mathbf{R} \sim \mathbf{I}$ and thus $\mathbf{k} \sim \mathbf{h}$, and the structure factors will therefore contribute only to themselves and not to other parts of R^*. This is particularly interesting in the case when the translational vectors are very close to integral fractions of the unit cell vectors. The translational shift will then lead to destructive interference between the X-rays scattered from the individual copies. As we noticed in the derivation of systematic absences or extinction rules in Chapter 5, perfect translation by fractions along unit cells generates corresponding zonal absences. The same happens when the NCS translation vectors are near integral fractions of the unit cell. For example, a near-perfect shift of $\mathbf{a}/2$ as in the above CAM example should generate weakening of the reflections in the (h, k, l) planes with $h \neq 2n$. The sequence of strong/weak/strong layers can be nicely seen in cross sections of reciprocal space as shown in Figure 11-8. With increasing resolution, the deviations from perfect translational periodicity become relatively more pronounced. When the lattice spacing d_h approaches the deviations, the extinction effect disappears, while in true crystallographic symmetry, the extinctions persist throughout the entire reciprocal space. As noted above, this doubled calmodulin cell can also be interpreted as a commensurately modulated superstructure with modulation vector $\mathbf{q} = 2\mathbf{a}^*$, with the weak $h \neq 2n$ reflections as commensurate satellite reflections.

As mentioned in the discussion of merohedral twinning in Chapter 8, translational NCS or pseudo-symmetry also affects the intensity distributions. Presence of a combination of compensating twinning and translational pseudo-symmetry can lead to near-normal intensity distribution statistics and obscure the detection of twinning.

The effect of translational NCS also manifests itself in native Patterson maps, where the map coefficients are the native intensities I_{obs} or F_{obs}^2. The atoms of each translational NCS copy are shifted by the same distance vector and although the signal from vectors between light atoms is weak, a distinct super-peak results

Figure 11-8 Translational NCS in reciprocal space. The presence of translational NCS with translational vectors very close to integer fractions of unit cell vectors manifests itself in systematic weakening of reflections, similar to extinctions from translational Bravais lattice centering indicating a commensurately modulated superstructure. In the depicted case, the NCS is nearly translational and the shift vector **T** is $a/2$, resulting in a sequence of strong/weak/strong planes hkl according to the reflection condition $h = 2n$. In cross section, the affected planes appear as weak lines (blue shade). Note that at high resolution the imperfectness of the NCS (in contrast to perfect crystallographic symmetry) leads to a weakening of the extinction effect. Structure factors from PDB entry 1up5.

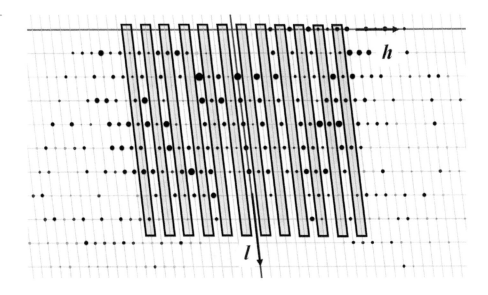

from the superposition of all collinear translational vectors. A strong peak at the NCS translation vector distance will thus be observable in the native Patterson map. Also in this case, the effect becomes obscured with increasing imperfection of the translational NCS. Figure 11-9 shows the strong translational peak at ~**a**/2 for our CAM example.

NCS axes parallel to crystallographic axes and native Patterson maps

A similar situation exists in a native Patterson map when a proper NCS axis is oriented parallel or very nearly parallel to a crystallographic axis. This is not an infrequent occurrence[16] and includes 2-fold NCS axes parallel to 2-, 4-, and 6-fold rotations or screw axes. We use as an example an F_{ab} antibody fragment structure (PDB entry 12e8)[11] with a proper 2-fold NCS axis nearly perfectly parallel to the single screw axis in $P2_1$. The native Patterson map again shows a strong peak in the Harker section and from the Harker vector $u, v, w = (2x, \frac{1}{2}, 2z)$ we can immediately determine the location of the axis (Figure 11-10). Even small deviations of a few degrees from parallel alignment of the NCS axis with the crystallographic axis lead to disappearance of the peak in the native Patterson map. A general explanation of how the real space scenario of two NCS related molecules with an axis parallel to the unique monoclinic axis gives rise to the peak in the Harker section of a native Patterson map is shown in Figure 11-11.

Figure 11-9 Translational NCS in Patterson space. Pure or nearly pure translational NCS manifests itself in native Patterson maps through peaks corresponding to the NCS vector. The map shows our example 1up5 with the two CAM molecules related by near translational symmetry with a translation vector of **T** ~ **a**/2. The remaining huge peaks at the origin of the primitive Patterson cell are the zero-distance self-vector peaks.

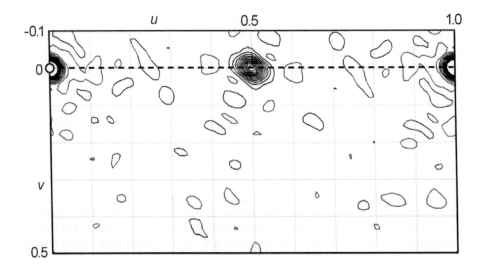

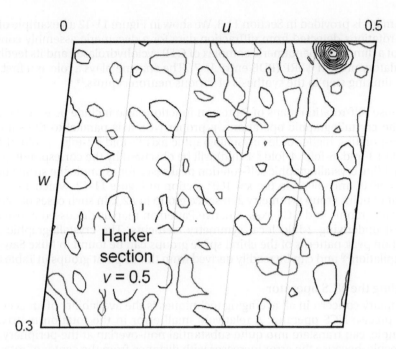

Figure 11-10 Native Patterson map for NCS axis parallel to crystallographic axis. A 2-fold NCS axis parallel or nearly parallel to a 2-, 4-, or 6-fold axis or screw axis will lead to a peak in the corresponding Harker section of the native Patterson map. The depicted map is a part of the Harker section u, $\frac{1}{2}$, w with Harker vector $(2x, \frac{1}{2}, 2z)$ in space group $P2_1$. The crystallographic 2_1 axis is perpendicular to the projection. The location of the parallel NCS axis can be read off the map as approximately $x = u/2 = 0.42/2$ and $z = w/2 = 0.02/2$. Application of the $P2_1$ symmetry operator $\bar{x}, y + \frac{1}{2}, \bar{z}$ leads to the NCS axis position $(-0.21, 0, -0.01)$ shown in Figure 11-2. Data from PDB entry 12e8.[11]

Proper NCS axes and Patterson self-rotation function

Proper non-crystallographic rotations are closed and cyclic point group operations, that is, perfect rotations about an n-fold axis *not* limited to the crystallographic values for n (n only 2, 3, 4, or 6). They manifest themselves as peaks in self-rotation functions of native Patterson maps. The values of the self-rotation function are presented in stereographic projections, that is, the rotation axis angle κ is plotted for fixed values of $360°/n$ (180°, 120°, 90°, 72°, 60°, etc.) as a function of the lateral (ϕ) and azimuthal angles (ψ) of the spherical coordinates in a stereographic projection. The explanation of Patterson self-rotation functions

Figure 11-11 NCS axis parallel to crystallographic axis. The real space scenario (left panel) of a dimer in space group $P2_1$ shows that in projection down the unique axis **b** the vectors between the dimeric subunits are collinear (but lie not in the paper plane, note the screw axes). In the Harker projection (right panel), these vectors give rise to the corresponding Harker vectors $u, v, w = (2x, \frac{1}{2}, 2z)$. = From these centrosymmetrically related Harker vectors, the location of the NCS axis can be determined as $(u/2, y, w/2)$ or $(-u/2, y, -w/2)$. Note that the two centrosymmetric Patterson peaks correspond to two different origin choices with respect to the NCS axis position.

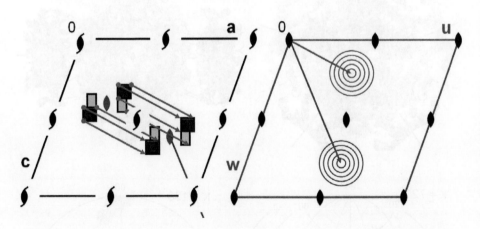

Box 11-5 Manifestation of NCS transformations in reciprocal space. Because NCS in real space also generates redundancies in reciprocal space, special NCS operations can be deduced directly from the experimental data. Proper rotational NCS manifests itself as peaks in maps of the Patterson self-rotation function in κ-sections corresponding to the rotation angle. Positions of 2-fold axes parallel to crystallographic 2-, 4-, or 6-fold rotation or screw axes that would otherwise be obscured by crystallographic symmetry in the Patterson self-rotation maps manifest themselves as peaks in native Patterson maps. Near-perfect translation-only NCS also shows itself through peaks in native Patterson maps.

and maps is provided in Section 11.3. We show in Figure 11-12 an example of the self-rotations detected from diffraction data for a decameric assembly consisting of a dimer of a pentameric complex of GTP cyclohydrolase I and its feedback regulatory protein GFRP (PDB entry 1is7). The enzyme plays a role as a first and rate-limiting step in the synthesis of various neuroreceptors.[19]

Because self-rotation maps of Patterson functions also include peaks generated by the crystallographic operations, a proper NCS axis parallel to the same or corresponding higher order crystallographic axis (2-fold ∥ 2-fold, 4-fold, 6-fold; 3-fold ∥ 3-fold, 6-fold; 4-fold ∥ 4-fold) will be obscured by the corresponding and coinciding crystallographic self-rotation peaks (see for example the strong peaks at $\phi = 90.0°$ and $-90.0°$ in the $\kappa = 180°$ section in Figure 11-12, which are to the result of point group symmetry 2 in space group in $P2_1$). In such cases of parallel crystallographic and NCS axes, a native Patterson map as discussed above may reveal underlying 2-fold local symmetry elements. The crystallographic self-rotation peak patterns of the chiral space groups can be found in Mike Sawaya's compilation,[20] and can be readily derived from the 12 point groups in Table 6-6.

Finding the NCS operator

A primary concern in all averaging techniques is the availability of an accurate and precise NCS operator. A relatively small error in the rotation matrix, for example, can translate into quite substantial non-overlap at the periphery of a molecule, because the error increases with distance from the center of rotation. Depending on the available information, we can obtain the NCS operator by a variety of methods.

Atomic model. In cases where a model (at least partial) is available, such as after molecular replacement, the superposition can be carried out in real space by any superposition program (Sidebar 11-2). The operators are then directly

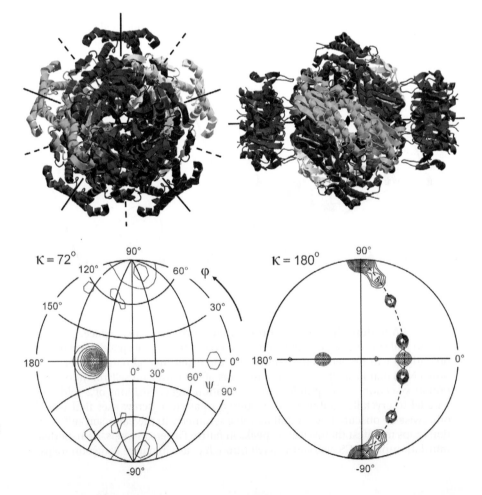

Figure 11-12 Proper NCS revealed in self-rotation maps. The Patterson self-rotation function of the decamer shown above is plotted in a stereographic projection (azimuthal angle ϕ, lateral angle ψ) in sections with different κ angles. The 5-fold local NCS axis reveals itself in a distinct peak at angles (180°, 43°, 72°) in the $\kappa = 72°$ section, and the five perpendicular 2-fold axes are distinctly visible on the dashed $\psi = 43°$ line in the $\kappa = 180°$ section. As the 2-folds "poke" through the molecule (see top panel), they are spaced every 36° symmetrically around the $\phi = 0°$ axis in the $\kappa = 180°$ section. The large peaks at $\phi = 90.0°$ and $-90.0°$ are the result of space group symmetry ($P2_1$), and the remaining spurious 2-fold peaks are probably due to additional local symmetry. The molecule itself has point group symmetry 52. The colors in the top panel have been selected differently in each panel to emphasize the 5- and 2-fold symmetry, respectively. The self-rotation functions were calculated with the CCP4 program *POLARRFN* and plotted with *PLTDEV*. A stereographic net as shown in the left panel can be plotted with the CCP4 program *STNET*. PDB entry 1is7.[19]

obtained and are fairly accurate. A minimal partial model can also be provided by the substructure heavy atoms, which generally also comply with the NCS. Complications can arise when there are not enough heavy atoms, there is different partial occupation in each NCS copy, or the atoms sit in special positions, but several programs such as *FINDNCS* or *PROFESSS* in the CCP4 suite or in the *SOLVE* package[21] are available to determine the NCS operator from the heavy atom substructure.

Real space molecular replacement. In the general real space case we might not have an atomic model available and real space molecular replacement methods can be used to overlap experimental electron density maps.[22] Electron density automatically implies phase information via map inversion, and real space molecular replacement methods employ domain rotation functions to determine the rotational component **R** and phased translation functions to obtain the translational part **T** of the NCS operator. These functions are further explained in Section 11.3. A frequent nuisance is that in initial electron density maps the full extent of the protein molecules is not known and parts of one or the other copy are generally outside of the unit cell. The electron density must then be examined over a large enough region in space so that all of the NCS related molecules are found intact and can be properly masked.

Self-rotation functions. In reciprocal space, proper NCS rotation operators can be determined without knowledge of a molecular model or phases from Patterson self-rotation functions (Section 11.3). If an initial electron density map is available, it can then be rotated according to the rotation determined by the self-rotation function, and the NCS translation is found by using the map as a search model in a phased translation function. In any case, the NCS operators and masks are progressively refined during the map averaging and phase extension process as discussed in Section 10-7.

Cross-correlation of self-rotation and cross-rotation. A useful program for the analysis of structures with NCS is *RFCORR* in the CCP4 suite, which cross-correlates peaks in the cross-rotation and self-rotation functions (Section 11.3). *RFCORR* uses the self-rotation function to filter out peaks in the cross-rotation function that are inconsistent with the NCS.[13] This reduces the number of noise peaks in the cross-rotation function that need to be tested with translation functions. The procedure works regardless of whether or not the NCS is proper, but is presently not employed by any of the automated MR scripts. One of the benefits of the self-rotation function is that it is inherently much more reliable than the cross-rotation function, because it only depends on the experimental data; the biggest source of error in the cross-rotations is of course the search model. The procedure works even better if the peaks are sharpened by using normalized amplitudes (*E*s) instead of structure factor amplitudes, which can be obtained from *ECALC* (CCP4).

> **Sidebar 11-2 Superposition of protein structure models.** Detecting 3-dimensional similarities between protein structures is a fundamental task in protein structure comparison and it is an active field of research in structural bioinformatics. In the case of 1:1 correspondence of the Cα-atoms, as is the situation between two identical symmetry related molecules in a crystal, the superposition is trivial, and can be readily solved by minimization of the coordinate deviation residual between the two molecules by least squares fitting. The situation is not so straightforward once the correspondence between the Cα-atoms of the molecules is not readily recognizable, and structure superposition becomes an *alignment* problem. A wide variety of 3-dimensional alignment methods have been developed, ranging from distance matrix alignment[23] (Dali) to combinatorial programs[24] (*CE*), local-global alignments[17] (*LGA*), and dynamic programming (*ProSup*,[25] *TopMatch*[26]). A comparison of some alignment programs can be found in a review by Levitt and co-workers.[27] The problem of recognizing 3-dimensional relationships becomes particularly difficult if similarities between structures extend over subunit boundaries, or in the case of domain swapping.

11.2   Phase improvement by NCS density averaging

A powerful opportunity for map improvement, generally applied in combination with solvent modification, is map averaging. When non-crystallographic, local symmetry (NCS) is present in a molecular structure, the asymmetric unit contains multiple copies of the molecule, and the diffraction pattern contains redundant information[1] as far as the molecular envelope is concerned. This is frequently the case: nearly half of all structures in the PDB are dimeric structures (80%) or structures with higher oligomerization state of the molecules.[2] In these cases, the initial experimentally phased electron density maps will contain the maps of two or more molecules, which can be averaged to improve the map quality. The theoretically expected (but not always achieved) improvement in the signal-to-noise ratio of the maps is proportional to $N^{1/2}$, where N is the number of independent molecular copies in the asymmetric unit. The strong dependence of improvement on the copy number explains why multi-copy map averaging is so powerful for highly symmetric virus structures.

The concept of NCS map averaging can be extended to maps from different crystals—either in different crystal forms, or different maps independently determined from the same crystal form. Multi-crystal or cross-crystal map averaging is elegantly implemented in the density modification program *DMMULTI* of the CCP4 suite.

Averaging in real space and in reciprocal space

It is intuitive that a simple arithmetic *N*-fold real space averaging of superimposed electron density maps will improve the signal-to-noise ratio in the map by $N^{1/2}$ by averaging out the noise and enhancing consistent features. Subsequent map inversion thus should give improved phases, particularly when applied simultaneously with solvent flattening. However, upon rotation of the electron density, the density grid points of the rotated map copy do not necessarily fall onto the grid points of the other copy. The density values at these grid points must be linearly[1] or non-linearly[28] interpolated. A corresponding real space map averaging scenario is shown in the left panel of Figure 11-13.

Reciprocal space averaging is not quite as simple as direct space averaging. As the molecules are related by general NCS operators that do not follow the crystallographic symmetry restrictions, the structure factors of one copy will not overlap with the structure factors of the rotated copy, that is, they do not map directly onto each other and fall on non-integer lattice points. The effect of each structure factor on its neighboring structure factors is therefore limited in reciprocal space, and determined by the so-called *G*-function or reciprocal space interference function. The *G*-function defines the reciprocal volume in which the structure factors are affected and will be of additional use in Section 11-3 in the context of molecular replacement. The reciprocal space scenario of a general NCS map rotation is shown in the right panel of Figure 11-13.

Real space averaging. To simplify the derivation of the averaging equations, we first define for each of the *N* NCS copies a real space mask function $g_i(\mathbf{x})$ (1 in the protein region and 0 outside) and subtract again the average solvent density from the entire electron density of the asymmetric unit, including the NCS related copies as shown in Figure 11-14. The initial map density at each grid point $\rho_i(\mathbf{x})$ is then given by

$$\rho(\mathbf{x}) = \rho_{\mathrm{OLD}}(\mathbf{x}) - \langle \rho_{\mathrm{S}} \rangle \tag{11-19}$$

Basic averaging of the density between *N* copies of the map is straightforward as

$$\langle \rho(\mathbf{x}) \rangle = \frac{1}{N} \sum_{i=1}^{N} \rho(\mathbf{x}_i) \tag{11-20}$$

Figure 11-13 NCS averaging in real space and reciprocal space. The left panel shows the real space scenario of a general near 2-fold NCS operation **Rx** + **T** of a masked map section (density A). To emphasize the grid mismatch, a coarse density grid is overlaid and marked with red dots on the molecule. The grid points rotating together with the density (density B) do not coincide with the original map grid, and the density values must be interpolated. The right panel shows a related reciprocal space scenario, where the corresponding rotation **R**$^\mathrm{T}$**x** of the reciprocal lattice leads to mapping of the rotated structure factors onto non-integral lattice points. The translational shift of the molecule does not affect the location of the reciprocal lattice points.

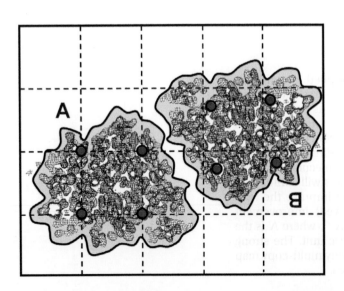

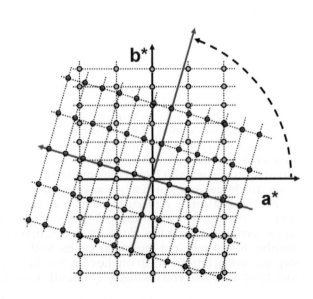

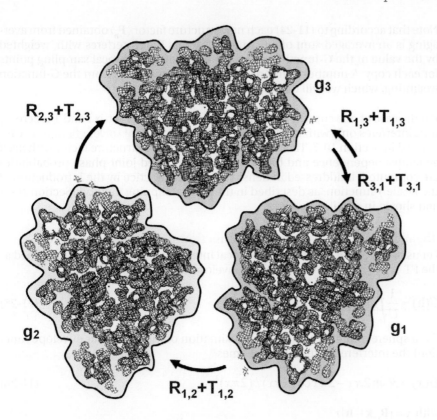

Figure 11-14 NCS averaging and solvent masks. Each molecule in the asymmetric unit is related to another NCS copy (including itself) by a transformation $\mathbf{x}_{i,j} = \mathbf{R}_{i,j}\mathbf{x} + \mathbf{T}_{i,j}$. Some of them are explicitly shown. Each molecule has a real space mask $\mathbf{g}_i$ defined around it, outside of which solvent density (white) is flattened. The drawing follows an illustration by Vellieux and Read.[29]

We now combine the averaging procedure with solvent flattening. We apply the first mask $g_i(\mathbf{x})$ to each density copy $\rho_i(\mathbf{x}_j)$ and average all points inside this mask. The process is repeated for each of the masks, and the expression for the averaged and masked protein density then becomes

$$\langle\rho(\mathbf{x})\rangle = \frac{1}{N}\sum_{i=1}^{N}\sum_{j=1}^{N}\rho(\mathbf{x}_{i,j})\cdot g_i(\mathbf{x}) \qquad (11\text{-}21)$$

where the coordinates $\mathbf{x}_{i,j}$ of each copy are given by the transformations $\mathbf{x}_{i,j} = \mathbf{R}_{i,j}\mathbf{x} + \mathbf{T}_{i,j}$ that relate the N map copies (Figure 11-14).

Reciprocal space averaging. To obtain the reciprocal space equivalent of this real space procedure, we have to Fourier transform both sides of Equation 11-21, and according to the convolution theorem, the product of the terms in the double sum becomes a convolution integral. The FT of the real space g-function is the reciprocal G-function, and because of the non-crystallographic relationships, G is a *general* function of the scattering vector $\mathbf{s}$, not of the discrete crystallographic lattice vector $\mathbf{h}$. $G(\mathbf{s})$ is therefore called the reciprocal space interference function, which we will meet again in molecular replacement.[30] The FT of the average density $\langle\rho(\mathbf{x})\rangle$ becomes the average structure factor $\langle\mathbf{F}_\mathbf{h}\rangle$ for the new map:

$$\langle\mathbf{F}_\mathbf{h}\rangle = \int_{ucell}\langle\rho(\mathbf{x})\rangle\exp(2\pi i\mathbf{h}\mathbf{x})d\mathbf{x} \qquad (11\text{-}22)$$

Substituting in the above equation for $\langle\rho(\mathbf{x})\rangle$ the expression in (11-21) yields

$$\langle\mathbf{F}_\mathbf{h}\rangle = \int_{ucell}\left(\frac{1}{N}\sum_{i=1}^{N}\sum_{j=1}^{N}\rho(\mathbf{x}_{i,j})\cdot g_i(\mathbf{x})\right)\exp(2\pi i\mathbf{h}\mathbf{x})d\mathbf{x} \qquad (11\text{-}23)$$

Replacing in the above Equation (11-23) $g_i(\mathbf{x})$ with its Fourier transform $U\cdot G_i(\mathbf{s})$ (where U is the sum of all protein mask volumes U_i) in Equation 11-23 and explicitly applying $\mathbf{x}_{i,j} = \mathbf{R}_{i,j}\mathbf{x} + \mathbf{T}_{i,j}$ finally leads to[31]

$$\langle\mathbf{F}_\mathbf{h}\rangle = \frac{U}{NV}\sum_{\mathbf{k}}\mathbf{F}_\mathbf{k}\sum_{i=1}^{N}\sum_{j=1}^{N}G_i(\mathbf{h} - \mathbf{R}_{i,j}^{\mathrm{T}}\mathbf{k})\cdot\exp(-2\pi i\mathbf{k}\mathbf{T}_{i,j}) \qquad (11\text{-}24)$$

Note that according to (11-24) each new structure factor $\langle \mathbf{F_h} \rangle$ obtained from averaging is an averaged sum of all structure factors $\mathbf{F_k}$ it interferes with, weighted by the value of the G-function at the corresponding reciprocal sampling points for each copy. A number of interesting consequences arise from the G-function weighting, which we will discuss in the next section.

For the case of one copy, $i,j = 1$ and $\mathbf{R}$ reduces to the unit matrix I and $\mathbf{T}$ is zero. $\mathbf{F_k}$ thus interferes only with itself and we obtain the classical solvent flattening case derived in Section 10-7. The fact that interference of a structure factor with itself generates dependence and bias when the associated joint phase probabilities are computed, is addressed in density averaging practice by the introduction of the Leslie γ-function, as described in the solvent flipping method (Section 10-7) and shown in detail elsewhere.[2]

The reciprocal space interference function G

Let us take a closer look at the reciprocal interference function G. It is defined as the FT of the real space masking or envelope function $g(\mathbf{x})$:

$$G(\mathbf{h}) = \frac{1}{U}\int_U g(\mathbf{x})\exp(2\pi i\mathbf{h}\mathbf{x})d\mathbf{x} \tag{11-25}$$

For a sphere of radius r (in first approximation of the molecular envelope function) the interference function becomes[30]

$$G(x) = 3(\sin 2\pi x - 2\pi x\cos 2\pi x)/(2\pi x)^3 \tag{11-26}$$

with $x = |\mathbf{R}_{i,j}\mathbf{k} - \mathbf{h}|r$

The G-function (11-26) is graphed in Figure 11-15 and shows that significant interference (influence of structure factors on each other) exists only within a narrow range around the origin of the function, that is, when the argument $x \ll 1$, or in other words, $\mathbf{k}$ must transform close to $\mathbf{h}$ for significant interference (which implies $\mathbf{R}_{i,j}^T\mathbf{h} \approx \mathbf{k}$). The practical consequences for us are twofold: first, given the limited effect of a reflection on its neighbors, phase extension to higher resolution in combination with averaging can only be carried out *in small steps*. At the surface of the resolution sphere (for which we have phases), only the narrow shell of new reflections that lie within the range of G from the outermost reflections can be added. Second, if we want to compute a correlation between two Patterson functions from differently oriented molecules, the number of terms that need to be evaluated is drastically reduced because only structure factors *within* the G-peak affect each other; this will be quite useful in molecular replacement.

Figure 11-15 The reciprocal space interference function G. The values of the axis are reciprocal space units given by $x = |\mathbf{R}_{i,j}\mathbf{k} - \mathbf{h}|r$ for each set of related reflections and imply that reflections that affect each other must lie close in reciprocal space. The practical implications are that phase extension must be extended successively in thin shells during averaging and that overlap of reciprocal functions for objects in different orientations is limited to neighboring reflections in reciprocal space. The latter allows computation of fast rotation functions in molecular replacement.

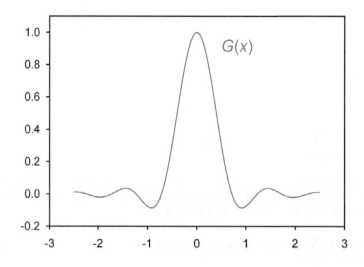

> **Box 11-6 Averaging in real and reciprocal space.** The redundancies in reciprocal space generated by real space NCS between molecules in the crystal can be exploited to improve the phases by electron density map averaging. NCS averaging procedures are generally combined with density modification and can be carried out in real space using the electron density maps or in reciprocal space using diffraction data. In both cases, NCS operations generate non-integer lattice points, which in real space are interpolated or in reciprocal space are determined by folding with the reciprocal space interference function *G*.

Examples of electron density map averaging

In the following we show two examples of map averaging. In the first one, we use only map averaging to further improve our already quite respectable electron density map obtained from the MAD phasing example in Chapter 10 before we submit it to automated model building in Chapter 12.

Basic map averaging

We will use the programs of the CCP4 suite for 2-fold averaging of the *SHELXE*-phased and density modified maps from our Se-MAD phasing experiment of the tesB dimer. The first step involves generating from the *SHELXE* ASCII *phs* file an *mtz* file, which the density modification and averaging program *DM*³ can read. The averaging program also needs the NCS operator, which we obtain from analyzing the symmetry of the heavy atom solution with the CCP4 program *PROFESSS*. We read the rotational part of the NCS operator in spherical angles and its translation part in Cartesian coordinates from the log file and enter it together with the remaining information into the *DM* GUI. Here is an interesting question for the observant reader: Does it matter for the operator search whether we use the correct heavy atom substructure enantiomorph or the inverted one? We also realize from the absence of a significant component in the **c**-direction in the translational part of the NCS operator that the 2-fold NCS axis must be nearly parallel to **c**. This explains why we cannot find a 2-fold axis in the self-rotation Patterson map: The NCS axis is obscured by the strong peaks present at $\kappa = 180°$ because of Patterson symmetry *mmm* for space group $C222_1$. The native Patterson map in Figure 11-16, however, shows a clear peak in Harker section $w = 0.5$ identifying the location of the axis.

The input for *DM* is straightforward; we select 'map averaging only' as the desired mode of operation and make sure the correct labels are assigned for the data columns of the *mtz* input file. Additional information includes the solvent content and number of monomers, and we select automated mask generation and mask updating. An idiosyncrasy of *DM* is that it needs an identity operator (zero rotation angles and translations) explicitly entered for the first molecule, followed by the spherical angles and translational component of the NCS operator relating the second molecule to the first copy. Because the map was already solvent modified and quite good, the gain in figure of merit (f.o.m.) that can be achieved by 2-fold averaging is modest (Figure 11-17). Nonetheless, even small map improvements will significantly improve the automated model building in the next step (Chapter 12).

Combined map averaging and density modification

An instructive case study demonstrating the power of map averaging combined with density modification is presented in our second example. A protein sample purified over a dextran-based affinity column was yielding only poor crystals diffracting to 6 Å at best for a year, when a new batch of protein suddenly yielded crystals diffracting to 2.5 Å. Soaking experiments were successful with Hg, and a Xe noble gas pressure derivatization experiment (Sidebar 3-15) yielded a Xe derivative. Anomalous diffraction data were collected from a highly isomorphous

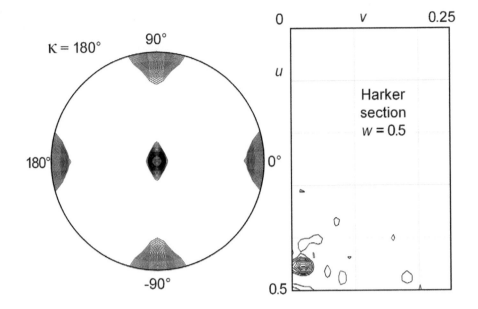

Figure 11-16 Patterson self-rotation map and Harker section of native Patterson map with NCS peak. The 2-fold NCS axis in the dimer of tesB in our MAD example from Chapter 10 is parallel to the **c**-axis and thus obscured by strong space group symmetry generated peaks in the $\kappa = 180°$ section of the Patterson self-rotation map (left panel). The NCS peak is, however, clearly visible in the Harker section $w = 0.5$ (perpendicular to **c**) in the native Patterson map (right panel).

Hg heavy atom derivative crystal, but the Xe noble gas derivative showed only limited isomorphism. SIRAS phases were calculated using *SHARP*[22] from the Hg derivative data which also served as the native data set. The rather poor phases were improved and extended to 2.7 Å by solvent flipping using *SOLOMON*[33] in the *SHARP* package[33] including a brief final run of *DM*[3] that included histogram matching (Section 10.7).

The resulting density modified phases were improved over the original Hg SIRAS phases, but the maps were still noisy and not traceable. To perform 2-fold map averaging, *DM* was supplied with the 2-fold NCS operator derived from heavy atom sites by *PROFESSS*, and *DM* was re-run in averaging and auto-masking mode. During combined NCS averaging and solvent modification, *DM* updated the initial solvent mask. The resulting averaged map had good connectivity and model building was initiated (Figure 11-18).

The anticlimax in this example is that it became rapidly obvious that the model did not match the expected sequence (which can be readily anchored to the Hg atoms bound to surface-exposed cysteines) and it turned out that the molecule was actually the dimer of maltodextrin phosphorylase (MP). Both MP and the sought-after protein have similar molecular weight and did not separate on the size exclusion column. MP likely bound to the dextrin resin of the column, and was released in sufficient amounts to actually form crystals. The incident emphasizes that in cases of doubt, quick N-terminal protein sequencing or mass spectroscopy can be quite helpful to confirm the nature of the material

Figure 11-17 Improvement of figure of merit by map averaging. The density-modified map of the tesB dimer obtained from *SHELXE* was 2-fold averaged using *DM*. As the map was already of high quality, the averaging gives only a small improvement in the figure of merit. Be also aware that solvent modified figures of merit can be inflated due to bias issues. The plot is one of the diagnostic log graphs output from *DM*.

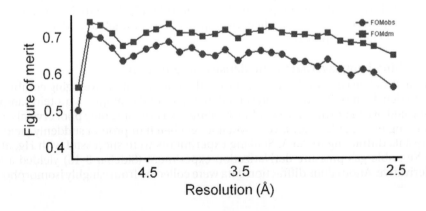

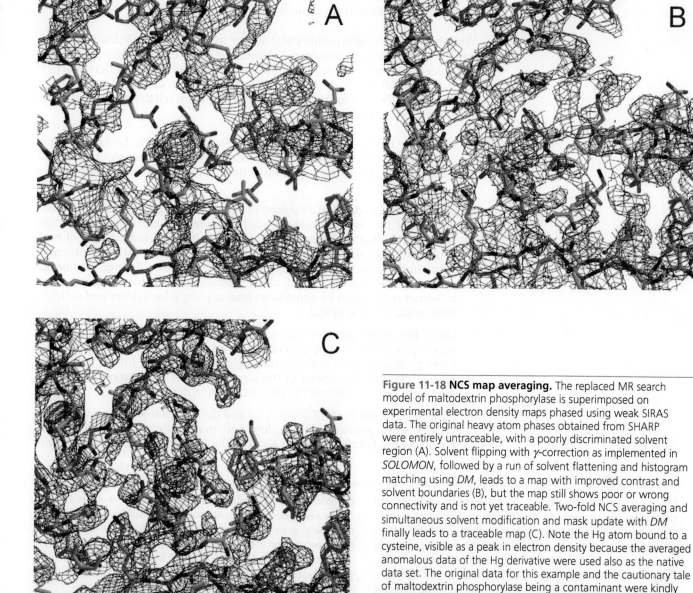

Figure 11-18 NCS map averaging. The replaced MR search model of maltodextrin phosphorylase is superimposed on experimental electron density maps phased using weak SIRAS data. The original heavy atom phases obtained from SHARP were entirely untraceable, with a poorly discriminated solvent region (A). Solvent flipping with γ-correction as implemented in *SOLOMON*, followed by a run of solvent flattening and histogram matching using *DM*, leads to a map with improved contrast and solvent boundaries (B), but the map still shows poor or wrong connectivity and is not yet traceable. Two-fold NCS averaging and simultaneous solvent modification and mask update with *DM* finally leads to a traceable map (C). Note the Hg atom bound to a cysteine, visible as a peak in electron density because the averaged anomalous data of the Hg derivative were used also as the native data set. The original data for this example and the cautionary tale of maltodextrin phosphorylase being a contaminant were kindly provided by Han Remaut, Institute of Structural Molecular Biology, School of Crystallography, Birkbeck College, London.

forming the crystals. The structure solved in minutes by molecular replacement with a model of the MP dimer.

11.3 Molecular replacement phasing

While experimental phasing by means of substructure determination has provided us with a general approach to determining protein structures, it comes at the cost and effort of either having to prepare isomorphous heavy atom derivatives, a Se-Met labeled protein, or very careful and redundant anomalous data collection for native S-SAD phasing. There is another method to obtain initial phases, which is conceptually simple but in practice not always trivial; it is molecular replacement (MR). The term *replacement*, first coined by Michael G. Rossmann,[34] is to be interpreted in the sense of re*locating* and not *substitution*: a known structure model, presumed to be similar to the sought-after unknown structure (generally based on sequence identity of ~30% or higher) is rotated and translated in the unit cell or asymmetric unit until the solution with the best fit between calculated diffraction data from the replaced model and the observed data from the unknown structure is obtained.

With the ever increasing number of structure models available in the PDB, MR is a method of increasing popularity.[35] Molecular replacement is also the method of choice if a known protein or mutant thereof crystallizes in a different crystal form. Co-crystallized drug–target complexes also often appear in crystal forms different from the apo-form, and MR allows a fast and reliable method of structure determination in those cases.

Given that we have collected all available information about the molecule and selected and prepared a proper search model,[36] which is not always a trivial task (see also Sidebar 11-5), the major tasks from a crystallographic point of view here are threefold:

- First, we need to develop a method to determine the proper orientation and location of the search probe—the actual molecular replacement structure solution. Although routine and highly automated for single copies of molecules with high sequence identity, the structure solution for cases with poor structural similarity or multiple copies of the same or different molecular subunits, often flexibly hinged, are non-trivial. In these cases, a heavy atom substructure solution by anomalous phasing might be quicker and deliver a more accurate final model.

- Given the often relatively flat solution landscape of the various possible search probe positions in difficult cases, we need to evaluate multiple solutions, preferably from multiple starting models, in order to find the best or most likely solution. Speed of the algorithms and automation of the programs are essential in order to find solutions in difficult cases.

- Finally, given the issue of phase bias and the fact that we have only model phases—of a generally incomplete and inaccurate model—we need to construct the Fourier map that gives us the least bias, and consider how to

---

Sidebar 11-3 Development of molecular replacement methods. About 80% of all protein structures are presently determined by molecular replacement (MR), where a structurally similar search model that has been correctly positioned in the unknown structure serves as initial phasing model. Interestingly, the methods used in MR were first developed in an attempt to use local, non-crystallographic symmetry (NCS) as a means for protein phasing given the difficulties experienced with the early heavy atom phasing experiments.[34] In addition, in the early 1960s there was certainly no indication that a large number of protein structures would ever be readily available to serve as search models. Instead, the second protein ever determined, hemoglobin, had already revealed that proteins can be composed of multiple similar subunits. It was also known that rhombohedral Zn-insulin, on which Dorothy Hodgkin worked from 1934, had two copies in the asymmetric unit (the structure was finally published in 1969,[41] after she had received the unshared 1964 Nobel Prize in Chemistry for her crystallographic work on cholesterol iodide, penicillin, and vitamin B_{12}).

The mathematical fundamentals of MR were fully laid out in 1962 by Michael Rossmann and David Blow,[30] largely in view of detecting subunits in structures to exploit the redundancies resulting from NCS for phasing. The first successful application of local 222-symmetry to improve the initial MIR phases of glyceraldehyde-3-phosphate dehydrogenase by real space averaging followed in 1973. A few years earlier fast rotation and translation functions providing a necessary means to handle MR searches in Patterson space efficiently were developed by Crowther[42] and Blow,[43] and in 1974 the first MR solution of the carboxypeptidase B using carboxypeptidase A as a search model succeeded.[44]

The MR method was continuously developed by many researchers, and improved programs were written while at the same time the PDB became populated with an increasing number of available search models. The first program geared toward automated structure solution, *AMoRe*,[45] was published in 1994, and evolutionary algorithms implemented in *EPMR*[46] and maximum likelihood molecular replacement functions[8] in *Phaser* are examples of recent developments allowing the determination of complex multi-subunit structures using molecular replacement.

Many of the original papers on molecular replacement can be found in the 1972 compilation by Michael Rossmann.[47] The 2001 and 2008 volumes of the CCP4 study weekend proceedings listed in the general reference section is dedicated to modern molecular replacement techniques. Recently, the phasing of crystal structures using molecular envelopes from electron microscopy by MR techniques[48] has become a further application of this powerful and popular structure determination technique.

minimize phase bias. We will therefore apply map reconstruction using σ_A-based maximum likelihood coefficients[37] that consider incompleteness and incorrectness in the phasing model. In addition, the severe phase bias practically always present in weak MR cases will require application of aggressive methods of phase bias minimization during the model rebuilding. Omit maps,[38] multi-model dummy atom refinement,[39] and combined procedures[40] assure that a minimally biased model can be obtained even when the initial MR solution was quite poor. We will discuss phase bias minimization methods in Chapter 12 during model building and refinement.

Placing a real space probe using reciprocal space data

If we have an atomic model or a phased map available, the relative orientation of the molecular models or maps can be determined by real space superposition and optimization of the overlap as discussed in Section 11.2. However, in the case of a MR problem, we have only reciprocal space amplitude data available to search for the unknown orientation and location of our molecule. The question is how to place and score a *real space probe* given only *reciprocal space data*.

- **Six-dimensional search methods.** The first idea that comes to mind is a procedure similar to the brute-force approach in heavy atom searches (Section 10.3), where we systematically place an atom into the asymmetric unit, and at each grid point compute the match to the experimental isomorphous or anomalous difference Patterson map. In a similar fashion, we could compute the structure factors for the probe located in different orientations at different grid positions in the unit cell and compute an R-value or correlation coefficient for each trial. The best solution would have the lowest R-value or highest correlation between observed and calculated structure factors.

  The problem with 6-dimensional searches lies largely in the technicalities of computation: in contrast to few atoms in a heavy atom search, we need to compute structure factors of an entire macromolecule in all possible orientations at each translational grid point. Such an exhaustive brute-force search in six dimensions (three rotational and three translational degrees of freedom) on a fine enough grid is impractically slow even with the commonly used FFT-based structure factor calculations (Sidebar 9-1). Nevertheless, using even faster structure factor calculations based on Fourier transform interpolation[49] and efficient evolutionary search algorithms,[46] stochastic search methods have become feasible given present computing power. Their benefit is that given enough trials (and time) such methods will eventually find the correct solution, provided that the problem has a solution at all. Poor data (as always) or unsuitable search probes generally lead to unsolvable MR problems.

- **Rotation–translation searches.** A computationally efficient approach is to break up the 6-dimensional search into a 3-dimensional rotation search and a subsequent 3-dimensional translation search once the correct rotation has been found. This reduces the computational effort from say 10^6 evaluations on a rather coarse $100 \times 100 \times 100$ point grid to $10^3 + 10^3$ evaluations—a substantial gain. The rotation search involves searching distance vector space for the best correlation of *intramolecular* vectors using Patterson methods, followed by fast translation searches for the best correlation with *intermolecular* Patterson vectors. The difficulty here is that the rotation solution with the highest Patterson correlation score is not always the correct one, and even a small rotational error combined with a correct translation generally does not give a good or clear solution. Nevertheless, the original Patterson-based rotation–translation search methods,[30] given the various fast routines that have been developed, are still present in the majority of methods in use today, and their target functions have been further improved by the introduction of maximum likelihood molecular replacement functions.[50]

As many of the basics and challenges of molecular replacement searches are the same in both 6-dimensional searches and the rotation–translation methods, we begin our discussion with the implementation of 6-dimensional search procedures.

Box 11-7 **Principles of molecular replacement.** Molecular replacement is a phasing technique that can be applied when a search model structurally similar to a molecule in the target crystal structure is available. Molecular replacement requires a search for the proper orientation and location of the search probe in the unknown crystal structure. The search for proper placement can be conducted in sequence by a deterministic 3-dimensional rotation search followed by a deterministic 3-dimensional translation search or in a single step by stochastic 6-dimensional placement searches.

Molecular replacement by multi-dimensional search

The principle of brute-force exhaustive searches is simple in nature: at each grid point in the asymmetric unit of the unknown cell we place the molecular search probe in varying orientations and compute a structure-factor based correlation score against the observed data. The scored solutions are ranked, with the best solution having the highest correlation coefficient. The problems are twofold:

- **Structure factor calculations:** The structure factor calculations—being a summation over all atoms for all structure factors—are notoriously slow even when conducted with conventional FFT-based algorithms, which poses a serious problem. The sensitivity of structure factor calculation to positional errors necessitates sampling on a fine grid. Sampling of a $100 \times 100 \times 100$ Å cell on a 1 Å grid, and at each grid point evaluating the orientation in independent $1°$ intervals requires $100^3 \times 360^3 = 4.7 \cdot 10^{13}$ relatively slow structure factor computations, which makes brute force exhaustive placement infeasible. However, super-fast structure factor computation algorithms based on Fourier transform interpolation algorithms (Sidebar 11-4) allow one to tackle the problem when combined with a more efficient search algorithm.

- **Efficient search algorithms:** Even with the fastest structure factor computations, we need a more efficient search algorithm than exhaustive searches, suitable for the complex, non-linear 6-dimensional molecular replacement search problem. Stochastic optimization algorithms such as evolutionary programming can overcome this challenge in multivariate searches.

Sidebar 11-4 Fast structure factor calculation by Fourier-transform-interpolation. The evaluation of correlation coefficients based on structure factors for each member of a population of 300 in each of 50 evolution cycles requires 15 000 complete structure factor calculations (each one for several thousand reflections and over several thousand atoms) per single run of evolutionary search. Even the commonly implemented, space group specific FFT-based structure factor calculations (Sidebar 9-1) are unacceptably slow for use in stochastic searches. However, as the search molecule in a molecular replacement search is rigid and its relative coordinates do not change (in contrast to structure refinement, where the atom positions are adjusted to achieve best fit to the data), a more rapid procedure has been devised, initially by Lattman[51] and Huber.[49] If the FT of the electron density is sampled on a fine enough grid (essentially providing near-continuous structure factors), then the structure factors of that same molecule in any orientation can be obtained by interpolation, and the translation can be accounted for by application of corresponding phase shifts (applied to all structure factors).

The practical implementation[46] of the procedure works as follows: First the structure factors are calculated for the molecule in a stationary box sufficiently large (about four times the dimensions of the molecule in each direction) to allow fine sampling of its transform. Instead of computing new structure factors at each position of the relocated search probe, the indices of the observed structure factors are rotated according to the current search probe rotation, and their values (non-integral in reference to the stationary box) are interpolated from the structure factors calculated for the molecule in the box. Then the phase shifts resulting from the current translation are added to the complex structure factors, and finally the structure factor contributions from all the molecules in the unit cell are combined to give the calculated structure factors. To score the placement of a probe, the correlation coefficient (11-27) between the computed structure factor amplitude of the placed probe molecule and the observed structure factors is the most robust measure.

Evolutionary search algorithms

An efficient class of optimization techniques that is generally suitable for complex searches is genetic algorithms, of which evolutionary algorithms are a subset. The name of these stochastic optimization procedures is derived from their conceptual similarity with biological selection: A starting random population competes for survival based on a selective process, and the surviving individuals continue to produce offspring. The population members that score highest in a stochastic tournament against an objective function defining the selection parameters remain in the population and restore the population to its original size with their offspring. Small random variations in the population in analogy to genetic mutations expand the search space around the most promising parameters defined by the parent population. Repetition of the procedure leads to an optimization process that samples the most promising regions in parameter space efficiently and comprehensively. True genetic algorithms also rely on crossover and combination of parent properties (similar to DNA shuffling introduced in Chapter 4), and given their heritage from DNA sequences, are based on discrete sampling. Evolutionary programming[52] in contrast does not employ crossover, which is less advantageous in the case of correlated parameters, but most of all can be applied to continuous, real-value variables, a significant advantage over discrete sampling: we do not need to restrict ourselves to specific grid point positions or rotation angle intervals in our molecular replacement search.

Implementation in _EPMR_. Evolutionary programming has been implemented successfully in the program _EPMR_ (Evolutionary Program for Molecular Replacement) by C. R. Kissinger and colleagues.[46] In _EPMR_, a starting population of 300 search models, evolving over 50 generations, is generated by assigning random variables to the six rigid-body parameters (three angles, three coordinates) that define the placement of each of the models. A fast Fourier transform interpolation-based structure factor calculation using data in the range of ~15–4 Å is performed for each model, and the linear correlation coefficient CC_F between model structure factors and observed structure factors is computed:

$$CC_F = \frac{\sum_{\mathbf{h}} \left| F_{obs} - \overline{F_{obs}} \right| \cdot \left| F_{calc} - \overline{F_{calc}} \right|}{\left(\sum_{\mathbf{h}} \left| F_{obs} - \overline{F_{obs}} \right|^2 \cdot \sum_{\mathbf{h}} \left| F_{calc} - \overline{F_{calc}} \right|^2 \right)^{1/2}} \qquad (11\text{-}27)$$

The stochastic tournament of the solutions against CC_F as an objective function then determines the winners in each cycle of evolution. A randomly selected, small number of individuals (2%) serves as a test set, against which all members of the population are scored. The population is ranked by the number of wins each member scored against the test set, and the top 50% of the solutions (decreasing to 20% at the end of the evolutionary search) are kept unmodified for the next scoring round. The advantage of a stochastic tournament compared with plain numerical ranking by CC_F is that as a non-deterministic ranking it preserves a certain chance that less fit members of the population survive; that is, it increases population diversity.

In the next round, the population is brought back to its initial size by the addition of new offspring with parameters varying randomly with a given variance from the parent parameters. The initial variance of the parameters is set to 6° for the rotation angles and to 3 Å for translations. This mutation variance is further optimized in a self-adaptive process, where the variance of the n parameters x_i for a new generation σ_i' is dependent on the variance of the parameter σ_i in the parent population

$$\sigma_i' = \sigma_i \exp\left(\frac{N(0,1)}{2^{1/2} n} + \frac{N_i(0,1)}{(2n)^{1/2}} \right) \qquad (11\text{-}28)$$

The term $N(0,1)$ indicates a normally distributed random number with zero mean and unit variance, and the indexed term indicates that a different random variable is used for each variable. The left term in the exponent thus affects all variables and the right term only affects individual variables. This self-adaptive parameter selection determines the optimal mutation sizes over the course of the evolutionary search.

After 50 cycles of evolution, the best solution is kept and the search model is subjected to a rigid body conjugate gradient refinement, in which the six placement parameters of the best solution are further optimized. The coordinates of the best solution are stored and a new run with a different set of random placements begins. After a selected number of runs, the ultimately best model is used to generate the first electron density map. The total number of runs necessary depends on the difficulty of the problem, and if a well-scoring solution with a CC_F above ~0.5 is found, the correctness of the solution is practically assured and the program terminates. Figure 11-19 illustrates the key steps of molecular replacement by evolutionary search as implemented in *EPMR*.

Multiple copies in the asymmetric unit. We have already discussed in a previous section that the presence of multiple copies in the asymmetric unit is a frequent occurrence in molecular structures. In this case we can either search with a complete oligomer, which is a viable option if the oligomer has strong intermolecular contacts and likely maintains the same conformation in the unknown cell. However, this is not an option if different molecules form a complex, if the oligomer is unknown, or if the relative orientation of the members of an oligomer is likely not the same in the search probe and the unknown model.

Figure 11-19 Molecular replacement by evolutionary search. Evolutionary algorithms are an efficient optimization technique implemented in multidimensional searches. A search model is placed at random into the unit cell, and evaluated in a stochastic tournament. The winners are kept, and a new generation with perturbations around the initial position regenerate the populations (inner loop). After 50 evolution cycles, the best model is kept, and a new run begins (outer loop). The best solution from all runs is finally used to generate a bias minimized map for model building.

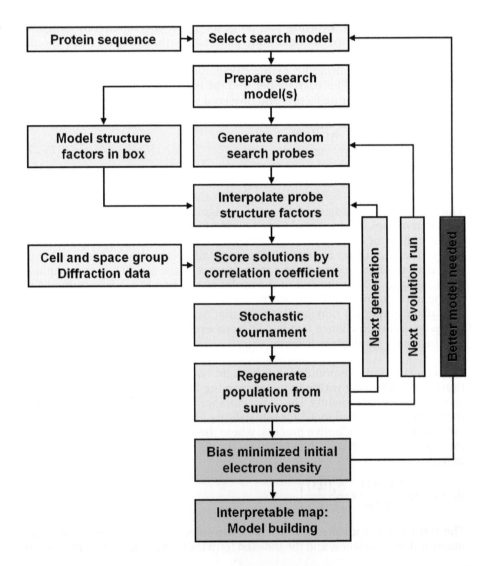

In this case we can, for example, first search for one monomer of a homo-dimer, and then fix the correctly placed first molecule and include its static contribution in the structure factor calculation while searching for the remaining molecule. As the probe contains only half of the scattering matter, the correlation coefficient in the first run of a sequential search will be less, and for a dimer, a correlation coefficient above ~0.3 probably indicates a correct solution. The more independent copies in the asymmetric unit need to be located, the more difficult the identification of correct partial solution generally becomes. We will discuss this problem again in the context of MR strategy development and maximum likelihood MR procedures.

Molecular replacement of a dimer structure with EPMR

In the following example we solve a not quite trivial molecular replacement problem involving a flexible dimer. We have available diffraction data of ferricytochrome c′ from the purple bacterium *Rhodobacter sphaeroides* (RS) in trigonal space group $P3_1$ (a, b = 48.10 Å, c = 115.9 Å) to 1.5 Å. The heme-binding bacterial cytochromes c′ occur primarily as 28 kDa homo-dimers and are of interest as they purportedly play a role in electron transport, and a large number of possible probe structures from other bacterial species are available from the PDB.[53] Sequence analysis revealed that the cytochrome c′ from *Rhodobacter capsulatus* (RC) has 41% sequence identity and probably constitutes a viable search probe. We cannot use an entire homo-dimer as a search probe in our problem, because the crossing angles between the interfacing four-helix bundles are different in most species, probably a result of variable dimer interfaces with a modest buried surface area between around 500 to 800 Å² per molecule. In addition, the anti-parallel four-helix bundles of each monomer are flexible themselves, and the (*a posteriori* determined) Cα coordinate r.m.s.d. of ~1.6 Å between the probe and the RS cytochrome monomer is relatively high—at the borderline of what can be expected to deliver a correct MR solution (Sidebar 11-5).

Model trimming. We execute some trimming of the model by removing all solvent atoms; those will be different in the target structure with the exception perhaps of a few structurally conserved water molecules. What about the heme moiety: should we leave it or truncate it? Given that this is a conserved region of the molecule, we might leave it in the probe. In addition, we are not executing an intramolecular distance vector based Patterson rotation search, and we therefore do not need to worry about possible strong intermolecular Fe–Fe distances (they are in fact all >20 Å apart), and the structure factor calculation in our stochastic search will be slightly more accurate with the Fe atoms and heme included.

We also know from the sequence alignment (Table 11-2) which residues are conserved and where gaps or insertions will probably be located. Instead of using a simple polyalanine model, we use the sequence alignment to truncate with the CCP4 program *CHAINSAW* non-identical residues past the first atom not in common with the corresponding residue in the target molecule. This truncation model in some cases delivers slightly better results than plain polyalanine truncation or replacement and side-chain modeling of residues.[54] The sequence alignment[55] also reveals a three-residue loop insertion in the target (our search model will have a gap there) and two single residue deletions, indicating regions where distinct structural differences between probe and target are likely.

Table 11-2 Pair-wise alignment between molecular search probe and target sequence. The total sequence identity is 41%, and one three-residue insertion and two single-residue deletions are present in the alignment. Identical residues are highlighted (*); conserved residues are indicated by a colon (:) and semi-conserved residues by a period (.).

```
Target      ADAEHVVEARKGYFSLVALEFGPLAAMAKGEMPYDAAAAKAHASDLVTLTKYDPSDLYAP 60
Probe 1cpq  ADTKEVLEAREAYFKSLG---GSMKAMTGVAKAFDAEEAAKVEAAKLEKILATDVAPLFPA 57
            **::..*:***:.**. :.  *.: **:   .:** ***..*:.* .:   * : *:..

Target      GTSADDVKG-TAAKAAIWQDADGFQAKGMAFFEAVAALEPAAGAGQ-KELAAAVGKVGGT 118
Probe 1cpq  GTSSTDLPGQTEAKAAIWANMDDFGAKGKAMHEAGGAVIAAANAGDGAAFGAALQKLGGT 117
            ***: *: * * ****** : *.* *** *:.** .*: .**.**:   :.**: *:***

Target      CKSCHDDFRVKR 130
Probe 1cpq  CKACHDDYREED 129
```

Sidebar 11-5 Divergence of sequence and structure of proteins. A strong correlation exists between the sequence identity and coordinate r.m.s.d. of related protein molecules. In 1986 Chothia and Lesk[56] compared the mutants of eight protein families providing 32 pairs of structures. The pair-wise Cα backbone r.m.s. coordinate deviation σ_r plotted against the sequence identity s, yields the following relation (graphed in Figure 11-20):

$$\sigma_r = 0.4 \cdot \exp\left(1.87(1-s)\right) \qquad (11\text{-}29)$$

Although the Chothia and Lesk study was limited to eight families of protein structures and to main chain atoms only, their equation is quite useful to roughly estimate the likelihood for success in a MR experiment. A widely accepted rule of thumb is that search models with a coordinate r.m.s.d. above ~1.5 Å from the target structure or

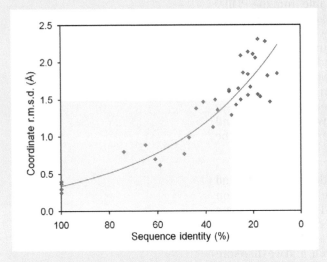

Figure 11-20 Correlation of sequence identity with coordinate r.m.s.d. The graph of Equation 11-29 is plotted together with experimental coordinate r.m.s.d. between pairs of structures with varying sequence identity. The green shaded area approximates the range where molecular replacement structure solution is generally successful.

with less than ~30% sequence identity rarely produce a viable MR solution.[57] From this rule it automatically follows that any atom in the correct position is a good atom, and every atom with a large deviation from the corresponding atom in the target molecule is undesirable. As a consequence, large flexible loops or flexible parts are generally trimmed from the search probe, as are parts that do not align in multiple sequence alignments. The study is also consistent with the experience that largely-sheet structures seem to be more sensitive to low sequence identity than mixed or largely-helical structures, which is probably a consequence of the higher percentage of conformationally flexible loop residues in sheet structures. Similarly, some reports indicate that polyalanine models—despite having lower total scattering contributions because the side chains are missing—have slightly higher scores than the full probes including side chains.[58] Various different truncation procedures exist[54] that can all be tried in multiple MR searches. It is also clear that the quality of the sequence alignment (Chapter 4) will affect the detection and choice of a suitable search model.[55]

In agreement with the 1.5 Å rule of thumb quoted above, below ~25–30% sequence identity the structural similarity between proteins enters a "twilight zone,"[59] where molecular replacement (and fold recognition or threading methods as well) generally fail. Some molecules are also extraordinarily flexible and even given high sequence identity, their coordinate r.m.s.d. may be unusually high. On the other hand, it must be recalled that structure (and function) is often more conserved than protein sequence, and in favorable cases when high functional similarity or evolutionary homology is evident, molecules with lower sequence identity may still provide viable search probes.

Maximum likelihood target functions for molecular replacement require *a priori* estimates for σ_A, a general measure for model incompleteness and coordinate error in maximum likelihood target functions. The relation (11-29) based on the sequence identity between MR search probe and actual protein provides the needed estimate for coordinate error (and the completeness of the search model relative to the probe is of course also known).

Multi-trial solution landscape. The evolutionary program *EPMR* needs only the search probe coordinates in PDB file format, the observed diffraction data of the unknown crystal, and its cell dimensions and space group as input. The default parameters for evolution and resolution cutoff (15–4 Å) work well, but we need to inform the program that we expect to find two copies of the search molecule in the asymmetric unit. We expect a relatively flat solution landscape and as a consequence, many necessary trials are required to find a good solution. In this particular case, discrimination between solutions was better with the entire search molecule including the complete heme group than with a polyalanine model. Given the relatively small size of the probe, computational speed is not of concern. We select at least 20 runs to increase our chance of finding a solution. The program provides a detailed listing of the progress for each generation of evolution, but for us the ranking of individual runs is most instructive.

Figure 11-21 shows the scores of 20 trial searches for the first and second molecule. The solution landscape for the first molecule is indeed relatively flat with poor discrimination between the trials. The best solution appears, by chance,

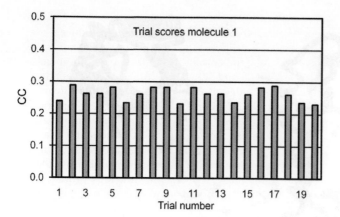
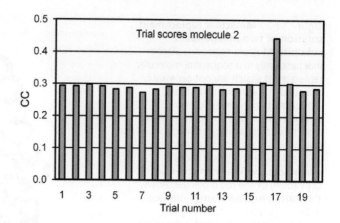

twice (trials 2 and 17, CC = 0.288, R = 0.535), but is only slightly better than the next best solutions. Repeated appearance of best solutions in weakly discriminated solution landscapes is in general a good sign in multi-solution methods, but it does not guarantee that the solution is correct. It is also evident that small changes in the probe molecule such as loop trimming may improve discrimination of trials in the solution landscape. Some programs such as *AMoRe*,[60] *Phaser*,[8] *COMO*,[61] *MOLREP*,[62] or *CNS*[63] provide for automated input of multiple search models, and/or adjust the models during the search process. During the search for the first molecule, the z-translation is held fixed at an arbitrary value, because the origin in polar space group $P3_1$ is floating. Once the first copy is placed, however, the relative position of the second copy must be determined in all three translational directions.

The second round of evolutionary search, with the first best solution kept fixed during the search and included in the structure factor calculation, shows a distinctly different solution behavior. Here the sharply distinct correct solution (CC = 0.444, R = 0.495) appears only once and well above the average solution, but relatively late, in trial 17. The late appearance of the correct solution in the evolutionary search runs emphasizes the general rule that in most multi-solution algorithms (compare the dispersive atom search in the VTA1 S-SAD phasing example in Section 10-8) only a sufficiently large number of trials assures that the correct solution—if it exists—is in fact found. Once the first molecule is placed correctly, the possible locations for the second placement are severely restricted, and the second solution is therefore more distinct.

Assembling the dimer. After a successful evolutionary search, the six placement parameters (θ_1, θ_2, θ_3, t_x, t_y, t_z) for both molecules are simultaneously rigid-body refined, which fine-adjusts the position further. At this point, we would construct a bias minimized map, generally from σ_A-based maximum likelihood coefficients (Section 12-1) or obtained from various other combinations of bias minimization methods (Section 12-3). It depends on chance whether the two molecules of the MR solution are in positions that represent the actual dimer, or have been placed in symmetry related positions that do not reflect a meaningful dimer (Figure 11-22). Inspection with a graphics program and/or computation of the buried surface area will in most cases help decide whether a crystal packing interface or the actual dimer interface is initially present. It is good practice (but not mandatory) to deposit the NCS related copies in an arrangement that represents the proposed oligomer. Although programs could automatically address this problem, automated "biomolecule" generation seems presently not implemented.

Other stochastic search procedures in addition to evolutionary programming have also been explored for MR problems. Similar to evolutionary programming, genetic algorithms[64] have also been implemented, and simulated annealing[65] (explained later in Chapter 12) also provides an efficient multivariate search as explored in the molecular replacement program *SOMoRe*.[66] One of the promising approaches explored in these implementations is a simultaneous, multidimensional search for multiple molecules[67] instead of the sequential approach

Figure 11-21 Solution scores of sequential 6-dimensional molecular replacement search. Both panels show the linear correlation coefficients of 20 independent runs each of a molecular replacement search using the evolutionary MR search program *EPMR*. The solution landscape of the search for the first molecule (top panel) is quite flat, with the correct solution (CC = 0.288) appearing twice in trials 2 and 17. In contrast, the solution for the second unit of the cytochrome dimer is quite sharp and well discriminated from the average solution (CC = 0.444), but it appears late in the trials.

Figure 11-22 Molecular replacement solution of two independent subunits. Multiple molecules placed independently in a sequential molecular replacement search are not necessarily arranged in a conformation that represents their native oligomeric arrangement. In the above example of a bacterial cytochrome c′ dimer, the two originally placed molecules (gray and red) show only weak packing interactions, while a symmetry-related copy (blue) of the red molecule shows the correct, extensive dimer interface with the gray monomer. It is good practice to deposit the molecules in the arrangement that represents the native oligomer. PDB entry 1gqa.[53]

we have used in the cytochrome example. So far it seems that the different implementations of stochastic programs for MR are very similar in convergence radius and performance, with the well-developed and robust evolutionary programming implementation in *EMPR* providing some practical advantage.

Box 11-8 Stochastic molecular replacement searches. Stochastic 6-dimensional molecular replacement searches rely on fast structure factor calculations and highly efficient search algorithms. Evolutionary programming algorithms are a special case of genetic algorithms and rely on competition of their population members for survival against a crystallographic selection criterion. Surviving members regenerate the population around their own parameter space with their offspring. After many generations, the fittest survivor is kept as the best solution. Many runs of evolutionary search may be necessary to find the correct solution, but the methods will eventually find the correct solution, if one exists, given model and data quality. Evolutionary molecular replacement search is implemented in the program *EPMR*.

Rotation–translation methods

Molecular replacement rotation–translation methods work distinctly differently from the multidimensional stochastic search techniques. They break up the search into two distinct phases: a 3-dimensional rotation search, determining the proper orientation of the probe, and subsequent determination of the correct location in the cell of the properly oriented probe in a 3-dimensional translation search. The methods are founded on the Patterson methods developed by Rossmann and Blow,[30] and they all employ fast rotation and translation functions of various sorts, and are also highly automated.

Searches in Patterson vector space

We recall that the Patterson function is an autocorrelation of the electron density, and that it can be calculated from the squared structure factor amplitudes (or intensities) serving as Fourier coefficients for the Patterson map construction. The Patterson space contains interatomic distance vectors between all pairs of molecules, self-vectors between equivalent atoms in symmetry related positions in Harker sections, and cross-vectors between atoms not related by crystallographic symmetry. As a consequence, the Patterson space unit cell is crowded with $N(N-1)$ peaks, which makes direct interpretation of the Patterson maps and thus structure determination of large light-atom-only structures impossible.

Despite the incredible crowding of a Patterson map from a protein molecule with many light atoms, the maps will in parts be characteristic for each molecule. Very large distances will in all likelihood be dominated by *intermolecular* vectors, which are different for each packing situation and thus not useful in orientation searches. In contrast, somewhat shorter *intramolecular* vectors that are characteristic of the relative locations of secondary structure elements to each other will be significant and distinct, while the spatial distribution of the ubiquitous distances smaller than the Cα–Cα distance of ~3.8 Å will be unspecific. In the search we will therefore use those low-resolution data that carry most of the characteristic distances, a common range being ~10–4 Å.

Patterson rotation searches

The distribution of the intramolecular distance vectors depends only on the orientation of the molecule, not on its translational position. This forms the basis of Patterson rotation searches illustrated in Figures 11-23 and 11-24. We can calculate a Patterson map of the search molecule in a sufficiently large box (generally a cube with the center of mass of the molecule placed at the origin or center), so that only intramolecular vectors are contributing to the Patterson map. To determine the orientation of the search molecule, we could rotate this map of intramolecular distance vectors in small increments over all angles, compute at each position its match with the observed Patterson map, and score the solutions by their correlation coefficient. This is the principle of a Patterson cross-rotation search illustrated in Figure 11-23. The procedure also can serve to determine the relative orientation of multiple NCS copies in the same asymmetric unit by rotating the native Patterson map onto itself, providing a Patterson self-rotation search (Figure 11-24).

Rotation functions

The real space formulation for the rotation (overlap) function $\mathfrak{R}(\mathbf{R})$ between Patterson maps or functions $P(\mathbf{r})$ from the target molecule t and the search probe s takes the form

$$\mathfrak{R}(\mathbf{R}) = \frac{1}{V} \int_{r_{min}}^{r_{max}} P_t(\mathbf{r}) \cdot P_s(\mathbf{R}^T\mathbf{r}) d^3\mathbf{r} \qquad (11\text{-}30)$$

Figure 11-23 Patterson cross-rotation search. Panel A shows a search molecule positioned in a *P*1 box that is selected to be large enough to contain only intramolecular vectors. The Patterson map containing the intramolecular distance vectors of the search molecule is shown in panel B. The target molecule in an as-yet-unknown orientation and location in its unit cell (C), and the corresponding vectors are shown in the next map (D). Intermolecular vectors to neighboring molecules are omitted. Note that the intramolecular vector map is independent from the location of the target molecule, but it depends on the orientation and is correspondingly rotated. By successively rotating the search map (blue dots) and calculating the overlap function at each orientation, we can find the correct rotation angle(s), in this case about 45° (panels E and F).

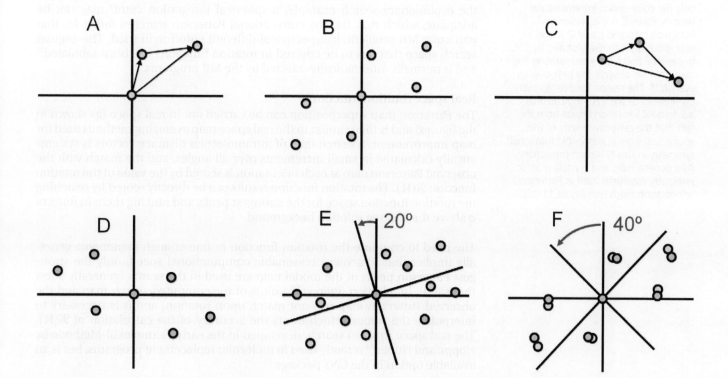

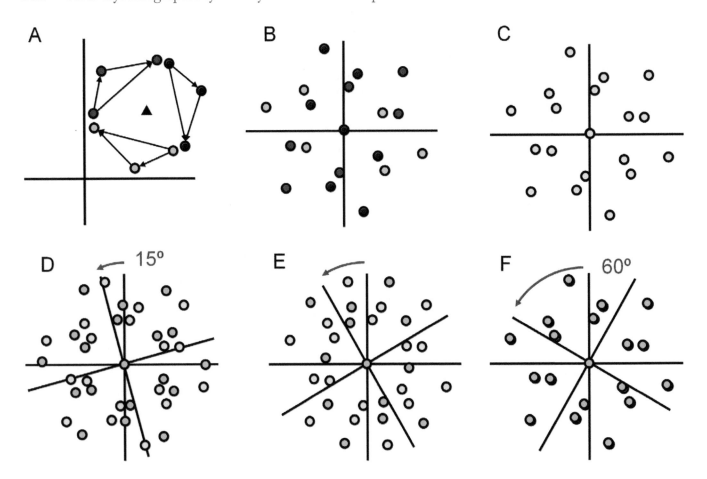

A

B

C

D ← 15°

E

F 60°

Figure 11-24 Patterson self-rotation search. Panel A shows a homo-trimer, where the molecules are related through a proper 3-fold NCS axis. Panel B is the corresponding Patterson map including only the color-coded intramolecular vectors. Panel C is the stationary Patterson map and panel D shows a superposition with itself rotated by 15°. In panels E and F, we keep rotating the map until we observe the best overlap every 60°. The reason for overlap every 60° instead of the 120° indicated by a 3-fold NCS rotation results from the fact that the centrosymmetry of the vector space generates 2-fold rotational symmetry in this Patterson projection. As a general rule, also crystallographic symmetry manifests itself in Patterson self-rotation maps (see Figure 11-16).

where **R** is the real space rotation matrix. The integration volume is selected so that the origin peak, short unspecific distances, as well as the long, mostly intermolecular vectors are excluded. A rule of thumb for the maximum integration radius is about 75% of the molecule diameter. With highly anisotropic molecules, such as the elongated four-helix bundle of cytochrome c′ used in the evolutionary search example, a spherical integration cutoff may not be adequate, which may be why conventional Patterson searches failed for that particular MR problem, irrespective of different cutoff radii tried. The angular search space that has to be covered in rotation functions has been tabulated[68] and is normally automatically selected by the MR programs.

Real space rotation functions

The Patterson map superposition can be carried out in real space (as shown in the figures) and is then similar to the real space map averaging methods used for map improvement. A search map of intramolecular distance vectors is systematically calculated in small increments over all angles, and the match with the observed Patterson map at each orientation is scored by the value of the rotation function $\mathfrak{R}(\mathbf{R})$. The rotation function results can be directly scored by searching the rotation function space for the strongest peaks and ranking them in units of σ above the average solution background.

The need to compute the rotation function in fine enough increments generally implies that to achieve reasonable computational speed, only the strongest Patterson peaks of the model map are used in the search (generally a few thousand). In addition, map grid points of the computed search map and the observed Patterson map will not match upon rotation, and it is necessary to interpolate the values, which limits the accuracy of the calculation of $\mathfrak{R}(\mathbf{R})$. The real space rotation search, developed in the early Faltmolekül-Methode by Hoppe and Huber,[69] is rarely used in molecular replacement programs, but is an available option in the *CNS* package.[70]

The fast rotation function

An elegant way to compute overlap of Patterson functions in reciprocal space has been devised by Rossmann and Blow,[30] and first put into general practice with fast rotation functions by Crowther.[42] We replace the Patterson function $P(\mathbf{r})$ with its Fourier transform

$$P(\mathbf{r}) = \sum_{\mathbf{h}} \frac{F^2(\mathbf{h})}{V} \exp(-2\pi i\mathbf{h}\mathbf{r}) \tag{11-31}$$

and substituting the Fourier summation (11-31) for P_t and P_s into (11-30) we obtain for $\mathfrak{R}(\mathbf{R})$ in reciprocal space

$$\mathfrak{R}(\mathbf{R}) = \sum_{\mathbf{h}} \sum_{\mathbf{k}} \frac{F_t^2(\mathbf{h})}{V_t} \frac{F_s^2(\mathbf{k})}{V_s} G(\mathbf{h} - \mathbf{k}\mathbf{R}^{\mathrm{T}}) \tag{11-32}$$

where G is again the reciprocal space interference function obtained by integration over the selected real space integration volume and plotted in Figure 11-15. As usual, the nested summation over the indices $\mathbf{h}$ of the target cell and $\mathbf{k}$ of the search cell, combined with the additional multiplication with $\mathbf{R}$ in the inner G-term, is notoriously slow. The remedy is to express the G-function in the form of spherical harmonic functions $Y_{l,m}$ and spherical Bessel functions J_l. Parameterized in Euler angles, the fast rotation function decouples the rotation from the index summation, and also keeps the reciprocal index contributions from target and search cell apart. In addition, the Crowther formulation of the fast rotation function allows the overlap function to be rapidly computed for each given angle value by a single 2-dimensional FFT.[15]

There are a number of clever implementations and extensions to the classical Crowther and Blow fast rotation function. One is to select the reference frame for the search probe so that the reference axis $\mathbf{Z}$ is parallel to the highest n-fold symmetry axis of the crystal cell, which limits the search around θ_1 to $0 \leq \theta_1 < 360/n$. The fast rotation functions implemented[71] in Jorge Navaza's AMoRe package[60] provide additional enhancements. Replacement of the classical Fourier–Bessel expansions with a numerical integration delivers higher numerical accuracy and the reduced rotation matrices are computed with a more stable recurrence relation. The net effect is that more of the higher term spherical harmonics can be included, which in turn results in a more accurate computation of the Patterson functions, particularly over large integration volumes.

Direct rotation function

An alternative to the fast rotation functions and real space searches, which all incorporate approximations in some form to achieve computational speed, is the option to directly compute the correlation between the stationary and rotated Patterson function as a direct rotation function or Patterson correlation function. The linear correlation coefficient CC_E between the squared normalized structure factor amplitudes (E^2) of the search model (E_{calc}) and the target (E_{obs})

$$CC_E = \frac{\sum_{\mathbf{h}} \left| E_{obs}^2 - \overline{E_{obs}^2} \right| \cdot \left| E_{calc}^2 - \overline{E_{calc}^2} \right|}{\left(\sum_{\mathbf{h}} \left| E_{obs}^2 - \overline{E_{obs}^2} \right|^2 \cdot \sum_{\mathbf{h}} \left| E_{calc}^2 - \overline{E_{calc}^2} \right|^2 \right)^{1/2}} \tag{11-33}$$

is a direct measure for the phase accuracy of a partial atomic model,[72] which is the situation in molecular replacement.

The search model is placed in a unit cell of the same dimensions as the target cell, and for each rotation of the model (not the map), the squared normalized structure factors are computed and correlated against the E^2 values for the target. The correlation uses all intra-molecular Patterson vectors of the search model, and no limitation of the integration radius is applied. However, in the Patterson correlation (PC) function the correlated values for E_{calc} are computed from the model and not from the Patterson vectors, and therefore E_{calc}^2 is a function of all distances, both intra- and inter-molecular. The rotational Patterson correlation search[70] as implemented in the *CNS* package[63] is therefore not always more accurate than other rotation functions,[13] but provides an alternate target function at the cost of about 10-fold slower computation. In *AMoRe*[60] the Patterson correlation is implemented differently; it computes structure factors from the Patterson function by transforming to spherical harmonics so that a radius cutoff can be applied to exclude the intermolecular vectors and computes correlation coefficients based on these modified structure factors.

Locked rotation functions

The more copies of a molecule we have to orient in a rotation search, the less each monomer will contribute to the total Patterson function, and solutions of the rotation functions will become increasingly weak and noisy. However, in some situations, we may have prior knowledge of the NCS operator relating the monomers. If the NCS operator is a proper one, the Patterson self-rotation function evaluated in spherical angles will show distinct peaks in sections of κ equal to the multiplicity of the NCS operation (as exemplified in Figure 11-12). In this case, we can apply to our advantage the fact that the NCS operator(s) relates two search probes to each other in a fixed relation, which imposes a powerful constraint on our search. This is particularly useful when multiple proper NCS operators intersect and form a closed group. Tetramers often form as dimers of dimers, which exhibit 222 point group symmetry, or icosahedral viruses exhibit 532 local point group symmetry. Another example is the decamer with 52 symmetry shown in Figure 11-12. The predetermined orientation of the NCS axes in these cases limits the search space and increases the signal-to-noise ratio of the solution. This is exploited in locked rotation functions. With $\mathbf{S}_n$ denoting one of the N proper NCS operations relating the molecules of an N-mer, any orientation $\mathbf{S}_n\mathbf{R}$ must also be a correct one. The locked rotation function $\mathfrak{R}_L(\mathbf{R})$ thus substitutes the averaged values of all orientations into the general rotation function (11-30) and becomes

$$\mathfrak{R}_L(\mathbf{R}) = \frac{1}{V} \int_{r_{min}}^{r_{max}} \left(\sum_{n=1}^{N} P_t(\mathbf{S}_n\mathbf{r}) / N \right) \cdot P_s(\mathbf{R}^T\mathbf{r}) d^3\mathbf{r} \qquad (11\text{-}34)$$

Both locked self-rotation functions and locked cross-rotation functions have been implemented, largely based on the work of Rossmann and Tong.[73] While the locked self-rotation function is mostly used in virus crystallography to determine the orientation of the virus particles (often located on a special position with related point group symmetry), the locked cross-rotation function is useful for the general molecule replacement problem. Instead of looking sequentially for say four copies of a 222 tetramer (with each monomer containing only 25% of the scattering mass), we can search for the orientation of the entire assembly in the cross rotation, which increases the sharpness of the rotation solution. In analogy, the subsequent translations can also be conducted with the complete assembly. This is done with locked translation functions described below.

Box 11-9 Patterson rotation searches. Patterson rotation searches are based on the overlap of Patterson maps. The orientation of the molecules is determined by the match of intramolecular Patterson vectors, calculated with a variety of direct or fast rotation functions. The relative orientation of multiple copies related by proper NCS in the asymmetric unit can be determined by rotating the native Patterson map onto itself. The resulting plot of the Patterson rotation function presented in a stereographic projection is called the self-rotation map. The overlap between search probe and target structure is determined from cross-rotation maps of the native Patterson function and the map of the search model placed in a large search box to avoid intermolecular vectors. If multiple copies in the subunit related by proper NCS are present, locked rotation functions are used. Multiple solutions represented by the highest peaks in the cross-rotation function are then selected for subsequent translation searches.

Patterson translation functions

Once the relative orientation of the search probe is determined, we must find its actual location in the unit cell by a translational Patterson search. The search can again be envisioned in real space by the overlap of Patterson functions, as Figure 11-25 illustrates. In this case, the *intermolecular* Patterson cross-vectors between the correctly oriented molecules in the unit cell are relevant.

Translational search space

The space that must be covered in a translation search depends on whether we are searching for the first (or only) copy in a unit cell, or for multiple copies. In the first case, we need to search only the Cheshire cell (see also Section 10.3)

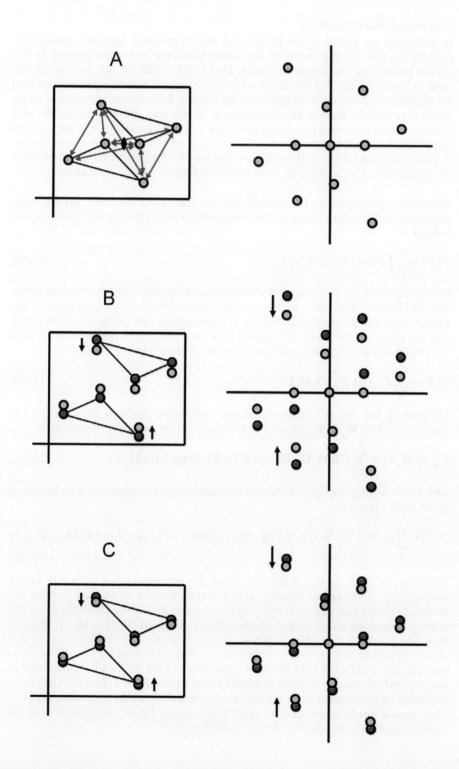

Figure 11-25 Patterson translation search. The target structure (green) with two molecules in a cell related by 2-fold crystallographic symmetry and its intermolecular cross-vectors and their corresponding Patterson map is shown in the top panels (A). If the molecules are properly oriented but not in the right location (panels B), the Patterson map will show the intermolecular vectors on translated positions. The displacement between the wrong and correct peaks is twice the displacement distance in the unit cell. If we translate the search probes (red) in the correct direction, the Patterson vectors will start to overlap until a peak is observed in the translation function when they match (C).

which in general is the minimum cell that needs to be covered to obtain unique search solutions.[74] It is obtained through normalization[75] of the space group by considering the possible multiple origins in the unit cell. As a consequence, the higher the symmetry of the space group, the larger the Cheshire cell that must be searched. Inversely, if no symmetry element is present as in $P1$, the Cheshire cell is a point and therefore the origin in $P1$ is arbitrary—we need no translation search and can place the molecule anywhere in the cell. A similar consideration holds for polar space groups, where the translation of the first search probe is necessary only in two dimensions; along the polar axis the translation is arbitrary, because no perpendicular symmetry elements fix the floating origin. Once the first molecule is fixed, however, a translational search for the remaining molecules must cover the entire unit cell unless the cell is Bravais-centered.

Fast translation functions

In principle we could again determine the translation function directly by computing the overlap between the intermolecular Patterson vectors of the search probe and the target structure. This looks trivial for our 2-dimensional case in Figure 11-25, but becomes relatively involved in three dimensions and for higher symmetry cells. In addition, we have to find the best overlap of intermolecular vectors against the background of the also present intramolecular vectors, which increases noise and the chance of wrong solutions. Crowther and Blow[43] have therefore developed a fast translation function (also known as the T_2 function) that subtracts the self-vectors from the Patterson map, and which is implemented in some form in most molecular replacement programs.

In analogy to the rotation function $\Re(\mathbf{R})$, the translation function $\Im(\mathbf{T})$ is again expressed in Patterson space as the convolution integral over the search space volume V:

$$\Im(\mathbf{T}) = \frac{1}{V}\int_V P_t(\mathbf{r}) \cdot P_s(\mathbf{r}-\mathbf{T})d^3\mathbf{r} \tag{11-35}$$

We can again apply the convolution theorem and replace the Patterson function with its Fourier transform, and the convolution integral becomes a discrete summation over the product of the squared structure factors of the oriented probe and the observed data. The translation function $\Im(\mathbf{T})$ expressed as a function of the positional vector $\mathbf{T}$ of the properly oriented search probe becomes

$$\Im(\mathbf{T}) = \sum_{\mathbf{h}} F_{obs}^2(\mathbf{h}) \cdot F_{calc}^2(\mathbf{h},\mathbf{R},\mathbf{T}) \tag{11-36}$$

Accounting for the G symmetry related molecules through crystallographic transformations $\mathbf{W}_g(\mathbf{M},\mathbf{t})$, the structure factor can be generally expressed as

$$\mathbf{F}_{calc}(\mathbf{h},\mathbf{R},\mathbf{T}) = \sum_{g=1}^{G} f(\mathbf{h}\mathbf{M}_g\mathbf{O}^{-1}\mathbf{R}\mathbf{O})\exp(2\pi i\mathbf{h}\mathbf{t}_g)\exp(2\pi i\mathbf{h}\mathbf{M}_g\mathbf{T}) \tag{11-37}$$

with $\mathbf{O}$ the orthogonalization matrix. Substituting this expression into Equation 11-36 finally leads to

$$\Im(\mathbf{T}) = \sum_{\mathbf{h}} m_{\mathbf{h}}\Delta F_{obs}^2(\mathbf{h})\sum_{g,g'} f(\mathbf{h}_g)^* f(\mathbf{h}_{g'})\exp\left(-2\pi i\mathbf{h}(\mathbf{t}_g - \mathbf{t}_{g'})\right)\exp\left(-2\pi i\mathbf{h}\left(\mathbf{M}_g - \mathbf{M}_{g'}\right)\mathbf{T}\right) \tag{11-38}$$

Here the $\Delta F_{obs}^2(\mathbf{h})$ is the difference from the corresponding structure factor average (i.e., the highest peaks), $f(\mathbf{h}_g)$ stands for the rotational part of the structure factor expression (11-37), and m accounts for the multiplicity of the corresponding structure factor. The search space, given by $(\mathbf{M}_g - \mathbf{M}_{g'})\mathbf{T}$, extends over the Cheshire cell. The function is real because the conjugate complex structure factors are included in the summation, indicated by f^* and the summation over g'. The full derivation of Expression 11-38 for $\Im(\mathbf{T})$ is presented in the papers by Navaza,[60] and the original Crowther and Blow T_2 function has been extended by Driessen et al. to more than one molecule in the asymmetric unit.[76] This translation function can be rapidly computed by FFT-based calculations without explicitly computing the structure factors.[77]

Phased and locked translation functions, overlap penalties

Expression 11-38 for the translation function readily allows the introduction of the structure factor contributions from already located subunits, which leads to a phased translation function $PT(\mathbf{T})$. The option to introduce phase information readily as an additional Fourier term into (11-38) is not limited to multiple search models; it can also come from externally supplied phases. As the translation function is in general the more difficult part of a MR search, even weak phases from a poor heavy atom solution that itself did not generate a traceable map may suffice to improve the translation function. Similar to the case of locked rotation functions, an entire NCS assembly with its members in a fixed relation can be used in the translation search by locked translation functions,[73] thereby increasing the signal-to-noise ratio of the solution.

An attempt to improve the translation function and its score by correlation coefficients by inclusion of a penalty for collisions between incorrectly placed molecules, through a steric overlap penalty function $O(\mathbf{T})$, was introduced by Harada and Lifchitz.[78] The total translation search function $T(\mathbf{T})$ is then provided by the quotient of the translation agreement function $TO(\mathbf{T})$ and the overlap penalty function:

$$T(\mathbf{T}) = \frac{TO(\mathbf{T})}{O(\mathbf{T})} \tag{11-39}$$

Given that search probes may have protruding loops or termini that are not present in the target structure, a certain amount of overlap may well be acceptable and the function must be properly weighted. The Harada function has been implemented in $TTFC$ of the CCP4 package, and it can be tried if the standard T_2-type functions fail.[79] More advanced packing functions that consider the expected deviations in regions of limited agreement between 3-dimensionally aligned search model ensembles are implemented in $Phaser$.

Scoring multiple rotation solutions by translation search

Common scoring functions for the translation function are the standard linear correlation coefficient (on I or F) and the R-value for F for the corresponding translation functions:

$$R_F = \frac{\sum_{\mathbf{h}} \left| F_{obs}(\mathbf{h}) - F_{calc}(\mathbf{h}, \mathbf{R}, \mathbf{T}) \right|}{\sum_{\mathbf{h}} F_{obs}(\mathbf{h})} \tag{11-40}$$

and

$$CC_I = \frac{\sum_{\mathbf{h}} \left| I_{obs}(\mathbf{h}) - \overline{I_{obs}(\mathbf{h})} \right| \cdot \left| I_{calc}(\mathbf{h}, \mathbf{R}, \mathbf{T}) - \overline{I_{calc}(\mathbf{h}, \mathbf{R}, \mathbf{T})} \right|}{\left(\sum_{\mathbf{h}} \left| I_{obs}(\mathbf{h}) - \overline{I_{obs}(\mathbf{h})} \right|^2 \cdot \sum_{\mathbf{h}} \left| I_{calc}c(\mathbf{h}, \mathbf{R}, \mathbf{T}) - \overline{I_{calc}(\mathbf{h}, \mathbf{R}, \mathbf{T})} \right|^2 \right)^{1/2}} \tag{11-41}$$

where the bar indicates the corresponding mean values.

The translation function provides an additional opportunity to score a set of possible rotation search solutions by evaluating them against the translation function. As the translation function defining the location of the molecule is the ultimate and final score in a MR search, a correct rotation solution should also score high in the translation functions. This *multi-solution approach* is taken in *AMoRe*, and is quite efficient in discriminating the correct rotation solution from a set of similar solutions.[60]

Improvement of initial solution by rigid body refinement

Once a MR solution has been found, it can be further optimized by rigid body refinement, in which the orientation and translation of the initially located molecule(s) as a whole is improved. The rigid body refinement can be applied to segments of a molecule which are presumed to be independent (such as groups of secondary structure) or to the different members in a multimeric assembly. Rigid body refinement is generally implemented in MR programs using the fast interpolation method for structure factors[49] described in Sidebar 11-4. The sum of residuals squared (SRS) between the calculated and observed structure factors is minimized (by least squares or other suitable algorithmic engines) as a function of the positional parameters defining $\mathbf{R}$ and $\mathbf{T}$, the overall B-factor, and a scale factor k:

$$SRS(\mathbf{R},\mathbf{T},B,k) = \sum_{\mathbf{h}} \left(F_{obs}(\mathbf{h}) \exp(-BS^2/4) - k \cdot F_{calc}(\mathbf{h},\mathbf{R},\mathbf{T}) \right)^2 \tag{11-42}$$

The rigid body refinement can affect the ranking of the initial solutions, particularly when they are weak and similar. This leads us to a related approach to improve the translation search, the Patterson correlation refinement.

Improved translation searches by PC refinement of orientations

The initial set of rotation searches can be further refined by a procedure called Patterson correlation refinement. As the name implies, the direct space Patterson correlation (PC) function serving as a score for the rotation searches as discussed above can be used as an optimization target.[80] PC refinement is implemented in the *CNS* package.[63] As in rigid body refinement, segments or groups of secondary structure are independently reoriented, thereby improving the quality and providing better discrimination of the possible orientation search solutions. This in turn increases the chance of obtaining a correct translation solution. The difference from rigid body refinement—which is applied to the final, rotated and translated solution—is that the PC refinement does not require symmetry relations between the molecules, and can therefore be applied to a search probe when only the orientation is known. Brunger has shown that initially quite poorly oriented molecules can be successfully reoriented using PC refinement.[81]

Box 11-10 Patterson translation searches. The location of the oriented search model in the target unit cell is determined by the match of intermolecular Patterson vectors, calculated with a variety of fast translation functions. The translation functions are very sensitive to small orientation errors of the search probe, therefore multiple top solutions of the rotation; search are evaluated in the translation search. During translation searches, the orientation of the molecules can be improved through Patterson correlation refinement, and the translation functions can also be used to evaluate a set of rotation solutions. If multiple copies in the subunit related by proper NCS are present, locked rotation functions are used. Packing functions introducing an overlap penalty also can improve the discrimination of solution translation searches. In successive multi-copy translation searches or when additional phase information is available, phased translation functions are used. The best solutions from the translation search are then optimized through rigid body refinement and re-ranked.

Probe selection and search strategies

Although molecular replacement is often straightforward in cases such as mutants or crystals of ligand-bound structures, where conformational changes are generally limited and local, MR can become quite challenging and difficult.[82] This is particularly true when low sequence identity approaching the twilight zone of ~20–30% between the search probe and target molecule is likely to imply a significant average r.m.s.d. (exceeding 1.5 Å) between their atomic backbone Cα atom coordinates.[55] The relation between sequence identity and structural similarity expressed as Cα-coordinate r.m.s.d. between probe and target has been estimated by Chothia and Lesk.[56] Their relation and its consequences which provide guidelines for trimming search models are described in Sidebar 11-5. As a general caveat, despite their undisputed value, relying solely on automated programs and their brute-force power while neglecting to use all available information to one's advantage may only waste time and computational resources instead of delivering a good MR solution.

NMR and homology models as MR search models

The large coordinate deviation between the models within an ensemble of models provided by NMR is also the main reason why NMR models with even 100% sequence identity have given mixed results for molecular replacement searches.[57] Nonetheless, in the absence of crystallographic search models, the carefully trimmed core regions of an averaged NMR model may provide an avenue to a successful MR structure solution. The ensembling procedure implemented in *Phaser* allows the computation of average structure factors for an ensemble of multiple, aligned models allowing one to make use of NMR models.

One of the objectives for the NIH Protein Structure Initiative[83] (PSI) was to populate the fold space so that for each newly discovered protein sequence an experimental model would be available close enough in sequence to allow homology modeling.[84] On the other hand, a potential use of computational models, particularly fold recognition (or threading) models, could be as search probes in molecular replacement. This of course requires that the model is close enough to the target structure so that it complies with the ~1.5 Å rule of thumb for the Cα-deviation.[57] The prospect presently is modest, as achievement of an average backbone deviation of less than 2 Å is still a fairly steep challenge[85] in computational modeling. Good theoretical models are generally obtained from structurally close experimental structure templates, which in that case might as well serve directly as MR probes. Nonetheless, consensus models or fragments with low coordinate deviation might occasionally prove useful in MR searches. The CCP4 suite provides a graphical interface to A. Sali's comparative structure modeling program *Modeller*[86] and includes some post-processing options to remove questionable parts. Another option for model improvement that occasionally benefits the MR solution is replacement of the original side chains to the correct target sequence with a modeling program like *SCRWL*.[87] A variety of options for trimming models and changing residues according to a provided sequence alignment also exist in CCP4 through the *CHAINSAW* module, which allows the testing of different truncation protocols for multi-model searches.[54]

Multiple molecules in the asymmetric unit

Another challenge for molecular replacement searches is the presence of multiple molecules in the asymmetric unit of the target crystal, either of the same kind in the form of a homo-oligomer, or in a multi-subunit complex of different proteins. The same situation is present when a large, flexibly hinged multi-domain molecule is broken up into its subunits and each of the domains must be located separately. In all these cases, the individual subunits contain only a correspondingly low fraction of the total scattering mass, which makes their detection in the large signal background of the remaining molecules that all contribute to Patterson functions or structure factors very difficult. The more individual copies are present, the more difficult the search becomes. However, once the first molecules are correctly placed, the situation gradually improves. Only in the case of proper closed group NCS symmetry do the locked search functions discussed above provide a means to exploit the multi-copy symmetry to advantage, although at least the rotational part of the NCS operator(s) must be known.

As a consequence of the small relative scattering mass of subunits, the solution landscape in sequential multi-copy searches is even less discriminating than in single-copy searches, and many equally possible solutions will appear, some even with differing rankings depending on the selected scoring function. A logical consequence is therefore to test as many possible solutions from the rotation search for as many different probes as reasonable and evaluate them all in the translation part of the search. Two such multi-solution approaches, combined with simultaneous positional improvement of orientation with Patterson correlations and rigid body refinement for translation searches, have been discussed above together with the corresponding search functions.

A strategy similar to locked translation functions is pursued in *MOLREP*,[62] where a multi-copy search model is constructed with the help of a special translation function from properly oriented monomers.[88] The entire pre-assembled multi-copy search model is then finally placed with a conventional fast translation function. The benefit is that the method does not require proper NCS to relate the search copies, and complexes of different molecules can also be assembled using this method.

A final note regarding the chance of recovering missing parts of correctly placed but incomplete models: Recovery of missing parts generally requires close enough starting phases (i.e., a largely correct partial model) and excellent data of sufficient resolution (at least 2.5 Å or much better) so that protocols such as *ARP/wARP*[89, 90] or its derivative *Shake&wARP*[40] can recover the electron density. Repeated cycling between real space rebuilding and reciprocal space refinement will be necessary to succeed and obtain gradual improvement of the maps. Even a maximum likelihood map based on the initial partial model will not readily recover the electron density of all the missing parts, particularly if the initial model is both incomplete and incorrect and therefore the phase error is high. Section 12.1 will provide more explanations why this is the case.

Automated multi-solution and multi-model searches

Automated multi-solution and multi-model searches are implemented in numerous molecular replacement programs and packages such as *MOLREP*[62] (which can be also used for placing a molecule in experimental electron density or electron microscopy envelopes), *AMoRe*, *Phaser*, *COMO*,[61] or *CNS*. The general strategy[82] is to take a group of top rotation solutions, either from the same model or additionally from multiple search probes, and process them all, subject to various intermediate refinement steps, in the translation search. The translation search itself is an excellent scoring function for the quality of the rotation search and a group of rotation solutions located around a promising translation maximum can all be explored in an attempt to further improve the contrast of the solutions. The specific strategies vary from program to program, but the general outline of a multi-model, multi-solution molecular replacement search is shown in Figure 11-26. The searches for multiple models can be conducted sequentially or in parallel. With computational power continuously and exponentially increasing (Moore's law),[91] combined multi-model, multi-solution searches are quite feasible, and have been implemented in most molecular replacement packages. *AMoRe* for example employs through the *CCP4i* graphical interface[92] a model database that is processed automatically during the execution of *AMoRe*.[60]

A wrapper program for molecular replacement that generates a number of different starting models and strategies is *MrBUMP*.[93, 94] Available as a part of CCP4, it interfaces to remote programs such as the multiple sequence alignment program

---

Box 11-11 Model selection and search strategies. Search models must have sufficient structural similarity to the target molecule. The Cα-coordinate deviation between model and target molecule can be estimated based on sequence identity by the Chothia–Lesk formula, and generally should not exceed ~1.5–2 Å. Regions with suspected flexibility and high coordinate deviation must be trimmed from the model. As a general rule, correctness is more important than completeness. The more subunits that must be located, the less total scattering mass is contained in the probe and the molecular replacement search becomes progressively more difficult. In the case of proper NCS, locked and phased search functions may improve the search. As many rotation solutions and differently modified models as practical should be evaluated in the translation searches to increase the chance for success in difficult cases. NMR models only occasionally, and homology models rarely, provide effective probes for MR searches. The method of assembling a probe from multiple models implemented in maximum likelihood MR searches may also improve the chances of success.

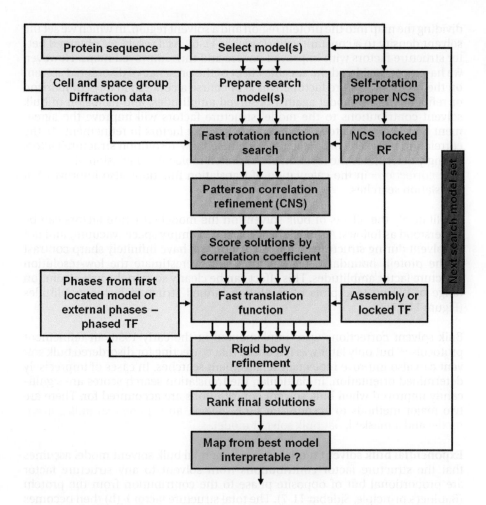

Figure 11-26 General multi-model and multi-solution MR search strategy. Owing to the poor contrast in the possible solutions, multiple models are subjected to the rotation search. The best solutions for each model are refined and multiple translational searches are conducted. The best translation solutions are rigid body refined and finally ranked. If more than one copy is in the asymmetric unit of the target crystal, the translation search is repeated until all copies are placed and again optimized in segmented rigid body refinement. Multimer assembling (*MOLREP*) and locked search functions allow exploitation of NCS. External phase information can also be used to increase the contrast in phased translation searches.

ClustalW,[95] and uses the alignment to trim the model and replace residues with *CHAINSAW,* similarly to the manual procedure we employed to prepare our cytochrome c′ search probe for molecular replacement. The search models are then passed on to *MOLREP,* and an ensemble model can be supplied to the maximum likelihood MR program *Phaser.* The external search options also allow connection to the structural domain database SCOP[96] and searching the quaternary structure server PQS[97] or its sucessor PISA for multimeric assemblies.

Another quite sophisticated program package within *CCP4i* for fully automated molecular replacement is *BALBES*[98] which automatically searches databases and manages the workflow. It uses *MOLREP* as the molecular replacement engine, *REFMAC* for refinement, and it will be linked with *ARP/wARP* for model building as well as with existing automatic experimental phasing procedures such as *CRANK*[99] and *Auto-Rickshaw,*[100] effectively providing a complete platform for automated structure solution.

A new *de novo* phasing method. A recent multi-solution hybrid phasing method[113] based on phase extension from multi-solution molecular fragment searches using the maximum likelihood program *Phaser* followed by density modification and fragment rebuilding with *SHELXE* is described in Sidebar 10-10. As it does not start from a specific probe molecule and uses only common secondary structure fragments and native data, it is in fact a *de novo* phasing method working at resolutions of 2 Å or better.

Bulk solvent corrections for model structure factors

At this point in our discourse, it is necessary to digress and address an issue we have so far ignored. In solvent flattening and solvent flipping techniques for phase improvement we have exploited the fact that the intermolecular space between proteins is filled with disordered solvent. The procedure consisted of dividing the map into the protein region and a solvent region, in which we set the solvent density to a constant value (Sidebar 11-6). As a result, we obtained better structure factors with improved phases and thus improved maps. However, we have not considered the specific effect of the presence of disordered solvent on the protein structure factors. This is of course a concern in any comparison or refinement of a model against observed amplitudes, and inclusion of bulk solvent contributions to the model structure factors will improve the agreement between calculated and observed structure factors in refinement. As the correlation between computed and calculated low-resolution structure factors strongly determines the location of the search molecules, inclusion of bulk solvent corrections in the calculation of translation functions also improves MR translation searches.

Qualitatively the effects of bulk solvent on the model structure factors can be understood as follows: A model that is placed in empty space (vacuum) and not in solvent during structure factor calculation will have infinitely sharp contrast at the protein boundary and therefore will overestimate the low-resolution structure factor amplitudes. This effect can be clearly seen in the low-resolution range of Wilson-type plots of resolution-binned structure factor amplitudes (Figure 11-27).

Bulk solvent corrections were implemented in the early 1990s in refinement protocols,[102] but only later was it realized that correcting for disordered bulk solvent can also improve molecular replacement searches. In cases of imprecisely determined orientation, in particular, the translation search scores are significantly improved when bulk solvent contributions are accounted for. There are two major methods to account for bulk solvent, an exponential bulk solvent model and a masked, flat bulk solvent model.

Exponential bulk solvent model. The exponential bulk solvent model assumes that the structure factor contributions from solvent to any structure factor are proportional but of opposite phase to the contribution from the protein (Babinet's principle, Sidebar 11-7). The total structure factor $\mathbf{F}_t(\mathbf{h})$ then becomes with $|\mathbf{S}| = S = 2\sin\theta/\lambda$

$$\mathbf{F}_t(\mathbf{h}) = \mathbf{F}_p(\mathbf{h})(1 - k_{sol}\exp(-B_{sol}S^2/4)) \tag{11-43}$$

Babinet's principle (Sidebar 11-7) holds only for uniformly scattering objects, and therefore the exponential model is in theory valid only for correction of very low-resolution reflections (> 15 Å).[103] The constant k_{sol} is the ratio of the mean

Figure 11-27 Effect of bulk solvent on structure factor amplitudes. Calculated structure factors of a model placed in vacuum (instead of solvent) overestimate the structure factors at low resolution (red graph). Application of bulk solvent correction greatly improves the match between the observed (green graph) and calculated low-resolution structure factors (blue graph). Model data were computed with *REFMAC*[101] with and without masked bulk solvent correction from PDB entry 2j9n.

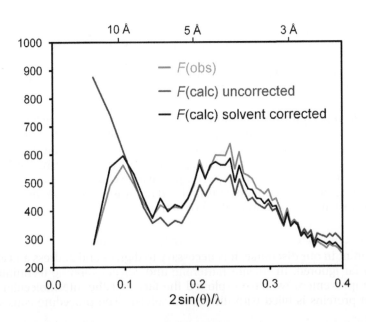

Sidebar 11-6 Estimating the electron density of solvent and protein.

Water. The molecular weight of water (H_2O), computes as 18.02 g mol^{-1}; 1 g of water thus contains 1/18.02 mol (0.0555 mol g^{-1}). Multiplying by the specific density of 1.0 g cm^{-3}, we find the molar volume of 1/18.02 mol of water per cm^3. Using Avogadro's number and the conversion from cm to Å (1 cm = 10^8 Å), we obtain the number of molecules per Å^3:

$$\frac{1}{18.02}\left[\frac{mol}{cm^3}\right] \cdot \frac{6.022 \cdot 10^{23}}{1}\left[\frac{molecule}{mol}\right] \cdot \frac{1}{10^{24}}\left[\frac{cm^3}{Å^3}\right] = 0.0334\left[\frac{molecule}{Å^3}\right]$$

Knowing that water has 10 e$^-$/molecule, we can easily finish the calculation:

$$0.0334\left[\frac{molecule}{Å^3}\right] \cdot 10\left[\frac{electrons}{molecule}\right] = 0.334\left[\frac{electrons}{Å^3}\right]$$

Note that the result of 0.334 e$^-$/Å^3 was derived for pure water. In a first approximation, for salt solutions with

higher density the value increases correspondingly up to around 0.40 e$^-$/Å^3. A compromise value commonly used by crystallographic programs is 0.375 e$^-$/Å^3.

Protein. A good approximation for the number of electrons in an average protein is 0.54 e$^-$/Dalton, or 0.54 e$^-$ mol g^{-1} (calculated by dividing the electron count from F_{000} by the molecular weight). Using the average specific density of 1.35 g cm^{-3} for a protein, we obtain for the average protein electron density 0.44 e$^-$/Å^3:

$$0.54\left[\frac{e^- mol}{g}\right] \cdot 1.35\left[\frac{g}{cm^3}\right] \cdot \frac{6.022 \cdot 10^{23}}{1}\left[\frac{molecule}{mol}\right] \cdot \frac{1}{10^{24}}\left[\frac{cm^3}{Å^3}\right] = 0.439\left[\frac{e^-}{Å^3}\right]$$

The ratio of the mean electron densities of water and protein used in exponential bulk solvent models, k_{sol}, is therefore 0.334/0.439 = 0.76.

---

electron density of solvent/protein (~0.76), but the exact value depends on the constitution of the solvent (Sidebar 11-6). The attenuation of $F_p(\mathbf{h})$ disappears rapidly at higher resolution because of the negative exponential B-factor term with B_{sol} of ~200 Å^2.

Flat bulk solvent model.

Flat bulk solvent model. A more realistic approximation for bulk solvent is the flat or mask solvent model, which uses a solvent mask distanced from the van der Waals surface of the protein by a gap of ~1.4 Å. The solvent mask is then extended back into this gap using a probe radius of ~0.8 Å, and the bulk solvent region is filled with continuous and uniform electron density. The complex structure factor contributions of protein and solvent are added to give the corrected complex model structure factors $\mathbf{F}_m$:

$$\mathbf{F}_m(\mathbf{h}) = \mathbf{F}_p(\mathbf{h}) + k_{sol}\mathbf{F}_{sol}(\mathbf{h})\exp(-B_{sol}S^2/4). \tag{11-44}$$

The parameters k_{sol} and B_{sol} are obtained by minimizing the residual between observed data $\mathbf{F}_o(\mathbf{h})$ and the solvent-corrected search model structure factors $\mathbf{F}_m(\mathbf{h})$:

$$SRS = \sum_{\mathbf{h}}\left|F_o(\mathbf{h}) - |\mathbf{F}_p(\mathbf{h}) + \mathbf{F}_{sol}(\mathbf{h}, k_{sol}, B_{sol})|\right|^2 \tag{11-45}$$

The parameters in the flat solvent model on average refine to values of $k_{sol} \sim 0.4$ and $B_{sol} \sim 45$ Å^2. This form of bulk solvent correction[104] is implemented in the refinement programs *CNS* and *REFMAC*, and has been adapted in a fast algorithm[105] to improve the translation target functions in *AMoRe* and *CNS*. In the maximum likelihood molecular replacement program *Phaser*, an exponential bulk solvent correction is incorporated in the likelihood functions through the sigma-A variance term (σ_A) together with the Chothia and Lesk estimate σ_r (Sidebar 11-5) for the model coordinate error:

$$\sigma_A = \left(\mathbf{F}_p\left[1 + k_{sol}\exp(-B_{sol}S^2/4)\right]\right)^{\frac{1}{2}}\exp\frac{2\pi}{3}S^2\sigma_r^2. \tag{11-46}$$

The improved estimate for σ_A given above is also implemented in the refinement program *REFMAC*.[101] That the inclusion of bulk solvent correction actually improves the fit between model and observed data and not just induces overfitting can be seen from the lower number for the cross-validation R-free value (Figure 11-28). In electron density maps, connectivity is improved and streaking is reduced when proper solvent corrections are applied.[102]

Sidebar 11-7 Babinet's principle of complementarity.

Babinet's principle states in simple words that a hole of a certain size has the same diffraction pattern as an object of the same size, but with opposite (ϕ + 180°) phases. In terms of crystallographic Fourier transforms, a flat electron density, where $\rho(\mathbf{x})$ is constant in the entire cell, has a transform $\mathbf{F}(\mathbf{h})$ that is zero except for $F(000)$. If we divide the cell into two parts, one representing the disordered solvent *sol* and the other one the protein *p*, then $\mathbf{F}_{sol}(\mathbf{h}) = -\mathbf{F}_p(\mathbf{h})$ because the contributions from the two regions must cancel out. As we can compute $\mathbf{F}_p(\mathbf{h})$ from the experimental map or model, we can estimate the solvent contribution as $-\mathbf{F}_p(\mathbf{h})$. By placing ordered atoms with very high B-factors (~200 Å^2) into the protein region and taking the opposite phase, we obtain an approximation for the solvent scattering. This is the basic exponential bulk solvent model expressed in Equation 11-43.

Figure 11-28 Lower R-free through bulk solvent correction. The decrease of R-free (the cross-validation R-factor) upon inclusion of bulk solvent correction demonstrates that bulk solvent correction does in fact provide a good description of low resolution data without over-fitting. R-free cross-validation values were computed with REFMAC[101] with and without masked bulk solvent correction from PDB entry 2j9n.

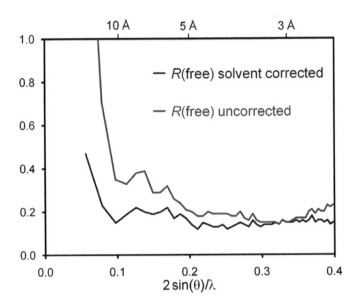

Sidebar 11-8 **Experimental data without solvent contribution?** A critical comment[106] regarding one of the crystal structures of a multi-unit complex of complement component C3, whose activation in the complement pathway to C3b is a key step, was published recently in the journal *Nature*. In addition to implausible data and refinement statistics as well as highly unusual packing characteristics of the particular model, the deposited data exhibit a highly unusual quality: It seems that there is no solvent contribution present in the observed data, that is, the observed high resolution structure factors do not show the suppression by bulk solvent contributions as expected in experimental data (evident in Figure 11-27). As a consequence, the R-values computed between data and model without solvent correction match and clearly lack the distinct increase in R-values that would be expected if the experimental data actually contained solvent contributions (compare Figure 11-28). Reprocessing the original diffraction image frames (which have not been made available) would help to clarify this contentious situation. Following a thorough investigation, it has been determined that twelve structures reported by the same individual are based on fabricated data, and removal of these entries from the PDB has been recommended.[114]

Maximum likelihood molecular replacement

We discussed in the experimental phasing section (Chapter 10) that maximum likelihood functions are more realistic target functions which allow explicit consideration of errors and incompleteness in models, two major realities in molecular replacement. Given that the solution landscape of MR searches is generally flat and possible solutions are poorly discriminated, it is reasonable to expect that maximum likelihood functions would also improve MR searches. Maximum likelihood functions have been implemented successfully in the molecular replacement and SAD phasing program *Phaser*[8] by Randy Read and colleagues. The idea of using ML targets in MR translation has been around since the late 1980s[107] and a general ML formalism was proposed by Bricogne in 1992.[108] However, the practical implementation of a full ML treatment of MR searches has proven less than trivial and has been implemented only recently in *Phaser* based on experience gained with its predecessors *BRUTE*[107] and *BEAST*.[50] To refresh the basics, we begin this section with a brief review of maximum likelihood functions (detailed in Section 7.9 or the highly recommended review by A. McCoy).[109]

A brief review of maximum likelihood functions

The data likelihood function *prob(data|model)* describes the probability that we will observe an experimental value given our present model. Following Fisher's "absolute criterion" and Bayes' theorem, a parameterized model optimized so that the likelihood function is maximal will also be the best model, given the data. For a multi-parametric model described by *data* (a set of discrete values or parameters), the model likelihood L will be a joint probability proportional to the product of the individual probabilities, generally expressed as a sum of the individual log-likelihood functions:

$$LL(model \mid \mathbf{data}) \propto \sum_i \ln\left[prob(data_i \mid model) \right] \tag{11-47}$$

In a crystallographic optimization problem, the model is generally described by a set of calculated structure factors $\mathbf{F}_{calc}$, and the data are the observed structure

Box 11-12 **Bulk solvent correction.** Bulk solvent corrections account for the overestimation of low resolution reflections in the calculation of model structure factors through disordered bulk solvent corrections. The flat or masked bulk solvent model is commonly implemented in refinement programs, and the molecular replacement program *Phaser* employs a basic exponential bulk solvent model.

factor amplitudes F_{obs}. The complex model structure factor contributions $\mathbf{f}_{calc}$ from each individual atom are subject to two types of errors; positional errors, which affect the phase, and errors in B-factor and occupancy, which affect the magnitude of the complex structure factor (Figure 7-17). Therefore, each complex structure factor (being the sum of all the atomic structure factors) is described by a complex, bivariate Gaussian distribution with its complex variance σ_Δ^2 centered on the vector $D \cdot \mathbf{F}_{calc}$ where the Luzzati factor D ($0 \le D \le 1$) is a measure of the coordinate error of the model (Figure 7-18).

The general conditional probability for observing a general acentric structure factor $\mathbf{F}_{obs}$ given a model structure factor $\mathbf{F}_{calc}$ is therefore given by a complex Gaussian likelihood function

$$prob(\mathbf{F}_{obs} \mid \mathbf{F}_{calc}) = \frac{1}{\pi \sigma_\Delta^2} \exp\left(-\frac{|\mathbf{F}_{obs} - D\mathbf{F}_{calc}|^2}{\sigma_\Delta^2} \right) \tag{11-48}$$

The obvious problem here is that we have expressed in (11-48) the conditional probability as a function of the complex structure factors, which is not what we obtain from our experiment—we have only the structure factor amplitudes F_{obs} available in the form of our experimental data. The phase angle can be marginalized or integrated out of the complex conditional probability, and depending on whether the structure factors have a centric or acentric distribution, and whether anything is known about the model or not (i.e., $\mathbf{F}_{calc} = 0$), we obtain the 1-dimensional probability distributions known as the Rice distribution (Sim distribution) R (acentric, with model)

$$prob(F_{obs} \mid F_{calc}) = \frac{2F_{obs}}{\sigma_\Delta^2} \cdot \exp\left(-\frac{F_{obs}^2 + D^2 F_{calc}^2}{\sigma_\Delta^2} \right) \cdot I_0\left(\frac{2F_{obs} D F_{calc}}{\sigma_\Delta^2} \right) = R(F_{obs}, F_{calc}, D, \sigma_\Delta^2) \tag{11-49}$$

the Woolfson distribution W (centric, with model)

$$prob(F_{obs} \mid F_{calc}) = \left(\frac{2}{\pi \sigma_\Delta^2} \right)^{1/2} \exp\left(-\frac{F_{obs}^2 + D^2 F_{calc}^2}{2\sigma_\Delta^2} \right) \cdot \cosh\left(\frac{F_{obs} D F_{calc}}{\sigma_\Delta^2} \right) = W(F_{obs}, F_{calc}, D, \sigma_\Delta^2) \tag{11-50}$$

or the centric (R_0) and acentric (W_0) Wilson distributions (in the case of no conditioning model) for the expectation values of structure factor amplitudes as derived in Chapter 7.

The above likelihood functions are applicable in cases where we directly compare calculated and observed structure factors of properly oriented models. This is the case in structure refinement and translation searches (or stochastic multi-dimensional searches), where the structure factor calculation includes the contributions from the neighboring symmetry related molecules of the entire unit cell. The situation is different for rotation searches, where the location of the search molecule is not known. A major challenge in the implementation of ML-MR therefore was the derivation of entirely new maximum likelihood rotation functions.

Maximum likelihood translation functions

To locate a properly oriented search molecule in the unit cell, we compare the calculated structure factor amplitudes computed from all symmetry related copies in the unit cell with the observed structure factor amplitudes. This scenario is shown for a general acentric likelihood translation function in Figure 11-29. The numerical value for the probability of observing an experimental structure amplitude F_{obs} given the complex structure factor $D\mathbf{F}_c$ of the model is given by integration (of the Rice function) over the F_{obs} phase circle intersecting with the 2-dimensional Gaussian.

Experimental errors in F_{obs} can be accounted for by inflating the variance, that is, adding to (the now 1-dimensional) variance σ_Δ^2 the estimated variance of the observations σ_F^2. For acentric reflections, the Rice function then becomes

$$prob(F_{obs} \mid F_{calc}) = \frac{2F_{obs}}{\sigma_\Delta^2 + \sigma_F^2} \cdot \exp\left(-\frac{F_{obs}^2 + D^2 F_{calc}^2}{\sigma_\Delta^2 + \sigma_F^2} \right) \cdot I_0\left(\frac{2F_{obs} D F_{calc}}{\sigma_\Delta^2 + \sigma_F^2} \right) \tag{11-51}$$

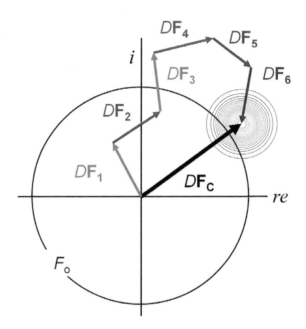

Figure 11-29 Maximum likelihood translation function. The structure factor contributions of six symmetry related molecules DF_1 to DF_6 contribute to the model structure factor DF_c, which has a 2-dimensional Gaussian variance indicated by the centroid contour lines. The magnitudes of the partial contributions remain the same during the translation search, but their phases change and therefore the total structure factor changes. The intersection of the centroid distribution with the observed structure factor magnitude circle with radius F_o yields represents the numerical value of the likelihood function which is obtained by integration over the phase circle. For acentric structure factors the value is given by the Rice distribution and for centrics by the Woolfson distribution.

Finally, if a likelihood function is expressed in normalized structure factor amplitudes E, the translation functions can be simplified by introducing a single parameter, σ_A, that accounts for both coordinate error and incompleteness in the model by replacing D with σ_A and the variance term with $1 - \sigma_A^2$. This yields for the acentric case the likelihood function for each structure factor:

$$prob(E_{obs} \mid E_{calc}) = \frac{2E_{obs}}{1-\sigma_A^2} \cdot \exp\left(-\frac{E_{obs}^2 + \sigma_A^2 E_{calc}^2}{1-\sigma_A^2}\right) \cdot I_0\left(\frac{2E_{obs}\sigma_A E_{calc}}{1-\sigma_A^2}\right) = R(E_{obs}, E_{calc}, \sigma_A^2)$$

(11-52)

The corresponding likelihood function for centric reflections can be derived similarly from the Woolfson distribution. The entire (log) maximum likelihood translation function (MLTF) is then as usual given by the sum of the logarithmic contributions:

$$\text{MLTF} = \sum_{\mathbf{h}}^{acent} \ln R(E_{obs}, E_{calc}, \sigma_A^2) + \sum_{\mathbf{h}}^{cent} \ln W(E_{obs}, E_{calc}, \sigma_A^2)$$

(11-53)

Phase information from already located probes can be readily incorporated into the MLTF. The F_{calc} contribution is then replaced by a combination of the fixed and the translational component contributions[8] which can be written as $F_\Phi = |D_m \mathbf{F}_m(\mathbf{T}) + D_f \mathbf{F}_f|$ where the subscripts m and f mean moving and fixed, respectively, and $\mathbf{T}$ is the corresponding translation vector for the moving search probe. The same considerations regarding search volume as discussed above for conventional translation functions apply.

Likelihood enhanced translation functions

In addition to the already slow structure factor calculations, a relatively complicated distribution needs to be calculated for the likelihood function of each reflection, which makes the computation of the brute force MLTF (11-53) derived above relatively slow. A more efficient way is to first search with a less accurate but faster function, and then use the more accurate MLTF function to re-score the top peaks. A number of fast likelihood enhanced translation functions (LETFs) are implemented in *Phaser*, derived from the first one or two terms of a Taylor expansion of the MLTF.[110] All these functions compute fast through a single FFT and are more sensitive than the commonly used linear correlation coefficients.

Anisotropy correction

The likelihood translation functions are susceptible to correlation in errors of the structure factor amplitudes, which are present if the crystal diffracts anisotropically, that is, with higher diffraction limits in certain directions. In this case, an anisotropy correction must be applied to both the structure factors and their variances to obtain corrected E and σ_E values. The six anisotropy parameters of the anisotropy tensor and a scale factor are refined to optimize the log-likelihood function (11-53). A detailed description of diffraction anisotropy corrections, which are generally combined with bulk solvent corrections, is provided in Chapter 12.

Maximum likelihood rotation functions

During an orientation search, we cannot compute the absolute phase of the structure factors, because we lack the position and translational relation to symmetry related molecules. A ML function based on the knowledge of the absolute phase of the search model, such as MLTF, cannot be used. However, we can calculate the *relative* phase relations between atoms within the search molecule and sum the corresponding atom structure factors to obtain the total amplitude, F_m. This is equivalent to computing the reciprocal space molecular transform of the molecule. We can also apply the space group operations of the target cell to generate all the s symmetry related orientations of the molecule, and compute their amplitudes $(F_m)_s$ (note that as always we cannot just add amplitudes to get the total structure factor amplitudes; we need a vector addition to accomplish this). The phase relation between these molecules will be random and the corresponding probability distribution will be a Wilson-like distribution, which for the acentric case is a Gaussian distribution centered around the origin $F = 0$. The corresponding log-likelihood function is denoted as Wilson-like maximum likelihood rotation function, or MLRF_0:

$$\text{MLRF}_0 = \sum_{\mathbf{h}}^{acent} \ln R_0(F_{obs}, \varepsilon\Sigma_W + \sigma_F^2) + \sum_{\mathbf{h}}^{cent} \ln W_0(F_{obs}, \varepsilon\Sigma_W + \sigma_F^2) \tag{11-54}$$

Note that this function depends only on F_{obs} and its total variance is defined by the measurement variance σ_F^2 and the Wilson-like variance Σ_W. The reflection redundancy is accounted for by the epsilon factor, ε.

The Wilson-like rotation function (11-54) can be improved by a more realistic assumption about the amplitudes of the molecular envelopes. As long as the positions of the symmetry related molecules are incorrect, one of the s symmetry related molecules will contribute more to the complex structure factors than the others. We assign this molecule's transform an arbitrary reference phase, and its contribution to the structure factors will be $\mathbf{F}_{big}$ (note that when all molecules are in their correct orientations, as in the case of a translation function, their contributions are equal in magnitude; compare Figure 11-29). The resulting distribution is centered on $D\mathbf{F}_{big}$ with the variance Σ_s depending on the distribution of the remaining structure factors:

$$prob(\mathbf{F}_{obs} \mid \mathbf{F}_{big}) = \frac{1}{\Sigma_s} \exp\left(-\frac{\left| \mathbf{F}_{obs} - D\mathbf{F}_{big} \right|^2}{\Sigma_s} \right) \tag{11-55}$$

Figure 11-30 visualizes the scenario described by probability distribution (11-55) and integrating out, in the usual fashion, the phase angle which we have introduced by selecting $\mathbf{F}_{big}$ as a phase reference leads in the centric case to a Sim-like probability distribution. Inflating the variance with the measurement error finally gives

$$prob(F_{obs} \mid \{Fm\}_{s \neq big}, F_{big}) = \frac{2F_{obs}}{\Sigma_s + \sigma_F^2} \cdot \exp\left(-\frac{F_{obs}^2 + D^2 F_{big}^2}{\Sigma_s + \sigma_F^2} \right) \cdot I_0\left(\frac{2F_{obs} DF_{big}}{\Sigma_s + \sigma_F^2} \right) \tag{11-56}$$

where the variance again depends on the distribution of the remaining structure factors.[50] The corresponding log-likelihood rotation functions MLRF is then again the sum of the individual log-likelihoods:

$$\text{MLRF} = \sum_{\mathbf{h}}^{acent} \ln R(F_{obs}, D_{big}F_{big}, \varepsilon\Sigma_s + \sigma_F^2) + \sum_{\mathbf{h}}^{cent} \ln W(F_{obs}, D_{big}F_{big}, \varepsilon\Sigma_s + \sigma_F^2) \tag{11-57}$$

Figure 11-30 Maximum likelihood rotation function. To derive the Sim-like maximum likelihood rotation function we separate the molecular transform that has the largest amplitude from the remaining components and assign it a random phase. The probability distribution is then centered on DF_{big}, with a variance determined by the distribution of the remaining components making up $\mathbf{F}_{rem}$. The numerical value of the likelihood function is obtained by integration over the phase circle. For centric reflections the distribution is described by the corresponding Woolfson function.

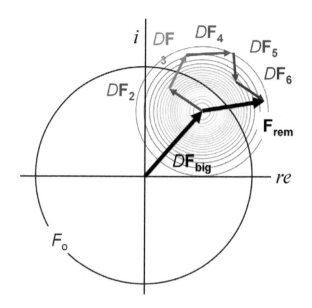

One of the interesting aspects of the maximum likelihood rotation function is that in contrast to the conventional Patterson rotation functions, the MLRF can account for partial information in the form of other models that have already been placed in the unit cell. Incorporating the amplitudes of known contributions modifies the variance, and thus how much of each structure factor remains to be accounted for by the current search model.

Just as with the translation functions, both the Sim- and the Wilson-like rotation functions are slow to compute, and corresponding likelihood-enhanced rotation functions (LERFs) have been derived.[111] They are again derived via Taylor series truncation and despite being approximations, they provide better discrimination that other, Crowther and Blow-based fast rotation functions. The strategy is to rapidly find a set of top solutions using a LERF, and to subsequently further discriminate the solutions by the slower but more accurate MLRF.

Variances, ensembling, and normal mode analysis

The likelihood functions critically depend on accurate variances, and thus accurate estimates for the expected coordinate error of the model. The Chothia and Lesk relation provides an estimate of the $C\alpha$-coordinate error σ_r based on sequence identity between probe and target molecule (Sidebar 11-5). The error estimate is further improved by accounting for the absence of modeled solvent through an exponential bulk solvent correction that is incorporated in the likelihood functions through the σ_A variance term together with the estimated $C\alpha$-coordinate error σ_r:

$$\sigma_A = \left(\mathbf{F}_p[1 + k_{sol}\exp(-B_{sol}S^2 / 4)]\right)^{\frac{1}{2}} \exp\frac{2\pi}{3}S^2\sigma_r^2 \tag{11-58}$$

If there is reason to suspect a higher coordinate deviation as a result of particular flexibility, the coordinate variance can be inflated by increasing the constant leading the exponent in the Chothia and Lesk estimate (11-29) up to 0.8. The availability of an *a priori* estimate of the coordinate deviation also allows the generation of an average structure factor from an ensemble of superimposed PDB structures (for example, different mutants or NMR models). The normalized ensemble structure factors E_{ens} then become the weighted average $E_{ens} = \Sigma w_i E_i$ and the ensemble variance σ_{ens} is derived from the normalized individual variances. The ensembling procedure is implemented in *Phaser*

Another option available in *Phaser* is normal mode perturbation of the search models. This is an interesting technique based on the fact that perturbation of a model along the lowest frequency normal mode will generate a model closer to the target in cases where there have been conformational changes.

> **Box 11-13 Molecular replacement with maximum likelihood functions.** Maximum likelihood functions allow for consideration of model incompleteness and model errors in the rotation and translation searches, thereby increasing the chance of success. Additional improvements such as bulk solvent correction can be included in the σ_A error estimate, and multiple model ensembling and automated multi-model search strategies are implemented in addition to 6-dimensional brute force searches in the maximum likelihood MR program *Phaser*.

Packing function

The success of a ML translation search is determined by the log-likelihood gain of a solution over the average solution. It is possible that this happens despite a clash between symmetry related molecules, and a packing function that constrains the solution space is implemented in *Phaser*. In addition to a hard packing function that is simply based on the number of contacts closer than 3Å between Cα atoms, the packing function considers probable deviations between probe and target structure. In ensemble searches, corresponding Cα atoms that differ by more than 3 Å between the models are removed from the packing evaluation, thereby accounting for likely differences between probe and target.

Automated MR strategies with *Phaser*

Most of the fundamentals such as model trimming and the increasingly flat solution landscape in the case of multi-subunit structures also apply to *Phaser*. The program excels in two major aspects: the general improvement of the searches resulting from advanced likelihood functions and algorithms; and secondly through the automation of a variety of molecular replacement tasks. There are four major automated tasks: search for multiple molecules in the asymmetric unit of the target structure; permutation of search order in the multi-unit searches (which can drastically increase the chance of finding a solution); handedness and screw axis periodicity ambiguities; and detecting the best model when a set of possible search models is given. Particularly for translation searches, the evaluation of multiple space group choices can be extremely helpful.

The automated multi-copy search employs a tree-search with pruning, and an entire recent review is dedicated to the strategies of solving protein complexes with *Phaser*.[112] Together with the excellent program documentation, examples, and the convenient implementation through the CCP4i graphical user interface, the program is quite user friendly despite its substantial internal complexity. The use of *Phaser* in a *de novo* phasing method based on multisolution fragent search followed by *SHELXE* phase extension with autotracing is discussed in Sidebar 10-10.

11.4   Next steps

Once we have properly placed a search model, we will have to generate a first electron density map and begin rebuilding the model. In contrast to the experimental maps, we have in general no independent and unbiased experimental phase information, and we rely entirely on the search model atom positions to

> **Box 11-14 Model bias in molecular replacement maps.** Because the phases for the electron density reconstruction originate entirely from the replaced search probe, electron density maps are highly susceptible to model bias and will reflect the model features, unless maps are constructed with methods that minimize phase bias.

phase the maps. As we recall from the previous chapters, the model phases will render our electron density seriously biased toward the model. This is a reason for caution, particularly at low resolution, where few data are available to overcome the strong model bias. We will therefore have to examine in the next chapter how to generate maps with minimum model bias to be able to effectively correct and rebuild our search model.

11.5   Key concepts

Unit cell contents

- Protein structures in general have a solvent content of ~25–80%, with proteins diffracting to high resolution tending to have lower solvent content
- Based on the distribution of the solvent content, the probability of the presence of multiple copies of a molecule in the asymmetric unit can be estimated indirectly from the Matthews coefficient and directly from the associated Matthews probabilities.
- The more molecules that can be accommodated in the asymmetric unit, the less discriminating the multiple, physically plausible copy numbers become.
- Biological evidence and further crystallographic evidence from Patterson self-rotation maps must complement the Matthews numbers in multi-copy situations.

Non-crystallographic symmetry operations

- When multiple copies of the same molecule are present in the asymmetric unit, they are related though local non-crystallographic symmetry (NCS).
- NCS can be classified according to the operations that relate the monomers. If the molecules are related though closed point group rotations, such as precise 2-, 3-, 4-, 5-, or 6-fold rotations, the NCS is termed *proper*.
- Any other NCS operation combining general rotations and translations is called *improper*.
- A transformations or mapping $\mathbf{W}(\mathbf{R},\mathbf{T})$ of an object onto another can be performed through a combination of rotation and translation.
- The translational part of the operation is determined by the translation vector $\mathbf{T}$.
- A rotation is described through the direction cosine matrix $\mathbf{R}$ which is a real square matrix whose transpose $\mathbf{R}^T$ equals its inverse $\mathbf{R}^{-1}$ (from which follows that $\mathbf{R}^T\mathbf{R} = \mathbf{I}$) and whose determinant det $\mathbf{R}$ equals +1.
- Any rotation can be described as a rotation round the single eigenvector $\mathbf{E}$ of the rotation matrix, the Euler axis, and the associated principal Euler angle.
- The rotation matrix can be expressed in various conventions in orthogonal coordinates through three Euler angles (θ_1, θ_4, θ_4) or in spherical coordinates with lateral, azimuthal, and axial rotations around the angles (ϕ, ψ, κ).
- The last rotation axis is the Euler axis, and κ is the associated principal Euler angle.
- To interpret a NCS operation in the crystallographic reference system, its operators need to be converted into crystallographic coordinates.

NCS in reciprocal space and NCS averaging

- Because NCS in real space also generates redundancies in reciprocal space, special NCS operations can be deduced directly from the experimental data.
- Proper rotational NCS manifests itself as peaks in maps of the Patterson self-rotation function in κ-sections corresponding to the rotation angle.
- Positions of 2-fold axes parallel to crystallographic 2-, 4-, or 6-fold rotation or screw axes that would be otherwise obscured by crystallographic symmetry in the Patterson self-rotation maps manifest themselves as peaks in native Patterson maps.
- Near-perfect translation-only NCS also shows itself through peaks in native Patterson maps.
- The redundancies in reciprocal space generated by real space NCS between molecules in the crystal can be exploited to improve the phases by electron density map averaging.

- NCS averaging procedures are generally combined with density modification and can be carried out in real space using the electron density maps or in reciprocal space using diffraction data.
- In both cases, NCS operations generate non-integer lattice points, which in real space are interpolated or in reciprocal space are determined by folding with the reciprocal space interference function G.

Molecular replacement

- Molecular replacement is a phasing technique that can be applied when a search model structurally similar to a molecule in the target crystal structure is available.
- Molecular replacement requires a search for the proper orientation and location of the search probe in the unknown crystal structure.
- The search for proper placement can be conducted in sequence by a deterministic 3-dimensional rotation search followed by a deterministic 3-dimensional translation search or in a single step by stochastic 6-dimensional placement searches.
- Stochastic 6-dimensional molecular replacement searches rely on fast structure factor calculations and highly efficient search algorithms.
- Evolutionary programming algorithms are a special case of genetic algorithms and rely on competition of their members for survival against a crystallographic selection criterion.
- Surviving members regenerate the population around their own parameter space with their offspring. After many generations, the fittest survivor is kept as the best solution.
- Many runs of a stochastic search may be necessary to find the correct solution, but the methods will eventually find the correct solution, if one exists, given model and data quality.
- Patterson rotation searches are based on the overlap of Patterson maps.
- The orientation of the molecules is determined by the match of intramolecular Patterson vectors, calculated with a variety of direct or fast rotation functions.
- The relative orientation of multiple copies related by proper NCS in the asymmetric unit can be determined by rotating the native Patterson map onto itself.
- The resulting plot of the Patterson rotation function presented in a stereographic projection is called the self-rotation map.
- The overlap between search probe and target structure is determined from cross-rotation maps of the native Patterson function and the map of the search model placed in a large search box to avoid intermolecular vectors.
- If multiple copies in the subunit related by proper NCS are present, locked rotation functions are used. Multiple solutions corresponding to the highest peaks in the cross-rotation function are then selected for subsequent translation searches.
- The location of the oriented search model in the target unit cell is determined by the match of intermolecular Patterson vectors, calculated with a variety of fast translation functions.
- The translation functions are very sensitive to small orientation errors of the search probe, therefore multiple top solutions of the rotation search are evaluated in the translation search.
- During translation searches, the orientation of the molecules can be improved through Patterson correlation refinement, and the translation functions can also be used to evaluate a set of rotation solutions.
- In successive multi-copy translation searches or when additional phase information is available, phased translation functions are used.
- The best solutions from the translation search are then optimized through rigid body refinement and re-ranked.
- Search models must have sufficient structural similarity to the target molecule. The Cα-coordinate deviation between model and target molecule can be estimated based on sequence identity by the Chothia–Lesk formula, and generally should not exceed ~1.5–2 Å.
- Regions with suspected flexibility and high coordinate deviation must be trimmed from the model. As a general rule, correctness is more important that

completeness. The more subunits that must be located, the less total scattering mass is contained in the probe and the molecular replacement search becomes progressively more difficult.

- In the case of proper NCS, locked and phased search functions may improve the search. As many rotation solutions and differently modified models as practical should be evaluated in the translation searches to increase the chance for success in difficult cases.
- NMR models only occasionally, and homology models rarely, provide effective probes for MR searches. The method of assembling a probe from multiple models implemented in maximum likelihood MR searches may also improve the chances for success.
- Bulk solvent corrections account for the overestimation of low resolution reflections in the calculation of model structure factors through disordered bulk solvent corrections.
- The flat or masked bulk solvent model is commonly implemented in refinement programs, and the molecular replacement program *Phaser* employs a basic exponential bulk solvent model.
- Maximum likelihood functions allow for consideration of model incompleteness and model errors in the rotation and translation searches, thereby increasing the chance of success.
- Additional improvements such as bulk solvent correction can be included in the σ_A error estimate, and multiple model ensembling and automated multimodel search strategies are implemented in addition to 6-dimensional brute force searches in the maximum likelihood MR program *Phaser*.
- Because the phases for the electron density reconstruction originate entirely from the replaced search probe, electron density maps are highly susceptible to model bias and will reflect the model features, unless maps are constructed with methods that minimize phase bias.

11.6   Additional reading

1.   Blow D (2002) *Outline of Crystallography for Biologists.* Oxford, UK: Oxford University Press.

2.   Cowtan KD, & Naismith JH (2001) Molecular replacement and its relatives – Proceedings of the CCP4 Study Weekend. *Acta Crystallogr.* D57(10).

3.   Murshudov GN, von Delft F, & Ballard C (2008) Molecular replacement – Proceedings of the CCP4 Study Weekend. *Acta Crystallogr.* D64(1).

4.   Rossmann MG, & Arnold E (Eds.) (2001) *International Tables for Crystallography, Volume F: Crystallography of*

Biological Macromolecules. Dordrecht, The Netherlands: Kluwer Academic Publishers.

5.   Carter CWJ, & Sweet RM (1997) Macromolecular crystallography. *Methods Enzymol.* 276, 277. London, UK: Academic Press.

6.   Rossmann MG (2001) Molecular replacement – historical background. *Acta Crystallogr.* D57, 1360–1366.

11.7   References

1.   Bricogne G (1974) Geometric sources of redundancy in intensity data and their use for phase determination. *Acta Crystallogr.* A30(3), 395–405.

2.   Kleywegt GJ, & Read RJ (1997) Not your average density. *Structure* 5, 1557–1569.

3.   Zhang KYJ, Cowtan KD, & Main P (2001) Phase improvement by iterative density modification. *International Tables for Crystallography F*, 311–324.

4.   Kantardjieff KA, & Rupp B (2003) Matthews coefficient

probabilities: Improved estimates for unit cell contents of proteins, DNA, and protein-nucleic acid complex crystals. *Protein Sci.* 12, 1865–1871.

5.   Matthews BW (1968) Solvent content of protein crystals. *J. Mol. Biol.* 33, 491–497.

6.   Matthews BW (1976) X-ray crystallographic studies of proteins. *Annu. Rev. Phys. Chem.* 27, 493–523.

7.   Quillin ML, & Matthews BW (2000) Accurate calculation of the density of proteins. *Acta Crystallogr.* D56, 791–794.

8. McCoy AJ, Grosse-Kunstleve RW, Adams PD, et al. (2007) Phaser crystallographic software. *J. Appl. Crystallogr.* 40(4), 658–674.

9. Adams PD, Afonine PV, Bunkoczi G, et al. (2010) PHENIX: a comprehensive Python-based system for macromolecular structure solution. *Acta Crystallogr.* D66(2), 213–221.

10. Stanfield RL, Zemla A, Wilson IA, et al. (2006) Antibody elbow angles are influenced by their light chain class. *J. Mol. Biol.* 357(5), 1566–1574.

11. Trakhanov S, Parkin S, Raffai R, et al. (1999) Structure of a monoclonal 2E8 Fab antibody fragment specific for the low-density lipoprotein-receptor binding region of apolipoprotein E refined at 1.9 Å. *Acta Crystallogr.* D55(1), 122–128.

12. Urzhumtseva LM, & Urzhumtsev AG (1997) Tcl/Tk-based programs. II. CONVROT: a program to recalculate different rotation descriptions. *J. Appl. Crystallogr.* 30(3), 402–410.

13. Tickle IJ (2007) personal communication.

14. Goldstein H, Poole C, & Safko J (2002) *Classical Mechanics.* Reading, MA: Addison Wesley.

15. Navaza J (2001) Rotation functions. *International Tables for Crystallography F*, 269–274.

16. Wang X, & Janin J (1993) Orientation of non-crystallographic symmetry axes in protein crystals. *Acta Crystallogr.* D49, 505–512.

17. Zemla A (2003) LGA: a method for finding 3D similarities in protein structures. *Nucleic Acids Res.* 31(13), 3370–3374.

18. Rupp B, Marshak DR, & Parkin S (1996) Crystallization and preliminary X-ray analysis of two new crystal forms of calmodulin. *Acta Crystallogr.* D52(2), 411–413.

19. Maita N, Okada K, Hatakeyama K, et al. (2002) Crystal structure of the stimulatory complex of GTP cyclohydrolase I and its feedback regulatory protein GFRP. *Proc. Natl. Acad. Sci. U.S.A.* 99(3), 1212–1217.

20. Sawaya MR (2007) Characterizing a crystal from an initial native dataset. In Doublie S (Ed.) *Macromolecular Crystallography Protocols.* Totowa, NJ: Humana Press.

21. Terwilliger T (2002) Rapid automatic NCS identification using heavy-atom substructures. *Acta Crystallogr.* D58(12), 2213–2215.

22. Vonrhein C, & Schultz GE (1998) Locating proper non-crystallographic symmetry in low-resolution electron-density maps with the program GETAX. *Acta Crystallogr.* D55, 225–229.

23. Holm L, & Sander C (1998) Touring protein fold space with Dali/FSSP. *Nucleic Acids Res.* 26(1), 316–319.

24. Bourne PE, & Weissig H (2003) *Structural Bioinformatics.* Hoboken, NJ: Wiley-Liss.

25. Kang F, & Sippl MJ (1996) Optimum superimposition of protein structures: ambiguities and implications. *Fold. Des.* 1(2), 123–132.

26. Sippl MJ, & Wiederstein M (2007) A note on difficult structure alignment problems. *Bioinformatics* 23, 426-427.

27. Kolodny R, Koehl P, & Levitt M (2005) Comprehensive evaluation of protein structure alignment methods: Scoring by geometric measures. *J. Mol. Biol.* 346(4), 1173.

28. Nordman C (1980) Procedures for detection and idealization of non-crystallographic symmetry with application to phase refinement of the satellite tobacco necrosis virus structure. *Acta Crystallogr.* A36(5), 747–754.

29. Vellieux FMD, & Read RJ (1997) Non-crystallographic symmetry averaging in phase refinement and extension. *Methods Enzymol.* 277, 18–53.

30. Rossmann M, & Blow D (1962) The detection of sub-units within the crystallographic asymmetric unit. *Acta Crystallogr.* A15, 24–51.

31. Arnold E, & Rossmann M (1986) Effect of errors, redundancy, and solvent content in the molecular replacement procedure for the structure determination of biological macromolecules. *Proc. Natl. Acad. Sci. U.S.A.* 83, 5489–5493.

32. Abrahams JL, & Leslie AGW (1996) Methods used in the structure determination of the bovine mitochondrial F1 ATPase. *Acta Crystallogr.* D52, 30–42.

33. Vonrhein C, Blanc E, Roversi P, et al. (2007) Automated stucture solution with autoSHARP. *Methods Mol. Biol.* 364(2), 215–253.

34. Rossmann MG (2001) Molecular replacement – historical background. *Acta Crystallogr.* D57, 1360–1366.

35. Evans P, & McCoy A (2008) An introduction to molecular replacement. *Acta Crystallogr.* D64, 1–10.

36. Dodson E (2008) The befores and afters of molecular replacement. *Acta Crystallogr.* D64, 17–24.

37. Read RJ (1986) Improved Fourier coefficients for maps using phases from partial structures with errors. *Acta Crystallogr.* A42, 140–149.

38. Hodel A, Kim S-H, & Brunger AT (1992) Model bias in macromolecular structures. *Acta Crystallogr.* D48, 851–858.

39. Perrakis A, Sixma TK, Wilson KS, et al. (1997) wARP: Improvement and extension of crystallographic phases by weighted averaging of multiple-refined dummy atomic models. *Acta Crystallogr.* D53, 448–455.

40. Reddy V, Swanson S, Sacchettini JC, et al. (2003) Effective electron density map improvement and structure validation on a Linux multi-CPU web cluster: The TB Structural Genomics Consortium Bias Removal Web Service. *Acta Crystallogr.* D59, 2200–2210.

41. Glusker JP (1994) Dorothy Crowfoot Hodgkin (1910–1994). *Protein Sci.* 3(12), 2465–2469.

42. Crowther R (1972) The fast rotation function. In Rossmann M (Ed.) *The Molecular Replacement Method.* New York, NY: Gordon and Breach Science Publishers.

43. Crowther R, & Blow DM (1967) A method of positioning a known molecule in an unknown crystal structure. *Acta Crystallogr.* 23, 544–548.

44. Lattman E (1985) Use of the rotation and translation functions. *Methods Enzymol.* 115, 55–77.

45. Navaza J (1994) AMoRe: an automated package for molecular replacement. *Acta Crystallogr.* A50(2), 157–163.

46. Kissinger CR, Gelhaar DK, & Fogel DB (1999) Rapid automated molecular replacement by evolutionary search. *Acta Crystallogr.* D55, 484–491.

47. Rossmann M (Ed.) (1972) *The Molecular Replacement Method.* New York, NY: Gordon and Breach Science Publishers.

48. Dodson E (2001) Using electron-microscopy images as a model for molecular replacement. *Acta Crystallogr.* D57(10), 1405–1409.

49. Huber R, & Schneider M (1985) A group refinement procedure in protein crystallography using Fourier transforms. *J. Appl. Crystallogr.* 18(3), 165–169.

50. Read RJ (2001) Pushing the boundaries of molecular replacement with maximum likelihood. *Acta Crystallogr.* D57, 1373–1382.

51. Lattman EE, & Love WE (1970) A rotational search procedure for detecting a known molecule in a crystal. *Acta Crystallogr.* B26(11), 1854–1857.

52. Fogel DB (1995) *Evolutionary Computation: Towards a New Philosophy of Machine Intelligence.* Piscataway, NJ: IEEE Press.

53. Ramirez L, Axelrod H, Herron S, et al. (2003) High resolution crystal structure of ferricytochrome c' from *Rhodobacter sphaeroides. J. Chem. Crystallogr.* 33, 413–424.

54. Schwarzenbacher R, Godzik A, Grzechnik SK, et al. (2004) The importance of alignment accuracy for molecular replacement. *Acta Crystallogr.* D60(7), 1229–1236.

55. Barton GJ (2008) Sequence alignment for molecular replacement. *Acta Crystallogr.* D64, 25–32.

56. Chothia C, & Lesk A (1986) The relation between the divergence of sequence and structure in proteins. *EMBO J.* 5(4), 823–826.

57. Chen YW, Dodson EJ, & Kleywegt GJ (2000) Does NMR mean "Not for Molecular Replacement"? Using NMR-based search models to solve protein structures. *Structure* 8, R213–R220.

58. Kissinger CR, Gehlhaar DK, Smith BA, et al. (2001) Molecular replacement by evolutionary search. *Acta Crystallogr.* D57(10), 1474–1479.

59. Rost B (1999) Twilight zone of protein sequence alignments. *Protein Eng.* 12(2), 85–94.

60. Navaza J (2001) Implementation of molecular replacement in AMoRe. *Acta Crystallogr.* D57, 1367–1372.

61. Jogl G, Tao X, Xu Y, et al. (2001) COMO: a program for combined molecular replacement. *Acta Crystallogr.* D57(8), 1127–1134.

62. Vagin A, & Teplyakov A (1997) MOLREP: an automated program for molecular replacement. *J. Appl. Crystallogr.* 30(6), 1022–1025.

63. Brunger AT, Adams PD, Clore GM, et al. (1998) Crystallography & NMR system: A new software suite for macromolecular structure determination. *Acta Crystallogr.* D54, 905–921.

64. Chang G, & Lewis M (1997) Molecular replacement using genetic algorithms. *Acta Crystallogr.* D53(3), 279–289.

65. Glykos NM, & Kokkinidis M (2000) A stochastic approach to molecular replacement. *Acta Crystallogr.* D56(2), 169–174.

66. Jamrog DC, Zhang Y, & Phillips GN, Jr. (2003) SOMoRe: a multi-dimensional search and optimization approach to molecular replacement. *Acta Crystallogr.* 59(2), 304–314.

67. Glykos NM, & Kokkinidis M (2003) Structure determination of a small protein through a 23-dimensional molecular-replacement search. *Acta Crystallogr.* D59(4), 709–718.

68. Rao S, Jih J-H, & Hartsuck J (1980) Rotation-function space groups. *Acta Crystallogr.* A36, 878–884.

69. Huber R (1965) Die automatisierte Faltmolekülmethode. *Acta Crystallogr.* 19(3), 353–356.

70. DeLano WL, & Brunger AT (1995) The direct rotation function: rotational Patterson correlation search applied to molecular replacement. *Acta Crystallogr.* D51(5), 740–748.

71. Navaza J (1993) On the computation of the fast rotation function. *Acta Crystallogr.* D49(6), 588–591.

72. Hauptman H (1982) On integrating the techniques of direct methods and isomorphous replacement. I. The theoretical basis. *Acta Crystallogr.* A38(3), 289–294.

73. Tong L (2001) How to take advantage of non-crystallographic symmetry in molecular replacement: "locked" rotation and translation functions. *Acta Crystallogr.* D57(10), 1383–1389.

74. Hirshfeld F (1968) Symmetry in the generation of trial structures. *Acta Crystallogr.* A24(2), 301–311.

75. Koch E, Fischer W, & Müller U (2002) Normalizers of space groups and their use in crystallography. *International Tables for Crystallography A*, 877–905.

76. Driessen HPC, Bax B, Slingsby C, et al. (1991) Structure of oligomeric βB2-crystallin: an application of the T2 translation function to an asymmetric unit containing two dimers. *Acta Crystallogr.* B47(6), 987–997.

77. Navaza J, & Vernoslova E (1995) On the fast translation functions for molecular replacement. *Acta Crystallogr.* A51(4), 445–449.

78. Harada Y, Lifchitz A, Berthou J, et al. (1981) A translation function combining packing and diffraction information: an application to lysozyme (high-temperature form). *Acta Crystallogr.* A37(3), 398–406.

79. Tickle IJ (1992) Fast Fourier translation functions. In Dodson EJ, Gover S, & Wolf W (Eds.) *Molecular Replacement – Proceedings of the 1992 CCP4 Study Weekend.* Daresbury, UK.

80. Grosse-Kunstleve RW, & Adams PD (2001) Patterson correlation methods: a review of molecular replacement with CNS. *Acta Crystallogr.* D57, 1390–1396.

81. Brunger AT (1997) Patterson correlation searches and refinement. *Methods Enzymol.* 276, 558–580.

82. Lebedev AA, Vagin AA, & Murshudov GN (2008) Model preparation in MOLREP and examples of model improvement using X-ray data. *Acta Crystallogr.* D64, 33–9.

83. Norvell JC, & Zapp-Machalek A (2000) Structural genomics programs at the US National Institute of General Medical Sciences. *Nat. Struct. Biol. Suppl.* 7, 931.

84. Hou J, Sims GE, Zhang C, et al. (2003) A global representation of the protein fold space. *Proc. Natl. Acad. Sci. U.S.A.* 100(5), 2386–2390.

85. Jones DT (2001) Evaluating the potential of using fold-recognition models for molecular replacement. *Acta Crystallogr.* D57, 1428–1434.

86. Marti-Renom MA, Stuart AC, Fiser A, et al. (2000) Comparative protein structure modeling of genes and genomes. *Annu. Rev. Biophys. Biomol. Struct.* 29, 291–325.

87. Canutescu AA, Shelenkov AA, & Dunbrack RL, Jr. (2003) A graph-theory algorithm for rapid protein side-chain prediction. *Protein Sci.* 12(9), 2001–2014.

88. Vagin A, & Teplyakov A (2000) An approach to multi-copy search in molecular replacement. *Acta Crystallogr.* 56(12), 1622–1624.

89. Perrakis A, Harkiolaki M, Wilson KS, et al. (2001) ARP/wARP and molecular replacement. *Acta Crystallogr.* D57, 1445–1450.

90. Cohen SX, Ben Jelloul M, Long F, et al. (2008) ARP/wARP and molecular replacement: the next generation. *Acta Crystallogr.* D64(1), 49–60.

91. Moore GE (1965) Cramming more components onto integrated circuits. *Electronics* 8(15), 2–5.

92. Potterton E, Briggs PJ, Turkenberg M, et al. (2003) A graphical user interface to the CCP4 program suite. *Acta Crystallogr.* D59, 1131–1137.

93. Keegan RM, & Winn MD (2007) Automated search-model discovery and preparation for structure solution by molecular replacement. *Acta Crystallogr.* D63, 447–457.

94. Keegan RM, & Winn MD (2008) MrBUMP: an automated pipeline for molecular replacement. *Acta Crystallogr.* D64, 119–24.

95. Larkin MA, Blackshields G, Brown NP, et al. (2007) ClustalW and ClustalX version 2.0. *Bioinformatics* 23(21), 2947–2948.

96. Murzin AG, Brenner SE, Hubbard T, et al. (1995) SCOP: A structural classification of proteins database for the investigation of sequences and structures. *J. Mol. Biol.* 247, 536–540.

97. Henrick K, & Thornton JM (1998) PQS: a protein quaternary structure file server. *Trends Biochem. Sci.* 23(9), 358–361.

98. Long F, Vagin AA, Young P, et al. (2008) BALBES: a molecular-replacement pipeline. *Acta Crystallogr.* D64, 125–32.

99. Ness SR, de Graaff RAG, Abrahams JP, et al. (2004) Crank: New methods for automated macromolecular crystal structure solution. *Structure* 12(10), 1753–1761.

100. Panjikar S, Parthasarathy V, Lamzin VS, et al. (2005) Auto-Rickshaw: an automated crystal structure determination platform as an efficient tool for the validation of an X-ray diffraction experiment. *Acta Crystallogr.* 61, 449–57.

101. Murshudov GN, Vagin AA, & Dodson ED (1997) Refinement of macromolecular structures by the maximum-likelihood method. *Acta Crystallogr.* D53, 240–255.

102. Kostrewa D (1997) Bulk solvent correction: Practical application and effects in reciprocal and real space. *CCP4 Newsl.Protein Crystallogr.* 34, 9–22.

103. Glykos NM, & Kokkinidis M (2000) On the distribution of the bulk-solvent correction parameters. *Acta Crystallogr.* D56(8), 1070–1072.

104. Jiang J-S, & Brunger AT (1994) Protein hydration observed by X-ray diffraction: Solvation properties of penicillopepsin and neuraminidase crystal structures. *J. Mol. Biol.* 243(1), 100–115.

105. Fokine A, Capitani G, Grutter MG, et al. (2003) Bulk-solvent correction for fast translation search in molecular replacement: service programs for AMoRe and CNS. *J. Appl. Crystallogr.* 36(2), 352–355.

106. Janssen JC, Read RJ, Brunger AT, et al. (2007) Crystallographic evidence for deviating C3b structure? *Nature* 448(7154), E1–E3.

107. Fujinaga M, & Read RJ (1987) Experiences with a new translation-function program. *J. Appl. Crystallogr.* 20, 517–521.

108. Bricogne G (1992) A statistical formulation of the molecular replacement and molecular averaging methods. In Wolf W, Dodson EJ, & Grover S (Eds.), *Proceedings of the CCP4 Study Weekend 1992, Molecular Replacement*. Daresbury, UK.

109. McCoy AJ, Storoni LC, & Read RJ (2004) Simple algorithm for a maximum-likelihood SAD function. *Acta Crystallogr.* D60(7), 1220–1228.

110. McCoy AJ, Grosse-Kunstleve RW, Storoni LC, et al. (2005) Likelihood-enhanced fast translation functions. *Acta Crystallogr.* D61(4), 458–464.

111. Storoni LC, McCoy AJ, & Read RJ (2004) Likelihood-enhanced fast rotation functions. *Acta Crystallogr.* D60(3), 432–438.

112. McCoy A (2007) Solving structures of protein complexes by molecular replacement with Phaser. *Acta Crystallogr.* D63(1), 32–41.

113. Rodríguez DD, Grosse C, Himmel S, et al. (2009) Crystallographic Ab Initio protein solution far below atomic resolution. *Nature Methods* 6(9), online publication August 16, 2009.

114. Baker EN, Dauter Z, Einspahr H, et al. (2010) In defence of our science - validation now! *Acta Crystallogr.* D66(2), 115.

Model building
and refinement

One should bear in mind that a macromolecular refinement against high resolution data is never finished, only abandoned.

George Sheldrick (2008) *Acta Crystallogr.*
D64, 112–122

Model building and refinement are intimately connected subjects. The step of building the initial model into electron density and local real space fitting of residues into electron density is repeatedly alternated with global restrained reciprocal space refinement of the positional parameters of the model. Related to model building is the reconstruction of minimally biased electron density maps, an issue that is particularly important to the rebuilding of molecular replacement models. In contrast to experimentally phased electron density, the initial electron density maps obtained from search model phases suffer from model bias.

We begin our discussion with an overview of the steps in model building and refinement, which soon leads to the question of how good or bad initial phases are. After a quite sobering analysis of our previous molecular replacement solution, we realize that incompleteness and errors must be accounted for in the map construction, and we reinvestigate in more detail the process of actually estimating the model errors though their likelihood estimate σ_A (sigma-A). From the estimate of σ_A, the widely and generally used maximum likelihood map coefficients for the construction of electron density maps are derived.

To remove the many small local conformation errors that remain after rebuilding, the model is subjected to global restrained reciprocal space refinement. The parameters of the model atoms (coordinates x, y, z, and in later stages also individual B-factors except in very low resolution structures) together with overall parameters such as scale factor and overall B-factors, bulk solvent corrections, and anisotropy corrections are refined against all experimental data. The target function to be minimized is the residual between the observed experimental structure factor amplitudes and the model structure factor amplitudes. Most macromolecular refinement programs have implemented maximum likelihood target functions, which are less susceptible to model errors and incompleteness than least squares target functions.

A brief investigation of the data-to-parameter ratio reveals the necessity of cross-validation as a safeguard against overparameterization. The low data-to-parameter ratio requires us to implement restraint functions, which we discuss in quite some detail, followed by the exploration of anisotropic atomic displacement and its implementation in translation-libration-screw (TLS) parameterization. We also realize the need for anisotropic scaling and examine the treatment of highly anisotropic data.

The implementation of maximum likelihood refinement targets forms a highly technical part of this section, which is followed by a discussion of the actual optimization algorithms used in the computations. In this section we also introduce simulated annealing, molecular dynamics, and torsion angle refinement, which are implemented in popular and powerful refinement programs.

The chapter closes with a section about the practice of model building and refinement. Model building requires some hands-on training in front of a computer graphics terminal. In addition, each of the model building programs varies greatly in the details of layout and available tools. We therefore restrict ourselves to a general description of important steps which we exercise by rebuilding the previous molecular replacement solution from Chapter 11 and the experimentally phased maps from Chapter 10. We also submit the MR model for automated rebuilding and demonstrate the use of automated *de novo* model building to obtain a first partial model of experimentally phased example structures.

12.1   Basics of model building and refinement

Model building—the construction of a stereochemically plausible atomic structure model that matches the electron density—and model refinement are conducted in tight connection. Building the initial model and real space fitting of the model into the electron density is repeatedly alternated with restrained reciprocal space refinement of the positional parameters of the model. Successful model building naturally depends on the quality of the electron density maps, an issue that is particularly important for the rebuilding of molecular replacement models. In contrast to experimentally phased electron density, the initial electron density maps obtained from molecular replacement search model phases suffer—often severely in the case of poor search models—from model bias.

At the present stage of practical work we have obtained an initial electron density map that hopefully contains traceable features and allows us to build a model of our structure into the map. In the case of an experimentally phased electron density map, the map will be completely empty and we need to build the entire model from scratch. Model building into experimentally phased maps is considered by many the most labor-intensive part of the structure determination. While much of the agony of early physical model building (Sidebar 12-1)

Sidebar 12-1 **Early tools for macromolecular model building.** Model building is accomplished these days nearly exclusively with graphical model building programs which provide numerous tools to place Cα-atoms, entire residues, or even structure fragments into the displayed electron density mesh. Many automated tracing and model building programs suitable for various map resolution ranges can build models or at least fragments, making initial model building much easier.

In the early days of protein crystallography, no graphics terminals were available and physical representations of electron density maps and structure models had to be built. Electron density maps were drawn in sections on transparent sheets and stacked on top of each other to obtain a 3-dimensional view of the electron density, and the first low resolution models of hemoglobin[1] were built at the MRC in Cambridge from stacked sections of balsa wood.[2] John Kendrew's first low-resolution model of myoglobin was kneaded from Plasticine[3] (a Play-doh-like

modeling clay), and is on exhibit in the Science Museum in London. Pictures of many of the early models can be found in the annotated collection of early crystallography literature by Richard Dickerson.[4]

A major step forward in 1968 was the Richards box,[5] a mechanical contraption including a semi-transparent mirror, allowing to assemble a model from prefabricated metal parts at a 2 cm/Å scale into the virtual image or reflection of the stacked map sections. Figure 12-1 shows a version of the Richards box which Brian Matthews and co-workers constructed to model one of the first enzyme structures, thermolysin, in 1972.[6]

With Byron's bender, invented around 1970, one could readily build low-resolution Cα-trace models from straight wire sections by dialing in appropriate torsion angles[8] (Figure 12-2).

As in other areas of macromolecular crystallography, true advances came to model building with the availability of computers powerful enough to compute map grids quickly and display them on graphics terminals. Early work at the MIT graphics laboratory and by Edgar Meyer and co-workers at Texas A&M University explored the principles of molecular graphics display and user interaction technology. With the Evans and Sutherland stereo graphics workstations (at a cost of $250 000 in 1985) a practical system became available. Parallel to the hardware developments,

the first popular computer program for model building, Alwyn Jones' *FRODO*[9] (Figure 12-3) paved the road for the modern model building programs in use today. An editorial in *Nature Structural Biology*[10] and the inaugural issue of *Protein Science*[11] contain additional interesting images of early model building efforts.

Figure 12-1 The Richards box used to build the model of thermolysin in 1972. Prior to the development of computer graphics, the "Richards Box," also known as "Fred's Folly," was used to build physical models of protein structures assembled from prefabricated parts. The panel above shows on the left side the wire model of the crystal structure of the cro-repressor assembled from "Kendrew parts" at a scale of 2 cm Å⁻¹ together with a Watson–Crick DNA model; there were no crystal structures of DNA available before 1979.[7] To the right of the model is a storage crate for the electron density sections plotted on clear plastic sheets, each suspended on a 3 by 3 feet square aluminum frame. A block of 11 map sections pulled out from the storage crate that represent the "active" part of the electron density map can be seen right of the model. A large, semi-

transparent mirror is mounted vertically between the model and the electron density map. A viewer standing at the extreme left in front of the model and looking toward the mirror would see the view photographed in the right panel. The electron density sections, visible through the mirror, are superimposed on the image of the model reflected from the face of the mirror. The physical model (out-of focus in the foreground) is assembled from the prefabricated Kendrew metal parts secured together by screw fasteners recognizable in the virtual image together with the electron density. In the original version introduced by Richards[5] the mirror was mounted at an angle of 45° to the map sections.Brian Matthews and Dale Tronrud (University of Oregon) kindly provided the photographs of the box.

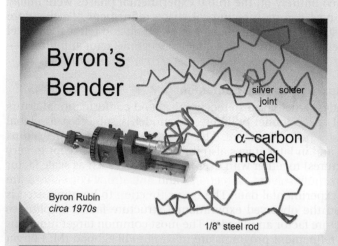

Figure 12-2 Byron's bender. The instrument, invented by Byron Rubin, allows dialing-in of Cα-backbone torsion angles and the bending of steel wires accordingly so that they form the Cα-backbone model of the protein structure. The annotated image of an early bender was kindly provided by Leonard Banaszak, University of Minnesota, Twin Cities.

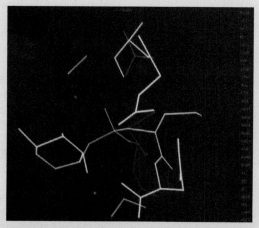

Figure 12-3 A screen shot of FRODO. The picture is a screen shot of a MMS-X CRT graphics system showing part of a thermolysin inhibitor displayed by the original 1978 version of FRODO,[9] which later became O,[12] a venerable graphics program by Alwyn Jones, still used in some laboratories today. Dale Tronrud kindly provided the picture taken in Brian Matthews' laboratory.

is now past, it is fair to say that even model building with modern computer graphics programs does require some practice. Model building skills improve greatly with experience, and there are numerous tools and semi-automated or fully automated model building programs available that now make the task much easier even for the aspiring protein crystallographer.

The model building and refinement process: an overview

Building a protein model into an empty electron density map is generally unproblematic when experimental electron density maps are reconstructed from reasonably accurate phases and high resolution data, largely because the side chains become distinct and the sequence can be readily locked into the electron density. At lower resolution, model building becomes more challenging for humans and machine algorithms alike. Electron density skeletonization methods ("bones") quickly provide an overview of the secondary structure elements that may be present. Manual model building usually starts with placement of Cα-atoms into detected side-chain branching points, which allows building stretches of a polyalanine backbone model. Very distinct residues such as tryptophan or phenylalanine, S–S bridges, or Se or heavy atom positions from the marker atom substructure make good anchor points from which the sequence can be assigned. Some loops will show traceable density, which allows the individually built secondary structure pieces to be connected and assembled into larger fragments, until as much of the model as can be reliably placed has been built. In fortunate cases with good experimental phases, even discrete solvent molecules may already be visible in the experimental electron density.[13]

Local real space refinement

Once contiguous backbone fragments are built, individual residues are fitted into the electron density by real space refinement and the geometry of entire backbone stretches is adjusted using real space geometry regularization tools that bring the model into more reasonable geometry. At this stage many errors, such as implausible backbone torsion angles, become obvious and the model must be rebuilt. Eventually, there will not be much that can be further improved by real space model building and refinement: Despite our very best efforts, the initial model will have numerous geometry errors. There will be bond lengths and angles substantially deviating from expectation values, implausible torsions, and various small errors that all are highly correlated and therefore cannot be reasonably fixed by further manual real space model rebuilding. This fact also explains why the very first refinement programs that worked solely in real space[14] and depended entirely on the initial experimental phases were finally abandoned for the now common reciprocal space refinement. We therefore need to change the strategy once the initial model is built and refine the model against the experimental diffraction data—that is, structure factor amplitudes—in reciprocal space.

Global reciprocal space restrained refinement

To correct the countless small local stereochemical and conformational errors remaining after real space model building, the model is subjected to global reciprocal space restrained refinement. The parameters of the model atoms (coordinates x, y, z, and in later stages also individual B-factors except in very low resolution structures) together with overall parameters such as scale factor and overall B-factors, bulk solvent corrections, and anisotropy corrections are refined against the experimental data. The target function to be minimized is the residual between the observed experimental structure factor amplitudes and the model structure factor amplitudes. The most common target functions in macromolecular refinement programs are maximum likelihood (ML) target functions which are generally less susceptible to non-random model errors and incompleteness than basic least squares (LSQ) target functions. A notable exception is the *SHELXL* program, which is based on a LSQ implementation that is best described as "likelihood by intuition" (Sidebar 12-5).

Cycling of real space and reciprocal space refinement. As the model improves during refinement, the available phases will improve as well. The next electron

> **Box 12-1 Real space model building and reciprocal space refinement.** The key to successful protein structure modeling is the cycling between local real space model building and model correction and global reciprocal space refinement. The molecular model is built in real space into electron density using computer graphics. The remaining and correlated geometry errors are corrected in restrained reciprocal space refinement by optimizing the fit between observed and calculated structure factor amplitudes. Successive rounds of rebuilding, error correction, and refinement are needed to obtain a good final protein model.

density map generally computed from maximum likelihood-based map coefficients will therefore be better and real space model building continues into the next round. More parts of the model can be built, and inspection of difference maps allows further corrections. The improved model is again refined, and the rebuilding–refinement cycles continue until no more significant improvements are possible and the model is as error-free as reasonably achievable (Figure 12-4).

Stereochemical restraints. Because we generally do not have enough diffraction data for entirely free or unrestrained positional refinement, we need to keep the reciprocal space refinement within the bounds of reasonable real space geometry by restraining the model stereochemistry so that the model features remain within universally valid expectations of stereochemistry (and all other laws of nature) during the refinement procedure.

Because model building and refinement are intimately intertwined, we must examine aspects of map quality relevant to model placement and understand the guiding principles of refinement before we can appreciate the proper model building and refinement practice. We therefore will discuss the basics of electron density map generation and of model refinement before we proceed to practical model building in Section 12.3.

How good or bad are initial phases?

In the case of an experimental electron density map, the initial phases are generated from experimental evidence based on measured protein structure factor amplitudes, combined with the marker atom substructure solution determined from experimental isomorphous derivative or anomalous difference data sets. Short of the defensible exclusion of unreasonably close heavy atom distances, generally valid structure factor probability distributions, and the fact that proteins and solvent have different densities, no prior assumptions went into the

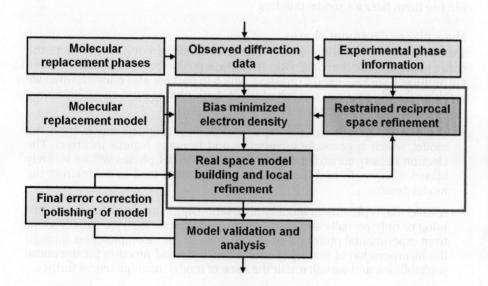

Figure 12-4 **Model building and refinement.** Model building takes place in real space by locally fitting residues of the model into electron density, while refinement in reciprocal space is based on global minimization of the observed and calculated structure factors against a likelihood target function. Local real space model building and global reciprocal space model refinement are alternated in multiple rounds (red box) until the model contains as few as possible remaining errors. In practically all cases except ultra-high atomic resolution, the reciprocal space refinement is restrained by prior knowledge of stereochemical geometry target values.

generation of the density-modified best phases used as Fourier coefficients to construct the initial electron density map. Therefore, an initial experimental electron density map will generally *not* contain any bias induced by our expectations about general protein stereochemistry or a specific protein structure model. The situation is certainly different for initial molecular replacement maps, where the phases are solely obtained from a model that is both incomplete and likely has substantial errors—massive model bias must be expected, because phases dominate the electron density reconstruction by Fourier methods (Sidebar 9-6).

Experimental electron density maps

In experimental electron density maps, the phase quality is generally expressed as a figure of merit of the probability-averaged best (centroid) structure factors $\mathbf{F}_{BEST}$ (Section 10-5):

$$\mathbf{F}_{BEST} = \frac{\sum_{\varphi=0}^{360} P(\varphi)\mathbf{F}(\varphi)}{\sum_{\varphi=0}^{360} P(\varphi)} \qquad (12\text{-}1)$$

The best phase angle $\varphi\alpha_{BEST}$ immediately results from $\mathbf{F}_{BEST}$ as

$$\varphi_{BEST} = \tan^{-1}\left(B_{BEST} / A_{BEST}\right) \qquad (12\text{-}2)$$

and we can express our degree of belief through a weight or figure of merit (m, w, or *fom* are frequently used symbols) in the range $0 \leq m \leq 1.0$, defined as the relative length of the resulting centroid vector $\mathbf{F}_{BEST}$

$$m = \frac{F_{BEST}}{F_P} \quad \text{or} \quad \mathbf{F}_{BEST} = mF_P \cdot \exp(i\varphi_{BEST}) \qquad (12\text{-}3)$$

By substituting Equation 12-3 back into the probability distribution we have shown in Section 10.2 that the figure of merit m is a direct measure for the mean phase error $\Delta\varphi$:

$$m = \langle \cos\Delta\varphi \rangle \qquad (12\text{-}4)$$

Although density modification programs in certain instances tend to inflate the figure of merit, a mean (resolution-dependent) *fom* of ~0.70 or better generally indicates a high probability of success for subsequent model building. There is a good chance that an automated model building program, properly selected for the resolution range, will deliver at least a partial model. Note that good experimental phases can result in much higher figures of merit and the model building will be correspondingly easier. We have seen already examples of quite respectable initial protein electron density maps in Chapters 10 and 11, and we will use them later for model building.

Molecular replacement phases

The situation is quite different for molecular replacement phases, where the entire phase information originates from the repositioned search model. There are multiple issues we need to address, which in part are also relevant when we combine the phases from our initial model with phases from an experimental map or rebuild a structure model in reciprocal space:

- The phases for a molecular replacement map originate solely from the model, which is generally incomplete and in many regions incorrect. The electron density construction of maps using model phases will be severely biased (Chapter 9) and the electron density maps tend to reconstruct the model density.

- A molecular replacement model will be seriously incomplete because of trimming or only partially successful multi-subunit solutions. We already know from experimental phasing that we can account for incompleteness through the incorporation of Sim weights into the likelihood function for the phase probabilities, and we will revisit the issue of model incompleteness further.

- A molecular replacement model will have large errors. The Luzzati D-factor provides a measure for the atomic coordinate deviation, and we have learned in Chapters 7 and 10 that model errors can be incorporated into the combined sigma-A (σ_A) measure together with the incompleteness of the model when expressing the likelihood functions in terms of normalized structure factors.

- As a result of low data-to-parameter ratio, unrestrained global reciprocal refinement of structure models will either be entirely unstable or yield physically improbable (i.e. nonsensical) models. We will address this through the introduction of stereochemical geometry restraints as (largely) independent prior information in the maximum likelihood refinement target functions.

The general solution to the problem of model bias is the use of maximum likelihood based functions for both map reconstruction and in the target functions for model refinement. Before we derive the maximum likelihood error model for the purpose of map reconstruction, we shall examine the quality of initial molecular replacement phases, which in turn provides a strong incentive to find ways to construct bias-minimized electron density.

Model errors and phase bias in molecular replacement

Let us investigate by way of a practical example how poor molecular replacement phases can be—in many cases, they are indeed quite abysmal. In Chapter 11 we solved the structure of the cytochrome c' dimer by molecular replacement. The MR solution by evolutionary search succeeded, although with a modest correlation coefficient CC_F on structure factors of only 0.44 after rigid body refinement. The initial overall R-value was correspondingly high at 0.50 for the data range of 15–4 Å used in solving the MR problem. As we are actually interested in how well the initial model compares with the final model, Figure 12-5 shows graphs of the properly scaled R-value, the phase angle error in degrees, and the corresponding mean figure of merit (m or fom) defined as $m = \langle \cos \Delta \varphi \rangle$ as a function of resolution.

The analysis of the graphs in Figure 12-5 is quite telling. Given that derived from Wilson statistics the expected R-value on F for an acentric random atom structure is 0.59,[15] we recognize that in the data range of 15 to 4 Å used for molecular replacement the R-value is only marginally better than random, and the corresponding average phase error of 60° is just reaching the interpretability limit. For higher resolution reflections, the R-value is essentially random, with correspondingly high phase errors and low fom values in the uninterpretable range far below ~0.5. We clearly see that the success of the molecular replacement solution actually hinged on *abysmally better than random* statistics.

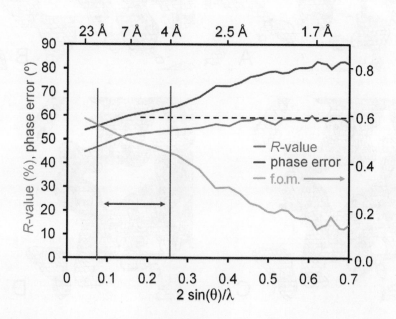

Figure 12-5 R-value, phase error, and figure of merit for a molecular replacement model. The R-value, phase error, and corresponding mean figure of merit are plotted as a function of resolution in bins of $2 \sin \theta / \lambda$ for an initially placed molecular replacement search probe relative to the final structure model. The resolution range of the data used for molecular replacement lies between the magenta boundaries. In the resolution range used for molecular replacement, the R-value is only marginally better than random (0.59, indicated by the dashed horizontal line), and the phase quality is in the lower range of what can reasonably be considered traceable. Beyond 4 Å, the phases approach random values; that is, they will only generate noise in the map. As a result of model phase bias, poor molecular replacement electron density maps will very much reflect the search model, and even ML maps may look deceptively good, despite significant deviations from the correct structure.

Box 12-2 **Phase quality and model phase bias.** Phase accuracy determines how well the reconstructed electron density will reflect the correct protein structure. While experimental electron density maps constructed from poor phases will be hard to interpret, the initial experimental map will not be biased toward any structure model. In contrast, when molecular replacement models are the sole source of phases, the electron density maps will be severely biased, and the map will reflect the model features. Therefore, bias minimization in the form of likelihood-based map coefficients, often combined with additional bias minimization measures, are applied before electron density map construction.

Given that our model phases φ_c are exactly the phases that according to Figure 12-5 do *not* represent the correct structure very well, this does not bode well for the expected appearance of a standard $(2F_o - F_c) \exp(i\varphi_c)$ map: The electron density map will be dreadfully noisy and what is discernible will largely resemble the model, with little room for improvement through model building in real space. As large positional errors are generally outside of the convergence radius of reciprocal space model refinement, just marginally fixing the model and then subjecting it to restrained refinement will only result in a refinement that stalls at high R-values, with no improvements visible in the resulting (highly biased) map. It is the *contrast* between regions of distinctly good and clearly bad fit between electron density and model that is indicative of a good map. Figure 12-6 shows various maps that suffer from decreasingly strong phase bias depending on the type of Fourier coefficients used in map reconstruction.

Likelihood-based error models

In order to obtain more realistic maps and to overcome the effects of model bias

Figure 12-6 Phase bias in initial molecular replacement electron density maps. The panels show electron density contoured at 1σ (blue) in the vicinity of a loop region where significant differences between an initial, only rigid body refined, molecular replacement model and a final crystal structure exist. The basic $2F_o - F_c$ map phased with the poor starting model shown in panel A is essentially useless; it is very noisy and there are no indications of how to rebuild the model. Panel B exemplifies the appearance of a biased map that largely reflects the model. Maximum likelihood maps are much clearer and bias minimized maps generated using more elaborate automated real space dummy atom placement alternated with unrestrained reciprocal space maximum likelihood refinement C show the most dramatic improvement and the sharpest contrast between correct and bad regions. Panel D shows a rebuilt model in the same map. The dummy atom method (described in the refinement section) generally works only at resolutions better than 2.5 Å and with good data—poor phases *and* bad data will under no circumstances give good maps.

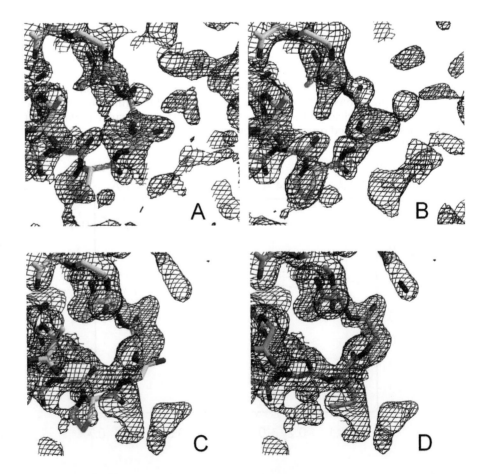

we need to construct the electron density with Fourier coefficients that account for the errors and incompleteness in the model. In principle such errors may affect any calculated model structure factors; the treatment of errors and incompleteness were discussed in Chapter 10 in the context of improving heavy atom refinement. In addition to model incompleteness and errors, in the refinement section we shall also consider how we can account for prior probabilities for our structure model, given that we know which stereochemistry a plausible model should comply with. Error modeling is particularly important for the early stages in structure model refinement, where we generally begin with incomplete models that additionally contain many quite substantial geometry errors.

In Chapters 7 and 10 we discussed how the Sim weights, which account for missing matter in a structure model, and the σ_A parameter, accounting for both incompleteness and incorrectness in structure models, are derived from maximum likelihood principles. We have not yet paid much attention to how they can be determined in practice and how the σ_A parameter can be used to reduce phase bias in map reconstruction.

Review of likelihood functions

The model likelihood function L describes the posterior probability of our model given the data. L is proportional to the experimentally accessible data likelihood function, which is the conditional probability that we will observe the experimental values given the model, $prob(data\,|\,model)$. Our protein structure model is described by a set of i calculated structure factors $\mathbf{F}_{calc}$, and the experimental observations for i structure factor amplitudes F_{obs}. The total likelihood function is then the product of the individual probabilities, generally expressed as a sum of the individual log-likelihoods:

$$LL(model\,|\,\mathbf{data}) \propto \sum_i \ln\left[prob(data_i\,|\,model)\right] \tag{12-5}$$

A protein structure model is commonly parameterized by the atomic coordinates, B-factors, and occupancy (which is rarely refined) for each individual atom. Both the coordinates and the B-factors will carry some error. The calculated complex model structure factor contributions from each atom $\mathbf{f}_{calc}$ are thus subject to two types of independent errors, *positional* error (expressed as real space coordinate error $\Delta\mathbf{r}$), which affects the *phase angle*, and error in B-factor and occupancy, which affects *amplitude*. For a single atom, the combined errors reduce the magnitude of the expected value for the observed structure factor. Averaging over the complex distribution of both phase- and position-error renders the expectation value for the partial structure factor somewhat shorter by a factor d as $d \cdot \mathbf{f}_{calc}$ (Figure 7-17). Each complex structure factor for the entire molecule (being the sum of all the atomic structure factor contributions $d \cdot \mathbf{f}_{calc}$) is then described by a complex, bivariate Gaussian distribution (Figure 7-18) with complex variance σ_Δ^2 centered on the vector $D \cdot \mathbf{F}_{calc}$ where the Luzzati[16] factor D $(0 \le D \le 1)$

$$D_\mathbf{h} = \langle\cos 2\pi(\Delta\mathbf{r}_j\mathbf{S})\rangle = \exp(-\pi^3\langle|\Delta\mathbf{r}_j|\rangle^2 S^2 / 4) \tag{12-6}$$

is a measure for the coordinate error of the model (Figure 7-18 illustrates the situation). The general conditional probability for observing a structure factor $\mathbf{F}_{obs}$ given structure factor $\mathbf{F}_{calc}$ from a model with errors is therefore given by a complex 2-dimensional Gaussian likelihood function in the acentric case, and a 1-dimensional Gaussian for centric reflections:

$$prob(\mathbf{F}_{obs}\,|\,\mathbf{F}_{calc}) = \begin{cases} \dfrac{1}{\pi\sigma_\Delta^2}\exp\left(-\dfrac{\left|\mathbf{F}_{obs} - D\mathbf{F}_{calc}\right|^2}{\sigma_\Delta^2}\right) & \text{(acentric)} \\[3ex] \dfrac{1}{2\pi\sigma_\Delta^2}\exp\left(-\dfrac{\left|\mathbf{F}_{obs} - D\mathbf{F}_{calc}\right|^2}{2\sigma_\Delta^2}\right) & \text{(centric)} \end{cases} \tag{12-7}$$

Positional errors and incompleteness in σ_A

In the late 1950s, G. A. Sim derived a structure factor probability distribution based on the proposition that a heavy atom model giving rise to the joint phase probabilities is incomplete. The total structure factor contribution of N atoms (which generate the diffraction pattern) can be formally split up into a contribution of P known atoms (our incomplete model, for which we can calculate the structure factors) and that of Q atoms, which we do not know anything about (but about which we can make some statistical assumptions):

$$\mathbf{F}_N = \mathbf{F}_P + \mathbf{F}_Q = \sum_{j=1}^{P} \mathbf{F}_j + \sum_{j=P+1}^{N} \mathbf{F}_j \tag{12-8}$$

Srinivasan[17] and Ramachandran[18] showed in the mid-1960s that the contributions from positional errors and missing atoms are formally equivalent, when the probability distributions (12-7) are expressed in terms of normalized structure factors $\mathbf{E}$ (Section 7.11). The Luzzati factor D and the variance or the errors σ_Δ^2 are anticorrelated; that is, if the positional errors are large, D will be small, and D and σ_Δ^2 can be combined into a single term σ_A. The values for σ_A thus varies between 0 (when no useful phase information comes from the model) and 1 (when the model is both correct and complete). Substituting $\sigma_A\mathbf{E}_{calc}$ for $D\mathbf{F}_{calc}$ and variance σ_Δ^2 with the corresponding term $1-\sigma_A$ and expressing the probability distribution in normalized structure factors, we obtain for the complex likelihood function:

$$prob(\mathbf{E}_{obs}\mid\mathbf{E}_{calc}) = \frac{1}{\pi(1-\sigma_A^2)}\exp\left(-\frac{\left|\mathbf{E}_{obs}-\sigma_A\mathbf{E}_{calc}\right|^2}{1-\sigma_A^2}\right) \quad\text{(acentric)} \tag{12-9}$$

$$prob(\mathbf{E}_{obs}\mid\mathbf{E}_{calc}) = \frac{1}{\left(2\pi(1-\sigma_A^2)\right)^{1/2}}\exp\left(-\frac{\left|\mathbf{E}_{obs}-\sigma_A\mathbf{E}_{calc}\right|^2}{2(1-\sigma_A^2)}\right) \quad\text{(centric)} \tag{12-10}$$

Because the conditional probabilities (12-9) and (12-10) are expressed as a function of the complex structure factors and not the structure factor amplitudes E_{obs} available from our experimental data, we must integrate out the phase as we showed explicitly in Chapter 7. The result of the marginalization of the phase is again a Rice or Sim distribution for the general case of acentric structure factors

$$prob(E_{obs}\mid E_{calc}) = \frac{2E_{obs}}{1-\sigma_A^2}\cdot\exp\left(-\frac{E_{obs}^2+\sigma_A^2 E_{calc}^2}{1-\sigma_A^2}\right)\cdot I_0\left(\frac{2\sigma_A E_{obs}E_{calc}}{1-\sigma_A^2}\right) \tag{12-11}$$

where I_0 is the modified Bessel function of order zero (Figure 7-16), and for centric structure factors a Woolfson distribution

$$prob(E_{obs}\mid E_{calc}) = \left(\frac{2}{\pi(1-\sigma_A^2)}\right)^{1/2}\exp\left(-\frac{E_{obs}^2+\sigma_A^2 E_{calc}^2}{2(1-\sigma_A^2)}\right)\cdot\cosh\left(\frac{\sigma_A E_{obs}E_{calc}}{1-\sigma_A^2}\right) \tag{12-12}$$

Estimating σ_A and figure of merit from experimental data

To use σ_A in practice we need to obtain an estimate for σ_A given the experimental data and the incomplete model. Sigma-A is defined as

$$\sigma_A = D(\Sigma_P/\Sigma_N)^{1/2} \tag{12-13}$$

where Σ_N is the variance of the structure factor probability distribution for the complete structure containing N atoms and Σ_P the variance of the structure factor distribution for the partial model. It can be obtained from the experimental structure factors, as shown in Chapter 7, from

$$\Sigma_N = \sum_{i=1}^{N} f_i^2 = \langle F_{obs}^2/\varepsilon\rangle \tag{12-14}$$

and the variance Σ_P is readily computed from summation over the scattering factors from the P known atoms in the partial model structure. As the Luzzati factor D is a function of the scattering vector $\mathbf{S}$ and thus resolution dependent

$$D_{\mathbf{h}} = \langle \cos 2\pi(\Delta\mathbf{r}_j\mathbf{S}) \rangle = \exp(-\pi^3 \langle |\Delta\mathbf{r}_j| \rangle^2 S^2 / 4) \tag{12-15}$$

σ_A needs to be calculated as a function of resolution. This can be achieved by maximizing the likelihood function of σ_A as suggested by Lunin and Urzhumtzev[19] and extended to both centric and acentric structure factors by Read.[20]

Computing σ_A. One way to compute the estimate of σ_A is through iterative solution of zero for the residual Q within ~20 resolution bins:

$$Q(\sigma_A) = \sum_{bin} w(\sigma_A - m E_{obs} E_{calc}) \tag{12-16}$$

where the weight w is 2 for acentric reflections and 1 for centric structure factors and m is the figure of merit. This requires in turn obtaining the figure of merit. In the absence of knowledge of the true model phases, for each reflection we need to use the estimated weights from the probability distributions which are simply the Sim weights for acentric and for centric reflections, respectively

$$m_a = \frac{I_1(X)}{I_0(X)} \quad \text{and} \quad m_c = \tanh\left(\frac{X}{2}\right) \tag{12-17}$$

with their argument X expressed in terms of σ_A:

$$X = \frac{2\sigma_A(\Sigma_P / \Sigma_N) F_{obs} F_{calc}}{\varepsilon\Sigma_P(1 - \sigma_A^2)} \tag{12-18}$$

The method of estimating σ_A described above is implemented in the program *SigmaA*[20] in the CCP4 suite.

An alternate method for estimating σ_A is through approximation by fitting a Wilson-like two-term exponential[21]

$$\sigma_A = \sigma_{A,0} \exp(-c_0 S^2 / 4)\left[(1 - \sigma_{A,1})\exp(c_1 S^2 / 4)\right] \tag{12-19}$$

to the experimental data. For a single partial structure, this fit needs only four parameters. The small number of parameters for this fit allows σ_A to be obtained from a fit against the cross-validation data that are excluded from refinement, which avoids bias and the resulting overestimation of σ_A. This form of σ_A estimate is implemented, for example, in the refinement program *REFMAC*.[22]

Estimating coordinate error and model completeness from σ_A

Having obtained an estimate of σ_A, we can in turn exploit its direct relation to the Luzzati D factor to estimate the coordinate error of our model based on the experimental data. With

$$\sigma_A = D(\Sigma_P / \Sigma_N)^{1/2} \tag{12-20}$$

and D a function of normally distributed coordinate errors[16]

$$D_{\mathbf{h}} = \exp(-\pi^3 \langle |\Delta\mathbf{r}_j| \rangle^2 S^2 / 4) \tag{12-21}$$

and taking the logarithm we obtain

$$\ln\sigma_A = \frac{1}{2}\ln(\Sigma_P / \Sigma_N) - \pi^3 \langle |\Delta\mathbf{r}| \rangle^2 S^2 / 4 \tag{12-22}$$

Function 12-22 can be fitted as a straight line against $S^2/4 = (\sin\theta/\lambda)^2$, whose intercept $\ln(\Sigma_P/\Sigma_N)$ provides an estimate for the overall completeness of the model, and from the slope $\pi^3 \langle |\Delta\mathbf{r}_j| \rangle^2$ the mean coordinate error $\langle |\Delta\mathbf{r}_j| \rangle$ can be calculated. This value for the mean coordinate error, among other and in practice better defined estimates such as the diffraction precision index (Section 12.2), is one of the overall coordinate r.m.s.d. estimates provided by maximum likelihood refinement programs. Because no bulk solvent contribution is included in this estimate of structure factor variances (Chapter 11), the straight line fit excludes low resolution reflections, while very high resolution reflections are omitted because of large measurement errors.

Generating bias minimized maximum likelihood maps

We finally need to reduce our results from maximum likelihood parameter estimation to practice and construct electron density maps.

Accounting for incompleteness. In electron density reconstruction with $F_{obs} \cdot \exp(i\varphi_{calc})$ coefficients, atoms missing from nearly complete models are displayed only at half the density level, and even less when the model is very incomplete.[23] Selecting $(2F_{obs} - F_{calc}) \cdot \exp(i\varphi_{calc})$ map coefficients brings these missing atoms back up to normal density levels. Main[24] has expanded this further by showing that $mF_{obs} \cdot \exp(i\varphi_{calc}) \simeq \frac{1}{2}F_{obs} + \frac{1}{2}F_{calc}$ (note the complex sum), and that, as a consequence, improved map coefficients can be obtained from a map using figure-of-merit weighted coefficients $(2mF_{obs} - F_{calc}) \cdot \exp(i\varphi_{calc})$ for general acentric reflections.

Accounting for incompleteness and errors. In addition to incompleteness, we also need to consider that the model has errors, and Read[20] has extended the above treatment to include errors and incompleteness in the map construction. We begin by applying the cosine rule to the complex difference δ between normalized acentric structure factors of the true model and calculated model $\mathbf{E}_{true} - \sigma_A \mathbf{E}_{calc}$:

$$\delta^2 = (\mathbf{E}_{true} - \sigma_A \mathbf{E}_{calc})^2 = E_{true}^2 + \sigma_A^2 E_{calc}^2 - 2\cos(\varphi_{true} - \varphi_{calc})\sigma_A E_{true} E_{calc} \qquad (12\text{-}23)$$

We now have introduced via the cosine (phase difference) term the figure of merit m and after expanding with $\exp(i\varphi_{calc})$ and some rearrangement[20] we obtain for the estimate of the true structure factor $\mathbf{E}_{true}$ the maximum likelihood (ML) coefficients

$$\mathbf{E}_{ML} = \mathbf{E}_{true} \simeq (2mE_{true} - \sigma_A E_{calc})\exp(i\varphi_{calc}) \qquad (12\text{-}24)$$

Having eliminated the unknown true phase, we can now use the model phases and substitute the observed amplitudes as our estimate of the true structure factor amplitudes. Converting back to structure factors yields

$$\mathbf{F}_{ML} = (\varepsilon\Sigma_N)^{1/2}(2mE_{obs} - \sigma_A E_{calc})\exp(i\varphi_{calc}) \qquad (12\text{-}25)$$

Finally, replacing σ_A with $\sigma_A = D(\Sigma_P / \Sigma_N)^{1/2}$ and using the definition of normalized structure factors (Chapter 7) $F_\mathbf{h} = (\varepsilon_\mathbf{h}\Sigma)^{1/2} E_\mathbf{h}$ we obtain

$$\mathbf{F}_{ML} = (2mF_{obs} - DF_{calc})\exp(i\varphi_{calc}) \qquad (12\text{-}26)$$

for the general acentric case. Following the same derivation, the result reduces to

$$\mathbf{F}_{ML} = mF_{obs}\exp(i\varphi_{calc}) \qquad (12\text{-}27)$$

for centric reflections.

Maximum likelihood map coefficients

The ML coefficients $\mathbf{F}_{ML}$ provide our best estimate $\mathbf{F}_{BEST}$ for the expectation value of the "observed" complex structure factors. Maps constructed from Fourier coefficients $(2mF_{obs} - DF_{calc}) \cdot \exp(i\varphi_{calc})$ for acentric reflections and $mF_{obs} \cdot \exp(i\varphi_{calc})$ for centric reflections have the minimal residual error and minimal phase bias achievable through maximum likelihood estimates of the structure factor amplitudes.[25] Such maps are generally used in model building, except in instances of very early models where more aggressive, iterative methods of bias removal such as automated dummy atom rebuilding and refinement and/or multi-model map averaging can be applied.[26] Although ML coefficients are the standard for map generation, they cannot overcome and correct severe phase bias in one single step of map generation, and multiple rounds of rebuilding and refinement will be necessary to arrive at the correct models.

Refinement programs generally either provide a pre-computed electron density map in a map format that model building programs can read, or include the option to list the map coefficients in a formatted reflection-list format (h, k, l, F_{wt}, φ_{wt}) that is suitable for internal Fourier reconstruction in model building programs such as *XtalView*[27] or write the CCP4 *mtz* format for *Coot*.[28]

There are two additional issues we need to consider when using actually measured F_{obs} amplitudes in the map reconstruction: (i) We may not have measured values for certain reflections, and (ii) the term $2mF_{obs} - DF_{calc}$ may actually become negative. Eliminating both types of reflections in the reconstruction would increase the noise in the maps.[22] Missing reflections with $F_{obs} = 0$ can be simply approximated by their expectation value DF_{calc} in the $2mF_{obs} - DF_{calc}$ ML maps, and are set to zero in ML difference maps. If the difference $2mF_{obs} - DF_{calc}$ becomes negative (i.e., the ML vector would point in the opposite direction), we use $|2mF_{obs} - DF_{calc}|$ as coefficients but must apply the opposite phase $\varphi_{wt} = \varphi_{calc} + 180°$ in the map calculation. For that reason, the map coefficients are commonly written with φ_{wt} instead of φ_{calc}, although φ_{wt} is not really a weighted phase. The forms of the ML Fourier coefficients are summarized in Box 12-3.

Note that the calculation of maximum likelihood map coefficients also avoids the necessity of applying arbitrary resolution cutoffs, with the associated risk

---

Box 12-3 Maximum likelihood map coefficients. The incorporation of model incompleteness and model errors in the σ_A (sigma-A) variance term for model structure factors leads to maximum likelihood map coefficients $\mathbf{F}_{ML} = (2mF_{obs} - DF_{calc}) \cdot \exp(\varphi_{wt})$ for acentric reflections and $\mathbf{F}_{ML} = mF_{obs} \cdot \exp(\varphi_{wt})$ for centric reflections. Electron density maps and difference maps constructed from maximum likelihood Fourier coefficients provide minimally biased electron density maps used as the standard in model building.

Summary. The maximum likelihood (ML) Fourier coefficients for ML map reconstruction

$$\rho_{ML}(xyz) = \sum_{\mathbf{h}=-\infty}^{+\infty} \mathbf{F}_{ML} \cdot \exp(-2\pi i \mathbf{h}\mathbf{x})$$

take the general form
$\mathbf{F}_{ML} = F_{wt} \exp(\varphi_{wt})$

Amplitude and phase terms of the ML coefficients for $2mF_{obs} - DF_{calc}$ ML maps:

$F_{wt} = |2mF_{obs} - DF_{calc}|$ with $\varphi_{wt} = \varphi_{calc} + 180°$ for $2mF_{obs} - DF_{calc} < 0$, else $\varphi_{wt} = \varphi_{calc}$

$F_{wt} = DF_{calc}$ if $F_{obs} = 0$

$F_{wt} = mF_{obs}$ for centric reflections.

For difference maps the amplitude and phase terms of the ML coefficients are
$F_{wt} = |mF_{obs} - DF_{calc}|$ with $\varphi_{wt} = \varphi_{calc} + 180°$ for $mF_{obs} - DF_{calc} < 0$, else $\varphi_{wt} = \varphi_{calc}$

$F_{wt} = 0$ if $F_{obs} = 0$

The argument for flipping the phase to $\varphi_{wt} = \varphi_{calc} + 180°$ when the amplitude part of the ML coefficients becomes negative is that amplitudes of complex numbers by definition are always positive.[29] In practice, omitting the absolute values and always setting $\varphi_{wt} = \varphi_{calc}$ works as well. Verify what your respective program (and version thereof) actually lists, particularly if you plan to use your own programs and ideas to further process the data—the tabulation above is for *REFMAC* 5.4.

of truncation effects (Chapter 9). Unreliable high resolution reflections will be weighted down accordingly so that they provide only minimal or zero contributions to the maps.

12.2   Refinement of macromolecular structure models

To correct the countless small local stereochemical and conformational errors remaining after real space model building, the model is subjected to global reciprocal space restrained refinement. The refinement procedure is conceptually easy to understand. The parameters describing structure model—generally the model coordinates, B-factors, and overall parameters such as scale factor and overall B-factors, bulk solvent corrections, and anisotropy corrections—are refined against all experimental data, so that we obtain a best fit between the observed structure factor amplitudes and the computed model structure factor amplitudes. The overall fit between diffraction data and model is numerically quantified by a global linear residual (the R-value) between the scaled structure factor amplitudes F_{obs} and F_{calc}:

$$R = \frac{\sum_{\mathbf{h}} \left| F_{obs} - F_{calc} \right|}{\sum_{\mathbf{h}} F_{obs}} \tag{12-28}$$

A single round of reciprocal space refinement usually consists of several internal cycles of parameter adjustment and optimization until the refinement reaches convergence, where no more parameter shifts occur.

Observations, models, target functions, and algorithms

A number of terms need to be defined and clearly distinguished when discussing reciprocal space refinement. The most relevant terms for our discussion are: the observations, the model and its parameters, the refinement target function, the optimization algorithm or refinement engine, and the refinement protocol.

Observations, constraints, and restraints. In the general context of optimization, observations are either specific experimental observations for our particular crystal structure (i.e., the measured structure factor amplitudes $F_{\mathbf{h}}$ and their standard uncertainties, σ_F), or they can be *universally valid* observations reflecting our knowledge about protein structure. The former specific observations are simply called the data, and the latter generally valid observations are restraints and constraints. Typical examples of restraints that are valid for all protein structures are bond lengths and bond angles, while constraints establish fixed relations between some model parameters. In the example of a surface-exposed side chain that is modeled in split conformations, the sum of partial occupancies is constrained to 1.0.

Model and parameterization. Our protein structure model is described by a mathematical model of p continuously variable parameters which we group into a parameter vector $\mathbf{p}$. The most obvious variable parameters that define each of the individual atoms our physical model is composed of are the atomic coordinate position x, y, z, B-factor, and in some instances, occupancy n. In addition to the specific atomic parameters, there are overall model parameters such as anisotropic scale factors and overall B-factors needed to bring calculated and observed data onto a common scale, and other parameterized improvements to our model such as bulk solvent corrections. The model can be described or *parameterized* in different terms; for example, basic isotropic B-factor parameterization can be replaced by a more complex and realistic translation-libration-screw model, or TLS parameterization. Another popular parameterization for low resolution models is torsion angle parameterization; the refinement procedure is then called torsion angle refinement.

Refinement target function. Obtaining the best fit between the experimental data and the model is equivalent to obtaining the minimum residual of a target

---

Sidebar 12-2 Local real space errors and global reciprocal space refinement. Given that we use the overall match of observed and calculated structure factors as a measure of our model quality, one point deserves immediate attention. Recall that the structure factor for each reflection $\mathbf{h}$ is a nonlinear function of each and every atom in the unit cell of the crystal structure:

$$\mathbf{F_h} = \sum_{j=1}^{atoms} f_j^B \cdot \exp(2\pi i \mathbf{h} \mathbf{x}_j). \tag{12-29}$$

The consequence of the summation in (12-29) over all atoms in the unit cell is that a local error that affects a handful of atoms will disperse itself over all the reflections in the diffraction pattern. The linear R-value

$$R = \sum_{\mathbf{h}} \left| F_{obs} - F_{calc} \right| / \sum_{\mathbf{h}} F_{obs} \tag{12-30}$$

is thus no more than a rough global quality indicator for a protein structure—it is only an overall measure that tells us nothing about which parts of the model are good and which are not. A second consequence of a local error being able to disperse itself over all reflections and the entrapment in false local minima is the tendency of a refined structure to cement in errors or bias from which the reciprocal space refinement cannot recover. Therefore, the repeated alternation between local real space rebuilding and global reciprocal space refinement is in general quite a successful combination. We have also discussed in Chapter 10 the improved performance of real space–reciprocal space cycling in dual space direct methods used for substructure solution.

function. This is generally a sum of squared residuals of a form specific to the target function. As shown in Section 7.9, the squared residual is the second, quadratic term of the Taylor series expansion of the corresponding probability distribution at its maximum value, and thus the residual is generally a squared term. The most basic target function is a least squares target, which is a special case of a maximum likelihood target when the variance of the error distribution is purely Gaussian. Knowing that we must consider errors and incompleteness of our model, we correctly expect that a maximum likelihood target function that accounts for errors and incompleteness of the model will give superior refinement results in such cases. Note, however, that the more correct and complete the model becomes, the more ML and conventional least squares as a special ML case converge. This is the reason why the versatile least squares program *SHELXL* performs so well for high resolution structure refinement.[30]

Assume that we have n discrete observations of a quality X_{obs}, each with a corresponding experimental variance σ^2_{obs}, and a corresponding mathematical model that predicts the function values X_{calc}. Let the model be defined by p continuously variable parameters which we group into a parameter vector $\mathbf{p}$. In a basic least squares target function, we assume that the model function is correct, and the LSQ residual Q_{LS} is then

$$Q_{LS} = \sum_{i=1}^{n} \frac{\left(X_{obs}(i) - X_{calc}(i,\mathbf{p}_{(1,..,p)})\right)^2}{\sigma^2_{obs}(i)} \tag{12-31}$$

In likelihood models we will also have assumptions about the model error, expressed as a variance term for the model—a typical measure of the model error is the sigma-A value as we have shown already. The residual Q_{ML} of a general maximum likelihood target function will therefore have the form

$$Q_{ML} = \sum_{i=1}^{n} \frac{\left(X_{obs}(i) - \langle X_{calc}(i,\mathbf{p}_{(1,..,p)})\rangle\right)^2}{\sigma^2_{obs}(i) + \sigma^2_{calc}(i,\mathbf{p}_{(1,..,p)})} \tag{12-32}$$

The brackets indicate that in the maximum likelihood case the calculated properties are expectation values defined by the corresponding likelihood functions (i.e., probability distributions), with an estimated error given by the variance $\sigma^2_{calc}(i,\mathbf{p}_{(1,...,p)})$ of the likelihood function. The important difference between the least squares residual and the likelihood formulation is the allowance of nonrandom errors for the proposed model.

Optimization algorithm. How we actually minimize our target function (or maximize its likelihood) depends to a degree on the choice of target function. There are various mathematical optimization algorithms available, ranging from gradient methods to stochastic optimization methods such as simulated annealing, the latter often applied together with torsion angle parameterization. The selection of refinement algorithm depends on suitability for the chosen refinement protocol and amongst other technical reasons is decided by numeric stability, availability of first and second derivatives of the function, convergence radius (Figure 12-7), computational speed, and ease of implementation. Simulated annealing in combination with torsion angle refinement, for example, has a large radius of convergence, at the price of slower speed and lower coordinate accuracy (see also Figure 12-25).

Refinement protocol. The term refinement protocol loosely describes the overall procedure used for refinement. This can, for example, be unrestrained refinement, restrained full-matrix least squares coordinate refinement, restrained maximum likelihood refinement, real-space dummy atom placement alternated with unrestrained maximum likelihood dummy atom refinement, or simulated annealing torsion angle refinement. Details about the procedure, such as how many model building rounds were iterated with reciprocal space refinement and other details, are also often provided under a more extensive description entitled "refinement protocol."

Figure 12-7 Local minima and radius of convergence. The figure visualizes the concept of trapping in a local minimum for a real space scenario. The $C\varepsilon_2$ atom of the misplaced Phe ring is trapped in the electron density of a water molecule, in which it happens incidentally to fit quite well. In such cases, a refinement program may not be able to proceed upwards over the "activation" barrier—or may allow only limited positional parameter shifts—that prevent the large movement of the entire ring out of the partial density until it snaps into the correct electron density. Increased ability to overcome local minima by allowing "upwards movement" during parameter search implies higher radius of convergence and higher probability to approach the global minimum, generally at the cost of more computation and lower accuracy.

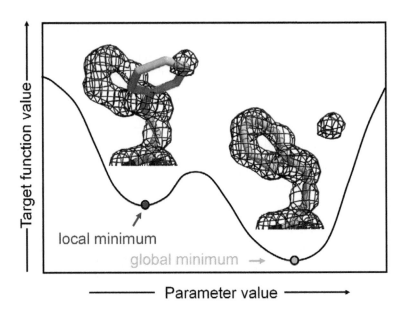

The data-to-parameter ratio

The concept of "refinement" requires that we actually have more experimental data points n available than adjustable model parameters p that need adjustment. The basic idea in refinement is that a system of p independent simultaneous equations is solved. This clearly cannot be done if $n < p$; there are either not enough data points or too many parameters, that is, the model is overparameterized and the problem is underdetermined. At the absolute determinacy limit, there must be at least $n = p$ observations, where a linearized system solves with a unique eigenvalue for each of the p parameters in our parameter vector **p**. Note that in this case the residual of the solution (not a "fit" anymore) is indeed zero.

The situation is most easily understood using the example of a linear fit. The function describing a linear model is the equation for a straight line

$$y = p_1 + p_2 x \tag{12-33}$$

which provides the values of the dependent variable y for the independent variable x as a function of two parameters, intercept (p_1) and slope (p_2). Leaving aside any error in the data, the refinement target function is then the basic least squares residual

Box 12-4 **Refinement basics.** During refinement the parameters describing a continuously parameterized model are adjusted so that the fit between discrete experimental observations and their computed values is optimized, as calculated from a target function. Observations can be experimental data specific to the given problem such as structure factor amplitudes, or general observations that are valid for all models. Stereochemical descriptors valid for all models such as bond angles, distances, torsions, chirality, and non-bonded interactions are incorporated as restraints to improved the data-to-parameter ratio of the refinement. The most accurate target functions are maximum likelihood target functions that account for non-random errors and incompleteness in the model. Various optimization algorithms can be used to achieve the best fit between parameterized model and all observations, which include measured data and restraints. The radius of convergence for an optimization algorithm describes its ability to escape local minima and approach the global minimum, generally with increased cost in terms of time and lower accuracy.

$$Q_{LS} = \sum_{i=1}^{n} \left(y_i - (p_1 + p_2 x_i) \right)^2 \qquad (12\text{-}34)$$

As long as we have more than two data points, we can fit a straight line through the points yielding an R-value that depends on how well the data are represented by the linear model. In the limiting case of two data points we obtain an exact solution with $R = 0$, and for the underdetermined case of a single data point, a line through it is not defined in orientation and no unique solution exists.

Model parameterization

A linear model may not be the best model describing the data, and crystallographic refinement is in fact highly *non-linear*, because no simple linear relationship exists between the structure factors and the model parameters; recall the basic structure factor formula (12-29). In the example in Figure 12-8 (describing the trajectory of a diffractometer thrown off a cliff) our experimental data (nine data points) are much better described by the physically reasonable model of a second order polynomial that has three parameters than they are by the two-parameter linear fit, as evidenced by a significantly lower R-value. The (patho)logical extension of this idea would be to add as many parameters as possible to the model description and at $n = p$ to obtain a perfect nine-parameter fit with an R-value of zero! At this point, however, we need to ask whether a model with so many additional parameters makes physically any more sense than the simple, three-parameter parabolic fit model (Sidebar 12-3). In the case of an object trajectory in a gravitational field and neglecting friction, any fit higher than second order is physically unreasonable, and higher order polynomial fits are out of the question. In such a simple case, the underlying laws of physics clearly delineate a reasonable *model parameterization* from an arbitrary *empirical fit*.

The need for cross-validation. The situation in the case of highly multivariate models is not as simple as in our previous picture. Consider for example spurious solvent density around the protein molecule in an electron density map. We

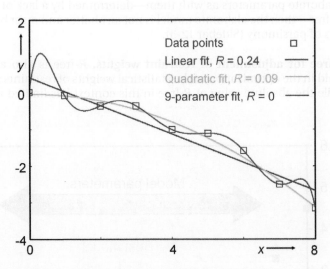

Figure 12-8 Fitting and over-fitting of a function. The data points are measurements of the drop y of a diffractometer pushed over a cliff, taken at constant time intervals x. The linear, 2-parameter model (red line) describes the data poorly, but a quadratic 3-parameter fit (green graph) clearly describes the data very well, as it represents the physically correct model of a parabolic function describing the trajectory of a dropping object. We can further improve the fit (but not the model) by adding more parameters, and a 9-parameter polynomial function (magenta) perfectly fits the data. Despite the perfect fit, the model is definitely nonsense, because the trajectory of the falling object takes upward turns, which is physically impossible. Following Bayesian reasoning the model, despite describing the data well, can be rejected based on a vanishingly small prior probability. In multi-parametric models such as crystal structures, over-fitting is unfortunately much less obvious, and cross-validation is a necessary practice.

Sidebar 12-3 **Numquam ponenda est pluralitas sine necessitate.*** As a general heuristic maxim (or rule of thumb), any explanation of an observation should make as few assumptions as possible, eliminating those that make no significant difference in the observable predictions of the explanatory hypothesis or model. When multiple competing theories—or models in our case—are equal in other respects, the principle recommends selecting the model that introduces the fewest assumptions and postulates the fewest parameters. This is particularly important to keep in mind when fitting poor and spurious electron density with complex ligands: Are they really justified based on crystallographic evidence and is there enough supporting evidence from other sources to justify their modeling? This basic idea of parsimony or the "law of succinctness" dates back to the English logician and Franciscan friar William of Ockham (1288–1347, Occam in Latin spelling) and is known as Occam's razor.

* Multitude must never be proposed without necessity.

can make our structure model more complicated (i.e., introduce more parameters) by building split residues and by indiscriminately placing water molecules into any piece of positive density. According to theory, the R-value will go down simply because we increased the number of refinement parameters, even if the physical model quality did not improve. How do we know whether the additional parameterization of our structure model made sense? This is indeed a universal problem in multivariate model optimization, and it is addressed through cross-validation.

Cross-validation and free R-value

Significance tests[31] and model comparison[32] are fundamental procedures in multivariate statistics and crystallography, and it is surprising that cross-validation was neglected for a long time in macromolecular crystallography. The principle of cross-validation is to set a small fraction of the experimental data aside—just enough to make the cross-validation statistic significant—and exclude them from any subsequent refinement. The agreement between fitted data and model is then computed separately for the working data set and for the excluded cross-validation data set—the "free" data set. The cross-validation R-value for the free data set is thus aptly termed R-free[33] while the R-value for the working data set is R-work:

$$R_{free} = \frac{\sum_{\mathbf{h} \in free} \left| F_{obs} - k F_{calc} \right|}{\sum_{\mathbf{h} \in free} F_{obs}} \text{ and } R_{work} = \frac{\sum_{\mathbf{h} \notin free} \left| F_{obs} - k F_{calc} \right|}{\sum_{\mathbf{h} \notin free} F_{obs}} \qquad (12\text{-}35)$$

Axel Brunger introduced the R-free value[33] and has shown that R-free is related to the mean phase error[34] and is therefore a measure for the phase accuracy and thus for model quality, in contrast to the working R-value. A change to the model that improves its description of physical reality will therefore also improve the fit to the excluded data, while purely cosmetic overparameterization will only lower R-work and not the cross-validation R-free (Figure 12-9). This can be loosely interpreted in terms of hypothesis testing: If the model refines as well without elaborate parameters as with them—determined by a lack of improvement in R-free—then the elaborate model is not any better and must be rejected on grounds of parsimony (Sidebar 12-3).

Use of R-free for adjustment of restraint weights. R-free is also a valuable objective aid in the optimization of the statistical weights of restraints and X-ray terms. While the absolute value of R-free in this context has limited relevance,

Figure 12-9 Cross-validation R-value (R-free). Before the first refinement steps, the experimental data are split into a small test set (~5% of reflections) that is never used in refinement and the working data set. After each successive round of model rebuilding and completion, the current model is refined to convergence and both R-work and R-free are plotted for the corresponding refinement run. Both R-values improve progressively as the model becomes more complete and more parameters are introduced. At a certain stage in refinement, the model will be optimal, and introduction of further parameters into the model (often by unjustified over-modeling of the discrete solvent or split side chain conformations) will not improve the model. At this point, R-free will stop improving, and with further over-fitting R-free will even start to increase again while R-work keeps dropping (exaggerated in the drawing; see Figure 12-41 for an actual example). Given proper weighting, the best model will be the model with the lowest R-free (or to be precise, the highest log-free-likelihood). The gap between R-work and R-free is only a secondary mark of over-fitting; it depends on a variety of parameters (Section 12.2), and observed values of the R-free:R-work ratio show a large variance (Figure 12-24).[35] Note that each individual reciprocal space refinement run itself must be allowed to reach convergence—stopping an individual refinement run when its R-free reaches a transient minimum is bad practice (Sidebar 12-6).

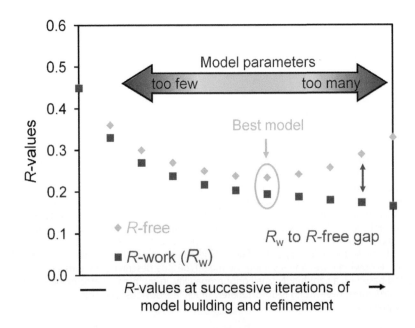

its relative minimum can be achieved through a proper selection of the weight between restraint residual and X-ray residual (revisited in the restraints section). The best model is in fact the one at the minimum negative log-free-likelihood ($-LL_{\text{free}}$), which is generally close to but not necessarily at the lowest R-free. Absolutely crucial is that refinements are allowed to proceed to convergence (where no more parameter shift occurs, see Sidebar 12-10) and not stopped arbitrarily when R-free or $-LL_{\text{free}}$ reaches a transient minimum. R-free (or better $-LL_{\text{free}}$) should only be actually minimized within the same refinement with respect to the weighting parameters (see subsection "Restraint weights and geometry r.m.s.d.").

Exclusion of cross-validation data does not significantly affect model quality. An often asked question is whether the absence of a certain amount of data reserved for the test set does not adversely affect the model refinement. As we have already shown in Chapter 8, the random omission of data does not affect map reconstruction up to considerable fractions of missing data. It can also be shown that based on a modified form of the diffraction precision index[36] (DPI, see subsection "Accuracy of protein structures") that the completeness C of the data affects the coordinate precision with $\sim C^{-5/6}$, so a 5% omission of data increases the coordinate uncertainty only by a factor of 1.04. For map reconstruction, all reflections including the test set reflections are generally used.

Amount of data needed for cross-validation. The significance of R-free is in practice solely dependent on the number of reflections used for its calculation, not on the size of the data set, molecule, or other specific descriptors.[37] The R-free set therefore has to be just large enough to provide a reliable R-free value. Brunger initially estimated[35] that the uncertainty in R-free is proportional to $(N_{ref})^{-1/2}$, which is reasonable to assume because this is how uncertainties vary with sample size. Tickle et al. finally showed[38] that the relative uncertainty in R_{free} is exactly equal to $(2N_{ref})^{-1/2}$ thus confirming Brunger's initial estimate and establishing the constant of proportionality as $2^{-1/2}$. Following this proportionality, ~ 1000 reflections are sufficient to obtain a better than 1% precision for an overall R-free in the 20–30% range.

Use of cross-validation data for σ_A estimation. In order to minimize bias, the *cross-validation* data set is also used to estimate the sigma-A coefficients (Section 12.1). Detailed analysis[39] shows that the optimal number of reflections depends to a degree on the model quality: The better the model, the more conservative the number of reflections selected for the cross-validation data set can be. Again, given a four-parameter fit (Equation 12-19) for the sigma-A coefficients, selecting about 1000–1500 reflections for the cross-validation data set is in any case sufficient to fit accurate sigma-A values for the purpose of model refinement.

Correlation between data. If reflections are not independent, as in the case of refinement against anomalous data, the Friedel mate $-h-k-l$ of an excluded reflection hkl must also be omitted from the working data set. In the case of NCS symmetry, a reflection hkl also affects its neighbors within the range of the reciprocal space interference function G (Figure 11-15). The effect, however, is generally weak enough not to generate appreciable correlation between the test set and working data set in practice,[35] except in cases where the NCS axis is precisely along rational directions in the crystal lattice.[40] A thorough discussion of bias in R-free introduced through complex cases of NCS has been provided by Chapman and colleagues,[41] and improper selection of cross-validation data in the presence of NCS may have been a contributing factor[42] to the problems with the MsbA and EmrE structures (Sidebar 12-8).

Another case of dependence between observations can exist in the case of highly isomorphous data sets, for example between a native apo-protein and its complex with a ligand in the same crystal form. If the model refined from the native data is used to phase the isomorphous complex data set, some memory of the previous refinement will persist even against the new data. Therefore, the R-free data set should contain the same reflections as the original data set; otherwise the R-free set will likely be biased toward the old model.

In some cases the *R*-free set may have been lost, conveniently "forgotten," or is otherwise not available. In this suboptimal case, an *a posteriori R*-free[35] can be computed if the model is perturbed by random coordinate shifts (shaking) or simulated annealing molecular dynamics and then re-refined against the new working data set.

Real space cross-validation

The selection of random reflections from a reciprocal data set for cross-validation has a complementary procedure in real space.[34] The electron density around a questionable area or that part of the model can be omitted from the calculation of model structure factors, and a so-called omit map based on the (back) calculated phases and the observed structure factor amplitudes is computed. The resulting map should then not be biased by the wrong model or density region as it was excluded from electron density reconstruction. Before the advent of ML map coefficients, these basic omit maps[43] were used as a means to reduce model bias, and modified sigma-A coefficients have also been suggested for omit map construction.[44] Omitting questionable parts from the electron density calculation (loops or ligands in poor density are a prime example) and checking whether they appear again is general practice, particularly in the case of molecular replacement models.

Despite omitting questionable parts of the model or map, the refined model itself can still maintain memory of the poor model parts trapped outside of its convergence radius, because its information is dispersed over all calculated structure factors (Sidebar 12-2). Shake-omit or simulated annealing-omit maps[45] are helpful to remove the remaining bias from the refined model. In the shake-omit method, the model is perturbed by introducing random coordinate errors and then re-refined in the absence of the questionable omitted part, and then the map is recalculated using ML coefficients (Figure 12-10). Molecular

Figure 12-10 Cross-validation in reciprocal space and in real space. A subset of unique reflections is set aside (the test or cross-validation set) before the model is refined and is excluded from any further refinement. The model is then refined against the working data set and the progress of the refinement is tracked against the test data set. In a similar fashion, the electron density or model in a questionable region can be removed, the model is again refined (often combined with a bias removal step), and the omitted region is inspected for new electron density. The figure layout follows an idea by A.T. Brunger.[34]

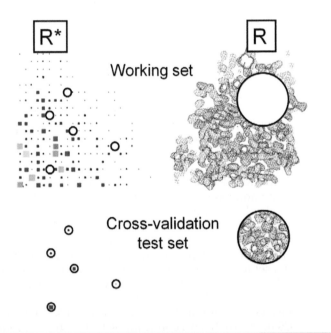

Working set

Cross-validation test set

Box 12-5 Cross-validation. Indiscriminate introduction of an increasing number of parameters into the model can lead to overparameterization, where the fit residual in form of the linear *R*-value still decreases, but the description of reality by the model structure does not improve. The evaluation of the residual against a data set excluded from refinement provides the cross-validation *R*-value or *R*-free. If parameters are introduced that do not improve the phase error of the model, *R*-free will not decrease any further or may even increase.

dynamics-simulated annealing refinement[46] is another option to reduce the cemented-in model bias prevalent particularly in low resolution models, and the resulting MD-SA maps will have reduced bias. The alternative real-space based dummy atom placement method of bias removal is described later in the practical refinement section.

Number of observations from experimental diffraction data

Given that the data-to-parameter ratio n/p is so important for a stable refinement, how many data points and parameters do we actually have available from the experimental observations for our macromolecular crystallographic refinement? Let us compute a quick estimate for our dimeric $P1$ calmodulin structure 1up5 (Figure 11-7). From the PDB file we find a total of 2567 atoms, including eight Ca atoms and about 230 water molecules, modeled as O atoms. Under the assumption that we will not refine any occupancies, we will need at least four parameters (x, y, z, and B) for each atom (10 108 parameters), and also set aside a few for anisotropic scaling and bulk solvent correction, totaling roughly 10 200 parameters for the crystal structure.

Next we need to compute the number of unique reflections in the reciprocal sphere for a given maximum resolution range. We can do this exactly using a program that calculates the structure factors to a given resolution such as *SFALL* from the CCP4 collection, or estimate it from the volume of unit cell (77 150 Å^3) and the Bravais and Laue symmetry of the crystal structure (Sidebar 12-4). In either case, the result is rather disturbing as far as the data-to-parameter ratio is concerned (Figure 12-11).

We clearly recognize from Figure 12-11 that positional and individual B-factor refinement based solely against experimental data—so-called unrestrained refinement—is underdetermined below about 2.5 Å resolution. Only at ultra-high resolution—providing an n/p ratio of about 10 or larger—does unrestrained refinement become viable. In the range in between, unrestrained refinement is at best numerically stable, but there are still not enough supporting data to ascertain that the refined model would even remotely display a reasonable stereochemistry representative of a real protein structure. However, we do have additional, very strong and well established prior information available about the molecular geometry that can be incorporated as restraints to stabilize (or regularize) the refinement.

Stereochemical constraints and restraints

There are in principle two avenues to improve the n/p ratio for our refinement: increase n, the number of observations, or decrease p, the number of parameters.

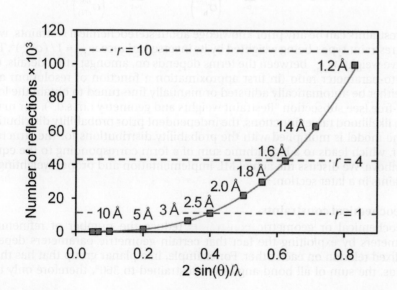

Figure 12-11 Data-to-parameter ratio for protein structures. The graph shows the number of reflections as a function of resolution (in units of $1/d$). The red dashed lines are drawn at numbers of reflections that correspond to a given data-to-parameter ratio r. The number of reflections approaches the number of refined parameters for positional and individual B-factor refinement around 2.5 Å. Below that resolution, unrestrained refinement is underdetermined, and only at atomic resolution does the redundancy of measured data become high enough ($r > 10$) that unrestrained refinement becomes even remotely conceivable. The redundancy levels are generally valid for proteins with a solvent content around 50%. Tighter packing means a smaller unit cell and thus fewer reflections compared with loose packing. For torsion angle only refinement, the n/p ratio is slightly better (Section 12.2); therefore it is often the only available refinement protocol for low resolution (below ~3.5 Å) structures.

Sidebar 12-4 Estimating the number of unique reflections. We can estimate the number of reflections available for refinement at a given resolution d_{min} as follows. The reciprocal sphere volume V_{RS} is given by the reciprocal space volume covered by the resolution sphere of radius d_{min}:

$$V_{RS} = \frac{4}{3}\pi r^3 = \frac{4}{3}\pi\left(\frac{1}{d_{min}}\right)^3 = \frac{4}{3}\pi\frac{1}{d_{min}^3} \qquad (12\text{-}36)$$

The total number of reflections in the reciprocal sphere volume N_{RS} then depends on the volume of reciprocal unit cell; the smaller the reciprocal cell (i.e., the larger the direct cell V_{UC}) the more reflections will be located in the reciprocal sphere. Thus, the total number of reflections in the entire reciprocal sphere is given by

$$N_{RS} = \frac{4}{3}\pi\frac{V_{UC}}{d_{min}^3} \qquad (12\text{-}37)$$

For refinement, the total number of unique reflections is relevant. Therefore, we need to correct this number by the fraction of reciprocal space that is covered according to Laue symmetry, and also consider systematic integral absences due to Bravais centering. The general point position multiplicity z in Table 6-6 provides this number, but we also need to consider that in most cases we refine against merged Friedel pairs, that is, the data in ½ of the reciprocal sphere, and thus divide by $2z$ to obtain the estimate for the number of unique reflections N_U for a given resolution d_{min}:

$$N_U \simeq \frac{2}{3z}\pi\frac{V_{UC}}{d_{min}^3} \qquad (12\text{-}38)$$

To be precise, we would need to apply the exact multiplicity for each reflection type, which is usually lower than the general reflection multiplicity. Therefore, our estimate is slightly pessimistic for high symmetry space groups. A web program for the estimate is available at http://www.ruppweb.org/

Both possibilities can be realized by incorporation of prior knowledge: an observation does not have to be an experimental observation (measured data) specific to the particular structure, but can be any type of general prior knowledge regarding molecular stereochemistry. Known stereochemistry can be exploited and implemented in macromolecular refinement in the form of geometric constraints and restraints.[47]

The first implementations of restrained macromolecular refinement were accomplished by Konnert[48] and Hendrickson[49] in the reciprocal space refinement program *PROLSQ*. Their restraint parameterization has been adopted with various enhancements in practically all subsequent macromolecular refinement programs. The general setup of a restrained target function as a least squares residual will be the sum of an X-ray term or X-ray residual Q_{xray} representing the specific experiment and a generic restraint geometry term or geometry residual Q_{geom} representing our independent prior knowledge:

$$Q_{total} = w_{xray}Q_{xray} + Q_{geom} = w_{xray}\sum_{\mathbf{h}}\frac{\left(F_{\mathbf{h}}^{obs} - F_{\mathbf{h}}^{calc}\right)^2}{\left(\sigma_{\mathbf{h}}^{obs}\right)^2} + \sum_{i=1}^{restraints}\frac{\left(r_i^{obs} - r_i^{calc}\right)^2}{\left(\sigma_{r(i)}^{obs}\right)^2} \qquad (12\text{-}39)$$

The restraints can be any prior knowledge about stereochemical restraints, with each restraint type r being weighted by its inverse variance $w_{r(i)} = 1/(\sigma_{r(i)}^{obs})^2$. The relative weight w_{xray} between the terms depends on, amongst other details, the data-to-parameter ratio (in first approximation a function of resolution) and can either be automatically adjusted or manually fine-tuned to obtain the lowest R-free (see subsection "Restraint weights and geometry r.m.s.d."). For maximum likelihood target functions, the independent prior probability distribution for the model is multiplied with the probability distributions for the structure factor, which leads to a logarithmic sum of a form corresponding to the equation above. We discuss the actual ML implementation and proper weighting of restraints in a later section.

Stereochemical constraints

Stereochemical or geometric constraints reduce the number of refinement parameters by exploiting the fact that certain geometric parameters depend in a fixed relation on each other. For example, in a planar group that has three ligands, the sum of all bond angles is constrained to 360°, therefore only two angles instead of three suffice to define the geometry. Another frequently appearing constraint is that the sum of partial occupancies of protein side chains must be 1.0. If one branch of a side chain is occupied at 0.5, then the occupancy of the other branch automatically must be 0.5 as well. Each of these constraints reduces the number of refinement parameters, but unfortunately, the number of "hard" constraints available in macromolecular refinement is limited, and does not increase our data-to-parameter ratio significantly.

Another implicit use of a geometric constraint is fixing of covalently bound hydrogen atoms in their computed, theoretical riding positions. Including them in the refinement does not therefore increase the number of parameters while improving the model structure factor calculation. Rigid body refinement can also be viewed as constrained refinement, with the distances between atoms in the molecule (or certain groups of the molecule) maintaining a fixed relation throughout translational and rotational shifts (see also Chapter 11). One of the first refinement programs that could also successfully handle nucleic acids was based on combined constrained–restrained least squares refinement (*CORELS*).[50, 51]

Bond distance, bond angle, and dihedral restraints

The alternative to decreasing the number of parameters via "hard" *constraints* is to improve the data-to-parameter ratio by adding stereochemical "soft" *restraints* serving as additional observations. We know from accurate and precise small molecule and peptide fragment structures (deposited in the Cambridge Structural Database, CSD),[52] that bond lengths and bond angles show distinct and relatively narrow distributions around their mean positions.[53] The standard uncertainties for the mean value of the observed target values also provide an approximate upper limit for the bond length and bond angle r.m.s.d. of our refined protein model.[54] A few common bond lengths are listed in Table 12-1.

Bond angles and dihedral angles can be expressed in the form of distance criteria between further distant atoms. Therefore, only one general restraint function may be used to describe the 1-2 (bond), 1-3 (angle), and 1-4 (dihedral) distance restraints, which greatly simplifies the restraint setup (Figure 12-12). Whether the restraints are implemented as 1-3 distance criteria or actual bond angles depends on the program. *SHELXL* use 1-3 distances while *REFMAC* and *TNT* as well as *XPLOR* and its successors *CNS* and *PHENIX* use direct angle implementation. Note that in this context, dihedral angle, in contrast to a torsion angle χ (which is also a dihedral angle) refers only to the 1-4 dihedrals in planar groups that can assume only one single value, while torsions may have up to six preferred angles in the range 0–2π.

To make the correspondence of the geometric restraint setup to the general SRS target clear, we express every restraint target value as an observation, and the

Peptide bond	Bond length (Å)	Single bond	Bond length (Å)
Cα−C	1.525 ± 0.026	C−C	1.540 ± 0.027
C−N	1.336 ± 0.023	C−N	1.489 ± 0.030
N−Cα	1.459 ± 0.020	C−O	1.420 ± 0.020
C=O	1.229 ± 0.019	C−S	1.807 ± 0.026
Cα−Cβ	1.530 ± 0.020	S−S (disulf.)	2.033 ± 0.016

Table 12-1 Common bond lengths. Mean values and their variance (expressed as σ or r.m.s.d.) for some of the most common 1-2 distances for covalent bonds and hydrogen bonds in protein structures. A complete tabulation including residue-specific distances, angels, dihedrals, and torsions can be found in Engh and Huber's recent compilation[55] in ITC Volume F. Refinement programs read restraint target values specific for each residue and each ligand molecule from restraint library files, which are subject to continuous empirical update.

Figure 12-12 Distance and torsion restraints. Bond distances, bond angles, and fixed dihedrals in planar groups are implemented in a common restraint function as specific 1-2, 1-3, and 1-4 distances (d_n). In contrast to dihedral angles d_n(1-4) restrained around one single value, torsions χ_n(1-4) can assume multiple preferred angles or minima. The special backbone torsions such as the ψ-torsion of the serine ψ_{Ser} depend on the secondary structure environment and are therefore not, or only weakly, restrained through energy or vdW terms, depending on the type of refinement parameterization. The absence of any strong restraints makes the backbone φ/ψ torsions quite useful for "geometric cross-validation" in the Ramachandran plots.

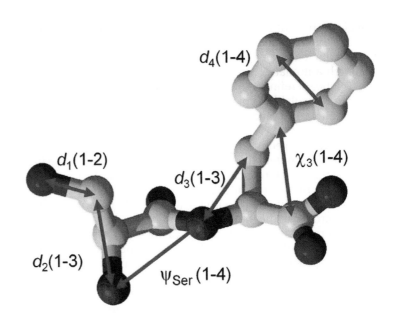

corresponding model value as the calculated term. For the distance restraints D we then obtain

$$Q_{\mathrm{D}} = \sum_{j=1}^{dist} w_{d(j)}(d_j^{obs} - d_j^{calc})^2 \tag{12-40}$$

where the summation extends over all j restrained 1-2, 1-3, and 1-4 distances. The individual restraint weights are determined by the inverse variance of the corresponding restraint type, $w_{d(j)} = 1/\sigma_{d(j)}^2$. This convention holds for all subsequent individual restraint weights.

Planarity and chirality restraints

The phenyl, indole, and imidazole rings of Phe, Tyr, His, and Trp, the guanidyl groups of Arg, the carboxyl and amide side chain groups of Asp, Glu, Asn, and Gln, and the backbone peptide bonds are all planar. Planar restraints are relatively strong and they are implemented as a deviation of the individual atoms from their common least squares plane (Figure 12-13). In the basic planar restraint function Q_{P}, the deviation from planarity for an atom j is expressed as the difference between the vector product $\mathbf{m}_k$ (the unit vector normal to the LSQ plane k) and $\mathbf{r}_{j,k}$, the atomic position vector of each atom j of plane k, and the difference to the scalar distance of each model atom to the LSQ plane, d_j:

$$Q_{\mathrm{P}} = \sum_{k=1}^{\substack{planes}} \sum_{j=1}^{\substack{coplanar \\ atoms}} w_{\mathrm{P}(j,k)}(\mathbf{m}_k\mathbf{r}_{j,k} - d_j^{calc})^2 \tag{12-41}$$

The drawback of this planar restraint implementation is that it defines and minimizes toward a LSQ plane about whose true location we actually do not know anything. More sophisticated implementations exist; *REFMAC*, for example, minimizes the eigenvalue of the product moment matrix of the atoms in the plane.[29]

The numeric values of the restraints discussed so far were *continuous*, and therefore can take any value. The situation is somewhat different for restraints that address discrete properties such as chirality—the chirality of a C_α center of any amino acid or the C_β in Thr or Ile is either correct or wrong. The problem of discrete values in restraints (recall that our parameter vector $\mathbf{p}$ determining the mathematical model has to be continuous) can be circumvented by computing a corresponding continuous value such as a chiral volume V_j, which is given by a scalar triple vector product $\mathbf{A} \cdot (\mathbf{B} \times \mathbf{C})$ originating at the central atom. With the smallest ligand pointed toward the observer and clockwise assignment of the vectors, the sign of the chiral volume is positive, and computes to about 2.5 Å^3.

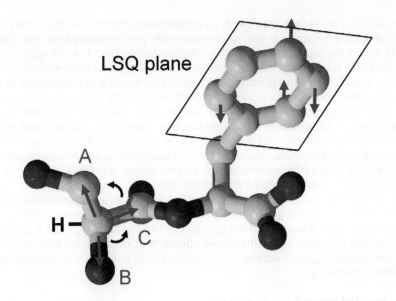

This is the case for L-amino acids with absolute (2S) configuration, where C_α is the center atom and H is the smallest ligand (Figure 12-13). The corresponding chiral residual Q_{CHIV} is given by

$$Q_{CHIV} = \sum_{i=1}^{\substack{chiral \\ centers}} w_{V(i)} (V_i^{obs} - V_i^{calc})^2 \qquad (12\text{-}42)$$

An alternative method of restraining chirality is by restraining the improper torsion (or improper dihedral or ζ-value) between Cα–N–C–Cβ to +34°.

It is important to realize that many of the higher order restraints such as planarity, dihedrals, or chiral volumes are redundant, and they introduce mutual, local dependence of the bond length and bond angle deviations. While this may make the geometry more stable and realistic,[56] it introduces correlations and complicates the assignment of proper weights. Improvement of restraint models[57] and weighting functions[54] in refinement is still an active field of research in computational crystallography.

Side chain versus backbone torsion angle restraints

Side chain torsion angles generally are much more variable than distance or dihedral restraints, and because of the larger variance, their statistical weight will be correspondingly lower. Long aliphatic side chains leading to the positively charged groups in lysine and arginine can assume a number of staggered conformations, while valine, for example, has a strong preference for a single Cα–Cβ torsion with a narrow torsion angle distribution. Side chain torsion angles are generally implemented sets of equal local minima with 1, 2, 3, or 4 periods. A torsion angle with a dihedral angle of 60° with a period of $n = 3$ for example, has minima at 60°, 180°, and 300°, that is, each $360/n$ degrees. The model value is restrained toward the nearest observed torsion angle minimum (therefore a wrongly built torsion is not likely to be corrected by conventional restrained refinement, and real space rebuilding is necessary). The torsion residual Q_T is defined as

$$Q_T = \sum_{t=1}^{\substack{torsions}} w_{\chi(t)} (\chi_t^{obs} - \chi_t^{calc})^2 \qquad (12\text{-}43)$$

Remember that as always *obs* refers to the ideal torsion values, and *calc* to the torsion of our structure model.

Backbone torsion restraints can affect validation. Backbone torsion angles (except the peptide bond torsion ω, which generally carries a medium-strong planarity restraint) assume different distinct ranges of values depending on the

Figure 12-13 Planarity and chiral volume restraints. The planarity restraints are expressed as a deviation of each atom from the least squares (LSQ) plane through its corresponding planar group. Chiral restraints (indicated for the Ser residue) are expressed as chiral volumes given by a scalar triple vector product **A** · (**B** × **C**) originating at the central atom. With the smallest ligand pointed toward the observer and clockwise assignment of the vectors, the chiral volume is about +2.5 Å³, as is the case for the L-α-amino acid Cα atoms. Note that different programs may use different definitions and atom naming conventions in their chiral volume calculation. *SHELXL* for example uses ASCII-alphabetic order of the atom names, and assigns positive chiral volumes for the L-amino acids. In the *REFMAC/PHENIX* monomer library the sign of the chiral volume is explicitly assigned for each atom, and with ligand order N CB C it is always negative for CA. If Cahn–Ingold–Prelog priority rules for absolute configuration (Sidebar 6-13) were strictly applied, cysteine would have R-configuration, all other chiral common amino acids would have S-configuration.

secondary structure environment, as evidenced in the Ramachandran plots.[58] The backbone torsions φ and ψ are normally not restrained to any specific angle values during refinement except for intramolecular 1-4 or higher van der Waals (vdW) repulsions, and can thus be considered as a geometric cross-validation set: If the strong geometric distance and bond angle restraints impose unwarranted good geometry on a grossly incorrectly built structure model, the torsion angles will be the critical point where the errors become obvious in the form of improbable backbone torsions, which the Ramachandran plot will clearly indicate. In certain cases, it might be acceptable to restrain the torsion angles of a low resolution model to the backbone torsions of a structurally similar high resolution model, which is possible in *XPLOR/CNS*, the flexible *TNT* package,[21] and the *PHENIX* project. However, any application of strong specific φ and ψ restraints obscures in principle the value of the Ramachandran plot as an independent geometric cross-validation tool to assess model quality. Nonetheless, the Ramachandran plot is hard to "fudge" even for a genuinely bad model: Kleywegt and Jones have shown that even under extremely high force constants applied during energy refinement, φ–ψ torsions remain robust against over-fitting.[58]

Non-bonded interaction restraints

Non-bonded interactions in protein structures are the major force responsible for intermolecular contacts, side-chain interactions, stabilization of secondary structure, and internal packing interactions in proteins. Non-bonded interactions are weak, and include hydrogen bonds, vdW forces, and ionic (charged) interactions. The latter are non-directional and long-range forces (see Figure 2-29), and are thus not specific enough to serve as geometric restraints. Hydrogen bonds are also multi-directional and only weakly attractive. Except as special hydrogen bond restraints in medium resolution refinement,[59] they are thus included in the repulsive terms of the vdW interactions, which increase rapidly below the equilibrium distance with $\sim r^{-12}$ and thus are excellent "anti-bumping" restraints. The implementation of the r^{-12} repulsive Lennard-Jones terms (Figure 12-18) is rather basic, either as standard quadratic residual or with its energy approximated as a repulsive distance term[49] (*REPEL* function) leading to the following residual:

$$Q_{\mathrm{NB}} = \sum_{n=1}^{\substack{non-\\bonded}} w_{d(n)} (d_n^{obs} - d_n^{calc})^4 \text{ for } d_{calc} < d_{obs}, \text{ else } 0 \qquad (12\text{-}44)$$

Here the weight is the inverse square of the variance of the non-bonded contact distance distribution, and the restraint does not have any effect as long as the model distance is larger than the target distance.

The target value d_{obs} is dependent on the type of interaction and only in second order on the involved atom type. The target values can be grouped into

---

Box 12-6 **Stereochemical restraints.** The known geometry target values for bond lengths, bond angles, and torsion angles as well as planarity of certain groups can be regarded as additional observations contributing to a higher data-to-parameter ratio. In addition, they constitute prior knowledge that keeps the molecular geometry in check with reality during restrained refinement. The geometry targets, chirality values, and non-bonded interactions are implemented as stereochemical restraints and incorporated into the target function generally in the form of a squared sum of residuals in addition to the structure factor amplitude residual. The structure factor amplitude residual is commonly called the X-ray term (or X-ray energy) and the restraint residuals the geometry- or chemical (energy) term. In terms of maximum posterior estimation, geometry target values and their variance define the prior probability of our model without consideration or knowledge of the experimental (diffraction) data.

single-torsion-separated pairs, multiple-atom-separated atoms, or hydrogen bonded atoms. This rather basic implementation of the repulsive vdW terms is one of the reasons they do not prevent Ramachandran outliers; therefore the Ramachandran plot remains a viable diagnostic tool for geometry cross-validation in restrained refinement.

Hydrogen atoms and riding restraints

A form of restraint that does not contribute parameters nor consume data is the addition of hydrogen atoms in riding positions. The locations of hydrogen atoms covalently bound to the refined N, C, and O atoms of amino acids and ligands are well known from basic stereochemistry and can thus be computed. As these hydrogen atoms move rigidly with the atoms they are attached to, they do not need to be refined and add no parameters to the refinement. Although the empirical stereochemical restraints implicitly include the contributions of hydrogen atoms, the inclusion of hydrogen atoms does improve the accuracy of the structure factor calculation and allows calculation of specific hydrogen bonded and specific non-bonded anti-bump restraints. The inclusion of riding hydrogen atoms generally leads to small but significant improvements in model quality and there is no reason not to include them in the refinement. As an example, we show below the refinement statistics listing (Table 12-2) from a refinement with and without added riding hydrogen atoms. In energy refinement, the hydrogen atoms are explicitly used in the calculation of pair-wise interaction potentials.

```
REMARK   3    OTHER REFINEMENT REMARKS: NULL
REMARK   3    FIT TO DATA USED IN REFINEMENT.
REMARK   3     CROSS-VALIDATION METHOD          : THROUGHOUT
REMARK   3     FREE R VALUE TEST SET SELECTION  : RANDOM
REMARK   3     R VALUE      (WORKING + TEST SET) : 0.20557
REMARK   3     R VALUE             (WORKING SET) : 0.20278
REMARK   3     FREE R VALUE                      : 0.25713
REMARK   3     FREE R VALUE TEST SET SIZE   (%) : 5.1
REMARK   3     FREE R VALUE TEST SET COUNT      :    941
REMARK   3
REMARK   3    ESTIMATED OVERALL COORDINATE ERROR.
REMARK   3     ESU BASED ON R VALUE                     (A): 0.323
REMARK   3     ESU BASED ON FREE R VALUE                (A): 0.236
REMARK   3     ESU BASED ON MAXIMUM LIKELIHOOD          (A): 0.157
REMARK   3     ESU FOR B VALUES BASED ON MAXIMUM LIKELIHOOD (A**2): 6.008
REMARK   3
REMARK   3   CORRELATION COEFFICIENTS.
REMARK   3     CORRELATION COEFFICIENT FO-FC       :    0.932
REMARK   3     CORRELATION COEFFICIENT FO-FC FREE  :    0.898

REMARK   3    OTHER REFINEMENT REMARKS:
REMARK   3    HYDROGENS HAVE BEEN ADDED IN THE RIDING POSITIONS
REMARK   3    FIT TO DATA USED IN REFINEMENT.
REMARK   3     CROSS-VALIDATION METHOD          : THROUGHOUT
REMARK   3     FREE R VALUE TEST SET SELECTION  : RANDOM
REMARK   3     R VALUE      (WORKING + TEST SET) : 0.20083
REMARK   3     R VALUE             (WORKING SET) : 0.19824
REMARK   3     FREE R VALUE                      : 0.24848
REMARK   3     FREE R VALUE TEST SET SIZE   (%) : 5.1
REMARK   3     FREE R VALUE TEST SET COUNT      :    941
REMARK   3
REMARK   3    ESTIMATED OVERALL COORDINATE ERROR.
REMARK   3     ESU BASED ON R VALUE                     (A): 0.316
REMARK   3     ESU BASED ON FREE R VALUE                (A): 0.228
REMARK   3     ESU BASED ON MAXIMUM LIKELIHOOD          (A): 0.156
REMARK   3     ESU FOR B VALUES BASED ON MAXIMUM LIKELIHOOD (A**2): 5.980
REMARK   3
REMARK   3   CORRELATION COEFFICIENTS.
REMARK   3     CORRELATION COEFFICIENT FO-FC       :    0.936
REMARK   3     CORRELATION COEFFICIENT FO-FC FREE  :    0.901
```

Table 12-2 Refinement statistics without and with hydrogen atoms added in riding positions. The top listing (yellow highlights) in the table contains an excerpt of the *REFMAC* refinement statistics in PDB format for a structure refined without riding hydrogen atoms, and the bottom listing (green highlights) is for one with riding hydrogen atoms. Note the improvement of the indices *R* and *R*-free as well as in the estimated overall coordinate uncertainty and the fit to the electron density expressed as correlation coefficient (FO-FC here refers to the correlation between F(obs) and F(calc), not "F(obs) minus F(calc)"). Both *R*-free and *R*-work drop by a similar amount (the gap between *R*-work and *R*-free even closes) and the model refined with riding hydrogen atoms is better than the one without.

Non-crystallographic symmetry restraints

In cases of non-crystallographic symmetry (NCS), multiple identical copies of a molecule occupy the asymmetric unit. As they are exposed to the same chemical environment during crystallization, observed differences between them can only be caused by the local packing environment or by genuine, probably dynamic, variation within the molecules. As a consequence, the backbone geometry in core regions will be very similar while side chains on the protein surface or loops may show larger deviations. Particularly for low resolution structures, where we lack observations, strong NCS restraints are very valuable during refinement and generally give lower R-free values. Side chains and loops are generally less restrained than the protein core, and for these regions individual NCS groups can be defined and restrained with correspondingly lower weights. For each group, the j atoms x_j in each copy k are related by the NCS operator $\mathbf{W}_k(\mathbf{R},\mathbf{T})$. The deviation from the mean position of all k related atoms in the group is then the minimization target. With substituting $\mathbf{x}'_{j,k} = \mathbf{R}_k \mathbf{x}_{j,k} + \mathbf{T}_k$ we obtain

$$Q_{\text{NCS}} = \sum_{j=1}^{atoms} w_{\text{NCS}(j)} \sum_{k=1}^{copies} \left(\mathbf{x}'_{j,k} - \langle \mathbf{x}'_{j,k} \rangle \right)^2. \tag{12-45}$$

The least squares refinement program *SHELXL*,[30] which is particularly useful for high resolution structure refinement, has a somewhat different NCS restraint implementation.[60] Instead of positions, it restrains corresponding 1-4 torsion angles between each sequence of four NCS related atoms. Depending on the individual protein, NCS restraints at resolution better than ~2.0 Å may not improve R-free by much, but it is always worth a try. In *XPLOR/CNS* and *Buster/TNT* it is also possible to completely constrain the NCS related structures, which can be useful in the early stages of refinement and at low resolution up to ~2.5 Å. In this case, only one of the molecules needs to be parameterized for refinement, and the other one is just generated from the refined copy by application of the NCS operators.

An interesting point is that the Chothia and Lesk equation (Sidebar 11-5) relating pair-wise Cα backbone r.m.s. coordinate deviation σ_r against the fractional sequence identity s

$$\sigma_r = 0.4 \cdot \exp\left(1.87(1-s)\right) \tag{12-46}$$

does not cross the ordinate at 0, but is offset at a value of 0.4 Å for sequence identity s of 1 or 100%. This rule was based on comparison of different structures refined by *different* individuals using *different* protocols and practices. In the case of NCS, the molecules are identical, and are refined by the *same* person and the *same* protocol and program, and share the same chemical environment (crystallization conditions). As a consequence, differences should in the NCS case originate only from genuine conformational differences. Extended analysis of high resolution NCS structures showed that average backbone Cα differences in fact exist in the range of 0.4–0.5 Å between identical NCS related molecules.[61] This number tends to be lower for lower resolution structures, but is in general above the estimated overall coordinate r.m.s.d. of the individual molecules. Thus, as Gerard Kleywegt put it,[61] NCS related molecules indeed "look more like distant cousins than like identical twins."

Experimental phase restraints

When starting from a new model built into a map obtained from experimental phases, independent and unbiased phase information is available. Particularly in the beginning of the refinement and at low resolution, it is good practice to include this unbiased phase information as a restraint into the X-ray term, resulting in phase restrained refinement. Phase information was first incorporated in energy refinement in *XPLOR/CNS* as a linear residual based on a square-well function with a weight corresponding to the figure of merit for each reflection.[64] More elegant is the implementation in maximum likelihood functions described in a later section. The phase information can be imported either as Hendrickson–Lattman coefficients (Chapter 10) or in the form of calculated

Sidebar 12-5 Macromolecular refinement with *SHELXL*.
The refinement program in George Sheldrick's *SHELX* program suite,[62] *SHELXL*, originally developed as a small-molecule refinement program, has been adapted for macromolecular refinement.[30] *SHELXL* is a least-squares-based refinement program, which makes it applicable for final stages of refinement—particularly high-resolution structures—where the models are generally complete and complex, and contain only minor errors. In this case, least squares and maximum likelihood targets become asymptotically equivalent. *SHELXL*'s small-molecule origin provides a highly flexible implementation of restraints, together with some other unique features. Proper restraint systems for correlated multiple side chains can be set up, and occupancy refinement of partially occupied ligands using only one common occupancy parameter is generally possible with data at ~1.5 Å or better, provided the *B*-factors are reasonably restrained. Other unique features of *SHELXL* include:

- the option to refine against intensities (F^2), not just structure factor amplitudes, which obviates the need for Bayesian estimates of *F* for negative intensities;

- refinement of twinned data, which is superior to refinement of de-twinned data, which generally gives poorer data, particularly with twin fractions approaching 0.5;

- conventional structure factor summation, which is slower but more accurate than FFT-based methods;

- choice of either full or blocked second derivative matrix optimization that includes relevant off-diagonal terms, and provision for inversion of this matrix, which provides accurate coordinate variances (but note that the matrix inversion is computationally intensive);

- a unique "likelihood by intuition" weighting scheme based on the goodness of fit that properly describes the variation of restraint weights with resolution as discussed in the subsection "Restraint setup and data-to-parameter ratio."

SHELXL also determines the Flack parameter[63] x (defined in the range $0 \le x \le 1$ as $F_{h,x}^2 = (1 - x)F_h^2 + xF_{-h}^2$). The Flack parameter indicates the absolute configuration during refinement against anomalous data. If the handedness is chosen correctly, x is close to 0; if the structure needs to be inverted, x is close to 1. This may give useful indications during heavy metal substructure refinement about substructure handedness. During refinement of protein structures (where the handedness is already established), an unexpected Flack parameter may indicate microscopic twinning.

phase and figure of merit. Programs that convert HL coefficients to *fom* presentation are available in CCP4 and other program suites.

Displacement parameter (*B*-factor) restraints

The *B*-factor or atomic displacement parameter is a general measure for the positional displacement of the scattering atoms including dynamic thermal motion, positional displacement resulting from multiple, partially occupied static conformations which cannot be resolved, as well as lattice disorder. In reciprocal space, the atomic displacement leads to additional attenuation of the scattering factor, with the real space equivalent of a broader electron density distribution.

The consequence of the inverse covariance or correlation of the displacement parameter and the occupancy is that in refinement we cannot distinguish low occupancy from high *B*-factor except in ultra-high-resolution structures (the only difference is a slightly different form of the fall-off in the scattering factor and therefore electron density shape; Figure 9-6). The problem of covariance largely affects solvent molecules and ligands, where partial occupancy is frequently the case. Unfortunately, based on its correlation with occupancy, the *B*-factor also absorbs any error in the structure when the refinement program, for whatever reason, does not appreciate scattering electrons at a given atomic position (or desires more scattering electrons). This leads directly to the need to (modestly) restrain the *B*-factors from refining to physically unreasonable values.

Box 12-7 Non-crystallographic symmetry restraints and phase restraints.
Geometric relations and redundancies between identical molecules in the asymmetric unit can be exploited through NCS restraints. Particularly at low resolution, NCS constraints or strong NCS restraints are an effective (and mandatory) means of stabilizing and improving the refinement. Experimental phase restraints are also an effective means in the early stages of model building to stabilize and improve the refinement.

If one atom is practically at rest, an adjacent atom covalently bound to it cannot excessively move around. There must be a restraint in the change of B-factor from one atom to the next one, a situation that is depicted in Figure 12-14. The implementation in the restraint function is generally a type of riding motion, expressed as the B-factor difference from the originating atom to the next target atom:

$$Q_B = \sum_{n=1}^{distance} w_{B(n)} \left(B_n^{origin} - B_n^{target} \right)^2 \tag{12-47}$$

Individual B-factor restraints. Depending on their location in the molecular structure, different types of bonds require different B-factor restraint weights. For example, atoms of solvent-exposed long side chains have more positional freedom and diverge much faster in B-factor than sequential backbone atoms. The relative weighting of the B-factor restraints is not trivial, and refinement programs differ in how they implement and weight these restraints. At the time of this writing, *REFMAC* tends to restrain side chains quite tightly given default values, and the B-factor variances are underestimated and often must be increased. In contrast, *CNS* has rather loose restraints, as evidenced by more rapidly increasing B-factors of flexible side chains in subsequent refinement cycles. For low-resolution structures, it is also possible to restrain the B-factors based on empirical data from high-resolution structures. Only high-resolution structures that have been refined without B-factor restraints (and thus do not introduce bias) can be used as reference data sets. Knowledge-based B-factor restraints[57] are presently implemented only in Dale Tronrud's *TNT* package.[21] The values for the weights (expressed in sigma) are around 5 Å^2 for main chain atoms and buried side chain atoms, and about 7–10 Å^2 on average for the exposed side chain atoms (the variance here is quite different depending on the residue type). These numbers may serve as a guideline for setting the B-factor restraint weights in other programs (subject to optimization using R-free or $-LL$(free)).

Anisotropic displacement factors. Long side chains have more freedom to swing perpendicular to bond directions than along the much more tightly restrained bond length. The consequence is an anisotropic, correlated movement of the atoms visualized in the right panel of Figure 12-14. We could (and in rare cases of high resolution better than ~1.4 Å we can) refine individual B-factors anisotropically. Unfortunately, the description of an anisotropic B-factor with an anisotropy ellipsoid consumes six parameters for the six unique elements U^{ij} of the symmetric anisotropy tensor **U** instead of one parameter for the isotropic B-factor (subsection "Anisotropic displacement and TLS parameterization"). We can rarely afford that many parameters, but it is clear that the isotropic B-factor model is a very poor description of reality, and one of the major reasons for the relatively high R-values around 22–18% we observe in medium resolution (2.5–1.5 Å) protein structures. Only at resolutions below ~1 Å and with careful refinement with *SHELXL*[30] does the R-value ever drop below 10%. The anisotropic displacement restraint function is quite complex, because the directional definition of the anisotropy ellipsoids requires implementation of directional restraints.[49] A compromise that can be applied at lower resolution is the option of a mixed mode available for B-factor restraints for example in *REFMAC* and *SHELXL*, where only atoms that are explicitly flagged anisotropic with an ANISO record in the PDB file are refined with anisotropic parameters and the rest is refined isotropically. This can be useful, for example, when a few heavy atoms that contribute significantly to the scattering show indications of anisotropic electron density.

A more efficient, non-trivial way to implement correlated anisotropic atomic motion in refinement is by TLS (translation-libration-screw) parameterization[65] described in the implementation section.

Restraint setup and data-to-parameter ratio

Let us now investigate how many observations we have gained through the implementation of restraint functions. This is not an entirely trivial exercise, because not all restraint types are independent (planar and chiral restraints are correlated with distances and angles), nor are they necessarily all active

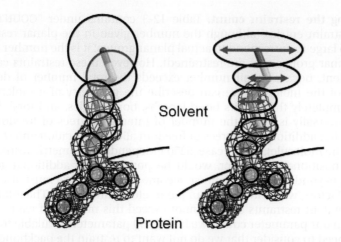

Solvent

Protein

Figure 12-14 Isotropic and anisotropic B-factors. The left panel shows electron density of a solvent-exposed Lys residue with isotropic B-factors of 5 Å^2 for the main chain atoms, successively increasing to 10, 20, 40, 80, and 160 Å^2. The radius of the circles approximates the r.m.s. isotropic displacement $\langle u_{iso}^2 \rangle^{1/2}$ of the atoms calculated from $\langle u_{iso}^2 \rangle = B_{iso}/8\pi^2$. However, it is very unrealistic that the Nζ atom of the lysine moves as much along the severely restrained bond direction (up and down) than perpendicular to it (to the left and right). It is much more likely that the whole chain undergoes some highly correlated, anisotropic movement as illustrated in the right panel, because motions involving torsions require significantly lower energy than is required to stretch covalent bonds. While the anisotropic displacement parameter description is much more realistic, it introduces excessive numbers of parameters into refinement—six instead of one per atom. Note that even the anisotropic parameterization is not entirely satisfactory; the actual motion of the Nζ atom probably lies within some crescent-shaped displacement 'body'.

(the repulsive vdW terms are only active for distances closer than the respective target value). We will examine an actual restraint setup, which is reported by most refinement programs. The relevant header sections of PDB file 2jn9, the small protein trypsin of 223 residues with about 1750 refined atoms, will serve as an example (on average there are eight protein atoms per residue, about 2000 atoms for 250 residues). With four positional parameters x, y, z, and B, per atom, some overall parameters and one or the other occupancy refined, we will need to adjust roughly 6800 parameters during refinement. From the CRYST1 record we estimate the number of unique reflections according to Sidebar 12-4 as 6800 reflections at 2.5 Å, 13 300 at 2.0 Å, and 31 500 at 1.5 Å.

Restraint setup. All refinement programs read the starting model coordinates, and a separate module generates the restraints from what the program perceives as the geometry of the molecule, given its library of stereochemical restraints. The program cannot set up restraints it does not recognize; therefore special links such as covalent bonds to specific modifications, ligands or metals, as well as the restraint description of entirely new entities not included in the library must be externally supplied. *REFMAC*, for example, reads an extensive library of molecular entities, collectively called "monomers," and special bonds can be restrained with a LINK statement.[66] Particular care must be taken to provide proper restraints for new ligands that do not yet have a valid dictionary entry and we discuss restraint generation for ligands separately in Section 13.4. Once all restraints are successfully set up, the refinement proceeds and the Table 12-3 of restraints was obtained at convergence of refinement.

```
CRYST1    53.924    56.695    66.054  90.00    90.00    90.00 P 21 21 21    4
REMARK   3 RMS DEVIATIONS FROM IDEAL VALUES          COUNT      RMS      WEIGHT
REMARK   3   BOND LENGTHS REFINED              (A):   1809 ;  0.010  ;  0.022
REMARK   3   BOND ANGLES REFINED          (DEGREES):   2451 ;  1.324  ;  1.964
REMARK   3   TORSION ANGLES, PERIOD 1 (DEGREES):        244 ;  6.466  ;  5.000
REMARK   3   TORSION ANGLES, PERIOD 2 (DEGREES):         59 ; 40.002  ; 25.932
REMARK   3   TORSION ANGLES, PERIOD 3 (DEGREES):        292 ; 11.833  ; 15.000
REMARK   3   TORSION ANGLES, PERIOD 4 (DEGREES):          2 ; 15.038  ; 15.000
REMARK   3   CHIRAL-CENTER RESTRAINTS      (A**3):       271 ;  0.086  ;  0.200
REMARK   3   GENERAL PLANES REFINED          (A):       2049 ;  0.006  ;  0.020
REMARK   3   NON-BONDED CONTACTS REFINED     (A):        308 ;  0.191  ;  0.200
REMARK   3   NON-BONDED TORSION REFINED      (A):        866 ;  0.174  ;  0.200
REMARK   3   H-BOND (X...Y) REFINED          (A):        131 ;  0.139  ;  0.200
REMARK   3   SYMMETRY VDW REFINED            (A):         41 ;  0.165  ;  0.200
REMARK   3   SYMMETRY H-BOND REFINED         (A):         25 ;  0.149  ;  0.200
REMARK   3
REMARK   3 ISOTROPIC THERMAL FACTOR RESTRAINTS. COUNT   RMS    WEIGHT
REMARK   3   MAIN-CHAIN BOND REFINED    (A**2):        1192 ;  1.739  ;  5.000
REMARK   3   MAIN-CHAIN ANGLE REFINED   (A**2):        1903 ;  2.691  ;  7.000
REMARK   3   SIDE-CHAIN BOND REFINED    (A**2):         646 ;  3.451  ;  9.000
REMARK   3   SIDE-CHAIN ANGLE REFINED   (A**2):         545 ;  4.886  ; 11.000
```

Table 12-3 Restraint table generated by REFMAC. For each implemented restraint type used in the refinement, count, r.m.s.d., and the weight (which is actually given as the variance) are listed. The table will also include entries for the restraints of explicitly refined hydrogen bond lengths. The overall bond and angle r.m.s. deviation from restraint targets of 0.01 Å and 1.3° is typical for a refinement in the resolution range of 2–1.5 Å. The restraints will contribute a substantial number of additional "observations," which make refinement at low resolution possible.

Estimating the restraint count. Table 12-3 contains under "COUNT" about 20 000 restraint entries (although the number given in the planar restraints is much too large to correspond to actual planar groups; it is the number of atoms in the planar groups that are restrained). However, these restraints cannot be independent, because their number exceeds the total number of degrees of freedom of the molecule. We can describe the geometry of a molecule of N atoms completely through its bond lengths, bond angles, and torsion angles, which universally is $3 \cdot N$, the number of internal degrees of freedom of the molecule (an additional 6 degrees of freedom allow for rotation and translation of the entire molecule). In our case, 5250 independent geometric restraints, one for each positional parameter, would be possible. Any additional geometric restraints beyond this number must become redundant. We must also account for the B-factors, which increases our parameter count to $4 \cdot N$. The total number of independent restraints again cannot exceed this number. If we include the B-factor in our parameter count, we have 7000 parameters available to restrain. Now we need to consider that we do not want to restrain the backbone torsions (except in very low resolution torsion angle refinements), which reduces the independent restraint count by two per residue. We thus can gain maximally ~6500 observations in the form of independent restraints. For a single molecule,

$$N(rest) = 4N(atom) - 2N(resi) - 6 \qquad (12\text{-}48)$$

How much each restraint contributes to the stability of the refinement is a non-trivial issue. To count as a full observation, a restraint would have to be independent of all other restraints. Bonds and angles are generally independent, with the exception of rings, where some dependence exists between bonds and angles. This dependence is relatively weak. Planarity restraints are orthogonal to bonds and angles and can be considered independent observations. Chiral restraints in contrast are not independent, because distances and angles already define the geometry around the central atom. A restraint between B-factors of atoms in a 1-2 (bond) distance certainly has an effect on how much the atoms in a 1-3 distance (angle) can deviate from each other. In addition, the actual effect that a restraint has as an observation also depends on the resolution. A bond length of 1.5 Å will be much less important in a 1.3 Å data set, where this distance is sampled by the diffraction experiment, compared with a lower resolution data set, where such data are absent.

Improvement in data-to-parameter ratio. The otherwise underdetermined refinement at 2.5 Å of ~7000 parameters with 6800 reflections becomes quite feasible with a total number of ~13 200 "observations" (6800 reflections and 6500 independent restraints). An empirical fit shows that the number of independent restraints is approximately ~$0.95 \times N$(parameters),[29] which is consistent with our estimate (12-48). In the case of refinement at 1.7 Å (13 300 reflections), we increase our number of observations to ~19 800, yielding a sufficient n/p of ~2.8. However, recall that this value for n/p is not really a true and strict statistic. The 1-2 distance restraints for example will have substantial impact, a "soft" torsion restraint much less. One way to quantify the impact of restraints on refinement in statistical estimates is through normalized restraint residuals.[67]

Can we attempt individual anisotropic B-factor refinement at 1.7 Å resolution? The parameter count immediately jumps from four per atom to nine per atom. What about a gain in restraints? Following Hirshfeld's rigid-bond postulate[68] only one anisotropic restraint per bond is physically justifiable, because for bonded atoms only the components of the **U** tensor in the bond direction must be equal. No restraints are imposed on the components perpendicular to the bond (although some programs apply weak similarity restraints for the remaining components), and we are therefore not gaining any additional independent restraints over the isotropic B case. We still have to fit ~18 000 parameters with ~13 300 + 6500 observations, which brings us back into a substantially over-parameterized situation with n/p barely reaching 1.1.

Although in practice anisotropic restraints can be made tighter than for the case of isotropic approximation where the rigid-bond postulate does not apply,[29] we still cannot justify full anisotropic refinement of individual B-factors in our 1.7 Å example. Nonetheless, although individual anisotropic B-factor refinement is not reasonable at this resolution, TLS parameterization or mixed isotropic/anisotropic parameterization, described in subsection "Anisotropic displacement and TLS parameterization," might allow a better dynamic model description that does not destabilize our refinement.

More data allow loose restraints. We also realize that given high resolution data, say 50 000 unique reflections, we have plenty of data available and many restraints actually become redundant. The 1.3 Å data certainly include information about atomic distances, for example 1.54 Å C–C distances, and we will be able to relax the geometric restraints. Now an interesting and perhaps counterintuitive phenomenon happens: the deviations from the target values can *increase* with a much lower penalty, and our model will actually show *true specific deviations* that might otherwise be suppressed by restraints.[69] One curse of the restraints is that the weaker the data, the more the model will have to—for the sake of refinement stability—comply with the stereochemical expectation values, and fine details regarding local deviations will be lost. In contrast, a substantial geometry deviation in a high resolution structure, confirmed by minimally biased electron density, does indicate an interesting feature, and is not just an outlier.

Restraint weights and geometry r.m.s.d.
How do we know how to properly weight the stereochemistry restraints versus the X-ray terms? Too high restraint weights will force our model to reflect just some ideal geometry and miss important local variation. On the other hand, too low a restraint weight may lead to bizarre geometry that does not reflect reality either. Here the cross-validation residual R-free comes to the rescue: As a measure of the phase error, for a given model, the point where R-free has its *relative* minimum will indicate the best choice of weights. We can thus adjust the X-ray weights w_a

$$Q_{total} = w_a Q_{xray} + Q_{geom} \qquad (12\text{-}49)$$

and observe the change of R-free and the overall bond length and angle deviations. In most cases between 2.5 and 1.5 Å resolution, R-free will reach a minimum, together with a reasonable distance and angle r.m.s. deviation from target values, in the range of ~0.020–0.007 Å and ~1.9–1.0°, respectively (see the discussion in the following paragraph for more on this subject). As the model improves, the *fom* and electron density correlation as well as the overall coordinate precision indices will also improve. In practice, for *REFMAC* a "matrix weight" w_a is defined that will range from 0.1 or even lower for ~2.8 Å resolution structures to ~0.4–0.5 in the 1.5 Å or better range. The default weight of 0.3 is seldom optimal. A value of $w = 0$ corresponds to geometry idealization (i.e., modeling), without consideration of the experimental data. *CNS* provides a script `optimize_wa.inp` for automated determination to optimize the overall weight w_a between X-ray and geometric terms.[70] *REFMAC* now provides a very basic (and not always justifiable, as discussed below) automated weight optimization that essentially adjusts the weight so that the average small molecule target r.m.s.d. values of 0.02 Å and 1.8° result.

Log(-free)-likelihood. While the R-free optimization is a fairly safe method to achieve proper overall restraint weights, it is not an absolute criterion. The use of R-free as a minimization target for optimal weights hinges on a number of assumptions; the most limiting one is that all measured diffraction data have the same uncertainty. Following the idea by Gérard Bricogne,[71] Ian Tickle has recently shown that the minimum of the negative log(-free)-likelihood (–LL-free) objectively indicates the optimal weights,[54] and it can be expected that such optimal weighting schemes will be implemented in automatic form in maximum likelihood refinement programs. Note also that optimization of the B-factor restraint weights is critical for obtaining the correct overall weight w_a between X-ray and geometric terms.[54] *CNS* also provides a combined X-ray and B-factor restraint optimization script `optimize_rweight.inp`. The optimal B-factor restraint weight is understandably different for individual structures (and probably locally as well). As a consequence, individual structures need

Sidebar 12-6 **The importance of convergence in refinement.** The optimization of parameters and weights in refinement requires that the refinement has converged. A common abuse of the refinement procedure, resulting from misguided efforts to obtain the lowest achievable *R*-free, is to stop the last refinement run at the cycle where *R*-free reaches a minimum, but the refinement has not yet converged (where *R*-free may rise again slightly). This is analogous to yanking the power cord out of an instrument once it shows a desired reading, and is absolutely bad practice. Refinements of models that still contain substantial errors tend to overshoot and reach convergence only after a number of small oscillations in parameter shift. Stopping the refinement prematurely is particularly tempting when a poorly modeled region repeatedly assumes bad geometry or over-modeled solvent atoms take off into the wild blue yonder of disordered solvent. Resist this temptation.

individual weights for the *B*-factor restraints. Unfortunately, they are rarely optimized, and in *REFMAC* the defaults (given as a σ of 2 Å^2 and 3 Å^2 for bonded main-chain and side-chain atoms) are much too tight for most structures. Better starting values are obtained from the empirical values[57] and are around 5 Å^2 and 7–10 Å^2, respectively.

For those who want to further explore the likelihood-based optimal estimation of weights, the recent work by Lebedev et al. is recommended reading.[72]

The geometry r.m.s.d. varies with resolution. It is also important to distinguish between the standard uncertainty of the mean value of a restraint target value, which is obtained from many *different* entries in the target structure libraries, and the overall bond length and bond angle r.m.s.d. as a statistic in the refinement of one *individual* structure. The restraint target standard uncertainty provides an upper limit[54] for a reasonable bond or angle r.m.s.d. from targets within the model (approximately 0.02 Å and 1.9°, respectively), but makes no further assumption where these values *de facto* should be in an optimal refinement of a protein structure.[73] Assume at one extreme a low-resolution torsion angle refinement with fixed bond lengths, where the bond length deviation from targets will be zero. At the other extreme, an unrestrained refinement at ultra-high resolution will reflect the bond length variation as observed in small molecules, around ~0.02 Å. In between the values will vary, depending on the properly selected overall restraint weight that minimizes *R*-free (or preferably –*LL*-free).[73] The notion that the r.m.s.d. from targets should be the same for all structures and that it should be close to small molecule values, as expressed in the validation program *WHAT-CHECK* and by some authors, is incorrect. In fact, refinements with *SHELXL* based on George Sheldrick's goodness-of-fit "likelihood by intuition" weighting show exactly the trend of increasing r.m.s. deviation from targets with increasing resolution![29]

Anisotropic displacement and TLS parameterization

As discussed earlier, anisotropic parameterization of atomic displacement provides a better description of reality than the isotropic *B*-factor parameterization. Unfortunately, the resolution of the data rarely allows stable refinement of the five additional parameters needed to describe each individual atomic anisotropic displacement parameter (ADP). Displacive attenuation of the atomic scattering factor is accounted for as a Debye–Waller factor $T_{\mathbf{h}}$ of the general form

$$T(\mathbf{h}) = \exp(-2\pi\langle(\mathbf{s}\cdot\mathbf{u})^2\rangle) \tag{12-50}$$

where $\mathbf{u}$ is the displacement vector of the atom.[74] In the isotropic case, the displacement vector $\mathbf{u}$ is replaced with the isotropic displacement amplitude u (in Å) related to the *B*-factor via $B = 8\pi^2\langle u^2\rangle$ and T becomes the standard Gaussian *B*-factor exponential describing the isotropic attenuation of the atomic scattering factor:

$$T(\mathbf{h}) = \exp(-B_{iso}(\sin\theta/\lambda)^2) \tag{12-51}$$

In the anisotropic case, the squaring of the product of displacement vector $\mathbf{u}$ and the reciprocal lattice vector $\mathbf{s}$ transforms the isotropic displacement into

Box 12-8 **Data-to-parameter ratio in macromolecular refinement.** The data-to-parameter ratio in protein structures is greatly increased through the introduction of stereochemical restraints. A protein of 2000 non-hydrogen atoms has about 8000 adjustable parameters and about the same number of restraints. At 2 Å about 15 000 to 25 000 unique reflections are observed, which yields a total data-to-parameter ratio of about 2–3 at 2 Å. Anisotropic *B*-factor refinement consumes five additional parameters per atom while gaining no additional restraint, and is generally not applied at resolutions lower than ~1.4 Å.

a second rank anisotropic displacement tensor $\mathbf{U}$ expressed as a 3×3 matrix. An anisotropy tensor essentially describes an ellipsoid with the three principal axes (Figure 12-15). Because of the symmetry of the ellipsoid, six components of the symmetric anisotropic displacement tensor suffice to describe the orientation and principal axes of the ellipsoid. The simple isotropic 1-dimensional Gaussian attenuation correction then becomes a trivariate Gaussian in space.

The anisotropic Debye–Waller factor can be defined[75] using the index vector $\mathbf{h}$ (instead of reciprocal scattering vector $\mathbf{s}$) as

$$T(\mathbf{h}) = \exp(-2\pi^2 \mathbf{h}^T \mathbf{U}^* \mathbf{h}) \tag{12-52}$$

Note that following definition (12-52), elements of $\mathbf{U}^*$ are dimensionless mean-square displacements (analogous to fractional coordinates), which simplifies their use in structure factor calculations without the need to explicitly use unit-cell parameters.

As the ellipsoid axes have a directional relation to the crystallographic axes, the general expression for the anisotropic temperature factor in Cartesian world coordinates $\mathbf{U}_W$ or $\mathbf{U}_{cart}$, with components U_W^{ij} in dimensions of Å^2, becomes quite complicated. We obtain $\mathbf{U}_W$ by orthogonalizing $\mathbf{U}^*$ with the orthogonalization matrix $\mathbf{O}$ (whose dimensions are in Å)

$$\mathbf{U}_W = \mathbf{O}\mathbf{U}^*\mathbf{O}^T \text{ and by inversion } \mathbf{U}^* = \mathbf{O}^{-1}\mathbf{U}_W(\mathbf{O}^{-1})^T \tag{12-53}$$

Symbolic expansion of $\mathbf{U}_W$ yields for the anisotropic displacement term

$$T(\mathbf{h}) = \exp(-2\pi^2(U_W^{11}h^2a^{*2} + U_W^{22}k^2b^{*2} + U_W^{33}k^2c^{*2} +$$
$$+2U_W^{12}hka^*b^*\cos\gamma^* + 2U_W^{13}hla^*c^*\cos\beta^* + 2U_W^{23}klb^*c^*\cos\alpha^*) \tag{12-54}$$

The tensor components U_W^{ij} are orthogonal mean square displacement amplitudes in Å^2 in the Cartesian world coordinate system. They can be approximated by an isotropic B-factor equivalent B_{eq}

$$B_{eq} = 8\pi^2 U_{eq} \text{ with } U_{eq} = \tfrac{1}{3}(U_W^{11} + U_W^{22} + U_W^{33}) = \tfrac{1}{3}\langle|\mathbf{u}|^2\rangle = \tfrac{1}{3}\langle\mathbf{u}\cdot\mathbf{u}\rangle \tag{12-55}$$

The isotropic equivalent B_{eq} is listed at the standard position for B_{iso} in the PDB records for each anisotropically refined atom followed by a separate line including the U_W^{ij} values in Å^2, multiplied by 10^4:

```
ATOM   3030  NZ  LYS  375      49.344 42.121 31.409  1.00 37.56         N
ANISOU 3030  NZ  LYS  375      3271   3788   7211  -1222   1450    421   N.
```

Unfortunately, the mmCIF format lists the anisotropy tensor components in dictionary item anisou_U in yet another format, and the conversion from $\mathbf{U}_{CIF}$ to $\mathbf{U}^*$ is given as follows:[76]

$$\mathbf{U}^* = \mathbf{A}^*\mathbf{U}_{CIF}(\mathbf{A}^*)^T \tag{12-56}$$

where $\mathbf{A}^*$ is the matrix that containing the reciprocal cell dimensions a^*, b^*, and c^* in the diagonal elements.

Anisotropic diffraction and anisotropic scaling

The overall translational periodicity in a crystal can be anisotropic, generally as a result of poor packing in a specific direction, and diffraction limits will be correspondingly anisotropic. Occasionally, this effect can be quite pronounced and needs to be taken into consideration during scaling and refinement. We can describe anisotropic displacive attenuation of the total scattering of the crystal by replacing the sphere of resolution in the Ewald construction with the transform of the real space displacement ellipsoid. Its transform is another ellipsoid, the resolution ellipsoid, with reciprocal principal axes. The diffraction limits will therefore be anisotropic as well.

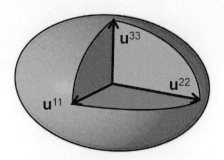

Figure 12-15 Anisotropic displacement ellipsoid. Anisotropic properties in general are described by a second rank anisotropy tensor $\mathbf{U}$ with components U^{ji}, which can be visualized as a property ellipsoid with principal axis vectors $\mathbf{u}^{ji}$. Individual anisotropic displacement parameters (ADPs) as well as correlated lattice disorder in a crystal structure manifesting itself as anisotropic diffraction, can be modeled by anisotropy ellipsoids.

Anisotropic scaling. Most refinement and scaling programs account for diffraction anisotropy by fitting anisotropic displacement as overall anisotropic B-factors,[74] so that the resolution of the scaled diffraction intensities becomes constant. The scale factor attenuation then becomes dependent on the direction of the reciprocal space vector $\mathbf{h}$, and expressed in terms of $\mathbf{U}^*$ we obtain for the anisotropic scale factor

$$k_{\mathbf{h}} = k_0 \exp(-2\pi^2 \mathbf{h}^\mathrm{T} \mathbf{U}^* \mathbf{h}) \tag{12-57}$$

Implementations of anisotropic displacement or scaling must consider that the space group symmetry imposes restraints on the anisotropic $\mathbf{U}$ tensor; its elements must remain invariant upon application of symmetry operations.

The anisotropy correction is generally implemented together with the bulk solvent correction. The masked bulk solvent model (Chapter 10) where structure factor contributions of protein and solvent are added to give the corrected complex model structure factors $\mathbf{F}_{calc}$

$$\mathbf{F}_{calc}(\mathbf{h}) = \mathbf{F}_{prot}(\mathbf{h}) + k_{sol}\mathbf{F}_{sol}(\mathbf{h})\exp(-B_{sol}S^2/4) \tag{12-58}$$

is further expanded by including the overall anisotropic scaling attenuation $k_{\mathbf{h}}$ given by the application of the above derived conversions as

$$k_{\mathbf{h}} = k_0 \exp\left(-\tfrac{1}{4}\mathbf{h}^\mathrm{T}\mathbf{O}^{-1}\mathbf{B}_\mathrm{W}\left(\mathbf{O}^{-1}\right)^\mathrm{T}\mathbf{h}\right). \tag{12-59}$$

For direct comparison with conventional B-factors, the anisotropy tensor is expressed in (12-59) as Cartesian $\mathbf{B}_\mathrm{W}$ obtained in Å² as $\mathbf{B}_\mathrm{W} = 8\pi^2 \mathbf{U}_\mathrm{W}$. The parameters k_{sol} and B_{sol} and $k_{\mathbf{h}}$ are fitted by minimizing the residual between observed data $F_{obs}(\mathbf{h})$ and the solvent-corrected, anisotropy-corrected search model structure factors $\mathbf{F}_{calc}(\mathbf{h})$:

$$SRS = \sum_{\mathbf{h}} \left| F_{obs}(\mathbf{h}) - \left| \mathbf{F}_{prot}(\mathbf{h}) + \mathbf{F}_{sol}(\mathbf{h}, k_{sol}, B_{sol}, k_{\mathbf{h}}) \right| \right|^2 \tag{12-60}$$

This type of anisotropy and bulk solvent correction is implemented in *CNS* and *REFMAC*. An even more elegant implementation[77] in *PHENIX*/*cctbx* avoids the least squares fit of the SRS by implementing the correction in the general maximum likelihood functions. This seems to further improve and stabilize the anisotropic bulk corrections, leading to significant improvements in R-free, although similar improvements have been achieved in *CNS* version 1.2 by introduction of a grid search in the k_{sol} bulk solvent parameter space.[78]

Anisotropic truncation. The anisotropic scaling procedure described above works fine as long as the anisotropy is not too severe: The anisotropic scaling essentially weights up the reflections in the poor direction and attenuates the good ones. We can clearly see that this is not optimal in cases of strong anisotropy: the very weak reflections will contain mostly noise and thus generate map noise, while strong high resolution reflections that carry detailed information are suppressed. In cases of severe anisotropy it is thus better to perform an elliptical truncation in the principal directions first, then anisotropically scale the data, and finally sharpen the data using a negative B-factor to restore the high resolution reflections that were down-scaled during the anisotropic scale[79] (Figure 12-16). Both map reconstruction and model refinement are impaired if extreme anisotropy (say 2.0 Å in a well-diffracting direction, 3.5 Å in the poor ones) is not properly corrected. A web service at UCLA determines the proper anisotropic truncation ellipsoid (http://www.doe-mbi.ucla.edu/~sawaya/anisoscale/). The sharpening of low resolution data through resolution-dependent, negative B-factor corrections has recently been proposed as a general method to improve map interpretation in low resolution structures.[80]

TLS description of correlated anisotropic motion

A realistically parameterized atomic model will have to account for the correlated, anisotropic motions of the atoms. The anisotropic displacement

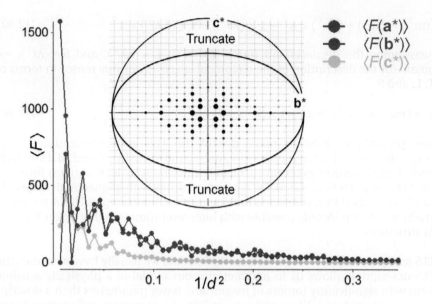

Figure 12-16 Highly anisotropic diffraction data. If lattice disorder varies anisotropically in a crystal, there will be anisotropy in reciprocal space as well and the diffraction limits will be anisotropic. The plot is an amplitude fall-off plot generated by *TRUNCATE* of the CCP4 suite. The graphs show the mean structure factor amplitudes in the reciprocal lattice directions against resolution. The fall-off in direction **c*** (green graph) is most pronounced and the reciprocal space plot confirms the assessment. In the case of such extreme diffraction anisotropy (insert), it is advisable to first truncate the data before applying anisotropic scaling. The principal axes of the truncation ellipsoid (insert) are ~1/1.8 Å, 1/1.8 Å, and 1/2.7 Å, respectively. Of ~10 000 reflections, about 4000 are discarded, the remaining reflections are anisotropically scaled, and finally a negative isotropic *B*-factor of ~–13 Å^2 is applied to restore the high resolution reflections that were attenuated by the anisotropic scaling. Note that in non-orthogonal crystal systems the always orthogonal truncation ellipsoid principal axes do not correspond directly to reciprocal axes. From the *TRUNCATE* manual: direction 2 = **b***; direction 3 = perpendicular to **a*** and **b***; direction 1 = perpendicular to **b*** and direction 3.

will have different sources, which all contribute to the total anisotropy. The following major four components are generally distinguished:[65]

$$U = U_{cryst} + U_{TLS} + U_{int} + U_{atom} \qquad (12\text{-}61)$$

U_{cryst} is a measure for the overall anisotropy resulting from lattice disorder, and is accounted for by the anisotropic scaling and bulk solvent correction as described above. The term U_{TLS} describes the three types of correlated, concerted rigid-body motions of individual groups of atoms. The group size may range from entities such as rigid phenyl rings to domains or even entire molecules in multi-subunit structures. In contrast to basic rigid body refinement, the translation-libration-screw (TLS) parameterization describes more complex, anisotropic motions. The TLS description was first introduced in small-molecule crystallography,[81] and was first implemented in 1985 in the macromolecular refinement program *RESTRAIN*.[82] Due to complexity in the setup and the difficulty in interpreting and visualizing the results, TLS has not been used much in practice, until the more recent implementations such as in *REFMAC*.[65]

The term U_{int} describes correlated intermolecular motions such as breathing modes that cannot be treated adequately by the rigid body TLS parameters and is presently not addressed in macromolecular refinement. Finally, U_{atom} describes remaining, local anisotropic atomic displacements. When TLS parameters are used, additional local displacement effects are absorbed in isotropic atomic *B*-factors refined *after* the TLS parameterization has stabilized.

The interesting part from a biological point of view is the anisotropic TLS movement of groups or domains, which may give clues to the dynamics of the molecule in solution. It is quite intuitive that two domains connected by a flexible linker may well display distinct motions, with corresponding functional implications.

TLS parameterization. Any displacement **u** of a point **x** from one position to another can be described by a combination of a rotational component **R** and a translational component **t**, as described in the NCS section of Chapter 11. Linearization of **u** = **t** + **Rx** with respect to the rotation term leads to **u** ≃ **t** + λ × **x** where the vector λ points in the direction of the rotation axis (Euler axis) defined by **R** with a magnitude corresponding to the rotation (similar to a quaternion representation). As the anisotropy tensor **U** is defined through the mean dyad product of the displacement vector ⟨**uu**^{T}⟩, we form the dyad product (column · row product) and average over the displacement terms:

$$\langle \mathbf{uu}^T \rangle = \langle \mathbf{tt}^T \rangle + \langle \mathbf{t\lambda}^T \rangle \times \mathbf{x}^T - \mathbf{x} \times \langle \mathbf{\lambda t}^T \rangle - \mathbf{x} \times \langle \mathbf{\lambda \lambda}^T \rangle \times \mathbf{x}^T \qquad (12\text{-}62)$$

Reassigning the displacement terms $\mathbf{T} = \langle \mathbf{tt}^T \rangle$, $\mathbf{L} = \langle \mathbf{\lambda \lambda}^T \rangle$, and $\mathbf{S} = \langle \mathbf{\lambda t}^T \rangle$, we obtain for the description of the reciprocal space anisotropy tensor in terms of $\mathbf{T}$, $\mathbf{L}$, and $\mathbf{S}$

$$\mathbf{U} = \langle \mathbf{uu}^T \rangle = \mathbf{T} + \mathbf{S}^T \times \mathbf{x}^T - \mathbf{x} \times \mathbf{S} - \mathbf{x} \times \mathbf{L} \times \mathbf{x}^T \qquad (12\text{-}63)$$

Two possibilities exist to make use of the TLS parameterization above. Originally it was used to *a posteriori* analyze individually refined ADPs in high resolution structures to reveal correlated movement by least squares fitting of the TLS parameters to the ADPs. A group of atoms with aligned ADP ellipsoids can be described by a concerted rocking motion, for example. This *a posteriori* analysis is generally only possible with high resolution (roughly better than 1.2 Å) structures.

TLS analysis. Even for the limited resolution data we usually have available, the TLS description allows us to parameterize our model in a physically sensible form with significantly (orders of magnitude) fewer parameters than a description with individual ADPs would require: The TLS parameterization contributes only 20 parameters per group. Which atoms form a correlated group can either be defined directly from knowledge about the structure (the hinged domains of a multi-domain structure each forming a TLS group would be good examples) or even from analysis of the isotropic *B*-factors. Various tools are available that allow the setting up of TLS groups and analysis and visualization of TLS groups. Basic tools such as *TLSANL*[83] convert the TLS tensors into the corresponding ADP description and provide principal axes of the **T** an **L** tensor for further analysis. Figure 12-17 presents three different visualizations of the anisotropic displacement and domain movements of calmodulin.

The most advanced analysis and visualization tools are available from Ethan Merritt's web site. *TLSview*,[84] available as a web service or a CCP4 plug-in, is a viewer that provides multiple modes of visualization and display of the TLS motions. *TLSMD*[85] is an analysis tool that partitions by graph-minimization a molecule into TLS groups based on the isotropic *B*-factors of the model. These TLS groups then can be used as input for refinement in *REFMAC*, and by tracking the improvements in *R*-free, the optimal number of TLS groups can be determined (Figure 12-17). The improvements in *R*-free are often substantial, and TLS clearly provides a much better description than basic isotropic displacement (*B*-factor) refinement, at a limited cost of 20 introduced parameters per refined TLS group. In addition to TLS group partitioning, the *TLSMD* web interface[86] provides excellent animated graphics for the visualization of the domain motions.

Implementation of restraints in least squares target functions

Stereochemical restraints were first implemented in the crystallographic refinement programs *PROLSQ* and *CORELS*, and have been adapted in similar form

---

Box 12-9 *B*-factor, anisotropic displacement parameterization, and TLS restraints. The most difficult task in the parameterization of macromolecular structure models is accounting for correlated dynamic or static displacement. Isotropic *B*-factors are inadequate to describe any correlated dynamic molecular movement, and anisotropic *B*-factors except at very high resolution lead to overparameterization of the model. A more realistic description of correlated molecular motion is by TLS (translation-libration-screw) parameterization of the anisotropic displacement of the model atoms.

Molecular and lattice packing anisotropy can also affect diffraction, and adequate correction by anisotropic scaling, or in severe cases, additional anisotropic resolution truncation is necessary.

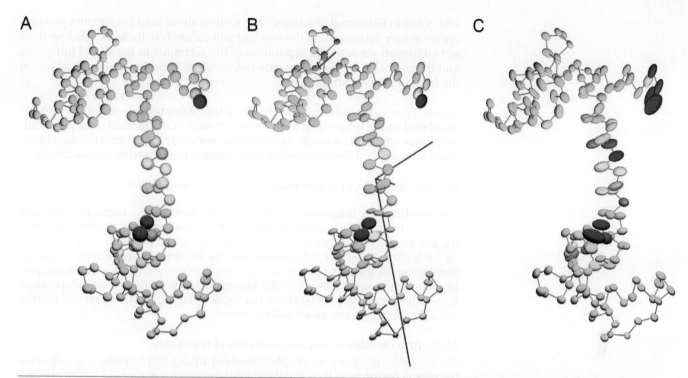

Figure 12-17 Three different representations of atomic displacement in the 1.0 Å resolution crystal structure of Ca²⁺-calmodulin (CaM). In each panel, the Cα atoms are shown connected by a gray line and the magnitude of the displacement is colored from blue (2 Å²) to red (25 Å²). Panel A shows a refined model with isotropic atomic displacements. The magnitude of the disorder at each atom is indicated by the radius and color of the isotropic displacement spheres, but no information about the anisotropy of motion and its directionality is available. In panel B, a three-domain translation-libration-screw (TLS) model that describes atomic mobility in terms of the rigid body displacements of a group of atoms was refined against the data. The principal axes of the libration tensor **L** for three groups are shown in red lines originating from the center of motion. The full TLS model contains additional information about rigid body translations and translation-libration correlations (the screw tensor) that are not shown here. The anisotropic displacement ellipsoids are computed from the TLS group description. In panel C, the directly refined individual atomic APD model from 1.0 Å data is shown. As the most parameter intensive of the three models, it is also the most detailed and accurate. A qualitative visual comparison of panels B and C shows that the TLS model describes the overall directional preferences of atomic mobility in CaM reasonably well. Figures and analysis courtesy of Mark Wilson, U. Nebraska. Animated analysis of the TLS motions is available from the *TLSMD*[86] server under PDB entry 1exr.[87]

into *SHELXL* and *TNT*, and finally, in modified form in maximum likelihood refinement programs such as *REFMAC* and *CNS*. The classical Hendrickson and Konnert incorporation of restraints into the least squares residual function (SRS)[49] is simply by summation of all the restraint residuals Q we derived above. The total refinement residual is then the sum of the X-ray term and all the geometric terms

$$Q_{total} = w_a Q_{xray} + Q_{geom} \tag{12-64}$$

with

$$Q_{geom} = Q_D + Q_P + Q_{CHIV} + Q_T + Q_{NB} + Q_B + Q_{NCS} \tag{12-65}$$

The individual residuals have been defined above. The relative weight w_a of the X-ray residual is adjusted as discussed so that minimum R-free or $-\log(-\text{free})$-likelihood is achieved with resolution-appropriate geometry target deviations.

Implementation of restraints in likelihood target functions

Particularly in early stages of macromolecular refinement, when models are necessarily incomplete and contain significant errors, the assumption of normal, random distribution of errors is not fulfilled. In this case, basic least squares target minimization gives different and less accurate results than the ML target functions. We are by now well aware of how to include model incompleteness and errors in likelihood functions—the section about map coefficients provides the necessary derivations of the maximum likelihood coefficients. What we have not addressed yet is the incorporation of the restraints in likelihood functions, and the provision to readily include independent experimental information in the form of phase restraints.

Bayes' theorem provides a direct means of implementing the restraints into our likelihood-based refinement as prior information in the form of prior probabilities. Bayes' theorem connects the posterior model likelihood with the data likelihood function and the prior probability through the following proportionality:

$$prob(model \mid data, I) \propto prob(data \mid model, I) \times prob(model \mid I) \tag{12-66}$$

The conditioning information I for the data likelihood is implicitly provided by the error model; we therefore omit it in the probability functions for brevity, but keep r for the prior restraint conditioning information as a reminder. As we include the prior information for the model, we are, strictly speaking, developing the formalism of maximum posterior refinement not just maximum likelihood refinement, which would assume a flat prior, that is, no independent knowledge (or assumptions) about our model. Technically this does not require any different treatment as we will see below.

Maximum likelihood and stereochemical restraints

The complete posterior model likelihood LM we actually maximize by adjusting the model parameters is as usual the joint probability expressed as the product of all individual likelihood functions for the structure factors, this time including the additional "observations" given by the stereochemical restraints:

$$LM(model \mid \mathbf{data}) \propto \prod_{\mathbf{h}}^{hkl} prob(data_{\mathbf{h}} \mid model) \times \prod_{r=1}^{restraints} prob(model \mid r) \tag{12-67}$$

The log-likelihood LL which we seek to maximize is then the sum of the logarithmic X-ray terms and the log of the restraint terms:

$$LL = \sum_{\mathbf{h}}^{data} \ln prob(data_{\mathbf{h}} \mid model) + \sum_{r}^{restaints} \ln prob(model \mid r) \tag{12-68}$$

which we abbreviate with split acentric and centric probabilities

$$LL = LL_{xray} + LP_{model} = \sum_{\mathbf{h}}^{acent} LL_{\mathbf{h}}^{a} + \sum_{\mathbf{h}}^{cent} LL_{\mathbf{h}}^{c} + LP_{model} \tag{12-69}$$

Our total target function for optimization of LM corresponds to the minimum of the –log-likelihood function:

$$\max\left[LM(model \mid \mathbf{data}) \right] = \min\left[-LL \right] = \min\left[-\left(LL_{Xray} + LP_{model} \right) \right] \tag{12-70}$$

Prior probability function for stereochemical restraints

The implementation of stereochemical restraints as a probability function is straightforward. Given a normal error distribution of the restraint expectation values, the individual stereochemical restraints r of a certain type b can be formulated as the Gaussian residual between the expectation value b_{obs} and the value calculated for the model b_{calc}:

$$q_b = w_b \left(b_{obs} - b_{calc} \right)^2 \tag{12-71}$$

The corresponding probability function for a given type of restraint b will be simply a Gaussian centered on the empirical mean or expectation value b_{obs} with its known empirical variance σ_b^2:

$$prob(b_{calc} \mid b_{obs}, \sigma_b^2) = \frac{1}{\left(2\pi\sigma_b^2 \right)^{1/2}} \exp\left(-\frac{(b_{obs} - b_{calc})^2}{2\sigma_b^2} \right) \tag{12-72}$$

The logarithm of distribution (12-72) then becomes

$$\ln prob(b_{calc} \mid b_{obs}, \sigma_b^2) = k_b - \left(\frac{(b_{obs} - b_{calc})^2}{2\sigma_b^2} \right) \text{ with } k_b = \ln \frac{1}{\left(2\pi\sigma_b^2\right)^{1/2}} \tag{12-73}$$

and the total contribution of all restraints to the log-likelihood function, the prior probability for the model LP, is written in general form

$$LP_{model} = \sum_b^{restr} \left(k_b - \frac{(b_{obs} - b_{calc})^2}{2\sigma_b^2} \right) \tag{12-74}$$

The evaluation of this term provides no further conceptual difficulties, and we need now to evaluate the X-ray terms, for which we already derived, in Section 12.1, the ML coefficients for the map calculation. We roughly follow the derivation given for the likelihood formalism as implemented in *REFMAC*,[22] except that for simplicity in notation we omit the separation into multiple partial models. Different parts of the structure may in practice need different treatment, but for the derivation this just adds more indices.

Likelihood function for the X-ray term

Starting with the general likelihood function for complex structure factors for a partial model with errors

$$prob(\mathbf{F}_{obs} \mid \mathbf{F}_{calc}) = \begin{cases} \dfrac{1}{\pi\sigma_\Delta^2} \exp\left(-\dfrac{\left| \mathbf{F}_{obs} - D\mathbf{F}_{calc} \right|^2}{\sigma_\Delta^2} \right) & \text{(acentric)} \\[2em] \dfrac{1}{2\pi\sigma_\Delta^2} \exp\left(-\dfrac{\left| \mathbf{F}_{obs} - D\mathbf{F}_{calc} \right|^2}{2\sigma_\Delta^2} \right) & \text{(centric)} \end{cases} \tag{12-75}$$

we convert to amplitudes by applying the cosine rule to the complex residual and marginalizing the phase as an explicit term:

$$prob(F_{obs} \mid F_{calc}) = \begin{cases} \dfrac{F_{obs}}{\pi\sigma_\Delta^2} \exp\left(-\dfrac{F_{obs}^2 + D^2 F_{calc}^2}{\sigma_\Delta^2} \right) \cdot \displaystyle\int_0^{2\pi} \exp\left(\dfrac{2F_{obs} DF_{calc}}{\sigma_\Delta^2} \cos\alpha_{calc} \right) d\alpha & \text{(a)} \\[2em] \left(\dfrac{1}{\pi\sigma_\Delta^2} \right)^{1/2} \exp\left(-\dfrac{F_{obs}^2 + D^2 F_{calc}^2}{2\sigma_\Delta^2} \right) \cdot \displaystyle\sum_{0,\pi}^{i} \exp\left(\dfrac{F_{obs} DF_{calc}}{\sigma_\Delta^2} \cos\alpha_{i,calc} \right) & \text{(c)} \end{cases} \tag{12-76}$$

After integrating out the phase and inflating the variance with the experimental errors σ_F we obtain the likelihood functions (I_0 is the modified Bessel function of order zero; see Figure 7-16):

$$prob(F_{obs} \mid F_{calc}) = \begin{cases} \dfrac{2F_{obs}}{2\sigma_F^2 + \sigma_\Delta^2} \exp\left(-\dfrac{F_{obs}^2 + D^2 F_{calc}^2}{2\sigma_F^2 + \sigma_\Delta^2} \right) \cdot I_0\left(\dfrac{2F_{obs} DF_{calc}}{2\sigma_F^2 + \sigma_\Delta^2} \right) & \text{(a)} \\[2em] \left(\dfrac{2}{\pi(\sigma_F^2 + \sigma_\Delta^2)} \right)^{1/2} \exp\left(-\dfrac{F_{obs}^2 + D^2 F_{calc}^2}{2(\sigma_F^2 + \sigma_\Delta^2)} \right) \cdot \cosh\left(\dfrac{F_{obs} DF_{calc}}{\sigma_F^2 + \sigma_\Delta^2} \right) & \text{(c)} \end{cases} \tag{12-77}$$

Now we finally combine the individual likelihood functions for centric and acentric reflections into the grand log-likelihood function for the X-ray term and take the logarithm, yielding

$$LL_{xray} = \sum_{\mathbf{h}}^{acent} LL_{\mathbf{h}}^a + \sum_{\mathbf{h}}^{cent} LL_{\mathbf{h}}^c \tag{12-78}$$

with

$$LL_{\mathbf{h}}^{a} = c_{a} + \ln F_{obs} + \ln(2\sigma_{F}^{2} + \sigma_{\Delta}^{2}) - \frac{F_{obs}^{2} + D^{2}F_{calc}^{2}}{\sigma_{\Delta}^{2}} + \ln I_{0}\left(\frac{2F_{obs}DF_{calc}}{\sigma_{\Delta}^{2}}\right) \quad \text{(acentric)}$$

$$LL_{\mathbf{h}}^{c} = c_{c} - \frac{1}{2}\ln(\sigma_{F}^{2} + \sigma_{\Delta}^{2}) - \frac{F_{obs}^{2} + D^{2}F_{calc}^{2}}{2(\sigma_{F}^{2} + \sigma_{\Delta}^{2})} + \ln\cosh\left(\frac{F_{obs}DF_{calc}}{\sigma_{F}^{2} + \sigma_{\Delta}^{2}}\right) \quad \text{(centric)}$$

$$(12\text{-}79)$$

Our total log-likelihood target function is now defined. What remains, before we can actually minimize the function, is to estimate our error model parameters σ_{Δ}^{2} and D.

Model error estimate and likelihood map coefficients

As described in Section 12.1, the model error estimate is made by expressing the likelihood functions in normalized structure factors and substituting the combined measure σ_{A}^{2} for both terms σ_{Δ}^{2} and D by simply replacing $\sigma_{A}E_{calc}$ for DF_{calc} and the variance σ_{Δ}^{2} with the corresponding term $1 - \sigma_{A}^{2}$.

Sigma-A itself is then estimated through approximation by fitting a Wilson-like two-term exponential[21]

$$\sigma_{A} = \sigma_{0}\exp(-c_{0}S^{2}/4)(1 - \sigma_{1}\exp(c_{1}S^{2}/4) \quad (12\text{-}80)$$

against the experimental cross-validation data set to avoid bias and overestimation of σ_{A}. This form of σ_{A} estimate is implemented in *REFMAC*[22] and in modified form in *XPLOR/CNS*. Once σ_{A} is estimated, the ML map coefficients are computed as derived in Section 12.1. The general forms of the coefficients are listed in Box 12-3.

Incorporating experimental phase restraints in likelihood functions

Incorporation of independent experimental phase information α_{obs} as additional refinement restraints to keep our model building efforts in check is useful in the early stages of refinement, particularly so at low resolution. The model tends in these cases to be quite poor and incomplete, so that even the ML error models are reaching the limits of their applicability. In the end stages of refinement, the model phases will become much better and will dominate the likelihood function. The experimental phase information therefore becomes less relevant. Experimental phase restraints can be elegantly implemented into ML target functions. As joint phase probabilities are simply obtained by the product of the phase probabilities,

$$prob(\alpha)_{comb} = prob(\alpha)_{ML} \cdot prob(\alpha)_{obs} \quad (12\text{-}81)$$

the logical point to incorporate it into the likelihood function is before we integrate the phase out.[88] We obtain

$$prob(F_{obs}|F_{calc}) =$$

$$= \begin{cases} \dfrac{F_{obs}}{\pi\sigma_{\Delta}^{2}} \cdot \exp\left(-\dfrac{F_{obs}^{2} + D^{2}F_{calc}^{2}}{\sigma_{\Delta}^{2}}\right) \cdot \displaystyle\int_{0}^{2\pi} prob(\alpha)_{obs}\exp\left(\dfrac{2F_{obs}DF_{calc}}{\sigma_{\Delta}^{2}}\cos(\alpha_{obs} - \alpha_{calc})\right)d\alpha \quad \text{(a)} \\[2em] \left(\dfrac{1}{\pi\sigma_{\Delta}^{2}}\right)^{1/2}\exp\left(-\dfrac{F_{obs}^{2} + D^{2}F_{calc}^{2}}{2\sigma_{\Delta}^{2}}\right) \cdot \displaystyle\sum_{\varphi,\varphi+\pi}^{i} prob(\alpha_{i})_{obs}\exp\left(\dfrac{F_{obs}DF_{calc}}{\sigma_{\Delta}^{2}}\cos(\alpha_{obs} - \alpha_{i,calc})\right) \quad \text{(c)} \end{cases}$$

$$(12\text{-}82)$$

The experimental phase probability information is generally provided in the "phase interchange" format of Hendrickson–Lattman coefficients

$$prob(\alpha|A,B,C,D) = K\exp(A\cos\alpha + B\sin\alpha + C\cos 2\alpha + D\sin 2\alpha) \quad (12\text{-}83)$$

and the likelihood function is expressed in the form of HL coefficients for phase combination. There are certain instabilities to address in this implementation,

> **Box 12-10 Restrained maximum likelihood refinement.** Maximum likelihood target functions that account for incompleteness and errors in the model are superior to basic least squares target functions, in the early, error-prone stages of the refinement. They are implemented in *REFMAC*, *Buster/TNT*, and *CNS* as well as the *PHENIX/cctbx* programs, together with all commonly used restraint functions including phase restraints, which are of advantage at low resolution or in the early stages of refinement.

and the experimental phases can be additionally blurred if they are less reliable.[89] The ML map coefficients for map construction provided by the programs are then of the form

$$\mathbf{F}_{ML} \simeq 2mF_{obs}\exp(i\varphi_{comb}) - DF_{calc}\exp(i\varphi_{calc}) \tag{12-84}$$

Non-stochastic optimization algorithms

Given that we have now formulated a likelihood function (or its log-likelihood) for the probability that our data are described by a parameterized model, we need to find the maximum of this function. In a single-parameter case, the maximum is simply the peak of the 1-dimensional likelihood function; for a two-parameter function, the peak corresponding to the highest contours in a 2-dimensional contour plot of the likelihood function; and so on. The brute-force, exhaustive approach to evaluating the function on a grid and simply checking which set of parameter values gives the highest peak works fine, as long as the number of parameters is low. We used this optimization method successfully when evaluating the molecular replacement rotation and translation function, where we had to determine only three parameters in each case (angles α, β, γ, or coordinates x, y, z, respectively). Now assume that we have a function with n parameters, and intend to evaluate the function on a relatively coarse 10-point grid in each dimension. Then we need to evaluate the function on each of the 10^n grid points. Given that we have to adjust several thousand parameters even in the simplest macromolecular refinement, the evaluation of a likelihood function on several 10^{1000} grid points will not finish in the lifetime of this universe.

Optimization as a general maximum likelihood procedure

The next idea to tackle the optimization procedure is based on the fact that the maximum of a likelihood function (or its logarithm) is the peak of its distribution, where the first derivative (the slope) of the function is zero (Figures 7-1, 7-8). Therefore, the peak of a log-likelihood function $LL(\mathbf{p})$ of n adjustable parameters p is the point where all partial derivatives are zero:

$$\frac{\partial LL(\mathbf{p}_0)}{\partial p_1} = 0, \frac{\partial LL(\mathbf{p}_0)}{\partial p_2} = 0, ..., \frac{\partial LL(\mathbf{p}_0)}{\partial p_n} = 0 \tag{12-85}$$

These n equations form a system of n simultaneous equations abbreviated with the *del* or *Nabla* symbol as $\nabla LL(\mathbf{p}_0) = 0$, also known as the gradient vector. The gradient vector of the log-likelihood at its maximum, defined by parameter vector $\mathbf{p}_0$, is zero.

For a linear system, the set of simultaneous equations $\nabla LL(\mathbf{p}_0) = 0$ can be solved in one shot by matrix inversion, and the resulting values of the parameter vector $\mathbf{p}$ are the optimal parameters $\mathbf{p}_0$. Unfortunately, the crystallographic likelihood functions—which include the model structure factor calculations for our refinement model—are anything but linear. In a general nonlinear case the partial derivatives will be unpleasantly complicated (and may not even exist in analytical form) and as a consequence the simultaneous equations will be difficult to solve.

Given the above complications, linearization of the problem might actually make it more tractable. Linearization will bring the likelihood function into a matrix equation form, $\nabla LL(\mathbf{p}_0) = \mathbf{H}\mathbf{p} + \mathbf{c}$, corresponding to a straight line "linear" equation $y = kx + d$. For the maximum value, $\nabla LL(\mathbf{p}_0) = 0$ and $\mathbf{H}\mathbf{p}_0 = -\mathbf{c}$ or $\mathbf{p}_0 = -\mathbf{H}^{-1}\mathbf{c}$. If the $n \times n$ matrix $\mathbf{H}$ and the vector $\mathbf{c}$ contain only constant values, the system can be solved by various linear equation solvers and matrix decomposition techniques. The matrix $-\mathbf{H}^{-1}$ is the covariance matrix $\sigma^2_{i,j}$, which contains the variances of each parameter $\sigma^2_{i,i}$ and the covariance between each parameter pair $\sigma^2_{i,j}$. If we solve the system by a method that also inverts the matrix, we can in fact obtain the standard uncertainty of all the atomic parameters. Unfortunately, full matrix inversion is a slow process of the order of $\sim n^3$ and is impractical in the course of a refinement. However, in selected cases inversion has been conducted once the refinements have converged, which provided valuable insight into estimated coordinate uncertainties (see subsection "Accuracy of protein structures").[38]

We can linearize our likelihood function by expanding it into a Taylor series. The Taylor expansion of $f(x)$ around x_0 is given by

$$f(x) = \sum_{n=0}^{\infty} \frac{1}{n!} \cdot \frac{\partial^n f(x_0)}{\partial x^n} \cdot (x - x_0)^n \cong f(x_0) + \frac{\partial f(x_0)}{\partial x}(x - x_0) + \frac{1}{2} \cdot \frac{\partial^2 f(x_0)}{\partial x^2}(x - x_0)^2 + \dots$$
(12-86)

Expanding the likelihood function into a multi-dimensional Taylor series is quite analogous, and terminating after the quadratic term we obtain

$$LL \cong LL(\mathbf{p}_0) + (\mathbf{p} - \mathbf{p}_0)^T \nabla LL(\mathbf{p}_0) + \frac{1}{2}(\mathbf{p} - \mathbf{p}_0)^T \cdot \nabla^2 LL(\mathbf{p}_0) \cdot (\mathbf{p} - \mathbf{p}_0)$$
(12-87)

where we use the notation $\nabla^2 LL(\mathbf{p}_0)$ for the second derivative Hessian matrix $\mathbf{H}$. Following our standard procedure, we can attempt to solve the problem directly by setting $\nabla LL(\mathbf{p}_0) = 0$, with $LL(\mathbf{p}_0)$ a (yet unknown) constant value that can be absorbed in the proportionality constant. The linearized LL then becomes

$$LL \simeq \frac{1}{2}(\mathbf{p} - \mathbf{p}_0)^T \cdot \nabla^2 LL(\mathbf{p}_0) \cdot (\mathbf{p} - \mathbf{p}_0)$$
(12-88)

Gradient descent methods

The system of equations describing LL now needs to be solved for $\mathbf{p}_0$. The inverse of the second derivative matrix is again a covariance matrix

$$\sigma^2_{i,j} = -\left(\nabla^2 LL(\mathbf{p}_0)\right)^{-1}$$
(12-89)

This is exactly analogous to the linear case $\mathbf{p}_0 = -\mathbf{H}^{-1}\mathbf{c}$ described above, and we can in principle use any of the discussed equation solvers to solve the simultaneous (least squares) equations. In practice, most direct linear equation solvers are too slow to be applicable in macromolecular crystallography, except for small problems. In addition, we do not need to solve the equations in a global, completely deterministic way, because we generally know an initial starting model and can start from its set of (positional) parameters. This makes the computation by *iterative* methods feasible—although at the price of possible entrapment in local minima.

As an alternative to the direct least squares solution, we can approximate the likelihood function instead of around the yet unknown solution $\mathbf{p}_0$, around a known starting parameter vector $\mathbf{p}_1$

$$LL \simeq LL(\mathbf{p}_1) + \nabla LL(\mathbf{p}_1)(\mathbf{p} - \mathbf{p}_1)^T + \frac{1}{2}(\mathbf{p} - \mathbf{p}_1)^T \cdot \nabla^2 LL(\mathbf{p}_1) \cdot (\mathbf{p} - \mathbf{p}_1) + \dots$$
(12-90)

and differentiate LL, which not unexpectedly gives a Taylor series of the derivatives

$$\nabla LL \simeq \nabla LL(\mathbf{p}_1) + \nabla^2 LL(\mathbf{p}_1)(\mathbf{p} - \mathbf{p}_1)^T + \dots$$
(12-91)

If we are close enough to the minimum, where $\nabla LL(\mathbf{p}_0) = 0$, we can formulate the difference $\mathbf{p} - \mathbf{p}_1$ as a parameter shift vector $\mathbf{s}$ and we can write for the parameter shift vector

$$\mathbf{s} = -\nabla LL(\mathbf{p}_1) \cdot \left(\nabla^2 LL(\mathbf{p}_1)\right)^{-1} \qquad (12\text{-}92)$$

Because we already have a current model with parameters $\mathbf{p}_1$, we can evaluate the first and second derivatives, and obtain the values for the shift vector components. Following this shift vector will bring us closer to the minimum, where we can repeat the procedure until we converge at $\mathbf{p}_0$ with $\mathbf{s} = 0$. In its basic form, this is essentially the Newton–Raphson algorithm. The gradient descent algorithms used in crystallographic refinement methods differ mostly in how the components of the derivative matrices providing the shift vector are evaluated. Note that the term cycle is used for each of the steps approaching the minimum until the refinement converges within a *single* refinement run, not to be mistaken with a *round* of real-space rebuilding and reciprocal-space refinement, with each individual refinement run requiring many cycles to reach convergence.

Full matrix. The full matrix minimization method evaluates each of the derivatives from the given model. This is particularly slow for the second derivative matrix, although the method converges faster than others when it is close to the minimum. For a good and complete high resolution model this will likely be the case, and in addition to the exact but slow direct structure factor summation, is the reason *SHELXL* is a good choice for such problems (in addition, when the model is final and accurate, maximum likelihood does not provide much advantage over LSQ, to which it asymptotically approaches for zero error, i.e. $\sigma_A = 1$ and thus $m = 1$ and $D = 1$ in the ML coefficients $2mF_o - DF_c$).

Sparse matrix. We can save some computational effort by concentrating on the second derivatives that have the highest impact on our shift vector. As $-\left(\nabla^2 LL(\mathbf{p})\right)^{-1}$ is a covariance matrix, only certain blocks of the matrix will contain large values, depending on the given LL function. For the positional parameters describing the model (from which F_{calc} in the X-ray terms is computed), the covariance is only strong if the electron density of the atoms is in close local proximity. Distant atoms will have little effect on each other's electron density. The situation is different for restraint functions: the parameters of the atoms that are restrained to another one will automatically have a high correlation and their parameters co-vary. Methods that use these blocked matrices, where the insignificant terms are ignored, are sparse matrix approximations. Programs that follow the classical Konnert–Hendrickson *PROLSQ* implementation such as *REFMAC* and *SHELXL* use a sparse matrix "conjugate gradient" algorithm (which is not identical to the Fletcher/Powell conjugate gradient algorithm used in function minimization). *XPLOR* uses a modified Powell algorithm (after the stochastic SA minimization stage), while *CNS* and *PHENIX* use a quasi-Newton LBGFS algorithm to approximate the Hessian matrix during coordinate refinement, and *TNT* uses a preconditioned conjugate gradient method.

Box 12-11 Optimization algorithms. Optimization algorithms are procedures that search for an optimum of a nonlinear, multi-parametric function. They can be roughly divided into analytic or deterministic procedures and stochastic procedures. Deterministic optimizations such as the gradient-based maximum likelihood methods are fast and work well when we are reasonably close to a correct model, at the price of becoming trapped in local minima. Stochastic procedures employ a random search that also allows movements away from local minima. They are slow but compensate for it with a large radius of convergence. Evolutionary programming as used in molecular replacement or simulated annealing in refinement is a stochastic optimization procedure. These are generally of advantage if we do not know the correct solution (MR) or are far from it (initial model refinement).

Deterministic optimizations can be classified depending on how they evaluate the second derivative matrix. They generally descend in several steps or cycles from a starting parameter set (model) downhill toward a hopefully—but not necessarily—global minimum.

Steepest descent and conjugate gradient. We can also ignore the second derivative matrix by treating the diagonal elements as identical constants. In these first-order minimization algorithms the shift vector is then given solely by the negative of the gradient matrix, and it will point downward. However, the downward path will not be optimal because of the omission of the second derivatives describing the curvature, and convergence will be slow. The benefit is that no numerical instabilities from ill-defined second derivatives can occur. The method is improved by approximating the second order terms by maintaining a portion of the previous descent vector to each following one. This essentially probes the change of the gradient, which is equivalent to the second derivative. A preconditioned conjugate gradient method therefore converges better than the gradient alone. The method can be further improved by using the diagonal elements of the curvature matrix in the first refinement round, as implemented in *TNT*.

Molecular dynamics, torsion angle refinement, and stochastic optimization

A somewhat different and alternative method of restrained refinement, introduced in the mid-seventies by Jack and Levitt,[89] is implemented in *XPLOR/CNS* and *PHENIX/cctbx*. Energy refinement is based on the interpretation of X-ray terms and geometric restraints as the sum of potential energy terms:

$$E_{total} = w_{xray} E_{xray} + E_{geom} \qquad (12\text{-}93)$$

The X-ray term can be implemented in different ways, either as basic quadratic amplitude potential, as a vector potential when phases are available, or as a maximum likelihood function. The geometric energy term is expressed in the form of empirical potentials, where the squared residuals such as bond lengths, angles, and chiral and planar restraints simply become a quadratic energy potential, and the torsion angles are a periodic potential with the number of minima corresponding to the period (the deviation of the calculated angle from target is in this case measured as phase difference δ). The non-bonded contact energy is given by the sum of all pair-wise Lennard-Jones type ($A/r^{12} - B/r^6$) potentials (Figure 12-18), and electrostatic terms are also included as pair-wise $1/r$ Coulomb terms with q_j the corresponding atom charge and ε the estimated local dielectric constant. The total geometric energy term then becomes

$$E_{geom} = \sum_{i=1}^{bonds} w_{d(i)} (d_i^{eq} - d_i^{calc})^2 + \sum_{j=1}^{angles} w_{\theta(j)} (\theta_j^{eq} - \theta_j^{calc})^2$$
$$+ \sum_{j=1}^{dihedrals} w_{\theta(j)} \left[1 + \cos(n\theta_j^{eq} - \delta_j^{calc}) \right] + \sum_{j=1}^{chi,plan} w_{w(j)} (\omega_j^{eq} - \omega_j^{calc})^2 \qquad (12\text{-}94)$$
$$+ \sum_{i<j}^{pairs} \left(\frac{A_{i,j}}{r_{i,j}^{12}} - \frac{B_{i,j}}{r_{i,j}^6} + \frac{q_i q_j}{\varepsilon r_{i,j}} \right)$$

Figure 12-18 Potential functions.
Two common basic empirical potential functions used in energy restraint functions are the Lennard-Jones potential of the general form $(r_0/r)^{-12} - (r_0/r)^{-6}$ and the quadratic potential of the general form $(d - d_0)^2$.

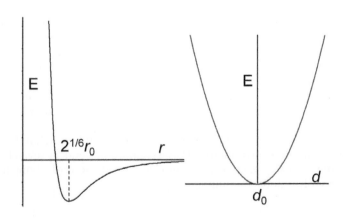

In practice, the electrostatic terms are often entirely omitted, and only the repulsive terms of the Lennard-Jones potential are implemented. The energy term in this form can be used to describe a molecular mechanics (or molecular dynamics, MD) force field, which can be minimized with efficient stochastic search algorithms, developed for molecular dynamics (MD calculations are essentially molecular energy (geometry) minimizations with zero weight for the X-ray term). The empirical (or "knowledge-based") potentials can be taken directly from MD programs such as *CHARMM*,[90] and are constantly updated by improved analysis.[91] In contrast to the gradient-based methods used in classical restrained or maximum likelihood refinement, where the search can proceed only down the gradient, the minimization algorithms developed in combination with MD also allow an uphill movement, and are therefore less susceptible to becoming trapped in local minima.

Simulated annealing

The most common method used to minimize potential energy functions is simulated annealing (SA), in which the model is perturbed and slowly returned to equilibrium following a certain annealing schedule $T_1 > T_2 > T_3 \ldots T(\text{final})$. The name results from the annealing technique used in materials science to relax stress in solids by heating the material up to a temperature where the thermal displacement of the atoms becomes large enough to overcome local energy minima, and allows the material to relax into an ordered minimum energy state.

In simulated annealing, model perturbation is performed by introducing coordinate variations, which increase the potential energy of the model. Following the Maxwell–Boltzmann distribution, the total energy of a system can be expressed as a temperature; for example at a room temperature of 293 K the thermal energy $E = k_B T$ of a system corresponds to a low value of ~0.6 kcal mol^{-1}. At each temperature the states of the system, defined by the adjustable parameters q_n, are populated corresponding to a Boltzmann distribution:

$$prob(q_1, q_2, \ldots, q_n) = \exp(-E(q_1, q_2, \ldots, q_n) / k_B T) \quad (12\text{-}95)$$

An ensemble of structure models with perturbed coordinates is now generated so that the temperature corresponds to several thousand degrees. The temperature has no physical meaning and only defines the parameter distribution of the starting model population.

Monte Carlo and molecular dynamics minimization

The selection parameter against which stochastic parameter moves are evaluated is the total system energy. There are two generally used algorithms to generate the change of parameters so that the system slowly returns to equilibrium while exploring the parameter space: Metropolis Monte Carlo searches (MMC) and molecular dynamics (MD) minimizations.

Metropolis Monte Carlo searches. In MMC, a random coordinate change that lowers the total energy is accepted right away and one that increases it is accepted with a certain probability $P = \exp(-\Delta E / k_B T)$. The random retrograde steps allow moving upwards against the gradient out of local minima, and the higher the temperature, the more likely it becomes that an energy barrier can be overcome. For $T = 0$, the MMC algorithm can proceed only downward, similar to a gradient descent method. The drawback of MMC is that many moves will be totally nonsensical and grossly violate strong restraints such as bond lengths and thus the moves have a high rejection rate. MMC optimization is presently used only in refinement of NMR models and in molecular replacement searches (*Queen of Spades*).[95]

Molecular dynamics. If parameter changes are expressed as generalized coordinates, the system can be described in terms of its temporal development by a system of Hamiltonian equations of motion:[96]

$$\frac{\delta H(p,q)}{\delta p_i} = \frac{\delta q_i}{\delta t} \quad \text{and} \quad \frac{\delta H(p,q)}{\delta q_i} = -\frac{\delta p_i}{\delta t} \quad (12\text{-}96)$$

Sidebar 12-7 Annealing of crystals. Actual annealing in real (crystal) space can be tracked through X-ray diffraction, and annealing of protein crystals by cycling between room temperature and liquid nitrogen temperature can sometimes relax the stress induced by cryoquenching. Successful annealing of protein crystals has in some cases led to increased resolution and reduced mosaicity.[92-94]

where p_i is the (generalized) momentum of its conjugate parameter q_i. In our minimization problem, the generalized coordinates are the degrees of freedom of our molecular model, and the equations are the MD equations of the system. Most of our parameters are atomic coordinates, where the MD equations assume the form of Newton's second law $F = m \cdot a$, where m is the mass of each atom and a the acceleration defined as $a = d^2x/dt^2$. We directly obtain a relation to potential energy and the time development of the system:

$$m_i \frac{\partial^2 \mathbf{r}_i}{\partial t^2} = F_i = -\frac{\partial E_{pot}}{\partial \mathbf{r}_i} \tag{12-97}$$

Simulated annealing. The system is initially described by the coordinates of the atoms and the potential energy. We can compute the force from the right side of Equation 12-97, and apply it to each atom for a short period of time, generally in the fs (femtosecond) range. Then we can calculate new coordinate positions. We repeat these steps while the molecular system relaxes into the energy minimum, and select the annealing schedule of decreasing temperatures so that at each step just enough momentum remains that the system can overcome local minima. A commonly used optimal annealing schedule is the so-called slow cooling protocol[97] implemented in *XPLOR* and *CNS*.

If the simulated annealing procedure is used without experimental restraints, we obtain information about the dynamic behavior of the molecule, which is applied in molecular modeling and molecular dynamics simulations of protein folding. In crystallographic refinement, the crystallographic energy term is added to the potential energy, and the relative weight is progressively adjusted during the refinement by estimating the σ_A values from the cross-validation data set.[37]

Simulated annealing torsion angle refinement

For crystallographic refinement, we can further adapt the SA procedure by using torsion angles (TA) as the adjustment parameters instead of directly adjusting the positional parameters. Torsion angle refinement requires significantly fewer parameters than coordinate refinement (the average parameter count for a residue decreases from 24 in coordinate parameterization to five torsion angles in addition to avoiding some of the weighting problems of coordinate SA refinement).[56] Torsion angle molecular dynamics (TA-MD) is therefore a particularly suitable protocol for low-resolution (generally denoting structures with a resolution of 3.5 Å or worse) refinement. In principle, the determinacy limit in the case of TA refinement is around 5.4 Å for a crystal with 50% solvent content, but the challenges in low resolution model building and refinement are formidable.[79] In the form of internal coordinate mechanics (ICM), the implementation of TA refinement has also been quite successful in molecular modeling and flexible protein ligand docking.[98] Crystallographic TA-MD requires higher annealing temperatures than required for coordinate-based MD because torsion angle changes are highly correlated across the molecule. As a consequence, TA-MD has on one hand a large radius of convergence; it can flip peptide bonds, for example. Torsion angle dynamics combined with exact solution of the equations of motion as implemented in *CNS* improves the convergence radius from a backbone atom coordinate r.m.s.d. of ~1.25 Å for coordinate refinement to ~1.7 Å for torsion angle refinement.[99] On the other hand, due to its stability, uncritically applied TA refinement can also produce unrealistic and divergent models (Sidebar 12-8). As one might expect, the combination of TA-MD with maximum likelihood targets[70] is the most powerful option. As a general strategy, once the model is in approximate shape, we can switch from torsion angle simulated annealing to the potential-based molecular dynamics simulated annealing and finally polish the model with gradient-based maximum likelihood coordinate refinement. Following this procedure prevents a model that has already improved from becoming unstable. Once large-scale errors are fixed by TA-MD and simulated annealing, the model is close enough to reality so that coordinate refinement and the gradient descent methods work well and safely.

Sidebar 12-8 With great power comes great responsibility. The introduction of crystallographic refinement by simulated annealing molecular dynamics by Axel Brunger, John Kuriyan, and Martin Karplus[46] has had significant impact on the refinement of macromolecular crystal structure models, particularly from low resolution data. The large radius of convergence of the procedure and its robust and elegant implementation in *XPLOR*, its successor *CNS*, and the *PHENIX* package has made simulated annealing a powerful and popular refinement technique.

Because *XPLOR* and *CNS* make it so easy to set up complex refinement protocols even for beginners, it is tempting to refine poor structure models (often molecular replacement models) right away with molecular dynamics and torsion angle refinement to (seldom global) convergence without much of the cumbersome model rebuilding. This is not a recommended strategy; inspection of the electron density and manual correction of remaining errors, followed by maximum likelihood energy (coordinate) refinement of likelihood target functions, is recommended and is also necessary practice in *XPLOR* and *CNS*.

Particularly hazardous is the uncritical application of multi-conformer refinement at low resolution. Multi-conformer refinement in principle provides a more accurate description of the dynamic behavior of the protein molecules, and its application in high-resolution structures (where sufficient data are available) is justified.[102] The original model of the ABC transporter,[103] however, was built into low resolution experimental electron density of wrong handedness (purportedly because of a swap of Friedel pairs in the data processing software, which had to give maps of inverted handedness). The subsequent refinement should not have yielded reasonable statistics and indeed stalled at high *R*-values, therefore raising questions about model correctness. Unfortunately, SA-MD multi-conformer refinement was uncritically used, and as a result the refinement statistics seem to have improved because of overparameterization (and incorrect selection of cross-validation data in the presence of NCS), and the error went unnoticed.[42] As a consequence, five high profile structures of the ABC membrane transporter that misled structural biologists and hindered progress for several years[104] had to be retracted.[105] The correct structure of a homolog was finally determined by R. J. Dawson and K. P. Locher in 2006.[106]

Multi-conformer refinement.
The fact that an ensemble of structures can be generated based on the system temperature allows two further extensions of the SA protocol. First, we can compute the X-ray residual from an averaged structure factor from an ensemble of models, which most likely describes the dynamic behavior of the molecule much better than the single isotropic *B*-factor. The procedure is termed time-averaged MD[100] or multi-conformer refinement. While this procedure has its justification in high-resolution structures, where such conformations are indeed observed[101] (and can subsequently also be stably refined by gradient methods), great care must be taken when multi-conformer refinement is used in low-resolution structures, where a great risk of overparameterization exists (Sidebar 12-8).

Dummy atom refinement protocols
We have already mentioned that repeated cycling between real space model rebuilding and reciprocal space global refinement is important to escape local minima and to arrive at optimal models. This method of frequent iteration of model building and refinement can be automated by the procedure of dummy

Box 12-12 Molecular dynamics, torsion angle refinement, and simulated annealing. Energy refinement of a molecular dynamics force field and torsion angle refinement are two parameterizations that are commonly used together with the stochastic optimization method of simulated annealing. In molecular dynamics the target function is parameterized in the form of potential energy terms and the development of the system is described by equations of motion. In torsion angle parameterization, the structure model is described by its torsion angles, which requires fewer parameters than coordinate parameterization. Both MD and TA parameterization are often combined with simulated annealing optimization, where the molecular system is perturbed and returns to equilibrium according to an optimized slow cooling protocol.

atom refinement. It was first suggested for phase improvement by Argawal and Isaacs[107] and for density modification by Bhat and Blow[108] and developed further by Lamzin, Wilson, Perrakis and co-workers[109] into a capable model building[110] and ligand-fitting program suite,[111] called *ARP/wARP*.[112]

Iterative dummy atom placement and refinement becomes possible as soon as discrete atoms are discernible in the electron density maps, including discrete water atoms in the first coordination shells. The atomicity limit is fairly relaxed in dummy atom placement, and can be applied as soon as the resolution of the data is better than about 2.5 Å. In addition to serving as a real space density modification technique, iterative dummy atom refinement has also been implemented for model rebuilding and is one of the most powerful model bias minimization techniques when combined with map averaging from multiple perturbed starting models (*Shake&wARP*).[26]

Automated water building

The application of automated model building is most easily understood in the case of solvent building. An incomplete model will have empty electron density blobs for as yet un-modeled solvent. We can search the maximum likelihood difference maps mF_o-DF_c for peaks and place water "atoms" into peaks above a certain density level and no closer than 2.3 Å to another atom (to a first approximation that would be the very closest possible hydrogen bond length). We then subject the resulting model with the new water atoms to restrained maximum likelihood refinement. If the atom was placed in "real" density, the refinement will most likely keep the atoms and the model will improve. If an atom was placed in non-genuine density or too close to a carbon atom, it will either drift away and/or end up with a high B-factor. We then construct the next, improved electron density map, and remove the atoms that did not survive and are no longer present in maximum likelihood $2mF_o-DF_c$ density, and place new ones in difference density that appeared in the improved discrete solvent model. We repeat this procedure, slowly reducing the levels of density and placing atoms in as the solvent model improves. Once no more new atoms can be placed, the discrete solvent is built and the model is ready for inspection.

Bias minimized map generation

In principle, by water placement we have performed a procedure that combines model completion with real space solvent density modification. The basic solvent building method (*arp_waters* in CCP4) worked well and was thus extended to use dummy atom placement and atom removal as a general method of model improvement. This time we use unrestrained maximum likelihood refinement, because the dummy water atoms in place of missing protein model atoms cannot, and need not, be restrained. After refinement, atoms in poor $2mF_o-DF_c$ density are removed, new atoms placed into empty difference density (this time with a minimum distance of 1.0 Å), and the atom placement/removal and refinement cycle is repeated about 30 times. In addition, different starting models can be created by perturbing the coordinates and randomly removing a certain number of atoms. The maps resulting from each of these models (six is a common count) will not be identical after a given number of rounds, but true features will likely appear in every map at significant density, while the noise features will be different in each map. The averaged map will then be much clearer and have significantly improved signal-to-noise ratio. In addition, entire questionable parts or ligands can be omitted from the starting model and when they are true, they will be rebuilt with dummy atoms and reappear in the map, while artifacts will be averaged out. This method is therefore ideal for powerful bias removal, because all bias minimization methods—coordinate perturbation and atom deletion to remove the model memory, maximum likelihood refinement to account for model errors and incompleteness, and solvent modification and map averaging for phase improvement—are used together. The only requirement is that the unrestrained refinement remains numerically stable, which is generally the case for resolutions higher than 2.5 Å given reasonably good data. This bias minimization method is implemented in the *Shake&wARP* protocol and service[26] (Figure 12-19).

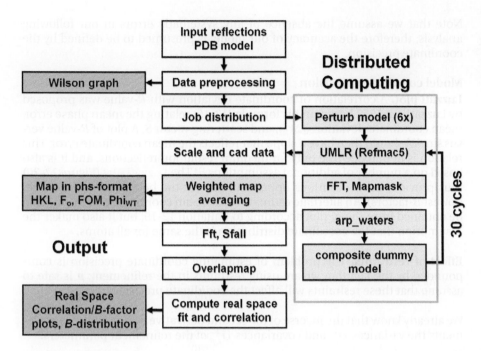

Distributed Computing

Output

Figure 12-19 **Flow diagram of the** ***Shake&wARP* procedure.** Six differently perturbed models are subjected to 30 rounds of atom placement/removal and unrestrained maximum likelihood refinement (UMLR). The best map is obtained by weighted averaging of the six individual maps. The method is a powerful bias removal and map reconstruction tool when data with resolution better than 2.5 Å are available, but it is slow and presently does not attempt to rebuild the protein model.

The next logical step in the development of dummy refinement protocols was to modify the procedure so that it builds actual protein models and not just provides maps from averaged dummy atom models. The *ARP/wARP* program allows one to do exactly that and it is implemented as a plug-in to the CCP4 suite or is available as a web service.[113] The program can be used to build into empty experimental electron density maps, or to correct and rebuild molecular replacement models. We will use the *ARP/wARP* program in the next section to rebuild our replaced cytochrome *c'* molecular replacement model.

Accuracy of protein structures

Let us assume we have arrived at a properly built and refined structure model. The question then is how accurately do the atomic coordinates of our model represent the true structure? As the coordinate error translates directly into a phase error, we expect that the refinement residual *R* will be to some degree a measure of the coordinate deviation, which in first approximation is true: the *R*-value for an acentric random atom model is 0.586 (0.828 centric) which has been derived by A. J. C. Wilson (of Wilson plot fame) based on intensity statistics[15] (see Sidebar 12-9). We therefore shall examine two questions:

- Given a certain *R*-value, how accurate are the coordinates of our protein model?

- What is the expected *R*-value and *R*-free for a given refinement (and thus also the *R*-free/*R* ratio)?

Box 12-13 **Dummy atom refinement.** Dummy atom placement and refinement is used for discrete solvent building, model completion, and phase improvement in general. Dummy atoms are placed in real space in difference electron density peaks, the new model is refined unrestrained in reciprocal space, and in the new map poorly positioned atoms are removed and new ones are placed again. About 30 rounds of repeated real space dummy atom placement and removal and reciprocal space refinement are conducted. Dummy atom refinement can be combined with multi-model map averaging where it forms the basis of bias minimization protocols and the automated model building program *ARP/wARP*.

Note that we assume the absence of any systematic errors in our following analysis, therefore the accuracy of our model is presumed to be defined by the coordinate precision.

Model coordinate precision

Luzzati plot. A correlation of coordinate deviation with R-value was proposed by Luzzati[16] in 1952 based on his formula (12-15) relating the mean phase error, mean coordinate deviation $\langle|\Delta\mathbf{r}_j|\rangle$, and scattering vector $\mathbf{S}$. A plot of R-value versus $\sin\theta/\lambda$ should then have a slope that reflects the mean coordinate error. This relation is in practice true only for medium resolution reflections, and it is also based on a number of additional assumptions.[114] The Luzzati plot (Figure 12-20) thus provides only a cautious upper estimate for the best parts of the model (lowest B-factors). An alternate estimate of the mean coordinate error $\langle|\Delta\mathbf{r}_j|\rangle$ can be obtained from the σ_A plot according to Equation 12-19, but it also makes the assumption that the coordinate distribution is the same for all atoms.

Effect of restraints. The problem of estimating coordinate precision is compounded by the fact that we are using restraints in the refinement; it is safe to assume that these restraints will affect the coordinate precision.

We already know that the inverse of the second derivative matrix contains as elements the variances $\sigma_{i,i}^2$ and covariances $\sigma_{i,j}^2$ of the refinement parameters:

$$\sigma_{i,j}^2 = -\left(\nabla^2 LL(\mathbf{p}_0)\right)^{-1} \tag{12-98}$$

Therefore, in the optimal case of unrestrained refinement possible at or beyond atomic resolution, we can invert the Hessian matrix $\mathbf{H}$ after the refinement has converged and directly obtain the coordinate precision *sans* restraints (inverting this matrix is a computationally challenging task[116] of order $\sim n^3$, which is why full inversion is not routinely applied in refinement and even when it is computed, only once the refinement has converged). For a number of high resolution structures, such matrix inversions of unrestrained refinements have been conducted and provide a baseline of coordinate precision values against which estimates can be tested.

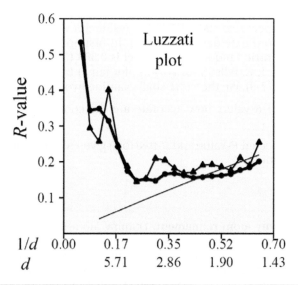

Figure 12-20 A Luzzati plot. In the Luzzati plot, shown here for historical interest more than for its actual usefulness compared with σ_A-estimates or the Cruickshank DPI, the high resolution part of the R-value (bold line, R; thin line, R-free) is fitted against a line through the origin. The slope then gives an upper estimate for the coordinate error,[16] in this case 0.134 Å. The more accurate DPI for this high resolution structure model is significantly lower at 0.082 Å. The plot is part of the output of *SFCHECK*,[115] a model validation program that also analyzes the data.

Sidebar 12-9 Why are protein structure models not perfect? Consider the following: A random structure (i.e., a completely wrong structure) has an R-value of 0.59 and a perfectly correct model should have an R-value of zero, if the data were also perfect. Given that the data generally have much better statistics and merge with much lower R-values (generally ~5%), our average resolution models with R-values of around 0.25–0.18 are thus about 1/3 "wrong," and most of it must be blamed on poor model parameterization. One of the weakest points in macromolecular structure refinement is certainly the lack of adequate description of correlated anisotropic displacements—that is, dynamic motion—present in the true molecular structure. The often dramatic improvements in R-free by TLS refinement, through better bulk solvent models, or with better restraint weight optimization indicate that there is still much room for improvement in model parameterization—a rich field for the theoretically inclined. On the other hand, high resolution models with carefully modeled multiple confirmations, discrete solvent, and anisotropic B-factors can refine to quite low R-values (see also Figure 12-23). One must remember that the diffraction data result from imperfect crystals, averaged over a large number of individual molecules, all subject to radiation damage—none of which is explicitly parameterized in our refinement model.

Most protein structures are refined in a resolution range where they depend more or less on the restraints, with the ratio of measured data to refinement parameters falling below 1 at about 2.5 Å. The refined coordinates will therefore in some complex form depend on the restraints. In comparisons of unrestrained refinement of the same protein model with restrained refinement, it turns out that the restraints strongly affect the atoms with high B-factor, while for average B-factors the restrained coordinate precision is more comparable with the unrestrained coordinate precision (Figure 12-21). It is not unexpected that stereochemical restraints prevent deviations for atoms with higher positional displacement. This is the curse of the restraints; any significant deviation will carry a penalty and only strong experimental evidence, that is, high resolution data with relaxed restraints, can overcome the imposed expectations. Note that a small positional variation of a low B-factor atom has a much greater effect on the structure factor calculation than a larger positional error of a high B-factor atom.

Diffraction precision index. Given that atoms with average B-factor are little affected by the restraints, in the absence of matrix inversion we can use an estimate for the coordinate precision derived without explicit dependence on specific restraints. Starting from a proven equation developed for small-molecule crystallography and inflating it by a factor of ~1.5 to account for the overall effect of restraints, D. W. J. Cruickshank derived the diffraction-component precision index[112] (DPI). The average coordinate uncertainty is described by the DPI, which for unrestrained refinement takes the form

$$\sigma(x, B_{ave}) = [N / (n_{obs} - n_p)]^{1/2} C^{-1/3} R d_{min} \qquad (12\text{-}99)$$

The estimated standard uncertainty of the coordinates therefore depends on the R-value, the number of non-hydrogen atoms N, the number of observations n_{obs}, the number of parameters describing the model n_p, and the fractional completeness C of the data up to the resolution limit d_{min}. This formula for the DPI is a good approximation for atoms with a B-factor close to the mean B-factor B_{ave}, and is independent of the contribution of restraints. If we are interested in the positional error and not in individual coordinate errors, we can form an isotropic average just as in the case of B-factors and obtain

$$\sigma(r, B_{ave}) = 3^{1/2} \sigma(x, B_{ave}) \qquad (12\text{-}100)$$

Figure 12-21 Positional uncertainties for atoms as a function of B-factors. Black data and graphs are for carbon, blue for nitrogen, and red for oxygen. The left panel shows the results from a matrix inversion for unrestrained refinement, and the right panel is for restrained refinement. For atoms with B-factors below 20 Å², the coordinate uncertainty is practically the same; only high B-factor atoms appear underestimated in restrained refinement. Modified figures from Cruickshank[114] reproduced with kind permission of the IUCr.

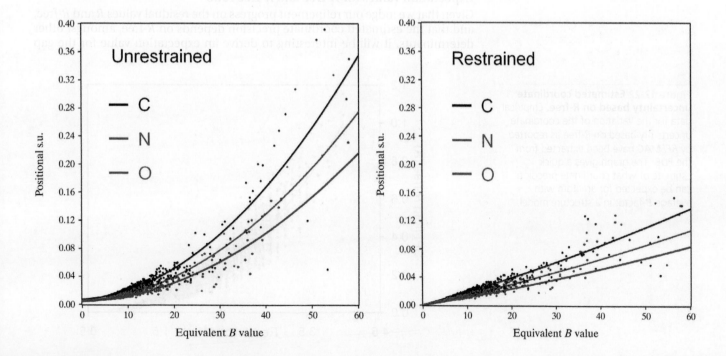

The DPI formula (12-99) requires modification when n_p approaches n_{obs}, where $\sigma(x,B_{ave})$ approaches infinity, that is, when there are fewer measured data than refined parameters. Based on empirical evidence, R is then replaced with R-free and the number of parameters is omitted. The positional DPI then becomes

$$\mathrm{DPI} = \sigma(r,B_{ave}) = 3^{1/2}(N/n_{obs})^{1/2}C^{-1/3}R_{free}d_{min} \qquad (12\text{-}101)$$

David Blow reformulated[36] the Cruickshank DPI estimates as a function of the Matthews coefficient V_M and incorporated the data-to-parameter ratio into the data completeness C, which then affects $\sigma(r)$ as $C^{-5/6}$. This shows, for example, that omission of the customary 5% of reflections for cross-validation increases coordinate uncertainty only by an acceptable 4% during refinement.

Fisher's information. An alternate estimate for the coordinate precision can be obtained from Fisher's information matrix, a statistical measure that includes information from the off-diagonal elements of the covariance matrix and the maximum likelihood function.[117] It is therefore not necessary to inflate the variance by a factor of ~1.5 as in the derivation of the macromolecular DPI. *REFMAC* also reports this maximum likelihood uncertainty:

```
REMARK   3   ESTIMATED OVERALL COORDINATE ERROR.
REMARK   3    ESU BASED ON R VALUE                    (A):   0.323
REMARK   3    ESU BASED ON FREE R VALUE               (A):   0.236
REMARK   3    ESU BASED ON MAXIMUM LIKELIHOOD         (A):   0.157
```

The above *REFMAC* listing for a 2.2 Å structure lists the estimated coordinate uncertainty based on R and R-free based on Equations 12-99 and 12-101, respectively, as well as the maximum likelihood estimate. At 2.2 Å, the R estimate is in a range where n_p approaches n_{obs}, and the R-free estimate is more realistic, but still inflated compared with the ML estimate. In practice, it is important to make sure the same method of coordinate error estimate has been used, preferably the free DPI or the likelihood estimate, when comparing the overall coordinate precision of different structures. The graph in Figure 12-22 shows the trend of the coordinate uncertainty extracted from the PDB. It provides a coarse guideline ignoring all other dependencies for the expected coordinate uncertainty given the resolution for a well-refined protein structure. Note that in the heavily underdetermined range above ~2.8 Å the name "diffraction-only coordinate precision" becomes somewhat optimistic, because refinement is then dominated by restraints.

Expectation values for *R*-free and *R*-free ratio

Given that we judge our refinement progress on the residual values R and R-free, and that the estimated coordinate precision depends on R-free, amongst other determinants, it will be interesting to derive an expectation value for the gap

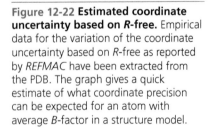

Figure 12-22 Estimated coordinate uncertainty based on *R*-free. Empirical data for the variation of the coordinate uncertainty based on *R*-free as reported by *REFMAC* have been extracted from the PDB. The graph gives a quick estimate of what coordinate precision can be expected for an atom with average *B*-factor in a structure model.

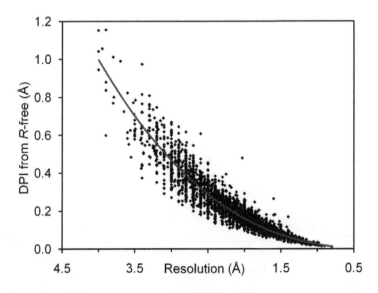

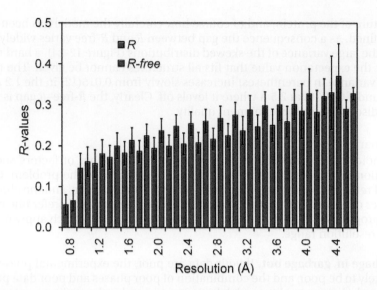

Figure 12-23 The variation of *R*-work and *R*-free extracted from the PDB. The mean values for each bin are plotted, with the error bars indicating the standard error serving as a rough indication for the spread. The mean resolution of a PDB entry is 2.1 Å. Many more such statistics have been compiled by Kleywegt and Jones.[118]

between the R and R-free values. We can obtain a first estimate for the ratio between R and R-free in unrestrained refinement from the two DPI expressions (12-99) and (12-101), which can only be both correct when the expectation value for R-free is given by the ratio

$$R_{free} / R = [n_{obs} / (n_{obs} - n_p)]^{1/2} \qquad (12\text{-}102)$$

A refined statistical derivation by Ian Tickle and co-workers[67] shows that the expected R-free ratio for restrained refinement is given by

$$R_{free} / R_{work} \simeq [(n_{obs} + d) / (n_{obs} - d)]^{1/2} \qquad (12\text{-}103)$$

where d is a more exact measure of the dependence on the number of parameters, the number of restraints n_r, and the value for the weighted sum of all restraint residuals Q_r as defined in the restraint section: $d = n_p - n_r + Q_r$. The problem here is that the residual Q_r is again obtained from inversion of the normal matrix, which is impractical in routine refinement.

In practice, the estimate of the R-free ratio allows us to judge whether our refinement behaves as expected or not. Too small a gap between R-work and R-free indicates that the refinement may still have room for additional improvement, while a very large gap raises suspicion about overparameterization or an incorrect or partly incorrect model. The actual values for R-work and R-free we expect for our refinements (Figure 12-23) depend on the quality of data, the resolution,

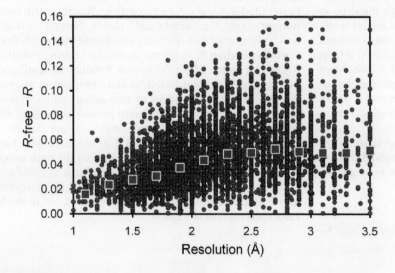

Figure 12-24 The variation of the difference *R*-free–*R* extracted from the PDB. The mean values for 0.2 Å wide bins are plotted in red symbols over the scatter plot of the R-free – R gap. The wide variance is clearly visible, making the gap a relatively uncertain discriminator for refinement quality.

the nature of the protein, and of course how carefully the model has been built and refined. As a consequence the gap between R and R-free varies widely and, given the large variance of the skewed distributions (Figure 12-24), a hard number for the expectation value that fits all structures cannot be given. The mean value (variance in parentheses) increases slowly from 0.015(12) in the 1.2 Å bin to around 0.047(20) at 2.5 Å where it levels off. Clearly, the R-free–R gap is not a sharp discriminator for refinement quality.

Summary of refinement programs and protocols

The choice of refinement program depends on a number of factors such as resolution or required parameterization and is specific to the problem. Often several refinement programs and protocols will work just fine, and the choice is a matter of convenience rather than any theoretical reason to prefer one or the other program. Nonetheless, a few fundamental rules exist which apply regardless of resolution range.

1. Garbage in, garbage out. If your data are poor, the experimental phases are also likely to be poor, and the combination of poor phases and poor data practically precludes any quality model. In the case of molecular replacement, a poor model equaling poor starting phases may be rebuilt into a good model only if the data are good and repeated rounds of rebuilding/refinement combined with *ARP/wARP* or other bias removal are patiently and often iterated.

2. Build and rebuild repeatedly. Starting from a poor model, just picking the method with largest convergence radius—generally simulated annealing, perhaps with torsion angle refinement—and expecting that the model will be close to final without rebuilding is futile. This is particularly tempting in molecular replacement, but it works rarely, if at all. Even the best reciprocal space refinement program will—for theoretical reasons rooted in the neglect of higher terms when linearizing the target functions—get stuck in local minima, and real space model rebuilding will be necessary. Ignoring this fact will lead to a wide gap between R and R-free, and generally suspect stereochemical quality. Another tangible benefit of frequent inspecting and rebuilding is that you get to know your molecule, a necessary requirement for the analysis of your structure. Serendipity is the mother of many *Nature* papers.

3. Monitor R-free. Always. The free R-value is your best safeguard against accidents and over-fitting. Anything you do to your model or refinement protocol

---

Sidebar 12-10 Hypothesis testing: which model is better? The magnitude of the R-values and the R-free/R ratio depend on specifics of refinement and data quality and therefore cannot be used as indicators of completed refinement. The ratio of the drop in R-work versus the drop in R-free upon introduction of a change in the model is still useful for judging whether the model improved in relation to the introduction of new refinement parameters: If both drop similarly, the change is meaningful; if R-free drops only a little compared with the drop in R-work, the change had little effect in terms of model improvement. If R-free remains stationary or increases, no useful modification to the model has been introduced.

A more rigorous statistical measure for the significance of improvement of model A compared with model B (serving as null-hypothesis) in terms of hypothesis testing is the (free)likelihood ratio[119] K:

$$K = \exp\left(LL_A(free) - LL_B(free)\right) \qquad (12\text{-}104)$$

The values of the log(-free)-likelihood $LL(free)$ can be

obtained from the refinement program output, and values for K can be roughly interpreted as a Bayes factor.[120] Values of $K < 1$ mean that model B is better than A (higher likelihood); $K > 1$ means that A is better than B. The higher K, the better model A; the values range from "barely worth mentioning" (1–3) via "substantial" (3–10), "strong" (10–30), and "very strong" (30–100) to > 100, where model A is decisively better than B. As the minor changes during polishing the model do not affect the overall R-values or likelihoods much, free-likelihood-based model comparison is most decisive in the early stages of refinement when making large changes, comparing different parameterizations, and for the adjustment of statistical weights.[54] Note that likelihood ratios are only meaningful if (i) identical data have been used in the model comparison, and (ii) both model refinements have converged. The usefulness of likelihood ratios as a general means of model testing are described in the literature[119, 121] and have been pointed out to me by Ethan Merritt and Ian Tickle.

that dramatically drops *R*-work and shows only little effect on *R*-free deserves great suspicion. Discard and try/build something else.

4. Monitor the overall geometry. Most model building programs allow you to compute a φ–ψ torsion angle plot on the fly, and its value in spotting errors is obvious. Less obvious is the fine-adjustment of overall weights (X-ray versus geometry) and specific stereochemistry and *B*-factor restraint weights. Even if this is automatically adjusted by the program, look at the stereochemistry statistics the refinement program provides. Proper fine adjustment of the restraint weights by minimizing *R*-free (or better –*LL*(free)) will give the correct bond length and angle restraint r.m.s. deviations for your refinement. Do not get distracted by squawks about "too tight" restraint target r.m.s. deviations made by geometry checking programs. Finally, if electron density supports a specific deviation, it is real. *Contra factum argumentum not facit*, no matter *WHAT-IF*.

5. Clean up your model. Many models are deposited with nuisance errors that are simply not necessary given the analysis and refinement tools currently available. This is not just a matter of cosmetics. As *every* atom in the unit cell contributes to *every* structure factor, the refinement of *all* atoms improves when the model is perfected. Overlooked errors include solvent (over)modeling, where water atoms are drifting into space during refinement and have no contacts to protein or other solvent molecules, recognizable often as waters with excessive *B*-factors. Reversed Gln, Asn, and His torsions that violate hydrogen bonding rules are another nuisance error. Analyze and flip them if the chemistry supports it, and re-refine. The fact that the PDB accepts your structure only means that it meets *minimum formal standards*, not that it is good.

6. Interpret validation reports with care. While you should polish your model to make it as error-free as possible, there is no need to get bent out of shape about every detail a well-meaning and overeager validation program flags as an "ERROR." As we explain in detail in the validation chapter, stereochemical model-only validation programs know only about target values, they do not have the evidence in the form of the electron density at hand; you do. The model you have built is based on data that represent an average of millions of molecules, collected over a period of time where changes resulting from exposure to intense ionizing radiation also has an effect on each of them. There will be parts in your structure that just cannot be reasonably modeled, and minor errors that fall into such regions where you lack interpretable electron density are nothing to be ashamed of. Remember George Sheldrick's famous quote[62] leading this chapter: *Macromolecular refinement against high resolution data is never finished, only abandoned*. More about the question of when model building and refinement is finished can be found in the last section of this chapter.

7. Avoid mental phase bias. High expectations, pressure, and the desire to see what one seeks can be hazardous, particularly when building ligands. Ligand binding sites have evolved to bind molecular entities, and there is nearly always something in the crystallization cocktail that is attracted by a binding site. As a consequence, most binding sites contain some spurious or obscure density that may lend itself to placing a ligand. As weakly binding ligands are often underoccupied, they generally have weaker density, making it easy to justify poor density fit and high *B*-factors. Global refinement statistics will be unaffected because of the small scattering mass of the ligand compared with the protein, and other chemical evidence should support the stereochemistry (or even existence) of a bound ligand. There are plenty of cautionary tales provided in Chapter 13. The model must make chemical and biological sense.

8. Select a suitable refinement protocol or program. Depending on the resolution or stage of refinement, some programs have certain advantages or unique features that can make them more suitable for one or the other resolution range or purpose. While restrained coordinate refinement with *REFMAC*, *TNT*, or *CNS* against maximum likelihood target functions yields good results in the general range of ~2.8 to ~1.4 Å, initial low resolution refinement given only a poor model may benefit from the increased convergence radius of simulated annealing–molecular dynamics and lower parameter count in torsion angle refinement available in *XPLOR*, *CNS*, or the *PHENIX* toolbox. For atomic high resolution structures, *SHELXL* provides a highly customizable restraint model and refinement against intensities (which obviates the ML estimate of *F*s for weak negative intensities), and refinement against *twinned* data (in contrast to *de-twinned* data, which generally become inferior during the de-twinning process; see Chapter 8). The absence of formal likelihood targets in *SHELXL* is generally not a problem, because of a smart weighting scheme applied and because high resolution models tend to be complete and relatively correct (provided that they have been properly rebuilt and refined beforehand by other ML refinement programs—high resolution alone is *not* a safeguard against model bias—see Figure 12-30). Table 12-4 lists refinement programs and their implemented parameterization, target functions, and minimization algorithms, while Figure 12-25 provides an overview of the properties of the minimization techniques.

Refinement program name	Refinement and restraint parameterization	Refinement target function	Minimization algorithm
REFMAC	*xyzB*, TLS, ADP	LS, ML, MLφ	SMG
CNS	*xyzB*, TA	MD, LS, ML, MLφ, LS-twin	SA, CG, QN-LBGFS
TNT	*xyzB*	LS	PCG
Buster/TNT	*xyzB*	ML, MLφ	PCG
SHELXL	*xyzB*, ADP	LS, LS-twin	SMG, FMG
PHENIX/cctbx	*xyzB*, TLS, ADP, IAS, (TA under development)	LS, ML, MLφ, LS-twin	SA, QN-LBGFS

Table 12-4 Macromolecular refinement programs. The following symbols are used: *xyzB*, coordinate refinement with Konnert–Hendrickson-type restraints; TLS, torsion-libration-screw parameterization of anisotropy; ADP, individual anisotropic displacement parameter refinement; TA, torsion angle parameterization; LS, least squares refinement target function; ML, maximum likelihood target function; MLφ, phase ML; LS-twin, least squares refinement target function for twinned data; MD, molecular dynamics potential function; SMG, sparse matrix gradient method; FMG, full matrix gradient; SA; simulated annealing; CG, conjugated (Powell) gradient; QN-LBGFS, quasi-Newton limited memory Broyden–Fletcher–Goldfarb–Shanno optimization; IAS, interatomic scattering (ultra-high-resolution refinement). The table is updated from Dale Tronrud's review of macromolecular refinement[56] with the latest information for the developing *PHENIX/cctbx* kindly provided by Paul Adams, LBNL.

Figure 12-25 Comparison of convergence, speed, and target function derivatives for refinement methods. Exploring a large parameter space in searches leads to a large radius of convergence at the price of computational speed. Use of first and second derivatives improves the definition of the downhill gradient in the minimization algorithm. The figure follows an idea of D. Tronrud.[56] MMC, Metropolis Monte Carlo; MD SA, molecular dynamics with simulated annealing.

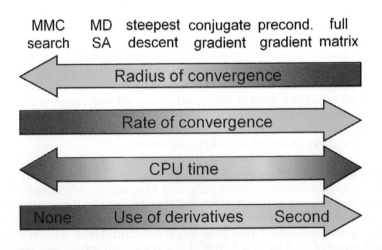

12.3   The practice of model building and refinement

One of the most exciting events in the course of a protein structure determination—on a par with appearance of the first crystals and a successful heavy atom substructure solution—is the first look at the initial electron density map. In the case of experimental phases, we are looking at an empty experimental electron density map, while in the case of molecular replacement we are looking at a more or less biased map, into which we can load the initial model.

In most cases where we have experimental phases available, the first attempt at model building will be to submit the map and a sequence file to a suitable program or server that builds at least a partial model into an experimental map. With good data and phases, the model building programs do a reasonable job. While the machine-built model often needs to be corrected, completed, and properly assembled into complete molecules, the entire model-building task is greatly simplified by the presence of an auto-built partial model.

Molecular replacement models can also be submitted to some servers and model building programs for correction and rebuilding. In this case, resolution below ~2.8 to 2.5 Å generally makes the task significantly harder with poor starting model phases, because the otherwise powerful real space–reciprocal space cycling dummy atom methods are unlikely to work and pattern recognition based methods must be used.

As model building is easier to learn when starting from an already placed model, we will first rebuild the model of our weak cytochrome c′ molecular replacement solution before we proceed to *de novo* model building. Note, however, that rebuilding a bad molecular replacement model may be more work and require more skills than polishing a decent model that has been built into good experimental density with an automated model-building program—another reason to phase experimentally should an MR solution turn out to be marginal. Automated model-building programs work quite reliably, but as always, good phases and good data are just as important for automated model-building programs as they are for successful manual model building.

The basics: manual rebuilding of a molecular replacement model

We begin with our poor molecular replacement solution of cytochrome c′ obtained from molecular replacement using *EPMR*. Given the good data to a resolution of 1.7 Å, we could use one of the dummy atom refinement protocols described above to generate a bias minimized map directly from the raw MR solution. Because of the incompleteness and degree of incorrectness of the model, ML maps generated from the raw model will still show bias. A first step of model improvement and refinement before we generate the first map will improve the maps and make model building easier.

The most common molecular graphics model building programs are *O*,[12] *XtalView*[27] (or its recent incarnation, *MIfit*),[122] and particularly *Coot*[28] (which is supported by CCP4 but is not actually part of the CCP4 suite). Each program provides a variety of different tools and each program has its own strengths and idiosyncrasies. Model building is best learned by practice and we will therefore not describe all programs in a comprehensive way, but by example illustrate some typical tools that are available in one form or another in these programs. It is also important to keep in mind that each model-building situation may require a different strategy, and there is no one-size-fits-all strategy. There are, however, good model building practices[123] that, exercised with necessary thoughtfulness, will lead to well-built models.

Correcting and refining the initial MR model

Side-chain replacement. We can make our life as model builders somewhat easier in two generally applicable ways: First, we can automatically replace the mutilated and incomplete residues with the correct ones using an automated

Box 12-14 **Model building.** Building a model into an empty map begins with the tracing of the backbone. This is aided by density skeletonization, followed by placement of Cα atoms into positions where side chains extend from the backbone. The sequence is docked from known atom positions from the heavy atom substructure solution or sequences of residues of characteristic shapes. The initial model is refined in reciprocal space with geometric restraints and phase restraints, and the next map is constructed from maximum likelihood coefficients. The model is then further completed and refined in subsequent rounds with increasing X-ray weights while tracking *R*-free and stereochemistry. Nuisance errors are removed after analysis in a polishing step.

Automated model-building programs greatly simplify model building, and auto-built models often only need to be completed and polished. Auto-building programs follow similar steps as manual model building and employ pattern recognition algorithms of various kinds to identify residues.

side-chain replacement program. Our search model of 1cpq was initially truncated by *CHAINSAW* so that any atoms beyond the last matching atom between non-identical residues were omitted. This was done in an attempt to improve the MR solution, and now that we have properly re-placed the model, we can repair all the mutilated residues according to the correct target sequence by submitting the model to a side-chain modeling program such as the *SCRWL* server.[124] The returned model now just contains unrefined side chains according to the sequence and in the most common torsions. Most model building programs also provide more or less convenient tools to replace residues.

Simulated annealing torsion angle dynamics. Second, because the replaced residues do not have correct torsions yet, we will subject the model initially to a refinement protocol with a large convergence radius to correct gross errors before we construct a map and rebuild. The present model will be far from the final solution and a first round of simulated annealing–torsion angle dynamics refinement—with its high radius of convergence—followed by positional energy refinement will improve the model geometry and provide a starting model from which we will construct a bias minimized map and begin model rebuilding.

We generate the molecular dynamics topology setup file with the *CNS* input program *generate_easy.inp*, and begin refinement with a round of rigid body refinement (which does not improve much because we already rigid-body refined the position with *EPMR*), combined with initial overall *B*-factor scaling (*rigid.inp*). The model is then ready for the simulated annealing–torsion angle refinement run, followed by energy-restrained coordinate refinement. We can edit the *anneal.inp* file from the *CNS* sample input collection, where we set the starting temperature to a conservative 4000 K, just enough to escape multiple local minima, and select the slow-cooling protocol. We select data up to 2.5 Å for the first refinement, and after a significant initial drop, the *R*-values finally stall as expected around 0.40/0.36.

Restrained ML refinement. At this point, the model will not improve any further without intervention in real space, and now it is time to generate the first

Box 12-15 **Improving poor initial MR models.** Rebuilding of very poor initial molecular replacement models can be aided by a first step of simulated annealing–torsion angle refinement. The large radius of convergence of SA-TA allows the necessary large corrections and escape from local minima. Also, before automated model rebuilding and correction, SA-TA can improve the amount and quality of model that is automatically rebuilt.

map and start rebuilding. As we will continue with regular geometry restrained maximum likelihood refinement with *REFMAC* for practice, we place the heme moieties back into the model (they were not included in the SA-TA refinement) which is done simply by copying the heme coordinates from the initially replaced model, and then adjusting the locations of the heme groups (momomer code HEC for the oxidized Fe^{3+} porphyrin) and make sure its covalent bonds to the Cys residues are properly restrained in refinement. We add a LINK statement to the coordinate file so that *REFMAC* applies the proper restraints:

```
LINK        SG   CYS A 119      1.840  CAB HEC A 131        CYS-HEC
LINK        SG   CYS A 122      1.840  CAC HEC A 131        CYS-HEC1
LINK        NE2  HIS A 123      1.950  FE  HEC A 131        HIS-HEC
LINK        SG   CYS D 119      1.840  CAB HEC B 131        CYS-HEC
LINK        SG   CYS D 122      1.840  CAC HEC B 131        CYS-HEC1
LINK        NE2  HIS D 123      1.950  FE  HEC B 131        HIS-HEC
```

Correcting register errors and loop rebuilding. To tighten the loose geometry we initially select a low X-ray weight of 0.15, and instruct *REFMAC* to write maximum likelihood map coefficients FWT and $\varphi(wt)$ in *XtalView* format. We generate and display the map, pick a slab range of about 6 Å, and step through the model residue by residue. We will find a number of typical small errors such as a Phe ring partially trapped in Fe density (similar to the situation in Figure 12-7), and many incorrect side-chain torsions and backbone deviations. Such errors are obvious and are easy to fix with real space electron density fitting tools.

Much more insidious are register errors, where the backbone is traced correctly but the side chains are wrong. Register errors can happen simply by wrong side chain assignment in experimental maps, or in molecular replacement situations like ours when a torsion angle refinement protocol with high convergence radius shifts entire stretches, which then snap out of register into a proper backbone conformation. Such large-range shifts do not happen with conventional coordinate refinement. While a register shift is relatively easy to notice in high-resolution electron density, they can be hard to detect in low-resolution structures,[125] and special validation tools have been developed to prevent such mishaps (Chapter 13).

In our case a register shift happened at the 3-residue insertion region beginning with Gly 22, and continues for 10 residues to where the density bulges out and can accommodate three additional residues (Figure 12-26). From then on, the model remains in sequence, and only two remaining loop regions need major rebuilding. We use a *baton tool* to add Cα-markers at fixed 3.84 Å distances for the three missing residues in the insertion region, and convert them into a polyalanine backbone and then replace and fit the correct side chains. After real space fitting of a stretch of residues into electron density, we locally refine the backbone geometry. The modern program *Coot* allows us to combine all these steps by simply instructing it to rebuild the whole range of residues with a certain sequence. We continue to rebuild and correct residues through one molecule, and superimpose the rebuilt first molecule onto the second monomer and replace it with the first model. This saves rebuilding the second monomer and allows application of NCS restraints. Inspection of the map shows that several stretches of the helices can be NCS restrained, and we pick corresponding groups of corresponding residues with medium backbone and weak side chain restraint weights. Loop regions and heme groups are excluded from the NCS restraints. We increase the X-ray weight to 0.3 in *REFMAC* to give more weight to the X-ray data and now include reflections to 1.8 Å. At this point the R-values are

Box 12-16 **Register errors.** In low resolution structures the backbone can be traced correctly, but the sequence may be shifted. Such register errors can be hard to detect from electron density shape alone and are usually detected by poor side-chain interactions or unusual local environment.

Figure 12-26 A three-residue register shift error. A shift of the sequence assignment with correct backbone conformation is a register error. In the depicted bias-minimized 1.8 Å electron density the incorrect residues Gly-Pro-Leu (A) are easily replaced with the correct residues Leu-Glu-Phe (B). Register shift is easy to detect in high resolution electron density, but not so in low resolution structures, where register shifts present one of the more insidious problems.

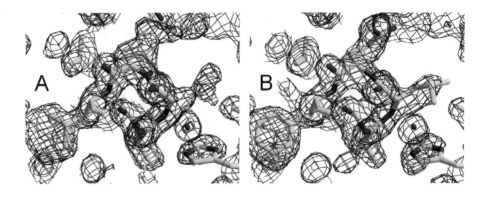

0.28 and 0.24, and the next map is again improved, and we begin to polish minor errors and build the discrete solvent as discussed in the following sections.

Building ordered solvent: the hydrogen bond network

The narrow boundary region between the protein and the solvent exhibits a specific and sometimes functionally relevant structure containing ordered and often conserved solvent molecules within one or two coordination shells. The solvent within this ~7 Å wide boundary region is generally well ordered in a crystal structure and often forms an integral part of the protein's solution structure.

If the resolution of X-ray data is better than ~2.5 Å, water molecules can be placed into discrete density blobs. The higher the resolution and the more rigid the protein surface, the easier this task is, and discrete water molecules, often arranged in extensive networks, can be detected. However, as we cannot discern the two hydrogen atoms protruding from the central oxygen, the water molecules are placed as "water atoms" in the form of a single oxygen atom. Based on the preferred directionality of hydrogen bonds and the H–O–H bond angle of 104.5°, chemical plausibility again needs to be employed to interpret the hydrogen bond network (Figures 12-27 and 12-28). Only very high resolution X-ray structures (generally 1.2 Å and higher) as well as neutron diffraction studies can explicitly reveal the position of the hydrogen atoms.

At this point in our rebuilding and refinement effort of the cytochrome dimer, many solvent atoms have become distinctly visible, and we automatically add water "atoms" with *arp_waters*, available as an option during *REFMAC* refinement. Alternatively we can employ a water placing function directly in the model building program. We inspect the waters, check for proper hydrogen bond donor–acceptor geometry (Figure 12-27), and remove any water molecules that drifted away during refinement and have excessive B-factors. The R-free and R values of the model now are 0.26 and 0.21, respectively.

Figure 12-27 An extended hydrogen bond network containing a number of typical hydrogen bonds. The green arrows in the schematic point toward the hydrogen accepting electron lone pairs. Note that (i) each bond has a proper donor–acceptor pair (which clearly defines the correct orientation of the Gln residue), and (ii) the different types of interactions: direct backbone–side chain, side chain–side chain, and water-mediated interactions between residues. Typical X–O–H angles are 120° for the sp2 hybridized orbitals of the -OH groups, and 104.5° for H–O–H in the nearly tetrahedrally coordinated water oxygen atom. The covalent O–H distance is 0.96 Å. The map is an omit map calculated without water molecules in the model.

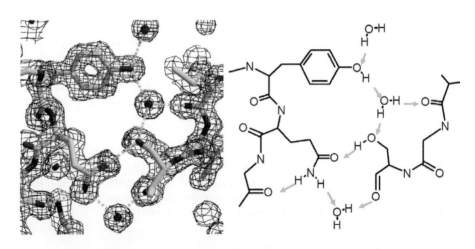

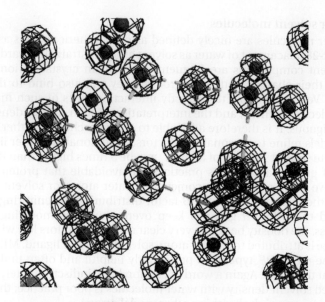

Figure 12-28 High resolution water network. Exceptionally well-defined water network in the intermolecular space of a high resolution (1.2 Å) structure (PDB entry 1bpi).[126] Note the typical, ice-like 5- and 6-membered water ring networks, with the central water atom possessing near-perfect tetrahedral arrangement of hydrogen bond partners.

Box 12-17 Solvent structure. One of the most common mistakes leading to overparameterization of the model is overbuilding of the solvent. Discrete water molecules should have hydrogen bonded contact(s) to other solvent molecules or to protein. Poorly placed waters tend to drift away during refinement because of lack of density and restraints and often end up far away from other molecules and with high *B*-factors.

Expected number of discrete water molecules. A commonly asked question is how many waters one should build in a protein structure. The answer is until all difference density peaks that can be plausibly assigned to water molecules are occupied. The emphasis is on *chemical plausibility*—as always, indiscriminately filling all unexplained density floating in the structure with water molecules is bad practice. An old rule of thumb by Alwyn Jones[123] states that about one water molecule per residue can be expected for the average structure. As one might assume, this number is lower for low-resolution structures where less discrete solvent molecules are expected, while many more waters may be placed into ordered solvent networks of high-resolution structures. The number of built water atoms binned by resolution is plotted in Figure 12-29 and confirms this trend.

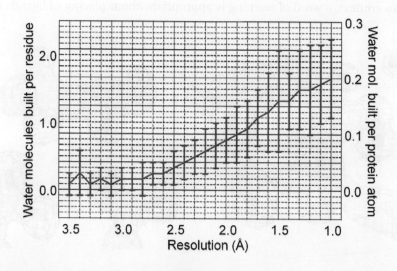

Figure 12-29 Number of discrete water solvent molecules built in protein structures. The graph shows the plausible trend that more discrete waters can be built in high-resolution structures than in low-resolution structures. Note that the linear part of the graph extrapolates to zero at about 2.9 Å, making one wonder what the water molecules in low resolution structures really might be. The graph has been redrawn with kind permission from a figure by Clemens Vonrhein, Global Phasing, Ltd.

Non-water solvent molecules

If the water molecules are nicely defined as clear spherical blobs as in Figure 12-27 or 12-28, placement of water as solvent atoms is straightforward. However, other solvent components are frequently present in crystallization cocktails, and given their charged and polar groups, they will also bind to the protein molecules. Very often, only the directly interacting parts of such moieties are visible in electron density, and the interpretation of such strange density can be quite ambiguous. It is therefore advisable to critically check whether such entities in models refine to reasonable B-factors. A "reasonable" upper limit would be a discrete solvent B-factor of about 2, maybe 3 times higher than the average B-factor of protein atoms. It is practically unavoidable that protein structure models have some of the more conjectural water atoms or solvent molecules located in the high-end tail of the B-factor distribution—monitoring B-factors, R-free, and the R-free/R gap will keep overenthusiastic modeling in check. Nonetheless, the density better be very clear and the B-factors low when special significance is attributed to a water atom, solvent entity, or ligand. ML difference maps of the $mF_o - DF_c$ type are a particularly helpful and often underappreciated help at this stage. Again a word of caution here: indiscriminately filling any unidentified solvent density with water molecules is poor practice; these waters then just serve as a sink absorbing other model errors!

Metal cations of Na, Ca, Mg, or Zn are quite ubiquitous in purification buffers and crystallization cocktails, and they are frequently found in the ordered solvent layer around proteins. They show typical, often octahedral, coordination geometries with water and negatively charged side chains or backbone oxygen atoms as partners, and the bond lengths are generally shorter than regular hydrogen bonds (Appendix A.11). From the level of electron density alone it is hard to tell precisely which kind of metal is present, and chemical plausibility and prior knowledge about the purification and crystallization conditions is once again necessary to make a convincing case for a specific assignment (Figure 12-30). Mg^{2+} ions are often found compensating the negative charges on the backbone phosphates of DNA, as shown in Figure 2-45.

As cations from the solvent environment can be readily detected, one would also expect to find their associated counteranions bound to the protein. Sulfate and phosphate, for example, have distinct tetrahedral shapes, which in addition to the higher electron density of S or P (and their anomalous signal) makes them easy to detect. Chloride ions (Figure 12-42), on the other hand, seem to be less frequently found in direct contact with positively charged protein surface residues. A partially occupied, spherical chloride ion is also difficult to distinguish from a well defined, low B-factor water molecule, and again plausibility of chemical environment and geometry are needed to support the assignment. Also the solvent part of a protein structure model must make sense.

Ligand density

In this context, a word of warning is appropriate about placing of ligands into

Figure 12-30 Solvent molecules in a protein structure. A metal ion (Mg^{2+}, left panel) and organic molecule (2-methyl-2,4-pentanediol, MPD, right panel) are bound to surface residues of concanavalin A. Note the distorted octahedral environment, which clearly indicates, together with shorter bond lengths (2.0–2.2 Å) that the coordinated central atom is not just another water molecule. MPD is a commonly used crystallization additive. Note the pentagonal arrangement formed by the hydrogen bonded atoms, similar to frequently observed arrangements involving ordered water molecules at protein surfaces. PDB entries 1gkb[127] and 1nls[128] (the MPD is modeled only as discrete water molecules in the deposited structure). Both maps are *Shake&wARP* omit maps computed without the solvent molecules.

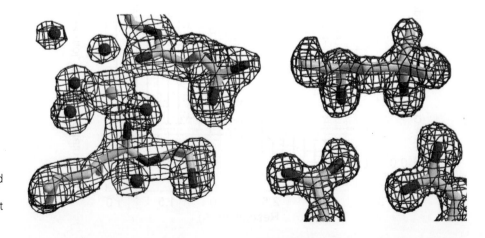

> **Box 12-18 Ligand density.** Binding sites have a tendency to attract various detritus from the crystallization cocktail, and will therefore often contain some weak, unidentifiable density that can be (wishfully) mistaken for desired ligand density. Plausible binding chemistry, ligand conformation, and independent evidence are necessary to avoid misinterpretation.

electron density: Binding sites have by nature specifically evolved to interact with a ligand, and the binding pockets are rich with opportunities for all types of interaction. Thus, the active site residues in binding pockets have a tendency to pick up whatever chemical detritus floats in the surrounding medium, particularly if it resembles the intended ligand. If a component of the crystallization cocktail is present in high concentration (such as buffer molecules, organic compounds, additives), this surrogate ligand does not even have to bind very well in order to yield some more or less defined unaccountable electron density in the binding site. It is very tempting to fill this mystery density with the ligand that one hoped to find in the structure (Figure 12-31). Clear electron density after rigorous bias removal and solid chemical plausibility are important points to verify that the density object is indeed the sought-after ligand.

The correct placement of ligands into electron density is a very important and unfortunately also often difficult and perilous task. While automated programs do a fairly good job searching for molecules in libraries, care must be taken to properly restrain the ligand geometry[130] and to assure plausible ligand stereochemistry and binding interactions with the target.[131] As drug–target complexes are particularly interesting and important, we will discuss ligand building and validation in Chapter 13 in detail.

Ambiguities in side-chain electron density: NQH flips

Well defined side-chain density generally allows reliable side chain placement based on the observed density shape. There is, however, some ambiguity that cannot be resolved by relying solely on electron density shape: The orientation of asparagine, glutamine, and histidine cannot be unequivocally assigned from electron density alone. In the amide residues, the difference of one single electron (equivalent to the scattering power of a hydrogen atom) between nitrogen ($7e^-$) and oxygen ($8e^-$) is normally not enough to distinguish the two. The same holds for the ambiguity of the orientation of the histidine ring, with C and N only one electron different in scattering power. Only with quality data and at high resolution (around ~1.2 Å and better) can N from O be reliably distinguished based on electron density levels alone.

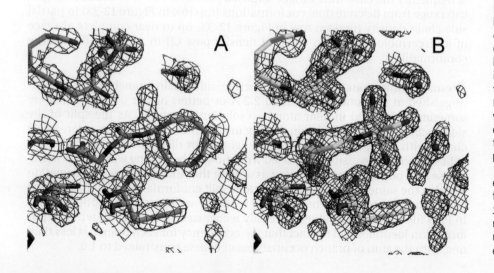

Figure 12-31 The sweetener that was bitter. The panels show 2.1 Å maps contoured at 1σ (blue) and 5σ (red). (A) A super-sweetener presumed to be bound to a taste receptor analog built into electron density map generated from *CNS* ML $2mF_o - DF_c$ coefficients. Poor density fit and a questionable vdW contact prompted critical re-examination of the model, and the fit and local chemistry proved to be perfect for TES buffer (2-(2-hydroxy-1,1-dihydroxymethyl-ethylamino)-ethanesulfonic acid), a component from the crystallization cocktail (B). The *Shake&wARP* map in (B) has slightly less noise and cleaner connectivity and clearly reveals the true nature of the ligand. PDB entry 1ynl.[129]

Figure 12-32 Ambiguity of glutamine side chain placement. Example of an incorrectly placed Gln side chain (A). The hydrogen bonding pattern is much more plausible in the correct, flipped position (B), where each hydrogen bond has a donor–acceptor pair.

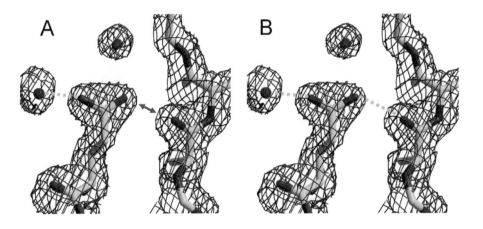

The proper conformation of Asn, Gln, and His needs to be assigned according to chemical plausibility, such as satisfying hydrogen bond networks, which is often straightforward in the case of Asn and Gln (Figure 12-32). The fact that His can assume different protonation states depending on its actual local pK_a makes the assignment of His side chain orientation often more subtle.

Surprisingly, a large fraction of protein crystal structures do not have correctly assigned Asn and Gln side chain amide conformation. This is even more surprising, as checking programs exist which flag questionable assignments. Recent checking programs based on self-consistent empirical potentials[132] or the *MolProbity* validation program[133] even return "flipped" coordinate files. Note that just swapping the atom names is not correct; the amide groups as well as the imidazole ring need to be rotated around the preceding torsion. In any case, because of the different bond lengths and angles, the swapped or flipped coordinates need to be re-refined again in the correct orientation. Improperly assigned NQH residues are more than a nuisance error. For purposes such as ligand docking and molecular modeling, the proper assignment of hydrogen bond donors and acceptors is crucial. It is not much effort to do it correctly the first time around and use *NQ-Flipper*[132] or *MolProbity*[133] during model building.

Side-chain torsions and multiple side-chain conformations

The local environment and intra- or inter-molecular contacts determine which of the possible favored conformers of side chains are actually assumed. Strong interactions with other residues or ligands can lead to strained conformations, but very unusual side-chain torsions certainly warrant some closer inspection. The fact that the local environment determines which torsion angle combinations are assumed also means that in a local environment that is ambivalent or fluctuating, multiple conformations of the same side chain are possible. This is frequently the case with surface-exposed residues. The conformational split can range from discrete dual conformations (top row in Figure 12-33) to partial side chain density (bottom row in Figure 12-33) up to near complete absence of a discernible side chain electron density past Cβ in multiple, fluctuating conformations.

Alternate conformations with two discrete side chain branches can be distinguished at resolutions of around 2.2 Å or better; higher discrete splits are sometimes observed in near atomic resolution structures. As the split of the side chain "dilutes" the electron density in each branch by $1/n$, disordered side chains with many n fluctuating positions are not distinct in electron density. Split side chains are usually modeled with the split beginning with the Cβ-atom, whose two positions are very close assuming that the backbone conformation remains the same. At very high resolution, split conformations that include the backbone are sometimes observed and modeled as discrete polypeptide chains. In PDB files, the split residues are easily recognized through the alternate conformation identifier and the fact that the occupancy for each branch is less than one, with the sum of branch occupancies of course constrained to 1.0.

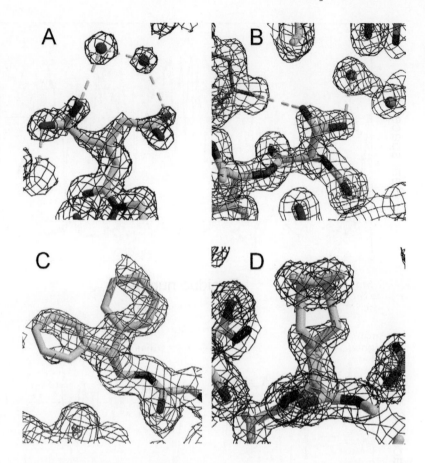

Figure 12-33 Split side chain conformations. Panel A shows the acidic residue Glu participating in a solvent hydrogen bond network, and the split serine residue (B) participates in intermolecular crystal contacts with its neighbor (purple Gln residue) in one conformer, and correlated hydrogen-bonds to a water molecule in the alternate conformation. The bottom panel row shows the hydrophobic residues Phe (C) and Met (D) located on the surface of a protein in split conformations, with only partial electron density visible. The resolution of the electron density maps is about 1.7 Å.

In our cytochrome c′ model, a few residues show distinctly split side chains, and we make sure these are excluded from the NCS restraints. At this point, we also add hydrogen atoms, and after restraint weight optimization the refinement converges at R-values of 0.233/0.195, somewhat better than in the original PDB submission.

Tracking progress through real space correlation plots

An extremely useful tool that quickly allows one to gain an overview of the model quality and to pinpoint local problem regions is a residue-by-residue plot of the real space correlation coefficient (RSCC).[134, 135] The RSCC is calculated between an "experimental" electron density map $\rho(\mathbf{r})_{obs}$ (computed from figure-of-merit weighted observed amplitudes and experimental or model phases) and the "calculated" model density map $\rho(\mathbf{r})_{calc}$ (computed from the model phases and amplitudes). A perfect fit has an RSCC of 1.0 with the real space correlation coefficient defined as

$$\text{RSCC} = \frac{\sum_{\mathbf{r}}(\rho(\mathbf{r})_{obs} - \overline{\rho(\mathbf{r})_{obs}}) \cdot (\rho(\mathbf{r})_{calc} - \overline{\rho(\mathbf{r})_{calc}})}{\left(\sum_{\mathbf{r}}(\rho(\mathbf{r})_{obs} - \overline{\rho(\mathbf{r})_{obs}})^2 \cdot \sum_{\mathbf{r}}(\rho(\mathbf{r})_{calc} - \overline{\rho(\mathbf{r})_{calc}})^2\right)^{1/2}} \tag{12-105}$$

The RSCC can be computed using the program *OVERLAPMAP* from CCP4, or the model and data can be submitted to the *Shake&wARP* service which returns a real space correlation plot (Figure 12-34). For models that are deposited in the PDB and have associated structure factors, the real space correlation plots can be downloaded from the EDS server in Uppsala,[136] which also provides the RSCC as a part of the structure analysis package.

Automated model building and improvement

In practical model building most of the time an automated model-building program will be used to build either an initial model into an experimental electron

Figure 12-34 Real space correlation coefficient (RSCC) plots. The progress of the rebuilding effort can be rapidly tracked by inspecting real space correlation plots, which measure the residue-by-residue fit of the model to the electron density. The top of each plot shows the RSCC in the range of 0 ≤ RSCC ≤ 1 (black graph); the bottom (blue graph) shows the B-factors of the model. Panel A shows the initial MR model, where the density fit is quite poor, and in fact around ~0.6 in practice is the lowest average RSCC value (black dotted line) where a model might be salvageable. Panel B shows the model fit after the simulated annealing–torsion angle refinement step, before individual B-factor refinement (note the flat B-factors at the Wilson-value of 18 Å^2). Panel C shows the fit of the final and polished model, which is quite excellent, with an average RSCC above 0.94. Note that RSCC and refined B-factor are often correlated because floppy residues that are hard to build generally have a high B-factor. Misplaced atoms also refine to higher B-factors and will show poor real space correlation.

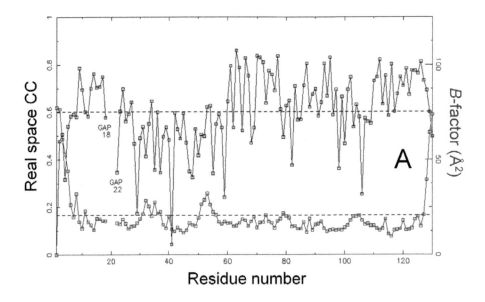

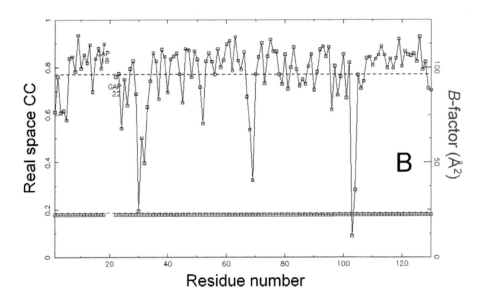

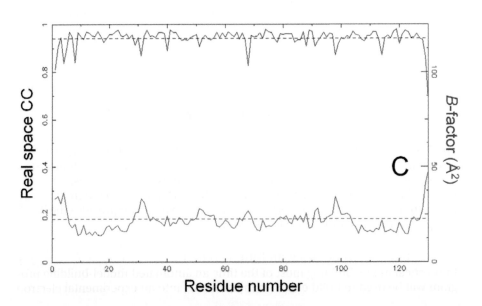

density map or to rebuild a molecular replacement model. New programs or extended features of existing programs are added to the crystallographer's toolbox frequently, and the following sections therefore provide only a snapshot of some of the presently available programs and the fundamentals of their functioning. We begin the section with an example of automated model rebuilding with *ARP/wARP*[137] and close the chapter with a brief discussion of a *de novo* building of a MAD structure into an experimental map with *TEXTAL*[138] and *ARP/wARP*. Both programs have the benefit of being available as well-functioning web services.

Automated model rebuilding with *ARP/wARP*

In our manual model building efforts, we began to rebuild the cytochrome c′ model and inspected the maps after the initial restrained refinement step with *REFMAC*. We had to manually rebuild a major register shift error and rebuild the model in two loop regions, which took considerable effort. We can try to save some labor by submitting the crudely refined model to the *ARP/wARP* program,[113] which is available as a separate plug-in for the CCP4 distribution or as a web service at the EMBL Hamburg.[137]

Automated rebuilding of the initial cytochrome MR model. The program *ARP/wARP* can correct MR models[139] and can also build into experimental electron density from scratch, with the limitation that the resolution should be 2.5 Å or better because of the dummy atom refinement steps at the core of *ARP/wARP*. For molecular replacement we submit our crude model from the *CNS* torsion angle refinement as input, together with the structure factor amplitude data file in CCP4 *mtz* format. We also provide the correct sequence file for the target molecule, and the number of identical units (2) in the asymmetric unit and the number of residues (260) we expect to be built. After an hour the job returns with a nearly complete model with $R = 0.25$ and R-free $= 0.30$ (without the heme groups, which *ARP/wARP* ignores, and the real improvement is thus even larger than anticipated from the R-values). The Cα coordinate r.m.s.d. between the automatically built *ARP/wARP* model and the final model is 0.33 Å, already quite low and in the range of what is expected between NCS related molecules. The register shift is completely rebuilt, and also gaps in the model are properly closed (Figure 12-35). What remains now is to fix one last proline-containing loop that *ARP/wARP* did not build (Figure 12-36) and to finish the model by fine-adjusting torsions, rebuilding solvent, and finally making sure all the remaining polishing steps such as checking NQH flips, close contacts, and perhaps split residues are properly accounted for.

Within an hour or two, we went from a first molecular replacement solution to a nearly finished model by using the right tools at the right place—one of the most important steps toward learning efficient crystallographic model building. For our cytochrome c′ example, the original manual model building, not so many years ago, took several days to complete.

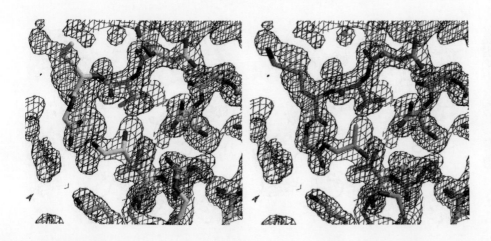

Figure 12-35 Automated loop rebuilding by *ARP/wARP*. Our starting model had a gap in a loop region (left panel). *ARP/wARP* rebuilt the loop and joined the ends properly. This worked well because we had good 1.7 Å data. The view is clipped and the loop actually continues out of the paper plane after the second glycine. Note the incorrect lysine torision typically for the minor errors that need to be manually adjusted during polishing of the final model.

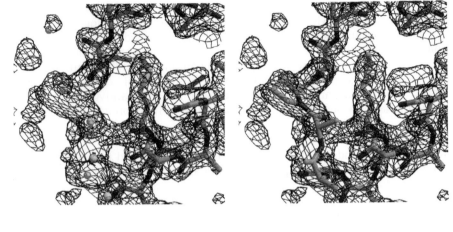

Figure 12-36 Manual finishing of a chain break. The starting model lacks a residue in a proline-containing loop. It remained unbuilt and needed to be corrected by hand (right panel). Note in the left panel the dummy atoms (cyan spheres) placed in the density by *ARP/wARP*. The program properly rebuilt the neighboring residues and broke the main chain, but it could not yet place the proline correctly and therefore could not close the gap.

Figure 12-37 Backbone Cα-trace and auto-built model in experimental electron density map. Panel A shows the trace of Cα-atoms in an experimental electron density map connected by 3.84 Å long batons (long brown sticks). The Cα-atoms are placed at positions where the side chains are believed to protrude from the modest quality 2.5 Å electron density. We picked as a starting point the Met residue, whose side chain position is known from the dispersive Se marker atom positions determined during heavy atom phasing. Panel B shows overlaid in purple the model that the model building program *TEXTAL* generated, which provided a good starting point for real space refinement and rebuilding. Panel C shows that the auto-built 2.5 Å model produced by *TEXTAL* (purple sticks) approximates the final model (yellow sticks) quite well in the displayed stretch around the Met residue and only minor corrections are necessary. The upper part of panel C reveals parts of the model, where larger tracing errors and deviations from the corrected model had to be corrected. Wrong initial backbone tracing across β-sheets because of incorrect strand-connecting density is hard to avoid in manual as well as in automated model building. The remedy is simple: better phases.

Model building into experimental maps

Compared with correcting the molecular replacement model, where we had to close only gaps of a few residues at most, we need a few more initial steps when building a protein model from scratch. Building a model into an empty experimental map is largely an exercise in pattern recognition, both for human model builders and for automated programs.[140]

A useful method for obtaining an overview of the molecule and visualizing the possible backbone trace particularly in noisy maps or maps of lower resolution is the skeletonization of the electron density.[141] It is essentially a density erosion technique, where the density is collapsed into a string ("bones") that most likely follows the backbone trace as well as distinct side chains. In medium resolution maps (approximately 3.0 Å) or better, the branching between side chains and the backbone is often clearly visible, and the oxygen atoms of the backbone form "knobby" extensions from the backbone. In this case, the placement of Cα-atoms is possible with little ambiguity at the intersections of the bones in the skeletonized map. "Baton" tools that place the next or preceding Cα-atom at a *trans* distance of 3.84 Å help in the building of the Cα backbone trace (Figure 12-37).

Once a marker in the form of the shape of a distinct residue combination or a known heavy atom position is found, the assignment of the sequence becomes possible, and the side chains can be built. The sequential position of the Se atoms located in a heavy atom search is always known, and the shape of the side chains upstream and downstream makes the assignment of a specific Met residue possible. The sequence is then locked in.

Even in the best initial models, at some loops or turns the trace will be lost. The first model thus consists of backbone fragments, some with clear sequence, some with ambiguity, and on occasion it requires some detective work find the correct connection of the stretches that complies with the sequence. Often these

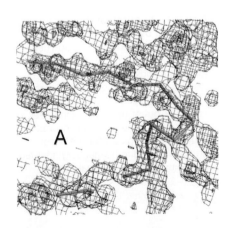

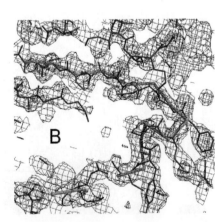

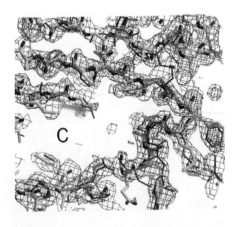

loops become progressively clearer as the phases improve during model refinement. While initially poor loop density that results from genuine disorder will not improve, loops that were intruded upon by tight density modification masks will come back, together with discrete solvent structure, in the course of refinement and rebuilding rounds.

At low resolution, the challenge to recognize certain regular patterns indicative of ordered secondary structure elements, such as helices and strands, increases. Some of the automated programs that build models from scratch in fact use pattern recognition methods to place residues or stretches of secondary structure fetched from libraries into electron density.[142] The risk of error increases with decreasing resolution: While complete nonsense leading to absurd geometry or impossible connections between segments is usually either evident or is flagged in interactive building programs, be aware of the insidious register shift error and improper connections between stretches of secondary structure.

Electron density of α-helices and β-sheets

In electron density, α-helices are usually easy to recognize. Looking down a helix, side chains point from the central backbone of the helix outward and to the right into the direction of the turn, and "droop down" in profile view toward the N-terminal of the chain. It is thus often possible to determine the direction of the helix quite well during initial model building. The cross section of α-helical density shows a tell-tale hole in the center of the helix, from which the residues point outward in the direction of the turn (Figure 12-38). Helical proteins tend to be relatively rigid with low intrinsic B-factors, and helical electron density is thus generally well defined; given a resolution of about 2.5 Å or better, the model is relatively easy to build in helical regions.

The tracing of β-sheets can be more challenging, largely because side chain and backbone carbonyl density can connect across the sheets and it can be hard to tell in which branch the backbone continues (Figure 12-39). In both helices and sheets, the direction of the chain can be identified by the fit of the carbonyl oxygen atoms—if the trace is reversed, the CO groups do not match

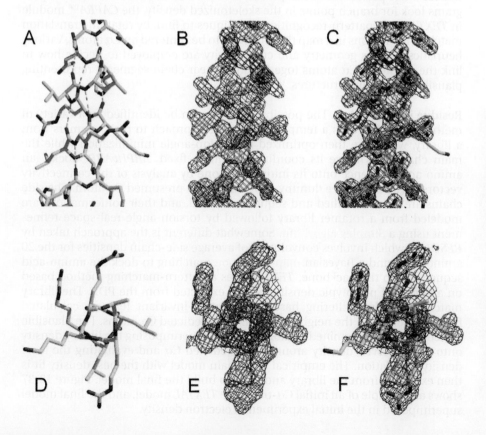

A   B   C

D   E   F

Figure 12-38 **Recognition of α-helical electron density.** Note the side chains "drooping" down toward the N-terminal end of the α-helix (top left panels), with the electron density somewhat reminiscent of a Christmas tree. A hole down the core of the α-helix (bottom panels) is generally visible at resolutions better than ~2.8 Å. Viewing along the helix from N-terminus to C-terminus, the density of the side chains protrudes outward and toward the right (bottom panels). Note in the cross section (D) that in projection the torsion angle of each following residue is a little tighter than 90°, thus at 100° only 3.6 residues (and not 4) are necessary to complete a full turn of the α-helix. Building secondary structure elements becomes considerably more challenging at low resolution (< 3.0–3.5 Å); for example, compare the low resolution electron density figures in a recent review.[143]

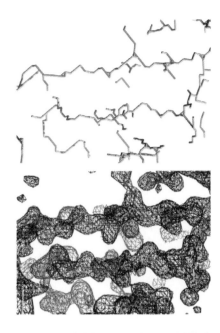

Figure 12-39 Electron density skeleton in β-sheet region of a 2.8 Å map. The skeletonized electron density (yellow "bones") (top) gives a quick overview of the backbone trace in an empty experimental electron density map as shown in the bottom panel together with the skeleton. Electron density of β-sheets has a tendency to connect across the strands, which makes building of sheet structures somewhat more challenging than building of helical sections.

the "knobby" extension typically protruding from the main chain density. In addition, hydrogen bonding will be poor and validation programs will likely flag unusual geometric features and environmental distributions of the residues. The risk of reverse tracing largely affects poor low resolution structures below ~3 Å.

Automated *de novo* model building

In addition to the *ARP/wARP* program we introduced for model completion in the previous section, various other automated model building programs exist. For automated model building the fundamental principle of garbage-in garbage-out holds as well: Poor starting phases will not give a good model. *RESOLVE*[144] and *TEXTAL*[138] are automated model-building programs that also work reasonably well for low resolution structures; *MAID*[145] and *ARP/wARP*[113] work well at resolutions above ~2.5 Å. Both *ARP/wARP*[137] and *TEXTAL*[138] are available as web services. *Buccaneer*[146] is based on the *FFFEAR* feature-recognition-based fragment building algorithm[147] and is available as a stand-alone version or in combination with the graphical model building program *Coot*.[28] Connecting loops between secondary structure elements are difficult to build because they are flexible and solvent-exposed, and generally show weak density. This is true for manual as well as automated building. The program *Loopy* for example is specially dedicated to building loops, and uses both electron density and a knowledge-based search to build conformationally plausible loops[148] (implemented in the 2009 *ARP/wARP* release).

Envelope identification and Cα placement. The first task during *de novo* model building is to identify which parts of the map form the actual molecule. As we have seen in Chapter 5 the asymmetric unit seldom contains an intact molecule, and the map must be extended across the asymmetric unit boundaries to cover the entire molecule. While this is intuitively grasped by the model builder during manual building, a program must determine the presumed molecular envelope; otherwise parts of the structure are built in disconnected fragments into different molecules, which naturally leads to gaps when building the model. Early model-building programs required manual assembly of the pieces into one molecule. The module *FINDMOL*[149] in *TEXTAL* for example, accomplishes this task. The placement of the Cα-atoms is the next task, and while some programs look for branch points in the skeletonized density, the *CAPRA*[150] module in *TEXTAL* uses pattern-recognition techniques to find, by rotation–translation matches, the regions in a map that are likely to be centered on Cα atoms. Various heuristics such as geometry and connectivity are employed to decide how to link the predicted Cα atoms together into linear chain segments representing plausible secondary structures.

Residue identification. The possible residues can be identified by a variety of methods. *MAID*[145] uses a template matching approach to pick rotamers from a library, which are then optimized by torsion-angle minimization while the main chain and hence its coordinates remain fixed. *ARP/wARP* "docks" an amino-acid sequence onto its initial backbone by analysis of the connectivity vectors between the free dummy atoms around a presumed Cα atom. The side chains are then identified and sequence docked, and their conformations are modeled from a rotamer library followed by torsion-angle real-space refinement using a Simplex algorithm. Somewhat different is the approach taken by *RESOLVE*, which involves convolution of average side-chain densities for the 20 amino acids and a Bayesian map-likelihood matching to dock the amino-acid sequence into the backbone. *TEXTAL* uses a pattern-matching method based on a library of prototypic density regions extracted from the PDB. The library matching involves filtering based on rotation-invariant features calculated from the density in the neighborhood of the predicted Cα atoms. The plausible matches are then examined in more detail by superimposing the library density onto the observed density around the predicted Cα and evaluating the local density correlation. The empirical side-chain model with the best density fit is then extracted from the library and used to build the final model. Figure 12-37 shows an example of an initial Cα-trace, the *TEXTAL* model, and the final model superimposed in the initial experimental electron density.

In between manual model building and fully automated model-building programs are a variety of tools for fragment and loop building, which look up empirical libraries. The tools either work from within model building programs such as *Coot*, *XtalView*, and *O* or are available as stand-alone modules such as *ESSENS*[151] and the FFT-based 6-dimensional fragment search *FFFEAR*.[152] These programs look for secondary structure elements by a process of template convolution, the former in a brute-force 6-dimensional approach and the latter by modified FFT methods.

Auto-building, correcting, and refining a MAD structure

In Chapter 11 we applied 2-fold density-averaging in real space to the experimental electron density map generated by *SHELXE*[153] from the tesB dimer Se-MAD data. We can now submit this 2.5 Å map to the *ARP/wARP* server at the EMBL.[137] As this map was not as outstanding and at such a high resolution as the VTA1 1.25 Å S-SAD map, we expect that there will be some polishing work left to perform after automated model building.

The submission procedure to the server is straightforward: we upload the *mtz* file written by *DM* and select the structure factor amplitudes (FP), their estimated error (SIGFP), the figure of merit (FOMDM), and phases (PHIDM) as input, not forgetting to instruct *ARP/wARP* to use the *R*-free flags and experimental phase restraints in the refinement. The remaining information includes the sequence (readily available in single letter format from PDBwiki.org), number of residues (570), and number of identical molecules (2) in the asymmetric unit. After about 1 hour the job finishes and we can download the results. A quick inspection shows that *ARP/wARP* did an outstanding job, and the model is nearly complete (Figure 12-40). We are only missing the N-terminal and C-terminal residues, and the flexible loops connecting the secondary structure elements: S2Q3, the loops S26 to V34 and L143 to R156 as well as the last residues H285 and N286.

Before we look at the model it is instructive to examine the diagnostic graphs *ARP/wARP* generated and which can be readily displayed with the CCP4i log-graphing utility. The graphs in Figure 12-41 shows consistent improvement in *R*-values and figure of merit with each building cycle, just as we would expect for successful manual model building. Note from the graphs that, in order to save time, the intermediate refinements do not have to go to convergence (the *R*-values still drop at the end of each cycle). This does not matter because the

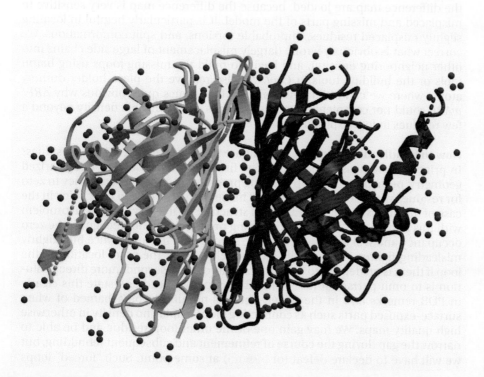

Figure 12-40 Auto-traced model built by *ARP/wARP*. The model of the tesB dimer built into the experimental 2.5 Å map is nearly complete; only terminal residues and a few flexible loop regions (indicated by the dotted connections) are missing. The red spheres are dummy atoms, which are used to fill any unmodeled strong density. Many of them are therefore not real solvent water molecules.

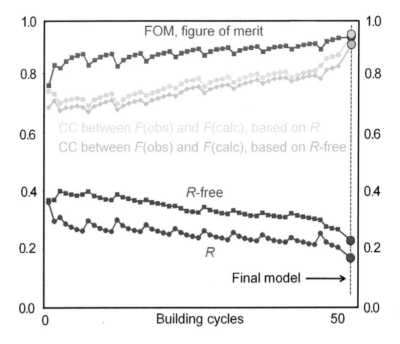

Figure 12-41 Model building and refinement progress. Typical trace of refinement indicators *R*, *R*-free, figure of merit, and correlation coefficient between *F*(obs) and *F*(calc). The circles at the right edge show the final values after completion of final manual rebuilding and polishing of the tesB (Se-Met) model.

model is rebuilt again in the next cycle, but for our final refinement and weight optimization, refinement to convergence will be crucial.

Although we expect no sequence docking problems, we first check whether the Se-Met residues display the strong density expected for the Se atom (Figure 12-42), which is reassuring. We immediately update in the auto-built model PDB file the MET residues to MSE, not forgetting to change the SD to SE (note that the elements are right adjusted; an example is listed below).

```
ATOM 4465  SD   MET B 282        6.576  67.108  52.680  1.00 6.44
ATOM 4465  SE   MSE B 282        6.576  67.108  52.680  1.00 6.44
```

Missing loops. We now load the updated model into our favorite model-building software, and walk through all the residues. Both the ML $2mF_o - DF_c$ map and the difference map are loaded, because the difference map is very sensitive to misplaced and missing parts of the model. It is particularly helpful in locating slightly misplaced residues, improbable torsions, and split conformations. We correct what is obviously wrong (largely misplacement of large side chains into other neighboring density), and begin to build the missing loops using baton tools or the building tools in *Coot*. We also remove the place holder dummy atoms where we can build new residues. It becomes quite obvious why *ARP/ wARP* could not complete the model: there is no discernible density beyond a few residues in four exposed loops.

How to handle missing parts. So what are we to do with the missing residues? In principle two possibilities exist. The first is to build a loop in a regularized geometry, perhaps with a loop building program, and set the occupancy to zero for residues with excessive *B*-factors. This seems to have been the choice in the case of the published native 1.9 Å tesB structure (PDB entry 1c8u). The problem with this approach is that most modeling programs will not recognize the zero occupancy and will display a complete model. This makes for nice but slightly misleading figures—we actually have no clue about the exact location of the loop if there is no density to support its location. The second, more direct, solution is to omit parts that just cannot be properly traced and state this clearly in PDB remarks and in the paper. There is nothing to be ashamed of when surface-exposed parts such as connecting loops display no density in otherwise high quality maps. We may gain one or the other loop residue and be able to narrow the gap during the course of refinement and subsequent rebuilding, but we will have to declare defeat (or tedium) at some point. Such "forced" loops without supporting experimental evidence also tend to provide an unrelenting source of φ–ψ violations, which then are significant errors (you find evidence for this problem clearly in the deposited 1.9 Å native structure 1c8u).

Discrete solvent atoms. While browsing through the remaining water molecules, we notice four highly coordinated "waters" with enormous density and *B*-factors as low as 2.0 (the lower limit most refinement programs permit). Such super-waters are often positively charged metal ions, but inspection of the distances to the nearest protein atoms shows (> 3 Å) that this hypothesis does not withstand scrutiny; metal–protein distances in coordination complexes are usually much closer, generally below hydrogen bond distance.[154, 155] Given that numerous backbone nitrogen atoms, as well as the positively charged guanidyl group of an arginine, are involved in contacts, it is possible that chlorine anions occupy these sites. The crystallization cocktail indeed included 2 M NaCl,[156] making the presence of discrete chlorine ions not unlikely (soaking crystals in high concentration Br⁻ or I⁻ solutions is in fact a common heavy atom derivatization procedure). The Cl⁻ ions indeed refine to a reasonable *B*-factor of 15–25 Å², close to the *B*-factors of the neighboring protein atoms (15–37 Å²) with typical Cl–protein coordination distances[157] of 3.1–3.6 Å (Appendix A.11).

Finishing the model. For the first round of refinement in *REFMAC* we select approximate X-ray weights based on experience (about 0.2–0.1 will work well in *REFMAC* for 2.5 Å resolution), but we are aware that we may have to further fine-adjust the X-ray weights at the end of the refinement based on the minimum of *R*-free or –*LL*(free). The same holds for *B*-factor weights; we increase the too-tight restraint sigmas to 4.0 Å² for the directly (1-2) bonded main chain atoms and 7.0 for the 1-3 bonded main chain atoms, and relax the corresponding side chain restraints to 9.0 and 11.0, respectively. Note that atomic *B*-factors are not normally distributed, and their distribution and variance is different for each structure. We exclude the potential loop regions from the medium NCS restraints. The resulting first model we immediately submit to *MolProbity* to obtain an error report and the NQH-rotamer-corrected PDB file, both of which we can read directly into *Coot*. The close contact violations reported by *MolProbity* can also be displayed in *Coot* and assist in visualizing the error-prone regions. Many clashes between atoms propagate to the surrounding areas, which often generates locally correlated error clusters. Fixing the root cause of the collisions often removes the remaining local errors in the next round of refinement. Note that simultaneous display of all maps and the clash contacts can result in a crowded display; selecting the proper subset of information is also a matter of personal preference—use what works best for you. Examples for some typical errors and their correction are given in Chapter 13.

With help of the difference map we correct a few remaining errors, rebuild one or two more residues into the traceable parts of the loops, and finally examine the remaining unoccupied density. We discover spurious elongated and fragmented

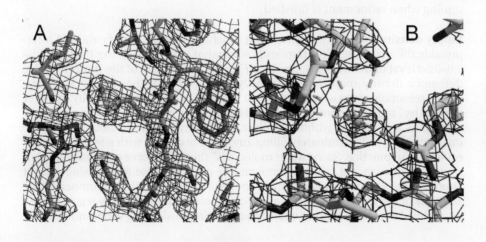

Figure 12-42 Se-Met residues and chloride ion in tesB structure. The sequence of the *ARP/wARP* auto-built 2.5 Å tesB model is properly assigned for 97% of the residues; the remainder are missing residues in disordered loops. Distinctly visible (A) is the high density for the Se-Met residues. Four "heavy waters" in the structure (B) are chloride ions based on the coordination distances > 3 Å, the environment, and the refinement results as detailed in the text.

density "sausages" and a few fragments of disjointed density left in intermolecular space which we cannot account for. We abstain from filling the unidentifiable features indiscriminately with solvent (in the native high resolution structure LDAO detergent molecules were modeled in different locations). The structure finally refines to an R-free/R of 22.9/17.7 with a clean Ramachandran plot and good stereochemistry. Both the bond length r.m.s.d. of 0.015 Å and the bond angle r.m.s.d. of 1.6° are also well within what is reasonable for a 2.5 Å structure with strong data. Upon comparison with the native structure we note that the backbone r.m.s.d. between our Se-Met structure and the native 1.9 Å crystal structure is only 0.35 Å. Given that we determined and refined the Se-Met structure with entirely different software and procedures, this is a nice result attesting to the reproducibility of crystal structure determination.

When is model building and refinement finished?

At some point in the building and refinement of a structure model you will need to decide when the time has come to stop. Short of deadlines or distressing supervisors there are different viewpoints to determine when refinement is finished.

From a task-based point of view, you might stop refinement as soon as the model is good enough to answer the question you posed in the first place when starting the structure determination. As an example take the search for a bound ligand in a drug target structure. This is most likely a molecular replacement case and if after some initial refinement the difference electron density shows nothing interpretable in the binding site, you are done. Such is often the case in industrial settings, and these models are generally not published.

The situation is different in most academic or thesis-related settings. Here your model and the diffraction data must be deposited and are subject to world-wide scrutiny by whoever feels like challenging your model-building prowess. Even if you are diligently employing the validation procedures described in the following chapter, there will always be minor imperfections where you fix one residue and all of sudden another minor irregularity surfaces. The major criteria for abandoning refinement and rebuilding are when:

- no more significant and interpretable difference density in $mF_o - DF_c$ maps remains;

- no more unexplained significant deviations from stereochemical target values and plausible stereochemistry remain (including bad contacts, flipped rotamers, etc.);

- the model makes chemical and biological sense;

or until you give up (see George Sheldrick's leading quote for this chapter, or as Randy Read phrased it: "Most people refine *ad tedium*, until they get bored"). Note that the bulleted statements above include a number of subjective qualifiers and are based on local measures, because global measures such as absolute values of R and R-free (or the level of boredom) are secondary at best in determining when refinement is finished.

Empty density. The practical judgment of what "no more significant and interpretable difference density" implies is somewhat subjective. Noise begins at the 0.8–0.5σ level or below in ML $2mF_o - DF_c$ maps. Inspection of the more sensitive difference density maps simultaneously displayed with the regular ML maps gives a reasonable idea where significant unexplained difference density (positive or negative) remains. Another problem is what to do with distinct solvent density that floats somewhere without contact to protein and/or without reasonable interpretation. Indiscriminately filling any obscure density with a few water molecules is bad practice. Such "water molecules" then only serve as a sink compensating for other model errors. In general, the effect of these few missing solvent atoms on the refinement is too small to make any perceptible difference in the accuracy of the remaining model and a comment in the PDB entry regarding the unmodeled density is a better choice than placing misleading water molecules.

Box 12-19 When to declare model building and refinement finished. The three major criteria for abandoning refinement and rebuilding are:

- no more significant and interpretable difference density in $mF_o - DF_c$ maps remains;
- no more unexplained significant deviations from stereochemical target values and plausible stereochemistry remain;
- the model makes chemical and biological sense.

Global measures such as absolute values of R and R-free (or the level of boredom) do not determine when refinement is finished.

Geometry outliers. Deviations from geometry targets are easily quantified because the expectation values and standard uncertainties are known. Following the normal distribution, a value deviating by more than 4σ occurs with a probability of less than 1/10000. Many checking programs therefore flag a geometry deviation of larger than 4σ; the PDB itself is more lenient and allows up to 6σ deviations without comment. As always, the electron density will determine if a significant deviation—particularly in backbone and side-chain torsion angles—is justified; improbable does not mean impossible. In Chapter 13 we will revisit the subject of geometry validation and introduce some of the analysis tools used concurrently with model building and refinement procedures in more detail. Note also that apparently minor errors such as close contacts, NQH flips, or waters without contact to protein or other solvent entities are "stereochemical anomalies" and deserve to be corrected.

Assembling "Table 2"

Once the refinement is considered complete and the remaining errors are corrected (and after "Table 1" has been prepared, hopefully as soon as the data were collected and memories were fresh), it is time to assemble the equally infamous "Table 2" summarizing global refinement statistics. The data items have been thoroughly discussed and can be found in the log files or harvesting files which the refinement programs generate. Several of those harvesting files can be directly uploaded to the PDB, and they contain much additional information beyond "Table 2" which is listed in the REMARK 3 section of the deposited PDB file.

The first block of data in Table 12-5 begins with an overall listing of which program was used and how many molecules and atoms were refined. A listing of overall and last resolution shell R-work and R-free follows. The R-values for the highest resolution shell are, as expected, higher than the overall values, but are still way below random (0.59). Given the resolution, the absolute values for R and R-free—which depend on many parameters as discussed—are representative of the expected average (Figure 12-23), and the gap between R and R-free is about 5%, again as expected (Figure 12-24). The coordinate deviations from target values are close to those expected for small molecule data (but not any higher) so the restraint weights were probably correctly adjusted for a good 2.5 Å structure. The structure is definitely not over-restrained, and the mean B-factors and their r.m.s.d. values are normal, so no scaling error or anomalies in overall atomic displacement are expected.

The Cruickshank diffraction-component precision index (DPI) based on R-free is 0.127, a good value for the given resolution (Figure 12-22). The next entry is the correlation coefficient between the observed and calculated structure factors as reported by the refinement program *REFMAC* based on R-free, and in the next line the actual real space correlation coefficient of a model map against a bias minimized map. Both numbers are above 0.9 and indicate that the overall map quality is good for this structure. The final overall backbone geometry is asserted by a Ramachandran plot, and looks excellent. Note, however, that this table of global quality indicators provides no assessment about the local quality of the structure. Local validation and its extraordinary relevance to structure interpretation will be discussed in the following chapter.

Table 12-5 Table of refinement statistics. Entry values have been explained in the refinement history of the example structure. Note that it is necessary to indicate what type of program was used to examine the backbone geometry, because the regions differ slightly depending on what empirical data set was used to establish the reference boundaries.

Refinement statistics, 1jia	
Refinement program	*REFMAC* 5.3.0026
Protein molecules in asymmetric unit	2
Residues refined	572
Non-hydrogen protein atoms refined	8950
Solvent waters refined	316
Chloride ions refined	4
Resolution range	74.95–2.50
No. of reflections used in refinement*	31 250 (2212)
Completeness (%)*	98.1 (95.3)
Reflections in cross-validation set (random, %)	5
R-value (working set)*	0.179 (0.214)
R-value (free)*	0.230 (0.297)
Overall *B*-factor (Wilson) ($Å^2$)	20.1
Mean *B*-value, main chain atoms ($Å^2$)	22.2
Mean *B*-value, side chain atoms ($Å^2$)	26.0
r.m.s.d. *B*-factors, bonded main chain atoms ($Å^2$)	3.4
r.m.s.d. *B*-factors, bonded side chain atoms ($Å^2$)	5.0
r.m.s.d. from bond length target values (Å)**	0.015
r.m.s.d. from bond angle target values (°)**	1.55
Coordinate precision, Cruickshank DPI (Å)	0.288
Coordinate precision, Cruickshank DPI, free (Å)	0.233
Coordinate precision, maximum likelihood (Å)	0.127
Correlation coefficient between F_o and F_c	0.94
Correlation coefficient between F_o and F_c (free)	0.90
Real space correlation coefficient	0.93
Ramachandran plot appearance*	
Favored regions (%)	100.0
Disallowed regions (%)	0.0

*Values in parentheses are for the highest resolution shell.
**Restraint target values from Engh and Huber, updated 2001.[55]   ***Regions as defined[158] in *MolProbity*.

12.4 Next steps

Model building, refinement, validation, and model correction are tightly connected and conducted in each round of rebuilding and refinement. In principle, there should be no need for separate validation after the refinement is declared finished (or abandoned, or a victim of tedium), although pressure and time constraints may prevent a thorough and timely validation. In this situation, surprises generally surface when the model is submitted to the PDB, which leads to a great deal of frustration and uncorrected "nuisance" errors. The next chapter will therefore outline in more detail some of the validation tools we (should have) extensively used during refinement and model building and how to interpret reports. The focus is on how to avoid the most frequently encountered omissions, errors, and difficulties in the first place—unfortunately there are plenty of cautionary tales available in the PDB.

12.5 Key concepts

Basics of model building and refinement

- The key to successful protein structure modeling is the cycling between local real space model building and model correction and global reciprocal space refinement.
- The molecular model is built in real space into electron density using computer graphics.
- The local geometry errors remaining after real space model building are corrected during restrained reciprocal space refinement by optimizing the fit between observed and calculated structure factor amplitudes.
- Successive rounds of rebuilding, error correction, and refinement are needed to obtain a good final protein model.
- While experimental electron density maps constructed from poor phases will be hard to interpret, an initial experimental map will not be biased toward any structure model.
- In contrast, when molecular replacement models are the sole source of phases, the electron density maps will be severely biased, and the map will reflect the model features.
- Bias minimization in the form of maximum likelihood map coefficients, often combined with additional bias minimization measures, are applied before electron density map construction.
- The incorporation of model incompleteness and model errors in the σ_A (sigma-A) variance term for model structure factors leads to maximum likelihood map coefficients $\mathbf{F}_{ML} = (2mF_{obs} - DF_{calc}) \cdot \exp(i\varphi_{wt})$ for acentric reflections and $\mathbf{F}_{ML} = (mF_{obs} \cdot \exp(i\varphi_{wt})$ for centric reflections.
- Electron density maps and difference maps constructed from maximum likelihood Fourier coefficients provide the standard, bias minimized maps used in model building.

Refinement

- During refinement the parameters describing a continuously parameterized model are adjusted so that the fit of discrete experimental observations to their computed values calculated by a target function is optimized.
- Observations can be experimental data specific to the given problem, such as structure factor amplitudes, or general observations that are valid for all models.
- Stereochemical descriptors valid for all models such as bond lengths, bond angles, torsion angles, chirality, and non-bonded interactions are incorporated as restraints to improve the observation-to-parameter ratio of the refinement.
- The most accurate target functions are maximum likelihood target functions that account for errors and incompleteness in the model.
- Various optimization algorithms can be used to achieve the best fit between parameterized model and all observations, which include measured data and restraints.
- The radius of convergence for an optimization algorithm describes its ability to escape local minima and approach the global minimum, generally with increased cost in time and lower accuracy.
- Indiscriminate introduction of an increasing number of parameters into the model can lead to overparameterization, where the refinement residual measured as linear R-value still decreases, but the description of reality, that is, the correct structure, does not improve.
- The evaluation of the residual against a data set excluded from refinement provides the cross-validation R-value or R-free. If parameters are introduced that do not improve the phase error of the model, R-free will not decrease any further or may even increase.
- Refined models carry some memory of omitted parts, which can be removed by slightly perturbing the coordinates and re-refining the model without the questionable part of the model.
- The known geometry target values for bond lengths, bond angles, and torsion angles as well as planarity of certain groups can be regarded as additional observations contributing to a higher data-to-parameter ratio.

- In addition, geometry targets constitute prior knowledge that keeps the molecular geometry in check with reality during restrained refinement.
- The geometry targets, chirality values, and non-bonded interactions are implemented as stereochemical restraints and incorporated into the target function generally in the form of squared sum of residuals in addition to the structure factor amplitude residual.
- The structure factor amplitude residual is commonly called the X-ray term (or X-ray energy) and the restraint residuals the chemical (energy) term.
- In terms of maximum posterior estimation, geometry target values and their variance define the prior probability of our model without consideration or knowledge of the experimental (diffraction) data.
- Geometric relations and redundancies between identical molecules in the asymmetric unit can be exploited through NCS restraints.
- Particularly at low resolution, strong NCS restraints are an effective means of stabilizing and improving the refinement.
- In the early stages of model building, experimental phase restraints are also an effective means to stabilize and improve the refinement.
- The data-to-parameter ratio in protein structures is greatly increased through the introduction of stereochemical restraints.
- A protein of 2000 non-hydrogen atoms has about 8000 adjustable parameters and about the same number of restraints.
- At 2 Å about 15 000 to 25 000 unique reflections are observed for a 2000 non-hydrogen atom protein, which yields a total data to parameter ratio of about 2–3 at 2 Å.
- Anisotropic B-factor refinement consumes five additional parameters per atom, and is generally not advisable at resolutions lower than 1.4 Å.
- The most difficult point in the parameterization of macromolecular structure models is accounting for correlated dynamic or static displacement.
- Isotropic B-factors are inadequate to describe any correlated dynamic molecular movement, and anisotropic B-factors, except at very high resolution, lead to overparameterization of the model.
- Molecular and lattice packing anisotropy can also affect diffraction, and adequate correction by anisotropic scaling, or in severe cases additional anisotropic resolution truncation, is necessary.
- Maximum likelihood target functions that account for incompleteness and errors in the model are superior to basic least squares target functions, particularly in the early, error-prone stages of the refinement.
- Maximum likelihood target functions are implemented in *REFMAC*, *Buster/TNT*, and *CNS* as well as the *PHENIX/cctbx* programs, together with all commonly used restraint functions including phase restraints, which is of advantage at low resolution or in the early stages of refinement.
- Optimization algorithms are procedures that search for an optimum of a non-linear, multi-parametric function.
- Optimization algorithms can be roughly divided into analytic or deterministic procedures and stochastic procedures.
- Deterministic optimizations such as the gradient-based maximum likelihood methods are fast and work well when we are reasonably close to a correct model, at the price of becoming trapped in local minima.
- Stochastic procedures employ a random search that also allows movements away from local minima. They are slow but compensate for it with a large radius of convergence.
- Evolutionary programming as used in molecular replacement or simulated annealing in refinement is a stochastic optimization procedure. This is generally of advantage if we do not know (MR) or are far from (initial model refinement) the correct solution.
- Deterministic optimizations can be classified depending on how they evaluate the second derivative matrix. They generally descend in several steps or cycles from a starting parameter set (model) downhill toward a hopefully—but not necessarily—global minimum.
- Energy refinement of a molecular dynamics force field and torsion angle refinement are two parameterizations that are commonly used together with the stochastic optimization method of simulated annealing.
- In molecular dynamics the target function is parameterized in the form of

potential energy terms and the development of the system is described by equations of motion. In torsion angle parameterization, the structure model is described by its torsion angles, which requires fewer parameters than coordinate parameterization.

- Both molecular dynamics and torsion angle parameterization are often combined with simulated annealing optimization, where the molecular system is perturbed and returns to equilibrium according to an optimized slow cooling protocol.
- Dummy atom placement and refinement is used for discrete solvent building, model completion, and phase improvement in general.
- Dummy atoms are placed in real space in difference electron density peaks, the new model is refined unrestrained in reciprocal space, and in the new map poorly positioned atoms are removed and new ones placed again.
- Dummy atom refinement can be combined with multi-model map averaging where it forms the basis of bias minimization protocols and the automated model building program *ARP/wARP*.

Model building and refinement practice

- Building of a model into an empty map begins with the tracing of the backbone.
- Tracing is aided by density skeletonization, followed by placement of Cα atoms into positions where side chains extend from the backbone.
- The sequence is docked from known atom positions from the heavy atom substructure solution or sequences of residues of characteristic shapes.
- The initial model is refined in reciprocal space with geometric restraints and phase restraints, and the next map is constructed from maximum likelihood coefficients.
- The model is then further completed and refined in subsequent rounds with increasing X-ray weights while tracking *R*-free and stereochemistry. Nuisance errors are removed after analysis in a polishing step.
- Automated model building programs greatly simplify model building, and auto-built models often only need to be completed and polished. Auto-building programs follow similar steps as manual model building and employ pattern recognition algorithms of various kinds to identify residues.
- Rebuilding of very poor initial molecular replacement models can be aided by a first step of torsion angle–simulated annealing (TA-SA) refinement.
- The large radius of convergence of TA-SA facilitates the necessary large corrections and escape from local minima. Also, before automated model rebuilding and correction, TA-SA can improve the amount and quality of the model that is automatically rebuilt.
- In low resolution structures the backbone can be traced correctly, but the sequence may be shifted. Such register errors can be hard to detect from electron density shape alone and are usually detected by poor side chain interactions or unusual environment.
- One of the most common mistakes leading to overparameterization of the model is overbuilding of the solvent. Discrete water molecules should have hydrogen bonded contact(s) to other solvent molecules or to protein.
- Poorly placed waters tend to drift away during refinement because of lack of density and restraints and often end up far away from other molecules and with high *B*-factors.
- Binding sites have a tendency to attract various detritus from the crystallization cocktail, and will therefore often contain some weak, unidentifiable density that can be (wishfully) mistaken for desired ligand density.
- Plausible binding chemistry, ligand conformation, and independent evidence are necessary to avoid misinterpretation.
- The three major criteria for abandoning refinement and rebuilding are: (i) no more significant and interpretable difference density in $mF_o - DF_c$ maps remains; (ii) no more unexplained significant deviations from stereochemical target values and from plausible stereochemistry remain; and (iii) the model makes chemical and biological sense.
- Global measures such as absolute values of *R* and *R*-free (or the level of boredom) do not determine when refinement is finished.

12.6 Additional reading

1. Noble M, & Perrakis A (2004) Model building and refinement – Proceedings of the 2004 CCP4 Study Weekend. *Acta Crystallogr.* D60(12).

2. Rossmann MG, & Arnold E (Eds.) (2001) *International Tables for Crystallography, Volume F: Crystallography of Biological Macromolecules.* Dordrecht, The Netherlands: Kluwer Academic Publishers.

3. Carter CWJ, & Sweet RM (1997) Macromolecular crystallography. *Methods Enzymol.* 276, 277. London, UK: Academic Press

4. Carter CW, & Sweet R (2003) Macromolecular crystallography. *Methods Enzymol.* 368, 374. London, UK: Academic Press.

5. Tronrud D (2004) Introduction to macromolecular refinement. *Acta Crystallogr.* D60(12), 2156–2168.

6. McCoy A (2004) Liking likelihood. *Acta Crystallogr.* D60(12), 2169–2183.

12.7 References

1. Cullis AF, Muirhead H, Perutz MF, et al. (1962) The structure of haemoglobin. IX. A three-dimensional Fourier synthesis at 5.5 Å resolution: description of the structure. *Proc. R. Soc. London* A265, 161–187.

2. Blundell TL, & Johnson LN (1976) *Protein Crystallography.* London, UK: Academic Press.

3. Perutz MF (1985) Early days of protein crystallography. *Methods Enzymol.* 114, 3–18.

4. Dickerson RE (2005) *Present at the Flood.* Sunderland, MA: Sinauer Associates.

5. Richards FM (1968) The matching of physical models to three-dimensional electron-density maps: A simple optical device. *J. Mol. Biol.* 37, 225–230.

6. Colman PM, Jansonius JN, & Matthews BW (1972) The structure of thermolysin: An electron density map at 2.3 Å resolution. *J. Mol. Biol.* 70, 701–724.

7. Wing R, Drew H, Takano T, et al. (1980) Crystal structure analysis of a complete turn of B-DNA. *Nature* 287(5784), 755.

8. Rubin B, & Richardson JS (1972) The simple construction of protein alpha-carbon models. *Biopolymers* 11(11), 2381–2385.

9. Jones T (1978) A graphics model building and refinement system for macromolecules. *J. Appl. Crystallogr.* 11(4), 268–272.

10. Editorial. (1997) String and sealing wax. *Nat. Struct. Biol.* 4(12), 961–964.

11. Dickerson RE (1992) A little ancient history. *Protein Sci.* 1(1), 182–186.

12. Jones TA (2004) Interactive electron-density map interpretation: from INTER to O. *Acta Crystallogr.* D60(12), 2115–2125.

13. Burling FT, Weis WI, Flaherty KM, et al. (1996) Direct observation of protein solvation and discrete disorder with experimental crystallographic phases. *Science* 271, 72–77.

14. Diamond R (1971) A real-space refinement procedure for proteins. *Acta Crystallogr.* A27, 436–452.

15. Wilson AJC (1950) Largest likely value for the reliability index. *Acta Crystallogr.* 3, 397–398.

16. Luzzati V (1952) Traitement statistique des erreurs dans la determination des structures cristallines. *Acta Crystallogr.* 5(6), 802–810.

17. Srinivasan R (1966) Weighting functions for use in the early stage of structure analysis when a part of the structure is known. *Acta Crystallogr.* 20(1), 143–144.

18. Srinivasan R, & Ramachandran GN (1966) Probability distribution connected with structure amplitudes of two related crystals. VI. On the significance of the parameter σA. *Acta Crystallogr.* 20(4), 570–571.

19. Lunin VY, & Urzhumtsev AG (1984) Improvement of protein phases by coarse model modification. *Acta Crystallogr.* A40(3), 269–277.

20. Read RJ (1986) Improved Fourier coefficients for maps using phases from partial structures with errors. *Acta Crystallogr.* A42, 140–149.

21. Tronrud DE (1997) TNT refinement package. *Methods Enzymol.* 277, 306–319.

22. Murshudov GN, Vagin AA, & Dodson ED (1997) Refinement of macromolecular structures by the maximum-likelihood method. *Acta Crystallogr.* D53, 240–255.

23. Luzzati V (1953) Resolution d'une structure cristalline lorsque les positions d'une partie des atomes sont connues: Traitement statistique. *Acta Crystallogr.* 6, 142–152.

24. Main P (1979) A theoretical comparison of the α, γ' and $2F_o - F_c$ syntheses. *Acta Crystallogr.* 35(5), 779–785.

25. Read RJ (2002) Model phases: Probabilities, bias and maps. *International Tables for Crystallography F*, 325–331.

26. Reddy V, Swanson S, Sacchettini JC, et al. (2003) Effective electron density map improvement and structure validation on a Linux multi-CPU web cluster: The TB Structural Genomics Consortium Bias Removal Web Service. *Acta Crystallogr.* D59, 2200–2210.

27. McRee DE (1999) XtalView/Xfit – a versatile program for manipulating atomic coordinates and electron density. *J. Struct. Biol.* 125, 156–165.

28. Emsley P, Lohkamp B, Scott WG, et al. (2010) Features and development of Coot. *Acta Crystallogr.* D66(4), 486–501.

29. Tickle IJ (2008) personal communication.

30. Sheldrick GM, & Schneider TR (1997) SHELXL: High resolution refinement. *Methods Enzymol.* 277, 319–343.

31. Hamilton WC (1965) Significance tests on the crystallographic R factor. *Acta Crystallogr.* 18, 502–510.

32. Prince E (1982) Comparison of the fits of two models to the same data. *Acta Crystallogr.* B38, 1099–1100.

33. Brunger AT (1992) Free R value: A novel statistical quantity for assessing the accuracy of crystal structures. *Nature* 355, 472–475.

34. Brunger AT (1993) Assessment of phase accuracy by cross-validation: The free R value. Methods and applications. *Acta Crystallogr.* D49, 24–36.

35. Kleywegt GJ, & Brunger AT (1996) Checking your imagination: applications of the free R value. *Structure* 4, 897–904.

36. Blow D (2002) Rearrangement of Cruickshank's formulae for the diffraction-component precision index. *Acta Crystallogr.* D58(5), 792–797.

37. Brunger AT (1997) Free R Value: Cross-validation in crystallography. *Methods Enzymol.* 277, 366–396.

38. Tickle IJ, Laskowski RA, & Moss DS (1998) Error estimates of protein structure coordinates and deviations from standard geometry by full-matrix refinement of γB- and βB2-crystallin. *Acta Crystallogr.* D54(2), 243–252.

39. Cowtan KD (2005) Likelihood weighting of partial structure factors using spline coefficients. *J. Appl. Crystallogr* (38), 193–198.

40. Tickle IJ, Laskowski RA, & Moss DS (1998) Rfree and the Rfree ratio. I. Derivation of expected values of cross-validation residuals used in macromolecular least-squares refinement. *Acta Crystallogr.* D54(4), 547–557.

41. Fabiola F, Korostelev A, & Chapman MS (2006) Bias in cross-validated free R factors: mitigation of the effects of non-crystallographic symmetry. *Acta Crystallogr.* D62(3), 227–238.

42. Jeffrey P (2009) Analysis of errors in the structure determination of MsbA. *Acta Crystallogr.* D65(2), 193–199.

43. Bhat TN (1988) Calculation of an OMIT map. *J. Appl. Crystallogr.* 21, 279–281.

44. Vellieux FMD, & Dijkstra BW (1997) Computation of Bhat's OMIT maps with different coefficients. *J. Appl. Crystallogr.* 30, 396–399.

45. Hodel A, Kim S-H, & Brunger AT (1992) Model bias in macromolecular structures. *Acta Crystallogr.* D48, 851–858.

46. Brunger AT, Kuriyan J, & Karplus M (1987) Crystallographic R factor refinement by molecular dynamics. *Science* 235, 458–460.

47. Evans PR (2007) An introduction to stereochemical restraints. *Acta Crystallogr.* D63, 58–61.

48. Konnert J (1976) A restrained-parameter structure-factor least-squares refinement procedure for large asymmetric units. *Acta Crystallogr.* A32(4), 614–617.

49. Hendrickson WA (1985) Stereochemically restrained refinement of macromolecular structures. *Methods Enzymol.* 115, 252–270.

50. Sussman JL, Holbrook SR, Church GM, et al. (1977) Structure-factor least-squares refinement procedure for macromolecular structure using constrained and restrained parameters. *Acta Crystallogr.* A33, 800–804.

51. Sussman JL (1985) Constrained-restrained least squares (CORELS) refinement of proteins and nucleic acids. *Methods Enzymol.* 115, 271–303.

52. Allen FH (2002) The Cambridge Structural Database: a quarter of a million crystal structures and rising. *Acta Crystallogr.* B, 58, 380–388.

53. Engh RA, & Huber R (1991) Accurate bond and angle parameters for X-ray structure refinement. *Acta Crystallogr.* A47, 392–400.

54. Tickle I (2007) Experimental determination of optimal root-mean-square deviations of macromolecular bond lengths and angles from their restrained ideal values. *Acta Crystallogr.* 63(12), 1274–1281.

55. Engh RA, & Huber R (2001) Structure quality and target parameters. In Rossmann MG & Arnold E (Eds.) *International Tables for Crystallography F.* Dordrecht/Boston/London: Kluwer Academic Publishers.

56. Tronrud D (2004) Introduction to macromolecular refinement. *Acta Crystallogr.* D60(12), 2156–2168.

57. Tronrud DE (1996) Knowledge-based B-factor restraints for the refinement of proteins. *J. Appl. Crystallogr.* 29(2), 100–104.

58. Kleywegt GJ, & Jones AT (1996) Phi/Psi-chology: Ramachandran revisited. *Structure,* 1395–1400.

59. Fabiola F, Bertram R, Korostelev A, et al. (2002) An improved hydrogen bond potential: Impact on medium resolution protein structures. *Protein Sci.* 11(6), 1415–1423.

60. Uson I, Pohl E, Schneider TR, et al. (1999) 1.7 Å structure of the stabilized REIv mutant T39K. Application of local NCS restraints. *Acta Crystallogr.* D55, 1158–1167.

61. Kleywegt G (1996) Use of non-crystallographic symmetry in protein structure refinement. *Acta Crystallogr.* D52(4), 842–857.

62. Sheldrick G (2008) A short history of SHELX. *Acta Crystallogr.* A64(1), 112–122.

63. Flack H (1983) On enantiomorph-polarity estimation. *Acta Crystallogr.* A39(6), 876–881.

64. Brunger AT (1988) Crystallographic refinement by simulated annealing: Application to a 2 Å resolution structure of aspartate aminotransferase. *J. Mol. Biol.* 203(3), 803–816.

65. Winn MD, Isupov MN, & Murshudov GN (2001) Use of TLS parameters to model anisotropic displacements in macromolecular refinement. *Acta Crystallogr.* D57, 122–223.

66. Vagin AA, Steiner RA, Lebedev AA, et al. (2004) Biological crystallography REFMAC5 dictionary: organization of prior chemical knowledge and guidelines for its use. *Acta Crystallogr.* D60, 2184–2195.

67. Tickle IJ, Laskowski RA, & Moss DS (2000) Rfree and the Rfree ratio. II. Calculation of the expected values and variances of cross-validation statistics in macromolecular least-squares refinement. *Acta Crystallogr.* D56(4), 442–450.

68. Hirshfeld F (1974) Can X-ray data distinguish bonding effects from vibrational smearing? *Acta Crystallogr.* A32, 239–244.

69. Jaskolski M, Gilski M, Dauter Z, et al. (2007) Stereochemical restraints revisited: how accurate are refinement targets and how much should protein structures be allowed to deviate from them? *Acta Crystallogr.* D63, 611–620.

70. Adams PD, Pannu NS, Read RJ, et al. (1997) Cross-validated maximum likelihood enhances crystallographic simulated annealing refinement. *Proc. Natl. Acad. Sci. U.S.A.* 94(10), 5018–5023.

71. Bricogne G (1997) Bayesian statistical viewpoint on structure determination: Basic concepts and examples. *Methods Enzymol.* 276, 361–423.

72. Lebedev AA, Tickle IJ, Laskowski RA, et al. (2003) Estimation of weights and validation: a marginal likelihood approach. *Acta Crystallogr.* D59(9), 1557–1566.

73. Tickle IJ (2007) personal communication.

74. Trueblood KN, Burgi HB, Burzlaff H, et al. (1996) Atomic displacement parameter nomenclature. Report of a Subcommittee on Atomic Displacement Parameter Nomenclature. *Acta Crystallogr.* A52(5), 770–781.

75. Murshudov GN, Vagin AA, Lebedev A, et al. (1999) Efficient anisotropic refinement of macromolecular structures using FFT. *Acta Crystallogr.* D55(1), 247–255.

76. Grosse-Kunstleve RW, & Adams PD (2002) On the handling of atomic anisotropic displacement parameters. *J. Appl. Crystallogr.* 35(4), 477–480.

77. Afonine PV, Grosse-Kunstleve RW, & Adams PD (2005) A robust bulk-solvent correction and anisotropic scaling procedure. *Acta Crystallogr.* D61, 850–855.

78. Brunger AT (2007) Version 1.2 of the Crystallography and NMR system. *Nat. Protoc.* 2(11), 2728–2733.

79. Strong M, Sawaya MR, Wang S, et al. (2006) Toward the structural genomics of complexes: Crystal structure of a PE/PPE protein complex from *Mycobacterium tuberculosis. Proc. Natl. Acad. Sci. U.S.A.* 103(21), 8060–8065.

80. Brunger AT, DeLaBarre B, Davies JM, et al. (2009) X-ray structure determination at low resolution. *Acta Crystallogr.* D65(2), 128–133.

81. Schomaker V, & Trueblood KN (1968) On the rigid-body motion of molecules in crystals. *Acta Crystallogr.* B24(1), 63–76.

82. Haneef I, Moss DS, Stanford MJ, et al. (1985) Restrained structure-factor least-squares refinement of protein structures using a vector processing computer. *Acta Crystallogr.* A41(5), 426–433.

83. Howlin B, Butler SA, Moss DS, et al. (1993) TLSANL: TLS parameter-analysis program for segmented anisotropic refinement of macromolecular structures. *J. Appl. Crystallogr.* 26(4), 622–624.

84. Painter J, & Merritt EA (2005) A molecular viewer for the analysis of TLS rigid-body motion in macromolecules. *Acta Crystallogr.* D61(4), 465–471.

85. Painter J, & Merritt EA (2006) Optimal description of a protein structure in terms of multiple groups undergoing TLS motion. *Acta Crystallogr.* D62(4), 439–450.

86. Painter J, & Merritt EA (2006) TLSMD web server for the generation of multi-group TLS models. *J. Appl. Crystallogr.* 39(1), 109–111.

87. Wilson MA, & Brunger AT (2000) The 1.0 Å crystal structure of Ca^{2+} bound calmodulin: an analysis of disorder and implications for functionally relevant plasticity. *J. Mol. Biol.* 301, 1237–1256.

88. Pannu NS, Murshudov GN, Dodson EJ, et al. (1998) Incorporation of prior phase information strengthens maximum likelihood structural refinement. *Acta Crystallogr.* D54, 1285–1294.

89. Jack A, & Levitt M (1978) Refinement of large structures by simultaneous minimization of energy and R factor. *Acta Crystallogr.* A34(6), 931–935.

90. Brooks BR, Bruccoleri RE, Olafson BE, et al. (1983) CHARMM: A program for macromolecular energy, minimization, and dynamics calculations. *J. Comput. Chem.* 4, 187–217.

91. Sippl MJ (1995) Knowledge-based potentials for proteins. *Curr. Opin. Struct. Biol.* 5, 229–235.

92. Hanson LB, Harp JM, & Bunick GJ (2003) The well-tempered protein crystal: Annealing macromolecular crystals. *Methods Enzymol.* 368, 217–235.

93. Hanson LB, Schall CA, & Bunick GJ (2003) New techniques in macromolecular cryocrystallography: macromolecular crystal annealing and cryogenic helium. *J. Struct. Biol.* 142(1), 77–87.

94. Kriminski S, Caylor CL, Nonato MC, et al. (2002) Flash-cooling and annealing of protein crystals. *Acta Crystallogr.* D58, 459–471.

95. Glykos NM, & Kokkinidis M (2000) A stochastic approach to molecular replacement. *Acta Crystallogr.* D56(2), 169–174.

96. Goldstein H, Poole C, & Safko J (2002) *Classical Mechanics.* 3rd edition, Pearson Education, Delhi, India.

97. Brunger AT, Krukowski A, & Erickson JW (1990) Slow-cooling protocols for crystallographic refinement by simulated annealing. *Acta Crystallogr.* A46(7), 585–593.

98. Abagyan R, Totrov M, & Kuznetsov D (1994) ICM: a new method for protein modeling and design. Applications to docking and structure prediction from the distorted native conformation. *J. Comput. Chem.* 15, 488–506.

99. Rice LM, & Brunger AT (1994) Torsion angle dynamics: reduces variable conformational sampling enhances crystallographic structure refinement. *Proteins: Struct, Funct, Genet.* 19, 277–290.

100. Gros P, vanGunsteren WF, & Hol WGJ (1990) Inclusion of thermal motion in crystallographic structures by restrained molecular dynamics. *Science* 249, 1149–1152.

101. Kuriyan J, Osapay K, Burley SK, et al. (1990) Exploration of disorder in protein structures by X-ray restrained molecular dynamics. *Proteins* 10, 340–358.

102. Levin EJ, Kondrashov DA, Wesenberg GE, et al. (2007) Ensemble refinement of protein crystal structures: validation and application. *Structure* 15(9), 1040–1052.

103. Chang G, & Roth CB (2001) Structure of MsbA from *E. coli:* A homolog of the multidrug resistance ATP binding cassette (ABC) transporters. *Science* 293(5536), 1793–1800.

104. Petsko GA (2007) And the second shall be first. *Genome Biol.* 8, 103–105.

105. Chang G, Roth CB, Reyes CL, et al. (2006) Retraction. *Science* 314(5807), 1875.

106. Dawson RJP, & Locher KP (2006) Structure of a bacterial multidrug ABC transporter. *Nature* 443(7108), 180–185.

107. Agarwal RC, & Isaacs NW (1977) Method for obtaining a high resolution protein map starting from a low resolution map *Proc. Natl. Acad. Sci. U.S.A.* 74(7), 2835–2839.

108. Bhat TN, & Blow DM (1982) A density-modification method for the improvement of poorly resolved protein electron-density maps. *Acta Crystallogr.* A38, 21–29.

109. Lamzin VS, & Wilson KS (1993) Automated refinement of protein models. *Acta Crystallogr.* D53, 448–455.

110. Morris R, Zwart P, Cohen S, et al. (2004) Breaking good resolutions with ARP/wARP. *J. Synchrotron Radiat.* 11(1), 56–59.

111. Evrard GX, Langer GG, Perrakis A, et al. (2007) Assessment of automatic ligand building in ARP/wARP. *Acta Crystallogr.* D63(1), 108–117.

112. Perrakis A, Sixma TK, Wilson KS, et al. (1997) wARP: Improvement and extension of crystallographic phases by weighted averaging of multiple-refined dummy atomic models. *Acta Crystallogr.* D53, 448–455.

113. Cohen SX, Morris RJ, Fernandez FJ, et al. (2004) Towards complete validated models in the next generation of ARP/wARP. *Acta Crystallogr.* D60(12), 2222–2229.

114. Cruickshank DWJ (1999) Remarks about protein structure precision. *Acta Crystallogr.* D55, 583–601.

115. Vaguine AA, Richelle J, & Wodak SJ (1999) SFCHECK: a unified set of procedures for evaluating the quality of macromolecular structure-factor data and their agreement with the atomic model. *Acta Crystallogr.* D55, 191–20.

116. Press WH, Flannery BP, Teulkolsky SA, et al. (1986) *Numerical Recipes.* Cambridge, UK: Cambridge University Press.

117. Steiner RA, Lebedev AA, & Murshudov GN (2003) Fisher's information in maximum-likelihood macromolecular crystallographic refinement. *Acta Crystallogr.* D59(12), 2114–2124.

118. Kleywegt GJ, & Jones TA (2002) Homo crystallographicus – quo vadis? *Structure* 10(4), 465–472.

119. Edwards AWF (1972) *Likelihood – An Account of the Statistical Concept of Likelihood and its Application to Scientific Inference.* Cambridge, UK: Cambridge University Press.

120. Goodman SN (1999) Towards evidence-based medical statistics. II: The Bayes factor. *Ann. Internal Med.* 120(12), 1005–1013.

121. Jaynes ET (2003) *Probability Theory: The Logic of Science.* Cambridge, UK: Cambridge University Press.

122. McRee D (2004) Differential evolution for protein crystallographic optimizations. *Acta Crystallogr.* D60(12 Part 1), 2276–2279.

123. Kleywegt GJ, & Jones TA (1997) Model building and refinement practice. *Methods Enzymol.* 277, 208–230.

124. Canutescu AA, Shelenkov AA, & Dunbrack RL, Jr. (2003) A graph-theory algorithm for rapid protein side-chain prediction. *Protein Sci.* 12(9), 2001–2014.

125. Kleywegt GJ, Hoier H, & Jones TA (1996) A re-evaluation of the crystal structure of chloromuconate cycloisomerase. *Acta Crystallogr.* D52(4), 858–863.

126. Parkin S, Rupp B, & Hope H (1996) Structure of bovine pancreatic trypsin inhibitor at 125 K: Definition of carboxyl-terminal residues Gly57 and Ala58. *Acta Crystallogr.* D52(1), 18–29.

127. Kantardjieff KA, Höchtl P, Segelke BW, et al. (2002) Concanavalin A in a dimeric crystal form: revisiting structural accuracy and molecular flexibility. *Acta Crystallogr.* D58, 735–743.

128. Deacon A, Gleichmann T, Kalb AJ, et al. (1997) The structure of concanavalin A and its bound solvent determined with small-molecule accuracy at 0.94 Å resolution. *J. Chem. Soc., Faraday Trans.* 93, 4305–4308.

129. Gokulan K, Khare S, Ronning D, et al. (2005) Co-crystal structures of NC6.8 Fab identify key interactions for high potency sweetener recognition: Implications for the design of synthetic sweeteners. *Biochemistry* 44, 9889–9898.

130. Kleywegt GJ, Henrick K, Dodson EJ, et al. (2004) Pound-wise but penny-foolish: How well do macromolecules fare in macromolecular refinement? *Structure* 11, 1051–1059.

131. Kleywegt GJ, & Harris MR (2007) ValLigURL: a server for ligand-structure comparison and validation. *Acta Crystallogr.* 63(8), 935–938.

132. Weichenberger CX, & Sippl MJ (2006) Self-consistent assignment of asparagine and glutamine amide rotamers in protein crystal structures. *Structure* 14(6), 967–972.

133. Chen VB, Arendall WB, III, Headd JJ, et al. (2010) MolProbity: all-atom structure validation for macromolecular crystallography. *Acta Crystallogr.* D66(1), 12–21.

134. Brändén CI, & Jones TA (1990) Between objectivity and subjectivity. *Nature* 343, 687–689.

135. Rupp B (2006) Real-space solution to the problem of full disclosure. *Nature* 444(7121), 817.

136. Kleywegt GJ, Harris MR, Zou J-Y, et al. (2004) The Uppsala Electron-Density Server. *Acta Crystallogr.* D60(12 Part 1), 2240–2249.

137. Cohen SX, Ben Jelloul M, Long F, et al. (2008) ARP/wARP and molecular replacement: the next generation. *Acta Crystallogr.* D64(1), 49–60.

138. Gopal K, McKee E, Romo T, et al. (2007) Crystallographic protein model-building on the web. *Bioinformatics* 23(3), 375–377.

139. Perrakis A, Harkiolaki M, Wilson KS, et al. (2001) ARP/wARP and molecular replacement. *Acta Crystallogr.* D57, 1445–1450.

140. Morris R (2004) Statistical pattern recognition for macromolecular crystallographers. *Acta Crystallogr.* D60, 2133–2143.

141. Greer J (1974) Three-dimensional pattern recognition: An approach to automated interpretation of electron density maps of proteins. *J. Mol. Biol.* 82, 279–288.

142. Kleywegt GJ, & Jones TA (2004) Databases in protein crystallography. *Acta Crystallogr.* D59, 1119–1131.

143. Karmali AM, Blundell TL, & Furnham N (2009) Model-building strategies for low-resolution X-ray crystallographic data. *Acta Crystallogr.* 65(2), 121–127.

144. Terwilliger TC (2004) SOLVE and RESOLVE: automated structure solution, density modification and model building. *J. Synchrotron Radiat.* 11(1), 49–52.

145. Levitt DG (2001) A new software routine that automates the fitting of protein X-ray crystallographic electron-density maps. *Acta Crystallogr.* D57, 1013–1019.

146. Cowtan K (2006) The Buccaneer software for automated model building. 1. Tracing protein chains. *Acta Crystallogr.* D62(9), 1002–1011.

147. Cowtan K (2008) Fitting molecular fragments into electron density. *Acta Crystallogr.* D64(1), 83–89.

148. Joosten K, Cohen SX, Emsley P, et al. (2008) A knowledge-driven approach for crystallographic protein model completion. *Acta Crystallogr.* 64(4), 416–424.

149. McKee EW, Kanbi LD, Childs KL, et al. (2005) FINDMOL: automated identification of macromolecules in electron-density maps. *Acta Crystallogr.* D61(11), 1514–1520.

150. Romo TD, Sacchettini JC, & Ioerger TR (2006) Improving amino-acid identification, fit and Cα prediction using the Simplex method in automated model building. *Acta Crystallogr.* D62(11), 1401–1406.

151. Kleywegt GJ, & Jones TA (1996) Efficient rebuilding of protein structures. *Acta Crystallogr.* D52(4), 829–832.

152. Cowtan KD (1998) Modified phased translation functions and their application to molecular-fragment location. *Acta Crystallogr.* D54, 750–756.

153. Sheldrick GM (2002) Macromolecular phasing with SHELXE. *Z. Kristallogr.* 217, 644–650.

154. Harding M (2006) Small revisions to predicted distances around metal sites in proteins. *Acta Crystallogr.* D62(6), 678–682.

155. Harding M (2004) The architecture of metal coordination groups in proteins. *Acta Crystallogr.* D60(5), 849–859.

156. Li J, Derewenda U, Dauter Z, et al. (2000) Crystal structure of the *Escherichia coli* thioesterase II, a homolog of the human Nef binding enzyme. *Nat. Struct. Biol.* 7(7), 555–559.

157. Backstrom S, Wolf-Watz M, Grundstrom C, et al. (2002) The RUNX1 Runt domain at 1.25A resolution: a structural switch and specifically bound chloride ions modulate DNA binding. *J. Mol. Biol.* 322(2), 259–272.

158. Lovell SC, Davis IW, Arendall WB, 3rd, et al. (2003) Structure validation by C-alpha geometry: phi, psi and C-beta deviation. *Proteins* 50(3), 437–450.

PART V
Making Sense of Your Structure

Structure validation, analysis, and presentation

> The scientist must be the judge of his own hypotheses, not the statistician.
> A. F. W. Edwards (1992) in *Likelihood – An account of the statistical concept of likelihood and its application to scientific inference*, p 34

Validation, analysis, and presentation are tightly connected subjects, and in practice validation is an intimate part of the model building and refinement stage in structure determination. While Chapter 13 is also useful as a stand-alone section for a class interested in the finer details of the analysis of a crystallographic protein structure model from a user's point of view, we treat validation and analysis here as an extension of improving our model during the structure refinement process.

We take a top-down view and begin with the emphasis on validation as a process that considers all available evidence and the full body of established prior knowledge. This naturally extends into the concept of hypothesis testing in general. Crystallographic models carry great persuasive power and the ultimate responsibility for proper validation of the model, for its realistic analysis, and for the plausibility of derived hypotheses rests with the crystallographer.

Starting with geometry validation, we emphasize the need to conduct local validation based on the plausibility of the model against the experimental evidence in the form of minimally biased electron density as well as the body of prior knowledge. It is important to understand that improbable deviations from expectation values are not necessarily impossible, but they demand strong experimental evidence in their support—unless they are simply errors that need to be corrected.

Much can be learned about proper validation, analysis, and hypothesis testing using instructive examples that contain errors of varying severity ranging from gross mis-tracing to what we termed nuisance errors. Large-scale errors such as reverse tracing, secondary structure connectivity errors, and register shifts nowadays can be generally detected and avoided given the validation tools that have been developed—often by those who made and also recognized these early errors. Given these tools, such gross errors seldom occur anymore except in cases when carelessness or strong desires reign to confirm—instead of critically testing—a model-based hypothesis.

In contrast, nuisance errors still unnecessarily prevail in protein structure models. Again, given the building and analysis tools available, they can be largely avoided and corrected. That small errors such as NQH flips, poor solvent modeling, or poor stereochemistry go beyond nuisance becomes painfully obvious when one attempts to use protein structure models with bound ligands for structure-guided drug discovery and design. The neglect of proper restraints and validation for the small (and not so small) molecule ligands in protein structures occasionally leads to absurd ligand stereochemistry and not infrequently to over-interpretation of "virtual" ligands in the structures. A brief section about binding pocket prediction, pocket analysis, and ligand docking and virtual ligand screening introduces analysis that adds value to a protein structure determination and provides further leads for new hypotheses and experiments.

After two case studies to which we apply basic hypothesis testing we finally examine B-factor distributions, a valuable tool to detect problems with B-factor restraint weights and potential scaling problems. The last validation section then briefly describes some general aspects of the PDB submission process.

Chapter 13 and our journey through the world of biomolecular structures concludes with some personal suggestions of how to present the (carefully analyzed and validated) fruits of one's labor. Ways and means of making a paper or presentation stand out are suggested together with some remarks on how to respond to, and make the best of, critical review comments.

13.1   Beyond mere geometry checking

As we have already stated *ad nauseam* in the refinement chapter, model validation should be (and in fact has to be) concurrent with model building—many errors are detected not by virtue of electron density alone but by support from assessments of the model geometry against expectations of reasonable stereochemistry. Model validation is essentially an appraisal of the model's plausibility (and any hypothesis derived from it) in face of both the experimental evidence (a bias minimized electron density map) and our prior knowledge of protein chemistry in the broadest sense. This includes not just basic geometry targets, but also local chemistry, environment, and compliance with all other known laws of physics and chemistry. The latter is particularly important when you are entering the dangerous regions of electron density interpretation in ligand binding sites and we therefore devote an entire section to ligand validation.

If you diligently build your model following the practices outlined in Chapter 12 and follow common sense, your structure model will be of good quality and it is very unlikely that it will contain any gross errors. How many small nuisance errors remain is essentially a question of your personal threshold of tedium—that is, when you abandon the polishing of the structure. The true risk in crystallography generally lies in over-interpretation of not-so-well defined regions of the electron density. Ligand molecules built into spurious or ambiguous density are a much higher risk factor than one or the other poor rotamer or close contact.

One fascinating observation is that almost all of the significant controversies of recent times—where clearly questionable structures have been published—could have been detected using the simple tools of inductive inference introduced in Chapter 7. Basic Bayesian reasoning almost always convincingly shows a low posterior probability for model correctness in such cases (which also sheds some doubt on the editorial and review process in these examples). We therefore begin this section with a top-down view by treating our model as a *hypothesis* that needs *testing* through critical validation.

Validation as hypothesis testing

We are at this point well aware, from the study of refinement techniques, that the crystallographic procedures leading to refined models are highly nonlinear and hard to parameterize, and the solution landscape is perilously spiked with local minima. In addition, the refinement is sometimes on the brink of being underdetermined. As a consequence, the resulting model is to a large degree dependent on—and therefore reflective of—our prior knowledge. It seems quite intuitive then that both the evidence supporting the model and the huge body of prior knowledge should be used for model validation. We know already that such a problem of inductive inference can be readily formulated as a Bayesian posterior probability. In simple words, the credibility of our model or hypothesis H will be the joint posterior probability PP (or posterior model likelihood) in the form of the joint probability of the data likelihood function L (based on the experimental data D) and all of the prior knowledge terms P_1, P_2, P_3 and so forth. The emphasis here is on *all*—every piece of established knowledge is fair game in the context of validation:

$$PP(H \mid D) \propto L(D \mid H) \times P_1(H) \times P_2(H) \times P_3(H)... \tag{13-1}$$

Equation 13-1 represents a joint probability and thus is subject to corresponding qualifiers. First, as pointed out in Chapter 7, the multiplied individual probabilities need to be independent. This is not always correct for the prior terms P_i for our model: A model that violates one fundamental law of physics will likely be incompatible with various other instances of established prior knowledge. In a strict quantitative sense this may be imperfect, but for the purpose of estimating the overall plausibility of the model, the joint probabilities are still a useful tool. On a similar note, for our qualitative assessment, we do not worry too much about the absolute scaling or weighting of the data likelihood versus the prior probabilities (we discussed the related problem of establishing proper restraint weights in several sections in the refinement chapter). We can just normalize the prior probabilities on a scale of 1 (highly probable, fully compatible) to zero (highly improbable, severely contradicts prior expectations). Any hypothesis must provide correspondingly strong supporting evidence in data likelihood terms. A model that contradicts prior knowledge will have to provide very convincing and solid experimental evidence to convince us to accept it as probable and eventually change our generally well established prior perceptions. Convincing and reproducible evidence is the means by which science eventually corrects it own misperceptions.[1]

Of course you are permitted to present additional biological or biochemical data in support of your model or hypothesis; they just go into an additional likelihood term. Nonetheless, there is still the need to critically check against all available knowledge. For example, you might have built a ligand of choice in acceptable geometry into some more or less featureless blob of density (possible), but it has no reasonable non-bonded contacts to the protein (improbable). The posterior probability will not be convincing, not even in view of a low R-value or great geometry for the rest of the otherwise superb structure: If the hypothesis is *local*, validation must be *local*. We will examine a few cases where such a critical analysis would have immediately raised questions and prevented major mishaps.

You may also recall that when you have two or more competing models or hypotheses based on exactly the same data (same data is important here), you can use a Bayes factor $PP(H_1 \mid D) / PP(H_2 \mid D)$ or a likelihood ratio[2] to get a feel for how competing hypotheses compare (Sidebars 12-10 and 13-1).

Model validation as a responsibility

One can approach the subject of model validation and analysis from the viewpoint of how to present the structure to a competent user or reader of a journal, and how to make sure that it withstands the scrutiny of a critical reviewer in the first round, and the merciless scrutiny of time in the foreseeable future; more or less as a form of career insurance. But more importantly, it is your responsibility as a crystallographer to deposit a structure that is as close to polished as reasonably achievable, given the tools available at the time of deposition.

Imagine the harm that a "hot" but flawed structure published in a high impact journal can cause to fellow researchers trying to obtain funding for a study based on preliminary results that contradict an incorrect structure[4] or flawed

Sidebar 13-1 Structure models form a basis for critical hypothesis testing. Given the associated effort and expense, protein structures are usually determined to test some general or specific biological hypothesis. The element of unplanned discovery—which has contributed at least as much to our knowledge of biomolecular structures[3] as strictly hypothesis driven studies—has unfortunately taken backstage and seems at present underappreciated and hardly fundable. This has led to some developments that are causes for concern, namely that structure determination is not always objectively conducted as a *test* for a hypothesis, but instead is sought as the ultimate *confirmation* of a hypothesis. This seemingly small difference between testing and confirming a hypothesis has widespread consequences. Practically all spectacular blunders in recent structure determinations originated from the desire to confirm a proposal and resulted in neglecting to critically investigate the underlying molecular structure models. Such can be described as mental model bias, which is quite insidious as no objective measure exists to determine its degree. As an example, the often unidentifiable electron density originating from the propensity of binding sites to accumulate whatever detritus is floating in the crystallization cocktail beckons to be filled with the desired ligand.

hypothesis developed from it. Crystallographic protein structure models of important drug targets or key molecules carry great weight and their pretty images purvey strong persuasive power. With great power comes great responsibility (Sidebar 12-8).

Having said this, getting bent out of shape over any minor deviation or complaint that a well-meaning checking program flags is not necessary. Sometimes checking programs use strong words like "ERROR" in bold letters or "disallowed." As these programs check only against expectation values, all this means is that an *improbable*, but not necessarily *impossible*, parameter value has been detected. There will be bona fide deviations from expectation values (which is all the geometry checking programs know) and if the electron density and chemical plausibility—the latter a sometimes underutilized help in such questions—support it, it is very likely still true.

Most of the nuisance errors mentioned previously can indeed be fixed; diligently working through the list of Ramachandran outliers, distance outliers, close contacts, NHQ flips, high B-factor waters (often waters without contact to any adjacent protein molecule) is generally rewarded by an improvement in R-free, a smaller R-free $-$ R gap, and better overall geometry—remember that every atom in the unit cell contributes to every structure factor and thus has a small but undeniable effect on the structure factor amplitude-based refinement residual. In the end, there is probably not a single big correction possible in your structure that reduces R-free by 2%, but there may be 100 smaller errors whose correction improves R-free and thus the model correctness by 0.02% each.

The most prevalent misconception by users and also new depositors is perhaps that once a structure is accepted by the PDB, it is also of sufficient quality and

---

Sidebar 13-2 Why crystallographers must deposit structure factor amplitudes—and should eventually deposit diffraction images. Deposition of measured structure factor amplitudes—although IUCr recommended practice[5] since 2000—was not a mandatory requirement for public PDB deposition of a structure model until February 2008. This is somewhat surprising given that full evaluation of deposited structure models is only possible on the basis of deposited structure factor amplitudes. If you do a reasonable job, there is no reason not to disclose your primary data, which constitutes an essential foundation of science. Without data your model cannot be falsified and, in Karl Popper's sense,[6] your work is not in compliance with fundamental scientific methodology—that is, it is on a par with pseudo-science and quackery. One can therefore argue that out of principle no protein structure model without supporting experimental data should ever be accepted for publication, and given the clear recommendation[5] by the International Union of Crystallography (IUCr) and even in earlier editorials on the subject,[7] deposition of structure factor amplitudes together with the model has finally become a requirement.

It is an interesting observation, quantified first by Gerard Kleywegt and Alwyn Jones,[8, 9] that crystallographic trade journals such as *Acta Crystallographica D* already had a very high voluntary deposition rate of structure factors (> 90%) while the high impact factor journals (often affectionately called the magazines or vanity journals) such as *Nature* and *Science* enjoyed a much lower experimental

data deposition rate. For example, up to 2002 only 38% of published structures in *Science* had deposited structure factors, and *Cell* and *Nature* fared not much better. There was also a (perhaps not surprising) correlation between poor R-values and the tendency not to submit structure factor amplitudes.[9] The deposition of structure factor amplitudes allows the re-refinement and improvement of models[10] once more advanced methods become available,[11] which contributes to better and more accurate empirical data extractable from the enormous but somewhat disorganized body of knowledge that the PDB comprises. A recent analysis of structure quality showed that the apparent lower quality of structures in high impact journals can be largely attributed to the larger size and complexity (generally resulting in lower resolution data) of a large fraction of structures published in these journals.[12]

There are also efforts underway to preserve the original diffraction images, because data processing routines also improve steadily, and reprocessing of original data at a later point may also lead to improved structure models (diffuse scattering or incommensurable satellite reflections,[13] for example, are just ignored). The large size of the image frames amounting to several GB for a single data set does generate some additional but not overwhelming challenges for original raw data preservation. The advantage of having original raw data available if questions of data provenance arise has been recently pointed out[14] during analysis of structure factor amplitude data giving rise to the unusual statistics discussed in Table 13-3.

Box 13-1 Model validation using all available means. Structure validation informs us about the probability of the features of our model. In cases of geometry outliers, a highly improbable conformation that has insufficient experimental support from electron density is very likely an error and needs to be corrected. In more subtle cases such as ligand binding or derived hypotheses based on model features, additional independent experimental support for the model as well as evaluation against the entire body of physicochemical knowledge should be sought. Validation is an integral process determining the plausibility of a model or hypothesis extending beyond simple geometry error checks.

nothing can be wrong. This is a grave mistake—the PDB presently does not validate your structure in great detail, it just makes sure that your structure complies with a minimum set of formal standards that make it acceptable and processable by the PDB. You still remain the final authority and responsible party regarding the correctness of your model and hypotheses.

Global versus local structure quality indicators

During our refinement and model building exercises we have realized that the certainty with which parts of a structure can be determined varies, and the reason lies foremost in the genuine plasticity of proteins. The reliability of protein structure models is a local property, about which the absolute values of global refinement indicators such as the linear refinement residuals R and R-free or coordinate r.m.s. deviations from target values reveal no detail.[15] Local definition of quality, however, becomes highly relevant for the accurate biological interpretation around active sites, mutations, and particularly for bound ligands. Substantial parts of a protein structure can be located in flexible regions or can be completely absent from the model[16] without any indications thereof evident in R or R-free, and even incorrectly traced models can be refined to unsuspicious R-values.[9] Intricately derived composite but still global quality indicators suffer equally from the same limitations.[17, 18]

While it should be obvious to the competent crystallographer that global descriptors are only a first quality indication and not proof of absence of gross indecencies, this is not necessarily clear to the users of crystallographic models. Over-reliance on (conveniently few) global indicators by the user community has unfortunately led to various inappropriate cosmetics such as targeting an R-value below 20% irrespective of the danger of over-fitting, or the mistaken interpretation of low r.m.s. deviations from restraint targets as a sign of crystallographic prowess. This is not the case and as discussed in Chapter 12, careful tracking of R-free during the refinement and the use of R-free or cross-validated log-likelihood functions[19] to properly adjust restraint weights have greatly reduced such deviant practices since the introduction and adoption of R-free in the early 1990s.

As the purpose of a structure model is nearly always to test a certain hypothesis of biological relevance, and most of these hypotheses are associated with specific locations such as binding sites, mutations, hinge regions, or otherwise specific regions of the protein structure, we clearly need to employ local measures in the evaluation of the reliability of the structure model. Practically all the validation tools we used during model building—or during the deposition process—make extensive use of prior knowledge of stereochemistry to evaluate the plausibility on a residue-by-residue basis as well as through related local measures such as close contact analysis or the analysis of the local chemical environment of residues. In the latter category, for example, fall NQH-flip checks, or analysis of the local packing propensities, or inside–outside distributions of residues.

It is necessary to realize that validation criteria that examine restrained properties such as basic stereochemical bond distances, angles, or torsions will

Sidebar 13-3 Z-values. A commonly reported measure for deviation from target values (or from distribution mean values) are the Z-values. They are a statistical standard score: Z represents the distance of the measured value from the distribution mean in units of the standard deviation. Z is negative when the observed value is below the mean, positive when it is above:

$$Z = \frac{x - \bar{x}}{\sigma_x} \qquad (13-2)$$

From the error function we can compute the probability that a value lies outside a given range from $-Z$ to Z: ± 1, 32%; ± 2, 5%; ± 3, 2%; ± 4, 0.01%; ± 5, 0.0001%. This means that for $Z = \pm 4$ only one in 10 000 observations will lie outside of 4σ. Recall that central limit theorem states the counterintuitive fact that the underlying sampled probability distributions do not have to be normal distributions themselves.

reflect these restraints according to their applied weights. They are still useful for examining outliers, but they need to be combined with evidence-based confirmation by examination of electron density (an important point discussed further below). As the restraint weights depend on specific particulars of the structure as well as on resolution, unconditional comments such as "the bond lengths are over-restrained" which are made for example by *WHATCHECK* when their r.m.s.d. Z-values fall below 0.5 are no reason for panic.[20] Nevertheless, such "squawks" can serve as a useful reminder to review how the overall restraint weights were adjusted during the course of the refinement (Chapter 12).

Irrespective of the usefulness of stereochemical or geometric validations, a complete evaluation of a protein structure must include specific evidence and cannot be solely based on generic prior knowledge. It is the irrefutable evidence of a clear, minimally biased electron density map that determines whether an outlier is the result of the inability to model a region of poor density correctly, or a true feature of the structure. It is clear that the more a feature deviates from prior knowledge, the stronger the support in its favor must be. It is also true that such points of higher local energy often do carry significance, either for specific local interactions at binding sites or perhaps as trigger points for other conformational rearrangement in response to environmental change.

The most powerful and complete analysis of a structure thus will include a combination of the real space electron density correlation with the location of geometric outliers.

13.2   Examining model stereochemistry

Geometric model validation compares model properties such as stereochemistry, local chemical environment, and packing propensity against their empirical expectation values based on prior knowledge. In many cases these target values have well-established uncertainties and the deviation from these values can be expressed in multiples of their estimated standard uncertainty, the so-called Z-values. Other programs may provide some specifically defined score such as the *clashscore* in *MolProbity*, the total number of close contacts that exceeds a certain threshold value,[21] which is obtained from reference data sets of high quality structure models. Many of the geometry checks are directly implemented in the model building programs and a great deal of validation and outlier detection already takes place during model building. *Coot* for example can read the score sheet of *MolProbity* and compute the clash spikes (Figure 13-3) which provides an easy way to step through and fix the errors of the model displayed in its electron density. Table 13-1 summarizes the checks implemented in common validation and model building programs.

The most popular macromolecular model geometry validation programs are *PROCHECK*, *WHATCHECK*, and *MolProbity*. Each of them provides the basic geometry checks and some extra information specific to each program. While *WHATCHECK* has unique and useful features, it is not for the faint hearted, as it has a tendency to exclaim "ERROR" in some instances when it is not quite appropriate. Presently the most convenient and complete of the geometry validation

Box 13-2 Local structure validation. Protein molecules are inherently flexible as a requirement for their molecular function. As a consequence, a model representing a protein structure will have well-defined parts and some parts that are less well defined in electron density. Therefore, only local validation including assessment of both geometry and electron density can give an accurate picture of the reliability of the structure model or any hypothesis based on local features of the model.

Property	Model building or validation program						
	Xtalview MI-Fit	Coot*	O/OOPS	PRO-CHECK	WHAT-CHECK	Mol-Probity	ERRAT Verify3d PROSA
Bond length, angles	x	x	x	x	x	x	
φ–ψ torsions	x	x	x	x	x	x	
Side chain torsions		x	x	x	x	x	
NCS φ–ψ torsions		x	x				
Peptide flips		x	x				
Clash score	x	x				x	
Cβ deviations		x				x	
NQH flips		x			x	x	
Local environment profiles, etc.							x

*Reads results table from *MolProbity*.

Table 13-1 Geometry validation in model building and validation programs. Practically all popular model building programs provide validation diagnostics during the model building process. The earlier and more diligently even small errors are corrected during model building, the less surprises will surface during the final validation upon coordinate and structure factor deposition to the PDB.

tools for the crystallographer is probably *MolProbity*, whose sensitive all-contact analysis in addition to basic checks and updated Ramachandran plot boundaries delivers a quite complete picture of the model geometry. The program also returns coordinates with hydrogen atoms added (needed for full contact analysis) and with the NQH residues flipped (which must be re-refined before deposition, of course). The squawk-list from *MolProbity* can also be uploaded into the model building program *Coot*, a very convenient feature.

PROCHECK stereochemistry analysis

The first comprehensive geometry validation program was *PROCHECK*,[22] originating from Janet Thornton's group. At the core of *PROCHECK* is analysis of deviations from target values,[23] which is presented in a graphically pleasing form in PostScript output (Figure 13-1). *PROCHECK* generates consistent atom label nomenclature (which may again get reversed by flipping torsions during model building). The analysis is quite self-explanatory. Deviations from targets are given in absolute numbers and as individual and combined *G*-factors, which are a log-likelihood measure of how far a value deviates from the expected target value. *PROCHECK* generates nice graphics that also compare the scores in resolution-dependent form against expectation values, and its target value database will likely be updated in the near future.

Geometry outliers are grouped into a number of categories (main chain bond lengths and angles; side chain bond lengths, angles, and torsions). By far the most common geometry deviations are side chain torsion angle outliers, which are the least restrained. The main chain torsion angles φ and ψ are generally unrestrained during refinement. Providing a particularly valuable validation aid, they are discussed in a later subsection.

MolProbity and all-atom contact analysis

A key feature in *MolProbity* is that it adds hydrogen atoms for all residues, and evaluates the so-called all-atom contacts. The addition of hydrogen atoms in turn allows analysis of local hydrogen bond networks, which enables optimal placement of movable hydrogen atoms such as rotatable hydroxyl groups. Analysis of hydrogen bond networks automatically lends itself to investigation of NQH side chain orientation and the PDB file with NQH residues flipped can

Figure 13-1 Presentation of stereochemistry outliers. The figure shows the observed r.m.s. deviations from restraint target values, in this case the non-(Pro, Gly) Cα–C–N bond angles. The values are distributed around the target value, with the ±4σ variance interval indicated by the dashed lines. For a value deviating by more than ±4σ, the probability of observing it is less than 1/10 000. Even with the leniency of ±6σ the PDB allows before listing the deviation in the actual PDB entry, the residues with bond angles at ~108° and ~126° certainly deserve close scrutiny. The figure is produced by *PROCHECK* as part of an extensive Postscript file graphically presenting the validation results.

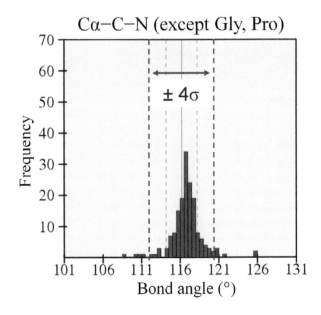

Figure 13-2 Close contact between ligand and a binding site residue. The ligand (2-dimethylamino-ethyl-diphosphate, bound to a Mg^{2+} ion) has an incorrect torsion between the preceding C and the –$N(CH_3)_2$ group (red circled regions). In addition to the poor fit of the model to electron density, a close contact between the ligand and the Oε of Glu 166 (see Table 13-2) is visible in the circled region in the left panel. In the 90° rotated view of the region in the right panel, the mismatch of the electron density becomes even more distinct. Flipping the leading C–N torsion angle will bring the dimethylamino group into the electron density while simultaneously eliminating the bad contact and generating a favorable hydrogen donor–acceptor pair between a carboxyl oxygen of Glu 166 and the N atom of the dimethylamino group.

be downloaded for re-refinement. The rapid increase of the repulsive potential in close contacts (Figure 12-18) makes all atom-contact analysis very sensitive, that is, a detected serious clash generally means that the position is indeed worth investigating. The geometric clashes are represented by dots in different colors (that turn into nasty spikes when contacts become offending, Figure 13-3) as generated by a rolling probe algorithm.[24] Investigation of these close contacts is made easy either through displaying the problem sites on the web page in interactive *Kinemage* displays or with a downloadable summary file that can be examined in the model building program *Coot* and corrected there with direct evidence of the (difference) electron density at the trouble spot. As an example we show here parts of the analysis of a ligand structure, revealing an NQH flip and an extremely close contact between ligand and protein target (Table 13-2 and Figure 13-2). Additional unique features of *MolProbity* are the analysis of

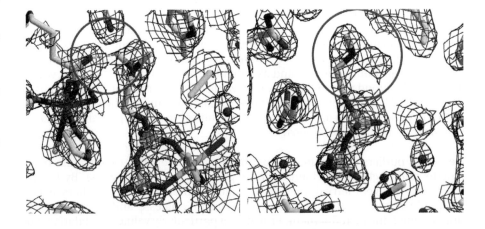

Box 13-3 Basic geometry validation. Geometry checks flag improbable values based on the deviation from expectation values, often in multiples of standard deviations in the form of *Z*-scores. The deviations can also be reported in percentile scores in relation to an empirical target distribution obtained from high quality reference structures.

Flip?	Chain	Res #	Res ID	Orig	Flip	Flip–Orig	Code	Explanation
	A	10	ASN	−6.8	−0.29	6.51	FLIP	Clear evidence for flip
	A	12	GLN	−1.1	−0.45	0.65	FLIP	Some evidence recommending flip
✓	A	64	ASN	2.6	5.8	3.2	FLIP	Clear evidence for flip
	A	76	ASN	−1.6	−0.025	1.575	FLIP	Some evidence recommending flip
✓	A	140	GLN	−2.4	−0.23	2.17	FLIP	Clear evidence for flip
✓	A	168	ASN	−4.9	0.61	5.51	FLIP	Clear evidence for flip
✓	B	64	ASN	5.9	8.1	2.2	FLIP	Clear evidence for flip
✓	B	76	ASN	−3.9	−0.25	3.65	FLIP	Clear evidence for flip

#	Res	High B	Clash > 0.4 Å	Ramachandran	Rotamer	Cβ deviation
B 116	GLU	22.08	1.322 Å OE1 with B1185 LIG C14	Favored (22.4%) General case / −147.6,138.6	57.7% (*mt-10*) χ angles: 274.1, 168, 179.3	0.115 Å

Table 13-2 Selected parts of *MolProbity* all-atom contact analysis. The top of the table is a selected part of the *MolProbity* web page output and shows examples for incorrect orientation of the Asn and Gln side chains. "Clear evidence" almost always is a correct assessment and examination of the other cases with "some evidence" is worthwhile. The table below shows a serious close contact between a Glu side chain oxygen and methyl group of the ligand. The situation in real space is illustrated in Figure 13-2.

Cβ-deviations in relation to the backbone (a sign that the subsequent side chain torsions may be suspect) and the sugar pucker analysis, which is of interest when an RNA/DNA structure or DNA/RNA–protein complex was determined.

Multiple side chain electron density and NQH flips

We already discussed that the orientation of asparagine, glutamine, and histidine cannot be unequivocally assigned from electron density alone, given the small difference in the number of electrons between C, N, and O. The proper conformation of Asn, Gln, and His needs to be assigned according to chemical plausibility, such as satisfying hydrogen bond networks, which is often straightforward in the case of Asn and Gln. The fact that His does have different protonation states depending on its actual local pK_a, often makes the assignment of His side chain orientation more subtle. It is not much effort to assign conformations correctly the first time around and use *NQ-Flipper*[25] or *MolProbity*[21] (both return flipped coordinates) during model building, but a large fraction of protein crystal structures still have incorrectly assigned Asn, Gln, or His side chain conformations.[26] In most cases, following the recommendations of the programs also improves the clash score (Figure 13-3).

However, there are subtle issues to be considered when split side chains are correlated, and these cases are not always correctly indicated by the NQH-flippers. Correlation between split side chains exists when one partial side chain can exist only when a neighboring or contacting side chain—or ligand—is in a specific partial conformation as well. These cases are not that rare and need to be examined carefully because most geometry validation programs do not analyze multiple conformations to the full extent necessary. Take for example a split side chain, combined with a correlated Gln flip in Figure 13-4. Correlated side chain splits can be clearly distinguished in high resolution structures and often involve partially occupied solvent molecules.

Backbone torsion angle validation

We discussed in Chapter 2 that the backbone conformation of a polypeptide is completely defined by the sequence of the torsion angles along its peptide

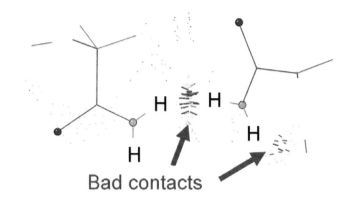

Figure 13-3 **Clash score improvement by Gln side chain flip.** The picture is a Kinemage[27] image as returned by the web server version of MolProbity.[21] The panel clearly shows close contacts between the NH atoms as indicated by the nasty red spikes. Note that both close contacts will disappear when the right residue is flipped.

Bad contacts

chain. The 2-dimensional scatter plot of the φ–ψ backbone torsion angle pairs for each residue, called the Ramachandran plot after its inventor,[28] shows preferred regions where the specific φ–ψ combinations cluster (Section 2.3 provides some theoretical background). These regions are equivalent to a 2-dimensional projection of the conformational energy surface, signifying the probability of finding a residue with a given torsion angle pair. The plots used in model validation are derived from empirical data and the boundaries of the most probable regions have changed over time, with larger data sets of high quality structures becoming available for analysis. While the first popular implementation of backbone torsion angle validation in the program *PROCHECK*[22] distinguished three regions with a quite "generously allowed" outer boundary,[23] recent reanalysis has shown that a tighter distribution with only a core region and an allowed region in addition to the improbable conformations (Figure 13-5) is appropriate and adequate.[29] In addition to separate treatment of glycines (the torsion angle distribution for glycine is centrosymmetric given the absence of a chiral Cα-center), individual analysis of the conformationally more limited prolines and pre-proline residues is also necessary. The latest versions of φ–ψ torsion angle distributions are implemented in *MolProbity*[21] and are also provided during PDB validation.

The φ–ψ torsion angle plot quickly shows the plausibility of the geometry of a protein backbone trace. In the process of assuming a minimum global energy state when the protein is folded, local strain can be induced on the protein backbone, and a few residues (approximately 1–2% in normal cases) can be expected to adopt an energetically less favorable conformation. For example, some residues such as catalytic serine residues in certain α/β hydrolases have been found

Figure 13-4 Ambiguity of side chain placement combined with split conformation. The glutamine side chain makes contact with a split serine side chain of a symmetry related molecule. To obtain an optimal donor–acceptor hydrogen bonding pattern it is necessary that the Gln residue is also split and flipped proportionally to the Ser occupancy, with occupancies in conformations A + B constrained to 1.0. *SHELXL*, which is particularly suitable for high resolution structure refinement, allows setting up and refining the correlated constraints necessary to describe such a situation correctly.

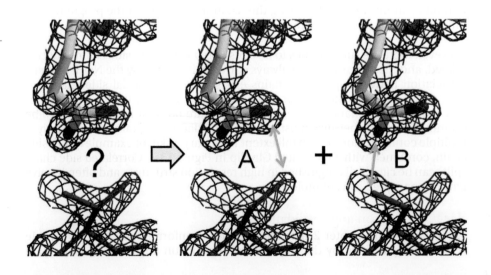

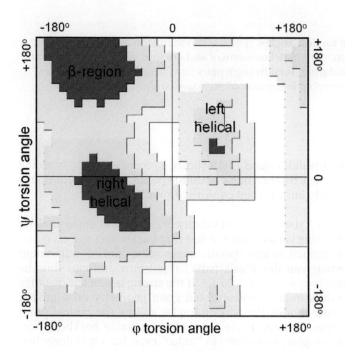

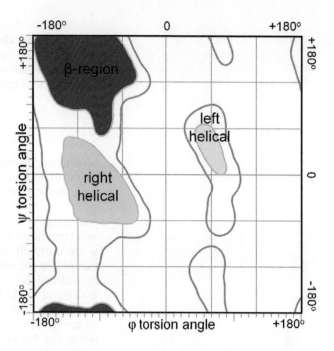

to be conserved in high energy conformations.[30] In any case, residues in extreme locations of conformational φ–ψ space deserve serious attention. The inspection of electron density around all Ramachandran outliers should be standard procedure. Most of the outliers may lie in regions of poor density, often loops, and are just difficult to model correctly—but an outlier in clear density may indicate an interesting "hot spot" in the structure.

Non-crystallographic symmetry. The φ–ψ torsion angle plots can also be used to check for proper application of NCS restraints. If NCS related molecules were properly restrained, the differences between the φ–ψ torsion of related residues should be small.[31] These multi-molecule "Kleywegt Plots" (Figure 13-6) are also implemented in *Coot*[32] and can be examined during model building. Another point that deserves attention is that a random coil peptide cannot have random φ–ψ torsions—the absence of a defined secondary structure in no way means that the close-contact repulsions that overwhelmingly determine the allowed

Figure 13-5 **Two popular presentations of the backbone torsion angle distribution.** The earlier *PROCHECK* boundaries (left panel) are somewhat more generous than the recently established empirical boundaries in *MolProbity*. In the *MolProbity* φ–ψ torsion angle plots, on average 98% of the residues are expected to lie within the core regions, and about 0.2% (only 1 in 500) outside the secondary boundary.

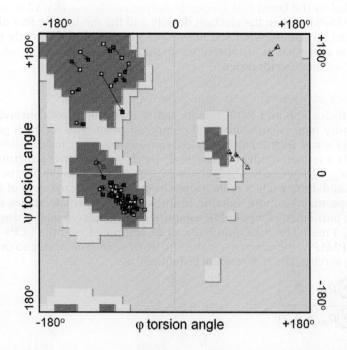

Figure 13-6 **Backbone torsion angle distribution for NCS related molecules.** Shown is the combined backbone torsion angle plot of two NCS related molecules of the cytochrome c′ dimer built in Chapter 12. The φ–ψ points of corresponding residues are connected with lines (triangles represent glycine residues). Except for one residue located in a flexible loop, no significant pair-wise differences between the torsion angle distributions of the two molecules are present, indicative of proper NCS restraint use (plot by *Coot*). PDB entry 1gqa.[33]

> **Box 13-4 Backbone torsion angles.** The backbone torsion angles φ and ψ are generally not restrained in model refinement and are thus an independent and reliable means of validation. Torsion angle pairs that fall outside of the "allowed" regions are highly improbable because of excessive conformational energy.

torsion angles become invalid. Figure 13-18 shows a recent example from the literature that purports the existence of a 41-residue peptide with exceptionally high (and thus exceedingly improbable) conformational energy.

As discussed in Chapter 12, the power and validity of the torsion angle analysis as a structure validation rule is based on the fact that backbone torsions φ and ψ are normally not restrained to any specific angle values during refinement except for intramolecular van der Waals (vdW) repulsions and can thus be considered as a geometric cross-validation set. If the strong geometric distance and bond angle restraints impose unwarranted good geometry on a grossly incorrectly built structure model, the torsion angles will be the critical point where the errors become obvious in the form of improbable backbone torsions. The Ramachandran plot seems hard to "fudge" even for a genuinely bad model: Kleywegt and Jones have demonstrated that even under extremely high force constants applied during energy refinement, φ–ψ torsions remain robust against over-refinement.[34] Note, however, that certain protocols in *CNS* or its successor suite *PHENIX*, for example, may allow restraining backbone torsion angles during (very) low resolution torsion angle refinement.

13.3 Electron density based model validation

When we correct the model during the real space building and rebuilding cycles, we actually already perform validation of the model in reference to the electron density. As the electron density is the ultimate evidence that crystallography provides, inclusion of local fit of the electron density in the validation process is the most direct means of obtaining an accurate picture about the local quality of the model. As electron density based validation is only possible when structure factor amplitudes are available, the importance of depositing structure factors cannot be overstated (Sidebar 13-2). The fastest way to identify problem regions is provided by the linear real space correlation coefficient (*RSCC*) or real space *R*-value (*RSR*) between the electron density and the model. The *RSR* plots were introduced by Carl Brändén and Alwyn Jones in 1990,[35] and recently the *RSCC* plots were suggested as the absolute minimum of crystallographic information required for manuscript review.[36]

Real space *R*-value and real space correlation coefficient

Both statistics *RSR* and *RSCC* require that a bias minimized "observed" electron density map is computed and compared point-wise for each grid point or density voxel with a calculated electron density map based solely on the model. As a refined model will hopefully be close to the true structure, a map from $2mF_{obs} - DF_{calc}$ maximum likelihood coefficients will suffice for the purpose of final validation, while a more aggressively bias minimized map of the *ARP/wARP* type may be more suitable in initial or intermediate stages of model building, particularly for poor MR solutions. The calculated map is simply a $F_{calc} \cdot \exp(i\varphi_{calc})$ map. The *RSCC* and *RSR* can be computed with the CCP4 program *OVERLAPMAP* or the programs in the *MAPMAN* suite, which are also used in the *EDS* web service. The *RSR* residual is defined[35] as

$$RSR = \frac{\sum_r \left| \rho_{obs} - \rho_{calc} \right|}{\sum_r \left| \rho_{obs} + \rho_{calc} \right|} \qquad (13\text{-}3)$$

where the grid points that are summed for each atom fall within a radius r which is given as 0.7 Å if the resolution d_{min} is less than 0.6 Å (an extraordinarily rare occurrence in protein structures and observed so far only for crambin)[37], 0.7 + $(d_{min} - 0.6)/3.0$ Å in the resolution range between 0.6 and 3.0 Å, and $0.5 \cdot d_{min}$ if the resolution is worse than 3.0 Å. This function accounts for the broader electron density peaks of each atom as a result of the larger overall B-factor in low resolution structures. The r.m.s. difference density plots provided in the model building program *Coot* are another, very sensitive measure allowing immediate inspection of local density fit already during model building.

The *RSCC* is defined (Chapter 7) as the linear correlation coefficient between observed and calculated electron density:

$$RSCC = \frac{\sum_{\mathbf{r}} (\rho(\mathbf{r})_{obs} - \overline{\rho(\mathbf{r})_{obs}}) \cdot (\rho(\mathbf{r})_{calc} - \overline{\rho(\mathbf{r})_{calc}})}{\left(\sum_{\mathbf{r}} (\rho(\mathbf{r})_{obs} - \overline{\rho(\mathbf{r})_{obs}})^2 \cdot \sum_{\mathbf{r}} (\rho(\mathbf{r})_{calc} - \overline{\rho(\mathbf{r})_{calc}})^2 \right)^{1/2}}$$

(13-4)

and has the benefit of being scale-independent. It therefore also works well when scaling of experimental to calculated data is problematic, as is the case in the initial stage of a molecular replacement search. The linear correlation coefficient also shows a good match for weak, high B-factor or low-occupancy atoms as long as they are in the right position. When the B-factor is graphed for each residue on the same plot, a quite complete picture of the fit of the model to the actual electron density can be quickly obtained.[38] As an example we show in Figure 13-7 the *RSCC* of a purported peptide ligand in a protease structure (the same peptide that displayed the highly improbable combination of φ–ψ backbone torsions discussed later and shown in Figure 13-18). This peptide has a mean B-factor of ~130 Å^2 compared with the mean B-factor of ~33 Å^2 for the protease, which does not bode well for the bound peptide's actual presence.[16]

The *EDS* server and *SFCHECK*

For deposited structure models that were accompanied with structure factors (about 80% of all entries), the Uppsala *EDS*[40] (Electron Density Server) web program provides the real space R-value and *RSCC* plots for amino acid residues and nucleic acids, together with a resolution-dependent empirical Z-value for the expected fit of each residue:

$$Z = (RSR_{obs} - RSR(d_{min})) / \sigma_{RSR}(d_{min})$$

(13-5)

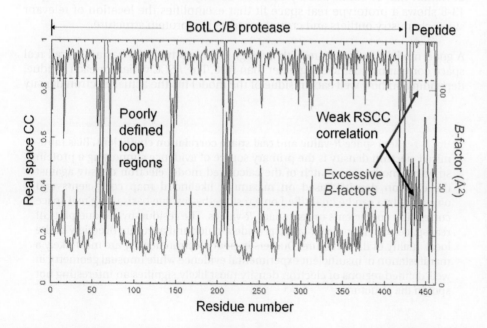

Figure 13-7 Real space correlation coefficient and *B*-factor plot. The superseded PDB entry 1f83 (downloadable from the book website) contains the model coordinates for the BotLC/B protease–synaptobrevin-II complex[39] refined at 2.0 Å. Shown in black (upper curve) is the residue-by-residue real space correlation coefficient; in blue (lower curve) the *B*-factors are plotted for each residue. The left part of the figure corresponds to the protease, which, with the exception of three loop regions is well defined. The synaptobrevin-II ligand peptide at the right figure edge, however, shows a very worrisome crossover between abysmal real space correlation and excessive *B*-factors. The plot was created by the *Shake&wARP* web service.[38] As an exercise, you may convince yourself that the superseding PDB entry 3g94 (in the meantime also retracted) suffers from the same problems.

where σ_{RSR} is the observed r.m.s. deviation of the *RSR* of the residue in question. As in the *RSCC* and *RSR* plots, a spike in the residue's *Z* score means that for some reason the fit is worse than expected, and it might be worthwhile examining it when polishing the model, or exercising some caution when interpreting the model in that region. The plots of the *EDS* server are linked to a Java-based web viewer, which allows immediate inspection of a questionable residue in the electron density map, a very useful feature.

Other useful data statistics the *EDS* server provides are a Wilson plot, the Padilla–Yeates plots providing statistics that detect the possibility of twinning (Chapter 8), a diffraction anisotropy plot (Chapter 12), and Ramachandran validation, as well as links to the PDB reports. It is expected that with the introduction of mandatory structure factor submission the deposition tools of the PDB—at least at the PDBe in Hinxton—will also provide real space validation during the course of deposition, a much needed extension of the PDB validation and deposition service.

Another useful program for structure factor based validation that can be run before deposition (and is also executed during the RCSB *ADIT* submission) is *SFCHECK*.[41] It provides the basic *R*-value, *R*-free, and overall correlation coefficient statistics, as well as publication quality Wilson plots, data completeness graphs, and Luzzati plots. A similar graphical representation of residue-by-residue real space fit statistics similar to those of *EDS* is also provided, although with slightly differently defined statistical descriptors.

Unexpected stereochemistry and real space validation

The real space correlation between the model and bias minimized electron density becomes particularly valuable when apparent stereochemistry deviations are correlated with good fit between density and model. Places where the electron density shows a good fit to a part of the model where geometry outliers are present indicate hot spots of unusual high energy conformations. That certain locations in a protein structure can have unusual conformations that do not conform to stereochemical expectation values clearly implies that model validation without considering the experimental evidence will not always provide correct answers, which has a far-reaching impact on the modeling of ligand docking with flexible local protein geometry. To our knowledge, no systematic correlation of geometry deviations against validated electron density has been attempted yet. Only the combination of both geometry validation and electron density validation can provide these potentially far-reaching insights. A geometry deviation in a region of poor electron density has little relevance, while a deviation that is confirmed by electron density fit deserves attention. Figure 13-8 shows a prototype real space fit that exemplifies the location of relevant stereochemistry outliers and crystal contacts in a protein structure.

A good model of a structure with average *B*-factors typically has an overall real space correlation coefficient greater than 0.90. The *RSCC* varies for each residue, depending on how well each residue of the model fits the actual electron density

---

Box 13-5 **Real space R-value and real space correlation coefficient.** Bias minimized electron density is the primary source of evidence supporting a protein structure model. The match of the calculated model electron density against the electron density based on maximum likelihood map coefficients and observed data can be quantified on a residue-by-residue basis using real space correlation coefficients or real space *R*-values. The residue-by-residue plots of real space correlation or *R*-value provide a quick and accurate overview of the local quality of the structure. Outliers in poor electron density are most likely a manifestation of insufficient experimental evidence while unusual geometry in well-defined regions of electron density most likely signifies an interesting hot spot in the structure.

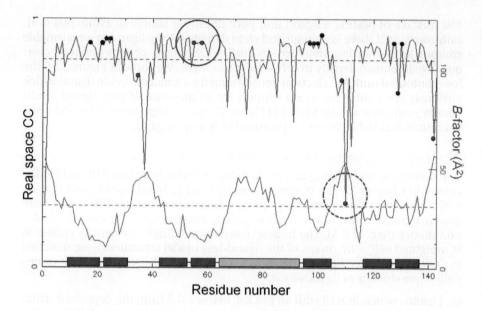

Figure 13-8 Real space correlation coefficients, plasticity, and location of stereochemistry deviations.
The real space correlation plot quickly provides an overview of the plasticity of the protein. Note the correspondence of high *B*-factors and density deviations in flexible regions of the molecule. The location of stereochemistry outliers in relation to the regions of intrinsic plasticity or rigidity (as indicated by good RS fit and low *B*-factors) is worth examining. If residues with unusually high backbone torsion angle energy (red dots) are located in regions of high real space density correlation (solid red circle), they deserve attention because they may carry some relevance. If they are located in regions of poor density and high molecular flexibility (dashed red circle), they are probably just reflections of our inability to correctly model the structure because of poor electron density. Crystal contacts are exemplified by blue dots. Regions where a protein forms crystal contacts may not necessarily be in a native conformation, which is relevant if contacts are in the vicinity of binding sites or have other significance associated with them.

map. A residue-by-residue plot of the real space correlation coefficient is thus a very powerful and quick measure to gain an overview of the local quality of a structure model.[36]

13.4 Validation of protein–ligand complex structures

The assessment of local quality around the binding sites that harbor small molecule ligands in complex structures is particularly important. These bound molecules are often important metabolite analogs, drug-like fragments, or drug leads, which serve as templates for ligand docking.[42] The conformation of ligands in protein–inhibitor complexes is often poorly defined to the point where it seriously affects the biological conclusion drawn.[16, 43] Such errors and inaccuracies also affect the confidence and accuracy of structure based virtual ligand screening (VLS) and the design of accurate protein ligand libraries and empirical parameters.

Ligand density validation
Building, refinement, and validation of ligands in structure models are more difficult than in the apo-protein molecule for a number of reasons:

- Even excellent stereochemical validation of the protein says nothing about the ligand, and the effect of either a well-built or an incorrectly built ligand on the *R*-value and other global refinement statistics will be negligible.

- Limited binding constants of ligands or lead compounds frequently cause binding sites to be only partially occupied, leading to proportionally weaker electron density.

- Only tightly bound parts of the ligand may have distinct density.

- Ligands are often present in multiple conformations, further reducing the electron density levels.

- Ligands can have higher internal symmetry than the asymmetric protein and thus can be located at special positions. In this case the ligand generally must have multiple conformations which may, in some refinement programs, require selectively turning off the packing vdW restraints so that the molecule does not collide with its own symmetry related parts.

- Because of the tendency of binding sites to scavenge all kind of detritus from the purification and crystallization environment, there will often be weak density of unknown origin, into which, regardless of its origin, an inappropriate ligand model can be placed.

The mistake of placing a ligand into poor density is simple to avoid: just critically examine if there is enough and clear density for the ligand at a reasonable contour level—not less than 1σ—in $2mF_o - DF_c$ maps, or appropriately clear positive difference density in $mF_o - DF_c$ omit maps. We show in Figure 13-9 the low-contoured omit ML electron density map for a small molecule ligand which obviously does not show much support for its presence. If your ligand (omit) density maps look similar to that of Figure 13-9, you might reconsider and resist your understandable desire (or pressure) to find the ligand.

Automated ligand building with *ARP/wARP*

The dummy atom building and refinement methods based on *ARP/wARP* discussed in Chapter 12 can of course also be used to build ligands into density while improving the overall map quality. Recent versions of *ARP/wARP* were used in a grand-scale project of refining all ligand structures[44] of suitable resolution (better than ~2.8 Å). The largest unaccounted difference density cluster in σ_A-weighted $mF_o - DF_c$ maps of the ligand-less model structures were matched to *CORINA*-idealized ligands in the *HIC-Up* database. The result was that essentially three clusters of ligands exist:

- Ligands which fit well with an r.m.s.d. below 1.0 Å from the deposited structure—roughly 25% of the models. In this case, the refinement following the ligand building will succeed and there is little doubt that these models are largely correct.

- The second cluster contains structures where the ligand is placed within 1.0 to 10 Å of the presumed position. In this group, at least the binding site is probably correctly identified, but the density apparently did not allow unambiguous automated ligand placement and refinement (~35% of test cases).

- In the remaining ~40% of cases, the deviation in location between auto-refined ligand and purported binding site lies between 10 and 100 Å, and it is safe to assume that even the binding site could not be verified based on electron density evidence.

As expected, the results were in agreement with the real space map correlation coefficient and essentially showed that only good density allows unambiguous fit of a ligand. The lesson to be learned is that although we have additional chemical plausibility and independent means at hand to place and validate a ligand, a large fraction of deposited ligand structures are not necessarily representing crystallographic evidence. As the ligand building module of *ARP/wARP*

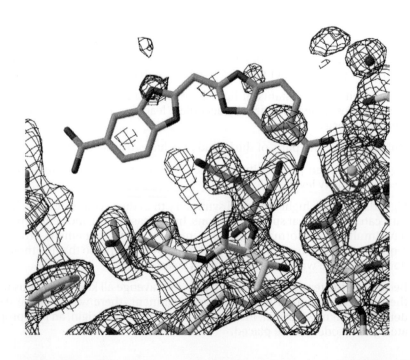

Figure 13-9 Absent ligand density.
A bias minimized ML omit $2mF_o - DF_c$ electron density map contoured at 0.8σ is shown for the ligand (a potential inhibitor) in a *Clostridium botulinum* serotype B neurotoxin light chain protease. The mean ligand B-factor is ~120 Å^2 while the mean B-factor of the protease is ~36 Å^2. Despite the low electron density contouring, and as expected from the high B-factors, no appreciable density for the ligand is present. Instead, even the catalytic water normally present in the apo-structure next to the Zn^{2+} site is visible in the minimally biased omit density. The two ring systems of the ligand are also in an unusual, perfectly flat conformation (180.0°), indicating lack of refinement. PDB entry 1fqh[43] (withdrawn).

is available as a CCP4 module, it also means that the procedure can be used as an unbiased validation of your own ligand placing—in case you are unsure and wish to acquire such independent, density-based conformation of your ligand placement.

Occupancy and *B*-factor adjustment

A second difficulty in ligand placing and refinement arises as a result of practically unavoidable under-occupation of the binding site (see Figure 3-40). Because of the actual ligand occupancy almost always being lower than 1.0, the refined *B*-factors of a ligand with occupancy of 1.0 will tend to be higher than the *B*-factors of the surrounding protein atoms. In the absence of independent refinement of the occupancy and *B*-factor, given the strong correlation between *B*-factor and occupancy, one can initially set the occupation to 1.0 and refine *B*. If the *B*-factor rises significantly above the *B*-factor of contacting protein atoms, the occupancy can be reduced until the *B*-factors of ligand atoms bound (and thus partly restrained in *B*) to the protein target are reasonably close to the protein *B*-factors. In the case of non-covalently bound atoms this can only be approximately estimated, but in many cases it is still a physically more reasonable representation than just assuming a high *B*-factor at full occupancy. Recall from Chapter 9 that in principle a distinct difference exists between attenuation of scattering by reduced occupancy versus attenuation through increased *B*-factor: The scattering factor curve of an atom and thus also its FT, the electron density, have different shapes for high *B*-factor and for low occupancy (Figure 9-6), but only at relatively high resolution (approximately 1.4–1.2 Å) is this difference distinct enough that independent (and accurate) simultaneous *B*-factor and occupancy refinement is possible. A very useful option for data with resolution of ~2 Å or better exists in *SHELXL* through refinement of one occupancy parameter for the whole ligand[45] (consuming only one *SHELXL* free variable), which generally gives reliable results if the *B*-factors are sensibly restrained as described in Chapter 12. This option is or will be available shortly in other refinement programs as well.

Geometry validation and restraint setup for ligands

Another frequent source of frustration is that there are no stereochemical restraints available in libraries for many known ligands and for new ligands. There are currently ~7500 hetero-compounds in the *HIC-Up* hetero-compound database[46] but significantly fewer are included in the standard restraint libraries. The ligands are then occasionally just placed into density and not properly refined[47] or they are poorly restrained and refine into implausible conformations. Figure 13-10 shows a "rogue's gallery" of ligands[48] that obviously were not kept in shape by reasonable stereochemical restraints.

The obvious problem that proper stereochemistry of ligands is often ignored—largely because errors in their conformation do not significantly affect global statistics—has been publicized recently,[47] and in view of the increasing importance of protein–ligand complex structures for structure-guided drug discovery and design, new tools have been developed that allow proper refinement and validation[49] of small-molecule ligands in protein structures.

Ligand restraint files. If you need to model a known ligand, a restraint file identified by the three-letter residue identifier of the ligand can be obtained from the hetero-compound dictionary file from the PDB or the more comprehensive *HIC-Up* (Hetero-compound Information Center, Uppsala) server,[48] which also contains numerous and valuable practical hints and cross-references for how to generate ligand restraint files for various refinement programs. Refinement programs periodically update their libraries according to the last official hetero-atom dictionary file posted on the PDB site and thus recognize the standard names. However, just because one can find the molecule in the ligand database does not mean the ligand geometry and/or restraint file are correct—verify, and if in doubt, ask a chemist (or determine the ligand crystal structure yourself).

If your ligand is new, you need to generate a new restraint file that your refinement program understands. For example, the monomer library sketcher program

SKETCHER provided with CCP4 generates a minimum description of the ligand from the stereochemical drawing you provide in *cif* format. The *LIBCHECK* program[54] (a part of the *REFMAC* distribution also included in the CCP4 suite) completes the restraint file and generates a new monomer library entry that is then

Figure 13-10 Stereochemical violations in ligand structures. (A) Bond distance and planarity violation, the N–C bond is shortened to 0.8 Å and forced planar, 4-piperidinopiperidine (1k4y); (B) planarity violation, the bond should be in the ring plane, 3-phenyl-propylamine (1tnk); (C) conjugated system not planar, the trans-polydiene "tail" of retinol (vitamin A) should be a conjugated, planar π-electron system (1hbp); (D) tetrahedral phosphate became a trigonal pyramid (1mzh); (E) a drug lead molecule with a nonplanar fluoride bond extending from the conjugated π-ring and a planar instead of tetrahedral sulfinyl group (1pme); (F) part of an FAD molecule, where one phosphate of the dinucleotide forms a tetragonal pyramid instead of a tetrahedron (1yy5); and (G), a completely disintegrated coenzyme A molecule with multiple and extreme stereochemistry violations (1q2c). These deviants have been unearthed by G. Kleywegt[48] using his ligand validation program *ValLigURL*.[49]

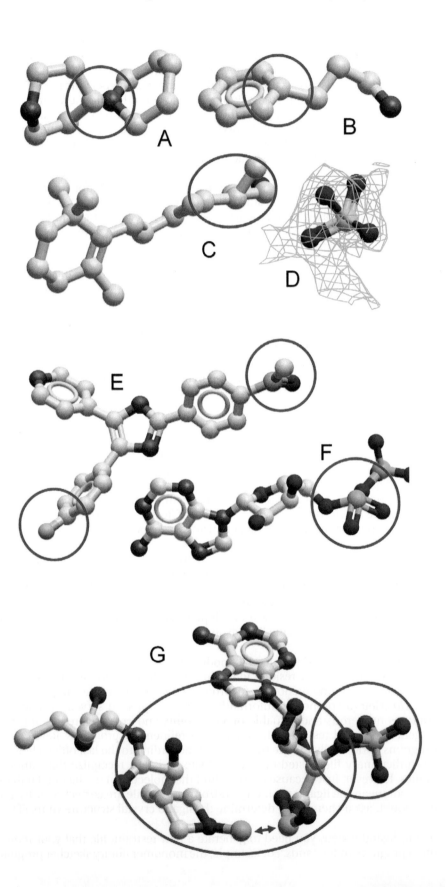

Sidebar 13-4 Crystallographic fragment screening. Proper identification of ligands and their binding characteristics plays an important role in crystallographic fragment screening,[50] which is employed in structure-guided drug discovery.[51] In fragment screening, a target protein is co-crystallized (or soaked) with multiple smaller building blocks (fragments) of drug-like molecules in each batch or drop. About 10 different fragment molecules of commonly no more than 300 Da and of different shape and properties are selected in each pool[52] (drop or batch). The small fragments generally will not bind as strongly as an optimized drug molecule and therefore are added in high concentration in order to achieve high enough occupancy to identify also weak binders with K_d values in the high mM range (see Figure 3-40). The difference in shape of the molecules in each pool simplifies identification of the bound fragment in electron density, and the specific interactions between the ligand and the target protein are analyzed. From various such identified fragments, larger drug-like molecules forming additional contacts can be assembled, resulting in greatly improved binding constants and likely improved selectivity (Figure 13-11). The benefits of fragment screening instead of screening against already drug-like molecules are larger coverage of chemical diversity, early establishment of druggability of the target, and more flexibility in anticipating ADME/T (absorption, distribution, metabolism, excretion, and toxicity) properties.[53] Fragment screening is also frequently conducted using NMR-based approaches.

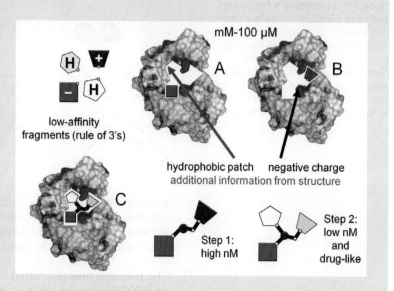

Figure 13-11 Crystallographic fragment screening. Various small, low-affinity molecule fragments with diverse properties (top left, H for hydrophobic, +/– indicating charged groups, and sometimes with functional groups suitable for combinatorial synthesis or *in situ* click chemistry) are soaked or co-crystallized with the target protein. Based on binding interaction analysis of various low-affinity complex structures (A and B) and additional structural analysis of the binding pocket, a new generation of leads is synthesized and the procedure is repeated until a drug-like, high-affinity molecule with no predicted adverse ADME/T properties is derived (C). *In vivo* studies in animals and ultimately clinical phase trials follow.

used in the refinement to restrain the new ligand. *SKETCHER* can be somewhat hard to use for beginners.

An alternative method of generating ligand restraint files is to submit a character-based description of the molecule generated with the web-based JME molecular editor to the *PRODRG* server,[55] an immensely useful tool that provides restraint files in many formats, such as *cif* files for *REFMAC*, topology parameter files for *CNS*, restraint setup records for *SHELXL*, and various other formats used in molecular modeling programs. *PRODRG* uses the *GROMOS87* force fields of the *GROMACS* molecular dynamics package[56] to energy-minimize the ligand geometry, and automatically generates the restraint file. A limitation of *PRODRG* is that only the most common elements (H, C, N, O, P, S, F, Cl, Br, and I) are included in the *GROMOS87* force field, and some such as Se, As, or metal cations are missing. The program is currently updated to a new force field.

Note that the restraint files can only be as good as the stereochemistry you have provided—particular attention needs to be paid to hybridization state, double bonds, planarity, and chirality. In case of doubt, let the friendly organic chemist next door examine your ligand sketch. Take, for example, the protease inhibitor benzamidine (Figure 13-12), where it is important to properly indicate that the aliphatic carbon atom is planar because of the resonance over the $NH_2-C=NH$ group. With the web-based JME sketcher we can readily draw the proper molecule, generate the restraint file, and refine the molecule we fitted

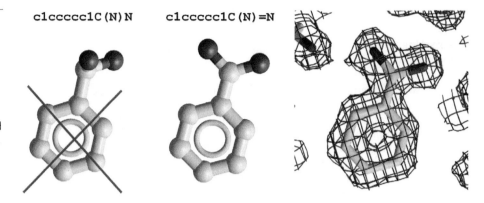

Figure 13-12 Correct molecular geometry is necessary for correct restraint files. Shown is the incorrect (left panel) and correct (center) conformation of the small molecule protease inhibitor benzamidine (benzyldiamine) with corresponding *Smiles* representation above. The right panel shows the fitted and refined ligand in 1.7 Å electron density. PDB entry 2j9n.

into the electron density (note that the torsion between the phenyl ring and the diamine group is different from 180° in the refined molecule). The JME editor also generates *SMILES*, another ASCII-character based description[57] of the molecule that is read as input by some molecular modeling programs.

The restraint files generated by automated setup programs such as *LIBCHECK* or *PRODRG* can be additionally edited and fine-tuned using a text editor. As an example the restraint file of benzamidine (Figure 13-12) is provided in Appendix A.9.

Once a previously known ligand has been refined, it can be compared with the conformations observed in other complex structures. Gerard Kleywegt's *ValLigURL*[49] web service (according to R. Laskowski "the absolutely most contrived acronym ever invented") is a very useful tool to examine whether the ligand has refined into an unusual and potentially suspicious conformation. Also provided are numerous links to additional tools for ligand building, validation, and database searches.

Chemical plausibility of fitted ligands and binding pocket analysis

Once the ligand molecule in a complex structure is properly fitted and refined, independent validation based on the plausibility of ligand binding chemistry and analysis of non-bonded contacts are conducted. It is quite common to display the ligand in its binding pocket, particularly when physicochemical properties of the binding pocket such as charge distribution (Figure 13-13) are mapped onto the pocket surface. Such figures are aesthetically pleasing and illustrate well the conformation and pose of the ligand in the binding site. Because the figures alone do not quantify the non-bonded contacts, they are generally supplemented with a table listing relevant protein–ligand interactions. Model building programs generally provide options to tabulate and visualize the nearest neighbor distances between ligand and protein target. An alternative, quite informative representation is provided by the *LIGPLOT* program,[58] which "flattens" the binding pocket and ligand into a schematic 2-dimensional representation and lists hydrogen bond contact distances directly in the diagram (panel C of Figure 13-13). Remember (Section 3.9) that a single hydrogen bond can change the binding constant of a ligand by nearly

> **Box 13-6 Ligand validation.** Non-covalently bound small molecule ligands, cofactors, or peptide ligands are the most difficult parts of the structure in terms of accurate modeling. Non-covalent binding sites with weak binding constants are out of principle partly occupied which generally leads to weaker electron density and becomes initially apparent in higher *B*-factors. Ligands require accurate restraint files for stable refinement, and chemical plausibility must accompany electron density analysis. Ambiguous density is particularly hazardous and tempting to model with a desired residue.

four orders of magnitude—the importance of proper binding site analysis for the purpose of structure-guided drug design cannot be overemphasized.

Binding pocket analysis. The 3-dimensional arrangement of structural features determines the physicochemical properties of the binding pocket. These properties can be generalized, and functional similarity can be identified, by

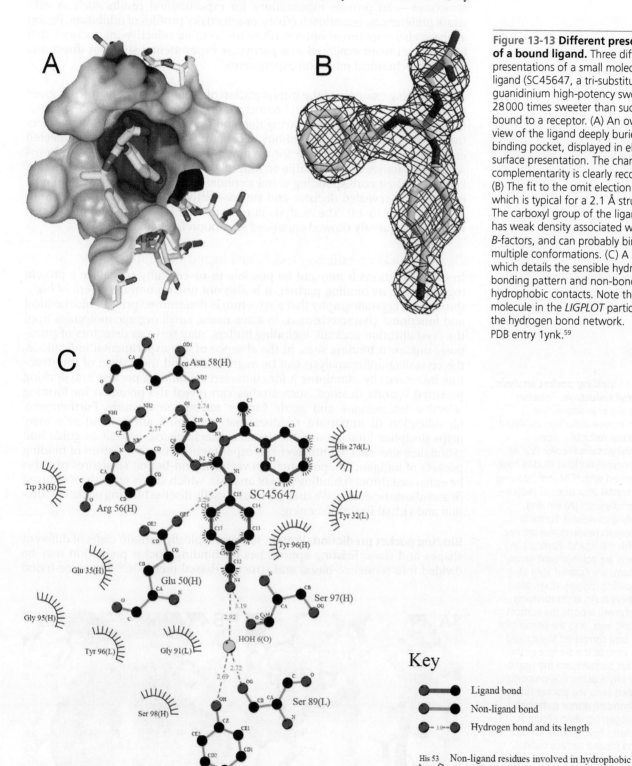

Figure 13-13 Different presentations of a bound ligand. Three different presentations of a small molecule ligand (SC45647, a tri-substituted guanidinium high-potency sweetener, 28 000 times sweeter than sucrose), bound to a receptor. (A) An overall view of the ligand deeply buried in the binding pocket, displayed in electrostatic surface presentation. The charge complementarity is clearly recognizable. (B) The fit to the omit election density, which is typical for a 2.1 Å structure. The carboxyl group of the ligand in (B) has weak density associated with high *B*-factors, and can probably bind in multiple conformations. (C) A *LIGPLOT*, which details the sensible hydrogen bonding pattern and non-bonded hydrophobic contacts. Note the water molecule in the *LIGPLOT* participating in the hydrogen bond network. PDB entry 1ynk.[59]

representing a binding pocket not by amino acids, but by the spatial arrangement of pseudo-functionalities such as hydrogen bond donor or acceptor potential.[60] Many computational programs that analyze binding pockets routinely use functional mappings to characterize the binding pocket. The comparative analysis of ligand binding pocket features and physicochemical properties—provided they are based on properly refined and validated protein structures—can provide explanations for experimental results such as substrate preferences, enantioselectivity, or selectivity profiles of inhibitors. Pocket analysis also may reveal opportunities for forming selective interactions that have not yet been exploited in a particular experimental study, or direct and predict biochemical mutation experiments.

An interesting example of the interpretation of binding pocket features is given in a study of the yeast carbonyl reductase from *Sporobolomyces salmonicolor* (SSCR),[61] prompted by the growing demand for reliable, efficient stereoselective, asymmetric synthesis methods. Biocatalysis with specifically engineered enzymes is an attractive tool for asymmetric synthesis: The SSCR carbonyl reductase catalyzes with versatile enantioselectivity the reduction of a variety of ketones to their corresponding chiral alcohols. Binding pocket analysis of residue contacts revealed decisive and stereoselective contacts with various substrates (Figure 13-14). The analysis in turn allowed rational design of mutants which subsequently showed enhanced enantiopreference.

Binding pocket prediction and virtual ligand screening

In some situations it may not be possible to successfully crystallize a protein together with its binding partner. It is also not uncommon in the era of high-throughput crystallography that a structure is determined prior to biochemical and functional characterization. In some cases, small organic molecules from the crystallization cocktail, including buffers, may serve as detectors of previously unknown binding sites. In the absence of any experimental indications, the crystallographic analysis can be augmented (and the impact of the structure increased) by examining it for characteristic binding pockets and docking potential ligands *in silico*. Such studies can reveal the potential for forming selective interactions and guide further experimental study. Furthermore, identification of important residues that determine affinity and selectivity helps decipher fundamental biochemical mechanisms, as well as guide and rationalize site-directed mutagenesis experiments. The comparison of binding pockets of unliganded apo-structures versus ligand-bound structures can also be enhanced through binding pocket analysis, which allows analysis of induced fit and allosteric effects. We shall therefore briefly discuss binding pocket prediction and virtual ligand screening.

Binding pocket prediction. Protein surfaces typically contain clefts of different shapes and sizes. Existing approaches for binding pocket prediction may be divided into sequence-based and structure-based methods.[63] Sequence-based

Figure 13-14 Binding pocket analysis in a carbonyl reductase. To better understand the enantioselective versatility in ketone reduction catalyzed by the carbonyl reductase from *Sporobolomyces salmonicolor* (SSCR), substrate–enzyme docking studies have been performed with ICM-Pro[62] starting with X-ray crystal structures of protein–cofactor complexes.[61] (A) Binding pocket analysis of residue contacts and areas reveals residues making key contacts with the ligand. Participating carbon atoms are colored light green, nitrogen atoms are colored light blue, and participating oxygen atoms are colored light red. An accompanying table (not shown) reports the contact area, exposed area, and the percentage of contact area compared to exposed area. (B) A view of the surface of the ligand pocket surrounding the ligand reveals key physicochemical properties, color-mapped onto the pocket surface: hydrogen bonding donor potential (blue), hydrogen bonding acceptor potential (red), hydrophobic surface (green), and neutral surface (white). Specific distances between atom types (hydrogen bonds for example) may also be monitored and displayed. Figures courtesy of Katherine Kantardjieff, W.M. Keck Foundation Center for Molecular Structure, CSU Fullerton.

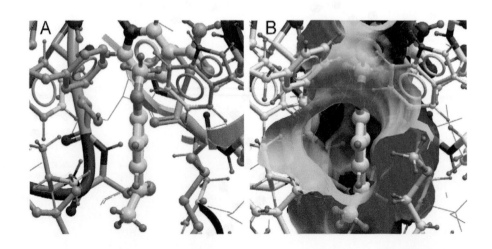

methods rely on the general assumption that the level of sequence similarity in a pair of proteins correlates with the structural similarity of their binding sites. As an example, the kinase families, important drug targets in oncogene signaling pathways, have been clustered by sequence similarity into target family landscapes.[64] Although kinases exhibit a diversity of structures, regulation modes, and substrate specificities, they share common structural features in their catalytic core, controlling the transfer of phosphate to their protein substrates, specifically serine, threonine, or tyrosine residues. Sequence-based pocket prediction methods fail, however, when the binding region itself is nonlocal in sequence. Thus, as a rule, computational programs that predict binding pockets use structure-based approaches to identify ligand-binding sites.

There are two general classes of methods for structure-based determination of binding pockets[65, 66] for drug-like molecules (which typically cover a molecular weight range of 300–700 Da): (i) geometry-based algorithms, which search for clefts in the molecule, and (ii) methods that by complementarity of shape and interactions dock and score actual molecules[67] or molecular fragments[68] (the latter are similar in concept to experimental fragment screening, Sidebar 13-4). The geometry-based algorithms allow a fast search, important if the location of the binding pockets is unknown or a large number of proteins have to be analyzed (as is the case in establishing the pocketome of an organism, the collection of all possible small molecule binding envelopes present in a cell).[69] Specific docking algorithms are important for virtual ligand screening applied in structure-based drug discovery.

Geometry-based methods are commonly enhanced by mapping general pseudo-functionalities such as hydrogen bonding potential, local charge, or hydrophobicity onto the pocket surface as described in the caption of Figure 13-14. A set of interaction surfaces is generated and stored for each atom in the ligand and the target receptor. The shape, size, and location of the interaction surfaces depend on the pseudofunctionalities mapped to stabilize the bound ligand. This sequence- and fold-independent spatial mapping of physicochemical properties instead of individual amino acid representations[60] not only decreases the size of the conformational search space, but also makes it much easier to identify functional similarity when comparing different binding pockets. As a caveat, binding pockets may be deceptively similar at the level of local physicochemical properties, despite the absence of a one-to-one correspondence of atom types. A new and fast computational algorithm, (PocketFinder)[69] employs pseudoproperties as well as Lennard-Jones potential-based shape analysis for accurate identification of the predicted ligand binding envelopes (rather than identifying only the binding pockets on the target protein). Sidebar 13-5 provides an example of pocket identification and analysis.

Ligand docking and virtual ligand screening. It is often the case that we know the molecular structure of a protein target and a ligand or putative drug, but not the structure of their complex. Ligand–target docking is the process by which the two molecules fit together in complementary fashion in 3-dimensional space through a series of torsional, rotational, and translational searches of the conformational space. Docking methods thus provide structural information that may be subsequently applied to the design of novel ligands that fit a particular pocket on a given target protein. Docking and screening methods may be divided into two types: (i) ligand docking/scoring protocols, which aim to predict for a given ligand molecule complementary of shape and interactions; and (ii) *de novo* design protocols, which incrementally combine fragments having predicted interactions with the binding pocket.

Most virtual ligand screening is receptor-based, beginning with a validated 3-dimensional structure of the protein target, preferably obtained crystallographically or by homology modeling (based on an experimental structure of similar sequence as template), and a library of small-molecule structures. Inherent methodological challenges lie in the speed and reliability with which ligands may be docked. Treating molecules as conformationally rigid bodies is computationally cheap, but useful only if the resulting conformational changes

Sidebar 13-5 Pocket identification and analysis: Cholinesterases. Cholinesterase inhibitors[70] have been used not only in chemical weapons (Sidebar 2-5), but also as therapeutics intended to slow the progression of Alzheimer's disease.[71] The severity of the disease parallels levels of the neurotransmitter acetylcholine in the brain under the control of two enzymes that hydrolyze acetylcholine, acetylcholinesterase (AChE) and butyrylcholinesterase (BChE). The use of cholinesterase inhibitors is predicated upon the assumption that inhibiting the activities of both enzymes will increase the concentration of acetylcholine in the brain.[72] Thus, widely prescribed inhibitors have been reversible inhibitors of both enzymes such as Rivastigmine®[71] However, because the activity of BChE increases with the progression of Alzheimer's disease while that of AChE decreases, it has been suggested that next generation cholinesterase inhibitor based drugs should include BChE selective inhibitors.[73]

The human cholinesterases share 51% sequence identity. Their 3-dimensional structures have been determined and their active sites have been characterized in the absence and presence of some inhibitors.[74, 75] The binding pockets in the two enzymes have subtly different physicochemical properties, which may be exploited in the design of selective BChE inhibitors.[76] In particular, the active site "gorge" in BChE is larger, with a higher degree of plasticity than that in AChE. These differences can be easily visualized by pocket-finding (Figure 13-15) and subsequent surface interaction mapping.

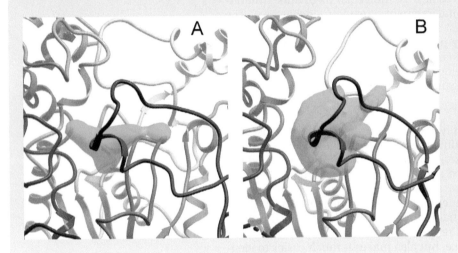

Figure 13-15 Cholinesterase active site binding pockets identified by ICM PocketFinder. (A) Active site binding pocket at the base of the active site gorge in human acetylcholinesterase; the pocket, shown in green, is shallow and narrow. (B) Active site binding pocket at the base of the active site gorge in human butyrylcholinesterase; the pocket, shown in green, is deeper and wider, which facilitates accommodation of bulkier functional groups. Figures courtesy of Katherine Kantardjieff, W.M. Keck Foundation Center for Molecular Structure, CSU Fullerton, rendered with ICM-Pro/PocketFinder.[69]

upon ligand binding to the target are small. Allowing molecular flexibility during the binding event increases the complexity of the conformational search. For example, if we attempt to dock a compound with 10 rotatable bonds, allowing a conformational search of three minima about each bond, this results in 3^{10} or 59 069 possible conformations to dock. If we extend this to six minima, we must dock $3.48 \cdot 10^9$ possible conformations. Accurate prediction of relative binding affinities must take into account electrostatic, electrodynamic, steric, and solvent-related forces. Ranking the resulting docked conformations may be based on a variety of scoring functions, such as force-field methods, knowledge-based potential of mean force, and interaction profiles.

Enrichment of libraries. Numerous docking programs are available, based on very different physicochemical approximations, which combine a docking engine with a scoring algorithm. Benchmarks for performance are based on their ability to accurately reproduce the conformations of small molecular weight ligands from experimental X-ray structures, propensity to predict binding free energies from the best-scored conformations, and capacity to discriminate known binders from randomly chosen molecules. As a general rule, virtual docking experiments do not provide a single answer, but a list *enriched* with promising candidates for subsequent validation verification by conventional assay-based or structure-based (NMR or X-ray crystallographic) high-throughput screening.

General caveats. The notion that the level of sequence similarity correlates with structural similarity of their binding sites is an oversimplification. The physicochemical properties imparted by binding site residues onto the cavity surface, as well as the 3-dimensional arrangement of interaction sites are neither entirely unambiguous, nor unchanging. For example, local variations in structure can have more distant effects, causing subtle shifts in interaction sites. Furthermore, a practical limitation for using only one crystal structure to examine protein–ligand interactions is *protein flexibility*. For these reasons, computational studies of ligand binding prefer to use bound complex structures for their starting coordinates whenever possible rather than the unbound target protein. The available conformational space is not completely represented without several crystal structures of the target proteins bound to different types of ligands.

13.5   Low resolution structures

Not without reason was the 2008 CCP4 study weekend[77] dedicated to low-resolution structure determination and validation. Given that low resolution structures ($d_{min} > 3.0$–3.5 Å) with their poor data-to-parameter ratio depend heavily on geometric restraints and that local density analysis is also impaired because the local side chain density is degenerate, low resolution structures pose their own challenges for validation. The global descriptors become even more insensitive than they already are: The structure is essentially held together by strong restraints, and deviations from geometry targets will be suppressed. Kleywegt and Jones[78] have shown that a willfully reverse-traced 3.0 Å model of cellular retinoic acid binding protein (CRABP) shows an entirely unsuspicious R-value and bond distance and normal bond angle r.m.s. deviations from restraint targets. However, R-free and the φ–ψ torsion angles are unacceptably poor for the reverse-traced model (Figure 13-16).

The reverse-tracing exercise drives the very important point across that the R-free value and the φ–ψ torsion angles are the most sensitive indicators of gross model errors. If either—or even worse, both—show suspicious behavior, it can be assured that something is severely wrong with the model. Insensitivity of the working R-value and geometry target deviations to errors also explain why in pre-R-free times it was much harder than today to detect partial reverse tracing or wrong connection of secondary structure elements in low resolution structures, and most of these errors appeared in pre-R-free times. Incidences such as the reverse tracing[79] of the 2.8 Å RuBisCO (ribulose-1,5-biphosphate carboxylase, a CO_2-fixing plant protein) small subunit model and its detection[80] thus prompted the involved crystallographers to develop complementary validation tools that are independent of the—at this resolution limited and highly restrained—geometric model descriptors. As an example, assume a packed helix that is traced with the residues shifted by two residues: hydrophobic residues that were packed inside are now exposed on the outside of the helix—the local packing score in regions with multiple hydrophobic residues sticking outside will then be rather unusual, and a number of additional empirical parameters describing the local environment preferences of side chains will also be unfavorable. Similarly, non-bonded contacts in an incorrectly traced structure will also be unusual and raise a warning flag.

Property profiling. The 3-dimensional profiling methods[81] (*verify3D*) and analysis of non-bonded interaction patterns[82] (*ERRAT*) are examples of such independent validation tools that examine the local environment of the protein chains. The reverse-traced RuBisCO subunit or the incorrect initial HIV protease model, for example, score poorly in both validation programs. Such complementary validation techniques are of course generally useful for the analysis of protein structure models—crystallographic, purely computational, and NMR models alike. The program *PROSA-II* was developed for the validation of theoretical models based on empirical threading potentials. It is also useful for low resolution crystallographic model evaluation.[83] An example of the improvement of low resolution structure with incorrectly connected secondary structure elements simply by collecting better data is illustrated in Figure 13-17.

Cα-atom-only backbone validation. A special case of low resolution models are Cα-only models, which sometimes can be found in the PDB (for example

Figure 13-16 Correct and reverse-traced CRABP model. To demonstrate the insensitivity of global descriptors to gross errors, the 3.0 Å model of cellular retinoic acid binding protein (CRABP, A) was purposefully traced backward (B) by Kleywegt and Jones.[78] The backbone traces are colored from N-terminal (blue) to C-terminal (red). The conventional *R*-value, bond distance and bond angle r.m.s. deviation from restraint targets, the overall *B*-factor, and the coordinate precision estimated from Luzzati plots are all unsuspicious and within expected limits for a 3.0 Å model. In contrast, the Ramachandran (φ–ψ torsion angle) plot and the linear cross-validation residual *R*-free (which is random (0.59) for the wrong model, and 0.32 for the correct model) clearly indicate problems. The output from the geometry validation program *MolProbity*[21] listed below the global descriptors also leaves no doubt that there are serious problems with the reverse-traced model (the lower the clash score and the higher its percentile range for comparable structures, the better).

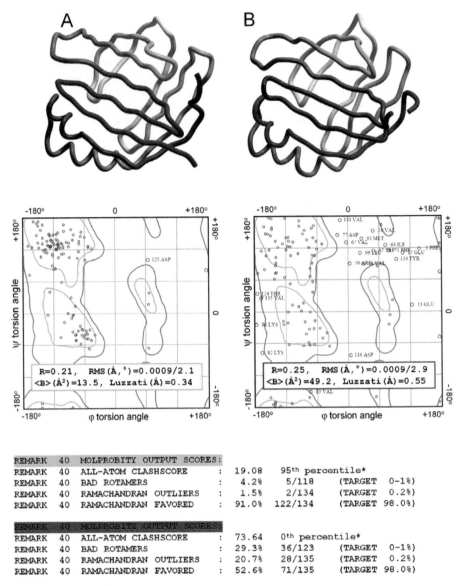

```
REMARK   40   MOLPROBITY OUTPUT SCORES:
REMARK   40   ALL-ATOM CLASHSCORE       :   19.08    95th percentile*
REMARK   40   BAD ROTAMERS              :    4.2%     5/118     (TARGET  0-1%)
REMARK   40   RAMACHANDRAN OUTLIERS     :    1.5%     2/134     (TARGET  0.2%)
REMARK   40   RAMACHANDRAN FAVORED      :   91.0%   122/134     (TARGET 98.0%)

REMARK   40   MOLPROBITY OUTPUT SCORES:
REMARK   40   ALL-ATOM CLASHSCORE       :   73.64     0th percentile*
REMARK   40   BAD ROTAMERS              :   29.3%    36/123     (TARGET  0-1%)
REMARK   40   RAMACHANDRAN OUTLIERS     :   20.7%    28/135     (TARGET  0.2%)
REMARK   40   RAMACHANDRAN FAVORED      :   52.6%    71/135     (TARGET 98.0%)
```

Clashscore is the number of serious steric overlaps (> 0.4 Å) per 1000 atoms.
* 100th percentile is the best among structures of comparable resolution; 0th percentile is the worst.

1pte shown in Figure 13-17, or the superseded PDB entry 1phy).[85] In the case of Cα-only models, in addition to obvious differences between *cis*- and *trans*-conformations (2.9 versus 3.8 Å Cα–Cα distance), pseudo-angles and torsions between Cα atoms are used for backbone validation similar to the familiar Ramachandran plot analysis.[86] Cα-only validation can also be useful in the very early stages of model building.[87]

Errors in trace direction, connection of side chains, and register shifts (Figure 12-26) are usually the result of the absence of distinct side chain density, which makes it difficult to clearly assign the sequence. Poor side chain density affects mostly low resolution structures. As noted before, heavy atom and dispersive atom positions—particularly engineered Se-Met, whose position in the sequence is known—are helpful to lock the sequence into place in the backbone trace. With a sufficiently critical mind and given the presently available validation tools that are already used during model building it is unlikely that a grossly wrong low resolution structure will pass muster—unless multiple warning signs are ignored[88]

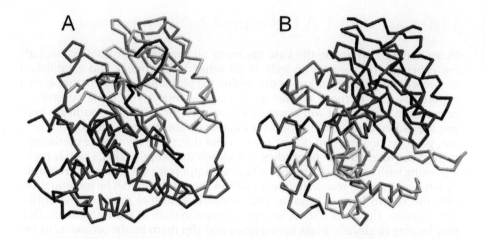

Figure 13-17 Incorrectly connected secondary structure elements. Comparison of the superseded initial Cα-backbone model (A, 2.8 Å) with the correct high resolution structure (B, 1.6 Å) shows that while helical secondary structure elements were partly correct, the connection as well as direction of the trace were incorrect in the low resolution structure. PDB entries 1pte and 3pte.[84]

> **Box 13-7 Low resolution structure validation.** Low resolution structures suffer from limited experimental evidence and require particularly careful validation against independent prior knowledge. Inside-outside distributions, 3-dimensional profiling, non-bonded pattern analysis, and empirical potentials can point out stretches of incorrectly traced or register-shifted residues.

(Sidebar 12-8). An instructive example of what potentially can go wrong is found in the Methods section of the publication describing the re-determination of the photoactive yellow protein[89] (PDB entry 2phy). A number of key points to consider in the case of low-resolution structures are summarized in Sidebar 13-6.

Sidebar 13-6 Building, refining, validating, and publishing low resolution structures. Low resolution structures (limiting resolution > ~3.5 Å) pose their own challenges because of the severely limited data-to-parameter ratio (Chapter 12). Even torsion-angle-only refinement reaches its determinacy limit, depending on solvent content, around 5 Å. Much useful guidance for low-resolution structure determination can be found in the 2008 CCP4 study weekend proceedings,[77] and a few key points as emphasized in a recent review[90] are summarized here.

Proper solvent correction. We already discussed in Chapter 12 the proper use of anisotropic scaling and sharpening procedures for low resolution data. Ultimately, an optimal anisotropic solvent correction model can be obtained by minimizing the *SRS*

$$SRS = \sum_{\mathbf{h}} \left(F_{obs} - F_{calc}(k, k_{sol}, B_{sol}, \mathbf{U}) \right)^2 \qquad (13\text{-}6)$$

with k the Wilson-type scale factor, k_{sol} and B_{sol} the bulk solvent correction parameters as discussed in Chapter 12, and $\mathbf{U}$ the overall displacement anisotropy tensor.

Sharpening of maps. As discussed in several places, the inversion of *B*-factor attenuation, as for example in normalized structure factors $\mathbf{E}$, is a general means of enhancing low resolution reflections and sharpening maps. Resolution dependent sharpening through a negative *B*-factor correction for the map coefficients has been applied successfully. Its primary effect is to enhance the "high resolution" part of the low-resolution data, thus enhancing the detail—that is, the side chain features—in a low resolution structure.

$$F^{\#} = F_{map} \cdot \exp\left(-B^{\#} (\sin\theta / \lambda)^2 \right) \qquad (13\text{-}7)$$

The sharpened map coefficients and (negative) *B*-correction are indicated by the sharp (#) sign. Because the conventional Wilson scaling becomes unreliable for low resolution structures, $B^{\#}$ is preferably determined via maximum likelihood procedures.[91] The invariable caveats are that the low resolution data have to be measured and treated accurately and that appropriate means of combating phase bias have been applied.

Use of high resolution domains or fragments whenever possible. The refinement of low resolution (complex) structures not surprisingly gives significantly better results when available high resolution starting models are placed into the low resolution electron density (see also Chapter 11 for real space molecular replacement methods).

Make your point carefully but assertively. Finally, some high impact journals recently experienced embarrassing retractions of low (and not so low) resolution structures (see Sidebar 12-8) and seem to have become somewhat hesitant to accept low resolution structures (see also Jim Naismith's comments at the end of this chapter). Nonetheless, a carefully phased and refined low-resolution structure can be extraordinarily informative within its limits. Relative location of domains, domain movements between apo- and complex forms, allosteric conformation changes upon ligand binding, and similar observations compatible with the resolution limit all can be valuable and highly publishable information (for an example see Figure 13-23)

13.6 Case study I: A highly improbable proposition

In Section 13.1 we made the case for using all available information, including experimental support as well as all other independent and established physicochemical information, to determine the probability of a structure-based hypothesis. The purpose of this section is that analysis of a highly unlikely proposition will prompt you to question your own models and hypotheses until you are either convinced beyond reasonable doubt that you are correct, or if not, you can clearly point out where you feel that there are more experiments required or where follow-up studies may clarify the propositions. There is nothing wrong with seeking or proposing further support for a hypothesis in a publication—but there is reason for great caution should the model be incompatible with the known laws of physics. The former is a normal process of scientific investigation, the latter almost always a demonstration of poor judgment. One may be able to cleverly mask deficiencies and slip them by the reviewers in the first round, but the inevitable scrutiny by other researchers in the field will eventually catch up with tricksters.

An interesting publication[39] describes the binding of a 41-residue target peptide to a protease, the light chain of *Clostridium botulinum* neurotoxin B (BotLC/B). Because of their catalytic function botulinum toxins are the most toxic substances known and thus of great relevance and good for a prime time story. After ingestion, the toxins enter the blood stream and are transported to the neuromuscular junction where they enter the presynaptic neuron and disrupt vesicle–membrane fusion by blocking the acetylcholine exocytosis pathway. Specifically, BotLC/B catalytically cleaves the synaptobrevin II (Sb2) peptide of the SNARE (soluble *N*-ethylmaleimide-sensitive-factor attachment receptor) complex[92] mediating vesicle exocytosis in motor neurons thus leading to the flaccid paralysis effects in botulinum poisoning.

Examination of prior probability. The experimental section of the publication reveals the following ligand soaking procedure. A 41-residue peptide (Sb2 residues 38–88) was soaked into the crystals as follows: "Crystals of the Sb2–BoNT/B-LC complex were obtained by soaking pre-formed BoNT/B-LC crystals in a solution containing 15% (v/v) ratio of 2,4-methylpentanediol to mother liquor and a 2.5 molar excess of the Sb2 peptide fragment. The apo crystals were soaked for approximately 5 s, followed by immediate freezing in liquid nitrogen."

This is of course a most interesting procedure because diffusion is a slow process, and we have learned in Chapter 3 (Figure 3-38) that even small molecule ligands take a long time—often hours—to diffuse through the solvent channels into preformed crystals. One can actually calculate the diffusion speed of such a peptide in random conformation from its radius of gyration and the second of Fick's laws. The diffusion distance of such a 41-residue, random conformer peptide in pure water within 5 s is less than the reported dimensions of the crystal, not to mention threading this peptide through solvent channels. We give the presence of this peptide throughout the crystal (a small portion of molecules located at the outside of the crystal might still be occupied by some peptide) therefore only a small probability based on simple physical chemistry considerations. The prior probability P_{phys}(Sb2) is a very small number, say about 5% or 0.05.

The next peculiar observation results from the analysis of the Ramachandran plot of the Sb2 peptide from the deposited coordinates (Figure 13-18). It seems that this random peptide also has near random backbone torsion angles, again a highly improbable proposition. As the Ramachandran plot is in essence a projection of an energy surface, this peptide must be sitting in the binding site in an exceptionally high internal conformational energy (like a wound-up spring). This scenario would require extremely strong non-bonded contacts to the protease and represents a highly improbable and so far not observed situation. P_{chem}(Sb2) is therefore very low as well, maybe 0.01. Remember: Extraordinary claims require extraordinary proof.

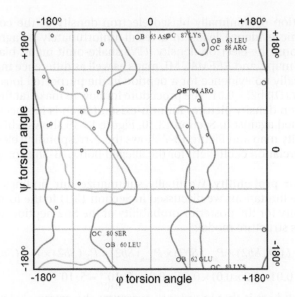

Figure 13-18 **Backbone torsion angle distribution for an improbable peptide.** *MolProbity* Ramachandran plot of a purported random comformation peptide with only 28% of residues in the favored region, 38% in additional allowed regions, and 34% total outliers. The conformational energy of this peptide is extraordinarily high, and it is highly improbable that it can exist in this conformation. In addition (and as a consequence of its improbable geometry), the *MolProbity* all-atom clash score,[21] a measure for serious steric overlaps, puts the peptide into the 0th percentile (worst) of all reference structure model entries. PDB entry 1f83.[94]

Equally condemning is the clash score for the peptide. According to *MolProbity*, it puts the Sb2 model into the 0th percentile of clash scores of comparable structures. We are generous and give P_{conf}(Sb2) a probability of 0.01.

The biochemists among us may also find that the direction of the peptide in the binding site is opposite to the well-established canonical binding direction of a peptide in Zn-proteases. While it has been shown that at low pH and high concentration at least the BotLC/A protease can be forced to digest itself in non-canonical binding mode,[93] non-canonical binding again is a strong claim that would require strong support. Perhaps P_{bio}(Sb2) might have a probability of 0.2.

In search of the extraordinary proof now required to convince us we first resort to the coordinate entry and observe that the mean *B*-factor for the peptide is ~130 Å², approximately four times the average *B*-factor for protease atoms. We can use these *B*-factors and calculate the relative scattering contribution and convince ourselves that there will not be much electron density visible in the map. This again is not conducive to proving the strong claim of the high conformational energy of the Sb2 peptide. We give the term P_{scatt}(Sb2) a correspondingly low value, 0.05.

Examination of primary evidence. So far our analysis has been solely based on prior knowledge determining the prior probability of the model. Although this does not bode well for the model and strong proof might require us to change a few fundamental laws of physics, fairness requires that we must now look at the experimental support, that is, the data likelihood function supporting the model. We fully recognize that nearly all scientific revolutions result from initially unexplainable and often contradictory experiments, that finally—when sufficiently supported by evidence—force us to change our theoretical beliefs and refine the underlying laws of nature.[1]

This is the point where the availability of experimental structure factor amplitudes becomes crucial: if we did not have the structure factor file available, we could never support or disprove the claimed proposition based on evidence. It is a fundamental principle of science that others should be able to come to the same conclusions given the data and proper procedures. In that strict sense, a structure model without structure factors cannot be falsified, and is, in the words of Karl Popper, on a par with quackery and pseudo-science.[6] It is amazing that it took until February 2008 to impose this fundamental requirement of making primary data (at least in the form of processed structure factor amplitudes) publicly available.

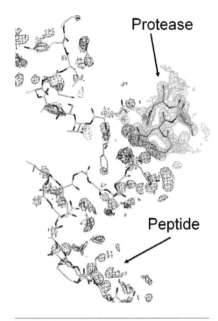

Protease

Peptide

Figure 13-19 Bias minimized electron density for an improbable peptide. The bias minimized electron density has been generated from deposited coordinates and structure factors by the *Shake&wARP* procedure described in Chapter 12 (electron density reconstructed by other omit- or maximum likelihood density reconstruction methods looks the same).[16, 95] The contour level (1.2σ) is the same for the peptide (blue) and the protease (green). The electron density is limited using a blob feature to within 4 Å of the models. Only noise density is visible around the peptide, while the protease density is clear. Compare the corresponding electron density real space correlation coefficient in Figure 13-7.

The examination of minimally biased electron density as the cornerstone of crystallographic evidence leads to a result unfortunately in agreement with first impressions: the electron density (*CNS* shake-omit maps, bias minimized *Shake&wARP* maps, and *REFMAC* ML maps as well as difference maps all agree) shows practically no evidence for a peptide in the purported locations (Figure 13-19). Thus, $L(D|\text{Sb2}) \sim 0.01$. The procedure for how to make (at first sight) convincing electron density pictures for non-existent ligands has been published[16] and is cautioned against in Section 13.10. Figure 13-19 shows a bias minimized electron density map and Figure 13-7 shows the corresponding electron density real space correlation coefficient for the alleged BotLC/B-Sb2 complex.

Joint posterior probability. We finally summarize our findings qualitatively (subject to the limitations we discussed in Section 13.1) in the form of a simple joint probability for the posterior probability of the Sb2 peptide actually being present in this structure model:

$$PP(\text{Sb2}|D) \approx L(D|\text{Sb2}) \times P_{phys}(\text{Sb2}) \times P_{chem}(\text{Sb2}) \times P_{bio}(\text{Sb2}) \times P_{conf}(\text{Sb2}) \times P_{scatt}(\text{Sb2})$$

$$\approx 0.01 \times 0.05 \times 0.01 \times 0.2 \times 0.01 \times 0.05 = 0.01 \times 5 \cdot 10^{-8} = 5 \cdot 10^{-10}$$

The clear result here is that in order to overcome the extremely low probability of the Sb2 peptide bound to the protease given the body of prior knowledge, we would have to provide at least an inversely huge supportive term to even put this hypothesis up for discussion. This is certainly not the case.

In summary, you are encouraged to judge your model and hypotheses derived from the model with utmost scrutiny against all available independent prior information. It does not really matter what value you assign to your respective evidence and probabilities, just try to be honest about your guess. This procedure will inevitably prevent you from embarrassment but will also provide you with the confidence to boldly propose unconventional new insights as soon as your evidence supports it (again, *all* supporting evidence is fair game too, not just the crystallographic electron density). After all, this is the stuff scientific revolutions are made of.

The structure of the BotLC/A serotype protease bound to the SNAP 25 peptide has finally been determined by M. Breidenbach and A. T. Brunger[95] who also confirmed the absence[16] of the Sb2 peptide in the BotLC/B complex based on the deposited data. Breidenbach and Brunger eliminated the problem of soaking a huge peptide into a crystal by co-crystallizing the target peptide with an inactive mutant of the protease. Another possibility is to link a non-hydrolyzable target peptide to the protease with a flexible linker and crystallize the "self-complex."

The original model (PDB entry 1f83) discussed above has recently been superseded by a new model with partially occupied ($n = 0.34$) Sb2 peptide (3g94) based on the same experimental data (and with phases of unspecified origin). As an exercise, you may compute your own biased-minimized electron density maps by whatever method you see fit and perform the same analysis on the new entry. Judge for yourself whether the new model (which in the meantime also has been retracted) had any merits.

13.7   Case study II: Perplexing features abound

A case study that drives across the point of checking all your tables for consistency is provided in the following. A report of a complex and flexible multidomain molecule that is proposed to undergo large conformational changes according to a supplemental movie has the supplemental data collection and refinement statistics table shown as Table 13-3.[96]

What strikes one immediately in Table 13-3 is the exceptionally low R_{merge} given the low signal-to-noise ratio $\langle I/\sigma(I)\rangle$ in the last resolution shell. For an $\langle I/\sigma(I)\rangle$

Data collection	
Space group	C2
Cell dimensions	
a, b, c (Å)	151.20, 142.70, 203.70
β (°)	98.9
Resolution (Å)	45.3–2.3
R_{merge}	0.07 (0.11)
Mean $I/\sigma I$	5.36 (1.32)
Completeness (%)	97.3
Refinement	
No. reflections	194 135
R_{work}/R_{free}	18.0/19.4

Table 13-3 Partial "Table 1" with interesting anomalies. Take a careful look at the data collection statistics for the last refinement shell (data in parentheses), and compare resolution, B-factor distribution (Figure 13-20), and the R-values for the refinement of a reportedly flexible molecule with disordered domains.

of 1.32, we would expect a corresponding R-merge in this shell of 0.8/1.32 = 0.61, and instead we find an exceptionally low R-merge for the last resolution shell of 0.11. This is highly improbable and deserves some explanation. Similarly, for a 2.3 Å structure of purported high flexibility, an R-free – R gap of 1.4% is highly remarkable—again not impossible, but quite improbable and deserving attention (see Figure 12-24). Finally, if we plot a B-factor distribution for both the main chain and side chain atoms, we find an exceptionally narrow distribution (Figure 13-20). This is highly improbable, particularly so in view of the claimed molecular rearrangement, missing residues, and the flexibility and partly disordered domains in the structure.

Further examination has revealed[14] that the structure factors reflect no bulk solvent attenuation (discussed in Sidebar 11-8) and that packing contacts exist in only two dimensions. As a result of all these irregularities, the origin of the structure factors has been called into question.[14] We are again faced with a highly improbable situation that would require strong experimental support, ultimately in the form of the original raw data, which have not been provided.

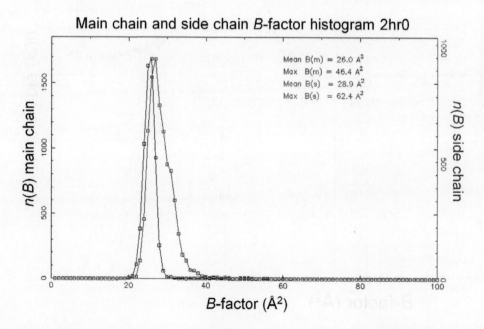

Main chain and side chain B-factor histogram 2hr0

Mean B(m) = 26.0 A²
Max B(m) = 46.4 A²
Mean B(s) = 28.9 A²
Max B(s) = 62.4 A²

Figure 13-20 Unusually narrow B-factor distribution. These B-factor distributions for main chain (black graph) and side chain (blue graph) atoms are extremely narrow and highly improbable for a 2.3 Å structure of a flexible molecule with missing side chains and entire missing domains. Compare this super-sharp distribution to a normal B-factor distribution in Figure 13-21. PDB entry 2hr0.[96]

Figure 13-21 *B*-factor distributions.
The top left panel shows a normal *B*-factor distribution for a relatively sturdy, compact globular molecule, 1.8 Å resolution. The top right panel reveals an arbitrary high *B*-factor cutoff set a 92 Å^2 in a 2.6 Å refinement. The reasons are unknown, but one might wonder whether many *B*-values over 100 Å^2 were considered a problem and an arbitrary cutoff was applied in the mistaken belief of correcting the problem. The bottom left panel shows a "really cool" 2.7 Å structure with many *B*-values at a lower cutoff value of 2.0 Å^2. The reasons are again unknown, and may indicate difficulty in overall scaling and/or *B*-factor restraint weighting. Uncertainty in scaling is also likely the problem for the most peculiar high-*B*-factor-only distribution shown for a "hot" low-resolution 3.9 Å structure in the bottom right panel. In all these cases, the resulting refinement is likely suboptimal. As a general rule, low resolution data can be hard to scale because of the uncertainties caused by pronounced deviations from linearity in the low-resolution part of the Wilson plot. Inspection of the Wilson plot is a highly recommended precaution and manual adjustments of the extrapolation range as well as grouped instead of individual *B*-factor refinement may be necessary (recall the data-to-parameter ratio discussion in Section 12.2).

Consistency of data tables. The cautionary tale presented here should serve as an encouragement to carefully check your data tables and make sure they make sense and are consistent both with model features (and derived hypothesis) and with prior knowledge. In the vast majority of cases you will by simple consistency checks prevent nuisance errors such as using refinement tables from intermediate stages, or depositing data collection statistics and structure factor files that do not belong to the final model. The *B*-factor distribution plots are also a quick means to assure that neither scaling nor *B*-factor restraints have produced any irregularities that defy basic physical chemistry (more in the next section).

13.8   Case study III: *B*-factor distribution plots

A very useful tool that prevents nuisance errors such as improperly set *B*-factor cutoffs and that also reveals major scaling or data problems is *B*-factor distribution plots. We already know from Chapter 12 that proper weighting of the restraint residuals is by no means trivial, and particularly the shape of the *B*-factor distribution is highly structure specific: some molecules are highly flexible, others tend to be more rigid. In any case the side chains show a higher variance, commonly manifest as a tail containing the surface-exposed residues in the distribution. As refinement programs depend largely on the less-restrained *B*-factor as a means to correct other problems that are not parameterized in the model, individual *B*-factors also absorb scaling anomalies, uncorrected anisotropy, and similar problems. Whenever the *B*-factor distributions deviate significantly from a physically reasonable distribution with a positive skew (tailing toward high *B*-factors, top left panel of Figure 13-21) it might be good idea to investigate further. In addition, abnormal cutoffs of *B*-factors that only serve cosmetic purposes or become necessary to keep the refinement from diverging become immediately obvious. Recall that there is nothing wrong with a few high *B*-factors as long as electron density, correlation plots, or difference density supports the existence of these atoms.

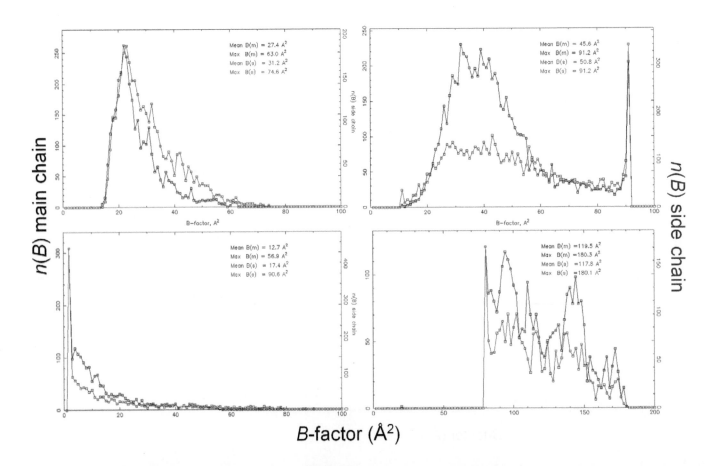

B-factor (Å^2)

13.9 Depositing the finished structure

Once you are satisfied with your structure and have pre-validated it during the polishing phase using the programs described above—*MolProbity* perhaps being the most comprehensive at this time, supported by *SFCHECK*[41]—you are ready to submit the coordinates and structure factors to the PDB. Deposition is possible via *ADIT* on the Rutgers site, and via *AutoDep* through the EBI in Hinxton, UK. The *ADIT* program has a useful pre-check feature that allows you to validate your structure using the PDB validation tools before you actually submit the coordinates and structure factors. This is a very useful step to resolve any final outstanding issues you may have overlooked, despite the ongoing validation you have diligently performed during refinement and model building.

In the near future, the PDB data deposition process[97] is expected to become more streamlined and consolidated with more validation tools recommended by a "Validation Task Force" team. Regardless of which deposition program you personally prefer, there are a number of items that you must have ready for the deposition process.

Deposition checklist

- **The coordinate file.** Common bookkeeping mistakes are non-compliance with residue nomenclature, missing TER records, or improperly adjusted element symbols. These are generally caused by peculiarities in model building or refinement programs. *PROCHECK* generates a new coordinate file that complies with proper residue nomenclature and the coordinate format pre-check feature during PDB submission also checks for format errors.

- **The sequence file.** The ATOM records will only contain the model atoms that could be modeled and missing residues, particularly in loops and N- and C-termini, are common. The SEQRES records, in contrast, should contain the sequence of the molecule *as crystallized*. In general, this will be the expressed construct sequence, subject to any known posttranslational modifications that may have taken place (inadvertent or planned digests, etc.).

- **The structure factor file.** The PDB accepts an increasing variety of structure factor file formats; most common are *mmCIF*, CCP4 *mtz*, *XPLOR/CNS*, and user generated *ASCII* (plain text) formats. Common mistakes are picking the wrong data file, importing/exporting wrong data columns (I instead of F, some intermediate kind of Fs instead of the actual experimental structure factor amplitudes $F(obs)$ refined against, etc.). Such errors can be prevented by good bookkeeping and use of *harvest files*, and are frequently detected by aberrant R-values during *SFCHECK*.

- **Log files or harvest files of scaling and merging.** This is perhaps the most cumbersome part of the PDB submission. There is usually a delay between data processing and the final refinement and data deposition. Now it really pays off if you heeded the reminders about good bookkeeping and project management and can indeed find the proper *log file* or *harvest file* generated by *Scalepack* or the CCP4 programs for example. At least, if you have already diligently prepared "Table 1 – Data collection statistics" when processing the data, you will find all the experimental data items needed for the PDB submission. A small application program *pdb_extract* provided by the PDB can be used to separately retrieve relevant information from coordinate file headers written by the refinement programs.

Deposition pre-check

The deposition pre-check provides a format check and a useful summary table of the most offensive geometry outliers according to the generous minimum standards of the PDB. While the summary report does not provide an excuse to ignore any full validation report, the deposition pre-check is a very useful tool helping to focus on the most serious issues and preventing you from getting lost in (and discouraged by) pages of full validation reports. To show a few of the more common errors, we run a pre-check of an intermediate stage of the tesB refinement from the Se-MAD data exercised in Chapters 10 and 12.

Coordinate format pre-check. Submitting the coordinate and structure factor files and selecting the nomenclature test shows a typical and easily fixed complaint about missing TER and END records. They usually originate from the model building program not writing these records:

```
ERROR: No 'TER' records after each polymer. A 'TER' card must be present after each polymer.
ERROR: File has no 'END' tokens. The 'END' record must be at the end of the file
```

Validation summary report. The validation report summary provides a number of useful checks; some of the more common ones we examine below:

```
CLOSE CONTACTS
    --------------

==> Close contacts in same asymmetric unit.  Distances smaller than 2.2
    Angstroms are considered as close contacts.

    Chain Atom    Res  Seq  Chain Atom    Res  Seq  Symm_Code        Distance
    ------------------------------------------------------------------------
      B  CZ    ARG 156 -     Z  O    HOH  34 ( 1,  5,  5,  5) Dist = 2.14
```

The table above informs us about an aberrant water molecule close to an Arg residue in the same asymmetric unit. It might have drifted during refinement, and inspection of the entire region around the poorly defined side chain is in order. Be aware that in the case of correlated partial occupancies (Figure 13-3) the validation report may misinterpret the geometry.

```
==> Close contacts based on crystal symmetry.  Distances smaller than 2.2
    Angstroms are considered as close contacts.

    Chain Atom    Res  Seq  Chain Atom    Res  Seq  Symm_Code        Distance
    ------------------------------------------------------------------------
      B  CD2   LEU 143 -     B  CD2  LEU 143 ( 3,  5,  5,  5) Dist = 1.83
```

The close contact above results from a collision of poorly defined residues in a loop region. We may have forgotten to display neighboring, symmetry related molecules or electron density when building the initial model.

```
The following table contains a list of the covalent bond angles
greater than 6 times standard deviation.

Deviation  Residue  Chain  Sequence  AT1  -  AT2  -  AT3   Bond     Dictionary  Standard
           Name     ID     Number                         Angle     Value       Deviation
    -----------------------------------------------------------------------------------
     -12.2   TYR      B      197      CB   -  CA   -  C    98.2      110.4       2.0
```

Six standard deviations is an exceptionally generous criterion for flagging outliers (probability of 1 in ~10^9 cases), and one can rest assured that there is positively something wrong. We probably mutilated this residue during real space refinement without regularization to a degree that the refinement program was not able to escape local minima despite the applied restraints. This error is also obvious and easy to fix.

```
==> The following peptide bonds deviate significantly from both cis
    and trans conformation. Trans is defined as 180 +/- 30 and cis
    is defined as 0 +/- 30 degrees.

    Residue  Chain  Sequence  Residue  Chain  Sequence        OMEGA
    SER      B      26        GLU      B      27             149.60
    GLU      B      27        ASP      B      28            -143.74
```

Here we are dealing with a poor backbone ω torsion angle to a point that the checking program even refuses to recognize these residues as either *cis*- or *trans*-peptides. These residues lie in a flexible loop that has not been properly rebuilt because it has practically no supporting electron density. These residues are absent in the final model. Note however that ω angles in proteins tend to have a relatively large standard deviation[98] of ~5.8° and if density supports it, even large deviations are justifiable (Figure 13-22).

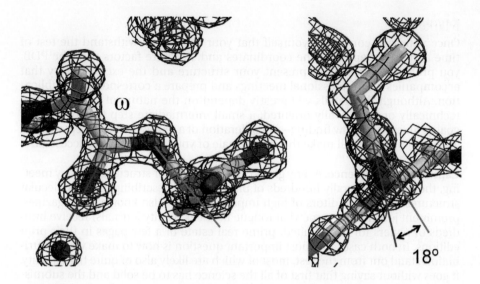

Figure 13-22 Omega backbone torsion angle. The depicted omega torsion angle deviates by 18° from planarity, which deviates 3–4σ from the target value but is well supported by electron density.

```
SOLVENT
-------

     The following solvent molecules are further than 3.5 Angstroms away from
     macromolecule atoms in the asymmetric unit that are available for
     hydrogen bonding. Solvent molecules in extended hydration shells
     separated by 3.5 Angstroms or less are not listed.

HETATM 9235  O   HOH Z  82     -0.606  84.408  43.158  1.00  47.51         O
HETATM 9274  O   HOH Z  96     11.269  43.986  67.578  1.00  44.39         O
HETATM 9529  O   HOH Z 200     36.482  71.269  41.317  1.00  41.00         O
HETATM 9547  O   HOH Z 208     16.673  72.245  40.539  1.00  34.55         O
HETATM 9628  O   HOH Z 242      7.215  65.898  39.749  1.00  35.15         O
HETATM 9694  O   HOH Z 269      6.683  76.293  40.639  1.00  47.70         O
HETATM 9733  O   HOH Z 291     30.632  32.474  51.412  1.00  27.73         O
```

These "water" molecules have no contact to either protein or other solvent molecules. They may have drifted away during refinement (because of lack of restraining electron density) or might have been placed into other unknown density as (inappropriate) placeholders. In this case, with *B*-factors not excessively high, they are likely located in some density; in fact they are left-over dummy-atoms from an *ARP*/*wARP* auto-building cycle placed in protein and ligand electron density that has not been properly rebuilt yet. Again, this is easily corrected by removing the waters and properly rebuilding the model.

Full validation report

The full validation report returned from the PDB via *ADIT* is essentially a *PROCHECK* report and a *SFCHECK* report of the structure factor analysis as discussed above. *AutoDep* at the EBI in Hinxton, UK, provides a similar check. The deposition programs may also relocate the waters that were associated with a symmetry related molecule so that they are closest to the deposited coordinates.

13.10 Publication and presentation

At this point you have successfully deposited a structure model that is well refined, free of any gross errors, and contains only a minimum of remaining nuisance errors. You know the model inside out and have developed an exciting structure-based hypothesis. The final act remaining is likely to present your tale to an audience such as a thesis committee or peers at a conference, and to publish your structure in a suitable journal. The presentation and publication of the structure model deserves as much thought and careful planning as the actual structure determination, and some (admittedly biased) advice will close our discourse.

Marketing

Once you have convinced yourself that your model will withstand the test of time and have deposited the coordinates and structure factors with the PDB, you probably will have to present your structure and the exciting story that accompanies it at professional meetings and prepare a corresponding publication. Although the details will greatly depend on the nature of your project—technically or biologically oriented; a small intermediate step toward a larger goal or an exciting new finding—consideration of a few generally accepted rules and good practice will make the grand finale of your project more successful.

Capturing the audience. At any given crystallography or structural biology meeting, there will be literally hundreds of presentations describing macromolecular structures. Similarly, editors of high impact journals (also known as magazines, prominent journals or, somewhat tongue in cheek, vanity journals) receive hundreds of papers for very limited, prime real estate of a few pages in their print editions. In both cases the most important question is how to make your contribution stand out from the rest, most of which are likely also of quite high quality. It goes without saying that first of all the science has to be solid and the submission should match the audience of the journal. Not every contribution is *Nature*, *Science*, or *Cell* material, and sending everything to those journals just wastes your time and that of the editor. Good papers will accumulate significant numbers of citations, regardless of the impact factor of the journal (on a personal note, if an institution judges your scientific potential solely by bean-counting impact points, you might reconsider applying there—the stress will eventually ruin you).

A few key elements can decide between successful acceptance for review or immediate editorial triage. A well-phrased title, a gripping abstract, and high quality illustrations are the most effective ways of making a paper or poster presentation stand out.

Title and abstract

The most critical parts of a publication from the marketing perspective are the title and the abstract. They are the first information a reader, editor, or reviewer absorbs, and if they are poorly crafted, they are probably also the last thing they read of your submission. It is amazing how poorly some paper titles are worded and how uninformative abstracts of structure papers can be. This is particularly relevant as the title and abstract form, together with keywords, the basis for indexing and search engines. Compare the following two fictitious titles:

1. Crystal structure of *Bacillus anthracis* alanine racemase at 2.0 Å bound to QED3019.

2. Structural basis for the development of a potent new anthrax inhibitor.

The first title gives technical information that the insider may appreciate, but the general audience (and the magazines) might find the second caption more News & Views worthy. While the title needs to be adapted to capture the interest of the intended audience, the art here is to find a balance between catering to marketing needs while still maintaining scientific honesty.

Similar considerations hold for the abstract. None of the readers of a general audience journal care in the abstract for cell parameters, space group, and crystallization conditions (unless this is the exciting part in a technical report or similar). The simple question here is: what are the objective and the key result of that study? Would the abstract allow an informed decision whether to keep reading the entire paper or not? Overselling here will backfire in the long run—it is likely that whoever reads your oversold stories will also eventually review some of your papers. You may rest assured that the vengeance for wasting readers' or reviewers' time will be served cold.

The art of scientific illustration

Captivating illustrations can convey a strong and convincing message to the viewers, provided they are based on a clear concept and are graphically well designed. This is where science meets art. There are many examples in this book (Sidebar 13-7 summarizes how the illustrations were made), and recent publications and cover illustrations will certainly provide excellent examples, some of truly stunning aesthetic appeal.

Overview versus detail. Figures and their captions are extremely important—they illustrate and support your textual descriptions and discussions in a persuasive and vivid form. The complexity of a large 3-dimensional structure almost always requires more than one figure. Composite overview illustrations such as a ribbon diagram or a surface charge rendering with emphasis on the location of the important regions of the molecule such as binding sites provide a first impression of the molecule. An insert of an important detail permits the viewer to rapidly gain orientation regarding the "business end" of the molecule. The discussion generally requires additional and quite detailed, topic-specific illustrations.

Electron density. As the experimental crystallographic evidence generally is (unbiased) electron density, showing a properly contoured section of relevant electron density together with the molecular model (hopefully inside density) of the relevant regions is justified. Very often the region of interest will be the target protein's binding site, harboring a ligand. The problem is often that despite proper slab selection and panning it can be difficult to find a view that is not obscured by some other part of density or model. Some programs therefore offer as a remedy so-called "blob" features, which allow the display of density only within a certain radius of the model atoms. This is where danger lurks. By lowering the electron density level to noise combined with a blob radius appropriate to the resolution, density can be displayed around any, however

Sidebar 13-7 The illustrations of *Biomolecular Crystallography*. I sincerely hope you found most figures in this book illustrative, informative and well designed, and that they were getting the intended point across. Some may have been confusing, poorly annotated, or not helpful at all, and I would surely like to hear about those, and also about your favorites. I thus briefly describe how most of the over 400 figures of *Biomolecular Crystallography* were generated and acknowledge the software used in their production. For graphics design, real estate usually counts, so I used dual HP LP3065 monitors with 2560×1600 pixel resolution each on an AMD quad core CPU workstation with 3 GB RAM running Windows XP. Raw images that contain electron density with few exceptions were rendered on an old Dell Latitude C840 laptop running RedHat Linux 9.

For images displaying electron density, I used Duncan McRee's venerable *XtalView/Xfit*,[99] which is somewhat outdated as far as complex model building tools are concerned, but generates excellent electron density images that can be automatically rendered with Ethan Merritt's versatile *Raster3d*[100] (*R3d*). A convenient interface window allows adjustment of all important *R3d* rendering parameters in *XtalView/Xfit*. Paul Emsley's *Coot*[32] provides more versatile model building tools, but presently I find the electron density rendering on the Windows version of *Coot* with the command line parameter adjustment for controlling the rendering less convenient than the *XtalView* interface.

The large majority of ribbon diagrams of structures were produced with Molsoft's *ICM Browser Pro*, a commercial viewer that is inexpensive and writes excellently rendered, high-resolution images, including sequence alignment figures, surface plots, and various other properties mapped onto grid meshes. *ICM Browser* (free, lacks some analytical tools of *Browser Pro*) also allows display of *iSee* datapacks[101] of annotated structure animations (used by the Oxford Structural Genomics Facility). A few images were rendered with *PovRay* (Persistence of Vision) out of Accelrys *Discovery Studio Visualizer*. Both *ICM Browser* and *DS Visualizer* provide embedding of ActiveX Control objects in Microsoft *PowerPoint*, which makes for great presentations. There are many other programs that are freely or cheaply available and can produce quality graphics, or even animation and movies, notably *PyMol* by Warren DeLano.[102] Another versatile program is *UCSF Chimera*[103]—you or your workgroup will certainly have their own favorites.

Basic functions were graphed with Microsoft *Excel* (the 2007 version is quite powerful and produces appealing charts), while more complicated functions were graphed with *MathCad* (which was also used for symbolic evaluation and checking of most computations). Graphic touch-up was performed with Adobe *Photoshop*, while for labeling and annotation most figures were simply transferred into *PowerPoint* and screen-captured. *PowerPoint* was also used to assemble composite images, flow charts, and schematic drawings.

For the remaining figures and those kindly provided by others, the credits and programs used (as far as known to me) are provided in the captions. Some sections of text and figures contain crystallographic symbols, for which I have used with kind permission Len Barbour's *Crystallography* OpenType font.

improbable, atomic model.[16] Note that such density will show tell-tale irregular or atypical borders. Similar considerations hold for touch-up with graphics programs: *Photoshop®* is not a density modification program. If you really need to use blob features or editing tools for clarity or for aesthetic reasons, it is a good idea to state why and to what level they had to be deployed. For example I had to remove neighboring density in some crowded images to make the features I wish to illustrate stand out clearly. This is acceptable for educational purposes, but certainly not unannounced and with the aim of bragging about "typical" clear electron density in an original publication.

Informative figure captions. One of the most common neglects apparent in publications is the lack of a succinct and still informative figure caption. The caption should lead with a statement sentence of what the figure illustrates, followed by a detailed enough description so that the viewer can appreciate the content and context without having to trace back to the text for the explanation. You cannot expect in a typeset publication that the figure will appear exactly where you want, and therefore the figure must have a self-contained explanation. On the other hand, it is perfectly acceptable to refer in the body text to the figure caption for a detail that would otherwise only interrupt the flow of the story.

Surviving and profiting from review

The central limit theorem (CLT) discussed in Chapter 7 informs us that regardless of the shape of the parent distribution, the sampling distribution will be a Gaussian normal distribution. From the CLT then follows that about half of all reviews you receive will be above average (review) quality and the other half will be below that average. Grouped in different percentiles (and according to informally polled experience) about 25% of all reviews are very valuable, where the reviewer obviously took great effort and time to examine the paper, and practically all comments are intended toward genuine improvement of your manuscript. In this category also fall critical and perhaps negatively perceived comments which may actually prevent you from publishing something really flawed—good reviews can save careers or at least public embarrassment.

The great bulk of comments in the 25th to 75th percentile contain some useful comments, and some less useful opinions. Ignore the opinion part and be grateful for the constructive elements. The remaining group contains reports that are not very helpful for improvement of manuscript quality at all. In this category also fall brief two-liners just recommending acceptance without any comment— do you really think that your manuscript is that perfect? A very small percentage also contains comments that are factually wrong, but those can be properly responded to. Harder are those rare reviews that contain personal attacks and plain stupid comments. Put them aside for a day or so instead of furiously responding and consider asking the editor to disregard the report because of blatant personal conflict and request additional alternative review. Some editors intercept such comments on their own and solicit alternate opinions.

A particularly difficult situation is immediate editorial rejection of a paper for technical reasons that you are convinced are not warranted. Based on his experience of an ultimately successful submission[104] of a low resolution (3.45 Å) structure of the open form of the *E. coli* small-conductance mechanosensitive channel MscS (Figure 13-23), Jim Naismith has compiled some thoughts and advice originally discussed and posted on the CCP4 bulletin board. He has allowed me to reproduce his insights, in his own words, below:

- **Do not be put off by the editor, argue for a technical review in advance of normal peer review.** After rejection without review, I sent a detailed technical document pointing out what the paper relied on from the structure (in my case, 10 Å movement of helices) and what the structure could reasonably be expected to show. I was also very clear what the structure could not show, side chain conformations and possible register errors. The editor sent this to an expert and the article then went to normal review. This strategy may be very useful for "magazine publishing" which we all decry but mostly aspire

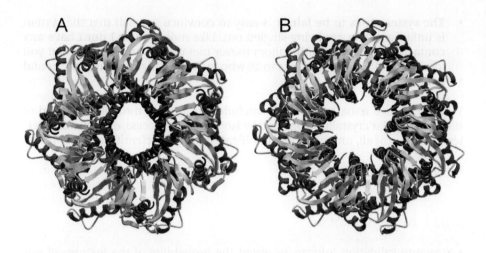

Figure 13-23 The small-conductance mechanosensitive channel in closed and open forms. Mechanosensitive channels open in response to membrane tension, allowing small osmotically active molecules and ions to cross the cell membrane, thus preventing hypo-osmotic shock of the cells. The heptameric MscS channel from *E. coli* is viewed in cross section from the cytoplasmic side of the membrane in closed[105] (A) and open[104] (B) form. Despite both models being derived from low resolution data (3.7 and 3.45 Å, respectively), the large movement of transmembrane helix TM3 (blue) pivoting around a glycine residue and opening the channel is distinctly visible. PDB entries 2oau and 2vv5.

to. These journals have had their fingers burned by general peer review for technically complex problems in low resolution structures (see Sidebars 10-6, 11-8, and 12-8 as well as a recent analysis).[88]

- **Try to find some way to establish unbiased electron density.** This was very helpful to me. I had not made it clear in my first draft that the helices were omitted from the molecular replacement search model, thus their location was unbiased. I made this clear in the revision and this helped convince. In my case I had 7-fold internal crystal averaging and 14-fold cross-crystal averaging. I did this with the unbiased phases and the resulting map was very clear. Clearly this (bias issue, BR) does not apply with heavy atom phasing.

- **Make the geometry very conservative.** I kept the model very conservative in geometry terms, tight NCS, low deviations, common rotamers. I found adding hydrogen atoms improved the refinement considerably in terms of geometry.

- **Sharpening.** Following the talks at the CCP4 workshop[77] (Axel Brunger mainly but others also) I sharpened my 14-fold averaged map. The results were very good, the register became much clearer.

- **Seek expert help.** In my case someone volunteered and kindly looked over the paper and told me what they as a reviewer would want to see. I explicitly asked them NOT to comment on the science of the paper or the correctness of the work, just what figures and data would be needed to convince them of the technical merits of this low resolution structure and that my inferences were not over-interpretation. I won't name somebody in case they get deluged although they are acknowledged in the paper. I did NOT use them as a shield with the editor though and did not mention them to the editor (except in the acknowledgment section that the editor will not read). Any mistakes are mine not theirs.

- **Provide supporting biology.** We did, but we made sure in the revised letter to the editor we made clear what the structure predicted, how we tested it and what the results were. Thus we changed the emphasis from what the structure "is" to what biology it predicts that can be tested.

- **Be respectful of the editor.** Probably the best piece of advice. It is very easy to get on ones high horse. However, the editor has a very difficult job. In our field there have been several high profile mistakes at low resolution and in many cases the authors will have been just as forceful as me (or you) in their conviction that in their case they are right. What's more, peer review did not spot or stop these efforts, so simply arguing for review did not help. I put myself in their shoes when I wrote back to them. I tried to make clear what we were uncertain about and what we were certain about. I avoided the blanket claims of "any fool would know this is right." That said, my own use of phrases of "we think this is correct," "appears convincing," in the original letter were taken as lack of belief in our work rather than qualification. What worked better was stating what we were sure was right and what we did not know.

- **The system tries to be fair.** It is easy to convince yourself that the system is unfair and you are being singled out. Like most people I don't have any contacts or influence with editors (never met one). You win some and you lose some. Once again, thanks to all who made such helpful suggestions and comments.

With these sage remarks we shall conclude our journey through the world of macromolecular crystallography and I wish you great success, exciting discoveries, and most of all, much fun and gratification in your scientific endeavors.

Bernhard Rupp

13.11 Key concepts

- Structure validation informs us about the probability of the features of our model.
- In cases of geometry outliers, a highly improbable conformation that has insufficient experimental support from electron density is very likely an error and needs to be corrected.
- In more subtle cases such as ligand binding or derived hypotheses based on model features, additional independent experimental support for the model as well as evaluation against the entire body of physicochemical knowledge should be sought.
- Validation is an integral process determining the plausibility of a model or hypothesis extending beyond simple geometry error checks.
- Protein molecules are inherently flexible as a requirement for their molecular function. As a consequence, a model representing a protein structure will have well-defined parts and some parts that are less well defined in electron density.
- Only local validation including assessment of both geometry and electron density can give an accurate picture of the reliability of the structure model or any hypothesis based on local features of the model.
- Geometry checks flag improbable values based on the deviation from expectation values, often in multiples of standard deviations in the form of Z-scores.
- The deviations can also be reported in percentile scores in relation to an empirical target distribution obtained from high quality reference structures.
- The backbone torsion angles φ and ψ are generally not restrained in model refinement, and are thus an independent and reliable means of validation.
- Torsion angle pairs that fall outside of the "allowed" regions are highly improbable because of excessive conformational energy.
- Bias minimized electron density is the primary source of evidence supporting a protein structure model.
- The match of the calculated model electron density against the electron density based on maximum likelihood map coefficients and observed data can be quantified on a residue-by-residue basis using real space correlation coefficients or real space R-values.
- The residue-by-residue plots of real space correlation or R-value provide a quick and accurate overview of the local quality of the structure.
- Outliers in poor electron density are most likely a manifestation of insufficient experimental evidence while unusual geometry in well-defined regions of electron density most likely signifies an interesting hot spot in the structure.
- Non-covalently bound small molecule ligands, cofactors, or peptide ligands are the most difficult parts of the structure in terms of accurate modeling.
- Non-covalent binding sites are out of principle partly occupied which often leads to weaker electron density which becomes apparent in higher B-factors during initial refinement.
- Ligands require accurate restraint files for stable refinement, and chemical plausibility must accompany electron density analysis.
- Ambiguous density is particularly hazardous and tempting to be modeled with a desired residue.
- Low resolution structures suffer from limited experimental evidence and

require particularly careful validation against independent prior knowledge.

- Inside–outside distributions, 3-dimensional profiling, non-bonded pattern analysis, and empirical potentials can indicate stretches of incorrectly traced or register-shifted residues.
- *B*-factor distribution plots can reveal abnormalities such as improper *B*-factor restraints, scaling errors, or artificial *B*-factor cutoffs.

13.12 Additional reading

1. Read RJ, & Kleywegt G (2009) Low-resolution structure determination and validation – Proceedings of the CCP4 study weekend 2008. *Acta Crystallogr.* D65.

2. Kleywegt GJ (2000) Validation of protein crystal structures. *Acta Crystallogr.* D56, 249–265.

3. Read RJ, & Kleywegt GJ (2009) Case-controlled structure validation. *Acta Crystallogr.* D65(2), 140–147.

4. Brunger AT (1997) Free *R* Value: Cross validation in Crystallography. *Methods Enzymol.* 277, 366–396.

5. EU 3-D Validation Network (1998) Who checks the checkers? Four validation tools applied to eight atomic resolution structures. *J. Mol. Biol.* 276, 417–436.

6. Rupp B (2010) Scientific inquiry, inference and critical reasoning in the macromolecular crystallography curriculum. *J. Appl. Crystallogr.* 43(5), 1242–1249.

13.13 References

1. Kuhn TS (1970) *The Structure of Scientific Revolutions.* Chicago, IL: University of Chicago Press.

2. Edwards AWF (1992) *Likelihood – An Account of the Statistical Concept of Likelihood and its Application to Scientific Inference.* Baltimore, MD: The Johns Hopkins University Press.

3. Perutz MF (2002) *I Wish I'd Made You Angry Earlier: Essays on Science, Scientists, and Humanity.* Oxford, UK: Oxford University Press.

4. Petsko GA (2007) And the second shall be first. *Genome Biol.* 8, 103–105.

5. Guss M (2000) Guidelines for the deposition and release of macromolecular coordinate and experimental data. *Acta Crystallogr.* D56(1), 2.

6. Popper K (2002) *The Logic of Scientific Discovery.* New York, NY: Routledge.

7. Baker EN, Blundell TL, Vijayan M, et al. (1996) Crystallographic data deposition. *Nature* 379, 202.

8. Kleywegt GJ, & Jones TA (2007) Experimental data for structure papers. *Science* 317, 194–195.

9. Kleywegt GJ, & Jones TA (2002) Homo crystallographicus – quo vadis? *Structure* 10(4), 465–472.

10. Joosten RP, Womack T, Vriend G, et al. (2009) Re-refinement from deposited X-ray data can deliver improved models for most PDB entries. *Acta Crystallogr.* D65(2), 176–185.

11. Joosten RP, & Vriend G (2007) PDB improvement starts with data deposition. *Science* 317, 195–196.

12. Read RJ, & Kleywegt GJ (2009) Case-controlled structure validation. *Acta Crystallogr.* D65(2), 140–147.

13. Lovelace JJ, Murphy CR, Daniels L, et al. (2008) Protein crystals can be incommensurately modulated. *J. Appl. Crystallogr.* 41, 600–605.

14. Janssen JC, Read RJ, Brunger AT, et al. (2007) Crystallographic evidence for deviating C3b structure? *Nature* 448, E1–E3.

15. Kleywegt GJ (1999) Validation of protein crystal structures. *Acta Crystallogr.* D56, 249–265.

16. Rupp B, & Segelke BW (2001) Questions about the structure of the botulinum neurotoxin B light chain in complex with a target peptide. *Nat. Struct. Biol.* 8, 643–664.

17. Badger J (2003) An evaluation of automated model-building procedures for protein crystallography. *Acta Crystallogr.* D59, 823–827.

18. Hoft RRW, Vriend G, Sander C, et al. (1996) Errors in protein structures. *Nature* 381, 272–272.

19. Abagyan R (1997) Protein structure prediction by global energy optimization. *Computer Simulation of Biomolecular Systems: Theoretical Experimental Applications, volume* 3, pp 363–394, Dordecht: Kluwer.

20. Tickle I (2007) Experimental determination of optimal root-mean-square deviations of macromolecular bond lengths and angles from their restrained ideal values. *Acta Crystallogr.* D63, 1274–1281.

21. Chen VB, Arendall WB, III, Headd JJ, et al. (2010) MolProbity: all-atom structure validation for macromolecular crystallography. *Acta Crystallogr.* D66(1), 12–21.

22. Laskowski RA, MacArthur MW, Moss DS, et al. (1993) PROCHECK: a program to check the stereochemical quality of protein structures. *J. Appl. Crystallogr.* 26(2), 283–291.

23. Morris AL, MacArthur MW, Hutchinson EG, et al. (1992) Stereochemical quality of protein structure coordinates. *Proteins: Struct. Fold. Des.* 12, 345–364.

24. Word JM, Lovell SC, LaBean TH, et al. (1999) Visualization and quantification of molecular goodness-of-fit: small probe contact dots with explicit hydrogens. *J. Mol. Biol.* 285, 1711–1733.

25. Weichenberger CX, & Sippl MJ (2006) Self-consistent assignment of asparagine and glutamine amide rotamers in protein crystal structures. *Structure* 14(6), 967–972.

26. Weichenberger C, Byzia P, & Sippl MJ (2008) Visualization of offending interactions in protein folds. *Bioinformatics* 24, 1206–1207.

27. Richardson DC, & Richardson JS (1992) The kinemage: a tool for scientific communication. *Protein Sci.* 1(1), 3–9.

28. Ramachandran GN, Ramakrishnan C, & Sasisekharan V (1963) Stereochemistry of polypeptide chain configurations. *J. Mol. Biol.* 7, 95–99.

29. Lovell SC, Davis IW, Arendall WB, 3rd, et al. (2003) Structure validation by C-alpha geometry: phi, psi and C-beta deviation. *Proteins* 50(3), 437–450.

30. Azizi AA, Gelpi E, Yang JW, et al. (2006) Mass spectrometric identification of serine hydrolase OVCA2 in the medulloblastoma cell line DAOY. *Cancer Lett.* 241(2), 235–249.

31. Kleywegt G (1996) Use of non-crystallographic symmetry in protein structure refinement. *Acta Crystallogr.* D52(4), 842–857.

32. Emsley P, & Cowtan K (2004) Coot: model-building tools for molecular graphics. *Acta Crystallogr.* D60, 2126–2132.

33. Ramirez L, Axelrod H, Herron S, et al. (2003) High resolution crystal structure of ferricytochrome c′ from *Rhodobacter sphaeroides*. *J. Chem. Crystallogr.* 33, 413–424.

34. Kleywegt GJ, & Jones AT (1996) Phi/Psi-chology: Ramachandran revisited. *Structure* 1395–1400.

35. Brändén CI, & Jones TA (1990) Between objectivity and subjectivity. *Nature* 343, 687–689.

36. Rupp B (2006) Real-space solution to the problem of full disclosure. *Nature* 444(7121), 817.

37. Jelsch C, Teeter MM, Lamzin V, et al. (2000) Accurate protein crystallography at ultra-high resolution: valence electron distribution in crambin. *Proc. Natl. Acad. Sci. U.S.A.* 97(7), 3171–3176.

38. Reddy V, Swanson S, Sacchettini JC, et al. (2003) Effective electron density map improvement and structure validation on a Linux multi-CPU web cluster: The TB Structural Genomics Consortium Bias Removal Web Service. *Acta Crystallogr.* D59, 2200–2210.

39. Hanson MA, & Stevens RC (2000) Cocrystal structure of synatptobrevin-II bound to botulinum neurotoxin type B at 2.0 Å resolution. *Nat. Struct. Biol.* 7(8), 687–692.

40. Kleywegt GJ, Harris MR, Zou J-Y, et al. (2004) The Uppsala Electron-Density Server. *Acta Crystallogr.* D60, 2240–2249.

41. Vaguine AA, Richelle J, & Wodak SJ (1999) SFCHECK: a unified set of procedures for evaluating the quality of macromolecular structure-factor data and their agreement with the atomic model. *Acta Crystallogr.* D55, 191–20.

42. Abagyan RA, & Totrov MM (2001) High-throughput docking for lead generation. *Curr. Opin. Chem. Biol.* 5(4), 375–382.

43. Hanson MA, Oost TK, Rich DH, et al. (2002) Structural basis for BABIM inhibition of botulinum neurotoxin type B protease (Correction). *J. Am. Chem. Soc.* 124(34), 10248.

44. Evrard GX, Langer GG, Perrakis A, et al. (2007) Assessment of automatic ligand building in ARP/wARP. *Acta Crystallogr.* D63(1), 108–117.

45. Sheldrick G (2008) A short history of SHELX. *Acta Crystallogr.* A64(1), 112–122.

46. Kleywegt GJ, & Jones TA (2004) Databases in protein crystallography. *Acta Crystallogr.* D59, 1119–1131

47. Kleywegt GJ, Henrick K, Dodson EJ, et al. (2004) Pound-wise but penny-foolish: How well do macromolecules fare in macromolecular refinement? *Structure* 11, 1051–1059.

48. Kleywegt G (2007) Crystallographic refinement of ligand complexes. *Acta Crystallogr.* D63(1), 94–100.

49. Kleywegt GJ, & Harris MR (2007) ValLigURL: a server for ligand-structure comparison and validation. *Acta Crystallogr.* 63(8), 935–938.

50. Rees DC, Congreve M, Murray CW, et al. (2004) Fragment-based lead discovery. *Nat. Rev. Drug Discovery.* 3(8), 660–672.

51. Blundell TL, Jhoti H, & Abell C (2002) High-throughput crystallography for lead discovery in drug design. *Nat. Rev. Drug Discovery.* 1(1), 45–54.

52. Congreve M, Carr R, Murray C, et al. (2003) A "rule of three" for fragment-based lead discovery? *Drug Discovery Today* 8(19), 876–877.

53. Hajduk PJ, & Greer J (2007) A decade of fragment-based drug design: strategic advances and lessons learned. *Nat. Rev. Drug Discovery.* 6, 211–219.

54. Vagin AA, Steiner RA, Lebedev AA, et al. (2004) Biological Crystallography REFMAC5 dictionary: organization of prior chemical knowledge and guidelines for its use. *Acta Crystallogr.* D60, 2184–2195.

55. Schüttelkopf AW, & van Aalten DMF (2004) PRODRG: a tool for high-throughput crystallography of protein-ligand complexes. *Acta Crystallogr.* D60(8), 1355–1363.

56. Van Der Spoel D, Lindahl E, Hess B, et al. (2005) GROMACS: Fast, flexible, and free. *J. Comput. Chem.* 26(16), 1701–1718.

57. Weininger D (1990) SMILES. 3. DEPICT. Graphical depiction of chemical structures. *J. Chem. Inf. Comput. Sci.* 30(3), 237–243.

58. Wallace AC, Laskowski RA, & Thornton JM (1995) LIGPLOT: a program to generate schematic diagrams of protein-ligand interactions. *Protein Eng.* 8(2), 127–134.

59. Gokulan K, Khare S, Ronning D, et al. (2005) Co-crystal structures of NC6.8 Fab identify key interactions for high potency sweetener recognition: Implications for the design of synthetic sweeteners. *Biochemistry* 44, 9889–9898.

60. Schmitt S, Hendlich M, & G K (2001) From structure to function: A new approach to detect functional similarity among proteins independent from sequence and fold homology. *Angew. Chem., Intl.* Ed. 40, 3141–3144.

61. Zhu D, Yang Y, Majcowicz S, et al. (2008) The enantioselectivity of a carbonyl reductase by substrate-enzyme docking-guided point mutation. *Org. Lett.* 10, 525–528.

62. Abagyan R, Totrov M, & Kuznetsov D (1994) ICM: a new method for protein modeling and design. Applications to docking and structure prediction from the distorted native conformation. *J. Comput. Chem.* 15, 488–506.

63. Laurie AT, & Jackson RM (2006) Methods for the prediction of protein-ligand binding sites for structure-based drug design and virtual ligand screening *Curr. Protein. Pept. Sci.* 7, 395–406.

64. Naumann T, & Matter H (2002) Structural classification of protein kinases using 3D molecular interaction field analysis of their ligand binding sites: target family landscapes. *J. Med. Chem.* 45, 2366–2378.

65. Campbell SJ, Gold ND, Jackson RM, et al. (2003) Ligand binding: functional site location, similarity and docking. *Curr. Opin. Struct. Biol.* 13, 389–395.

66. Sotriffer C, & Klebe G (2002) Identification and mapping of small-molecule binding sites in proteins: computational tools for structure-based drug design. *Farmaco* 57, 243–251.

67. Cummings MD, DesJarlais RL, Gibbs AC, et al. (2005) Comparison of automated docking programs as virtual screening tools. *J. Med. Chem.* 48(4), 962–976.

68. Schneider F, & Fechner U (2005) Computer-based de novo design of drug-like molecules. *Nat. Rev. Drug Discovery* 4, 649–663.

69. An J, Totrov M, & Abagyan R (2005) Pocketome via comprehensive identification and classification of ligand binding envelopes. *Mol. Cell. Proteomics* 4, 752–761.

70. Patocka J (1998) Acetylcholinesterase inhibitors – From nervous gas to Alzheimer's disease therapeutics. *Chem. Listy* 92, 1016–1019.

71. Grossberg T (2003) Cholinesterase inhibitors for the treatment of Alzheimer's disease: getting on and staying on. *Curr. Ther. Res.* 64, 216–235.

72. Giacobini E (2002) Acetyl- and butyrylcholinesterase inhibition by rivastigmine in cerebrospinal fluid of patients with Alzheimer's disease correlates with cognitive benefit. *J. Neural Transm.* 109, 1053–1065.

73. Greig NH, Utsuki T, Yu Q, et al. (2001) A new therapeutic target in Alzheimer's disease treatment: attention to butyrylcholinesterase. *Curr. Med. Res. Opin.* 17, 159–165.

74. Nicolet Y, Lockridge O, Masson P, et al. (2003) Crystal structure of human butyrylcholinesterase and of its complexes with substrates and products. *J. Biol. Chem.* 278, 41141–41147.

75. Kryger G, Harel M, Giles K, et al. (2000) Structures of recombinant wild-type human acetylcholinesterase and of its E202Q mutant as complexes with fasciculin-II, a "three-finger" polypeptide. *Acta Crystallogr.* D56, 1385–1394.

76. Law K-S, Acey RA, Smith CR, et al. (2007) Dialkyl phenyl phosphates as novel selective inhibitors of butyrylcholinesterase. *Biochem. Biophys. Res. Commun.* 355, 371–378.

77. Read RJ, & Kleywegt G (2009) Low-resolution structure determination and validation – Proceedings of the CCP4 Study Weekend 2008. *Acta Crystallogr.* D65.

78. Kleywegt GJ, & Jones TA (1995) Where freedom is given, liberties are taken. *Structure* 3, 535–540.

79. Chapman MS, Suh SW, Curmi PMG, et al. (1988) Tertiary structure of plant RuBisCO: domains and their contacts. *Science* 241, 71–74.

80. Knight S, Andersson I, & Brändén C (1988) Reexamination of the three-dimensional structure of the small subunit of RuBisCo from higher plants. *Science* 244, 702–705.

81. Lüthy R, Bowie JU, & Eisenberg D (1992) Assessment of protein models with three-dimensional profiles. *Nature* 356, 83–85.

82. Colovos C, & Yeates TO (1993) Verification of protein structures: Patterns of nonbonded atomic interactions. *Protein Sci.* 2(9), 1511–1519.

83. Sippl MJ (1993) Recognition of errors in three-dimensional structures of proteins. *Proteins* 17, 355–362.

84. Kelly JA, & Kuzin AP (1995) The refined crystallographic structure of a DD-peptidase penicillin-target enzyme at 1.6 Å resolution. *J. Mol. Biol.* 254, 223–236.

85. McRee DE, Tainer JA, Meyer TE, et al. (1989) Crystallographic structure of a photoreceptor protein at 2.4 Å resolution. *Proc. Natl. Acad. Sci. U.S.A.* 86(17), 6533–6537.

86. Kleywegt GJ (1997) Validation of protein models from Cα coordinates alone. *J. Mol. Biol.* 273, 371–376.

87. Ioerger TR, & Sacchettini JC (2002) Automatic modeling of protein backbones in electron density maps via prediction of Cα coordinates. *Acta Crystallogr.* D58, 2043–2054.

88. Jeffrey P (2009) Analysis of errors in the structure determination of MsbA. *Acta Crystallogr.* D65(2), 193–199.

89. Borgstahl GE, Williams DR, & Getzoff ED (1995) 1.4 Å structure of photoactive yellow protein, a cytosolic photoreceptor: unusual fold, active site, and chromophore. *Biochemistry* 34(19), 6278–6387.

90. Brunger AT, DeLaBarre B, Davies JM, et al. (2009) X-ray structure determination at low resolution. *Acta Crystallogr.* D65(2), 128–133.

91. Popov AN, & Bourenkov GP (2003) Choice of data-collection parameters based on statistic modelling. *Acta Crystallogr.* D59(7), 1145–1153.

92. Sutton RB, Fasshauer D, Jahn R, et al. (1998) Crystal structure of a SNARE complex involved in synaptic exocytosis at 2.4 Å resolution. *Nature* 395, 347–353.

93. Segelke B, Knapp M, Kadhkodayan S, et al. (2004) Crystal structure of *C. botulinum* neurotoxin protease in a product bound state: Evidence for non-canonical zinc protease activity. *Proc. Natl. Acad. Sci. U.S.A.* 101, 6888–6893.

94. Hanson MA, & Stevens RC (2001) Response to Rupp and Segelke. *Nat. Struct. Biol.* 8(8), 664.

95. Breidenbach MA, & Brunger AT (2004) Substrate recognition strategy for botulinum neurotoxin serotype A. *Nature* 432(7019), 925–929.

96. Ajees AA, Gunasekaran K, Volanakis JE, et al. (2006) The structure of complement C3b provides insights into complement activation and regulation. *Nature* 444, 221–225.

97. Dutta S, Burkhardt K, Swaminathan GJ, et al. (2008) Data deposition and annotation at the Worldwide Protein Data Bank. In Kobe B, Guss M, & Huber T (Eds.) *Structural Proteomics: High-Throughput Methods*. New York, NY: Humana Press/Springer.

98. Jaskolski M, Gilski M, Dauter Z, et al. (2007) Stereochemical restraints revisited: how accurate are refinement targets and how much should protein structures be allowed to deviate from them? *Acta Crystallogr.* D63, 611–620.

99. McRee DE (1999) XtalView/Xfit – a versatile program for manipulating atomic coordinates and electron density. *J. Struct. Biol.* 125, 156–165.

100. Merritt EA, & Bacon DJ (1997) Raster3D: Photorealistic molecular graphics. *Methods Enzymol.* 277, 505–524.

101. Abagyan R, Lee WH, Raush E, et al. (2006) Disseminating structural genomics data to the public: from a data dump to an animated story. *Trends Biochem. Sci.* 31, 76–78.

102. DeLano WL (2008) *The PyMOL Molecular Graphics System*. DeLano Scientific, Palo Alto, CA.

103. Pettersen EF, Goddard TD, Huang CC, et al. (2004) UCSF Chimera – a visualization system for exploratory research and analysis. *J. Comput. Chem.* 25(13), 1605–1612.

104. Wang W, Black SS, Edwards MD, et al. (2008) The structure of an open form of an *E. coli* mechanosensitive channel at 3.45 Å resolution. *Science* 321, 1179–1183.

105. Bass RB, Strop P, Barclay M, et al. (2002) Crystal structure of *Escherichia coli* MscS, a voltage-modulated and mechanosensitive channel. *Science* 298, 1582–1587.

Appendix

The Appendix contains supplemental material and information helpful for basic crystallographic computing tasks. Explanation of the PDB file records is provided, and a small linear algebra refresher follows, limited to what is needed to follow or to verify most of the derivations provided in the main text. Also useful are a reminder of the cosine rule and the proof of Euler's equation by MacLaurin series expansion. The explicit proof that electron density is real is provided as an exercise. Various useful tables and definitions follow.

A.1 Coordinate file formats

The protein structure file is essentially a collection of atomic coordinate records, preceded by some model-specific information. The two most common file formats are the PDB file, a fixed-length, 80 character per line, key-worded record format descending from the FORTRAN language standards, and a more modern, variable length record format called the *macromolecular Crystallographic Information File*, or mmCIF. Additional translators between the formats and to extensible mark-up language (XML) formats are available from the PDB and EBI sites.

Despite its shortcomings, the fixed record PDB file format is persistent, largely because it is quite easy to read. Unfortunately, efforts to overcome its limitations have led to some rogue extensions, particularly in columns 73–80 (once the so-called FORTRAN comment columns). The PDB file format is fully described at the PDB web site and we will limit the description in this section to what is needed in order to start viewing crystallographic structure models based on PDB files.

The PDB file header section

A PDB file is identified by a four-character PDB identification code. The first character is a number, which was originally intended to indicate revisions to the file by incrementing the numbers, followed by three alphanumeric characters. The PDB file has two main parts: a header section and the actual atom coordinate records.

The header begins with the title section, containing title, compound name, source, and biographic data, followed by a REMARK section in free text format. The REMARK section often contains rather important auxiliary information, such as warnings about poorly defined regions, location of *cis*-peptides, the biological oligomerization state, and crystallization conditions. The free-text format regrettably is suboptimal for automated data retrieval, but in any case, the remark section is worth reading.

Primary and secondary structure information. The primary structure records contain a sequence database reference and the SEQRES record, which provides the complete amino acid sequence of the molecule that has been crystallized. This does not necessarily mean that all the residues were also actually observed in the structure and listed in the coordinate section—parts of a structure can be disordered and absent from electron density. Additional records contain sequence advisories if the used construct deviates from the published genetic sequence. After an optional section for the description of *heterogens* (which are simply non-standard amino acids, ligands, ions, solvent molecules, etc.) which might be present in the coordinate section, a listing of secondary structure elements, S–S bonds, and special sites follows. The secondary structure is usually assigned automatically by the DSSP program during deposition, but the description can be overridden by the depositor of the structure.

The CRYST1 and SCALE records

Important crystallographic information is contained in the CRYST1 record, which contains unit cell dimensions, space group, and the number of molecular monomers in the unit cell (Z). Note that this number is not the general position multiplicity M or z as listed in the space group tables, but considers the oligomerization state. For example, for space group $P2_12_12_1$, $M = 4$, and for a dimer forming the asymmetric unit, the value for Z listed in the CRYST1 record is 8.

Information in the CRYST1 record is necessary to properly create the symmetry related molecules that form the unit cell. The origin shift record ORIGX is rarely used, and practically always contains the identity transformation matrix. The next record is the scale record, SCALE. It contains the elements of the deorthogonalization matrix, which converts the orthonormal Cartesian coordinates of the subsequent ATOM coordinate records into fractional crystallographic coordinates. This transformation can actually be created from the CRYST1 record as discussed in Chapter 5. The SCALE record is thus somewhat redundant and due to its fixed format limited in accuracy for large unit cells. Cartesian coordinates must be converted into fractional crystallographic coordinates before the symmetry operators defined by the space group in the CRYST1 record can be applied to the molecule to create the remaining symmetry related molecules in the unit cell. Coordinate conversion and symmetry operations are explained in great detail in Chapter 5. If non-crystallographic symmetry between the molecules listed in the coordinate section exists, MTRIX records containing the non-crystallographic Cartesian transformations $\mathbf{W}(\mathbf{R},\mathbf{T})$ relating the molecules may follow. Cartesian transformations are extensively discussed in Chapter 11.

The coordinate section of the PDB file

The coordinate record section is the largest part of the PDB file. All atoms of each residue that could be built or modeled into the electron density are listed in the ATOM records (or HETATM records with the same format for heterogens, i.e. all atoms not belonging to the 20 standard amino acids). These records are the minimum information that must be read by any program in order to display the molecule. The record format is thus very important and is explicitly provided in Table A-1.

Table A-1 Key for the format of the ATOM record of a PDB file.

COLUMNS	DATA TYPE	FIELD	DEFINITION
1 - 6	Record name	"ATOM "	
7 - 11	Integer	serial	Atom serial number
13 - 16	Atom	name	Atom name, right-adjusted
17	Character	altLoc	Alternate location indicator
18 - 20	Residue name	resName	Residue name
22	Character	chainID	Chain identifier
23 - 26	Integer	resSeq	Residue sequence number
27	AChar	iCode	Code for insertion of residues
31 - 38	Real(8.3)	x	Orthogonal coordinates for X in Å
39 - 46	Real(8.3)	y	Orthogonal coordinates for Y in Å
47 - 54	Real(8.3)	z	Orthogonal coordinates for Z in
55 - 60	Real(6.2)	occupancy	Occupancy
61 - 66	Real(6.2)	tempFactor	Temperature factor
73 - 76	LString(4)	segID	Segment identifier, left-adjusted
77 - 78	LString(2)	element	Element symbol, right-adjusted
79 - 80	LString(2)	charge	Charge of the atom

Atom name: column 13,14: element symbol, right adjusted
          Column 15,16: Atom position label

As an example we interpret actual ATOM records of one residue in the coordinate section of a protein structure model (Table A-2), beginning with the first highlighted record.

Table A-2 PDB coordinate records for residue valine 25 in a protein structure model.

```
          1         2         3         4         5         6         7         8
1234567890123456789012345678901234567890123456789012345678901234567890123456789 0
ATOM    145  N   VAL A  25      32.433  16.336  57.540  1.00 11.92      A1   N
ATOM    146  CA  VAL A  25      31.132  16.439  58.160  1.00 11.85      A1   C
ATOM    147  C   VAL A  25      30.447  15.105  58.363  1.00 12.34      A1   C
ATOM    148  O   VAL A  25      29.520  15.059  59.174  1.00 15.65      A1   O
ATOM    149  CB AVAL A  25      30.385  17.437  57.230  0.50 13.88      A1   C
ATOM    150  CB BVAL A  25      30.166  17.399  57.373  0.50 15.41      A1   C
ATOM    151  CG1AVAL A  25      28.870  17.401  57.336  0.50 12.64      A1   C
ATOM    152  CG1BVAL A  25      30.805  18.788  57.449  0.50 15.11      A1   C
ATOM    153  CG2AVAL A  25      30.835  18.826  57.661  0.50 13.58      A1   C
ATOM    154  CG2BVAL A  25      29.909  16.996  55.922  0.50 13.25      A1   C
```

According to the key in Table A-1, the atom coordinates belong to residue valine 25 of peptide chain A. The blue highlighted record is for atom 146, the Cα atom (note that element symbols are right adjusted, i.e. a calcium atom would be "CA " instead of " CA ") of valine A 25 (i.e. valine 25 of chain A). The following three decimal numbers are the *Cartesian coordinates* (X, Y, Z) of that atom. Cartesian coordinates in Å are used instead of dimensionless fractional crystallographic coordinates, because even basic graphics display programs can then generate the molecule without any additional crystallographic unit cell and space group information. The drawback is, as we already mentioned, that programs that cannot interpret and apply crystallographic information are only able to display the molecule but cannot display symmetry- and packing-related molecules. Consequently, possibly important effects and artifacts of crystal packing and assembly geometry cannot be examined (and in fact are often overlooked or ignored for that reason).

The following two decimal numbers in the record are the *occupancy* (n) and the *B*-factor or *isotropic displacement parameter*. The occupancy for the covalently connected atoms in a protein chain is usually one, but it can be lower for ligands located in incompletely occupied binding sites or for other heterogens such as metal ions or solvent components that are not covalently attached to the protein chain. Residue side chains can also partially occupy alternate or multiple conformations. In high-resolution structures, it is sometimes possible to distinguish two distinct side chain conformations, and the residue is built as a "split" residue. This is most common in solvent-exposed side chains. The second highlighted ATOM record, Cβ of valine 25, carries an alternate indicator (CB A), and the occupation of the branch A of the side chain is set to 0.5. Note that the branch B then also has to have occupation of 0.5, because the sum of occupancies is constrained to 1.0. Also notice that the residue splits beginning with the Cβ atom. It is very common that only the side chain can be modeled in clearly split conformations. Only in very high resolution structures is it sometimes possible to build two distinctly split models for backbone atoms.

The last of the five decimal number columns in each ATOM record deserves special attention, as it carries some of the otherwise absent information about dynamic behavior or disorder. The individual *B*-factor, or isotropic atomic displacement parameter, is a measure of the probability that an atom is present at the location given by the coordinates, and is related to the *mean square displacement* of the atom from its mean position (thus its dimensions are Å²). The origin of the displacement can be manifold, including thermal vibrations, dynamic disorder, multiple but not resolved conformations, or simply an attempt by the refinement program to weigh down scattering contributions from that specific atom (we discuss *B*-factors in great detail in Chapter 12 dealing with protein structure refinement). Clearly, atoms in the rigid core of the protein will have considerably lower *B*-factors than long, flexible side chains of solvent-exposed residues. Nonetheless, extended regions of very high B-factors are a sure indication that the structure—for whatever reason—is less well defined in that region. Once the B-factors exceed 100 Å², there is certainly reason to examine closer and to be slightly suspicious (although some programs or users artificially limit the B-factors to 99 Å²). Such attempts manifest themselves in a pathological B-factor histogram, as explained in Chapter 13. The PDB file ends with bookkeeping records MASTER and END.

A.2   Basic linear algebra

The following provides a short overview of elements of linear algebra as used in various chapters. A more formal introduction can be found in the corresponding chapters on crystallographic computing in *Fundamentals of Crystallography*[1] or the booklet *Vectors and Tensors in Crystallography*.[2]

Let **a** be a general simple column vector in a rectilinear, 3-dimensional coordinate system with Cartesian components a_i:

$$\mathbf{a} = \begin{pmatrix} a_1 \\ a_2 \\ a_3 \end{pmatrix} \tag{A-1}$$

The sum of two vectors **a** and **b** is obtained by addition of the components, and the difference vector by subtraction of the components:

$$\mathbf{a} \pm \mathbf{b} = \begin{pmatrix} a_1 \pm b_1 \\ a_2 \pm b_2 \\ a_3 \pm b_3 \end{pmatrix} \tag{A-2}$$

The dot (or inner, scalar) product of two column vectors **a**, **b** is defined as

$$\mathbf{a} \cdot \mathbf{b} = ab\cos\gamma \tag{A-3}$$

where γ is the angle enclosed between the vectors and a and b represent the norm (or length, magnitude) of **a** and **b**, respectively:

$$a = |\mathbf{a}| = \left(\sum_i a_i^2\right)^{1/2} \tag{A-4}$$

The cross (or vector) product of two column vectors **a**, **b** defines a third vector **c** perpendicular to the plane spanned by **a** and **b**

$$\mathbf{c} = \mathbf{a} \times \mathbf{b} \tag{A-5}$$

with its norm defined by

$$c = |\mathbf{c}| = |\mathbf{a} \times \mathbf{b}| = ab\sin\gamma \tag{A-6}$$

Equations A-3 and A-6 can be used to compute the angle between two vectors such as crystallographic and/or other axes.

Matrix representations

A 2-dimensional $m \times n$ matrix **A** is an array filled with elements a_{ij} where the row index i runs from $1 \le i \le m$ and the column index j runs from $1 \le j \le n$. The $m \times n$ matrix **A** has m rows and n columns:

$$\mathbf{A} = \begin{pmatrix} a_{11} & \cdots & a_{1n} \\ \vdots & & \vdots \\ a_{m1} & \cdots & a_{mn} \end{pmatrix} \tag{A-7}$$

The most common case in basic computations involving coordinate transformations and symmetry operations is a square 3×3 matrix. Matrices of higher dimensionality than 2 are used in advanced crystallographic computing, largely to achieve compactness of notation.

Vectors can be written in matrix notation: a column vector is an $m \times 1$ matrix, and a row vector is a $1 \times n$ matrix. The most common case for m and n is 3. We denote the format of a matrix by [row $\times$ column].

Sum, difference, and multiplication of matrices

The sum or difference of matrices is formed by component-wise addition or subtraction, and thus is only defined if the matrices have the same dimensions n and m. The vector operation (A-2) may serve as an example for the sum or difference of two 3×1 (column) matrices.

Multiplication of two matrices $\mathbf{A}$ and $\mathbf{B}$

$$\mathbf{A}\,\mathbf{B} = \mathbf{C} \tag{A-8}$$

requires that the number of columns n of $\mathbf{A}$ equals the number m of rows of $\mathbf{B}$. The result of a multiplication of a $u \times v$ matrix with a $v \times w$ matrix is a $u \times w$ matrix $\mathbf{C}$. The elements of $\mathbf{C}$ can be readily (but lengthily) obtained by evaluating

$$c_{ij} = a_{i1}b_{1j} + a_{i2}b_{2j} + a_{i3}b_{3j} + \cdots + a_{iv}b_{vj} \tag{A-9}$$

Note that matrix multiplication (and thus vector multiplication) is not commutative, that is $\mathbf{AB} \neq \mathbf{BA}$. Matrix multiplication is commutative only with respect to its inverse (next section). Matrix multiplication is associative:

$$(\mathbf{AB})\mathbf{C} = \mathbf{A}(\mathbf{BC}) \tag{A-10}$$

Transpose, inversion, determinant, and trace of a matrix

An $m \times n$ matrix has a transpose $m \times n$ matrix $\mathbf{A}^{\mathrm{T}}$. The most commonly encountered transpose matrices in crystallography are transposes of square matrices, where the elements are simply swapped across the diagonal of the matrix:

$$\mathbf{A}^{\mathrm{T}} = \begin{pmatrix} a_{11} & a_{12} & a_{13} \\ a_{21} & a_{22} & a_{23} \\ a_{31} & a_{32} & a_{33} \end{pmatrix}^{\mathrm{T}} = \begin{pmatrix} a_{11} & a_{21} & a_{31} \\ a_{12} & a_{22} & a_{32} \\ a_{13} & a_{23} & a_{33} \end{pmatrix} \tag{A-11}$$

It immediately follows that $\mathbf{A} = \mathbf{A}^{\mathrm{T}}$ for a symmetric matrix with $a_{ij} = a_{ji}$ or a diagonal matrix ($a_{ij} = 0$ for $i \neq j$). Matrices of certain physical properties can be described as symmetric square matrices, called tensors, for example the anisotropic displacement tensor $\mathbf{U}$ or the metric tensor $\mathbf{G}$.

A square matrix $n \times n$ may have an inverse $\mathbf{A}^{-1}$, which if it exists is unique. If a matrix is invertible, then

$$\mathbf{AA}^{-1} = \mathbf{A}^{-1}\mathbf{A} = \mathbf{I} \tag{A-12}$$

where $\mathbf{I}$ is the identity matrix having only 1 for all its diagonal elements. The inverse of a transpose is also the transpose of the inverse:

$$\left(\mathbf{A}^{\mathrm{T}}\right)^{-1} = \left(\mathbf{A}^{-1}\right)^{\mathrm{T}} \tag{A-13}$$

For orthogonal matrices $\mathbf{G}$, the transpose is also its inverse:

$$\mathbf{GG}^{\mathrm{T}} = \mathbf{G}^{\mathrm{T}}\mathbf{G} = \mathbf{I} \tag{A-14}$$

The manual computation of inverse matrices is lengthy, given in any linear algebra text, and shall not concern us here. It is of interest that inversion is a process of the order n^3, which means that the inversion of large matrices is a *computationally expensive task* and thus used only—if at all—at the end of macromolecular refinement calculations.

Each square matrix $\mathbf{A}$ has the equivalent of a vector norm, its determinant $\det(\mathbf{A})$. Its computation is again a lengthy but standard procedure. Determinants are useful, for example, in preventing undesirable handedness changes during symmetry operations, where the determinant of an operator that generates the enantiomorph of a motif is negative.

Finally, the trace of a matrix tr(**M**) is the sum of its diagonal elements. Note that the trace is preserved during basis transformation.

Direction cosine matrix. The determinant of a proper direction cosine matrix describing a rotation is 1 and the trace equals $1 + 2\cos\kappa$, which is practical to compute the principal Euler angle.

Vector operations as matrix operations

From the generally applicable matrix operation rules it follows that vector operations can be written as matrix operations, which is commonly used in crystallography largely for compactness of notation.

Based on the requirement for matching of column and row numbers of the operands, we immediately find that the vector multiplication rule requires that we represent the first vector as a $1 \times n$ vector and the second one as an $n \times 1$ vector, that is, the dot product of a row vector with a column vector. It further follows that the resulting product is a 1×1 matrix (i.e. a single scalar value). Finally, to obtain a row vector from our standard column vector, we need to transpose for example a 3×1 column vector **a** into a 1×3 row vector $\mathbf{a}^T$. We write in matrix notation for the scalar vector product of **a** and **x**:

$$\mathbf{r} = \mathbf{a}^T\mathbf{x} = \begin{pmatrix} a_1 & a_2 & a_3 \end{pmatrix} \cdot \begin{pmatrix} x_1 \\ x_2 \\ x_3 \end{pmatrix} = a_1x_1 + a_2x_2 + a_3x_3 = \begin{bmatrix} 1 \times 1 \end{bmatrix} \tag{A-15}$$

In most cases of simple vector dot products, the transposition symbol is omitted; for example, in the exponential phase term of the structure factor equations we generally write $\exp(2\pi i \mathbf{h}\mathbf{x})$ instead of the more accurate $\exp(2\pi i \mathbf{h}^T\mathbf{x})$. Here **h** is the (dimensionless) reciprocal lattice index vector, and **x** is the (dimensionless) crystallographic fractional coordinate vector.

We can expand the case to a general square transformation matrix **M**:

$$\mathbf{r} = \mathbf{a}^T\mathbf{M} = \begin{pmatrix} a_1 & a_2 & a_3 \end{pmatrix} \cdot \begin{pmatrix} m_{11} & m_{12} & m_{13} \\ m_{21} & m_{22} & m_{23} \\ m_{31} & m_{32} & m_{33} \end{pmatrix} \text{ and } \mathbf{r} = \mathbf{M}^T\mathbf{a} = \begin{pmatrix} m_{11} & m_{21} & m_{31} \\ m_{12} & m_{22} & m_{32} \\ m_{13} & m_{23} & m_{33} \end{pmatrix} \cdot \begin{pmatrix} a_1 \\ a_2 \\ a_3 \end{pmatrix}$$
$$\tag{A-16}$$

Note that the result is the same, except that one time it is in the form of a row vector (row · matrix = row) and the other time in the form of a column vector (matrix · column = column).

Vectors are often transformed by matrices to obtain new vectors. The relation to the matrix presentation of linear systems of equations is obvious:

$$\mathbf{b} = \mathbf{R}\mathbf{x} = \begin{array}{l} b_1 = r_{11}x_1 + r_{12}x_2 + r_{13}x_3 \\ b_2 = r_{21}x_1 + r_{22}x_2 + r_{23}x_3 \\ b_3 = r_{31}x_1 + r_{32}x_2 + r_{33}x_3 \end{array} \tag{A-17}$$

where both **x** and **b** are column vectors. An example is the application of a point group symmetry operator **R** (written as a rotation matrix) to a fractional coordinate vector **x** in real space generating a new, symmetry related fractional coordinate vector **x′**:

$$\mathbf{x}' = \begin{pmatrix} x' \\ y' \\ z' \end{pmatrix} = \mathbf{R}\mathbf{x} = \begin{pmatrix} r_{11} & r_{12} & r_{13} \\ r_{21} & r_{22} & r_{23} \\ r_{31} & r_{32} & r_{33} \end{pmatrix} \cdot \begin{pmatrix} x \\ y \\ z \end{pmatrix} = \begin{pmatrix} r_{11}x + r_{12}y + r_{13}z \\ r_{21}x + r_{22}y + r_{23}z \\ r_{31}x + r_{32}y + r_{33}z \end{pmatrix} = \begin{bmatrix} 3 \\ \times \\ 1 \end{bmatrix} \tag{A-18}$$

If we swap the sequence of the matrix multiplication to $\mathbf{b} = \mathbf{xR}$ without transposing the matrix the result will differ from $\mathbf{b} = \mathbf{Rx}$ because matrix multiplication is not commutative. Furthermore, the column vector $\mathbf{x}$ must be transposed and becomes a $[1 \times 3]$ row vector (and consequently the result $\mathbf{b}$ will also be a row vector). Although it is occasionally omitted for brevity of notation, it is sometimes useful to keep the transposition notation $\mathbf{b}^T = \mathbf{x}^T\mathbf{R}$ as a reminder that the components of the result vector $\mathbf{b}^T$ must now be formed differently than in the case of the (column) result vector for $\mathbf{Rx}$—note the transposed matrix element indices in the example below.

Transformations of the form $\mathbf{b}^T = \mathbf{x}^T\mathbf{R}$ (or $\mathbf{b} = \mathbf{R}^T\mathbf{x}$) are generally found in operations in reciprocal space, when for example a real space symmetry operator $\mathbf{R}$ is applied to the reciprocal lattice index vector $\mathbf{h}$:

$$\left(\mathbf{h}'\right)^T = \begin{pmatrix} h' & k' & l' \end{pmatrix} = \mathbf{h}^T\mathbf{R} = \begin{pmatrix} h & k & l \end{pmatrix} \cdot \begin{pmatrix} r_{11} & r_{12} & r_{13} \\ r_{21} & r_{22} & r_{23} \\ r_{31} & r_{32} & r_{33} \end{pmatrix} = \tag{A-19}$$

$$= \begin{pmatrix} r_{11}h + r_{21}k + r_{31}l & r_{12}h + r_{22}k + r_{23}l & r_{13}h + r_{23}k + r_{33}l \end{pmatrix} = \begin{bmatrix} 1 & \times & 3 \end{bmatrix}$$

Vector elements can be vectors themselves, and the same rules apply. Example: A positional vector $\mathbf{r}$ in a crystallographic basis $[0, \mathbf{a}, \mathbf{b}, \mathbf{c}]$ and with fractional coordinates $\mathbf{x}$ is defined by a scalar product as follows:

$$\mathbf{r} = \begin{pmatrix} \mathbf{a} & \mathbf{b} & \mathbf{c} \end{pmatrix} \cdot \begin{pmatrix} x \\ y \\ z \end{pmatrix} = (x\mathbf{a} + y\mathbf{b} + z\mathbf{c}) \tag{A-20}$$

with the equivalent notation of $\mathbf{A}$ being a diagonal matrix containing the basis vectors

$$\mathbf{r} = \mathbf{A}^T\mathbf{x} = \begin{pmatrix} \mathbf{a} & 0 & 0 \\ 0 & \mathbf{b} & 0 \\ 0 & 0 & \mathbf{c} \end{pmatrix} \begin{pmatrix} x \\ y \\ z \end{pmatrix} = \begin{pmatrix} x\mathbf{a} \\ y\mathbf{b} \\ z\mathbf{c} \end{pmatrix} \tag{A-21}$$

providing the result in a column vector. Even if the matrix is diagonal and $\mathbf{A}^T = \mathbf{A}$, we keep the transposition symbol in matrix notation as a reminder that it is a representation of a column vector. In full matrix representation, we obtain the same result, except this time in form of the diagonal elements of the result matrix:

$$\mathbf{r} = \mathbf{A}^T\mathbf{X} = \begin{pmatrix} \mathbf{a} & 0 & 0 \\ 0 & \mathbf{b} & 0 \\ 0 & 0 & \mathbf{c} \end{pmatrix} \begin{pmatrix} x & 0 & 0 \\ 0 & y & 0 \\ 0 & 0 & z \end{pmatrix} = \begin{pmatrix} x\mathbf{a} & 0 & 0 \\ 0 & y\mathbf{b} & 0 \\ 0 & 0 & z\mathbf{c} \end{pmatrix} \tag{A-22}$$

In the most general case, the matrix representation(s) may contain non-diagonal elements, for example when a different reference basis has been chosen for the basis vectors. Example:

$$\begin{pmatrix} \mathbf{a} \\ \mathbf{b} \\ \mathbf{c} \end{pmatrix} = \mathbf{A} = \begin{pmatrix} a_1 & a_2 & a_3 \\ b_1 & b_2 & b_3 \\ c_1 & c_2 & c_3 \end{pmatrix}$$

$$\text{and} \begin{pmatrix} \mathbf{a} \\ \mathbf{b} \\ \mathbf{c} \end{pmatrix}^T = \begin{pmatrix} \mathbf{a} & \mathbf{b} & \mathbf{c} \end{pmatrix} = \mathbf{A}^T = \begin{pmatrix} a_1 & a_2 & a_3 \\ b_1 & b_2 & b_3 \\ c_1 & c_2 & c_3 \end{pmatrix}^T = \begin{pmatrix} a_1 & b_1 & c_1 \\ a_2 & b_2 & c_2 \\ a_3 & b_3 & c_3 \end{pmatrix} \tag{A-23}$$

The result of $\mathbf{A}^T\mathbf{X}$ then is a product matrix built from elements as described in the matrix multiplication section.

A.3   Crystallographic coordinate system and fractional coordinates

In order to describe molecular objects in space, which is the ultimate goal of crystallography, we need to have a precise way to locate their components (such as the individual atoms comprising a protein) in 3-dimensional space. In direct space, which is where crystallographers and molecule alike reside, we use a rectilinear (as opposed to curvilinear) crystallographic coordinate system. This system is spanned by a right-handed set of three unit cell vectors, denoted as [0, **a**, **b**, **c**]. The system [0, **a**, **b**, **c**] forms the basis of the crystallographic coordinate system, which is not necessarily orthogonal. Only for the orthorhombic, tetragonal, and cubic crystal systems is the crystallographic basis also an orthogonal basis. Crystallographic computations involving symmetry operations are generally conducted in the crystallographic basis using fractional coordinates, while metric calculations and graphic display are generally based on transformations into orthogonal, Cartesian world coordinates.

A point in a crystallographic coordinate system [0, **a**, **b**, **c**] is described by its crystallographic position vector **r** through the fractional crystallographic coordinates x, y, z contained in a column vector **x**:

$$\mathbf{x} = \begin{pmatrix} x \\ y \\ x \end{pmatrix} \tag{A-24}$$

The scalar product of the basis row vector and the fractional coordinate column vector provides the crystallographic position vector **r**

$$\mathbf{r} = \begin{pmatrix} \mathbf{a} & \mathbf{b} & \mathbf{c} \end{pmatrix} \cdot \begin{pmatrix} x \\ y \\ z \end{pmatrix} = (x\mathbf{a} + y\mathbf{b} + z\mathbf{c}) = \mathbf{A}^T \cdot \mathbf{x} \tag{A-25}$$

with components $x\mathbf{a}$, $y\mathbf{b}$, and $z\mathbf{c}$ in the corresponding crystallographic directions.

Direction cosines

The direction cosines of a general vector $\mathbf{r} = (x\mathbf{a} + y\mathbf{b} + z\mathbf{c})$ in basis [0, **a**, **b**, **c**] are the cosines of the angles between the vector and the three basis vectors:

$$\cos\alpha = \frac{\mathbf{r} \cdot \mathbf{a}}{|\mathbf{r}|}, \quad \cos\beta = \frac{\mathbf{r} \cdot \mathbf{b}}{|\mathbf{r}|}, \quad \cos\gamma = \frac{\mathbf{r} \cdot \mathbf{c}}{|\mathbf{r}|} \tag{A-26}$$

and

$$\cos^2\alpha + \cos^2\beta + \cos^2\gamma = 1 \tag{A-27}$$

The metric tensor **G**, bond lengths, and bond angles

The dot product of a column vector and its row vector is the dyad product (also known as outer product, or tensor product) which yields a matrix. Following (A-23), the dyad product of **A** is the metric tensor **G**:

$$\mathbf{G} = \mathbf{A}\mathbf{A}^T = \begin{pmatrix} \mathbf{a} \\ \mathbf{b} \\ \mathbf{c} \end{pmatrix} \cdot \begin{pmatrix} \mathbf{a} & \mathbf{b} & \mathbf{c} \end{pmatrix} = \begin{pmatrix} \mathbf{a} \cdot \mathbf{a} & \mathbf{b} \cdot \mathbf{a} & \mathbf{c} \cdot \mathbf{a} \\ \mathbf{a} \cdot \mathbf{b} & \mathbf{b} \cdot \mathbf{b} & \mathbf{c} \cdot \mathbf{b} \\ \mathbf{a} \cdot \mathbf{c} & \mathbf{b} \cdot \mathbf{c} & \mathbf{c} \cdot \mathbf{c} \end{pmatrix} \tag{A-28}$$

The metric tensor **G** is useful for computing distances, angles, and volumes. It is generally computed as a first preparatory step in most crystallographic programs. Its dimensions are Å². Substituting the symbolic evaluations of the dot products into **G** provides a number of practical metrics. The determinant of **G** is the square of the unit cell volume:

$$\det\mathbf{G} = G = V_{UC}^2 = (abc)^2(1 - \cos^2\alpha - \cos^2\beta - \cos^2\gamma + 2\cos\alpha\cos\beta\cos\gamma) \tag{A-29}$$

It can be readily seen that for an orthogonal system where all angles are 90° and the direction cosines are zero, the cell volume computes as $V = abc$. As the determinant of a 3×3 matrix contains triple vector products of its elements, the dimensions of $\det(\mathbf{G})$ are $Å^6$ and the dimensional analysis checks.

The scalar product between two identical crystallographic position vectors $\mathbf{r}$ defines the square of the norm given via $\mathbf{G}$ as

$$r^2 = \mathbf{r} \cdot \mathbf{r} = \mathbf{X}^T \mathbf{G} \mathbf{X} = \begin{pmatrix} x & y & z \end{pmatrix} \mathbf{G} \begin{pmatrix} x \\ y \\ z \end{pmatrix} \tag{A-30}$$

which enumerates to

$$r^2 = (xa)^2 + (yb)^2 + (zc)^2 + 2xyab\cos\gamma + 2xzac\cos\beta + 2yzbc\cos\alpha \tag{A-31}$$

Again, it can be easily seen that for an orthogonal system the cosine terms disappear.

For two points defined by position vectors $\mathbf{r}_1$ and $\mathbf{r}_2$ we apply the same formula $\mathbf{X}^T \mathbf{G} \mathbf{X}$ to the difference vector $\mathbf{r}_{12}$ to obtain the distance between points 1 and 2:

$$r_{12}^2 = \mathbf{r}_{12} \cdot \mathbf{r}_{12} = \mathbf{X}_{12}^T \mathbf{G} \mathbf{X}_{12} = \begin{pmatrix} x_2 - x_1 & y_2 - y_1 & z_2 - z_1 \end{pmatrix} \mathbf{G} \begin{pmatrix} x_2 - x_1 \\ y_2 - y_1 \\ z_2 - z_1 \end{pmatrix} \tag{A-32}$$

The bond angles between three atoms defined by vectors position vectors $\mathbf{r}_1$, $\mathbf{r}_2$, and $\mathbf{r}_3$ and atom 1 at the center compute as

$$\cos\theta = \frac{\mathbf{X}_{12}^T \mathbf{G} \mathbf{X}_{13}}{r_{12} \cdot r_{13}} \tag{A-33}$$

Basis transformations

The general transformation from one basis $[0, \mathbf{a}, \mathbf{b}, \mathbf{c}]$ to a new one $[0, \mathbf{a}', \mathbf{b}', \mathbf{c}']$ sharing the same origin can be described using a transformation matrix $\mathbf{P}$ with elements p_{ij} that is formed by expressing the new system axes in terms of the old axes:

$$\mathbf{a}' = p_{11}\mathbf{a} + p_{12}\mathbf{b} + p_{13}\mathbf{c}$$
$$\mathbf{b}' = p_{21}\mathbf{a} + p_{22}\mathbf{b} + p_{23}\mathbf{c} \tag{A-34}$$
$$\mathbf{c}' = p_{31}\mathbf{a} + p_{32}\mathbf{b} + p_{33}\mathbf{c}$$

or in compact matrix notation

$$\mathbf{A}' = \mathbf{P}\mathbf{A} \text{ and the inverse } \mathbf{A} = \mathbf{P}^{-1}\mathbf{A}'$$

where matrix $\mathbf{A}$ contains the basis vector components of the respective systems as a row vectors. Basis transformations allow the crystallographic data to be presented in whatever system fits a particular task best; examples are fractional crystallographic coordinates for symmetry operations or re-indexing; Cartesian world coordinates for metric calculations and display; and reciprocal coordinates (reciprocal space) in the processing of diffraction data.

Re-indexing of diffraction data

The transformation matrix $\mathbf{P}$ takes different forms depending on the type of basis transformation. In re-indexing from reciprocal basis $[0, \mathbf{h}, \mathbf{k}, \mathbf{l}]$ to a new one $[0, \mathbf{h}', \mathbf{k}', \mathbf{l}']$, $\mathbf{P}$ is generally defined by fractions or multiples of the original reciprocal axes:

$$\mathbf{h}' = p_{11}\mathbf{h} + p_{12}\mathbf{k} + p_{13}\mathbf{l}$$
$$\mathbf{k}' = p_{21}\mathbf{h} + p_{22}\mathbf{k} + p_{23}\mathbf{l} \tag{A-35}$$
$$\mathbf{l}' = p_{31}\mathbf{h} + p_{32}\mathbf{k} + p_{33}\mathbf{l}$$

or in short matrix notation,

$\mathbf{H}' = \mathbf{PH}$ and the inverse operation $\mathbf{H} = \mathbf{P}^{-1}\mathbf{H}'$ (A-36)

The determinant of $\mathbf{P}$ then has a number of interesting properties:

- If $\det(\mathbf{P}) < 0$, the system changed handedness (generally undesirable).

- If $\det(\mathbf{P}) = 0$, the chosen new basis vectors are not linearly independent (and do not form a proper basis).

- The magnitude $|\det(\mathbf{P})|$ provides the ratio of the reciprocal unit cell volumes, V'/V.

Coordinate system transformations

A common basis transformation is from fractional crystallographic coordinates into orthonormal Cartesian world coordinates. Atomic position coordinates (X, Y, Z) in a PDB ATOM or HETATM record

```
ATOM    367  O    VAL A  47     -22.742  -1.823  28.183  1.00 23.68
-------------------------------- -- X ------ Y ----- Z -------------
```

are listed in a right-handed, Cartesian (orthonormal) coordinate system [0, $\mathbf{X}$, $\mathbf{Y}$, $\mathbf{Z}$] in Å while the fractional crystallographic positions are described in [0, $\mathbf{a}$, $\mathbf{b}$, $\mathbf{c}$] by a dimensionless fractional coordinate vector $\mathbf{x}$.

The general basis transformation $\mathbf{A}' = \mathbf{PA}$ and the inverse $\mathbf{A} = \mathbf{P}^{-1}\mathbf{A}'$ between basis [0, $\mathbf{X}$, $\mathbf{Y}$, $\mathbf{Z}$] and [0, $\mathbf{a}$, $\mathbf{b}$, $\mathbf{c}$] sharing a common origin therefore must accomplish two tasks: The basis orthogonalization provided by the fractional orthogonalization matrix $\mathbf{Q}$, which is then expanded into the Cartesian orthogonalization matrix $\mathbf{O}$ by multiplication with the unit cell matrix $\mathbf{A}^\mathrm{T}$ containing the unit cell vectors $\mathbf{a}$, $\mathbf{b}$, $\mathbf{c}$ in its columns:

$\mathbf{O} = \mathbf{A}^\mathrm{T}\mathbf{Q}$ and $\mathbf{O}^{-1} = (\mathbf{A}^\mathrm{T})^{-1}\mathbf{Q}^{-1}$ (A-37)

There are multiple choices of $\mathbf{O}$ depending on the order and selection of the axis rotations. The following convention is followed by PDB (and most crystallographic programs):

- Cartesian axis $\mathbf{X}$ is collinear with crystallographic axis $\mathbf{a}$.

- $\mathbf{Y}$ is collinear with $(\mathbf{a} \times \mathbf{b}) \times \mathbf{X}$.

- $\mathbf{Z}$ is collinear with $(\mathbf{a} \times \mathbf{b})$ (or with $\mathbf{c}^*$).

Using the above convention

$$\mathbf{O} = \mathbf{A}^\mathrm{T}\mathbf{Q} = \begin{pmatrix} a & b\cos\gamma & c\cos\beta \\ 0 & b\sin\gamma & \dfrac{c(\cos\alpha - \cos\beta\cos\gamma)}{\sin\gamma} \\ 0 & 0 & \dfrac{V}{ab\sin\gamma} \end{pmatrix}$$ (A-38)

with

$$V = abc(1 - \cos^2\alpha - \cos^2\beta - \cos^2\gamma + 2\cos\alpha\cos\beta\cos\gamma)^{1/2}$$ (A-39)

or

$$V = \det(\mathbf{O})$$ (A-40)

The deorthogonalization matrix $\mathbf{O}^{-1}$ can be obtained by inversion of $\mathbf{O}$ following Cramer's rule via the adjunct matrices

$$O^{-1} = \frac{adj(O)}{det(O)} \tag{A-41}$$

and reduces to

$$O^{-1} = \begin{pmatrix} \dfrac{1}{a} & -\dfrac{\cos\gamma}{a\sin\gamma} & \left(\dfrac{b\cos\gamma\,c(\cos\alpha - \cos\beta\cos\gamma)}{\sin\gamma} - bc\cos\beta\sin\gamma\right)\dfrac{1}{V} \\[2ex] 0 & \dfrac{1}{b\sin\gamma} & -\dfrac{ac(\cos\alpha - \cos\beta\cos\gamma)}{V\sin\gamma} \\[2ex] 0 & 0 & \dfrac{ab\sin\gamma}{V} \end{pmatrix} \tag{A-42}$$

The relation of O with the metric tensor G is distinctly visible and is given by $G = OO^T$, and with $\det(G) = V^2$ it follows that $\det(O) = V$.

Reciprocal space metric relations

The third important basis transformation is from direct space R into reciprocal space R^*. The crystallographic coordinate system $[0, a, b, c]$ with units of Å spanning a direct lattice described by a metric tensor G can be transformed into a reciprocal system $[0, a^*, b^*, c^*]$ spanning a reciprocal lattice described by a reciprocal tensor G^*. The relation of the axis is defined by the dyad product:

$$\begin{pmatrix} a^* \\ b^* \\ c^* \end{pmatrix} (a \quad b \quad c) = \begin{pmatrix} aa^* & ba^* & ca^* \\ ab^* & bb^* & cb^* \\ ac^* & bc^* & cc^* \end{pmatrix} = I \tag{A-43}$$

It follows from relation (A-43) above that

- a^* is normal to plane (b, c), b^* is normal to (a, c), and c^* is normal to (a, b); that is, all scalar products between normal and reciprocal basis vectors are zero except

- $a^* \cdot a = b^* \cdot b = c^* \cdot c = 1.$

Further,

$$G^* = G^{-1} \text{ and } \begin{pmatrix} a^* \\ b^* \\ c^* \end{pmatrix} = G^* \cdot \begin{pmatrix} a \\ b \\ c \end{pmatrix}. \tag{A-44}$$

It directly follows from relation (A-44) that $V^* = 1/V$, and the following metric relations between lattices can be readily derived:[1]

$$a^* = \frac{bc\sin\alpha}{V}, \quad b^* = \frac{ac\sin\beta}{V}, \quad c^* = \frac{ab\sin\gamma}{V}$$

$$\sin\alpha^* = \frac{V}{abc\sin\beta\sin\gamma}, \quad \cos\alpha^* = \frac{\cos\beta\cos\gamma - \cos\alpha}{\sin\beta\sin\gamma}$$

$$\sin\beta^* = \frac{V}{abc\sin\alpha\sin\gamma}, \quad \cos\beta^* = \frac{\cos\alpha\cos\gamma - \cos\beta}{\sin\alpha\sin\gamma}$$

$$\sin\gamma^* = \frac{V}{abc\sin\alpha\sin\beta}, \quad \cos\gamma^* = \frac{\cos\alpha\cos\beta - \cos\gamma}{\sin\alpha\sin\beta}$$

$$V = abc\sin\alpha\sin\beta\sin\gamma^* = abc\sin\alpha\sin\beta^*\sin\gamma = 1/V^* \tag{A-45}$$

Reciprocal lattice vectors

Let $[0, a, b, c]$ be the basis of a direct lattice and $[0, a^*, b^*, c^*]$ the basis of the corresponding reciprocal lattice (common origin). With h, k, l the Miller indices of a set of lattice planes (hkl), a reciprocal vector

$$\mathbf{d}^*_{hkl} = h\mathbf{a}^* + k\mathbf{b}^* + l\mathbf{c}^* = (\mathbf{a}^* \quad \mathbf{b}^* \quad \mathbf{c}^*)\begin{pmatrix} h \\ k \\ l \end{pmatrix} = (\mathbf{A}^*)^{\mathrm{T}}\mathbf{h} \tag{A-46}$$

is perpendicular to the set of lattice planes (hkl). The corresponding direct space vector $\mathbf{d}_{hkl}$ is the interplanar distance vector from origin to lattice plane (hkl) and is collinear with $\mathbf{d}^*_{hkl}$ extending from the origin to the reciprocal lattice point hkl. The metric relations in terms of reciprocal lattice can be derived from the reciprocal metric tensor $\mathbf{G}^* = \mathbf{G}^{-1}$ in the same fashion as for the direct space, and the general relation in terms of reciprocal space becomes

$$d^*_{hkl} = \frac{1}{d_{hkl}} = \left(h^2 a^{*2} + k^2 b^{*2} + l^2 c^{*2} + 2hka^*b^*\cos\gamma^* + 2hla^*c^*\cos\beta^* + 2klb^*c^*\cos\alpha^*\right)^{1/2} \tag{A-47}$$

The metric relations in direct space can be readily computed from the fact that in the Bragg case the scattering vector $\mathbf{S}$ equals the reciprocal lattice vector $\mathbf{d}^*_{\mathbf{h}}$:

$$d^*_{\mathbf{h}} = \frac{1}{d_{\mathbf{h}}} = \left|(\mathbf{O}^{-1})^{\mathrm{T}}\mathbf{h}\right| \tag{A-48}$$

Proof. The phase angle is independent of the coordinate basis and is given by $\mathbf{h}^{\mathrm{T}}\mathbf{x} = \mathbf{S}^{\mathrm{T}}\mathbf{r}$. We can set $\mathbf{r} = \mathbf{O}\mathbf{x}$ which is the orthogonalized coordinate vector obtained from fractional crystallographic coordinate vector $\mathbf{x}$.

Substituting $\mathbf{r} = \mathbf{O}\mathbf{x}$ into $\mathbf{S}^{\mathrm{T}}\mathbf{r}$ provides

$$\mathbf{h}^{\mathrm{T}}\mathbf{x} = \mathbf{S}^{\mathrm{T}}\mathbf{O}\mathbf{x}$$

$$\mathbf{h}^{\mathrm{T}} = \mathbf{S}^{\mathrm{T}}\mathbf{O} \tag{A-49}$$

$$\mathbf{h} = \mathbf{O}^{\mathrm{T}}\mathbf{S}$$

from which it immediately follows that

$$\mathbf{S} = \left(\mathbf{O}^{\mathrm{T}}\right)^{-1}\mathbf{h} = \left(\mathbf{O}^{-1}\right)^{\mathrm{T}}\mathbf{h} \tag{A-50}$$

Substituting into $|\mathbf{S}| = S = d^*_{\mathbf{h}}$ we obtain the final result

$$|\mathbf{S}| = S = d^*_{\mathbf{h}} = \frac{1}{d_{\mathbf{h}}} = \frac{2\sin\theta}{\lambda} = \left|(\mathbf{O}^{-1})^{\mathrm{T}}\mathbf{h}\right| \tag{A-51}$$

and thus

$$\left(\frac{1}{d_{\mathbf{h}}}\right)^2 = 4\left(\frac{\sin\theta}{\lambda}\right)^2 = \left|(\mathbf{O}^{-1})^{\mathrm{T}}\mathbf{h}\right|^2 \tag{A-52}$$

The results of enumeration of (A-52) for the orthogonal, hexagonal, and triclinic system are given below. The remaining cases can be readily obtained from these equations:

$$\frac{1}{d^2} = \frac{h^2}{a^2} + \frac{k^2}{b^2} + \frac{l^2}{c^2} \quad \text{(orthogonal)}$$

$$\frac{1}{d^2} = \frac{4}{3}\cdot\frac{h^2 + k^2 + hk}{a^2} + \frac{l^2}{c^2} \quad \text{(hexagonal)}$$

$$\frac{1}{d^2} = \frac{\dfrac{h^2}{a^2}\sin^2 a + \dfrac{k^2}{b^2}\sin^2\beta + \dfrac{l^2}{c^2}\sin^2\gamma + \\ + \dfrac{2kl}{bc}(\cos\beta\cos\gamma - \cos\alpha) + \\ + \dfrac{2hl}{ac}(\cos\alpha\cos\gamma - \cos\beta) + \dfrac{2hk}{ab}(\cos\alpha\cos\beta - \cos\gamma)}{(1 - \cos^2\alpha - \cos^2\beta - \cos^2\gamma + 2\cos\alpha\cos\beta\cos\gamma)} \quad \text{(triclinic)} \tag{A-53}$$

A.4 Cosine rule

Cosine rule. For any arbitrary triangle the cosine rule (Figure A-1) relates the sides as follows:

$$a^2 = b^2 + c^2 - 2bc\cos\alpha \tag{A-54}$$

A.5 Proof of Euler's formula

Euler's formula

$$e^{i\varphi} = \cos\varphi + i\sin\varphi \tag{A-55}$$

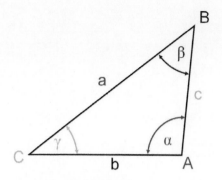

Figure A-1 The cosine rule: $a^2 = b^2 + c^2 - 2bc\cos\alpha$.

is used extensively in any structure factor related calculations. From (A-55) directly follows the Euler Identity, considered the perhaps most beautiful equation which includes five fundamental numbers: the transcendental numbers e and π as well as the elementary integers 1 and 0, and the fundamental complex number i:

$$e^{i\pi} + 1 = 0 \tag{A-56}$$

Proof. We expand the three terms in (A-55) via Taylor expansion around $f(0)$ into a MacLaurin series:

$$f(x) = \sum_{k=0}^{\infty} \frac{\partial^k f(0)}{\partial x^k} \cdot x^k \cdot \frac{1}{k!} \tag{A-57}$$

We use the basic derivatives (indicated by the superscript $'$)

$$(e^x)' = e^x, \ (\sin x)' = \cos x, \ (\cos x)' = -\sin x \tag{A-58}$$

to calculate all three series

$$e^{ix} = \sum_{k=0}^{\infty} (e^{ix})_{(0)}^{k'} \cdot x^k \cdot \frac{1}{k!} = 1 + i\frac{x}{1!} - \frac{x^2}{2!} - i\frac{x^3}{3!} + \frac{x^4}{4!} + i\frac{x^5}{5!} - \ldots - \ldots + \ldots + \ldots$$

$$\sin(x) = \frac{x}{1!} - \frac{x^3}{3!} + \frac{x^5}{5!} - \ldots + \ldots$$

$$\cos(x) = 1 - \frac{x^2}{2!} + \frac{x^4}{4!} - \frac{x^6}{6!} + \ldots - \ldots \tag{A-59}$$

Observing the alternating exponentials and grouping real and imaginary terms we immediately obtain the result

$$e^{ix} = \cos x + i\sin x \tag{A-60}$$

A.6 Electron density is real

As an exercise in practical computing with complex numbers and trigonometric functions, we give an explicit arithmetic proof that the electron density is real.

We can group the structure factors in the electron density summation

$$\rho(\mathbf{x}) = \frac{1}{V} \sum_{\mathbf{h}=-\infty}^{+\infty} \mathbf{F}(\mathbf{h}) \cdot \exp(-2\pi i \mathbf{h}\mathbf{x}) \tag{A-61}$$

into centrosymmetric pairs, and evaluate the expression for each pair. Written in our short-hand notation, for each pair $\mathbf{F_h} + \mathbf{F_{-h}}$ we can write

$$\mathbf{F_h} \cdot \exp\left(-2\pi i(\mathbf{h}\cdot\mathbf{x})\right) + \mathbf{F_{-h}} \cdot \exp\left(-2\pi i(-\mathbf{h}\cdot\mathbf{x})\right)$$

$$= \mathbf{F_h} \cdot \exp\left(-2\pi i(\mathbf{h}\cdot\mathbf{x})\right) + \mathbf{F_{-h}} \cdot \exp\left(2\pi i(\mathbf{h}\cdot\mathbf{x})\right) \tag{A-62}$$

Setting now $2\pi(\mathbf{h}\cdot\mathbf{x}) = \varphi_r$, using Euler's formula, and substituting the factorization of the complex structure factor $\mathbf{F_h}$ into real and imaginary terms A and B, we obtain for each structure factor pair

$$(A_h + iB_h)\cdot\exp(-i\varphi_r) + (A_h - iB_h)\cdot\exp(i\varphi_r)$$
$$= (A_h + iB_h)(\cos(-\varphi_r) + i\sin(-\varphi_r)) + (A_h - iB_h)(\cos\varphi_r + i\sin\varphi_r)$$
$$= (A_h + iB_h)(\cos\varphi_r - i\sin\varphi_r) + (A_h - iB_h)(\cos\varphi_r + i\sin\varphi_r)$$
$$= A_h\cos\varphi_r + iB_h\cos\varphi_r - A_h i\sin\varphi_r + B_h\sin\varphi_r$$
$$+ A_h\cos\varphi_r - iB_h\cos\varphi_r + A_h i\sin\varphi_r + B_h\sin\varphi_r$$
$$= 2(A_h\cos\varphi_r + B_h\sin\varphi_r) \tag{A-63}$$

We can see that all the imaginary terms have been cancelled out. As the expression above holds for each pair $\mathbf{F_h} + \mathbf{F_{-h}}$, the electron density now can be summed twice over half of the reflections and becomes explicitly real:

$$\rho(\mathbf{x}) = \frac{2}{V}\sum_{h=0}^{+\infty}\sum_{k=-\infty}^{+\infty}\sum_{l=-\infty}^{+\infty} A_h\cos\varphi_r + B_h\sin\varphi_r \tag{A-64}$$

We can again make this formula explicit in F_h and its associated phase α_h by substituting $A_h = F_h\cos\alpha_h$ and $B_h = F_h\sin\alpha_h$ back into Equation A-64, which gives

$$\rho(\mathbf{x}) = \frac{2}{V}\sum_{h=0}^{+\infty}\sum_{k=-\infty}^{+\infty}\sum_{l=-\infty}^{+\infty} F_h\cos\alpha_h\cos\varphi_r + F_h\sin\alpha_h\sin\varphi_r$$

$$= \frac{2}{V}\sum_{h=0}^{+\infty}\sum_{k=-\infty}^{+\infty}\sum_{l=-\infty}^{+\infty} F_h(\cos\alpha_h\cos\varphi_r + \sin\alpha_h\sin\varphi_r) \tag{A-65}$$

Applying the basic trigonometric addition theorem

$$\cos(\alpha - \beta) = \cos\alpha\cos\beta + \sin\alpha\sin\beta \tag{A-66}$$

we finally obtain

$$\rho(\mathbf{x}) = \frac{2}{V}\sum_{h=0}^{+\infty}\sum_{k=-\infty}^{+\infty}\sum_{l=-\infty}^{+\infty} F_h\cos(\varphi_r - \alpha_h) \tag{A-67}$$

which is again real and in a useful form for actual calculation of the electron density transform. Make sure to clearly distinguish between φ_r and α_h: $\varphi_r = 2\pi(\mathbf{h}\cdot\mathbf{x})$ contains here the fractional coordinate vector $\mathbf{x}$ of each grid point in the density reconstruction $\rho(\mathbf{x}) = \rho(xyz)$ with components xyz, while α_h is the phase angle of each reflection $\mathbf{h}$ and contains the fractional atom coordinate vector $\mathbf{x}_j$ for each atom j.

A.7   Direction cosine matrix in spherical coordinates

The direction cosine matrix describing a rotation in spherical coordinates is defined by

$$\mathbf{R} = \mathbf{R}_Z(\varphi)\mathbf{R}_Y(\psi)\mathbf{R}_Z(\kappa)\mathbf{R}_Y(-\psi)\mathbf{R}_Z(-\varphi) \tag{A-68}$$

Using the substitution for the Euler vector components

$$\mathbf{E} = \begin{pmatrix} l \\ m \\ n \end{pmatrix} = \begin{pmatrix} \sin\psi\cos\varphi \\ \sin\psi\sin\varphi \\ \cos\psi \end{pmatrix} \tag{A-69}$$

the matrix product (A-68) evaluates symbolically to

$$\mathbf{R} = \begin{pmatrix} l^2 + (m^2 + n^2)\cos\kappa & lm(1-\cos\kappa) - n\sin\kappa & nl(1-\cos\kappa) + m\sin\kappa \\ lm(1-\cos\kappa) + n\sin\kappa & m^2 + (m^2 + l^2)\cos\kappa & mn(1-\cos\kappa) - l\sin\kappa \\ nl(1-\cos\kappa) - m\sin\kappa & mn(1-\cos\kappa) + l\sin\kappa & n^2 + (l^2 + m^2)\cos\kappa \end{pmatrix} \tag{A-70}$$

A.8 Crystallographic restriction theorem

By symbolic evaluation of Equation A-70 one can shown that the trace of a rotation matrix (a DCM) is $1 + 2\cos\kappa$, independent of any basis. In a crystallographic lattice basis, the limitation exists that the rotation must map lattice points onto lattice points, and therefore crystallographic rotations $\mathbf{R}_C$ in a crystallographic basis always contain ones (positive or negative) or zeroes as matrix elements, and the trace must also be integral. Therefore $1 + 2\cos\kappa$ must be integral and the only possible solutions are rotation matrices with integers $-1 \le n \le 3$; for -1: $\cos\kappa = -1$, $180°$; for 0: $\cos\kappa = -\frac{1}{2}$, $120°$; for 1: $\cos\kappa = 0$, $90°$ for 2: $\cos\kappa = \frac{1}{2}$, $60°$; for 3: $\cos\kappa = 1$, $0°$; for $n \le -2$ and $n \ge 4$ no solutions exist.

As the trace is preserved during basis transformation, we can transform every crystallographic rotation $\mathbf{R}_C$ by deorthogonalization

$$\mathbf{R}_W = \mathbf{O}\mathbf{R}_C\mathbf{O}^{-1} \tag{A-71}$$

into a Cartesian (world coordinate) DCM $\mathbf{R}_W$, and would obtain the same result.

A.9 Anatomy of a ligand restraint file

The restraint files generated by the automated setup programs such as *LIBCHECK* or *PRODRG* can be additionally edited and fine tuned using a text editor. The example in Table A-3 lists the restraint file for benzamidine (benzyldiamine) in CIF format.

```
#
data_comp_list
loop_
_chem_comp.id
_chem_comp.three_letter_code
_chem_comp.name
_chem_comp.group
_chem_comp.number_atoms_all
_chem_comp.number_atoms_nh
_chem_comp.desc_level
BEN        .    'BENZYLDIAMINE              ' non-polymer   18   9  .
#
data_comp_BEN
#
loop_
_chem_comp_atom.comp_id
_chem_comp_atom.atom_id
_chem_comp_atom.type_symbol
_chem_comp_atom.type_energy
_chem_comp_atom.partial_charge
   BEN           N2      N     NH2       0.000
   BEN           C       C     C         0.000
   BEN           N1      N     NH2       0.000
   BEN           C1      C     CR6       0.000
   BEN           C2      C     CR16      0.000
   BEN           C3      C     CR16      0.000
   BEN           C4      C     CR16      0.000
   BEN           C5      C     CR16      0.000
   BEN           C6      C     CR16      0.000
loop_
_chem_comp_tree.comp_id
_chem_comp_tree.atom_id
_chem_comp_tree.atom_back
_chem_comp_tree.atom_forward
_chem_comp_tree.connect_type
   BEN        N2      n/a      C       START
   BEN        C       N2       C1      .
   BEN        C1      C        C2      .
   BEN        C2      C1       C3      .
   BEN        C3      C2       C4      .
   BEN        C4      C3       C5      .
   BEN        C5      C4       C6      .
   BEN        C6      C5       H6      END
```

continued

```
loop_
_chem_comp_bond.comp_id
_chem_comp_bond.atom_id_1
_chem_comp_bond.atom_id_2
_chem_comp_bond.type
_chem_comp_bond.value_dist
_chem_comp_bond.value_dist_esd
 BEN       C       N2        coval      1.320    0.020
 BEN       N1      C         coval      1.320    0.020
 BEN       C1      C         coval      1.400    0.020
 BEN       C1      C6        coval      1.390    0.020
 BEN       C2      C1        coval      1.390    0.020
 BEN       C3      C2        coval      1.390    0.020
 BEN       C4      C3        coval      1.390    0.020
 BEN       C5      C4        coval      1.390    0.020
 BEN       C6      C5        coval      1.390    0.020
loop_
_chem_comp_angle.comp_id
_chem_comp_angle.atom_id_1
_chem_comp_angle.atom_id_2
_chem_comp_angle.atom_id_3
_chem_comp_angle.value_angle
_chem_comp_angle.value_angle_esd
 BEN       N2      C       N1      118.000    3.000
 BEN       N2      C       C1      121.000    3.000
 BEN       N1      C       C1      121.000    3.000
 BEN       C       C1      C2      120.000    3.000
 BEN       C       C1      C6      120.000    3.000
 BEN       C2      C1      C6      120.000    3.000
 BEN       C1      C2      C3      120.000    3.000
 BEN       C2      C3      C4      120.000    3.000
 BEN       C3      C4      C5      120.000    3.000
 BEN       C4      C5      C6      120.000    3.000
 BEN       C5      C6      C1      120.000    3.000
loop_
_chem_comp_tor.comp_id
_chem_comp_tor.id
_chem_comp_tor.atom_id_1
_chem_comp_tor.atom_id_2
_chem_comp_tor.atom_id_3
_chem_comp_tor.atom_id_4
_chem_comp_tor.value_angle
_chem_comp_tor.value_angle_esd
_chem_comp_tor.period
 BEN    var_2    N2    C     C1    C2    169.238    20.000    1
 BEN    var_3    N2    C     C1    C6    -10.638    20.000    1
 BEN    CONST_1  C     C1    C2    C3    180.000     0.000    0
 BEN    CONST_2  C1    C2    C3    C4      0.000     0.000    0
 BEN    CONST_3  C2    C3    C4    C5      0.000     0.000    0
 BEN    CONST_4  C3    C4    C5    C6      0.000     0.000    0
 BEN    CONST_5  C4    C5    C6    C1      0.000     0.000    0
loop_
_chem_comp_plane_atom.comp_id
_chem_comp_plane_atom.plane_id
_chem_comp_plane_atom.atom_id
_chem_comp_plane_atom.dist_esd
 BEN     plan-1   C1       0.020
 BEN     plan-1   C2       0.020
 BEN     plan-1   C6       0.020
 BEN     plan-1   C        0.020
 BEN     plan-1   C3       0.020
 BEN     plan-1   C4       0.020
 BEN     plan-1   C5       0.020
 BEN     plan-2   C        0.020
 BEN     plan-2   C1       0.020
 BEN     plan-2   N1       0.020
 BEN     plan-2   N2       0.020
```

Table A-3 Ligand restraint file for benzamidine in CIF format.

A.10 The 44 indexing possibilities and possible space groups

No	PENALTY	LATT	a	b	c	alpha	beta	gamma	Possible spacegroups
44	999	mI	86.42	176.89	38.83	102.7	116.7	60.8	C2 (transformed from I2)
43	888	oF	86.42	86.40	176.89	106.9	119.2	53.4	F222
42	888	tI	109.14	86.41	86.42	78.4	50.9	50.8	I4,I41,I422,I4122
41	888	tI	86.41	86.42	109.14	50.9	50.8	78.4	I4,I41,I422,I4122
40	888	oI	86.41	86.42	109.14	50.9	50.8	78.4	I222,I212121
39	887	mC	86.40	86.42	86.41	78.4	101.7	53.4	C2
38	780	hR	176.88	109.22	38.83	90.0	77.3	108.0	H3,H32 (hexagonal settings of R3 and R32)
			115.83	77.19	77.21	90.0	177.7	177.6	R3,R32 (primitive rhombohedral cell)
37	668	cF	115.84	115.83	115.90	96.4	96.4	140.8	F23,F432,F4132
36	668	cI	86.41	86.42	109.14	50.9	50.8	78.4	I23,I213,I432,I4132
35	668	hR	38.83	86.40	247.15	102.0	99.0	116.7	H3,H32 (hexagonal settings of R3 and R32)
			77.21	86.41	109.14	50.8	50.8	50.8	R3,R32 (primitive rhombohedral cell)
34	556	tI	38.83	77.19	176.89	64.2	77.3	90.0	I4,I41,I422,I4122
33	556	oI	38.83	77.19	176.89	115.8	102.7	90.0	I222,I212121
32	444	hP	38.83	77.19	77.21	90.0	90.0	90.0	P3,P31,P32,P312,P321,P3112,P3121,P3212,P3221
									P6,P61,P65,P62,P64,P63,P622,P6122,P6522,P6222,P6422,P6322
31	444	hP	77.19	77.21	38.83	90.0	90.0	90.0	P3,P31,P32,P312,P321,P3112,P3121,P3212,P3221
									P6,P61,P65,P62,P64,P63,P622,P6122,P6522,P6222,P6422,P6322
30	444	oC	77.19	172.58	38.83	90.0	90.0	116.5	C222,C2221
29	443	mC	77.19	172.58	38.83	90.0	90.0	63.5	C2
28	443	mC	172.58	77.19	38.83	90.0	90.0	63.5	C2
27	335	tI	109.14	115.90	38.83	70.5	90.0	90.0	I4,I41,I422,I4122
26	335	oF	38.83	159.19	159.21	86.6	104.1	104.1	F222
25	333	hR	86.40	86.43	115.83	116.4	63.6	101.6	H3,H32 (hexagonal settings of R3 and R32)
			77.19	38.83	77.21	90.0	90.1	90.0	R3,R32 (primitive rhombohedral cell)
24	332	hR	86.40	86.41	115.93	116.4	63.6	101.7	H3,H32 (hexagonal settings of R3 and R32)
			77.19	38.83	77.21	90.0	89.9	90.0	R3,R32 (primitive rhombohedral cell)
23	332	oC	86.42	86.40	77.21	90.0	90.1	53.4	C222,C2221
22	332	mC	86.42	86.40	77.21	90.0	90.1	53.4	C2
21	332	tP	38.83	77.19	77.21	90.0	90.0	90.0	P4,P41,P42,P43,P422,P4212,P4122,P41212,P4222,P42212,P4322,P43212
20	332	mC	86.42	86.40	77.21	90.0	90.1	126.6	C2
19	332	cP	38.83	77.19	77.21	90.0	90.0	90.0	P23,P213,P432,P4232,P4332,P4132
18	225	oI	38.83	109.14	115.90	90.0	70.5	90.0	I222,I212121
17	224	mC	159.19	38.83	109.14	90.0	133.3	75.9	C2
16	113	oC	38.83	159.19	77.21	90.0	90.0	104.1	C222,C2221
15	113	oC	38.83	159.21	77.19	90.0	90.0	104.1	C222,C2221
14	113	mC	38.83	159.21	77.19	90.0	90.0	75.9	C2
13	113	mC	38.83	159.19	77.21	90.0	90.0	75.9	C2
12	112	mC	159.19	38.83	77.21	90.0	90.0	75.9	C2
11	112	mC	159.21	38.83	77.19	90.0	90.0	75.9	C2
10	1	tP	77.19	77.21	38.83	90.0	90.0	90.0	P4,P41,P42,P43,P422,P4212,P4122,P41212,P4222,P42212,P4322,P43212
9	1	oP	38.83	77.19	77.21	90.0	90.0	90.0	P222,P2221,P21212,P212121
8	0	mP	77.19	38.83	77.21	90.0	90.0	90.0	P2,P21
7	0	mC	109.14	109.22	38.83	90.0	90.0	90.0	C2
6	0	aP	38.83	77.19	77.21	90.0	90.0	90.0	P1
5	0	oC	109.14	109.22	38.83	90.0	90.0	90.0	C222,C2221
4	0	mC	109.22	109.14	38.83	90.0	90.0	90.0	C2
3	0	mP	38.83	77.21	77.19	90.0	90.0	90.0	P2,P21
2	0	mP	38.83	77.19	77.21	90.0	90.0	90.0	P2,P21
1	0	aP	38.83	77.19	77.21	90.0	90.0	90.0	P1
No	PENALTY	LATT	a	b	c	alpha	beta	gamma	Possible spacegroups

Table A-4 The 44 indexing possibilities and their possible space groups. The data are for a tetragonal structure: note the distinct drop of the penalty function for the correct tetragonal primitive indexing (No. 10). For primitive lattices, a stands for anorthic, that is, P1; the remaining lattice symbols are self-explanatory.

A.11 Protein–metal coordination and distances

Distance (Å) between coordinated metal, water molecules, and specific types of protein ligands						
Metal	H$_2$O	Monodentate carboxylate	Bidentate carboxylate	Main chain carbonyl	Histidine imidazole	Cysteine thiolate
Na	2.42(19)	2.41(11)	2.8	2.37(6)	–	–
Mg	2.09(8)	2.08(8)	2.5	2.26(23)	–	–
K	2.82(14)	2.82(13)	3.2	2.67(10)	–	–
Ca	2.40(10)	2.33(7)	2.8	2.36(10)	–	–
Mn	2.22(2)	2.12(5)	2.5	2.19(5)	2.21(10)	2.35(4)
Fe	2.17(8)	2.10(6)	2.5	2.04(6)	2.16(10)	2.30(5)
Co	2.09(3)	2.05(6)	2.5	2.08(5)	2.14(10)	2.25(10)
Cu	2.13(22)	1.96(4)	2.5	2.04(14)	2.02(10)	2.15(9)
Zn	2.06(13)	2.01(9)	2.5	2.07(5)	2.03(5)	2.31(10)
For Cl$^-$ ions, coordination distances are 3.1–3.7 Å, backbone N often involved; also positively charged residues						

Table A-5. Table of coordination distances of metal coordination groups in proteins. Values are based partly on distances in the Cambridge Structural Database (CSD), and partly on those in protein structures determined at or near atomic resolution as described by M. Harding.[3,4] The provided variances serve as a guidance only, because the chemical environment may vary depending on each specific case. The following cautionary comments are taken from M. Harding's web site: Interactions may have a varying degree of covalent/electrostatic nature (e.g. Na, K), and coordination numbers and distances can vary due to Jahn–Teller distortions (e.g. in the case of Cu or Co). In addition some cases are based on a limited number of observations. For asparagines and glutamine, M–O distances are similar to monodentate carboxylates, perhaps very slightly longer. For serine and threonine, expect M–O distances between those for water and for monodentate carboxylate. For tyrosine, expect M–O distances significantly shorter (by ~0.1 Å) than for monodentate carboxylate. For distances to non-protein O, N, or S the values for O of asparagine, N of histidine, or S of cysteine may be used. The distances for bidentate ligands can vary greatly, and the values are upper limits based on a single metal coordinated by both oxygen atoms of the same carboxyl group. Data are reproduced with kind permission from the IUCr and M. Harding.

A.12 Selected physical constants and conversions

Common symbol	Name	Common units	
h	Planck constant	4.135 667 33(10) × 10^{-15} eV · s	6.626 068 96(33) × 10^{-34} J · s
k, k_B	Boltzmann constant	8.617 343(15) × 10^{-5} eV · K^{-1}	1.380 6504(24) × 10^{-23} J · K^{-1}
N_A	Avogadro number	6.022 141 79(30) × 10^{23} mol^{-1}	
c	Speed of light	299,792,458 m · s^{-1}	299 792.5 · 10^4 cm · s^{-1}
R	Gas constant	5.189 4856 × 10^{19} eV · K^{-1} · mol^{-1}	8.314 472(15) J · K^{-1} · mol^{-1}
Cb	Coulomb, e$^-$ charge	1 C	1 A · s
Da	Atomic mass unit	931.494028(23) MeV · c^{-2}	1.660 538 782(83) × 10^{-24} g
Energy conversions			
E	Energy	1 eV = 8065.5 cm^{-1}	1 J = 0.23901 cal
		1 eV = 1.602 176 53 × 10^{-19} J	1 J = 1 kg · m^2 · s^{-2}
		λ (Å) = 12397.639/E(eV)	

Table A-6. Selected physical constants and conversions.

A.13 References

1. Giacovazzo C, Monaco HL, Viterbo D, et al. (2002) *Fundamentals of Crystallography*. Oxford, UK: Oxford Science Publications.

2. Sands DE (1995) *Vectors and Tensors in Crystallography*. New York, NY: Dover Publications, Inc.

3. Harding M (2004) The architecture of metal coordination groups in proteins. *Acta Crystallogr.* D60(5), 849–859.

4. Harding M (2006) Small revisions to predicted distances around metal sites in proteins. *Acta Crystallogr.* D62(6), 678–682.

Table of notation

Crystal geometry

$\mathbf{x}$	fractional coordinate vector (x, y, z), in crystallographic system $[0,\mathbf{a},\mathbf{b},\mathbf{c}]$
$\mathbf{X}$	position vector in Cartesian space (X, Y, Z) in Cartesian system $[0,\mathbf{A},\mathbf{B},\mathbf{C}]$
$\mathbf{r}$	a general real space position vector in crystallographic system $[0,\mathbf{a},\mathbf{b},\mathbf{c}]$
$\mathbf{r}^*$	a general reciprocal lattice vector in reciprocal system $[0, \mathbf{a}^*, \mathbf{b}^*, \mathbf{c}^*]$
$\mathbf{A}^*$	reciprocal unit cell basis matrix, contains $\mathbf{a}^*, \mathbf{b}^*, \mathbf{c}^*$ in row vectors
$\mathbf{d}_\mathbf{h}^*$	reciprocal lattice vector with directional indices h, k, l, perpendicular to plane (hkl), $\mathbf{d}_\mathbf{h}^* = (\mathbf{A}^*)^{\mathrm{T}}\mathbf{h}$
$\mathbf{Q}, \mathbf{Q}^{-1}$	fractional orthogonalization matrix, fractional deorthogonalization matrix
$\mathbf{O}, \mathbf{O}^{-1}$	orthogonalization matrix, deorthogonalization matrix, $\mathbf{O} = \mathbf{A}^{\mathrm{T}}\mathbf{Q}$, $\mathbf{O}^{-1} = \mathbf{A}^{\mathrm{T}}\mathbf{Q}^{-1}$
$\mathbf{I}$	unit vector, unit matrix
$\mathbf{R}$	3×3 rotation matrix, directional cosine matrix, DCM
$\mathbf{R}^{-1}, \mathbf{R}^{\mathrm{T}}$	inverse of $\mathbf{R}$, transpose of $\mathbf{R}$, with $\mathbf{R}^{-1}\mathbf{R}^{\mathrm{T}} = \mathbf{I}$
$\mathbf{T}$	translation vector
$\mathbf{W}$	crystallographic symmetry transformation, augmented 4×4 matrix
$\mathbf{W(R,T)}$	general Cartesian transformation, with operations $\mathbf{R}$ and $\mathbf{T}$
$\theta_1, \theta_2, \theta_3$	Euler angles determining components of $\mathbf{R}$
φ, ϕ, κ	sperical angles determining components of $\mathbf{R}$
$\mathbf{E}$	Euler axis, describes rotation axis about the principal Euler angle κ
κ	principal Euler angle, rotation about $\mathbf{E}$

Structure factors

$\mathbf{x}, \mathbf{x}_j$	fractional atomic coordinate vector (x, y, z) ; of atom j
$|\mathbf{x}|, x$	norm or magnitude of $\mathbf{x}$
$\mathbf{S}, \mathbf{s}_0, \mathbf{s}_1$	general scattering vector; incoming and diffracted wave vector
$\mathbf{d}_\mathbf{h}^*$	reciprocal lattice vector, identical to scattering vector in Bragg case $\mathbf{S} = \mathbf{d}_\mathbf{h}^* = (\mathbf{A}^*)^{\mathrm{T}}\mathbf{h}; |\mathbf{d}_\mathbf{h}^*| = 1/d = 2\sin\theta_\mathbf{h}/\lambda; |(\mathbf{O}^{-1})^{\mathrm{T}}\mathbf{h}|^2/4 = |\mathbf{S}|^2/4 = (\sin\theta/\lambda)^2$
$\mathbf{h}$	vector containing Miller indices h, k, l, of lattice plane (hkl); indices of reciprocal lattice vector $\mathbf{d}_\mathbf{h}^*$

$I, I_h, I(\mathbf{h})$	intensity of reflection; of reciprocal index vector $\mathbf{h}$
$\rho, \rho_j, \rho_r, \rho(\mathbf{r})$	electron density; of atom j; at real space position or grid point $\mathbf{r}$
f, f_j, f_j^0	wavelength independent atomic scattering factor; of atom j; the Fourier transform of electron density ρ_j
f^B, f_j^B	B-factor corrected wavelength independent atomic scattering factor
$\mathbf{F}, \mathbf{F_h}, \mathbf{F(h)}, \mathbf{F}_j$	complex structure factor; of reflection with reciprocal lattice vector $\mathbf{h}$; of atom j $$\mathbf{F_h} = \sum_{j=1}^{N} f_j(\mathbf{h}) \cdot \exp(2\pi i \mathbf{h} \mathbf{x}_j) = F_h \cdot \exp(i\varphi_h) \text{ for a structure of } N \text{ atoms}$$
$F, F_h, F(\mathbf{h}), F_j$	structure factor magnitude or amplitude; of reflection with reciprocal index vector $\mathbf{h}$; of atom j
$\alpha, \varphi, \alpha_h, \varphi_h$	phase angle; of structure factor $\mathbf{F_h}$

Displacement parameters, B-factors

$T(\mathbf{h}), T_j(\mathbf{h})$	temperature factor, Debye–Waller factor; for atom j isotropic: $T_j(\mathbf{h}) = \exp(-B_j(\sin\theta / \lambda)^2)$
B_{iso}, B_j, u, u_j	isotropic B-factor; of atom j; isotropic displacement parameter in Å; of atom j; $B = 8\pi^2 \langle u^2 \rangle$
$\mathbf{U}^*, \mathbf{U}_j^*, \mathbf{u}, \mathbf{u}_j$	reciprocal anisotropic displacement tensor; of atom j atomic displacement vector, of atom j
$\mathbf{U}_{cart}, \mathbf{U}_W, U_W^{ij}$	Cartesian anisotropic displacement tensor; components thereof

Statistics, likelihood

$\bar{x}$	arithmetic mean, unweighted average; $\bar{x} = \dfrac{1}{N}\sum_{j=1}^{N} x_j$
$\langle x \rangle$	expectation value, mean; of random variable
σ_x^2	variance of variable x: $\sigma_x^2 = \langle (x - \langle x \rangle)^2 \rangle = \langle x^2 \rangle - \langle x \rangle^2$
σ_x	estimated standard error, standard deviation, standard uncertainty
r.m.s.d.	root mean square deviation
w_i	statistical weight of observation i: $w_i = 1/\sigma_i^2$
x_{best}	weighted average, best value $$x_{best} = \sum_{i=1}^{n} w_i x_i / \sum_{i=1}^{n} w_i, \text{ with } \sigma_{best} = \left(\sum_{i=1}^{n} w_i\right)^{-1/2}$$
$\langle \mathbf{F} \rangle, \mathbf{F}_{BEST}$	expectation value probability weighted average (best value), of structure factor
$\mathbf{E}, \mathbf{E_h}, \mathbf{E(h)}, \mathbf{E}_j$	normalized structure factor; of reciprocal lattice vector $\mathbf{h}$; of atom j $\mathbf{E_h} = \dfrac{\mathbf{F_h}}{(\varepsilon_h \Sigma_N)^{1/2}}$
$E, E_h, E(\mathbf{h}), E_j$	magnitude of normalized structure factor; of reflection $\mathbf{h}$; of atom j $$E_h = \frac{F_h}{\langle F_h^2 \rangle^{1/2}} = \frac{F_h}{(\varepsilon_h \Sigma_N)^{1/2}}$$

σ_N^2, Σ_N	variance of a structure factor or of its expectation value; sample of size N:

$$\Sigma_N = \sum_{j=1}^{N} f_j^2 = \frac{1}{\varepsilon_{\mathbf{h}}} \langle F^2 \rangle$$

$\varepsilon, \varepsilon_{\mathbf{h}}$	statistical weight of reflection with index $\mathbf{h}$, depending on reflection zone
$m, m_{\mathbf{h}}$, f.o.m.	figure of merit, statistical weight; of structure factor with index $\mathbf{h}$

defined as mean phase error: $m = \langle \cos(\varphi_T - \varphi_{calc}) \rangle$ with φ_T the "true" phase;

in likelihood function for acentric reflections: $m_a = \dfrac{I_1(X)}{I_0(X)}$

in likelihood function for centric reflections: $m_c = \tanh\left(\dfrac{X}{2}\right)$.

I_0, I_1	modified (hyperbolic) Bessel functions of order zero; of order one
X	argument of Bessel functions I or hyperbolic functions in likelihood function for a partial

structure without errors: $X = \dfrac{2 F_{obs} F_{calc}}{\varepsilon \Sigma_Q}$

for a partial structure with errors: $X = \dfrac{2 \sigma_A (\Sigma_P / \Sigma_N) F_{obs} F_{calc}}{\varepsilon \Sigma_P (1 - \sigma_A^2)}$

$D, D_{\mathbf{h}}$	Luzzati D factor; for each structure factor as function of real space coordinate error $\Delta \mathbf{r}$:

$D_{\mathbf{h}} = \langle \cos 2\pi(\Delta \mathbf{r}_j \mathbf{S}) \rangle = \exp(-\pi^3 \langle |\Delta \mathbf{r}_j| \rangle^2 S^2 / 4)$;

as function of σ_A below

σ_A	Sigma-A, variance-term accounting for error and incompleteness in LF

$\sigma_A = D(\Sigma_P / \Sigma_N)^{1/2} = D F_{\mathbf{h}} / E_{\mathbf{h}}$

σ_Δ^2	complex variance of structure factor when atoms have positional errors $\Delta \mathbf{r}$:

$\sigma_\Delta^2 = 1 - \sigma_A^2$

∇	vector differential operator, Nabla operator, del operator

Refinement

$\mathbf{F}_{obs}, \mathbf{F}_o, \mathbf{F}(obs)$	true complex structure factors of a crystal structure
$F_{obs}, F_o, F(obs)$	magnitude of experimentally observed structure factors
$\mathbf{F}_{calc}, \mathbf{F}_c, \mathbf{F}(calc)$	calculated complex structure factors of any model structure, also partial
$F_{calc}, F_c, F(calc)$	calculated structure factor magnitudes of any model structure, also partial

Glossary

Bold terms within definitions denote other Glossary terms. B = biological meaning or definition, C = crystallographic context, M = mathematical context. Numbers in parentheses next to the entry refer to defining equations.

A

absent reflections – *see* **systematic absences**

absolute configuration
Absolute configuration is defined by the sequence of the ligands around a chiral center (S = *sinister*, left; R = *rectus*, right) following the **Cahn–Ingold–Prelog priority rules**. The absolute configuration of a molecule can be determined by **anomalous** X-ray **diffraction** experiments.

absolute scale – *see* **Wilson scaling**

absorption edge – *see* **X-ray absorption edge**

absorption edge scan – *see* **X-ray excitation scan**

accuracy
A measure of how close an experimental mean value or **expectation value** is to the true value. *Compare* **precision**.

acentric molecule
Molecule that possesses no center of **inversion**.

acentric structure factor (or reflection)
General structure factor that lacks **centrosymmetry** and has no **phase restricted reflections**.

acetyl (–COCH$_3$)
Chemical group derived from acetic acid (CH$_3$COOH). Acetyl groups are important in metabolism and are also added covalently to some proteins as a **posttranslational modification**.

acetylcholine (ACh)
Neurotransmitter that functions at cholinergic synapses. Found in the brain and peripheral nervous system. The neurotransmitter at vertebrate neuromuscular junctions.

achiral
Not **chiral**

acid
A proton donor. Substance that releases **protons** (H$^+$) when dissolved in water, forming **hydronium ions** (H$_3$O$^+$) and lowering the pH.

active site
Region of an enzyme surface to which a substrate molecule binds in order to undergo a catalyzed reaction.

acyl group (–CO–R)
Functional group derived from a carboxylic acid.

additive (C)
An ion or molecule, added to the crystallization cocktail aiming to alter or improve protein crystallization behavior.

ADME, ADME/T
Acronym in pharmacology for **a**bsorption, **d**istribution, **m**etabolism, **e**xcretion (and **t**oxicity), which describes the disposition of a pharmaceutical compound within an organism.

affinity
The strength of binding of a molecule to its ligand at a single binding site.

affinity capture
Enrichment of a protein construct with an **affinity tag** by batch methods or **affinity chromatography**.

affinity chromatography
Type of chromatography in which the protein mixture to be purified is passed over an affinity matrix to which specific ligands for the required protein are attached, so that the protein is retained on the matrix.

affinity constant (association constant) (K_a)
Measure of the strength of binding of the components in a complex. For components A and B and a binding equilibrium A + B $\leftrightarrow$ AB, the association constant is given by [AB]/[A][B]; the tighter the binding between A and B, the larger the association constant. K_a is the inverse of the **dissociation constant** K_d.

affinity tag
Small **fusion peptide**, linked to a **protein construct**, causing it to binding to an affinity matrix. *See also* **affinity chromatography**.

agonic (C) – *see* **triclinic**

alanine
Small L-α-amino acid with –CH$_3$ side chain, somewhat hydrophobic.

alkyl group (C$_n$H$_{2n+1}$)
General term for a group of covalently linked carbon and hydrogen atoms such as methyl (–CH$_3$) or ethyl (–CH$_2$CH$_3$) groups. Usually exist as part of larger organic molecules. On their own they form extremely reactive free radicals.

allele
One of several alternative forms of a gene. In a diploid cell each gene will typically have two alleles, occupying the corresponding position (locus) on homologous chromosomes.

allostery (adjective **allosteric**)
Change in a protein's conformation brought about by the binding of a regulatory ligand (at a site other than the protein's **active site**), or by covalent modification. The change in **conformation** alters the activity of the protein and can form the basis of directed movement.

alpha helix (α-helix)
Common folding pattern in proteins, in which a linear sequence of amino acids folds into a right-handed helix stabilized by internal hydrogen bonding between backbone atoms. Figure 2-11.

ambiguity – *see* **substructure handedness, phase ambiguity**

amide
Molecule containing a **carbonyl group** linked to an **amine**.

amine
Chemical group containing nitrogen and hydrogen, positively charged at neutral pH.

amino acid
Organic molecule containing both an **amino group** and a **carboxyl group**. **Proteinogenic amino acids** serving as building blocks of proteins are L-α-amino acids, where the **amino** and **carboxyl** groups linked to the same Cα carbon atom in 2S **absolute configuration**. *See also* **CORN rule**.

amino group (–NH$_2$)
Weakly basic functional group derived from ammonia (NH$_3$) in which one or more hydrogen atoms are replaced by another atom. In aqueous solution the amino group can accept a proton and carry a positive charge (–NH$_3$$^+$).

amino terminus
First amino acid residue in a polypeptide chain, with free NH$_2$ group, generally **protonated** at physiological pH, or decorated with posttranslationally modified amino group.

amorphous (C)
Non-crystalline, vitreous, of no defined shape.

amphipathic
Having both hydrophobic and hydrophilic regions, as in a **phospholipid** or a **detergent** molecule.

amphiphile
Small **detergent**-like molecule used in **membrane protein** crystallization.

amplitude
In a transversal wave, the magnitude at the peak of the wave. In the case of X-rays, the maximum of the electromagnetic field vector. *See also* **structure factor amplitude**.

angstrom (Å)
Unit of length, non-SI, but conveniently of the same order of magnitude as atom size. $1 Å = 10^{-8} cm = 10^{-10} m = 0.1 nm$.

anion
Negatively charged ion.

anisotropic
Different physical properties in different directions, such as refractive index, different **diffraction limits**, or **anisotropic displacement parameters** of atoms.

anisotropic *B*-factor – *see* anisotropic displacement parameters

anisotropic displacement ellipsoid
Graphical representation of the anisotropic atomic displacement in the form of an ellipsoid with principal axes the diagonal elements of the **anisotropic displacement tensor**.

anisotropic displacement parameter refinement
Also known as anisotropic *B*-factor refinement, requires six adjustable parameters instead of one in **isotropic** *B*-factor [Q1] refinement. Due to increased number of parameters, generally only possible at resolution of ~1.4 Å or higher. Can be accomplished in groups by **TLS refinement** at lower resolution.

anisotropic displacement parameters
Define the anisotropic displacement of an atom from its mean position, expressed as the six parameters defining the **anisotropic displacement tensor**.

anisotropic displacement tensor
Symmetric matrix $\mathbf{U} = \langle \mathbf{uu}^T \rangle$, defined by the mean **dyad product** of the displacement vector **u**. Can be graphically represented as an anisotropy ellipsoid, with its principal axes given by the diagonal elements of **U**.

anisotropic overall *B*-factor
Applied during scaling of anisotropic diffraction data, that is, an overall scattering attenuation ellipsoid instead of an **isotropic overall** *B*-factor.

anisotropic scaling
Applied to account for anisotropic diffraction limits in the form of **anisotropic overall** *B*-factor. In extreme cases, combined with anisotropic data truncation.

anisotropic scattering tensor
In the presence of polarization anisotropy as used in anisotropic anomalous scattering, the scalar anomalous scattering factors become tensors.

anisotropy tensor
The **dyad product** of an anisotropic property vector, for example **anisotropic displacement tensor U**.

annealing
A process of relaxing stress or high energy states in a material by temperature cycling. Also used to describe minimization procedure used in crystallographic **refinement**.

anomalous difference Patterson map
Patterson map formed from $\left(F_{PA}^+ - F_{PA}^- \right)^2$ as Fourier coefficients.

anomalous differences
The anomalous intensity differences $\Delta I = I^+ - I^-$ or differences in structure factor amplitudes $\Delta F = F^+ - F^-$ between members of a **Bijvoet pair**.

anomalous diffraction
Change of intensities between certain related reflections (**Bijvoet pairs**) due to presence of **anomalous scattering** contributions, caused by anomalous X-ray dispersion.

anomalous diffraction phasing – *see* anomalous phasing

anomalous diffraction ratio (6-64)

anomalous dispersion – *see* anomalous scattering

anomalous phasing
Experimental phasing methods that exploit differences in intensities between certain related reflections (**Bijvoet pairs**) due to presence of **anomalous scattering** contributions resulting from phasing marker atoms. *See also* **SAD, MAD, S-SAD**.

anomalous scattering
In the vicinity of an absorption edge the **atomic scattering factor** becomes wavelength- or energy-dependent (anomalous dispersion). The anomalous contributions are a real part f' and an imaginary part $i \cdot f''$ rendering the scattering factor complex. The complex contributions change both phase and amplitude of a general acentric structure factor, leading to a breakdown of **Friedel's law** and generating **anomalous differences** between the members of acentric **Friedel** and **Bijvoet pairs**.

anomalous scattering factor (6-56)
Wavelength-dependent anomalous contributions with real part f' and an imaginary part $i \cdot f''$ rendering the atomic scattering factor complex, thereby introducing additional phase shift and anomalous intensity differences in Bijvoet pairs.

anomalous signal
Colloquial term for **anomalous scattering** contributions in anomalous (difference) diffraction data.

antibiotic
Substance such as penicillin or streptomycin that is toxic to microorganisms. Often a product of a particular microorganism or plant. Used as selective marker in **cloning**.

antibody (immunoglobulin, Ig)
Protein produced by B cells in response to a foreign molecule or invading microorganism. Binds tightly to the foreign molecule or cell, inactivating it or marking it for destruction by phagocytosis or complement-induced lysis. Molecule composed of a heavy and light chain, forming an F_c and two F_{ab} fragments upon enzymatic digestion. Used in crystallization for **antibody scaffolding**.

antibody scaffolding
Utilization of **antibody** fragments for co-crystallization of (membrane) proteins in order to mediate a 3-dimensional network of intermolecular interactions ("scaffold") leading to the formation of a crystal.

anti-freeze – *see* cryoprotectant

antigen
A molecule that can induce an adaptive immune response or that can bind to an **antibody** or T-cell receptor.

apo-protein, apo-enzyme
Protein not bound to its ligands or complex partners.

archaea
Archaea or archaebacteria are single-celled organisms without a nucleus, superficially similar to **bacteria**. At a molecular level, more closely related to bacteria in metabolic machinery, but more similar to **eukaryotes** in genetic machinery. Archaea and bacteria together make up the **prokaryotes**.

Argand diagram
Graphical representation of **complex numbers** in a 2-dimensional diagram, with the real axis horizontal and the imaginary axis vertical.

arginine
Positively charged L-α-amino acid, side chain $-(CH_2)_3NHC(NH_2)_2^+$ with planar conjugated guanidinium group.

ARP/wARP
Dummy atom refinement and model building program, available on the web.

asparagine
Polar L-α-amino acid with $-CH_2CONH_2$ side chain. Hydrogen bond donor.

aspartate, aspartic acid
Charged, acidic L-α-amino acid with $-CH_2COO^-$ side chain. Hydrogen bond acceptor.

association constant – *see* **affinity constant**

associativity (M)
Associativity implies that $a(bc) = (ab)c$ (which does not automatically apply **commutativity**, that is, $ab \neq ba$).

asymmetric
Not possessing any symmetry, **chiral**.

asymmetric unit (cell)
Smallest part of a crystal structure, containing the **motif**, from which upon application of **space group symmetry operations** the complete unit cell of the crystal is generated.

atomic coordinates (C)
Define the positions of the atoms in a crystal structure in reference to a common origin. Can be in **Cartesian coordinate** form (world coordinates) or **crystallographic** or **fractional coordinates**.

atomic displacement parameter
Also called *B*-factor, a measure related to the atomic displacement (vector) of an atom from its mean position. The nature of atomic displacement is variable; can be thermal vibration about the equilibrium position, static disorder of atomic position in different unit cells, or dynamic effects of group movement. Atomic displacement can be accounted for in **isotropic** parameterizations (overall or individual refinement) or **anisotropic** parameterization (individual or group **TLS refinement**).

atomic force microscope (AFM)
The atomic force microscope is a high-resolution type of scanning probe microscope, with a resolution of fractions of a nanometer, more than 1000 times better than the optical diffraction limit for a light microscope.

atomic resolution (C)
Diffraction limit beyond which individual atoms can be discerned in electron density, commonly meaning better than ~1.2 Å.

atomic scattering factor, *f*
Measure for the X-ray scattering power of an atom, equal to the number of **electrons** at zero scattering angle, and decreasing rapidly with scattering angle. The atomic scattering factor is the **Fourier transform** of the atomic electron density, commonly approximated by a spherically symmetric superposition of **Gaussian** functions. Most common is the approximation by nine **Cromer–Mann coefficients**. Only at ultra-high **resolution** (0.8 Å and better) the anisotropy of the electron density due to valence electrons (charge density deformation) needs to be considered. The units of the scattering factor are electrons.

ATP (adenosine 5′-triphosphate)
Nucleoside triphosphate composed of adenine, ribose, and three phosphate groups. The principal carrier of chemical energy in cells. The terminal phosphate groups are highly reactive in the sense that their hydrolysis, or transfer to another molecule, takes place with the release of a large amount of free energy.

autoconvolution – *see* **autocorrelation**

autocorrelation (9-25)
Convolution of the same function with its inverse; for example, the **Patterson function** is a autocorrelation of the electron density, $P(\mathbf{u}) = \rho(\mathbf{r}) \otimes \rho(-\mathbf{r})$

autoinduction
Elegant method to induce expression in diauxic *E. coli* when switching from glucose to lactose metabolism.

Avogadro's number (Loschmidt's number)
Number of molecules in a mole of material, $N_A = 6.022142 \times 10^{23}$ mol^{-1}.

B

B-factor – *see* **atomic displacement parameter**

B-factor distribution
A positively skewed distribution that is specific for each molecule and can indicate abnormal *B*-factor restraints or scaling.

B-factor restraints
Restraints placed on the difference in displacement parameters (*B*-factors) of adjacent atoms.

B-factor scaling – *see* **Wilson scaling**

Babinet's principle
States that the diffraction pattern of an object is the same as that of a hole of the same size and shape as the object, but with opposite phase.

backbone atoms – *see* **main chain atoms**

bacteriophage (phage)
Any virus that infects **bacteria**. Phages were the first organisms used to analyze the molecular basis of genetics, and are now widely used as cloning vectors.

bacteriorhodopsin
Pigmented protein found in the plasma membrane of a salt-loving archaean, *Halobacterium salinarium* (*Halobacterium halobium*). Pumps protons out of the cell in response to light. Among the first **integral membrane proteins** whose 3-dimensional structure was determined. Figure 3-45.

bacterium (plural bacteria) (eubacterium)
Member of the domain bacteria, one of the three main branches of the tree of life (archaea, bacteria, and **eukaryotes**). Bacteria and archaea both lack a distinct nuclear compartment, and together comprise the **prokaryotes**.

baculovirus
Virus that exclusively infects insect cells; used as corresponding expression vector.

ball and stick model
A model of a molecular structure in which the covalent bonds are represented as sticks, with the atoms at the intersection of the bond sticks represented as balls. *Compare* **stick model**.

bandwidth
Generally the region between a lower and upper frequency or energy cut-off selected from a continuous spectrum. In spectroscopy also the **full width at half maximum** of an emission line.

base
A substance that can reduce the number of **protons** in solution, either by accepting H⁺ ions directly, or by releasing OH⁻ ions, which then combine with H⁺ to form H_2O. The purines and pyrimidines in DNA and RNA are organic nitrogenous bases and are often referred to simply as bases.

base excision repair
DNA repair pathway in which single faulty bases are removed from the DNA helix and replaced. *Compare* **nucleotide excision repair**.

base pair (Watson–Crick base pair)
Two nucleotides in an RNA or DNA molecule that are held together by hydrogen bonds—for example, G paired with C, and A paired with T or U.

basic (alkaline)
Having the properties of a base.

basis (C)
The *n* linearly independent vectors with a common origin that span an *n*-dimensional space.

batch crystallization

Direct combination of protein solution and crystallization cocktail. Often conducted under paraffin oil without occurrence of significant **vapor diffusion**, or under a mixture of paraffin oil and silicone oil which is semi-permeable to water.

Bayes factor (12-104)

Confidence measure expressing the relative merits of competing hypotheses, similar to **likelihood ratios**, but with consideration of prior (model) odds. Sidebar 12-10.

Bayes' theorem (7-17)

Law of formal inductive inference: the **posterior probability** of a model or hypothesis is proportional to its **data likelihood function** times its **prior probability**.

Bayesian inference (reasoning)

Form of inductive inference, where prior probabilities are explicitly used in the formulation of joint probabilities and likelihood functions.

Bayesian posterior – *see* **posterior probability**

bending magnet

Located between the straight sections of a synchrotron storage ring, strong bending (electro) magnets are used to deflect the electrons and steer them on a closed circular path. At the same time, due to the acceleration they experience, the electrons emit polychromatic synchrotron radiation, used for X-ray diffraction experiments. Figure 8-6.

Bessel functions

Set of ubiquitous functions obtained as solutions of the Bessel differential equation. Modified Bessel functions of the first kind $I_n(x)$ of order n are exponentially increasing functions obtained during the marginalization of nuisance variables from complex probability distribution functions.

beta-bulge

Distortion caused in β-sheets by insertion of an extra residue, perturbing the regular hydrogen bonding pattern.

beta-hairpin turn

Short secondary structure element with distinct n to $n + 3$ hydrogen bonding pattern connecting two antiparallel β-strands.

beta-sheet (β-sheet)

Common structural motif in proteins in which different strands of the polypeptide chain run either parallel or antiparallel alongside each other, joined together by hydrogen bonding between atoms of the polypeptide backbones.

bias (C) – *see* **model bias**

bias (M)

Tendency of a result toward a certain outcome, often implying non-randomness in statistical sampling or dependence on hidden parameters.

bidentate

Ligand or additive molecule in crystallization with two functional groups.

bifurcated β-sheet

A β-**sheet** structure where a strand is shared and connects two sheets in different orientation. Figure 2-16.

bifurcated hydrogen bond

Hydrogen bond in which a hydrogen atom is shared between two hydrogen bond acceptors.

Bijvoet difference

The anomalous intensity differences $\Delta I = I^+ - I^-$ or differences in structure factor amplitudes $\Delta F = F^+ - F^-$ between members of a **Bijvoet pair**.

Bijvoet pair

Each of the m symmetry related equivalents $\mathbf{h}\mathbf{R}_m$ of reflection $\mathbf{h}$ (or F^+) has a **Friedel mate**, $(-\mathbf{h})\mathbf{R}_m$ (or F^-). These reflections F^+ and F^- form Bijvoet pairs. Similarly, starting from the Friedel mate $-\mathbf{h}$ (or F^-) of a reflection $\mathbf{h}$ and then applying the m point group

followed by **inversion** generates $-(-\mathbf{h})\mathbf{R}_m$ reflections (or F^+), and the pairs F^+ and F^- are again Bijvoet mates. In the general acentric case, **anomalous differences** exist between Bijvoet pairs.

binding pocket

Deep cleft or depression on the surface of a protein molecule that interacts with another molecule through noncovalent bonding.

binding site

Region on the surface of one molecule (usually a protein or nucleic acid) that interacts with another molecule through noncovalent bonding.

birefringence

Splitting of light into two components with different planes of polarization as a result of different refractive index in optically **anisotropic** media, such as protein crystals, where it can be detected with the aid of a polarizing microscope.

bitopic membrane protein

Protein passing through the lipid membrane of a cell with one or more transmembrane helix, connecting an extracellular ectodomain and the intracellular cytosolic domain, which in the case of a single transmembrane helix often can be expressed separately. *Compare* **monotopic membrane protein**.

body-centered

A lattice or unit cell with centering vector $I = (½ ½ ½)$. Possible in **tetragonal** (*tI*), **orthorhombic** (*oI*), and **cubic** (*cI*) lattices.

Bragg equation

Fundamental relation between lattice spacing $d(hkl)$ of reflecting lattice planes hkl of a crystal structure and the diffraction angle of a X-ray reflection: $d(hkl) = n\lambda / (2 \sin \theta)$.

Bragg reflection

An X-ray diffraction spot or discrete reflection fulfilling the **Bragg equation**. *Compare* **thermal diffuse scattering**.

Bravais centering – *see* **Bravais translations**

Bravais lattices

The combination of lattice symmetry with requirements for translational periodicity generates in addition to the six **primitive** lattices additional eight **centered lattice** types, collectively termed the 14 Bravais lattices.

Bravais symbols – *see* **Bravais translations**

Bravais translations

The translational centering vectors for the eight centered **Bravais lattices** (Figure 5-30). The centering vectors are $(½ ½ 0)$ for C-centering; $(½ ½ 0)$, $(½ 0 ½)$, and $(0 ½ ½)$ for F-centering; $(½ ½ ½)$ for I-centering; and $(⅔ ⅓ ⅓)$, $(⅓ ⅔ ⅔)$ for R-centering. Table 6-6.

Bremsstrahlung

Continuous part of the X-ray emission spectrum emanated by a laboratory **X-ray generator**. Figure 8-2.

brilliance

A measure for **photon flux** that takes the beam divergence into account, defined as the number of photons per second that pass through an area of 1 mm² with a given divergence measured in millirad: $B_\lambda = \text{photons} \cdot s^{-1} \cdot mm^{-2} \cdot mrad^{-2}$.

buffer

Solution of weak acid or weak base that maintains (buffers) the pH when small quantities of acid or base are added, or when solution is diluted. *See also* **crystallization cocktail**.

bulk solvent correction

Disordered solvent attenuates low resolution Bragg reflections, which is accounted for by bulk solvent correction.

C

C-centering

Lattice centering with (Bravais) centering vector $C = (½ ½ 0)$. Possible in monoclinic (*mC*) and orthorhombic (*oC*) lattices. Figure 5-30.

C-terminus – *see* **carboxyl terminus**

Cahn–Ingold–Prelog priority rules

Rules for assignment of ligand priority in establishing the **absolute configuration** of a **chiral center**. In first order, following the size of the ligand, for α-amino acids the sequence is from Cα to H, N, C, Cβ. *See also* **CORN rule**.

calmodulin

Ubiquitous intracellular Ca^{2+}-binding protein that undergoes a large conformation change when it binds Ca^{2+}, allowing it to regulate the activity of many target proteins. In its activated (Ca^{2+}-bound) form, it is also called Ca^{2+}/calmodulin.

calorie

Unit of energy, equal to 4.184 joules (J). One calorie (cal) is the amount of heat needed to raise the temperature of 1 gram of water by 1°C.

cancer

Disease featuring abnormal and improperly controlled cell division resulting in invasive growths, or tumors, that may spread (metastasize) throughout the body.

capsid

Protein coat of a virus, formed by the self-assembly of one or more types of protein subunit into a geometrically regular structure, frequently of icosahedral symmetry. Figure 2-2.

carbohydrate

General term for sugars and related compounds containing carbon, hydrogen, and oxygen, usually with the empirical formula $(CH_2O)_n$. Structural elements in **glycoproteins**.

carbonyl group (>C=O)

Carbon atom linked to an oxygen atom by a double bond.

carboxyl group (–COOH)

Carbon atom linked both to an oxygen atom by a double bond and to a hydroxyl group. Molecules containing a carboxyl group are weak carboxylic acids.

carboxyl terminus

Last amino acid residue in a polypeptide chain, with free COOH group, **deprotonated** under physiological pH.

Cartesian coordinates (basis)

An **orthogonal** coordinate basis [0, **X**, **Y**, **Z**] in **real space**, with dimensions suitable to the object to be represented (Å for molecules); world coordinates.

Cauchy distribution – *see* **Lorentz–Cauchy distribution**

CCP4 (Collaborative Computing Project 4)

A major collection of popular crystallographic programs bundled with a graphical user interface (CCP4i) and supported by a large user community.

cDNA

DNA molecule made as a copy of mRNA and therefore lacking the introns that are present in eukaryotic genomic DNA.

cell (B)

The cell is the structural and functional unit of all known living organisms, separated by a membrane from its environment.

cell (C) – *see* **unit cell**

cell-free system

Fractionated cell homogenate that retains the protein synthesis function of the intact cell.

cell line

Population of cells of plant or animal origin capable of dividing indefinitely in culture.

cell parameters – *see* **unit cell parameters**

center of mass

Defined as mass-weighted average, of i atoms in a molecule,

$$\mathbf{C} = \sum_i \mathbf{r}_i m_i \Big/ \sum_i m_i .$$

centered lattice

The eight centered **Bravais lattices** are *mC, oI, oC, oF, tI, hR, cI,* and *cF*.

central limit theorem

States that the distribution obtained by random sampling from almost any unimodal parent probability distribution will be a **normal distribution**, irrespective of the shape of the parent distribution.

central moment (*n*th central moment)

The *n*th moment of a distribution about the mean is defined as $\mu_n = E(x - x_0)^n$ with the **expectation value** E defined as the integral over the distribution. The first central moment is zero, the second moment about x_0 is the **variance** σ_x^2, the third is the **skewness**, and the fourth is the **kurtosis**, a measure how "squat" or "peaky" a distribution is. *Compare* **raw moment**.

centric Patterson projections – *see* **Harker sections**

centric reflections

Centric reflections are **phase restricted reflections** resulting from scattering by centrosymmetrically related atoms, that is, related by the point group symmetry of the crystal structure. They are located in centric zones of reciprocal space, and have phases of φ or φ + 180°, often (in the absence of transitional symmetry elements) but not necessarily 0° or 180°. **Bijvoet pairs** of centric reflections cannot carry any **anomalous difference** signal even in the presence of **anomalous scattering** contributions, because they also fulfill the symmetry relations of the corresponding space (point) group. Not all space groups have centric reflections; for example, in *P*3 none of the three point group operators **R** applied to **h** generates a reflection **hR** that is also a Friedel mate –**h** of itself.

centroid phases (10-43)

Phases defined by the centroid or weighted average of the possible phase probabilities, obtained by integration over all phase angles.

centrosymmetry

Centrosymmetry relates any object with positional coordinate vector **x** to an inverted object with coordinate vector –**x**. Centrosymmetry (**inversion**) changes handedness of the motif and is therefore not possible in native protein crystal structures.

change of handedness – *see* **substructure handedness**

channel (membrane channel)

Transmembrane protein complex that allows inorganic ions or other small molecules to diffuse passively across the lipid bilayer **membrane**.

chaotropic

A chaotropic reagent (chaotrope) is a substance which disrupts the 3-dimensional structure in proteins, DNA, or RNA and denatures them. Chaotropic agents interfere with stabilizing intramolecular interactions mediated by noncovalent forces such as hydrogen bonds, van der Waals forces, and hydrophobic effects.

chaperone (molecular chaperone)

Protein that facilitates the proper folding of other proteins, or helps them avoid misfolding, for example heat shock proteins (Hsp).

characteristic radiation – *see* **X-ray fluorescence**

charge, charged

The unit of charge is the charge of one electron, 1.602176487(40) × 10^{-19} Coulomb (Cb). Charged **residues** carry a positive or negative charge and are involved in charged or ionic interactions, also called salt bridges.

charge coupled device (CCD)

Imaging device built from a semiconductor array that transforms radiation into electric signal.

charge flipping

Substructure solution method related to **solvent flipping**, where beyond a certain threshold value the low electron density is subtracted from the high electron density, new structure factors are calculated, and the process is cycled.

chemical shift
Term used in spectroscopic techniques for energy shift of absorption signal (X-ray absorption) or resonance signal (NMR) due to a specific local chemical environment.

Cheshire cell
The limits of the Cheshire cell determine the space that has to be searched during translational searches.

Cheshire group
The different possible origin and **enantiomorph** choices combined with the symmetry operators of a space group generate the complete set of operators of the so called Cheshire groups. Cheshire groups can be visualized as the symmetry groups of the symmetry operators, and play a role in crystallographic search procedures. They are formally derived as the Euclidean normalizers of the space groups.

chi-square, χ^2 (7-35)
Sum of normalized residuals squared, used in **refinement** target functions and statistical χ^2 tests.

chi-square test (7-36)
The reduced χ^2 is employed in the statistical χ^2 tests deciding whether a distribution is a **normal distribution** ($\chi^2 = 1$). Used to adjust error models, for example during data processing, to reflect a normal distribution.

chiral center, asymmetric center
Atom (generally sp^3 hybridized carbon) with four different, nonplanar ligands. Cannot undergo **inversion** or **mirror operation** without changing **handedness**.

chiral, chirality
A molecule that does not possess inversion or mirror symmetry is chiral. In proteins chirality results from the presence of **asymmetric** chiral Cα carbon atoms with four different covalently bound groups. **Isoleucine** and **threonine** have a second chiral center at Cβ; their **absolute configuration** is (2S, 3R). Chirality is associated with optical activity.

chiral space group
Space groups that can act on a **chiral** motif, that is, contain no **mirror operations** or **inversions**. Sixty-five of the 230 space groups are chiral (Table 6-6).

chiral volume
A scalar triple vector product $\mathbf{A} \cdot (\mathbf{B} \times \mathbf{C})$ originating at the central atom with the vector order following the **Cahn–Ingold–Prelog priority rules**. For L-amino acids with absolute (2S) configuration, the sign of the chiral volume is invariably positive, and computes to about 2.5 Å^3.

chromatography
Broad class of biochemical techniques in which a mixture of substances is separated by charge, size, hydrophobicity, noncovalent binding affinities, or some other property by allowing the mixture to partition between a moving phase and a stationary phase.

circular dichroism (spectroscopy)
Circular dichroism is observed when optically active matter absorbs left and right hand circular polarized light slightly differently. Used to experimentally determine secondary structure composition of a protein.

cis
On the same or near side. Compare *trans*.

clathrate
A local cage of ordered water molecules, similar in structure to that of ice, around a molecule such as a protein, whose formation leads to a loss of **entropy**.

clone, cloning
Population of identical individuals (cells or organisms) formed by repeated (asexual) division from a common ancestor. Creation of multiple copies of a gene by growing a clone of carrier cells (such as *E. coli*) into which the gene has been introduced, and from which it can be recovered, by **recombinant DNA** techniques.

cloning vector
Small DNA molecule, usually a circular plasmid, which is used to carry the fragment of DNA to be cloned into the recipient cell, and which enables the DNA fragment to be replicated.

close-packed lattice
The face-centered cubic lattice (*cF*) and hexagonal close-packed lattice (*hP*) achieve this same highest possible packing density for spheres with a coordination number of 12 and a packing density of ~74%.

closure requirement (M)
Requirement that operations on members of a set forming a mathematical **group** under these operations are again group members.

closure residual (C)
The difference between observed and calculated **heavy atom derivative** structure factor amplitudes, $F_{PH}(obs) - F_{PH}(calc)$.

CNS (C)
A versatile high-level programming language and program package for macromolecular refinement (Crystallography and **NMR** System) by Axel Brünger and colleagues. Unique feature **simulated annealing torsion angle refinement**. *CNS* also includes all aspects of phasing (including MAD, SAD, SIR/MIR) and density modification. **Cross-validation** is used throughout the program.

co-crystallization
Crystallization of a complex of a protein and a binding partner from an incubated solution of both components. *Compare* **soaking**.

codon
Sequence of three nucleotides in a DNA or mRNA molecule that represents the instruction for incorporation of a specific amino acid into a growing polypeptide chain.

coenzyme
Small molecule tightly associated with an enzyme that participates in the reaction that the enzyme catalyzes, often by forming a covalent bond to the substrate. Examples include biotin, PLP, NAD, and coenzyme A.

cofactor
Inorganic ion or coenzyme required for an enzyme's activity.

coherence
Constant phase relation in two or more waves or photon wave packets. Normal light sources and X-rays are incoherent sources of radiation; lasers are coherent radiation sources.

coherence length (C)
Distance over which the electromagnetic scattering process is coherent.

coherent scattering
Scattering process during which defined phase relations are maintained.

coiled-coil
Stable rod-like protein structure formed by two or more α-helices coiled around each other.

collimator
Device that passively limits divergence of X-rays through entrance and exit slits.

colony PCR
Direct screening of bacterial colonies by **polymerase chain reaction** to screen for correct DNA vector constructs.

commensurately modulated structure
Crystal structure in which motifs vary in a systematic fashion compatible with the periodic translation period of the original lattice. Occurs in the form of **translational NCS** and manifests itself in Patterson space as well as in reciprocal space as a superstructure pattern. *Compare* **incommensurately modulated structure**.

commutativity (M)

Commutativity implies that the outcome of an operation is independent of the order of elements, that is, *ab* = *ba*.

competent cell (B)

Competence is the ability of a cell to take up extracellular DNA from its environment.

complex (B)

Molecules bound to each other, commonly a protein and ion, small molecule, ligand, DNA or other protein molecule, most frequently formed by noncovalent interactions.

complex conjugate (M)

The complex conjugate of a **complex number** is obtained by changing the sign of its imaginary part; in the case of a complex **structure factor** by changing the sign of the **phase angle**.

complex number

A complex number or function has a real part and an imaginary part, with the imaginary unit *i* defined as the square root of –1. They are conveniently represented in an **Argand diagram**.

complex structure factor

The resultant electromagnetic wave describing a diffracted X-ray, expressed as a complex number.

composite omit map – *see* **omit map**

Compton scattering

Inelastic scattering process involving electrons and electromagnetic radiation, generally leading to loss of energy of scattered radiation or particle.

conditional probability

A probability or probability function that depends on certain conditioning parameters or **conditioning information**. For example, the probability of the outcome of a certain face of a die depends on the number of faces the die has.

conditioning information – *see* **conditional probability**

conditions limiting reflections – *see* **systematic absences**

confidence interval (7-10)

Region or range of a probability function within which an observation will fall with a given probability, expressed in standard deviations or percent probability.

conformation

The spatial arrangement of atoms in a molecule with respect to bonds, angles, and torsions. *See also* **absolute configuration**.

consensus sequence

Average or most typical form of a sequence that is reproduced with minor variations in a group of related DNA, RNA, or protein sequences. Indicates the nucleotide or amino acid most often found at each position. Preservation of a sequence generally implies that it is functionally important.

constraints

Mathematical relation between two or more parameters in a multivariate model.

contours – *see* **electron density maps**

convergence (C)

Convergence is approached or reached when, in an iterative procedure such as **macromolecular refinement**, the refinement parameter shifts become negligible.

convergence radius

Measure for parameter deviation from which a global minimum can be reached by a given optimization method. About 1.7 Å coordinate r.m.s.d. from target for torsion angle parameterization, 1.2 Å for coordinate refinement.

convolution (9-24)

Operation involving two functions, defined by a **convolution integral** and similar to probing or point-wise multiplying one function with the other.

convolution integral (9-24)

convolution theorem – *see* **Fourier convolution theorem**

coordinate systems

Spanned by a set of **basis** vectors; most commonly **crystallographic coordinate** systems with **fractional coordinates** and the **Cartesian coordinate** system with coordinates in Å. Can be transformed into each other by affine similarity transformation using the **orthogonalization matrix** and its inverse, the deorthogonalization matrix.

CORN rule

The CORN rule is a practical aid for determining the configuration of **chiral** Cα centers of **amino acids**. When the central Cα atom is viewed with the H atom pointing toward the observer (or out of the paper plane), the ligand sequence reads "CO-R-N" in a clockwise rotation. Figure 2-5.

correlation coefficient (7-53)

A scale-factor-independent measure for the agreement between observed and calculated quantities. *See also* **real space correlation coefficient**.

COSY (correlation spectroscopy)

2-dimensional homonuclear **NMR** technique indicating which hydrogen atoms are spin-coupled through bonds. *Compare* **NOESY**.

covalent bond

Strong, directional chemical bond in which atoms share electrons.

covariance (matrix)

Pair-wise measure for how one parameter or variable in a multivariate function depends on another parameter. The covariance matrix contains the variance of each single parameter in its diagonal.

critical energy, of synchrotron storage ring

Mode of the energy spectrum emitted from a synchrotron (8-1).

Cromer–Mann coefficients (6-15)

The coefficients in the commonly used isotropic nine-parameter Cromer–Mann approximation of the **atomic scattering factor**.

cross-crystal map averaging

Used to improve electron density maps by averaging and density modification of density from the same molecule in different crystal forms.

cross-Fourier map (10-16)

Difference map based on difference amplitudes of native and second derivative and SIR phases of first derivative, used to maintain origin choice between substructures.

cross-Patterson vector

Interatomic distance vector between atoms of different molecules in a crystal structure. Generally longer than **self-Patterson vectors** between atoms in the same molecule.

cross-product – *see* **vector product**

cross-rotation – *see* **Patterson cross-rotation**

cross-section

Quantifies the probability of a certain particle–particle or particle–wave interaction such as X-ray or neutron scattering or absorption.

cross-validation

Statistical method to avoid overparameterization of a model by excluding a subset of data from refinement against which a cross-validation statistic such as **R-free** is computed.

cross-validation R-value, R-free (12-35)

R-value calculated for the cross-validation data excluded from refinement.

cryobuffer

A suitable buffered solution of a **cryoprotectant**.

cryocrystallography

Method of collecting diffraction data at **cryogenic** temperatures, largely to minimize **radiation damage** in the case of protein crystals. Generally requires **cryoprotection** of the protein crystals and/or **flash cooling**.

cryogenic

Very low temperature (below 123 K or −150 °C, −238 °F). *See also* **cryoprotection**.

cryoprotectant

A substance or solution that prevents the formation of crystalline ice, in common use anti-freeze. Ice formation (taking place in **mother liquor** surrounding a harvested protein crystal) almost invariably degrades or destroys the crystal. Substances acting as cryoprotectants are salts, organic precipitants, PEGs, glycerol, sucrose solutions, and others.

cryoprotection

Prevention of ice formation in the **mother liquor** surrounding harvested protein crystals during cooling, because ice formation almost always degrades or destroys protein crystals. The crystals are cryoprotected through one or more of (i) significant amounts of **cryoprotectants** already in the mother liquor, (ii) swiping or soaking them in cryoprotectant, (iii) **flash cooling** in liquid nitrogen.

crystal

A periodic assembly of identical objects such as atoms or molecules.

crystal class – *see* **crystallographic point groups**

crystal contacts

Intermolecular contacts that connect symmetry related molecules (including translationally related molecules from neighboring cells) in a **crystal structure**.

crystal faces

Distinct lattice planes forming and defining the macroscopic shape of the crystal.

crystal form

Different **crystal structures** formed by the same material. *Compare* **crystal habit**.

crystal habit

Different appearance of the same **crystal form** of the same material. Shape, isotropy, development of crystal faces, as well as diffraction properties can differ for different crystal habits.

crystal lattice

A periodic construct generated by translation of unit cells along the **lattice vectors**.

crystal structure

A specific arrangement of molecules in a **crystal**, defined by unit cell dimensions, space group, and molecular arrangement. Same **space group** does not mean same crystal structure, even for the same material. *See also* **crystal form**.

crystal system

The seven crystals systems arise from the six **lattice types** under consideration of minimal internal symmetry, which splits the hexagonal lattice into a **trigonal** (only 3-fold internal symmetry) and a **hexagonal** (full 6-fold internal symmetry) system.

crystallization cocktail

Mixture of various **precipitants**, **buffers**, and **additives** used to reduce protein solubility and achieve **supersaturation**.

crystallization diagram

A pseudo-binary **phase diagram** of protein concentration versus precipitant concentration, depicting protein solubility and phase relations of a protein solution together with a depiction of tentative information about kinetic phenomena such as nucleation.

crystallographic coordinates – *see* **fractional coordinates**

crystallographic fragment screening

A method of experimental ligand screening in which small molecules—fragments of drug-like substances—are soaked or co-crystallized with target molecules. From the binding fragments, drug-like molecules are designed in an iterative process.

crystallographic point group

Point groups are mathematical **groups** of operations that keep the origin fixed (e.g. all symmetry operations intersect at the origin). There is an infinite number of 3-dimensional point groups but only 32 point groups comply with the crystallographic translation restrictions. Only eleven of the crystallographic point groups are enantiomorphic, that is, allowed for **chiral** motifs such as protein molecules. Table 6-6.

crystallographic restriction theorem

Requirement that all lattice points need to superimpose upon each other when a symmetry operation is applied. Follows from properties of the **direction cosine matrix** defining a rotation.

crystallographic rotation axis

An n-fold rotation axis ($n = 2, 3, 4, 6$) in specific orientation (parallel, perpendicular, along space diagonal) to a **unit cell vector**. The orientation of the axes can be found in the **unit cell projections** depicted for each **space group** in the International Tables for Crystallography, Volume A.

crystallographic rotations

Crystallographic rotations are **proper** n-fold **symmetry operations** limited by crystallographic translational periodicity restrictions to n ($n = 2, 3, 4, 6$) successive rotations by $360/n$ degrees around a **crystallographic rotation axis** in a specific orientation (parallel, perpendicular, along space diagonal) to a unit cell vector.

crystallographic screw axis

An n-fold rotation axis n_m combining n successive **rotations** by $360/n°$ parallel to a unit cell vector **k** with m successive translations of $\mathbf{k}m/n$ along the rotation axis where $n = 2,3,4,6$ and $m = 1$ to $n-1$. For example, a 4_3 screw axis parallel to unit cell axis **c** generates 3 additional **symmetry equivalent** copies by clockwise rotations by $m \times 90°$ plus m translations by $\mathbf{c}m/n$. Figures 5-32, 5-33, 5-34.

cubic

Lattice or crystal structure with cell parameters $a = b = c$, $\alpha = \beta = \gamma = 90°$, with primitive ($cP$), I-centered (cI), and F-centered (cF) lattice types. Minimal internal symmetry four 3-fold rotation axes along the space diagonals.

cumulative distribution function

Integral over a probability distribution function.

curvature (M)

Second **derivative**, or change of **slope**, of a function. At a maximum of a function the curvature is negative; at the minimum, positive.

cyclic group (C)

A space group that is generated by successions of a single generator, such as a **proper rotation**.

cysteine

Polar L-α-amino acid containing reactive thiol group in side chain $-CH_2SH$. Forms disulfide bonds to adjacent cysteines upon oxidation.

cytochrome

Colored heme-containing protein that transfers electrons during respiration, drug metabolism (cytochrome P-450), and photosynthesis.

cytosol

Contents of the main compartment of the cellular cytoplasm, excluding membrane-bounded organelles such as endoplasmic reticulum and mitochondria.

D

dalton

Unified atomic mass unit, Da. One twelfth of the mass of a ^{12}C atom at rest and approximately equal to the mass of a hydrogen atom (1.66054×10^{-24} g). 1 Da = $1/N_A$ grams where N_A is **Avogadro's number**.

data (C)

Experimental **observations** (structure factor amplitudes or intensities).

data likelihood (function)

Term frequently used for the probability function $prob(data|model, I)$, also known as the sampling probability.

data-to-parameter ratio

A measure for the **determinacy** of a system; with a data-to-parameter ratio > 1 the system is overdetermined, but for sensible parameter refinement, multiple **observations** per adjustable parameter are required.

de Broglie relation

Relates mass m and energy (or **wavelength**) of particles. For particles moving at light speed c, the wavelength is $\lambda = h/mc$ and the frequency $v = mc^2/h$ with h the Planck constant. *See also* **energy–wavelength conversion**.

Debye effect

Deviations from linearity in **Wilson plot** for low-resolution data, due to correlation between atomic distances.

Debye–Waller factor (6-16)

The entire scattering vector dependent exponential term bearing the **atomic displacement parameter** or ***B*-factor** in the (negative) exponent.

deductive inference (logic)

An argument is deductive when the truth of the conclusion follows necessarily as a logical consequence of the premises, that is, its corresponding conditional is a necessary truth. *Compare* **inductive inference**.

denaturation

Dramatic change in conformation of a protein caused by heating or by exposure to chaotropic reagents, usually resulting in the loss of biological function.

density averaging

An averaging procedure whereby electron density of multiple molecules in the asymmetric unit related by **non-crystallographic symmetry** is averaged, generally combined with solvent flipping. Very powerful with high NCS—highly symmetrical virus capsids can be phased starting from a spherical shell as a model.

density modification

Methods for phase improvement and phase extension that utilize the distinctly different properties of protein versus disordered solvent. **Solvent flattening**, **solvent flipping**, and the **sphere of influence algorithm** (often combined with electron **density averaging**) are common density modification procedures.

deorthogonalization matrix

Inverse of the orthogonalization matrix.

deoxyribonucleic acid – *see* **DNA**

deoxyribose

The five-carbon monosaccharide component of DNA. Differs from **ribose** in having H at the C-2 position rather than OH. $C_5H_{10}O_4$.

derivative (C)

A derivative crystal contains (heavy) atoms used as phasing marker atoms.

derivative (M)

A measure of how a mathematical function changes as its parameters change. The first derivative of a function at a given point defines its **slope** (tangent), the second derivative the **curvature** (i.e. the change of the slope).

derivative-difference Fourier map (10-17)

Difference Fourier map use to identify additional low-occupancy sites.

derivative Patterson map

A Patterson map based on coefficients $(F_{PA1} - F_{PA2})^2$ used (historically) to maintain origin choice between derivatives.

detergent

Organic molecules that are amphiphilic, meaning they contain both hydrophobic groups (their "tails") and hydrophilic groups (their "heads"). Therefore, they are soluble in both organic solvents and water. They form micelles at concentrations above the critical micelle concentration. Used in crystallization as an additive and to keep (membrane proteins) in solution.

determinacy (limit)

The limiting resolution defined by a data-to-parameter ratio of unity below which a system becomes undetermined. Around 4.5 Å for torsion angle refinement, ~3.5–3.0 Å for restrained refinement.

determinant (M)

The determinant of a matrix describing a **linearly independent** set of vectors (or set of equations) is nonzero. In a 3-dimensional **basis** system, right-**handedness** implies a positive determinant; **inversion** of **handedness** changes the sign of the determinant.

deviate – *see* **residual**

dialysis

Exchange of small molecules, water, or ions across a membrane separating two systems. Process follows the chemical gradient.

dielectric constant

Relative static permittivity, a measure for the electrostatic flux density or polarity of a solvent. For example, water (highly polar) has a dielectric constant of ~80 while *n*-hexane (nonpolar) has a dielectric constant of 1.9 at 20°C.

difference electron density map

An **electron density map** constructed from **Fourier coefficients** based on the difference of observed and calculated **structure factor amplitudes**. Missing parts of the protein model show up as positive difference density, incorrectly placed parts as negative difference density.

difference Patterson map

A **Patterson map** constructed from **Fourier coefficients** based on the difference between **isomorphous difference** intensities or **anomalous difference** intensities. Reveals the interatomic distances between heavy or anomalous **marker atoms** and is used for marker atom **substructure** solution.

differentiation (M)

The process of finding a **derivative** of a function is called differentiation.

diffraction

Scattering discrete directions of electromagnetic radiation or particles by periodic assemblies.

diffraction-component precision index (12-101)

Measure for the coordinate precision based on *R*-free, data-to-parameter ratio, completeness, and resolution.

diffraction limit

The extent in solid angle to which diffracted X-rays can be observed. Generally expressed as a function of the (smallest) interplanar **lattice spacing**, and termed (highest) **resolution**, d_{min}. A direct measure for the information content of a diffraction pattern (Figures 1-6 and 9-8). Lattice spacing and **diffraction** angle are related through the **Bragg equation**. High resolution means small interplanar *d*-spacing.

diffraction image, pattern

A 2-dimensional image of diffracted X-rays or **X-ray reflections**, generally recorded on an area detector.

diffraction pattern symmetry – *see* **Laue symmetry**

diffraction spots – *see* **X-ray reflections**

diffractometer

Instrument to generate X-rays and record diffracted X-rays. Minimally requires an X-ray source, X-ray optics, a goniostat, and an X-ray detector.

diffuse scattering – *see* **thermal diffuse scattering**

diffusion

Net drift of molecules in the direction of lower concentration due to random movement. A slow process in solution and even more so in solvent channels of crystals.

dihedral angle

The dihedral angle between two planes ABC and BCD of four consecutive atoms ABCD. **Torsion angles** are a special class of dihedral angles.

dimer

Molecular complex of two monomers (molecules). See also **homodimer**.

dimethyl sulfoxide, DMSO

A highly polar aprotic solvent, $(CH_3)_2SO$, that dissolves both polar and nonpolar compounds and is miscible in a wide range of organic solvents as well as water. Highly concentrated solutions of various small molecule ligand molecules in DMSO are used to introduce ligands with poor water solubility into crystallization drops for ligand **soaking**.

dipole interactions

Uncharged interactions between molecules, regions of molecules, or side chains that carry a dipole moment.

dipole moment

Represented by a dipole vector **P** and caused by charge separation or non-uniform charge distribution.

direct methods

Methods of crystallographic phase determination exploiting relations between strong normalized structure factors with certain index relationships. Routinely used in small molecule crystallography; in macromolecular crystallography predominantly for **substructure** solution.

direct rotation function

A rotation function based on the direct correlation between observed Patterson map and search model map. *See also* **Patterson correlation function**.

direct space – *see* **real space**

directed evolution

Method of random mutagenesis under selective pressure for specific desired properties such as solubility or crystallizability of the gene product.

direction cosine matrix

Matrix describing an arbitrary rotation through successive rotations about the three **Eulerian angles**. The determinant of a proper **direction cosine** matrix describing a rotation is 1 and the trace equals $1 + 2\cos\kappa$, which is practical to compute the **principal Euler angle**.

direction cosines (A-26)

dispersion

The change of a physical property such as polarizability, refractive index, or atomic scattering factor with wavelength. Passing through an **X-ray absorption edge**, the **atomic scattering factor** changes rapidly with X-ray wavelength and becomes complex, leading to anomalous dispersion or **anomalous scattering**.

dispersive atom, marker

An atom that provides **anomalous scattering** contributions, used for anomalous phasing methods.

dispersive differences

Dispersive differences exist between the diffracted intensities of pairs of the same reflection recorded at two different wavelengths.

displacement parameter – *see* **atomic displacement parameter**

dissociation constant (K_d)

Measure of the tendency of a complex to dissociate. For components A and B and the binding equilibrium A + B ↔ AB, the dissociation constant is given by [A][B]/[AB]; the tighter the binding between A and B, the smaller the dissociation constant. The dissociation constant K_d is the reciprocal of K_a, the **affinity constant**.

disulfide bond (–S–S–)

Covalent linkage formed between two **thiol** groups on cysteines,

target bond length 2.03 Å. For extracellular proteins, a common way of joining two proteins together or linking different parts of the same protein. Formed in the endoplasmic reticulum of eukaryotic cells. Under reducing conditions during overexpression in bacteria, S–S bonds generally do not form, and special bacterial strains for the production of disulfide bond containing proteins have been developed.

DNA (deoxyribonucleic acid)

Polynucleotide formed from covalently linked deoxyribonucleotide units. The store of hereditary information within a cell and the carrier of this information from generation to generation. Common canonical form is B-DNA (Figure 2-42). A-DNA is a dehydrated form, and Z-DNA is a third form, not observed *in vivo*.

DNA ligase

Enzyme that joins the ends of two strands of DNA together with a covalent bond to make a continuous DNA strand.

DNA polymerase

Enzyme that synthesizes DNA by joining nucleotides together using a DNA template as a guide. Thermostable, high-fidelity proofreading polymerases are used for **polymerase chain reaction**.

DNA shuffling

A form of synthetic, *in vitro* gene recombination, often combined with **directed evolution**.

domain (C)

Region of crystal containing a certain well-defined orientation of molecules, separated from other domains by domain boundaries. *See also* **mosaicity**.

domain (protein domain)

Portion of a protein that has an independently folding **tertiary structure**, often with a specific function. Larger proteins are generally composed of several domains, each connected to the next by short flexible regions.

domain rotation function

Cross-rotation function that uses phase information from NCS related electron density domains to determine NCS rotations.

domain swapping

Occurs in dimeric proteins when one part or domain of one molecule exchanges position with the same part or domain of the other molecule.

domain truncation

Separating a multi-domain protein into smaller, individually expressed domains and modifying the domain boundaries by recombinant DNA techniques or biochemical proteolysis.

dose mode

Data collection mode where the total incident radiation over each exposure increment (frame) is constant.

dot product – *see* **scalar product**

double helix

The 3-dimensional structure of DNA, in which two antiparallel DNA chains, held together by hydrogen bonding between the bases, are wound into a helix. Figure 2–44.

dual space (direct) methods

Substructure solution methods, Patterson-seeded in the case of *SHELXD*, that cycle between reciprocal space tangent refinement and direct space peak picking.

dual space methods

General term for iterative crystallographic methods that cycle between reciprocal space and real space; such as in direct methods or dummy atom placement.

dummy atom placement (refinement)

Dual space method for phase improvement and model building based on iterative placement, refinement, and removal of dummy atoms. Implemented in *ARP/wARP*.

duplex DNA
Double-stranded DNA.

dyad (C)
Name for 2-fold **rotation axis**.

dyad product (M, A-43)
Square matrix obtained by multiplication of column vector with row vector.

dynamic light scattering, PCS, QELS
Single angle light scattering method for particle size determination that measures the scattered photon correlation. Also called photon correlation spectroscopy (PCS) or quasi-elastic light scattering (QELS).

E

E. coli – see Escherichia coli

E-squared minus one, $\langle |E^2-1|\rangle$ *– see* **mean *E*-squared minus one.**

E-values *– see* **normalized structure factors**

edge scan *– see* **X-ray excitation scan**

effective resolution
Resolution d_{min} of diffraction data, corrected for completeness C: $d_{eff} = d_{min} \cdot C^{-1/3}$.

eigenvalue
Scalar value λ associated with a non-zero **eigenvector** of a linear matrix equation in the form of an **eigenvalue equation**.

eigenvalue equation
Linear system of equations $\mathbf{Ax} = \lambda\mathbf{x}$ where $\mathbf{A}$ is a matrix describing a linear transformation (such as the **direction cosine matrix** of a **rotation**) with $\mathbf{x}$ a nonzero **eigenvector** and λ the scalar **eigenvalue**.

eigenvector (eigenfunction)
Solution vector (function) of linear matrix equation, associated with a given **eigenvalue**.

elastic scattering
Scattering process in which no energy transfer occurs and **wavelength** or energy is maintained. The discrete Bragg scattering exploited in X-ray diffraction is elastic.

electromagnetic radiation
Described in dual picture as either (i) transversal electromagnetic wave with electromagnetic field vector and magnetic field vector perpendicular to each other and to the propagation direction; or (ii) an uncharged **photon** particle with no rest mass of energy $E = h\upsilon$ with h the Planck constant and υ the frequency. See Box 6-1 and **energy–wavelength conversion**.

electromagnetic wave *– see* **electromagnetic radiation**

electron
Subatomic particle that carries a negative electric charge. Electrons, together with atomic nuclei made of protons and neutrons, make up the atoms. Electrons scatter X-rays, and the **Fourier transform** of the **diffraction** patterns in **reciprocal space** therefore yields the **electron density** of the scattering object in **real space**.

electron density
The **Fourier transform** of experimentally determined **structure factor amplitudes** (proportional to the square root of diffraction spot intensities) plus additional **phases** (from phasing experiments) as **Fourier coefficients** generates the electron density of the diffracting objects (molecules).

electron density map
A representation of the **electron density** in the form of a 2-dimensional contour plot or a 3-dimensional grid connecting points or voxels of equal electron density. The level of density is often expressed in terms of **standard deviations** (σ) above the mean density. A high density level is represented for example by 5σ contours, while a low density level is at a 1σ level.

electrophoresis
Technique for separating molecules (typically proteins or nucleic acids) on the basis of their speed of migration through a porous medium when subjected to a strong electric field.

electroporation
Method for introducing DNA into cells, especially bacteria, in which a brief electric shock makes the cell membrane temporarily permeable to the foreign DNA.

empirical potentials
Expression of deviation from stereochemical target values as an energy potential function.

enantiomorph
One of the two partners in an enantiomorphic pair having opposite **chirality** or handedness.

enantiomorph ambiguity
Both **enantiomorphs** of a substructure produce the same **Patterson map**.

enantiomorphic pairs of space group
Twenty-two of the 65 **chiral space groups** with **screw axes** $3_1, 3_2; 4_1, 4_3; 6_1, 6_5; 6_2, 6_4$ form 11 enantiomorphic pairs. The pairs are $P3_1$-$P3_2$, $P3_121$-$P3_221$, $P3_112$-$P3_212$, $P4_1$-$P4_3$, $P4_122$-$P4_322$, $P4_12_12$-$P4_32_12$, $P6_1$-$P6_5$, $P6_2$-$P6_4$, $P6_122$-$P6_522$, $P6_222$-$P6_422$, and cubic $P4_132$-$P4_332$. When the handedness of the (sub)structure is changed, the screw axis must also be changed to the other **enantiomorph**. See also **substructure handedness**.

endoglycosidase
Enzyme that cleaves **oligosaccharides** *within* the polysaccharide chain; used for enzymatic deglycosylation of **glycoproteins**.

endonuclease
Enzyme that cleaves nucleic acids *within* the polynucleotide chain. *Compare* **exonuclease**.

endoplasmic reticulum (ER)
Labyrinthine membrane-bounded compartment in the cytoplasm of eukaryotic cells, where lipids' membrane-bound proteins, and secretory proteins are synthesized.

energy refinement
Refinement in which the X-ray and restraint SRS residuals are expressed as potential energy terms. Minimized commonly by simulated annealing.

energy–wavelength conversion
Energy in electronvolts (eV) = 12397.639 / wavelength in Å.

enthalpy, *H*
Change in enthalpy is the heat transfer ΔH during a chemical or physical process. *Compare* **free energy**.

entropy, *S*
The change in entropy ΔS is a measure of the change in order or degrees of freedom during a chemical or physical process. ΔS is negative for a process in which order is created, as in formation of an ordered crystal. The associated energy term is $-T\Delta S$. *Compare* **free energy**.

enzyme
Protein that catalyzes a specific chemical reaction.

enzyme-coupled receptor
A major type of cell-surface receptor (membrane protein) that has a cytoplasmic domain that either has enzymatic activity or is associated with an intracellular enzyme. In either case, the enzymatic activity is stimulated by an extracellular ligand binding to the receptor. Typical example GPCRs, G-protein coupled receptors.

epitaxy
Growth of crystals on a crystalline substrate acting as nucleation site with a matching or specific relation of cell dimensions between growing crystal and crystalline substrate.

epitope
Specific region (antigenic determinant) of an antigen that binds to an **antibody** or a T-cell receptor.

epitope-binding regions
Hypervariable complementarity determining regions (CDRs) of

an antibody binding to antigen.

epsilon-factor (epsilon-zone)

A correction factor applied to **intensity** statistics accounting for overlap in reciprocal space of certain symmetry related reflections.

equipoints – *see* **equivalent positions**

equivalent positions

Positions in a crystal structure that are symmetry related and generated by the space group symmetry operations.

equivalent reflections

X-ray diffraction spots or structure factor amplitudes related by reciprocal space symmetry.

error function, erf(x)

Sigmoid function obtained by integration from 0 to x of squared negative exponential functions, such as the normal distribution.

***Escherichia coli* (*E. coli*, E.C.)**

Rod-like bacterium normally found in the colon of humans and other mammals. Engineered strains are widely used as heterologous overexpression hosts.

eucentric point

Unique point in a diffractometer setup where all goniostat axes as well as the X-ray beam and optical centering microscope axis intersect.

eukaryote (eucaryote)

Organism composed of one or more cells that have a distinct nucleus. Member of one of the three main divisions of the living world, the other two being bacteria and archaea. *Compare* **prokaryote**.

Euler angle – *see* **principal Euler angle**

Euler axis

The Euler axis is the single, unique axis around which any general **rotation** is defined. It is obtained as the **eigenvector** belonging to the real **eigenvalue** of 1 when solving the linear matrix equation represented by the **direction cosine matrix R** describing the rotation. The **principal Euler angle** defines the single rotation angle around the Euler axis, and is equivalent to the rotation angle κ in **spherical coordinate** representation.

Eulerian angles

The three angles used in the **direction cosine matrix** to describe a rotation. Multiple conventions exist to define these three angles (Sidebar 11-1). Different from **principal Euler angle**.

Euler's formula (6-5; Section A.5)

evolutionary programming

Genetic algorithm that does not apply crossover and combination of parent properties but works well in continuous cases.

Ewald sphere

A sphere with radius of $1/\lambda$, with the diffracting crystal in its center, used to visualize diffraction geometry. Whenever a **reciprocal lattice point** intersects the Ewald sphere, the diffraction conditions in the form of the **Laue** and **Bragg equations** are fulfilled.

EXAFS – extended X-ray absorption fine structure

Region of the X-ray absorption spectrum above the absorption edge. Typical feature are EXAFS wiggles representing a Fourier transform of nearest neighbor geometry and distances.

excluded volume effect

General effect in macromolecular solutions increasing the effective concentration of macromolecules, because one part of a macromolecular chain cannot occupy space that is already occupied by another part of the same molecule.

exon

Segment of a eukaryotic gene that consists of a sequence of nucleotides that will be represented in mRNA or other mature RNA molecule. In protein-coding genes, exons encode the amino acids in the protein. An exon is usually adjacent to a noncoding DNA segment called an **intron**.

exonuclease

Enzyme that cleaves nucleotides one at a time from the ends of polynucleotides. *Compare* **endonuclease**.

expectation value (7-42)

In general terms, the integral of a random variable with respect to its probability measure. For discrete random variables this is equivalent to the weighted sum of the possible (or observed) discrete values forming the discrete probability distribution. For continuous random variables with a **probability density function** it is the probability density-weighted integral of the function values. Annotated also as $E(x)$ or $\langle x \rangle$.

experimental design

Formal statistical design of an experiment or study so that unbiased and valid statistical analysis of the outcome is possible.

experimental phasing

Encompasses all crystallographic **phasing methods** that provide phase information independent of any starting model. *Compare* **molecular replacement**.

exponential bulk solvent model

Bulk solvent correction valid for very low resolution reflections only, based on **Babinet's principle**.

expression (B)

Process by which information from a **gene** is used in the synthesis of a functional gene product, generally a **protein**. Used also for **overexpression** of proteins in various **expression hosts**.

expression host

Organism in which **heterologous overexpression** of a protein takes place. Most commonly bacteria, yeasts, insect cells, and cell lines of higher organisms.

expression vector

A **virus** or **plasmid** that carries a DNA sequence into a suitable **expression host** cell and there directs the synthesis of the protein encoded by the DNA sequence.

extensive parameters

In thermodynamics, extensive parameters are those explicitly depending on the amount of material. *Compare* **intensive parameters**.

extinctions (C) – *see* **systematic absences**

extrema (M)

Maxima or minima of a function or probability distribution.

F

face centered

Lattice or unit cell centering with (Bravais) centering vector F = (½ ½ 0), (½ 0 ½), and (0 ½ ½). Possible in orthorhombic (*oF*) and cubic (*cF*) lattices.

factorial design

A type of **experimental design** balancing the occurrence of possible factors (reagents, pH, drop size/ratio), their factor levels, and their combinations during the sampling process.

fast Fourier transform (FFT)

Method to rapidly compute Fourier transforms on $n = 2^k$ grid points.

fast rotation function (11-32)

A function calculating the overlap of Patterson functions in reciprocal space rapidly by expressing the **reciprocal space interference function** in spherical functions, with specially parameterized Euler angles. Developed originally by Crowther and Blow, and implemented in modern variants in various molecular replacement packages.

fast translation function (11-38)

A Patterson **translation function** that computes fast due to subtracting intermolecular vectors from the **Patterson map**, developed originally by Crowther and Blow, and implemented in modern variants in various molecular replacement packages.

fatty acid
Carboxylic acid with a long hydrocarbon tail.

fermentation
Anaerobic energy-yielding metabolic pathway.

figure of merit, fom, m
A statistic for the probability of a **phase angle** to be correct, expressed as mean phase error $m = \langle \cos(\Delta\varphi) \rangle$.

fine slicing
Method of recording diffraction images in small rotation increments, generally in the range of 0.1 to 0.3°. Allows 3-dimensional diffraction spot profile fitting.

fission yeast
Common name for the yeast model organism *Schizosaccharomyces pombe*. It divides to give two equal-sized cells.

Flack parameter
The Flack parameter x is defined in the range $0 \le x \le 1$ as $F_{\mathbf{h},x}^2 = (1-x)F_{\mathbf{h}}^2 + xF_{-\mathbf{h}}^2$. If the handedness is chosen correctly, x is close to 0; if the structure needs to be inverted, x is close to 1. This can give useful indications during heavy atom **substructure** refinement about substructure **handedness**. During refinement of protein structures strange Flack parameters may indicate **microscopic twinning**.

flash cooling
Rapid cooling of protein crystals by direct transfer into a **cryogenic** environment (generally **liquid nitrogen**) that prevents ice formation in the **mother liquor** surrounding a **harvested** protein crystal. The mother liquor solidifies in **vitreous** or **amorphous** form during flash cooling. Generally requires some form of **cryoprotection**.

flat bulk solvent model
Bulk solvent correction that generates a solvent mask around the protein and then fills the solvent regions with continuous density.

fluorescence
Emission of radiation by atoms when excited electrons return to their energy ground state, generally within nanoseconds. *Compare* **phosphorescence**.

fluorescence scan – *see* **X-ray excitation scan**

flux (C) – **see photon flux**

fold
A specific 3-dimensional arrangement of **secondary structure** elements of a protein forming a specific and stable **tertiary structure** that belongs to a certain fold family of structurally similar proteins.

form factor (C) – *see* **atomic scattering factor**

Fourier coefficients
The coefficients in a **Fourier integral** or series. In crystallography, the **complex structure factor**; or the **structure factor amplitude** and its associated **phase angle**.

Fourier convolution theorem (9-15)
States that the Fourier transform (FT) of a **convolution** of two functions equals the product of the individual FTs.

Fourier indexing
Indexing method that exploits **Fourier transforms** of the diffraction spots in certain directions to assign lattice vectors.

Fourier integral, series (9-2)
A complex integral or infinite complex series expressing a function as sum of cosine and sine terms or their equivalent complex exponential representation.

Fourier synthesis, summation (9-13, 9-14)
Procedure of carrying out the Fourier integration or Fourier summation. In crystallography, the result is either complex structure factors in reciprocal space or electron density in real space, which are Fourier transforms of each other.

Fourier transform (9-5)
A bijective transformation of a function from one domain into its reciprocal domain by means of Fourier integration or summation. In crystallography, the transformation of **complex structure factors** in **reciprocal space** into **electron density** in **real space** and *vice versa*.

Fourier transform interpolation
Fastest method to calculate structure factors, primarily used, for example, in **stochastic search** algorithms.

Fourier truncation ripples
Result around (heavy) atoms when truncated (phased) data are used for Fourier reconstruction of electron density.

fractional coordinates
Position of a point in a crystal structure expressed in dimensionless fractions x, y, z of the **unit cell vectors a, b, c**.

fragment screening – *see* **crystallographic fragment screening**

frame – *see* **diffraction image**

free energy, G
Also Gibbs energy, the driving force for chemical reactions. The change in free energy ΔG must be negative for a reaction to spontaneously occur. At the equilibrium, the free energy is zero. It is the sum of an enthalpic term and an entropic term, $\Delta G = \Delta H - T \Delta S$.

free-interface diffusion
Crystallization technique where the crystallization cocktail and the protein solution freely diffuse against each other. Carried out in capillaries or in **microfluidic** chips.

free likelihood ratio
Measure for statistical significance of differences between models. *See also* **Bayes factor**.

Free Lunch
A method of **phase extension** initially replacing unobserved high resolution reflections with those calculated by George Sheldrick's **sphere of influence algorithm** (Sidebar 10-10).

free R-value – *see* **cross-validation R-value** (12-35)

free radical
Atom or molecule which is extremely reactive due to at least one unpaired electron. Responsible for intracellular DNA damage *in vivo* and in part for the radiation damage to protein crystals during exposure to X-rays.

frequency
Common symbol υ, related to energy by $E = h\upsilon$ with h the Planck constant. *See also* **wavelength–energy conversion**.

Friedel pair
The reflection **h** and its Friedel opposite or mate **–h** form a Friedel pair. Friedel pairs are also **Bijvoet pairs**, but not all Bijvoet pairs are Friedel pairs. For general acentric reflections, **anomalous differences** exist between members of a Bijvoet pair.

Friedel's law
States that in the absence of **anomalous scattering** contributions, $F_{\mathbf{h}} = F_{-\mathbf{h}}$. The **structure factor amplitudes** (or measured intensities) of reflections with conjugate phase (i.e. the members of a Friedel pair) are identical. In the presence of anomalous signal, Friedel's law breaks down, and the relation $F_{\mathbf{h}} = F_{-\mathbf{h}}$ remains true only for **centric reflections**.

full matrix refinement (minimization)
Minimization algorithm in which all first and second derivative matrix elements are evaluated.

full reflection
An X-ray reflection or reciprocal lattice point that appears entirely on a single diffraction image or frame.

full width at half maximum (FWHM)
The width across a peak at half of its maximum. Particularly useful for cases such as Lorentz or Cauchy (spectral) lines where the variance is undefined. Figure 7-3.

function (M)

Arithmetic expression describing the dependence between two or more quantities, one or more of which are known (independent variables) and the others which are generated. Example: $y = f(x) = x^2$ means that y, a dependent variable, which is a function of the independent variable x, is given by the square of x.

fungi

Eukaryotic organisms that include the yeasts, molds, and mushrooms. Many plant diseases and a relatively small number of animal diseases are caused by fungi.

fusion protein, peptide

Engineered protein that combines two or more normally separate polypeptides, expressed from recombinant gene. Most commonly **affinity tags** or solubility enhancers.

G

G-function – *see* **reciprocal space interference function**

G-protein-coupled receptor (GPCR)

A cell-surface receptor with seven transmembrane helices that, when activated by its extracellular ligand, activates a G-protein, which in turn, activates either an enzyme or ion channel in the plasma membrane.

ganglioside

Any glycolipid having one or more sialic acid residues in its structure. Found in the plasma membrane of eukaryotic cells and especially abundant in nerve cells.

Gaussian normal distribution (7-24)

A ubiquitous **probability distribution function**, due to the **central limit theorem**. The normal distribution is symmetric about its maximum representing the mean or **expectation value**, with its **variance** defined as a squared residual. Figure 7-3 and 7-6.

gel-shift assay (gel-mobility shift assay)

Technique for (i) detecting proteins bound to a specific DNA sequence by the fact that the bound protein slows down the migration of the DNA fragment through a gel during gel electrophoresis, or (ii) detecting ligands or heavy atoms bound to proteins by slowing down protein migration in a native, non-denaturing SDS-PAGE gel.

gene

Region of DNA that is transcribed as a single unit and carries information for a discrete hereditary characteristic, usually corresponding to (i) a single **protein** (or set of related proteins generated by variant post-transcriptional processing), or (ii) a single RNA (or set of closely related RNAs).

general position (C)

A point position in a crystal structure with **fractional coordinates** x, y, z that is not located on any **symmetry element**.

generator (C)

A symmetry operation that is one of the one to three symmetry operations combined with the lattice translations into a **space group**.

genetic algorithms

Stochastic search or optimization methods that generate and recombine solutions from random starting values following the principles of genetic evolution. *Compare* **evolutionary algorithms**.

genetic engineering – *see* **recombinant DNA methods**

geometry term

Sum of residuals squared term (SRS) for the deviations of model geometry from independently determined geometry target values.

GIGO principle

Universal scientific principle stating that in any given system, garbage in equals garbage out.

glide plane

A combination of **mirror operation** with translation. Not allowed for **chiral** motifs.

glutamate, glutamic acid

Charged, acidic L-α-amino acid with $-CH_2CH_2COO^-$ side chain. Hydrogen bond acceptor.

glutamine

Polar L-α-amino acid with $-CH_2CH_2CONH_2$ side chain. Hydrogen bond donor.

glutaraldehyde

A dialdehyde $HOC-(CH_2)_3-COH$ used as a cross-linking reagent for protein crystals.

glycine

Smallest, non-chiral α-amino acid with a second H atom instead of a side chain branch.

glycoprotein

Any **glycosylated** protein with one or more oligosaccharide chains covalently linked to amino-acid side chains. Most secreted proteins and most proteins exposed on the outer surface of the plasma membrane are glycoproteins. *See also* **glycosylation**.

glycosylation

Covalent attachment of **carbohydrate** moieties, most frequently to Asn (N-glycosylation) or Ser and Thr (O-glycosylation). Can cause problems in crystallization due to conformational and chemical inhomogeneity. A **posttranslational modification** that does not occur in bacterial **expression hosts**.

goniometer – *see* **goniostat**

goniometer head

Small precision instrument allowing translational adjustment of the crystal for centering, mounted with standardized thread on the **goniostat**. Some eucentric goniometer heads also have adjustable arcs.

goniostat

Precision device allowing orienting a crystal so that it remains centered in the **eucentric point** of a diffractometer. Goniostats can be simple single-axis instruments or complex multi-axis (multi-circle) instruments.

gradient (descent) minimization (optimization)

Multivariate optimization algorithms that use in various implementations the first derivative to determine parameter shifts.

green fluorescent protein

A β-barrel structure harboring a chromophore, used in many variants as a fluorescence label.

grid screen

An exhaustive screening protocol varying two experimental parameters in a 2-dimensional matrix (grid) design.

group (M)

An abstract algebraic structure, consisting of a set (of objects or numbers, for example) and some operators acting on the members of the set. The conditions for set and operators forming a group are (i) **closure**—any operation can only generate a member of the group, (ii) identity—one and only one **identity operation** exists, (iii) **inversion**—in the mathematical sense that reversal of the operation generates the original object, and (iv) **associativity**—that $a(bc) = (ab)c$ (which does not automatically apply **commutativity**, that is, $ab \neq ba$). **Cyclic groups** are commutative.

guanidinium, guanidyl

Chemical group $-NHC(NH_2)_2^+$, derived from guanidine, in side chain of arginine.

H

habit (C) – *see* **crystal habit**

handedness (M)

Refers to the relative orientation of **basis** vectors. In the 3-dimensional case, the basis $[0, \mathbf{a}, \mathbf{b}, \mathbf{c}]$ is right-handed if the vectors $\mathbf{a}$, $\mathbf{b}$, $\mathbf{c}$ follow the sequence of thumb, index finger, middle finger of the right hand. Figure 5-25.

handedness (C)

Refers to one of the two configurations of a **chiral center**, chiral molecule, or **asymmetric** atomic assembly or marker atom **substructure**. *See also* **chirality**.

handedness ambiguity – *see* **substructure handedness**

hanging drop (C)

Crystallization method by **vapor diffusion** where a drop of protein solution plus precipitate is placed over a well solution of **precipitate** solution. *See also* **sitting drop**.

Harker diagram

Graphic representation of the **phasing equations**, where the **structure factor amplitudes** are represented as circles.

Harker sections

Certain sections (space-group-dependent **centric projections**) of a **Patterson map**, which reveal interatomic distance vectors between symmetry related atoms. Used in marker atom **substructure** solutions for **experimental phasing**.

harvesting (C)

The process of removing a crystal from the growth solution or **mother liquor**, most frequently with some form of harvesting loop.

heat shock protein (Hsp, stress-response protein)

Large family of highly conserved molecular chaperone proteins, so named because they are synthesized in increased amounts in response to an elevated temperature or other stressful treatment. Hsps have important roles in aiding correct protein folding or refolding.

Heaviside function

A unit step function, with Fourier transform of $\sin(x)/x$.

heavy atom

An atom of a heavy element (generally a heavy metal ion soaked into a protein crystal) used as marker atom in **experimental phasing**. Most heavy atoms also are **dispersive markers** providing **anomalous signal** for phasing.

heavy atom derivative (crystal)

Obtained by soaking **heavy atoms** into protein crystals or by co-crystallizing a protein with heavy atom compounds. In order to be useful for **experimental phasing**, the heavy atom derivative crystal needs to be **isomorphous** to the native crystal.

heavy atom refinement

Minimization of **sum of residuals squared** $(F_{PH}(obs) - F_{PH}(calc))^2$ by adjusting the parameters of the **heavy atom substructure**, generally by **maximum likelihood** methods, in order to obtain improved protein phases. Can be omitted if the substructure solution is (i) correct and (ii) complete, frequently the case for **dual space** substructure solution **methods** such as *SHELXD* or *Shake and Bake*.

heavy chain (H chain)

The larger of the two types of polypeptide chain in an immunoglobulin molecule, extending into the F_c fragment.

helix–loop–helix

DNA-binding structural motif present in many gene regulatory proteins, consisting of a short alpha helix connected by a flexible loop to a second, longer alpha helix. Distinct from the helix–turn–helix motif.

helix–turn–helix

DNA-binding structural motif present in many gene regulatory proteins, consisting of two alpha helices held at a fixed angle and connected by a short chain of amino acids, constituting the turn. Proteins containing this motif frequently form symmetric dimers and bind to DNA sequences that are themselves similar and arranged symmetrically.

hemihedral twinning

Merohedral twin with a single **twin operator**. If the twin fraction of a single twin operator is 0.5, it forms a perfect hemihedral twin with exactly equal amounts of each domain orientation (Figure 8-18). A corresponding projection of the diffraction pattern acquires additional perfect symmetry.

Hendrickson–Lattman coefficients (10-82)

Used for phase combination where each phase probability is expressed in terms of $\cos\varphi$, $\sin\varphi$, $\cos 2\varphi$, and $\sin 2\varphi$ and coefficients *A, B, C, D* (plus a normalization constant *K*) whose form depends on the particular **phase probability distribution**.

Hermann–Mauguin symbol – *see* **space group symbol**

Hermitian matrix (M)

A Hermitian matrix (or self-adjoint matrix) is a **square matrix** with complex elements which is equal to its own conjugate **transpose**. **Eigenvalues** of a Hermitian matrix are real with **orthogonal eigenvectors**. Physical observables obtained via Hermitian operators are real.

Hessian matrix

The Hessian (matrix) is the square matrix of second-order partial derivatives of a function; that is, it describes the local curvature of a multivariate function.

heterodimer

Molecular complex, **heterooligomer**, composed of two different **monomers**; protein complex composed of two different **polypeptide** chains. *Compare* **homodimer**.

heterogeneous nucleation – *see* **nucleation**

heterologous overexpression

Expression of high levels of a protein in a host system different from the native organism.

heterooligomer

Molecular complex composed of multiple different **monomers**; protein complex composed of multiple different **polypeptide** chains. *Compare* **homooligomer**.

heteropolymer

Macromolecule assembled from a subset of different building blocks or **monomers**. Proteins are heteropolymers made of **proteinogenic amino acids**. *Compare* **homopolymer**.

hexagonal

Lattice or crystal structure with cell parameters $a = b$, c, $\alpha = \beta = 90°$, $\gamma = 120°$, with primitive (*hP*) lattice type and 6-fold internal minimal symmetry.

high energy remote data set

Anomalous diffraction data collected about 100 eV or so above the absorption edge. Intended to maximize dispersive signal to **inflection data set**.

high resolution

Higher resolution means higher diffraction angle, equivalent to smaller sampled interplanar *d*-spacing d_{min}, implying finer detail discernible in the molecular structure. Figure 1-6.

histidine

Aromatic amino acid with imidazole ring in side chain $-CH_2(C_3N_2H_3)^+$, weakly protonated at physiological pH.

histogram matching (C)

A **solvent modification** method providing phase improvement and **phase extension** based on matching the initial electron density distribution to that observed for actual protein structures.

homeodomain

DNA-binding domain that defines a class of gene regulatory proteins important in development of higher organisms.

homodimer

Molecular complex, **homooligomer**, composed of two identical **monomers**; protein complex composed of two identical **polypeptide** chains. *Compare* **heterodimer**.

homogeneous nucleation – *see* **nucleation**

homolog

One of two or more genes that are similar in sequence as a result of derivation from the same ancestral gene. The term covers both **orthologs** and **paralogs**. Homology often but not exclusively indicates structural similarity.

homology modeling
Method of comparative modeling based on experimental structure template of a related structurally similar model as identified by sequence alignment.

homooligomer
Molecular complex composed of multiple identical **monomers**; protein complex composed of multiple identical **polypeptide** chains. *Compare* **heterooligomer**.

homopolymer
Macromolecule assembled from a subset of identical building blocks. *Compare* **heteropolymer**.

homozygous
Individual who carries two identical alleles of a gene affecting a given trait on the two corresponding homologous chromosomes.

HSQC
Heteronuclear single-quantum coherence, 2-dimensional NMR technique, separating backbone amides according to their ^{1}H and ^{15}N resonance frequencies.

hybridoma
Cell line used in the production of monoclonal **antibodies**, obtained by fusing antibody-secreting B cells with cells of a B-lymphocyte tumor.

hydrogen bond
The most important bond type in biological systems and processes. Ubiquitous noncovalent, semi-directional bond in which an electropositive hydrogen atom is partially shared by two electro-negative atoms, one acting as a hydrogen bond donor and the other as hydrogen bond acceptor. *See also* **bifurcated hydrogen bond**.

hydrodynamic radius
Radius of a macromolecule in solution determined by the Stokes–Einstein relation (4-4).

hydronium ion
Solvated **proton**, H_3O^+. The form generally taken by protons in aqueous solution.

hydrophilic
Dissolving readily in water. Literally, "water loving."

hydrophobic (lipophilic)
Not dissolving readily in water. Literally, "water hating."

hydrophobic force
Force exerted by the hydrogen-bonded network of water molecules that brings two nonpolar surfaces together by excluding water between them.

hydrophobic moment
The hydrophobic moment reflects the periodicity of hydrophobicity of a peptide, as measured per residue for a specified angle of rotation.

hydroxyl
Chemical group consisting of a hydrogen atom linked to an oxygen (–OH), as in an alcohol.

hygroscopic
A hygroscopic substance attracts water molecules from the surrounding environment.

I

I-centered – *see* **body centered**

identity matrix
An identity matrix has unity values in the diagonal elements and zeros elsewhere.

identity operation
An operation that maps an object onto itself; described by an **identity matrix**.

imaging plate
Area detector that stores X-ray photon energy in a **phosphorescent** material, read out later optically by laser excitation.

improper non-crystallographic symmetry (NCS)
General non-crystallographic symmetry that is both rotational and translational, does not comply with closed **group** limitations. *Compare* **proper non-crystallographic symmetry**.

inclusion body
Inclusion bodies are dense proteinaceous aggregates that form in the cytoplasm, notably when the protein folding machinery becomes overwhelmed. Inclusion bodies generally contain misfolded proteins.

incommensurately modulated structure
Crystal structure in which motifs vary in a systematic fashion out-of-phase or incommensurate with the periodic translation period of the original lattice. Manifests itself in reciprocal space through **satellite reflections** next to the Bragg peaks of the parent cell. *Compare* **commensurately modulated structure**.

indexing
The consistent assignment of three linearly independent basis vectors to a reciprocal or real lattice, generally by **Fourier indexing** methods.

indexing possibilities
The combination of the 14 **Bravais lattice** types and axis permutation generates 44 possible indexing choices. The highest symmetry indexing with the smallest penalty score is generally the correct choice.

inducible promoter
A regulatory DNA sequence that allows expression of an associated gene to be switched on by a particular molecular or physical stimulus.

inductive inference (logic)
An argument is inductive when the truth of the conclusion is given with a certain probability based on experimental or prior knowledge, that is, its corresponding conditional is a likelihood (function). *Compare* **deductive inference**.

inelastic scattering
Scattering process where the scattered radiation or particle loses or gains energy during the scattering process.

inflating the variance
Method to increase the variance of a **likelihood function** in order to implicitly account for additional (model) errors.

inflection data set
Anomalous diffraction data collected at the absorption edge inflection point, corresponding to the negative peak in the **Kramers–Kronig transform** of the **absorption edge** scan.

insertion device
Magnetic instrument that is inserted into the straight sections of an **electron synchrotron storage ring**, used to generate intense **synchrotron** X-ray **radiation**. **Wigglers** and **undulators** are insertion devices.

inside–outside distribution
Distribution of propensity of residues to be located either at the surface of protein or in the hydrophobic core of a protein. Useful in low-resolution X-ray structure validation and validation of predicted structures.

insulin
Polypeptide hormone that is secreted by B-cells in the pancreas regulating glucose metabolism in animals.

integral extinctions – *see* **systematic absences**

integral membrane protein
Membrane protein that is embedded in the lipid bilayer with transmembrane helices so that it does not have independently stable or expressible ectodomains or cytosolic domains.

integrating (C)
Reading of detector pixels and combining them to raw reflection intensities.

intensity, of diffracted X-rays (6-49, 6-52)

The intensity of an X-ray reflection or diffraction spot is proportional to the square of the corresponding structure factor amplitude.

intensity distribution

Probability distribution functions for diffraction intensities, commonly expressed in normalized form as squared normalized structure factors (E^2-values). *See also* **normalized structure factors** (E).

intensive parameters

In thermodynamics, intensive parameters are those not depending on the amount of material in the system, such as pH or temperature. *Compare* **extensive parameters**.

interatomic distance vectors

Distances between atoms in a structure (cross-vectors) and symmetry related atoms (self-vectors) in a crystallographic unit cell, comprising the Patterson space. *See also* **Patterson function**.

interference function – *see* **reciprocal space interference function**

interferon (IFN)

Member of a class of cytokines secreted by virus-infected cells and certain types of activated T cells. Interferons induce antiviral responses, inhibiting viral replication and stimulating macrophages and natural killer cells to kill virus-infected cells.

internal symmetry

The symmetry of a **lattice** as defined by the internal arrangement of motifs, defining the **crystal system**.

interplanar distance vector – *see* **lattice spacing**

interplanar spacing – *see* **lattice spacing**

intron

Noncoding region of a eukaryotic gene that is transcribed into an RNA molecule but is then excised by RNA splicing during production of the mRNA or other functional RNA.

inverse probability – *see* **likelihood**

inversion (C)

Symmetry operation that inverts the position vector, that is, **x** becomes –**x**. Inversion changes the handedness of a **substructure** or the **chirality** of a molecule.

inversion (M)

Operation that generates a matrix M^{-1} so that $MM^{-1} = I$, the **identity matrix**.

inversion axis

A crystallographic **symmetry operation** combining **rotation** with **inversion**, not allowed on **chiral** motifs such as protein molecules.

inversion center

The unique point through which a structure is inverted. Generally an origin of a unit cell, but three exceptions to the inversion about the origin exist: In chiral space groups $I4_1$, $I4_122$, and $F4_132$, the origin is not located on the enantiomorph axis, and the center of inversion does not coincide with the origin. Their inversion operators are $(-x, \frac{1}{2}-y, -z)$, $(-x, \frac{1}{2}-y, \frac{1}{4}-z)$, $(\frac{1}{4}-x, \frac{1}{4}-y, \frac{1}{4}-z)$, respectively. *See also* **substructure handedness**.

ion

An atom that has either gained or lost electrons to acquire a negative (anion) or positive (cation) charge.

ion channel

Transmembrane protein complex that forms a water-filled channel across the lipid bilayer through which specific inorganic ions can diffuse down their electrochemical gradients.

ionizing radiation

High energy electromagnetic or corpuscular radiation causing bond breakage and formation of **free radicals**.

isoelectric focusing

A technique for separating different proteins by their electric charge, on 2-dimensional gels for analytical purposes; also possible on preparative scale.

isoelectric point, pI

The pH value at which the net charge (the sum of all local charges) of a protein is zero. Also the point where protein **solubility** is minimal, and the protein does not migrate in an electric field.

isomer

Molecule formed from the same atoms and connectivities as another but having a different 3-dimensional conformation.

isoleucine

A **hydrophobic** L-α-amino acid with an asymmetrically branched aliphatic side chain –$CH(CH_3)CH_2CH_3$ and a second chiral center at Cβ, with absolute configuration (2S, 3R).

isomerase

Enzyme that catalyzes the rearrangement of bonds within a single molecule.

isomorphous (C)

Of the same crystal structure. Native structure and **heavy atom derivative** structure need to be isomorphous to generate **isomorphous difference** intensities useful for phasing. **Anomalous** and **dispersive difference** intensities are inherently isomorphous, subject to **radiation damage**. *See also* **RIP**.

isomorphous difference

Isomorphous differences exist between the diffracted intensities of a native protein and a **heavy atom derivative**.

isomorphous difference Patterson map

Patterson map formed from isomorphous difference coefficients $(F_{PA} - F_P)^2$.

isomorphous replacement

A method of experimental phase determination based on determination of a heavy atom marker **substructure**. Requires both a native crystal and one or more **isomorphous derivative crystal(s)**. Based on the fact that the complex structure factors (but not the **amplitudes**) are additive: $\mathbf{F}_{PH} = \mathbf{F}_P + \mathbf{F}_H$, or split into amplitude and phase term, $F_{PH} \cdot \exp(i\varphi_{PH}) = F_P \cdot \exp(i\varphi_P) + F_H \cdot \exp(i\varphi_H)$. F_P and F_{PH} as well as $\mathbf{F}_H$ (both F_H and φ_H) are known from the substructure solution. P, protein; PH, heavy atom derivative; H, heavy atom.

isotope

One of a number of forms of an element differing in atomic weight, that is, having the same number of protons and electrons but different neutron number. Isotopes are chemically equivalent with the exception of kinetic isotope effects caused by different atomic mass.

isotropic

Directionally independent, uniform.

isotropic displacement parameter – *see* **atomic displacement parameter**

isotropic overall *B*-factor – *see* **overall *B*-factor**

J

joint probability

The product of two or more, generally conditional, probabilities.

joule

Standard unit of energy in the meter–kilogram–second system. One joule is the energy delivered in 1 second by a 1-watt power source. Equal to 0.23901 calories.

K

kinase

Enzyme that catalyzes the addition of phosphate groups to molecules.

Kramers–Kronig transform

A general formalism that relates the real part of any analytic complex function to its imaginary part. Useful for physical response functions such as an X-ray absorption spectrum.

*k*th raw moment – *see* raw moment

kurtosis

The fourth **central moment** of a distribution; a measure how "squat" or "peaky" a distribution is.

L

Lambert–Beer absorption law (6-54)

Basic exponential law for the absorption of electromagnetic radiation.

lattice

An infinite, periodic 2-dimensional or 3-dimensional mathematical construct defined by three unit lattice vectors.

lattice parameters

The dimensions (norm, magnitude) *a*, *b*, *c*, of the **lattice** vectors **a**, **b**, **c**, and the enclosed angles α, β, and γ. Identical to **unit cell parameters**.

lattice planes

Sets of planes in a real lattice [0, **a**, **b**, **c**] spanned by three **lattice points** with indices (*u***a**, *v***b**, *w***c**) where *u*, *v*, *w*, are integers and known as direct **Weiss indices**. Lattice planes are more commonly described or indexed by reciprocal **Miller indices** *h*, *k*, *l* corresponding to a reciprocal lattice point **r*** = (*h***a*** + *k***b*** + *l***c***).

lattice point

Any point **r** = (*u***a** + *v***b** + *w***c**) with *u*, *v*, *w* integers in a lattice with basis [0, **a**, **b**, **c**]. Centered lattices possess additional lattice points corresponding to their **Bravais translations**. *Compare* **reciprocal lattice point**.

lattice spacing

The shortest distance d(*hkl*) (d$_h$) between **lattice planes** in direct space indexed with reciprocal **Miller indices** *hkl*. The interplanar distance vector **d**$_h$ is collinear with reciprocal lattice vector **d***$_h$ and reciprocal to it in **norm** or length. *See also* **resolution**.

lattice symbols

Symbols used for the 14 **Bravais lattices**. Table 6-6.

lattice types – *see* **Bravais lattices**

lattice vector

A vector from the origin (or other lattice point) to a **lattice point**.

Lattman angles

A special combination of **Eulerian angles** which prevents correlation problems at certain angle combinations.

Laue diffraction

Diffraction method that uses a broad **bandwidth** of X-ray energies instead of monochromatic X-rays. More reflections fulfill the diffraction condition at the same time as the **Ewald sphere** now becomes a Ewald shell with a thickness corresponding to the inverse of the bandwidth limits. Allows **time-resolved X-ray diffraction** experiments for the elucidation of enzyme mechanism and transport reaction.

Laue equations (6-20)

Set of three basic equations defining diffraction conditions in three dimensions.

Laue group (class)

Derived from the crystallographic **point groups** by addition of **inversion**, generating 12 distinct Laue groups (point group 32 splits into two Laue groups depending on the 2-fold axis orientation as defined by the corresponding **trigonal** space groups).

Laue symmetry

Point group symmetry plus **inversion**; the symmetry of **reciprocal space**. Important for efficient data collection strategies. Table 6-6.

LBFGS

A quasi-Newton optimization algorithm (limited-memory Broyden–Fletcher–Goldfarb–Shanno). LBFGS is particularly well suited for optimization problems with a large number of dimensions, as it does not explicitly form or store the Hessian (second derivative) matrix.

lead optimization

Improvement of desirable properties such as target binding and specificity of a drug lead compound by chemical modification.

least squares minimization – *see* **least squares refinement**

least squares refinement

General refinement method that minimizes the **sum of residuals squared** (SRS; hence the name) between observed and calculated data by adjusting model parameters. In crystallographic refinement, the squared differences between observed and calculated structure factor amplitudes are minimized by adjusting positional parameters (coordinates) and **B-factors** of the model atoms. Least squares methods are a special case of **maximum likelihood** methods applicable for complete models with only random errors.

Lennard-Jones potential

Potential energy curve of the basic form A*r*$^{-12}$ – B*r*$^{-6}$. Figure 12-18.

leucine

Hydrophobic L-α-amino acid with a branched aliphatic –$CH_2CH(CH_3)_2$ side chain.

leucine zipper

Structural motif in DNA-binding proteins in which two alpha helices from separate proteins are joined together in a coiled-coil (rather like a zipper), forming a protein dimer. Figure 2–47.

libration

From Latin *librare* "to balance, to sway." Rocking motion of a group of atoms along an arc, used in the **TLS** parameterization of molecular motion.

ligand

Any molecule that binds to a specific site on a protein or other molecule.

ligand docking

Molecular modeling technique that finds and scores the orientation and conformation (pose) of a ligand molecule.

ligase

Enzyme that ligates two molecules in an energy-dependent process. DNA ligase joins two DNA molecules together end-to-end through phosphodiester bond formation.

light scattering methods

Non-destructive optical methods for conformational analysis of macromolecules. **Dynamic light scattering** and **static light scattering** methods provide information about hydrodynamic properties, conformational state, oligomerization state, and molecular mass.

likelihood

The probability of random variables in an experiment with known outcomes, that is, the estimate or adjustment of model parameters based on known outcomes. In other words, likelihood is the "inverse" of basic probability, which is the prediction of outcomes given the parameters. *Compare* **probability**.

likelihood by intuition

A number of chemically and physically reasonable common-sense assumptions incorporated in George Sheldrick's successful (non-maximum likelihood) programs.

likelihood enhanced translation function (LETF)

Translation function that is faster than the full maximum likelihood translation function, used in initial stages of translation search.

likelihood function (*L*)

A probability function describing the probability of variables or parameters of a model given certain experimental outcomes or data.

likelihood ratio

Ratio of (posterior model) likelihoods of two competing hypotheses, based on the *same* data. See also **Bayes factor**.

linear merging *R*-value (8-15)

linear regression

Modeling or fitting of relationship between one or more independent variables and one dependent (response) variable by a linear **least squares** regression function. A linear regression with one independent variable represents a straight line. Linear regression does not imply a straight line, but refers to the linearity of the regression function.

linear residual, deviate (7-28)

The difference between two values, often between observed and calculated values or observed and **expectation values**.

linearly independent

A set of vectors (or equations) is linearly independent if none of them can be written as a linear combination of finitely many other vectors (or equations) in the set. The **determinant** of a matrix describing a linearly independent set of vectors (or set of equations) is nonzero.

lipid cubic phase

The lipid mono-olein forms a complex phase system with water. One of the phases in the mono-olein/water system is a bilayered, cubic phase containing 50–80% lipid as well as interconnected solvent channels. Bacterial rhodopsins, halorhodopsins, and photosynthetic reaction centers have been crystallized in lipid cubic phases.

liquid nitrogen, LN$_2$

Colorless clear **cryogenic** liquid with density of 0.807 g·ml^{-1} at its boiling point of 77 K (–196°C; –321°F), produced by fractional distillation of liquid air. Used to cool **protein crystals** to cryogenic temperatures during X-ray exposure in order to minimize **radiation damage**.

local minimum

A parameter combination in a multi-dimensional solution landscape that is not the global (absolute) minimum.

local pK_a

Low **solvent polarity** raises the **pK_a** of an acid residue, because it destabilizes the ionized form. Conversely, if a basic lysine residue is buried in a hydrophobic environment, its pK_a will be lower.

local refinement – *see* **real space refinement**

local scaling

Scaling method where data sets are scaled together in local groups instead of simple overall **Wilson scaling**.

locked rotation function (11-34)

A self-rotation or cross-rotation function used for searching for multiple molecules, in which the NCS relation between the molecules is known and held fixed.

locked translation function

A **translation function** used for searching for multiple molecules, in which NCS between the molecules or already positioned molecules or parts thereof are accounted for.

log-likelihood (*LL*)

The natural log$_e$ (ln) of a likelihood function *L*. Reduces multiplication of likelihood functions to sum of log-likelihood functions when forming **joint probabilities**.

log-likelihood gain (*LLG*)

The *LLG* is a measure for the improvement of the model during refinement or when comparing different models against the same data. In crystallographic model comparison, the *LLG* is commonly computed as the difference between the log-likelihood of the model and the log-likelihood calculated from a Wilson distribution, thus measuring how much better the data can be predicted with a particular model than with a random distribution of the same atoms.

loop (B)

Sequence of residues of varying length, connecting secondary structure elements, with no specific hydrogen binding pattern and thus not a secondary structure element. Often disordered in crystal structures. *Compare* **turn**.

Lorentz–Cauchy distribution

The form of a spectral line, similar—in shape only—to a **Gaussian normal distribution**, but with unpleasant properties such as undefined mean and infinite variance.

Lorentz exclusion region – *see* **Lorentz factor**

Lorentz factor (6-51)

A correction factor for raw diffraction intensities taking into account the different amount of time it takes for a **reciprocal lattice point** to pass through the **Ewald sphere**. The shallower (tangential) the intersection with the Ewald sphere, the longer the reciprocal lattice point remains in **reflection condition** during the recording of a rotation image. Points along the rotation axis can remain in reflection conditions too long to be properly corrected, and fall into the **Lorentz exclusion region**. Often combined with the **polarization factor**.

low energy remote data set

Anomalous diffraction data collected about 100 eV or so below the absorption edge. Sometimes used for additional dispersive signal relative to **inflection data set**, but carries no **anomalous signal** for the edge atom.

low resolution structure

Structure refined against data with resolution of less than ~3.5 Å (approximately the **determinacy limit** for restrained coordinate refinement, therefore **torsion angle** refinement with fewer adjustable parameters is commonly used at low resolution).

lunes

Elliptic regions of reflections recorded on an area detector during a rotation increment.

Luzzati *D*-factor

A factor by which the amplitude of a model structure factor is reduced due to positional (atom coordinate) and scattering factor errors.

Luzzati plot

A graph of *R*-value plotted versus $\sin\theta/\lambda$ where the slope in its linear region provides an upper estimate of the mean coordinate error.

lysine

Positively charged L-α-amino acid, long side chain $-(CH_2)_4NH_3^+$ with high surface entropy.

lysine methylation

Mild chemical methylation of surface lysine residues in proteins, aiming to improve crystallization.

lysozyme

Enzyme that catalyzes the cutting of polysaccharide chains in the cell walls of bacteria. A hardy perennial of protein crystallization, for practice and physical chemistry studies, with atypically high **supersaturation** levels.

M

macromolecular refinement

Conducted as restrained global reciprocal space refinement against experimental structure factor amplitudes or local real space refinement against electron density.

macroscopic twinning

Intergrowth of crystal in multiple regions or domains that are highly misaligned and generally can be visually distinguished and sometimes mechanically separated.

MAD – *see* **multi-wavelength anomalous diffraction phasing**

magic angle

Angle between the base of a cube and its space diagonal of 54.74°, $(\tan^{-1}\sqrt{2})$.

magnitude (M)

Length or **norm** of a vector; amplitude of the electric field vector.

main chain atoms

The atoms N, Cα, C, and the carbonyl oxygen O of each residue form the main chain (backbone) of a polypeptide.

main chain torsion angle plot
Representation of pairs of φ–ψ torsion angles of each residue in an energy contour plot, representing a potential energy surface. Repulsive **van der Waals interactions** limit the probable torsion angles to certain regions, typical for the **secondary structure** in which the residues partake. As the main chain torsion angles are generally not restrained during refinement, the torsion angle analysis provides valuable stereochemical **cross-validation**. Also known as Ramachandran plot, named after its inventor.

main chain torsion angles
The **torsion angles** phi (φ), psi (ψ), and omega (ω) around N and Cα of a residue, Cα and C of a residue, and around C of one residue and N of the subsequent residue in a polypeptide chain, respectively.

map
A graphical representation of a 2-dimensional or 3-dimensional **function** where points of equal value are connected with 2-dimensional contour lines or a 3-dimensional mesh or grid. *See also* **electron density map**, **Patterson map**.

map averaging
Procedure of averaging electron density, often combined with **solvent modification** for phase improvement.

map contrast and connectivity
Measures for electron density map quality. Contrast between protein region and solvent region, as well as connectivity within the protein region, should be high.

map inversion
Computation by Fourier transformation of complex structure factors from electron density.

map-likelihood phasing
Procedure of density modification based on the adjustment of the initial electron density map likelihood toward the likelihood function of a proper protein map.

marginalization
Elimination of a **nuisance variable** from a **probability distribution** by integrating it out.

marker atom
Atoms incorporated in a protein structure or crystal providing the source of **isomorphous differences** or **anomalous differences** in diffraction data. Common markers are **heavy metal** ions, native or soaked/co-crystallized with proteins; Se in **selenomethionine** labeled proteins; or native sulfur in case of **S-SAD phasing**.

marker atom substructure – *see* **substructure**

masked bulk solvent model – *see* **flat bulk solvent model**

mass absorption coefficient
Measure for how much X-rays are absorbed by a certain element, in $cm^2\, g^{-1}$.

mass spectrometry (MS)
Technique for identifying compounds on the basis of their mass-to-charge ratio. Common uses are identification of proteins, determination of molecular mass, sequencing polypeptides, or hydrogen–deuterium exchange mass spectroscopy.

mathematical group – *see* **group**

matrix operators
Representation of a symmetry operation as a combined rotation and translation matrix. *See also* **symbolic operators**.

matrix weight
Term used by *REFMAC* for the **weight** of the X-ray term, w_a in *CNS* terminology.

Matthews coefficient
Ratio of asymmetric unit volume over protein molecular weight, on average 2.5 Å^3/Da^{-1}; related to solvent content (11-1).

Matthews probability
Conditional probability that the **asymmetric unit** of a crystal

structure harbors a homooligomer, given the diffraction limit or **resolution** of the data.

maximum entropy method
Special method of direct phase determination based on maximum (informational) entropy distributions of atoms in a structure, can use prior information.

maximum likelihood – *see* **maximum likelihood principle**

maximum likelihood coefficients
Fourier coefficients for electron density reconstruction derived from **maximum likelihood functions** based on **sigma-A**.

maximum likelihood function
A joint probability distribution that can accommodate incompleteness and various errors (as well as prior information) of the model. *Compare* **maximum posterior refinement**.

maximum likelihood method
Any optimization method that maximizes the **data likelihood function** in order to obtain the proportional **model likelihood**.

maximum likelihood principle
States that the best estimate for any parameter is the value that maximizes a **likelihood function**, that is, the joint probability of experimental outcomes given the model parameters (as well as prior information).

maximum likelihood rotation function MLRF (11-57)
Rotation function based on conditional Rice and Woolfson distributions accounting for relative model phases and data errors.

maximum likelihood translation function MLTF (11-53)
Translation function based on Rice and Woolfson distributions, accounting for model incompleteness and errors.

maximum posterior refinement
A maximum likelihood method that incorporates prior probability (commonly in the form of stereochemical restraints) in the maximum likelihood target function. Sometimes called maximum *a posteriori* refinement or regularized maximum likelihood refinement.

mean (M)
Expectation value of a **probability distribution**, in simple cases the arithmetic average or a weighted average.

mean absolute error (7-38)

mean E-squared minus one $\langle |E^2 - 1| \rangle$
Mean absolute of normalized structure factor squared minus one, indicating whether a structure is centrosymmetric (0.968) or acentric (0.736). Can be less for twinned structures.

median (M)
The median of a **probability distribution** $P(x)$ is the value of x where the integral over the distribution (area) above and below x are equal. *Compare* **mean**.

membrane
The fluid lipid bilayer plus associated proteins that encloses all cells and, in eukaryotic cells, many organelles as well.

membrane protein
Protein associated or anchored with, or extending across, lipid membrane of cells. Commonly used for transmembrane proteins that have one or more **transmembrane helices**. *See also* **integral membrane protein**.

merging (C)
Combining various measurements of identical or symmetry related reflections into one (unique) data set.

merging R-value – *see* **linear merging R-value (8-15)**

merohedral twinning
Growth of domains in different orientations that can be precisely described by a symmetry operation compatible with, or extending, the crystal symmetry. Reflections from the distinctly oriented **domains** perfectly superimpose and the diffraction pattern looks unsuspicious and normal. The symmetry operator is called the

twin operator. Merohedral twinning can be detected by deviations from expected **intensity distributions**. *See also* **hemihedral twinning**.

messenger RNA (mRNA)
RNA molecule that specifies the amino acid sequence of a protein. Produced in eukaryotes by processing of an RNA molecule made by RNA polymerase as a complementary copy of DNA. It is translated into protein in the translation process catalyzed by **ribosomes**.

metabolic inhibition
Suppression of Met biosynthesis through inhibition of aspartokinase by excess Ile, Lys, and Thr in the medium. Used to overexpress **selenomethionine** labeled proteins.

metastable
State of a system where upon disturbance or **nucleation** it returns under phase separation to equilibrium.

methionine
A hydrophobic amino acid with a $-CH_2CH_2SCH_3$ side chain. *See also* **seleno-methionine**.

methyl group
Hydrophobic $-CH_3$ group.

methylene group
Hydrophobic $-CH_2-$ group.

metric tensor, G
Dyad product of the basis vectors; used in molecular geometry claculations.

Metropolis–Monte Carlo algorithm
Stochastic optimization procedure based on random parameter changes; accepts a certain fraction of uphill movements.

microfluidics
The precise control and manipulation of fluids that are geometrically constrained to a small, typically sub-millimeter, scale. Microfluidic chips made of transparent silicone rubber are used in protein crystallization experiments and require very little material.

microseeding
Introduction of microscopic, external **nucleation** seeds into crystallization drop by **streak seeding** or other methods.

Mie ratio
The ratio of particle radius over **wavelength** determining the characteristics of a scattering process. *See also* **Rayleigh scattering**.

Miller indices
Integer number triple *h, k, l* written as reciprocal index vector **h** specifying sets of parallel, equidistant 3-dimensional **lattice planes** (Figure 5-46). Derived from direct **Weiss indices** by inversion and normalization with smallest common denominator.

MIR – *see* **multiple isomorphous replacement**

MIRAS
Multiple isomorphous replacement with anomalous signal, uses orthogonal anomalous phasing to improve **phase probabilities**.

mirror operation
A **symmetry operation** that generates a mirror image of an object. Not allowed for **chiral** motifs such as protein molecules, because mirror operations change the **handedness** of the motif.

mitogen, mitogenic
Extracellular signal molecule that stimulates cells to proliferate.

mode (M)
Peak value of a probability distribution, most probable value. The **mean** and the **mode** are equal only in monomodal symmetric distributions.

model (C)
The physical model of atoms representing the crystal structure. The same physical model can be differently parameterized, that is, represented by a different *mathematical model*.

model (M)
The mathematical set of adjustable parameters, describing a *physical* model.

model bias
Tendency of calculated **phases** to dominate the electron density reconstruction, notably in case of **molecular replacement** phasing.

model building (C)
Process of manually or by means of automated programs placing the atomic model of the protein structure into the **electron density**.

model likelihood (function) – *see* **posterior probability**

modulated structure (C)
Crystal structure with additional periodicity superimposed over crystal lattice period. Can be **commensurate** or **incommensurate** with the lattice period.

molecular dynamics (MD)
Potential energy minimization based on empirical potential parameterization and minimization of force equations. Can include X-ray terms or NOEs for experimental structure refinement.

molecular envelope – *see* **scattering envelope**

molecular replacement
Phasing method that uses the calculated phases from a correctly placed, structurally similar model to phase an unknown structure. Commonly broken up into a 3-dimensional rotational and subsequent 3-dimensional translational search. Can be based on 6-dimensional **stochastic searches**, **Patterson search** methods, and modern **maximum likelihood**-based search functions.

monochromator
Selects a narrow energy band of X-rays. Commonly one or more crystal monochromators separating X-rays by Bragg reflection. Focusing multi-layer X-ray mirrors can replace/enhance monochromators.

monoclinic
Lattice or crystal structure with cell parameters $a \neq b \neq c$, $\alpha = \gamma = 90°$, $\beta \neq 90°$, with primitive (*mP*) and *C*-centered (*mC*) lattice types. Minimal internal symmetry a 2-fold axis parallel to **unique axis b**.

monoclonal antibody
Antibody secreted by a **hybridoma** cell line. Because the hybridoma is generated by the fusion of a single B cell with a single tumor cell, each hybridoma produces **antibodies** that are all identical.

monomer (B)
Single molecular entity, component of **oligomer**.

monomer (C)
Molecular entity for which a **stereochemical restraint** file exists in a restraint file library (**CCP4**).

monotopic membrane protein
Protein associated or anchored in the lipid membrane of a cell with a simple tail or membrane anchor. *Compare* **bitopic membrane protein**.

morphology (C)
Shape and form of a crystal. *See also* **polymorphism**.

mosaic crystal
Single crystal consisting of multiple, slightly misaligned blocks or **domains**.

mosaicity
The degree of misalignment between **domains** of a **mosaic crystal**.

mother liquor
The liquid medium surrounding a growing protein crystal and filling its **solvent channels**.

motif (B)
Element of structure or pattern that recurs in many contexts, particularly a small structural pattern, often with specific function, that can be recognized in a variety of proteins.

motif (C)
Physical contents of the asymmetric unit which makes up the physical crystal structure upon application of **space group symmetry operations**. Can be atoms, parts of a molecule, an entire molecule, or an **oligomer**.

mtz file
Binary data format used by the CCP4 suite and other programs.

multi-conformer refinement
Refinement protocol that captures the dynamic nature of molecules by refining multiple models using time-averaged molecular dynamics.

multimodal
A function or probability distribution with multiple peaks or local extrema.

multiple conformations
Two (or rarely more) discrete conformations of the same molecule or part thereof, often observed in case of surface-exposed residues or ligand molecules.

multiple isomorphous replacement (MIR)
Method of **isomorphous replacement** which uses several heavy atom derivatives to resolve the **phase ambiguity**.

multiplicity of space group (C)
Multiplicity of the **point group** times the multiplicity of the **Bravais translation** (and times 2 in case of a **centrosymmetric** space group). Equal to the number of **asymmetric units** in the unit cell.

multi-solution methods
Methods for (sub)structure solution that generate and rank multiple possible solutions.

multivariate
A function or probability distribution depending on multiple variables or parameters.

multi-wavelength anomalous diffraction phasing (MAD)
Anomalous phasing technique utilizing both **anomalous** and **dispersive differences**.

mutagen, mutagenic
A mutagen is a physical or chemical agent that changes the genetic material of an organism, generating mutations. As many mutations cause cancer, mutagens are typically also carcinogens.

N

N-terminus – *see* **amino terminus**

native Patterson map
A map of the Patterson function computed from the intensities of a native protein structure as Fourier coefficients, compared to a difference Patterson map.

NCS – *see* **non-crystallographic symmetry**

neutron
Subatomic particle with no net electric charge carrying a magnetic moment and a mass slightly larger than that of a **proton**. Neutrons are produced in nuclear fission reactors or **spallation neutron sources**. For **neutron diffraction** they are moderated (thermal neutrons) so that their energy is equivalent to a wavelength on the order of 1 Å. Kinetic energy, speed, and **wavelength** of the neutron are related through the **de Broglie relation**.

neutron diffraction
Technique using thermal **neutrons** instead of X-rays for diffraction on crystals. Advantage is that the neutron scattering factor for hydrogen and deuterium is different and of the same order as for heavy atoms, therefore hydrogen atoms can be located and hydrogen–deuterium exchange studied.

nitrogen
Colorless, odorless, tasteless, and mostly inert diatomic gas at standard conditions, constituting 78% by volume of the earth's atmosphere. *See also* **liquid nitrogen**.

NMR – *see* **nuclear magnetic resonance spectroscopy**

NOESY
Nuclear Overhauser effect (NOE) spectroscopy, 2-dimensional NMR technique for structure determination of macromolecular structures. NOE resonances are observed through space, not through bonds. *Compare* **COSY**.

non-centric reflections – *see* **acentric reflections**

non-commutativity (M)
Outcome of operation is dependent on the order of elements, that is, $ab \neq ba$.

non-crystallographic symmetry (NCS)
Presence of more than one copy of a motif in the **asymmetric unit**. Frequently the case; nearly half of all structures in the PDB are **dimers** or higher **oligomers** state of the molecules. Molecules may be related through a general combination of rotation and translation; NCS is local and not limited to crystallographic **symmetry operations**. *See also* **proper non-crystallographic symmetry, translational non-crystallographic symmetry**. Figures 5-19, 11-2, 11-12.

non-merohedral twinning
Non-merohedral twinning can occur if cell dimensions of a crystal are such that they either *match* or are *reasonably close* in two dimensions when domains are specifically oriented to each other. In this case, the crystals look visually unsuspicious, but the diffraction pattern in three dimensions will reveal two different interpenetrating lattices. If the spots are reasonably resolved and one orientation dominates, the major component can often be indexed and integrated separately.

norm (M)
Euclidean norm: length, magnitude, of a vector, square root of sum of squared vector components.

normal distribution – *see* **Gaussian normal distribution**

normal equations
An $n \times p$ system of linear equations used in the setup of least squares minimization, with n the number of observations and p the number of model parameters.

normalized structure factor distribution
Probability distribution functions for **normalized structure factor** (or **structure factor amplitude**) to have a certain value. Can be unconditional (centric and acentric **Wilson distribution**), or conditional on a model (centric **Woolfson**, acentric **Sim** or **Rice distribution**).

normalized structure factors (7-109)
Structure factor amplitudes E_h reflecting the scattering of point atoms, adjusted for B-factor attenuation, **atomic scattering factor** attenuation, and **epsilon-factor**. *Compare* **unitary structure factors**.

normalized sum of residuals squared – *see* **chi-square**

*n*th central moment – *see* **central moment**

nuclear magnetic resonance spectroscopy (NMR)
Spectroscopic technique for solution structure determination that exploits the resonance of magnetic states of spin-carrying nuclei induced by radio frequency **electromagnetic radiation**.

nucleation (C)
Formation of sites from which phase separation proceeds, either due to homogeneous formation of nucleation sites at high supersaturation or heterogeneous introduction of nuclei (e.g. foreign matter or nucleation **seeding**).

nucleic acid
RNA or DNA, a macromolecule consisting of a chain of nucleotides joined together by phosphodiester bonds.

nucleophile
A **nucleophile** (literally "nucleus lover") is a chemical entity that seeks to bind to its reaction partner (the electrophile) by donating bonding electrons.

nucleoside
Purine or pyrimidine base covalently linked to a ribose or deoxyribose sugar.

nucleosome
Beadlike structure in eukaryotic chromatin, composed of a short

length of DNA wrapped around an octameric core of histone proteins. The fundamental structural unit of chromatin.

nucleotide

Nucleoside with one or more phosphate groups joined in ester linkages to the sugar moiety. DNA and RNA are polymers of nucleotides. Figure 2-43.

nucleotide excision repair

DNA repair pathway in which entire nucleotides are removed from the DNA helix and replaced. *Compare* **base excision repair**.

nuisance variable

A parameter whose value is generally not known but which appears in a **probability distribution**. It can often be eliminated from the probability distribution by integrating it out (**marginalization**).

Nyquist theorem

If a discrete function $F(x)$ contains no terms higher than N (corresponding to $1/d_{min}$) its transform is completely determined by representing it by a series of points spaced $1/2N$ (that is $d_{min}/2$) apart.

O

oblique (C)

A lattice or crystal system that is not **orthogonal**, also specifically identifying a 2-dimensional plane lattice with $a \neq b$, $\gamma \neq 90°$ or $120°$.

observations (C)

Any experimental **data** point such as measured data or knowledge-based **restraints**. In restrained refinement, experimental data (structure factor amplitudes or intensities) as well as independently established model restraints.

obverse (C)

Definition of normal or standard assignment of crystallographic axes; primitive **rhombohedral** cells are indexed as rhombohedrally centered **hexagonal** cells in obverse setting when $R = (^2/_3\ ^1/_3\ ^1/_3)$, $(^1/_3\ ^2/_3\ ^2/_3)$.

occupancy – *see* **occupancy factor**

occupancy factor

Fraction $0 \leq n \leq 1$ of atoms or molecules that actually occupy a crystallographic position, often less than one for ligands, depending on the **dissociation constant**.

occupancy refinement

Used in substructure solution to improve the solution. A sharp drop in occupancy after the expected number of marker atoms is indicative of a correct solution.

oligomer

Complex made of a number of subunits. *See also* **heterooligomer**, **homooligomer**.

oligosaccharide

Short linear or branched chain of covalently linked sugars.

omega backbone torsion angle

The backbone **torsion angle** defining the torsion around the C atom of one **residue** and N atom of the subsequent residue in a polypeptide chain. Restrained to 180°. Figure 2-7.

omega-loop – *see* **loop**

omit map

Electron density reconstruction from **Fourier coefficients** in whose computation a questionable part of the model has been omitted. Used as means of reducing **model bias**, optionally in combination with coordinate perturbation and/or **simulated annealing**. To generate composite omit maps, a different block of model is omitted each time the map is calculated, and the final composite map is averaged from the individual maps.

optical transforms

Fourier transforms of objects obtained experimentally by diffraction with visible light, used as a means of trial-and-error structure determination in the very early days of crystallography, and still useful for educational purposes.

optimization algorithm

Optimization algorithms are procedures that search for an optimum of a nonlinear, multi-parametric function. Deterministic optimizations such as the gradient-based **maximum likelihood methods** are fast and work well when we are reasonably close to a correct model, at the price of becoming trapped in local minima. Stochastic procedures employ a random search that also allows movements away from local minima. They are slow but compensate for it with a large convergence **radius**.

optimization experiment

A crystallization experiment that follows initial crystallization trials, with the objective of narrowing down the parameter space and adjusting initial conditions to obtain improved crystals.

origin (C)

The point from which the three **unit cell vectors** in a crystal structure can originate. Many space groups allow multiple origins.

orthogonal

A system whose **basis** vectors are perpendicular to each other.

orthogonalization matrix

Describes a 3×3 similarity matrix used to transform a **crystallographic coordinate basis** into an **orthogonal Cartesian basis**. Its inverse is the deorthogonalization matrix.

orthologs

Genes or proteins from different species that are similar in sequence because they are descendants of the same gene in the last common ancestor of those species. *Compare* **paralogs**.

orthonormal

An **orthogonal** system whose basis are **unit vectors**.

orthorhombic

Lattice or crystal structure with cell parameters $a \neq b \neq c$, $\alpha = \beta = \gamma = 90°$, with primitive (*oP*), *F*-centered (*oF*), or *I*-centered (*oI*) lattice types. Minimal symmetry three non-intersecting 2-fold rotation axes.

oscillation method – obsolet; *see* **rotation method**

outlier detection

Statistical method to flag data that exceed a certain deviation from the expectation value.

overall *B*-factor

Isotropic overall *B*-factor, obtained by **Wilson scaling** of diffraction data. During subsequent refinement, generally replaced by **anisotropic** overall displacement tensor.

overexpression

High expression levels of a protein in the native host or in other organism (heterologous expression).

overfitting

Introduction of too many parameters into mathematical refinement model, often by adding too many waters or other obscure features into the structure model. Controlled by **cross-validation**.

P

p53

Tumor suppressor gene found mutated in about half of human cancers. Encodes a gene regulatory protein that is activated by damage to DNA and is involved in blocking further progression through the cell cycle. Figure 2–47.

palindromic sequence

Nucleotide sequence that is identical to its complementary

strand when each is read in the same chemical direction, for example GATC.

paralogs
Genes or proteins that are similar in sequence because they are the result of a gene duplication event occurring in an ancestral organism. *Compare* **orthologs**.

parameter
Variable in a mathematical model.

parameter estimation (M)
Process of estimating parameters of a model based on experimental data. An estimator takes data as input and provides model parameters as output. Common estimators in crystallography are least squares, maximum likelihood, or maximum *a posteriori* estimators.

Parseval's theorem
States that the mean square of a complex function $F(x)$ and the mean square of its Fourier transform are proportional. Sidebar 9-5.

partial derivative
Derivative of a multi-parametric function with respect to one specific parameter.

partial occupancy – *see* **occupancy factor**

partial reflection, partial
An X-ray deflection or reciprocal lattice point extending over multiple diffraction images or frames is a "partial" on each frame.

partial structure (C)
Molecular structure where a part is unknown, accounted for with **Sim weights**, or if model errors are also included, with the **sigma-A** parameter.

Patterson correlation function
A scoring function based on the correlation coefficient between two Patterson maps.

Patterson correlation refinement
Refinement of atom or molecule positions by optimizing the correlation between observed and calculated Patterson functions. Requires only correct orientation, translation invariant.

Patterson cross-rotation (function)
Used in **molecular replacement** Patterson rotation searches, based on the overlap of **cross-Patterson vectors**.

Patterson cross-vectors
Interatomic distance vectors between different atoms in the asymmetric unit in a Patterson map. Not the same as **cross-Patterson vectors**.

Patterson function (9-29)
The Patterson function is the **autocorrelation** of the **electron density**, which is equivalent to the Fourier transform of the squared **structure factor amplitudes**, or the reflection intensities. The Patterson function has its maxima at the interatomic distance vectors between all atoms in the unit cell. It is inherently **centrosymmetric** and its symmetry is the **space group** symmetry plus **inversion**. It is used for (i) **substructure** solution, (ii) **molecular replacement** searches, and (iii) detection of translational NCS parallel to unit cell axes and proper NCS axes parallel to crystallographic axes via **native Patterson maps**.

Patterson map
A contoured 2- or 3-dimensional representation of the **Patterson function**.

Patterson minimum function
A robust scoring function for the correctness of a trial substructure; used in *SHELXD*.

Patterson search
A search algorithm that is based on the **Patterson function**, matching **interatomic distance vectors** of the search probe and those obtained from experimental data. Can be a **Patterson self-rotation** search, a **Patterson cross-rotation** search, or a Patterson translation search.

Patterson seeding
Method to pick starting phases (atoms) for **direct methods** from Patterson superposition, implemented in substructure solution program *SHELXD*.

Patterson self-rotation (function)
Used in Patterson searches, based on the overlap of **self-Patterson vectors**. *See also* **Patterson self-rotation map**.

Patterson self-rotation map
A map containing in **stereographic projection** the peaks of the **Patterson self-rotation function**, which reveal the orientation of proper non-crystallographic n-fold symmetry axes in the $\kappa = 360/n$ degree section. Figure 11-12.

Patterson self-vectors
Interatomic distance vectors between the same atom and its symmetry mates in a Patterson map. Not the same as **self-Patterson vectors**.

Patterson superposition methods – *see* **Patterson vector methods**

Patterson symmetry
Space group symmetry plus inversion.

Patterson vector methods
Patterson function-based search method that uses atom pairs or trial structures to locate (heavy) atom positions.

PCR – *see* **polymerase chain reaction**

peak data set
Anomalous diffraction data collected at the absorption edge maximum defined by an **absorption edge** scan.

PEG – *see* **poly(ethylene glycol)**

PEG-MME – *see* **poly(ethylene glycol) monomethyl ether**

pegylation
Attachment of drug molecules or enzymes to PEG for better delivery and release properties.

peptide – *see* **polypeptide**

peptide bond
Chemical bond between the **carbonyl group** of one **amino acid** and the **amino group** of a second amino acid—a special form of amide linkage. Peptide bonds link amino acids into long macromolecular **protein** chains. Almost always in *trans*-**conformation**, only one in 1000 peptide bonds assumes the less favorable *cis*-**conformation** for non-**proline** residues.

perfect hemihedry – *see* **hemihedral twinning**

phase ambiguity
The phasing equations based on a single substructure always have two solutions for the phase angle. The ambiguity can be broken by (i) additional independent substructures (isomorphous derivatives), (ii) additional orthogonal **anomalous** data, or (iii) **density modification** techniques.

phase angle
The angle by which the phase of a **complex structure factor** or reflection is shifted from a reference phase, commonly zero. Readily computed for known structures, but requires **experimental phasing** for unknown structures.

phase bias – *see* **model bias**

phase combination (10-41)
Weighted averaging of experimental and/or model phase probabilities from different sources, yielding best phases from **joint** phase **probabilities** with **phase probability distributions** frequently expressed using **Hendrickson–Lattman coefficients**.

phase diagram
A diagram depicting the phase relations in a multi-component system. *See also* **crystallization diagram**.

phase extension
Procedure which provides phases for (high resolution) reflections, for which no initial experimental phases from the marker

atom substructure solution are available. Solvent flattening, solvent flipping, sphere of influence, histogram matching and, Free Lunch are examples for common phase improvement and phase extension methods.

phase probability distributions
The probability that an experimentally determined phase angle has at each angle in the range 0 to 2π rad. Determined by solving the phasing equations, often in the form of specific **maximum likelihood functions**.

phase problem
As the intensity of diffracted X-rays is given by the product of the **complex structure factor F** and its **complex conjugate F***, it is **Hermitian** and the result is thus real, as required for a physical observable. This means that phase information is lost during the measurement process, creating the phase problem. The phases must be separately supplied by a variety of **phasing methods**. *See also* **experimental phasing**.

phase restrained refinement
Reciprocal space refinement that includes the independent experimental phase information as a restraint for the structure factors in the likelihood target function.

phase restricted reflections
Centric reflections are phase restricted.

phase separation
Occurs when a **metastable**, **supersaturated** system returns to equilibrium.

phased translation function
Translation function that uses electron density maps and thus phase information to determine translation vectors of properly oriented maps or models.

phases – *see* **phase angle**

phasing
Colloquial term for establishing the **phase angles** associated with measured **structure factor amplitudes**. Necessary procedure to solve the crystallographic **phase problem**. *See also* **experimental phasing**.

phasing circles
Circles in a **Harker diagram**.

phasing marker – *see* **marker atom**

phasing methods
See Table 10-1.

phasing power (10-58)
An occasionally found statistic defined for each derivative as the sum of heavy atom or anomalous contributions divided by the sum of **closure residuals**.

phenylalanine
Hydrophobic, aromatic L-α-amino acid with an aromatic $-CH_2C_6H_5$ side chain.

phi backbone torsion angle
The backbone **torsion angle** for the torsion around N and Cα of a **residue**. Figure 2-7.

phi–psi torsion angle plot – *see* **main chain torsion angle plot**

phospholipid
Most common category of lipids used to construct biomembranes. Generally composed of two fatty acid tails linked through glycerol (or sphingosine) phosphate to one of a variety of polar head groups.

phosphorescence
Photoluminescence process where the absorbed energy generates a state of higher spin multiplicity. Once the energy is trapped in the triplet state, transition back to the lower singlet energy states is forbidden and thus slow, sometimes taking hours. *See also* **imaging plate**.

photon
Uncharged particle with no rest mass, of energy $E = h\upsilon$ with h the Planck constant and υ the frequency. *See also* **energy–wavelength conversion**.

photon flux
Number of **photons** per second that pass through an area of 1 mm² Φ_λ = photons·s⁻¹·mm⁻². Dimensions depend on the energy unit selected for the photon.

physiological pH
Commonly the pH of blood, ~ pH 7.4 Varies for different cellular environments.

pI – *see* **isoelectric point**

pi-helix
A rare, loosely packed helical secondary structure element with backbone hydrogen bonds from residue n to $n + 5$.

pi-stacking
An uncharged dipole interaction between π-electron systems of aromatic or **arginine** residues.

pK_a value
The negative logarithm of the acid dissociation constant, $-\log_{10}(K_a)$, with $K_a = [H^+]\cdot[A^-]/[HA]$.

plain rotation
A rotation (axis, operation) that does not possess any additional symmetry elements such as translation (in **screw axes**) or inversion (in **inversion axes**).

plane equation (M)
A plane can be defined by a point with coordinate vector **x** on the plane and the normal vector **h** to the plane, $\mathbf{h}\cdot\mathbf{x} = n$.

plane group (C)
Set of **symmetry operations** forming a mathematical **group**, giving rise to the general 17 (five chiral) different arrangements of motifs in a plane (2-dimensional) periodic structure.

plasmid
Small circular extrachromosomal DNA molecule that replicates independently of the genome. Modified plasmids are used extensively as vectors for DNA cloning.

pocketome
The pocketome of an organism is the collection of all possible small molecule binding envelopes present in its cells.

point group
Mathematical **group** of operations that keep the origin fixed at one point and comply with the rules for mathematical groups. *See also* **crystallographic point groups**.

Poisson distribution (7-20)
A one-parametric, discrete **probability distribution function** describing for example random counting errors.

polar
In the electrical sense, describes a structure (for example, a chemical bond, chemical group, molecule, solvent) with positive charge concentrated toward one end and negative charge toward the other as a result of anisotropic electron distribution. Polar molecules are likely to be soluble in water. See also **dielectric constant**.

polar coordinates
Define any point in a plane through a distance from origin (or pole) and the azimuth (or polar) angle.

polar solvent
Solvent consisting of polar molecules.

polar space group
Space group in which the origin is arbitrarily located on a crystallographic axis. Table 6-6.

polarizability
Proportionality constant quantifying the interaction of the electric field vector and the resulting induced electric dipole moment.

polarization (C)
Describes the orientation of the *electric field vector* relative to the propagation direction of the wave. The oscillations may be fixed in

a single direction (linear polarization), or the oscillation direction may rotate as the wave travels (circular or elliptical polarization).

polarization factor (6-50)
A correction factor for raw diffraction intensities taking into account the effects of instrument- and diffraction-geometry dependent **polarization** of scattered X-rays. Often combined with the **Lorentz factor**.

polychromatic
Radiation that contains photons with widely varying energy, also termed "white radiation."

polycrystalline
Material that contains a randomly oriented sample of many small microcrystals.

poly(ethylene glycol) (PEG)
A common precipitate for protein crystallization, reducing protein **solubility** by competing for solvent water molecules (excluded volume effect). PEGs are organic polyalcohols $HO-(O-CH_2-CH_2)_n-OH$ of varying chain lengths between ~200 and ~15 000 Da average molecular weight. PEGs above 1 kDa are solid white powders, and freshly prepared aqueous solutions are used as precipitants.

poly(ethylene glycol) monomethyl ether (PEG-MME)
Same as **poly(ethylene glycol)**, except PEG-MMEs have a CH_3O- (methoxy) group on one terminal $CH_3O-(O-CH_2-CH_2)_n-OH$.

polyion, polyionic
Molecule that carries multiple charges; proteins are polyionic macromolecules or polyionic **heteropolymers**.

polymer
Macromolecule made of many monomers.

polymerase
Enzyme that catalyzes polymerization reactions such as the synthesis of DNA and RNA.

polymerase chain reaction
Technique for amplifying specific regions of DNA by the use of sequence-specific **primers** and multiple cycles of DNA synthesis, each cycle being followed by a brief heat treatment to separate complementary strands.

polymorphism (B)
Refers to genes with two or more alleles that coexist at high frequency in a population.

polymorphism (C)
Refers to different **crystal forms** (morphologies) of the same material. *Compare* **habit**.

polypeptide
A linear **heteropolymer** built by combination of different proteinogenic **amino acids**. The term peptide is generally applied to shorter stretches or molecules, while larger polypeptides (> ~50 residues) folding into distinct **tertiary** structures are called **proteins**.

polyproline-II helix
A left-handed, helical secondary structure element consisting of *trans*-proline or *trans*-proline-rich stretches with backbone hydrogen bonds from residue n to $n + 2$. Long PP-II helices containing enzymatically Cγ-hydroxylated prolines are building blocks of the twisted collagen fibers.

polytopic membrane protein
Integral membrane protein passing through the lipid membrane of a cell with multiple transmembrane helices and having multiple functional domains, commonly in the ectodomain, which sometimes can be expressed separately. *Compare* **monotopic membrane protein**.

pose (of a ligand)
The orientation of the ligand relative to the receptor including the specific conformation of the ligand when bound to the receptor.

posterior probability
Joint probability function describing the **model likelihood** $prob(model|data, I)$ as the product of **data likelihood function** $prob(data|model, I)$ and **prior probability** $prob(model|I)$. *See also* **Bayes' theorem**.

posttranslational modification (PTM)
Covalent modifications to the expressed protein, commonly carried out by other enzymes. **Glycosylation**, N-terminal acetylation, phosphorylation, and myristilation are common PTMs.

powder diffraction
Diffraction method applied to polycrystalline material. Powerful for chemical characterization, limited in case of *de novo* structure determination.

precession camera
An ingenious, once popular, mechanical contraption invented by Martin Buerger that provides an undistorted projection of the reciprocal lattice onto film or detector.

precipitant
A primary component of the **crystallization cocktail** that reduces the protein solubility.

precipitate (C)
Describes, in the context of protein crystallization, a reagent that reduces the **solubility** of proteins.

precision
High precision indicates that measurements of a quantity have a narrow distribution (small variance) about their mean. An experiment can be precise but have low **accuracy**. Figure 7-4.

precision-indicating merging R-value (8-17)

preconditioned conjugate gradient minimization
Iterative multivariate minimization algorithm that approximates the second derivatives to improve the gradient descent path.

primary structure
The linear sequence of the amino acid **residues** in the protein chain, beginning at the **amino-terminus**.

primer
Oligonucleotide that pairs with a template DNA strand and promotes the synthesis of a new complementary strand by a polymerase.

primitive (C)
Non-centered **lattice**, **space group**, **crystal structure**; the six non-centered primitive **Bravais lattices** are aP, mP, oP, tP, hP, and cP. Primitive does not mean simple.

principal Euler angle
The matrix equation represented by the **direction cosine matrix** $\mathbf{R}$ describing a rotation can be solved and yields the three **eigenvalues**. The **eigenvector** belonging to the real eigenvalue 1 is the **Euler axis**, with the principal Euler angle defining the single rotation angle around the Euler axis $\mathbf{E}$. Equivalent to the rotation angle κ in **spherical coordinate** representation.

prior probability
Probability distribution function describing the probability of a hypothesis or model prior to considering the specific experimental data of the present experiment. *See also* **Bayes' theorem**.

probability
The probability of a random variable is its *relative frequency of occurrence*, that is, it allows us to predict unknown outcomes based on known parameters. In general terms, also the **likelihood** of a certain outcome of an experiment, that is, the estimate of unknown parameters based on known outcomes. *Compare* **likelihood**.

probability density function
A continuous, integrable **probability distribution function**.

probability distribution function
General term for the probability of occurrence $P(x)$ of a random variable for given values of x.

projection (M)
Linear transformation described by matrix $\mathbf{P}$ fulfilling $\mathbf{P} = \mathbf{P}^2$. Used for example to project 3-dimensional objects (molecules or maps) onto a 2-dimensional plane.

prokaryote (procaryote)
Single-celled microorganism whose cells lack a well-defined, membrane-enclosed nucleus, either a bacterium or archaeon. Prokaryotes generally lack the capability of **posttranslational modifications**, have no dedicated translocation machinery, and their reducing cytoplasm often prevents disulfide formation when overexpressing eukaryotic (mammalian) proteins in prokaryotic hosts.

proline
Cyclic, hydrophobic, proteinogenic L-α-amino acid. The imide bonds between a non-proline residue and a subsequent proline are more frequently (~6%) found in *cis*-conformation than for two non-proline residues (0.1%). For the cyclic amino acid proline, the *cis*- and *trans*-conformations are geometrically very similar, with comparable energies differing by only a few kcal mol^{-1}. Figure 2-13.

proper non-crystallographic symmetry (NCS)
Non-crystallographic symmetry that is rotational only and close to a **point group** (closed group) operation. In contrast to the translation periodicity restrictions of crystallographic point groups, also 5-fold and higher NCS rotation axes are allowed. Proper NCS manifests itself in **Patterson self-rotation maps**. Proper crystallographic NCS axes parallel or very near parallel to crystallographic axes that are otherwise obscured in self-rotation maps can manifest themselves in **native Patterson maps**. Figures 11-12, 13-23. Note that proper has a different meaning here than in **proper symmetry operation**.

proper rotation – *see* **plain rotation**

proper symmetry operation
A proper symmetry operation does not change the properties of the **motif** it is acting upon. Only **plain rotations** and **screw axes** are allowed crystallographic symmetry operations for **asymmetric (chiral)** motifs such as protein molecules.

proportional counter
Detector where a photon ionizes a counting gas, generating multiple electrons pulled toward the anode and measured as an electric pulse.

protease (proteinase, proteolytic enzyme)
Enzyme that degrades proteins by hydrolyzing specific peptide bonds between amino acids.

protein
A linear **heteropolymer** built by combination of 20 different common **proteinogenic amino acids**. The term protein is generally used for longer molecules, while shorter stretches (< ~50 residues) not necessarily folding into distinct **tertiary structures** are termed **polypeptide**. Proteins are the major macromolecular constituent of cells.

protein construct
A protein or fragment thereof with a specific sequence, generally engineered and overexpressed, often linked with **affinity tags**, fluorescent probes, and so forth.

Protein Data Bank (PDB)
A curated, world-wide repository of all experimentally determined protein structures, also providing a multitude of online structure analysis tools and database access.

protein engineering
General term for modifying, mostly at the genetic level by **recombinant DNA** techniques, the **sequence** and properties of a protein.

protein stock solution
Aqueous, weakly buffered solution of protein that contains additional components as required to stabilize the protein, plus unknown components acquired and carried through during purification.

proteinogenic amino acids
Proteinogenic amino acids serve as building blocks of **proteins** and are L-α-**amino acids**. There are 20 common proteinogenic amino acids, plus **selenocysteine** and **pyrrolysine**.

proton
Positively charged subatomic particle that forms part of the atomic nucleus. Equivalent to a hydrogen atom stripped of its **electron**, forms in solution a **hydronium ion**.

protonation
Loss of electron and acquisition of a positive charge in aqueous medium.

protozoa
Free-living or parasitic, nonphotosynthetic, single-celled, motile eukaryotic organisms, such as *Paramecium* and *Amoeba*. Free-living protozoa feed on bacteria or other microorganisms.

pseudo-symmetry
Non-crystallographic symmetry with rotation axes parallel to crystallographic axes and/or translations with integer fractions of unit cell vectors that generate higher apparent symmetry and abnormal intensity distributions.

psi backbone torsion angle
The backbone **torsion angle** for the torsion around Cα and C of a **residue**. Figure 2-7.

pyrrolysine
A rare **proteinogenic amino acid**, found only in methanogenic **archaea** in **enzymes** that are part of their methane-producing metabolism.

Q

quasi-elastic light scattering – *see* **dynamic light scattering**

quaternary structure
The 3-dimensional arrangement of more than one polypeptide chain (subunits) in a molecular complex.

quaternion
Quaternions form a 4-dimensional normed division algebra over the real numbers. Used as 4×4 matrices in the description of certain 3-dimensional rotations and superpositions.

R

R-anom (8-18)
The **linear merging R-value** for anomalous pairs of reflections.

R-cryst
Commonly the **linear merging R-value** for reflections from different crystals.

R-Cullis (10-56)
Various linear residuals once used for tracking **heavy atom refinement**.

R-delta (10-4)
Generic difference data merging *R*-value.

R-factor – obsolete *see* **R-value**

R-free – *see* **cross-validation R-value** (12-35)

R-free/R ratio (12-103)

R-int
Commonly the **linear merging R-value** for identical reflections; based on intensities.

R-Kraut (10-57)
R-value based on linear **closure residual**.

R-meas – *see* **redundancy-independent merging R-value** (8-16)

R-merge – *see* **linear merging R-value** (8-15)

R-pim – *see* **precision-indicating merging R-value** (8-17)

R-rim – *see* **redundancy-independent merging R-value** (8-16)

R-sigma (8-25)

R-sym
Commonly the **linear merging R-value** for symmetry related reflections.

R-value, R_F-value (7-51)

The normalized **linear residual** between observed and calculated structure factor amplitudes. Expectation value for non-centrosymmetric random structure ~0.59, for centrosymmetric case ~0.83.

R-work (12-35)

R-value for all reflections not belonging to the **cross-validation** data set. *See also* **cross-validation R-value (R-free)**.

racemic mixture, racemate

Mixture of equal parts of **enantiomorphs**.

radiation damage

Resulting from interaction of high-energy **ionizing radiation** such as X-rays, causing bond breakage and formation of **free radicals**. Destroys protein crystals and is routinely counteracted by keeping crystals at **cryogenic** temperature during exposure.

radiation damage-induced phasing (RIP)

An unusual phasing technique that exploits the intensity differences between data recorded from a crystal at different stages of **radiation damage**.

radical

Reactive species featuring unpaired electrons.

radius of convergence – *see* **convergence radius**

radius of gyration

The root mean square distance of the atoms from the **center of mass** of a molecule.

Ramachandran plot – *see* **main chain torsion angle plot**

random error

Deviations from the mean following a **normal distribution**.

random variable

A variable that can take any value in a given interval depending on its probability distribution, independent of past or future observations.

raw moment, *k*th raw moment

The moments of a distribution about the origin, statistical descriptors used in intensity statistics and defined as $\langle x^k \rangle$ for the *k*th raw moment. *Compare* **central moment**.

Rayleigh scattering

Elastic scattering electromagnetic radiation by particles much smaller than the **wavelength** of the light, that is, with a **Mie ratio** $x_M \ll 1$.

real space

The 3-dimensional space in which crystallographers and physical objects reside.

real space correlation coefficient (plot), RSCC (7-54)

A scale-factor-independent measure, plotted residue-by-residue, for the agreement between observed (bias minimized) and calculated (model) electron density.

real space molecular replacement

Molecular replacement methods to determine **NCS** operators, applied to electron density maps by **domain rotation functions** and **phased translation functions**.

real space R-value (13-3)

Linear residual between observed and calculated electron density.

real space refinement

Local refinement of the positional parameters of the structure model against experimental or **bias** minimized electron density.

receptor

Any protein inside the cell or at the cell surface that binds a specific signal molecule (ligand) and initiates a response in the cell.

reciprocal lattice

A periodic lattice in **reciprocal space** R^* with basis [0, $\mathbf{a}^*$, $\mathbf{b}^*$, $\mathbf{c}^*$] consisting of **reciprocal lattice points** and reciprocal to real space (crystal lattice) [0, $\mathbf{a}$, $\mathbf{b}$, $\mathbf{c}$]. *See also* **reciprocal space**.

reciprocal lattice point

Any point $\mathbf{r}^* = (h\mathbf{a}^* + k\mathbf{b}^* + l\mathbf{c}^*)$ with *h, k, l* integers in a lattice with basis [0, $\mathbf{a}^*$, $\mathbf{b}^*$, $\mathbf{c}^*$] reciprocal to basis [0, $\mathbf{a}$, $\mathbf{b}$, $\mathbf{c}$]. **Reciprocal lattice** points represent sets of parallel equidistant lattice planes with **Miller indices** *hkl* in a real space lattice or crystal structure.

reciprocal lattice vector

A vector from the origin to a **reciprocal lattice point** (*hkl*), collinear with the normal vector and interplanar distance vector $\mathbf{d}(hkl)$ of lattice plane *hkl*.

reciprocal space

A space R^* related to a real space R by the reciprocity conditions of the basis vectors so that the **dyad product** of their **basis vectors** forms the **identity matrix**, that is, in three dimensions $\mathbf{a}^* \cdot \mathbf{a} = \mathbf{b}^* \cdot \mathbf{b} = \mathbf{c}^* \cdot \mathbf{c} = 1$. The diffraction space spanned by the **reciprocal lattice** with basis [0, $\mathbf{a}^*$, $\mathbf{b}^*$, $\mathbf{c}^*$] is a reciprocal to the **real space** with basis [0, $\mathbf{a}$, $\mathbf{b}$, $\mathbf{c}$] and *vice versa*. Real space and reciprocal space share the same origin. Objects or functions are mapped from one space or domain into its reciprocal space or domain by **Fourier transforms**.

reciprocal space interference (G-)function

Function that determines how reciprocal lattice points (structure factors) calculated from different regions of direct space affect each other (11-26).

reciprocal space refinement

Nonlinear refinement procedure in which the minimization target function is based on structure factor amplitudes, which are elements of reciprocal space (i.e. diffraction space). *See also* **restrained reciprocal space refinement**.

recombinant DNA

Any DNA molecule formed by joining DNA segments from different sources.

recombinant DNA methods, technology

Collection of techniques by which DNA segments from different sources are combined to make a new DNA, often called a recombinant DNA. Recombinant DNAs are widely used in the **cloning** of genes, in the genetic modification of organisms, and in production of proteins by overexpression in an **expression host** or **cell-free system**.

rectangular lattice

A 2-dimensional plane lattice with $a \neq b$, $\gamma = 90°$.

redundancy-independent merging R-value (8-16)

refinement

General term for iterative adjustment of variable **parameters** of a model so that the fit between model and observed data is optimized by minimizing the sum of squared residuals of a refinement target function, generally a **maximum likelihood** residual. Different refinement protocols and **optimization algorithms** exist. *See also* **restrained reciprocal space refinement**.

reflection conditions – *see* **systematic absences**

reflections (reflexions) – *see* **X-ray reflection**

refractive index

Measure for how much the propagation speed of a photon is reduced in a medium compared to vacuum. In most materials proportional to square root of the **dielectric constant**, and approaching unity in all materials for **X-ray** radiation.

register error

Model building error in which the main chain conformation is correct, but the sequence assignment and with it the side chains are shifted.

re-indexing

Assignment of new unit cell axes using a basis transformation in the form of a re-indexing matrix.

residual, deviate (7-28)

The difference between two values, often between observed and calculated values or observed and **expectation values**.

residual index – *see* **R-value** (7-51)

residue

Amino acid **monomer** within a **polypeptide** chain; **side chain** plus peptide **main chain atoms** (Figure 2-19).

resolution (C)

The limit in **diffraction angle** up to which X-rays diffracted by a crystal can be detected. Expressed in smallest interplanar **lattice spacing** d_{min} sampled, approximately equal (slightly larger than) to the distance at which objects can be distinguished in an electron density map (Sidebar 9-4). The higher the resolution, the smaller the lattice spacing, and the more detailed the **electron density** map. Lattice spacing and diffraction angle are related through the **Bragg equation**. *Compare* **effective resolution**.

resolution sphere (C)

A sphere (or ellipsoid in the anisotropic case) with radius of $1/d_{min}$ containing the actually observed reflections intersecting the **Ewald sphere**, drawn centered at the reciprocal lattice origin in a Ewald construction.

resonance

Maximal oscillation of a system in response to a stimulus, such as atomic electrons excited by the electric field vector of electromagnetic radiation. A region in the spectrum where absorption and anomalous dispersion effects occur.

restrained reciprocal space refinement

Refinement against structure factor amplitudes where the **sum of residuals squared** includes additional residual terms for the deviations of the model geometry from **stereochemical restraint target values**. *See also* **restraint weights**.

restraint target values – *see* **stereochemical restraint target values**

restraint weights

The inverse variance of the respective restraint target distribution.

restraints – *see* **stereochemical restraints**

restriction nuclease (restriction enzyme)

Enzyme that can cleave a DNA molecule at any site where a specific short sequence of nucleotides occurs. Important for recombinant DNA technology.

rhombohedral

A lattice or unit cell that can be either indexed rhombohedrally primitive with $a=b=c$, $\alpha=\beta=\gamma \neq 90°$, or preferably indexed in **obverse hexagonal** setting as a rhombohedrally centered **Bravais lattice** or cell.

rhombohedral centering

A rhombohedral cell can be described as a triple-sized trigonal cell with rhombohedral R-centering vectors $(2/3\ 1/3\ 1/3)$ and $(1/3\ 2/3\ 2/3)$, lattice type hR.

ribosome

Particle composed of ribosomal RNAs and ribosomal proteins that catalyzes protein synthesis using information provided by mRNA (Figure 2-1).

ribozyme

RNA with catalytic activity.

Rice distribution (7-148)

A special case of non-central χ-distribution, the form of the **Sim distribution** and the general, conditional **acentric** structure factor probability distribution.

riding positions

The calculated positions of hydrogen atoms in a molecular structure.

rigid body refinement

Refinement of the position of entire molecules or molecular groups treated as rigid units in a crystal structure. Often used subsequent to **molecular replacement** searches in early steps of macromolecular refinement.

RIP – *see* **radiation damage-induced phasing**

r.m.s.d., RMSD (7-33)

Root mean square deviation, square root of the **variance** of a distribution.

RMS-Z score (7-37)

Normalized statistical standard score, equals unity when the sample distribution has the same variance as the target distribution, smaller than 1 for tighter distributions, larger for more loose distributions. *See also* **Z-score.**

RNA (ribonucleic acid)

Polymer formed from covalently linked ribonucleotide monomers. Different functions, such as messenger RNA, ribosomal RNA, transfer RNA. Forms secondary structures that sometimes cause problems in the translation phase during heterologous overexpression.

rotation

Linear transformation or operation by which an object is rotated around a given **rotation axis**. Every rotation can be described by a rotation by one **principal Euler angle** around the **Euler axis**. Expressed as linear matrix equation $\mathbf{x'} = \mathbf{Rx}$ with rotation matrix $\mathbf{R}$ the **direction cosine matrix**.

rotation axis

Axis around which a general rotation is performed. *Compare* **crystallographic rotation axis**.

rotation camera – *see* **rotation method**

rotation function

Rotation functions are used to find the relative orientation of a search molecule in an unknown crystal structure. They can be based on the Patterson rotation function or can be implemented as **maximum likelihood rotation functions**.

rotation matrix

Description of a rotation operation by a 3×3 matrix with special properties. *See* **direction cosine matrix**.

rotation method

Simplest method of X-ray data collection by exposing a crystal to X-rays while it is rotated in small increments around a single axis. During each rotation increment one **diffraction image** or **frame** is recorded.

rotation search – *see* **rotation function**

roto-inversion – *see* **inversion axis**

A combination of crystallographic **rotation axis** with **inversion**. This improper symmetry operation is not allowed in protein crystals.

roto-translation – *see* **crystallographic screw axis**

S

SAD – *see* **single-wavelength anomalous diffraction phasing**

salt bridge, ion pair

Charged, non-directional (ionic) interaction between two side chains, for example Glu-Asn.

salting in

Increasing protein solubility by increasing salt concentration, in low-salt regime of protein solubility diagram.

salting out

Exponential decrease of protein solubility due to increasing salt concentration, in high-salt regime of protein solubility diagram.

salvage strategies

Methods applied to modify protein after the fact that crystallization did not succeed, such as **lysine methylation** or enzymatic digests.

sampling probability – *see* **data likelihood**

satellite reflections

Weak systematic reflections next to Bragg reflections, often indicative of an **incommensurately modulated structure**.

Sayre's equation (10-21, 10-24)

Convolution of **structure factors** in **reciprocal space**, equivalent to squaring of **electron density** in **real space**. Used together with **structure invariants** in **direct methods** of phase determination.

SBS standard

The Society for Biomolecular Screening (SBS) provides standards for labware and plastics assuring that their form factors are compatible with SBS standard compliant instruments and robotics.

scalar product (M)

The scalar (or inner, dot) product of two column vectors **a**, **b** is the defined as $\mathbf{a} \cdot \mathbf{b} = ab \cos \gamma$ where γ is the angle enclosed between the vectors and a and b the **norm** of **a** and **b**.

scale factor (overall, linear)

Brings observed diffraction intensities or structure factor amplitudes onto a common, absolute scale with calculated intensities or structure factor amplitudes. Obtained initially from a **Wilson plot**.

SCALE records

Records in PDB file containing the **deorthogonalization matrix**.

scaling

Procedure of bringing various data sets onto a common relative or absolute scale. *See also* **Wilson scaling**.

scattering diagram

A diagram representing the change of wave vectors (momentum transfer) during a scattering process.

scattering envelope

Commonly used term for the scattering function of an entire (single) molecule.

scattering factor – *see* **atomic scattering factor**

scattering probability

Probability distribution function describing the spatial distribution of scattered waves or particles. *See also* **atomic scattering factor**.

screw axis – *see* **crystallographic screw axis**

SDS-PAGE (sodium dodecyl sulfate–polyacrylamide gel electrophoresis)

Type of electrophoresis used to separate proteins by size. The protein mixture to be separated is treated with a harsh negatively charged detergent (SDS) and with a reducing agent (β-mercaptoethanol), before being run through a polyacrylamide gel. The detergent and reducing agent unfold the proteins, free them from association with other molecules, and separate the polypeptide subunits. *See also* **electrophoresis**.

Se-Met labeling

Introduction of Se-Met either by overexpression in methionine synthesis deficient (Met⁻) host cells or by **metabolic inhibition** in Se-Met augmented medium.

second virial coefficient

A measure for the thermodynamic excess energy, measuring pair-wise intermolecular interactions in a solution, available via osmotic pressure or **light scattering methods**.

secondary structure

Sequence of structural elements in a macromolecule with regular local folding patterns, defined in proteins by a specific pattern of **hydrogen bonds** between **main chain atoms** of the protein chain. Most common secondary structure elementsare α-**helices**, β-**sheets**, and various **turns** and **loops** connecting them.

seeding (C)

Introduction of external crystallization nuclei in **supersaturated** protein solution, in order to induce **nucleation** and phase separation.

self-adjoint matrix – *see* **Hermitian matrix**

self-Patterson vectors

Interatomic distance vectors between atoms of the same molecule in a crystal structure. They are typical for a molecule irrespective of the crystal structure it is in, and can therefore be used in Patterson rotation searches for determining the orientation of a molecule.

self-rotation map – *see* **Patterson self-rotation map**

self-vectors – *see* **Patterson self-vectors**

selenocysteine

A rare **proteinogenic amino acid**, isomorphous to L-α-**cysteine** but with Se instead of S, found in certain redox enzymes and hydrogenases.

selenomethionine

Isomorphous to L-α-**methionine** but with Se instead of S, introduced by overexpression in medium augmented with Se-Met instead of native methionine. Used as a source of isomorphous **anomalous signal** for **anomalous phasing**. Figure 2-21. *See also* **Se-Met labeling**.

semi-invariants – *see* **structure semi-invariants**

sequence – *see* **primary structure**

sequence identity

The percentage of identical residues between two aligned protein sequences.

sequence similarity

The percentage of identical and functionally similar residues between two aligned protein sequences, based on a similarity or substitution matrix.

serial extinctions

Systematic absences that extend along a specific direction in the reciprocal lattice, along a reciprocal axis. The 11 **enantiomorphic pairs of space groups** display the same serial extinctions ($l = 3n$ only for $P3_1$ and $P3_2$, for example) and cannot be distinguished by extinction rules (Table 6-6) alone.

serine

A **polar** L-α-amino acid with a short hydroxyl side chain $-CH_2OH$, common target for phosphorylation by kinases.

serine protease

Type of protease that has a reactive serine in the active site.

Shake and Bake (SnB)

A general purpose dual-space direct method program for structure solution, also used for marker atom **substructure** solution.

Shake and wARP

A slow but powerful, iterative **model bias** minimization and model completion protocol based on ARP/wARP combining coordinate perturbation, **omit** techniques, unrestrained maximum likelihood **dummy atom refinement**, and multi-model **map averaging** to generate bias minimized electron density maps. Generally requires data with **resolution** of ~2.5 Å or better.

shape complementarity

Fit of molecules to **receptor** based on shape giving rise to **van der Waals interactions** (in addition to specific interactions) which contribute a significant part of the binding energy of drug–receptor interactions

sharpening

Sharpening implies elimination of the attenuation of the structure factor magnitude with increasing scattering angle $\sin \theta / \lambda$ by assumption of point atoms resulting in **normalized structure factor** amplitudes E_h. Maps generated from these E_h as coefficients are sharpened. In practice, somewhat less extreme sharpening (sharpening also leads to increased **truncation ripples**) is desirable, and "mixed" structure factor amplitudes of the form $(E_h^3 \cdot F_h)^{1/2}$ are commonly used.

Sheldrick's rule

States that a resolution of 1.2 Å or better is required for *ab initio* structure determination of proteins by direct methods.

SHELX programs

Suite of fast and robust programs by George Sheldrick for general and macromolecular crystallography, including experimental phasing, density modification, and refinement.

side chain

Amino acid **monomer** or **residue** sans **main chain atoms**, beginning with Cβ. Figure 2-19.

side chain torsions

The set of torsion angles in an amino acid side chain.

sigma-A, σ_A

A **maximum likelihood**-derived statistical **weight** (variance) accounting for both *incompleteness* and errors in **structure models**, thus providing more accurate structure factors and thus better Fourier coefficients for **electron density map** reconstruction. Generally estimated from a 2-parameter, scattering angle dependent exponential fit (12-19) using the **cross-validation** data set in order to minimize bias.

sigma cutoff

Truncation of data below a certain signal-to-noise ratio, commonly applied to high resolution shells. Necessary for **difference data**, but not in maximum likelihood refinement.

signal sequence

Short continuous sequence of amino acids that determines the final location of a protein in the cell, for example the N-terminal signal peptides directing nascent secretory and transmembrane proteins to the endoplasmic reticulum.

signal-to-noise ratio (SNR)

The fraction of the measured signal divided by its standard deviation or noise level.

signaling cascade, pathway

Sequence of linked intracellular reactions, typically involving multiple amplification steps in a relay chain, triggered by an activated cell-surface **receptor**.

Sim distribution (7-127)

A probability distribution function accounting for incompleteness in structure models (partial structure), similar in shape to the **Rice function**.

Sim weights (7-127)

Statistical **weights** accounting for incompleteness in structure models (partial structure), expressed in terms of modified Bessel functions of the first kind. Figure 7-16.

similarity transformation

Transformation of type $R_w = AR_C A^{-1}$ between basis systems W and C.

Simplex (Nelder–Mead) algorithm

Non-linear numerical optimization algorithm that does not need derivatives.

simulated annealing

A general technique of energy refinement by successively minimizing the potential energy of a perturbed model that slowly returns to equilibrium following a certain *annealing schedule* $T_1 > T_2 > T_3 \ldots T$(final). The X-ray term as well as the stereochemistry of the model are expressed as potential energy terms. During the cooling the model relaxes into a minimum energy. High **radius of convergence**, particularly when combined with **torsion angle refinement**.

simultaneous equations

A system of n equations with multiple parameters, not necessarily linear.

single anomalous diffraction (SAD) phasing

Phasing technique that utilizes a single **anomalous difference** data set. Needs to resolve two ambiguities: **substructure handedness** and **phase angle ambiguity** in phasing equations, generally by means of **density modification**.

single isomorphous replacement (SIR)

Phasing technique that utilizes a single **isomorphous derivative**. Needs to resolve two ambiguities: **substructure handedness** and **phase angle ambiguity** in phasing equations, generally by means of **density modification** or addition of orthogonal **anomalous signal** (SIRAS).

single isomorphous replacement with anomalous scattering (SIRAS)

Phasing method that uses the additional **anomalous signal** of the heavy marker atom to break the **phase angle ambiguity** inherent in **single isomorphous replacement** (SIR).

SIR – *see* **single isomorphous replacement**

SIRAS – *see* **single isomorphous replacement with anomalous scattering**

site-directed mutagenesis

Technique by which a mutation can be made at a particular site in DNA. Sidebar 4-3.

sitting drop (C)

Crystallization method by **vapor diffusion** where a drop of protein solution plus precipitate sitting on a shelf, post, or depression equilibrates against a well solution of **precipitate** solution. Most common method in high-throughput format. *See also* **hanging drop**.

skew (skewness)

The third **central moment** of a distribution defining the asymmetry of the distribution.

slope (M)

The slope of a **function** at a given point is given by its first **derivative**. At **extrema** of a function the slope or first derivative is zero. The **curvature** is negative for maxima, positive for minima of a function.

slow cooling protocol

Temperature schedule for simulated annealing optimization in molecular dynamics refinement.

small angle scattering

Scattering experiment on amorphous or liquid sample in which the scattering is limited to forward scattering representing coarse features of the molecular **scattering envelope**.

soaking

Incorporation of a heavy atom, ligand, or other (small) binding partners into an already grown crystal by adding the substance to the mother liquor.

solubility, of proteins

Protein solubility is affected by various reagents such as salts determining ionic strength, small alcohols affecting dielectric constant of the solvent, **poly(ethylene glycol)s** (PEGs) competing for solvent interactions, and net and local charge as affected by pH.

solubility diagram

A diagram of protein concentration versus precipitant concentration depicting the **solubility line** (solubility limit). *See also* **crystallization diagram**.

solubility line

Line in a phase diagram indicating maximum (protein) solubility and separating the single-phase (protein) solution from a two-phase region (or region of metastability).

solvent (C)

In protein crystallization, generally referring to disordered **mother liquor** in **solvent channels** and intermolecular voids in the crystal structure. Contains all components of the **crystallization cocktail** plus all components present in the **protein stock solution**.

solvent channels

Contiguous channels in protein crystals filled with disordered mother liquor, often along **plain rotation axes** in crystal structures.

solvent content

Percentage of crystals structure occupied by disordered solvent (crystallization **mother liquor**) obtained from **Matthews probabilities**.

solvent flattening

An iterative **solvent modification** method providing phase improvement and **phase extension** based on setting the solvent density to constant value.

solvent flipping

An iterative **solvent modification** method providing phase improvement and **phase extension** that is robust against bias from the protein map. Similar in concept to **charge flipping**.

solvent mask
Mask of connected real space grid points delineating the protein molecule (density) from the disordered solvent region.

solvent modification
General term encompassing methods for phase improvement and phase extension

solvent polarity
Solvents consisting of polar molecules such as water have a high polarity; aliphatic molecules such as hexane have a low polarity.

space group
Mathematical **group** of symmetry operations, with one to three **generators**, giving rise to the general 230 (65 chiral) different arrangements of motifs in a 3-dimensional periodic crystal structure.

space group enantiomorph – *see* **enantiomorphic pairs of space group**

space group symbol
Symbolic representation of the **space group** symmetry and its generators, commonly in the form of the **Hermann–Mauguin symbol** including the **Bravais symbol** and one to three **generators** of the group. The alternative Schönflies notation and Hall notation are less common in macromolecular crystallography.

spallation neutron source (SNS)
When a high-energy **proton** is accelerated into a target made of heavy elements such as mercury or uranium, spallation particles including **neutrons** are produced. For every proton striking the nucleus, 20 to 30 neutrons are expelled. The Oak Ridge SNS is presently the most powerful neutron source in the world.

sparse matrix (C)
A crystallization experiment with a given combination of reagents and reagent levels, based on prior experience, but not following a rational **experimental design**.

sparse matrix (M)
A **sparse matrix** is a matrix populated primarily with zeros. Similar to a scientific meeting.

sparse matrix refinement
Minimization algorithm in which only the largest second derivative matrix elements are evaluated.

special position (C)
A crystallographic position in a unit cell that is located on a symmetry element or intersection of symmetry elements. A motif on such a position needs to have at minimum the same symmetry as the position where it is located. Asymmetric protein molecules cannot be located on symmetry elements, only on **general positions**. Heavy atoms or solvent molecules can be found on special positions.

sphere of influence algorithm
A method of **density modification** that uses a fuzzy cross-over region instead of a sharp solvent mask to define the solvent–protein boundary.

spherical coordinates
Spherical coordinates describe a rotation in terms of an azimuthal angle φ (horizontal rotation), a lateral angle ψ (up/down rotation), and a rotation κ around the **principal Euler axis** defined by the rotations about φ and ψ. Figure 11-5.

spherical harmonics
Ubiquitous functions $Y_{l,m}$ that are the angular part of orthogonal solutions to the Laplace differential equation, represented in a system of spherical coordinates.

spherulites
Term from mineralogy indicating spherical clusters of needle-like crystals that originate from a **nucleation** core in the center of the spherulite.

split conformation, side chain
Two discrete conformations of the same molecule or part thereof, often observed in surface-exposed residues or ligand molecules.

square lattice (C)
A 2-dimensional plane lattice with $a = b$, $\gamma = 90°$.

square matrix (M)
An $n \times n$ matrix.

squaring method
Basis of **direct methods**; **convolution** (**autocorrelation**) of electron density in direct space is equivalent to squaring of structure factors.

Src (Src protein family)
Family of cytoplasmic tyrosine kinases (pronounced "sark") that associate with the cytoplasmic domains of some enzyme-linked cell-surface receptors (for example, the T-cell antigen receptor) that lack intrinsic **tyrosine kinase** activity. They transmit a signal onwards by phosphorylating the receptor itself and specific intracellular signaling proteins on tyrosines.

S-SAD – *see* **sulfur single anomalous diffraction phasing**

standard deviation (standard uncertainty)
Square root of **variance**, measure for width of a probability distribution.

static light scattering, MALS
Multi-angle light scattering (MALS) method for molecular weight determination. Also provides the second thermodynamic virial coefficient as a measure for intermolecular interaction.

statistical weight
A statistical measure for the reliability of observations or computed parameters. Examples are **variance**, **sigma-A**, and **restraint weights**.

steepest descent minimization
Iterative multivariate minimization algorithm that uses only the first derivatives or gradients to determine the downward path.

stereochemical restraint target values
Expectation values and **variance** of stereochemical parameters such as bond lengths and angles, torsion angles, planarity of groups and rings, and chiral volumes.

stereochemical restraints
Provide a means of keeping the model stereochemistry in physically reasonable bounds in macromolecular refinement because the experimental **data-to-parameter ratio** is generally too low for unrestrained refinement. Most restraints are implemented as sum of residuals squared reflecting model deviations from stereochemical restraint target values.

stereographic projection
A representation of a sphere on a plane, using a stereographic net (Figure 11-12). Historically used for experimental determination of **crystal class** and frequently in the representation of a rotation, particularly **Patterson self-rotations**, in **spherical coordinates** (lateral angle ψ, azimuthal angle φ) in sections with different κ-angles (**principal Euler angles**).

stick model
A model of a molecular structure in which the covalent bonds are represented as sticks, with the atoms located at the intersection of the bond sticks. *Compare* **ball and stick model**.

stochastic search (optimization) methods
Search or optimization algorithms that use random starting parameters combined with efficient algorithms such as evolutionary programming or Metropolis-Monte Carlo to improve the parameters.

streak seeding
Introduction of **nucleation** seeds by swishing a fiber with **microseeds** through a protein crystallization drop.

structure amplitude – *see* **structure factor amplitude**

structure factor
A complex function $\mathbf{F}(hkl)$ (interpreted as a transversal electromagnetic wave) whose magnitude $F(hkl)$ is proportional to the square root of the observed **intensity** of a diffraction spot or reflection with **reciprocal lattice** indices hkl. $\mathbf{F}(hkl)$ is the sum of

all atomic scattering contributions in a **unit cell** in the direction defined by *hkl*.

structure factor amplitude

The norm, magnitude, or amplitude of the complex structure factor **F**, corresponding to the **amplitude** to the diffracted electromagnetic X-ray wave.

structure factor distribution

Probability distribution function showing the probability for structure factors (or structure factor amplitudes) to have a certain value. Can be unconditional (centric and acentric **Wilson distribution**), or conditional on a model (centric **Woolfson**, acentric **Sim**, or **Rice distribution**).

structure guided drug design

A method of rational drug design where receptor–drug interactions are improved and designed based on knowledge of the receptor, ligand, or a receptor–ligand complex structure.

structure invariants

Structure invariants are **triplet relations** independent of the origin of the structure. Used in **direct methods** phasing. *Compare* **structure semi-invariants**.

structure model – *see* **model** (C)

structure semi-invariants

Structure semi-invariants are **triplet relations** invariant in relation to positions in the cell with the same point group (i.e. the structure factors have same amplitude but different phase). Used in **direct methods** phasing.

substrate analog

Molecule similar to a natural substrate, often a non-hydrolyzable or non-processable form of the substrate.

substructure

A partial assembly of atoms of a structure, generally referring to the **marker atom** substructure in **experimental phasing**.

substructure handedness

Due to symmetry of the **Patterson space**, the **substructure** handedness is generally undefined. When the substructure handedness is chnaged during phasing, the **screw axis** of an **enantiomorphic pairs of space group** must also be changed to the other enantiomorph of the pair. There are three exceptions to the **inversion** about the origin: In **chiral space groups** $I4_1$, $I4_122$, and $F4_132$, the origin is not located on the enantiomorph axis, and the center of inversion does not coincide with the origin. Their inversion operators are $(-x, \frac{1}{2}-y, -z)$, $(-x, \frac{1}{2}-y, \frac{1}{4}-z)$, $(\frac{1}{4}-x, \frac{1}{4}-y, \frac{1}{4}-z)$, respectively.

suicide substrate

A substrate, ligand, or drug that binds covalently to its **receptor**.

sulfhydryl (thiol)

Reactive chemical group –SH containing sulfur and hydrogen; present in the amino acid cysteine and other molecules. Two sulfhydryls can join to produce a **disulfide bond**. Often covalently modified and subject to **radiation damage**.

sulfur single anomalous diffraction (S-SAD) phasing

Phasing technique that utilizes a single **anomalous difference** data set resulting from weak anomalous signal of native sulfur atoms. Needs to resolve two ambiguities: **substructure handedness** and **phase angle ambiguity** in phasing equations, generally by means of **density modification**.

sum of residuals squared (SRS) (7-34)

Quantity minimized during optimization procedures.

superposition

Minimization of RMSD between corresponding atoms of similar structures. Problem is the detection of the correspondence at low similarity.

supersaturation

Thermodyamically **metastable** state reached by increasing the protein concentration beyond its equilibrium solubility limit.

superstructure – *see* **commensurately modulated structure**

surface entropy reduction

A method that potentially increases probability of crystal formation by reducing the adverse surface entropy term by replacing high-entropy residues such as lysine with smaller and more rigid residues.

symbolic operators (C)

Representation of a symmetry operation not through combined rotation and translation matrix but by symbols representing the coordinates of the object after application of the respective operation. *See also* **matrix operators**.

symmetric matrix

Matrix that is identical with its **transpose**.

symmetry

Arrangement of like objects in a defined fashion.

symmetry averaging – *see* **density averaging**

symmetry axis

The axis around which a **rotation** or **roto-translation** (screw operation) occurs.

symmetry center – *see* **inversion** (C)

symmetry element – *see* **symmetry operation**

symmetry equivalents

Identical symmetry related points (equipoints), objects, or reflections.

symmetry operation

An affine **transformation** or operation acting on an object or **motif** that generates identical copies of the object, that is, the physical properties of the object must remain invariant with respect to any symmetry operation. Represented as a matrix equation $\mathbf{x}' = \mathbf{Rx} + \mathbf{T}$.

symmetry-related molecules

Identical copies of a given molecule that are related by **symmetry operations**.

symmetry-related reflections

Reflections that are symmetry related in reciprocal space as a result of crystal symmetry. Symmetry related reflections have invariably identical reflections and defined phase relations if the symmetry operation includes translational elements, $\varphi_{hR} = \varphi_h - 2\pi \mathbf{hT}$.

synchrotron (storage ring, source)

A machine that keeps electrons circling at relativistic speeds, providing a source of tunable high energy **electromagnetic radiation** (**synchrotron radiation**) ranging from far UV to hard X-rays (Figure 8-5).

synchrotron radiation

High energy electromagnetic radiation emitted from a synchrotron storage ring by **bending magnets** and **insertion devices** such as **wigglers** and **undulators**. The advantage of synchrotron X-ray sources is that from wide bandwidth of the emitted **brilliant** and polychromatic radiation, narrow **X-ray photon** energies can be selected by suitable X-ray optics and **monochromators**. Tunable X-rays are necessary for **multi-wavelength anomalous diffraction** (MAD) experiments and for optimal selection of wavelength in **anomalous diffraction** experiments in general.

synergistic

Term used to describe a situation where different entities cooperate or interact toward a final outcome.

systematic absences

Reflections that have invariably zero intensity due to the presence of translational symmetry elements (in **screw axes** and **Bravais translations**). Screw axes cause **serial extinctions** and Bravais translations cause **zonal** or **integral extinctions**.

systematic error

Non-random error in measurement affecting a deviation of the **mean** or **expectation value** from the unknown true value. *See also* **accuracy**.

T

tangent formula (10-28)

Relates phase angles between sets of certain **normalized structure factors**. It is used for **phase extension** and phase refinement in **direct methods**.

target values – *see* **stereochemical restraint target values**

Taylor series (7-70)

Expansion of a function in a series of derivative terms.

tedium (Latin *taedium*: disgust, boredom)

Occurring during late stages of macromolecular **model building** and **refinement**, ultimately terminating macromolecular refinement.

temperature factor – *see* **Debye–Waller factor**

template convolution

Template matching technique for finding small parts of an image or electron density which match a template image or density.

tensor

Usually a second rank tensor or 3×3 matrix describing an anisotropic physical property, obtained as the dyad product of an anisotropic property vector, for example anisotropic displacement.

tertiary structure

Complex 3-dimensional structure of a folded polymer chain, especially a protein or RNA molecule.

tetrad (C)

Name for 4-fold **rotation axis**.

tetragonal

Lattice or crystal structure with cell parameters $a = b \neq c$, $\alpha = \beta = \gamma = 90°$, with primitive (*tP*) or *I*-centered (*tI*) lattice type and 4-fold internal minimal symmetry.

TEV protease

Very specific and relatively efficient protease of the tobacco etch virus, used commonly in His-tagged variants for tag removal.

thermal diffuse scattering (TDS)

Thermal diffuse scattering is the intensity that is scattered by crystals outside of discrete Bragg reflections due to deviations from the perfect order such as thermal motion and various static or dynamic defects of **atomic displacements**. Diffuse scattering features depend on the correlation functions of the deviations.

thermal ellipsoid – *see* **anisotropic displacement ellipsoid**

thiol – *see* **sulfhydryl**

Thomson scattering

Thomson scattering is the elastic scattering of **electromagnetic radiation** by a charged particle such as an **electron**.

three-ten helix (3_{10} **helix**)

A rare, tightly packed helical secondary structure element with backbone hydrogen bonds from residue n to $n + 3$.

threonine

A **polar** L-α-amino acid with asymmetrically branched hydroxyl side chain $-CH(OH)CH_3$, absolute configuration (2S, 3R), common target for phosphorylation by kinases.

time-resolved X-ray diffraction

Rapid collection of data of a crystal or molecules triggered to assume a specific conformational state. Used in exploration of enzyme or electron transport kinetics.

TLS refinement

Macromolecular **refinement** protocol in which the displacement of whole groups of atoms or parts of the molecule such as domains are parameterized as a combination of **translation**, **libration**, and **screw** motion. Improved representation of molecular flexibility and mobility, requiring only 20 parameters per TLS group, therefore usable also at much lower resolution than individual **anisotropic *B*-factor** refinement.

topoisomerase (DNA topoisomerase)

Enzyme that binds to DNA and reversibly breaks a phosphodiester bond in one or both strands. Topoisomerase I creates transient single-strand breaks, allowing the double helix to swivel and relieving superhelical tension. Topoisomerase II creates transient double-strand breaks, allowing one double helix to pass through another and thus resolving tangles.

torsion angle

The relatively weakly restrained side-chain and main-chain dihedral angles are called torsion angles as compared with general **dihedral angles**.

torsion angle refinement (parameterization)

Refinement protocol in which instead of **atomic coordinates** the **torsion angles** are refined. Requires fewer parameters and has a large **radius of convergence**. Used for refinement of **low resolution structures** and in initial stages of refinement of **molecular replacement** models.

trace (M)

Sum of diagonal elements of a matrix.

trans

On the other or opposite side. *Compare cis*.

transcription (DNA transcription)

Copying of one strand of DNA into a complementary RNA sequence by the enzyme RNA polymerase.

transfection

Introduction of a foreign DNA molecule into a host cell. Usually followed by expression of one or more genes in the newly introduced DNA.

transform – *see* **Fourier transform**

transformation (B)

Insertion of new DNA (e.g. a plasmid) into a cell or organism, such as into competent *E. coli*.

transformation, affine (M,C)

Function or mapping in vector space that can be expressed as a **linear transformation** plus a translation. A symmetry operation containing a rotation plus a translation is an affine transformation.

transformation, linear (M,C)

Function or linear mapping in vector space that can be expressed as a matrix and preserves vector addition and scalar multiplication. Scaling and rotations are linear transformations.

translation (C)

Linear **transformation**, mapping an object in the same orientation onto a different position, $\mathbf{x}' = \mathbf{x} + \mathbf{T}$.

translation (RNA translation)

Process by which the sequence of **nucleotides** in a **messenger RNA** (mRNA) molecule directs the combination of amino acids into a protein. Takes place at and is catalyzed by **ribosomes**.

translation function

Function used to determine the location of correctly oriented search probe molecule in the unit cell. The search must cover the **Cheshire cell**, and can be a Patterson-based fast translation function or a maximum likelihood translation function.

translation-libration-screw parameterization – *see* **TLS refinement**

translational non-crystallographic symmetry (NCS)

Non-crystallographic symmetry that is solely or largely translational does not comply with closed **group** limitations. Translational NCS parallel or near parallel and in integer fractional relation to unit cell vectors manifests itself as **commensurately modulated superstructures** and in **native Patterson maps**.

transmembrane helix

A hydrophobic helix of a membrane protein that extends across the cell membrane in contact with the hydrophobic tails of the phospholipid molecules.

transmembrane protein

Integral membrane protein that extends through the lipid bilayer, generally with functional part present on both sides of the membrane.

transpose (M)
A matrix **M** has a transpose $\mathbf{M}^T$ with diagonally swapped elements.

triad (C)
Name for 3-fold **rotation axis**.

triclinic
Lattice or crystal structure in which all cell parameters are different, **lattice type** aP, agonic, with no internal minimum symmetry.

trigonal
Lattice or crystal structure with cell parameters $a = b, c, \alpha = \beta = 90°$, $\gamma = 120°$, with primitive (hP) or R-centered (hR) lattice type and 3-fold internal minimal symmetry.

triple vector product
Defines the volume V of a parallelepiped (or **unit cell**) spanned by vectors **a**, **b**, **c** in a right-handed system as positive: $V = \mathbf{a} \cdot (\mathbf{b} \times \mathbf{c})$.

triplet relations (10-26)
Fundamental to direct methods, stating that if all reflections in the reflection triplet F_h, F_k, and F_{h-k} are very large, then the phase relation $\varphi_{-h} + \varphi_k + \varphi_{h-k} \simeq 0$ is fulfilled.

truncation ripples – *see* **Fourier truncation ripples**

tryptophan
Largest, aromatic and polar L-α-amino acid with indole ring in side chain $-CH_2(C_8NH_6)$; can act as hydrogen bond donor and participate in π-stacking interactions.

turn
Short secondary structure elements with a distinct hydrogen bonding pattern that facilitate tight changes of direction in the polypeptide chain. *Compare* **loop**.

twin operator
Symmetry operator that relates orientation of twinned domains in a twinned crystal.

tyrosine
Aromatic, polar L-α-amino acid with side chain $-CH_2(C_6H_4)OH$ that structurally participates in hydrogen bonds and π-stacking interactions.

tyrosine kinase
Enzyme that phosphorylates the **tyrosine** residues of certain proteins.

U

undulator
Insertion device consisting of a periodic structure of magnets alternating along its length. Electrons traversing the periodic magnet structure are forced to undergo oscillations and therefore radiate. The X-rays produced in an undulator are very intense with narrow energy **bandwidth** and are collimated in the orbit plane of the synchrotron ring.

uniaxial system – *see* **unique axis**

unimodal
A function or probability distribution with a single peak or extremum.

unique axis (C)
Special single axis parallel to minimal symmetry element, **b** in monoclinic, **c** in trigonal, tetragonal, and hexagonal systems.

unique reflections
The reduced set of recorded reflections belonging to one and the same asymmetric unit of reciprocal space.

unit cell
The transitionally repeating building blocks of a crystal, comprising a **unit lattice** filled with motifs according to the **space group** symmetry of the crystal.

unit cell parameters
The dimensions (norm, magnitude) a, b, c, of the **unit cell** vectors

a, **b**, **c**, and the enclosed angles α, β, and γ.

unit cell vector
The three vectors **a**, **b**, **c**, enclosing angles α, β, and γ that form the **basis** of the **crystal system**.

unit cell volume – *see* **triple vector product**

unit lattice
Mathematical construct spanned by three lattice vectors **a**, **b**, **c**, enclosing angles α, β, and γ that form the **basis** of the lattice.

unit vector
A **vector** in a vector space whose length (**norm**) is the unit length, 1.

unitary structure factor (7-108)
Structure factor amplitudes $0 \leq U(\mathbf{h}) \leq 1$ reflecting the scattering of point atoms, that is, adjusted for *B*-**factor** attenuation and **atomic scattering factor** attenuation, but not **epsilon-factor** corrected. *Compare* **normalized structure factors**.

univariant
Function or probability distribution depending on a single variable.

unrestrained refinement
Refinement that uses diffraction data only and does not incorporate knowledge-based restraints. Rarely possible in macromolecular refinement, requires atomic resolution data.

usage bias
In the absence of a count for negative results, success rates (e.g. percentage successful trials) cannot be properly quantified, and anything that is used more frequently will likely deliver more positive results regardless of its actual success rate.

V

valine
Hydrophobic L-α-amino acid with a $-CH(CH_3)_2$ side chain.

van der Waals (vdW) interaction
Interatomic interaction with potential curve following the **Lennard-Jones potential** (Figure 12-18). Strong and rapidly increasing repulsive component once atoms come closer than the van der Waals contact distance, but only weakly attractive force for distances larger than the van der Waals contact distance. Van der Waals repulsions determine the **main chain torsion angle** distribution, giving rise to typical secondary structures such as helices and sheets.

vapor diffusion
A method of crystallization in a closed system where the protein solubility is decreased by exchange of water vapor from a drop of protein plus precipitant into a **hygroscopic** well solution. Used in **hanging drop** and **sitting drop** setup.

variance (7-29)
Fundamental and universal statistical measure for the width of a **probability distribution** $\sigma_n^2 = \langle (n - \langle n \rangle)^2 \rangle = \langle n^2 \rangle - \langle n \rangle^2$.

variance, of sample distribution (7-32)

vector (B) – *see* **expression vector**

vector (C)
Entity endowed with both length and direction, represented as a 1-dimensional matrix. Can be a row or column vector.

vector product (M)
The vector product or cross-product $\mathbf{a} \times \mathbf{b}$ of two column vectors **a**, **b** defines a third vector **c** perpendicular to the plane spanned by **a** and **b** with its **norm** defined by $c = |\mathbf{c}|\ |\mathbf{a} \times \mathbf{b}| = ab \sin \gamma$.

virtual ligand screening
In-silico method of screening small molecules by means of **ligand docking**.

virus
Particle consisting of nucleic acid (RNA, retrovirus, or DNA) enclosed in a protein **capsid** and capable of replicating within a host cell and spreading from cell to cell.

vitreous
Amorphous, non-crystalline, glass-like.

W

Watson–Crick base pair – *see* **base pair**

wave vector
Propagation vector of **electromagnetic wave**, vector (cross) product of perpendicular electric field vector and magnetic field vector, with magnitude inverse to the **wavelength** or directly proportional to the **energy** of the wave.

wavelength
The extent of a transversal wave between two points of maximum amplitude (or any repeating point). Inversely related to energy or **frequency**. *See* **wavelength-energy conversion**.

wavelength–energy conversion
Wavelength in Å = 12397.639/Energy in electronvolts (eV).

wedge
A sector of reciprocal space.

weight (C) – *see* **statistical weight**

weighted average (7-55)
Variance-corrected average (mean) value.

Weiss indices – *see* **lattice planes**

Weissenberg camera
An ingenious mechanical device of early crystallography to record the diffraction spots on circular film so that reciprocal layer lines are separated on typical curves. Still used in some detectors in conjunction with special cylindrical image plates.

white line
An additional peak on and at the **X-ray absorption edge**, resulting from electronic transitions into unoccupied higher states.

white radiation
Polychromatic radiation, radiation of a broad energy **bandwidth**, such as **Bremsstrahlung** or synchrotron radiation from a **bending magnet** or **wiggler**.

wiggler
Insertion device consisting of a periodic series of magnets in a Halbach array design to laterally deflect or wiggle the electron beam inside a **synchrotron storage ring**, thereby generating **brilliant** X-ray radiation of a broad energy **bandwidth**.

Wilson distribution
The unconditional probability distribution for structure factor amplitudes or intensities, different for centric and acentric reflections.

Wilson-like maximum likelihood rotation function (11-54)
Basic likelihood rotation function based on the unconditional Wilson structure factor distribution.

Wilson plot (7-121)
The logarithm of mean observed intensity over mean absolute intensity, binned in resolution shells and plotted against a function of $1/d^2$. The intercept of a regression line though the linear high resolution part of the graph provides the linear **scale factor** k as the intercept and the **isotropic overall B-factor** from the slope. Figure 7-15.

Wilson scaling
Used to bring diffraction data onto a common absolute scale, based on the total scattering estimated for the unit cell from the atomic scattering factor contributions. Requires a **scale factor** and an initial **overall B-factor** accounting for attenuation of scattering by **atomic displacement**, obtained from a **Wilson plot**.

Woolfson distribution (7-148)
The conditional structure factor probability distribution for **centric** reflections.

world coordinates
The **Cartesian coordinates** of **real space** harboring crystallographers and their objects.

X

X-ray(s)
Term coined by Wilhelm Conrad Röntgen for ionizing radiation in the keV energy range (Ångstrom wavelength range).

X-ray absorption edge
Above certain element-specific energies X-ray absorption rapidly increases, forming an edge in the absorption spectrum. Absorption edges have a fine structure, and in their vicinity atomic scattering factors contain significant, wavelength-dependent **anomalous scattering** contributions.

X-ray absorption near edge structure
Region of the X-ray absorption spectrum in the close vicinity of an absorption edge. Typical feature are **white lines**.

X-ray crystallography
Fundamental technique for the exploration of the atomic and molecular structure of matter. Based on discrete **diffraction** of X-rays by periodically ordered assembly of atoms or molecules, that is, a single **crystal** or crystalline matter in general.

X-ray diffraction
Process of interaction of **X-rays** (high energy **electromagnetic radiation**) with the **electrons** of periodically arranged atoms and molecules in **crystals**.

X-ray emission line – *see* **X-ray fluorescence**

X-ray excitation scan
An absorption edge scan obtained by measuring the intensity of the fluorescence excitation (instead of the **X-ray absorption** directly).

X-ray fluorescence
Emitted from materials when an electron from a higher shell fills a hole (generated by high energy X-ray or electron bombardment) in a tightly bound core electron shell. In **X-ray generators**, typically the strong K-shell fluorescence emission lines are used. Figure 8-2.

X-ray generator
Instrument that generates X-rays by bombarding water-cooled anode material (such as Cr, Cu, or Mo) with high energy electrons of ~40–100 keV. Generates continuous **Bremsstrahlung** as a result of electron deceleration, superimposed with about 100–1000 times as intense **characteristic X-ray fluorescence** emission lines with energies specific for the anode material.

X-ray optics
General term for devices to affect properties and geometry of X-rays, such as monochromators, filters, X-ray mirrors, and collimators.

X-ray reflection
Diffraction spot recorded on a film or detector resulting from diffracted X-rays, interpreted as reflections of X-rays on crystal **lattice planes** with **interplanar spacing** $d(hkl)$ related to diffraction (reflection) angle by the **Bragg equation**.

X-ray scattering
Fundamental process of interaction of **electrons** with **X-rays**. Term also used for describing experimental techniques that do not require crystalline matter. *Compare* **X-ray diffraction**.

X-ray sources
X-ray radiation generating devices, generally by (i) acceleration of electrons (**synchrotron** radiation and **Bremsstrahlung**), (ii) characteristic **X-ray fluorescence** (laboratory **X-ray generators**), (iii) radioactive decay (flood field sources), or (iv) inverse Compton effect.

X-ray term
The residual in a refinement that depends on the experimental diffraction data only.

X-ray weights (12-49)
Weight given the X-ray residual relative to the geometry restraint residual in restrained reciprocal space refinement.

XANES – *see* **X-ray absorption near edge structure**

xenobiotic

A chemical which is found in an organism but which is not normally produced by it or expected to be present in it, for example drug molecules. *See also* **cytochrome** P-450.

XPLOR

A versatile high level programming language and program package for macromolecular refinement, superseded by *CNS* (Crystallography and NMR System) by Axel Brunger and colleagues. Unique feature simulated annealing torsion angle refinement.

Y

yeast

Common name for several families of unicellular fungi. Among the simplest of eukaryotes, used as efficient expression hosts for eukaryotic proteins. Generally capable of most posttranslational modifications, secretion, and disulfide formation.

Yeates–Padilla plots

Cumulative probability plot of pair-wise local intensity relations of reflections, revealing (also perfect) **hemihedral twinning**.

Z

Z-score

Normalized residual, statistical standard score measuring the deviation of values from their **expectation value** in units of **standard deviations** of their sampling distribution. *See also* **RMS-Z-score**.

zinc finger

DNA-binding structural motif present in many gene regulatory proteins. All zinc finger motifs incorporate one or more Zn^{2+} ions. Figure 2-46.

zonal absences, extinctions

Systematic absences caused by translations in certain planes; for example, C-centering causes zonal absences for reflections $hk \neq 2n$ (odd).

zone (C)

Set of planes in a lattice whose *intersections* are parallel; often used as a general term for a set of parallel planes.

zone axis

Direction of the intersection of a **zone** of lattice planes hkl, indicated by square brackets [hkl]. Example: all lattice planes $hk0$ have a zone axis [001], and their reflections lie in the $hk0$ reciprocal lattice plane.

Index

For Product Safety Concerns and Information please contact our EU
representative GPSR@taylorandfrancis.com Taylor & Francis Verlag GmbH,
Kaufingerstraße 24, 80331 München, Germany

Printed and bound by CPI Group (UK) Ltd, Croydon, CR0 4YY

01/05/2025

01858594-0001